REVISED EDITION

WORLD GEOGRAPHY TODAY

HOLT, RINEHART AND WINSTON
Harcourt Brace & Company
Austin • New York • Orlando • Atlanta • San Francisco • Boston • Dallas • Toronto • London

The Authors

Dr. Robert J. Sager is Professor and Chair of Earth Sciences at Pierce College in Tacoma, Washington. Dr. Sager received his BS in geology and geography and MS in geography from the University of Wisconsin and holds a JD in international law from Western State University College of Law. He is the coauthor of several geography textbooks and has written many articles and educational media programs on the geography of the Pacific. Dr. Sager has received several National Science Foundation study grants and was a recipient of the University of Texas NISOD National Teaching Excellence Award. A founding member of the Southern California Geographic Alliance, Dr. Sager is vice president of the Association of Washington Geographers and a member of the Washington Geographic Alliance.

Dr. David M. Helgren is Director of the Center for Geographic Education at San Jose State University in California, where he also teaches in the Department of Geography and Environmental Studies. Dr. Helgren received his PhD in geography from the University of Chicago. He is the coauthor of several geography textbooks and has written many articles on the geography of Africa. Awards from the National Geographic Society, the National Science Foundation, and the L.S.B. Leakey Foundation have supported his many field research projects. Dr. Helgren is a former president of the California Geographical Society and a founder of the Northern California Geographic Alliance.

Director: John Lawyer

Project Editor: Rachel Guichard Tandy

Managing Editors: Greg Metz, Jim Eckel

Editorial Staff: Anne Norman, Daniel M. Quinn, Deborah J. Holtzman, Valerie Larson, *Department Secretary*

Editorial Permissions: Janet Harrington

Design and Production: Pun Nio, *Senior Art Director;* Sylvia Harrington, *Designer;* Donna McKennon, *Designer;* Sally Bess, Janet Brooks, Tonia Klingensmith, Anne Wright, *Design Staff;* Donna Lewis, *Production Coordinator;* Carol Martin, *Electronic Publishing Manager;* Debra Saleny, *Photo Research Manager;* Tim Taylor, *Senior Photo Researcher;* Shelley Boyd, Angi Cartwright, Sabrina Hirst, Mitzi Markese, *Photo Staff*

Art Credits: **GeoSystems, an R.R. Donnelley & Sons Company:** Text Maps, Atlas and Regional Maps; **Don Collins:** 53, 54 56; **John Edwards and Associates:** S13, 11, 12, 13, 14, 16, 19, 20, 21, 24, 26, 36, 38, 39, 45, 46, 47, 48, 144, 145, 228, 352-353, 382, 448-449, 502, 512-513; **David Griffin:** 317, 595, 623; **Joe LeMonnier:** 163, 167, 211, 289, 302, 339, 343, 345, 363, 525, 575, 609; **Mike Obersham:** locator maps; **Ortelius Design:** 207, 293; **Precision Graphics:** S10, S11, S12, S13, S17, 68, 75, 83, 84, 91, 116, 142, 194, 196, 306, 404, 501, 508, 589; **Pronk and Associates:** 427, 486; Other: **BB & K:** Cover design; **Amy Wasserman:** Cities of the World, Themes in Geography collage

Acknowledgments appear on pages 660–662, which are extensions of the copyright page.

Printed in the United States of America
ISBN 0-03-016802-3
3 4 5 6 7 032 99 98 97

Consultants and Reviewers

Robert E. Gabler
Professor of Geography
Western Illinois University
Macomb, Illinois

Truman Hartshorn
Professor of Geography
Georgia State University
Atlanta, Georgia

Mohammad Hemmasi
Professor of Geography
University of North Dakota
Grand Forks, North Dakota

Robert B. Kent
Professor of Geography and Planning
University of Akron
Akron, Ohio

Christopher T. King
Social Studies Department Chairman
Walton High School
DeFuniak Springs, Florida

Michael Kort
Professor of Social Science
Boston University
Boston, Massachusetts

Jane Young Leatherman
Social Studies Teacher
Durant High School
Durant, Oklahoma

Nancy Lehmann-Carssow
Social Studies Teacher
Lanier High School
Austin, Texas

James F. Marran
Chair Social Studies Department, Emeritus
New Trier High School
Winnetka, Illinois

Todd O'Banion
Social Studies Teacher
Scott County High School
Georgetown, Kentucky

Roger Sass
Social Studies Teacher
Kelvyn Park High School
Chicago, Illinois

Ron Scholten
Social Studies Department Chair
Providence High School
Burbank, California

Earl P. Scott
Professor of Geography
University of Minnesota
Minneapolis, Minnesota

Catharine Stoner
Social Studies Teacher
Wentzville High School
Wentzville, Missouri

Bill Takizawa
Professor of Geography
San Jose State University
San Jose, California

Kenneth Wood
Social Studies Teacher
Appleton West High School
Appleton, Wisconsin

Table of Contents

UNIT 3

The United States and Canada 102

UNIT 4

Middle and South America 184

UNIT 5

UNIT 9

East and Southeast Asia 490

UNIT 10

South Asia

UNIT 11

The Pacific World and Antarctica

List of Maps

To The Student

World Geography Today is a guide to learning about the constantly changing world in which you live. Through the chapter text and special features, this textbook is meant to help you understand the study of geography and make sense of our complex world. Hundreds of photographs of people and places will transport you around the world on your journey to learning about other cultures. Full-color diagrams will take you deep beneath the earth's crust, through the canals of a working lock, and to the top of some of the world's highest mountains. Locator maps and globes put the area under study in perspective for you. A geography dictionary in each chapter introduces new words and ideas to help you express your geographic knowledge. And focus and review sections will guide you in your study.

While the emphasis of ***World Geography Today*** is global, the concepts of geography found in it apply to the place where you live as well. The text seeks to make geography relevant to you by making the connection between regional and global issues and your own community.

Organization of *World Geography Today*

It is helpful to think of this book as having three parts. The first part is the front matter. The pages you are reading now are part of the front matter, as are the Table of Contents and the Student Atlas, a collection of handy reference maps of the world and its continents.

The second part of ***World Geography Today*** is devoted to text describing the study of geography and the world's major regions. The text is divided into 11 units and 50 chapters.

The third part of your textbook is the back matter. It includes an Appendix with statistical information on and flags of the world's countries. The back matter also includes a Glossary, which lists all the vocabulary words that appear in boldface type throughout the book.

One World, Many Voices

Literature

"1911: Machu Picchu
The Last Sanctuary of the Incas

isn't dead; it only sleeps. For centuries the Urubamba River, foaming and roaring, has exhaled its potent breath against these sacred stones, covering them with a blanket of dense jungle to guard their sleep. Thus has the last bastion of the Incas, the last foothold of the Indian kings of Peru, been kept secret.

Among snow mountains which appear on no maps, a North American archeologist, Hiram Bingham, stumbles upon Machu Picchu. A child of the region leads him by the hand over precipices to the lofty throne veiled by clouds and greenery. There, Bingham finds the white stones still alive beneath the verdure [green vegetation], and reveals them, awakened, to the world."

Translated by Cedric Belfrage

© MARCELO ISARRUALDE

Eduardo Galeano (1940–) is an Uruguayan writer who is the author of *OPEN VEINS OF LATIN AMERICA, THE BOOK OF EMBRACES, WE SAY NO,* and the three-volume history of North and South America, *MEMORY OF FIRE.* He has worked as a journalist and political cartoonist.

In this excerpt from *MEMORY OF FIRE,* Galeano writes about the discovery of a great Incan city located near the peak of Machu Picchu. Until 1911, only Native Americans living high in the Andes Mountains were aware of its location or knew what remained of the ancient city.

Interpreting Literature

1. What elements of physical geography kept the Incan city hidden for centuries?
2. Examine the photographs of Machu Picchu on this page and on page 185. What purposes do you think the "white stones" mentioned in this passage served in this city?

For the Record

HIRAM BINGHAM COLLECTION, COURTESY OF THE NATIONAL GEOGRAPHIC SOCIETY AND YALE UNIVERSITY, PEABODY MUSEUM OF NATURAL HISTORY

Hiram Bingham (1875–1956) was an archaeologist and historian. Bingham led an expedition to find the ancient Incan capital city. He chanced upon the site after crossing the rugged Andes Mountains and following the course of the Urubamba River through dense jungle terrain. The excerpt below describes his amazement at the physical beauty of the Urubamba canyon.

"It will be remembered that it was in July 1911, that I began the search for the last Inca capital. . . . I had entered the marvelous canyon of the Urubamba below the Inca fortress of Salapunco near Torontoy.

Here the river escapes from the cold plateau by tearing its way through gigantic mountains of granite. The road runs through a land of matchless charm. . . . In the variety of its charms and the power of its spell, I know of no place in the world which can compare with it. Not only has it great snow peaks looming above the clouds more than two miles overhead; gigantic precipices of many-colored granite rising sheer for thousands of feet above the foaming, glistening, roaring rapids, it has also, in striking contrast, orchids and tree ferns, the delectable [pleasing] beauty of luxurious vegetation, and the mysterious witchery of the jungle. One is drawn irresistibly onward by ever-recurring surprises through a deep, winding gorge, turning and twisting past overhanging cliffs of incredible height."

Analyzing Primary Sources

1. What words does Bingham use to show that the Urubamba was not a calm river along the course that he followed?
2. What aspects of the physical geography were most surprising to Bingham?

186 *Unit 4* | *Unit 4* 187

Last, the back matter includes the Index. The Index is your guide to the hundreds of topics discussed in the book, with page references for each topic.

Special Features

You must have certain skills to understand the changes that are affecting our world. In the ***Skills Handbook*** starting on page S1, you will learn important skills that will help you understand the impact of these changes. In addition, skill lessons throughout the book will give you additional tools for your study as well as opportunities to practice the skills you have acquired.

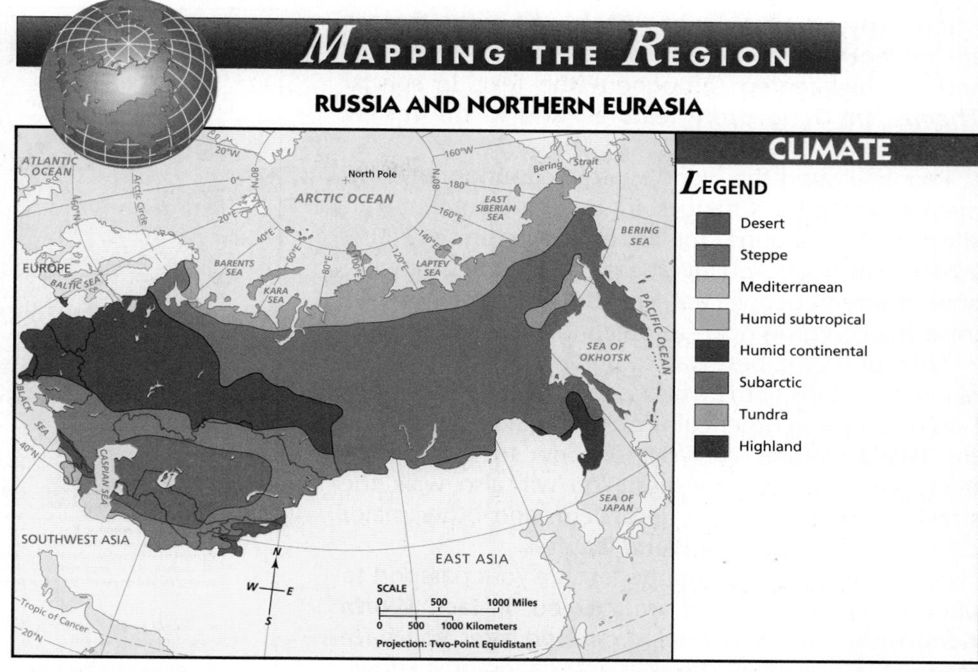

MAPPING THE REGION
RUSSIA AND NORTHERN EURASIA
CLIMATE
LEGEND
Desert
Steppe
Mediterranean
Humid subtropical
Humid continental
Subarctic
Tundra
Highland
ATLANTIC OCEAN
North Pole
ARCTIC OCEAN
EAST SIBERIAN SEA
Bering Strait
BERING SEA
EUROPE
BALTIC SEA
BARENTS SEA
KARA SEA
LAPTEV SEA
SEA OF OKHOTSK
PACIFIC OCEAN
BLACK SEA
CASPIAN SEA
SEA OF JAPAN
SOUTHWEST ASIA
EAST ASIA
Tropic of Cancer
Arctic Circle
SCALE
0 500 1000 Miles
0 500 1000 Kilometers
Projection: Two-Point Equidistant

Literature can help you in your study of geography by painting vivid pictures of faraway places. In the beginning of each unit of the text, you will find a feature called ***One World, Many Voices***. These literature passages and firsthand accounts describe the geography of various places around the world. You might be familiar with some of the authors of these passages; other authors will be new to you. All will help you understand how different people see the world around them.

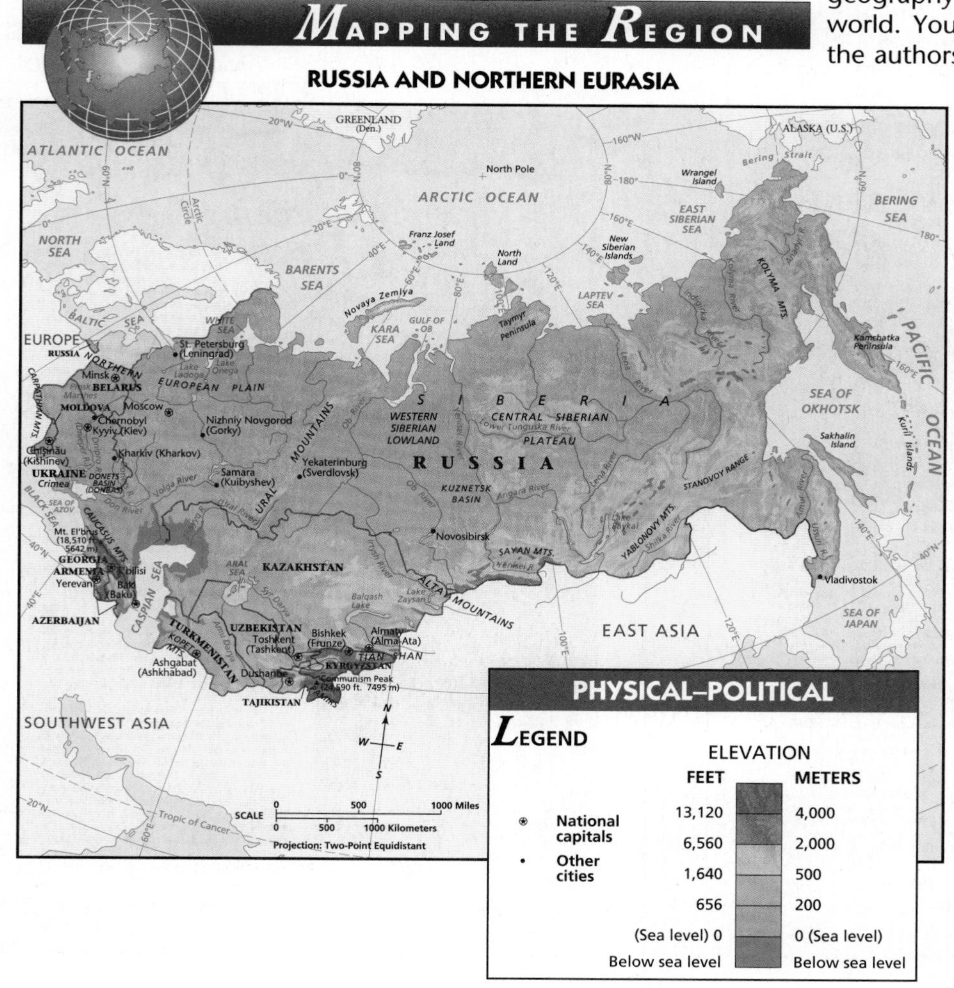

MAPPING THE REGION
RUSSIA AND NORTHERN EURASIA
GREENLAND (Den.)
ALASKA (U.S.)
ATLANTIC OCEAN
North Pole
ARCTIC OCEAN
Wrangel Island
EAST SIBERIAN SEA
BERING SEA
NORTH SEA
Franz Josef Land
North Land
New Siberian Islands
BARENTS SEA
Novaya Zemlya
KARA SEA
GULF OF OB
Taymyr Peninsula
LAPTEV SEA
BALTIC SEA
WHITE SEA
EUROPE
RUSSIA
St. Petersburg (Leningrad)
NORTHERN EUROPEAN PLAIN
Minsk
BELARUS
MOLDOVA
Moscow
Chernobyl
Kyyiv (Kiev)
Nizhniy Novgorod (Gorky)
Kharkiv (Kharkov)
Chişinău (Kishinev)
UKRAINE
Crimea
Samara (Kuibyshev)
URAL MOUNTAINS
Yekaterinburg (Sverdlovsk)
SIBERIA
WESTERN SIBERIAN LOWLAND
CENTRAL SIBERIAN PLATEAU
RUSSIA
KUZNETSK BASIN
Novosibirsk
SAYAN MTS.
YABLONOVY MTS.
STANOVOY RANGE
KOLYMA MTS.
Kamchatka Peninsula
SEA OF OKHOTSK
Sakhalin Island
Kuril Islands
PACIFIC OCEAN
Vladivostok
SEA OF JAPAN
BLACK SEA
SEA OF AZOV
Mt. El'brus (18,510 ft. 5642 m)
CAUCASUS MTS.
GEORGIA
Tbilisi
ARMENIA
Yerevan
Baku
AZERBAIJAN
CASPIAN SEA
ARAL SEA
KAZAKHSTAN
Balqash Lake
Lake Zaysan
ALTAI MOUNTAINS
UZBEKISTAN
Toshkent (Tashkent)
Bishkek (Frunze)
Almaty (Alma-Ata)
TIAN SHAN
KYRGYZSTAN
TURKMENISTAN
KOPET MTS.
Ashgabat (Ashkhabad)
Dushanbe
Communism Peak (24,590 ft. 7495 m)
TAJIKISTAN
EAST ASIA
SOUTHWEST ASIA
Tropic of Cancer
SCALE
0 500 1000 Miles
0 500 1000 Kilometers
Projection: Two-Point Equidistant
PHYSICAL–POLITICAL
LEGEND
National capitals
Other cities
ELEVATION
FEET METERS
13,120 4,000
6,560 2,000
1,640 500
656 200
(Sea level) 0 0 (Sea level)
Below sea level Below sea level

To help you learn about all aspects of the world's regions, each unit has a series of physical–political, climate, population, and economy maps called ***Mapping the Region***. These full-color maps will help you compare characteristics within and between world regions. As you will learn, however, regions are not fixed areas. Just as the world is constantly changing, so are the ways in which it is divided. Special ***Regions in Transition*** features highlight areas of the world that are undergoing significant changes.

You also will learn about the themes that tie the study of geography together. These themes are movement, human-environment interaction, location, place, and region. These Five Themes of Geography help us answer the questions of where, why, and how in geography. All of the themes are evident in the changes taking place in the world and are highlighted throughout the text, in special ***Themes in Geography*** features, and in the review activities.

People around the world share a common environment—the earth. Changes in our environment can affect people around the globe. Each unit's ***Planet Watch*** feature lets you investigate case studies showing how different people are working in different ways toward a common goal—protecting the environment.

Sometimes the best way to understand change and different regions of the world is to learn about the daily lives of people in other cultures. The ***News from Around the World*** features are your window to the lives of teenagers in other countries. You will also walk the streets of Sydney, Tokyo, Jerusalem, and other major cities in the ***Cities of the World*** features.

All of these features and the text are your passport to understanding the world around you. In fact, ***World Geography Today*** will help you long after you have read the last chapter. The skills and geographic knowledge you are about to learn will help you understand and adapt to future changes in our world.

NEWS FROM AROUND THE WORLD *Russia*

Teenagers in Russia Have More Free Time

When Russia was part of the Soviet Union, teenagers spent much of their free time in government-sponsored groups. Almost all children joined the Young Pioneers when they reached nine years of age. At age 14, those who did not join Komsomol (the Young Communist League) were considered disloyal to the government and could not expect to attend a university or receive good jobs as adults.

Now that Russia is no longer under communism, few teenagers join political groups. Some Young Pioneer groups still exist, although the organization suffers from lack of funds. Those attending the summer camps run by the Young Pioneers have noticed big changes. No longer intent on promoting communism, Young Pioneer groups have become far less rigid. Boys and girls used to resemble young soldiers as they saluted and marched with red scarves secured around their necks. "There are no songs, no marching, no red ties anymore," said Artyom V. Chibisov, a 13-year-old boy from Moscow. One young woman had a practical reason for not missing the scarves. "It means less ironing," she said.

As groups such as the Young Pioneers and Komsomol weaken or disappear, young Russians are left with more free time and are seeking new ways to spend it. Although Russian teenagers rarely worked at after-school jobs, Russia's economic crisis offers them even less chance to do so now. Many spend their free time socializing with friends.

Moscow teenagers who enjoy playing rollerball, like those pictured above, arrive each day in Gorky Park to meet friends and practice their skills. Other teens have become swept up in Western music, spending so much time in music-related activities that they are called *rokry* (rockers) or *metallisty* (heavy metal fans).

Many Russian teens, however, have other interests besides sports and music. Some join large groups that work to clean up or conserve the environment. Others remain in the International Clubs that schools organized as the old Soviet Union became more open. In these clubs, teenagers plan parties and meetings and correspond with pen pals in other nations.

progress has been slow. Stable political relationships have yet to be created.

Russia's greatest resource is its people, who are well educated. They have survived the challenges of living in a country characterized by isolation and a harsh environment. During the Soviet era, government officials controlled the lives of the people. Now the people can make their own decisions and determine their own futures. The Russian people have much to offer themselves and the world.

4 SECTION REVIEW

1. What is significant about the Tatarstan Republic?
2. Name two political challenges facing the Russian Federation.
3. **Critical Thinking** **List** ways in which a free-market economy and a democratic government might benefit the Russian Federation.

348 *Unit 6*

PLANET WATCH

Women Tree Planters of Kenya

Avocado, mango, and acacia trees thrive in the Kenyan highlands, where not long ago farm workers could not even find shade.

Ten million young trees scattered throughout the country offer new hope to Kenyan farmers and families. The trees were planted not by foresters or government workers but by the many women who farm the lands. "We are planting trees to ensure our own survival," says Wangari Maathai, founder of Kenya's Green Belt Movement. The organization encourages tree planting in Kenya's rural communities as a way of fighting deforestation and desertification.

Many of the trees in the highlands of **Kenya** have been cleared for crops.

Deforestation

Like many other developing nations, Kenya is a country mainly of farmers and herders. The most productive farmlands are in the highlands, an area once green with trees and lush vegetation. Over the past century, however, much of the land has been stripped. Today, only about 10 percent of the original forests remain. Many trees were cut for firewood, which farm families use in open-hearth cooking.

One by one, the trees began to disappear from the highlands. Without tree roots to hold the soil in place, the land eroded. Erosion of banks along streams and rivers then threatened the water supplies.

The land was losing its ability to sustain the region's growing population as well as its fertility. Farmers moved down to the savannas in search of better land. These plains are home to nomadic peoples and much of Kenya's wildlife.

Wangari Maathai recognized what was happening to her country. The daughter of a farm worker, she is a biologist and the first woman in Kenya to receive a PhD. "When I would visit the village where I was born," she says, "I saw whole forests had been cleared for cultivation and timber. People were moving onto hilly slopes and riverbeds and marginal areas that were only bush when I was a child. Springs were drying up." Maathai was shocked to find children suffering from malnutrition. "[M]y community was supposed to be a rich, coffee-growing area." Instead of eating nutritious, traditional foods, such as beans and corn, she explains, people were relying on refined foods such as rice because they need less cooking—and thus less firewood.

The Green Belt Movement

On June 5, 1977, in honor of World Environment Day, Maathai and a few supporters planted seven trees in Nairobi, Kenya's capital. This small tree planting began the Green Belt

456 *Unit 8*

CITIES OF THE WORLD

Jerusalem

The capital of ancient Israel 3,000 years ago, Jerusalem is the capital of modern Israel today. The city's past and present merge in its skyline of ancient ruins, medieval towers, and modern high-rise buildings.

Perhaps more than in any other city, the threads of human history wind through Jerusalem's streets.

Jerusalem is of central importance to people of the Jewish faith. For centuries, Jews living outside the city have turned toward Jerusalem in prayer three times each day. Hundreds of millions of non-Jews also care deeply about the city. Christians cherish it as the place where Jesus lived, preached, and was crucified. Muslims honor the site in Jerusalem where they believe the prophet Muhammad rose to heaven.

An Ancient History

Because it is sacred to so many, several groups have fought for control of Jerusalem in its 5,000-year history. It became the center of Jewish life in 1003 B.C., when King David made it his capital. Thirty-three years later, David's son Solomon built his magnificent Temple there. Although the Babylonians conquered Jerusalem and destroyed the Temple in 586 B.C., the Israelites recovered the city and built a second Temple.

Jerusalem was under Roman control when Jesus lived and preached there. A generation later, in A.D. 66, the Jews revolted against Roman rule. The Romans crushed the revolt, however, and in A.D. 70 destroyed the second Temple. The Temple's only remnant, known today as the Western Wall, remains Judaism's most sacred religious shrine.

After the Romans were finally driven out, Jerusalem came under the control of a series of groups, including the Byzantines, Arabs, Christian Crusaders, Egyptians, and Turks. In the late seventh century, the Arabs built the Dome of the Rock and the al-Aqsa Mosque where the Jewish Temple once stood. Later, during the sixteenth century, the Turks constructed the high stone walls that still surround what today is called the Old City.

Visitors view the Children's Memorial, part of the Holocaust Memorial, or Yad Vashem, in **Jerusalem**. Israel was founded in 1948, three years after the end of World War II and the defeat of Nazi Germany.

Modern Jerusalem

The modern history of Jerusalem begins in the 1860s, when the city's growing Jewish population built the first neighborhoods outside the city's walls. Jerusalem continued to spread outward, especially after the British wrestled the city from the Turks in 1917.

When modern Israel formed in 1948, another round in the struggle for Jerusalem began. Neighboring Arab states attacked Israel, and Jordan occupied the Old City. Israel recovered the western part of the city in 1949 and made it the country's capital. The eastern part, including the Old City, remained under Jordanian control.

Jerusalem was reunified when Israel defeated Jordan, Egypt, and Syria in the Six-Day War in 1967. Today, the Old City is part of a metropolis of more than 500,000 people. Within

392 *Unit 7*

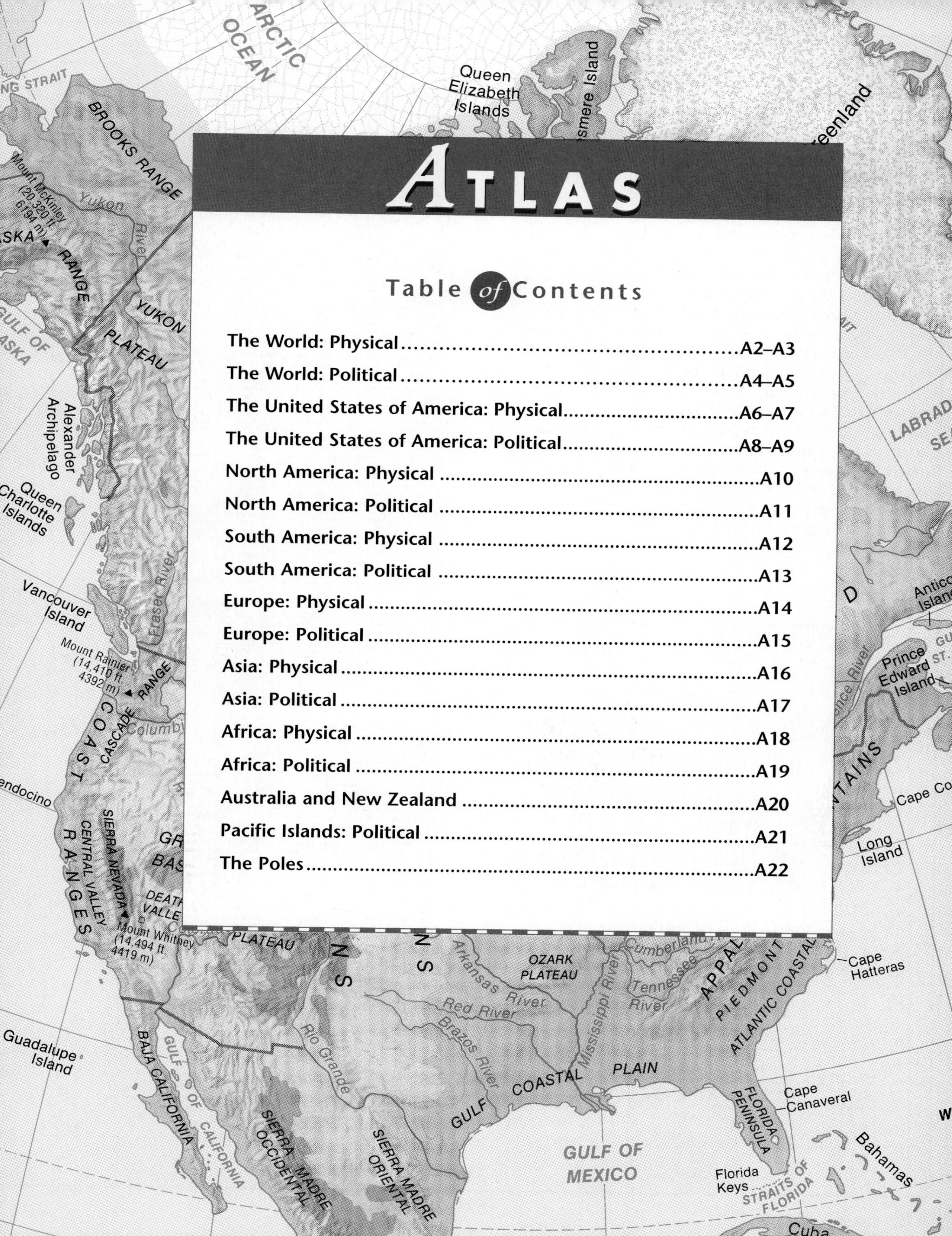
ARCTIC OCEAN
Queen Elizabeth Islands
BROOKS RANGE
Mount McKinley (20,320 ft. 6194 m)
Yukon River
YUKON PLATEAU
Alexander Archipelago
Queen Charlotte Islands
Fraser River
Vancouver Island
Mount Rainier (14,410 ft. 4392 m)
CASCADE RANGE
COAST RANGES
SIERRA NEVADA
CENTRAL VALLEY
Mount Whitney (14,494 ft. 4419 m)
PLATEAU
OZARK PLATEAU
Arkansas River
Red River
Brazos River
Rio Grande
Mississippi River
Tennessee River
PIEDMONT
ATLANTIC COASTAL PLAIN
GULF COASTAL PLAIN
Cape Hatteras
Guadalupe Island
BAJA CALIFORNIA
GULF OF CALIFORNIA
SIERRA MADRE OCCIDENTAL
SIERRA MADRE ORIENTAL
GULF OF MEXICO
FLORIDA PENINSULA
Cape Canaveral
Florida Keys
STRAITS OF FLORIDA
Bahamas
Cuba
Prince Edward Island
Long Island

ATLAS

Table of Contents

The World: Physical

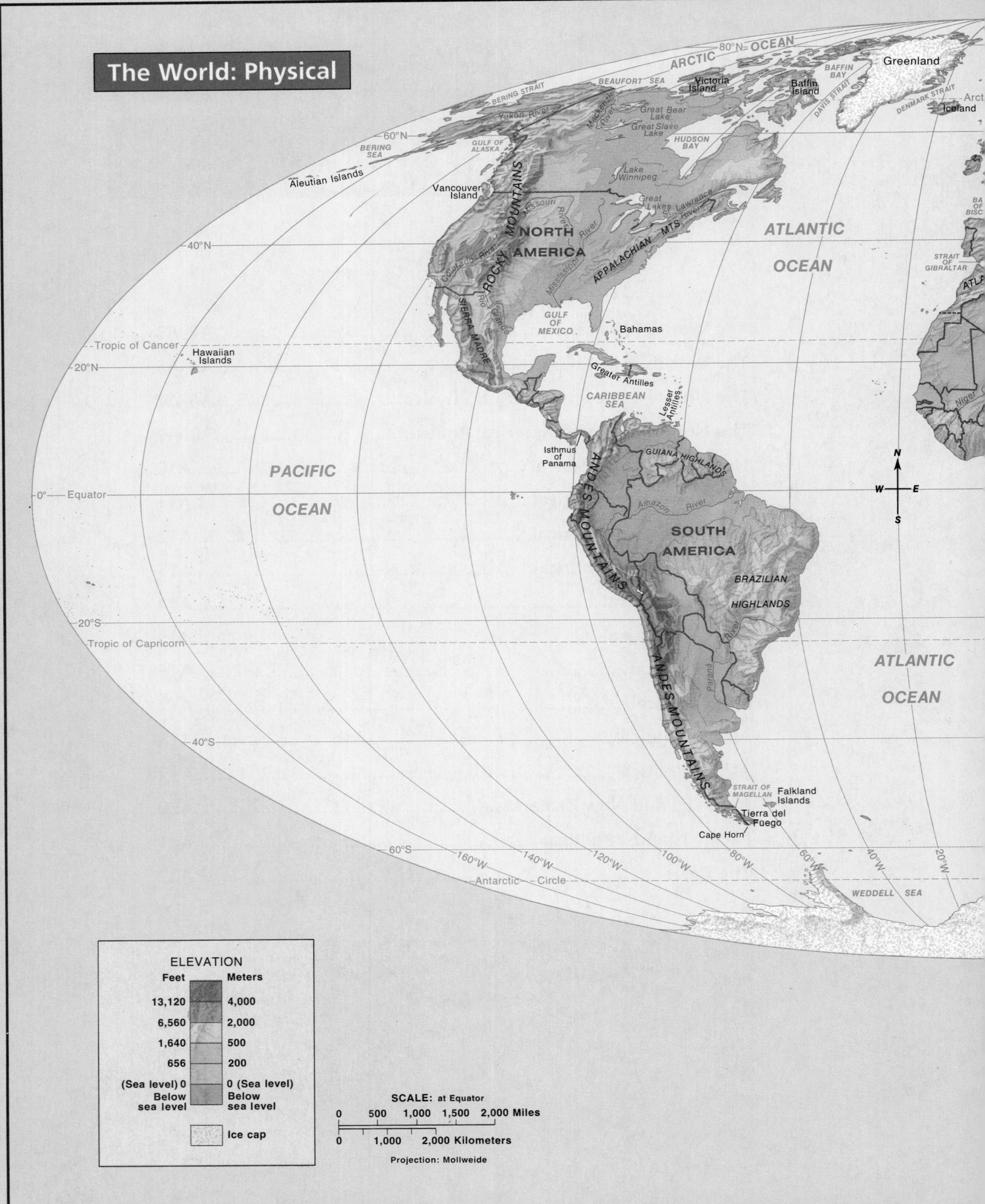

ARCTIC
80°N
OCEAN
BERING STRAIT
BEAUFORT SEA
Victoria Island
BAFFIN BAY
Baffin Island
Greenland
DAVIS STRAIT
DENMARK STRAIT
Iceland
Yukon River
Mackenzie River
Great Bear Lake
Great Slave Lake
HUDSON BAY
60°N
BERING SEA
GULF OF ALASKA
Aleutian Islands
Lake Winnipeg
Vancouver Island
ROCKY MOUNTAINS
Missouri River
Great Lakes
St. Lawrence River
NORTH AMERICA
APPALACHIAN MTS.
Mississippi River
Colorado River
Rio Grande
40°N
ATLANTIC
OCEAN
STRAIT OF GIBRALTAR
SIERRA MADRE
GULF OF MEXICO
Bahamas
Tropic of Cancer
Hawaiian Islands
20°N
Greater Antilles
CARIBBEAN SEA
Lesser Antilles
Niger
Isthmus of Panama
GUIANA HIGHLANDS
PACIFIC
OCEAN
0°
Equator
ANDES MOUNTAINS
Amazon River
SOUTH AMERICA
BRAZILIAN
HIGHLANDS
N
W
E
S
20°S
Tropic of Capricorn
Paraná River
ATLANTIC
OCEAN
40°S
STRAIT OF MAGELLAN
Falkland Islands
Tierra del Fuego
Cape Horn
60°S
160°W
140°W
120°W
100°W
80°W
60°W
40°W
20°W
Antarctic Circle
WEDDELL SEA
ELEVATION
Feet
Meters
13,120
4,000
6,560
2,000
1,640
500
656
200
(Sea level) 0
0 (Sea level)
Below sea level
Below sea level
Ice cap
SCALE: at Equator
0
500
1,000
1,500
2,000 Miles
0
1,000
2,000 Kilometers
Projection: Mollweide

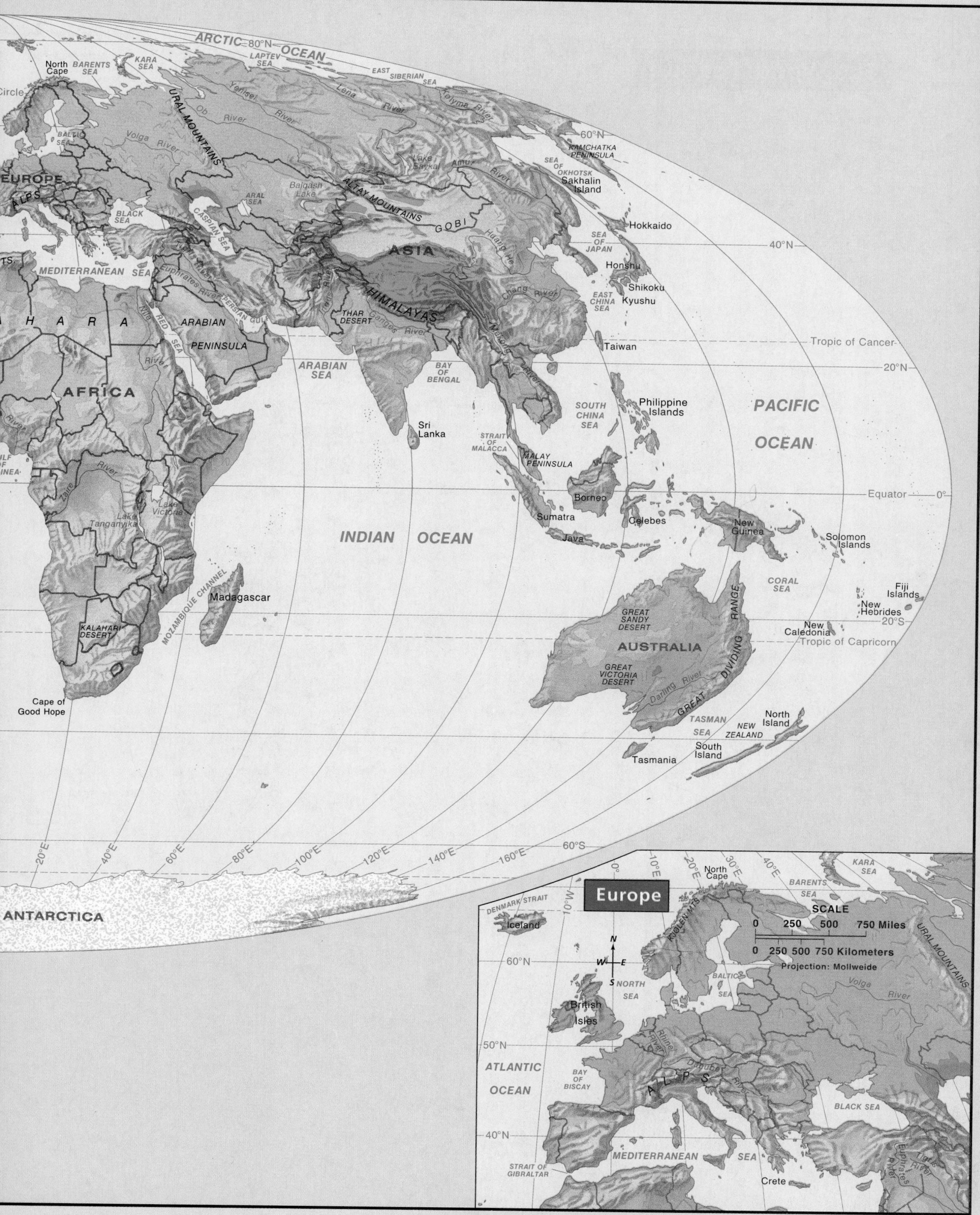
ARCTIC OCEAN
80°N
LAPTEV SEA
KARA SEA
BARENTS SEA
North Cape
EAST SIBERIAN SEA
Yenisei River
Lena River
Kolyma River
Ob River
URAL MOUNTAINS
Volga River
BALTIC SEA
60°N
KAMCHATKA PENINSULA
SEA OF OKHOTSK
Sakhalin Island
Lake Baykal
Amur River
EUROPE
ALPS
Balqash Lake
ARAL SEA
ALTAY MOUNTAINS
GOBI
BLACK SEA
CASPIAN SEA
Hokkaido
SEA OF JAPAN
40°N
ASIA
Huang He
MEDITERRANEAN SEA
Euphrates River
Honshu
Shikoku
Kyushu
EAST CHINA SEA
HIMALAYAS
Chang River
PERSIAN GULF
THAR DESERT
Ganges River
Indus
Nile
RED SEA
ARABIAN PENINSULA
Taiwan
Tropic of Cancer
20°N
ARABIAN SEA
BAY OF BENGAL
Mekong River
AFRICA
SOUTH CHINA SEA
Philippine Islands
PACIFIC OCEAN
Sri Lanka
STRAIT OF MALACCA
MALAY PENINSULA
Borneo
Equator
0°
Zaire
Lake Victoria
Lake Tanganyika
Sumatra
Celebes
New Guinea
INDIAN OCEAN
Java
Solomon Islands
CORAL SEA
Fiji Islands
MOZAMBIQUE CHANNEL
Madagascar
GREAT SANDY DESERT
New Hebrides
KALAHARI DESERT
New Caledonia
20°S
Tropic of Capricorn
AUSTRALIA
GREAT DIVIDING RANGE
GREAT VICTORIA DESERT
Darling River
Cape of Good Hope
TASMAN SEA
NEW ZEALAND
North Island
South Island
Tasmania
60°S
20°E
40°E
60°E
80°E
100°E
120°E
140°E
160°E
ANTARCTICA
Europe
10°W
0°
10°E
20°E
30°E
40°E
DENMARK STRAIT
Iceland
KARA SEA
BARENTS SEA
North Cape
KJOLEN MTS.
SCALE
0 250 500 750 Miles
0 250 500 750 Kilometers
Projection: Mollweide
N
W
E
S
NORTH SEA
BALTIC SEA
URAL MOUNTAINS
Volga River
British Isles
Rhine River
50°N
ATLANTIC OCEAN
BAY OF BISCAY
ALPS
Danube River
BLACK SEA
40°N
MEDITERRANEAN SEA
STRAIT OF GIBRALTAR
Crete
Tigris River
Euphrates River

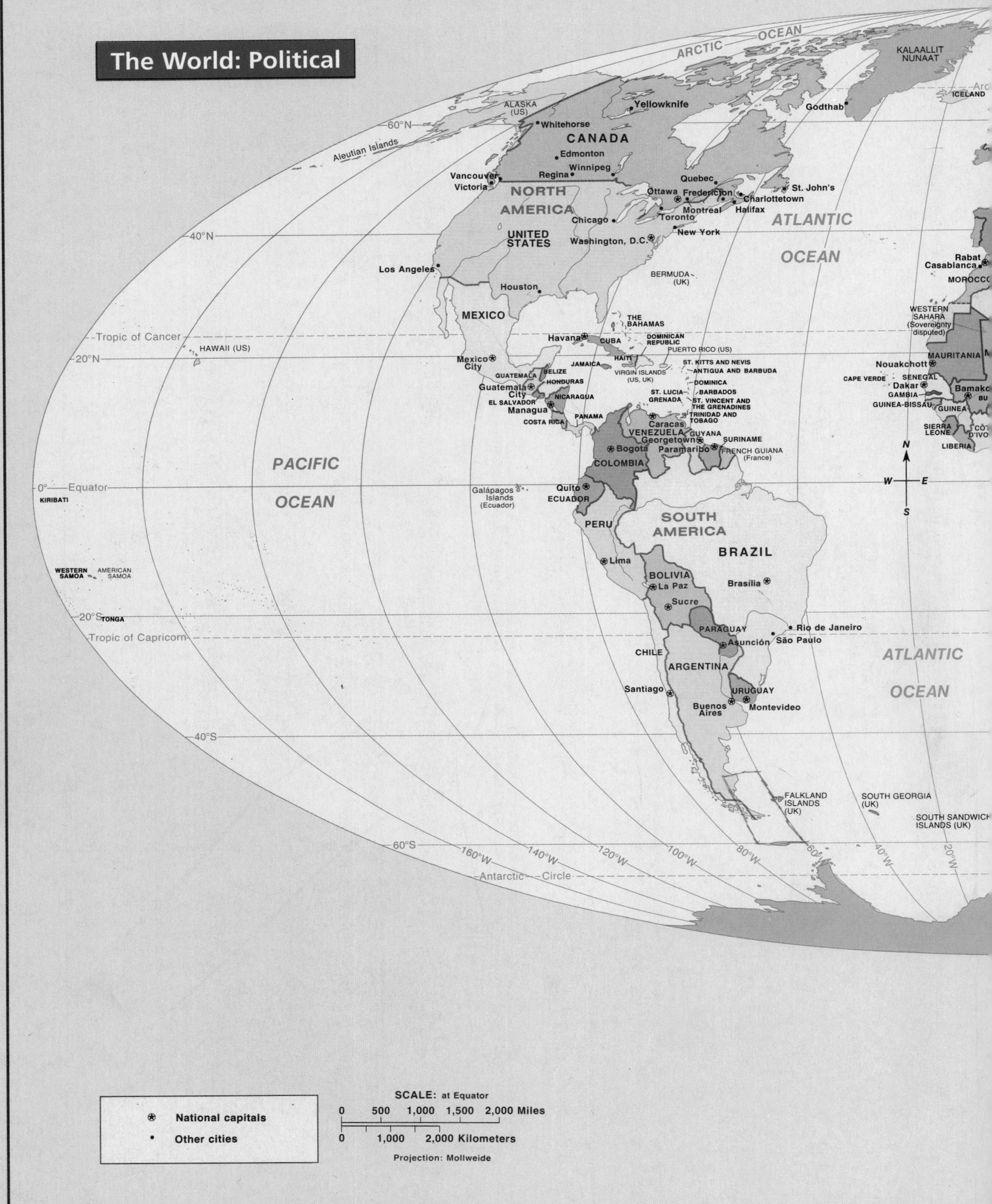
The World: Political
ARCTIC
OCEAN
KALAALLIT NUNAAT
ICELAND
ALASKA (US)
Yellowknife
Godthab
Whitehorse
CANADA
Aleutian Islands
60°N
Edmonton
Regina
Winnipeg
Vancouver
Victoria
Quebec
St. John's
NORTH AMERICA
Ottawa
Fredericton
Charlottetown
Montreal
Halifax
Toronto
Chicago
ATLANTIC
OCEAN
40°N
UNITED STATES
Washington, D.C.
New York
Los Angeles
Houston
BERMUDA (UK)
Rabat
Casablanca
MOROCCO
MEXICO
THE BAHAMAS
WESTERN SAHARA (Sovereignty disputed)
Tropic of Cancer
Havana
CUBA
DOMINICAN REPUBLIC
HAWAII (US)
PUERTO RICO (US)
20°N
Mexico City
HAITI
ST. KITTS AND NEVIS
MAURITANIA
JAMAICA
ANTIGUA AND BARBUDA
Nouakchott
BELIZE
VIRGIN ISLANDS (US, UK)
GUATEMALA
CAPE VERDE
SENEGAL
Dakar
HONDURAS
DOMINICA
Guatemala City
GAMBIA
Bamako
ST. LUCIA
BARBADOS
EL SALVADOR
NICARAGUA
GRENADA
ST. VINCENT AND THE GRENADINES
GUINEA-BISSAU
GUINEA
Managua
TRINIDAD AND TOBAGO
PANAMA
COSTA RICA
Caracas
SIERRA LEONE
VENEZUELA
GUYANA
Georgetown
SURINAME
LIBERIA
PACIFIC
Bogotá
Paramaribo
FRENCH GUIANA (France)
N
COLOMBIA
W
E
S
0°
Equator
Galápagos Islands (Ecuador)
Quito
KIRIBATI
OCEAN
ECUADOR
SOUTH AMERICA
PERU
BRAZIL
Lima
WESTERN SAMOA
AMERICAN SAMOA
BOLIVIA
La Paz
Brasília
Sucre
20°S
TONGA
PARAGUAY
Rio de Janeiro
Tropic of Capricorn
Asunción
São Paulo
CHILE
ATLANTIC
ARGENTINA
OCEAN
URUGUAY
Santiago
Buenos Aires
Montevideo
40°S
FALKLAND ISLANDS (UK)
SOUTH GEORGIA (UK)
SOUTH SANDWICH ISLANDS (UK)
60°S
160°W
140°W
120°W
100°W
80°W
60°W
40°W
20°W
Antarctic Circle
National capitals
Other cities
SCALE: at Equator
0
500
1,000
1,500
2,000 Miles
0
1,000
2,000 Kilometers
Projection: Mollweide

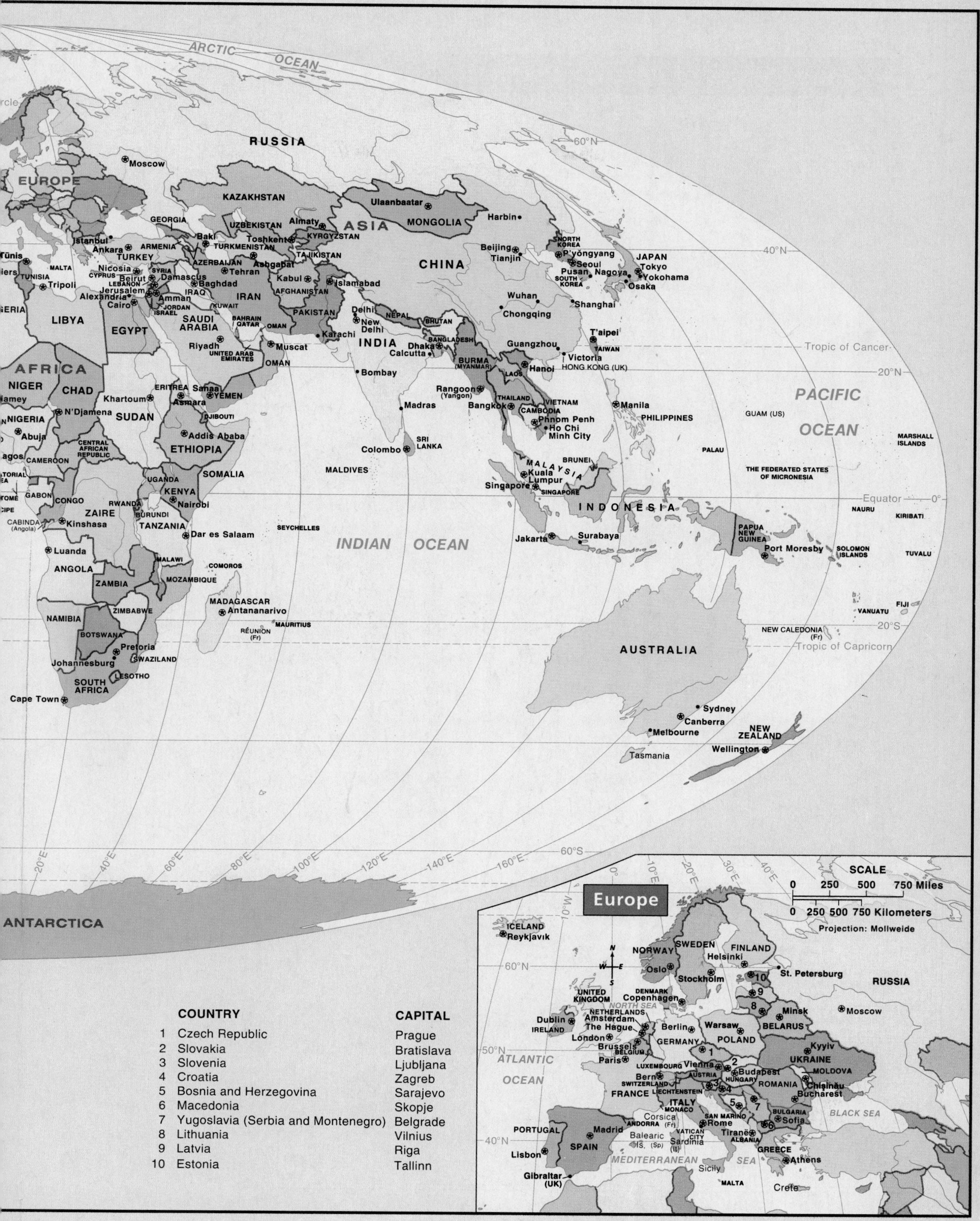
ARCTIC OCEAN
RUSSIA
Moscow
EUROPE
KAZAKHSTAN
GEORGIA
UZBEKISTAN
Almaty
ASIA
MONGOLIA
Ulaanbaatar
Harbin
KYRGYZSTAN
Toshkent
Baki
ARMENIA
TURKMENISTAN
Istanbul
Ankara
TURKEY
AZERBAIJAN
Ashgabat
TAJIKISTAN
CHINA
Beijing
Tianjin
NORTH KOREA
P'yŏngyang
Seoul
SOUTH KOREA
Pusan
JAPAN
Tokyo
Yokohama
Nagoya
Osaka
Tunis
MALTA
TUNISIA
Tripoli
Nicosia
CYPRUS
SYRIA
Beirut
LEBANON
Damascus
Jerusalem
Tehran
Baghdad
IRAQ
IRAN
Kabul
AFGHANISTAN
Islamabad
Alexandria
Cairo
Amman
JORDAN
ISRAEL
KUWAIT
Wuhan
Shanghai
Chongqing
LIBYA
EGYPT
SAUDI ARABIA
BAHRAIN
QATAR
OMAN
PAKISTAN
Delhi
New Delhi
NEPAL
BHUTAN
Riyadh
UNITED ARAB EMIRATES
Muscat
Karachi
INDIA
BANGLADESH
Dhaka
Calcutta
Guangzhou
Victoria
HONG KONG (UK)
T'aipei
TAIWAN
Tropic of Cancer
AFRICA
NIGER
CHAD
ERITREA
Sanaa
YEMEN
Asmara
Bombay
BURMA (MYANMAR)
LAOS
Hanoi
Khartoum
SUDAN
N'Djamena
NIGERIA
Abuja
DJIBOUTI
Rangoon (Yangon)
Bangkok
THAILAND
VIETNAM
CAMBODIA
Phnom Penh
Ho Chi Minh City
Madras
Manila
PHILIPPINES
GUAM (US)
PACIFIC OCEAN
CENTRAL AFRICAN REPUBLIC
Addis Ababa
ETHIOPIA
CAMEROON
SRI LANKA
Colombo
PALAU
MARSHALL ISLANDS
SOMALIA
MALDIVES
BRUNEI
MALAYSIA
Kuala Lumpur
Singapore
SINGAPORE
THE FEDERATED STATES OF MICRONESIA
UGANDA
KENYA
Nairobi
GABON
CONGO
RWANDA
BURUNDI
ZAIRE
Kinshasa
CABINDA (Angola)
TANZANIA
Dar es Salaam
SEYCHELLES
INDONESIA
Equator
NAURU
KIRIBATI
Jakarta
Surabaya
PAPUA NEW GUINEA
Port Moresby
SOLOMON ISLANDS
TUVALU
INDIAN OCEAN
Luanda
ANGOLA
MALAWI
COMOROS
ZAMBIA
MOZAMBIQUE
ZIMBABWE
MADAGASCAR
Antananarivo
NAMIBIA
MAURITIUS
RÉUNION (Fr)
BOTSWANA
Pretoria
SWAZILAND
Johannesburg
LESOTHO
SOUTH AFRICA
Cape Town
VANUATU
FIJI
NEW CALEDONIA (Fr)
Tropic of Capricorn
AUSTRALIA
Sydney
Canberra
Melbourne
Tasmania
NEW ZEALAND
Wellington
ANTARCTICA
60°N
40°N
20°N
0°
20°S
60°S
20°E
40°E
60°E
80°E
100°E
120°E
140°E
160°E
Europe
SCALE
0 250 500 750 Miles
0 250 500 750 Kilometers
Projection: Mollweide
ICELAND
Reykjavík
NORWAY
SWEDEN
FINLAND
Helsinki
Oslo
Stockholm
St. Petersburg
RUSSIA
DENMARK
Copenhagen
UNITED KINGDOM
NORTH SEA
NETHERLANDS
Amsterdam
The Hague
Minsk
BELARUS
Moscow
Dublin
IRELAND
London
Berlin
Warsaw
POLAND
GERMANY
Brussels
BELGIUM
Kyyiv
UKRAINE
Paris
LUXEMBOURG
Vienna
Budapest
HUNGARY
AUSTRIA
MOLDOVA
ATLANTIC OCEAN
Bern
SWITZERLAND
LIECHTENSTEIN
ROMANIA
Chisinău
Bucharest
FRANCE
ITALY
MONACO
BULGARIA
Sofia
BLACK SEA
PORTUGAL
Madrid
SPAIN
ANDORRA
Corsica (Fr)
SAN MARINO
Rome
VATICAN CITY
Tirane
ALBANIA
Balearic Is. (Sp)
Sardinia (It)
Lisbon
GREECE
Athens
MEDITERRANEAN SEA
Sicily
Gibraltar (UK)
MALTA
Crete
COUNTRY | CAPITAL
1 Czech Republic | Prague
2 Slovakia | Bratislava
3 Slovenia | Ljubljana
4 Croatia | Zagreb
5 Bosnia and Herzegovina | Sarajevo
6 Macedonia | Skopje
7 Yugoslavia (Serbia and Montenegro) | Belgrade
8 Lithuania | Vilnius
9 Latvia | Riga
10 Estonia | Tallinn

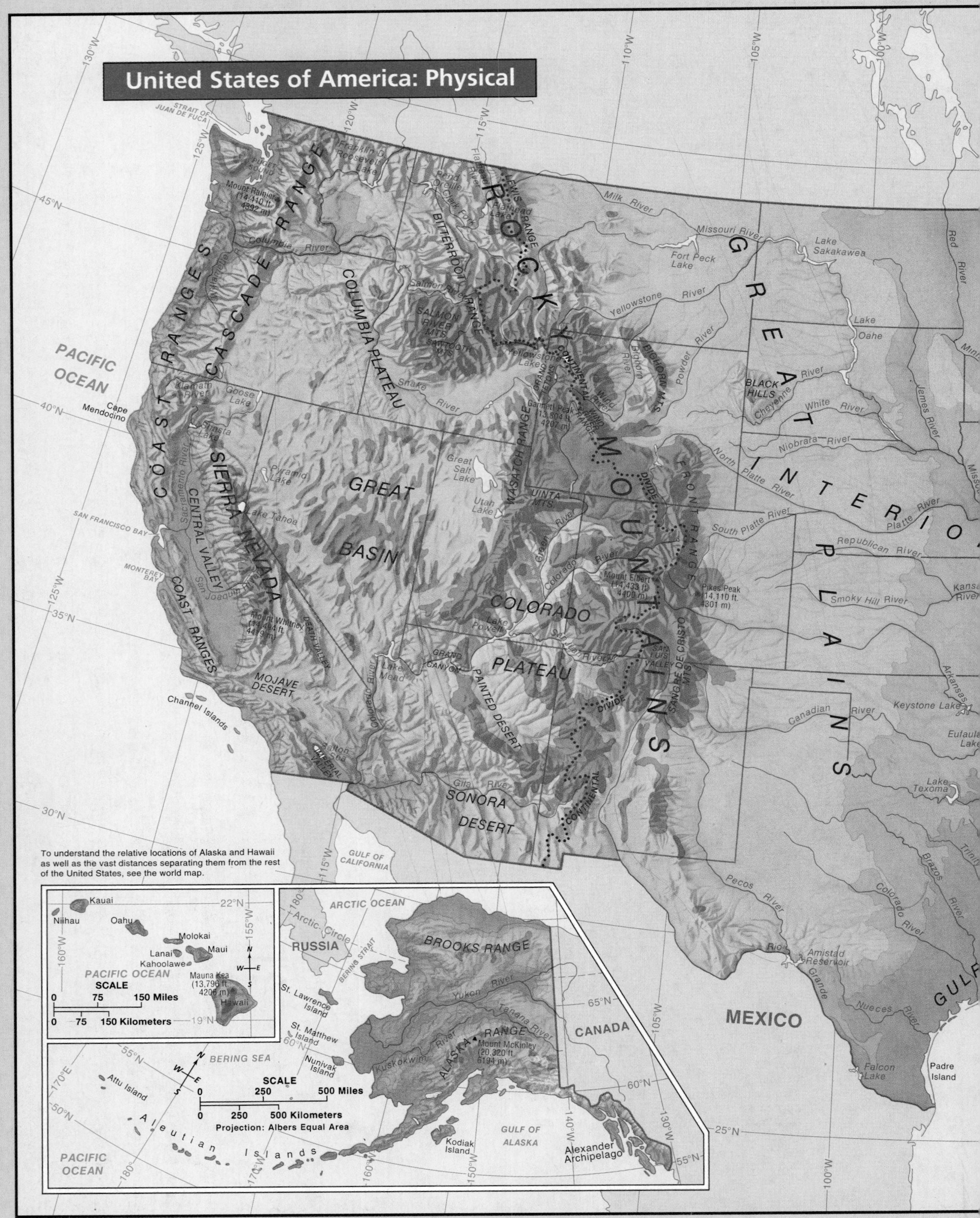

United States of America: Physical
PACIFIC OCEAN
STRAIT OF JUAN DE FUCA
PUGET SOUND
Mount Rainier (14,410 ft. 4392 m)
COAST RANGES
CASCADE RANGE
Columbia River
Franklin D. Roosevelt Lake
Cape Mendocino
Klamath River
Goose Lake
Shasta Lake
SIERRA NEVADA
Sacramento River
CENTRAL VALLEY
Pyramid Lake
Lake Tahoe
SAN FRANCISCO BAY
MONTEREY BAY
San Joaquin River
COAST RANGES
Mount Whitney (14,494 ft. 4418 m)
DEATH VALLEY
MOJAVE DESERT
Channel Islands
Salton Sea
IMPERIAL VALLEY
COLUMBIA PLATEAU
Snake River
SALMON RIVER MTS.
SAWTOOTH MTS.
BITTERROOT RANGE
Salmon River
Clark Fork
Flathead Lake
LEWIS RANGE
GREAT BASIN
Great Salt Lake
Utah Lake
WASATCH RANGE
UINTA MTS.
ROCKY MOUNTAINS
CONTINENTAL DIVIDE
Yellowstone Lake
GRAND TETONS
Gannett Peak (13,804 ft. 4207 m)
WIND RIVER RANGE
BIGHORN MTS.
Bighorn River
Powder River
Milk River
Missouri River
Fort Peck Lake
Lake Sakakawea
Yellowstone River
Lake Oahe
BLACK HILLS
Cheyenne River
White River
James River
Niobrara River
North Platte River
Platte River
South Platte River
Republican River
Smoky Hill River
Kansas River
Arkansas River
FRONT RANGE
Mount Elbert (14,433 ft. 4400 m)
Pikes Peak (14,110 ft. 4301 m)
SAN LUIS VALLEY
SANGRE DE CRISTO MTS.
Green River
Colorado River
COLORADO PLATEAU
Lake Powell
San Juan River
GRAND CANYON
Lake Mead
PAINTED DESERT
Gila River
SONORA DESERT
GULF OF CALIFORNIA
GREAT INTERIOR PLAINS
Red River
Canadian River
Keystone Lake
Eufaula Lake
Lake Texoma
Trinity
Brazos River
Colorado River
Pecos River
Rio Grande
Amistad Reservoir
Nueces River
Falcon Lake
Padre Island
MEXICO
GULF
To understand the relative locations of Alaska and Hawaii as well as the vast distances separating them from the rest of the United States, see the world map.
Kauai
Niihau
Oahu
Molokai
Lanai
Maui
Kahoolawe
Mauna Kea (13,796 ft. 4205 m)
Hawaii
PACIFIC OCEAN
SCALE
0 75 150 Miles
0 75 150 Kilometers
ARCTIC OCEAN
Arctic Circle
RUSSIA
BERING STRAIT
BROOKS RANGE
Yukon River
St. Lawrence Island
St. Matthew Island
Nunivak Island
Tanana River
ALASKA RANGE
Mount McKinley (20,320 ft. 6194 m)
Kuskokwim River
CANADA
BERING SEA
SCALE
0 250 500 Miles
0 250 500 Kilometers
Projection: Albers Equal Area
Attu Island
Aleutian Islands
PACIFIC OCEAN
Kodiak Island
GULF OF ALASKA
Alexander Archipelago

CANADA
ATLANTIC OCEAN
GULF OF MEXICO
THE BAHAMAS
CUBA
Lake Superior
Isle Royale
MESABI RANGE
Lake Huron
Lake Michigan
Lake Ontario
Lake Erie
Finger Lakes
ADIRONDACK MTS.
CATSKILL MTS.
GREEN MTS.
WHITE MTS.
LONGFELLOW MTS.
Lake Champlain
St. Lawrence River
St. Lawrence Seaway
St. John River
Cape Cod
LONG ISLAND SOUND
Long Island
DELAWARE BAY
CHESAPEAKE BAY
PAMLICO SOUND
Cape Hatteras
PLATEAU
ALLEGHENY
APPALACHIAN
MOUNTAINS
BLUE RIDGE MOUNTAINS
GREAT SMOKY MTS.
CUMBERLAND PLATEAU
PIEDMONT
ATLANTIC COASTAL PLAIN
COASTAL PLAIN
PLAINS
Lake of the Ozarks
OUACHITA MTS.
Mississippi River
Wisconsin River
Des Moines River
Illinois River
Wabash River
Ohio River
Scioto River
Allegheny River
Monongahela River
Kanawha River
Susquehanna River
Delaware River
Hudson R.
Connecticut River
Potomac River
James River
Roanoke River
Cumberland River
Tennessee River
Lake Barkley
Kentucky Lake
White River
Tombigbee River
Pearl River
Alabama River
Coosa River
Chattahoochee River
Oconee River
Savannah River
Altamaha River
Sea Islands
Okefenokee Swamp
Chandeleur Islands
Mississippi Delta
FLORIDA PENINSULA
Cape Canaveral
Lake Okeechobee
The Everglades
Cape Sable
Florida Keys
STRAITS OF FLORIDA
ELEVATION
Feet
Meters
13,120
4,000
6,560
2,000
1,640
500
656
200
(Sea level) 0
0 (Sea level)
Below sea level
Below sea level
Ice cap
SCALE
0
250
500 Miles
0
250
500 Kilometers
Projection: Albers Equal Area
N
S
E
W
90°W
85°W
80°W
75°W
70°W
65°W
60°W
50°N
45°N
40°N
35°N
30°N
25°N

United States of America: Political
WASHINGTON
OREGON
IDAHO
MONTANA
NORTH DAKOTA
SOUTH DAKOTA
WYOMING
NEBRASKA
NEVADA
UTAH
COLORADO
KANSAS
CALIFORNIA
ARIZONA
NEW MEXICO
OKLAHOMA
TEXAS
MEXICO
PACIFIC OCEAN
Seattle
Tacoma
Olympia
Spokane
Portland
Salem
Eugene
Boise
Pocatello
Helena
Billings
Bismarck
Fargo
Pierre
Sioux Falls
Casper
Cheyenne
Omaha
Lincoln
Reno
Carson City
Salt Lake City
Provo
Sacramento
Concord
Berkeley
Oakland
San Francisco
Hayward
Fremont
San Jose
Sunnyvale
Stockton
Modesto
Fresno
Bakersfield
Las Vegas
Lakewood
Aurora
Denver
Colorado Springs
Topeka
Wichita
Glendale
Pasadena
Oxnard
Los Angeles
Pomona
Inglewood
San Bernardino
Torrance
Ontario
Riverside
Fullerton
Long Beach
Anaheim
Santa Ana
Garden Grove
Huntington Beach
San Diego
Phoenix
Scottsdale
Mesa
Tucson
Santa Fe
Albuquerque
Amarillo
Oklahoma City
Lubbock
El Paso
Odessa
Abilene
Irving
Fort Worth
Arlington
Waco
Austin
San Antonio
Laredo
Corpus Christi
Padre Island
Strait of Juan de Fuca
Puget Sound
Franklin D. Roosevelt Lake
Pend Oreille
Flathead Lake
Columbia River
Missouri River
Fort Peck Lake
Lake Sakakawea
Yellowstone River
Lake Oahe
Yellowstone Lake
Snake River
Goose Lake
Shasta Lake
Cape Mendocino
Sacramento River
San Francisco Bay
Monterey Bay
San Joaquin R.
Pyramid Lake
Lake Tahoe
Great Salt Lake
Utah Lake
Green River
Platte River
Lake Powell
Lake Mead
Colorado River
Salton Sea
Channel Islands
Gila River
Canadian River
Arkansas River
Keystone Lake
Eufaula Lake
Pecos River
Rio Grande
Amistad Reservoir
Brazos River
Colorado River
Gulf of California
130°W
125°W
120°W
115°W
110°W
105°W
100°W
45°N
40°N
35°N
30°N
25°N
To understand the relative locations of Alaska and Hawaii as well as the vast distances separating them from the rest of the United States, see the world map.
Kauai
Niihau
Oahu
Honolulu
Molokai
Maui
Lanai
Kahoolawe
HAWAII
Hawaii
PACIFIC OCEAN
SCALE
0 75 150 Miles
0 75 150 Kilometers
22°N
19°N
160°W
155°W
ARCTIC OCEAN
Arctic Circle
RUSSIA
BERING STRAIT
Nome
St. Lawrence Island
St. Matthew Island
Nunivak Island
BERING SEA
Yukon River
Fairbanks
ALASKA
CANADA
Anchorage
Juneau
Kodiak Island
GULF OF ALASKA
Alexander Archipelago
Attu Island
Aleutian Islands
PACIFIC OCEAN
SCALE
0 250 500 Miles
0 250 500 Kilometers
Projection: Albers Equal Area
65°N
60°N
55°N
50°N
180°
170°E
174°W
160°W
150°W
140°W
130°W

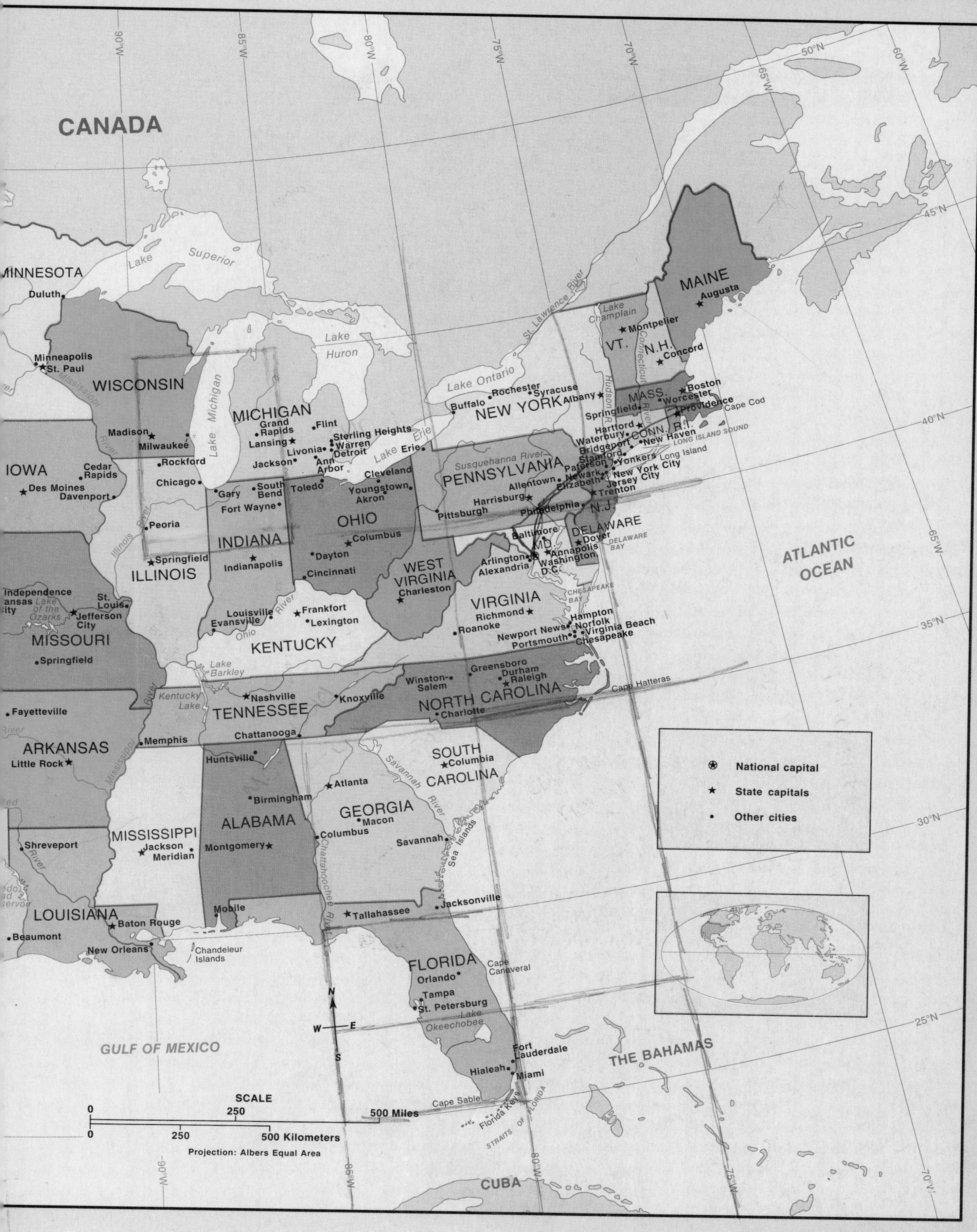

CANADA
ATLANTIC OCEAN
GULF OF MEXICO
THE BAHAMAS
CUBA
National capital
State capitals
Other cities
SCALE
0
250
500 Miles
0
250
500 Kilometers
Projection: Albers Equal Area
MAINE
Augusta
VT.
Montpelier
N.H.
Concord
MASS.
Boston
Worcester
Springfield
R.I.
Providence
CONN.
Hartford
Waterbury
New Haven
Bridgeport
Stamford
Cape Cod
NEW YORK
Albany
Buffalo
Rochester
Syracuse
Yonkers
New York City
Long Island
PENNSYLVANIA
Harrisburg
Philadelphia
Pittsburgh
Allentown
Erie
N.J.
Paterson
Newark
Elizabeth
Jersey City
Trenton
DELAWARE
Dover
MD.
Baltimore
Annapolis
Washington, D.C.
Arlington
Alexandria
VIRGINIA
Richmond
Roanoke
Hampton
Norfolk
Newport News
Portsmouth
Virginia Beach
Chesapeake
WEST VIRGINIA
Charleston
OHIO
Columbus
Cleveland
Toledo
Youngstown
Akron
Dayton
Cincinnati
MICHIGAN
Lansing
Grand Rapids
Flint
Sterling Heights
Warren
Detroit
Livonia
Ann Arbor
Jackson
INDIANA
Indianapolis
Gary
South Bend
Fort Wayne
Evansville
ILLINOIS
Springfield
Chicago
Rockford
Peoria
WISCONSIN
Madison
Milwaukee
MINNESOTA
Duluth
Minneapolis
St. Paul
IOWA
Des Moines
Cedar Rapids
Davenport
MISSOURI
Jefferson City
St. Louis
Independence
Springfield
KENTUCKY
Frankfort
Louisville
Lexington
TENNESSEE
Nashville
Knoxville
Chattanooga
Memphis
NORTH CAROLINA
Raleigh
Greensboro
Durham
Winston-Salem
Charlotte
Cape Hatteras
SOUTH CAROLINA
Columbia
GEORGIA
Atlanta
Macon
Columbus
Savannah
Sea Islands
ALABAMA
Montgomery
Birmingham
Huntsville
Mobile
MISSISSIPPI
Jackson
Meridian
ARKANSAS
Little Rock
Fayetteville
LOUISIANA
Baton Rouge
New Orleans
Shreveport
Beaumont
Chandeleur Islands
FLORIDA
Tallahassee
Jacksonville
Orlando
Tampa
St. Petersburg
Cape Canaveral
Fort Lauderdale
Hialeah
Miami
Cape Sable
Florida Keys
STRAITS OF FLORIDA
Lake Superior
Lake Michigan
Lake Huron
Lake Erie
Lake Ontario
Lake Champlain
Lake Okeechobee
Lake of the Ozarks
Lake Barkley
Kentucky Lake
St. Lawrence River
Hudson R.
Connecticut River
Susquehanna River
Ohio River
Mississippi River
Illinois River
Savannah River
Chattahoochee River
CHESAPEAKE BAY
DELAWARE BAY
LONG ISLAND SOUND
50°N
45°N
40°N
35°N
30°N
25°N
90°W
85°W
80°W
75°W
70°W
65°W
60°W
N
S
E
W

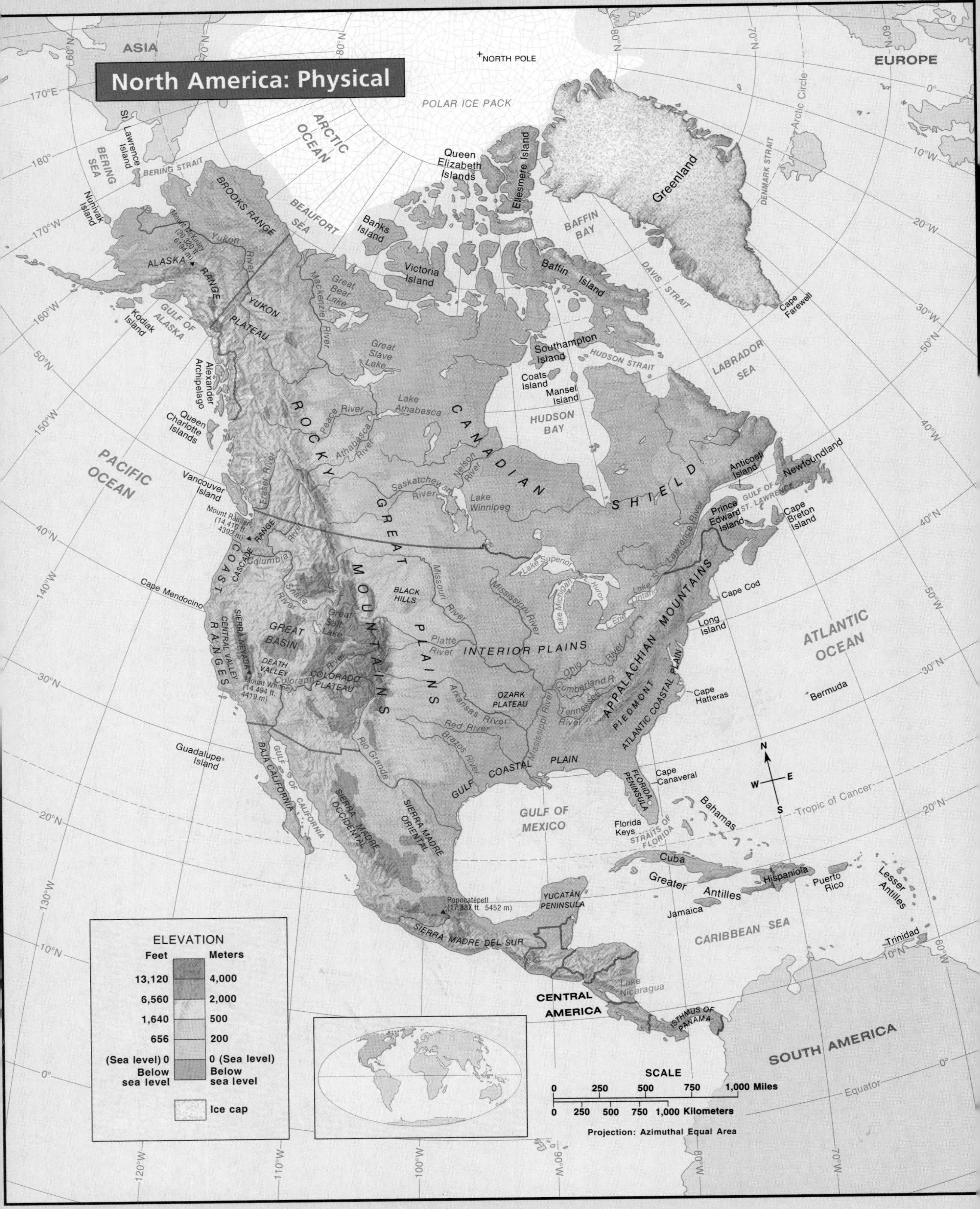
North America: Physical
ASIA
EUROPE
NORTH POLE
POLAR ICE PACK
ARCTIC OCEAN
BERING SEA
BERING STRAIT
St. Lawrence Island
Nunivak Island
BEAUFORT SEA
Queen Elizabeth Islands
Ellesmere Island
Greenland
Banks Island
Victoria Island
BAFFIN BAY
Baffin Island
DAVIS STRAIT
DENMARK STRAIT
Arctic Circle
Cape Farewell
BROOKS RANGE
Mount McKinley (20,320 ft. 6194 m)
ALASKA RANGE
Yukon River
YUKON PLATEAU
GULF OF ALASKA
Kodiak Island
Mackenzie River
Great Bear Lake
Great Slave Lake
Southampton Island
HUDSON STRAIT
LABRADOR SEA
Coats Island
Mansel Island
HUDSON BAY
Alexander Archipelago
Queen Charlotte Islands
ROCKY
Peace River
Lake Athabasca
Athabasca River
CANADIAN SHIELD
Anticosti Island
Newfoundland
GULF OF ST. LAWRENCE
Prince Edward Island
Cape Breton Island
PACIFIC OCEAN
Vancouver Island
Fraser River
Nelson River
Saskatchewan River
Lake Winnipeg
Mount Rainier (14,410 ft. 4392 m)
CASCADE RANGE
COAST RANGES
Columbia River
GREAT PLAINS
MOUNTAINS
Lake Superior
St. Lawrence River
Lake Michigan
Lake Huron
Lake Ontario
Lake Erie
Cape Cod
Cape Mendocino
Snake River
BLACK HILLS
Missouri River
Mississippi River
APPALACHIAN MOUNTAINS
Long Island
SIERRA NEVADA
CENTRAL VALLEY
GREAT BASIN
Great Salt Lake
Platte River
INTERIOR PLAINS
Ohio River
ATLANTIC OCEAN
DEATH VALLEY
Mount Whitney (14,494 ft. 4419 m)
Colorado River
COLORADO PLATEAU
Cumberland R.
Tennessee River
PIEDMONT
ATLANTIC COASTAL PLAIN
Cape Hatteras
Bermuda
OZARK PLATEAU
Arkansas River
Red River
Brazos River
Rio Grande
Guadalupe Island
BAJA CALIFORNIA
GULF OF CALIFORNIA
GULF COASTAL PLAIN
FLORIDA PENINSULA
Cape Canaveral
N
E
W
S
Tropic of Cancer
Bahamas
SIERRA MADRE OCCIDENTAL
SIERRA MADRE ORIENTAL
GULF OF MEXICO
Florida Keys
STRAITS OF FLORIDA
Cuba
Greater Antilles
Hispaniola
Puerto Rico
Lesser Antilles
Jamaica
Popocatépetl (17,887 ft. 5452 m)
YUCATÁN PENINSULA
SIERRA MADRE DEL SUR
CARIBBEAN SEA
Trinidad
Lake Nicaragua
CENTRAL AMERICA
ISTHMUS OF PANAMA
SOUTH AMERICA
Equator
ELEVATION
Feet
Meters
13,120
4,000
6,560
2,000
1,640
500
656
200
(Sea level) 0
0 (Sea level)
Below sea level
Below sea level
Ice cap
SCALE
0 250 500 750 1,000 Miles
0 250 500 750 1,000 Kilometers
Projection: Azimuthal Equal Area
170°E
180°
170°W
160°W
150°W
140°W
130°W
120°W
110°W
100°W
90°W
80°W
70°W
60°W
60°N
70°N
80°N
50°N
40°N
30°N
20°N
10°N
0°
10°W
20°W
30°W
40°W
50°W

North America: Political

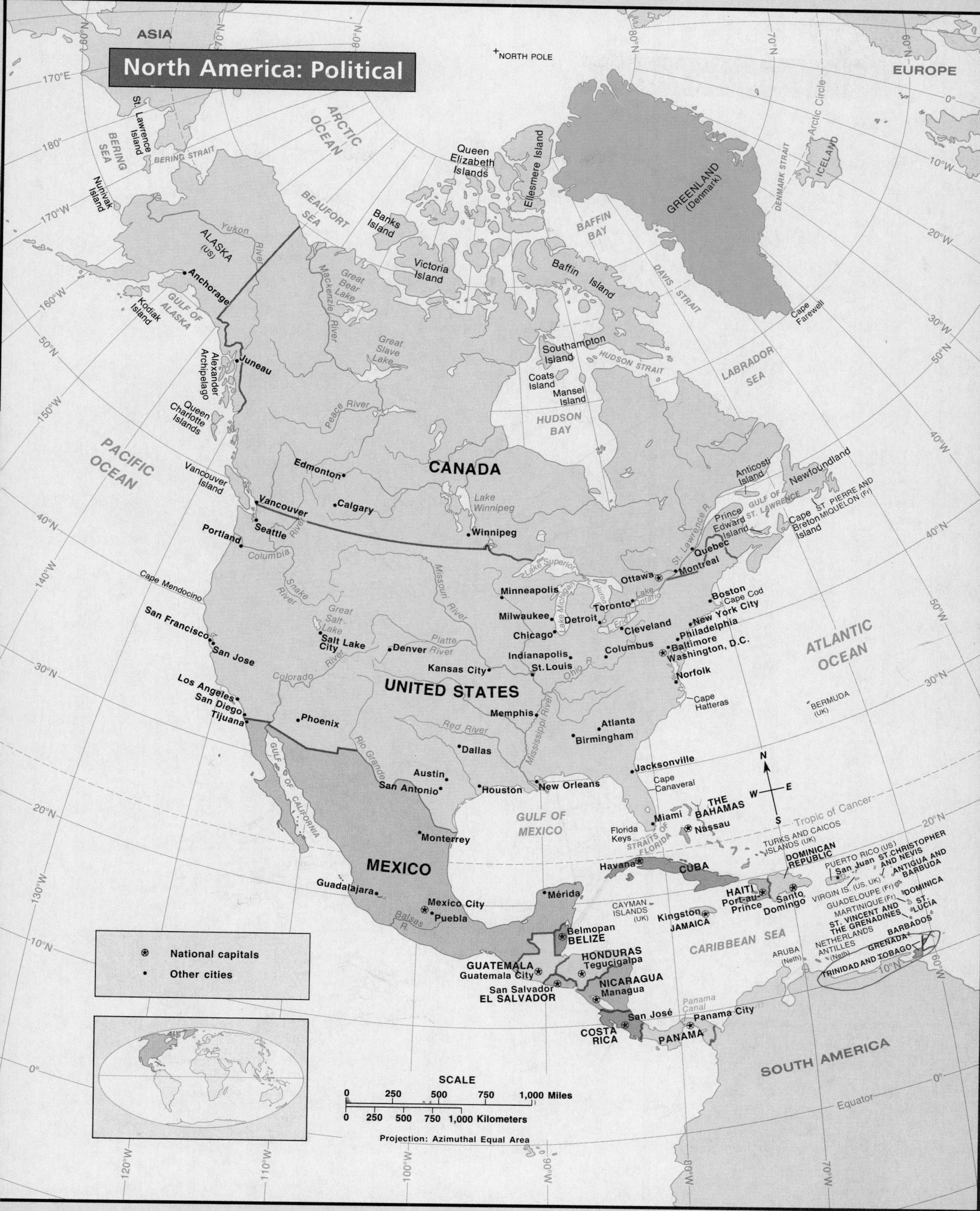

ASIA
NORTH POLE
EUROPE
ARCTIC OCEAN
BERING SEA
BERING STRAIT
St. Lawrence Island
Nunivak Island
Queen Elizabeth Islands
Ellesmere Island
GREENLAND (Denmark)
ICELAND
Arctic Circle
DENMARK STRAIT
BEAUFORT SEA
Banks Island
Victoria Island
BAFFIN BAY
Baffin Island
DAVIS STRAIT
Cape Farewell
ALASKA (US)
Yukon River
Anchorage
GULF OF ALASKA
Kodiak Island
Great Bear Lake
Mackenzie River
Great Slave Lake
Southampton Island
HUDSON STRAIT
Coats Island
Mansel Island
LABRADOR SEA
HUDSON BAY
Alexander Archipelago
Juneau
Queen Charlotte Islands
Peace River
PACIFIC OCEAN
Vancouver Island
Edmonton
CANADA
Anticosti Island
Newfoundland
GULF OF ST. LAWRENCE
ST. PIERRE AND MIQUELON (Fr)
Cape Breton Island
Prince Edward Island
St. Lawrence R.
Lake Winnipeg
Calgary
Vancouver
Seattle
Winnipeg
Portland
Columbia River
Quebec
Montreal
Ottawa
Lake Superior
Lake Huron
Lake Michigan
Lake Ontario
Lake Erie
Cape Mendocino
Snake River
Missouri River
Minneapolis
Toronto
Boston
Cape Cod
Great Salt Lake
Milwaukee
Detroit
New York City
San Francisco
Salt Lake City
Platte River
Chicago
Cleveland
Philadelphia
Baltimore
San Jose
Denver
Indianapolis
Columbus
Washington, D.C.
ATLANTIC OCEAN
Kansas City
St. Louis
Ohio R.
Norfolk
Colorado River
UNITED STATES
Los Angeles
San Diego
Tijuana
Cape Hatteras
BERMUDA (UK)
Memphis
Phoenix
Mississippi River
Atlanta
Red River
Birmingham
Dallas
Rio Grande
Jacksonville
GULF OF CALIFORNIA
Austin
San Antonio
Houston
New Orleans
Cape Canaveral
N
W
E
S
THE BAHAMAS
Miami
Nassau
Tropic of Cancer
GULF OF MEXICO
Florida Keys
STRAITS OF FLORIDA
Monterrey
TURKS AND CAICOS ISLANDS (UK)
DOMINICAN REPUBLIC
PUERTO RICO (US)
San Juan
ST. CHRISTOPHER AND NEVIS
ANTIGUA AND BARBUDA
MEXICO
Havana
CUBA
VIRGIN IS. (US, UK)
GUADELOUPE (Fr)
DOMINICA
Guadalajara
Mérida
HAITI
Port-au-Prince
Santo Domingo
MARTINIQUE (Fr)
ST. LUCIA
Mexico City
Puebla
CAYMAN ISLANDS (UK)
Kingston
JAMAICA
ST. VINCENT AND THE GRENADINES
BARBADOS
Balsas R.
Belmopan
BELIZE
NETHERLANDS ANTILLES (Neth)
GRENADA
CARIBBEAN SEA
ARUBA (Neth)
TRINIDAD AND TOBAGO
HONDURAS
Tegucigalpa
GUATEMALA
Guatemala City
NICARAGUA
Managua
San Salvador
EL SALVADOR
Panama Canal
San José
Panama City
COSTA RICA
PANAMA
SOUTH AMERICA
Equator
National capitals
Other cities
SCALE
0
250
500
750
1,000 Miles
0
250
500
750
1,000 Kilometers
Projection: Azimuthal Equal Area
170°E
180°
170°W
160°W
150°W
140°W
130°W
120°W
110°W
100°W
90°W
80°W
70°W
60°W
50°N
40°N
30°N
20°N
10°N
0°
60°N
70°N
80°N
10°W
20°W
30°W

South America: Physical
CARIBBEAN SEA
CENTRAL AMERICA
Margarita Island
Tobago
Trinidad
Panama Canal
Lake Maracaibo
Orinoco River Delta
LLANOS
Orinoco River
Meta River
GULF OF PANAMA
Angel Falls
GUIANA
HIGHLANDS
ATLANTIC OCEAN
Mount Tolima (18,425 ft. 5616 m)
Devil's Island
Cape Orange
Malpelo Island
Amazon River Delta
Río Negro
Caqueta River
Galápagos Islands
Equator
Mount Chimborazo (20,561 ft. 6267 m)
Japura River
AMAZON
BASIN
Amazon River
GULF OF GUAYAQUIL
ANDES MOUNTAINS
Marañon River
Jurua River
Purus River
Madeira River
Tapajós River
Xingu River
Tocantins River
Parnaíba River
Mount Huascarán (22,205 ft. 6768 m)
Ucayali River
BRAZILIAN
HIGHLANDS
MATO GROSSO PLATEAU
Beni River
Mamore River
Araguaia River
São Francisco River
PACIFIC OCEAN
Lake Titicaca
Ancohuma Peak (20,958 ft. 6388 m)
Lake Poopó
Pilcomayo River
ATACAMA DESERT
CHACO
BRAZILIAN PLATEAU
ANDES
Paraguay River
Paraná River
Salado River
Uruguay River
Tropic of Capricorn
San Ambrosio Island
San Félix Island
Mount Aconcagua (22,834 ft. 6960 m)
Juan Fernández Islands
RÍO DE LA PLATA
PAMPAS
ATLANTIC OCEAN
Salado River
Colorado River
ELEVATION
Feet Meters
13,120 4,000
6,560 2,000
1,640 500
656 200
(Sea level) 0 0 (Sea level)
Below sea level Below sea level
GULF OF SAN MATÍAS
Chiloé Island
Chonos Archipelago
MOUNTAINS
PATAGONIA
GULF OF SAN JORGE
Cape Tres Puntas
BAHÍA GRANDE
SCALE
0 250 500 750 1,000 Miles
0 250 500 750 1,000 Kilometers
Projection: Azimuthal Equal Area
STRAIT OF MAGELLAN
Falkland Islands
Tierra del Fuego
Cape Horn
South Georgia Islands
20°N 10°N 0° 10°S 20°S 30°S 40°S 50°S
100°W 90°W 80°W 70°W 60°W 50°W 40°W 30°W 20°W
N S E W

South America: Political

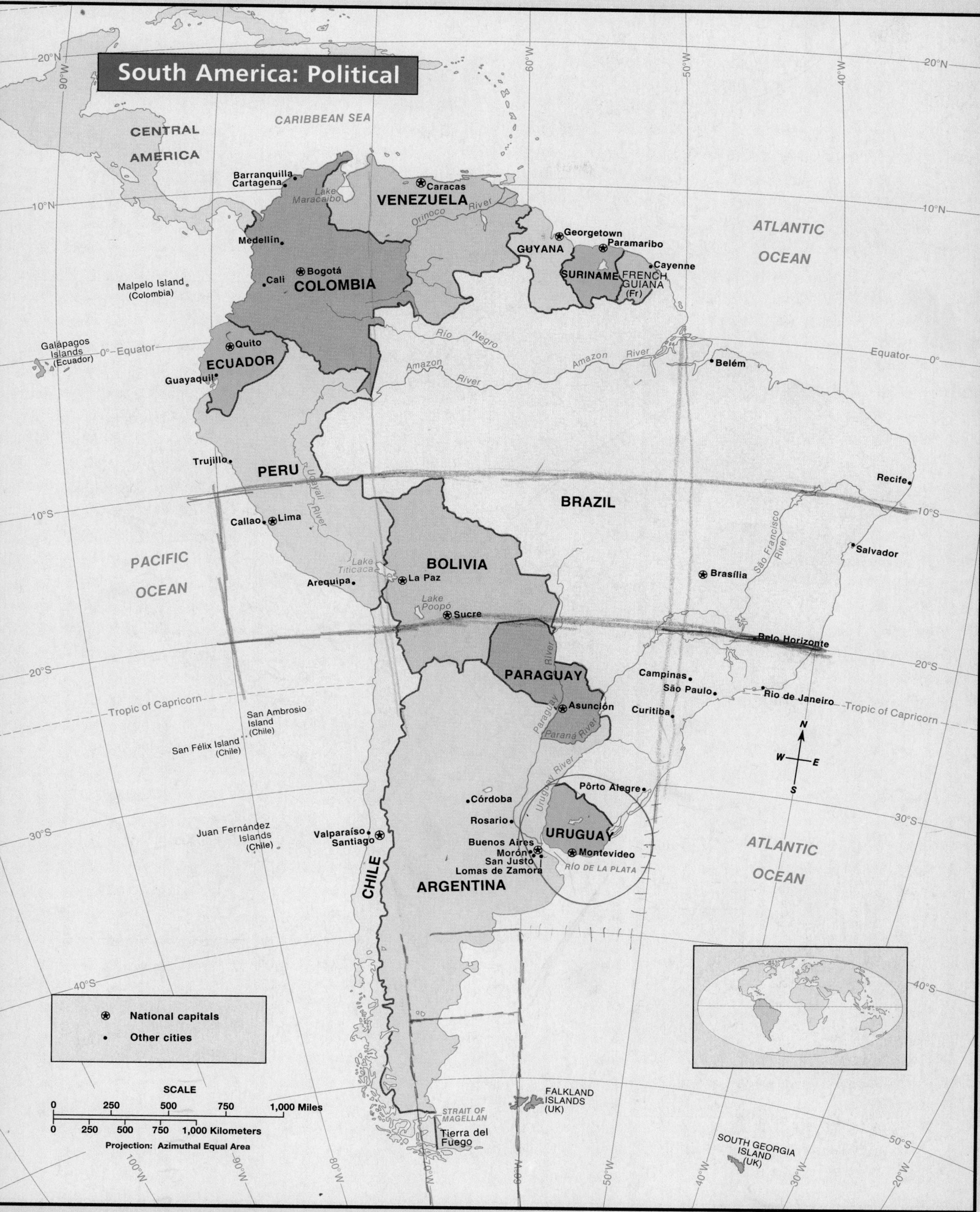

CARIBBEAN SEA
CENTRAL AMERICA
Barranquilla
Cartagena
Lake Maracaibo
Caracas
VENEZUELA
Orinoco River
Medellín
Georgetown
Paramaribo
GUYANA
SURINAME
FRENCH GUIANA (Fr)
Cayenne
ATLANTIC OCEAN
Bogotá
Cali
COLOMBIA
Malpelo Island (Colombia)
Galápagos Islands (Ecuador)
Equator
Quito
ECUADOR
Guayaquil
Río Negro
Amazon River
Belém
Trujillo
PERU
Ucayali River
Recife
BRAZIL
Callao
Lima
PACIFIC OCEAN
Lake Titicaca
BOLIVIA
La Paz
Arequipa
Lake Poopó
Sucre
Brasília
São Francisco River
Salvador
Belo Horizonte
PARAGUAY
Campinas
São Paulo
Rio de Janeiro
Tropic of Capricorn
Asunción
Curitiba
Paraguay River
Paraná River
San Ambrosio Island (Chile)
San Félix Island (Chile)
Uruguay River
Pôrto Alegre
Córdoba
Rosario
URUGUAY
Juan Fernández Islands (Chile)
Valparaíso
Santiago
Buenos Aires
Morón
San Justo
Lomas de Zamora
Montevideo
RÍO DE LA PLATA
CHILE
ARGENTINA
National capitals
Other cities
SCALE
0 250 500 750 1,000 Miles
0 250 500 750 1,000 Kilometers
Projection: Azimuthal Equal Area
FALKLAND ISLANDS (UK)
STRAIT OF MAGELLAN
Tierra del Fuego
SOUTH GEORGIA ISLAND (UK)
20°N
10°N
0°
10°S
20°S
30°S
40°S
50°S
100°W
90°W
80°W
70°W
60°W
50°W
40°W
30°W
20°W
N
S
E
W

Europe: Physical

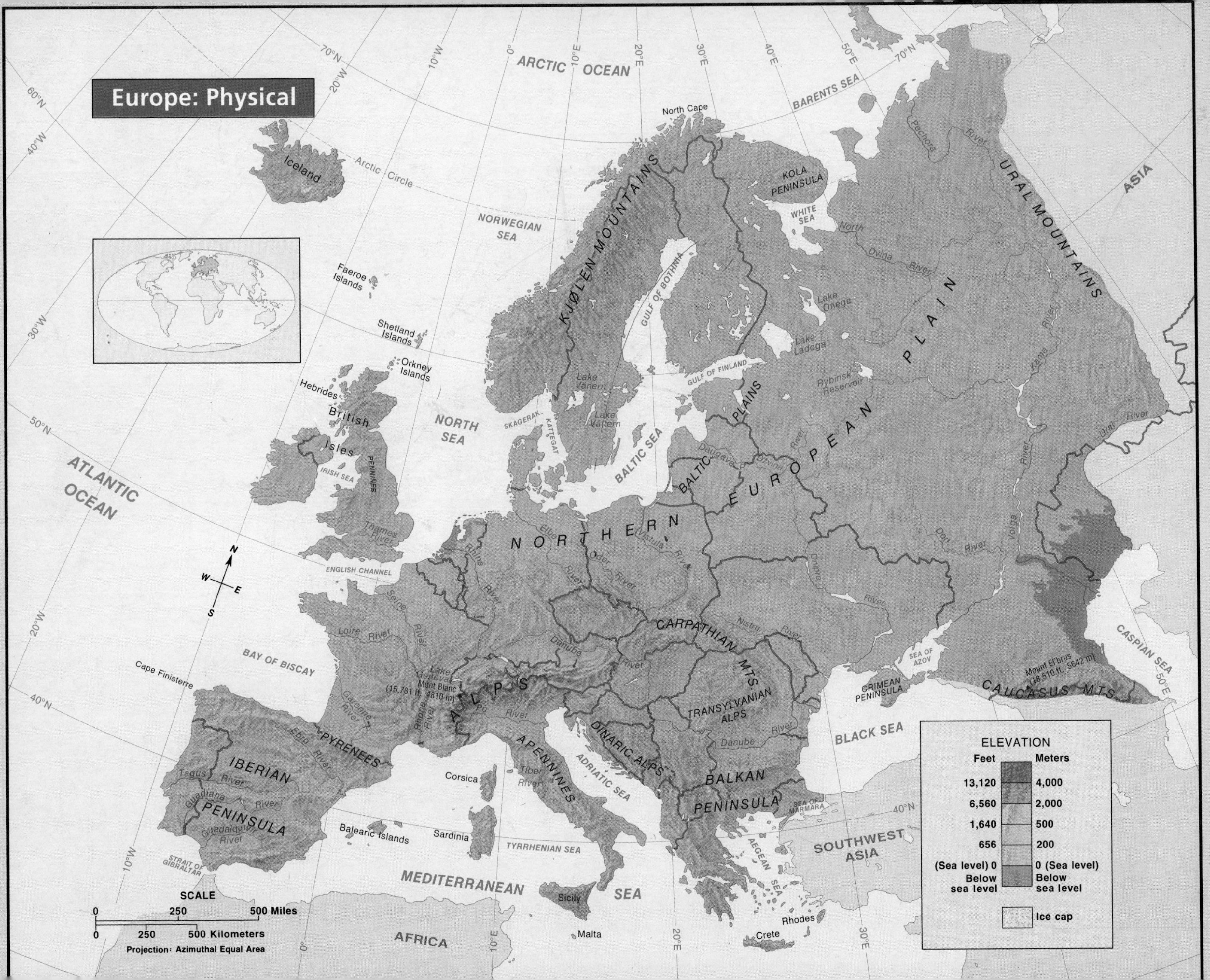

ARCTIC OCEAN
BARENTS SEA
North Cape
Iceland
Arctic Circle
NORWEGIAN SEA
KOLA PENINSULA
WHITE SEA
URAL MOUNTAINS
ASIA
Pechora River
North Dvina River
KJØLEN MOUNTAINS
GULF OF BOTHNIA
Faeroe Islands
Shetland Islands
Orkney Islands
Hebrides
British Isles
IRISH SEA
PENNINES
Thames River
NORTH SEA
SKAGERAK
KATTEGAT
Lake Vänern
Lake Vättern
BALTIC SEA
GULF OF FINLAND
Lake Onega
Lake Ladoga
Rybinsk Reservoir
Kama River
Ural River
Volga River
Don River
PLAINS
BALTIC
NORTHERN EUROPEAN PLAIN
Daugava R.
Dzvina River
ATLANTIC OCEAN
ENGLISH CHANNEL
Elbe River
Rhine River
Oder River
Vistula River
Dnipro River
Seine River
Loire River
Danube
CARPATHIAN MTS.
Nistru River
SEA OF AZOV
CRIMEAN PENINSULA
CASPIAN SEA
Mount El'brus (18,510 ft., 5642 m)
CAUCASUS MTS.
BAY OF BISCAY
Cape Finisterre
Lake Geneva
Mont Blanc (15,781 ft., 4810 m)
ALPS
Po River
Rhone River
Garonne River
TRANSYLVANIAN ALPS
Danube River
BLACK SEA
PYRENEES
Ebro River
IBERIAN PENINSULA
Tagus River
Guadiana River
Guadalquivir River
STRAIT OF GIBRALTAR
Corsica
Tiber River
APENNINES
DINARIC ALPS
ADRIATIC SEA
BALKAN PENINSULA
SEA OF MARMARA
Balearic Islands
Sardinia
TYRRHENIAN SEA
AEGEAN SEA
SOUTHWEST ASIA
MEDITERRANEAN SEA
Sicily
Malta
Rhodes
Crete
AFRICA
N
W
E
S
SCALE
0
250
500 Miles
0
250
500 Kilometers
Projection: Azimuthal Equal Area
ELEVATION
Feet
Meters
13,120
4,000
6,560
2,000
1,640
500
656
200
(Sea level) 0
0 (Sea level)
Below sea level
Below sea level
Ice cap
70°N
60°N
50°N
40°N
40°W
30°W
20°W
10°W
0°
10°E
20°E
30°E
40°E
50°E

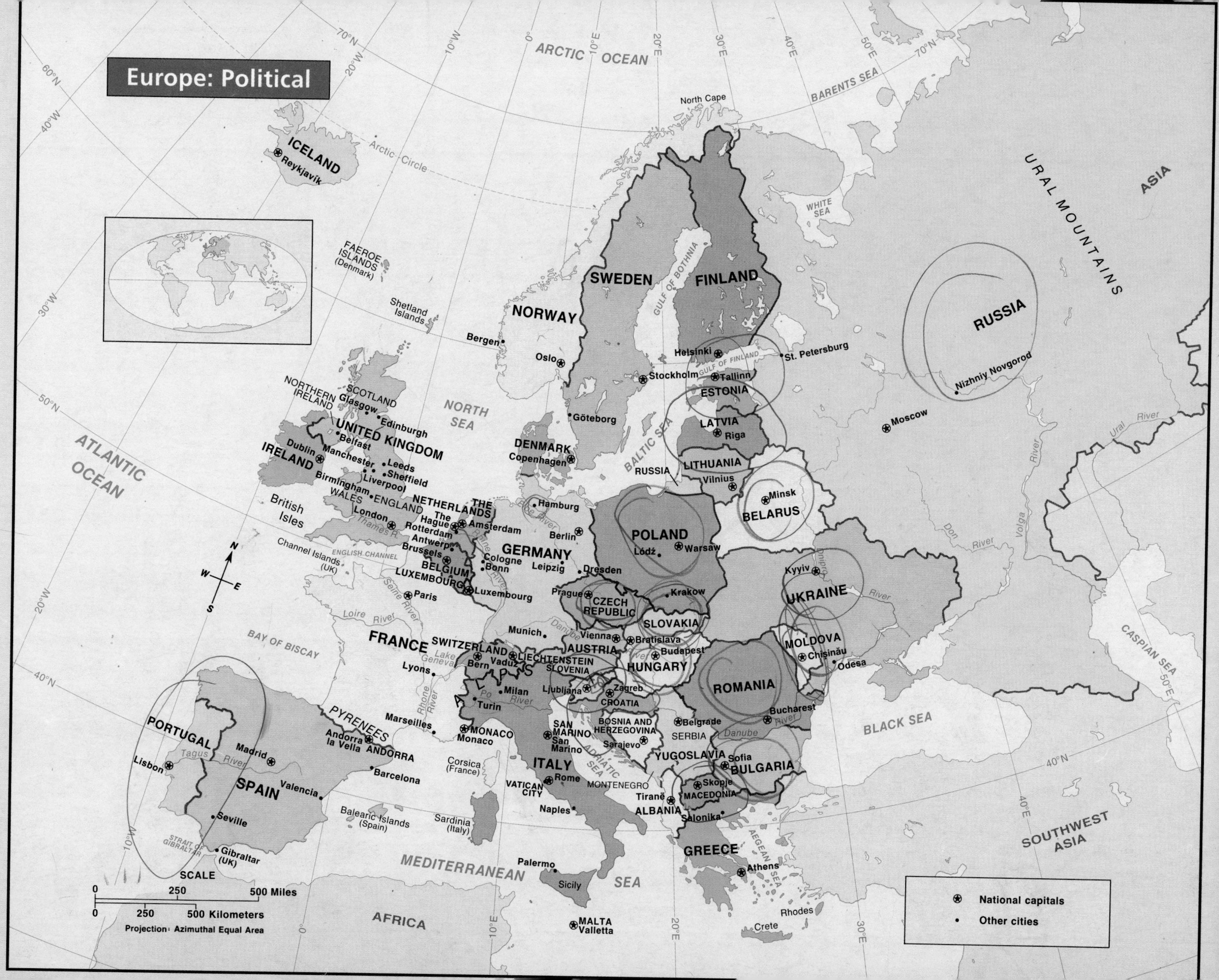

Europe: Political
ARCTIC OCEAN
BARENTS SEA
North Cape
Arctic Circle
ICELAND
Reykjavík
FAEROE ISLANDS (Denmark)
Shetland Islands
WHITE SEA
URAL MOUNTAINS
ASIA
RUSSIA
Nizhniy Novgorod
Moscow
St. Petersburg
SWEDEN
FINLAND
GULF OF BOTHNIA
NORWAY
Bergen
Oslo
Stockholm
Helsinki
GULF OF FINLAND
Tallinn
ESTONIA
LATVIA
Riga
Göteborg
NORTH SEA
BALTIC SEA
RUSSIA
LITHUANIA
Vilnius
Minsk
BELARUS
DENMARK
Copenhagen
NORTHERN IRELAND
SCOTLAND
Glasgow
Edinburgh
UNITED KINGDOM
Belfast
IRELAND
Dublin
Manchester
Leeds
Sheffield
Liverpool
Birmingham
WALES
ENGLAND
London
Thames R.
British Isles
ATLANTIC OCEAN
THE NETHERLANDS
The Hague
Amsterdam
Rotterdam
Antwerp
Brussels
BELGIUM
LUXEMBOURG
Luxembourg
Channel Islands (UK)
ENGLISH CHANNEL
Hamburg
Elbe River
Berlin
GERMANY
Cologne
Bonn
Rhine River
Leipzig
Dresden
POLAND
Lódź
Warsaw
Krakow
Prague
CZECH REPUBLIC
SLOVAKIA
Paris
Seine River
Loire River
BAY OF BISCAY
FRANCE
Munich
Danube
Vienna
Bratislava
Budapest
AUSTRIA
SWITZERLAND
Lake Geneva
Bern
Vaduz
LIECHTENSTEIN
SLOVENIA
HUNGARY
Lyons
Ljubljana
Zagreb
CROATIA
ALPS
Milan
Po River
Turin
Rhone River
Marseilles
MONACO
Monaco
SAN MARINO
San Marino
BOSNIA AND HERZEGOVINA
Sarajevo
Belgrade
SERBIA
YUGOSLAVIA
ADRIATIC SEA
MONTENEGRO
Tiranë
ALBANIA
Skopje
MACEDONIA
Salonika
Sofia
BULGARIA
Danube River
Bucharest
ROMANIA
MOLDOVA
Chişinău
Odesa
UKRAINE
Kyiv
Dnipro River
Don River
Volga River
Ural River
CASPIAN SEA
BLACK SEA
SOUTHWEST ASIA
GREECE
Athens
AEGEAN SEA
Rhodes
Crete
PYRENEES
Andorra la Vella
ANDORRA
Barcelona
PORTUGAL
Lisbon
Tagus River
Madrid
SPAIN
Valencia
Seville
STRAIT OF GIBRALTAR
Gibraltar (UK)
Balearic Islands (Spain)
Corsica (France)
Sardinia (Italy)
ITALY
Rome
VATICAN CITY
Naples
Palermo
Sicily
MEDITERRANEAN SEA
MALTA
Valletta
AFRICA
SCALE
0 250 500 Miles
0 250 500 Kilometers
Projection: Azimuthal Equal Area
N E S W
National capitals
Other cities
70°N
60°N
50°N
40°N
40°W
30°W
20°W
10°W
0°
10°E
20°E
30°E
40°E
50°E

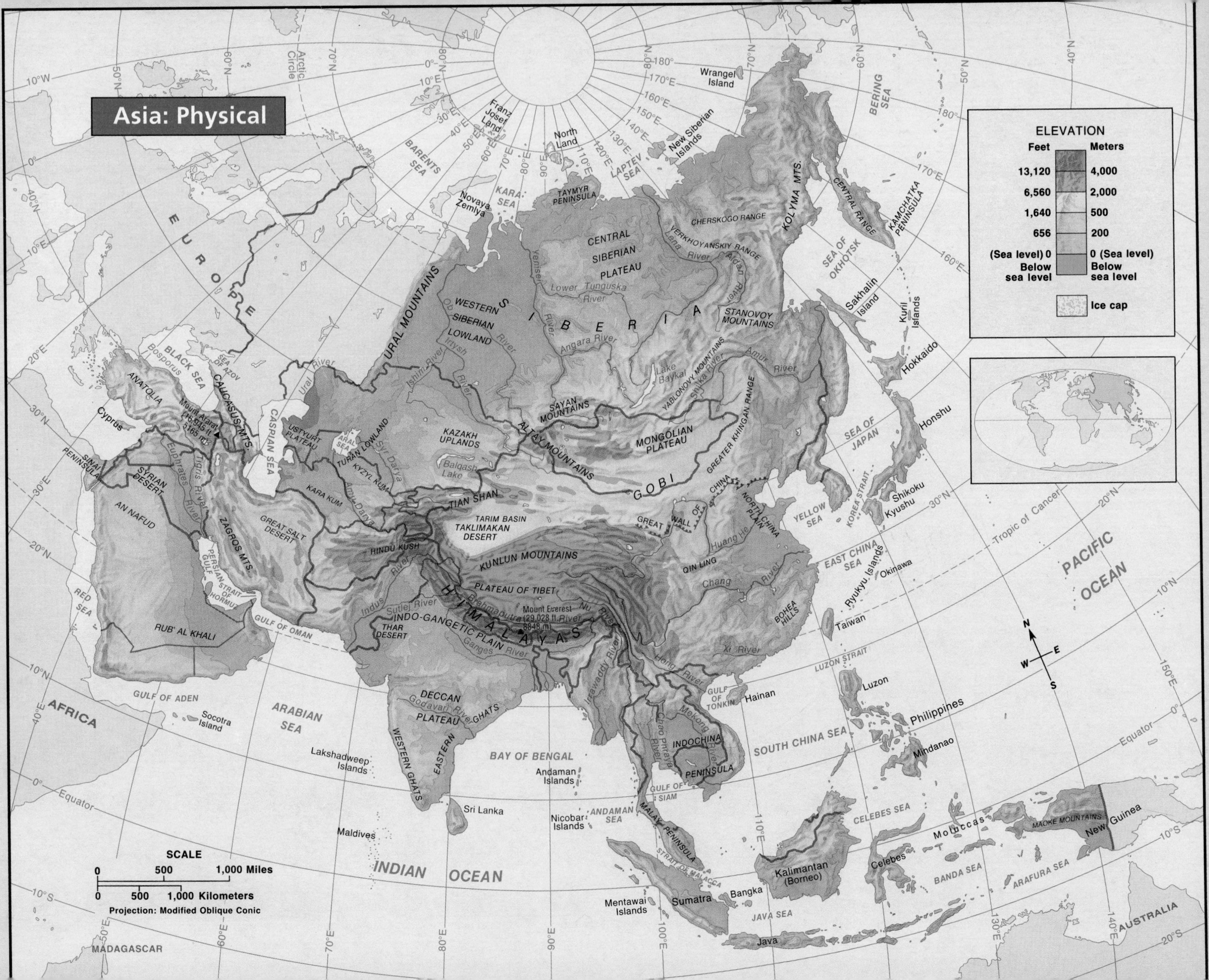
Asia: Physical
ELEVATION
Feet Meters
13,120 4,000
6,560 2,000
1,640 500
656 200
(Sea level) 0 0 (Sea level)
Below sea level Below sea level
Ice cap
SCALE
0 500 1,000 Miles
0 500 1,000 Kilometers
Projection: Modified Oblique Conic
EUROPE
AFRICA
AUSTRALIA
MADAGASCAR
SIBERIA
PACIFIC OCEAN
INDIAN OCEAN
BARENTS SEA
KARA SEA
LAPTEV SEA
BERING SEA
SEA OF OKHOTSK
SEA OF JAPAN
YELLOW SEA
EAST CHINA SEA
SOUTH CHINA SEA
CELEBES SEA
BANDA SEA
ARAFURA SEA
JAVA SEA
ANDAMAN SEA
BAY OF BENGAL
ARABIAN SEA
GULF OF ADEN
RED SEA
PERSIAN GULF
STRAIT OF HORMUZ
GULF OF OMAN
CASPIAN SEA
ARAL SEA
BLACK SEA
SEA OF AZOV
Bosporus
KOREA STRAIT
LUZON STRAIT
GULF OF TONKIN
GULF OF SIAM
STRAIT OF MALACCA
URAL MOUNTAINS
CAUCASUS MTS.
ZAGROS MTS.
HINDU KUSH
HIMALAYAS
KUNLUN MOUNTAINS
TIAN SHAN
ALTAY MOUNTAINS
SAYAN MOUNTAINS
YABLONOVY MOUNTAINS
STANOVOY MOUNTAINS
GREATER KHINGAN RANGE
VERKHOYANSKIY RANGE
CHERSKOGO RANGE
KOLYMA MTS.
CENTRAL RANGE
QIN LING
BOHEA HILLS
MAOKE MOUNTAINS
WESTERN GHATS
EASTERN GHATS
WESTERN SIBERIAN LOWLAND
CENTRAL SIBERIAN PLATEAU
KAZAKH UPLANDS
TURAN LOWLAND
USTYURT PLATEAU
KYZYL KUM
KARA KUM
GREAT SALT DESERT
TARIM BASIN
TAKLIMAKAN DESERT
PLATEAU OF TIBET
MONGOLIAN PLATEAU
GOBI
NORTH CHINA PLAIN
INDO-GANGETIC PLAIN
THAR DESERT
DECCAN PLATEAU
INDOCHINA PENINSULA
MALAY PENINSULA
ANATOLIA
SYRIAN DESERT
AN NAFUD
RUB' AL KHALI
SINAI PENINSULA
KAMCHATKA PENINSULA
TAYMYR PENINSULA
GREAT WALL OF CHINA
Mount Everest (29,028 ft. 8848 m)
Mount Ararat (16,945 ft. 5165 m)
Ob River
Irtysh River
Ishim River
Yenisei River
Lower Tunguska River
Angara River
Lena River
Aldan River
Amur River
Shilka River
Lake Baykal
Balqash Lake
Syr Darya
Amu Darya
Ural River
Tigris River
Euphrates River
Indus River
Sutlej River
Ganges River
Brahmaputra River
Godavari River
Nu River
Irrawaddy River
Mekong River
Chao Phraya River
Hong River
Xi River
Chang
Huang He
Novaya Zemlya
Franz Josef Land
North Land
New Siberian Islands
Wrangel Island
Sakhalin Island
Kuril Islands
Hokkaido
Honshu
Shikoku
Kyushu
Okinawa
Ryukyu Islands
Taiwan
Hainan
Luzon
Philippines
Mindanao
Celebes
Moluccas
New Guinea
Kalimantan (Borneo)
Java
Sumatra
Bangka
Mentawai Islands
Nicobar Islands
Andaman Islands
Sri Lanka
Maldives
Lakshadweep Islands
Socotra Island
Cyprus
Arctic Circle
Tropic of Cancer
Equator
N
S
E
W

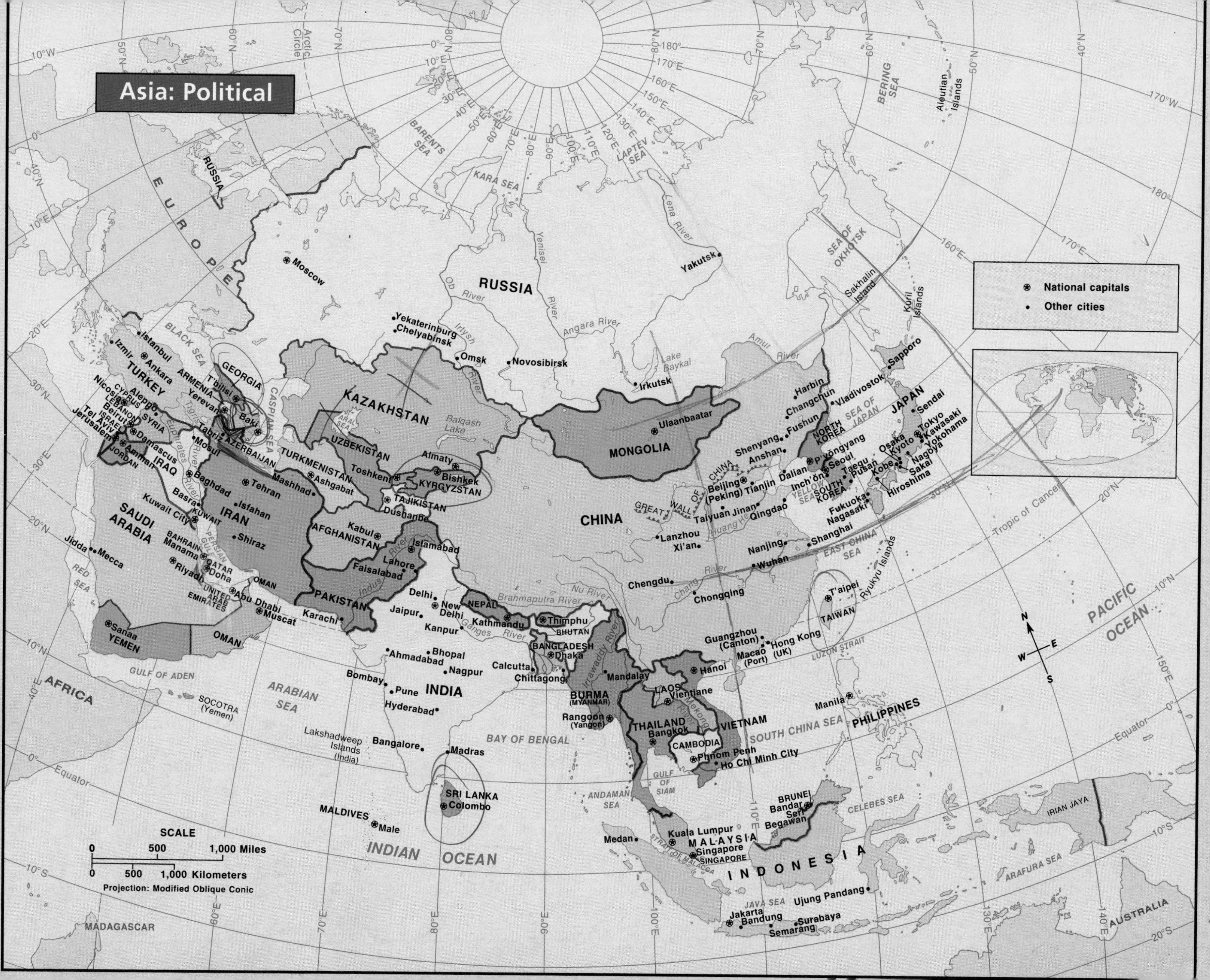

Asia: Political
National capitals
Other cities
SCALE
0 500 1,000 Miles
0 500 1,000 Kilometers
Projection: Modified Oblique Conic
PACIFIC OCEAN
INDIAN OCEAN
ARABIAN SEA
BAY OF BENGAL
SOUTH CHINA SEA
EAST CHINA SEA
YELLOW SEA
SEA OF JAPAN
SEA OF OKHOTSK
BERING SEA
LAPTEV SEA
KARA SEA
BARENTS SEA
CASPIAN SEA
BLACK SEA
RED SEA
PERSIAN GULF
GULF OF ADEN
GULF OF SIAM
ANDAMAN SEA
JAVA SEA
CELEBES SEA
ARAFURA SEA
STRAIT OF MALACCA
LUZON STRAIT
Tropic of Cancer
Equator
Arctic Circle
EUROPE
AFRICA
AUSTRALIA
MADAGASCAR
RUSSIA
KAZAKHSTAN
UZBEKISTAN
TURKMENISTAN
KYRGYZSTAN
TAJIKISTAN
AFGHANISTAN
PAKISTAN
INDIA
NEPAL
BHUTAN
BANGLADESH
BURMA (MYANMAR)
THAILAND
LAOS
CAMBODIA
VIETNAM
MALAYSIA
SINGAPORE
BRUNEI
INDONESIA
PHILIPPINES
TAIWAN
CHINA
MONGOLIA
NORTH KOREA
SOUTH KOREA
JAPAN
SRI LANKA
MALDIVES
IRAN
IRAQ
SYRIA
LEBANON
ISRAEL
JORDAN
CYPRUS
TURKEY
GEORGIA
ARMENIA
AZERBAIJAN
KUWAIT
BAHRAIN
QATAR
UNITED ARAB EMIRATES
OMAN
YEMEN
SAUDI ARABIA
IRIAN JAYA
Moscow
Yekaterinburg
Chelyabinsk
Omsk
Novosibirsk
Irkutsk
Yakutsk
Vladivostok
Ulaanbaatar
Beijing (Peking)
Tianjin
Shanghai
Nanjing
Wuhan
Chongqing
Chengdu
Lanzhou
Xi'an
Taiyuan
Jinan
Qingdao
Shenyang
Anshan
Dalian
Harbin
Changchun
Fushun
Guangzhou (Canton)
Hong Kong (UK)
Macao (Port)
P'yongyang
Seoul
Inch'ŏn
Taegu
Pusan
Tokyo
Yokohama
Kawasaki
Nagoya
Kyoto
Osaka
Kobe
Sakai
Hiroshima
Fukuoka
Nagasaki
Sendai
Sapporo
T'aipei
Manila
Hanoi
Ho Chi Minh City
Phnom Penh
Vientiane
Bangkok
Rangoon (Yangon)
Mandalay
Kuala Lumpur
Singapore
Bandar Seri Begawan
Jakarta
Bandung
Semarang
Surabaya
Ujung Pandang
Medan
Dhaka
Chittagong
Thimphu
Kathmandu
Calcutta
Nagpur
Bhopal
Delhi
New Delhi
Kanpur
Jaipur
Ahmadabad
Bombay
Pune
Hyderabad
Bangalore
Madras
Colombo
Male
Islamabad
Lahore
Faisalabad
Karachi
Kabul
Dushanbe
Toshkent
Bishkek
Almaty
Ashgabat
Tehran
Mashhad
Isfahan
Shiraz
Tabriz
Baghdad
Mosul
Basra
Kuwait City
Manama
Doha
Abu Dhabi
Muscat
Riyadh
Mecca
Jidda
Sanaa
Damascus
Aleppo
Beirut
Nicosia
Amman
Tel Aviv
Jerusalem
Ankara
Istanbul
Izmir
T'bilisi
Yerevan
Baku
Lakshadweep Islands (India)
SOCOTRA (Yemen)
Kuril Islands
Aleutian Islands
Sakhalin Island
Ryukyu Islands
Lena River
Yenisei River
Angara River
Ob River
Irtysh River
Amur River
Huang He
Chang Jiang
Mekong River
Irrawaddy River
Brahmaputra River
Ganges River
Indus River
Nu River
Tigris River
Euphrates River
Lake Baykal
Balqash Lake
ARAL SEA
GREAT WALL OF CHINA

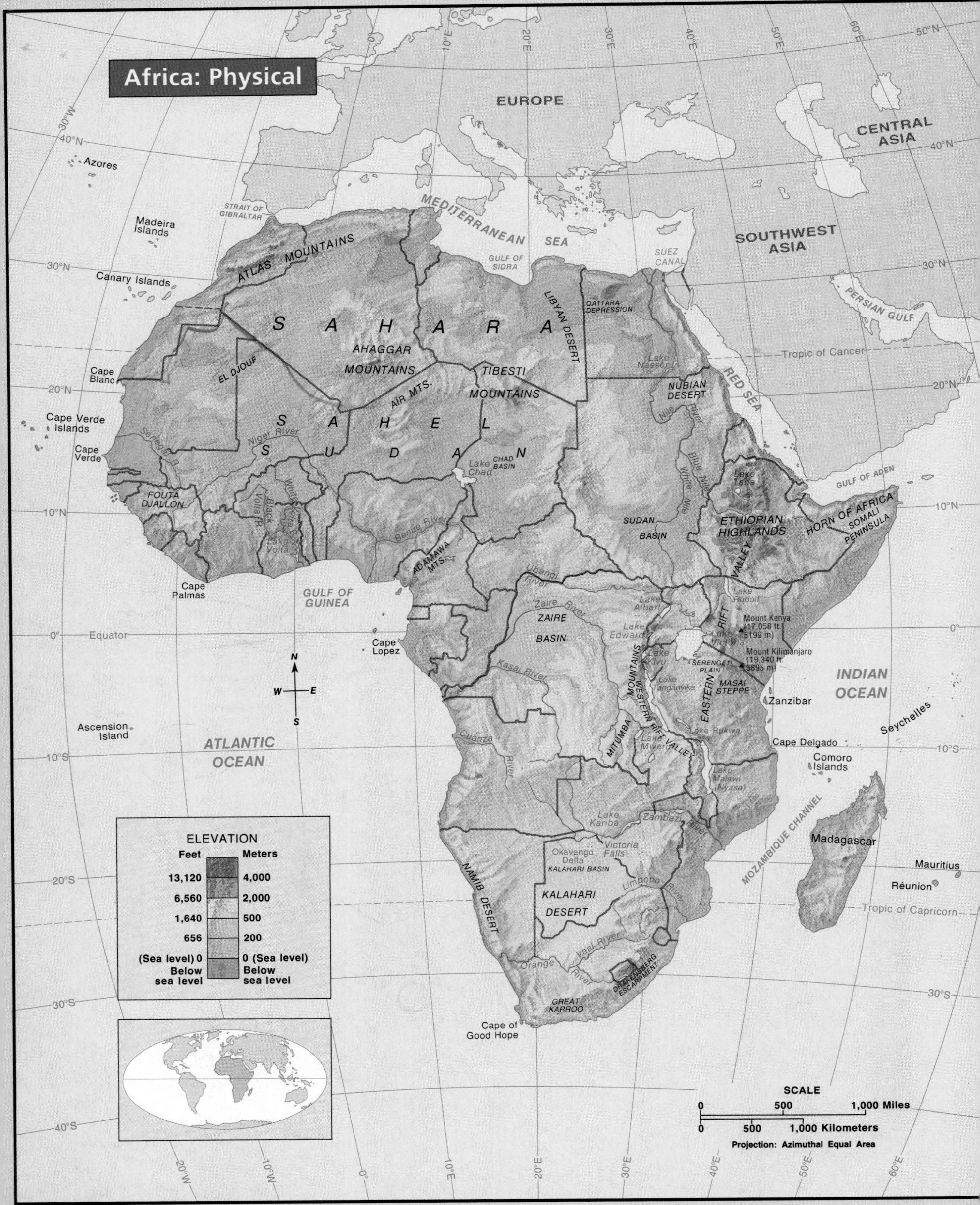

Africa: Physical
EUROPE
CENTRAL ASIA
Azores
Madeira Islands
Canary Islands
STRAIT OF GIBRALTAR
MEDITERRANEAN SEA
GULF OF SIDRA
SUEZ CANAL
SOUTHWEST ASIA
PERSIAN GULF
ATLAS MOUNTAINS
SAHARA
AHAGGAR MOUNTAINS
TIBESTI MOUNTAINS
AÏR MTS.
EL DJOUF
LIBYAN DESERT
QATTARA DEPRESSION
Lake Nasser
NUBIAN DESERT
RED SEA
Tropic of Cancer
Cape Blanc
Cape Verde Islands
Cape Verde
SAHEL
SUDAN
Senegal R.
Niger River
Lake Chad
CHAD BASIN
Nile River
Blue Nile
White Nile
Lake Tana
GULF OF ADEN
ETHIOPIAN HIGHLANDS
HORN OF AFRICA
SOMALI PENINSULA
FOUTA DJALLON
Volta R.
Black Volta
White Volta
Lake Volta
Benue River
ADAMAWA MTS.
SUDAN BASIN
RIFT VALLEY
Cape Palmas
GULF OF GUINEA
Ubangi River
Lake Albert
Lake Rudolf
Zaire River
ZAIRE BASIN
Lake Edward
Mount Kenya (17,058 ft. 5199 m)
Lake Victoria
Mount Kilimanjaro (19,340 ft. 5895 m)
Equator
Cape Lopez
Lake Kivu
SERENGETI PLAIN
MASAI STEPPE
Kasai River
Lake Tanganyika
MITUMBA MOUNTAINS
WESTERN RIFT VALLEY
EASTERN
INDIAN OCEAN
Zanzibar
Ascension Island
ATLANTIC OCEAN
Cuanza River
Lake Mweru
Lake Rukwa
Cape Delgado
Seychelles
Comoro Islands
Lake Malawi (Nyasa)
Lake Kariba
Zambezi River
MOZAMBIQUE CHANNEL
Madagascar
Victoria Falls
Okavango Delta
KALAHARI BASIN
KALAHARI DESERT
NAMIB DESERT
Limpopo River
Mauritius
Réunion
Tropic of Capricorn
Vaal River
Orange River
DRAKENSBERG ESCARPMENT
GREAT KARROO
Cape of Good Hope
ELEVATION
Feet
Meters
13,120
4,000
6,560
2,000
1,640
500
656
200
(Sea level) 0
0 (Sea level)
Below sea level
Below sea level
SCALE
0
500
1,000 Miles
0
500
1,000 Kilometers
Projection: Azimuthal Equal Area
N
S
E
W
50°N
40°N
30°N
20°N
10°N
0°
10°S
20°S
30°S
40°S
30°W
20°W
10°W
0°
10°E
20°E
30°E
40°E
50°E
60°E

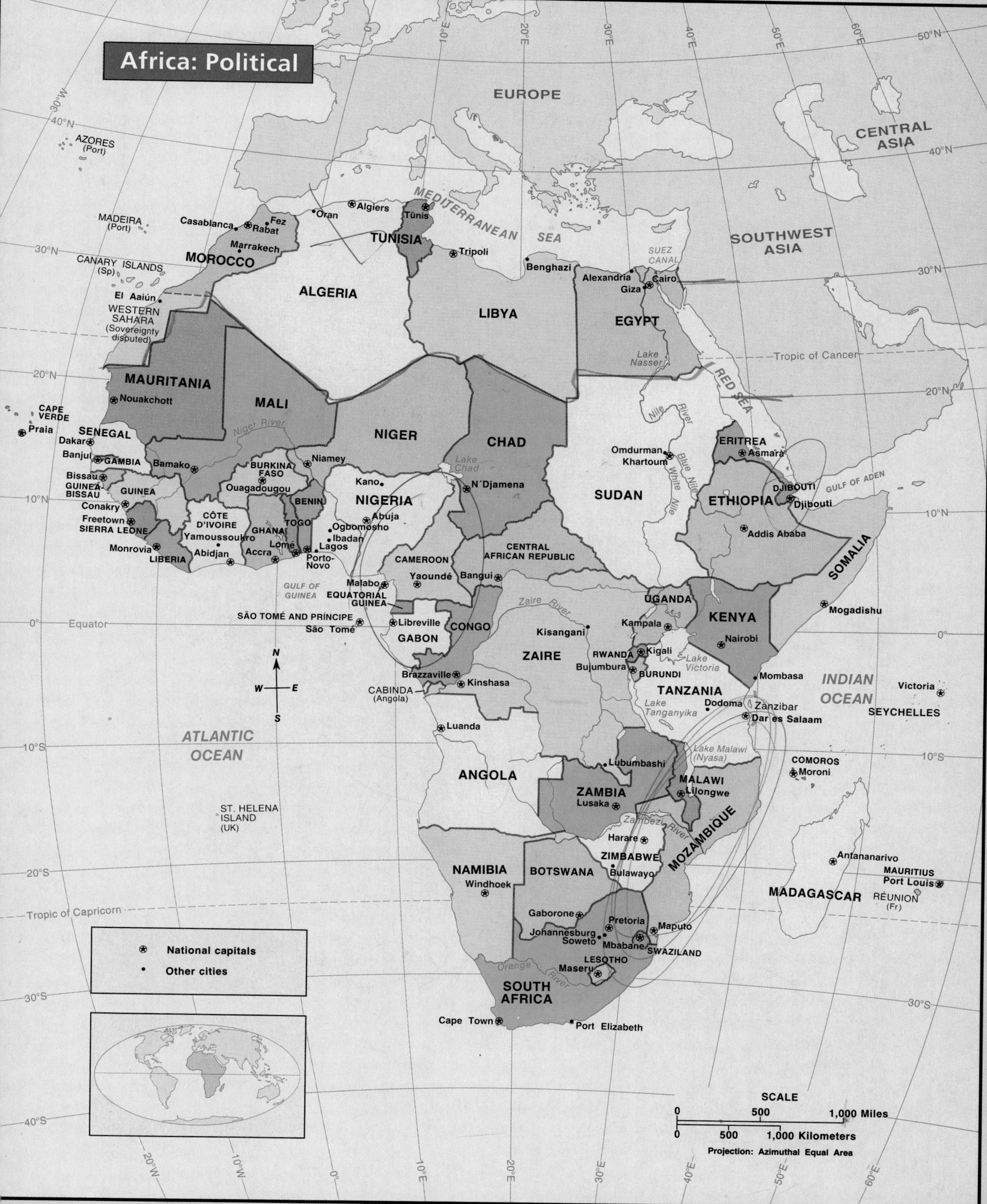
Africa: Political
EUROPE
CENTRAL ASIA
SOUTHWEST ASIA
AZORES (Port)
MADEIRA (Port)
CANARY ISLANDS (Sp)
MEDITERRANEAN SEA
SUEZ CANAL
RED SEA
GULF OF ADEN
Tropic of Cancer
Equator
Tropic of Capricorn
ATLANTIC OCEAN
INDIAN OCEAN
GULF OF GUINEA
MOROCCO
Casablanca
Fez
Rabat
Marrakech
El Aaiún
WESTERN SAHARA (Sovereignty disputed)
ALGERIA
Algiers
Oran
TUNISIA
Tunis
LIBYA
Tripoli
Benghazi
EGYPT
Alexandria
Cairo
Giza
Lake Nasser
MAURITANIA
Nouakchott
MALI
Bamako
Niger River
NIGER
Niamey
CHAD
N'Djamena
Lake Chad
SUDAN
Omdurman
Khartoum
Blue Nile
White Nile
Nile River
ERITREA
Asmara
DJIBOUTI
Djibouti
ETHIOPIA
Addis Ababa
SOMALIA
Mogadishu
CAPE VERDE
Praia
SENEGAL
Dakar
Banjul
GAMBIA
Bissau
GUINEA-BISSAU
GUINEA
Conakry
Freetown
SIERRA LEONE
Monrovia
LIBERIA
CÔTE D'IVOIRE
Yamoussoukro
Abidjan
BURKINA FASO
Ouagadougou
GHANA
Accra
TOGO
Lomé
BENIN
Porto-Novo
NIGERIA
Kano
Abuja
Ogbomosho
Ibadan
Lagos
CAMEROON
Yaoundé
CENTRAL AFRICAN REPUBLIC
Bangui
Malabo
EQUATORIAL GUINEA
SÃO TOMÉ AND PRÍNCIPE
São Tomé
Libreville
GABON
CONGO
Brazzaville
Kinshasa
CABINDA (Angola)
ZAIRE
Zaire River
Kisangani
Lubumbashi
UGANDA
Kampala
KENYA
Nairobi
Mombasa
RWANDA
Kigali
Bujumbura
BURUNDI
Lake Victoria
TANZANIA
Dodoma
Zanzibar
Dar es Salaam
Lake Tanganyika
Lake Malawi (Nyasa)
MALAWI
Lilongwe
ANGOLA
Luanda
ZAMBIA
Lusaka
Zambezi River
ZIMBABWE
Harare
Bulawayo
MOZAMBIQUE
Maputo
NAMIBIA
Windhoek
BOTSWANA
Gaborone
Pretoria
Johannesburg
Soweto
Mbabane
SWAZILAND
LESOTHO
Maseru
Orange River
SOUTH AFRICA
Cape Town
Port Elizabeth
ST. HELENA ISLAND (UK)
SEYCHELLES
Victoria
COMOROS
Moroni
MADAGASCAR
Antananarivo
MAURITIUS
Port Louis
RÉUNION (Fr)
N
W
E
S
National capitals
Other cities
SCALE
0
500
1,000 Miles
0
500
1,000 Kilometers
Projection: Azimuthal Equal Area
50°N
40°N
30°N
20°N
10°N
0°
10°S
20°S
30°S
40°S
30°W
20°W
10°W
0°
10°E
20°E
30°E
40°E
50°E
60°E

Australia and New Zealand

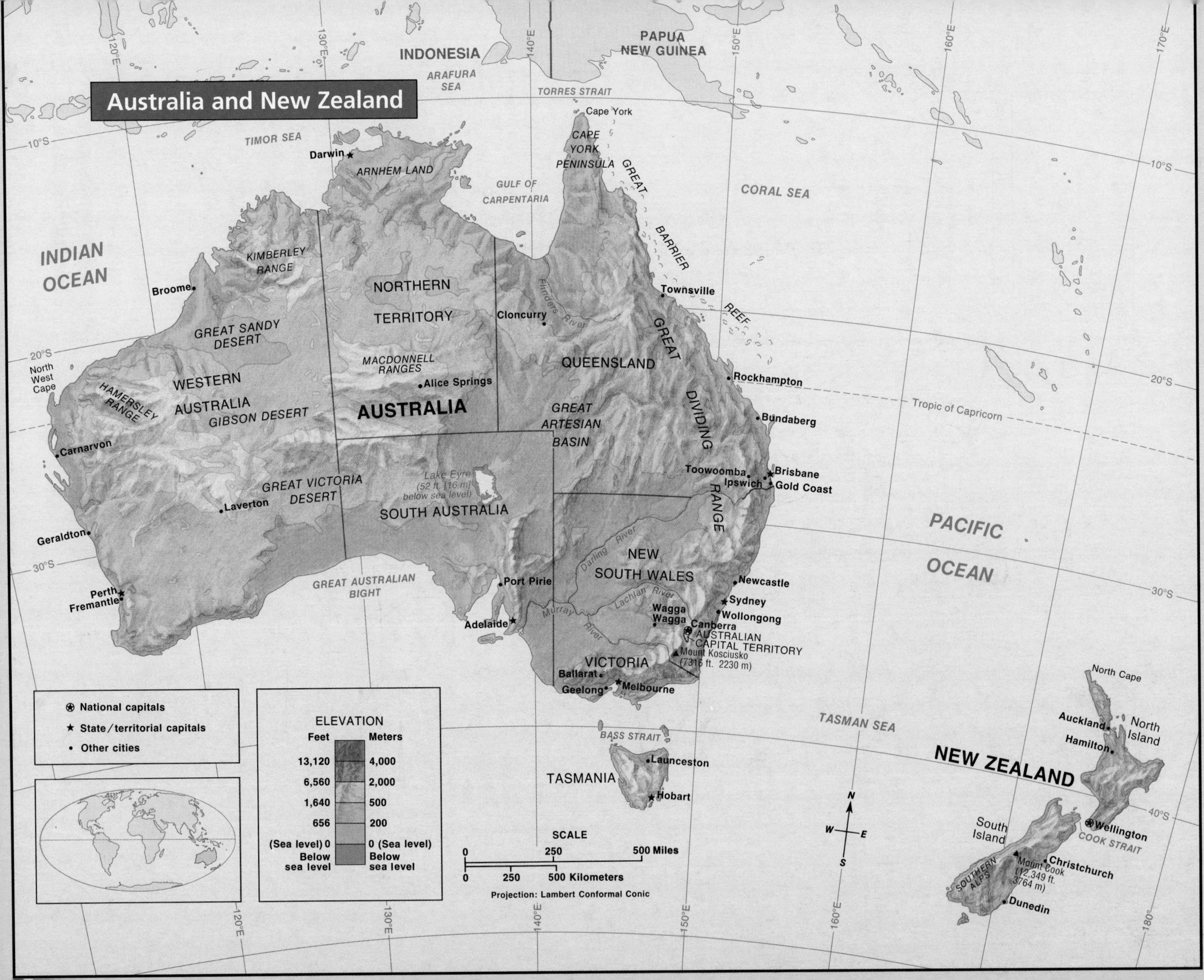

INDONESIA
PAPUA NEW GUINEA
ARAFURA SEA
TORRES STRAIT
Cape York
TIMOR SEA
Darwin
ARNHEM LAND
CAPE YORK PENINSULA
GULF OF CARPENTARIA
GREAT BARRIER REEF
CORAL SEA
INDIAN OCEAN
KIMBERLEY RANGE
Broome
NORTHERN TERRITORY
Cloncurry
Flinders River
Townsville
GREAT SANDY DESERT
MACDONNELL RANGES
Alice Springs
QUEENSLAND
North West Cape
HAMERSLEY RANGE
WESTERN AUSTRALIA
GIBSON DESERT
AUSTRALIA
GREAT ARTESIAN BASIN
GREAT DIVIDING RANGE
Rockhampton
Tropic of Capricorn
Bundaberg
Carnarvon
Lake Eyre (52 ft. [16 m] below sea level)
GREAT VICTORIA DESERT
Laverton
SOUTH AUSTRALIA
Toowoomba
Ipswich
Brisbane
Gold Coast
Geraldton
Darling River
NEW SOUTH WALES
PACIFIC OCEAN
Perth
Fremantle
GREAT AUSTRALIAN BIGHT
Port Pirie
Newcastle
Sydney
Lachlan River
Murray River
Wagga Wagga
Wollongong
Adelaide
Canberra
AUSTRALIAN CAPITAL TERRITORY
Mount Kosciusko (7316 ft. 2230 m)
VICTORIA
Ballarat
Melbourne
Geelong
North Cape
National capitals
State/territorial capitals
Other cities
ELEVATION
Feet
Meters
13,120
4,000
6,560
2,000
1,640
500
656
200
(Sea level) 0
0 (Sea level)
Below sea level
Below sea level
TASMAN SEA
BASS STRAIT
Auckland
North Island
Hamilton
Launceston
NEW ZEALAND
TASMANIA
Hobart
N
W
E
S
SCALE
0
250
500 Miles
0
250
500 Kilometers
Projection: Lambert Conformal Conic
South Island
Wellington
COOK STRAIT
SOUTHERN ALPS
Mount Cook (12,349 ft. 3764 m)
Christchurch
Dunedin
10°S
20°S
30°S
40°S
120°E
130°E
140°E
150°E
160°E
170°E
180°

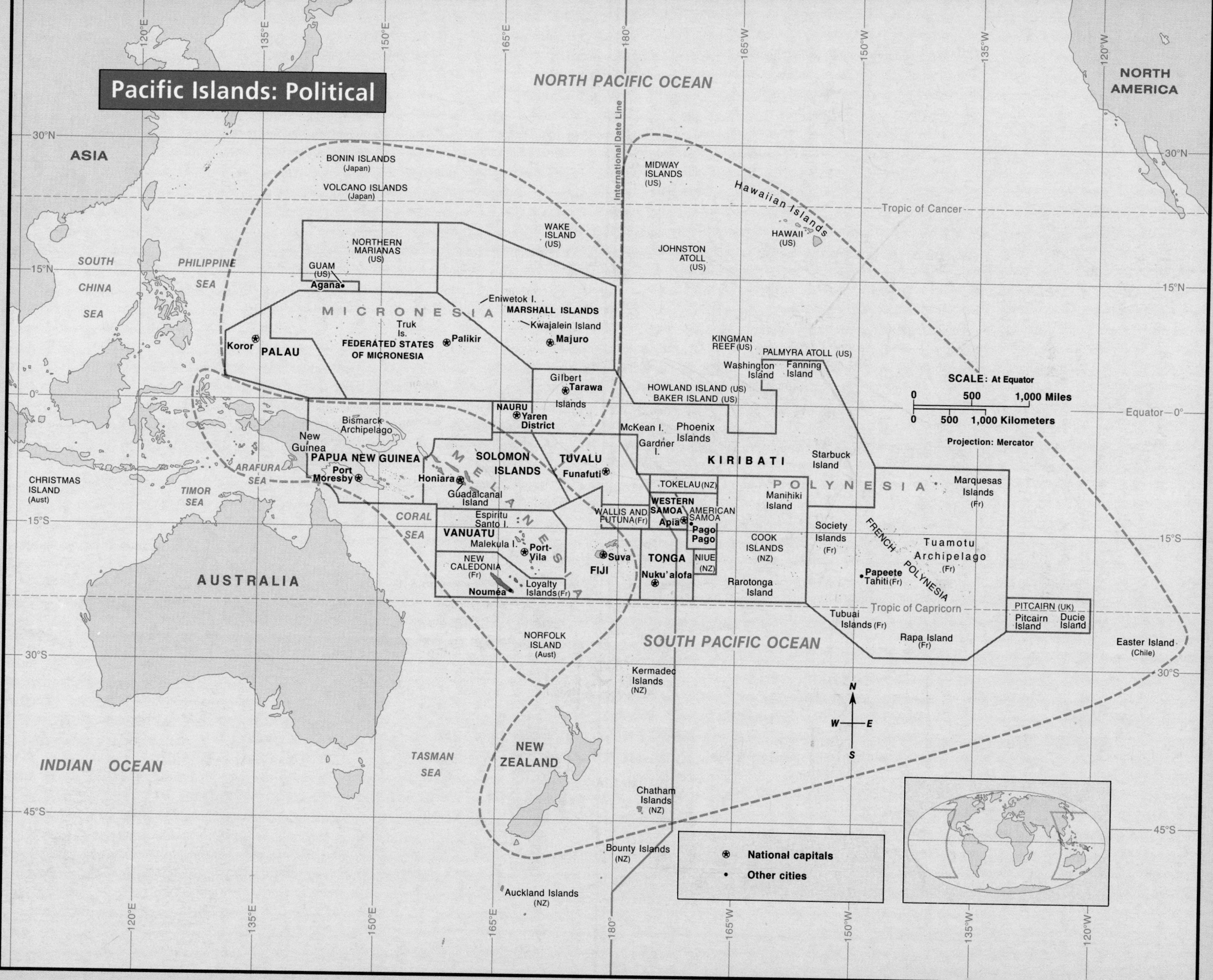
Pacific Islands: Political
NORTH PACIFIC OCEAN
NORTH AMERICA
ASIA
SOUTH CHINA SEA
PHILIPPINE SEA
BONIN ISLANDS (Japan)
VOLCANO ISLANDS (Japan)
MIDWAY ISLANDS (US)
Hawaiian Islands
HAWAII (US)
International Date Line
Tropic of Cancer
NORTHERN MARIANAS (US)
GUAM (US)
Agana
WAKE ISLAND (US)
JOHNSTON ATOLL (US)
MICRONESIA
Eniwetok I.
MARSHALL ISLANDS
Kwajalein Island
Majuro
Truk Is.
FEDERATED STATES OF MICRONESIA
Palikir
Koror
PALAU
KINGMAN REEF (US)
PALMYRA ATOLL (US)
Washington Island
Fanning Island
Gilbert Islands
Tarawa
HOWLAND ISLAND (US)
BAKER ISLAND (US)
SCALE: At Equator
0 500 1,000 Miles
0 500 1,000 Kilometers
Projection: Mercator
Equator
NAURU
Yaren District
Bismarck Archipelago
New Guinea
PAPUA NEW GUINEA
Port Moresby
McKean I.
Phoenix Islands
Gardner I.
KIRIBATI
Starbuck Island
SOLOMON ISLANDS
Honiara
Guadalcanal Island
TUVALU
Funafuti
MELANESIA
ARAFURA SEA
TIMOR SEA
CHRISTMAS ISLAND (Aust)
CORAL SEA
TOKELAU (NZ)
POLYNESIA
Marquesas Islands (Fr)
Manihiki Island
WESTERN SAMOA
Apia
AMERICAN SAMOA
Pago Pago
WALLIS AND FUTUNA (Fr)
Espiritu Santo I.
VANUATU
Malekula I.
Port-Vila
NEW CALEDONIA (Fr)
Nouméa
Loyalty Islands (Fr)
Suva
FIJI
TONGA
Nuku'alofa
NIUE (NZ)
COOK ISLANDS (NZ)
Rarotonga Island
Society Islands (Fr)
FRENCH POLYNESIA
Tuamotu Archipelago (Fr)
Papeete
Tahiti (Fr)
AUSTRALIA
Tropic of Capricorn
Tubuai Islands (Fr)
PITCAIRN (UK)
Pitcairn Island
Ducie Island
Rapa Island (Fr)
Easter Island (Chile)
NORFOLK ISLAND (Aust)
SOUTH PACIFIC OCEAN
Kermadec Islands (NZ)
N
E
S
W
INDIAN OCEAN
TASMAN SEA
NEW ZEALAND
Chatham Islands (NZ)
Bounty Islands (NZ)
Auckland Islands (NZ)
National capitals
Other cities
30°N
15°N
0°
15°S
30°S
45°S
120°E
135°E
150°E
165°E
180°
165°W
150°W
135°W
120°W

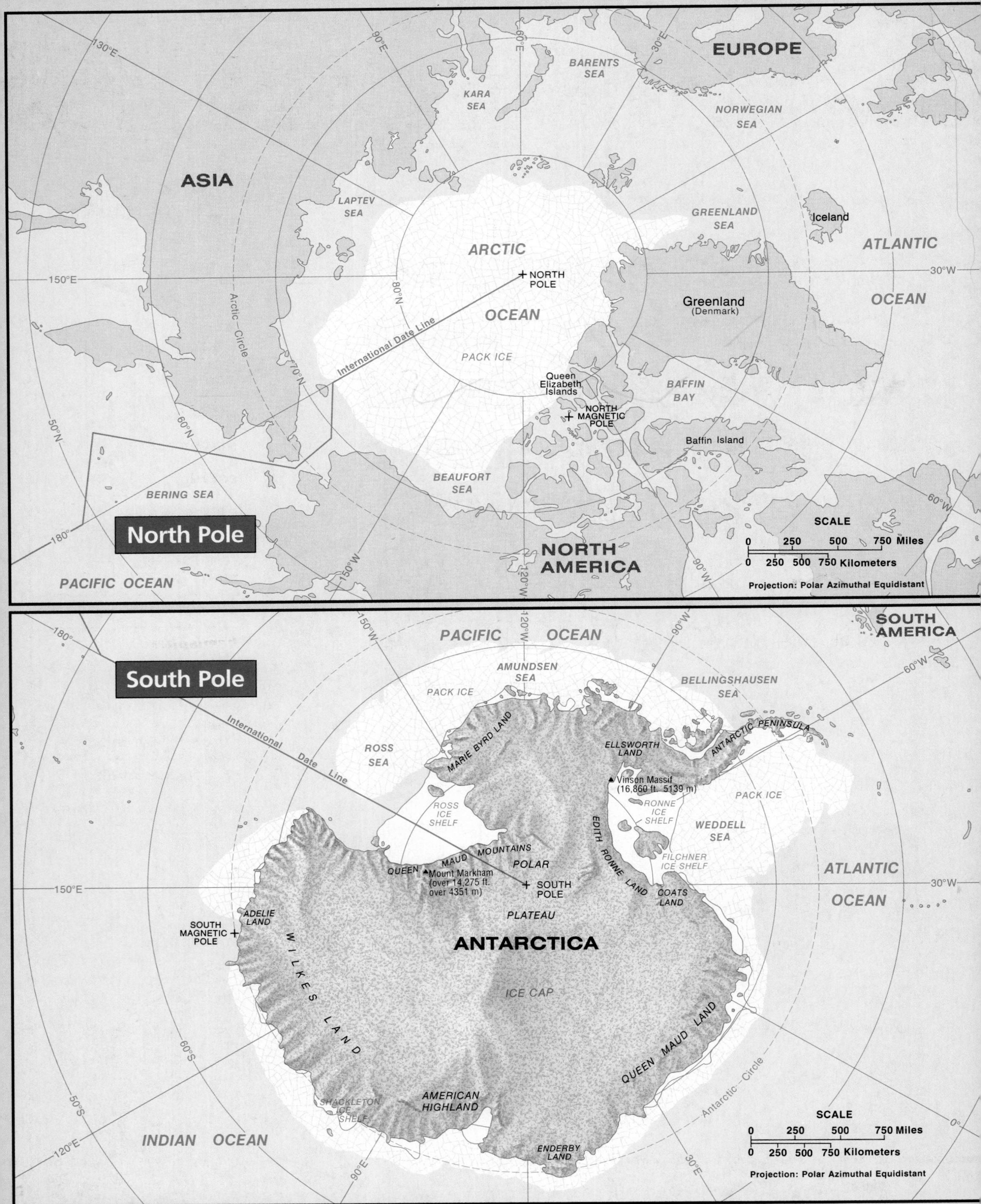
North Pole
EUROPE
BARENTS SEA
KARA SEA
NORWEGIAN SEA
ASIA
LAPTEV SEA
GREENLAND SEA
Iceland
ARCTIC OCEAN
NORTH POLE
ATLANTIC OCEAN
Greenland (Denmark)
International Date Line
Arctic Circle
PACK ICE
Queen Elizabeth Islands
BAFFIN BAY
NORTH MAGNETIC POLE
Baffin Island
BEAUFORT SEA
BERING SEA
NORTH AMERICA
PACIFIC OCEAN
130°E
90°E
60°E
30°E
150°E
30°W
80°N
70°N
60°N
50°N
180°
150°W
120°W
90°W
60°W
SCALE
0 250 500 750 Miles
0 250 500 750 Kilometers
Projection: Polar Azimuthal Equidistant
South Pole
PACIFIC OCEAN
SOUTH AMERICA
AMUNDSEN SEA
BELLINGSHAUSEN SEA
PACK ICE
ROSS SEA
MARIE BYRD LAND
ELLSWORTH LAND
ANTARCTIC PENINSULA
International Date Line
Vinson Massif (16,860 ft. 5139 m)
ROSS ICE SHELF
RONNE ICE SHELF
PACK ICE
WEDDELL SEA
EDITH RONNE LAND
QUEEN MAUD MOUNTAINS
POLAR PLATEAU
FILCHNER ICE SHELF
Mount Markham (over 14,275 ft. over 4351 m)
SOUTH POLE
ATLANTIC OCEAN
COATS LAND
ADELIE LAND
SOUTH MAGNETIC POLE
ANTARCTICA
WILKES LAND
ICE CAP
QUEEN MAUD LAND
Antarctic Circle
SHACKLETON ICE SHELF
AMERICAN HIGHLAND
ENDERBY LAND
INDIAN OCEAN
180°
150°W
120°W
90°W
60°W
150°E
30°W
120°E
90°E
30°E
0°
60°S
50°S
SCALE
0 250 500 750 Miles
0 250 500 750 Kilometers
Projection: Polar Azimuthal Equidistant

SKILLS HANDBOOK

CONTENTS

Studying geography requires the ability to understand and use various tools. This Skills Handbook explains how to use maps, charts, and other graphics to help you learn about geography and the various regions of the world. Throughout this textbook, you will have the opportunity to improve these skills and build upon them.

GEOGRAPHY DICTIONARY

globe
continent
island
ocean
grid
latitude
longitude
equator
parallel
prime meridian
meridian
degree
hemisphere
map
atlas
map projection
great-circle route
directional indicator
scale
legend
compass rose
mental map

MAPPING THE EARTH

The Globe

A globe is a scale model of the earth. It is useful for looking at the entire earth or at large areas of the earth's surface. The earth's land surface is organized into seven large landmasses, called **continents**, which are pictured in the four maps in Figure 1. Landmasses smaller than continents and completely surrounded by water are called **islands**. Geographers also organize the earth's water surface into parts, the largest of which is the world **ocean**. Geographers divide the world ocean into four oceans: the Pacific Ocean, the Atlantic Ocean, the Indian Ocean, and the Arctic Ocean. Lakes and seas are smaller bodies of water.

Figure 2 is a diagram of a globe. The pattern of lines that circle the earth in east-west and north-south directions is called a **grid**. The intersection of these imaginary lines helps us find the location of places on the earth. Some mapmakers label the lines with letters and numbers. The grid on many maps and globes, however, is made up of lines of **latitude** and **longitude**.

Lines of latitude are drawn in an east-west direction and measure distance north and south of the **equator**. The equator is an imaginary line that circles the globe halfway between the North Pole and the South Pole. Lines of latitude are called **parallels** because they are

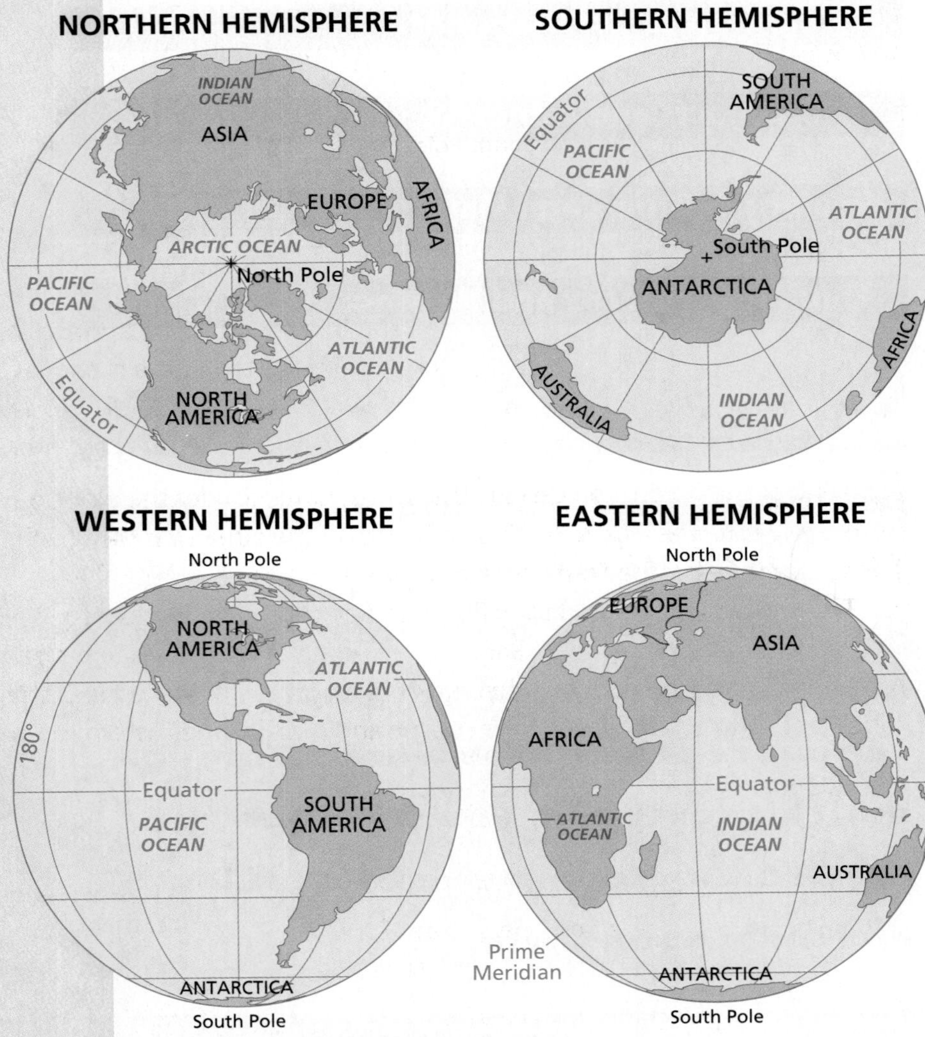

NORTHERN HEMISPHERE
INDIAN OCEAN
ASIA
EUROPE
AFRICA
ARCTIC OCEAN
North Pole
PACIFIC OCEAN
ATLANTIC OCEAN
Equator
NORTH AMERICA
SOUTHERN HEMISPHERE
SOUTH AMERICA
Equator
PACIFIC OCEAN
ATLANTIC OCEAN
South Pole
ANTARCTICA
AFRICA
AUSTRALIA
INDIAN OCEAN
WESTERN HEMISPHERE
North Pole
NORTH AMERICA
ATLANTIC OCEAN
180°
Equator
SOUTH AMERICA
PACIFIC OCEAN
ANTARCTICA
South Pole
EASTERN HEMISPHERE
North Pole
EUROPE
ASIA
AFRICA
Equator
ATLANTIC OCEAN
INDIAN OCEAN
AUSTRALIA
Prime Meridian
ANTARCTICA
South Pole

▲ **Figure 1: The hemispheres**

always parallel to the equator. Parallels north of the equator are labeled with an *N*, and those south are labeled with an *S*.

Lines of longitude are drawn in a north-south direction and measure distance east and west of the **prime meridian**. The prime meridian is an imaginary line that runs through Greenwich, England, from the North Pole to the South Pole. Lines of longitude are called **meridians**.

▼ **Figure 2: Globe**

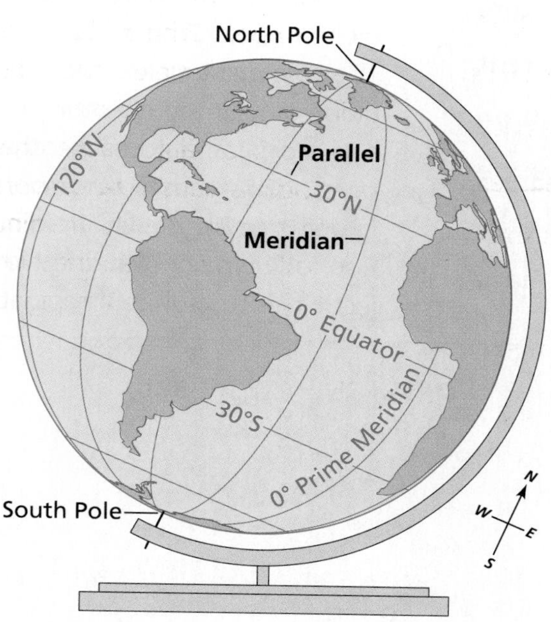
North Pole
120°W
Parallel
30°N
Meridian
0° Equator
30°S
0° Prime Meridian
South Pole
N
W
E
S

Parallels measure distance from the equator, and meridians from the prime meridian, in **degrees**. The symbol for degrees is °. Degrees are further divided into minutes, for which the symbol is ′. There are 60 minutes in a degree.

Lines of latitude range from 0°, for locations on the equator, to 90°N or 90°S, for locations at the North Pole or South Pole. Lines of longitude range from 0° on the prime meridian to 180° on a meridian in the mid-Pacific Ocean. Meridians west of the prime meridian to 180° are labeled with a *W*. Those east of the prime meridian to 180° are labeled with an *E*.

Looking at the globe, you can see that the equator divides the globe into two halves, or **hemispheres**. See **Figure 1**. The half north of the equator is the Northern Hemisphere. The southern half is the Southern Hemisphere.

The prime meridian and the 180° meridian divide the world into the Eastern Hemisphere and the Western Hemisphere. Because the prime meridian separates parts of Europe and Africa into two different hemispheres, some mapmakers divide the Eastern and Western hemispheres at 20° W. This places all of Europe and Africa in the Eastern Hemisphere.

YOUR TURN

1. Look at the Student Atlas map on page A4. What islands are located near the intersection of latitude 20° N and longitude 160° W?
2. Name the four hemispheres. In which hemispheres is the United States located?
3. Name the continents of the world.
4. Name the oceans of the world.

MAP-MAKING

A map is a flat diagram of all or part of the earth's surface. An **atlas** is an organized collection of maps in one book. **M**apmakers have different ways of presenting a round earth on flat maps. These different ways are called **map projections**. Because the earth is round, all flat maps have some distortion. Some flat maps distort size, especially at high latitudes. Those maps, however, might be useful because they show true direction and shape. Some maps, called equal-area maps, show size in true proportions but distort shapes. **M**apmakers must choose the type of map projection that is best for their purposes. Many map projections are one of three kinds: cylindrical, conic, or flat-plane.

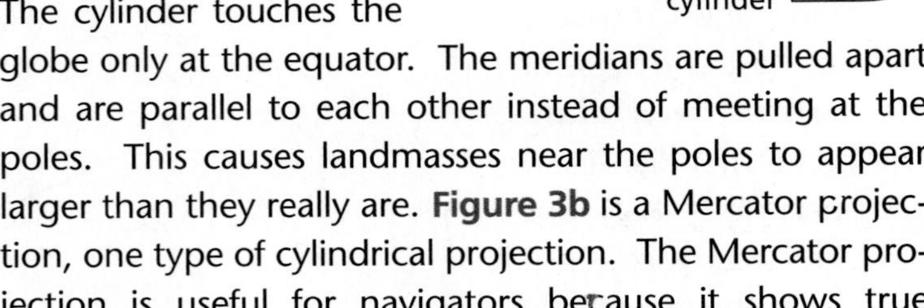

Paper cylinder

▶ **Figure 3a: Paper cylinder**

Cylindrical projections are designed from a cylinder wrapped around the globe. See **Figure 3a**. The cylinder touches the globe only at the equator. The meridians are pulled apart and are parallel to each other instead of meeting at the poles. This causes landmasses near the poles to appear larger than they really are. **Figure 3b** is a Mercator projection, one type of cylindrical projection. The Mercator projection is useful for navigators because it shows true direction and shape. The Mercator projection for world maps, however, emphasizes the Northern Hemisphere. Africa and South America are shown to be smaller than they really are.

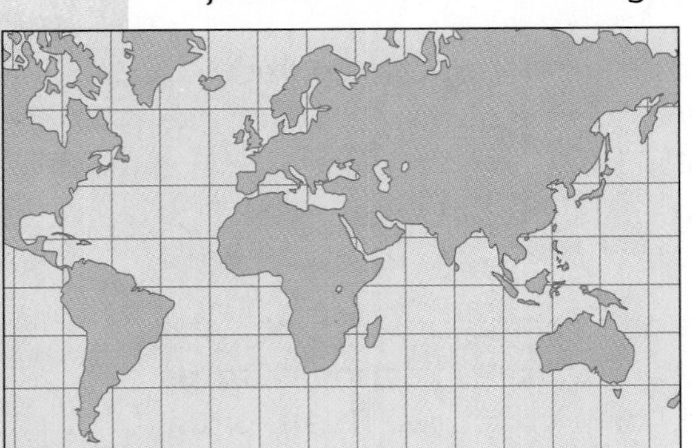

▲ **Figure 3b: Mercator projection**

▼ **Figure 4a: Paper cone**

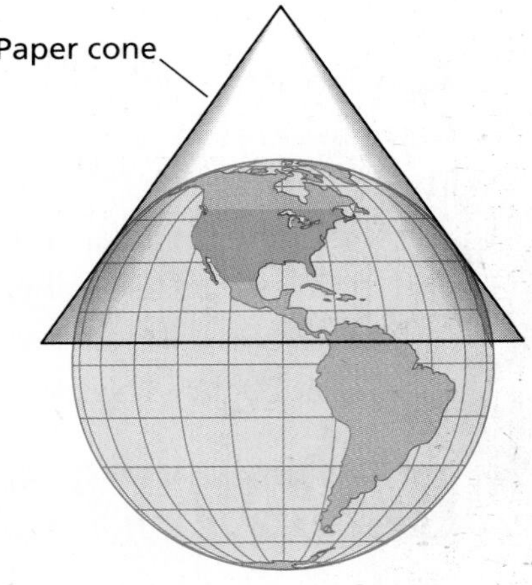

Paper cone

Conic projections are designed from a cone placed over the globe. See **Figure 4a**. A conic projection is most accurate along the lines of latitude where it touches the globe. It retains almost true shape and size. Conic projections are most useful for areas that have long east-west dimensions, such as the United States. See the map in **Figure 4b**.

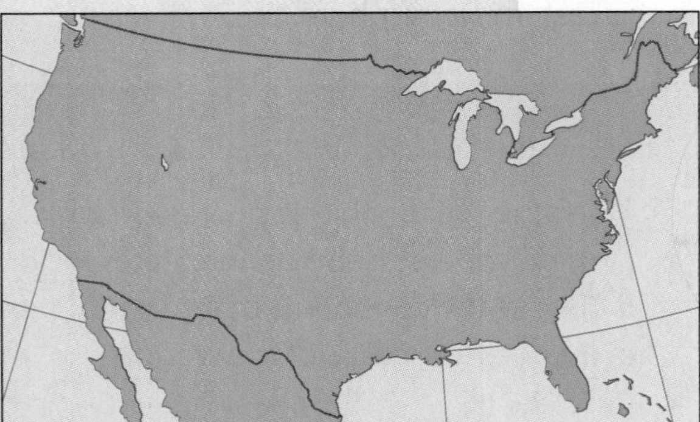

◀ **Figure 4b: Conic projection**

PANAMA
GULF OF PANAMA
VENEZUELA
Orinoco
Caroni R.
Angel Falls
Georgetown

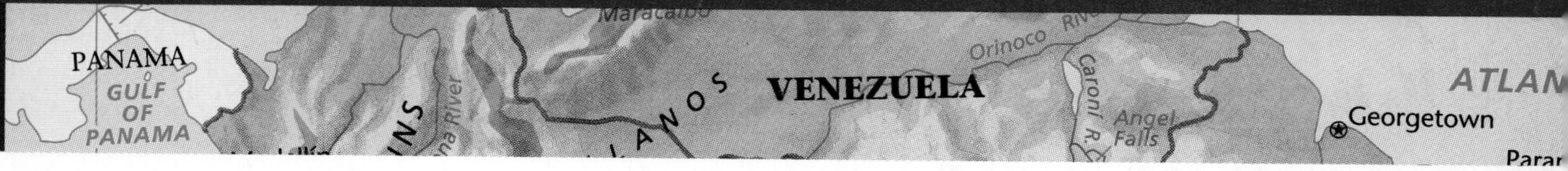

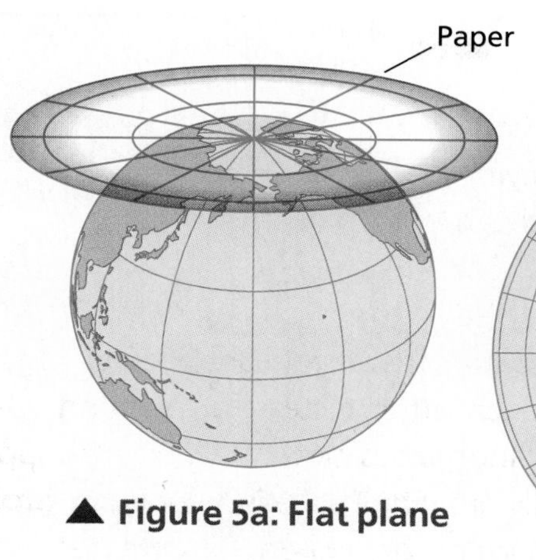

Paper

▲ **Figure 5a: Flat plane**

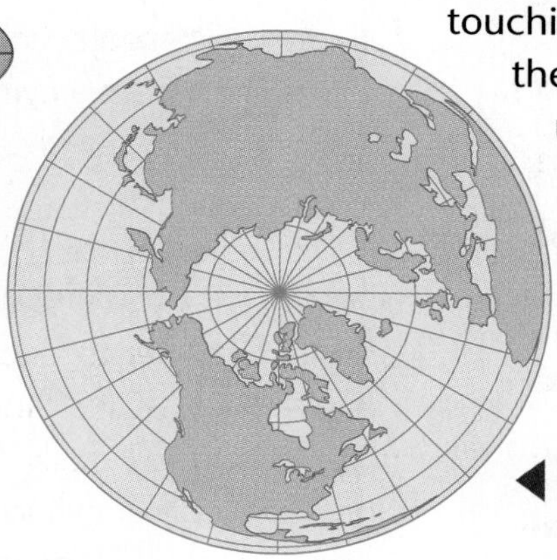

◀ **Figure 5b: Flat-plane projection**

Flat-plane projections are designed from a plane touching the globe at one point, such as at the North Pole or South Pole. See **Figures 5a** and **5b**. A flat-plane projection is useful for showing true direction for airplane pilots and ship navigators. It also shows true area, but it distorts true shape.

The Robinson projection is a compromise between size and shape distortions. It often is used for world maps, such as the map on pages 28–29. The minor distortions in size at high latitudes on Robinson projections are balanced by realistic shapes at the middle and low latitudes.

Drawing a straight line on a flat map will not show the shortest route between two locations. Remember, maps represent a round world on a flat plane. The shortest route between any two points on the earth is a **great-circle route**. See **Figures 6a** and **6b**. Any imaginary line that divides the earth into equal parts is a great circle. The equator is a great circle. Airplanes and ships navigate along great-circle routes.

▼ **Figure 6a: Great-circle route**

NORTH AMERICA
Great-circle route
EUROPE
40°N
Philadelphia
Lisbon
ATLANTIC OCEAN
AFRICA
PACIFIC OCEAN
SOUTH AMERICA

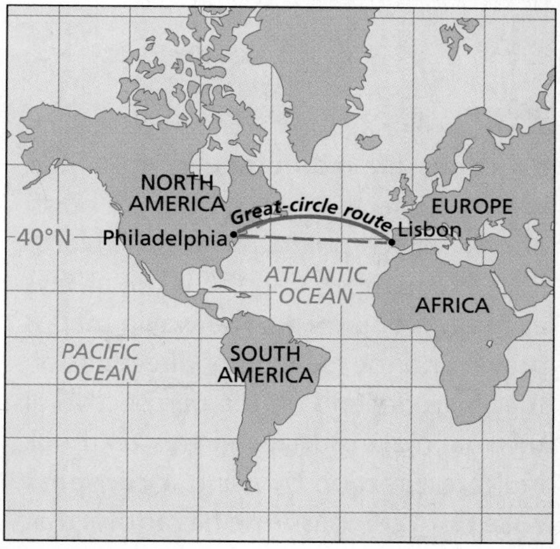

MERCATOR PROJECTION

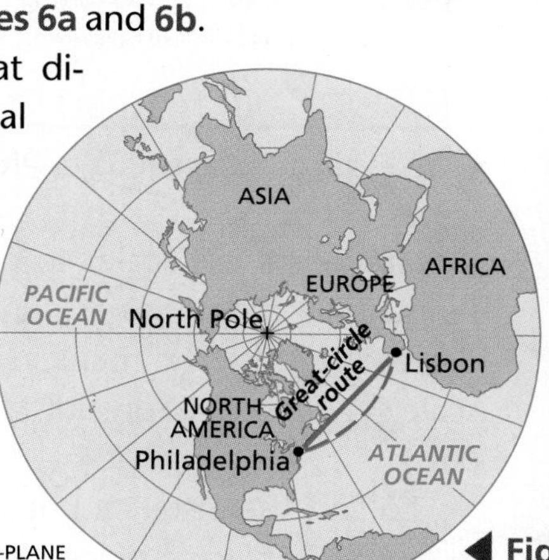

ASIA
AFRICA
EUROPE
PACIFIC OCEAN
North Pole
Great-circle route
Lisbon
NORTH AMERICA
Philadelphia
ATLANTIC OCEAN

FLAT-PLANE PROJECTION

◀ **Figure 6b: Great-circle route**

YOUR TURN

1. Name three major kinds of map projections.
2. Why is a Robinson projection often used for world maps?
3. What kind of projection is a Mercator map?

MAP ESSENTIALS

In some ways, maps are like messages sent out in code. Mapmakers provide certain elements that help us translate these codes to understand the information, or message, they are presenting about a particular part of the world. Almost all maps have several common elements: **directional indicators**, **scales**, and **legends**, or keys. **Figure 7**, a map of East Asia, has all three elements.

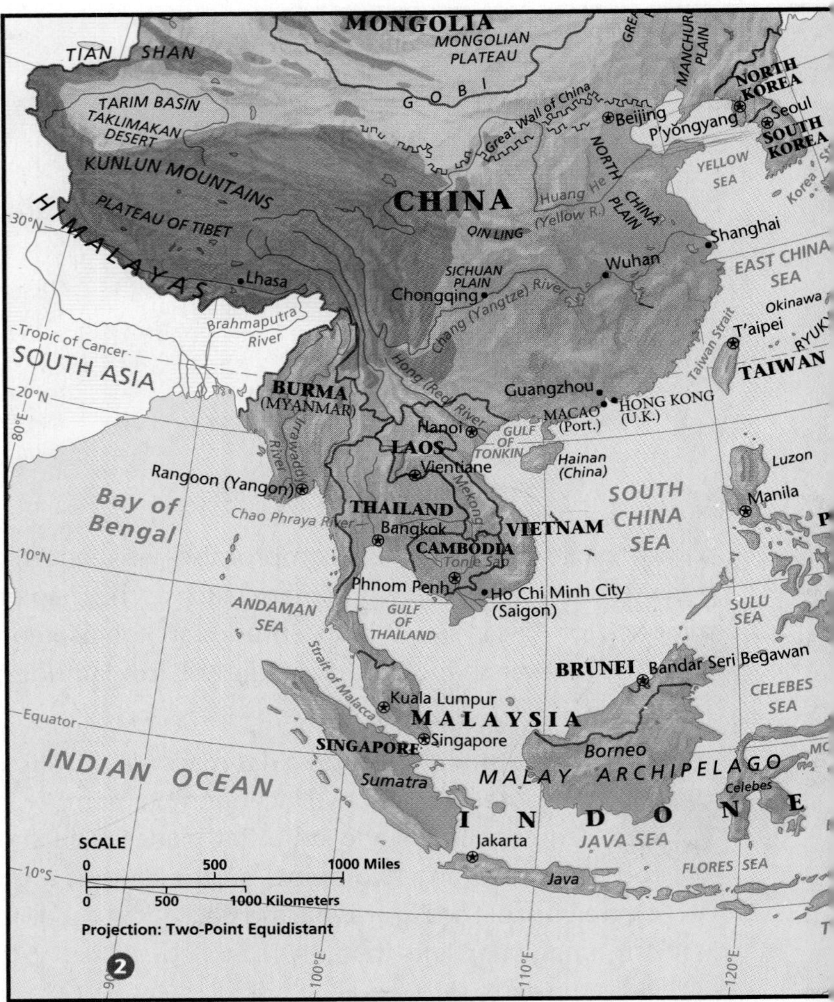

MONGOLIA
MONGOLIAN PLATEAU
TIAN SHAN
GOBI
TARIM BASIN
TAKLIMAKAN DESERT
KUNLUN MOUNTAINS
PLATEAU OF TIBET
HIMALAYAS
Great Wall of China
Beijing
P'yŏngyang
NORTH KOREA
Seoul
SOUTH KOREA
YELLOW SEA
NORTH CHINA PLAIN
CHINA
Huang He (Yellow R.)
QIN LING
Shanghai
EAST CHINA SEA
SICHUAN PLAIN
Chongqing
Wuhan
Chang (Yangtze) River
Lhasa
Brahmaputra River
Okinawa
T'aipei
TAIWAN
Taiwan Strait
Tropic of Cancer
SOUTH ASIA
30°N
20°N
10°N
Equator
10°S
80°E
100°E
110°E
120°E
BURMA (MYANMAR)
Irrawaddy River
Hong (Red) River
Guangzhou
HONG KONG (U.K.)
MACAO (Port.)
Hanoi
GULF OF TONKIN
LAOS
Vientiane
Hainan (China)
Luzon
Manila
Rangoon (Yangon)
Bay of Bengal
Chao Phraya River
THAILAND
Bangkok
Mekong
CAMBODIA
Tonle Sap
VIETNAM
SOUTH CHINA SEA
Phnom Penh
Ho Chi Minh City (Saigon)
ANDAMAN SEA
GULF OF THAILAND
SULU SEA
BRUNEI
Bandar Seri Begawan
Strait of Malacca
Kuala Lumpur
MALAYSIA
CELEBES SEA
SINGAPORE
Singapore
Borneo
INDIAN OCEAN
Sumatra
MALAY ARCHIPELAGO
Celebes
INDONE
Jakarta
JAVA SEA
FLORES SEA
Java
SCALE
0 500 1000 Miles
0 500 1000 Kilometers
Projection: Two-Point Equidistant
2

▲ **Figure 7: East Asia—Physical–Political**

A directional indicator shows which directions are north, south, east, and west. Some mapmakers use a "north arrow," which points toward the North Pole. Remember, "north" is not always at the top of a map. The way a map is drawn and the location of directions on that map depend on the perspective of the mapmaker. Maps in this textbook indicate direction by using a **compass rose ❶**. A compass rose has arrows that point to all four principal directions, as shown in **Figure 7**.

Mapmakers use scales to represent distances between points on a map. Scales may appear on maps in several different forms. The maps in this textbook provide a line scale ❷. The scales give distances in miles and kilometers (km).

To find the distance between two points on the map in **Figure 7**, place a piece of paper so that the edge connects the two points. Mark the location of each point on the paper with a line or dot. Then, compare the distance between the two dots with the map's line scale. The number on the top of the scale gives the distance in miles. The number on the bottom gives the distance in kilometers. Because the distances are given in intervals, you will have to approximate the actual distance on the scale.

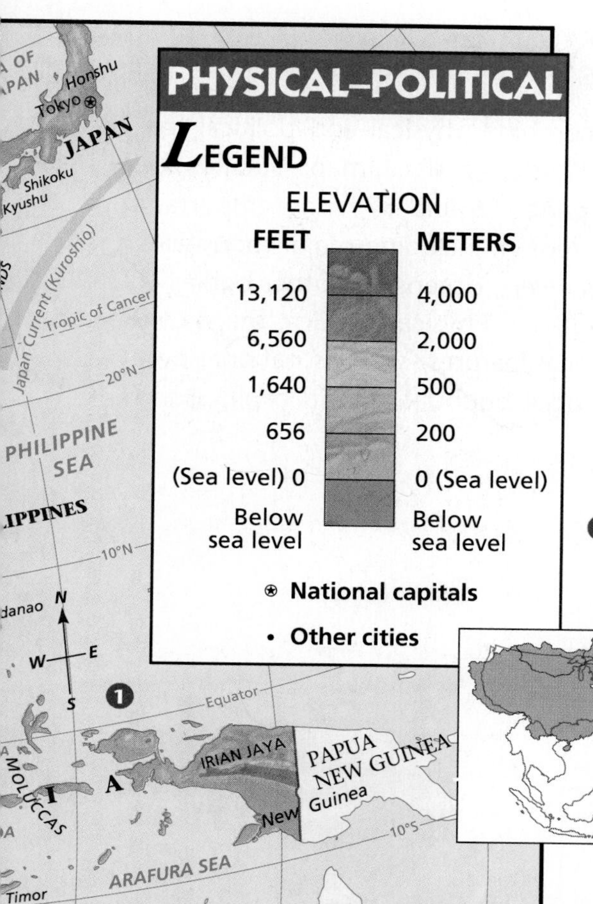
PHYSICAL–POLITICAL
LEGEND
ELEVATION
FEET
METERS
13,120
4,000
6,560
2,000
1,640
500
656
200
(Sea level) 0
0 (Sea level)
Below sea level
Below sea level
National capitals
Other cities
Honshu
Tokyo
JAPAN
Shikoku
Kyushu
Japan Current (Kuroshio)
Tropic of Cancer
20°N
PHILIPPINE SEA
10°N
N
W
E
S
Equator
IRIAN JAYA
PAPUA NEW GUINEA
New Guinea
MOLUCCAS
10°S
ARAFURA SEA
Timor
AUSTRALIA
130°E
140°E
150°E

❶ ❸

The legend ❸, or key, explains what the symbols on the map represent. Point symbols are used to specify the location of things, such as cities, that do not take up much space on a large-scale map. Some legends, such as the one in **Figure 7**, show which colors represent certain elevations. Other maps might have legends with symbols or colors that represent things such as roads, economic resources, land use, population density, and climate.

Size comparison of China, Taiwan, and Mongolia to the contiguous United States

❹

Physical–political maps at the beginning of each regional chapter have size comparison maps ❹. An outline of the mainland 48 U.S. states (not including Alaska and Hawaii) is compared to the area under study in that chapter. These size comparison maps help you understand the size of the areas you are studying in relation to the size of the United States.

Inset maps are sometimes used to show a small part of a larger map. Mapmakers also use inset maps to show areas that are far away from the areas shown on the main map. Maps of the United States, for example, often include inset maps of Alaska and Hawaii. (See the map on page 107.) Those two states are too far from the other 48 states to accurately represent the true distance on the main map. Subject areas in inset maps can be drawn to a scale different from the scale used on the main map.

YOUR TURN

Look at the Student Atlas map on pages A4 and A5.

1. Locate the compass rose. What country is directly west of Madagascar in Africa?
2. What island country is located southeast of India?
3. Locate the distance scale. Using the inset map, find the approximate distance in miles and kilometers from Oslo, Norway, to Stockholm, Sweden.
4. Identify the capital of Brazil. Identify the other cities shown in Brazil.

WORKING WITH MAPS

The Atlas at the front of this textbook includes two kinds of maps: physical and political. At the beginning of most units in this textbook, you will find four kinds of maps. These physical–political, climate, population, and economic maps provide different kinds of information about the region you will study in that unit. These maps are accompanied by questions. Some questions ask you to show how the information on each of the maps might be related.

Mapmakers often combine physical and political features into one map. Physical–political maps, such as the one in **Figure 7** on pages S6 and S7, show important physical features in a region, including major mountains and mountain ranges, rivers, oceans and other bodies of water, deserts, and plains. Physical–political maps also show important political features, such as national borders, state and provincial boundaries, and capitals and other main cities.

▼ **Figure 8: East Asia — Climate**

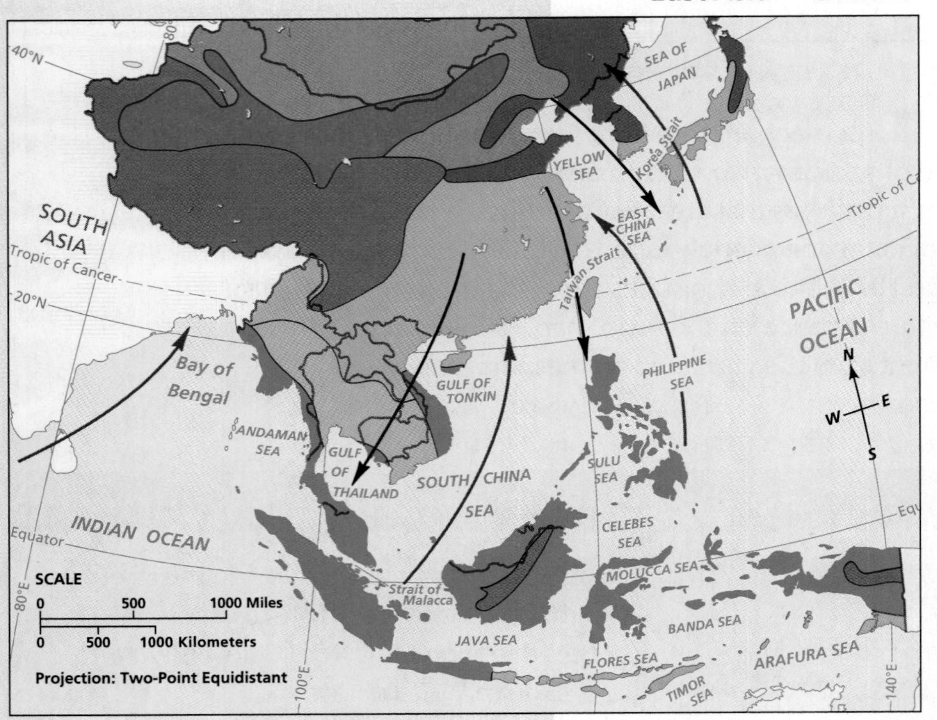

40°N
80°E
SEA OF JAPAN
Korea Strait
YELLOW SEA
EAST CHINA SEA
Tropic of Cancer
SOUTH ASIA
Tropic of Cancer
20°N
Taiwan Strait
PACIFIC OCEAN
Bay of Bengal
GULF OF TONKIN
PHILIPPINE SEA
N
W
E
S
ANDAMAN SEA
GULF OF THAILAND
SOUTH CHINA SEA
SULU SEA
INDIAN OCEAN
Equator
CELEBES SEA
MOLUCCA SEA
SCALE
0 500 1000 Miles
0 500 1000 Kilometers
Projection: Two-Point Equidistant
Strait of Malacca
JAVA SEA
BANDA SEA
FLORES SEA
ARAFURA SEA
TIMOR SEA
100°E
140°E
80°E

Mapmakers use climate maps to show dominant weather patterns in certain areas. Climate maps throughout this textbook use color to show the various climate regions of the world. See **Figure 8**. Colors that identify climate types are found in a legend that accompanies each map. Boundaries between climate regions do not indicate an abrupt change in dominant weather conditions between two climate regions. Instead, boundaries approximate areas of gradual change between two climate regions.

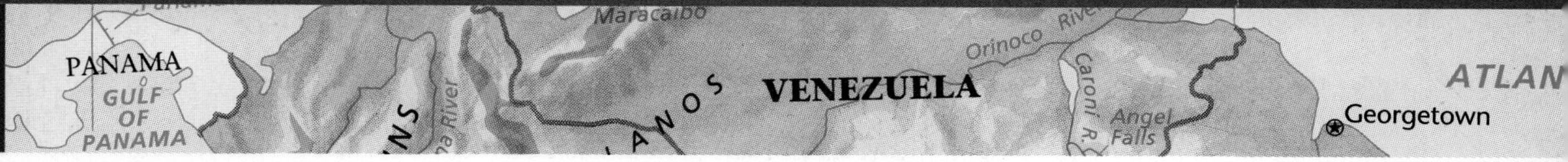

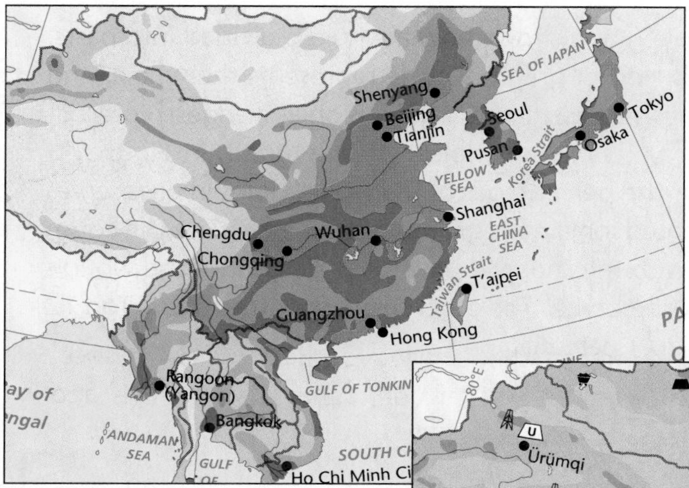

▲ **Figure 9: East Asia—Population**

Population maps show where people live in a particular region and how crowded, or densely populated, regions are. Population maps throughout this textbook use color to show population density. See **Figure 9**. Each color represents a certain number of people living within a square mile or square kilometer. The population maps also use symbols to show metropolitan areas with populations of a particular size. These symbols and the color categories are identified in a legend.

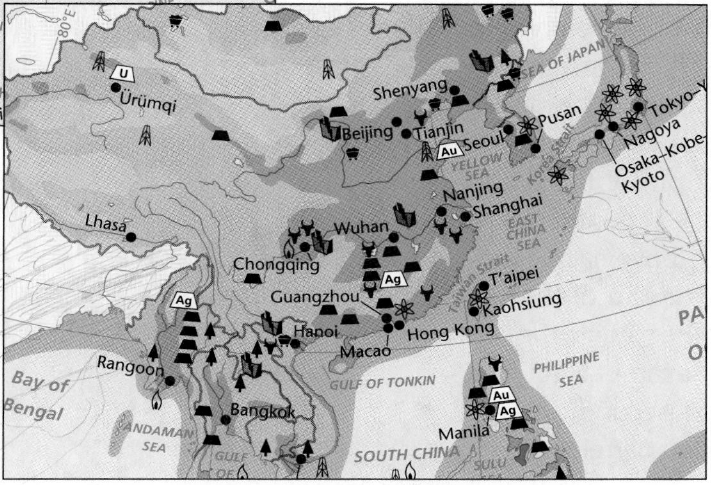

Economic maps show the important resources of a region. See **Figure 10**. Symbols and colors are used to show information about economic development, such as where industry is located or where farming is most common. The meanings of each symbol and color are shown in a legend.

◀ **Figure 10: East Asia—Economy**

Some kinds of maps are not found in a textbook. **Mental maps** are those maps that we see in our minds. We use mental maps to help us make sense of the world around us. A person can, for example, visualize the community he or she lives in: the location and kinds of buildings there, the activities that take place there, and the surrounding environment. Similarly, mental maps help people structure what they know about any other place: the location and relative size of countries and continents, for example. As people learn more about places, they add more details to their mental maps of these places. Throughout this textbook, you will learn more about the details people use in constructing and using their mental maps.

YOUR TURN

1. How many climate types are shown on the climate map on page 108?
2. In the economy map on page 110, what color represents commercial farming?
3. Use your mental map of your school's neighborhood to write a short paragraph describing the route you take to get to school. Then compare your paragraph with a partner's, and discuss how your mental map differs from that of your partner.

USING GRAPHS AND DIAGRAMS

Bar graphs are a visual way to present information. The bar graph in **Figure 11** compares the amount of oil reserves in the largest oil-producing countries. Various amounts of oil reserves in billions of barrels are listed on the left side of the graph. Along the bottom of the graph are the names of the 12 countries with the most oil reserves. Above each country is a vertical bar. The top of the bar corresponds to a number along the left side of the graph. For example, Iraq has 100 billion barrels of oil, second to Saudi Arabia's nearly 260 billion barrels.

A pie graph shows how a whole is divided into parts. In this kind of graph, a circle represents the whole, and wedges represent the parts. Bigger wedges represent larger parts of the whole. The pie graph in **Figure 11** shows the percentages of the world's known oil reserves estimated for various groups of countries. The five countries with the most oil have 64.4 percent of the world's known reserves. The other seven countries of the Top 12 have 26.1 percent.

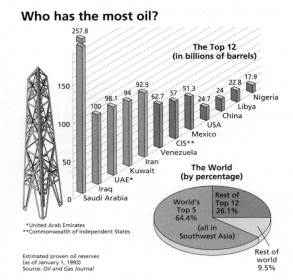

Who has the most oil?

The Top 12 (in billions of barrels)

Saudi Arabia 257.8; Iraq 100; UAE* 98.1; Kuwait 94; Iran 92.9; Venezuela 62.7; CIS** 57; Mexico 51.3; USA 24.7; China 24; Libya 22.8; Nigeria 17.9

The World (by percentage)

World's Top 5 64.4% (all in Southwest Asia); Rest of Top 12 26.1%; Rest of world 9.5%

*United Arab Emirates
**Commonwealth of Independent States

Estimated proven oil reserves (as of January 1, 1993)
Source: *Oil and Gas Journal*

▲ **Figure 11: Reading bar and pie graphs**

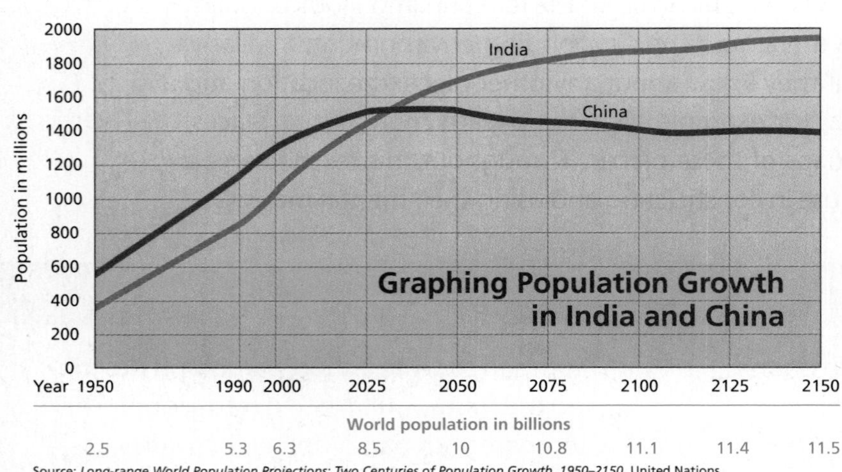

Graphing Population Growth in India and China
India
China
Population in millions
2000
1800
1600
1400
1200
1000
800
600
400
200
0
Year 1950
1990 2000
2025
2050
2075
2100
2125
2150
World population in billions
2.5
5.3 6.3
8.5
10
10.8
11.1
11.4
11.5

Source: *Long-range World Population Projections: Two Centuries of Population Growth, 1950–2150*, United Nations

▲ **Figure 12: Reading a line graph**

Often, line graphs are used to show such things as trends, comparisons, and size. The line graph in **Figure 12** shows the population growth of India and China over time. The information on the left shows the number of people in millions. The years being studied are listed along the bottom. Lines connect points that show the population in millions at each year under study. This line graph projects population growth far into the future.

A climograph compares temperatures and precipitation amounts in different latitudes and locations. See **Figure 13**. Along the left side of the graph is a range of average temperatures. Letters representing each month of the year are listed along the bottom of the graph. Along the right side is a range of average precipitation totals. The climograph uses elements of line and bar graphs. A red line shows the average high temperature in a particular month. The blue line shows the average low temperature each month. The vertical bars along the bottom indicate the average precipitation amount each month.

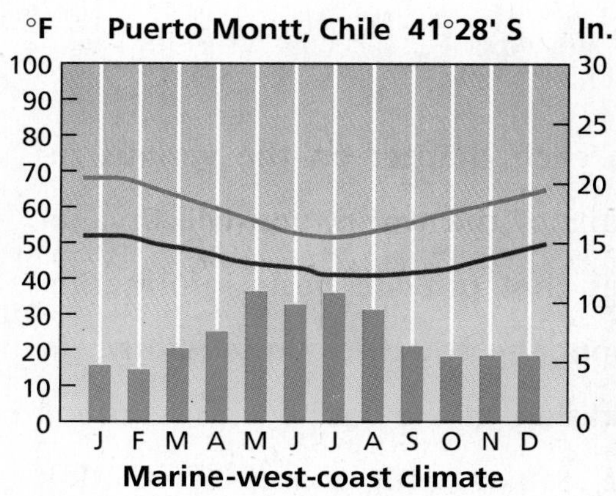

°F
Puerto Montt, Chile 41°28' S
In.
100
90
80
70
60
50
40
30
20
10
0
30
25
20
15
10
5
0
J F M A M J J A S O N D
Marine-west-coast climate

▲ **Figure 13: Reading a climograph**

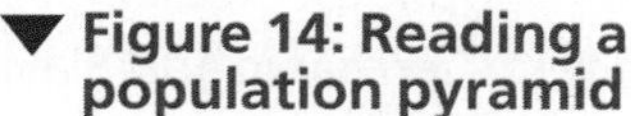

▼ **Figure 14: Reading a population pyramid**

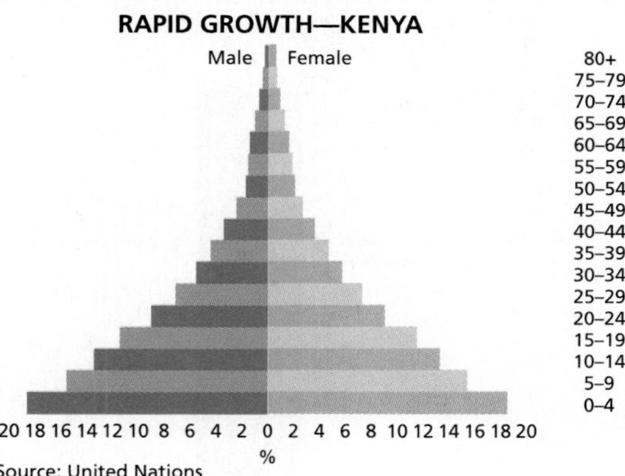

RAPID GROWTH—KENYA
Male
Female
80+
75–79
70–74
65–69
60–64
55–59
50–54
45–49
40–44
35–39
30–34
25–29
20–24
15–19
10–14
5–9
0–4
20 18 16 14 12 10 8 6 4 2 0 2 4 6 8 10 12 14 16 18 20
%
Source: United Nations

Population pyramids show the percentages of males and females by age group in a particular country or region. Population pyramids are split into two sides, one for male and one for female. Along the bottom are numbers that show the percentage of the male or female population that is represented by the age groups. The age groups are listed on the side of the population pyramid. The wider the base of a country's pyramid, the younger the population of that country.

Population pyramids help us understand economic, cultural, and historical patterns of a place or region. Look at the pyramid in **Figure 14**. The wide pyramid base for Kenya indicates a young, rapidly growing population. The pyramid shows that a little over 19 percent of Kenya's female population is made up of females under the age of 4. Nearly 20 percent of the country's male population is made up of males under the age of 4.

YOUR TURN

1. Look at the bar graph in Figure 11. Which countries shown in this graph have less than 50 billion barrels of oil reserves?
2. Study the pie graph in Figure 11. What percentage of the world's proven oil reserves is found in the Top 5 countries?
3. Refer to the line graph in Figure 12. What is the projected population of India by 2150?
4. Look at the climograph in Figure 13. When are average high and average low temperatures coolest in Puerto Montt?

USING CHARTS AND TABLES

In each chapter on the various regions of the world, you will find tables that provide basic information about the countries under study. In addition, at the back of this textbook are tables with more information about each country of the world.

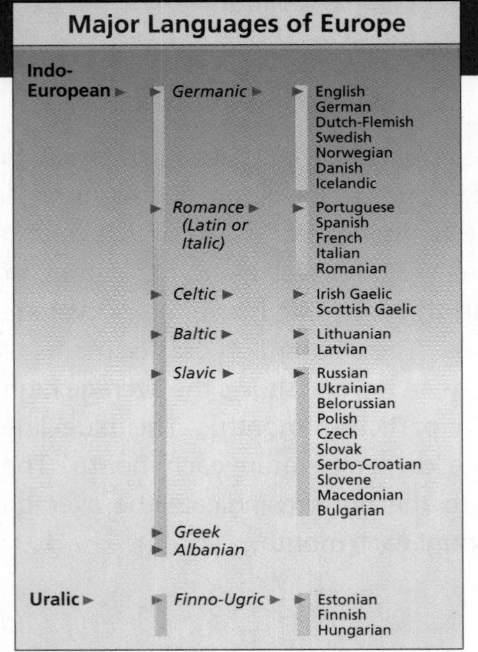

Major Languages of Europe
Indo-European
Germanic
English
German
Dutch-Flemish
Swedish
Norwegian
Danish
Icelandic
Romance (Latin or Italic)
Portuguese
Spanish
French
Italian
Romanian
Celtic
Irish Gaelic
Scottish Gaelic
Baltic
Lithuanian
Latvian
Slavic
Russian
Ukrainian
Belorussian
Polish
Czech
Slovak
Serbo-Croatian
Slovene
Macedonian
Bulgarian
Greek
Albanian
Uralic
Finno-Ugric
Estonian
Finnish
Hungarian

◀ **Figure 16: Reading a chart**

Not all charts take the form of tables. The illustration in **Figure 16** shows major languages in Europe. Related languages are grouped together as part of two language families: Indo-European and Uralic. English, Swedish, and Danish are all Germanic languages. Russian, Polish, and Czech are all Slavic languages. Germanic and Slavic languages are all part of the Indo-European language family.

The countries of Atlantic South America are listed on the left in the table in **Figure 15**. You can match statistical information on the right with the name of each country listed on the left. The categories of information are listed across the top of the table.

FACTS IN BRIEF **Atlantic South America**

COUNTRY POPULATION (1994)	LIFE EXPECTANCY (1994)	LITERACY RATE (1990)	PER CAPITA GDP (1993)
ARGENTINA 33,912,994	68, male 75, female	95%	$5,500
BRAZIL 158,739,257	57, male 67, female	81%	$5,000
PARAGUAY 5,213,772	72, male 75, female	90%	$3,000
URUGUAY 3,198,910	71, male 77, female	96%	$6,000
UNITED STATES 260,713,585	73, male 79, female	97% (1991)	$24,700

Source: *The World Factbook 1994*, Central Intelligence Agency

▲ **Figure 15: Reading a table**

Figure 17: Reading a time line ▶

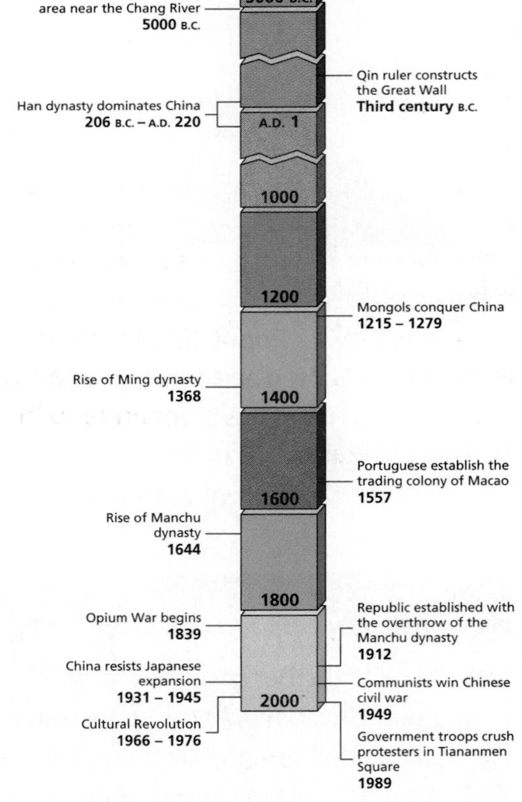

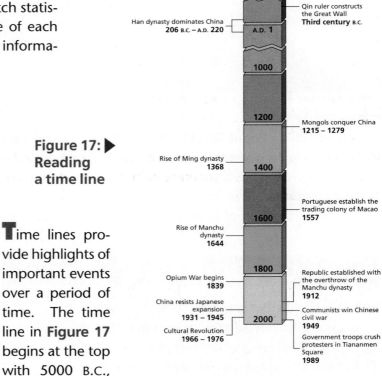

5000 B.C.
First rice farmers cultivate the area near the Chang River 5000 B.C.
Qin ruler constructs the Great Wall Third century B.C.
Han dynasty dominates China 206 B.C. – A.D. 220
A.D. 1
1000
1200
Mongols conquer China 1215 – 1279
Rise of Ming dynasty 1368
1400
Portuguese establish the trading colony of Macao 1557
1600
Rise of Manchu dynasty 1644
1800
Opium War begins 1839
Republic established with the overthrow of the Manchu dynasty 1912
China resists Japanese expansion 1931 – 1945
Communists win Chinese civil war 1949
2000
Cultural Revolution 1966 – 1976
Government troops crush protesters in Tiananmen Square 1989

Time lines provide highlights of important events over a period of time. The time line in **Figure 17** begins at the top with 5000 B.C., when rice was first cultivated by people in present-day China. The time line highlights important events that have shaped the human and political geography of China.

Flowcharts are visual guides that explain different processes. They lead the reader from one step to the next, sometimes providing both illustrations and text. The flowchart in **Figure 18** shows the different steps involved in harvesting cacao and preparing it for use by consumers. Among other things, chocolate and cocoa are made from cacao. The flowchart takes you through the steps of harvesting and processing cacao. Captions help guide you through the four such flowcharts in this textbook.

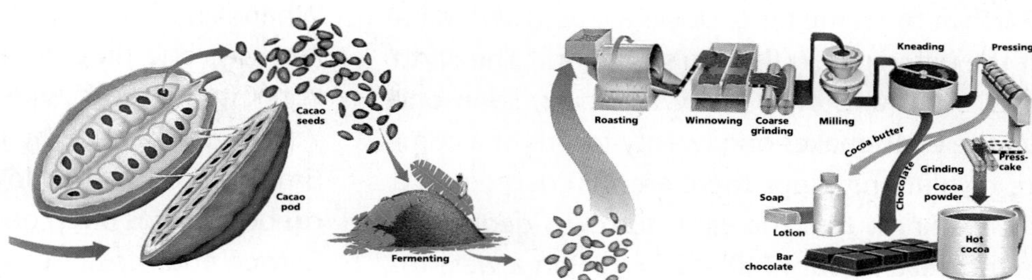

Cacao seeds
Cacao pod
Fermenting
Roasting
Winnowing
Coarse grinding
Milling
Kneading
Pressing
Cocoa butter
Chocolate
Grinding
Press-cake
Cocoa powder
Soap
Lotion
Bar chocolate
Hot cocoa

▲ **Figure 18: Reading a flowchart**

Graphic organizers can help you understand certain ideas and concepts. For example, the word web in **Figure 19** helps you think about the sources of water in the hydrosphere, which is discussed in Chapter 2, "The Earth in Space." In this word web, one source goes on each line. Depending on the subject, a word web can continue to branch out as you gather more and more details.

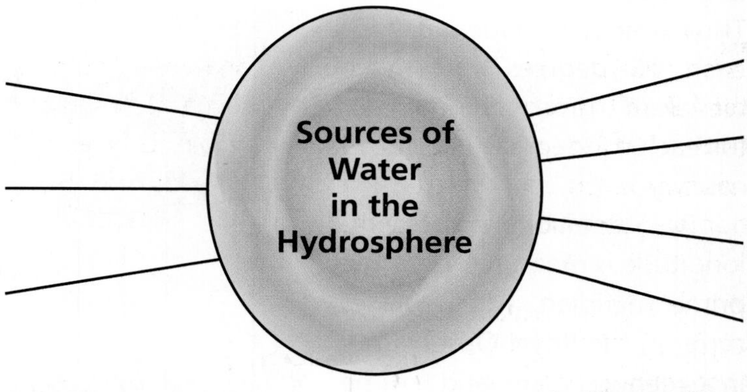

Sources of Water in the Hydrosphere

▲ **Figure 19: Word web**

YOUR TURN

1. Look at the statistical table for Atlantic South American countries in Figure 15. Identify the Atlantic South American country with the highest literacy rate.
2. Look at the language chart in Figure 16. Name the two languages that are part of the Celtic language group.
3. Look at the China time line in Figure 17. Identify two important events in China's history between 1200 and 1400.

READING A TIME-ZONE MAP

Since the sun is not directly overhead everywhere on Earth at the same time, clocks are adjusted to reflect the difference in the sun's position. The earth rotates on its axis once every 24 hours, so in one hour, the earth makes one twenty-fourth of a complete revolution. Since there are 360 degrees in a circle, we know that the earth turns 15 degrees of longitude each hour (360° ÷ 24 = 15°) in a west-to-east direction. Therefore, if a place on Earth has the sun directly overhead at this moment (noon), then a place 15 degrees to the west will have the sun directly overhead one hour from now. The earth will have rotated 15 degrees during that hour. As a result, the earth is divided into 24 time zones. Thus, time is an hour earlier for each 15 degrees you move westward. Time is an hour later for each 15 degrees you move eastward.

By international agreement, longitude is measured from the prime meridian, which passes through the Royal Observatory in Greenwich, England. Time also is measured from Greenwich and is called Greenwich mean time (GMT). For each time zone east of the prime meridian, clocks must be set one hour ahead of GMT. For each time zone west of Greenwich, clocks are set back one hour from GMT. When it is noon in London, it is 1:00 P.M. in Oslo, Norway, one time zone east, and 7 A.M. in New York City, five time zones west.

As you can see by looking at the map above, time zones do not follow meridians exactly. Political boundaries often determine time-zone lines. In Europe and Africa, for example, time zones follow national boundaries. The mainland 48 U.S. states,

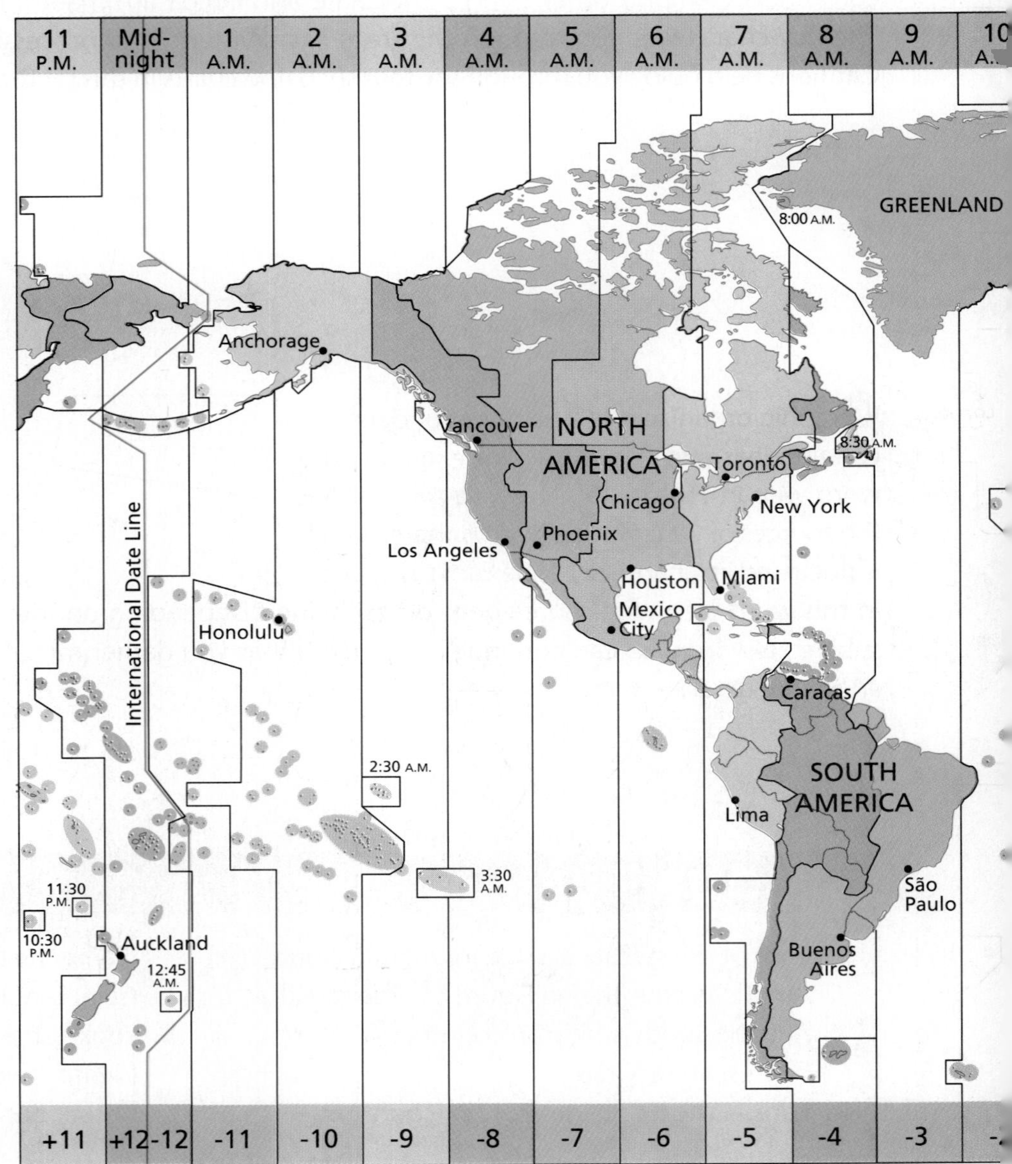
WORLD TIME ZONES
11 P.M.
Mid-night
1 A.M.
2 A.M.
3 A.M.
4 A.M.
5 A.M.
6 A.M.
7 A.M.
8 A.M.
9 A.M.
GREENLAND
8:00 A.M.
Anchorage
Vancouver
NORTH AMERICA
8:30 A.M.
Toronto
Chicago
New York
Phoenix
Los Angeles
Houston
Miami
Mexico City
International Date Line
Honolulu
Caracas
2:30 A.M.
SOUTH AMERICA
Lima
3:30 A.M.
São Paulo
11:30 P.M.
10:30 P.M.
Auckland
12:45 A.M.
Buenos Aires
+11
+12
-12
-11
-10
-9
-8
-7
-6
-5
-4
-3

meanwhile, are divided into four major time zones: Eastern, Central, Mountain, and Pacific. Alaska and Hawaii are in separate time zones to the west.

Some countries have made changes in their time zones. For example, most of the United States has daylight savings time in the summer in order to have more evening hours of daylight.

The international date line is a north-south line that runs through the Pacific Ocean. It is located at 180°, although it sometimes varies from that meridian to avoid dividing countries. At 180°, the time is 12 hours from Greenwich time. There is a time difference of 24 hours between the two sides of the 180° meridian. The 180° meridian is called the international date line because when you cross it, the date and day change. As you cross the date line from the west to the east, you gain a day. If you travel from east to west, you lose a day.

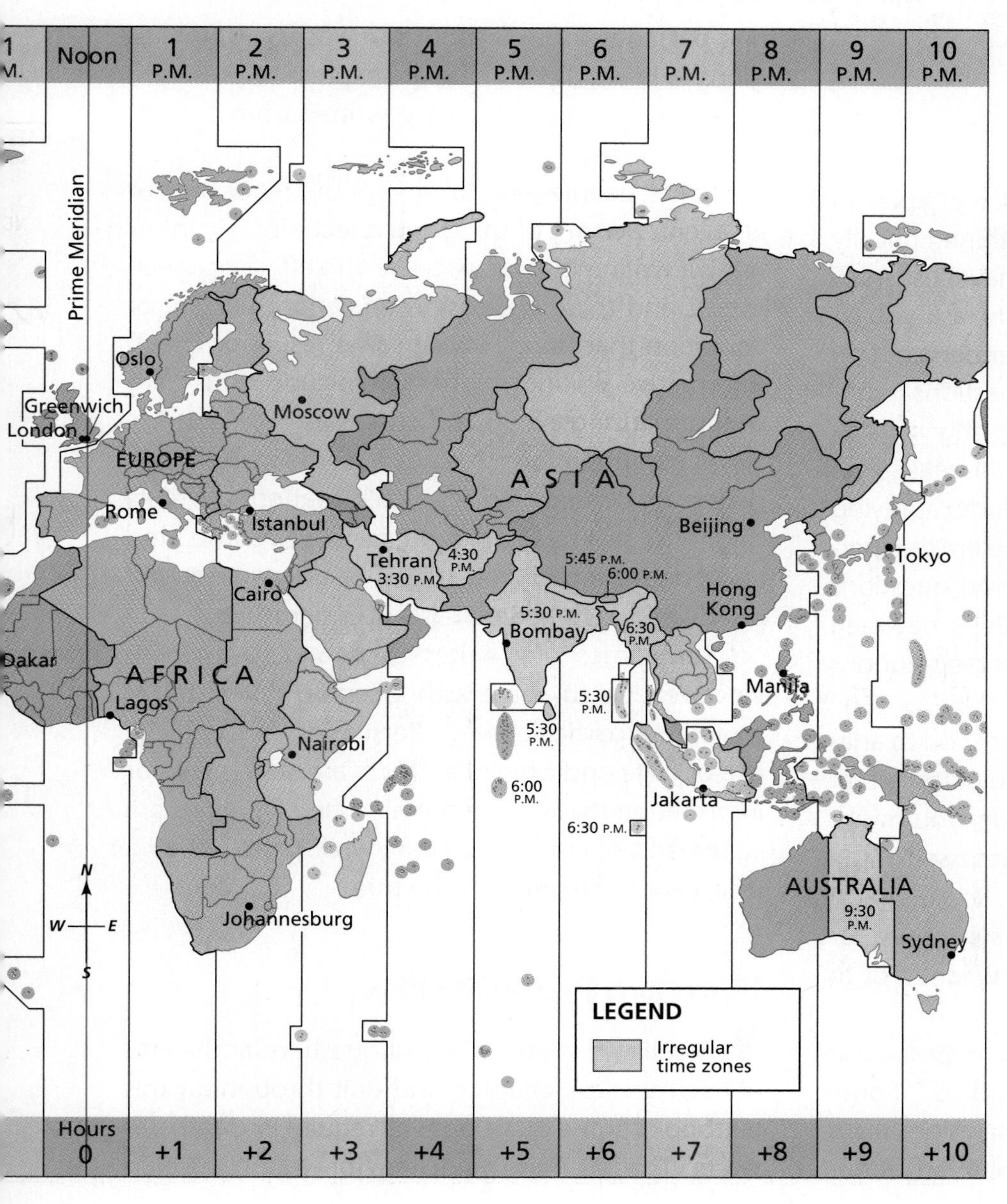

Noon
1 P.M.
2 P.M.
3 P.M.
4 P.M.
5 P.M.
6 P.M.
7 P.M.
8 P.M.
9 P.M.
10 P.M.
Prime Meridian
Oslo
Greenwich
London
Moscow
EUROPE
Rome
Istanbul
ASIA
Beijing
Tokyo
Tehran
3:30 P.M.
4:30 P.M.
5:45 P.M.
6:00 P.M.
Hong Kong
Cairo
5:30 P.M.
Bombay
6:30 P.M.
Dakar
AFRICA
Manila
Lagos
5:30 P.M.
Nairobi
6:00 P.M.
Jakarta
6:30 P.M.
AUSTRALIA
9:30 P.M.
Sydney
Johannesburg
N
W
E
S
LEGEND
Irregular time zones
Hours
0
+1
+2
+3
+4
+5
+6
+7
+8
+9
+10

YOUR TURN

1. In which time zone do you live? Check your time now. What time is it in New York?
2. How many hours behind New York is Anchorage, Alaska?
3. How many time zones are there in Africa?
4. If it is 9 A.M. in the middle of Greenland, what time is it in São Paulo?

THINKING CRITICALLY AND WRITING

The study of geography requires more than analyzing and understanding information provided by tools like maps and graphs. Throughout this textbook, you will be given opportunities to think critically and write about information you have studied.

THINKING CRITICALLY

At the end of each section and each chapter in this textbook, you will be asked to think critically about some of the information you have just studied. Critical thinking questions might ask you to demonstrate your knowledge or understanding about something in the text. Such questions might ask you to tackle such tasks as evaluating, analyzing, or interpreting information from the text.

Some critical thinking skills include applying what you have learned to solve problems or perform certain tasks. For example, some questions might ask you to examine relationships between pieces of information and to seek possible solutions to problems.

Other critical thinking questions ask you to analyze— to break down information into parts so that it is easier to understand. For example, you might be asked to show how some things discussed in the text are related. On the other hand, some critical thinking questions might ask you to synthesize, or pull information together, to form new ideas, or to develop additional information.

Finally, thinking critically also means evaluating or judging information you have studied. Some questions might ask you to draw conclusions from pieces of information, to give an opinion about something you have read, or to challenge or verify the truth of information you read.

PURPOSES FOR WRITING

Writers have many different reasons for writing. In your study of geography, you might write to accomplish many different tasks.

Sometimes writers *express* their own personal feelings or thoughts about a topic or event. At the beginning of most units in this textbook, you will find passages that tell about some aspect of the regions to be studied. These are expressive passages, some prose and some poetry, written from personal experiences.

Some people write to *inform* the reader about an event, person, or thing. One technique in informative writing is to show cause and effect between events and things. Cause is an action, person, or condition that brings about some result, or effect. Informative writing might also include describing or summarizing events and processes.

In aiming to inform readers, some authors also define or *classify* things or information. Writers might use several techniques, including comparing and contrasting characteristics or advantages and disadvantages of different items under study.

Sometimes one writes to *persuade*—to convince readers to agree with a certain statement or to act in a particular way. Persuasion relies on the use of facts and opinions. Facts are statements or information that can be proven true. Opinions are statements or information that, while perhaps being true, cannot be proven to be true.

WRITING PARAGRAPHS

You will find various kinds of questions at the end of each section, chapter, and unit throughout this textbook. Some questions will require in-depth answers. The following guidelines for writing will help

The Writing Process

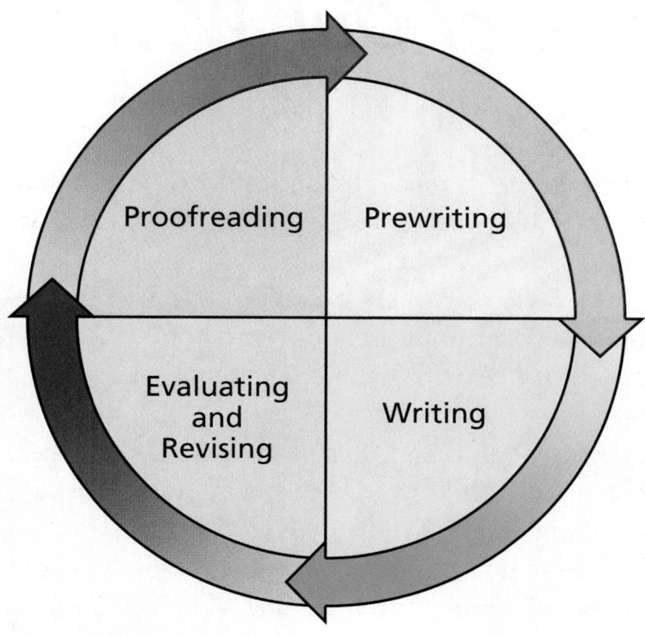
Proofreading
Prewriting
Evaluating and Revising
Writing

you structure your answers so that they clearly express your thoughts.

Prewriting

Prewriting is the process of thinking about and planning what to write. Prewriting includes gathering and organizing information into a clear plan. Writers also use the prewriting stage to identify the audience and purpose for what is to be written.

Often, writers must conduct research to get the information they need. Research can include finding primary and secondary sources. Primary sources are original information provided by first-hand accounts. Secondary sources pull information from other works. Secondary sources include encyclopedias, newspaper and magazine articles, and biographies.

Writing a Draft

After you have gathered and arranged your information, you are ready to begin writing. Many paragraphs are structured in the following way:

The Topic Sentence The topic sentence states the main idea of the paragraph. Putting the main idea into the form of a topic sentence helps keep the paragraph focused.

The Body The body of a paragraph develops and supports the main idea. Writers use a variety of information, including facts, opinions, and examples, to support the main idea.

The Conclusion The conclusion summarizes the writer's main points or restates the main idea.

Evaluating and Revising

Read over your paragraph and make sure you have clearly expressed what you want to say. Sometimes it helps to read your paragraph aloud or ask someone else to read it. Revise parts of your paragraph that are not clear or that stray from your main idea. You might want to add, cut, replace, or reorder sentences to make your paragraph clearer.

Proofreading

Before you write your final draft, read over your paragraph and correct any spelling, grammar, and punctuation errors.

Writing Your Final Version

After you have revised and corrected your draft, neatly rewrite your paragraph. Make sure your final version is clean and free of mistakes. The appearance of your final version can affect how your writing is perceived and understood.

YOUR TURN

1. What are the steps in the writing process?
2. What elements are found in many paragraphs?

UNIT 1

GEOGRAPHY: *A View of Our World*

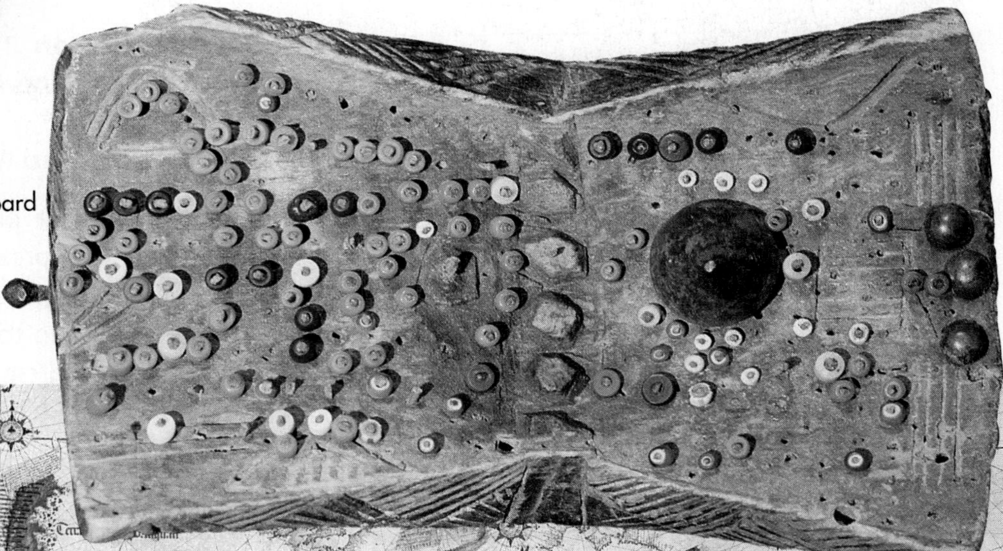

African memory board

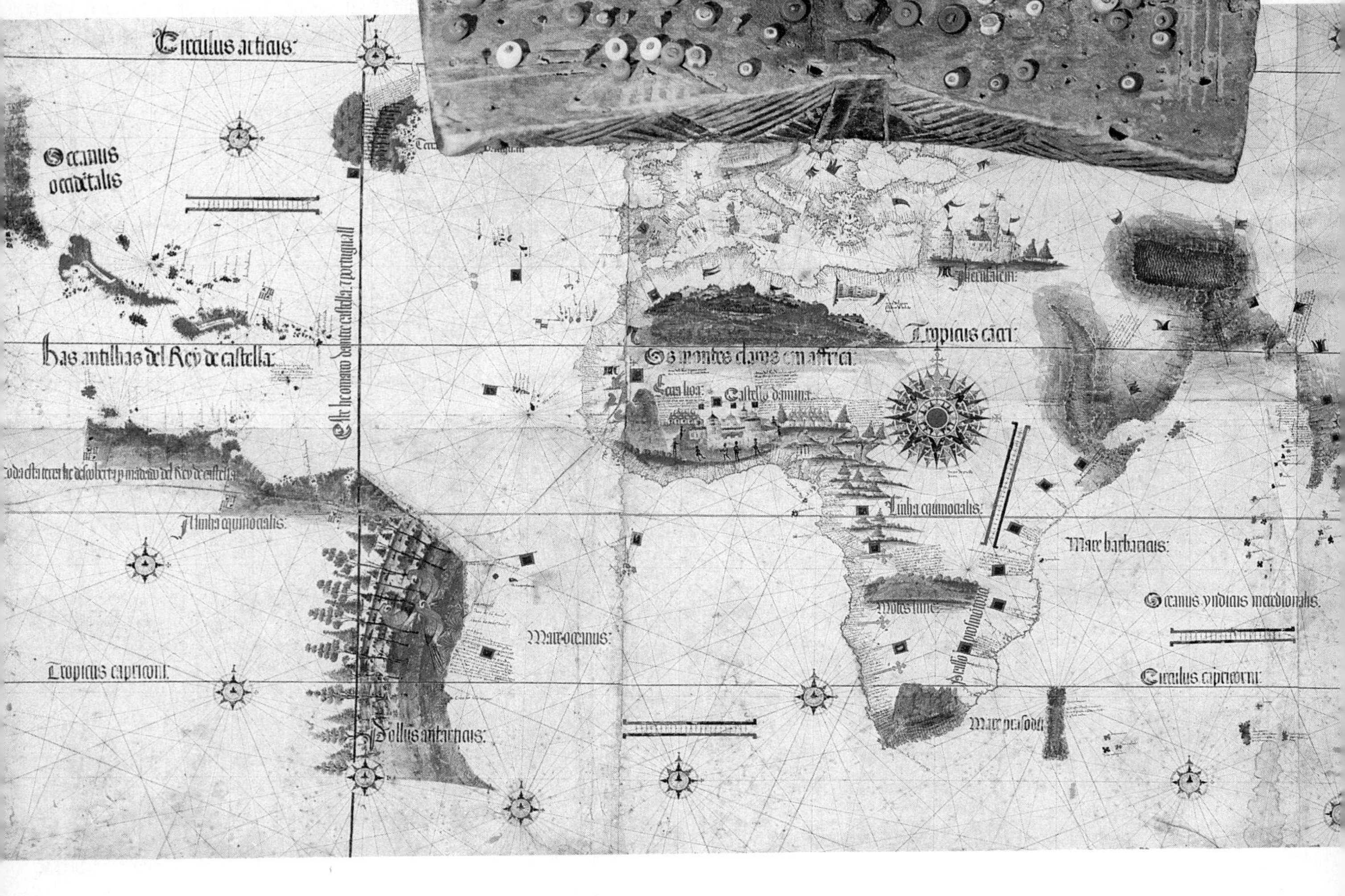

How people view their world is reflected in how they represent it on maps. Some nineteenth-century Africans used a memory board (top, opposite) to locate important sites and record information about their society. The Cantino map (bottom, opposite) shows how little the Europeans in 1502 knew of what they called the New World. The fifteenth-century Korean map (bottom of this page), with an oversized China and Korea, reflected an East Asian view of the world. Today, we can see our world through computers and in great detail, as in this three-dimensional view of the Sphinx in Egypt (below).

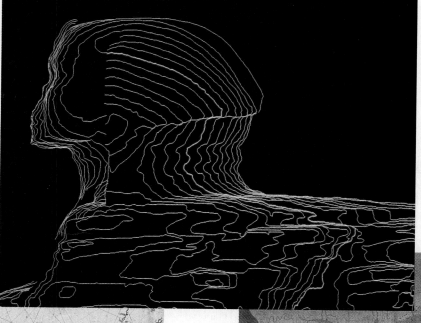

UNIT 1 CHAPTERS

Computer-generated image of the Sphinx

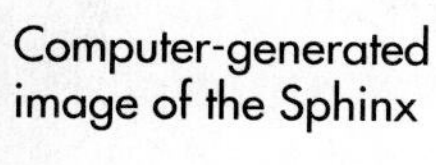
Comercial agriculture

Fifteenth-century Korean map

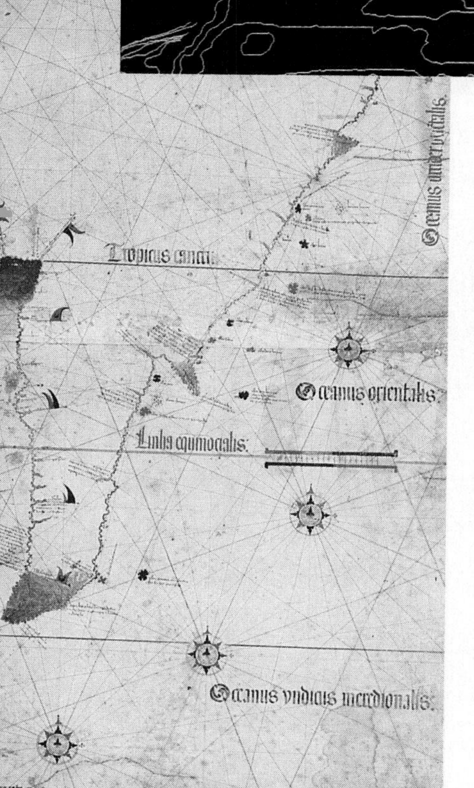

Cantino world map

CHAPTER 1

The Geographer's World

> "Once in his life a man [person] ought to concentrate his mind upon the remembered earth, I believe. He ought to give himself up to a particular landscape in his experience, to look at it from as many angles as he can, to wonder about it, to dwell upon it."
>
> *N. Scott Momaday*

GEOGRAPHY DICTIONARY

site
absolute location
situation
relative location
drought
spatial interaction
cartography
topographic
meteorology

What does the location of your school or city have in common with an African farmer's method of raising crops or herding animals? What does the location and design of a shopping mall have in common with national parks and tourism? The answer is space. The use of space and how people interact with their surroundings is the focus of geographers, and this focus gives geographers a special way of looking at the world.

In its broadest sense, geography is about discovery of the world around us, because people everywhere make use of space and interact with their environment and with each other. Thus, we use geography every day in finding our way to school, going on trips, or reading about how people live in other countries. We think geographically every time we decide what to do, where to go, how to dress, and even what to eat.

Today, all of the world's peoples are linked. Events far away influence us all, and we share the world's resources, products, and ideas. Geographers study places where products were designed and made. They study how and where products and resources are transported. And they study the people around the world who are involved in these and countless other activities.

Sunrise near Blue Ridge Parkway, **North Carolina**

1 GEOGRAPHERS LOOK AT THE WORLD

FOCUS

- *What are some topics that geographers study?*
- *What are the Five Themes of Geography?*

Understanding the World Nearby Shopping malls make fascinating geography. For many Americans, a mall is much more than a place to shop. It is a place to socialize with friends and to share news. But for geographers, the building of a mall is a serious study in land use.

Location is a key characteristic of a shopping mall. A mall needs a physical **site**—an exact, **absolute location**. So that the foundations of the buildings and the parking lots will function well, the land should be stable and well drained. A site also must have a good **situation**, or **relative location**. Shoppers should not have to travel too far from their homes, and the route to the mall should be fairly simple. For these reasons, malls often are located at exits along highways. This situation means that customers and merchandise can get to the mall easily.

Place also is a key characteristic of a shopping mall. The shopping mall must be an interesting and pleasant place. Much time and money is spent deciding how the mall should look and feel.

FIVE THEMES OF GEOGRAPHY
- LOCATION
- PLACE
- HUMAN-ENVIRONMENT INTERACTION
- MOVEMENT
- REGION

FIVE THEMES OF GEOGRAPHY

LOCATION

Every place on Earth has a location. Location is defined in terms of absolute and relative location.

Absolute location: *the exact spot on Earth where something is found*

Example: Niamey, the capital of Niger, is located at 13° 31′ north latitude and 2° 07′ east longitude.

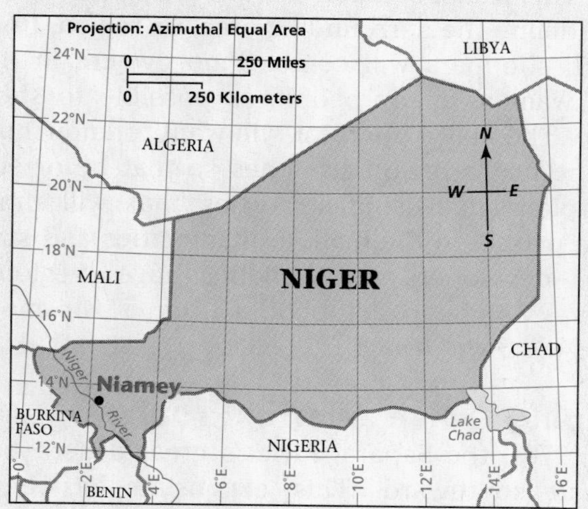

Projection: Azimuthal Equal Area
0
250 Miles
0
250 Kilometers
LIBYA
ALGERIA
NIGER
MALI
CHAD
Niamey
Niger River
BURKINA FASO
NIGERIA
Lake Chad
BENIN
N
W
E
S
24°N
22°N
20°N
18°N
16°N
14°N
12°N
0°
2°E
4°E
6°E
8°E
10°E
12°E
14°E
16°E

Relative location: *the position of a place in relation to other places*

Example: Yosemite National Park is located north of Los Angeles, California, and east of San Francisco, California.

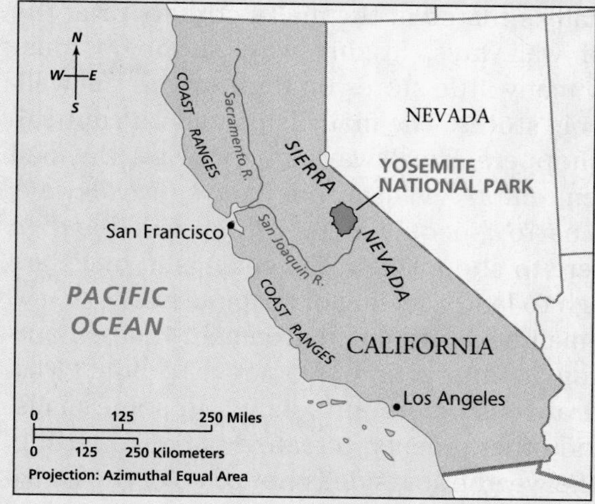

N
W
E
S
COAST RANGES
Sacramento R.
SIERRA NEVADA
NEVADA
YOSEMITE NATIONAL PARK
San Francisco
San Joaquin R.
PACIFIC OCEAN
COAST RANGES
CALIFORNIA
Los Angeles
0
125
250 Miles
0
125
250 Kilometers
Projection: Azimuthal Equal Area

Your Turn

1. Use an atlas to find the absolute location of your city or town.

2. Write a sentence describing the relative location of your home.

PLACE

Every place on Earth has special characteristics that make it different from every other place.

Every place can be described in terms of its physical characteristics, including weather, land features, plants, and animals.

Example: The Sahel in Africa is hot and dry. Overgrazing by animals and the search for firewood by humans has stripped the Sahel of vegetation.

Every place can be described in terms of its human features. These features include a place's peoples, their cultures, and their ideas.

Example: Yosemite National Park has several hotels and campgrounds to accommodate tourists.

Your Turn

Describe the physical and human characteristics of your school.

This study includes selecting the positions of individual stores. For example, the biggest stores are usually situated at the mall's corners or at the ends of walkways. In this way, shoppers must walk by many little stores on their way to the well-known big stores. The mall also should be appealing to shoppers. If walkways are too wide, the mall will seem empty even on busy days. If walkways are too narrow, people might feel crowded and go elsewhere to shop. Often, the ceilings in malls are very high to lend a feeling of spaciousness.

A mall becomes an environment that encourages human interaction. Attractive product displays draw customers into shops. In some malls, trees and other greenery create a park-like setting. When the weather outside is hot or cold, the mall's pleasant, controlled environment attracts people.

The movement of people and merchandise is also part of the geography of the shopping mall. Trucks, railways, airplanes, and ships move goods from distant places to mall stores. Nearby roads and highways lead to and from shoppers' homes.

The transportation routes to the mall will help define the surrounding area, or region, from which shoppers will come. Shop owners in the mall want as many people as possible to shop there. The mall's operator will want to know how far the mall's region stretches so that he or she can plan advertising strategies that will draw more people to the mall. In many cities and suburbs, developers are building huge "regional malls," which attract people from 50 or more miles (80.5 km) away.

Helping Solve World Problems Since the 1970s, the Sahara, Africa's great desert, has spread southward. This expansion has been caused by **drought** (a long, dry period with no rain) and the overgrazing of land by livestock. The countries of Mauritania, Mali, and Niger have been severely affected, and the place characteristics of the Sahel have changed. Lands once

FIVE THEMES OF GEOGRAPHY

HUMAN-ENVIRONMENT INTERACTION

People are constantly interacting with their surroundings. This geographic theme is called human-environment interaction.

People adapt to their environment.

Example: People who live in the Sahel wrap their heads with heavy material to keep the hot sun and sand out of their faces.

People change their environment.

Example: The developer of a shopping mall chooses to bulldoze the land to level it for buildings and parking lots.

Your Turn

Give one example of how you have adapted to your environment and one example of how you have changed your environment.

covered with brush and grass are now bare. The movement of people and animals away from the desert has created overcrowded conditions in nearby environments. Geographers are working with governments and international agencies to help solve this problem.

Geographers may use satellite images and aerial photos to locate the exact boundaries of Africa's overgrazed regions. Teams of researchers, including geographers, then travel to these places for on-site surveys. They study the human-environment interaction in the region. Some investigate how the local economy has declined. Others research the crisis' impact on families and the extent of hunger. Still others check on the health of livestock and soil and water quality. Deterioration of the natural environment is then studied in detail.

Meanwhile, other geographers travel to regions where the drought has been less severe and where lands have not been overgrazed. The drought and non-drought regions can be compared in order to make recommendations for helping the victims of the drought. Perhaps roads need improvements so that emergency supplies can be moved quickly to all parts of the region.

Solving the world's problems depends increasingly on understanding local and regional geography. Solutions to problems in richer countries may not always work in poorer countries.

Preserving Natural Environments One of the most rapidly growing industries in the world is tourism. In the United States, where national parks are popular destinations, the National Park Service is struggling to maintain the natural beauty of the parks as the number of visitors continues to increase.

Yosemite in California's Sierra Nevada is one of the country's most popular national parks. Yosemite's relative location to San Francisco and Los Angeles is convenient for visitors. Tourists

MOVEMENT

People, goods, and ideas move continuously. The movement of people, goods, and ideas is called **spatial interaction**.

People meet their needs either by traveling to other places or by trading with people in other places.

Example: Open markets in the Sahel provide centers where herders and farmers can trade their goods.

As the amount of movement and kinds of transportation change, so do other geographic features.

Example: The movement of tourists into Yosemite National Park has created a need for a shuttlebus service.

Your Turn

Give one example of how the amount of movement has changed geographic features in your community.

traveling to California's coastal cities often add Yosemite to their travel plans.

Yosemite's place characteristics include beautiful mountain scenery. Tourists can see breathtaking vistas of steep mountains, green forests, and waterfalls. Hiking trails wind throughout the park. Yosemite Valley also has lovely hotels and camping sites. The park draws visitors from throughout the United States and many foreign countries. Many signs at park facilities are printed in several languages, including Spanish, French, German, and Japanese.

During the 1960s and 1970s, thousands of tourists crowded into the park in a day. On holidays, park employees could expect 40,000 people to visit. The interaction between Yosemite's natural environment and its many visitors created problems. Many of the hiking trails had to be paved. The hotels and campgrounds were overflowing. Air pollution from the movement of automobiles in and out of the park clouded the air and limited the views. A steady procession of garbage trucks carried the valley's waste away. Rather than guide visitors through Yosemite's natural wonders, park rangers directed traffic.

Solving the park's transportation problems was important to the future of Yosemite. Geographers and other planners had to make difficult decisions about the building of more roads and hotels. Automobile access to the park has been limited, and shuttlebuses now take tourists into and out of Yosemite Valley. As a result, air pollution has decreased. Preserving the beauty and wildlife of Yosemite and managing tourism in the park are ongoing challenges. The effort to protect Yosemite is an excellent example of geographers working with other professionals to solve environmental problems.

REGION

A region is an area defined by common characteristics.

Regions can be defined by more than one characteristic.

Example: The Sahel can be defined as a dry grassland region in Africa because of its physical characteristics. Many people in the Sahel speak Arabic, so it also can be defined as a region by language. And the Sahel can be defined as a region by religion because many people are Muslims, or members of the Islamic faith. Shown at right is a mud mosque in West Africa.

To learn more about how geographers classify regions, read the special feature "What's in a Region?" on pages 100–101.

Your Turn

Name three different regions in which you live.

The Five Themes of Geography There are a number of common geographic themes in the case studies you have just read about. These common themes are location, place, human-environment interaction, movement, and region. Together these themes are known as the Five Themes of Geography.

One or more of the themes can be found in the study of any topic in geography. As you study the major world regions in the other parts of this textbook, it will be helpful to use the Five Themes of Geography to organize your thinking. Refer to the charts in this section throughout your study of *World Geography Today.*

1 SECTION REVIEW

1. How can geographers help solve local and regional problems? Use examples from this chapter.
2. Name the Five Themes of Geography and give an example of each.
3. **Critical Thinking** **Apply** the Five Themes of Geography to the drought problem in Africa.

2 THE WIDE WORLD OF GEOGRAPHY

FOCUS

- *What is the difference between human geography and physical geography?*
- *What are some professions that involve geography?*

Branches of Geography By now you know that geography covers a wide range of topics. Often, it is divided into different branches. The two main branches of geography are human and physical geography, and each of these branches can be further divided into many subjects.

Human geography is the study of how people and their activities vary from place to place. It includes political, economic, and cultural factors such as cities and countries, farms and factories, and ways of life. Geographers are unique social scientists in that they borrow elements of many subject areas to conduct their studies. Geographers who work in human geography might specialize in historical, political, economic, medical, population, or urban geography.

Physical geography is the study of how the earth's natural features vary from place to place. It includes the study of plains and mountains, weather and climates, and plants and animals. Specialty areas of physical geography include the study of water resources, the study of the distribution of plants and animals, and the study of the earth's surface. The interactions of people with water, the weather, soil and minerals, and plants and animals are links between human geography and physical geography.

In this textbook, both human and physical geography are used to study the world's regions. Geographers use regional geography to divide our world into smaller topics for easier study. In this way, different parts of the world can be compared and analyzed. The boundaries of regions might be rivers, mountain ranges, or oceans. Or they might be highways or lines on a map defined by a government. The Five Themes of Geography will serve as a guide to your study of geography and the connections between the earth and its people.

Geography as a Profession Most professions require an understanding of geography. For example, a grocery store manager must know the region of his or her store and what types of products to stock. A physician must understand the physical and social environments of her or his patients. A news reporter may travel all over the world writing articles about people and events. There are, however, several professions that rely on specially trained geographers.

Cartographers are people who make maps. **Cartography** is the branch of geography that studies maps and mapmaking. Although many maps are drawn by hand, computers are revolutionizing map production. Computers store data from old maps as well as satellite images, photographs, and data from surveys and reports. The cartographer then "draws" the map on a computer screen and sends it to a printer that produces a copy almost instantly. Cartographers work for companies that publish maps, atlases, newspapers, magazines, and books. They also are employed by nearly all government agencies. The largest employer of cartographers in the United States has been the United States Geological Survey (USGS), which produces all the **topographic** (surface feature) maps for the United States.

You see geographers at work whenever you watch the nightly news. The person providing information about the weather is called a meteorologist. **Meteorology** is the field of geography that specializes in weather and weather forecasting.

Geographers are taking on an increasingly important role in studying the environment. They are able to research, map, and analyze environmental data. They may gather information about pollution, water quality, hazardous waste, endangered species of plants and animals, the loss of rain forests, and many other environmental issues.

Geographers also make decisions about managing places, guiding the transport of products, and encouraging regional development. Many geographers who work in environmental management investigate land-use practices and advise how lands should be used.

Businesses hire geographers, too. They decide where new stores should be located, schedule routes for airlines and trucking companies, and help identify new markets for businesses. Geographers also work at United Nations offices all around the world. They help organizations such as the Red Cross and the World Bank. Geographers also become experts on different peoples, cultures, and regions of the world.

Geography teachers help develop an informed public. Many people seem to know little about their own country and even less about places far away. In a closely linked world, we must all learn as much as we can about locations, places, human-environment interaction, movement, and regions.

Geographic knowledge also is necessary for good citizenship. People need to be a positive influence on their government's policies. Should we allow suburbs to be built over farmland? Where should garbage and poisonous chemicals be disposed of? How can we encourage economic progress in a poor region? These kinds of questions are geographic. Helping the government and its citizens find answers to these questions is the job of geographers.

2 SECTION REVIEW

1. List three specialty areas of human geography.
2. List three specialty areas of physical geography.
3. **Critical Thinking** **Explain** how geography is useful in everyday life.

CHAPTER 1 CHECK

Cairo, Egypt

Reviewing the Main Ideas

1. Geography is the study of the world, both nearby and far away. Its focus is spatial. Geography can help solve almost any problem that has to do with space and how people use space.
2. There are five fundamental themes that provide an organizational framework for geography: location, place, human-environment interaction, movement, and region.
3. Human geography and physical geography are the two main branches of geography. An understanding of geography is important for many professions and for good citizenship.

Building a Vocabulary

1. Discuss the difference between a site and a situation, and between an absolute location and a relative location. How are the terms related?
2. What is a drought?
3. Define *spatial interaction*. To which geographic theme is it related?

Recalling and Reviewing

1. Give an example of a world problem that geography can help solve.
2. Which of the Five Themes of Geography might you use in planning a trip to a destination of your choice, and how would you use them?
3. What is the difference between human geography and physical geography? Name one profession that uses mostly human geography, one profession that uses mostly physical geography, and one profession that uses both.

Thinking Critically

1. Think of a major news event you have heard about recently in the newspaper or on television. How is geography involved in that event?
2. What are some professions that involve geography that are not mentioned in this chapter? Consider the jobs of your friends and family, and discuss how geography is a part of them.

Learning About Your Local Geography

Individual Project

Many of the review exercises in this textbook will ask you about your "community," and it will be helpful for you to decide in advance what your community is, geographically. Is your community the area around your home, your school, neither, or both? Is your community a street block or rural route, a neighborhood, a whole town? What are the boundaries of your community? Who lives in your community? Write two or three sentences defining your community. You may want to include your definition in your individual portfolio.

Using the Five Themes of Geography

In a list or chart, apply the Five Themes of Geography to a movie you have seen recently—any movie of any kind. In what **locations** did the movie take place? What characteristics did the **places** in the movie have? What evidence of **human-environment interaction** did the movie give? What kinds of **movement** were shown? And finally, how did the movie apply the concept of **region**?

CHAPTER 2

The Earth in Space

> "The sight of stars always sets me dreaming just as naïvely as those black dots on a map set me dreaming of towns and villages. Why should those points of light in the firmament, I wonder, be less accessible than the dark ones on the map of France?"
>
> *Vincent van Gogh*

GEOGRAPHY DICTIONARY

solar system
planet
satellite
rotation
revolution
tropics
polar region
solstice
Tropic of Capricorn
Antarctic Circle
Arctic Circle
Tropic of Cancer
equinox
earth system
atmosphere
lithosphere
hydrosphere
biosphere

If you look at the sky on a clear night, you can see thousands of stars. With a telescope, you can see millions more. Beyond the telescope's view are billions more. All these stars are part of the universe, made up of all existing things, including space and Earth. Scientists estimate that the universe is 15 to 20 billion years old. Since its birth, according to scientists, the universe has been expanding continuously and is unimaginable in size.

The largest objects in space are stars, which generally are grouped together in huge, rotating collections called galaxies. Many objects that look like individual stars to the naked eye are really billions of stars in a galaxy far, far away. One galaxy, the Milky Way, appears as a bright milky streak across the night sky. Many individual stars in the Milky Way are visible because it is the galaxy in which we live. The sun is actually a medium-size star near the edge of the Milky Way.

When you look up at the night sky, do you ever wonder if there are places with life besides Earth? Scientists wonder this and continue to search for clues. The vast universe holds unlimited possibilities for future discovery.

Earth and Moon

The Solar System

Uranus
Neptune
Mercury
Venus
Jupiter
Pluto
Sun
Earth
Saturn
Mars

The sun and nine known planets make up part of our solar system. Which two of the nine planets are closest to Earth?

THE SUN, THE EARTH, AND THE MOON

FOCUS

- *What is the solar system?*
- *What factors determine the amount of solar energy that the earth receives?*

The Solar System The sun is the star that is the center of our **solar system**. The nine **planets**, including Earth, that move around the sun are part of the solar system. Planets are spheres, or ball-like objects, that orbit around a star. Jupiter is the largest planet, followed by ringed Saturn. Earth is one of the smaller planets and is fifth in size. It is the third nearest planet to the sun. (See the diagram on this page.)

The Moon Moons also are part of the solar system. A moon orbits around a planet. A body that orbits a larger body is called a **satellite**. Moons are natural satellites. Some planets have many moons—Saturn has 17. Others, such as Venus, have none. Earth has one large moon whose diameter is a little more than one-fourth that of Earth. On a clear night, we often can see the barren volcanic surface of the moon.

The moon, the earth, and the sun exert forces on each other that influence physical processes on Earth such as the movement of the ocean known as tides. Tides are complex, but, generally, tides occur because of the gravitational pull of the moon and sun, and the force produced by the rotation of the earth and moon together. (See the diagram on page 12.)

The Sun The sun is small compared to some other stars. Yet, when compared to Earth, the sun is enormous. For instance, the diameter of the earth is about 8,000 miles (12,872 km). The diameter of the sun is about 865,000 miles (1,391,785 km). The sun is more than 100 times the diameter of Earth and more than 400 times the diameter of

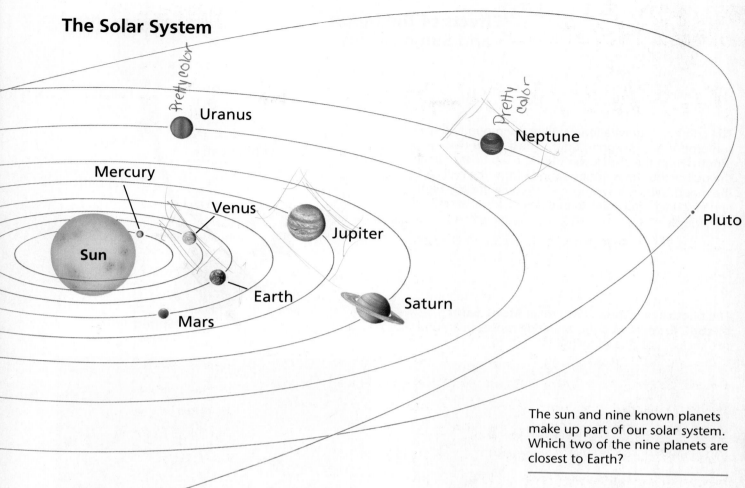

Sun

Effects of the Moon and Sun on Tides

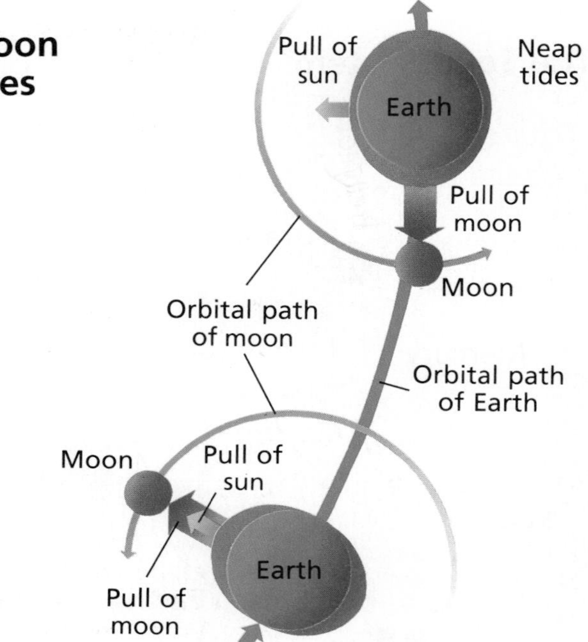
Pull of sun
Neap tides
Earth
Pull of moon
Moon
Orbital path of moon
Orbital path of Earth
Moon
Pull of sun
Earth
Pull of moon
Spring tides

The combined gravitational pull of the moon and the sun produces high spring tides, which occur twice a month. High tides are lower during neap tides, when the pull of the sun is at right angles from the pull of the moon. Why is knowing the schedule of daily high and low tides important to people in coastal areas?

The photo at left shows low tide at **Mont Saint Michel, France**. At right, high tide moves in around the ancient abbey and town.

the moon. The sun looks much larger and brighter than other stars because it is so much closer to the earth. Even though the earth averages 93 million miles (150 million km) from the sun, the next nearest star, Proxima Centauri, is about 25 trillion miles (40 trillion km) from Earth.

Solar Energy Almost all of the earth's energy comes from the rays of the sun. This energy makes life as we know it possible. The amount of solar energy received plays a major role in determining weather, vegetation, animal life, and the ways of life of people around the world. It influences the clothes people wear, the types of homes they live in, the food they grow and eat, and even the types of sports they play.

Three factors control the amount of solar energy that falls on different parts of the earth. These are the earth's rotation, revolution, and tilt.

MOVEMENT **The Earth's Rotation** Think of Earth as having an imaginary rod running through it from the North Pole to the South Pole. This rod represents the earth's axis, and the planet spins around on it. One complete spin of Earth on its axis is one **rotation**, which takes 24 hours. We see the effects of the earth's eastward rotation as the sun "rising" in the east and "setting" in the west. To us, it appears that the sun is moving, while actually it is Earth rotating on its axis.

Solar energy strikes only the half of the earth that is facing the sun. If Earth did not rotate

Angle of Sun's Rays Hitting Earth

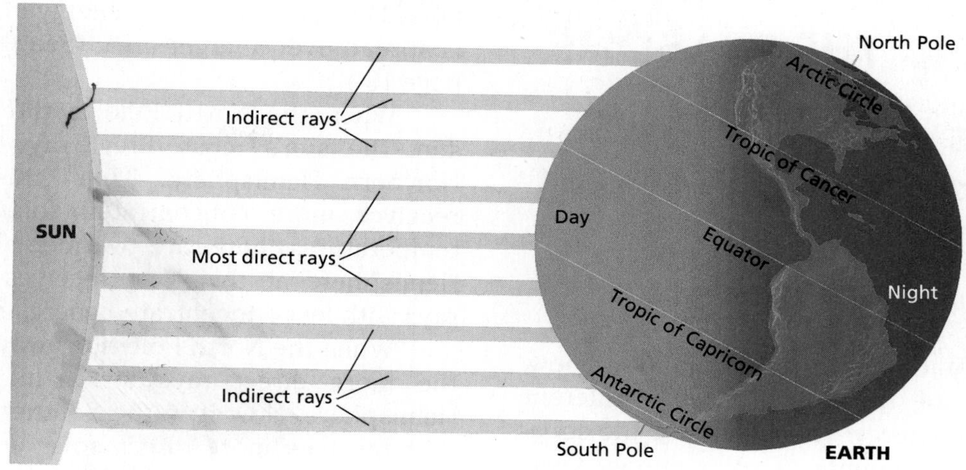
North Pole
Arctic Circle
Tropic of Cancer
Indirect rays
SUN
Day
Equator
Most direct rays
Night
Tropic of Capricorn
Indirect rays
Antarctic Circle
South Pole
EARTH

The tilt of the earth's axis and the position of the earth on its orbit determine where the sun's rays most directly strike the planet. Why is the area along the equator always warmer than the polar regions?

on its axis, only the half facing the sun would receive the sun's energy. That side of the earth would be extremely hot. The half of the earth facing away from the sun would always be dark and extremely cold. The earth's rotation makes it possible for the earth's surface to be exposed for a time to the warming effects of daylight and to the cooling effects of darkness. The earth's rotation creates day and night.

The Earth's Revolution In addition to rotating on its axis, the earth revolves around the sun. The earth makes one nearly circular orbit, or one **revolution**, every 365¼ days (one Earth year). Each time you celebrate your birthday, you have just completed another orbit around the sun. For convenience, our calendars have 365 days in one year. To account for the one-fourth day gained each year, an extra day (February 29) is added to the calendar every fourth year. This year, one day longer than the previous three, is called a leap year.

The Earth's Tilt The earth's axis is always pointed toward the same spot in the sky. The north polar axis always points to a star known as the North Star. The position of the axis is fixed in respect to the North Star, but not to our sun. The earth's orbit lies on a plane that runs from the center of the sun to the center of the earth. The earth's axis is tilted 23½° from the perpendicular (90°) to the plane of its orbit. Consequently, as the earth revolves around the sun, the North Pole points at times toward the sun and at times away from the sun.

When the North Pole leans toward the sun, the North Pole is in constant sunlight, and the Northern Hemisphere receives more of the sun's energy. At this time, the South Pole is in complete darkness, and the Southern Hemisphere receives less of the sun's energy. When the North Pole leans away from the sun and is in complete darkness, the region around the South Pole is bathed in constant sunlight.

What happens when the poles are not leaning toward or away from the sun? If the earth's axis pointed straight up and down all the time in relation to the sun, every day would be the same length. Each day would consist of 12 hours of daylight and 12 hours of darkness. This would be true all over the earth throughout the year. This is not the case, however, since the earth's axis is tilted. During the year, the tilt causes some places to be exposed longer to daylight while others are exposed longer to darkness.

SECTION REVIEW

1. What objects are included in the solar system?
2. What is the difference between rotation and revolution, as applied to Earth?
3. **Critical Thinking** **Analyze** the factors affecting the amount of solar energy that Earth receives.

SOLAR ENERGY AND THE SEASONS

FOCUS

- *What is the relationship between latitude and the amount of solar energy that the earth receives?*
- *What causes the seasons?*

Solar Energy and Latitude As you know, different parts of the earth receive different amounts of solar energy. The areas near the equator receive great amounts of solar energy year-round. These areas are warm all the time. We call these warm, low-latitude areas near the equator the **tropics**. Other areas of the world get very little solar energy. These areas are at high latitudes and are cold most of the time. Because they surround the North and South poles, we call these areas the **polar regions**. In some areas, the amounts of solar energy vary greatly during the year, and they may be warm or cool depending on the time of year. These areas lie between the low-latitude tropics and the high-latitude polar regions and are called the middle-latitude regions.

The amount of energy that a place receives depends on the angle at which the sun's rays strike the earth. Direct, vertical solar rays are much stronger than angled rays in heating the earth's surface. This is because the amount of energy in a direct ray is concentrated on a small area. The same amount of energy in an angled ray is spread over a larger area. (See the diagram on page 13.)

When the North Pole is tilted toward the sun, the sun's most direct rays strike in the Northern Hemisphere. Thus, this hemisphere receives more concentrated solar energy, and temperatures here are warmer. The Southern Hemisphere at this time receives more angled rays with less concentrated energy and is cooler.

When the North Pole tilts away from the sun, the more direct rays strike in the Southern Hemisphere. Now it is the Southern Hemisphere that receives more direct solar energy and has warmer temperatures. At this time, the Northern Hemisphere receives less energy, and its temperatures are cooler.

The Seasons We know these times of greater or lesser heat as the seasons. In each hemisphere, the days are longer, and the sun's energy is stronger, during the summer. In the winter, the days are shorter, and the sun's energy is weaker. During spring and fall, when daylight and darkness are of equal length in both hemispheres, the sun's energy is more evenly distributed. It is the tilt of the earth's axis that causes the Northern and Southern hemispheres to have opposite seasons on the same dates.

The Solstices. At two moments during the year, the earth's poles point toward or away from the sun more than at any other time. The time that the earth's poles point at their greatest angle toward or away from the sun is called a **solstice**. Solstices occur around December 21 and around June 21.

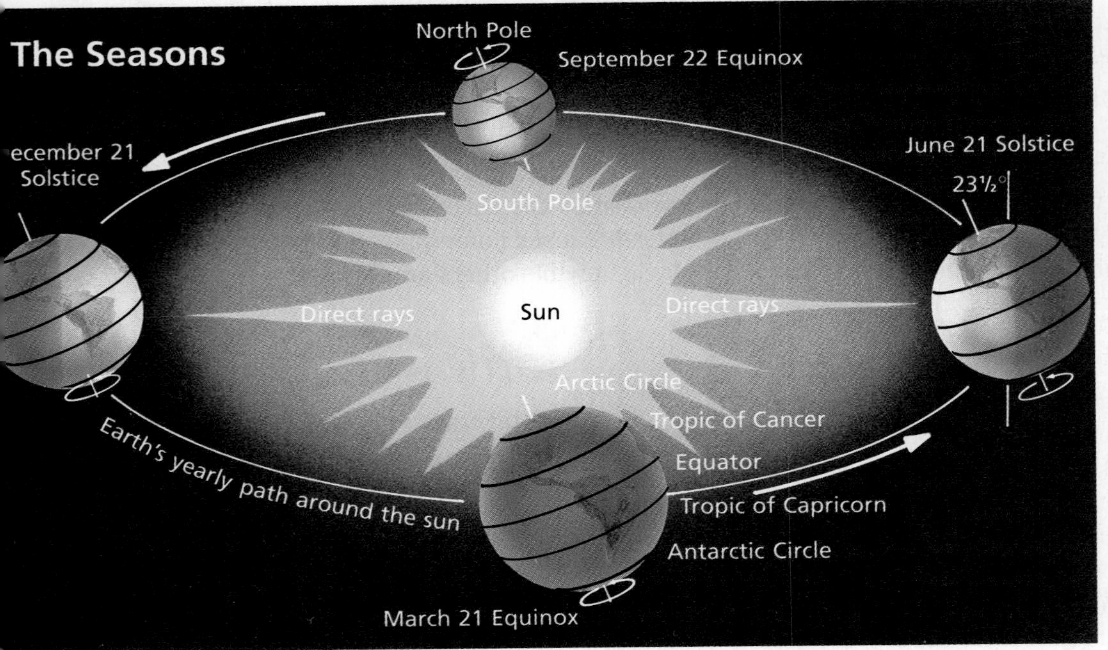

The Seasons
North Pole
September 22 Equinox
ecember 21
Solstice
June 21 Solstice
23½°
South Pole
Direct rays
Sun
Direct rays
Arctic Circle
Tropic of Cancer
Equator
Tropic of Capricorn
Antarctic Circle
Earth's yearly path around the sun
March 21 Equinox

As the earth revolves around the sun, the tilt of the poles toward and away from the sun brings about seasonal change. When are both the North and the South poles at a 90° angle from the sun?

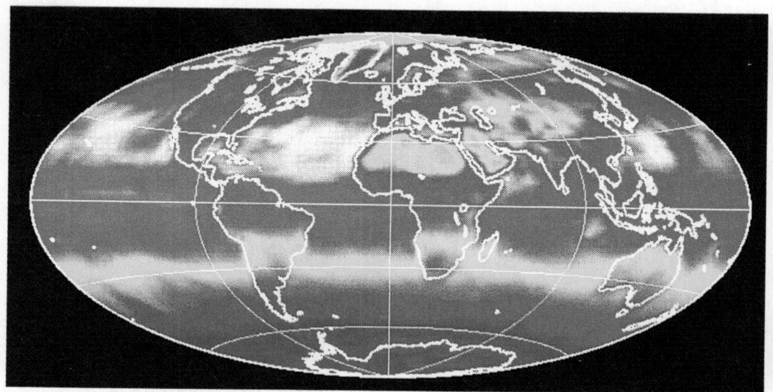

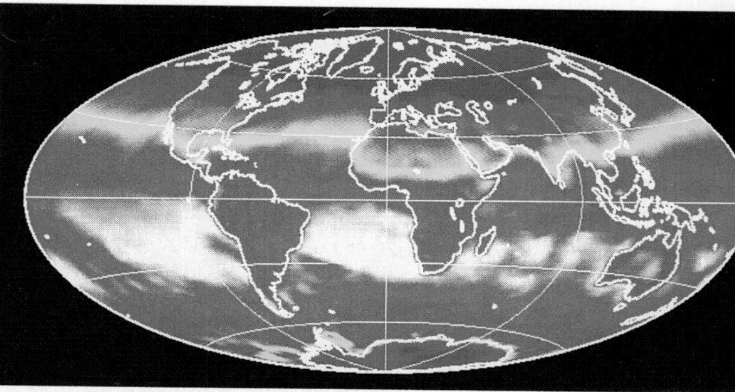

Because surface temperatures are cooler there, areas represented by blue and green colors radiate less heat than do red areas in these infrared images of Earth. The image at the left, taken in July, shows blue and green colors limited mostly to the Southern Hemisphere, where it is winter. The image at the right, taken in January, shows much of the Northern Hemisphere deep in winter.

In the Northern Hemisphere, the day in December on which the solstice occurs is the shortest day of the year and the first day of winter. In the Southern Hemisphere on that same day, it is the longest day of the year and the first day of summer.

During the December solstice, the sun's most direct rays of energy strike the earth in the Southern Hemisphere, along a parallel 23½° south of the equator. This parallel is called the **Tropic of Capricorn**. The South Pole is tilted toward the sun and receives constant sunlight. All areas located south of the **Antarctic Circle** have 24 hours of daylight. The Antarctic Circle is located at 66½° south of the equator. On this same date, the area around the North Pole experiences constant darkness and is very cold. The parallel beyond which no sunlight shines on this day is known as the **Arctic Circle**. It is located at 66½° north of the equator.

During the June solstice, the Northern Hemisphere experiences the longest day of the year, and it is the first day of summer. On this day, the North Pole is pointed at its greatest angle toward the sun. The direct rays of the sun are at their most northerly position, striking the earth most directly at a line 23½° north of the equator. This line is called the **Tropic of Cancer**. If you were to travel to Australia on the June solstice, it would be the first day of winter and the shortest day of the year.

During the June solstice in the Northern Hemisphere, the sun never sets north of the Arctic Circle. Daylight lasts 24 hours. The opposite occurs during this time south of the Antarctic Circle, where darkness lasts 24 hours.

The Equinoxes. An **equinox**, which means "equal night" in Latin, occurs twice a year when the earth's poles are not pointed toward or away from the sun. At this time, the direct rays of the sun are striking the equator, and both poles are at a 90° angle from the sun. Both hemispheres receive an equal amount of sunlight—exactly 12 hours each. If you were to travel anywhere in the world during an equinox, daylight and darkness would last the same length of time.

Equinoxes occur around March 21 and around September 22. In the Northern Hemisphere, the March equinox signals the beginning of spring. To people living in the Southern Hemisphere, however, the March equinox signals the beginning of fall. The opposite situation occurs around September 22.

Days between the solstices and equinoxes gradually become longer or shorter and warmer or cooler, depending on the hemisphere in which you live. This cycle is repeated each year, creating the four seasons.

2 SECTION REVIEW

1. Is the amount of solar energy that the earth receives greater in the low latitudes or in the high latitudes? Why?
2. What is the difference between a solstice and an equinox?
3. **Critical Thinking** **Estimate** the chances that Australia will get a lot of snow next January.

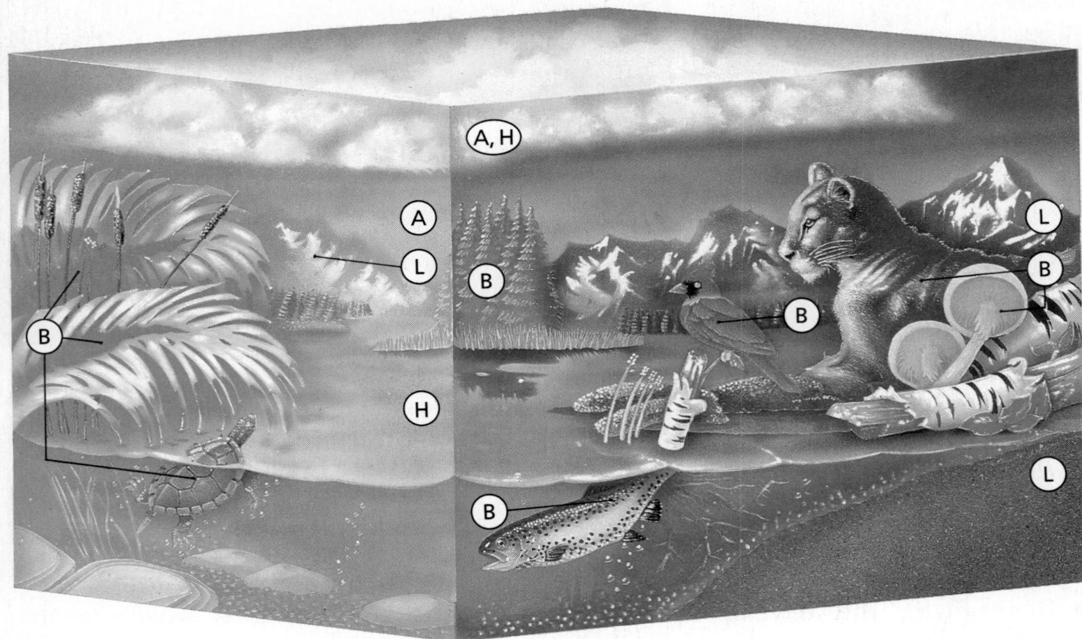
A, H
A
L
B
B
B
H
B
L
B
L
B
A = Atmosphere
B = Biosphere
L = Lithosphere
H = Hydrosphere

The Earth System

The interactions of the atmosphere, the lithosphere, the hydrosphere, and the biosphere make up the earth system. Which elements of this illustration are part of the hydrosphere?

3 THE EARTH SYSTEM

FOCUS

- *What are the four parts of the earth system?*
- *Why is Earth the only planet in our solar system to support life as we know it?*

Earth is a complex planet. Of all of our solar system's planets, Earth is unique in that it is the only planet that supports life as we know it. Geographers call the interactions of elements on and around the earth the **earth system**. The earth system can be divided into four physical systems that work together to create our physical environment: the atmosphere, the lithosphere, the hydrosphere, and the biosphere.

The **atmosphere** is the layer of gases that surrounds the earth. This mixture of moving gases provides the air you breathe and protects the earth from the sun's intense harmful radiation. The **lithosphere** is the surface of the planet that forms the continents and the ocean floor. All the water of the planet—that in the oceans, on the land, and in the atmosphere—form the **hydrosphere**. (The prefix *hydro-* means "water.") The part of the earth in which all of the planet's plant and animal life exists is called the **biosphere**.

These systems are interrelated, so we cannot draw strict dividing lines between them. The hydrosphere supplies humans with water and also serves as a home to animal and plant life. The hydrosphere affects the lithosphere when, for example, a stream flows over rock, causing the rock to be worn down. Soil can be examined as part of the lithosphere, biosphere, or hydrosphere.

As far as scientists have determined, no other planet in the solar system has four physical systems. Earth's nearest neighbors, Venus and Mars, each have an atmosphere and a lithosphere. But because of extreme heat or cold, these planets lack supplies of water. Without water, there is no life or biosphere. In our solar system, only Earth has the necessary four physical systems that work together to support life.

Is there another planet in a faraway galaxy with just the right temperatures to allow for the formation of water and the existence of life? We have yet to discover it.

3 SECTION REVIEW

1. What do we call the part of the earth system that forms the surface of the earth?
2. Why is it that Venus and Mars cannot support life as we know it?
3. **Critical Thinking** **Agree or disagree** with the following statement: "The biosphere is Earth's most important physical system."

CHAPTER 2 CHECK

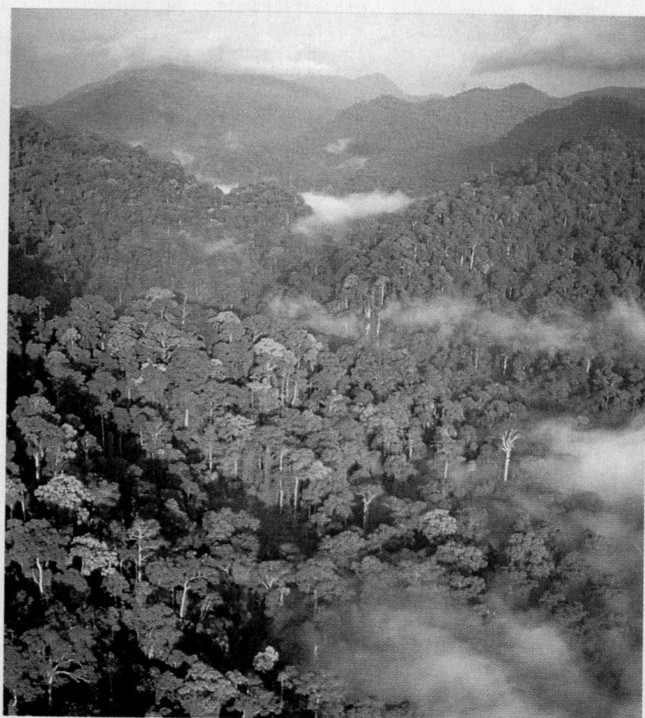

Rain forest in **Borneo**

Reviewing the Main Ideas

1. The rotation of the earth on its axis makes it possible for the earth's surface to be exposed for a time to the sun's energy. This creates day and night.
2. The sun is a star at the center of our solar system. The earth makes one revolution around the sun every 365¼ days, or one year.
3. The earth's axis points toward the North Star at all times. Therefore, as the earth revolves, it leans toward or away from the sun at different positions in its orbit, causing the seasons.
4. The earth is a unique and complex planet. The physical environment of the earth is made up of four interrelated physical systems.

Building a Vocabulary

1. What is a satellite? Give an example of a natural satellite.
2. Define *polar regions* and *tropics*. What do we call the regions that lie between them?
3. Define *solstice*. How do the winter and summer solstices differ from the fall and spring equinoxes?
4. What is the earth system?

Recalling and Reviewing

1. What factors determine the amount of solar energy that the earth receives from the sun?
2. Describe the position of the earth in relation to the sun during winter in the Northern Hemisphere.
3. What are the four physical systems that make up the physical environment on Earth?

Thinking Critically

1. Would life on Earth be possible if the earth did not rotate on its axis? Why or why not?
2. If you lived in an area near the equator, would you experience all of the seasons? Explain.

Learning About Your Local Geography

Cooperative Project

With your group, take a walk around your immediate school area. Find examples of the four physical systems that make up the physical environment. Present your information in an illustration or chart.

Using the Five Themes of Geography

Carefully examine the photographs on this page. Using the photographs as examples, write two or three sentences that explain how the geographic theme of **location** relates to the amount of solar energy that a place receives.

Antarctica

CHAPTER 3

The Earth's Atmosphere and Climates

"Then Poseidon drove together the clouds and stirred up the sea. In his hands the Trident; he excited violent squalls, among all the winds. With dense clouds he covered earth and sea; and night advanced from the sky. Together broke the east wind, the south wind and the stormy west wind. And the north winds from cold lands came, driving huge waves."

From Homer's *Odyssey*

The earth's atmosphere, a thin layer of gases, is made up mostly of nitrogen and oxygen. The atmosphere is necessary for human life because it contains the oxygen we breathe. The condition of the atmosphere varies constantly from day to day, season to season, and year to year. **Weather** is the condition of the atmosphere at a given place and time.

Weather conditions in an area averaged over a long period of time are called **climate**. Climate includes the expected, or average, weather events, as well as rarer events such as devastating storms. Climate also includes long-term global changes such as ice ages.

In this chapter, you will learn how weather and climate are affected by the sun's energy, air pressure, and the global circulation systems. Both weather and climate also are influenced by physical features such as oceans and mountains. Today, there is concern that human activities also may produce changes in the earth's weather and climate.

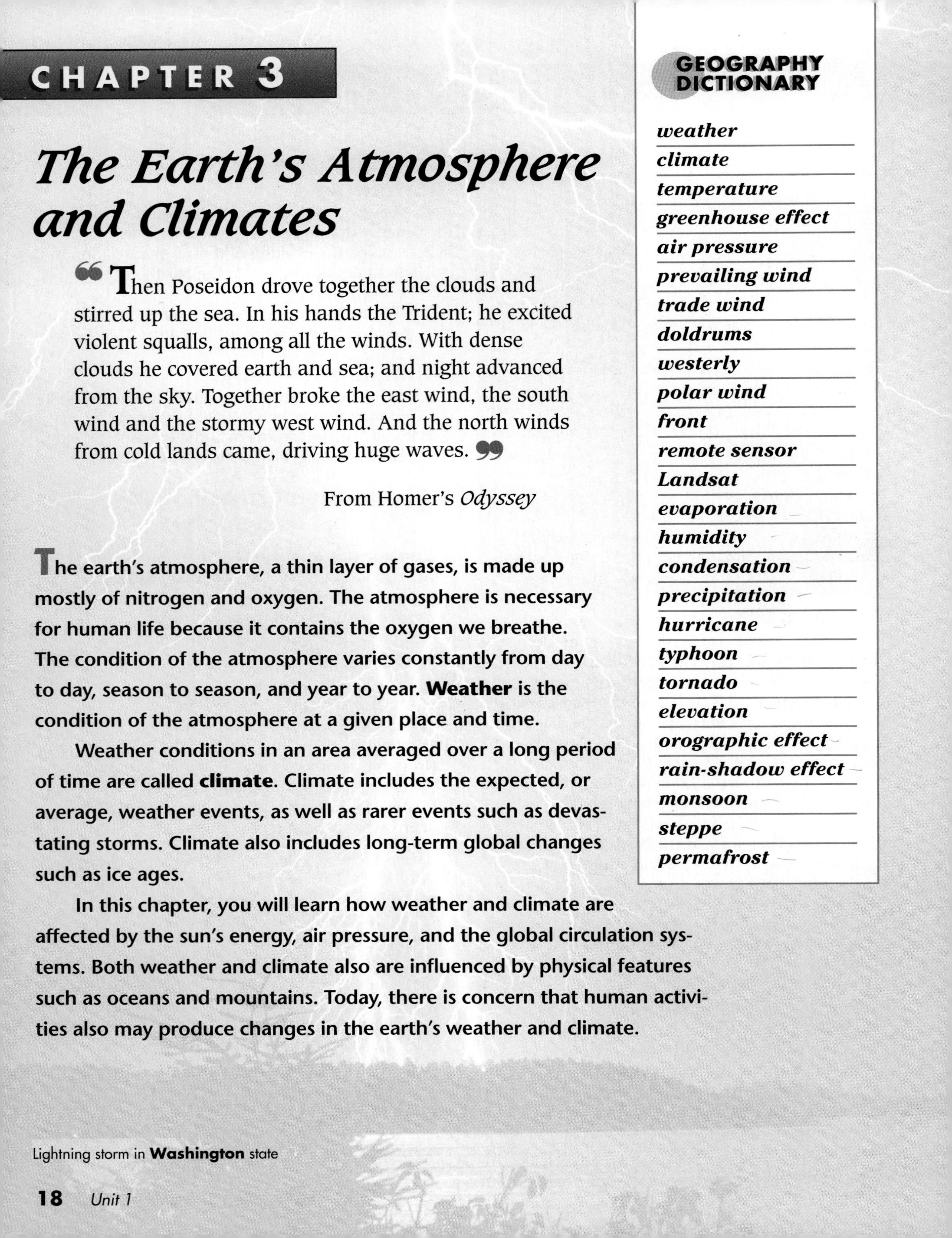

GEOGRAPHY DICTIONARY

weather
climate
temperature
greenhouse effect
air pressure
prevailing wind
trade wind
doldrums
westerly
polar wind
front
remote sensor
Landsat
evaporation
humidity
condensation
precipitation
hurricane
typhoon
tornado
elevation
orographic effect
rain-shadow effect
monsoon
steppe
permafrost

Lightning storm in **Washington** state

GLOBAL ENERGY SYSTEMS

FOCUS

- *What causes high- and low-pressure areas?*
- *How do winds and ocean currents help the earth maintain an energy balance?*

Every day, a small portion of the sun's total energy reaches the earth. Some of that energy is reflected directly back to space. A small amount stays in the atmosphere, where it is absorbed by atmospheric gases. Much of the sun's energy, however, is absorbed by the earth's surface—its oceans and its continents.

Once the sunlight is absorbed, it is changed into heat energy. For example, if you stand in direct sunlight, you become warm. What you feel is the sunlight changing into heat energy. Just as you warm up by absorbing the sunlight, the earth's surface and the atmosphere also warm up when sunlight is changed into heat. The measurement of this heat in the atmosphere is called **temperature**.

The atmosphere traps energy in much the same way that a greenhouse does. The glass of the greenhouse allows the sunlight to pass through it. But it traps and delays the escape of energy after the sunlight has been changed into heat. Like the greenhouse, the earth's atmosphere allows sunlight to pass through it. When the sunlight is changed into heat energy by the earth's surface, it is trapped by the atmosphere, and the earth is kept warm. This process is called the **greenhouse effect**.

This mirage in **Colombia** is created when sunlight heats the ground and the air above it. Rays of sunlight are bent as they pass through the layers of hot air rising from the desert ground. From a distance, you see what appears to be a pool of water.

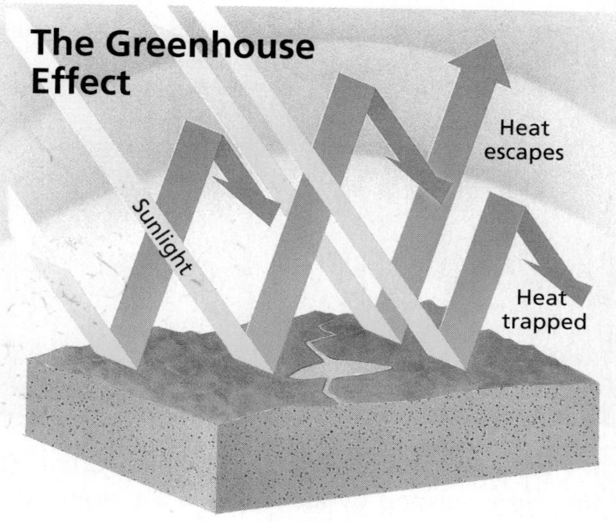
The Greenhouse Effect
Sunlight
Heat escapes
Heat trapped

Light from the sun passes through the atmosphere and is changed into heat energy by the earth's surface. Most heat energy later escapes into space. What would happen if too much heat energy remained trapped and could not escape into space?

Why doesn't the earth get warmer and warmer, until it is so hot that we cannot survive? It is because all heat energy trapped by the atmosphere eventually escapes back into space. Most of the heat is lost to space at night and during the winter season. This is how Earth maintains an energy balance.

As you have learned, the amount of solar energy received by the earth varies from place to place. The equator receives much more solar energy than do the North and South poles. In the tropics, more solar energy is received than is lost to space. In contrast, the polar regions receive less energy than they lose to space. An imbalance appears to exist. We might expect the tropics to reach boiling temperatures and the polar regions to become colder and colder. Yet, this does not happen.

The areas near the equator do not continuously get warmer and the poles get colder because major exchanges of heat energy take place between them. Excess heat energy is moved from the tropics to the polar areas via the atmospheric winds and ocean currents, and cold air and water are moved from the polar areas to the tropics to be reheated. These systems of global-energy exchange balance temperature extremes between latitudes.

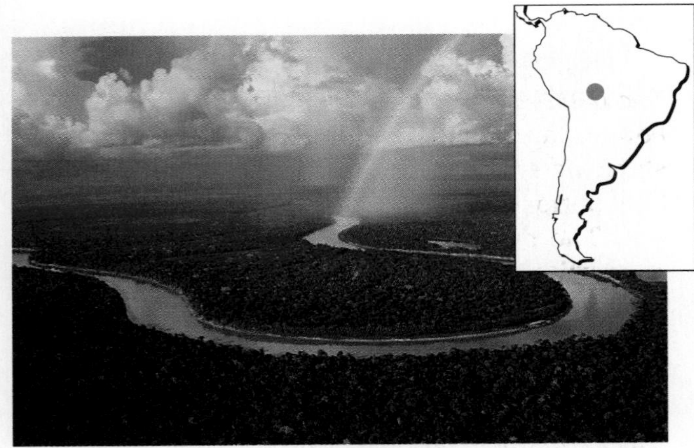

Rapidly rising warm air creates unstable weather conditions over the Juruá River near the equator in **Brazil**.

Air Pressure and Wind The measurement of the force exerted by air is called **air pressure**. Air pressure is important in understanding climates because it creates the winds and ocean currents that make up the systems of global-energy exchange. Unequal heating of the earth's surface causes air-pressure differences. When air is warmed, it expands, becomes lighter, and rises. This creates low-pressure areas, which tend to bring unstable conditions. That is, the air rises and cools. As it cools, the water vapor in the air forms clouds. These clouds may bring rain or storms. Persistent low-pressure areas are found along the equator and at about 60° north and south latitudes.

Cold air is dense and heavy and tends to sink. This action produces high-pressure areas, which tend to be stable. As the cold, heavy air sinks, it heats up. This generally causes clear, dry weather. Persistent high-pressure areas are found at the poles and at about 30° north and south latitudes.

Wind is the horizontal motion of air between areas of different pressure. Wind always flows from high- to low-pressure areas. For example, when air is released from a tire, it flows from inside the tire (high-pressure area) to outside (low-pressure area). In the atmosphere, the seashore illustrates this point. On a sunny day, as the land heats more quickly than the ocean, warm air rises over the land and creates a low-pressure area. Cooler air from over the ocean (high-pressure area) then blows onshore to replace the rising air over the land (low-pressure area). This cooler sea breeze makes the beach very comfortable, even on a warm summer day. During the night, the land cools more rapidly than the ocean. The air then moves in the opposite direction. Then, the cooler land breeze flows offshore from the land (high-pressure area) to replace the warmer, rising air over the ocean surface (low-pressure area). (See the illustration on this page.)

On a global scale, there are four major air-pressure zones: (1) equatorial low pressure, (2) subtropical high pressure, (3) subpolar low pressure, and (4) polar high pressure. (See the diagram on page 21.) At the equator, intense heating by the direct rays of the sun causes warm air to rise, forming the equatorial low-pressure zone. This rising air then flows toward the poles in the upper atmosphere. At about 30° north and south latitudes, the warm air that was rising from the equator now cools and sinks back to the surface. The air piles up, causing the very stable subtropical high-pressure zones. At the poles, the intense cooling causes a continuous sinking of very cold air. This cold, sinking air forms the polar high-

Sea and Land Breezes

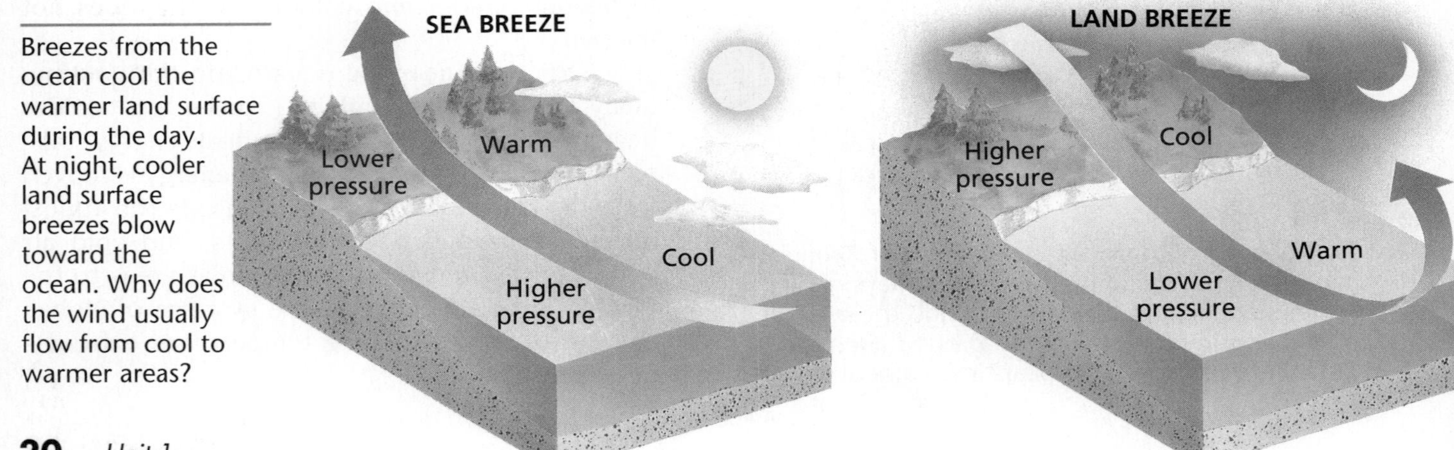
SEA BREEZE
Lower pressure
Warm
Cool
Higher pressure
LAND BREEZE
Higher pressure
Cool
Warm
Lower pressure

Breezes from the ocean cool the warmer land surface during the day. At night, cooler land surface breezes blow toward the ocean. Why does the wind usually flow from cool to warmer areas?

pressure zones. At about 60° north and south latitudes are the subpolar low-pressure zones. They are formed between the high-pressure areas of the polar zones and the subtropical zones. Air flows from these high-pressure zones toward the subpolar low-pressure zones.

MOVEMENT **Global Wind Belts** Winds move heat and cold across the earth between the high- and low-pressure areas. This movement of heat and cold helps the earth maintain an energy balance. Some parts of the world have winds that usually blow from the same direction. These persistent winds are known as the **prevailing winds**, or global wind belts. These winds blow from areas of high pressure toward areas of low pressure.

The **trade winds** blow from the subtropical high-pressure zone toward the equatorial low-pressure zone. European explorers used the northeast trade winds to sail from Europe to America. Have you ever wondered why Columbus landed in the West Indies instead of along the East Coast of the United States? Look at the arrows showing the northeast trade winds on the diagram on this page. Then examine the world map in the Atlas on pages A4–A5. Notice how these winds would guide a sailing ship from Spain across the Atlantic.

Not all areas of the world are in wind belts. The calm areas with no prevailing winds along the equator are called the **doldrums**. Long ago, sailing ships were sometimes caught in the doldrums for long periods of time.

The middle latitudes are dominated by the **westerlies**. After Europeans settled in North America, mariners sailed with the northeast trade winds to America and returned to Europe with the westerlies. Most of the United States is located in the path of the westerlies. These winds carry weather patterns and storms across the nation from west to east.

In the high latitudes, the winds are more variable but mainly come from the east. These are the **polar winds**. Here, cold masses of air flow out of the high-pressure zones of the frozen Arctic and Antarctic regions. They bring cold conditions into the middle latitudes. Between 40° and 60° latitude, the warm westerlies meet the cold polar winds. This meeting zone of warmer subpolar air and cold polar air is called a polar **front**. A front occurs when two air masses with very different temperatures and amounts of moisture meet. Fronts usually cause stormy weather.

There also are prevailing winds that blow in the upper atmosphere, high above the surface of the earth. Though we do not feel them directly,

Text continues on page 24.

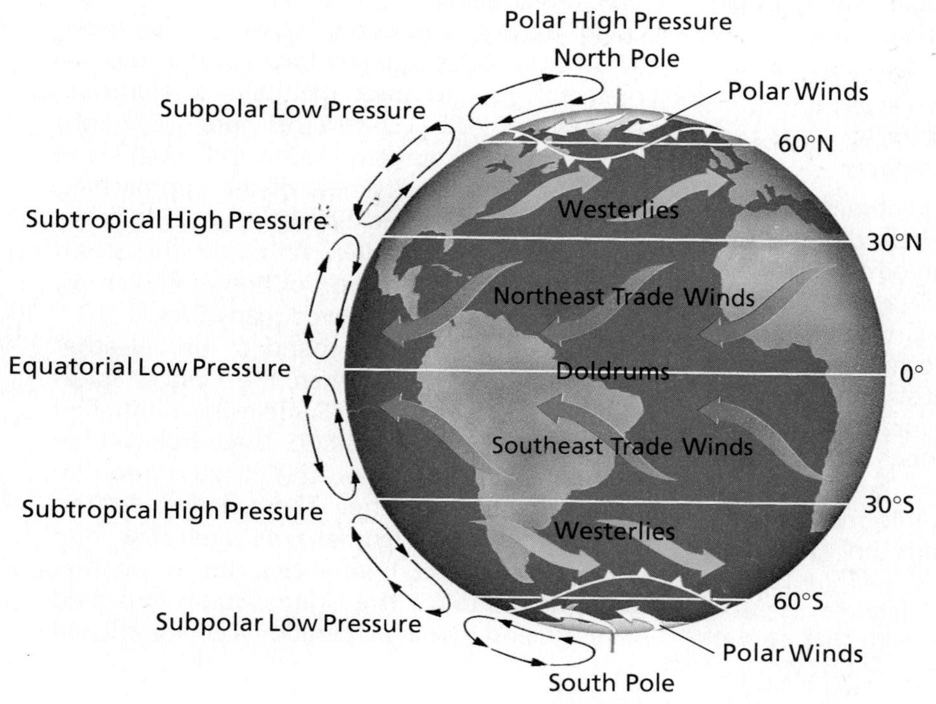

Polar High Pressure
North Pole
Polar Winds
Subpolar Low Pressure
60°N
Subtropical High Pressure
Westerlies
30°N
Northeast Trade Winds
Equatorial Low Pressure
Doldrums
0°
Southeast Trade Winds
Subtropical High Pressure
30°S
Westerlies
60°S
Subpolar Low Pressure
Polar Winds
South Pole
Polar High Pressure

Global Pressure and Wind Systems

Winds between the earth's high- and low-pressure areas help regulate the globe's energy balance. What happens when warm westerlies come into contact with cold polar winds?

Watching Over Earth

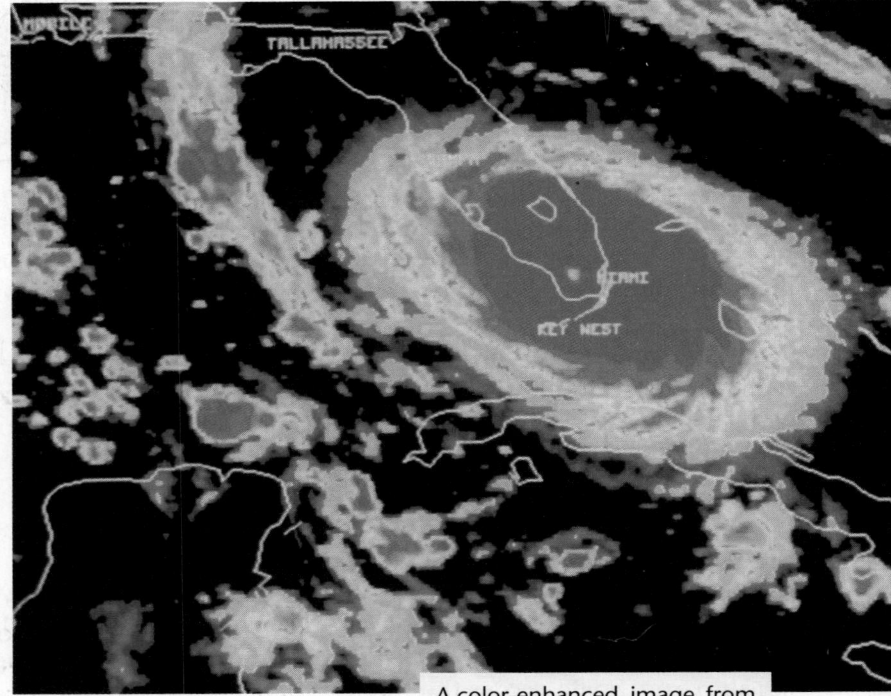
TALLAHASSEE
MIAMI
KEY WEST
A color-enhanced image from a weather satellite catches Hurricane Andrew as it bears down on **Florida** in 1992.

As you read in Chapter 2, a satellite is a body that orbits a larger body. Some satellites, such as the moon, are natural satellites. There are also human-made satellites, which are becoming more important to the study of geography. Satellites collect and send back information about the earth.

Satellites use different types of **remote sensors** to collect information. Remote sensors are devices attached to satellites that act as "eyes in the sky." These instruments allow scientists to keep track of physical changes that are occurring almost anytime and anywhere on the earth.

A common type of remote sensor is a camera that records light. Satellite cameras take photographs that are broken down into information to be sent back to Earth. Computers on Earth then arrange the information back into images. Satellite cameras mainly record visible light and sometimes infrared light, which is not visible to the human eye. Infrared sensors detect heat given off by clouds, water, and land.

Satellites depend primarily on the earth's gravity to power their movement once they are in orbit.

Tracking the Weather

For centuries, people have looked to the skies for information about the weather. Only in recent times, though, has technology allowed scientists to make accurate forecasts. Part of this new technology is the weather satellite. Today, meteorologists monitor the atmosphere with information gathered from observation stations on Earth and from satellites.

Images collected by the satellites' sensors may show, for example, cloud swirls that warn of a hurricane or typhoon. Or they may reveal dense cloud masses, indicating heavy rain or snow. Weather satellites supplied the data that allowed meteorologists to track the path of Hurricane Andrew, whose 135-mile-per-hour (217-kph) winds battered southern Florida and Louisiana in 1992. Meteorologists knew of the approaching hurricane's power and warned people to leave the area or find shelter. Although the storm caused massive destruction of homes and crops, the advanced warnings saved many lives.

Scientists have depended on weather satellites since 1960, when the United States launched the first TIROS (Television InfraRed Observation Satellite). TIROS 1 was followed by other satellites, all of which collected information about the weather. These early satellites, however, did not remain over one area long enough to detect hourly changes in weather patterns. In 1975, the United States launched GOES (Geostationary Operational Environmental

Satellite), the first geostationary satellite. A geostationary satellite orbits at about 22,300 miles (35,890 km) over the equator. At that height, the earth's gravity has exactly enough force to make the satellite move at a speed that matches the earth's rotation. Thus the satellite remains over the same portion of the earth and can gather information continuously from the same area.

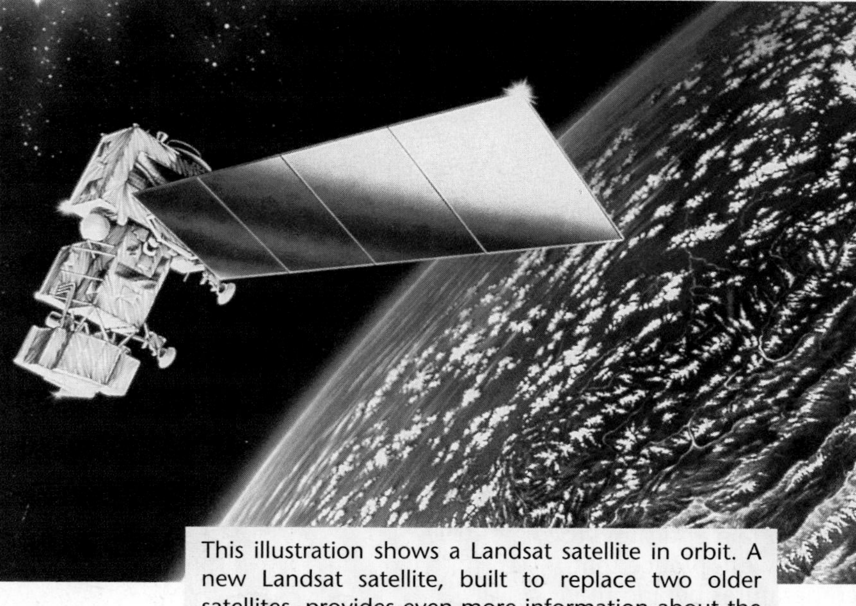

This illustration shows a Landsat satellite in orbit. A new Landsat satellite, built to replace two older satellites, provides even more information about the size and shape of surface features.

Eyeing the Environment

While weather satellites send scientists information about clouds, storms, and oceans, **Landsat** satellites reveal the condition of the environment. Landsat satellites usually orbit around the North and South poles and carry sensors that scan the same area about every 18 days. They detect changes in the earth's surface, such as changes in vegetation and moisture. These satellites view Earth in both visible and infrared light.

After the information gathered by Landsat satellites is processed, each color of the Landsat pictures show a different type of surface feature of the earth. Plants, for example, might appear as bright red. Blue might represent bare soil. Because Landsat photographs do not show true color, they are often called false-color images.

Environmental scientists can use Landsat images to gather information about air pollution, pests that threaten forests and farmlands, or algae that might pollute lakes. They can also monitor human activities such as urban growth or the cutting and burning of rain forests. In addition, Landsat satellites have been used to track forest fires, crop damage, the paths of oil spills, the movement of herds of animals, and ground movement caused by earthquakes.

The colors in this satellite image of **New York City** are used to distinguish between different kinds of surface features. To the right of the Hudson River (the dark band in the middle of the photograph) is Manhattan. Central Park is the orange rectangle in the middle of the island.

Geographic Information Systems

Photos from satellites also are used in mapmaking. In fact, some maps are not drawn but are made up of pieced-together satellite photos. Such maps are often used in geographic information systems, or GISs, which use computers to combine the maps with other data. These systems allow people to observe the spatial patterns of a wide variety of phenomena. For example, a GIS might show what roads are subject to flooding.

Weather satellites, Landsat satellites, and GISs have revolutionized the way scientists and geographers view our world. Satellites can help save lives, money, and the environment.

Your Turn

Think of ways in which nations and individuals can make use of Landsat images to protect the environment. **Write** a paragraph describing how the images might be used to highlight an environmental issue that is of concern to you. **Explain** how you and others might use the information to provide solutions to the problem.

Reading a Weather Map

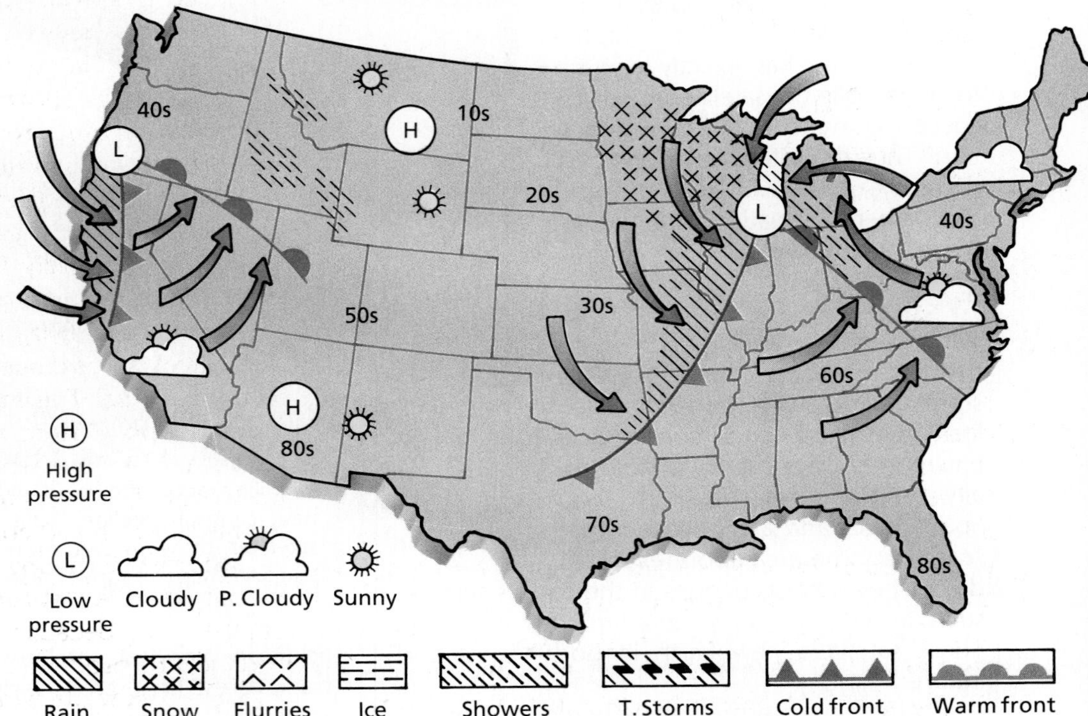
40s
10s
20s
40s
30s
50s
60s
80s
70s
80s
H
L
High pressure
Low pressure
Cloudy
P. Cloudy
Sunny
Rain
Snow
Flurries
Ice
Showers
T. Storms
Cold front
Warm front

MAP STUDY Weather maps show climate conditions as they currently exist or as they are forecast for a particular time period. Most weather maps have legends that explain what the symbols on the map mean. This map shows a cold front sweeping through the central United States. A low-pressure system is at the center of a storm bringing rain and snow to the Midwest. Notice that temperatures behind the cold front are considerably cooler than those ahead of the front.

these strong winds move energy, storms, and major weather patterns. These high-speed, westerly moving winds are called jet streams. When an airplane experiences turbulence at high altitude, it has crossed the part of a jet stream's path where there are sudden changes in wind speed and direction.

MOVEMENT **Ocean Circulation** The oceans are the most important solar-energy collectors. Most of the sun's energy that reaches the earth's surface is absorbed by seawater. Water heats and cools much more slowly than land. Therefore, areas near oceans, such as islands and coastlines, do not experience great temperature extremes. Large land areas, especially the centers of large continents, have the most extreme temperature ranges.

As the prevailing winds blow across the surface of the ocean, they set in motion the ocean currents. Ocean currents are giant rivers of seawater flowing at the surface of the oceans. The ocean currents generally flow in circular paths. (See the map on pages 28–29.) Warm currents carry heated water from the low latitudes into the cooler high latitudes. Cool currents return cooled water from the higher latitudes to be rewarmed in the lower latitudes. The ocean currents are one of the major ways in which heat is moved between the warm tropical regions and the cold polar regions. This movement of heat helps the earth maintain its energy balance.

The oceans do much more than move heat. They are a major source of oxygen for the air. Oxygen is produced by tiny ocean plants, which release it into the atmosphere. The oceans also are the major source of moisture for the atmosphere. You will learn more about the oceans in Chapter 4.

SECTION REVIEW

1. What causes differences in air pressure? How does air behave differently in a low-pressure area than in a high-pressure area?
2. How do global wind patterns and global ocean-current patterns help the earth maintain an energy balance?
3. **Critical Thinking** **Evaluate** the importance of the atmosphere to Earth's energy systems.

ATMOSPHERIC EFFECTS

FOCUS

- *What causes precipitation?*
- *How is elevation related to weather and climate?*

Water and the Atmosphere Water vapor is an important gas in our atmosphere. Most water vapor in the air is evaporated from the oceans. **Evaporation** is the process by which water is changed from a liquid to a gas. The remainder of the water vapor in the air comes from lakes, plants, and the soil. Without water vapor, there could be no clouds, rain, or storms. The invisible water vapor stores the heat energy necessary to bring about changes in the atmosphere.

The amount of water vapor in the air is called **humidity**. You feel warmer on a humid, muggy day than on a dry day with the same temperature. As perspiration evaporates, it cools your skin. Perspiration will evaporate much more slowly on a humid day, however, because if much moisture is already in the air, there is less room for the air to absorb the moisture evaporating from your skin.

The higher the temperature, the more water vapor the air can hold. When air cools, it will reach a temperature at which it can hold no more water vapor. At this point, **condensation** will occur. Condensation is the process by which water vapor changes from a gas into liquid droplets. You most often see condensation as the formation of clouds, fog, and dew. If the condensed droplets become large enough, they fall as **precipitation**. Rain, snow, sleet, and hail are the four common forms of precipitation. Precipitation is not evenly distributed throughout the year. It may also vary from year to year. Too much precipitation in a short time period may cause floods, and too little precipitation may result in a drought.

Storms The most sudden and violent weather event is a storm, which occurs when energy stored in the water vapor of the atmosphere is released. The large amount of energy released from a storm is shown by the violence and damage that sometimes results.

Storms occur under low-pressure conditions during which rising, unstable air is present. In the middle latitudes, these storms travel with the upper-level, westerly flow and are called middle-latitude storms, or extratropical storms. They can be as large as 1,000 miles (1,609 km) in diameter and travel across oceans and continents during their life cycle. The middle-latitude storms occur along the polar front that forms a boundary line between warm, wet tropical air and cold, dry polar air. Moving fronts occur within the middle-latitude storms as warm and cold air clash.

Storms in the low latitudes differ from those of the middle latitudes. Tropical storms are smaller and, because only warm air is present, they do not produce fronts. They move from the east with the trade winds, bringing rain and wind. (See the photographs on this page.)

Severe tropical storms sometimes form over warm ocean waters. Called **hurricanes** or **typhoons**, these revolving storms bring violent winds, heavy rain, and dangerously high seas. Hurricanes are most destructive to islands and coastal areas. Sometimes they follow warm ocean currents into the middle latitudes, such as along the Gulf and Atlantic coasts of the United States.

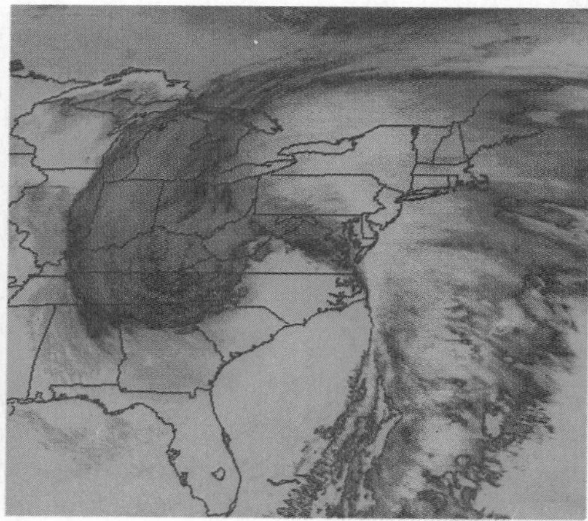

Hurricane Bob (left), churning toward the Northeastern **United States** in 1991, and the "Blizzard of 1993" (right) are examples of tropical and extratropical storms that affect weather in the United States.

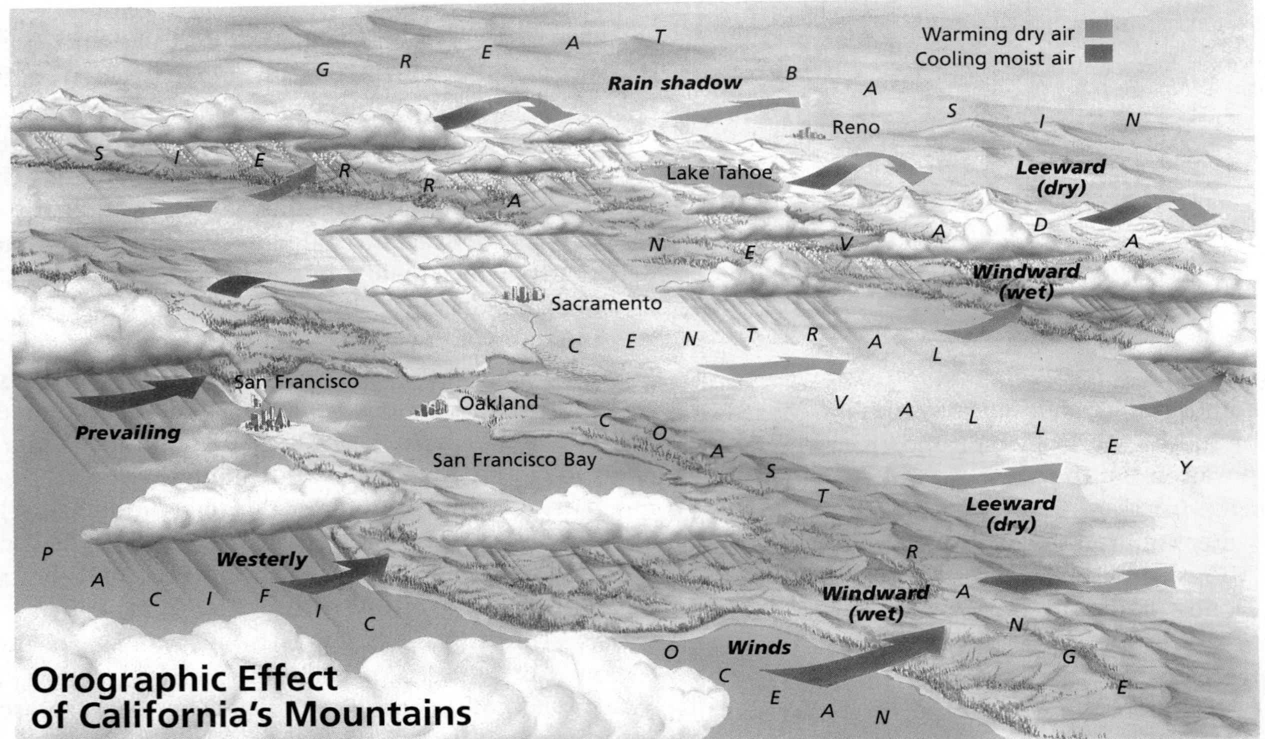
Warming dry air
Cooling moist air
G R E A T B A S I N
Rain shadow
Reno
S I E R R A
Lake Tahoe
N E V A D A
Leeward
(dry)
Windward
(wet)
Sacramento
C E N T R A L V A L L E Y
San Francisco
Oakland
Prevailing
San Francisco Bay
C O A S T R A N G E
Leeward
(dry)
P A C I F I C O C E A N
Westerly
Windward
(wet)
Winds
Orographic Effect
of California's Mountains

As moist air from the ocean moves up the windward side of a mountain, it cools. The water vapor condenses and falls in the form of rain or snow. The descending drier air that moves down the leeward side brings almost no precipitation to lands in the rain shadow of the mountains. This illustration shows that the air that has traveled over two mountain ranges and the Central Valley does not bring much precipitation to the Great Basin desert of the western United States.

Tornadoes are the smallest but most violent of storms. These small, twisting spirals of air can destroy almost anything in their path. They generally form along fronts in middle-latitude storms and are most common in the southeastern and central United States. In fact, the United States has more tornadoes than any other nation.

A thunderstorm, the most common type of storm, is any storm with lightning and thunder present. At any moment, about 2,000 thunderstorms are taking place on Earth. Thunderstorms form most often in the hot, unstable equatorial regions and along fronts in middle-latitude storms. Lightning, which carries a dangerous electrical charge, kills or injures hundreds of people each year. Thunderstorms may drop great amounts of rainfall in a short period of time, producing flash floods.

LOCATION Elevation and Temperature

Landforms, particularly high mountains, have a major effect on weather. **Elevation**, or height on the earth's surface above or below sea level, has the greatest impact. An increase in elevation causes a lowering of the temperature. For every 1,000 feet (305 m) up the side of the mountain, the air temperature cools about 3.6°F (2°C). The 19,340-foot (5,895-m) elevation of Kilimanjaro (kil-uh-muhn-JAHR-oh), in Africa, shows us this effect. Even though the mountain is on the equator, its twin peaks are always covered with ice and snow.

Orographic Effect Another effect of elevation on weather and climate is the **orographic** (ohr-uh-GRAF-ik) **effect**. When moist air flowing from the ocean meets a mountain barrier, the air is forced to rise. The higher the air rises, the more it cools. As the rising air cools and condensation begins, clouds form and precipitation occurs in the form of rain or snow. As a result, the side of the mountain facing the wind receives a great deal of moisture. This side of the mountain is called the windward side.

The side of the mountain facing away from the wind is called the leeward side. As the air moves down the leeward side, it warms and becomes drier. Deserts commonly form on the leeward side of mountains. This is called the **rain-shadow effect** because areas in a rain shadow do not receive much rain, just as areas in the shade do not receive direct sun. For example, because of the orographic effect of the mountains in the western United States, rain-shadow deserts are located on the leeward side of these mountains. (See the diagram on page 26.)

SECTION REVIEW

1. What is the relationship between evaporation, condensation, and precipitation?
2. How do elevation and the orographic effect influence land and climate?
3. **Critical Thinking** **Rate** the danger of different types of storms.

GLOBAL CLIMATES

FOCUS

- *What are the major influences on climate?*
- *Where are the major global climate regions located?*

People living in advanced industrial nations such as the United States have used technology to control the effects of climate on their daily lives. Such societies have invested billions of dollars to cool and heat homes and offices, to build water delivery systems for farms and cities, and to construct transportation systems that can operate in good weather and bad. But for four out of every five people on Earth, climate continues to have an immediate impact on every aspect of daily existence. The clothes people wear, the places they live, and the foods they eat all are influenced by climate.

The major global climate types are classified mainly by temperature and precipitation differences, both of which are influenced by latitude. That is, latitude has the most control over which regions are hot or cold, wet or dry. To a slightly lesser degree, a region's location on land and its elevation also affect climate.

Because the vegetation found in a place is a response to its climate, vegetation is sometimes used to classify climate. Some climate types are named for the dominant vegetation of the region, which you will learn more about in Chapter 6. Refer to the climate map and chart on pages 28–29 as you read the rest of this chapter.

LOCATION **Low-Latitude Climates** The areas close to the equator mostly have warm temperatures and a large amount of rainfall year-round. These wet, hot areas have a humid-tropical climate. People who live in the humid tropics never experience winter or even cool weather.

Because the equator is constantly being heated by the sun's rays, warm air is always rising in the humid tropics. This continuous rising of warm, unstable air brings almost daily thunderstorms and heavy rainfall. The combination of continuous warm temperatures and heavy rainfall creates ideal conditions for plant growth, and dense tropical rain forests thrive here.

In some tropical areas, especially India and Southeast Asia, the rain is concentrated in one very wet season. During the summer months, moist air flows into these parts of Asia from the warm ocean (high-pressure area) to the hotter land (low-pressure area), bringing heavy rains. During the winter, dry air flows off the cooling continent (high-pressure area) to the warm oceans (low-pressure area), bringing dry conditions to the area. The wind that blows from the same direction for months at a definite season of the year is called a **monsoon**.

Text continues on page 30.

Residents of **Bombay, India**, wade through the high waters caused by the wet monsoon. From where do monsoon winds blow during the wet monsoon season in India?

The World's Climate Regions

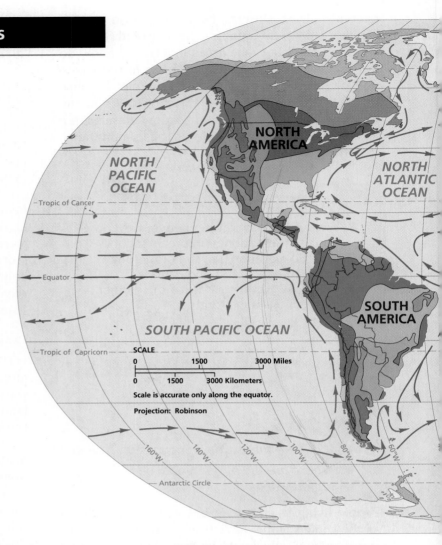

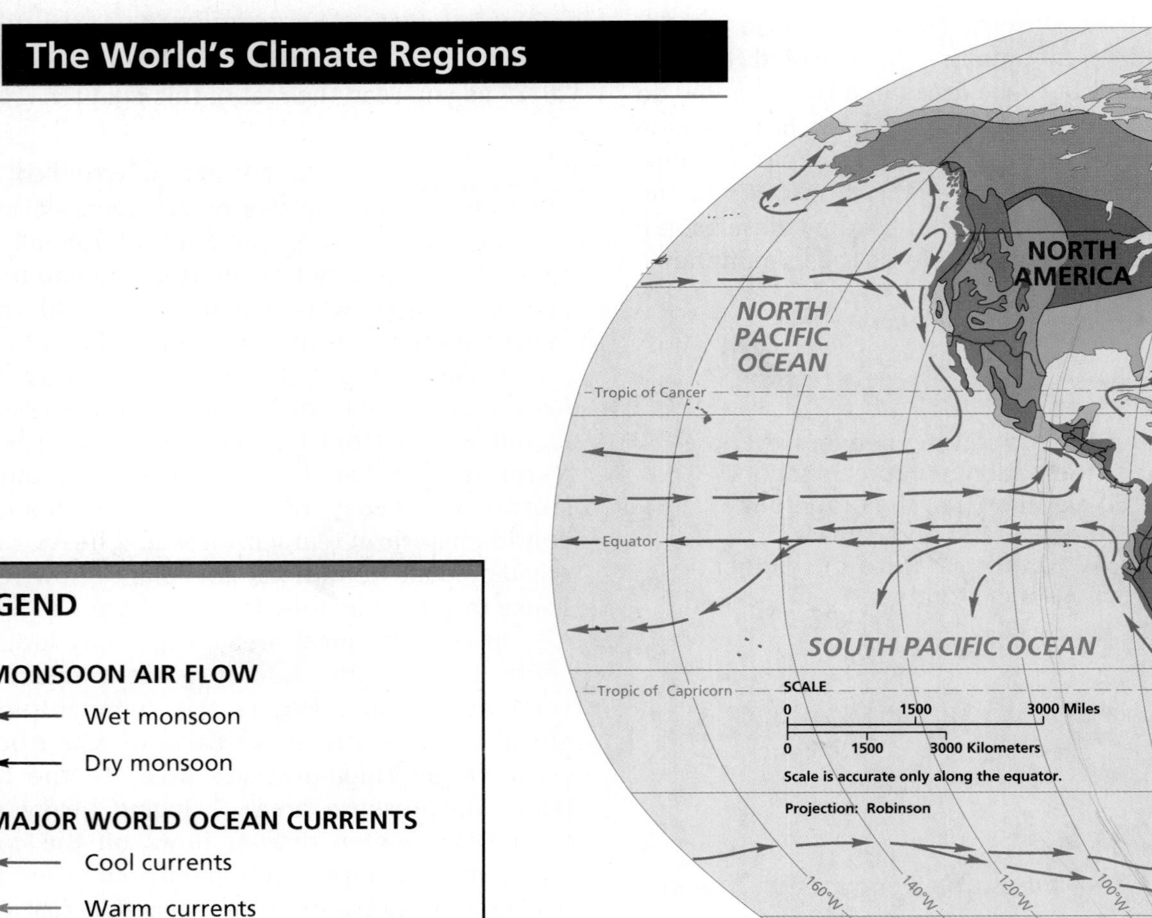
NORTH AMERICA
NORTH PACIFIC OCEAN
NORTH ATLANTIC OCEAN
SOUTH AMERICA
SOUTH PACIFIC OCEAN
Tropic of Cancer
Equator
Tropic of Capricorn
Antarctic Circle
SCALE
0
1500
3000 Miles
0
1500
3000 Kilometers
Scale is accurate only along the equator.
Projection: Robinson
160°W
140°W
120°W
100°W
80°W
60°W
LEGEND
MONSOON AIR FLOW
Wet monsoon
Dry monsoon
MAJOR WORLD OCEAN CURRENTS
Cool currents
Warm currents

	Climate	Geographic Distribution	Major Weather Patterns	Vegetation
Low Latitudes	HUMID TROPICAL	along Equator; particularly equatorial South America, Zaire Basin in Africa, Southeast Asia	warm and rainy year-round, with rain totaling anywhere from 65 to more than 450 in. (165–1,143 cm) annually; typical temperatures are 90°–95°F (32°–35°C) during the day and 65°–70°F (18°–21°C) at night	tropical rain forest
	TROPICAL SAVANNA	between humid tropics and deserts; tropical regions of Africa, South and Central America, southern and Southeast Asia, Australia	warm all year; distinct rainy and dry seasons; precipitation during the summer of at least 20 in. (51 cm) and in some locations exceeding 150 in. (380 cm); summer temperatures average 90°F (32°C) during the day and 70°F (21°C) at night; typical winter temperatures are 75°–80°F (24°–27°C) during the day and 55°–60°F (13°–16°C) at night	tropical grassland with scattered trees
Dry/Semiarid	DESERT	centered along 30° latitude; some middle-latitude deserts in interior of large continents and along western coasts; particularly Saharan Africa, southwest Asia, central and western Australia, southwestern North America	arid; precipitation of less than 10 in. (25 cm) annually; sunny and hot in the tropics and sunny with great temperature ranges in middle latitudes; typical summer temperatures for lower-latitude deserts are 110°–115°F (43°–46°C) during the day and 60°–65°F (16°–18°C) at night, while winter temperatures average 80°F (27°C) during the day and 45°F (7°C) at night; in middle latitudes, the hottest month averages 70°F (24°C)	sparse drought-resistant plants; many barren, rocky, or sandy areas
	STEPPE	generally bordering deserts and interiors of large continents; particularly northern and southern Africa, interior western North America, central and interior Asia and Australia, southern South America	semiarid; about 10–20 in. (25–51 cm) of precipitation annually; hot summers and cooler winters with wide temperature ranges similar to desert temperatures	grassland; few trees
Middle Latitudes	MEDITERRANEAN	west coasts in middle latitudes; particularly southern Europe, part of southwest Asia, north-western Africa, California, southwestern Aus-tralia, central Chile, southwestern South Africa	dry, sunny, warm summers and mild, wetter winters; precipitation averages 15–20 in. (38–51 cm) annually; typical temperatures are 75°–80°F (24–27°C) on summer days; the average winter temperature is 50°F (10°C)	scrub woodland and grassland
	HUMID SUBTROPICAL	east coasts in middle latitudes; particularly southeastern United States, eastern Asia, central southern Europe, southeastern parts of South America, South Africa, and Australia	hot, humid summers and mild, humid winters; precipitation year-round; coastal areas are in the paths of hurricanes and typhoons; precipitation averages 40 in. (102 cm) annually; typical temperatures are 75°–90°F (24°–32°C) in summer and 45°–50°F (7°–10°C) in winter	mixed forest

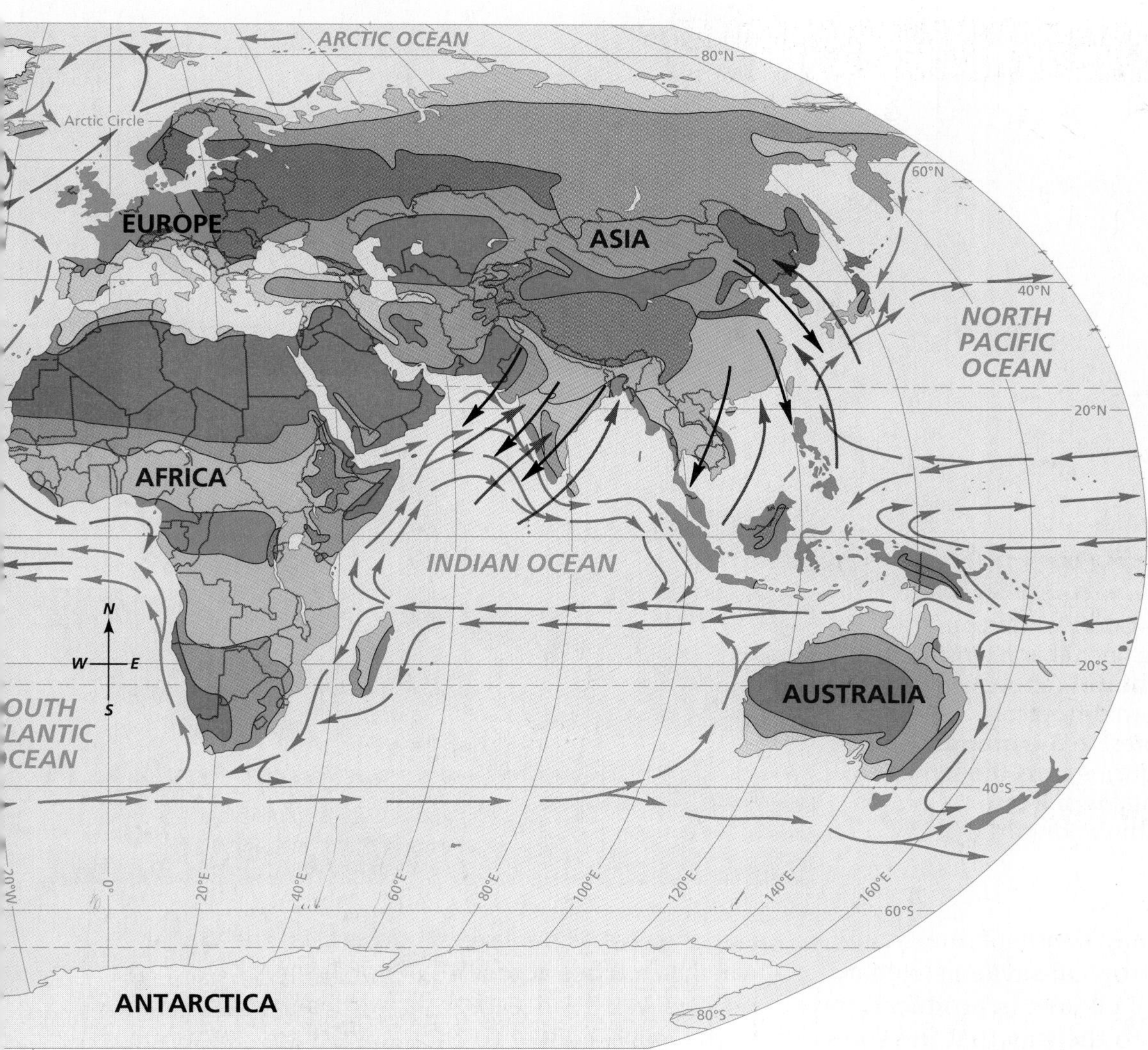

ARCTIC OCEAN
80°N
Arctic Circle
60°N
EUROPE
ASIA
40°N
NORTH PACIFIC OCEAN
20°N
AFRICA
INDIAN OCEAN
N
W
E
S
20°S
AUSTRALIA
40°S
20°W
0°
20°E
40°E
60°E
80°E
100°E
120°E
140°E
160°E
60°S
ANTARCTICA
80°S

	Climate	Geographic Distribution	Major Weather Patterns	Vegetation
Middle Latitudes	MARINE WEST COAST	west coasts in upper-middle latitudes; particularly northwestern Europe and North America, southwestern South America, central southern South Africa, southeastern Australia, New Zealand	cloudy, mild summers and cool, rainy winters; strong ocean influence; precipitation averages 20–60 in. (51–152 cm) annually, with some coastal mountains receiving more than 200 in. (508 cm); average temperature in hottest month usually is between 60°F and 70°F (16°–21°C); average temperature in coolest month usually is above 32°F (0°C)	temperate evergreen forest
	HUMID CONTINENTAL	east coasts and interiors of upper-middle-latitude continents; particularly northeastern North America, northern and eastern Europe, northeastern Asia	four distinct seasons; long, cold winters and short, warm summers; precipitation amounts vary, usually 20–50 in. or more (51–127 cm) annually; average summer temperature is 75°F (24°C); average winter temperature is below freezing	mixed forest
High Latitudes	SUBARCTIC	higher latitudes of interior and east coasts of continents; particularly northern parts of North America, Europe, and Asia	extremes of temperature; long, cold winters and short, warm summers; low precipitation amounts all year; precipitation averages 5–15 in. (13–38 cm) in summer; temperatures in warmest month average 60°F (16°C), but can warm to 90°F (32°C); winter temperatures average below 0°F (–18°C)	northern evergreen forest
	TUNDRA	high-latitude coasts; particularly far northern parts of North America, Europe, and Asia, Antarctic Peninsula, subantarctic islands	cold all year; very long, cold winters and very short, cool summers; low precipitation amounts; precipitation average is 5–15 in. (13–38 cm) annually; warmest month averages 40°F (4°C); coolest month averages a little below 0°F (–18°C)	moss, lichens, low shrubs; permafrost bogs in summer
	ICE CAP	polar regions; particularly Antarctica, Greenland, Arctic Basin islands	freezing cold; snow and ice year-round; precipitation averages less than 10 in. (25 cm) annually; average temperatures in warmest month are not higher than freezing	no vegetation
	HIGHLAND	high mountain regions, particularly western parts of North and South America, eastern parts of Asia and Africa, southern and central Europe and Asia	greatly varied temperatures and precipitation amounts over short distances as elevation changes	forest to tundra vegetation, depending on elevation

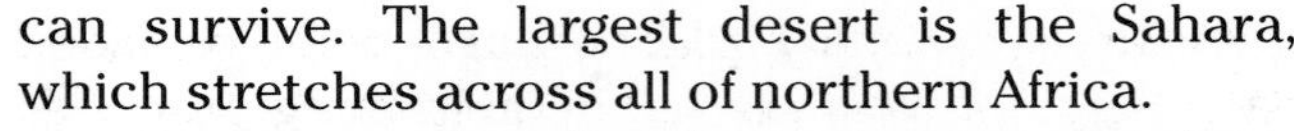

Climates have shaped the world's landscapes and created regions with distinct characteristics. Much of bitterly cold **Greenland** (above) is covered by an ice cap. A humid-tropical climate nurtures this rain forest in **Suriname, South America** (upper right). Across the same continent, the Andes Mountains of **Peru** (right) provide a climate with conditions that vary by elevation.

Just to the north and south of the humid-tropical climate is the tropical-savanna climate. This wet- and dry-tropical climate is produced by the seasonal change due to the way that the sun's rays strike the areas north and south of the equator. For example, during high-sun season (summer), the sun's rays strike most directly. This high sun increases the temperature and causes low pressure and unstable, rising air that produces heavy rainfall. During the low-sun season (winter), the opposite occurs. As the direct solar rays move to the opposite hemisphere, the subtropical high-pressure zone moves into the area. This causes stable, cool, sinking air and a dry season.

LOCATION **Dry Climates** Though their temperatures may vary greatly, all dry climate regions share aridity, or low annual rainfall. The two types of dry climates are the arid desert climate and the semiarid steppe climate.

Most of the desert climate areas are centered at about 30° north and south of the equator. The dryness is caused by the subtropical high-pressure zone, which brings stable, sinking, dry air all year. Very little rain is produced, and few plants can survive. The largest desert is the Sahara, which stretches across all of northern Africa.

As you read earlier, large deserts also form in the rain shadow to the leeward side of mountains. An example is the Great Basin of the United States. Other deserts are far into the interior of continents, away from moisture-bearing winds. These deserts include the Gobi and Taklimakan deserts of Asia, and the deserts of the dry interior of Australia. These desert regions are blocked from rain by mountain ranges and great distances across large continents. They experience temperature extremes ranging from extremely cold temperatures during winter to hot, scorching temperatures during summer.

Small and very dry deserts are found along the west coasts of continents. Here, the cool ocean currents cause stable conditions, and it may not rain for many years. Examples of dry coastal deserts are found along the west coasts of South America, southwestern Africa, and Mexico.

The semiarid **steppe** climate is a transition area between the arid deserts and the more humid climates. The term *steppe* refers to short-grass vegetation. Areas with steppe climates

Dunes and hardy vegetation mark a desert-climate region in **West Texas** (below). Sunbathers take advantage of a Mediterranean climate in **Nice, France** (right).

receive more moisture than the deserts but less than the more humid areas. Steppe climates generally support grasslands, and trees are rare except along river banks. Today, the natural grasslands in many places have been replaced with fields of food grains. Poor farming practices and overgrazing have turned some natural grasslands into human-made deserts.

LOCATION Middle-Latitude Climates As is the case with low-latitude climates, there are several types of middle-latitude climates. Located between 30° and 40° latitude, the Mediterranean climate is confined mainly to the coastal areas of southern Europe and the west coasts of continents with cool ocean currents. Mediterranean climates usually do not extend far inland beyond mountain ranges. The stable, sinking air of the subtropical high-pressure zone causes long, sunny, dry summers. During the mild winter, however, cool middle-latitude storms enter the region with the westerlies and bring needed rains.

The humid-subtropical climate is much more widespread than the Mediterranean climate. This climate is found on the eastern side of continents with warm ocean currents. The humid-subtropical climate is greatly affected by moist air flowing off the warm ocean waters. Summers are hot and humid. Winters are mild, with occasional frost and some snow. The warm ocean currents cause areas with the humid-subtropical climate to be struck by occasional hurricanes or typhoons.

A climate type influenced mainly by oceans is called the marine-west-coast climate. This climate is generally found on the west coasts of continents in the upper middle latitudes. Temperatures are mild all year. Middle-latitude storms traveling across the oceans in the westerlies bring most of the rainfall to areas with this climate. Winters are foggy, cloudy, and rainy, but summers can be warm and sunny. This climate supports dense evergreen forests in the Northwest United States. The marine-west-coast climate is most widespread in northwestern Europe where the absence of mountain ranges along the coast allows the cool, moist ocean air to spread far into the interior of the continent.

The humid-continental climate is found in latitudes that are subject to periodic invasions of both warm and cold air. This climate has the most changeable weather conditions and experiences four distinct seasons. Its midcontinental locations are responsible for extreme differences between summer and winter temperatures. Because this climate type is situated along the polar front, middle-latitude storms bring rain throughout much of the year and bring snow in winter. Precipitation is heavy enough to support forests.

Ireland is often called the Emerald Isle for its green countryside. The country is dominated by a marine-west-coast climate.

Reindeer and humans have adapted to the cold of the tundra in northern **Siberia**.

LOCATION **High-Latitude Climates** Located in the high latitudes, the subarctic climate is centered above 50° north latitude. This climate has long, dark, cold winters, with temperatures staying well below freezing for half of the year. During the short summers, however, very warm temperatures can occur. The subarctic climate has the greatest annual temperature ranges in the world. Though severe, the subarctic climate supports vast evergreen forests.

The subarctic climate region is very large, extending across northern North America, Europe, and Asia. The Southern Hemisphere lacks this climate type because there is no land at these latitudes.

Another climate with a long winter is the tundra climate. Temperatures are above freezing only during the short summers. The tundra climate takes its name from the only vegetation that can survive there. Tundra vegetation is made up of small, hardy plants such as mosses, lichens (LIE-kuhnz), herbs, and low shrubs. The climate is so severe that no trees grow.

Water below the tundra surface remains frozen throughout the year. This condition is called **permafrost**. Although the melting winter snow creates swamps and bogs on the surface during the summer, permafrost makes it difficult for the water to seep into the ground. These wet areas do, however, support great numbers of insects and birds during the short summer.

The polar ice-cap climate has cold temperatures all year, and snowfall is likely year-round. Life is almost impossible in polar ice-cap climates. Only animals and plants that can live in the icy polar seas survive here.

LOCATION **Highland Climates** As you learned earlier, an increase in elevation causes a decrease in temperature. Highland climates are determined by elevation rather than by latitude. Mountain areas of the world can have a variety of climate types in a very small area.

The lowest elevations of a mountain will have a climate and vegetation similar to that of the surrounding area. As you climb higher up the mountain, the climate conditions change. Temperatures and air pressure are lower. The cooler temperatures limit the vegetation that can grow, especially trees. At the highest elevations, climate conditions are similar to those of the polar ice-cap climate. Here, temperatures remain below freezing, and ice and snow are always present.

3 SECTION REVIEW

1. How does the latitude of a place influence its climate?
2. Give an example of a climate type found in each of the following climate regions: low-latitude, dry, middle-latitude, and high-latitude.
3. **Critical Thinking** **Compare and contrast** desert climates and tundra climates.

CHAPTER 3 CHECK

Space shuttle photo of large thunderstorm east of **Madagascar**

Reviewing the Main Ideas

1. The earth maintains an energy balance by absorbing and releasing heat from the sun. Differences in the heating of the earth's surface create differences in air pressure. These differences create the global wind belts and global ocean currents that move heat and cold across the earth, helping the earth maintain its energy balance.
2. Water vapor is an important gas in the atmosphere. Condensation occurs when water vapor changes from a gas to a liquid. If the condensed droplets become large enough, they fall to the earth as precipitation. Storms occur when the energy stored in water vapor is released.
3. Generally, the global climate regions are classified by differences in temperature and precipitation. Major factors that influence climate include latitude, elevation, and location on land.

Building a Vocabulary

1. What is the difference between weather and climate?
2. Define *air pressure*. How do differences in air pressure affect climate?
3. What are the prevailing winds called at the low, middle, and high latitudes? What are the doldrums?
4. What is permafrost, and where is it found?

Recalling and Reviewing

1. What important role do the oceans play in the global heat or energy exchange system?
2. Where does water vapor come from? How does its form change when the air temperature rises or falls?
3. What climate types are found in the low, middle, and high latitudes?
4. Where are most of the desert climate areas located? Why?

Thinking Critically

1. Imagine that the year is 2050 and you live in a highly technological society. What technologies might be in place to protect you from the climate?
2. Why do you think that the humid-continental climate is not found in the Southern Hemisphere?

Learning About Your Local Geography

Cooperative Project

Watch a local television news show. Take careful notes of the type of information presented by the weatherperson. Then, track the weather in your area for one week. With your group, prepare a weather report using daily figures, charts, and graphs. Present your weather report to the rest of the class.

Using the Five Themes of Geography

An aspect of **human-environment interaction** is the way in which people are affected by the climate in which they live. Make a list of as many factors as possible that describe how the climate affects your daily life. Next to each factor, describe how the use of technology has changed the impact of climate on your life.

CHAPTER 4

The Water Planet

"I never cease to wonder at the tenacity of water—its ability to make its way through various strata of rock, zigzagging, backtracking, finding space, cunningly discovering faults and fissures in the mountain, and sometimes traveling underground for great distances before emerging into the open. Of course, there's no stopping water. For no matter how tiny that little trickle, it has to go somewhere!"

Ruskin Bond

GEOGRAPHY DICTIONARY

irrigation
industrialization
hydroelectricity
hydrologic cycle
transpiration
evapotranspiration
headwaters
tributary
watershed
estuary
wetlands
groundwater
water table
aquifer
continental shelf

If you looked at Earth from a spaceship, you might wonder why we do not call our planet *Water*. Almost three-fourths of the earth's surface is covered with water. Earth is the only planet in the solar system that has water, and it is sometimes called the "water planet." The oceans, water frozen as ice, lakes and rivers and other surface water, water found underground, water vapor in the atmosphere, and water in all living things make up the hydrosphere.

Water is the most precious of Earth's natural resources. It is essential for all known life, especially for cell growth. All living things, plant and animal, are made up mainly of water. In fact, your body is about 70 percent water.

Though plentiful on a global scale, fresh water is a scarce resource throughout much of the world. Lack of water limits a country's ability to develop economically and to produce enough food to feed its people. In poor countries, many people have access only to unclean, unsafe water. About 30 percent of the world's population lives in areas with severe water problems.

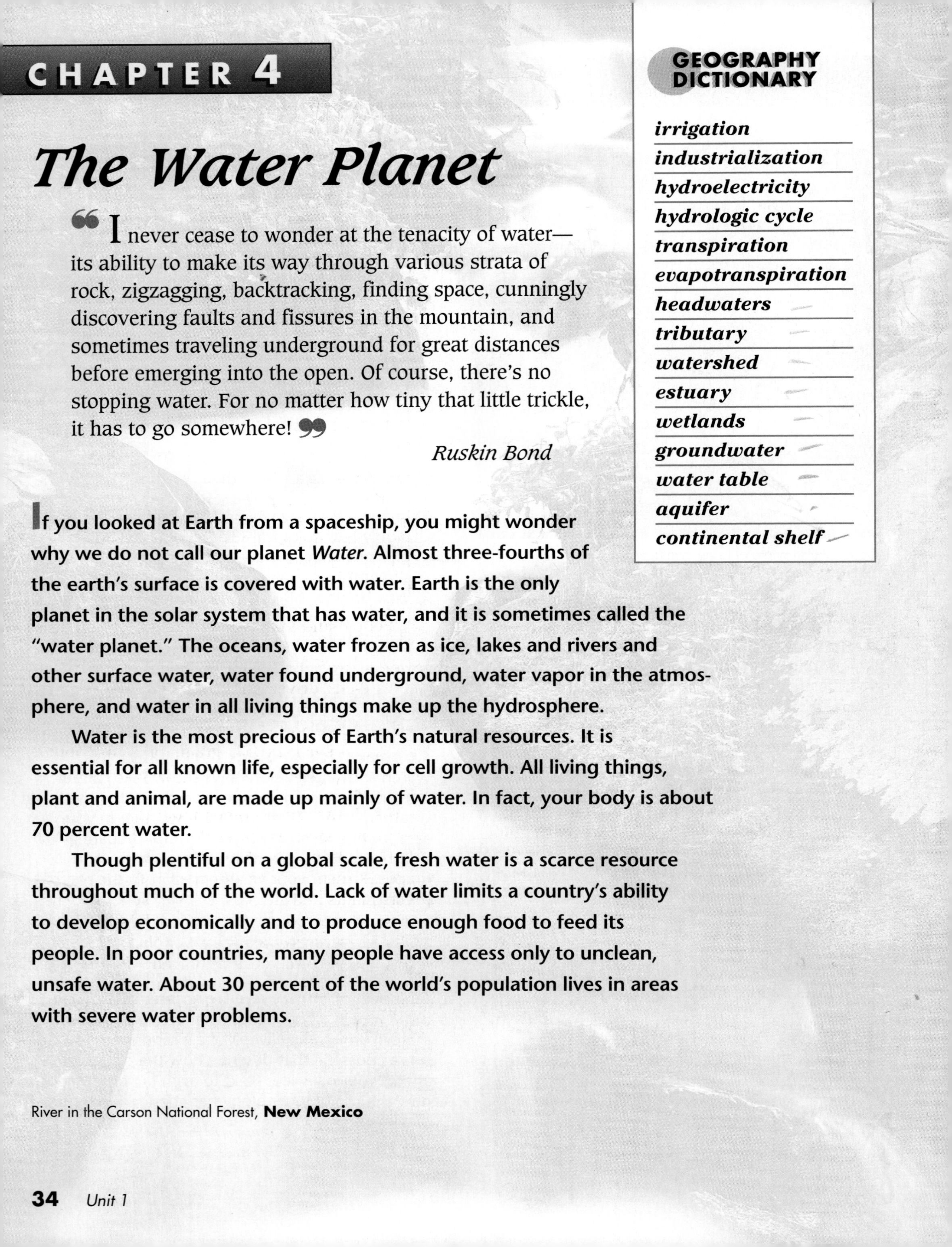

River in the Carson National Forest, **New Mexico**

The Grand Coulee Dam towers above the town of **Coulee Dam, Washington**. This dam is part of a system of U.S. and Canadian dams on the Columbia River that provides water for hydroelectric power and irrigation. The Grand Coulee Dam produces more electricity than any other hydroelectric dam in the United States.

THE GEOGRAPHY OF WATER

FOCUS

- *Why is water essential to life on Earth?*
- *What characteristics make water a unique substance?*
- *How does water circulate among the parts of the hydrosphere?*

People and Water Water is one of our most important resources. Humans cannot survive for more than about a week without it. How much water do you use each day in your home? In modern society, we often take our water supply for granted. While you may simply turn a faucet to get a glass of water, people in other parts of the world may spend hours each day walking to a community well or stream just for a pail of water.

Water is essential for agriculture, and in dry climates, **irrigation** is necessary for growing crops. Irrigation is the watering of land through pipes, ditches, or canals. In the United States, few crops could be grown in areas of Texas, California, and Arizona if farmers could not get irrigation water. Yet, these states are among the nation's biggest agricultural producers. Millions of dollars are spent each year moving water to these farming regions. In countries with little technological development, people are dependent on rains and careful water conservation to irrigate their fields.

Industry also could not function without water. Water is an ingredient in the manufacture of many products and serves as a coolant and lubricant for many machines. Countries that are beginning to industrialize can do so only where water is available. **Industrialization** is the large-scale manufacturing of goods using machine power.

People and industry also rely on water as a valuable power source. **Hydroelectricity**, a major power source, is produced when water stored behind dams drives engines to produce electric power. Our oceans, rivers, and lakes also provide a world transportation network. To transport bulk cargo such as oil, grain, or automobiles, shipping by water is a preferred method because of low cost.

As the world becomes more densely populated, our water resources become more vulnerable. Pollution from city, farm, and industrial wastes is dumped into our bodies of water. This affects plants and animals, and our sources of fish and other seafood are threatened. In some areas, we have overused our water supplies, and water is becoming scarce. It is important to remember that our quality of life depends upon water, and we must learn to conserve and keep clean our most vital resource.

Characteristics of Water Water is a unique substance with three important characteristics. First, it is the only substance that can exist as a liquid, solid, or gas within the earth's temperature range. This means that water can be transported and stored nearly anywhere on Earth. Second, water has the ability to dissolve almost anything over time, including the hardest rocks. This allows mineral matter to be dissolved and moved by water in the oceans, rivers and streams, and underground.

Third, water heats and cools very slowly compared to other Earth materials. This characteristic enables water to moderate Earth's air and surface temperatures. Without water, Earth's temperatures probably would be too extreme to support life.

Geographic Distribution Although water is abundant on Earth, it is not evenly distributed. Because we depend on water for survival, all people must live in a place where they have some access to water. As a result, access to and control of water has been the cause of many conflicts throughout world history.

The majority of the hydrosphere is in the oceans, which contain roughly 97 percent of Earth's water. Of the remaining 3 percent, more than 2 percent is found frozen in the polar and mountain ice sheets. Thus, more than 99 percent of the earth's hydrosphere is either salty or frozen.

As a result, less than one percent of Earth's water is available as a freshwater resource. This very small amount is found underground or in lakes, rivers, and streams. Even smaller, but very essential, amounts are found in the atmosphere and biosphere.

MOVEMENT The Hydrologic Cycle The amount of water on Earth is constant, yet it is always changing form and location. The circulation of water among parts of the hydrosphere is called the **hydrologic cycle**. Water is cycled endlessly through the oceans, ice sheets, lithosphere, atmosphere, and all living things in the biosphere. (See the diagram on this page.)

The hydrologic cycle is driven by solar energy. Most water enters the atmosphere after being heated by the sun's energy and evaporating from the ocean. It becomes water vapor and rises into the air. As it rises, it cools and condenses, changing from a gas to tiny liquid droplets. These droplets join to form clouds, and if the droplets become heavy enough, they will fall back onto the ocean as precipitation. This completes the cycle of evaporation, condensation, and precipitation.

The hydrologic cycle becomes more complex, however, when precipitation falls on the continents and islands. Some precipitation falls as snow on high mountains or polar areas and accumulates as ice. Some flows off the land through rivers and streams, possibly to be stored in lakes, and eventually returns to the ocean.

The Hydrologic Cycle

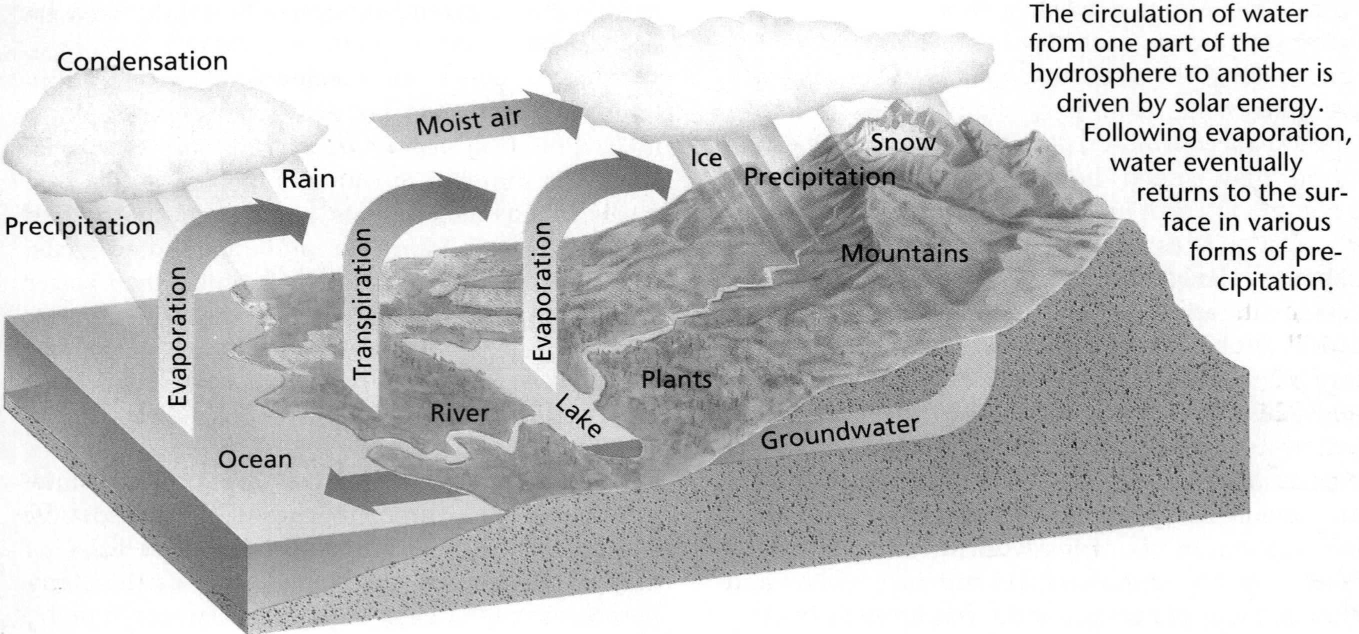
Condensation
Moist air
Rain
Precipitation
Evaporation
Transpiration
Evaporation
Ice
Snow
Precipitation
Mountains
Plants
River
Lake
Groundwater
Ocean

The circulation of water from one part of the hydrosphere to another is driven by solar energy. Following evaporation, water eventually returns to the surface in various forms of precipitation.

The ground absorbs some water by taking it into the air spaces in the soil and rock, where it can be used by plants. Here, the water moves underground to the ocean or is returned to the atmosphere by evaporation from the land or by a process called **transpiration**. Transpiration occurs when plants give off water vapor through their leaves. The evaporation of water from the ground combined with the transpiration of water by the plants is called **evapotranspiration**. This is a measure of the total loss of water from the land to the atmosphere.

Although water is constantly being renewed within the hydrologic cycle, its rate of renewal is very slow. The best technology in the world cannot increase the rate at which water is renewed in the global water cycle.

SECTION REVIEW

1. How is water essential to agriculture and industry?
2. What is the importance of each of the three characteristics of water?
3. **Critical Thinking** **Analyze** cause-and-effect relationships in the hydrologic cycle.

WATER RESOURCES

FOCUS

- *What are the various types of surface water found on Earth?*
- *What is groundwater, and why is it important?*
- *What are the characteristics of seawater?*

Surface Water As precipitation falls on the continents and islands, it flows down the slopes of hills and mountains toward the lowlands and coasts. The first and smallest streams to form from this runoff are called **headwaters**. As these headwaters join, they form larger streams, and farther downstream, they eventually form rivers. Any smaller stream or river that flows into a larger stream or river is a **tributary**. In the United States, major rivers such as the Ohio and Missouri are tributaries of the Mississippi River.

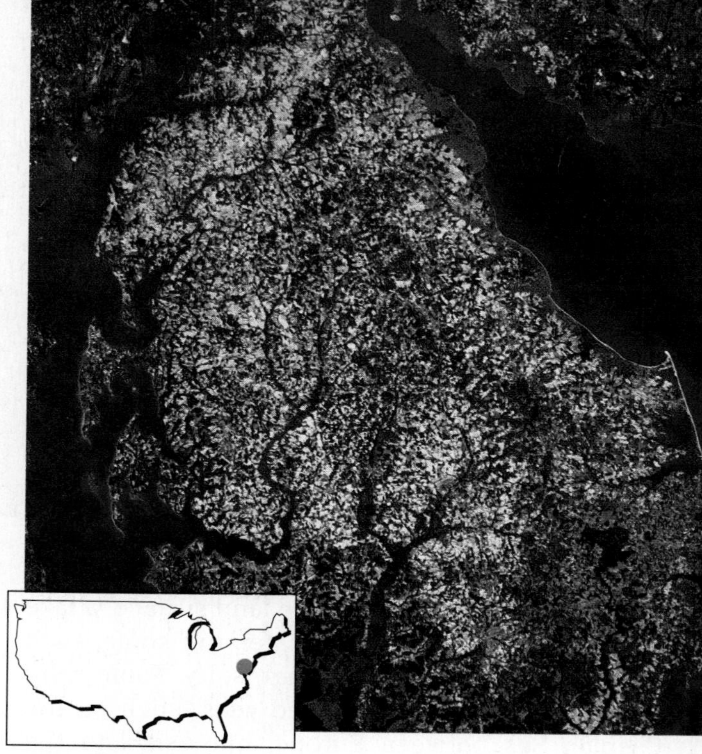

The **Delmarva Peninsula**, on which **Delaware** and part of **Maryland** and **Virginia** sit, is drained by a number of rivers, including the Choptank River (center) and the Nanticoke River (bottom center). Both rivers empty into Chesapeake Bay, to the left of the peninsula. Delaware Bay is to the upper right of the peninsula.

An area of land that is drained by a river and its many tributaries is a **watershed**, or drainage basin. Very large rivers, such as the Amazon in South America or the Mississippi in the United States, may flow for thousands of miles and be joined by hundreds of tributaries. Their drainage basins might extend over 1 million square miles (2.6 million sq. km) and drain a major portion of a continent. Rivers are a valuable water resource, providing water for agriculture, electricity, transportation, and cities. When a river is polluted from the clearing of land or by the disposal of waste materials, the effects may be felt along the entire river system.

Where rivers meet an inlet of the sea, an **estuary** may form. In this semi-enclosed coastal body of water, seawater and fresh water mix. Estuaries are particularly important because they are rich in fish and shellfish. Good examples of large estuaries are Chesapeake Bay on the Atlantic coast of the United States and Puget Sound on the Pacific coast. As rivers carry fresh water to the coasts, they also bring essential minerals that were removed from the land. These minerals contain nutrients necessary to ocean life.

These caribou are among the many species that move through and live in the wetlands of **Alaska**. The advantages and disadvantages of economic development in the wetlands of Alaska, as elsewhere, are sharply debated.

When runoff is slowed or stopped inland and water fills a depression on the land surface, a lake forms. Most lakes are freshwater, but some, such as Great Salt Lake in Utah, are salty. Some salty lakes were incorrectly named seas, such as the Caspian Sea between Europe and Asia and the Dead Sea in Southwest Asia. Unlike seas, however, lakes are totally surrounded by land, are usually not at sea level, and do not exchange water with the oceans.

Many lakes, such as the Great Lakes along the United States–Canadian border, were formed by glaciers. The Great Lakes is the largest freshwater lake system in the world. Like the United States and Canada, Russia, Sweden, and Finland have thousands of lakes left from the last ice age. Other lakes, such as Crater Lake in Oregon, have formed in erupted volcanoes. Lakes are an important resource for our water supply, fish, and recreation. In addition, large lakes have a moderating effect on the local climate, allowing for a long fruit-growing season in some areas.

Other important surface waters are **wetlands**. These are land areas that become flooded for at least part of the year. Some wetlands are continuously flooded, while others change with the seasons. Wetlands contain some of the earth's most productive land. Wetlands include coastal marshes, wooded swamps, tundra bogs, and river bottomlands. Wetlands support huge populations of fish, shellfish, and birds, and also filter pollutants from water. One of the best-known wetlands in the United States is the Everglades in southern Florida. Almost half of Alaska is wetlands, and that state holds about 70 percent of the wetlands remaining in the United States. More than half of U.S. wetlands have been destroyed by draining, paving, and filling for farmland, housing, ports, and industry.

Groundwater While the water resources in rivers, lakes, and wetlands are visible to us, there

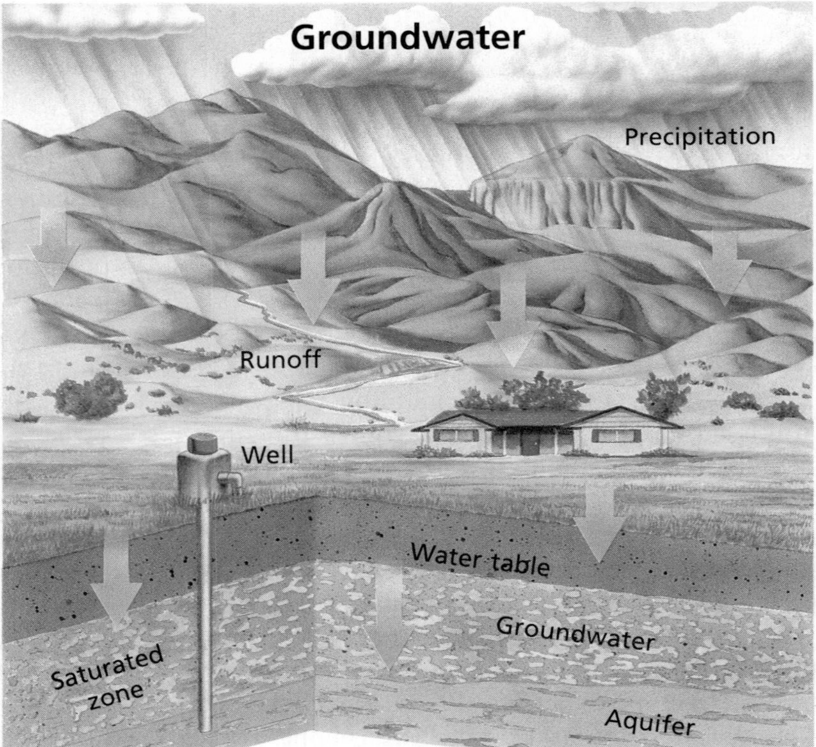
Groundwater
Precipitation
Runoff
Well
Water table
Groundwater
Saturated zone
Aquifer

Water sometimes exists underground in regions where surface water is scarce. How do people gain access to groundwater?

is a large freshwater resource beneath the earth's surface. This **groundwater** is found in tiny spaces between soil and rock grains. Groundwater is an important resource, especially in dry climates where surface water is scarce.

The major source of groundwater is precipitation. Water from rain or melting snow sinks into the ground and slowly seeps downward. The water stops when there is no more space between the rock grains or when all the spaces are saturated with water. The top of this saturated zone is called the **water table**. To tap this groundwater, we would have to drill a well to a depth below the water table.

The depth of the water table varies. In areas with a wet climate, the water table will be near the surface. In desert areas, it may be hundreds of feet below the surface or it may not even exist. The water table follows the land and will rise under hills and drop under valleys. It will vary from year to year as rainfall patterns change. If a region suffers a drought, the water table falls. Too many wells in a region can also lower the water table.

Like rivers, groundwater can travel long distances but at a much slower rate. Generally, groundwater flows through a rock layer called an **aquifer**. In an aquifer, water can be stored in and carried through porous rock and then released for use. Aquifers may be quite large and may measure hundreds of miles across. The Ogallala Aquifer stretches from Texas to Nebraska and is a major source of water for the irrigation of farms in the Great Plains subregion.

The Oceans As you have learned, most of the earth's water is not found on land. Oceans, the largest bodies of salt water, hold more than 97 percent of the earth's water and cover about 71 percent of the planet's surface. Actually, oceans are one continuous global body of water surrounding all the continents. For geographic identification, this global ocean is divided into four separate oceans: the Pacific, the Atlantic, the Indian, and the Arctic.

The Pacific Ocean is the largest geographic feature on Earth and covers an area larger than all the continents combined. The Atlantic and Indian oceans are each about half the size of the Pacific Ocean. The Arctic Ocean is much smaller than the other three oceans and is located near the North Pole.

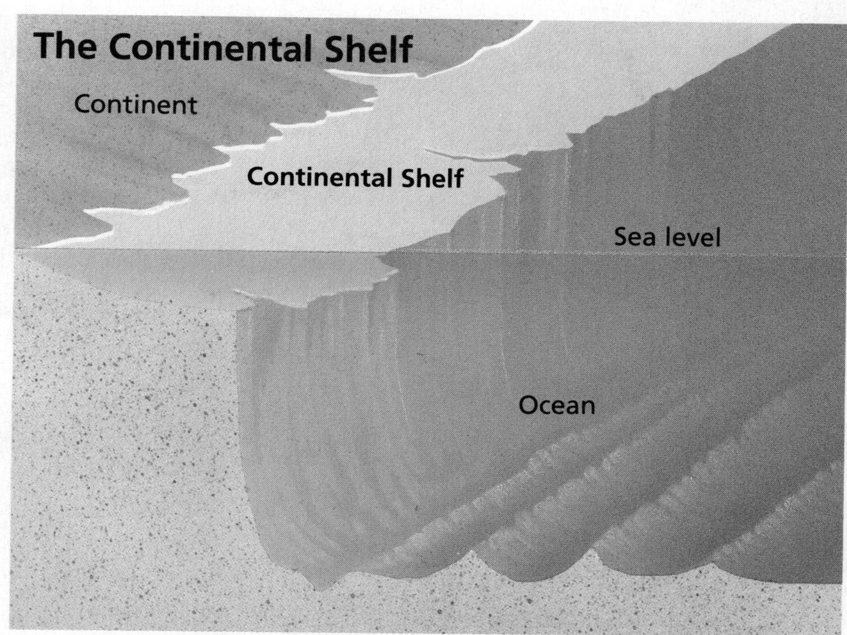
The Continental Shelf
Continent
Continental Shelf
Sea level
Ocean

The continental shelf is the shallowest part of the ocean and slopes gently away from the continents. Much of the mud and sand from rivers is deposited onto the continental shelf.

Seas are smaller saltwater bodies connected to the oceans. Examples include the Mediterranean Sea, the Coral Sea, and the Caribbean Sea. Saltwater bodies near coasts that extend into land are sometimes called gulfs, such as the Gulf of Mexico, the Gulf of Alaska, and the Persian Gulf in Southwest Asia. A few large saltwater bays, such as Asia's Bay of Bengal and Canada's Hudson Bay, are actually seas, not bays. Bays are inlets of seas.

The oceans average about 13,000 feet (3,962 m) in depth. There is a variety of features at the ocean bottom, and the depth varies greatly. Flat plains are found in the deep ocean basins, while ridges rise high above them. The greatest ocean depth is in the Mariana Trench, located in the North Pacific Ocean. Its dark, cold waters reach more than 36,000 feet (10,973 m) below sea level.

The most shallow part of the ocean is the **continental shelf**, which slopes gently from the continents. Some parts of the continental shelf were actually land thousands of years ago during the last ice age. As the ice melted, the sea level rose, flooding the continental shelf. If the massive ice sheets of Antarctica and Greenland were to completely melt, low areas far inland would flood.

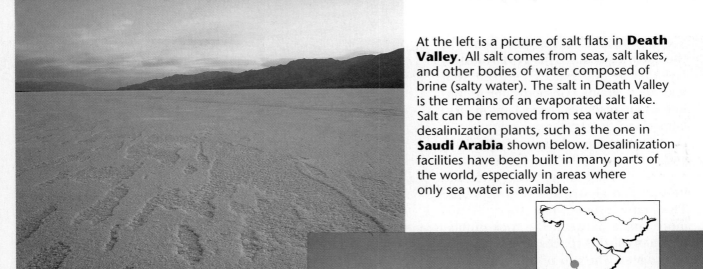

At the left is a picture of salt flats in **Death Valley**. All salt comes from seas, salt lakes, and other bodies of water composed of brine (salty water). The salt in Death Valley is the remains of an evaporated salt lake. Salt can be removed from sea water at desalinization plants, such as the one in **Saudi Arabia** shown below. Desalinization facilities have been built in many parts of the world, especially in areas where only sea water is available.

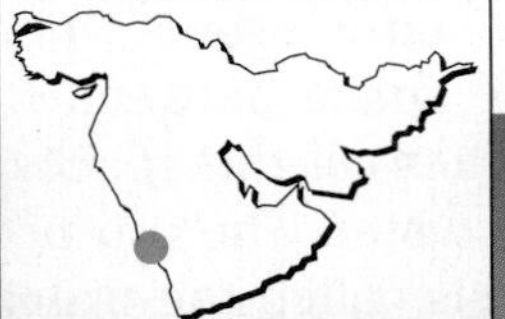

Seawater and Its Characteristics

Because water has the unique ability to dissolve almost anything, it contains the materials it dissolves. Thus, seawater contains every element known on Earth, even gold. Most of these dissolved materials are in such small quantities that it is not practical to "mine" seawater. You may have wondered why you taste salt when you swim in the ocean. Salts dissolve easily and are the most common material found in seawater. In fact, about 3.5 percent of seawater is dissolved salts. Because of the salt content, we cannot use the oceans as a resource for drinking water or for agriculture. It is possible to remove salt from seawater, but the process is expensive.

Because water is slow to heat up and cool down, the oceans do not have the temperature extremes of land. On a daily basis, the temperature of ocean water changes less than one degree, while the temperature of land between day and night will vary greatly. The seasonal changes in the ocean also are less than the land's seasonal changes, and the oceans help prevent the climates along the coasts from being too warm or too cool. Because sunlight does not penetrate very deeply into the oceans, only the upper portion of the water is heated.

Like the atmosphere, the ocean is always moving and circulating. The winds push the ocean currents and cause waves, and the waves and currents mix the surface waters. The gravitational pull of the moon and sun causes the tides to rise and fall, which in turn causes many currents, especially in the coastal areas. Even the deepest ocean waters move slowly as a result of varying temperatures or the amount of salt in the water. Though the oceans appear huge, their life-producing regions are limited mainly to the shallow near-shore areas, as most seawater is deep, dark, and cold. Much of the world ocean remains unexplored.

SECTION REVIEW

1. What environmental benefits are provided by rivers, estuaries, lakes, and wetlands?
2. In what climate type is groundwater especially important? Why?
3. **Critical Thinking** **Explain** why the Arctic Ocean freezes but the Pacific Ocean does not.

CHAPTER 4 CHECK

Iceberg near **Antarctica**

Reviewing the Main Ideas

1. Water is one of the most important natural resources on Earth. Life as we know it depends on water. Besides its personal uses, water is used for irrigation, as a power source, and in industry.
2. Water has three important characteristics that make it a unique substance: it can exist as a solid, liquid, or gas; it can dissolve almost anything; and it heats and cools very slowly.
3. Water resources on Earth include surface water, such as rivers, lakes, and wetlands; groundwater found beneath the surface of the earth; and seawater.

Building a Vocabulary

1. Define *irrigation*. Name three states in the United States that depend upon irrigation.
2. How is hydroelectricity produced?
3. What is the term used to describe the circulation of water among parts of the hydrosphere?
4. What are the differences among headwaters, tributaries, estuaries, and wetlands?
5. What is a watershed?

Recalling and Reviewing

1. Explain the importance of water to agriculture and to industry.
2. Why are rivers and estuaries important to ocean life?
3. What is the major source of groundwater?
4. Into what four major oceans do geographers divide the global ocean?

Thinking Critically

1. Describe three ways in which life would change if the world's water resources were reduced.
2. Design a poster promoting the protection of wetlands.

Learning About Your Local Geography

Individual Project

Conduct research to find out about water resources in your area. Where does the water in your home come from? Who monitors the water quality? What is the depth of the water table in your area? Is irrigation important? Is there a hydroelectric plant nearby? Organize your information in a written report or in a visual presentation. You may want to include your work in your individual portfolio.

Using the Five Themes of Geography

Write a paragraph about how the geographic themes of **place, human-environment interaction,** and **movement** apply to irrigation. What place characteristics make irrigation necessary, and how does irrigation change places? How do humans change the environment with irrigation? How does water move in irrigation? From where does it come, and to where does it go?

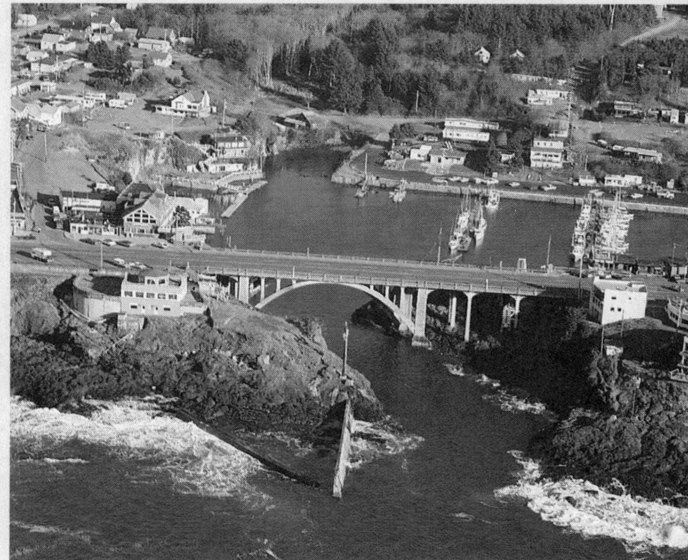
Depoe Bay, Oregon

CHAPTER 5

Global Landforms

"There are few more barren places on earth than the plains surrounding a volcano in the aftermath of its eruption. Black tides of lava lie spilt over its flanks like slag from a furnace. Their momentum has gone but they still creak and boulders still tumble as the flow settles. Steam hisses between the blocks of lava, caking the mouths of the vents with yellow sulphur. Pools of liquid mud, grey, yellow or blue, boiled by the subsiding heat from far below, bubble creamily. Otherwise all is still."

David Attenborough

GEOGRAPHY DICTIONARY

erosion
rock weathering
sediment
glacier
sand dune
lava
volcano
fold
fault
earthquake
landform
plain
plateau
plate tectonics
mid-ocean ridge
abyssal plain
subduction
trench
relief
floodplain
alluvial fan
delta

During a heavy rain, water running in street gutters carries leaves, twigs, litter, and small rocks. If you were to strain the water, you would discover small particles of soil as well. This running rainwater is wearing away the land instead of soaking into it. Wearing away of land is called **erosion.** Water, as well as wind, waves, and ice, causes erosion. Like sandpaper, these agents of erosion smooth the earth. They also transport rock and soil, depositing these particles in new locations.

Other forces deep below the surface of the earth also are at work. These forces sometimes raise masses of rock, building mountains. They may even bend and break the continents. Usually, the forces shaping the earth's surface work only gradually. Sometimes, however, due to a great flood or a volcanic eruption, sudden changes occur. Together, the forces working below and above the earth's surface are constantly changing the world around us.

Hawaii Volcanoes National Park, **Hawaii**

1 FORCES SHAPING THE LAND

FOCUS

- *What external forces cause changes in the earth's surface?*
- *What internal forces cause changes in the earth's surface?*

Rock Weathering At the earth's surface, rocks break and decay. The process that breaks up rocks and causes them to decay is called **rock weathering**. Weathering often is slow and almost impossible to detect, but even the hardest rocks eventually will break into smaller fragments. Given enough time, rocks can dissolve like salt in a pan of water. Some rock weathering is like tooth decay. Chemicals in the air and water pass through rock, slowly eating away at it. Other rock weathering is more like chipping or breaking a tooth. When a forest fire moves over the land, heat can cause rocks to crack. Rocks also can break when water passing through freezes into ice. Similarly, a tree root can pry a rock apart. Rock weathering also is the first step in soil formation.

Weathering breaks rock into smaller particles of mud, sand, or gravel called **sediment**. The weathered sediment is subject to movement due to gravity or removal by erosion. For example, sediment may simply roll or slide downhill in a rockfall or landslide. Sediment also may be eroded and carried away by water, wind, or even moving ice.

Water Whether running in streams and rivers or pounding the beach, water is the most powerful force of erosion and transportation. Eventually, water will carve through the hardest rock. Rivers sometimes carry sediment long distances. Erosion on a hillside can begin as a tiny channel narrower than a pencil. As the water flows farther downhill, the channel might grow wider and deeper, forming a gully—a scar on the land that shows where erosion is particularly active. Rivers also erode and shape the valleys into which they flow. When a river cuts downward for a long period of time, a steep-sided canyon can result. Even in desert regions, rains can erode the bare surfaces between the sparse plants.

Glaciers Thick masses of ice that glide slowly across the earth's surface are called **glaciers**. There are two kinds of glaciers—sheet glaciers and mountain glaciers. Sheet glaciers, or ice sheets, cover large areas. Due to the force of gravity, they flow outward from great domes of ice where snow accumulates. The thick ice in the center of the ice sheet presses downward with great force. The edges of the glacier, which are usually far away from the center, creep outward. Glaciers have the power to level anything in their path. They can move rocks bigger than houses long distances. Ice sheets also can act like

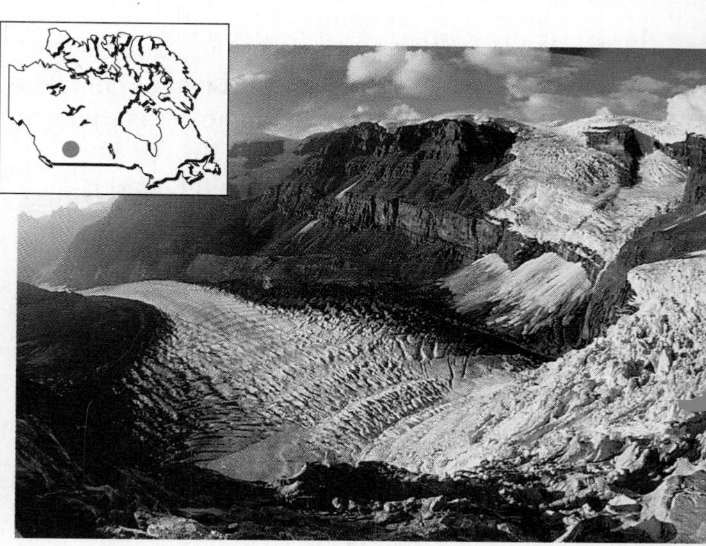

Water and ice are powerful shapers of land. At the left is Utah's **Bryce Canyon**, whose colorful and unusual rock formations were created by centuries of weathering by water and ice. Above, a glacier has carved a path through the Rocky Mountains of **Alberta, Canada**.

Erosion by the Colorado River has created **Marble Canyon** (left), part of Grand Canyon National Park in Arizona. South of Marble Canyon, the Colorado River has cut a gorge as much as one mile (1.6 km) deep and 18 miles (29 km) wide in the Grand Canyon.

blankets, however, protecting the ground underneath from severe cold and other eroding forces.

Today, ice sheets up to 10,000 feet (3,048 m) thick cover most of Antarctica and Greenland. Geographers have discovered that during the last 2 million years there have been many periods when great ice sheets spread across North America and Europe. During these ice ages, glaciers spread as far south as St. Louis, New York City, and Moscow. The last ice age ended about 10,000 years ago. During each ice age, climates changed all across the world. Some areas that are forests today were deserts during the ice ages.

Mountain glaciers are smaller and much more common than ice sheets. Flowing downhill like slow rivers of ice, mountain glaciers are found in high mountain valleys all around the world. Today, there are more than 1,000 mountain glaciers in the Rocky Mountains, the Cascade Mountains, and the Sierra Nevada in the western United States. The ice of mountain glaciers forms from surplus winter snow that does not melt during the summer. This snow is the result of low temperatures caused by high elevation. Mountain glaciers create sharp peaks and U-shaped valleys. During the ice ages, mountain glaciers were more common and much larger than they are today. Usually, they extended farther down the mountainsides, creating much of today's spectacular mountain scenery.

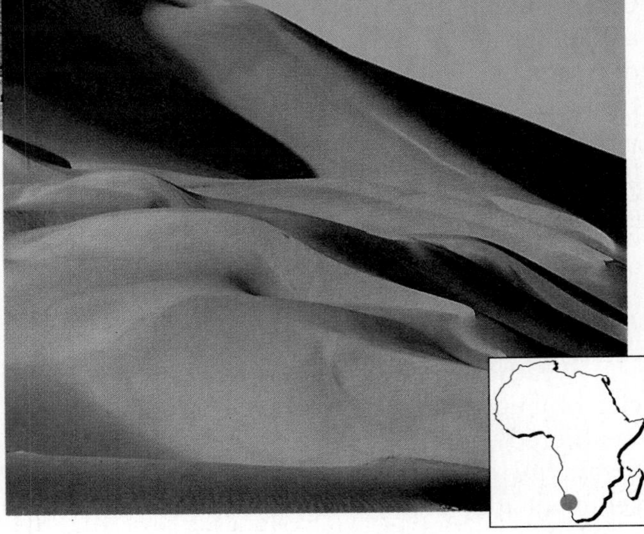

Wind plays a large part in shaping the land. Wind-blown sand deposits create sand dunes, such as those in Africa's **Namib Desert**.

MOVEMENT **Wind** Another force that erodes and transports sediment is wind. Because plants protect land surfaces from erosion, wind erosion is greatest where there are the fewest plants—in deserts, along beaches, and on dry, bare ground where people and animals have removed the plants. Wind erosion works in two ways. First, wind wears down hard rocks by blowing sand and other particles against them, often leaving smoothed and polished surfaces. Wind also changes the land by blowing sand and dust from one place to another. Hills of wind-deposited sand are called **sand dunes**. Often, sand dunes are near where the sand was eroded, such as along an ocean beach or beside a

The Interior of the Earth

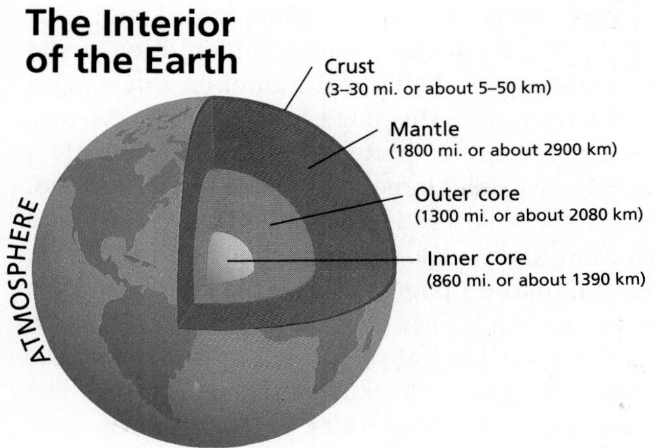
Crust
(3–30 mi. or about 5–50 km)
Mantle
(1800 mi. or about 2900 km)
Outer core
(1300 mi. or about 2080 km)
Inner core
(860 mi. or about 1390 km)
ATMOSPHERE

Temperatures within the earth increase with depth. In the inner core, temperatures may reach as high as 12,000°F (6900°C). How far below the surface is the earth's inner core?

dry stream bed. Sometimes dust particles in deserts are lifted high into the atmosphere and carried great distances. Dust from the Sahara, the great desert in northern Africa, often is transported across the Atlantic to islands in the Caribbean.

Inside the Earth If erosion were the only force shaping the earth's surface, the world around us would have been worn smooth long ago. But as erosion wears down the land, forces inside the earth continue to push up the surface and break it apart.

Scientists have found that our planet is composed of four distinct layers. (See the diagram above.) In the center is the solid inner core, and the next layer is the liquid outer core. The minerals of the core areas are rich in iron. Surrounding the two cores is the mantle, a mostly solid layer. A thin, rigid crust covers the mantle. If the earth were an apple, the crust would be as thin as the apple's skin. The gases of the atmosphere surround the four layers of the solid earth. The world of people and all other living things is located in a narrow zone at the top of the crust and the bottom of the atmosphere.

Forces Beneath the Crust The earth's surface is bent and broken by heat and by rock movements in the crust and upper mantle. Heat currents travel slowly upward from the core to the crust. These heat currents partially melt the upper parts of the mantle. Nearby parts of the lower crust may be melted as well. The melted, liquid rock, or magma, from within the earth that spills out on the earth's surface is called **lava**.

The crust above the rising heat currents usually rises. Lava may break through and flow over the surface, filling valleys and building the land upward. The opening in the earth's crust through which lava flows is called a **volcano**. (The term *volcano* also is used to refer to the surface feature created by the molten lava.) On the other hand, the crust above the cooler parts of the mantle may sink, creating broad depressions, or low spots, on the earth's surface.

Forces working within the earth not only create lava flows and volcanoes but also bend and break rock. When rock layers are bent, the result is a **fold**. When rock layers break and move, the result is a **fault**. (See the diagram below.) The shock waves, or vibrations, caused by movement along a fault are called **earthquakes**. Whether rocks fold or fault depends on how hard the rocks are and how sudden and strong the forces are. Many of the world's great mountain ranges show much evidence of folding and faulting.

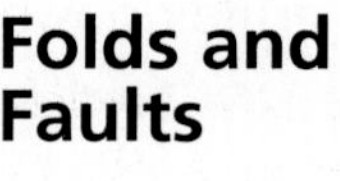
Folds and Faults

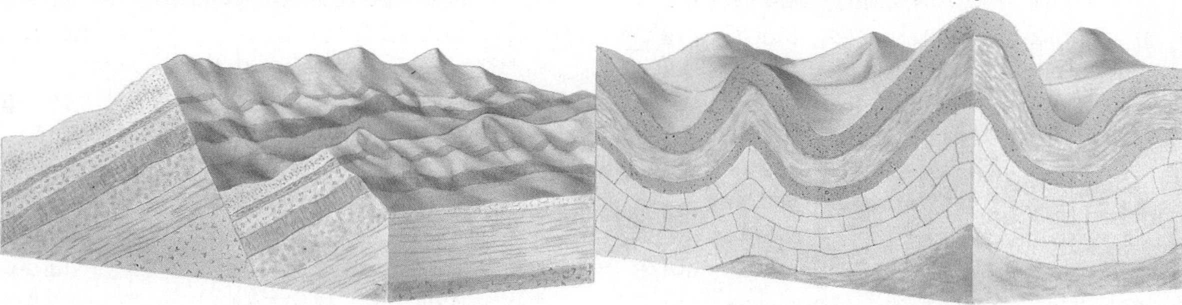
Mountains formed by faults
Mountains formed by folds

Many large mountain ranges are formed by folding, which occurs when two plates collide and their edges wrinkle, as shown above. Fault-block mountains are formed when parts of the earth's crust are pushed up along a fracture line, or fault.

The weight and hardness of rocks also play a role in shaping the earth's crust. Hard rocks resist erosion and often remain as high ground. In contrast, the erosion of weaker rocks usually produces lowlands. Over the long history of the earth, the lightest rocks have risen to the crust. They now form the continents. If an area of a continent is undergoing rapid erosion, its weight becomes less and less. As a result, the eroded area will steadily rise. Similarly, areas that are being continually buried by the deposit of sediment will tend to sink. Scientists believe these forces that change the surface of the earth have been at work for billions of years.

SECTION REVIEW

1. What processes erode land?
2. What occurs when rock is bent or broken?
3. **Critical Thinking** **Assess** the extent to which human activities cause changes in the earth's surface.

LARGE LANDFORMS AND PLATE TECTONICS

FOCUS

- *What is the theory of plate tectonics?*
- *What are the three types of plate boundaries?*

Plates The shapes on the earth's surface are called **landforms**. Examples of landforms include hills, valleys, nearly flat areas called **plains**, and elevated flatlands called **plateaus**. Landforms vary in size from the smallest indentation made by a raindrop to the largest mountain range. (See the illustration on this page.)

The theory of **plate tectonics** helps explain how the world's large landforms were formed. This theory views the earth's crust as divided into more than a dozen rigid, slow-moving plates. (See the map on page 47.) The plates shift only a few inches each year. Scientists have determined that the processes of plate tectonics usually take millions of years to cause major changes. The plates move like giant rafts cruising slowly over the upper mantle. The less heavy rocks of the continents ride along like passengers on the plate surfaces. Some plates are huge. For example, the North American plate begins under the middle of the Atlantic Ocean and crosses Arctic Canada before reaching its distant edge in northern Japan. Other plates are much smaller. One plate lies totally beneath the Caribbean Sea. The thickness of the plates also varies. They can be more than 60 miles (97 km) thick under the mountains of continents. But they are only about 6 miles (10 km) thick at sites below the oceans where the plates are moving apart.

MOVEMENT **Plate Boundaries** The processes of folding and faulting along plate boundaries build mountains and long valleys. Volcanoes are common, too. In fact, the area bordering the Pacific plate (those areas that lie along the Pacific Ocean) is often called the "Ring of Fire" because it experiences frequent earthquake and volcanic activity. These active regions along the Ring of Fire contrast sharply with the usually inactive interiors of plates. Where plate interiors also are continent interiors, erosion steadily grinds down the land, resulting in wide plains.

Landforms

Many of the landforms shown here can be found together in various regions of the earth. Which landforms can you identify where you live?

Glacier
Mountains
Volcano
Alluvial fan
Lake
Canyon
Plateau
Valley
Plain
Isthmus
River
Hills
Desert
Peninsula
Delta
Sandy beach
Floodplain
Strait
Coastline
Island

Plate Tectonics

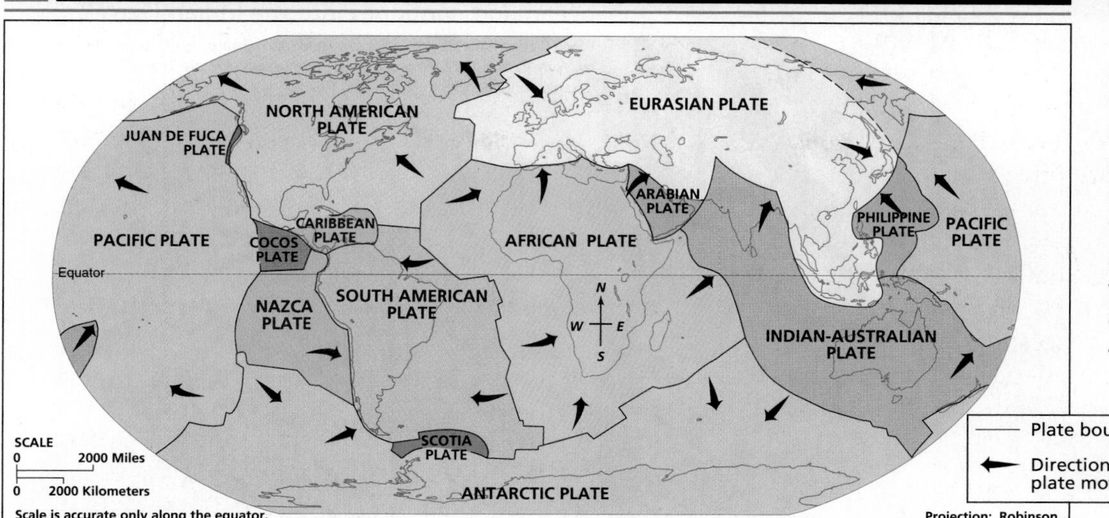
NORTH AMERICAN PLATE
EURASIAN PLATE
JUAN DE FUCA PLATE
ARABIAN PLATE
CARIBBEAN PLATE
PHILIPPINE PLATE
PACIFIC PLATE
COCOS PLATE
AFRICAN PLATE
PACIFIC PLATE
Equator
SOUTH AMERICAN PLATE
NAZCA PLATE
N
W
E
S
INDIAN-AUSTRALIAN PLATE
SCALE
0 2000 Miles
0 2000 Kilometers
Scale is accurate only along the equator.
SCOTIA PLATE
ANTARCTIC PLATE
Plate boundaries
Direction of plate movement
Projection: Robinson

MAP STUDY Scientists believe that the earth's crust is divided into more than a dozen rigid, slowly moving plates. This map shows the location of the plates. The arrows indicate the direction in which they are drifting. What happens where plates meet each other?

There are three types of plate boundaries: those where the plates are moving away from each other, those where plates are pushing against each other, and those where the plates are sliding past each other. Because the crust is thickest and hardest to break where there are continents, most plate boundaries are under the ocean floors. (See the illustration on this page.)

Along the boundaries where plates are moving away from each other, the seafloor spreads, and hot, molten rock moves upward from below to fill the opening. This movement happens because the upper mantle beneath this area of the crust is above heat currents rising from the interior. As a result, the crust is lifted upward, and on the ocean floor a chain of mountains, called a **mid-ocean ridge**, is formed. The mid-ocean ridge system reaches into every ocean floor, extending more than 40,000 miles (64,360 km) in total. (See the map on page 48.) With no wind and rain to erode them, the mid-ocean ridges remain steep and jagged. Occasionally, a mid-ocean ridge sticks up above the ocean, forming an island. Iceland is the largest of these. Away from the ridges, the ocean floors have large, flat areas. These **abyssal** (uh-BIS-al) **plains** are the earth's flattest and smoothest features.

Along the boundaries where plates collide, a heavier plate may dive under the other plate and sink into the upper mantle. This process occurs above the cooler parts of the mantle and is called **subduction**. This sinking creates a **trench**, a long valley on the ocean floor. Trenches can be more than 600 miles long, 60 miles wide, and 7 miles deep (965 km long, 97 km wide, 11 km deep). They are the oceans' deepest places. Trenches often have a row of volcanic islands on one side. When one of the colliding plates has a continent on it,

Movement at Plate Boundaries

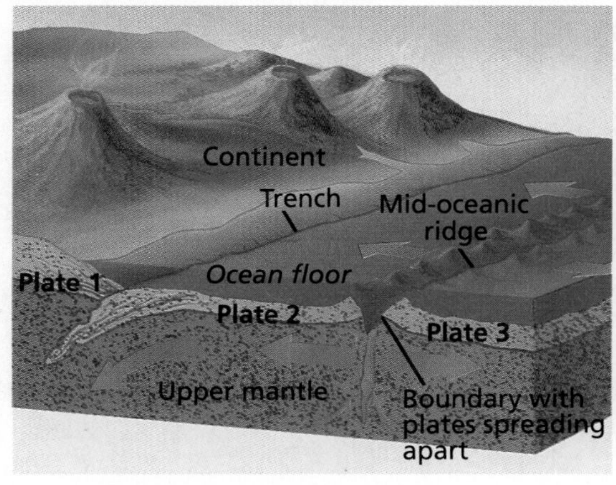
Continent
Trench
Mid-oceanic ridge
Ocean floor
Plate 1
Plate 2
Plate 3
Upper mantle
Boundary with plates spreading apart

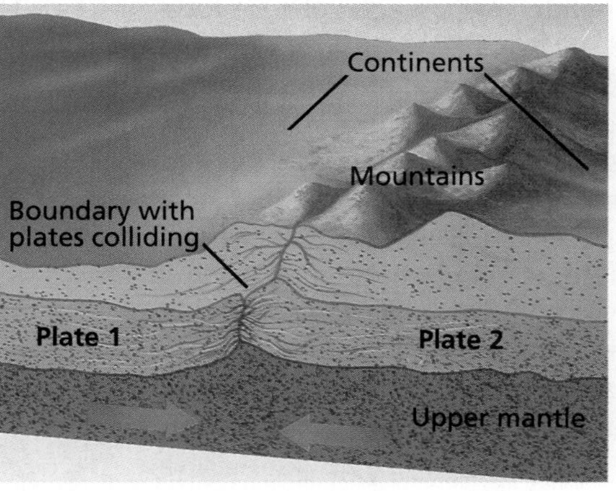
Continents
Mountains
Boundary with plates colliding
Plate 1
Plate 2
Upper mantle

The movement of tectonic plates and the plate boundaries between them create many of the large landforms on the earth. How does the study of plate boundaries explain how some mountains are formed?

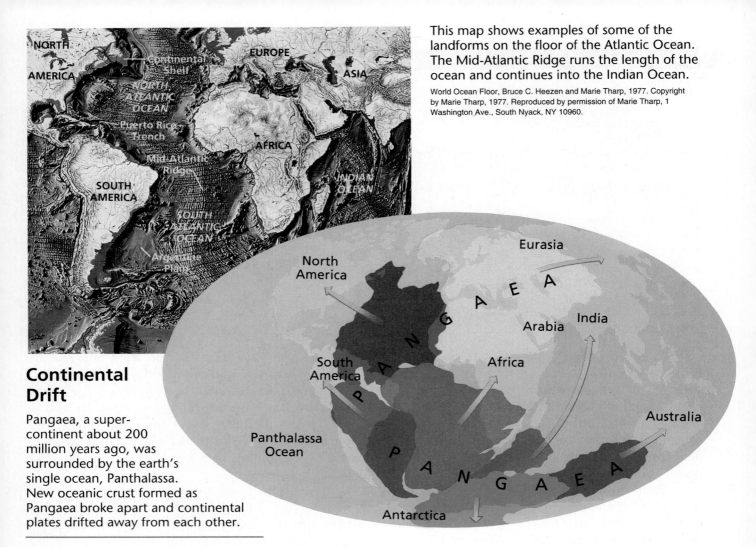

This map shows examples of some of the landforms on the floor of the Atlantic Ocean. The Mid-Atlantic Ridge runs the length of the ocean and continues into the Indian Ocean.

Continental Drift

Pangaea, a supercontinent about 200 million years ago, was surrounded by the earth's single ocean, Panthalassa. New oceanic crust formed as Pangaea broke apart and continental plates drifted away from each other.

the lighter rocks of the continent do not sink. Instead, they crumple and form a mountain range. (See the diagram on page 47.) An example is the Andes Mountains of South America. Sometimes two plates, each carrying a continent, collide. When this happens, no trench forms as the lighter rocks of the continents move together, building huge mountains. The Himalayas between India and China are an example of this.

When plates slide past each other, long creases are formed in the earth's surface, and low mountains may result. Most of these plate boundaries are under the oceans, but the most famous example on land is the San Andreas Fault system in California. There, the Pacific plate slips and jerks northward against the western side of the North American plate, rocking cities such as Los Angeles and San Francisco with earthquakes.

Scientists use the theory of plate tectonics to explain the long history of the earth's surface. They have determined that about 200 million years ago, today's continents were linked as one supercontinent called Pangaea (pan-JEE-uh). (See the diagram on this page.) Pangaea then broke into two smaller supercontinents named Laurasia and Gondwanaland. These then split further into the modern continents, which continue to drift slowly across the world. The theory of continental drift helps explain the apparent fit in coastal shape and rock types between Africa and South America.

SECTION REVIEW

1. How does the theory of plate tectonics explain the formation of the earth's large landforms?
2. What landforms are formed at each of the three types of plate boundaries?
3. **Critical Thinking** **Determine** which type of plate boundary is likely to have the greatest effect on humans.

3 LANDFORM DEVELOPMENT

FOCUS

- *What are primary and secondary landforms?*
- *How are landforms classified?*

Primary and Secondary Landforms Landforms are the result of the struggle between the forces working above and below the earth's surface. Primary landforms are large masses of rock raised by volcanic eruptions and other forces that uplift the earth's surface. A good example is the island of Surtsey, which appeared in 1963. It is the top of a volcano on the mid-ocean ridge south of Iceland. In just a few months, land existed where only waves had washed before.

Yet, even as a primary landform is being formed, the forces of erosion are at work to produce secondary landforms. As soon as the island of Surtsey appeared, the forces of erosion became evident. Ocean waves battered the growing pile of volcanic ash, and sediment washed into the sea. During its first year, the island shrank and grew as the forces of the volcano and the ocean battled each other. Finally, lava flowing out of the volcano hardened into a tough crust over the pile of ash.

Surtsey still exists, but its form is steadily changing. The rain, wind, and pounding surf have worn its edges smooth. Gulleys and valleys are being cut into the island, and beaches have appeared. Thus, secondary landforms develop from primary landforms due to erosion.

Volcanic forces created the island of **Surtsey** near **Iceland**. Wind and water are continuously eroding the face of the island.

Rain and wind wear down the surface of a primary landform and remove a little more of it each year. For this reason, older secondary landforms are usually smoother, lower, and more rounded than younger ones. Older landforms usually have less **relief**, or smaller differences in elevation, than younger ones. The older age of a landform is usually evident in hillsides that are not very steep and in wide valleys.

Classifying Landforms Another way to understand landforms is to divide them into three categories, based on how they were formed. One kind of landform is made of rock and has a thin layer of weathered sediment and soil at the surface. Its surface is slowly being lowered by the forces of erosion. Most valleys, plains, and plateaus were formed this way. Another kind of landform is formed by sediment deposited by water, wind, or ice. A sand dune in a desert is an example of this kind of landform. Another example is a **floodplain**. A floodplain is a landform of level ground built by sediment deposited by a river or stream. The third landform is that caused by tectonic activity, the uplifting of the earth's crust. While erosion and deposition change landforms, tectonic activity creates them. In fact, the word *tectonic* comes from the Greek word *tekton,* meaning "builder." Tectonic activity formed landforms of high elevation, such as the Himalayas, and also deeply depressed landforms, such as Death Valley in California.

The terrain in any area is usually a jigsaw puzzle of these three kinds of landforms. For example, a mountain is formed by tectonic activity. The agents of erosion may then form deep valleys in the mountain. The sediment eroded from the mountain may be deposited by a stream on the plain at the mountain's base. This fan-shaped deposit of sediment is called an **alluvial fan**. Later, mud and sand might erode from the alluvial fan and be carried to a river mouth. If it is deposited there, a **delta** may form. Still later, waves might carry the sediment to the ocean floor. Eventually, the sediment could be pulled into a trench and crushed to form a new mountain range.

Landforms and People The size, shape, and location of major and minor landforms have influenced human settlement throughout history. Just as important has been the ability of people to

Elevation Map: Guadalcanal

LEGEND
ELEVATION

FEET	METERS
6,000	1,830
4,000	1,220
2,000	610
1,000	305
0	0

Honiara
Mataniko R.
Lunga
Tenaru R.
PROFILE LINE
Guadalcanal
Mount Makarakomburu (8028 ft. 2447 m)
CORAL SEA
SCALE 0 30 Miles 0 30 Kilometers
Projection: Lambert Conformal Conic

Contour Map: Guadalcanal

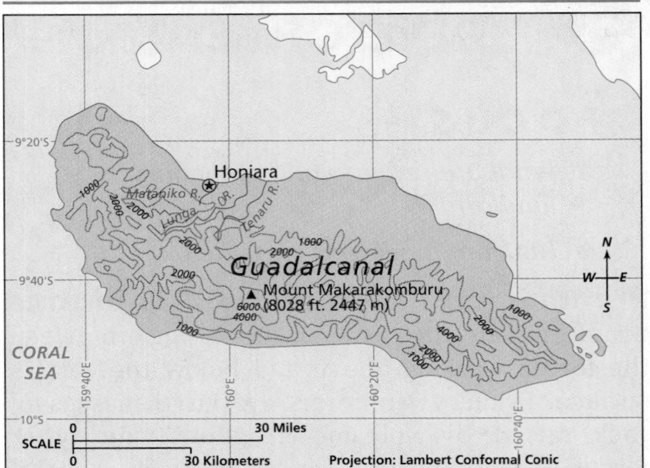

Honiara
Mataniko R.
Lunga
Tenaru R.
Guadalcanal
Mount Makarakomburu
(8028 ft. 2447 m)
CORAL SEA
30 Miles
SCALE
30 Kilometers
Projection: Lambert Conformal Conic

MAP STUDY

Elevation is the height of the land above sea level. An elevation map, such as the one for Guadalcanal (above, left) uses color to show how high landforms are. The illustration to the right is a cross-sectional diagram of Guadalcanal. This kind of diagram shows a region's landscape even more clearly because it is a side view, or profile, of a specific region. The diagram here is a profile of Guadalcanal along a line from Point A to Point B in the elevation map.

Vertical and horizontal distances are calculated differently on cross-sectional diagrams. The vertical distance (the height of Mount Makarakomburu, for example) is exaggerated when compared to the horizontal distance between Point A and Point B. This technique is called vertical exaggeration. If the vertical scale were not exaggerated, even tall mountains would appear as small bumps on a cross-sectional diagram.

Contour maps, such as the one above right, also show elevation. Instead of colors, map makers use lines that connect points of equal elevation. Elevation levels are written along the lines. The closer together the lines are, the steeper the terrain in that area. For example, on this contour map, the terrain to the south of Mount Makarakomburu is steeper than the terrain on the northwestern end of the island.

Elevation Profile: Guadalcanal

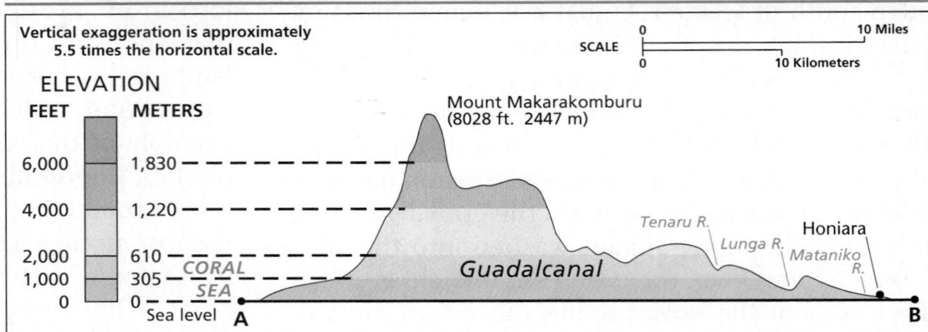

Vertical exaggeration is approximately 5.5 times the horizontal scale.
SCALE 0 10 Miles 0 10 Kilometers
ELEVATION
FEET METERS
6,000 1,830
4,000 1,220
2,000 610
1,000 305
0 0
Sea level
CORAL SEA
Mount Makarakomburu
(8028 ft. 2447 m)
Guadalcanal
Tenaru R.
Lunga R.
Honiara
Mataniko R.
A
B

adapt to living on, and changing, the earth's landforms. Such adaptation may be as simple as a farmer plowing around the side of a hill to prevent erosion. Or human adaptation may involve technology, such as explosives, that makes it possible to build dams across rivers or to carve highways on the sides of mountains.

Theories about how the continents formed may seem to have little to do with our daily lives. But stories in the news about mud slides destroying roads or villages, or lava from a volcano threatening a nearby town, remind us that the earth is constantly changing. Today, no less than thousands of years ago, people face the challenge every day of adapting to those changes.

SECTION REVIEW

1. How do primary and secondary landforms differ? Give one example of each.
2. How is Surtsey a good example of a landform that is both primary and secondary?
3. **Critical Thinking** **Classify** landforms into three categories other than primary and secondary.

CHAPTER 5 CHECK

Coastal plain of **Bangladesh**

Reviewing the Main Ideas

1. The earth's form is constantly being changed by forces inside the earth and on the earth's surface.
2. The forces of running water, waves, glaciers, and wind erode the land. Other influences, such as heat and plate movement, are constantly building new landforms.
3. Primary landforms are those made of large masses of rock that formed as a result of volcanic eruptions and other uplifting forces. Secondary landforms are the result of erosive forces working on primary landforms.

Building a Vocabulary

1. What is the difference between folding and faulting?
2. What is a mid-ocean ridge? How is it formed?
3. What is the term used to describe the differences in elevation of a landform?

Recalling and Reviewing

1. In what ways does erosion of the earth's surface occur? Explain.
2. Explain how the forces within the earth prevent the earth's surface from becoming totally smooth due to erosion.
3. Describe the theory of plate tectonics, and explain how the continents and major continental landforms were made.

Thinking Critically

1. According to the theory of plate tectonics, the great mountain ranges of the world are still forming, although some are more stable than others. What might happen to the Alps as the African plate continues to move under the Eurasian plate?
2. What could you do to prepare for a predicted earthquake? Make a list of items for a survival kit.

Learning About Your Local Geography

Individual Project

Draw a cross-sectional diagram of the landforms found in your area. (Refer to the diagram on page 46.) Label the diagram, and include brief explanations of how each landform might have been formed. You may want to include your diagram in your individual portfolio.

Using the Five Themes of Geography

People changing and adapting to the environment falls under the geographic theme of **human-environment interaction**. Make two lists to show this interaction in your area. In one list, write descriptions of local landforms and the forces of change taking place, such as erosion, weathering, earthquakes, and so on. In the other list, write ways in which people have made changes in the local environment to better suit their needs or to adapt to natural forces of change.

Mitre Peak near **Milford Sound, New Zealand**

CHAPTER 6

Patterns of Life

"The forests of today are very much the same, in essence, as those that developed soon after the appearance of the flowering plants, fifty million years ago. Then, as now, there were jungles in Asia, dank rain forests in Africa and South America, and cool verdant woods in Europe. . . . Everywhere, leaves sprouted; season after season, century after century, they offered an ever-renewing, inexhaustible supply of food for any animal able to gather and digest them."

David Attenborough

GEOGRAPHY DICTIONARY

photosynthesis
food chain
plant community
ecosystem
plant succession
climax community
humus
soil horizon
leaching
coral reef
biome
deciduous forest
coniferous forest
mixed forest
savanna
prairie
extinct

As you learned in Chapter 2, the biosphere contains all living things on Earth. Plants make up most of the biosphere and are the basis for almost all other life forms. In one way or another, all animals depend upon plants for their food supply. Some animals eat only plants, while others eat the animals that eat the plants. Still others, including humans, eat both plants and animals. Plants also are a major source of oxygen for the earth's atmosphere.

Soil provides the nutrients, or nourishment, that plants need to grow and to survive. Soil also supports the food crops necessary to feed the human population. This chapter will help you understand the interrelationships of plants, animals, soil, water, and climate.

Rain forest of Olympic National Park, **Washington** state

THE WORLD OF PLANTS AND SOIL

FOCUS

- *What is a plant community?*
- *What determines the type of soil a location will have?*
- *What is the marine food chain?*

Cacti and other hardy plants make up this desert plant community. Why do you think a desert plant community is not very complex?

Primary Production Every plant is a small factory helping to support life on Earth. These factories convert sunlight into chemical energy through a process called **photosynthesis**. Plants use the sun's energy as fuel to change carbon dioxide from the air, and water and nutrients from the soil, into food. Plants' roots obtain the nutrients from the water, minerals, and gases in the soil. Oxygen is produced as a result of this process and is released, along with water vapor, into the atmosphere. Some of the food produced is used to keep the plant alive, while the rest is used for growth.

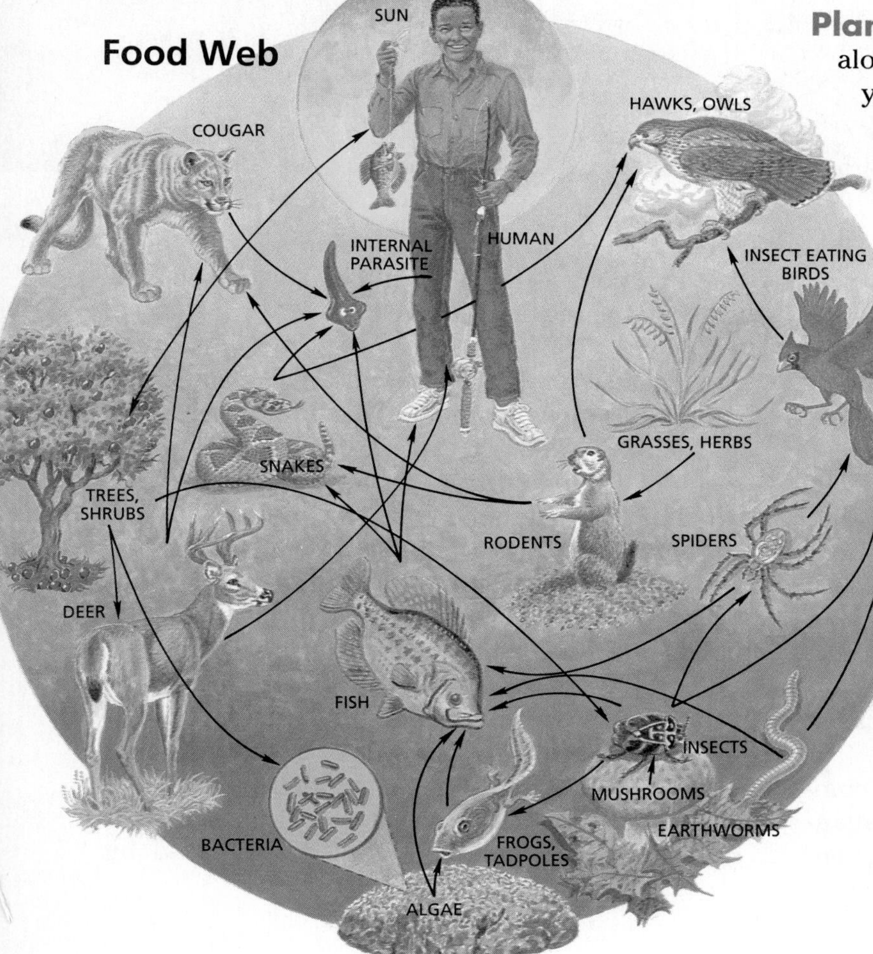
Food Web
SUN
COUGAR
HAWKS, OWLS
INTERNAL PARASITE
HUMAN
INSECT EATING BIRDS
GRASSES, HERBS
SNAKES
TREES, SHRUBS
RODENTS
SPIDERS
DEER
FISH
INSECTS
MUSHROOMS
EARTHWORMS
BACTERIA
FROGS, TADPOLES
ALGAE

Plants are at the bottom of the **food chain** and support animal life. A food chain is a series of organisms in which energy is passed along through living things. Some animals eat plants, and these animals are eaten by other, usually larger, animals. At each stage of the food chain, there are fewer living things. Only a few animals, such as lions, hawks, and humans, are at the top of the food chain.

Plant Growth Generally, plants do not live alone, since they are not independent. Just as you live in a society, surrounded by friends, family, and other people, plants live in groups. Such a group is called a **plant community**. In the humid-tropical climates, these communities are quite large, and thousands of different kinds of plants are found here. The many lush tropical plants in such a community support great numbers of animals, from millions of insects that inhabit the forest floor to birds and monkeys in the treetops. In the desert climate, the plant community is less complex. Here,

Food chains rarely occur alone in nature. More common are food webs, which are interlocking networks of food chains. This food web includes tiny organisms such as parasites and bacteria as well as human beings and other large mammals.

Difficult conditions after a forest fire mean that the first plants to grow back must be very hardy. In this illustration, which plants are the first to grow back?

Forest Succession After a Fire

A. Forest Fire in Progress

B. Early Plant Growth

C. Middle Stage

D. Forest Recovered

there may be only a few highly specialized plants, such as cacti, and a few specialized animals, such as lizards. Specialized plants and animals are those that have adapted to a particular environment.

Each plant evolves in such a way that it plays a special role in its community. Different types of plants require different amounts of sunlight, water, and nutrients from the soil. Some plants, such as grass, have shallow roots that capture water and nutrients close to the soil's surface. Other plants, such as trees, send their roots deep into the ground. Some plants survive in soil that is poor in nutrients, while others require rich soil.

Large trees need a great deal of direct sunlight in order to make food. Ferns, however, can survive with very little sunlight. Only the plants whose requirements for growth are fulfilled in a community will survive there. Variations in sunlight, temperature, water, and soil help explain the geographic distribution of forests or grasslands in different regions of the world.

Nearly all plants depend on other plants to grow. For example, trees provide the shade that keeps ferns growing on the forest floor from receiving too much sunlight. Trees also provide support for clinging vines so that the vines can capture needed sunlight at the top of the trees. Individual types of plants work to aid their own growth and, often, the growth of other plants in their community.

This interdependence of living things combined with the physical environment these organisms inhabit forms an **ecosystem**. Thus, an ecosystem is a community of plants and animals that functions as a unit with the climate, soil, and water of the region.

Plant Succession The cluster of plants that constitutes a community does not develop by accident. Rather, the plant community is the result of orderly change in a geographic area over a long period of time.

Imagine an area just after a forest fire. The first plants to grow will be very hardy, and because there are no trees to give shade, the plants must be able to endure direct sunlight. They will grow rapidly when rain falls on the ash and soil. These plants will help prevent erosion by holding the soil in place, and they also will provide shelter for the seeds of other plants.

Grass and weeds may cover the ground for some time, but soon they will be replaced by taller plants. Often, these newer plants are shrubs and small trees. Their seeds may have blown in from a distance by the wind or have been carried by animals such as squirrels, bats, and birds. When these seeds are dropped, a new tree may sprout. When the new tree is tall enough and has enough leaves, it will shade the ground. This shade will then make it difficult for the earlier sun-loving plants to survive. Shade-loving plants, such as ferns, will then cover the forest floor. After many years, the area may again look as it did before the fire. (See the diagram on page 54.)

The process by which one group of plants replaces another is called **plant succession**. Each group of plants grows best in one set of conditions. When these conditions change, the growth of a new plant community begins.

The succession process can be controlled by people, but this is not easy to do. If you decide that you want only grass to grow in your yard, you can prevent other plant communities from developing. The grass must be cut frequently to keep tree seedlings from sprouting. Some trees might have to be trimmed so they do not block needed sunlight from the grass. If insects or a lack of fertilizer causes bare spots to appear on your lawn, weeds quickly will fill the open spaces. Time, effort, and money must be spent to prevent succession from changing your grassy lawn into a weed patch. Without your involvement, a new plant community, similar to the original plant community, may in time grow back.

INTERACTION **Climax Community** For each set of climate and soil conditions in a geographic area and for any given time, there is a plant community that is best suited to use the resources there. Only after a long process of plant succession does a stable plant community develop. This last stage of succession is called the **climax community**.

Changes in the environment can alter a climax community. It might be destroyed by natural events such as forest fires, storms, volcanic eruptions, and plant disease. If any of these occur, the succession cycle will start over again. Humans, too, keep climax communities from continuing by burning and clearing the land for farming, by logging, by grazing their animals, and by introducing foreign plants. For these reasons, true climax communities are seldom found today in many areas of the world.

Factors of Soil Formation As you have learned, soil supplies the nutrients plants need to grow and survive, which helps provide the food we need to live. But what is soil? If you dig a shovel into the ground, what do you bring up? Soil is made up primarily of minerals from the weathering of rocks. Most of these minerals are from the rocks found just below the soil layer. Sometimes, weathered rock material is brought to an area by rivers, wind, or glaciers.

Decayed plant or animal matter, called **humus**, also is an important ingredient of soil. Bacteria and insects help break down the humus, and they produce necessary space between the soil particles for gases and water. Weathered rock material, organic matter such as humus, gases, and water must exist together in order for soil to provide plants with the necessities of life.

Soil differs from place to place, and there are hundreds of soil types. Climate is the major factor that determines the type of soil a location will have. A region's climate type controls the amount of sunlight, moisture, and natural vegetation that influence the soil's formation. The second major control is the rock type that provides minerals for the soil. One region may have dark, thick soil with much humus, while another has only a thin layer of sandy soil. Other factors, such as the slope of the land, human activities, and time for soil development, also are important.

Soil Horizons Soil develops very slowly, taking hundreds or even thousands of years to form. As it develops, it forms distinct layers, called **soil horizons**. Most soil has three layers, called the A, B, and C horizons. (See the diagram on page 56.)

The top layer, or A horizon, is called topsoil. This is where decaying plant or animal material accumulates and forms humus. This horizon is very active and has much bacterial and insect activity as well as most of the plant roots. The middle layer, or B horizon, is called subsoil. Only deep plant roots penetrate the B horizon. The C horizon is made up of the broken rock pieces that form the parent material for the soil.

Because climate is the major control of soil formation, there are similarities among soils in the same climate region. For instance, in the humid-tropical climate, warm temperatures and heavy rainfall support lush forests. These tropical rain forests, in turn, supply great amounts of humus to the soil below.

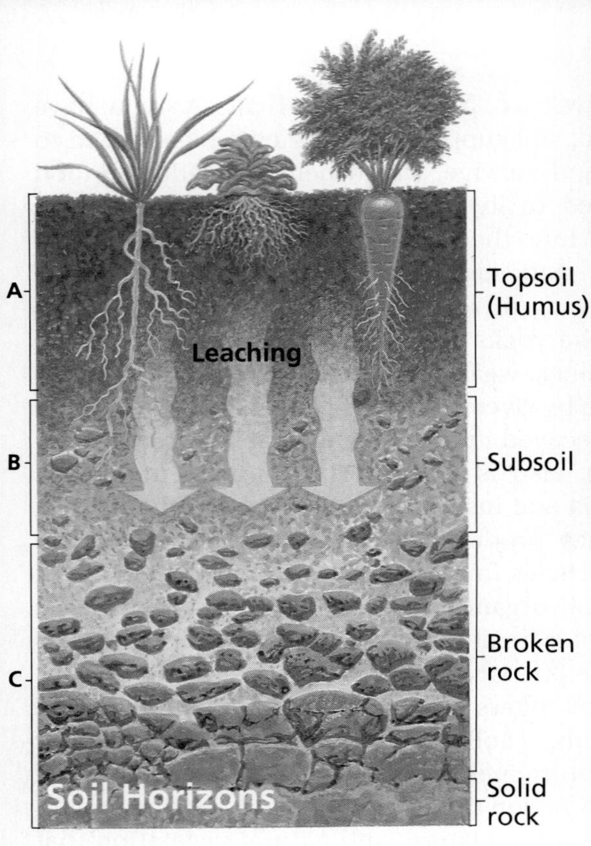
A
B
C
Leaching
Topsoil (Humus)
Subsoil
Broken rock
Solid rock
Soil Horizons

Most soils have three layers, as shown to the left. Only deep plant roots penetrate the B horizon. Leaching occurs when nutrients are washed downward by heavy rainfall.

Fish are found in abundance near this coral reef in the South Pacific Ocean. Why do you think coral reefs have such rich marine ecosystems?

At the same time, humid-tropical soil retains little of this humus due to a process called **leaching**. Leaching occurs when the nutrients necessary for plant growth are washed downward out of the topsoil by rainfall. The nutrients then cannot be reached by the plant roots. Fortunately, the tropical rain forest continues to produce humus as fast as the rain washes the nutrients away. An area that supports such a lush forest would seem ideal to clear for farming. Just a few years after the trees are removed, however, the soil will lose its nutrients because of the constant leaching, and only scrub plants will grow. Thus, the plants and soil of an ecosystem are interrelated.

Marine Ecosystems Like ecosystems on land, marine (sea) ecosystems depend on sunlight for life. Drifting marine plants convert sunlight and nutrients from seawater into growth. They also produce oxygen for the atmosphere. These sea plants form the bottom of the marine food chain. Small drifting sea animals eat the sea plants, and these small sea animals are then eaten by larger sea animals. At the top of the marine food chain are fish, marine birds, and marine mammals such as seals and whales.

Most marine life is concentrated near the water's surface where sunlight is available to support the food chain, and near coastlines where nutrients are brought to the sea by rivers. The richest marine ecosystems, with many kinds of animals and plants, are coastal marshes and **coral reefs**. A coral reef is a ridge found in warm, tropical, shallow water close to shore. It is made of coral, which is a rocky limestone material formed of the hard skeletons of the coral polyp. Coral polyps are tiny marine animals, usually smaller than the tip of a ballpoint pen. They live in colonies that are anchored to the shallow ocean floor. As old coral polyps die, their limestone skeletons serve as a platform on which new corals grow.

Not all of the ocean is rich in marine life. In fact, much of the ocean is a biological desert—the open ocean far from coastlines as well as the cold, dark depths.

SECTION REVIEW

1. What roles do plants play in a plant community?
2. What is the major factor in soil formation?
3. **Critical Thinking** **Compare** the food chain on land to the marine food chain.

The World's Basic Biomes

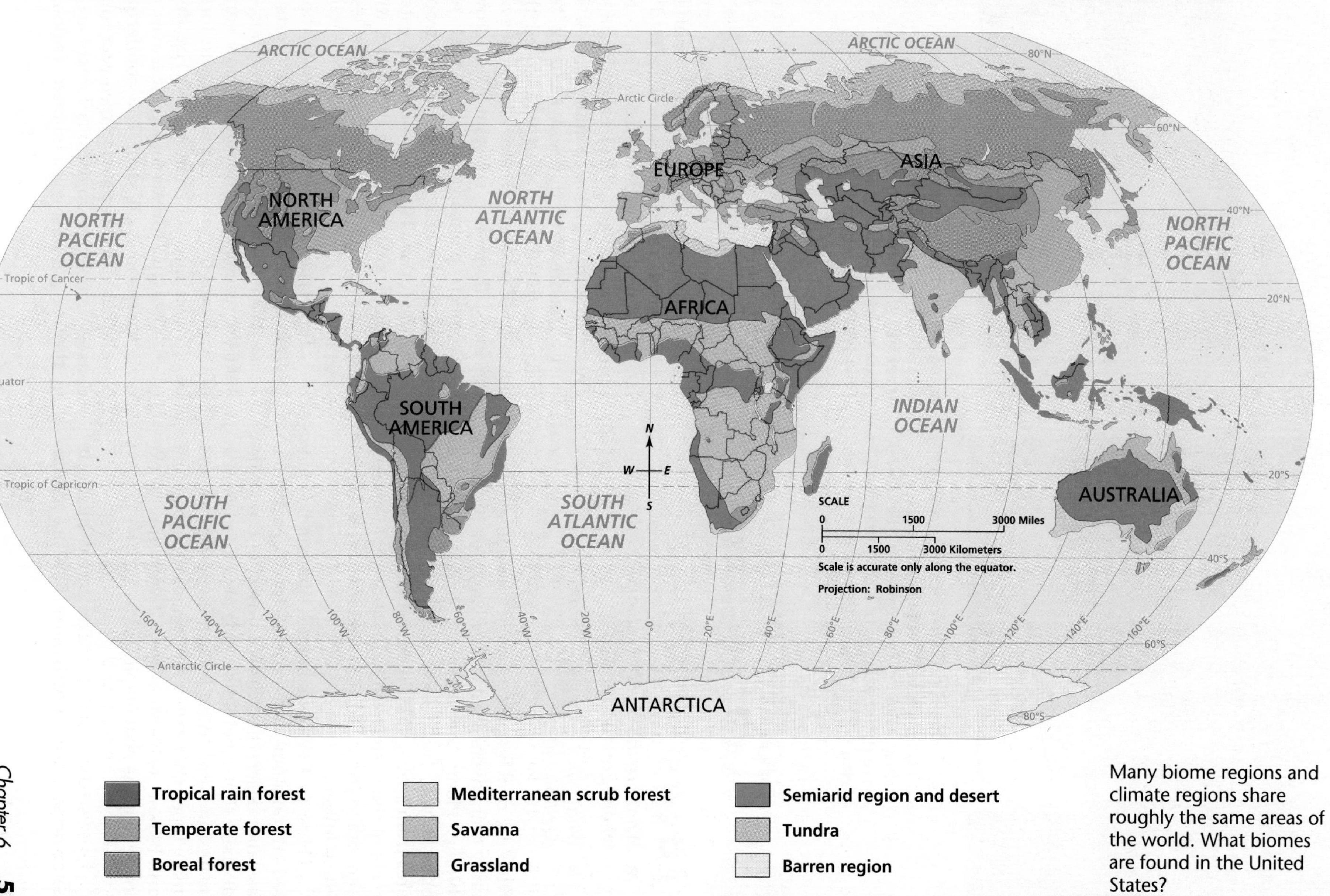
ARCTIC OCEAN
ARCTIC OCEAN
NORTH PACIFIC OCEAN
NORTH AMERICA
NORTH ATLANTIC OCEAN
EUROPE
ASIA
NORTH PACIFIC OCEAN
AFRICA
SOUTH AMERICA
INDIAN OCEAN
AUSTRALIA
SOUTH PACIFIC OCEAN
SOUTH ATLANTIC OCEAN
ANTARCTICA
Arctic Circle
Tropic of Cancer
Equator
Tropic of Capricorn
Antarctic Circle
80°N
60°N
40°N
20°N
20°S
40°S
60°S
80°S
160°W
140°W
120°W
100°W
80°W
60°W
40°W
20°W
0°
20°E
40°E
60°E
80°E
100°E
120°E
140°E
160°E
N
S
E
W
SCALE
0
1500
3000 Miles
0
1500
3000 Kilometers
Scale is accurate only along the equator.
Projection: Robinson
Tropical rain forest
Temperate forest
Boreal forest
Mediterranean scrub forest
Savanna
Grassland
Semiarid region and desert
Tundra
Barren region

Many biome regions and climate regions share roughly the same areas of the world. What biomes are found in the United States?

Deciduous and coniferous trees color autumn in eastern **North America** (above). Giraffes and scattered trees tower above the grasslands of a savanna biome in **Africa** (above, right). Clockwise from top left on the next page is an example of a grasslands biome in central **North America**, of summer vegetation in the tundra of **Alaska**, and of the desert biome of the Patagonia region in **Argentina**.

2 BIOMES: WORLD PLANT REGIONS

FOCUS

- *What are the five basic biomes, and where are they located?*
- *What types of vegetation are found in each biome?*

What Is a Biome? A plant and animal community that covers a very large land area is called a **biome**. Plants are the most visible part of a biome. If you could look down on the United States from space, you would see the biomes. The forests of the eastern United States would appear green, while the deserts of the Southwest would be light brown.

There are five basic biomes that can be identified throughout the world: forest, savanna, grassland, desert, and tundra. Areas of the world that share the same biome type also have similar climates and soil types. In fact, some biomes and climate types share the same names. Compare the World Biome map on page 57 with the World Climate Regions map on pages 28–29.

PLACE **Forest Biome** The forest biome is tree covered. Within this biome, there are a number of different climax forest communities. Each forest community is determined by the types of trees growing there. In addition to trees, the forest biome supports a great variety of living things from the treetops to beneath the soil surface.

Tropical rain forests are located in the humid-tropical climate regions. Here, temperatures are warm, and water is abundant all year. These conditions allow plants to grow continuously. Hundreds of different kinds of tall trees make up the tropical rain forest, and the thick umbrella of treetops creates continuous shade on the forest floor. The constant plant growth provides much humus to replace that which is removed by the heavy rains from the tropical soil below the forest. The tropical rain forests are the most complex ecosystem in the world, with thousands of kinds of plants and animals, especially insects.

Temperate forests are located in the middle-latitude climates. Generally, these forests are less dense than tropical rain forests. Thus, enough sunlight reaches the ground to allow shrubs and other plants to grow. The temperate forests can be divided into different types. The **deciduous** (di-SIJ-uh-wuhs) **forests** lose their leaves during a certain season each year. The **coniferous** (koh-NIF-uh-ruhs) **forests** remain green year-round. The densest temperate coniferous forests are located in the marine-west-coast climate regions. Some

areas are **mixed forests** where deciduous and coniferous forests blend. These are located primarily in the humid-subtropical and humid-continental climates of the middle latitudes.

Boreal (BOHR-ee-uhl) forests consist mainly of coniferous trees. Because these needle-bearing trees can live under very cold conditions, boreal forests are located in the subarctic climate regions. The floors of these forests build up a layer of tree needles, and due to the long winters, decay is slow. Boreal forests stretch in a broad band across subarctic North America, Europe, and Asia. *Boreal* means "northern," and these forests are not found in the Southern Hemisphere.

Like the Mediterranean climate region, the Mediterranean scrub forests cover a small percentage of the earth's land surface along middle-latitude coasts. Because of the long, sunny, dry summers, most of the vegetation of this forest type is made up of short trees and shrubs. This drought-resistant vegetation has leathery leaves and thick bark that help the plants survive the long summers. Forest and brush fires are frequent during the dry season. Most of the original Mediterranean scrub forests have been lost to clearing for farming, the growth of cities, and overgrazing by herds of domestic animals.

PLACE Savanna Biome The **savanna** biome is a tropical grassland with scattered trees and shrubs. Usually, the savanna is located in areas with distinct wet and dry seasons. The savanna biome occupies tropical regions between the wet, tropical rain forests and the arid deserts. This biome supports great herds of grazing animals and many predators, such as those in East Africa. In Asia and Africa, the savanna has increased in area at the expense of forests. Clearing forests for logging and farming and overgrazing have caused the savanna to spread into what once were forested regions.

PLACE Grassland Biome Middle-latitude grasslands are located between the temperate forests and the desert biomes. Because of the semiarid climate, these grasslands usually do not have tree cover, except along rivers. There are two types of middle-latitude grasslands. Those parts that are wetter and closer to the forest biome support a tall grassland cover called **prairie**. The deep fertile soil of the prairie is particularly well suited for growing grains, and the world's largest wheat- and corn-growing areas are found in these prairie regions. Because of agricultural expansion, little original prairie grassland remains.

The drier desert margins of the middle-latitude grasslands support the short-grass vegetation called steppe. Though the steppe soil is rich, lower rainfall amounts make this grassland less productive than prairie. Great herds of animals once roamed the steppes of the world. But today, for example, the vast grasslands and buffalo herds of the North American Great Plains have been replaced with wheat fields and cattle.

PLACE **Desert Biome** Desert biomes generally are found in the low and middle latitudes in the desert climate regions. Desert biome plants survive by using very little water or by storing water. Cacti are examples of desert plants that store water. Other desert plants, such as sagebrush, have tiny leaves with waxy surfaces and large root systems. They can withstand hot sunny days without letting very much water escape into the air.

Some desert plants actually avoid the dry conditions. These plants have short growing cycles and seeds that can survive underground for several years. When the rare rains come, the seeds sprout and the desert is covered with flowers. These plants die within a few weeks, but not before spreading seeds for the next rains. Because there is so little moisture in the desert biome, plants grow far apart, produce almost no humus for soil formation, and produce limited food for animals.

PLACE **Tundra Biome** Like the desert biome, the tundra biome supports little plant growth and has no trees. Only a few small specialized plants can grow in the cold climate conditions. The permafrost keeps plant roots from penetrating the soil. During the long winter, the plant life is frozen, and only a few strong animals survive. During the short summer, however, the tundra thaws and comes alive with mosses, lichens, low shrubs, and small flowering plants. Animals from other regions arrive to feed on the fresh plant growth and millions of insects. Birds migrate thousands of miles to feed and nest during this brief growing season. The largest areas of tundra are found in the Arctic regions of North America, Europe, and Asia. Smaller areas are located in the Antarctic and at high mountain elevations.

PLACE **Barren Regions** In the far polar regions, ice and snow remain all year, and it is impossible for plants to grow. These areas with no plant life or soil are called barren regions. The largest barren regions are the ice-cap climate regions of Antarctica and Greenland. Here, any animal life must depend upon the ocean for food, as the land provides none. Many high mountain peaks in other areas of the world also are barren of plant and animal life. Here, elevation causes ice-cap climate conditions outside the polar regions. The cold temperatures, severe winds, snow, and glaciers make it impossible for plants and most animals to survive.

INTERACTION People and the Biosphere

All living things—plants, animals, and humans—are interrelated within the biosphere. That is, events in one place in the biosphere can affect plants and animals in another part far away. Natural changes in the biosphere take place all the time. For example, a volcanic eruption may destroy a whole ecosystem around a mountain. Then, plants and animals slowly return to the area and repopulate it.

Changes that last for a long period of time may have a greater impact. When an ice age replaces warmer conditions, many plants and animals die. This long-term destruction may cause some plants and animals to become **extinct**, which means that they cease to exist.

Human actions also can have an effect on all parts of the biosphere. As you have learned, some of the biomes have changed size as a result of human activities. The tropical rain forests, for example, are becoming smaller as humans cut the trees for timber and clear the land for farms. Because of the loss of the forest, the soil loses its nutrients. Although the farms may be abandoned and the loggers might move elsewhere, the forest cannot return because the necessary soil was changed. Many kinds of rain-forest plants and animals, some not yet discovered, may become extinct.

While the forest biomes are becoming smaller, other biomes are expanding due to human activities, especially the savanna and desert biomes. If our biosphere is completely changed, what will life be like in the future? Humans have a special responsibility to handle carefully the resources on which all life depends.

2 SECTION REVIEW

1. Where is each of the five world biomes found?
2. Which biomes support little plant life?
3. **Critical Thinking** **Explain** why there are many types of forests.

CHAPTER 6 CHECK

Yellowstone National Park one year after the 1988 fire

Reviewing the Main Ideas

1. Life on Earth depends on the earth's plant, animal, and soil resources. Plants are at the bottom of the food chain and support animal life.
2. Soil provides the nutrients plants need to grow and survive. It also supports the food humans grow for survival. Soil is composed primarily of minerals from the weathering of rocks.
3. A plant and animal community that covers a large area of the earth's surface is a biome. There are five basic world biomes.

Building a Vocabulary

1. How is a plant community related to an ecosystem?
2. Describe the process of plant succession.
3. How is a coral reef formed?
4. Define the term *extinct*. What causes a living thing to become extinct?

Recalling and Reviewing

1. Would animals be able to survive on Earth if there were no plants? Why or why not?
2. Describe the stages of either a food chain on land or a marine food chain.
3. What are the five biomes, and what kinds of vegetation would you find in each?

Thinking Critically

1. In what biome type do you live? Give two examples of other places in the world that have the same type of biome as the one in which you live.
2. Predict what will happen to the biosphere if the forest biomes continue to get smaller.

Learning About Your Local Geography

Cooperative Project

With your group, investigate the soil in or near your schoolyard. Is it rocky, sandy, thick with humus, or hard like clay? Draw a cross-sectional diagram of what you think the soil horizon near your school might look like. On the diagram, label the climate type that determines the soil in your region.

Using the Five Themes of Geography

In a paragraph, explain how **human-environment interaction** has resulted in changes to the size of biomes.

Coral reef of **Mili Atoll, Marshall Islands**

 **SKILLS**

Using Sketch Maps

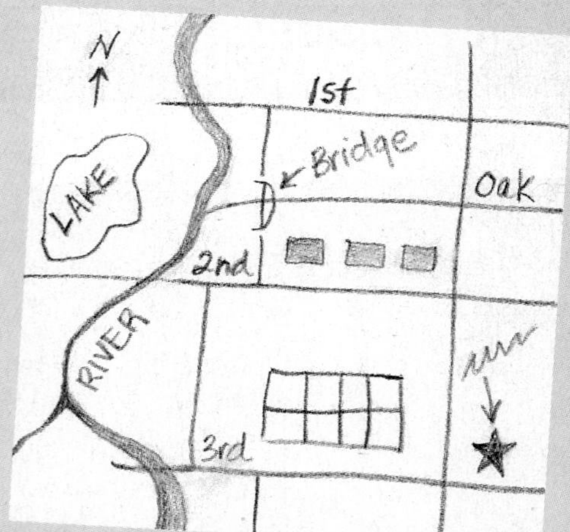
N
1st
Bridge
Oak
LAKE
2nd
RIVER
3rd

As you have read, maps are an important tool of geographers. But you may occasionally find it useful to make your own maps. Sketching your own map can be a valuable study aid, giving you a better grasp of where places are located. You can study a map in a book or atlas for hours and still not know where things are. If you draw your own map based on what you see in the book, however, you will probably have a better understanding of the subject of the map. You will also be able to remember it better.

Of course, you will not always have access to other people's maps, and sometimes you will have to draw your own map truly "from scratch." You might be having a discussion with a friend about a geographical area during which it will be helpful for you to sketch a map quickly. Drawing your own maps might help you to understand events you read about in the newspaper or learn about on television. Sketch maps also have immediate, practical uses. For instance, someone might ask you for directions to a nearby shopping mall, football field, or museum. You could give directions orally, by saying "go down Main Street and make a left at the first traffic light," but it would probably be more helpful to the person if you drew a map.

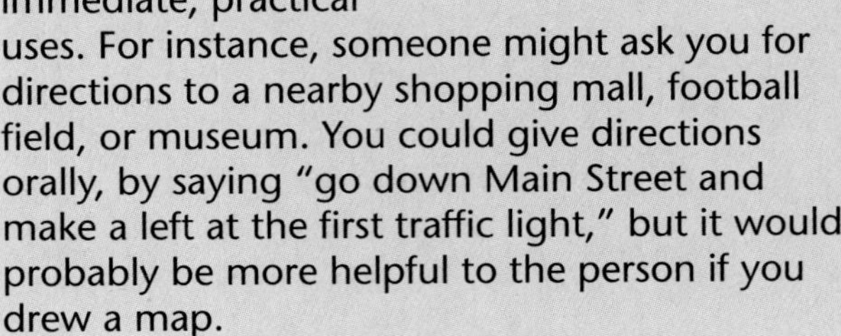

How to Draw a Sketch Map

1. Decide what region your map will show. Choose boundaries so that you do not sketch more than you need to.
2. Determine how much space you will need for your map. Things that are the same size as each other in reality should be about the same size as each other on your map.
3. Decide on and note the orientation of your map. Most maps use a directional indicator. On most maps, north is "up."
4. Select reference points so that viewers of your map can quickly and easily figure out what they are looking at. For a sketch map of the world, reference points might be the equator and prime meridian. For a map of your community, a major street or river might be your reference point, or you might include a grid labeled by numbers and letters. Even maps of shopping malls indicate reference points—usually by saying "You Are Here" next to a dot or an asterisk.
5. Decide how much detail your map will show. The larger the area you want to represent, the less detail you will need. For example, a map of the world will not need names of streets, but a map of your community will.
6. You are ready to begin sketching. First, sketch general shapes, such as the continents if you are drawing a world map. If you do not know or cannot remember exact shapes, you can use circles, rectangles, and triangles.
7. Now, fill in more details as they occur to you—names of places, major land features, and so on.
8. Do not spend more than an hour working on your map, and do not try to make it perfect or overly detailed. Many useful maps can be sketched in just a few minutes.

PRACTICING THE SKILL

1. Select any map in this book and copy the main features of the area onto a piece of paper. Do not trace the map.
2. Make a map of your classroom, indicating the location of pieces of furniture, windows, closets, doors, and so on. Use a grid system to divide the map into sections.
3. Suppose a visitor to your community approached you right now and asked you for directions to a nearby library. Sketch a map for the visitor, indicating how to get from your present location to the library. What reference points will be most helpful?

UNIT 2

PEOPLE and GEOGRAPHY

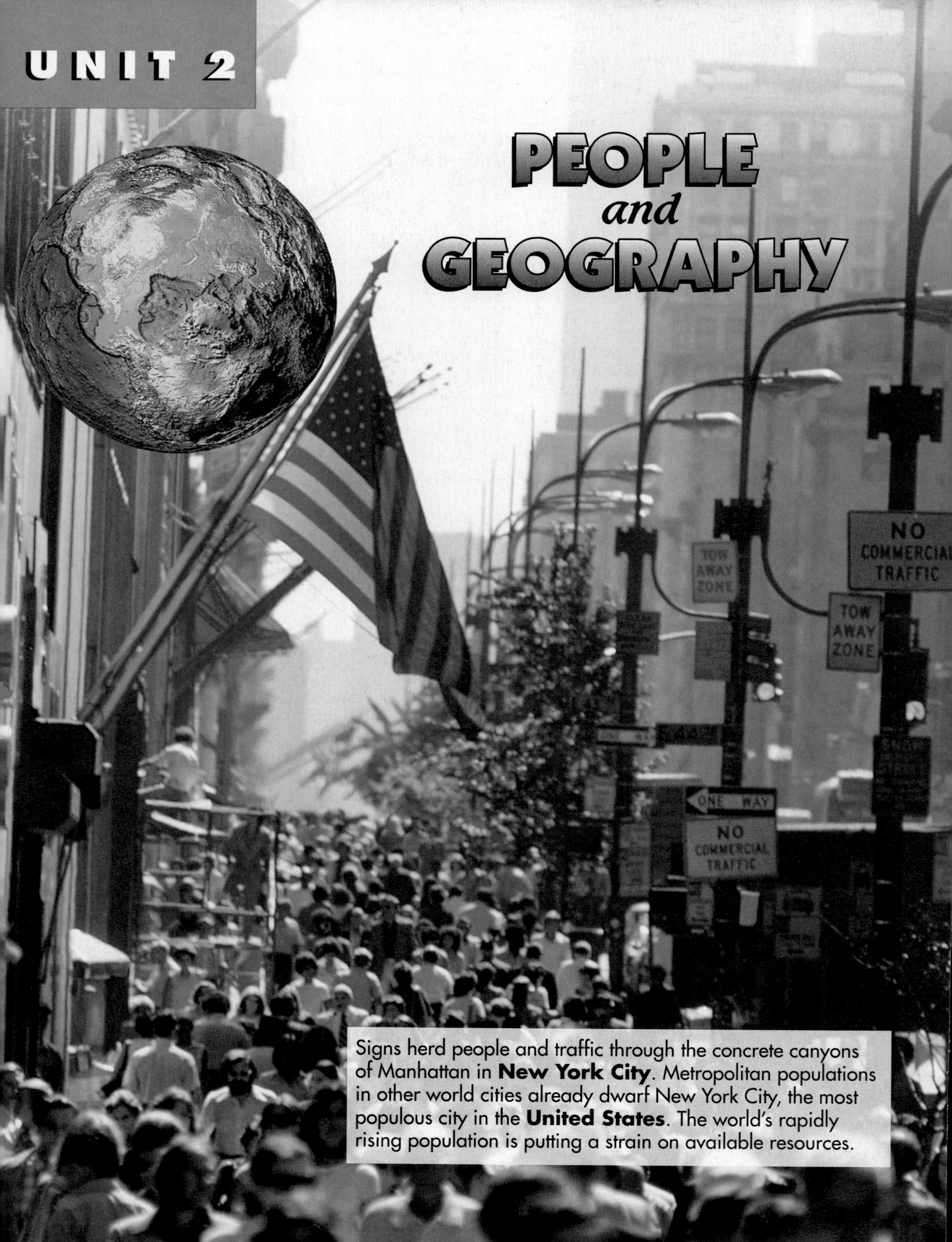
NO
COMMERCIAL
TRAFFIC
TOW
AWAY
ZONE
TOW
AWAY
ZONE
ONE WAY
NO
COMMERCIAL
TRAFFIC

Signs herd people and traffic through the concrete canyons of Manhattan in **New York City**. Metropolitan populations in other world cities already dwarf New York City, the most populous city in the **United States**. The world's rapidly rising population is putting a strain on available resources.

Information, a valuable resource, is communicated partly through this telephone network control center in **New York**. Modern technology can help people put resources to work so that societies can provide for their members.

UNIT 2 CHAPTERS

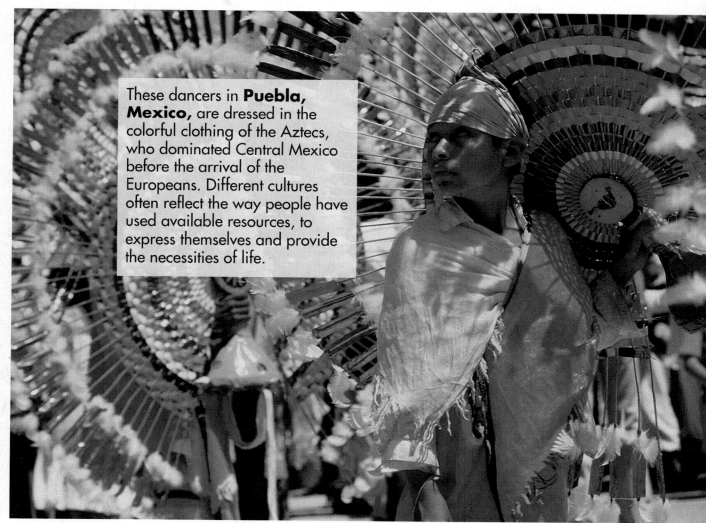

These dancers in **Puebla, Mexico,** are dressed in the colorful clothing of the Aztecs, who dominated Central Mexico before the arrival of the Europeans. Different cultures often reflect the way people have used available resources, to express themselves and provide the necessities of life.

CHAPTER 7

Global Cultures

> "Our most important task is to develop a spirit of democratic pluralism, a culture that respects the other person, whether he's a neighbor or the people across the water. But to have the strength for that kind of respect, one must be confident in one's own heritage and national personality."
>
> *Mongi Bousnina*

GEOGRAPHY DICTIONARY

- *culture*
- *culture trait*
- *culture region*
- *ethnic group*
- *innovation*
- *diffusion*
- *acculturation*
- *domestication*
- *subsistence agriculture*
- *commercial agriculture*
- *urbanization*
- *culture hearth*
- *nationalism*
- *fundamentalism*
- *totalitarian government*
- *democratic government*
- *tariff*
- *quota*

In Unit 1, you learned about the physical geography of our world and the forces at work that continue to shape the earth's surface. Landforms, climate, water resources, and vegetation are key features of our environment. Much of the world around us, however, has been formed by people. Cities and villages, farm fields and grocery stores, and highways and airports have been created to meet our ever-changing needs.

People have divided the earth into nearly 200 countries, which are home to people of varying languages, religions, economic systems, and political views. The study of the world's people—how they live and how their activities vary from place to place—is called cultural or human geography.

1993 State of the Union Address before the U.S. Congress

HUMAN GEOGRAPHY

FOCUS

- *What elements make up a culture?*
- *What is a culture region? Why do cultures change?*

What Is Culture? The key term in understanding human geography is **culture**. Culture includes all the features of a society's way of life. Culture is learned, passed down from generation to generation through teaching, example, and imitation. Culture includes language, religion, government, economics, food, clothing, architecture, and family life. It includes all of a society's shared values, beliefs, institutions, and technologies—or skills, tools, and abilities. Culture gives roles to individuals and establishes relationships within groups.

To better understand culture, we can divide it into activities and behaviors that people repeatedly practice. These activities and behaviors are called **culture traits**. There are many kinds of culture traits. Some, such as learning to read and do arithmetic, are much the same around the world even though various cultures use different alphabets or symbols. Many culture traits, however, vary from place to place. For example, a typical U.S. teenager usually eats with a knife, fork, and spoon. Chinese teenagers, however, eat with chopsticks, and Malaysian teenagers have been taught to eat with their fingers. Each of these learned culture traits is considered the best method in its own culture.

Culture traits often are linked together. For example, Islam and Judaism are two religions that do not allow the eating of pork. Therefore, in Southwest Asia, pigs usually are not raised by either Islamic or Jewish farmers. Islam also does not allow alcoholic beverages, while Judaism permits wine. As a result, grapes grown on an Islamic farm are eaten fresh or are dried into raisins, while the grapes grown on a Jewish farm may be made into wine. These examples show that the culture traits of religion and farming are linked.

Culture traits also change through time. Sometimes, these changes are as simple as new clothing styles. For example, ask your parents or other older adults what clothing was fashionable when they were teenagers. Often, changes in culture traits through time are more complex. In the United States, for example, changes in transportation have changed shopping habits. People once rode horse carts or streetcars to downtown or central areas to do their shopping. Over time, these modes of transportation were replaced by buses and automobiles. Today, highway networks and mass transit systems take shoppers away from city centers to huge shopping malls in the suburbs.

REGION **Culture Regions** An area that has many shared culture traits is called a **culture region**. One country may be a single culture region. France and Japan are two examples.

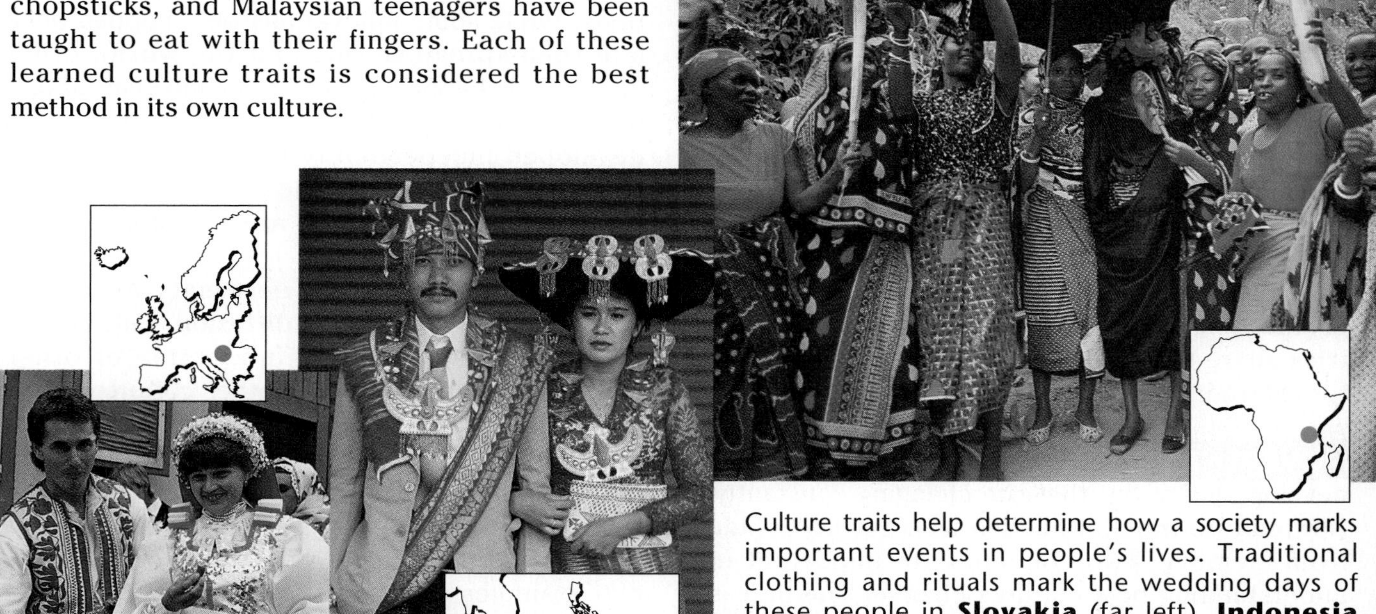

Culture traits help determine how a society marks important events in people's lives. Traditional clothing and rituals mark the wedding days of these people in **Slovakia** (far left), **Indonesia** (middle), and **Tanzania** (above).

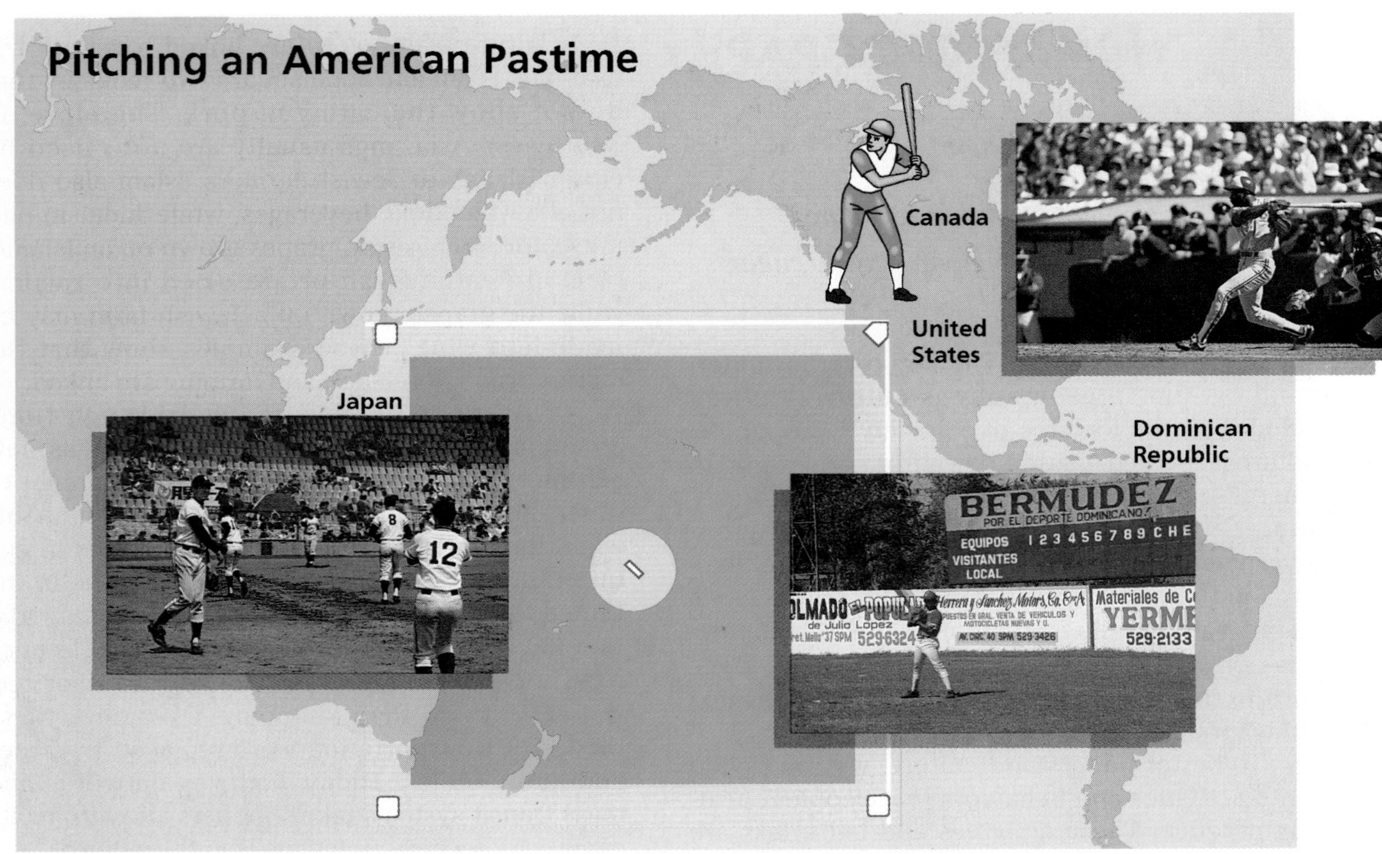
Pitching an American Pastime
Canada
United States
Japan
Dominican Republic
BERMUDEZ
POR EL DEPORTE DOMINICANO
EQUIPOS 1 2 3 4 5 6 7 8 9 C H E
VISITANTES
LOCAL
Materiales de C
YERM
529-2133

Baseball might be as American as apple pie, but fans around the world are evidence of its wide diffusion from the Yankee bullpen. About 16 percent of players on Major League Baseball 1993 spring rosters were born outside the United States. Many of those players were from Middle and South America.

Sometimes, though, political boundaries cut across culture regions. As a result, some countries contain several culture regions. This is particularly true in the countries of Africa. In countries such as Nigeria and Zaire, many **ethnic groups** representing different culture regions live within the boundaries of one country. An ethnic group is a population that shares a common cultural background.

A culture region also can be made up of several countries. An example is Central America, which includes Guatemala, Honduras, Nicaragua, El Salvador, Costa Rica, and Panama. These countries share similar languages, religions, and historical traditions.

MOVEMENT Culture Change Cultures develop slowly, but they are changing constantly. New culture traits are added as older traits fade from memory. Two concepts are important in understanding how cultures change. The first is **innovation**. People are always thinking of new ways to do things. Only those new ideas that are useful and valuable will last, however. New ideas that are accepted into a culture are called innovations. While some innovations only happen once and then spread throughout the world, other innovations happen in more than one place and at different times. Boat building, for example, was developed independently by people all around the world. Though the boat designs differ in detail, boats appear in the artwork of all ancient civilizations.

The second concept necessary for understanding culture change is **diffusion**. Cultural diffusion happens when an innovation or other culture trait spreads into another culture region, which then adopts that element into its own culture. For example, while jazz music partly has its roots in African music, it took hold as an art form in New Orleans and now has spread throughout the world. Video games and compact discs are two examples of innovations that have diffused into other culture regions during your lifetime.

Culture traits diffuse to new regions in two ways. First, an innovation might gradually spread from its source. Imagine a ripple of water spreading outward in a pond when a rock is thrown into the water. In the United States, fashions spread in this way from a center of innovation such as Los Angeles, California. Likewise, country music diffuses from Nashville, Tennessee, to the rest of the country. Second, culture traits spread when people move from a region and carry the innovation with them to a new area. America's 500-year period of settlement by Europe, Africa, and other areas of the world had a strong impact on American culture. For example, European colonists brought the Christian religion to the Americas, and Europeans and Africans brought new foods and cooking methods.

When one culture changes a great deal through its meeting with another culture, the process is called **acculturation**. The adoption of U.S. blue jeans and fast foods by teenagers throughout the world is an example of acculturation. Acculturation is usually a two-way process in which each culture changes the other. For example, while U.S. fast-food restaurants have spread hamburgers and fried chicken worldwide, Chinese, Mexican, Japanese, and Italian foods have spread across the United States.

One of the best examples of cultural diffusion is the spread of languages. Geographers study the spread of languages to trace patterns of political, economic, and religious expansion of cultural groups over time. Language is probably the most important culture trait. People who cannot talk with each other have a difficult time sharing ideas. Often, the spread of innovations is accompanied by the language used to describe them. As U.S. popular culture diffuses around the world, so does the English language.

SECTION REVIEW

1. List eight parts of society that are influenced by culture.
2. What happens when different culture regions come into contact with each other?
3. **Critical Thinking** **Assess** whether an individual can belong to more than one cultural group.

PEOPLE SHAPE THE LAND

FOCUS

- *In what ways have people acquired the resources they need to live?*
- *Why do cities develop and grow? What purposes does industrialization serve, and what problems does it create?*

All people have the same basic needs: food, clothing, and shelter. Food, water, and the materials needed for daily life are basic resources. To live, people must gather these resources from their environment. All of the resources we need cannot be found in one place, however. People must travel and trade to obtain resources from other regions. How people have used the land to acquire their resources has greatly affected Earth's human geography. Throughout history, new discoveries and their diffusion have altered places dramatically.

MOVEMENT Hunting and Gathering

Hunting and gathering was the main way of life for most of human history. Hunter-gatherers live by foraging, or collecting foods from the surrounding land. To be successful, hunter-gatherers must know their environments very well.

In early times, hunter-gatherers often roamed across large areas. They moved their camps with the seasons, in search of different plants and animals throughout the year. There are few hunter-gatherers left today. They survive only in isolated areas where farming is difficult because of harsh climates or poor soils. Among these survivors are the San (Bushmen) of the Kalahari Desert in Africa and the Aborigines of the Australian deserts. Small groups of hunter-gatherers also live in the humid-tropical forests of South America. The Inuit (Eskimo) who hunt and fish along the Arctic shores of North America and Siberia are hunter-gatherers as well.

Even though hunter-gatherers often have simple tools and limited technology, they all have complex social customs and religious beliefs. Most hunter-gatherers live without the modern products other cultures have come to rely on.

Text continues on page 72.

GLOBAL CONNECTIONS

Building a World Culture

Thousands of Russians jam this stadium in **Moscow** for a music festival. American rock music is increasingly popular in the republics of the former Soviet Union.

In the past few decades, modern technology has brought different peoples into closer contact than ever before. Jet airplanes and high-speed trains carry immigrants, tourists, and travelers quickly from nation to nation. Television, radio, and motion pictures have turned the world into one giant theater. Music television is available in more than 70 countries, bringing together performers and audiences from around the world.

Blue jeans are popular with consumers around the world, including these young shoppers in **Tokyo, Japan**. U.S. companies exported more than 90 million pairs of pants made from blue denim to other countries in 1994.

An Emerging World Culture

Throughout the world, cultures are mixing and mingling to form a new "world culture." Although they may live oceans apart, people are exposed to common music, food, clothing, and ideas. In this new world culture, music plays a central role. Boris Grebenshikov, a well-known popular musician in Russia, recalls that he first heard rock music on the Voice of America, a U.S. radio station that broadcast into the Soviet Union. The freedom and energy expressed in the music, he says, "changed my life, and the lives of my friends, forever." For Grebenshikov and young people like him, rock music became a means of expressing their individuality and of defying authority—even the authority of the Communist government.

Modern transportation and communication have bridged the geographical obstacles that once separated countries and cultures.

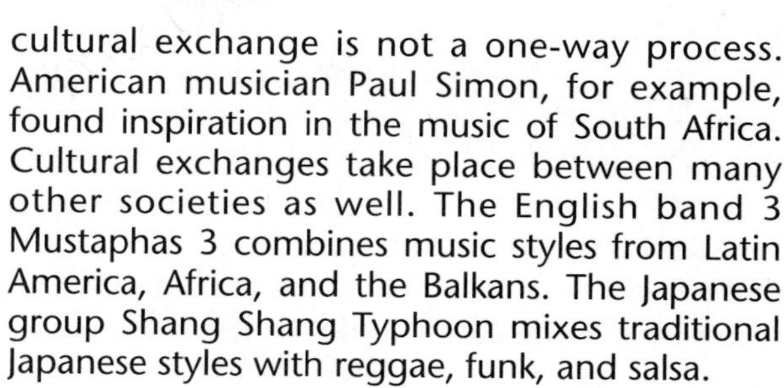
Maulana Karenga teaches these African American children in **Los Angeles** about *Kwanzaa*, an annual celebration of black culture. Karenga developed *Kwanzaa* in the United States in 1966. The word is part of a phrase that means "first fruits" in the African language of Swahili. Based on a traditional African festival, *Kwanzaa* is an example of how cultures from around the world help shape U.S. culture.

In the same way that rock music represents the music of the new world culture, blue jeans represent the clothing. Blue jeans are so sought after in Russia, for example, that jean-clad tourists often are asked if they would be willing to sell the pants they are wearing. In some countries, even used jeans sell at high prices.

The food of the new world culture, like the clothing, is casual and convenient. Fast-food outlets have sprung up around the world. In Moscow, people form lines stretching entire city blocks to sample the menu at a new hamburger franchise. In Ho Chi Minh City, Vietnam, people flock to the *Ca-li-pho-nia Ham-bu-gó*, the "California Hamburger."

The American Influence

Rock music, blue jeans, and fast food are all products of American popular culture. People in less prosperous nations view these products as symbols of a wealthy nation, able to produce many consumer goods. Yet some critics question this emphasis on consumer items. They worry that people may value wealth and material possessions more than tradition, family, community, and spirituality.

In contrast, those who defend American popular culture say that it reflects important ideals such as democracy. They point out that rock music, blue jeans, and fast food are available to many people rather than to a privileged few. Others say that American culture reflects the important ideal of diversity. American culture is molded by a variety of ethnic groups and appeals to different peoples around the world.

Although American popular culture is important in shaping the new global culture, cultural exchange is not a one-way process. American musician Paul Simon, for example, found inspiration in the music of South Africa. Cultural exchanges take place between many other societies as well. The English band 3 Mustaphas 3 combines music styles from Latin America, Africa, and the Balkans. The Japanese group Shang Shang Typhoon mixes traditional Japanese styles with reggae, funk, and salsa.

A World of Bridges, Not of Walls

Around the world, cultural ideas are crossing borders and barriers that once separated people. One symbol from the past is the Berlin Wall, which divided Germany and separated Communist and Western societies. Today, a new symbol is needed for the bonds of shared tastes and ideas being forged by the emerging world culture.

One such symbol of world culture was offered by Gianni de Michelis, chairman of the World Arts Council. "Walls," de Michelis said, "are the architectural emblem of the age we have just left; bridges of the age ahead."

Your Turn

Do you agree with critics of a world culture or those who believe a world culture brings people together? **Imagine** you are a delegate to the United Nations who serves on a commission concerned with promoting cultural exchange. **Write** a letter reflecting your opinion about a world culture.

This satellite image shows ordered crop patterns near the Salton Sea in **California** contrasting with the irregular farm plots found across the border in **Mexico**. More efficient cultivation techniques make for more productive land on the northern side of the border.

INTERACTION **Agriculture** Discovering how to raise animals and how to plant crops were perhaps the two most important innovations in human history. The development of agriculture (farming) changed how people interacted with their environments. Agriculture appeared when hunter-gatherers studied a plant or animal so closely that they found out how to grow or tame it. This innovation is called **domestication**. Geographers believe agriculture first appeared about 10,000 years ago in several areas of southwestern Asia.

Farming is a more reliable source of food than hunting and gathering. To grow food, however, the earth's natural vegetation had to be altered. As agriculture spread, forests were cleared. Waters were diverted to irrigate drylands. Food now came from fields and pastures, and people did not have to move with the seasons. Agriculture permitted people to grow more food year after year.

Farming also encouraged village life. Because people moved less often, they could live in larger groups. They shared innovations easily, and social life became more complex. People began forming governments to create social order, and religions became more structured. Large areas of land were then devoted to agriculture. More food led to increased populations.

Many people still make their living by agriculture. Two types of agriculture are common today. In traditional or **subsistence agriculture**, people grow food on small farms mostly for their own families. Some surplus food may be traded or sold. Perhaps one or two crops are raised to sell for cash at a nearby market. In subsistence agriculture, tools are basic. The farm's family members and the farm animals are the main source of labor. Subsistence agriculture is common today in the world's poorer countries, where it may be a town or village's main economic activity.

In **commercial agriculture**— the growing of crops for sale—farms are large and use modern technology. High-yield seeds, industrial fertilizers, and powerful machinery are typical. Farm animals are raised using scientific methods. Commercial agriculture requires fewer people, yet the yields far exceed the needs of the farmers. In the United States, only a few million farmers produce the food on commercial farms for more than 260 million people. A large amount remains to sell to other countries. Commercial agriculture is practiced throughout the world's wealthier countries.

City Life Hunter-gatherers lived in small camps, and the first farmers lived in villages surrounded by fields. Once food was easier to grow, people had more time to work at other things. Some workers specialized in making tools or clothing. Others worked in government and religion or became traders and merchants. These types of work are key characteristics of cities. Cities first appeared more than 5,000 years ago. Growth in the proportion of people living in towns and cities is called **urbanization**.

The development of cities signaled a change in human geography. People in cities lived close together, and communication was easier. As a result, new ideas and technologies could be shared and tested quickly. The first cities became centers of innovations. Regions of important new ideas and developments are called **culture hearths**. For example, Ancient Mesopotamia (mehs-uh-puh-TAY-mee-uh), where writing developed, is a culture hearth. Culture hearths developed in many areas of the world and at different times. Cultural diffusion became more rapid as trade carried goods and new ideas from city to city.

Cities continue to play a major role in world geography as the source of most innovations.

Early World Culture Hearths

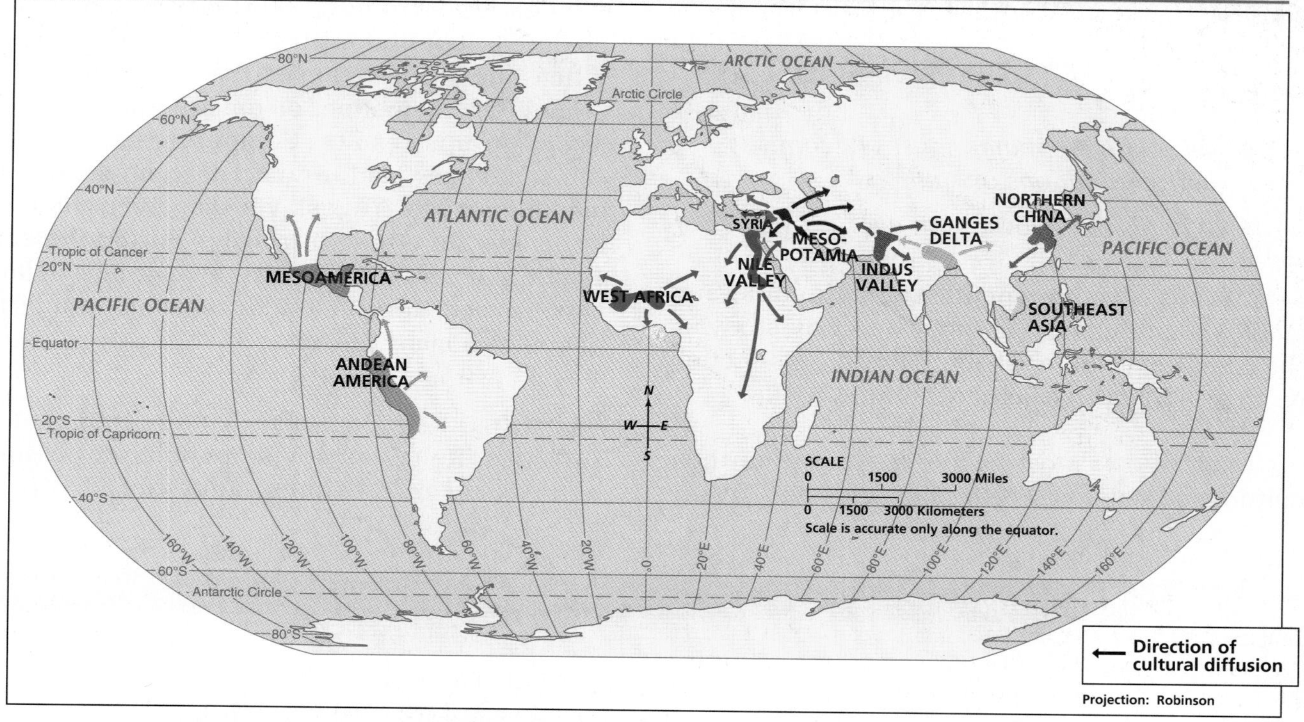

ARCTIC OCEAN
Arctic Circle
ATLANTIC OCEAN
PACIFIC OCEAN
INDIAN OCEAN
Tropic of Cancer
Equator
Tropic of Capricorn
Antarctic Circle
MESOAMERICA
ANDEAN AMERICA
WEST AFRICA
SYRIA
NILE VALLEY
MESO-POTAMIA
INDUS VALLEY
GANGES DELTA
NORTHERN CHINA
SOUTHEAST ASIA
SCALE
0 1500 3000 Miles
0 1500 3000 Kilometers
Scale is accurate only along the equator.
Direction of cultural diffusion
Projection: Robinson

Early culture hearths were scattered across the world and diffused ideas and inventions to other regions. To what region did some ideas and inventions from Mesoamerica diffuse?

Cities are centers of manufacturing, communications, education, government services, and the arts. Television, radio, and computer networks, along with transportation facilities, link major cities throughout the world.

Industrialization In the early cities, workers used simple tools to make clothing and other goods by hand. Such goods were custom-made and carefully crafted, but they were scarce and costly. Industrialization allowed large amounts of standardized goods to be made at a lower cost. It also dramatically altered the earth's physical and human geography.

Industrialization spread rapidly in western Europe during the eighteenth and nineteenth centuries. In world history, this period of time was known as the Industrial Revolution. Factories were clustered near large rivers where running water provided power. Later, coal and petroleum provided even more efficient sources of power and freed industrial sites from having to be located near water. As a result, most factories sprang up in cities where large pools of labor were available. In turn, industrialization led to increased urbanization as people from the countryside moved into the cities to find jobs in factories.

Industrialization continues to spread around the world. New factories are being built in countries where only a few years ago subsistence farming was the main economic activity. These countries hope that industrialization will improve their economies. Changing to an industrial economy can be expensive, however, and requires resources such as water, fuel, and low-cost labor. In addition, industrialization creates pollution and industrial waste and depletes natural resources.

2 SECTION REVIEW

1. Why was the development of agriculture an important innovation?
2. Why do urbanization and industrialization often accompany each other?
3. **Critical Thinking** **Imagine** that a previously isolated hunter-gatherer group suddenly came into contact with an industrial society. What might be the consequences of this meeting?

3 CULTURE AND WORLD EVENTS

FOCUS

- *What are the major causes of conflicts between nations or cultures?*
- *What is the information age?*

Sometimes, disagreements among people and among countries can be understood by studying the countries' different culture traits. This section examines the main causes of cultural conflict.

Nationalism Most people feel proud of their culture and country. Feelings of pride and loyalty for one's country are called **nationalism**. Nationalism often is expressed in patriotic songs, symbols, and writings. Unfortunately, one group's pride can conflict with that of another group. When one group believes that it is better than others, discrimination, or unfairness, toward different groups results. Competing nationalisms can sometimes lead to war. For example, the former country of Yugoslavia dissolved violently into separate ethnic territories during the early 1990s. And conflicts among ethnic groups in India have caused the country to redefine its internal boundaries many times.

Religion Religion is a key culture trait in many societies. Religions advise people on personal behavior, sacred times, and sacred places.

World Religions

ARCTIC OCEAN
Arctic Circle
ATLANTIC OCEAN
PACIFIC OCEAN
INDIAN OCEAN
PACIFIC OCEAN
80°N
60°N
40°N
Tropic of Cancer
20°N
Equator
20°S
Tropic of Capricorn
40°S
60°S
Antarctic Circle
80°S
160°W 140°W 120°W 100°W 80°W 60°W 40°W 20°W 0° 20°E 40°E 60°E 80°E 100°E 120°E 140°E 160°E
N
W
E
S
SCALE
0 1500 3000 Miles
0 1500 3000 Kilometers
Scale is accurate only along the equator.

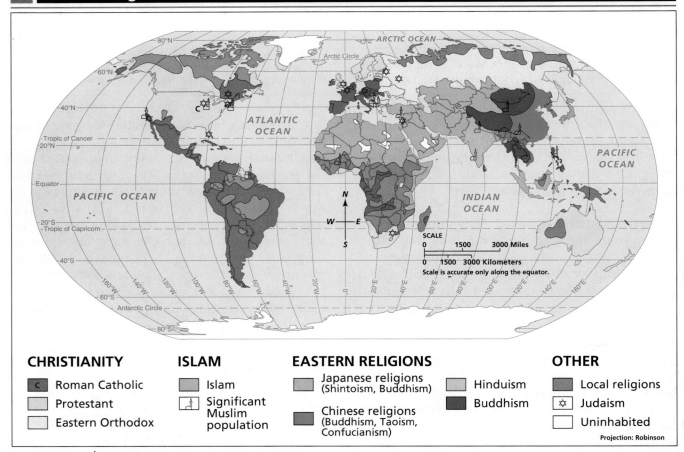

CHRISTIANITY
- C Roman Catholic
- Protestant
- Eastern Orthodox

ISLAM
- Islam
- Significant Muslim population

EASTERN RELIGIONS
- Japanese religions (Shintoism, Buddhism)
- Chinese religions (Buddhism, Taoism, Confucianism)
- Hinduism
- Buddhism

OTHER
- Local religions
- Judaism
- Uninhabited

Projection: Robinson

MAP STUDY This map shows which religions are dominant in various parts of the world. The followers of most major religions are found in more than one region of the world. Significant numbers of followers of Judaism, for example, can be found in Southwest Asia, Europe, and North America. Which religions have places of worship in or near your community?

People also may receive a sense of personal fulfillment from their religion.

Conflicts emerge when people of one religion believe that theirs is the one true religion. When people do not tolerate other religions, disputes can lead to unrest and even war. Because each side believes its cause is holy, religious conflicts can be bitter and long lasting.

There are many modern examples of religious conflict. In Sri Lanka, the Buddhist Sinhalese have warred with the Hindu Tamils. In Northern Ireland, Roman Catholics and Protestants have battled each other. In Southwest Asia, religious differences among Jews, Muslims, and Christians have caused bitter conflicts.

Traditional Versus Modern Values

Some cultures adapt to change easily, while other cultures do not. Changes that appear to threaten traditional religious practices or the traditional roles of men and women may be resisted. In some cases, religious **fundamentalism** has been a direct reaction to the spread of Western culture and values throughout the world. Fundamentalism is a movement that stresses the strict following of basic traditional principles. An example is Islamic fundamentalism now spreading in Southwest Asia and North Africa.

Conflicts over values arise from several causes. Often, young people accept innovations that older generations reject, such as popular music and new styles of clothes. People who live in rural areas sometimes resist new technology, while city dwellers welcome it.

Politics How a group of people should be governed has been a source of conflict throughout world history. Some countries are governed by one person and a few advisers. In these **totalitarian governments**, only a few people decide what is best for everyone. In other countries, everyone has a voice in their government. These are **democratic governments**. Human rights often are violated by totalitarian governments. In contrast, democratic governments promote human rights and value individual freedom.

In countries with many political parties, conflicts may arise over which party should govern the country. Even when democratic elections are held, sometimes the losing party or parties may not honor the election. Conflicts also appear over different ideas about the basis for a country's laws. For example, should laws be based on religious rules or on the ideas of political philosophers or elected officials?

The United Nations is a worldwide organization that tries to settle problems among and within countries. Representatives from countries can work with the United Nations to solve problems without violence. UN peacekeeping forces are on duty in many places. In the early 1990s, UN forces aided the United States in its effort to stop civil war and feed starving people in Somalia. UN forces are not always successful. In the former Yugoslavia, UN forces at times were blockaded from providing food and medical relief.

Economics The world's resources are not evenly distributed. Some countries have many raw materials and successful industries. Others have only a few. These inequalities may encourage a poor but well-armed country to invade its richer neighbor. In the past, wealthy countries have colonized other lands to acquire valuable resources.

Sometimes, economic conflicts between countries are nonviolent struggles. The conflicts may appear

Minimum Voting Ages (in selected countries)

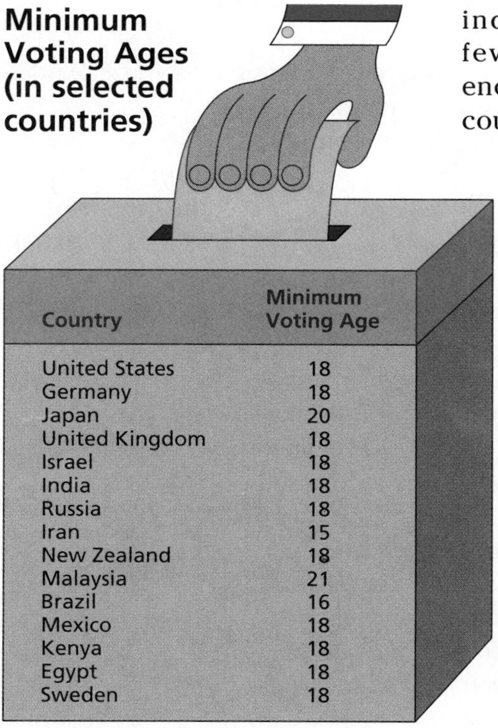

Country	Minimum Voting Age
United States	18
Germany	18
Japan	20
United Kingdom	18
Israel	18
India	18
Russia	18
Iran	15
New Zealand	18
Malaysia	21
Brazil	16
Mexico	18
Kenya	18
Egypt	18
Sweden	18

Source: *The World Factbook 1994,* Central Intelligence Agency

This chart provides a random sample of legal voting ages in various countries. As of 1994, Iran had one of the lowest legal voting ages in the world. Teenagers in the United States can legally vote at age 18.

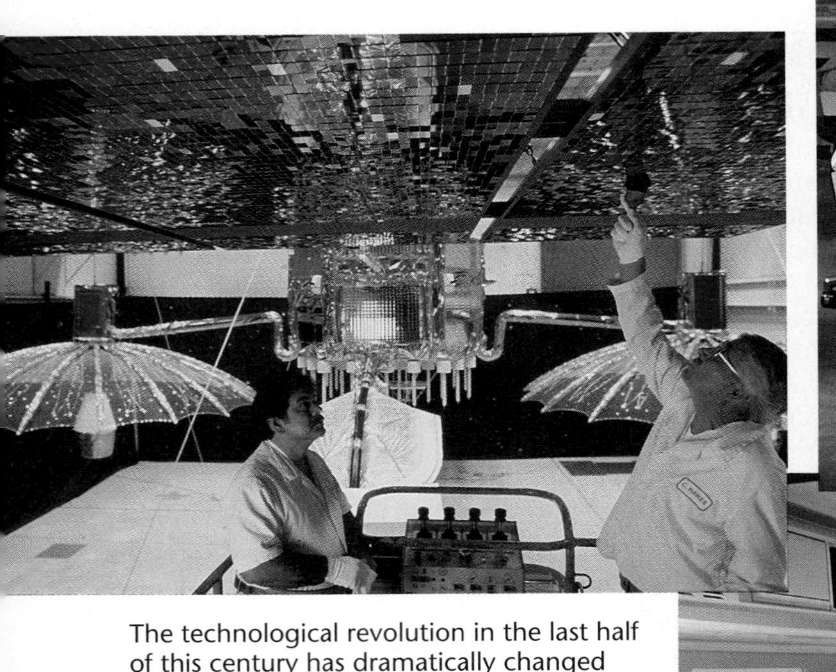

The technological revolution in the last half of this century has dramatically changed the way we exchange information. The telecommunications satellite (above), telephone bank (above, right), and high school computer lab (right) all are tools of the information age.

over trading rules. For example, countries often establish **tariffs**, which are taxes placed on imports and exports. A similar tool is a **quota**, which limits the amount of a particular good that can be imported. Tariffs and quotas protect a country's industries from foreign competition. They can, however, cause ill will among the countries in competition with each other.

Issues in Human Geography In spite of cultural conflicts throughout the world, the earth's people are linked closely. Innovations in high technology, particularly computers and satellites, are permitting rapid and easy communications. Information now travels great distances almost instantly. Many people around the world watch the same television programs, eat the same foods, and speak the same language.

In this "information age," our definitions of resources and what is of economic value are changing. For example, fewer raw materials are needed in the information age. Fuels such as oil, coal, and natural gas are used little in moving information. Instead, computer chips are made from silicon, an element common in sand.

As a result of changes brought about by the information age, jobs in traditional industries are disappearing. Workers must retrain for new jobs in the information industry. Today, the world's richest countries are those with educated work forces. Clearly, the skills you are learning in school are more important than ever before.

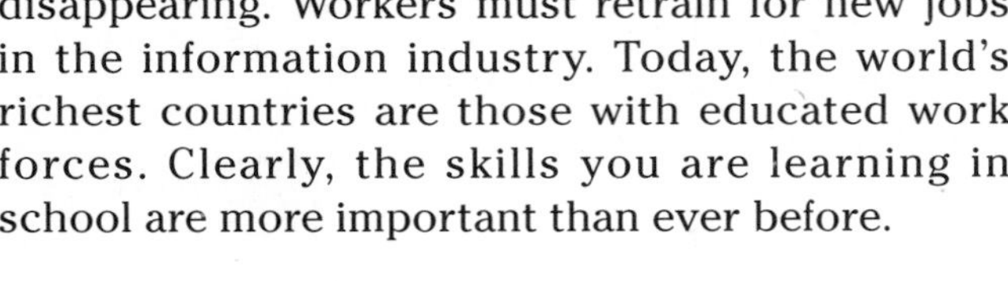

3 SECTION REVIEW

1. List and explain four sources of world conflicts.
2. How does your daily life reflect the information age? Give examples.
3. **Critical Thinking** **Decide** whether the arrival of the information age will increase or decrease the number of world conflicts. Explain your answer.

CHAPTER 7 CHECK

Modern farming in **Kansas**

Reviewing the Main Ideas

1. Culture includes all aspects of a society's way of life. A culture is made up of many culture traits, and a culture region is an area where many culture traits are shared. Cultures change through innovation and diffusion.
2. All people have the same basic needs for food, shelter, and clothing. Throughout history, the innovations of agriculture, urbanization, and industrialization have affected how our basic needs are met.
3. World conflicts are caused by many factors, including nationalism, differences in religious and political beliefs, different values, and the uneven distribution of resources. The current era, in which geographical distance no longer slows world communication, is called the information age.

Building a Vocabulary

1. What is an ethnic group? Name three ethnic groups who live in the United States.
2. What is the term used to describe the change of one culture through its meeting with another culture?
3. What is the difference between subsistence agriculture and commercial agriculture?
4. Compare and contrast totalitarian governments and democratic governments.
5. What are tariffs? What is a quota? Why are they used?

Recalling and Reviewing

1. How do innovation and diffusion cause cultures to change?
2. How was the development of agriculture related to the growth of cities?
3. Why is the present era called the information age?

Thinking Critically

1. Some people believe they are not part of any culture. Do you think it is possible for someone not to be part of any culture? Explain.
2. In your opinion, what one factor is most likely to cause conflicts between countries or cultures? Support your answer with examples.

Learning About Your Local Geography

Cooperative Project

On a sheet of paper, write down at least two cultural, ethnic, or other types of groups of which you, as an individual, consider yourself to be a part. As a class, make a master list of all the groups mentioned on the individual lists. Discuss your findings. Are many different groups listed, or just a few? What are some of the various ways people define themselves? What can you learn from people who define themselves differently than you do?

Using the Five Themes of Geography

From pictures in magazines, newspapers, and other materials, make a collage showing human **place** characteristics of various culture **regions**. Show evidence of **movement** between various culture regions. Make sure you have permission before cutting up any materials. If you do not have access to materials from which to cut pictures, draw your own illustrations.

CHAPTER 8

Global Economics and Global Population

> "Whether we look at population, environment, or development, the next 10 years will be critical for our future. The decisions we make or don't take will widen or narrow our options for a century to come. They could decide the fate of the earth as a home for human beings."
>
> *Nafis Sadik*

In the past hour, about 10,000 babies were born. Indeed, the world's population is fast approaching 6 billion people. In fact, we are living during the period of greatest population increase in human history. Can our planet continue to support this ever-increasing global population? The question is an urgent one for the more than 1 billion people around the world who will go to bed hungry tonight.

Economic progress varies among the world's countries and from place to place within individual countries. A country's level of economic development also changes over time. For example, in its earliest years, the United States was a poor country of farmers. Today, it is the richest country in the world. Yet pockets of poverty still exist within its borders. All countries, regardless of their level of economic development, wish to improve their economies. This chapter explores the important link between global economics and global population.

GEOGRAPHY DICTIONARY

- *primary industry*
- *secondary industry*
- *tertiary industry*
- *quaternary industry*
- *gross national product (GNP)*
- *gross domestic product (GDP)*
- *literacy*
- *infrastructure*
- *telecommunications*
- *developed country*
- *developing country*
- *foreign aid*
- *multinational company*
- *free enterprise*
- *capitalism*
- *market economy*
- *command economy*
- *communism*
- *demography*
- *birthrate*
- *death rate*
- *emigrant*
- *immigrant*

New York City street festival

1 GLOBAL ECONOMICS

FOCUS

- *What are the categories of economic activities?*
- *How do geographers measure varying levels of economic development?*

These photographs of the dairy business show four industries at work. Dairy farming (top), a primary industry, provides the milk needed for processing into cheese (middle, left), a secondary industry. A grocery store, a tertiary industry, offers a customer a selection from the dairy case (middle, right). A specialist, part of a quaternary industry, conducts research to find ways to improve dairy production and quality (bottom).

Economic Geography The branch of geography concerned with how people use Earth's resources, how they earn a living, and how products are distributed is called economic geography. People's economic activities can be divided into four categories—primary (first order), secondary (second order), tertiary (third order), and quaternary (fourth order) industries.

Primary industries include agriculture, forestry, and mining. These activities directly use natural resources and raw materials and are found at or near sources of those materials. Agriculture and forestry are found in regions with suitable soils and climates. Mines are located where rich mineral deposits have been found.

Secondary industries take goods made in primary industries and change the goods into products that are useful to consumers. Secondary industries involve manufacturing. For example, a dairy takes a raw material (raw milk) and processes it into products such as ice cream and butter. Often, a product will go through several stages before it becomes a finished good. Many factors determine the locations of manufacturing industries. Some industries, such as dairies and bakeries, are located near markets because their products may spoil if they must be transported too far. Modern refrigeration and improved roads, however, have made longer hauls possible. Other industries are located where special manufacturing needs can be met easily. These may be needs for cheap power sources, skilled workers, or specialized transportation facilities.

Tertiary industries are service industries. These industries provide services to primary and secondary industries, to communities, and to individual consumers. Service industries include stores, restaurants, banking, insurance, transportation, and government agencies. Services usually are located close to where people live.

The use of computers and electronic communications is changing some service industries. Today, consumers can buy a wide variety of catalog items over the telephone. Computers allow some people to work at home, which gives them more flexibility in deciding where they want to live. A worker living in Jackson Hole, Wyoming, for example, may work for a firm in New York City.

Some geographers include this movement and processing of information in a fourth category of economic activity. **Quaternary industries**, a subdivision of tertiary industries, are performed by professionals having specialized skills or knowledge. Information research, management, and administration are quaternary industries. Your school principal is an example of a worker involved in a quaternary industry.

Economic Indicators As economic activities vary, so do economic conditions. Accurately measuring levels of economic development is difficult, but geographers can use several means to compare data and make generalizations. It is important

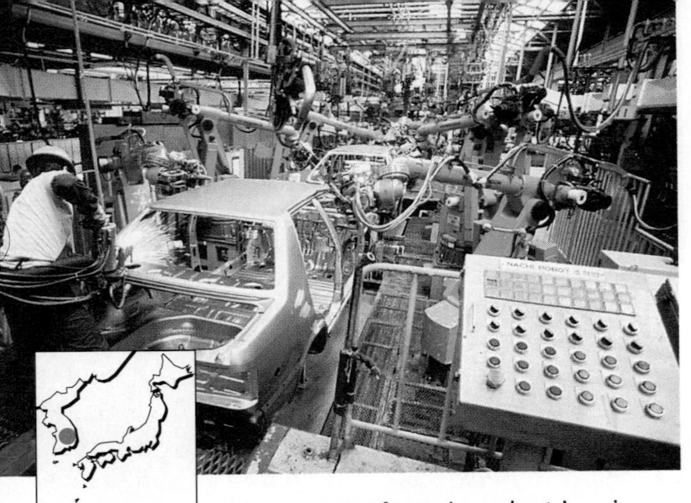

This car manufacturing plant is evidence of the great improvements in the economy of **South Korea** since the Korean War. Manufacturing, a secondary industry, is an important part of South Korea's economic growth. Automobiles are among South Korea's most important exports.

to remember that a country's economic level says little about its cultural values and contributions.

One measure of a country's economic development is **gross national product** or **GNP**. Gross national product is the total value of goods and services produced by a country in a year, both inside and outside the country. Some geographers use **gross domestic product (GDP)** instead of gross national product. Gross domestic product includes only those goods and services produced within a country and does not include income earned outside that country. This number becomes more useful when it is divided by the number of people in that country. This figure tells us the per capita GDP, which then can be used to compare the economic level among countries.

Another way to describe levels of economic development is to identify the extent to which a country has industrialized. The world's poorest countries have no industrial activity. The next poorest countries are industrializing, which means they are in the process of developing manufacturing and industry. Some countries are beginning to move out of the industrial stage into a postindustrial stage. In a postindustrial country, more people are employed in tertiary and quaternary industries than in primary and secondary industries. This trend, however, has resulted in industrial decline in some regions, as experienced in the Northeastern and Midwestern United States. The full impact of this trend will not be known for some time.

Other measures of economic development are literacy, infrastructure, and telecommunications. **Literacy** is the ability to read or write. In general, the higher a country's literacy rate, or the higher the number of people who can read or write, the more economically developed the country. **Infrastructure** includes roads, bridges, transportation facilities, power plants, water supply, sewage treatment facilities, and government buildings that are necessary to build industries and to move goods in and out of a country quickly and easily. **Telecommunications** is electronically transmitted communication. Today, a rapid, efficient telecommunications network is important for developing high-technology industries.

SECTION REVIEW

1. Why do different kinds of economic activities take place in different areas?
2. List three different ways of measuring a country's economic development.
3. **Critical Thinking** **Explain** the relationship between literacy rates and economic development.

DEVELOPED AND DEVELOPING COUNTRIES

FOCUS

- *How do developed countries and developing countries differ?*
- *What are the various types of economic and political systems?*

Categorizing the world's countries by economic level identifies the world's richest and poorest countries and those that lie in between. The world's wealthy countries are called **developed countries**. These include many of the countries in Europe, the United States, Canada, Japan, Australia, Singapore, and others.

Developed countries share a number of common characteristics. Among these are a good educational system, widely available health care, and many manufacturing and service industries. Developed countries are industrialized and participate in international trade. People mostly live in cities. The few people who work as farmers use modern technology. In developed countries, most people have access to a telephone and are part of the global network formed by modern telecommunications. The wealthiest of the developed nations have per capita GDPs above $19,000.

About three-quarters of the world's people live in the poorer countries, which often are called **developing countries**. Many of the people in developing countries live by subsistence farming in rural areas. These countries have few manufacturing and service industries to provide jobs. Though capital cities and other major cities may be large, poverty and unemployment are widespread. Many people migrate, or move, from rural areas to cities. Many city dwellers are underemployed, working only part time.

In developing countries, health services are limited and schools often are crowded. The literacy rate is usually low. Modern telecommunications are not usually found outside of major cities. Exporting agricultural products and minerals to developed countries often is an important source of income for developing countries. Some of the world's poorest countries have per capita GDPs of less than $600.

Between the world's richest and poorest countries are middle-income countries such as Argentina, Mexico, Venezuela, Brazil, Hungary, South Korea, and Malaysia. These countries combine characteristics of developed and developing countries. Some middle-income countries, such as South Korea and Taiwan, are newly industrialized. Their per capita incomes are growing rapidly and soon will be in the developed category. Other countries, such as Argentina, have remained in the middle category for decades.

Choices in Economic Development How can a poor country improve its economy? If it has many natural resources such as oil fields, mineral deposits, or valuable crops, then development of these resources can lead to economic progress. Income from economic activity can be invested in education and infrastructure.

For a developing country with few natural resources, the economic choices often are very limited. Sometimes the governments of developed countries help developing countries with loans and gifts of money. Some of this **foreign aid** goes to ease human suffering.

Developed countries often debate how to help developing countries. Some wealthy countries provide little help. They argue that there are too few resources to share and that by helping others their own economic growth will suffer. Other developed countries give aid only to countries with rich natural resources or a low rate of population growth. Still other countries give aid only to those countries with which they are on close political terms.

Multinational companies build factories in many poor countries around the world. These companies are businesses with activities in many countries. Multinational firms seek raw materials and labor at low cost. Multinational companies provide jobs and often transfer technology from the developed countries to developing countries. Multinational companies sometimes are criticized, however, for harming the natural environment or changing the cultures of developing countries.

Politics and Economics Economic development is of concern to all governments. What kinds of economic and political ideas lead to economic progress?

The world's economically developed countries share two important characteristics. The first is an economic system guided by **free enterprise**. Under free enterprise, prices are determined mostly through competition in which buyers and sellers are free to choose what and when to sell and buy. As a result, goods that people want are produced in large amounts and are sold at fair prices. Under free enterprise, people are free to move in search of higher wages and better working conditions.

Free enterprise is the basis of **capitalism**, an economic system in which resources, industries, and businesses are owned by private individuals. Capitalism is also called a **market economy**. In a market economy, consumers determine what is to be bought or sold by buying or not buying certain goods and services. The term *market* refers to a system in which goods and services are exchanged freely.

The second important characteristic of the developed countries is democracy. People must have personal freedom to make free enterprise work. During the past few years, all of the countries whose economies have improved also have expanded free enterprise and democratic ideas.

In contrast to economies based on free enterprise are **command economies**. In a command economy, the government determines wages, the kinds and amounts of goods produced, and the prices of goods. Communist countries have command economies. **Communism** is an economic and political system in which the government owns or controls almost all of the means of production. These definitions are generalizations in today's world, however. Since the breakup of the Soviet Union, the once-Communist countries of Eastern Europe and the former Soviet Union have been struggling to reform and improve their economic systems.

The world's developing economies often are a mix of command economies and free enterprise. India, for example, is a democracy, but much of its economy is controlled by the government. Though China has greatly expanded free enterprise, democracy in the country has been put down, sometimes violently.

Many developing countries' economies are hindered by poor government. In some countries, few people are trained in government and business. In other countries, much of the wealth is controlled by a small number of people who pressure the government to pass laws favorable to them. Such governments are often unstable. As the developing countries turn toward free enterprise and democracy, their economic conditions can be expected to improve.

SECTION REVIEW

1. What determines whether a country is considered developed or developing?
2. What is the main difference between capitalism and communism?
3. **Critical Thinking** **Formulate** your own ideal political and economic system.

POPULATION GEOGRAPHY

FOCUS

- *How is the world's population distributed, and why? Why has the population grown over time?*
- *What factors will affect the size of future populations?*

PLACE **Global Population Density** The geography of the human population is uneven. Some regions are densely populated, while others have many fewer people. Some regions have aging populations, while others have mostly children. Population geography is the study of the variations and changes in a population's makeup, distribution, movement, and relationship to its environment. **Demography** is the study of the human population that emphasizes statistical characteristics, or facts and figures. It is used to compare population distribution, population density, trends in population, and so on.

It is easy to understand why some parts of the world have fewer people. These regions usually have harsh climates, and farming is difficult. In contrast, understanding patterns of dense population settlement is more complex.

There are four large regions of dense human settlement. These are eastern Asia, particularly eastern China; south Asia, particularly India and Bangladesh; Western Europe; and eastern North America. A few smaller areas of dense population include southeastern Brazil, Nigeria, and southern California. Clearly, regions with environments favorable to farming usually are densely settled. The fertile coastal lowlands and river valleys of China, India, and Egypt have been heavily populated for more than 2,000 years. Other factors also lead to population growth and settlement in regions, however. The populations of Western European and eastern North American cities increased dramatically when these regions industrialized.

Population Growth Archaeologists believe the world's population at the end of the last ice age was less than 10 million. With the development of agriculture, populations grew rapidly. Two thousand years ago, the world's population numbered about 250 million, growing only gradually during the next centuries. In periods of prosperity, populations rose, only to decline during periods of war or widespread outbreak of disease. By about A.D. 1650, world population finally had doubled to 500 million. Soon after, populations began to increase steadily. With modern health practices, people began to live longer. More babies lived to become adults. Food supplies increased, and cities grew rapidly.

By 1850, the world's population had doubled to 1 billion. Since then, we have witnessed explosive population growth. World population passed the 2 billion mark around 1940, doubling to 4 billion in 1975. In the year 2000, the population of the world will be 6 to 7 billion. Demographers predict that if this trend continues, world population will surpass 10 billion in the next century.

Stages of Population Growth What factors lead to rapid population growth? Demographers explain that a country experiences three stages of population growth as its economy develops. As a model, geographers use Western Europe, which went through industrialization and urbanization beginning in the mid-1700s. Of course, not all countries follow this model, but it is useful to study how and why population growth in a region changes in relation to economic development.

Geographers use birthrates and death rates to explain the three stages of population growth. **Birthrate** is the number of births per 1,000 people in a given year. **Death rate** is the number of deaths per 1,000 people in a given year. If a country's birthrate exceeds its death rate, the population will increase. If the number of deaths is more than the number of births, the country's population will decrease.

During the first stage of population growth of a country, the birthrate and the death rate are high but equal. Though there are many births, widespread disease takes its toll on the population. As health care and safe water become more widespread, the death rate declines steadily while the birthrate remains high. In this second stage, family sizes increase, and more children than before survive to adulthood. Population growth is rapid. In the third stage, the birthrate drops, and most children live to adulthood. Family sizes decrease as more people decide to have fewer children. The birthrate and death rate again become about the same, though now at lower levels. The country's rate of population growth again becomes low. The population now is much higher than in the first stage, however.

World Population Growth

Population in billions
12
11
10
9
8
7
6
5
4
3
2
1
0
A.D.1 100 200 300 400 500 600 700 800 900 1000 1100 1200 1300 1400 1500 1600 1700 1800 1900 2000 2100
Year
Source: Population Reference Bureau, Inc.

The world's population has increased rapidly since the late 1700s. This sign reminds residents of **New Delhi, India,** of their country's birthrate and population.

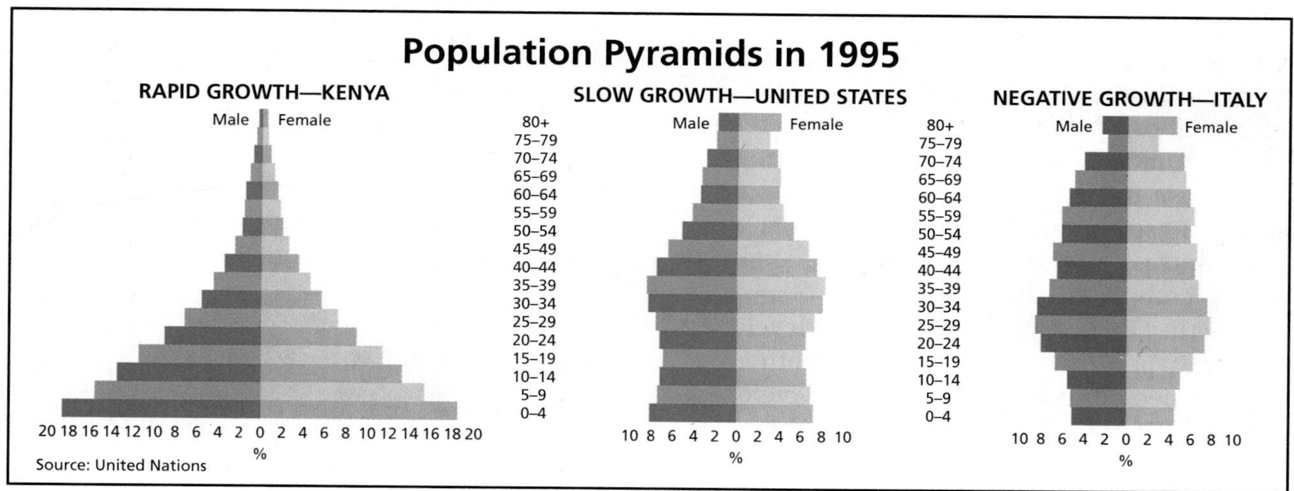
Population Pyramids in 1995
RAPID GROWTH—KENYA
Male
Female
20 18 16 14 12 10 8 6 4 2 0 2 4 6 8 10 12 14 16 18 20
%
Source: United Nations
SLOW GROWTH—UNITED STATES
80+
75–79
70–74
65–69
60–64
55–59
50–54
45–49
40–44
35–39
30–34
25–29
20–24
15–19
10–14
5–9
0–4
Male
Female
10 8 6 4 2 0 2 4 6 8 10
%
NEGATIVE GROWTH—ITALY
80+
75–79
70–74
65–69
60–64
55–59
50–54
45–49
40–44
35–39
30–34
25–29
20–24
15–19
10–14
5–9
0–4
Male
Female
10 8 6 4 2 0 2 4 6 8 10
%

These population pyramids compare three countries' population distributions by age and sex. Italy, with a relatively high percentage of older residents, is an example of a negative-growth country. The population of Kenya is rapidly growing, leaving the country with a larger percentage of younger residents. The bulge in the 20–40 age range in the United States pyramid is a result of the post–World War II baby boom. Can you predict what the U.S. pyramid will look like in the year 2020?

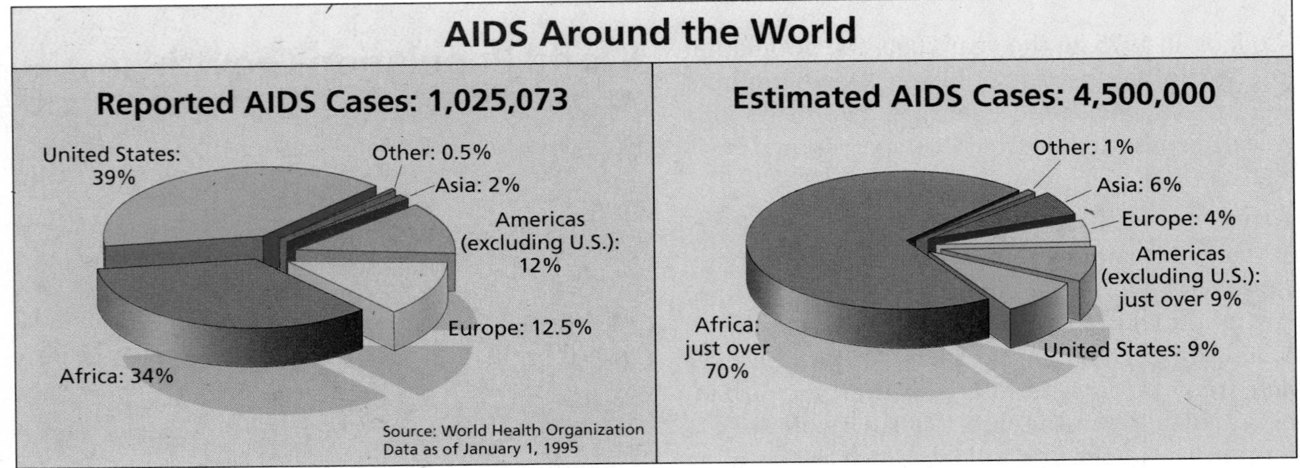
AIDS Around the World
Reported AIDS Cases: 1,025,073
United States: 39%
Other: 0.5%
Asia: 2%
Americas (excluding U.S.): 12%
Europe: 12.5%
Africa: 34%
Source: World Health Organization
Data as of January 1, 1995
Estimated AIDS Cases: 4,500,000
Other: 1%
Asia: 6%
Europe: 4%
Americas (excluding U.S.): just over 9%
United States: 9%
Africa: just over 70%

Some demographers believe projections for population growth and life expectancy rates must be revised for some parts of the world because of the toll taken by the disease AIDS. These pie graphs show the worldwide distribution of reported and estimated AIDS cases from the beginning of the pandemic in the late 1970s or early 1980s to the end of 1994. (Diseases become pandemic when they are widespread in many locations at the same time.) The World Health Organization estimated that by the end of 1994, about 19.5 million people, including 1.5 million children, had become infected with the virus that causes AIDS. By the year 2000, the total number of infected individuals has been projected at between 30 and 40 million, with the vast majority of infections taking place in developing countries.

A country's population growth or decline also depends upon the number of its emigrants and immigrants. An **emigrant** is someone who moves out of a country. An **immigrant** is someone who moves into a country. If a country has more immigrants than emigrants, its population will increase.

Population change and economic success are related. In all the world's developed countries, the period of rapid population growth has ended. These countries are experiencing low death rates and low birthrates. People in these countries live mostly in cities and die at an old age. Women have the opportunity to work outside the home, and many choose to have smaller families. The populations of these countries are changing little.

In contrast, the world's developing countries have not completed the three stages of population growth. Often, with aid from the developed countries to improve public health, death rates have dropped dramatically among developing countries. Though birthrates have begun to decline, they still exceed the death rates. In these countries, large families are often important to the family economically. Children are needed to work around the home or farm. International organizations have begun providing education and information to these countries about controlling population growth.

Future Growth Because the world's developing countries are home to a high number of children who will themselves have children, rapid population growth will continue. Providing for this still-expanding population will mean increases in food production, industrial development, and education within the developing regions.

An issue with an uncertain effect on future population growth or decline is the spread of disease. Of particular concern is AIDS (Acquired Immune Deficiency Syndrome) and the virus that causes it (HIV). According to the UN World Health Organization Global Programme on AIDS, "With every day that passes, 5000 more people worldwide become infected with the virus." Many more cases go unreported or undiagnosed.

3 SECTION REVIEW

1. Why are some areas more densely populated than others? What factors explain why populations change in size?
2. What are some of the challenges future populations will face?
3. **Critical Thinking** **Predict** what would happen if world population exceeded world food production. Would growth still continue?

CHAPTER 8 CHECK

Mall of America in **Bloomington, Minnesota**

Reviewing the Main Ideas

1. Economic activities can be divided into four categories: primary, secondary, tertiary, and quaternary. Geographers have several different ways of measuring economic development.
2. The countries of the world can be grouped according to their economic conditions. Most of the world's people live in developing countries. Developed countries usually have an economic system based on free enterprise, whereas the economic systems of developing countries often are a mix of command and market economies.
3. Some regions of the world are densely populated, while other regions are sparsely populated. The world's population has increased dramatically over the past few centuries.

Building a Vocabulary

1. What is the difference between gross national product and gross domestic product?
2. Why is infrastructure necessary for industrialization?
3. Define *telecommunications*.
4. What do we call the study of the statistical characteristics of the human population?
5. What is the difference between an emigrant and an immigrant?

Recalling and Reviewing

1. List the four categories of economic activities, and give an example of a job or profession in each category.
2. What advantages do people in developed countries often have that many people in developing countries do not have?
3. What are some of the problems faced by developing countries in their efforts to improve their economies?
4. Describe the three stages of population growth.

Thinking Critically

1. If a country has a very high per capita GDP, what can you tell about how much money any given individual in that country might have?
2. Do you think that developed countries should try to help developing countries? Explain.

Learning About Your Local Geography

Individual Project

Collect information from an almanac or statistical abstract on how your state's population is distributed. What percentage of your state's population lives in urban or metropolitan areas, and what percentage lives in rural or nonmetropolitan areas? Construct a pie graph to present your information. You may want to include your graph in your individual portfolio.

Using the Five Themes of Geography

In this chapter, you have learned that the world can be divided into developed, developing, and middle-income **regions**. Make a chart with three columns, one for each of these regions. Fill in the chart with the human **place** characteristics that often are associated with each of these regions.

CHAPTER 9

Resources and Environmental Change

> "Now it seems as though our mother planet is telling us, "My children, my dear children, behave in a more harmonious way. My children, please take care of me.""
>
> *The Dalai Lama*

GEOGRAPHY DICTIONARY

nonrenewable resource
renewable resource
crop rotation
soil salinization
desertification
reforestation
deforestation
sustained yield use
desalinization
photochemical smog
acid rain
ozone
fossil fuel
petrochemical
geothermal energy

The world's almost 6 billion people draw from a common pool of Earth's natural resources for survival. Yet, resource depletion and pollution do not know geographic boundaries. An oil spill in the ocean can destroy fisheries along a coast many miles away. Air pollution, carried by winds, can destroy forests and lakes hundreds of miles away. The destruction of forests in one nation may cause severe flooding and soil erosion to a nation downstream. Soil erosion may decrease crop yields and increase costs of food, or even lead to starvation.

The earth's resources are divided into two major categories: nonrenewable and renewable. **Nonrenewable resources** are those that are not replaced by natural processes or are replaced at extremely slow rates. Examples of nonrenewable resources include oil, coal, and minerals. **Renewable resources** are those that can be replaced by the earth's natural processes. These include solar energy, soil, water, and forests. Even renewable resources, however, can be lost if we are careless or wasteful.

The population of the United States makes up only 5 percent of the world's population, yet we use 25 percent of its resources and produce 50 percent of its solid waste. We have a special responsibility to look after our planet's natural resources.

Hikers in the rain forest of **Borneo**

SOIL AND FORESTS

FOCUS

- *Why is soil conservation important to agriculture?*
- *What are the causes and effects of deforestation?*

Soil Throughout history, people have developed ways to conserve and enrich the soil. "Return to the soil what you get from it" has been the rule for thousands of years in China. The world's farmers know that the soil is like a bank. It is impossible to continue making withdrawals without making deposits.

The terraced rice paddy, developed in Asia thousands of years ago, is an example of how farmers learned to preserve the soils and store water even on the steepest slopes. Many of today's farmers also are aware of the importance of soil conservation. They use contour plowing, which involves plowing and planting crops in rows across the slope of the land to prevent runoff from eroding topsoil. Farmers also reduce wind erosion by planting rows of trees along the edges of fields to block the wind.

The fertility of the soil can be preserved in several ways. Use of natural or chemical fertilizers is one; **crop rotation** is another. In crop rotation, a field planted one year with corn, a soil-depleting crop, may be planted the next year with alfalfa or some other soil-enriching crop.

Many dry regions that depend upon irrigation are suffering from **soil salinization**, or salt buildup in the soil. The evaporation of irrigation water leaves behind salt. High salt content becomes destructive to crops and eventually destroys the soil's productivity. Soil salinization has reached problem levels in areas of the southwestern United States, Egypt, and India.

Many nations have lost their precious soil to **desertification**, a process by which the loss of plant cover and soil eventually leads to desert-like conditions. In developing countries, desertification is primarily the result of overgrazing and the removal of trees over many years. Climatic change also is involved. The condition spreads as people move their herds from overgrazed areas to vegetated areas, and the process repeats itself. Thousands of people have died from starvation in regions of Africa as fertile grasslands have turned to desert.

These terraced rice paddies in the **Philippines** help prevent soil erosion in the hilly terrain. The terraces keep rainwater from carrying soil down the hillsides.

In developed countries, such as the United States, loss of farmland is mainly the result of suburban and urban expansion across some of the best farmland. Today, because the world's growing population must be fed, our soil resource is more valuable than ever.

Forests Forests, if properly used and cared for, are valuable renewable resources. The first settlers in the United States found a land of seemingly endless forests. As the nation's population grew following European settlement, many forests were cut to provide timber for fuel, homes, and industry. New trees were not planted to replace those that were destroyed.

It was not until the early 1900s that the federal government took steps to conserve our forests. The government declared certain areas national forests to control the cutting of the trees and to preserve wildlife and recreation in those areas. In national forests that are logged, or cut for lumber, **reforestation** is required.

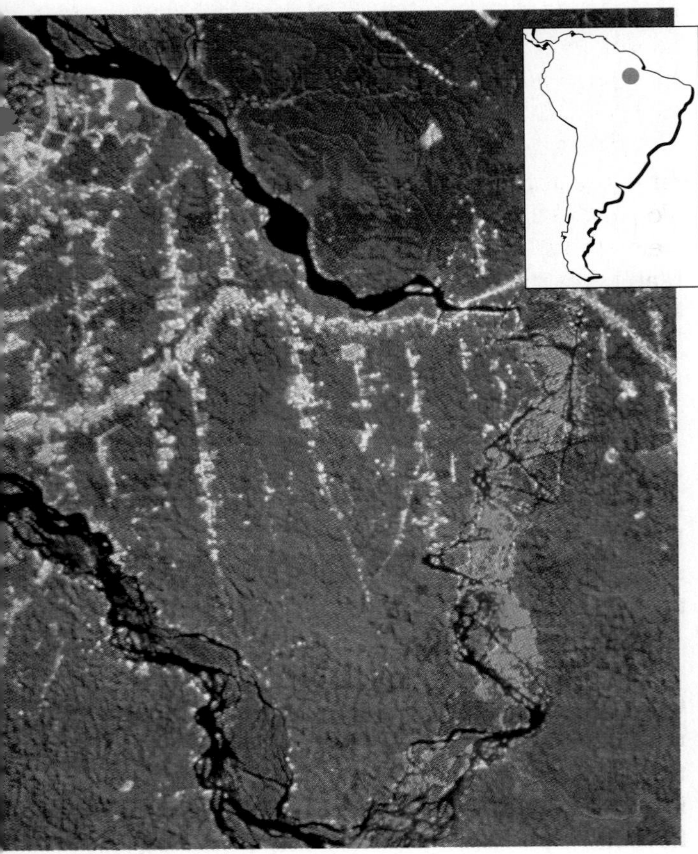

The blue and white areas in this satellite image show large paths that have been cleared through the rain forest of **Brazil.** The thick forest vegetation shows as red. Deforestation is a danger to the world's overall environment, to local wildlife, and to the native peoples whose livelihoods depend on the rain forests.

Reforestation is the process by which forests are renewed through the planting of seeds or young trees. Other nations, such as Canada, Sweden, Germany, Finland, and New Zealand, also practice reforestation. In many areas, the original-growth forests have been lost and replaced with tree farms.

In the tropical rain forests of Asia, Africa, and Latin America, **deforestation**, or the clearing of forests, is occurring at a rapid rate. In some nations, including Bangladesh and Haiti, all the original rain forests have been destroyed. Other nations, such as Costa Rica and Australia, are trying hard to preserve what is left of their original rain forests.

One-third of the earth's remaining rain forests are located in Brazil, Zaire, and Indonesia. Before the rain forests vanish completely, it is hoped that these countries will realize the consequences of losing the trees as well as the soil, wildlife, and medical benefits of forest plants.

The loss of the rain forests is mainly due to the logging of tropical hardwoods, clearing for farmland and pastures, and cutting for fuel. Charcoal from wood is the main source of energy for cooking and heating in most developing countries. In fact, more than 2 billion people rely on wood as their main source of fuel. Logging companies in these countries also continue to cut vast stretches of rain forest for lumber, offering no plans for reforestation. Government officials in developing countries defend these actions by saying that cutting the rain forests provides jobs and income to invest in their economies.

In the middle latitudes, the native temperate forests also are being logged at a rapid rate. This is especially true in northwestern North America, where U.S. and Canadian lumber companies clear-cut, or remove all trees from, large areas of the forests for domestic lumber and export to Japan and other countries. Though much of the logging is on replanted tree farms, there is continued cutting of original old-growth forest, a resource that can never be replaced.

We must recognize the importance of preserving original-growth forests. By practicing **sustained yield use**, countries can take products from the forest such as rubber, fruits, nuts, and medicinal plants without destroying the trees. Sustained yield use of a resource ensures that the resource is well managed and remains renewable. When the forests are logged or cut for fuel, reforestation must be practiced. Forests are an extremely important part of the biosphere because they produce oxygen and absorb carbon dioxide, helping maintain the proper ratio of gases in the lower atmosphere.

1 SECTION REVIEW

1. How can farmers practice soil conservation and preserve soil fertility?
2. Are forests always considered renewable resources? Why or why not?
3. **Critical Thinking** **Discuss** advantages and disadvantages of practicing sustained yield use in forests.

WATER AND AIR

FOCUS

- *What effect does a growing world population have on water quality?*
- *What might be changing the earth's climate?*

INTERACTION **Water Quantity** With the rapidly growing world population, maintaining water quantity and quality are serious challenges in many areas of the world. In the western United States, Australia, and Russia, expensive water transfer systems with dams, reservoirs, and canals transport water long distances. For example, most of California's cities and agriculture are dependent on water that is transferred hundreds of miles across desert and mountains.

In many arid parts of the world, deep wells are the only source of water. As water demands increase, the water table drops, forcing the drilling of deeper wells and further depleting the aquifers.

To solve the problem of inadequate water supplies, some countries rely on **desalinization** plants, which change salty seawater into fresh water. But desalinization is costly, especially for the energy that must be generated to produce fresh water. Presently, two-thirds of the desalinization plants are in Southwest Asia, where water is expensive and energy is cheap. Desalinization also is used where there is no alternative, such as in the Florida Keys and in the islands of the Caribbean.

A laboratory technician checks the water quality at a water treatment plant in **Michigan**. Most people in the United States get their water from public water supply systems, which use treatment plants to purify the water.

Drought has caused water shortages in Africa, Australia, and parts of the United States. To survive droughts and meet the continued demand for more water, people need to conserve water. Water conservation must be practiced on farms, in factories, in cities, and in our homes.

Flood Control Although water shortages affect many countries, floods also create hazards. As natural disasters, floods rank high in the taking of human life. In fact, floods have killed millions of people over the centuries. The Huang He (Yellow River) of northern China is known as "China's sorrow." The Huang He has drowned 1 million people in a single flood.

Dams are built to control flood waters. After the building of the Aswān High Dam in Egypt, floods from the Nile River were minimized. As a result, however, the agricultural fields along the riverbanks were no longer enriched by fertile sediments deposited by the floodwaters. Egyptian farmers now must use chemical fertilizers to replenish their soils.

INTERACTION **Water Quality** As the population of the world continues to grow along with the expansion of agriculture and industry, the byproducts of human activity also increase. As a result, impure water supplies can endanger human life. This is particularly true in developing nations that are densely populated. There, when sewage comes into contact with the drinking-water supply, diseases may spread.

Sewage is not the major source of water pollution in industrialized nations, which have sewage-treatment plants. In these countries, water is contaminated by industrial chemicals, fertilizers, pesticides, metals, and oil. Farms also pollute rivers and streams with the runoff of pesticides and chemical fertilizers, as do homes with the runoff from lawns.

The oceans are the final dumping ground for some human wastes. Some pollution occurs by accident, as when an oil tanker sinks. In other cases, pollution is caused by the direct dumping of dangerous toxic-chemical wastes and radioactive materials. Most marine pollution comes from rivers, which wash all types of human waste to the

Text continues on page 92.

PLANET WATCH

What Can *You* Do?

Persuading Industry

A ninth-grade biology student in Maine learns that tuna nets have been needlessly capturing and killing hundreds of thousands of dolphins each year. He and 75 other students write to tuna company executives, pleading that they stop the practice. As a result of public pressure such as this, within months the nation's three leading tuna companies pledge to stop killing dolphins.

Persuading Government

A group of New Hampshire tenth graders become concerned that their lunch trays are made of styrofoam. The students write and distribute a pamphlet discussing the environmental problems that result from making, using, and disposing of styrofoam. Shortly afterwards, the local school board bans styrofoam trays, and the town council decides to make the government offices and schools styrofoam free.

People around the world are choosing to tackle the earth's environmental problems. As the examples above show, kids play a major role in this effort. When you take positive steps to protect the environment and Earth's resources, your actions benefit the entire planet. Environmentalists call this "thinking globally, acting locally." Many of the most important accomplishments are a result of people adjusting everyday behaviors in simple, small ways.

Reduce, Reuse, Recycle

It is estimated that Americans generate enough trash each year to fill a line of trash trucks that would stack halfway to the moon. You yourself may have thrown away 60 pounds (27 kg) of plastic last year alone.

Have you ever thought about what happens to all that trash? While putting a piece of trash in a garbage can may seem like getting rid of the item once and for all, really it is just being moved to another place. An astounding 80 percent of all trash goes straight to landfills where waste is buried between layers of earth. Much of the trash will sit there for hundreds of years.

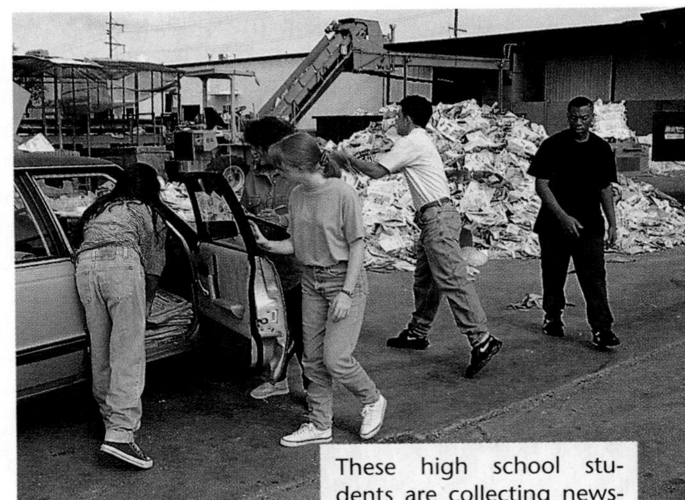
These high school students are collecting newspapers for recycling. Some cities have residential collection programs for used paper, glass, and plastic.

This **New York** landfill provides a feast for hungry seagulls. For humans, the landfill is a warning about the burdens on our environment. How much trash do you think you throw away in one week?

The earth is running out of space for all this garbage. To cut down on the amount of trash piling up around us, one major thing you can do is follow a simple strategy called "Reduce, reuse, recycle."

1. ***Reduce*** the amounts of disposable goods, water, and energy you use. Not only will this help the environment, it will also save money. For example, write on both sides of notebook paper, so that you will use only half as

much. If you go shopping, bring a cloth bag in which to carry goods home. While brushing your teeth or washing dishes, turn the water off while you are not actually using it. Turn the lights off when you leave a room.

2. ***Reuse*** products you already have instead of throwing them out and buying new ones. For instance, if you take your lunch to school in a brown bag, fold up the bag and use it again the next day. If you go on a picnic and use plastic forks and spoons, do not throw them out. Take them home, wash them, and use them again for your next picnic.

3. ***Recycle*** items that you can no longer reuse, and use recycled products. Most areas now have recycling centers or even residential pick-up for items such as cans, glass bottles and jars, newspapers, cardboard, and plastic. Set aside a place in your home and school to put goods to be recycled. Even if everyone recycled all the time, however, it would do little good if no one ever used products made from recycled materials! Today, many such products are available. Even some notebook paper is made from recycled paper.

Should you wish to take your efforts a little further, you can become involved with environmental groups that share your concerns. Such organizations can make your voice heard and can bring people together to make an even greater difference.

SCHOOL BUS

Some high school students spend a state-sponsored field trip cleaning a Texas beach. About 8,000 volunteers picked up nearly 160 tons (145 metric tons) of trash from Texas beaches on one day in 1995. Many times that amount of trash has been collected since the start of the program in 1987.

Here are some of the many organizations committed to protecting Earth and its resources:

America the Beautiful Fund
Free Seeds for a Green Earth Program
219 Shoreham Bldg.
Washington, DC 20005

American Forests Association
Global Releaf Program
P.O. Box 2000
Washington, DC 20013

The Children's Rainforest, U.S.
P.O. Box 936
Lewiston, ME 04243–0936

Citizens Clearinghouse for Hazardous Waste
P.O. Box 6806
Falls Church, VA 22040

Earth Day USA
P.O. Box 470
Peterborough, NH 03458

EarthSavers
Dept. ESC
1400 16th St., N.W.
Washington, DC 20036

Educators for Social Responsibility
23 Garden St.
Cambridge, MA 02138

Greenpeace
1436 U St., N.W.
Washington, DC 20009

Kids Against Pollution (KAP)
Tenakill School
275 High St.
Closter, NJ 07624

Kids for a Clean Environment (Kids FACE)
P.O. Box 158254
Nashville, TN 37215

National Recycling Coalition
1727 King St., Suite 105
Alexandria, VA 22314–2720

National Wildlife Federation
1400 16th St.
Washington, DC 20036–2266

Renew America
1400 16th St., N.W., Suite 710
Washington, DC 20036

Sierra Club
730 Polk St.
San Francisco, CA 94109

Student Environmental Action Coalition
P.O. Box 1168
Chapel Hill, NC 27514

Toxic Avengers
c/o El Puente
211 South 4th St.
Brooklyn, NY 11211

World Resources Institute
1709 New York Ave., N.W.
Washington, DC 20006

Your Turn

Organize a clean-up of your school grounds. With several classmates, pick up any trash you find and put it in the garbage. Make a list of five items you pick up. **Explain** how each item on your list might have been reduced (not used at all), reused, or recycled.

sea. Such pollution can contaminate fish and shellfish and can make people sick or even poison them.

Some water, such as that of rapidly flowing rivers, recovers quickly from pollution unless chemicals are present. Water that circulates slowly, in deep lakes or groundwater, may stay polluted for centuries.

Air Pollution Though it is often taken for granted, air also is a natural resource. It is an essential gas that we breathe, it protects us from harmful radiation, and it helps maintain the temperature balance of the planet. Global winds also link ecosystems in remote regions of the planet. For example, dust from Africa travels to the Amazonian rain forests, where it may help replace nutrients in the soil. Although we may have an endless supply of air, our atmosphere is subject to change from human pollution.

Air pollution, or smog, is a serious problem facing many large urban and industrial areas. *Smog* originally meant "smoke plus fog," but it now refers to both industrial air pollution and **photochemical smog**. Photochemical smog is produced when sunlight interacts with exhaust gases to produce visible and dangerous air pollution.

Automobiles release exhaust fumes, and factories pour smoke and fumes into the atmosphere. Some cities, such as Mexico City and Los Angeles, have become infamous for their smog. In many developed countries, clean air laws and modern technology have slowed or reduced air pollution. In their rush to industrialize, however, most developing nations face increasing smog problems in their cities.

Acids released by industrial smokestacks combine with water vapor in the atmosphere, causing **acid rain**. These acid droplets can harm lakes, forests, and human health. Often, the industries of one nation produce the pollutants that cause acid rain in another nation hundreds of miles downwind. Industries in Germany and in Poland may harm lakes and forests in Sweden and in Finland.

Global Air Pollution Air pollution is a global problem that knows no political boundaries and has the potential to change the earth's climate. In the upper atmosphere is a region known as the **ozone** layer. Ozone (O_3) is a gas formed from an interaction between oxygen and sunlight. The ozone layer protects life beneath it by filtering out dangerous ultraviolet solar radiation. Recent evidence collected from satellites, high-flying research aircraft, and weather balloons indicates that the ozone layer has thinned over the Antarctic. Here, there is a "hole" in the ozone layer. As this ozone hole has grown, skin cancer rates in the Southern Hemisphere have risen.

Scientists also are studying the upper atmosphere over the Arctic region for evidence of an ozone hole. Should an Arctic ozone hole exist, skin cancer rates in North America and Europe are likely to increase. Scientists also are very concerned about the effect of the loss of ozone protection on the rich Arctic and Antarctic waters, where the food chain may be damaged by increased ultra-

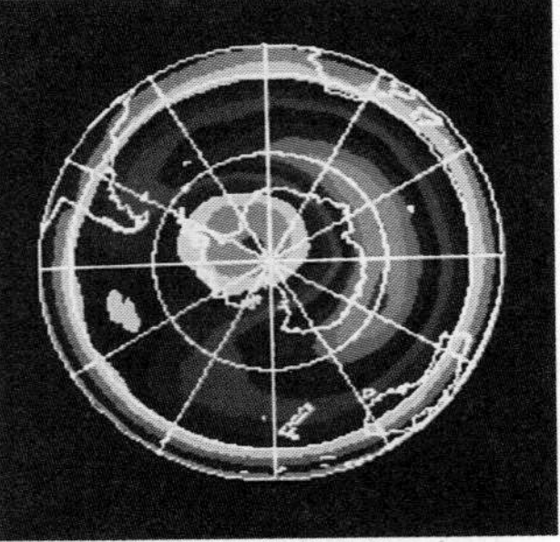

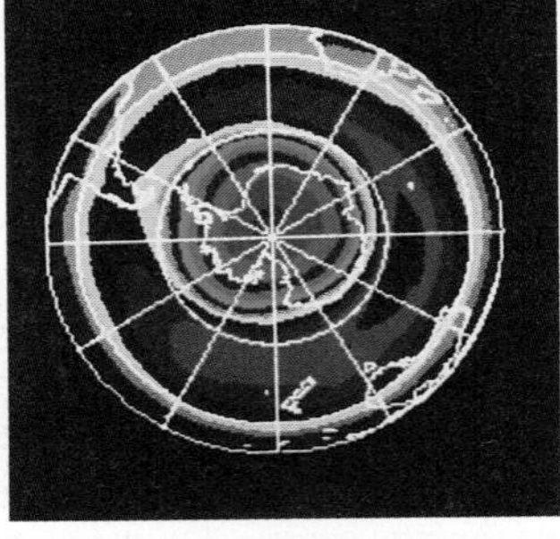

Russian factories belch pollution into the air around **Murmansk** (above). Scientists are worried that some pollutants being pumped into the atmosphere are damaging the earth's ozone layer. In the satellite images at the right, blue shades show areas with lower ozone levels than areas with brown and yellow shades. The image at the bottom shows much less ozone over Antarctica in 1987 than there was eight years earlier (top).

violet radiation. Increased ultraviolet radiation may damage or kill plants and bacteria. Plants and bacteria are essential to continued life on Earth.

A group of chemicals called CFCs (chlorofluorocarbons) apparently rise in the atmosphere and set off chemical reactions that destroy the ozone layer. CFCs are released from air conditioners, refrigerators, liquid cleaners, and plastic foams. Concerned nations, including the United States, Canada, and some European countries, have agreed to decrease the manufacture of products that release CFCs. Unfortunately, some developing nations find that substitutes for CFCs are too expensive to produce.

Global warming caused by the greenhouse effect may be a further threat to the atmosphere. The greenhouse effect is the result of an apparent buildup of carbon dioxide (CO_2) in the lower atmosphere. This buildup traps heat and causes a long-term warming of the planet. Many scientists have noted that massive coal burning causes increased heat to be trapped in the atmosphere.

When the earth has natural fluctuations in carbon dioxide, the climate changes. Ice ages occurred when carbon dioxide content was low. Warmer periods of the earth's history occurred when carbon dioxide had increased in the atmosphere. Natural conditions such as volcanic activity and forest fires release great amounts of carbon dioxide. Humans also produce carbon dioxide by burning fuels. An increase in carbon dioxide in the atmosphere over the past two centuries coincides with industrial development.

Although we cannot be sure of the effects of global warming and climate change, some scientists warn that sea levels will rise, some types of vegetation will die, and entire ecosystems will be destroyed. Scientists say two ways to protect ourselves from the greenhouse effect are to decrease the burning of fuels that pollute and to stop clearing the world's forests.

2 SECTION REVIEW

1. How are water quantity and water quality affected by a growing world population?
2. Why is there concern that the earth's climate might be changing?
3. **Critical Thinking** **Decide** whether clean air is a renewable or nonrenewable resource, and explain your answer.

3 ENERGY RESOURCES

FOCUS

- *What fossil fuels are used as major energy sources?*
- *Why is the use of nuclear energy as a power source declining?*
- *What are some renewable energy sources?*

Fossil Fuels Coal, petroleum, and natural gas are known as **fossil fuels**. Fossil fuels are believed to have been formed slowly from the buried remains of prehistoric plants and animals. Coal originates from ancient swamps and bogs; petroleum and natural gas were formed from ancient marine plants and animals.

Since the Industrial Revolution, great amounts of fossil-fuel energy have been needed for factories, transportation, and farms. During the previous two centuries, coal was the major resource. In this century, petroleum and natural gas have replaced coal as the preferred energy sources. In the next century, as petroleum and gas supplies decline, what will be the major energy resource?

Coal is still an important source of energy. Most of it is mined and burned to produce electricity. Along with iron ore, it is also used to make steel, a basic material for all industrial nations. Coal is the source of many chemical products as well, such as synthetic rubber, synthetic fabrics, plastics, and paints.

Because coal is a solid fuel, it has certain disadvantages, and burning it causes acid rain and air pollution. Through expensive liquification and gasification processes, coal can be changed into a cleaner-burning liquid fuel or gas. The nations with the greatest coal reserves include the United States, Russia, China, and Australia.

The main use of petroleum is for liquid fuels such as gasoline, diesel, and jet fuel. Petroleum also provides fuel oils for heat and electrical generation. Many chemicals also are produced from petroleum. These **petrochemicals** include chemical fertilizers, pesticides, food additives, explosives, and medicines. No nation can be an industrial power without a dependable source of petroleum and a petrochemical industry. Petroleum is a

Nuclear power plant

Natural gas flame

Coal mine

Offshore oil rig

nonrenewable resource, however, and the world's supply may last only another 50 years or so. About one-fourth of the world's petroleum comes from beneath coastal waters, and this percentage may increase as the reserves on land are depleted.

The largest petroleum reserves are found in Saudi Arabia and in the neighboring Persian Gulf countries. In the United States, the largest oil reserves are in Alaska, Texas, and Louisiana. Oil is shipped mainly by sea in large tankers, and it is the leading commodity traded in the world today. Oil and oil products also are a major cause of air and marine pollution.

The use of natural gas is growing rapidly. Natural gas often is found in the same wells with oil, but it may also be found separately. The largest reserves of natural gas are in Russia and the Persian Gulf nations.

Natural gas is the cleanest-burning fossil fuel and can be shipped great distances through pipelines. In fact, there are more than 1 million miles of natural-gas pipeline in the United States. Natural gas drilled in Texas heats homes in Massachusetts, and people cooking with gas stoves in California receive natural gas from New Mexico. The United States has abundant supplies of natural gas. Like petroleum reserves, however, they are limited and may become more expensive and scarce over the next several decades.

Nuclear Energy The world's first nuclear power plant began producing electricity in Britain in 1956. Today, there are more than 400 nuclear power plants scattered around the world in more than 25 nations. The United States has just over 100 operating nuclear power plants, which produce about 20 percent of the nation's electricity. In France, nuclear power plants produce about three-fourths of the electricity. In other nations, such as South Korea and Hungary, nuclear power plants produce almost half of the electricity. Japan depends heavily on nuclear energy.

Originally, nuclear power was seen as a clean, inexpensive power source. Today, however, the nuclear power industry faces two serious problems: nuclear accidents and nuclear waste. In 1986, a nuclear accident occurred at Chernobyl, in the former Soviet Union. Many people died from radiation, and thousands of others who were contaminated will have shorter lives. Many square miles of land around Chernobyl were evacuated, and radiation spread over much of Europe. The accident was devastating to agriculture in the region.

All nuclear power plants produce nuclear waste, and exposure to these radioactive substances is dangerous. Radioactive wastes decay slowly, which means that they must be stored and monitored for thousands of years. The best storage method developed so far is to bury the waste in sealed containers deep underground in solid rock. Transporting these waste products to storage facilities involves a risk of accidents.

Since the Chernobyl accident, public opposition to expansion of nuclear power has grown. Some nations, such as Sweden and Germany, are phasing out existing power plants. The United States, meanwhile, foresees almost no growth in this power source. A few nations, including Denmark and New Zealand, have declared themselves "nuclear free" and have no intention of ever building nuclear power plants.

Renewable Energy Sources Flowing water is an important energy resource. Hydroelectric

Geothermal power plant

Hydroelectric dam

Wind turbines

Solar power panels

power, or hydropower, produced by the force of running water is a renewable and relatively pollution-free source of electrical energy.

Development of hydropower requires rivers with a steady flow, drops in elevation, and the construction of dams, which are very expensive. In some areas, such as the Columbia River in the northwestern United States, numerous dams provide great amounts of electricity. Overall, 10 percent of the United States' electricity is produced by hydropower. Russia, Canada, New Zealand, and Norway also produce large amounts of hydroelectricity. Unfortunately, hydropower can change the natural flow of rivers, flooding vast areas under reservoirs and interfering with the migration of fish, such as salmon. It also depletes farmland.

Another source of renewable energy is underground heat, or **geothermal energy**. Geothermal energy originates in the hot interior of the planet and escapes to the surface through hot springs and steam vents. The escaping steam can be used to generate electricity. Today, geothermal plants produce electricity in countries all over the world, including Ethiopia, Kenya, Nicaragua, India, Indonesia, the United States, Japan, New Zealand, Italy, Mexico, and Iceland.

A third renewable energy source is the sun. Solar energy involves capturing the sun's light energy and converting it into heat or electricity. Many homes in sunny climates now have solar panels that collect sunshine and convert it into heat for hot water. An experimental solar electric plant is now operating in the desert of southern California. If this plant is successful, many more may be built in the future.

For centuries, wind power has been used to grind grain and pump water from wells with windmills and to power sailing ships. Wind also can be harnessed to produce electricity by the use of machines called aerogenerators. On a small scale, aerogenerators may be used on tropical islands to generate electricity for a school. On a much larger scale, "wind farms" with hundreds of aerogenerators can produce electricity for much larger areas. California, for example, supplies 1 percent of its electricity needs from wind power.

In the air, in oceans, underground, and in living things, Earth has virtually unlimited energy resources. Many of these energy sources are nonrenewable, and they may be in short supply in the future. We must continue to seek and improve alternative energy sources and use them more efficiently. Only then can we maintain our energy supply and also reduce environmental hazards.

3 SECTION REVIEW

1. What are the main uses of coal, petroleum, and natural gas?
2. What are the positive and negative aspects of nuclear energy as a power source?
3. **Critical Thinking** **Imagine** what might happen if we do not seek alternative, renewable energy sources.

4 POPULATION AND RESOURCES

FOCUS

- *Why can a rapid population growth rate be a problem for a nation?*
- *What are some consequences of uneven resource distribution?*

Population Growth As you have learned, the world's population is not evenly distributed. Extreme climates, rugged terrain, and lack of water and fertile soil prevent many areas from being populated. Most of the productive regions of the world today, however, are populated.

Population growth rates are important to the future of the world. Some nations, such as Kenya and Pakistan, have rapid growth rates. Many industrial nations have slower growth rates. These include the United States, Canada, Russia, Australia, New Zealand, and most European countries. Some nations, such as China and Indonesia, are trying hard to control population growth.

Why would a nation want to control population growth? Control of population growth would contribute to a higher standard of living for its citizens. Nations must provide jobs, education, roads, and medical care for its people. A slower growth rate will put less strain on the limited resources of that nation. The nations with the highest growth rates are generally the poorest nations.

In contrast, some nations, such as Singapore, Thailand, Japan, and several European countries, have been so successful at controlling population that they are now concerned about low birthrates. With low birthrates, there are fewer young people entering the work force. This may lower productivity and consumer spending. At the same time, the percentage of elderly people is growing. This may cause an economic strain because of the increased cost of health care.

Balancing Population and Resources The present population of the world is more than 5.5 billion. If current growth rates continue, it is estimated that the world's population will be 8 billion by 2020. More than 1 million people are added to the world every five days. Can the earth support this many people?

Natural resources, like people, are not distributed evenly. Some regions have surplus water, food, or energy resources. Other areas have shortages. Uneven distribution explains why we need world trade, but it also leads to world conflict as nations try to secure needed resources. The people of the world must share and use the planet's resources wisely.

Many people are optimistic about the future. They believe that as the earth's population increases, scientists will discover ways to produce new kinds of foods. Farmers will increase crop yields through new methods, more food and water will be taken from the sea, and high-rise buildings will permit more people to live in less land space.

Yet, other people question whether we have already reached the point of limitation for important resources and food production. Will substitutes be found for many of our dwindling natural resources? Can expanding food production keep up with the population growth? Can we continue to pollute the environment without harming our planet, our nation, and ourselves? Will countries continue to trade valuable resources? Will the richer nations share their wealth and resources with the less fortunate nations?

It is important to remember that people sometimes have conflicting interests. Politicians, economists, environmentalists, scientists, wealthy people, and poor people all have different sets of priorities. A better understanding of the world's geography will help us face these challenging issues. Turn to the "Planet Watch" feature on pages 90–91 for information about how you can help protect the environment and conserve our resources.

4 SECTION REVIEW

1. Why do some nations try to practice population control?
2. What human activities occur due to the unequal distribution of resources?
3. **Critical Thinking** **Evaluate** which would be of greater concern to a country: a high growth rate or a low growth rate.

CHAPTER 9 CHECK

A nuclear power plant towering over the French countryside

Reviewing the Main Ideas

1. With an ever-growing population, the world faces a serious problem of resource depletion. Air, water, forests, soil, and fossil fuels are some of the natural resources people must work together to preserve and use responsibly.
2. Human activities, such as industrialization, urbanization, and the burning of fossil fuels, have caused serious environmental pollution. Global warming, caused by the trapping of greenhouse gases in the atmosphere, may result in climate change.
3. The unequal and uneven distribution of resources results in world trade but also in world conflict. In an interdependent world, we must all work together to secure the future health of our shared planet.

Building a Vocabulary

1. Define *renewable* and *nonrenewable resources.*
2. How do soil salinization and desertification threaten the world's soil?
3. What is the difference between reforestation and deforestation?
4. Define *desalinization.* Why is this process not always practical?
5. What is acid rain? Why is it of concern to nations other than the one producing it?

Recalling and Reviewing

1. What are two methods farmers use to prevent soil erosion? How do farmers protect the soil's fertility?
2. Forests are renewable resources, but it is still necessary to conserve them. Why?
3. Why is the recent discovery of a hole in the ozone layer a cause for concern?
4. How does the burning of fossil fuels contribute to the greenhouse effect?
5. Why are some nations concerned about the success of their attempts to control population?

Thinking Critically

1. What is the difference between environmental problems in developing and developed nations? How is pollution related to economic development?
2. The United States currently imports 50 percent of its oil. By the year 2010, the United States may be importing as much as 75 percent of its oil. What might be the consequences of this increase?

Learning About Your Local Geography

Individual Project

Find out about any future plans for development in your community. Then, make inquiries about the kinds of environmental studies that are completed before a development plan is approved by city or county officials. Present your information in a brief report accompanied by a flowchart.

Using the Five Themes of Geography

Find out how your home is heated and/or air conditioned, and write a paragraph about how the geographic themes of **region, movement,** and **human-environment interaction** play a role. How does the region you live in affect your home's energy needs and the forms of energy that are available? How does the energy move from its place of origin to your home? What consequences might the forms of energy you use have for the environment? What could you do to reduce the amount of energy you use?

SKILLS

Analyzing Points of View

Have you ever had a disagreement with a friend in which you were both convinced that you were right and the other was wrong? Sometimes it is difficult to get the facts on a particular topic, and sometimes the facts seem to contradict each other. As a member of our global society, you will need to learn how to understand the various, and often opposing, points of view held by geographers. Ultimately, you will have to make decisions about which opinions you agree with and what actions you will take to support your opinions.

A topic on which many geographic experts have different points of view is the greenhouse effect. Andrew Revkin, the author of a book on global warming, believes that global warming could have some very serious, negative consequences. He states:

"If levels of carbon dioxide continue to rise at the current rate, there is a significant chance that disruptive climate shifts will occur within the lifetimes of children born today. The list of potential consequences is familiar: coastal flooding, searing droughts, wars over shrinking water supplies, accelerated extinction of species."

In contrast, Robert C. Balling, the author of another book on global warming, believes that if the earth is warming, it is not warming much, and small increases in the earth's temperature might even be good:

"There is a large amount of . . . evidence suggesting that . . . the highly touted greenhouse disaster is most improbable. The empirical evidence suggests that our future will see a rise in temperature of approximately 1.0°C (1.8°F) . . . the earth will probably be wetter, cloudier, with substantial increases in soil moisture; droughts may diminish in frequency, duration, and intensity. . . . In the case of the greenhouse effect, we can find legitimate benefits from our environmental impact."

This difference of opinion poses difficult questions. Is global warming a threat large enough to justify spending a large amount of money trying to control it? If so, who should take the primary responsibility for this control? These debates are far from being resolved.

PRACTICING THE SKILL

Another topic over which geographers differ is population growth. As you know, the earth's human population is growing rapidly. Not everyone, however, agrees on what effects this rapid growth will have. Read the two quotations below, and then answer the questions that follow.

"There are limits to the rates at which [the] human population . . . can use materials and energy, and there are limits to the rates at which wastes can be emitted without harm. . . . Human society is now using resources and producing wastes at rates that are not sustainable. . . . Even with much more efficient institutions and technologies, the limits of the earth's ability to support population . . . are close at hand."

Meadows, Meadows,
and Randers,
Population Researchers

"The effect of higher population density actually seems to be positive. . . . The central benefit of more people in a more developed world is that there are more . . . people to invent new ideas. . . . If population had not increased about 8,000 years ago and made hunting and gathering become less productive, we would be having wild roots, rabbits, and berries for lunch."

Julian Simon, Economist

1. What do the authors of each quotation think about population growth? About what do the two quotations disagree?

2. What are the implications of each point of view for how governments should think about and make decisions on population growth?

3. What additional information might help you decide which quotation you agree with? How might you find this information?

BUILDING YOUR PORTFOLIO

Individually or in a group, complete one of the following projects to show your understanding of the geography concepts involved. Both projects are aimed at class presentations. Your report may also take the form of a community education project. You will need to speak with your teacher about specific activities.

A *When Different Cultures Meet*

From the point of view of many Americans, an important example of different cultures coming into contact has been the meeting of Native Americans with Europeans and European Americans. This contact was not a one-time event—it has lasted several hundred years. These cultures are still interacting today.

Research the results of the meeting of one or more groups of Native Americans with one or more groups of European Americans. Prepare an oral report, supported by written and visual materials, about your findings. To complete your report, think about the following questions.

1. How were Native American cultures different from European cultures? Demonstrate how Native American cultures differed from European cultures, and how these cultural differences caused conflicts. Focus on three or four culture traits. Use illustrations, with captions, to present this information.
2. How did Native American populations change in size after contact with Europeans? What factors were responsible for these changes? In a graph or chart, show how the size of Native American populations has changed over time, from the first meeting with Europeans to today.
3. Did Native Americans and Europeans have similar views about the physical environment? How might different views about the use of the environment cause problems for people who want to share the same environment? Summarize your findings in a few statements.

Organize the information you have gathered, and present your information to the rest of the class. Conclude with a few statements giving your reactions to what you have learned. Do you think there is any way that conflict between European and Native American cultures might have been avoided?

B *The Geography of Industry*

As you learned in this unit, many countries have industrialized or are industrializing. Investigate industrial development and its effects on a local level. Select a city or large town near where you live. For that city or town, identify one neighborhood or community that contains much industry, and one neighborhood or community in which there is very little industry. Prepare an oral report, supported by written and visual materials, comparing the two communities. To complete your report, think about the following questions.

1. Do the two communities have similar ethnic compositions? Make two bar graphs. One should show the proportions of people of various ethnic groups living in the industrial community. The other one should show the proportions of people of various ethnic groups living in the nonindustrial community.
2. What are the economic activities of people in each community? Make two lists, one showing the most common occupations of people who live in the industrial community, and one showing the most common occupations of people who live in the nonindustrial community.
3. How do the environmental problems faced by the people in each community differ? On a poster, illustrate the environmental hazards in the industrial community. Write a few sentences on the poster about how these hazards compare to those in the nonindustrial community.

Organize the information you have gathered, and present your information to the rest of the class. As you do so, suggest some possible reasons for any differences you have discovered between the two communities. Also, discuss the implications of your findings for the city government and for the members of both communities.

What's in a Region?

Think for a moment about where you live, where your school is located, and where you usually shop. These places are all part of your neighborhood. In geographic terms, your neighborhood can be called a **region**. A region is an area that has more common characteristics than differences. Each region has location, area, and unique descriptive elements.

Geographers classify places by region. Dividing the world into manageable parts allows geographers to compare one region to another, or to study aspects of a single region.

Try to list the many regions in which you live. You live on a continent, in a country, in a state, in a county, and in a city or town. Your zip code, your telephone area code, and your school district are regions. All of these regions can be mapped.

Regions often are organized into smaller regions called **subregions**. For example, Africa is an enormous region that can be divided into several subregions: North Africa, West Africa, East Africa, Central Africa, and Southern Africa. Each of these subregions can be further organized into smaller subregions. Your hometown may be organized into several subregions, such as fire, police, and school districts.

Cultural characteristics, such as religion, help define regions. These Buddhist monks are praying in **Thailand**. Buddhism is a major religion in **Southeast Asia**.

High mountain ranges, such as the Alps here in northern **Italy**, are found in many regions of the world. Ways of life, terrain, and climates in mountainous regions may differ greatly from those in regions with lower elevations.

Regional Characteristics

There are limitless characteristics for forming and mapping regions. The four most common types of determining characteristics are physical, political, economic, and cultural. Physical regions are based on the natural features of the earth, such as continents, seas, landforms, and climates. Political regions are based on the nations of the world and their subregions, such as states, provinces, and cities. Economic regions are defined by activities such as agriculture and other industries, or they may center around transportation systems and world trade routes. Cultural regions may classify areas by language, religion, or ethnic groups.

Regional Boundaries

All regions have boundaries, or borders. Boundaries indicate where the characteristics of one region meet the differing characteristics of a neighboring region. The boundaries of some regions are distinct and can be shown as lines on a map. For example, look at the world political map on pages A4–A5. Notice the boundaries that separate countries.

Other boundaries may also appear on maps. For example, turn to the United States physical map on pages A6–A7. Coastlines,

Regions of the World

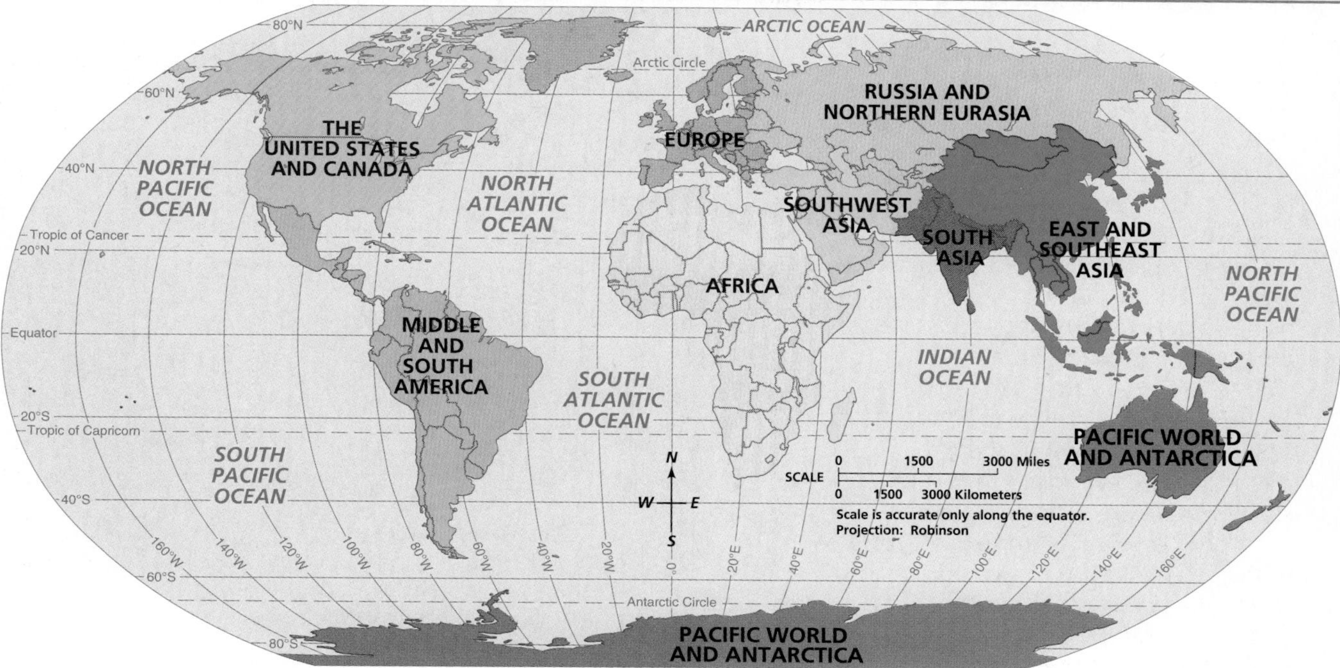
ARCTIC OCEAN
Arctic Circle
80°N
60°N
40°N
Tropic of Cancer
20°N
Equator
20°S
Tropic of Capricorn
40°S
60°S
80°S
Antarctic Circle
THE UNITED STATES AND CANADA
RUSSIA AND NORTHERN EURASIA
EUROPE
SOUTHWEST ASIA
SOUTH ASIA
EAST AND SOUTHEAST ASIA
AFRICA
MIDDLE AND SOUTH AMERICA
PACIFIC WORLD AND ANTARCTICA
PACIFIC WORLD AND ANTARCTICA
NORTH PACIFIC OCEAN
NORTH ATLANTIC OCEAN
NORTH PACIFIC OCEAN
SOUTH ATLANTIC OCEAN
INDIAN OCEAN
SOUTH PACIFIC OCEAN
N
W
E
S
SCALE
0 1500 3000 Miles
0 1500 3000 Kilometers
Scale is accurate only along the equator.
Projection: Robinson
160°W
140°W
120°W
100°W
80°W
60°W
40°W
20°W
0°
20°E
40°E
60°E
80°E
100°E
120°E
140°E
160°E

In dividing the world into nine regions, the authors of this textbook took into account physical, cultural, political, and economic factors. As you read this textbook, think about how you would have divided the world into regions.

mountain ranges, and rivers show regional boundaries. Some regional boundaries are less clear, however. Areas called **transition zones** exist when different characteristics such as language, religion, and economic activity overlap into the same area. For example, when a city's suburbs are expanding into a rural area, a transition zone forms. During this period, it is difficult to determine where the exact boundary between the rural and urban areas lies.

Types of Regions

Geographers identify two types of regions. The first type is a **uniform region**—an area distinguished by one or more common characteristics. The country of Mexico is a uniform region because its people share a common government, language, and culture.

The second type, a **nodal region**, is an area distinguished by the movement or activities that take place in and around it. A node is the region's center point or point of activity. Paris, France, is at the center of a nodal region defined by the distances that goods, services, and workers travel to and from the city each day. A shopping center, an airport, or a seaport also can be a region's node.

Often, uniform and nodal regions overlap to form complex world regions. In Units 3 through 11 of this textbook, nine regions form the basis of our geographic study of the world. (See the map above.) Each of these major world regions has unifying characteristics based on physical location and on historical, cultural, economic, and political elements.

Economic activities, such as oil production here in **Saudi Arabia**, are important characteristics of the world's regions.

UNIT 3

The UNITED STATES and CANADA

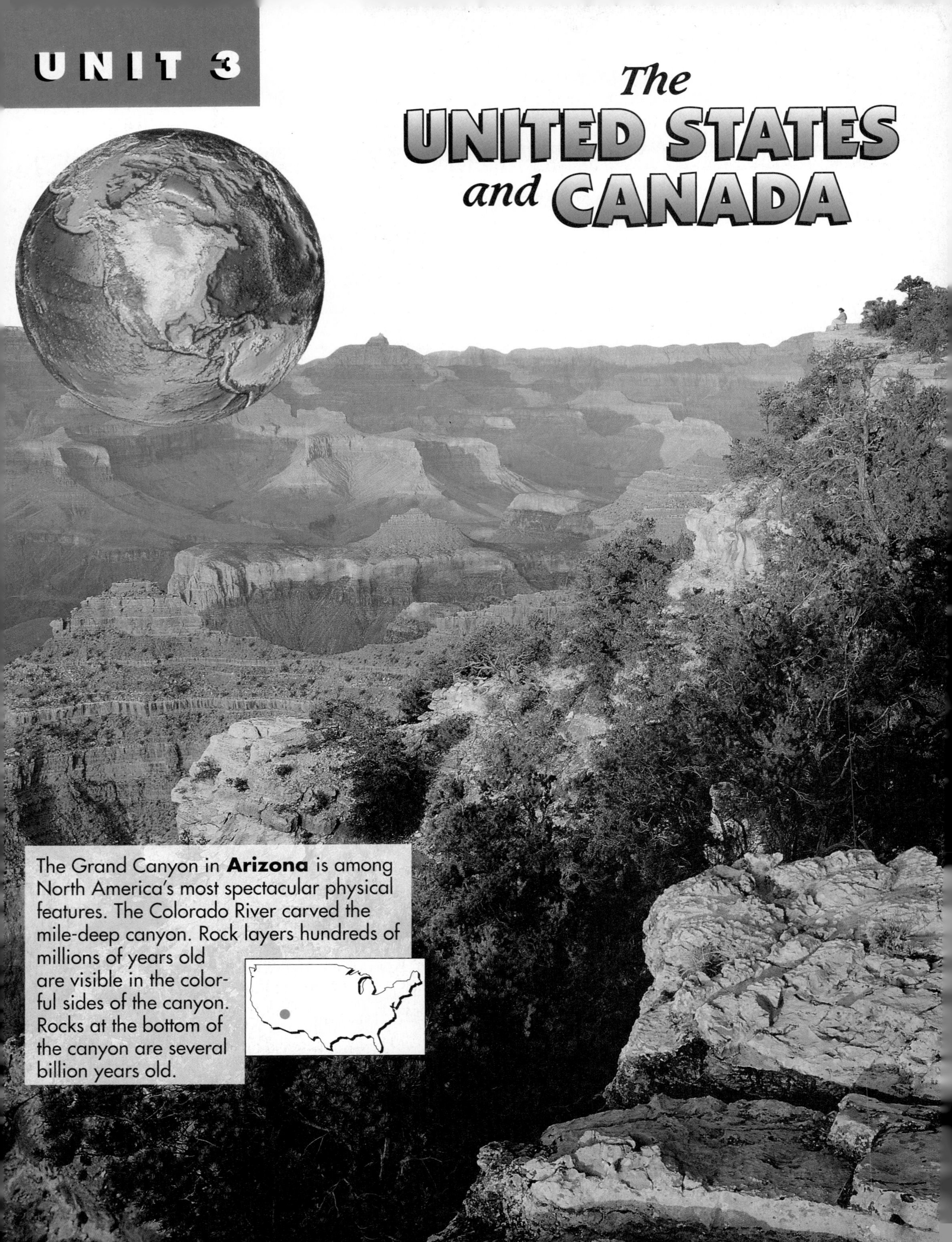

The Grand Canyon in **Arizona** is among North America's most spectacular physical features. The Colorado River carved the mile-deep canyon. Rock layers hundreds of millions of years old are visible in the colorful sides of the canyon. Rocks at the bottom of the canyon are several billion years old.

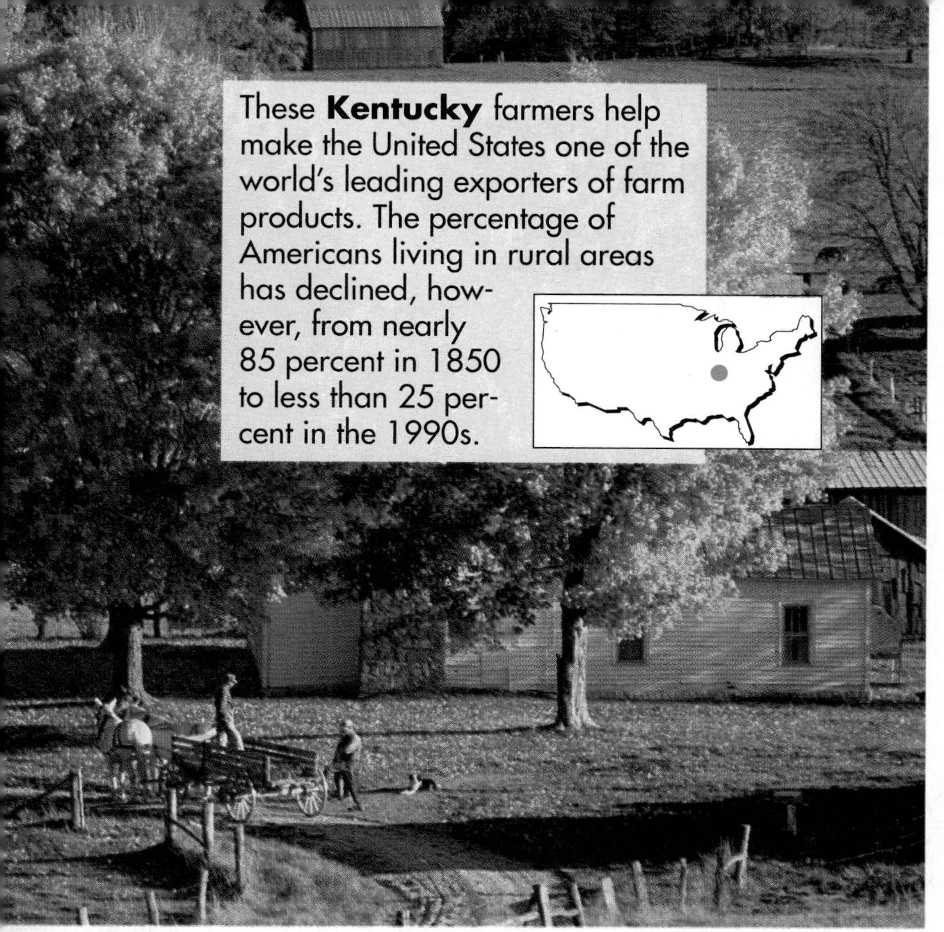

These **Kentucky** farmers help make the United States one of the world's leading exporters of farm products. The percentage of Americans living in rural areas has declined, however, from nearly 85 percent in 1850 to less than 25 percent in the 1990s.

UNIT 3 CHAPTERS

Olympic Stadium in **Montreal, Canada,** is home to the Expos Major League baseball team. More of the Montreal metropolitan area, located in Quebec, can be seen in the background. Montreal is Canada's second largest urban area and is one of the world's largest French-speaking cities.

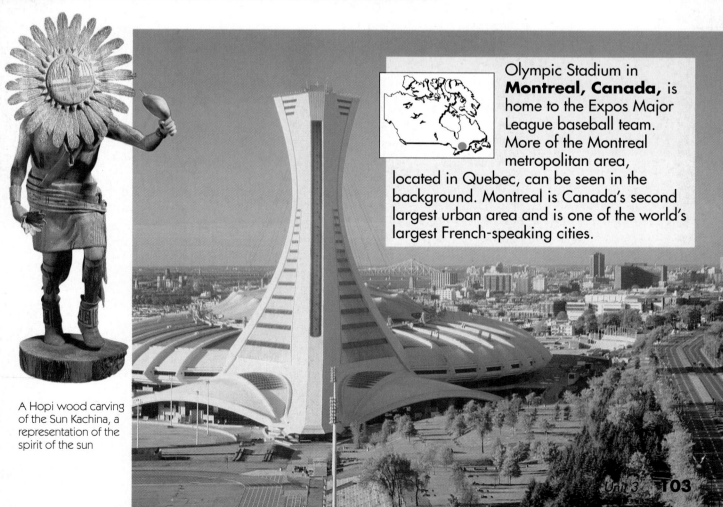

A Hopi wood carving of the Sun Kachina, a representation of the spirit of the sun

One World, Many Voices

Literature

Wendy Rose (1948–) was born in California. She began her writing career in 1967. Many of her published books of poetry reflect her Native American heritage and emphasize environmental themes. "Before I learned how to write I expressed myself through making up songs. It has always been a tradition of my people [the Hopi of Arizona] to celebrate everything in life with song. For all occasions, both happy and sad, I sang my feelings."

Loo-Wit*

The way they do
this old woman
no longer cares
what we think
but spits
her black tobacco
any which way
stretching
full length
from her bumpy bed.
Finally up
she sprinkles
ash on the snow,
cold buttes
promise nothing
but the walk
of winter.
Centuries of cedar
have bound her
to earth,
huckleberry ropes
lay prickly
on her neck.
Around her
machinery growls,
snarls and ploughs
great patches of her skin
She crouches in the north,
her trembling
the source
of dawn.

Light appears
with the shudder
of her slopes,
the movement
of her arm.
Blackberries unravel,
stones dislodge;
it's not as if
they were not warned.

She was sleeping
but she heard the boot scrape, the creaking floor,
felt the pull of the blanket
from her thin shoulder.
With one hand free she finds her
weapons
and raises them high; clearing the
twigs from her throat
she sings, she sings, shaking the
sky
like a blanket about her
Loo-wit sings and sings and sings!

****Loo-Wit is a name that Native Americans of the Cowlitz group gave to Mount St. Helens, a volcano in Washington state. Mount St. Helens erupted violently in 1980 after being inactive for more than 120 years.***

Interpreting Literature

1. Why is this image of an elderly woman awakening from her sleep appropriate for a volcano such as Mount St. Helens?
2. How does Loo-Wit warn of the impending eruption? What action by Loo-Wit represents the eruption itself?

FOR THE RECORD

Thomas Jefferson, third president of the United States, played an important role in acquiring the vast territory known as the Louisiana Purchase. Bought from France in 1803, this land extended the boundary of the new nation from the Mississippi River to present-day Montana. Jefferson commissioned young army officers Meriwether Lewis and William Clark to lead an exploration team across this region. A letter of instructions to Lewis reveals the president's keen interest in the geography of the territory.

To Meriwether Lewis, esquire, Captain of the 1st regiment of infantry of the United States of America:

The object of your mission is to explore the Missouri river, & such principal stream of it, as, by its course & communication with the waters of the Pacific Ocean, may offer the most direct & practicable water communication across this continent, for the purposes of commerce.

Beginning at the mouth of the Missouri, you will take observations of latitude & longitude, at all remarkable points on the river, & especially at the mouths of rivers, at rapids, at islands & other places. . . . Other object worthy of notice will be

the soil & face of the country, its growth & vegetable productions . . .

the animals of the country . . .

the mineral productions of every kind . . .

Volcanic appearances.

climate as characterized by the thermometer, by the proportion of rainy, cloudy & clear days, by lightning, hail, snow, ice, by the access and recess of frost, by the winds prevailing at different seasons.

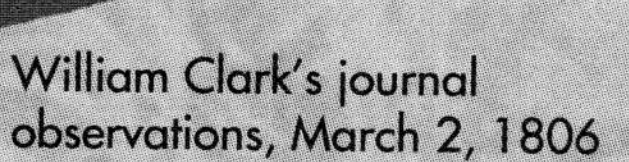

Missouri Historical Society Archives

William Clark's journal observations, March 2, 1806

Analyzing Primary Sources

1. What did President Jefferson consider the most important purpose of the Lewis and Clark expedition?
2. What elements of physical geography in the Louisiana Purchase territory were of special interest to Jefferson?

CHAPTER 10

Introducing the United States

GEOGRAPHY DICTIONARY

piedmont
Fall Line
basin
contiguous
Gulf Stream
seaboard
plantation
edge city
ghetto
trade deficit

SEAL OF THE PRESIDENT OF THE UNITED STATES
E PLURIBUS UNUM

Presidential Seal of the **United States**

The United States and Canada share the majority of the North American continent as well as the world's longest peaceful boundary. The United States shares a southern boundary with Mexico.

The United States consists of 50 states and often is referred to as America or the USA. It features a federal system in which the national government shares power with the states. The United States is a democracy, a form of government in which all citizens have a voice. Regarded as a "land of opportunity" throughout its history, the United States continues to attract millions of immigrants from around the world.

The United States ranks fourth in the world in land area and ranks third in population. The country faces oceans and seas in three directions, making its location ideal for international trade. With the breakup of the Soviet Union in the early 1990s, the United States remains the world's single undisputed military superpower.

This unit explores the diverse and complex geography of the United States and Canada. Chapters 10 through 15 examine the United States; Chapter 16 discusses Canada.

The U.S.–Canadian border between **New York** and **Quebec**

MAPPING THE REGION

THE UNITED STATES AND CANADA

RUSSIA
BERING SEA
Aleutian Islands
Bering Strait
Arctic Circle
ARCTIC OCEAN
BEAUFORT SEA
BROOKS RANGE
ALASKA (U.S.)
Mount McKinley (20,320 ft. 6194 m)
ALASKA RANGE
Yukon River
GULF OF ALASKA
YUKON TERRITORY
Mount Logan (19,850 ft. 6050 m)
Banks Island
Victoria Island
AMUNDSEN GULF
Queen Elizabeth Islands
Prince of Wales Island
Somerset Island
Baffin Island
Baffin Bay
GREENLAND (Den.)
ICELAND
Davis Strait
NUNAVUT
Prince Charles Island
Southampton Island
Coats Island
Mansel Island
Hudson Strait
LABRADOR SEA
NORTHWEST TERRITORIES
Great Bear Lake
Great Slave Lake
Mackenzie River
Liard River
CANADA
Lake Athabasca
Hudson Bay
Belcher Islands
James Bay
Akimiski I.
NEWFOUNDLAND
Newfoundland Island
Anticosti I.
GULF OF ST. LAWRENCE
PRINCE EDWARD ISLAND
Cape Breton Island
NOVA SCOTIA
NEW BRUNSWICK
QUEBEC
St. Lawrence River
CANADIAN SHIELD
Nelson River
Saskatchewan River
MANITOBA
SASKATCHEWAN
ALBERTA
Athabasca R.
Peace River
BRITISH COLUMBIA
Queen Charlotte Islands
COAST RANGES
ROCKY MOUNTAINS
GREAT PLAINS
Fraser River
Vancouver Island
Edmonton
Calgary
Vancouver
Lake Winnipeg
Winnipeg
Lake Nipigon
ONTARIO
Quebec
Montreal
Ottawa
Toronto
Seattle
Mount Rainier (14,410 ft. 4392 m)
Mount St. Helens (8364 ft. 2549 m)
CASCADE RANGE
WASH.
Columbia River
COLUMBIA PLATEAU
OREG.
IDAHO
Snake River
MONT.
Missouri River
N. DAK.
S. DAK.
WYO.
MINN.
WIS.
MICH.
Lake Superior
Lake Michigan
Lake Huron
Lake Ontario
Lake Erie
Niagara Falls
Mississippi R.
UNITED STATES
Mount Shasta (14,162 ft. 4317 m)
COAST RANGES
CENTRAL VALLEY
SIERRA NEVADA
CALIF.
San Francisco
GREAT BASIN
NEV.
Great Salt Lake
UTAH
Colorado River
N. Platte
S. Platte
NEBR.
IOWA
Chicago
Detroit
INTERIOR PLAINS
ILL.
IND.
OHIO
Ohio River
PA.
N.Y.
VT.
N.H.
MAINE
MASS.
Boston
R.I.
CONN.
Connecticut River
New York
Philadelphia
N.J.
DEL.
MD.
Washington, D.C.
Chesapeake Bay
W. VA.
VA.
ALLEGHENY PLATEAU
APPALACHIAN MOUNTAINS
CUMBERLAND PLATEAU
PIEDMONT
ATLANTIC COASTAL PLAIN
Denver
COLO.
Arkansas River
KANS.
MO.
OZARK PLATEAU
KY.
TENN.
N.C.
S.C.
Atlanta
GA.
ALA.
MISS.
Mississippi River
ARK.
OKLA.
Red River
DEATH VALLEY
COLORADO PLATEAU
MOJAVE DESERT
Los Angeles
San Diego
ARIZ.
Phoenix
Rio Grande
N. MEX.
Pecos River
TEXAS
Dallas
San Antonio
Houston
LA.
GULF COASTAL PLAIN
FLA.
Miami
Florida Keys
Straits of Florida
GULF OF MEXICO
GULF OF CALIFORNIA
PACIFIC OCEAN
ATLANTIC OCEAN
BERMUDA (U.K.)
BAHAMAS
Tropic of Cancer
CUBA
HAITI
DOMINICAN REPUBLIC
PUERTO RICO (U.S.)
MEXICO
BELIZE
GUATEMALA
HONDURAS
EL SALVADOR
NICARAGUA
COSTA RICA
PANAMA

Niihau
Kauai
Kaula
PACIFIC OCEAN
Oahu
Molokai
HAWAII
Lanai
Maui
Kahoolawe
Hawaii

SCALE
0 100 200 Miles
0 100 200 Kilometers
Projection: Albers Equal Area

To understand the relative location of Hawaii as well as the vast distance separating it from the rest of the United States, see the map on page A4.

SCALE
0 500 1000 Miles
0 500 1000 Kilometers
Projection: Azimuthal Equal Area

PHYSICAL–POLITICAL

LEGEND

Ice caps
⍟ National capitals
• Other cities

ELEVATION

FEET	METERS
13,120	4,000
6,560	2,000
1,640	500
656	200
(Sea level) 0	0 (Sea level)
Below sea level	Below sea level

MAPPING THE REGION

THE UNITED STATES AND CANADA

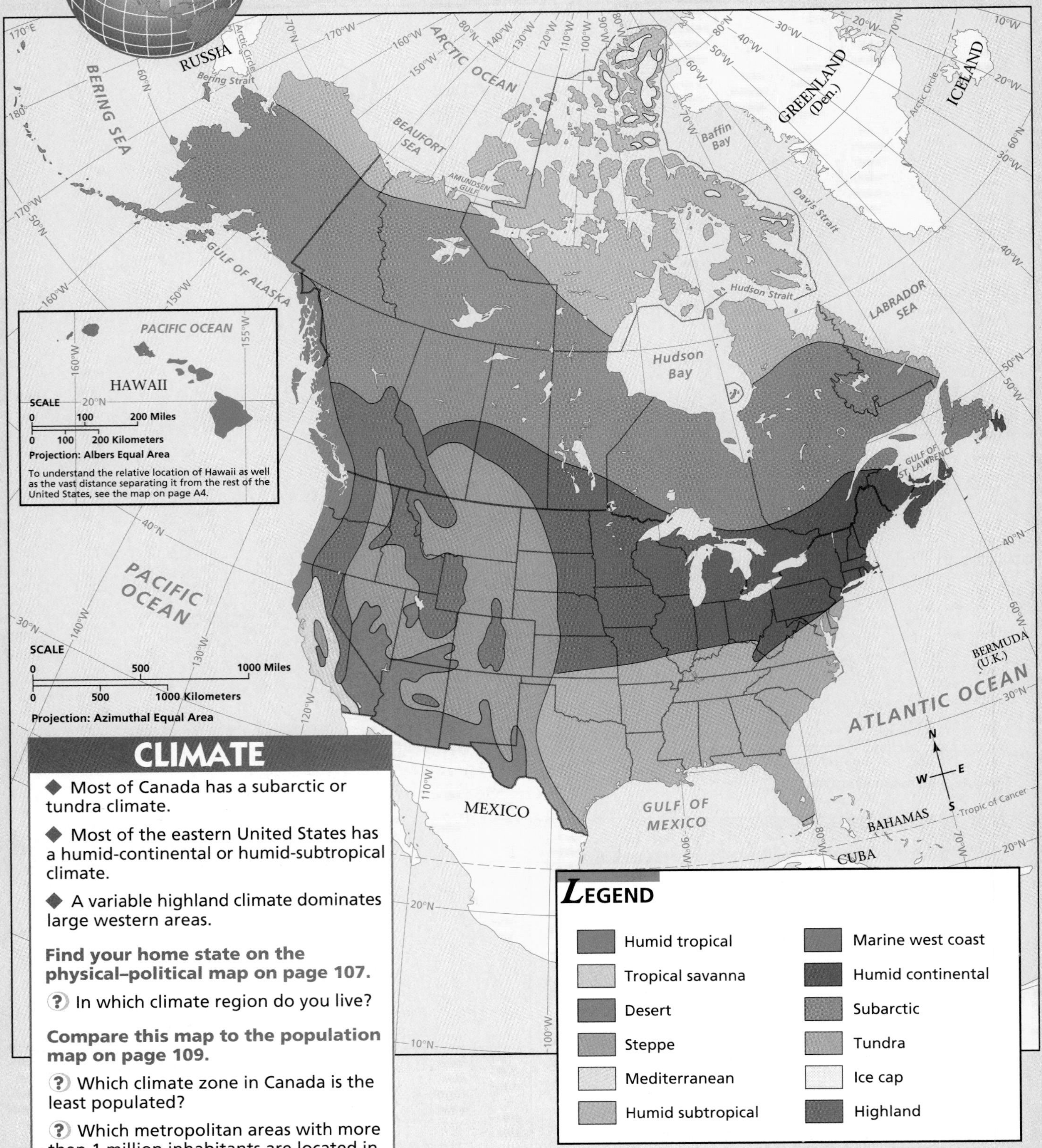

PACIFIC OCEAN

HAWAII

SCALE

0 100 200 Miles

0 100 200 Kilometers

Projection: Albers Equal Area

To understand the relative location of Hawaii as well as the vast distance separating it from the rest of the United States, see the map on page A4.

SCALE

0 500 1000 Miles

0 500 1000 Kilometers

Projection: Azimuthal Equal Area

RUSSIA

BERING SEA

ARCTIC OCEAN

BEAUFORT SEA

AMUNDSEN GULF

GULF OF ALASKA

GREENLAND (Den.)

ICELAND

Baffin Bay

Davis Strait

Hudson Strait

LABRADOR SEA

Hudson Bay

GULF OF ST. LAWRENCE

PACIFIC OCEAN

ATLANTIC OCEAN

BERMUDA (U.K.)

MEXICO

GULF OF MEXICO

BAHAMAS

CUBA

Tropic of Cancer

CLIMATE

◆ Most of Canada has a subarctic or tundra climate.

◆ Most of the eastern United States has a humid-continental or humid-subtropical climate.

◆ A variable highland climate dominates large western areas.

Find your home state on the physical–political map on page 107.

? In which climate region do you live?

Compare this map to the population map on page 109.

? Which climate zone in Canada is the least populated?

? Which metropolitan areas with more than 1 million inhabitants are located in a desert climate?

LEGEND

- Humid tropical
- Tropical savanna
- Desert
- Steppe
- Mediterranean
- Humid subtropical
- Marine west coast
- Humid continental
- Subarctic
- Tundra
- Ice cap
- Highland

MAPPING THE REGION

THE UNITED STATES AND CANADA

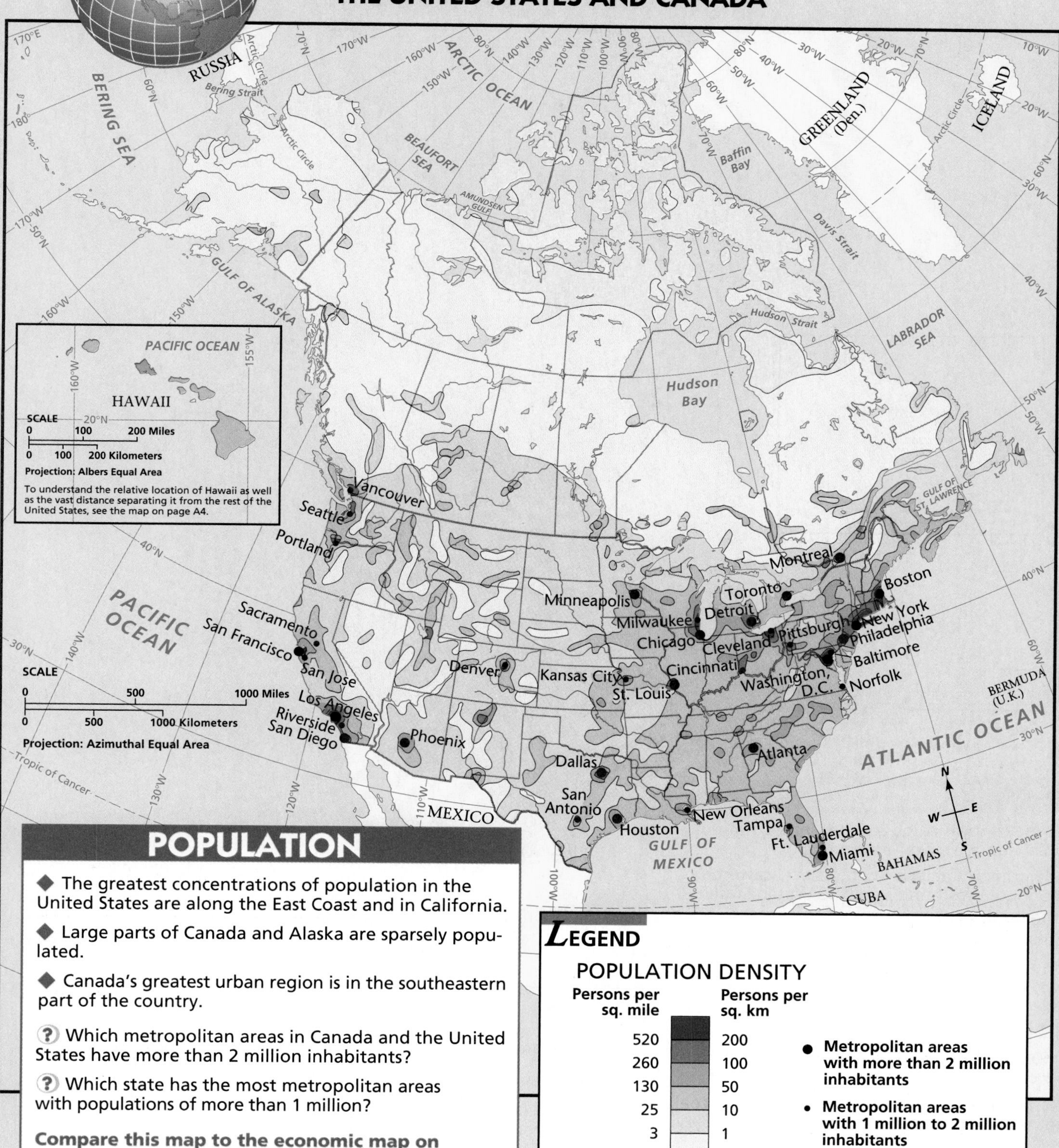

RUSSIA
BERING SEA
Bering Strait
Arctic Circle
ARCTIC OCEAN
BEAUFORT SEA
AMUNDSEN GULF
GREENLAND (Den.)
ICELAND
Baffin Bay
Davis Strait
Hudson Strait
LABRADOR SEA
Hudson Bay
GULF OF ALASKA
GULF OF ST. LAWRENCE
PACIFIC OCEAN
HAWAII
SCALE
0 100 200 Miles
0 100 200 Kilometers
Projection: Albers Equal Area
To understand the relative location of Hawaii as well as the vast distance separating it from the rest of the United States, see the map on page A4.
Vancouver
Seattle
Portland
Sacramento
San Francisco
San Jose
Los Angeles
Riverside
San Diego
Phoenix
Denver
Minneapolis
Milwaukee
Chicago
Detroit
Toronto
Montreal
Boston
New York
Philadelphia
Baltimore
Washington, D.C.
Norfolk
Pittsburgh
Cleveland
Cincinnati
Kansas City
St. Louis
Dallas
San Antonio
Houston
New Orleans
Atlanta
Tampa
Ft. Lauderdale
Miami
MEXICO
GULF OF MEXICO
BAHAMAS
CUBA
BERMUDA (U.K.)
ATLANTIC OCEAN
Tropic of Cancer
SCALE
0 500 1000 Miles
0 500 1000 Kilometers
Projection: Azimuthal Equal Area
N
S
E
W
LEGEND
POPULATION DENSITY
Persons per sq. mile
Persons per sq. km
520 200
260 100
130 50
25 10
3 1
0 0
Metropolitan areas with more than 2 million inhabitants
Metropolitan areas with 1 million to 2 million inhabitants

POPULATION

◆ The greatest concentrations of population in the United States are along the East Coast and in California.

◆ Large parts of Canada and Alaska are sparsely populated.

◆ Canada's greatest urban region is in the southeastern part of the country.

? Which metropolitan areas in Canada and the United States have more than 2 million inhabitants?

? Which state has the most metropolitan areas with populations of more than 1 million?

Compare this map to the economic map on page 110.

? Name some economic activities found in the more sparsely populated areas of the western United States.

MAPPING THE REGION

THE UNITED STATES AND CANADA

HAWAII

SCALE
0 100 200 Miles
0 100 200 Kilometers
Projection: Albers Equal Area

To understand the relative location of Hawaii as well as the vast distance separating it from the rest of the United States, see the map on page A4.

SCALE
0 500 1000 Miles
0 500 1000 Kilometers
Projection: Azimuthal Equal Area

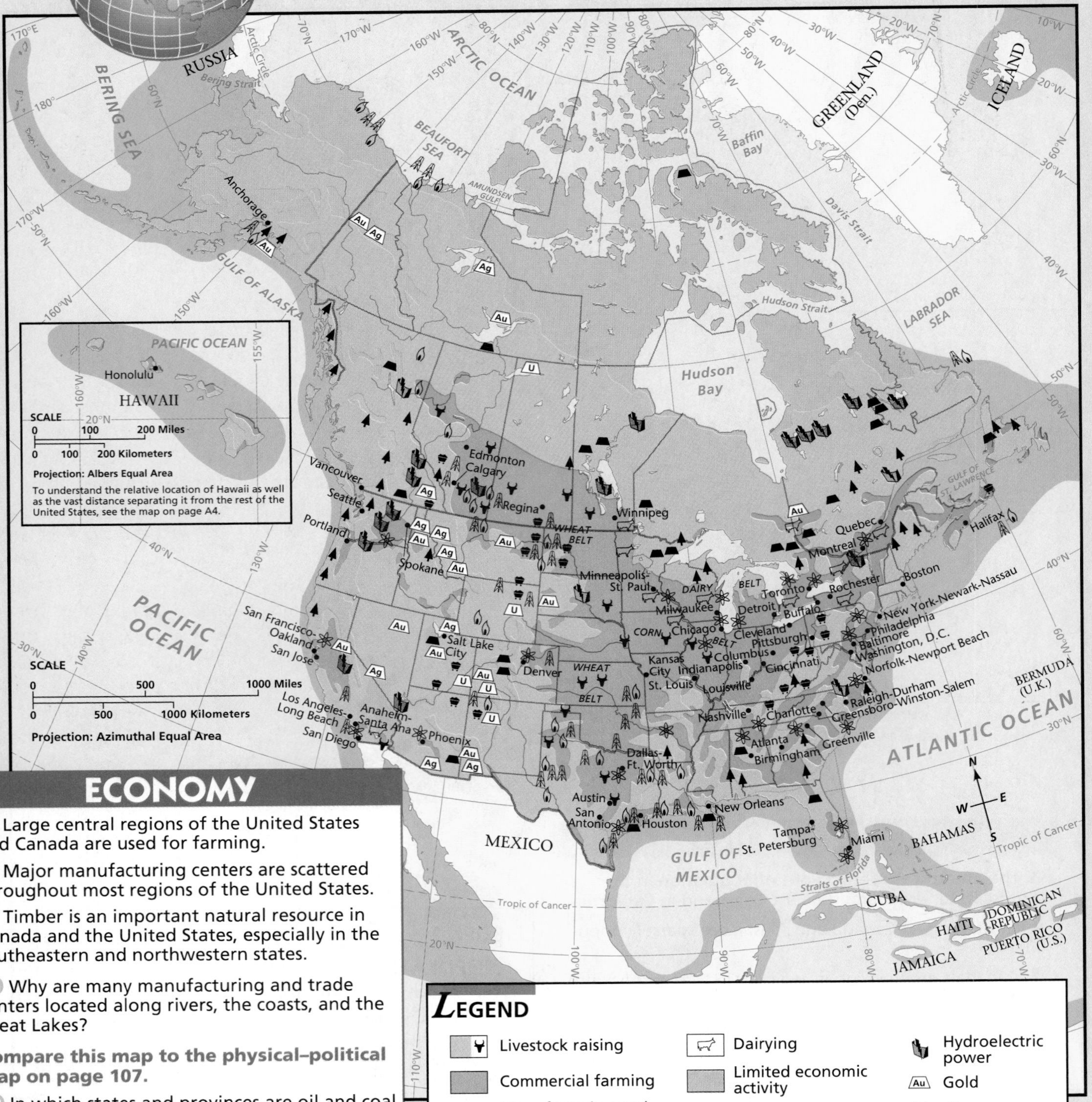

ECONOMY

◆ Large central regions of the United States and Canada are used for farming.

◆ Major manufacturing centers are scattered throughout most regions of the United States.

◆ Timber is an important natural resource in Canada and the United States, especially in the southeastern and northwestern states.

? Why are many manufacturing and trade centers located along rivers, the coasts, and the Great Lakes?

Compare this map to the physical–political map on page 107.

? In which states and provinces are oil and coal production important economic activities?

Compare this map to the climate map on page 108.

? What climates dominate the livestock-raising areas in the western United States?

LEGEND

- Livestock raising
- Commercial farming
- Manufacturing and trade
- Commercial fishing
- Forests
- Dairying
- Limited economic activity
- Coal
- Natural gas
- Oil
- Major nuclear power plants
- Hydroelectric power
- Au Gold
- Ag Silver
- U Uranium
- Other minerals
- Timber

A blue, misty haze gives the Great Smoky Mountains their name. Shown here in **North Carolina**, the mountain range is part of the Appalachian Mountains.

PHYSICAL GEOGRAPHY

FOCUS

- *What are the major landform regions in the United States?*
- *What different climates are found in the United States?*

REGION **Landforms** The United States features some of the world's most spectacular scenery. Its landforms vary from vast, flat plains to high mountains and plateaus to volcanic islands. (See the map on page 107.)

The landforms of the eastern half of the United States are older than those of the western half. The mountains have eroded, and rolling hills and flatlands dominate most of the region. In contrast, the western United States has a younger landscape. Steep mountains, active volcanoes, deep canyons, high plateaus, and wide plains are common here.

The region known as the Coastal Plain stretches along the Atlantic Ocean and Gulf of Mexico from New York to the border with Mexico. This low region lies close to sea level and rises only gradually inland. The Coastal Plain is narrowest along the Atlantic coast, widens to the south in the Carolinas and Georgia, and includes all of Florida. Along the Gulf of Mexico, it widens considerably and includes all of the lower Mississippi River valley as far north as southern Illinois.

The Piedmont is a rolling plateau region inland from the Coastal Plain. A **piedmont** is an area at or near the foot of a mountain region. The Piedmont begins in New Jersey and stretches as far south as Alabama. Many of America's first cities, including Philadelphia, Pennsylvania, and Richmond, Virginia, were established along the **Fall Line**, where the Coastal Plain meets the Piedmont. Here, rivers plunge over rapids and waterfalls from the hard rock underlying the Piedmont to the softer rocks underneath the Coastal Plain. These rivers provided waterpower for the early towns and cities.

The Appalachian Mountains rise to the west and north of the Piedmont. The Appalachians were formed more than 300 million years ago when, according to the theory of plate tectonics, eastern North America collided with Africa. The mountains' peaks have been lowered and smoothed by erosion. The Appalachians stretch from Maine to Alabama and are divided into several ranges. Their highest peaks rise to more than 6,000 feet (1,829 m) in North Carolina's Blue Ridge Mountains and New Hampshire's White Mountains. A major portion of the Appalachians is made up of a series of parallel ridges and valleys. Well-known features in the Appalachians are the Blue Ridge, Catskill, and Pocono mountains.

FACTS IN BRIEF United States

COUNTRY POPULATION (1994)	LIFE EXPECTANCY (1994)	LITERACY RATE (1991)	PER CAPITA GDP (1993)
UNITED STATES 260,713,585	73, male 79, female	97%	$24,700

Source: *The World Factbook 1994,* Central Intelligence Agency

The United States has a higher gross domestic product (GDP) per capita than Canada. Canadians, however, have a longer life expectancy than Americans. See the chart on page 174.

Physiographic Regions of the United States and Canada

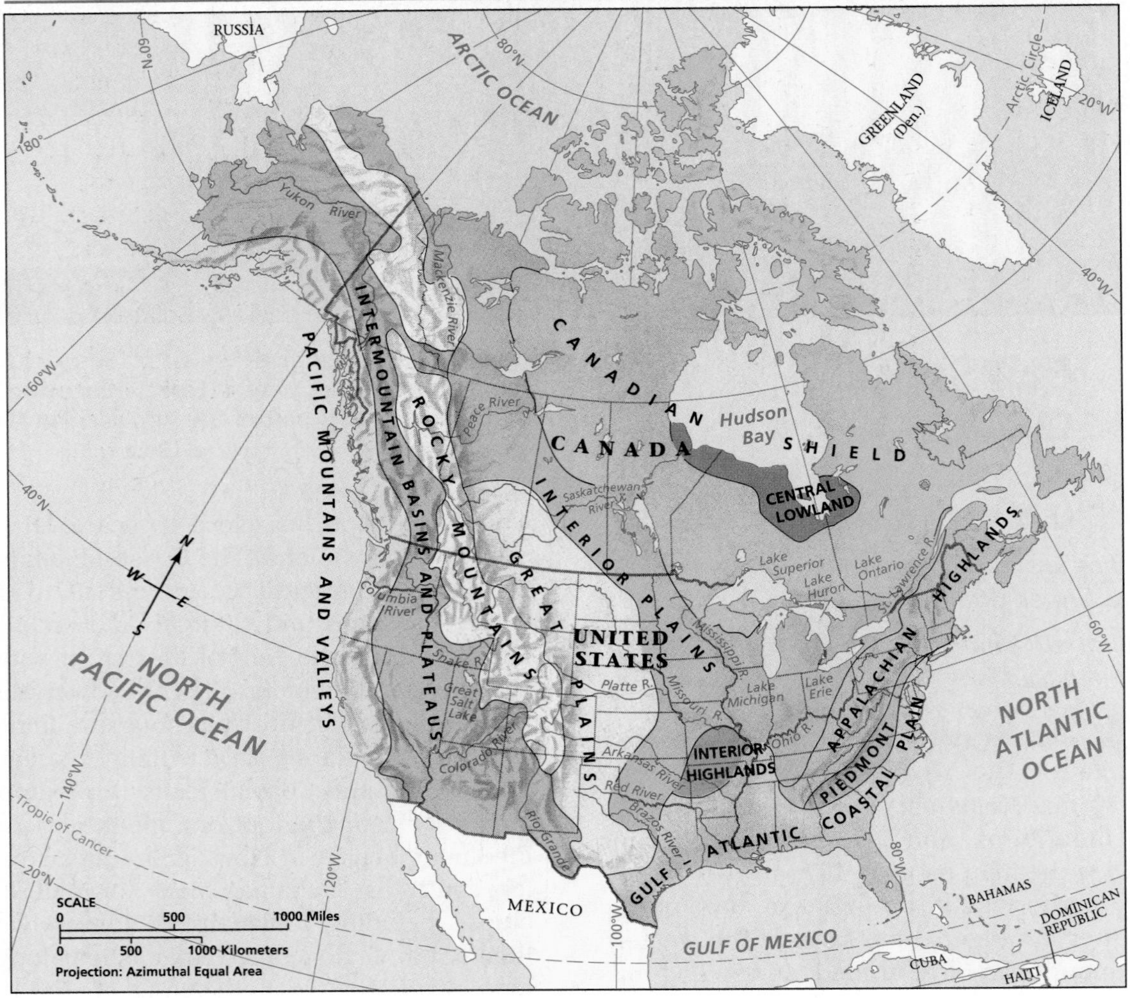

MAP STUDY The colors in this map designate the physiographic regions of the United States and Canada. A physiographic region, or province, has the same general types of landforms throughout. In which physiographic region do you live?

The Ozark Plateau, although separated from the Appalachians, also is an old, eroded upland. It lies in Missouri, Arkansas, and Oklahoma.

There are many fertile river valleys in the Appalachians. Among these are the Connecticut, Hudson, and Shenandoah valleys. In the western part of the highlands, a long plateau slopes toward the plains of the Midwest. This is known as the Allegheny Plateau in the northeast and the Cumberland Plateau farther south.

Between the Appalachians and the Rocky Mountains is a vast area called the Interior Plains. The Mississippi River and its many tributaries drain most of the region. The northern Interior Plains, north of the Ohio and Missouri rivers, were covered by glaciers during the last ice age. The great ice sheets changed the landscape. Today, this area has rolling hills, thousands of lakes (including the five Great Lakes), river systems, and a wide variety of productive soil types.

The Great Plains stretch from Canada to Mexico in a north-south direction and from approximately 100° west longitude to the foot of the Rocky Mountains in an east-west direction. Although the Great Plains are the flattest portion of the Interior Plains, they are not as low in elevation. At the eastern edge of the Rocky Mountains, the Great Plains reach elevations of more than 5,000 feet (1,524 m).

The Rocky Mountains extend north and south from Canada to New Mexico and from the Great

Plains west to the Intermountain region. Many of the highest peaks reach more than 14,000 feet (4,267 m) above sea level. Their jagged peaks and deep valleys are evidence of glacial action. The Rocky Mountains are not a single range but a series of several ranges separated by high plains and valleys.

The Intermountain region is located between the Rocky Mountains on the east and the Cascade and Sierra Nevada ranges on the west. Because rivers in this region (except the Colorado River) never reach the ocean, much of the region is called the Great Basin. A **basin** is a low area of land, generally surrounded by mountains. The rugged landscape of the Intermountain region consists mainly of high plateaus with deep canyons, isolated mountain ranges, and desert basins. The lowest point on the continent of North America, Death Valley in California, lies at the western edge of the Great Basin. Death Valley drops 282 feet (86 m) below sea level.

The Pacific coast region is characterized by two major mountain systems and a series of valleys between the mountains that extend in a north-south direction. On the eastern edges of this region are the Sierra Nevada and Cascade ranges. The Sierra Nevada is a giant faulted mountain range that runs almost the length of California. It is the longest and highest range in the 48 **contiguous** states. Contiguous states are those that border each other as a single unit. Only Alaska and Hawaii are not contiguous states. To the north of the Sierra Nevada, in northern California, Oregon, and Washington, are the Cascade Mountains. This range consists of a series of high volcanoes, such as Mount Rainier, Mount Hood, Mount Shasta, and Mount St. Helens.

Along the Pacific Ocean are the rugged Coast Ranges that stretch from California to Canada. Between the Coast Ranges and the Sierra Nevada and Cascades are three fertile valleys: the Puget Sound Lowland, the Willamette Valley, and the Central Valley of California.

Mount St. Helens in **Washington** state is one of a number of volcanoes in the western United States. What is the cause of the volcanic activity in the western part of North America?

The rugged western United States is part of the Pacific Ring of Fire, with its active volcanoes and earthquake faults. This is where, according to the theory of plate tectonics, the North American plate is colliding with the Pacific plate. All of the continental United States is on the North American plate except for coastal California, which is located on the Pacific plate. This part of California is separated from the North American plate along the San Andreas Fault, where major earthquakes occur.

The two westernmost states, Alaska and Hawaii, also are geologically active. All of the Hawaiian Islands are a result of huge volcanoes that have grown from the seafloor. Southeast and south-central Alaska is very mountainous. North America's highest peak, Mount McKinley in the Alaska Range, is 20,320 feet (6,194 m) in elevation. Except for the Brooks Range, northern and interior Alaska have flat and hilly terrains. The volcanic Aleutian (uh-LOO-shuhn) Islands extend into the Pacific from Alaska.

REGION **Climate** The 48 contiguous states are located in the middle latitudes and have a variety of climates. The climates of the United States are influenced mainly by this middle-latitude location, a large land area, high mountain ranges, westerly winds, and ocean currents. The United States has 11 climate types, the greatest variety of climates of any nation. (See the map on page 108.)

A humid-continental climate covers the northeastern quarter of the nation. It stretches westward from the Atlantic coast to approximately 100° west longitude. This climate region has four distinct seasons, including a warm, humid summer and a cold, snowy winter. The nearness of the Great Lakes and the Atlantic Ocean moderates temperatures slightly and is a source of increased precipitation.

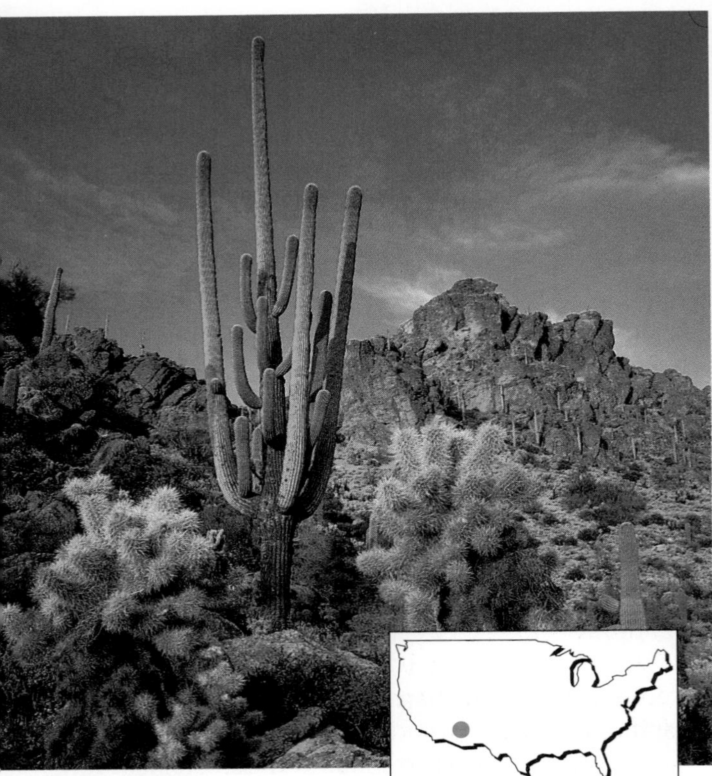

These teddy bear cholla (CHOL-yah) (foreground) and saguaro cacti near **Tucson, Arizona**, are native to the desert climate of the southwestern United States.

Lightning arcs over the Space Needle and downtown **Seattle, Washington**. Seattle is located in a wet, marine-west-coast climate.

A humid-subtropical climate is found in the southeastern quarter of the United States. This region stretches from the Atlantic and Gulf of Mexico coasts to about 100° west longitude. The climate is influenced by the warm waters of the Gulf of Mexico and by the warm **Gulf Stream** current in the Atlantic Ocean. The Gulf Stream is an ocean current that moves warm tropical water northward. Summers are hot and humid, and winters are mild. Rainfall is distributed fairly evenly throughout the year, and thunderstorms are common. Some snow and frost may occur in winter. The very southern tip of Florida extends beyond the humid-subtropical region. This area has a tropical-savanna climate.

West of 100° west longitude is the steppe climate of the Great Plains. This semiarid climate supports vast grasslands but has few trees. Air from the deserts, Arctic Canada, and the Gulf of Mexico invade the region. Summers are hot, while winters can be very cold. The Great Plains region is subject to violent weather, including thunderstorms, hailstorms, tornadoes, and dust storms. Droughts are also a major concern here.

The interior western states have a variety of climates because of the mountainous terrain. The Rocky Mountains have highland climates, with temperatures and precipitation dependent upon elevation and exposure. The Intermountain region, located between the major mountain areas of the west, has mainly desert and steppe climates due to its rain-shadow location.

The Pacific coast region has two climates, marine west coast and Mediterranean. The mild marine-west-coast climate dominates the Pacific Northwest. It has cool, cloudy, wet winters and mild, sunny summers. The Mediterranean climate bathes southern and central California. This climate is known for its long, sunny, dry summers and its mild, wet winters.

Hawaii is the southernmost state and the only one that is fully within the tropics. Because the Hawaiian Islands fall in the easterly trade wind belt, they are wetter on the windward, eastern sides where they have a humid-tropical climate. On the leeward, western slopes there is a drier tropical-savanna climate.

Alaska is the northernmost state and the only one that extends into the subarctic and tundra climates. It actually has three climates: a narrow strip of marine-west-coast climate along the southeast coast, a subarctic climate in the interior, and a tundra climate along the north and west coasts. The subarctic climate area is forested and has severe seasonal changes. Winters are cold and long, while summers are short and relatively warm. The tundra climate area is cold all year. It has no trees, and only small plants grow.

SECTION REVIEW

1. Name four major mountain ranges in the United States. Where is each one located?
2. Name and describe the climate region in which you live.
3. **Critical Thinking** **Explain** why the United States has such a wide variety of landform regions and climate types.

HUMAN GEOGRAPHY

FOCUS

- *Over time, what groups of people have settled in what is now the United States?*
- *What are the characteristics of the population of the United States today, and how is the population distributed?*

MOVEMENT **Early Settlement** The first inhabitants of North America were Native Americans, who probably arrived from Asia between 10,000 and 40,000 years ago. The first Europeans to arrive were Viking explorers about A.D. 1100. By this time, several million Native Americans lived throughout the North American continent.

The Spanish arrived about 500 years ago, claiming mainly the southern part of the present-day United States from Florida to California. They were soon followed by the French, who explored major river systems and claimed the interior of North America. The British, however, became the major influence during the nation's early history, establishing 13 colonies along the Atlantic **seaboard**. A seaboard is a land area near the ocean. Among the earliest settlers along the eastern seaboard were Africans, who first arrived in Jamestown in 1619.

Beginning in the 1600s, millions of Africans were brought to the American colonies and sold as slaves to work on southern **plantations**. A plantation is a large farm that concentrates on one major crop, such as cotton or tobacco. Although the slave trade with Africa ended in 1808, the tragic practice of slavery did not end completely until after the Civil War.

After gaining independence from Britain, the United States grew as immigrants arrived, mainly from Europe. By the mid-1800s, the United States stretched from the Atlantic to the Pacific coast and had formed its present borders with Canada and Mexico. During this time, as the nation's growing population gradually moved westward, more states were added to the original 13. Diseases and warfare with settlers killed many Native Americans. The surviving populations were driven farther west, where they were eventually forced onto reservations.

Some archaeologists believe the Great Serpent Mound—constructed near present-day **Cincinnati, Ohio**, by Native Americans—was used as a burial site and for defense.

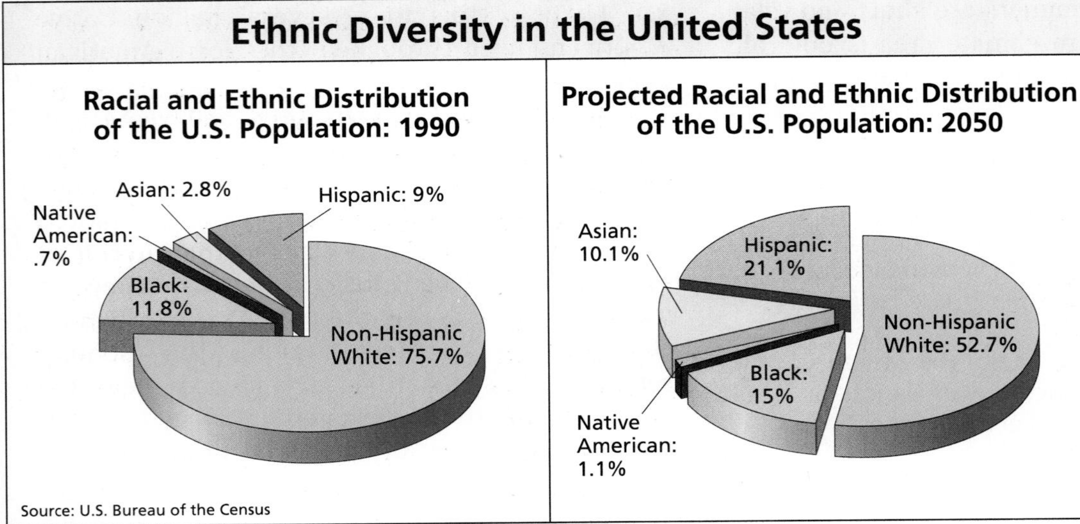
Ethnic Diversity in the United States
Racial and Ethnic Distribution of the U.S. Population: 1990
Asian: 2.8%
Hispanic: 9%
Native American: .7%
Black: 11.8%
Non-Hispanic White: 75.7%
Projected Racial and Ethnic Distribution of the U.S. Population: 2050
Asian: 10.1%
Hispanic: 21.1%
Non-Hispanic White: 52.7%
Black: 15%
Native American: 1.1%
Source: U.S. Bureau of the Census

These pie graphs show the racial distribution of the U.S. population in 1990 and the projected distribution for 2050. The percentage of Americans of Hispanic (Latino) origin and of Asian Americans is projected to increase dramatically by 2050. The figures for Hispanic origin include all those who identified themselves as Hispanic in surveys conducted by the U.S. Bureau of the Census.

Population Diversity Today, the United States has the most culturally diverse population in the world. It is about 75 percent European American (also called non-Hispanic white by the U.S. Bureau of the Census), 12 percent African American, 9 percent Hispanic (Latino), 3 percent Asian, and 1 percent Native American (Indian, Inuit, and Hawaiian).

Europeans were the dominant immigrant group in America until the 1960s. Recent immigration, however, has been mostly from Middle and South America and Asia. In the past few decades, millions of Hispanics (Latinos) have immigrated to the United States, especially from Mexico, Puerto Rico, Cuba, and Central America. Early Asian immigrants included Chinese laborers brought to the West Coast between 1849 and 1882. Since the 1960s, Asians have arrived from many countries, including the Philippines, Vietnam, Korea, China, and India. In recent years, many people have immigrated to the United States illegally. Most come to seek jobs or to escape political instability in their home countries.

The United States is predominantly Christian in religion, though nearly every world religion is represented here. The population is about one-half Protestant and one-third Roman Catholic. There also are significant numbers of Jews, Mormons, and Muslims.

The U.S. population, which ranks third in the world after China and India, is only about 5 percent of the total world population. With low birthrates by world standards, the population of the United States will form an even smaller percentage in the future. Low birthrates mean that there are fewer young people, and high health standards mean that the nation's population is living longer. In the early 1990s, half of Americans were older than 33. It is estimated that by 2050 half of Americans will be over 40.

Population Distribution The densest settlement in the early years of the United States was along the eastern seaboard, especially in the northeastern quarter of the country. Today, the greatest concentrations of population are still along the East Coast and in California, the most populous state. As the United States has developed from an industrial society into a technological society over the past few decades, many people have moved from the northern states to the Sunbelt states, particularly California, Texas, Florida, and Arizona. In recent years, some of the most rapid population growth has been in the less populated states of the Intermountain West and Pacific Northwest. (See the map on page 109.)

The United States is a highly urbanized nation. Its three largest metropolitan areas are New York, Los Angeles, and Chicago. Most U.S. metropolitan areas contain large central cities and ever-growing suburbs. Most commuters now drive from suburb to suburb instead of from suburb to city. **Edge cities**, or suburbs with large employment and commercial centers, are common.

This suburbanization has left many of America's inner cities abandoned and in decay. These areas have become pockets of urban poverty with high crime rates, substandard schools, and little economic opportunity. Many areas are **ghettos,** parts of cities where minority groups are concentrated because of economic pressure and social discrimination.

SECTION REVIEW

1. List six major groups of people who have inhabited what is now the United States.
2. What is meant by suburbanization?
3. **Critical Thinking** **Analyze** the changes occurring in the makeup and distribution of the population of the United States.

ECONOMIC GEOGRAPHY

FOCUS

- *How does the United States compare to other nations of the world in resources, industry, communications, and trade?*
- *What political, economic, and domestic challenges does the United States face?*

Resources Rich natural resources and a productive population have made the United States an industrial giant and the world's leading agricultural nation. Even though less than 3 percent of the population farms, the nation easily feeds its people and is a major food exporter. U.S. farmers produce the world's widest variety of agricultural products, including huge supplies of corn, wheat, and soybeans. Much of this success is due to the nation's wide variety of climates and rich soils.

Water also has been important to the nation's economic development. Many rivers, including the Tennessee, Colorado, and Columbia, are a source of irrigation and hydroelectricity. The Great Lakes and rivers such as the Mississippi, Ohio, Hudson, and Columbia are excellent inland water routes. The waters of the nearby continental shelf provide rich ocean resources.

Forests are another valuable natural resource to the country. One-third of the United States is forested, with the major commercial forests located in the southeastern United States and in the Pacific Northwest. Early logging depleted the forests of the Northeast and the Midwest. Much of today's logging, both for domestic use and export, takes place on private tree farms and in national forests.

The United States is rich in energy and mineral resources. Almost 30 percent of the world's coal reserves are in the United States, which is a major coal exporter. Most of the coal is mined in the Appalachian and Rocky mountains and on the Interior Plains.

The United States also is a major oil producer, although it uses more than it produces. More than one-third of its needed oil must be imported. Most domestic oil production takes place on the Gulf Coast of Texas and Louisiana, and in California and Alaska.

INTERACTION **Industrial Growth** During the nineteenth century, the United States experienced rapid industrialization. America's early steel industry was centered in Midwest cities and was dependent on the region's coal and iron ore. The nation's infrastructure was a major boost to U.S. industrial leadership. The early development of waterways and railroads helped industry spread into the interior of the country.

Free enterprise encouraged innovation, inventions, and competition. Henry Ford's mass automobile production is a good example of American

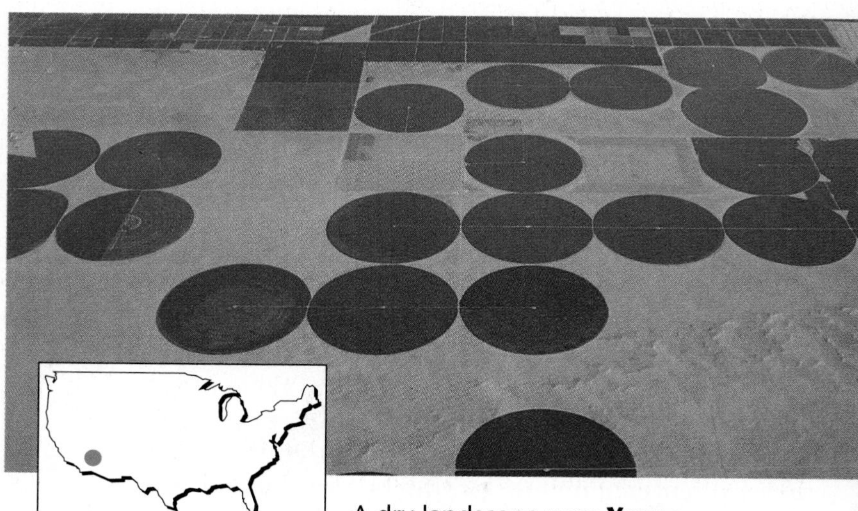

A dry landscape near **Yuma, Arizona**, has been transformed into productive farmland by irrigation systems. Much of the western and southwestern United States has a climate that is too dry to support agriculture without irrigation.

The world's most mobile people help make U. S. airports the world's busiest. In one year, more than 45 million passengers pass through the terminals of **Los Angeles International Airport**, shown here.

innovation. The United States was the first nation to develop this method of production. American industry has also been a leader in industrial research, which led to the development of products such as plastics, nylon, aircraft, computers, and computer software.

Today, American companies are world leaders in fields such as aerospace, computer technology, pharmaceuticals (medicinal drugs), chemicals, and construction engineering. The United States also is the leader in the entertainment industry (movies, television, and music) and the fast-food industry. Service industries, especially banking, education, food services, health services, retail sales, tourism, and communications, are the most rapidly growing area of the U.S. economy.

With the world's largest highway system and more than 135 million automobiles, Americans are the most mobile people in the world. The nation also has a vast aviation network. Many Americans, in homes and offices, have access to the world's most developed telecommunications network. Televisions, computers, fax machines, and telephones are a part of everyday life.

International Trade The United States is a world leader in international trade. For the past 20 years, however, the nation has had a **trade deficit**. A trade deficit occurs when the value of a nation's imports exceeds that of its exports. Although the United States exports aircraft, computers, entertainment products, and food, it imports (in value) more oil, automobiles, and electronic consumer products.

The United States is a major part of the global economy, and imports and exports will continue to grow. An important challenge for the nation is to reverse the trade deficit and continue innovation in the face of economic competition.

The United States has worked with other countries to lower barriers to world trade. For example, the United States approved the North American Free-Trade Agreement (NAFTA) with Canada and Mexico in 1993. Also, the United States and other countries finished work on the new General Agreement on Tariffs and Trade (GATT) in 1994. These agreements lowered tariffs and other trade barriers.

Issues The United States is a wealthy and democratic nation. Its future lies in solving problems created by a changing political and economic world. Major issues in foreign affairs include trade and the U.S. role in world conflicts and affairs. The United States also faces domestic challenges such as a high national debt, poverty, crime, urban decay, and pollution. With the most productive, innovative, and diverse people in the world, the United States is well equipped to confront the challenges of the twenty-first century.

3 SECTION REVIEW

1. Describe three characteristics of the United States that have helped it become a world leader.
2. Name two international and two domestic challenges facing the United States.
3. **Critical Thinking** **Assess** the ability of the United States to remain a world leader over the next 50 years.

CHAPTER 10 CHECK

Loading grain onto a freighter in **Duluth, Minnesota**

Reviewing the Main Ideas

1. The United States has a wide range of landforms, including plains, mountains, plateaus, and volcanic islands. The United States also has many climate types, from humid tropical in Hawaii to desert in the western United States to tundra in Alaska.
2. Many different groups of people have lived in what is now the United States. Today, the population is ethnically and religiously diverse, and it is aging. It is shifting toward southern and western states and out of cities into suburbs as well.
3. The United States is rich in natural and human resources and is a world leader in industrialization and international trade. Even so, the United States faces many challenges.

Building a Vocabulary

1. Define *piedmont* and *Fall Line.*
2. Which two states are not contiguous with the other states?
3. What is the Gulf Stream? Which region's climate is influenced by it?
4. What is a plantation? Who provided the labor for early U.S. plantations?

Recalling and Reviewing

1. Which parts of the United States are the most geologically active? Why?
2. From where do most recent immigrants to the United States come? For what reasons are they immigrating to the United States?
3. If the United States is a major oil producer, why must oil be imported?
4. Which U.S. industries employ the most people?

Thinking Critically

1. Do you think the presence of tariffs would serve to increase or decrease a trade deficit? Explain.
2. Of all of the challenges facing the United States, which one do you think is the most urgent, and why? List some things you might do to help the United States confront this challenge.

Learning About Your Local Geography

Individual Project

How has the average age of your community's population changed in recent years? Request information from a community elementary school about student enrollments over the past 20 years. Has the number of students in the school increased, decreased, or stayed the same? Graph your data. What can you conclude from your data about the average age of your community's population? Write a caption for your graph, suggesting reasons for any shifts. You may want to include your graph in your individual portfolio.

Using the Five Themes of Geography

Create a graphic organizer to show how the geographic themes of **location**, **place**, and **movement** have interacted to make the United States a world leader in many respects. What advantages does the United States have because of its location? What physical and human place characteristics have enabled the United States to make the most of its location? How has movement played a role in helping the United States use both its location and its place characteristics?

A Mexican-style rodeo near **Austin, Texas**

CHAPTER 11

The Northeastern United States

"There are roughly three New Yorks. There is, first, the New York of the man or woman who was born here, who takes the city for granted and accepts its size and turbulence as natural and inevitable. Second, there is the New York of the commuter—the city that is devoured by locusts each day and spat out each night. Third, there is the New York of the person who was born somewhere else and came to New York in quest of something."

E. B. White

GEOGRAPHY DICTIONARY

Megalopolis
textile
granite
peninsula
moraine
second-growth forest
biotechnology
truck farm
bituminous coal
coke
anthracite coal
borough
break-of-bulk center

The Northeast is the most populated and urbanized region of the United States. It has the nation's greatest concentration of factories, banks, universities, and transportation centers. The region is home to New York City, the nation's largest city, and Washington, D.C., the nation's capital.

This chapter organizes the Northeastern United States into two subregions that are overlapped by a third subregion. The New England states are located in the far northeastern portion of the country. The Middle Atlantic states stretch eastward from the Appalachians to the Atlantic coast. Megalopolis (megh-uh-LAHP-uh-luhs) is a giant urban area that extends along the eastern seaboard. This subregion overlaps the New England and Middle Atlantic states.

Subway tokens for the **New York City** Transit Authority

The Statue of Liberty and **New York City**

NORTHEASTERN UNITED STATES: PHYSICAL–POLITICAL

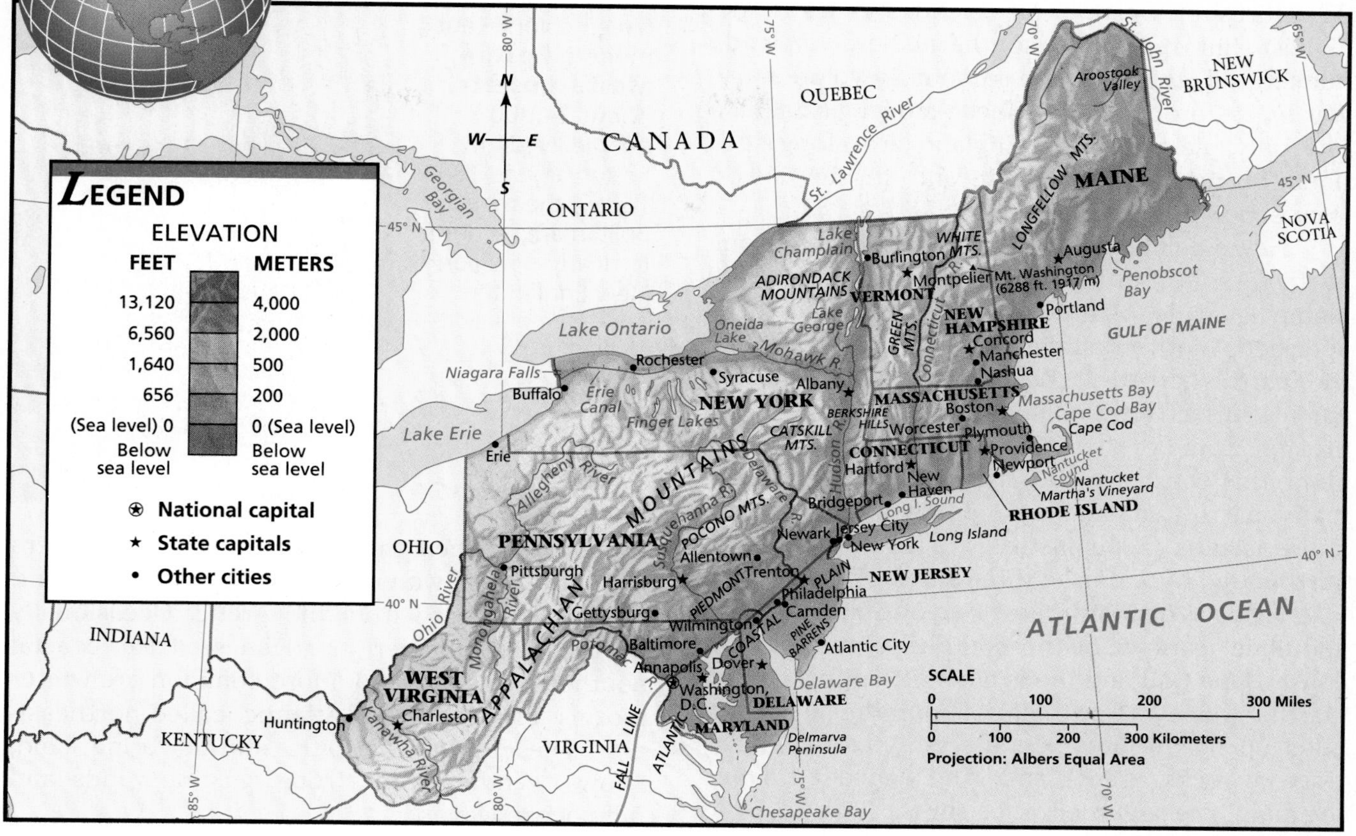

LEGEND

ELEVATION

FEET	METERS
13,120	4,000
6,560	2,000
1,640	500
656	200
(Sea level) 0	0 (Sea level)
Below sea level	Below sea level

⊛ National capital
★ State capitals
• Other cities

SCALE: 0–100–200–300 Miles; 0–100–200–300 Kilometers
Projection: Albers Equal Area

NEW ENGLAND

FOCUS

- *How have New England's landforms and climates affected the region's economy?*
- *Why is New England's economy changing, and in what ways is it changing?*

The New England states include Maine, New Hampshire, Vermont, Massachusetts, Rhode Island, and Connecticut. The population of this region is about 5 percent of the nation's total.

New England has played a major role in U.S. history. The Pilgrims arrived at Plymouth, on Massachusetts Bay, in 1620. They soon were followed by other colonists, who traded furs, farmed, fished, and logged the forests. In the eighteenth century, New England became the center of the American Revolution.

During the nineteenth century, New England became the first industrialized region in North America. Its oldest industries, including **textiles** (cloth products), timber, and shoes, are being replaced by industries centered around high technology and finance. These new industries are concentrated mainly in southern New England.

REGION Landforms The New England states are famous for their scenic beauty and variety of landforms. (See the map above.) The Appalachian Mountains cross much of northern New England. The major ranges are the White Mountains in New Hampshire, the Green Mountains in Vermont, and the Longfellow Mountains in Maine. The highest peak, Mount Washington in the White Mountains, reaches an elevation of 6,288 feet (1,917 m). Because of glacial erosion during the last ice age, northern New England has thousands of lakes and thin, rocky soils.

Southern New England primarily is a glaciated, hilly region. As the great ice sheets pushed their way to the Atlantic coast, they deposited rock material, forming hundreds of hills. The Berkshires of western Massachusetts form the highest hill region in southern New England. Bunker Hill, the site of a famous Revolutionary War battle, probably is the best-known glacial hill. The only plains in New England are along sections of the Atlantic coast and in the lower Connecticut River valley.

New England's coastline varies greatly from north to south. In the north, the scenic coast of Maine is made of eroded **granite**. Granite is a speckled, hard, crystalline rock formed deep in the earth's crust. It becomes exposed at the earth's surface after millions of years of uplift and erosion. Some of the world's greatest tides occur along this coast, which is rugged and rocky with many narrow inlets, bays, **peninsulas**, and offshore islands. A peninsula is a landform that is surrounded by water on three sides.

Glacial deposits are responsible for the shoreline features of the southern New England coast. Cape Cod and the islands of Nantucket and Martha's Vineyard are glacial **moraine** materials piled up by the huge ice sheets. A moraine is a ridge of rocks, gravel, and sand deposited along the margins of a glacier or ice sheet.

Climate New England has a humid-continental climate. Each year, autumn's brightly colored, changing leaves attract tourists, while cold, snowy winters are inviting to winter sports enthusiasts. Fog is common along the coast, especially in summer, when the warm Gulf Stream waters meet the cool waters of the Labrador Current, which flow south from the Arctic. Occasionally during late summer, hurricanes strike coastal southern New England. More common are winter storms from the North Atlantic, called northeasters. These middle-latitude storms bring cold, snowy weather with strong, gusting winds and high ocean waves.

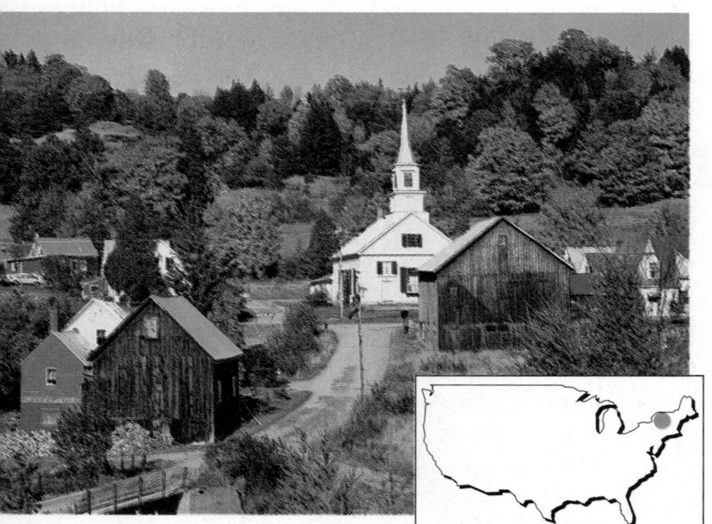

Autumn explodes in color around this **Vermont** town. Look at the biome map on page 57. Which biomes are dominant in the Northeast United States?

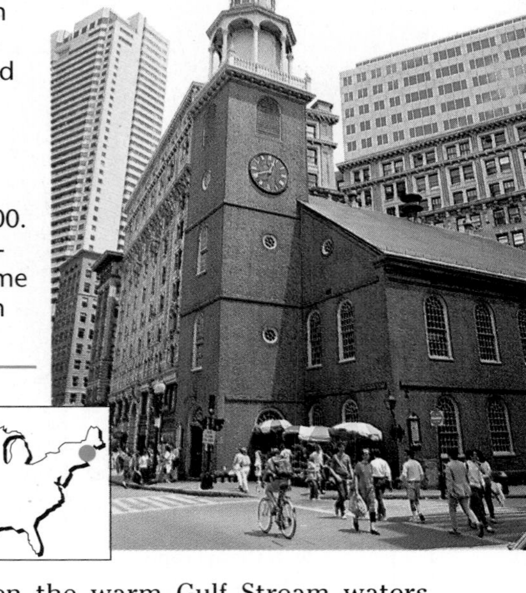

The Old South Meeting House, built in 1729, is a link between colonial and modern **Boston, Massachusetts**. About 24,000 people lived in Boston around 1800. Today, the metropolitan area is home to nearly 3 million New Englanders.

Agriculture Because of the region's rocky terrain and short growing season, agriculture in New England is quite limited. The Connecticut River valley is the leading agricultural area. The swampy lowlands of Cape Cod grow more than half of all the cranberries produced in the United States. The Aroostook (uh-ROOS-tuhk) Valley in northern Maine produces about 10 percent of the country's potatoes.

New England's cool, moist climate and hilly pastureland make dairy farming the region's most important agricultural activity. Poultry farms also can be found. Farming in New England has always been a small-scale operation and has been in decline for more than a century. Many farms have been abandoned, while others have been bought by urban dwellers to use as second homes or retirement retreats.

Resources The shallow waters off New England's coast are among the world's most productive fishing grounds. Where the cool Labrador Current flows over the shallow continental shelf, fish are plentiful. Cod and lobster are the region's

These traps, called lobster pots, are tools of the trade for lobster catchers of **Bernard, Maine**. Lobstering is an important part of the economy of coastal New England.

most valuable seafoods. The most productive waters have been around Georges Bank, an underwater plateau off the Massachusetts coast. After years of overfishing and declining fish stocks, however, the U.S. government ordered restrictions on fishing in some areas around Georges Bank in 1994.

Softwood forests, especially tall white pines, were an important natural resource for New England's early shipbuilding industry. Nearly all forests in New England are now **second-growth forests**. These are the trees that cover an area after the original forest has been removed. New England supports a pulpwood industry in Maine.

New England's few minerals cannot support a mining industry. The region's only major mining resource is building stone, especially Vermont granite and marble.

INTERACTION **Industry** New England became the nation's first industrial area. Swift streams provided power for the mills, and the relatively large local population provided a ready labor source. Boston served as a gateway city, where raw materials were imported and finished products could be shipped out by sea or by land. Many textile-mill towns and shoe factories sprang up along New England's rivers.

By the twentieth century, these industries had declined. Cheaper labor and more efficient factories could be found in the southern states and overseas. To remain competitive, New England has developed new industries and modernized its remaining traditional industries. For example, the region now makes textile and footwear products geared for outdoor and sporting activities.

New England long has been a national leader in higher education. More than 250 colleges and universities are located here, including Harvard, Yale, and Massachusetts Institute of Technology (MIT). This leadership in education made it possible for the region to become a leader in high-technology industries, such as electronics and computers. With the end of the Cold War, the region's economy has worked to recover from cuts in defense spending, which once provided thousands of jobs.

With some of the nation's leading hospitals and medical schools, New England's future economic growth may also be in **biotechnology**. Biotechnology is the application of biology to industrial processes, such as pharmaceutical and genetic engineering research and the manufacture of products from that research. In addition, New England is the center for numerous service industries. Many banks, investment houses, and insurance companies have their headquarters here.

Tourism also is important to New England's economy. Mountains, beaches, and historic cities and towns attract tourists from around the world. Popular ski resorts are scattered throughout the region's scenic, forested mountains.

Issues New England's future growth must be balanced against the need to protect the region's environment. Increased industrialization and urbanization, especially in southern New England, will place increasing pressure on the surrounding rural areas. Tourism and dense coastal development may also strain the region's resources. Controlling acid rain and other forms of pollution remains a major challenge.

SECTION REVIEW

1. Why was New England better suited for industrial growth than for agriculture?
2. What industries will dominate New England's economy in the near future? Why?
3. **Critical Thinking** **Suggest** possible origins of the name "New England." Do you think the name is appropriate for the region today?

THE MIDDLE ATLANTIC STATES

FOCUS

- *What are the major landform regions and climate types of the Middle Atlantic states?*
- *How were resources and location once important to the Middle Atlantic states, and why are they less important today?*

REGION The Middle Atlantic states include New York, Pennsylvania, New Jersey, Delaware, Maryland, and West Virginia. The nation's capital, Washington, District of Columbia, (or Washington, D.C.) also is included in this region. About one-fifth of the U.S. population lives here, making it one of the world's most industrialized and urbanized areas.

The Middle Atlantic states have played an important role in the economic development of the United States. Vast coal deposits, together with important land and water routes, helped the region emerge as an industrial giant during the nineteenth and early twentieth centuries. This industrialization attracted many immigrants to the region. Today, the Middle Atlantic region is where many of the nation's economic and political decisions are made.

Landforms Three landform regions cross the Middle Atlantic states: the Coastal Plain, the Piedmont, and the Appalachian Mountains. (See the map on page 121.) The Coastal Plain stretches across all of the Middle Atlantic states except West Virginia. This flat plain does not rise much above sea level. Long Island and Chesapeake Bay are both part of the Coastal Plain.

Chesapeake Bay, fed by the Susquehanna River, is the largest and most productive estuary on the Atlantic coast. As the sea level rose after the ice-age glaciers melted, the sea drowned the lower river valleys and formed the bay. Chesapeake Bay is a major supplier of oysters and crabs and has several important seaports.

Inland from the Coastal Plain is the Piedmont region, which gently slopes down from the Appalachians to meet the Coastal Plain. Because of the available waterpower at the Fall Line, many cities developed here.

The northern part of the Appalachian Mountains crosses the Middle Atlantic States region in a northeast-southwest direction. Several major rivers cut through some of the Appalachian ranges here on their way to the Atlantic. These include the Potomac, Susquehanna, Delaware, and Hudson rivers.

Climate The Middle Atlantic states have two major climate types: a humid-continental climate in the north and a humid-subtropical climate in the south. The dividing line between the two climates stretches west from the Atlantic coast south of New York City.

Summers in both climate regions can be very hot and humid. Winds passing over the warm Gulf Stream and the Gulf of Mexico bring hot, humid air inland. During winter, cold periods occur when arctic air from the north enters the region. Occasional snowstorms can cause serious transportation problems for the major cities. Farther

FACTS IN BRIEF — States of the Northeastern United States

STATE POPULATION (1993)	POP. RANK (1993)	AREA (1990)	AREA RANK (1990)	PER CAPITA INCOME (1993)
CONNECTICUT 3,277,000	27	5,544 sq. mi. 14,359 sq. km	48	$28,110
DELAWARE 700,000	46	2,397 sq. mi. 6,208 sq. km	49	$21,481
MAINE 1,239,000	39	33,741 sq. mi. 87,389 sq. km	39	$18,895
MARYLAND 4,965,000	19	12,297 sq. mi. 31,489 sq. km	42	$24,044
MASSACHUSETTS 6,012,000	13	9,241 sq. mi. 23,934 sq. km	45	$24,563
NEW HAMPSHIRE 1,125,000	41	9,283 sq. mi. 24,043 sq. km	44	$22,659
NEW JERSEY 7,879,000	9	8,215 sq. mi. 21,277 sq. km	46	$26,967
NEW YORK 18,197,000	2	53,989 sq. mi. 139,832 sq. km	27	$24,623
PENNSYLVANIA 12,048,000	5	45,759 sq. mi. 118,516 sq. km	33	$21,351
RHODE ISLAND 1,000,000	43	1,231 sq. mi. 3,188 sq. km	50	$21,096
VERMONT 576,000	49	9,615 sq. mi. 24,903 sq. km	43	$19,467
WEST VIRGINIA 1,820,000	35	24,232 sq. mi. 62,761 sq. km	41	$16,209

Rankings are for all 50 states.
Source: *Statistical Abstract of the United States 1994*

Which are the two most populous states of the Northeastern United States?

inland and northward in the Middle Atlantic states, winters are colder and snowfall can be heavy, especially in the Appalachians and in upstate New York. Like the New England coast, the Middle Atlantic coast is subject to hurricanes and northeasters, which can cause severe damage to coastal communities.

Agriculture Agricultural production in the Middle Atlantic states supplies the region's huge population. Fresh fruits and vegetables are produced on farms on Long Island and in New Jersey, Delaware, and Maryland. These farms are sometimes called **truck farms** because their products can be trucked easily to city markets. Some areas of the region specialize in certain crops. Potatoes are grown in New Jersey and on Long Island. Vineyards are found in the Finger Lakes area of New York and on Long Island. Other farms in the region produce milk, poultry, and grains.

During recent years, there has been a decrease in the amount of farmland in the Middle Atlantic states. Part of this loss has resulted from urban expansion into rural areas. Valuable farmland is being bought for housing and industry, while many farms in more remote rural areas have been abandoned or combined into larger agricultural operations.

Snow blankets **Westchester County** north of New York City. Winters generally are colder farther inland from coastal areas.

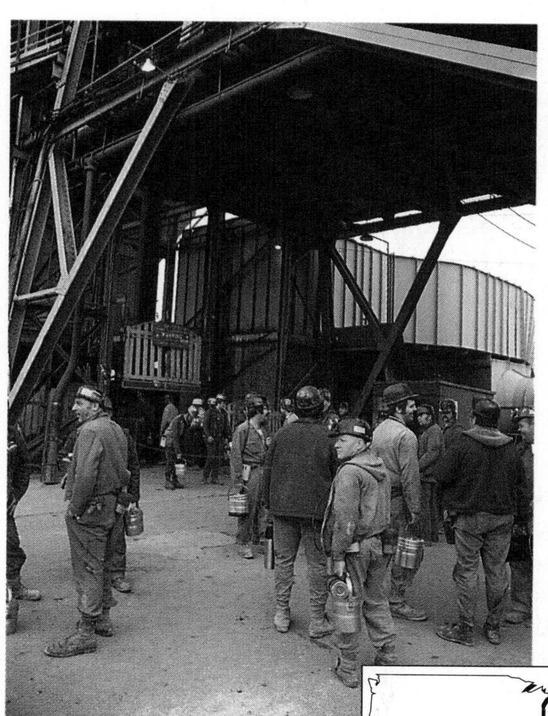

These **Pennsylvania** miners are waiting to enter one of the many coal mines in the Middle Atlantic states.

Resources In this highly industrialized region, coal has been the key natural resource. One of the largest **bituminous coal**, or soft coal, areas in the world is found in West Virginia, the western counties of Pennsylvania, and bordering areas in Ohio and Kentucky. Bituminous coal is used to produce electricity and chemicals. It also is used in the production of **coke**, or coking coal, which is used in blast furnaces to purify iron ore for making steel. Bituminous coal mining became the basis for industrialization in the Middle Atlantic states.

Deposits of **anthracite coal**, or hard coal, are found in eastern Pennsylvania. Anthracite is a higher grade coal than bituminous, but because it is more expensive to mine, anthracite mining has declined.

The Appalachian coal-producing region has suffered economic hardships over the past several decades. This area has some of the highest unemployment and poverty rates in the nation. Thousands of people have left the region, seeking jobs elsewhere. The coal industry has left behind abandoned mine debris and polluted water.

The Middle Atlantic states have several forests and wilderness areas. The Adirondack and Catskill mountains in New York, the Pine Barrens in New Jersey, and many mountain ridges in Pennsylvania and West Virginia remain forested and thinly populated. These forests are not major wood producers but instead serve as wildlife reserves and as recreation areas for the huge urban populations nearby.

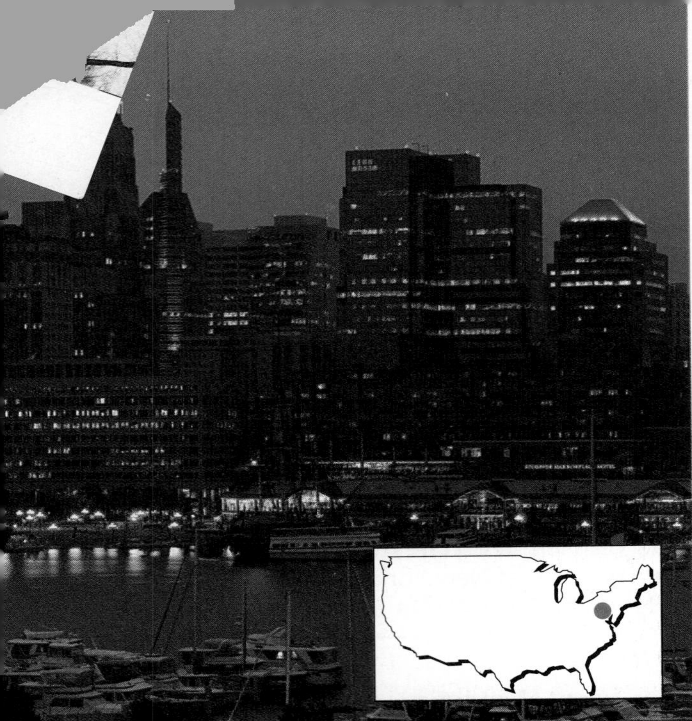

Office towers, hotels, shops, and apartments have replaced the decay that once haunted the Inner Harbor of **Baltimore, Maryland**. Many old industrial cities of the Northeastern United States are faced with the task of urban renewal.

INTERACTION Industry and Urban Growth

The Middle Atlantic states have benefited from their transportation access to national and world markets. This access along with a large labor force helped maintain the region's early industrial growth. As canals, roads, and railroads were built, the Middle Atlantic states found it easy to obtain raw materials from almost every part of the country and the world. Today, nearly every kind of manufacturing and service industry can be found in the region.

Industrial activity spread to several areas, such as upstate New York and western Pennsylvania, outside the region's major cities. New York's industries are located chiefly along the Hudson and Mohawk river valleys. This is the route of the Erie Canal, which was the first route to connect trade between the Atlantic Ocean and the Great Lakes. The canal was completed in 1825 and soon attracted industrial development in several cities. Albany, New York's state capital, is located near the junction of the Hudson and Mohawk rivers. Buffalo, the state's second largest city, is an industrial port on Lake Erie.

Pittsburgh, Pennsylvania, the largest industrial city in the Appalachians, also is the nation's largest inland harbor. It began as the French-built Fort Duquesne in 1754, where the Allegheny and Monongahela (muh-nahn-guh-HEE-luh) rivers join to form the Ohio River. In 1758, the fort was captured by the British and renamed Fort Pitt. Situated in the coal-producing region of western Pennsylvania, Pittsburgh became a major steel producer. As a result, it also became one of the nation's most polluted cities. The Pittsburgh urban area suffered economic problems and population losses with the decline of the steel industry. In recent years, new high-tech industries have helped renew the city, and it now has cleaner air.

The tourism industry is vital to many cities and towns throughout the Middle Atlantic States region. The major beach destinations include eastern Long Island, the Delmarva Peninsula, and the New Jersey shore, especially Atlantic City. Tourists also flock to the scenery of the Blue Ridge, Catskill, Adirondack, and Pocono mountains, and Niagara Falls. Countless historic and battlefield sites, such as Gettysburg, also are popular.

Issues The Middle Atlantic states played a key role in the industrialization of America. The region, however, now faces industrial decline as the nation moves from manufacturing to service and information industries. Since the 1970s, the Middle Atlantic states have experienced unemployment and migration to the Sunbelt. The industrial decline has affected most severely the small communities that depended on traditional industries such as coal mining, textiles, and steel. In contrast, the region's large cities have a greater economic diversity and major ports, which provide employment opportunities and a foundation for future growth.

2 SECTION REVIEW

1. What are the characteristics of the Middle Atlantic states' various landform regions? their climate types?
2. How did the location of the Middle Atlantic states help the region make the most of its resources? Why does the region now face economic decline?
3. **Critical Thinking** **Analyze** advantages and disadvantages of a plan to make the region's wilderness areas more accessible to tourists.

MEGALOPOLIS

FOCUS

- *What are the major cities of Megalopolis? What characteristics make each city important?*
- *What challenges do the cities of Megalopolis face?*

REGION The ongoing commercial and industrial growth of the Atlantic seaboard has led to the development of a continuous urban corridor. This corridor, called Megalopolis, was named by a French geographer in the 1960s. Gradually, the five major nodal cities of the Northeastern United States—Boston, New York, Philadelphia, Baltimore, and Washington, D.C.—spread toward each other. (See the map on page 121.) Excluding Washington, D.C., the original growth was due to the cities' good port sites, which linked the nation's interior to world trade routes.

The boundaries of Megalopolis continue to change and expand. Expansion has been most rapid as new highways allow commuters easy access to the cities. Subways, high-speed rails, and airline shuttles offer commuters an alternative to the automobile. In fact, this nodal region is connected by some of the densest railroad and interstate highway networks in the nation. The region has some of the world's busiest seaports and most crowded airports, as well.

Between these major cities are expanding suburbs and edge cities. Some farms and even forested land still exist within Megalopolis, however. Preserving the rural character within the expanding urban sprawl is a major challenge for the region.

Boston Megalopolis is anchored in the north by Boston, the capital of Massachusetts and New England's largest industrial, educational, financial, and commercial center. The city was founded as a British harbor in 1630, and its main commercial activities were connected with the sea. Boston Harbor has been modernized, and the waterfront and marketplace areas have been renovated. Much of Boston's recent growth has centered around highways called "electronics parkways." This term refers to the high-technology industries established there.

MOVEMENT **New York** South of Boston is New York City. Situated at the mouth of the Hudson River, the city has an excellent natural harbor with several bays and islands. It was originally settled by the Dutch, who named it New Amsterdam. The British seized the city in 1664 and renamed it New York. Today, it is the central node of Megalopolis. New York City is one of the world's most important cities and the site of the United Nations headquarters.

New York City is a major seaport and rail center and has three major airports. As the nation's commercial and financial center, the city is the site of stock exchanges, corporate headquarters, major banks, law firms, publishers, and advertising agencies. It is also the heart of the nation's garment industry and is a world fashion center. New York City serves as the nation's center for theater, art, and entertainment as well, and is the headquarters for the country's major television networks.

The city is divided into five **boroughs**, or administrative units. These are Manhattan, Staten Island, Brooklyn, the Bronx, and Queens. The commercial heart of New York City is Manhattan Island. Because New York City always has been a landing point for immigrants to the United States, it is a city of great ethnic diversity.

The importance of movement in a place's economic growth is clearly shown by New York City. The city surpassed its rivals, Philadelphia and Boston, in population and commerce when the Erie Canal connected it to the Great Lakes in 1825. As a result of the canal, New York City experienced tremendous growth and benefited from its role as a **break-of-bulk center**. A break-of-bulk center is a place where shipments of goods are moved from one mode of transportation to another, such as from train to ship.

The city's urban area has spread east across Long Island, north into Connecticut, and west across the Hudson River into New Jersey. The three-state region is connected by interstate highways, toll roads, a large subway and rail system, bridges, several underwater tunnels, and a ferry system.

Philadelphia Philadelphia is the fifth largest urban area in the United States and is an important port and industrial center. Located on the lower Delaware River, it has access to the Atlantic through Delaware Bay. Philadelphia was planned

Text continues on page 130.

Racing with the Sun

Each Memorial Day weekend, hundreds of thousands of people gather to watch the Indianapolis 500, a world-famous, high-speed automobile race. Noise shatters the quiet of a lazy holiday, fuel flows, and exhaust fumes pollute the air.

As cars circle the track in Indianapolis, Indiana, each year, another, quieter race is taking place. The drivers are younger; some are high-school students. One year, drivers raced from Albany, New York, to the finish line in Plymouth, Massachusetts. Although the cars reached speeds of almost 70 miles per hour (113 kph), they spent five days traveling the 250 miles (402 km) between the two cities, a snail's pace compared to the "Indy 500."

> *The goal of the creators of the Tour de Sol is to encourage the development and everyday use of solar-charged electric cars.*

Solar-Charged Cars

The name *Tour de Sol* offers a clue to what the annual race is about. *Sol* means "sun" in Latin, and the young drivers rely on energy from the sun. Lift the hood of a Tour de Sol car, and you will not find a methanol or gasoline engine. Instead, you might see an electric motor and batteries. The batteries are connected to solar cells that convert sunlight into electric current. One full day of sunlight can store up to 50 miles (80 km) of power. When a Tour de Sol car needs a recharge, the driver makes a pit stop and utilizes the sun, rather than gasoline, to provide energy.

A winning car in one Tour de Sol, the *Electric Hilltopper* would never have been mistaken for one of the sleek cars in the Indianapolis 500. Owned by a high-school teacher, the car once was destined for the junkyard. But some enterprising students saw a new use for the tired automobile.

The students worked for nine months. They removed the engine and other parts to make the car as light as possible. Then they added an electric motor and 16 batteries. The *Electric Hilltopper* set a Tour de Sol record, covering more than 100 miles (161 km) before needing its batteries recharged. "The car ran like a dream," one proud student beamed. "We never had to take a tool out of the toolbox."

Although fun and competition are all part of the Tour de Sol and similar races, such as Sunrayce '95 from Indianapolis to Colorado, the race has a serious purpose. The creators of these races believe that an alternative to the gasoline engine must be found. Their mission is to encourage the development and everyday use of solar-charged electric cars.

Gasoline engines are responsible for much of the air pollution that threatens our environment. Exhaust fumes help to produce the thick clouds of smog that cause watery eyes, clogged throats, and perhaps even lung cancer. Gases emitted from automobiles also contribute to acid rain and global warming.

VIRGINIA TECH

Solar car exhibition in **Putnam, Connecticut**

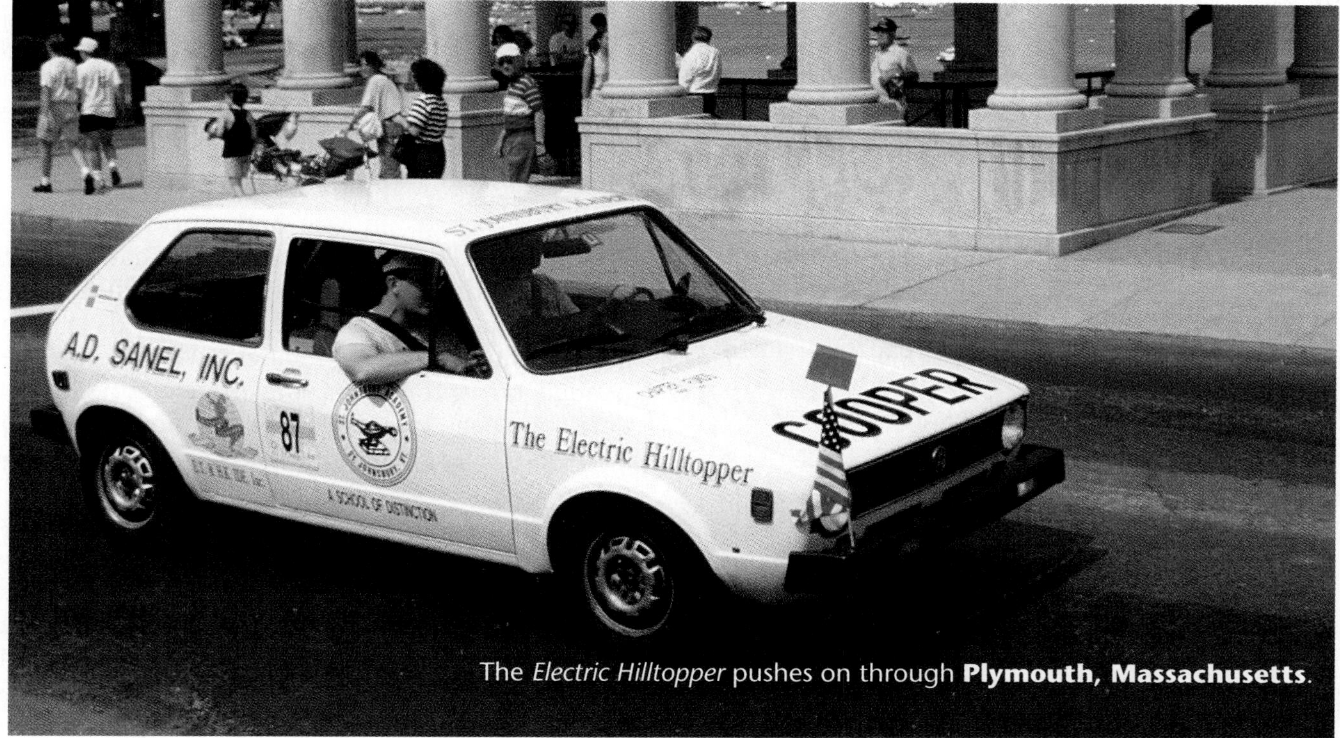
A.D. SANEL, INC.
87
The Electric Hilltopper
COOPER
A SCHOOL OF DISTINCTION
The *Electric Hilltopper* pushes on through **Plymouth, Massachusetts**.

A Renewable Resource

Unlike petroleum products, the sun provides a clean and inexpensive energy source. Better yet, solar energy is renewable—there is an endless supply. In contrast, petroleum is a fossil fuel, formed millions of years ago from microscopic marine plants and animals. Once pumped from the ground, petroleum takes millions of years to replace.

People have always made use of solar energy to meet their needs. In the southwestern United States, for example, many Native Americans built their homes of adobe brick, or sun-dried clay and mud. The adobe kept the homes cool in summer and warm in winter. Today, people living in sunny climates can place solar panels on roofs in order to heat water for homes and office buildings. Automobile manufacturers, meanwhile, are experimenting with solar and electric cars.

One Tour de Sol winner hopes to find marketable success with his solar-powered car. He claims that his efficient design will allow people to commute to work without using a drop of gasoline. The car would reach a speed of 60 miles per hour (97 kph) and travel 200 miles (322 km) before needing a recharge. Stressing other advantages, one solar-car enthusiast says, "If you're sitting in a traffic jam in the sun, this car isn't using power, it's gaining power. There's no pollution, no noise. That's the most amazing part."

Solar energy can power more than cars. Solar panels on the roof collect sunlight to heat water for this **Vermont** home. Satellites and other spacecraft also use solar panels for power.

Your Turn

Write an eyewitness account of a Tour de Sol race. In your account, **describe** one of the cars, and explain the role of the race in protecting the environment. You may also want to **invent** a solar-powered machine. Write a paragraph describing your invention.

by William Penn in the late 1600s and was the nation's most important city until the early 1800s. It is where the Declaration of Independence was signed in 1776.

Today, Philadelphia is a major manufacturing, pharmaceutical, banking, and educational center. Though many older sections of the city have suffered urban decay, downtown rebuilding programs have helped reverse this trend. The Philadelphia urban area has spread west farther into Pennsylvania, east beyond Camden, New Jersey, and south along the Delaware River.

Baltimore Baltimore, Maryland, is a seaport on the western shore of Chesapeake Bay. Founded along the Fall Line in 1729, it grew as a major port because of its rail connections to interior coal mines, steel mills, and farms. Its main industries are large oil refineries and steel mills. The city is also home to Johns Hopkins Hospital and University. A redevelopment program has helped rebuild the inner harbor area as well as some of the poorer neighborhoods.

Washington, D.C. Anchoring Megalopolis in the south is Washington, D.C. The nation's capital occupies a unique place among American cities. It is a federal district, located outside the authority of any state. There has long been an active movement to make the district the 51st state.

The District of Columbia was founded for one reason—to be the seat of the nation's government. The original city was designed in the late 1700s by French engineer and architect Pierre L'Enfant (lahn-FAHNT). Later work on the city was done by Benjamin Banneker, an African American mathematician, astronomer, and surveyor.

Washington, D.C., is a city mainly of politicians, lawyers, journalists, diplomats, military personnel, and tourists. In fact, tourism has become the city's second major source of income after government employment. Each year, millions of tourists are attracted to the city's many sites, including the U.S. Capitol, the White House, the Smithsonian Institution, the Vietnam Veterans Memorial, and hundreds of other historical and political landmarks.

Thousands of people who work in Washington, D.C., commute daily from their homes in suburbs and edge cities outside the boundaries of the district. The expanding Washington, D.C., urban area has grown so large that it has spilled over into Maryland and the southern state of Virginia. The southern boundary of Megalopolis now is approaching Richmond, Virginia.

Issues Changes in U.S. and world economies over the past 30 years have had a dramatic impact on Megalopolis. As traditional manufacturing has declined, factory jobs have disappeared. In turn, neighborhoods in large cities, which once depended on factory jobs, have declined. Substandard housing, crowded schools, and a lack of services have been problems for all the major urban areas of the region.

Meanwhile, new businesses in service industries and high technology have opened in cities on the edge of large urban centers. These edge cities around Boston, New York, Philadelphia, Baltimore, and Washington, D.C., have experienced explosive growth. As the population has spread out, new problems such as long commutes and traffic jams have become everyday challenges. Farmland and open space are being used for housing developments, business centers, and recreation areas.

Air pollution and disposal of sewage and toxic waste materials are vital concerns for Megalopolis. Indeed, pollution threatens rich ecosystems in the region, such as Chesapeake Bay. Environmental awareness by corporations and by the general public, though, is beginning to improve the situation. For example, the Hudson River, once polluted from the Adirondacks to New York Harbor, is recovering. The river now supports a healthy fish population, and people once again are making their homes nearby. Protecting its precious water resources, especially coastal bays and rivers, is a major challenge for the entire Northeastern United States.

3 SECTION REVIEW

1. List the major cities of Megalopolis. Next to each city, write down two of that city's important features.
2. What are two challenges faced by Megalopolis?
3. **Critical Thinking** **Predict** what effects a major strike of railroad workers might have on Megalopolis.

CHAPTER 11 CHECK

Row houses in **Boston, Massachusetts**

Reviewing the Main Ideas

1. In the nineteenth century, New England was well suited for industrial development because of its landforms, climates, and resources. The region's current industries are based increasingly on high technology.
2. Because of access to major land and water routes and resources such as coal, the Middle Atlantic states played a key role in the industrialization of the United States. Today, many of the nation's economic and political decisions are made in the region.
3. Megalopolis, a region overlapping the New England and Middle Atlantic states, is a continuous urban area along the eastern seaboard. The five major cities of the region are important industrial, commercial, financial, educational, and entertainment centers. They also face many economic and environmental challenges.

Building a Vocabulary

1. What is granite? What is a moraine? What natural forces have made granite and moraine materials common along the New England coastline?
2. Why are most of New England's forests second-growth forests?
3. What is biotechnology?
4. Why are some farms called truck farms?
5. What is the term for New York's five administrative units?

Recalling and Reviewing

1. How does New England's agriculture differ from the agriculture of the Middle Atlantic states? What accounts for the differences?
2. What makes each of the Northeastern United States subregions attractive to tourists? List three features of each subregion that attract tourists.
3. What environmental challenges are faced by New England, the Middle Atlantic states, and the cities of Megalopolis?

Thinking Critically

1. Imagine that you were a visitor to the Northeastern United States in the early nineteenth century. How would the region look in contrast to the way it looks today? How do you think it will look 100 years from today?
2. Evaluate the following statement as true or false: "Regions are defined solely on the basis of location; therefore, any individual place cannot be part of more than one region." Explain your evaluation, using examples from the chapter.

Learning About Your Local Geography

Cooperative Project

With your group, research the extent to which a nearby city or town has expanded over the past 20 years. Look for the growth of housing developments, office buildings, shopping centers, and malls. Locate or draw two maps, one of the city or town 20 years ago, and one of the city or town today.

Using the Five Themes of Geography

Construct a flowchart linking the two themes of **location** and **region** in the Northeastern United States. How did the location of the region and its cities affect its growth? How, in turn, did the growth of the region influence how some geographers divide the Northeast into subregions?

CHAPTER 12

The Southern United States

GEOGRAPHY DICTIONARY

barrier island

bayou

levee

lock

tenant farmer

diversify

bilingual

"Never saw anything more beautiful than these trees [live oaks]. . . . They are the first objects that attract one's attention here. They are large, noble trees with small glossy green leaves. Their great beauty consists in the long bearded moss with which every branch is heavily draped. This moss is singularly beautiful, and gives a solemn almost funereal [gloomy] aspect to the trees."

Charlotte Forten

Fiddle from **Louisiana**

The Southern United States stretches in a great arc from Virginia to Texas. The states of this region are Virginia, North Carolina, South Carolina, Georgia, Kentucky, Tennessee, Alabama, Florida, Mississippi, Arkansas, Louisiana, and Texas. The largest state in the region, Texas, shares a long international border with Mexico.

Historically, the region was rural and agricultural. The majority of the population lived on small farms, but enslaved African Americans provided the labor for large cotton, rice, and tobacco plantations in the region. After the Civil War and the end of slavery, many African Americans as well as whites migrated to northern cities in search of factory jobs.

Since the 1960s, there has been a large movement of people back to the South from the North. In addition, thousands of people have moved to the region from the Caribbean, Mexico, and Middle and South America.

Louisiana live oak trees draped in Spanish moss

SOUTHERN UNITED STATES: PHYSICAL–POLITICAL

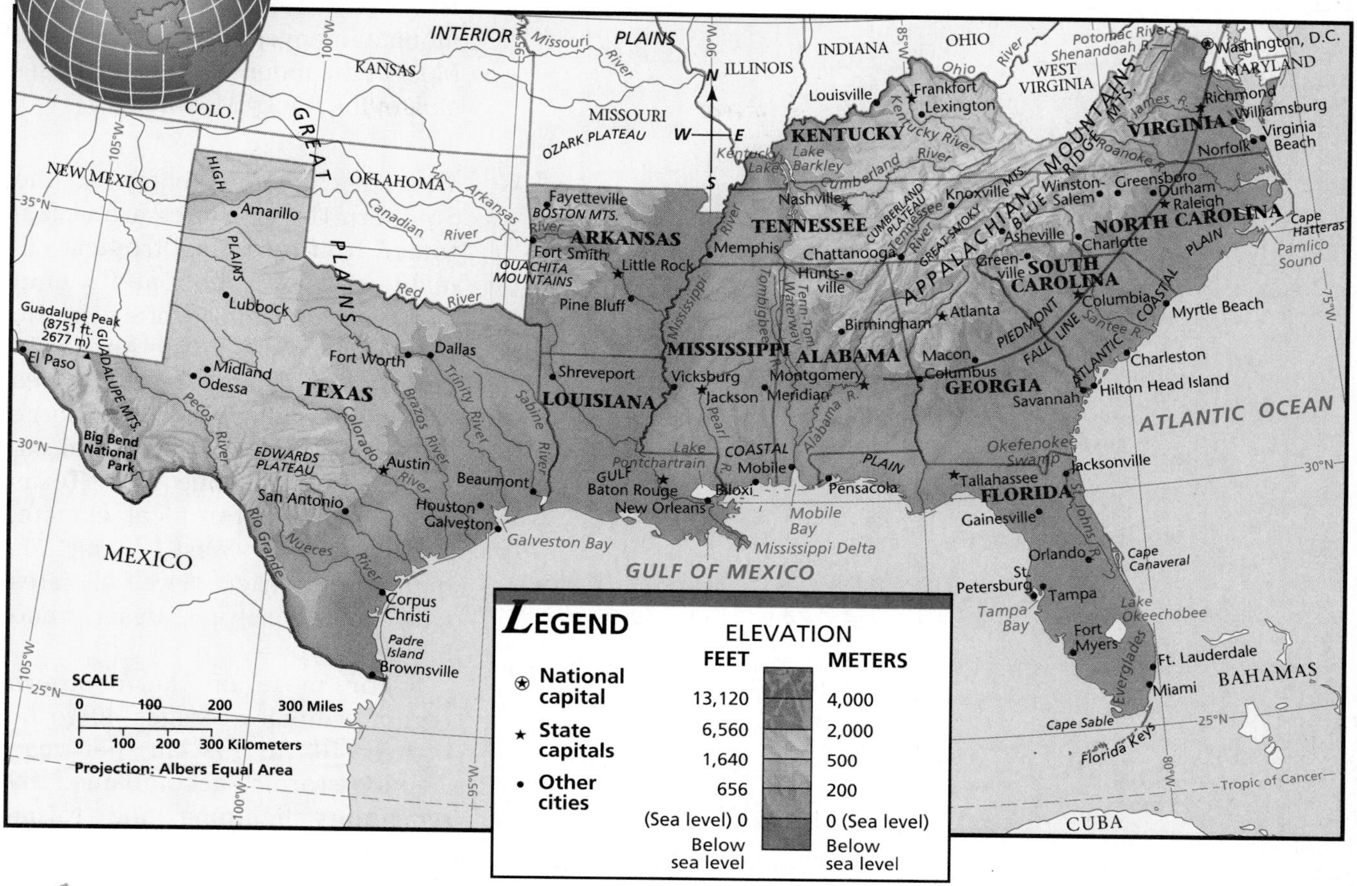

LEGEND

- ⊛ National capital
- ★ State capitals
- • Other cities

ELEVATION

FEET	METERS
13,120	4,000
6,560	2,000
1,640	500
656	200
(Sea level) 0	0 (Sea level)
Below sea level	Below sea level

SCALE: 0–100–200–300 Miles; 0–100–200–300 Kilometers

Projection: Albers Equal Area

PHYSICAL GEOGRAPHY

FOCUS

- *What landform regions are found in the Southern United States?*
- *What are the characteristics of climates in the southern states?*
- *Why are river systems important to the southern region?*

REGION Landforms The dominant landform region of the Southern United States is the Coastal Plain. Miles of sandy beaches mark where the plain meets the waters of the Atlantic and the Gulf of Mexico. Inland from the Coastal Plain are the other major landform regions—the Piedmont, the southern Appalachian and Ozark highlands, and the Interior Plains. West Texas extends into the Great Plains. (See the map above.)

The Coastal Plain stretches inland from the Atlantic Ocean and the Gulf of Mexico. Most of this low area has a shoreline bordered by **barrier islands**. A barrier island is a long, narrow, sandy island separated from the mainland by a shallow lagoon or wetland.

Lying just above sea level, the Coastal Plain holds the nation's largest wetlands. These include the Everglades in Florida, the Okefenokee Swamp in Georgia, the Mississippi Delta in Louisiana, and hundreds of coastal marsh areas behind the barrier islands. The Coastal Plain also contains the Mississippi Delta, the lower Mississippi Valley, and all of the Florida peninsula. The plain then stretches inland to the Fall Line, where it meets the Piedmont. The rolling hills of the Piedmont rise inland in central Virginia, in the Carolinas, and in northern Georgia.

Adding to the beauty of the region are the southern Appalachian Mountains. These include the Blue Ridge and Great Smoky mountains and the Cumberland Plateau. The Appalachians cross the western part of Virginia and the Carolinas, the

Natural Regions of Texas

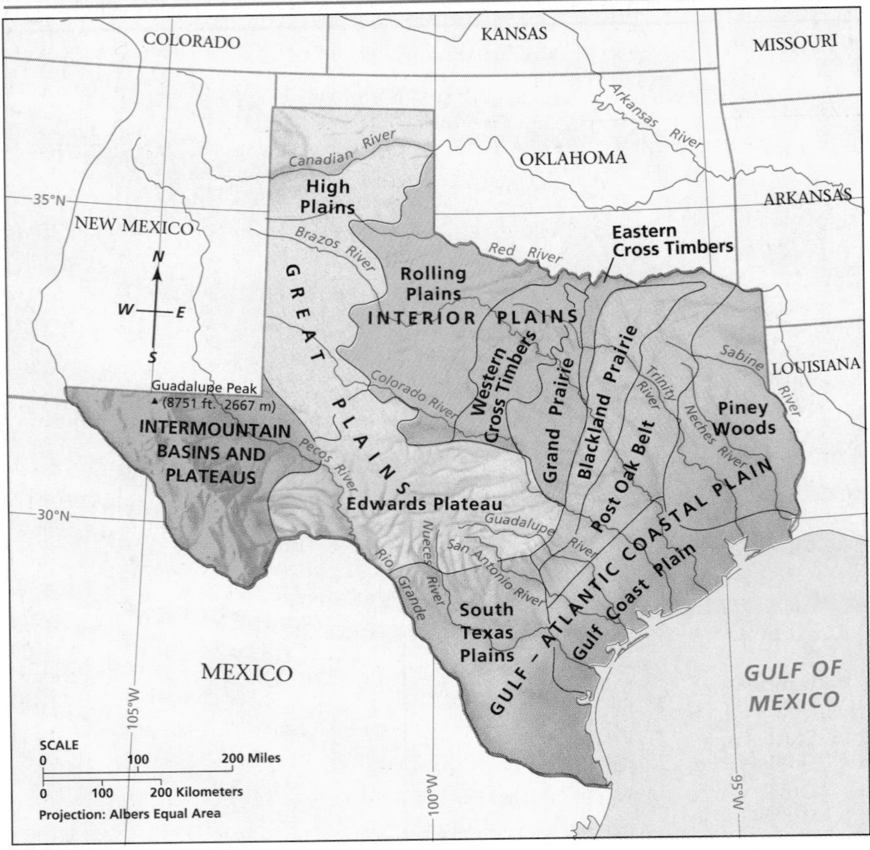
COLORADO
KANSAS
MISSOURI
OKLAHOMA
ARKANSAS
NEW MEXICO
LOUISIANA
MEXICO
GULF OF MEXICO
35°N
30°N
105°W
100°W
95°W
Canadian River
Arkansas River
Red River
Brazos River
Colorado River
Pecos River
Rio Grande
Nueces River
San Antonio River
Guadalupe River
Trinity River
Neches River
Sabine River
High Plains
Rolling Plains
INTERIOR PLAINS
GREAT PLAINS
Eastern Cross Timbers
Western Cross Timbers
Grand Prairie
Blackland Prairie
Post Oak Belt
Piney Woods
Edwards Plateau
South Texas Plains
Gulf Coast Plain
GULF – ATLANTIC COASTAL PLAIN
INTERMOUNTAIN BASINS AND PLATEAUS
Guadalupe Peak
(8751 ft. 2667 m)
N
S
E
W
SCALE
0 100 200 Miles
0 100 200 Kilometers
Projection: Albers Equal Area

MAP STUDY The large state of Texas has a variety of landforms, climates, and biomes. Texas can be divided into natural regions, which are part of the four major physiographic regions in Texas. See the map on page 112 for the physiographic regions of the United States and Canada. Environmental conditions are similar within each of the state's natural regions.

eastern portions of Kentucky and Tennessee, and northern Georgia and Alabama.

The Ozark Plateau lies mainly in Arkansas but extends into Oklahoma and Missouri. This ancient plateau region has eroded into rugged hill country. The Arkansas River forms a valley through the center of the Ozarks.

Central Kentucky and Tennessee and portions of Texas are part of the Interior Plains, while most of central and western Texas is part of the Great Plains. The northwest part of Texas' Great Plains region is called the High Plains, where elevations reach between 2,000 and 5,000 feet (610 and 1,524 m). South of the High Plains is the Edwards Plateau. This hilly area features many springs where groundwater reaches the surface.

Eastern Texas has the most in common with the rest of the southern states. Most of eastern Texas lies in the Coastal Plain and is lowland with gently rolling hills. Southwest Texas is a region of mountain ranges and desert basins. Most of the mountains are an extension of the Rocky Mountains.

Climate People living in the Southern United States are accustomed to a humid-subtropical climate with long, hot, and humid summers and mild winters. In the higher elevations of the Appalachians and Ozarks, temperatures are cooler. Because of its large size, Texas has a more varied climate pattern. While East Texas has a humid-subtropical climate, drought-prone West Texas is higher and drier. Here, climates vary among steppe, desert, and highland types.

For most of the southern region, rainfall averages 40 to 60 inches (102 to 152 cm) per year. Thunderstorms, accompanied by dangerous lightning and tornadoes, bring much rain to the region. In the Appalachians, annual rainfall totals may reach more than 100 inches (254 cm), and some areas experience occasional winter snowstorms. Invasions of cold arctic air, called "northers," have caused crop damage as far south as southern Texas and Florida.

During late summer and early fall, hurricanes from the Atlantic Ocean or Gulf of Mexico may strike coastal areas. In 1992, Hurricane Andrew cut a path of destruction across south Florida and Louisiana. The worst natural disaster in the United States was the 1900 hurricane that flattened Galveston, Texas, and claimed more than 6,000 lives. Cities like Galveston that are located on barrier islands are threatened by hurricanes and severe coastal erosion.

INTERACTION **River Systems** The Mississippi is one of the largest river systems in the world. Along with its tributaries, which include the Ohio and Missouri rivers in the north and the Arkansas, Tennessee, and Red rivers in the south, the Mississippi drains nearly half of the continental United States.

Sediment is carried by the river and deposited at its mouth, forming the large Mississippi Delta. The delta is located in Louisiana, where the river empties into the Gulf of Mexico. The delta's land is low and swampy and cut by small, sluggish streams called **bayous**.

The Mississippi has often flooded its banks and caused much destruction to the areas along its floodplain. Yet, the river serves as a means of transportation, and the floods are a source of fertile new soil. Authorities have tried to lessen flood danger and damage from the Mississippi with better flood forecasting, dams along the river's upper tributaries, and an improved system of **levees** (LEHV-eez). A levee is a ridge of earth along a riverbank that hinders flooding.

The Tennessee River originates in the Appalachians and later joins the Ohio River. In 1933, the federal government formed the Tennessee Valley Authority (TVA). The TVA built dams, reservoirs, and canals in the Tennessee River valley to control flooding, slow soil erosion, allow barge navigation, produce electricity, and increase economic development.

The newest river project is the Tenn-Tom Waterway. Partly as a result of the construction of **locks** on the Tombigbee River, the Tennessee River now connects the Ohio River to the port of Mobile, Alabama, on the Gulf of Mexico. A lock is a part of a waterway enclosed by gates at each end. It is used to raise or lower boats as they pass from one water level to another. These channel and lock systems allow barge transportation to the Gulf ports and provide flood control and hydroelectricity.

The lower Rio Grande forms the international boundary between the United States and Mexico. The irrigation waters from this river are important to the productive farms along this borderland.

SECTION REVIEW

1. Describe three of the landform regions of the Southern United States.
2. What kinds of storms is the southern region likely to experience? What damage can these storms cause?
3. **Critical Thinking** **Assess** the importance of river control techniques to the Southern United States.

ECONOMIC GEOGRAPHY

FOCUS

- *Why and how has agriculture been changing in the Southern United States?*
- *What resources and industries are important to the region's economy?*

Agriculture The climate, rainfall, and soil on the Coastal Plain are well suited to growing cotton. Cotton became the leading commercial crop of the South in the years before the Civil War. After the end of slavery, many former slaves and some whites became **tenant farmers**. A tenant farmer is one who rents a plot of land on which to grow crops.

From the 1930s to 1980s, cotton production in the region fell. Pests, such as the boll weevil, foreign competition, expensive mechanization, and synthetic materials all contributed to this decline. Since then, cotton demand has increased, but now most of the nation's cotton is grown on large irrigated and mechanized farms farther west. Today, the leading cotton-producing state is Texas.

Tobacco was introduced to the early settlers by Native Americans. Tobacco is grown mainly on the Coastal Plain and the Piedmont. The farming of tobacco is now mechanized, and the United States remains one of the world's largest producers and exporters of this profitable crop. Health concerns related to tobacco products have forced most tobacco farmers to **diversify**, or to produce a variety of crops.

Today, the Southern United States leads the nation in the production of citrus fruits. Oranges, lemons, and grapefruits are grown in central Florida and in the lower Rio Grande valley of Texas. Florida produces about 75 percent of the nation's oranges and grapefruits.

Many agricultural areas of the Southern United States specialize in certain crops. These crops include winter vegetables from the lower Rio Grande valley; peaches from Georgia; sugarcane from Louisiana, Florida, and Texas; rice from Arkansas and Mississippi; and peanuts and pecans from Georgia and Alabama.

Soybeans have become one of the region's most valuable crops as well, particularly in the

FACTS IN BRIEF States of the Southern United States

STATE POPULATION (1993)	POP. RANK (1993)	AREA (1990)	AREA RANK (1990)	PER CAPITA INCOME (1993)
ALABAMA 4,187,000	22	52,237 sq. mi. 135,294 sq. km	30	$17,234
ARKANSAS 2,424,000	33	53,182 sq. mi. 137,741 sq. km	28	$16,143
FLORIDA 13,679,000	4	59,988 sq. mi. 155,369 sq. km	23	$20,857
GEORGIA 6,917,000	11	58,977 sq. mi. 152,750 sq. km	24	$19,278
KENTUCKY 3,789,000	24	40,411 sq. mi. 104,664 sq. km	37	$17,173
LOUISIANA 4,295,000	21	49,650 sq. mi. 128,594 sq. km	31	$16,667
MISSISSIPPI 2,643,000	31	48,286 sq. mi. 125,061 sq. km	32	$14,894
NORTH CAROLINA 6,945,000	10	52,672 sq. mi. 136,420 sq. km	29	$18,702
SOUTH CAROLINA 3,643,000	25	31,189 sq. mi. 80,780 sq. km	40	$16,923
TENNESSEE 5,099,000	17	42,145 sq. mi. 109,156 sq. km	36	$18,434
TEXAS 18,031,000	3	267,277 sq. mi. 692,247 sq. km	2	$19,189
VIRGINIA 6,491,000	12	42,326 sq. mi. 109,624 sq. km	35	$21,634

Rankings are for all 50 states.
Source: *Statistical Abstract of the United States 1994*

Which two states in the Southern United States are the third and fourth most populous in the country?

lower Mississippi Valley with its rich alluvial soil. Corn, for livestock feed, also has become more important. Florida, Georgia, and Texas are major beef-cattle producers. The poultry industry is important in Arkansas, Georgia, and Mississippi.

Resources The coastal waters of the Gulf of Mexico and the Atlantic are rich in ocean life. Though all the southern coastal states have fishing industries, Louisiana and Texas are the leaders. The shallow waters around the Mississippi Delta produce great quantities of oysters, shrimp, and other seafood.

Before the region filled with settlers, nearly all of the Southern United States was forested. Today, second-growth timber covers more than half the area. The southern states generate about one-third of the nation's commercial forest production. Southern pine is plentiful on the Coastal Plain, and hardwoods are common in the Mississippi Valley and on the Piedmont. The softwood forests and tree farms support a paper and pulp industry.

The major mineral and energy resources of the Southern United States include coal, sulfur, salt, phosphates, oil, and natural gas. Coal and iron-ore deposits supplied the steel mills of Birmingham, Alabama, in the late 1800s. A century later, the iron has been mined out and production is down. The city now depends more on service industries, although it still produces metal products.

Texas and Louisiana lead the nation in the production of sulfur and are also important salt-mining states. Florida, meanwhile, has the largest phosphate-mining industry in the nation. Much of the product, which is used to make fertilizer, is exported to Japan and other Asian nations.

Oil is found in West Texas and in coastal Texas and Louisiana. In fact, these two states produce about one-third of the country's crude oil. Much of the production is from offshore oil rigs in the continental shelf waters of the Gulf of Mexico.

With low oil prices on the world market during the 1980s and early 1990s, Texas and Louisiana both experienced high unemployment rates and a loss in government income. The oil and gas industries here suffer from "boom and bust" periods, causing ups and downs in the regional economy.

Industry The Southern United States now contains about 90 percent of the U.S. textile industry. This industry is heavily concentrated on the Piedmont of Georgia, in the Carolinas, and in Virginia. Most of the large cotton and synthetic-fiber mills are located in cities in North Carolina and Virginia.

Oil refineries are concentrated along the Texas Gulf Coast, producing gasoline, kerosene, and lubricating oils. Associated petrochemical products include plastic, synthetic rubber, paint, tar, fiber, asphalt, explosives, fertilizer, and insecticide. Most petrochemical plants lie in the Houston, Texas, area and along the lower Mississippi in the area around Baton Rouge, Louisiana.

Many new industries related to high technology, publishing, and aerospace have located in the Southern United States. Cities that boast new

A shrimper clears the day's catch from the hold of a shrimping boat along the **Louisiana** coast. The Gulf Coast is the richest shrimp-producing region in the United States.

industries and research facilities include Austin, Texas; Tampa and Orlando, Florida; Raleigh, North Carolina; and Nashville, Tennessee. The region's low labor costs, transportation access, and relatively inexpensive land have also drawn many foreign companies.

Tourism is a major industry in many southern states, particularly Florida. The state's warm winters attract millions of visitors to beach resorts, the Florida Keys, and Everglades National Park. Coastal resorts such as Padre Island in Texas and Hilton Head and Myrtle Beach in South Carolina attract tourists to their sandy beach locations. Other major destinations in the region include scenic areas of the Ozarks and southern Appalachians, especially Great Smoky Mountains National Park in Tennessee.

SECTION REVIEW

1. What crops have been important in the Southern United States?
2. What industries are found in the southern region?
3. **Critical Thinking** **Predict** what might happen to the economies of Texas and Louisiana when their oil resources are depleted. How might these states prepare?

URBAN GEOGRAPHY AND REGIONAL ISSUES

FOCUS

- *What are the major cities of the Southern United States?*
- *What challenges does the region face?*

The Southern United States is home to some of the nation's oldest cities, including St. Augustine, Florida, founded by Spanish settlers in 1565. Today, the region's metropolitan areas are among the fastest growing in the country as more and more people move to the Sunbelt states.

Atlanta Georgia's largest city, Atlanta is the major transportation hub of the Southern United States. Atlanta's selection as host of the 1996 Olympics is evidence of the city's importance as a major commercial center. The metropolitan area, with nearly 3 million people, is one of the largest urban regions in the southeastern United States.

Miami The rapid growth along the southeast coast of Florida started with tourism and retirement communities. Today, Miami is a major center for transportation, especially with its close connections to the Caribbean, Mexico, and Central America. Electronics, aerospace, and medical technology industries also are established here. Much of Miami's recent population growth is the result of immigration from Cuba, Nicaragua, and other Central American countries.

Tampa and St. Petersburg Located on Florida's Gulf Coast, this Sunbelt metropolitan area grew rapidly during the 1980s. Much of the growth was due to a large migration of retirees from the northern states, which fueled a construction boom. Tampa's port is the export harbor for products from Florida's phosphate mines and citrus farms.

New Orleans Located on the Mississippi River about 100 miles (161 km) from the Gulf of Mexico, New Orleans was founded by the French in 1718. Once a riverboat port, New Orleans now is a major oil-tanker port. Its main industries include petroleum, chemicals, and metal. Its famous jazz

Danny Barker and The Jazz Hounds

Jazz musicians such as these are popular performers in **New Orleans**. African Americans made New Orleans the birthplace of a brand of music known as Dixieland jazz.

and annual Mardi Gras festivals help make tourism a major source of income as well. Much of New Orleans is below sea level, so serious drainage problems occur during storms. Pollution threatens the city's water supply.

Houston Houston is the nation's fourth largest metropolitan area, its third busiest port, and Texas' largest city. The city was founded in 1836 and is named after Sam Houston, who helped gain Texas' independence from Mexico that same year. Its recently diversified economy includes a large medical complex and service and convention industries. Houston's many refineries and petrochemical and oil-equipment companies reflect its location near the local oil fields.

Dallas-Fort Worth Area About 250 miles (402 km) north of Houston is the Dallas-Fort Worth metropolitan area, known locally as "the Metroplex." This large urban area has grown around two nodes that are 30 miles (48 km) apart. Dallas was founded in 1841 as a railroad shipping stop for cattle and cotton. Today, it is a large cosmopolitan city and center for electronics, banking, insurance, and fashion. To the west is Fort Worth, an early cattle-marketing center. Today, it is a center for petroleum, grain, and aerospace, though the city retains its large cattle stockyards. Between Dallas and Fort Worth are many high-tech industries and recreational facilities, as well as one of the nation's busiest airports.

San Antonio Located in hilly south central Texas, San Antonio has experienced rapid growth. It was founded as a Spanish mission town in 1718 and is one of the region's oldest cities. It is home to food processing, electronics, and aerospace industries, and several military bases. The city is also Texas' major tourist destination. San Antonio is a **bilingual** city with half of its population of Hispanic (mainly Mexican American) origin. A bilingual person has the ability to speak two languages.

INTERACTION **Issues** Although economic progress in the southern Sunbelt cities and coastal areas has been rapid, several southern states suffer the lowest per capita incomes, health standards, and literacy rates in the United States. High unemployment and poverty rates plague Appalachia, part of which is in the Middle Atlantic states, and the lower Mississippi Valley and Delta area.

Rapid growth in urban areas has contributed to traffic congestion, air pollution, ghetto growth, and housing shortages. Newcomers have strained public facilities such as schools, water supplies, and public transportation. Thousands of people have settled on the barrier islands. Such high population densities put a strain on the already-limited roads and bridges. Future hurricanes may endanger the lives and property of the many people who live here.

Pollution and new drainage systems have caused environmental damage to the region's wetlands. The Florida Everglades, a giant "river of grass," has shrunk as natural drainage systems have been altered by increased farming demands. Lower water levels also expose the Everglades and its wildlife to increased dangers from hurricanes and from fires during periods of drought.

3 SECTION REVIEW

1. Which cities of the Southern United States are transportation centers or ports? To which cities is the electronics industry important?
2. What challenges exist in the Southern United States as a result of growth?
3. **Critical Thinking** **Evaluate** the following statement: "Miami and San Antonio both have large Spanish-speaking populations. Therefore, cultures in the two cities are similar."

CHAPTER 12 CHECK

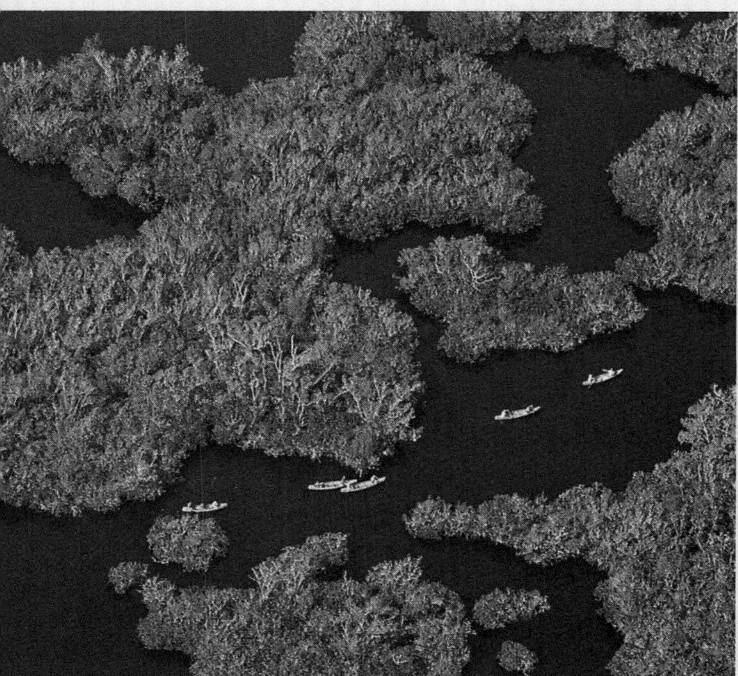

Boating in the **Florida Everglades**

Reviewing the Main Ideas

1. The Southern United States has a variety of landforms, ranging from plains to mountains. Most of the region has a humid-subtropical climate. The Mississippi River, one of the largest river systems in the world, has been critical in the development of the economy of the Southern United States.
2. Cotton and tobacco traditionally have been important crops in the southern region, but fruits, vegetables, rice, soybeans, and corn have become increasingly important. Ocean, forest, and mineral resources are also significant. Industries in the region are based on textiles, oil, technology, publishing, aerospace, and tourism.
3. Some of the largest urban areas of the Southern United States are Atlanta, Miami, Tampa and St. Petersburg, New Orleans, Houston, Dallas-Fort Worth, and San Antonio. Rapid growth has presented the region with a number of challenges.

Building a Vocabulary

1. What is a barrier island?
2. What is the term for a small stream that cuts through the swampy land of a delta?
3. How are levees and locks used to control rivers and river systems?
4. Why might farmers diversify their crops?

Recalling and Reviewing

1. Compare and contrast the landform regions found in Texas with the landform regions found in other parts of the Southern United States. What accounts for the differences?
2. What are some of the economic challenges that the southern states have faced in recent years?
3. What groups of people are migrating to the southern region?

Thinking Critically

1. What precautions might people who live in the southern region take against hurricanes and floods? Design an evacuation plan for the residents of barrier islands in the case of a severe hurricane.
2. Imagine that you are the president of a major corporation based in the Northeastern United States, and you are considering moving your corporate headquarters to the southern region. What factors should you consider in making your decision?

Learning About Your Local Geography

Individual Project

Investigate the last major storm that caused damage to your community. Interview long-term residents of your community to determine when the storm occurred. Next, go to the library and look through old newspapers from the time of the storm. Create illustrations showing damage done by the storm. Write captions for your illustrations. You may want to include your illustrations in your individual portfolio.

Using the Five Themes of Geography

Write three to five statements discussing the themes of **movement** and **place** in the Southern United States. What place characteristics encourage movement into and within the region? What place characteristics allow such movement to occur?

CHAPTER 13

The Midwestern United States

GEOGRAPHY DICTIONARY

dredge

channel

township and range system

Corn Belt

Dairy Belt

"I came to know and care for a landscape that few who are not midwesterners ever call beautiful. Travelers, whether in the air or on the ground, usually see the Middle West less as a destination than as a place to pass through. Only after a long while does one appreciate that the very plainness of the countryside is its beauty. . . . When people speak . . . of the American heartland, this is one of the places they mean."

William Cronon

The Midwest includes the eight states of Ohio, Indiana, Illinois, Michigan, Wisconsin, Minnesota, Iowa, and Missouri. As the nation's agricultural and industrial center, the region has great economic importance. Most of the nation's corn and much of its meat and dairy products come from the Midwest. Access to the Atlantic Ocean through the Great Lakes–St. Lawrence Seaway and to the Gulf of Mexico through the Mississippi River system make the region well situated for international trade.

Butter churn

Prior to European exploration, the Midwest was home to Native Americans who hunted and farmed in the region. By the 1840s, most of the Native American lands were taken by European settlers. As transportation routes from the East Coast developed, thousands more settlers moved to the Midwest, attracted to the promising farmlands and growing cities. In the late 1800s, African Americans, many of whom were former slaves, migrated from the southern states seeking jobs in the new Midwestern cities.

A dairy farm near **Baraboo, Wisconsin**

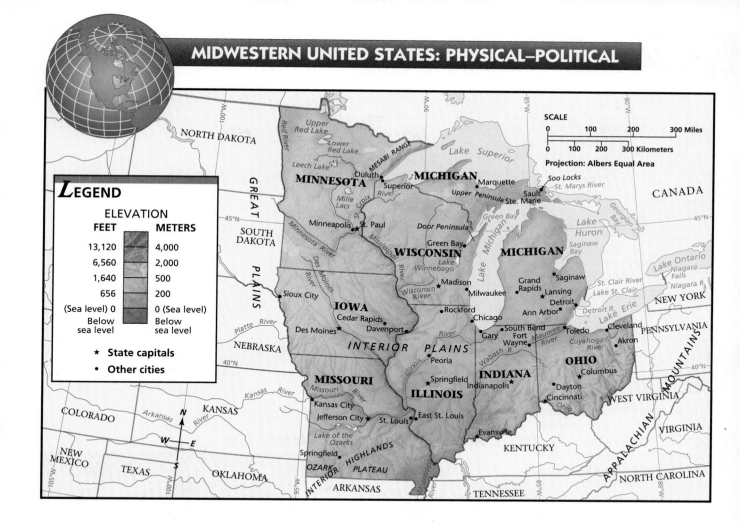

MIDWESTERN UNITED STATES: PHYSICAL–POLITICAL

LEGEND

ELEVATION

FEET	METERS
13,120	4,000
6,560	2,000
1,640	500
656	200
(Sea level) 0	0 (Sea level)
Below sea level	Below sea level

★ State capitals

• Other cities

SCALE

Projection: Albers Equal Area

1 PHYSICAL GEOGRAPHY

FOCUS

- *What are the characteristics of landforms and climates found in the Midwest?*
- *What lakes and rivers are found in the region, and why are they important?*

REGION **Landforms and Climate** The Midwest lies almost entirely within the Interior Plains, a region of flat plains and low hills. More varied landforms are found in the eroded hills of the ancient Canadian Shield, which extends south from Lake Superior. The most rugged areas are in the Ozark Plateau, which stretches into southern Missouri. (See the map above.)

During the last ice age, huge ice sheets covered the Midwest as far south as the Missouri and Ohio rivers. The northern part of the region was eroded glacially, leaving behind ice-scoured rocks, thin soils, and thousands of lakes. Minnesota alone has more than 10,000 lakes. Most of the southern portion of the Midwest has thick, rich soils that were deposited by glaciers.

The entire Midwest region has a humid-continental climate with four distinct seasons. Generally, the northern Midwest states along the Great Lakes have cool, short summers with severe winters. The southern Midwest states along the Ohio and Missouri river valleys have longer summers filled with hot, humid days. While winters tend to be milder to the south, all of the Midwest region experiences cold arctic air and snow in winter. The region receives adequate rainfall for agriculture, though summer droughts occasionally occur. The entire region is subject to severe thunderstorms and tornadoes.

The Great Lakes The Great Lakes are the largest freshwater lake system in the world and contain one-fifth of the world's fresh surface

The Great Lakes and St. Lawrence Seaway

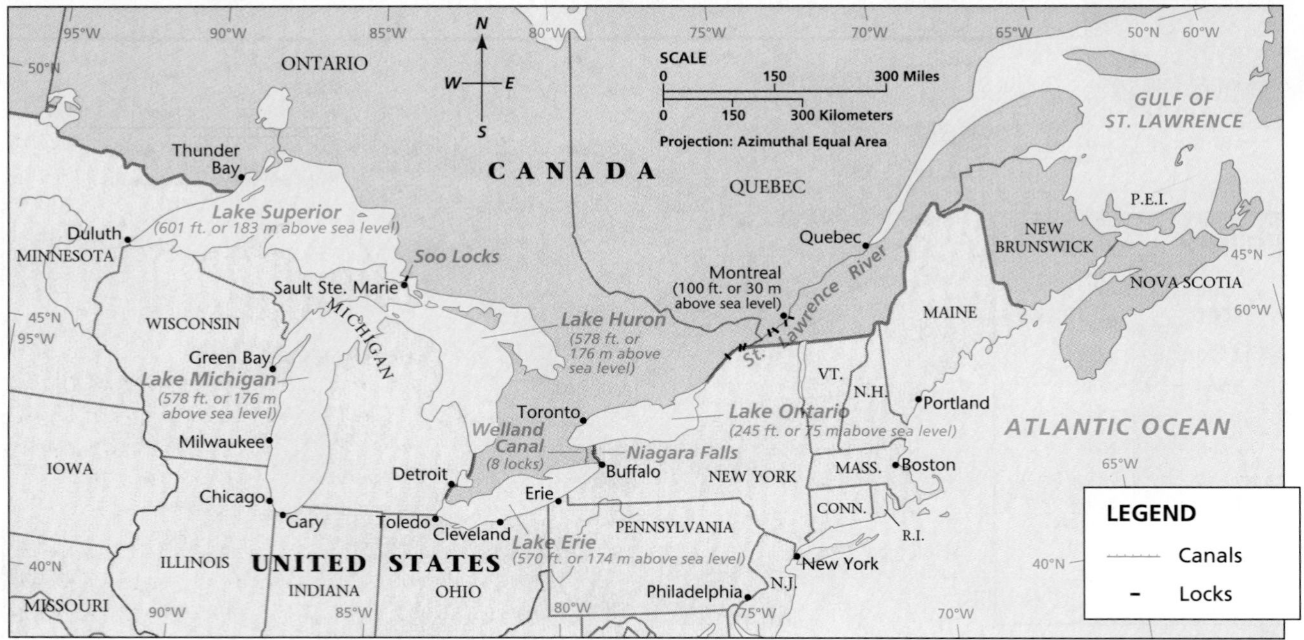
ONTARIO
CANADA
QUEBEC
SCALE
0 150 300 Miles
0 150 300 Kilometers
Projection: Azimuthal Equal Area
GULF OF ST. LAWRENCE
Thunder Bay
Lake Superior
(601 ft. or 183 m above sea level)
Duluth
MINNESOTA
Soo Locks
Sault Ste. Marie
MICHIGAN
WISCONSIN
Green Bay
Lake Michigan
(578 ft. or 176 m above sea level)
Milwaukee
IOWA
Chicago
Gary
ILLINOIS
MISSOURI
INDIANA
OHIO
UNITED STATES
Detroit
Toledo
Cleveland
Erie
Lake Erie
(570 ft. or 174 m above sea level)
Lake Huron
(578 ft. or 176 m above sea level)
Toronto
Welland Canal
(8 locks)
Niagara Falls
Buffalo
Lake Ontario
(245 ft. or 75 m above sea level)
NEW YORK
PENNSYLVANIA
Philadelphia
N.J.
New York
CONN.
R.I.
MASS.
Boston
VT.
N.H.
Portland
MAINE
Montreal
(100 ft. or 30 m above sea level)
St. Lawrence River
Quebec
NEW BRUNSWICK
P.E.I.
NOVA SCOTIA
ATLANTIC OCEAN
LEGEND
Canals
Locks

The St. Lawrence Seaway permits the passage of ships between ports on the Great Lakes and the Atlantic Ocean. The waterway includes canals and locks that allow ships to move from one water level to another. The eastern end of Lake Erie, at the western end of the waterway, is about 570 feet (174 m) above sea level. Farther east, Montreal lies at about 100 feet (30 m) above sea level. Ships moving from one water level to another enter a lock. The water level inside the lock is raised or lowered to match the water level in the portion of the waterway ahead of the ship.

Moving Through a Canal Lock

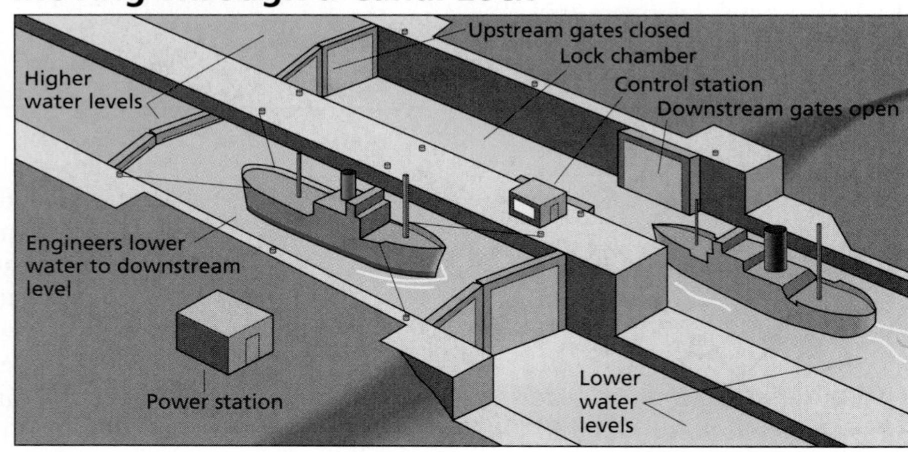
Upstream gates closed
Lock chamber
Control station
Downstream gates open
Higher water levels
Engineers lower water to downstream level
Power station
Lower water levels

water. They include lakes Superior, Michigan, Huron, Erie, and Ontario. All of the Great Lakes were carved by ice sheets during the last ice age. All but Lake Michigan are shared with Canada.

Lake Superior is the largest and deepest of the lakes and has the highest lake-level elevation. It is connected to neighboring lakes Michigan and Huron by the St. Marys River. They, in turn, are connected to Lake Erie via the St. Clair River, Lake St. Clair, and the Detroit River. Lake Erie is connected to the lowest lake, Lake Ontario, by the Niagara River, which flows over Niagara Falls. Lake Ontario flows into the St. Lawrence River, which empties into the Atlantic Ocean. Locks between the lakes allow for interlake shipping.

The St. Lawrence Seaway, completed in 1959, is one of the world's greatest inland navigation and transportation systems. It allows oceangoing ships from the Atlantic to enter the St. Lawrence River in eastern Canada and travel into the heartland of North America as far as Chicago, Illinois, or Duluth, Minnesota. Ice during the severe winters causes the seaway to close for several months each year.

FACTS IN BRIEF States of the Midwestern United States

STATE POPULATION (1993)	POP. RANK (1993)	AREA (1990)	AREA RANK (1990)	PER CAPITA INCOME (1993)
ILLINOIS 11,697,000	6	57,918 sq. mi. 150,008 sq. km	25	$22,582
INDIANA 5,713,000	14	36,420 sq. mi. 94,328 sq. km	38	$19,203
IOWA 2,814,000	30	56,276 sq. mi. 145,755 sq. km	26	$18,315
MICHIGAN 9,478,000	8	96,705 sq. mi. 250,466 sq. km	11	$20,453
MINNESOTA 4,517,000	20	86,943 sq. mi. 225,182 sq. km	12	$21,063
MISSOURI 5,234,000	16	69,709 sq. mi. 180,546 sq. km	22	$19,463
OHIO 11,091,000	7	44,828 sq. mi. 116,105 sq. km	34	$19,688
WISCONSIN 5,038,000	18	80,374 sq. mi. 208,169 sq. km	16	$19,811

Rankings are for all 50 states.
Source: *Statistical Abstract of the United States 1994*

Which state had the highest per capita income in 1993 in the Midwestern United States?

MOVEMENT The Upper Mississippi River The drainage patterns of the Midwest were formed by the melting of the great ice sheets. Today, the upper Mississippi River system drains the region. The major tributaries include the Ohio, Missouri, and Illinois rivers. These rivers have been **dredged** and **channeled**, which means they have been dug out and straightened. The dredging and channeling allows barge traffic to move between the Midwest and the Gulf of Mexico. In 1993, two months of heavy rainfall swelled the upper Mississippi and its tributaries, flooding much of the Midwest.

SECTION REVIEW

1. Within which landform region is most of the Midwest located? within which climate region?
2. Name the Great Lakes and the major rivers of the Midwest. How have the rivers been modified to allow for shipping?
3. **Critical Thinking** **Imagine** that the last ice age had not occurred. How might the physical geography of the Midwest be different?

AGRICULTURE

FOCUS

- *What patterns of farm ownership and land use are found in the Midwest?*
- *What agricultural products are grown in the region? How are they important on a global scale?*

Hot, humid summers, fertile soils, level land, and agricultural technology have made the Midwest one of the most productive farming regions in the world. It is the United States' leading producer of dairy products, pork, corn, soybeans, and many other vegetables.

From an airplane, the land of the Midwest appears to be divided into squares. The settlement pattern is one in which fields, roads, and towns all are aligned according to the **township and range system**. The federal government started this survey system in the late 1780s to divide and sell land northwest of the Ohio River. Each township was divided into 36 sections, each one-mile square. During the late 1800s, in order to settle the land, the government offered homesteaders free land if they would farm it for at least five years.

Many Midwestern farms have been family owned for generations and are productive, highly mechanized, and well maintained. Changing economic conditions over the past two decades, however, have led to a decrease in the number of family-owned farms and an increase in the number of large, corporate-owned farms. These changes have affected many of the Midwest's small rural towns that serve farmers and their families. As the number of family farms drops, these towns, especially those located away from major highways, suffer economic decline.

In a world with an ever-growing population, the agricultural yields of the Midwest are among America's most important exports. The Midwest, with its productive soils and efficient farmers, will maintain its global importance.

The Corn Belt Corn originated in Mexico (called "maize" there) and was introduced to the early European settlers by Native Americans. Although corn is grown in every Midwestern state, it is concentrated in the **Corn Belt**, the core of which stretches from Ohio to Iowa. The Corn

Belt also extends into southern Wisconsin and Minnesota, northern Missouri, and the eastern portions of the Great Plains states of Nebraska and South Dakota. (See the map on page 141.) Iowa and Illinois are the country's two leading corn producers.

The Corn Belt is often called the Feed and Grain Belt because most of the corn grown in the Midwest is used to feed livestock. Many Corn Belt farmers raise beef cattle and hogs. In fact, Iowa and Illinois are the leading pork-producing states. The animals are fed corn to fatten them before they are sent to market. Many western ranches send their cattle to the Midwest to be fattened in feedlots before marketing the cattle.

A small percentage of the corn grown in the Midwest is consumed as fresh sweetcorn, as canned and frozen corn, in breakfast cereals, and in cooking oils. By-products of corn, such as corn oil, cornstarch, and corn syrup, are found in many other store products, including soft drinks, instant coffee, and cake mix. Even some plastic grocery bags are derived from corn.

A significant change in the Corn Belt in recent decades has been the expansion of soybean production. Soybeans originated in China and were introduced into the Midwest after World War II. By 1980, the United States was the world's leading soybean producer, with the Midwest its major growing region. Soybeans are commonly used to produce margarine, vegetable oil, and bean curd (tofu). Almost half of the soybeans produced in the United States are exported, mainly to East Asia.

Soil loses its fertility if fields are planted with the same crop year after year. To prevent this loss, Midwestern farmers practice crop rotation. Crops such as soybeans and clover are substituted for corn in some years. The soybeans and clover help return the necessary mineral balance to the soil. Commercial fertilizers also keep production levels high.

The Dairy Belt The **Dairy Belt** is located north of the Corn Belt, where the summers are cooler and soils become rocky and less fertile. This region covers almost all of Wisconsin and most of Michigan and Minnesota. The Dairy Belt also extends into Iowa and into some Northeastern states. Known as "America's Dairyland," Wisconsin produces more milk, butter, and cheese than any other state. Products from Dairy Belt farms are sold in nearby Midwestern urban areas and are shipped by refrigerated trucks to the rest of the country.

Much of the land in the Dairy Belt is devoted to pasture and to the cultivation of the hay, oats, and corn used to feed dairy cattle. Midwestern dairy farms are efficient and use scientific breeding methods to improve their herds. Strict laws regulate the dairy industry to ensure that dairy products are safe and pure. In recent years, the Midwest's dairy farmers have had problems because of large dairy product surpluses, low prices, and the changing American diet.

Many dairy farms have woodlots, which can be used to earn extra income from the sale of firewood, maple syrup, Christmas trees, and wood pulp. Because of the warming effect of the Great Lakes on nearby land, specialized fruit-growing

Corn: From Field to Consumer

- Most ears of corn in the United States are about nine inches long. The kernels, arranged in rows on the ear, are edible seeds.
- In the United States, corn usually is harvested anytime from August to late October.
- The corn is dried, stored in bins, and then sold or used on the farm for livestock feed.

Combine harvesting

areas are located around the Great Lakes, especially in western Michigan and the Door Peninsula in Wisconsin. Farmers grow cherries, cranberries, apples, and other fruits.

SECTION REVIEW

1. How have farm ownership patterns been changing in the Midwest? How do farmers keep the soil productive?
2. What products come from the Corn Belt, and where do they go?
3. **Critical Thinking** **Discuss** the relevance of science and technology to agriculture in the Midwest.

RESOURCES AND INDUSTRY

FOCUS

- *What are the primary industries of the Midwest?*
- *What challenges does the region face? How have these challenges been addressed?*

Forests After the major forests of the Northeastern states were logged, the lumberjacks moved into Michigan, Wisconsin, and Minnesota. Today, the region's virgin pine forests are gone, although many second-growth forests are taking their place. Lumbering is no longer a major industry. Instead, the forests of northern Wisconsin and Minnesota and of Michigan's Upper Peninsula now serve as major recreational areas. The region's forests, lakes, streams, and hills attract millions of tourists, who come to camp, hunt, fish, and ski. The Ozarks of southern Missouri remain largely forested and serve as a tourist center. The forests here also provide valuable wildlife and watershed areas. Recreation and tourism have become vital to the Midwest and join agriculture and industry as a cornerstone of the region's economy.

Industry The Midwest has been an important industrial region for most of the twentieth century. Mining interests grew with discoveries of lead in Wisconsin, copper in Michigan, and iron in Minnesota. The early settlement and growth of the region's many industrial cities were based in part on the cities' locations on or near important transportation routes. Key industries include food processing and the manufacturing of iron, steel, automobiles, and machinery.

Some industrial cities of the Midwest grew as ports on the Great Lakes. Here, they had access to the coal deposits of the Appalachian Mountains, coal mines in Illinois, Indiana, and Ohio, and the iron-ore deposits of upper Michigan and northern Wisconsin and Minnesota. The iron ore was shipped out of the Lake Superior ports of Duluth and Superior to industrial cities such as Gary, Milwaukee, Detroit, Toledo, and Cleveland.

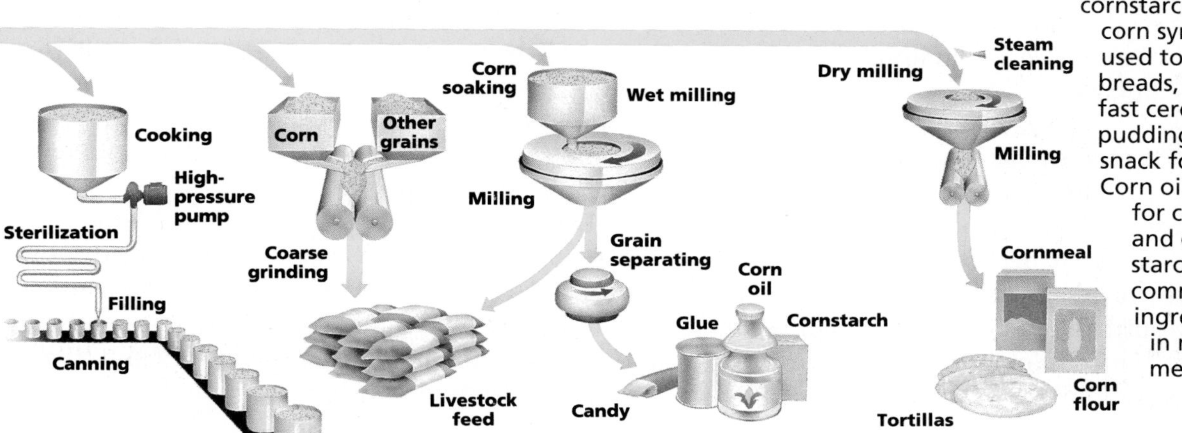

Cooking
High-pressure pump
Sterilization
Filling
Canning
Corn
Other grains
Coarse grinding
Livestock feed
Corn soaking
Wet milling
Milling
Grain separating
Corn oil
Glue
Cornstarch
Candy
Dry milling
Steam cleaning
Milling
Cornmeal
Corn flour
Tortillas

• Corn can be processed in a variety of ways. Some corn is cooked and canned, or ground and used for livestock feed. Corn also might be wet-milled or dry-milled, after which various grain parts are used to make different products.
• Corn by-products, such as cornstarch and corn syrup, are used to make breads, breakfast cereals, puddings, and snack foods. Corn oil is used for cooking, and cornstarch is a common ingredient in many medicines.

The 125-mile-long (201-km) Lake of the Ozarks in southern **Missouri** is a popular destination for vacationers. The hydroelectric dam that forms the lake on the Osage River provides power for the St. Louis area.

Today, iron is mined only in the Mesabi (muh-SAHB-ee) Range in northern Minnesota and in Michigan and Missouri. After many years of mining, the high-grade ores have become exhausted. The area, however, does have a plentiful supply of low-grade iron ore called taconite (TAK-uh-nite).

In many ways, the Great Lakes are the economic lifeline of the Midwest. They furnish inexpensive transportation for bulk cargo materials. Because of the transport connections provided by the St. Lawrence Seaway, the products of the Midwest can reach and compete more easily in world markets.

The industrialized Ohio River valley stretches across southern Ohio, Indiana, and Illinois. Barges carry coal and other bulk goods to the Ohio Valley cities. The Ohio River empties into the Mississippi River and eventually the Gulf of Mexico.

The Ohio Valley cities produce a variety of goods, including steel, plastics, glass, fertilizers, textiles, and machinery. The largest Midwest industrial cities on the Ohio River are Cincinnati, Ohio, and Evansville, Indiana. Tributaries of the Ohio connect to Dayton and Columbus, Ohio, and to Indianapolis, Indiana.

INTERACTION Issues The Midwest's manufacturing industries have experienced the same decline as those in the Northeastern states. Traditional industries based on coal, iron, steel, and automobiles have suffered from foreign competition. Many outdated factories with high labor costs and pollution problems have closed, which has increased unemployment in the region. Some factories, especially in the Ohio Valley, burn local coal, which has caused acid rain in the Northeast and in eastern Canada. The high cost of cleaning up these industries may lead to further closings.

Some midwesterners have migrated to the Sunbelt states in the South and West to seek employment. Even so, the Midwest's central location, excellent transportation network, and experienced labor force have strong potential for industrial redevelopment. An example of this redevelopment is the automobile industry. American automobile manufacturers have renovated factories, and Japanese automakers have built assembly plants in states such as Ohio and Indiana. Other areas have added new industries and now produce high-technology engineering products, such as precision machine tools and automated machines.

Of major concern to the people of the Midwest and to Canada's neighboring provinces is the water quality of the Great Lakes. The people of this region depend on these lakes not only for transportation but also for drinking water, industrial water, fishing, and recreation. All the Great Lakes have suffered from numerous types of pollution, including pesticides, industrial metals, and toxic wastes, and from erosion from construction and dredging.

Lake Erie, however, once considered the most polluted body of water in the world, now sustains a variety of plants and animals. Stricter pollution laws have made many of the rivers in the Midwest, including the industrial Ohio River, cleaner today than they have been in decades.

SECTION REVIEW

1. What industries are supported by the Midwest's forest resources? List three other industries of the Midwest.
2. Why has the Midwest faced industrial decline in recent years? Is redevelopment a possibility? Why or why not?
3. **Critical Thinking** **Assess** the ability of industrial corporations to reduce labor costs without cutting workers' salaries and without increasing unemployment.

URBAN GEOGRAPHY

FOCUS

- *How is the interstate highway system significant to the Midwest?*
- *What are the major cities of the Midwestern United States? What are some characteristics of each?*

MOVEMENT The Interstate Highways

The United States is crisscrossed by a network of more than 45,000 miles (72,405 km) of multilane interstate highways. This network links major cities from coast to coast. Interstate highways have a simple purpose: to move people and goods from city to city.

Before the late 1950s, roads linking major U.S. cities passed through many small towns. The building of the interstate system changed this situation because the new highways bypassed towns and villages. Travelers benefited by saving time. By pulling business away from towns, however, the interstate system caused economic decline in some bypassed towns.

As the transportation center of the United States, the Midwest relies on the interstate system as well as other forms of transportation. The region is supported by an intricate network of railroads, rivers, and highways, which connect the Midwest's major metropolitan areas.

Many large cities, such as **Chicago,** rely on rail transportation to help ease heavy traffic on clogged highways.

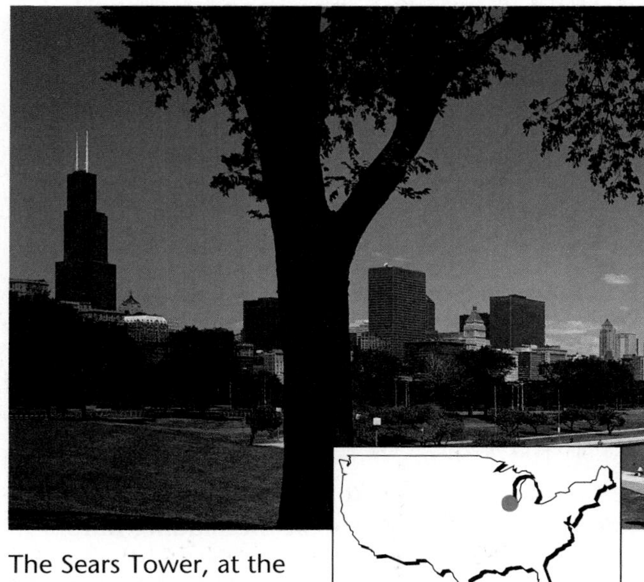

The Sears Tower, at the far left, in downtown **Chicago** is the world's tallest building. Chicago is the core of the third largest metropolitan area in the United States.

Chicago The largest of the Midwest's industrial and commercial centers and the nation's third largest metropolitan area is Chicago, Illinois. Chicago was originally the site of Fort Dearborn and became a city in 1837. Its location at the southern tip of Lake Michigan and its network of major railroads caused Chicago to grow rapidly and to become the nation's transportation center. In the late 1800s, the city swelled with immigrants who were attracted by jobs in Chicago's steel mills, meat-packing houses, and grain elevators.

Today, Chicago is the busiest port on the Great Lakes and has the world's busiest airport. Its downtown area, called "the Loop," is the site of several of the world's tallest buildings. The Chicago area manufactures steel, electronic equipment, machinery, clothing, and food products. The city is primarily a business center, however. Chicago's universities, theaters, zoos, and museums make the city the cultural capital of the Midwest.

Detroit Located on the Detroit River between Lake Erie and Lake St. Clair, Detroit was established as a French fur-trading fort in 1701. Detroit, Michigan, grew to become the Midwest's second largest city and in 1994 was the nation's seventh largest. Here, Henry Ford built his first automobile factory, which produced its first car in 1908. Since then, Detroit has been the leading automobile producer in the nation. Recent cutbacks in the auto industry have led to high unemployment rates and a population decline in the city.

St. Louis St. Louis, Missouri, is located on the west bank of the Mississippi River, just south of its junction with the Missouri and Illinois rivers. The city began as a French fur-trading post in 1764. It later became the center for pioneers heading west and the nation's leading riverboat port. St. Louis remains a transportation hub, but it is mainly an industrial city. The urban area produces food products, aircraft, beer, chemicals, machinery, and clothing. The city's waterfront district has undergone urban renewal.

The Twin Cities Minneapolis and St. Paul, Minnesota, called the Twin Cities, are located where the Minnesota River joins the upper Mississippi River. Both cities serve as a major distribution center for the agricultural produce of the Midwest and the Great Plains. They also serve as the retail center of the upper Midwest. An enormous enclosed shopping mall sprawls in the Twin Cities urban area. The cities are leaders in food processing and machinery and electronics manufacturing. St. Paul is the state capital.

Cleveland The major port and manufacturing city on Lake Erie is Cleveland, Ohio. It was established at the mouth of the Cuyahoga River in 1796. In 1834, Cleveland was connected to the busy Ohio River by canal and grew rapidly. Besides being a coal and iron-ore port, Cleveland has a variety of industries, including steel, clothing, paint, and chemical. After years of battling industrial decline and pollution, the city is making a comeback as a high-technology center.

Kansas City The sister cities of Kansas City, Missouri, and Kansas City, Kansas, are located at the junction of the Missouri and Kansas rivers. This site was the starting point of the Oregon Trail and became a pioneer supply town. By 1869, Kansas City, Missouri, was one of the nation's major railroad centers and was the site of grain elevators, stockyards, and meat-packing plants. Today, it is a major financial and retail center. Both cities serve the Midwest and Great Plains agricultural areas.

The Gateway Arch rises over **St. Louis, Missouri**, to a height of 630 feet (192 m).

4 SECTION REVIEW

1. How has the interstate system caused decline in some Midwestern towns?
2. Name four cities of the Midwest. How is each city's relative location to bodies of water significant?
3. **Critical Thinking** **Determine** whether the major cities of the Midwest form a subregion similar to the Northeast's Megalopolis. Support your answer.

CHAPTER 13 CHECK

The Soo Locks of **Sault Ste. Marie, Michigan**

Reviewing the Main Ideas

1. Landform regions of the Midwest include the Interior Plains, the Canadian Shield, the Appalachian Mountains, and the Ozark Plateau. The Midwest has a humid-continental climate and contains the Great Lakes and many important rivers that are key to the region's economy.
2. The Midwest, with the Corn Belt and the Dairy Belt, is one of the world's most productive agricultural regions. Farms are owned by families and corporations, and farmers use various methods to keep soils fertile.
3. Industries of the Midwest include tourism, food processing, plastics, glass, fertilizers, textiles, and the manufacturing of steel, automobiles, and machinery. Although the region's industries have been undergoing a period of decline, there is potential for redevelopment.
4. Many Midwestern cities, including Chicago, Detroit, St. Louis, the Twin Cities, Cleveland, and Kansas City, are located on key shipping routes and are important industrial and transportation centers.

Building a Vocabulary

1. If a river has been dredged and channeled, what has happened to it? What purposes do dredging and channeling serve?
2. In which states is the Corn Belt located? Where is the Dairy Belt located?

Recalling and Reviewing

1. What landform and climate characteristics make the Midwest suitable for agriculture?
2. Discuss the following statement: "Without its lake and river systems, the Midwest would be neither a major agricultural region nor a major industrial region."
3. How do corporations play a role in the economy of the Midwestern region?
4. How are Midwestern towns and cities changing in response to economic conditions?

Thinking Critically

1. How do product surpluses and low prices present problems for farmers? Suggest some solutions for these problems.
2. Using examples from the chapter, explain how pollution created by one country can affect other countries. Who do you think should bear the responsibility for pollution that crosses international boundaries?

Learning About Your Local Geography

Cooperative Project

With your group, conduct interviews to find out how goods that are locally grown or produced are transported to their destinations and what some of those destinations are. Make a chart showing the major transportation routes in and out of your area.

Using the Five Themes of Geography

Write a paragraph about how the themes of **location** and **region** are evident in the Midwestern United States. How has the region's location affected its development? Into what subregions might you divide the Midwestern region?

CHAPTER 14

The Interior West

"Being in the midst of these mountains was different from anything we'd ever known. Inside the Rockies even the colors of things were different. Instead of the orange and red skies that seemed to cover the earth on the plains and in West Texas, the sky was so blue you had to wear sunglasses to look into it. . . . Slashes of white rock cut upward into the blue sky that dared you to try to look to the end of it."

Peter and Barbara Jenkins

GEOGRAPHY DICTIONARY

badlands
chinook
Wheat Belt
shelterbelt
center pivot irrigation
strip mining
land reclamation
Continental Divide
tree line
ghost town
aqueduct
toponym

In books, in movies, and on television, the Interior West of the United States has often been romanticized as the "Old West." To geographers, this region generally encompasses most of the drier and mountainous areas of the western United States.

The Interior West states include North Dakota, South Dakota, Nebraska, Kansas, Oklahoma, Montana, Wyoming, Colorado, Idaho, Utah, Nevada, New Mexico, and Arizona. These states can be grouped into three subregions: the Great Plains, the Rocky Mountains, and the Intermountain West.

Water is the vital resource for the entire Interior West region. Irrigation agriculture is the major consumer of this precious resource. Rapid urban development and the coal-mining industry are making competitive claims to the water, however. In some areas, underground water supplies are dangerously low. Future growth in the Interior West will require that communities balance water supply, preservation of the environment, and economic growth.

A painted cowhide dance shield of the Native American Dakota people

The Rocky Mountains in the White River National Forest of **Colorado**

INTERIOR WEST: PHYSICAL–POLITICAL

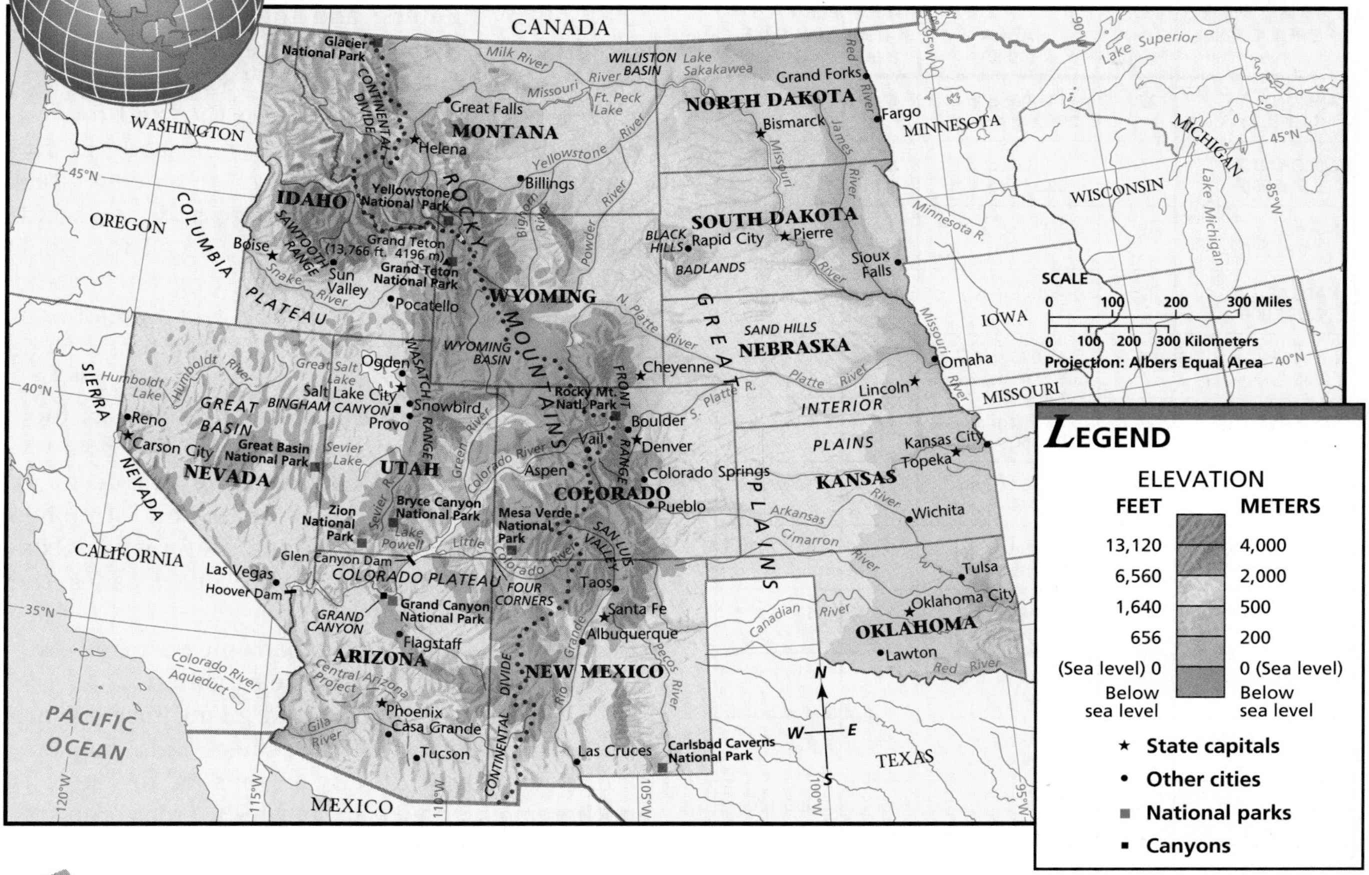

THE GREAT PLAINS

FOCUS

- *What are the characteristics of the Great Plains' landforms and climate?*
- *What economic activities have taken place in the Great Plains?*
- *What challenges do the region's residents face?*

REGION **Landforms** The Great Plains lie in the middle of the United States between the Interior Plains on the east and the Rocky Mountains on the west. The region includes Oklahoma, Kansas, Nebraska, North Dakota, South Dakota, and parts of New Mexico, Colorado, Wyoming, Montana, and Texas. (See the map above.)

The Great Plains were formed by sediment that eroded from mountains and was deposited by rivers. The plains rise gradually from east to west and resemble a huge, tilted table. Although the Great Plains are known for their flat, sweeping horizons, a few areas feature more varied landforms. The Sand Hills of Nebraska are ancient, grass-covered sand dunes. In the Dakotas, rugged areas of soft, sedimentary rock called **badlands** are found. Badlands are lands that have been eroded by wind and water into small gullies and have been left without vegetation and soil.

Climate A steppe climate dominates the region. The Great Plains are semiarid and become drier toward the west. Temperatures can be extreme, with winters in some areas reaching to -40°F (-40°C) and summers reaching above 100°F (38°C). In winter, blizzards occur, bringing cold winds with blowing snow. In summer, hot, moist air from the Gulf of Mexico brings severe thunderstorms and destructive tornadoes.

Droughts are the major climate hazard of the region and occur every few decades. During drought periods, dust storms may cover huge areas. In addition, strong, dry mountain winds,

FACTS IN BRIEF States of the Interior West

STATE POPULATION (1993)	POP. RANK (1993)	AREA (1990)	AREA RANK (1990)	PER CAPITA INCOME (1993)
ARIZONA 3,936,000	23	114,006 sq. mi. 295,276 sq. km	6	$18,121
COLORADO 3,566,000	26	104,100 sq. mi. 269,619 sq. km	8	$21,564
IDAHO 1,099,000	42	83,574 sq. mi. 216,457 sq. km	14	$17,646
KANSAS 2,531,000	32	82,282 sq. mi. 213,110 sq. km	15	$20,139
MONTANA 839,000	44	147,046 sq. mi. 380,849 sq. km	4	$17,322
NEBRASKA 1,607,000	37	77,359 sq. mi. 200,360 sq. km	17	$19,726
NEVADA 1,389,000	38	110,567 sq. mi. 286,369 sq. km	7	$22,729
NEW MEXICO 1,616,000	36	121,598 sq. mi. 314,939 sq. km	5	$16,297
NORTH DAKOTA 635,000	47	70,704 sq. mi. 183,123 sq. km	19	$17,488
OKLAHOMA 3,231,000	28	69,903 sq. mi. 181,049 sq. km	21	$17,020
SOUTH DAKOTA 715,000	45	77,121 sq. mi. 199,743 sq. km	18	$17,666
UTAH 1,860,000	34	84,904 sq. mi. 219,901 sq. km	13	$16,180
WYOMING 470,000	50	97,819 sq. mi. 253,351 sq. km	9	$19,539

Rankings are for all 50 states.
Source: *Statistical Abstract of the United States 1994*

How many states of the Interior West are among the 10 largest states in area in the country? Which state has the smallest population of all 50 states?

called **chinooks**, descend from the Rockies and affect the western portions of the Great Plains.

The vegetation of the Great Plains is prairie and steppe grasses, shrubs, and sagebrush. The wetter eastern plains originally were covered with continuous tall-grass prairie. The drier High Plains had shorter steppe grasses. Today, most of the original grasslands have been changed by agriculture and grazing.

The End of the Open Range Because of the lack of moisture, most of the Great Plains has been better suited to grazing than to farming. During the "cowboy era" of the mid-1800s, herds of cattle and flocks of sheep roamed the open range of the Great Plains. They replaced the huge buffalo herds hunted by the Native Americans prior to the arrival of the railroads and settlers. Within a short time, the buffalo herds had been killed off by hunters, and the Native Americans were forced onto government reservations.

The Homestead Act of 1862 granted 160 acres to settlers willing to farm the land for five years, causing rapid population growth in the region. Eventually, barbed wire for fencing, steel plows that cut through the tough sod, and windmill pumps for wells ended the era of the open range. Fences divided the plains into ranches and farms. Today, livestock raising is still a major economic activity, but it often is combined with wheat farming.

Wheat is one of the most valuable U.S. food crops and an important export for the country. Nearly all Americans eat wheat in one form or another every day. The **Wheat Belt** stretches across the Dakotas, Montana, Nebraska, Oklahoma, Colorado, and Texas. In recent decades, the Wheat Belt has extended westward into drier steppe grasslands. This expansion occurred with the development of large-scale irrigation.

Hailstorms, insects, and plant disease are constant challenges in the Wheat Belt. Soil erosion, particularly during periods of drought, is another major concern. In windy areas, rows of trees, called **shelterbelts**, block the wind and protect the soil.

Irrigation Agriculture The western part of the region generally receives less than 20 inches (50 cm) of rainfall per year, making irrigation nec-

A storm rolls over these lonely structures on the Great Plains of **Wyoming**. To see the area covered by the Great Plains, see the map of the physiographic regions of the United States and Canada on page 112.

essary for crop growth. Underground water supplies are pumped for use in the fields. The largest source of water under the Great Plains is the Ogallala Aquifer, which stretches from South Dakota to Texas. For many years, this aquifer supplied all the farmers' water needs. With increased irrigation and wasteful water practices, however, the water table began to drop. Today, the depletion rate has slowed as a result of more efficient irrigation methods.

A number of rivers cross the Great Plains. In the past, the rivers and tributaries were full in the spring, but in the summer the water slowed to a trickle. Today, major dam-building projects control the flow of these river systems and make farming less risky.

From the air, much of the irrigated Great Plains appears as a series of green circles. This pattern is caused by **center pivot irrigation**, which employs long sprinkler systems mounted on huge wheels that rotate slowly. Center pivot irrigation has been introduced throughout the western United States. Major irrigation crops of the Great Plains include corn, sugar beets, and alfalfa, which is used for cattle feed.

Farming Issues Although the Great Plains region is productive agriculturally, the farmers are challenged continually by climate and world markets. During the 1930s, the combination of a drought and the Great Depression meant disaster for thousands of farmers on the plains.

Today, farmers also are concerned about the price of wheat on world markets—low prices can be as disastrous to them as drought. Political decisions, such as how much the United States will sell to the largest wheat importers, China and Russia, influence prices. During the 1980s, a surplus of wheat caused prices to drop. Thousands of farms were abandoned, and many rural plains towns began shrinking. States such as North Dakota actually lost population. Recently, some of the region's small towns have been undergoing renewal.

Resources and Industry Both coal and oil have been found beneath the Great Plains. Oil production on the plains has been mainly in Oklahoma, Wyoming, Colorado, and eastern New Mexico. The largest coal deposits are in Wyoming. The coal, buried just beneath the surface, can be removed by **strip mining**. In strip mining, soil and rock are stripped away by large machines to get at the coal. The law now requires that mining companies use **land reclamation** methods, so that the land is usable after the mining is finished. To reclaim the land, the topsoil is saved, and the land surface is returned to its original contours. If land reclamation is not practiced, serious erosion occurs, and an ugly pit remains.

Located at the foot of the Rockies, Denver is the dominant city and economic center of the Great Plains and of the adjoining Rocky Mountains region. It is also the capital of Colorado. Unfortunately, Denver's rapid growth has caused severe air pollution.

Workers have dug this pit to mine coal near **Sheridan, Wyoming**.

1 SECTION REVIEW

1. Describe the dominant landform and dominant climate of the Great Plains region.
2. What economic activities take place in the region, and how has the environment been modified?
3. **Critical Thinking** **Explain** how international concerns play a role in the region.

THE ROCKY MOUNTAINS

FOCUS

- *What are the physical features of the Rocky Mountains?*
- *What economic activities have been important in the region?*

REGION **Physical Geography** West of the Great Plains are the Rocky Mountains. The Rockies have rich natural resources and great scenic beauty. They stretch from Canada through Idaho, Montana, Wyoming, Utah, Colorado, and New Mexico. The Rockies are actually a series of mountain ranges separated by passes and wide valleys. The crest of the Rocky Mountains forms the **Continental Divide**, which divides the major river systems of North America into those that flow eastward and those that flow westward. Snow-capped peaks more than 12,000 feet (3,658 m) high are numerous in the Rockies.

The highland climates of the Rockies are variable, depending upon elevation. Semiarid grasslands usually are found at the foot of the mountains, while most of the slopes are forested. The forests capture the winter snowfall and form the source for the rivers that flow across the Great Plains and Intermountain West states. The highest mountain peaks reach above the **tree line** to a zone of tundra, rock, and ice. The tree line is the elevation above which trees cannot grow.

Economic Geography Most of the Rocky Mountains region is still wilderness or forest land. Several highways and railroads cross the mountain passes, but no major cities have developed in the Rockies. Most of the residents of the region live in small towns that are dependent on mining, ranching, forestry, or tourism.

Early prospectors struck major veins of gold and silver in Montana, Idaho, Colorado, Utah, and New Mexico. Utah, Montana, and New Mexico are leading copper-producing states. Lead and many other ores also are found in the Rocky Mountains. Though mining communities are scattered throughout the region, many have become **ghost towns**. Ghost towns are towns that once flourished but are deserted today, usually because of the depletion of a nearby resource such as gold or silver. Oil also has been found on the western slopes of the Rockies. In the northern Rockies of Idaho and Montana, logging supports lumber and pulp mills. Grazing of sheep and cattle is important in some Rocky Mountain areas. Farming is limited, however.

One of the most valuable industries in the Rocky Mountains region is tourism. Popular Rocky Mountain ski areas include Aspen, Vail, Sun Valley, and Taos (TAUS). The federal government has set aside large scenic areas in the region as national parks. Among these are Yellowstone, Grand Teton, Rocky Mountain, and Glacier national parks. Millions of tourists come to these parks to camp, hike, mountain climb, and fish. The largest landowner in the Rocky Mountains is the federal government. Private businesses must acquire permits to mine, log, graze cattle, or build ski resorts on these government lands.

A skier takes advantage of the snow in Taos Ski Valley in **New Mexico**. Each year tourists flock to the slopes of ski resorts that dot the Rocky Mountains.

SECTION REVIEW

1. Why do the Rockies have several vegetation zones, and what are they?
2. Why has mining activity decreased in the region? What is one of the region's most valuable economic activities today?
3. **Critical Thinking** **Predict** what might happen in the Rocky Mountains if the federal government sold off its lands here.

3 INTERMOUNTAIN WEST

FOCUS

- *What is the physical and economic geography of the Intermountain West?*
- *How is the population distributed in the Intermountain West?*

REGION **Physical Geography** Between the Rocky Mountains and the Sierra Nevada and Cascade ranges lies the vast Intermountain West. The Intermountain West includes all of Arizona, Nevada, and Utah, southern Idaho, and the western portions of Colorado and New Mexico. The Intermountain West region can be further divided into three landform regions: the Basin and Range region (also known as the Great Basin), the Colorado Plateau, and the Columbia Plateau.

The Basin and Range region is centered in Nevada. It is a region of desert basins separated by faulted mountains. The central portion of this area is called the Great Basin because its river waters do not reach the sea. Instead, the rivers flow into low basins and evaporate, leaving behind dry lake beds or salt flats.

The Colorado Plateau is a beautiful uplifted area of horizontal rock layers that have been eroded by wind and water. This region is known for its deep canyons, the largest of which is the Grand Canyon of the Colorado River. The Columbia Plateau lies in the northern part of the Intermountain West. It is centered in southern Idaho but extends into eastern Washington and Oregon. The Snake River has cut deep canyons into the region's plateaus.

The climate of the Intermountain West region varies. Generally, the southern part, in Arizona, New Mexico, and Nevada, is desert. The northern part, in Utah and southern Idaho, is mainly steppe. Some of the low regions, especially the Great Basin, experience the rain-shadow effect and receive almost no rain at all.

Agriculture While only a small portion of the Intermountain West is used as farmland, farming is an important part of the economy. In the warmer and drier south, all farming requires irrigation. The Colorado River is the lifeblood of the southwestern states as it flows more than 1,450 miles (2,333 km) from the Rocky Mountains to the Gulf of California. By agreement, several states and Mexico share the Colorado River's valuable waters. Dams, including giant Hoover Dam and Glen Canyon Dam, now control the once wild river.

Today, agriculture in all of these dry states is possible because of the Colorado River basin projects. Arizona and southern California have **aqueduct**, or canal, systems to carry water hundreds of miles from the Colorado River. The desert city of Las Vegas, Nevada, also receives most of its water from the Colorado River. As the most regulated and tapped river in the United States, the Colorado is virtually dry by the time it reaches the Gulf of California.

Other Resources Mining is vital to the economy of the Intermountain West. The most plentiful ores are copper, gold, silver, lead, zinc, and uranium. The Basin and Range area in Arizona and Utah is the richest copper-mining area in the United States. Coal is mined in Utah, Arizona, and New Mexico. Mining development has come under scrutiny, however, because of its impact on water and air quality.

The Intermountain West is a favorite tourist destination, particularly Arizona and Utah. Grand Canyon National Park in northwestern Arizona attracts millions of tourists each year. The canyon is more than one mile (1.6 km) deep, carved by the Colorado River over millions of years. Other heavily visited national parks in the Intermountain West are Zion Canyon and Bryce Canyon in Utah, Mesa Verde in southwestern Colorado, and Great Basin in Nevada.

Urban Geography The Intermountain West began gaining population after World War II when the federal government built dams and water projects, military bases, and major highways here. The widespread use of air conditioning made the region even more attractive. Recent growth here has been the most rapid in the nation.

People and industry are attracted to the area because of the wide-open spaces and warm, sunny climate. Cities have grown rapidly, especially Phoenix, Tucson, and Las Vegas. Many retirement communities have sprung up in the middle of the desert. The growing population requires large quantities of water—a limited resource.

Phoenix, the capital of Arizona, is the largest city in the Intermountain West. It contains more

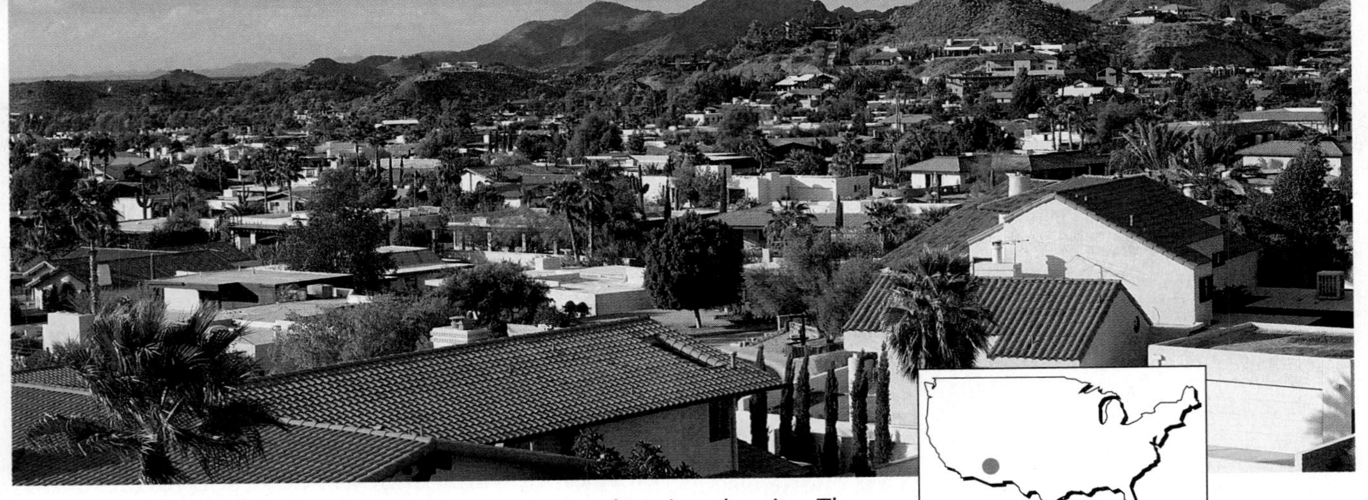

Phoenix rises toward the mountains that ring the city. The population of this rapidly growing city increased by nearly 400,000 from 1970 to 1990.

than half of the population of Arizona in its expanding urban area. Like Denver, Phoenix is experiencing environmental problems related to rapid population growth. Today, the state has an urban-based economy that is dependent on industry, tourism, and retirement communities.

The second largest urban area in the Intermountain West is Salt Lake City, the capital of Utah. The area has recently become a center for computer and communications industries. More than 65 percent of Utah's population lives in Salt Lake City and other nearby cities.

Salt Lake City is the headquarters of The Church of Jesus Christ of Latter-day Saints, whose members are called Mormons. Mormon pioneers, led by Brigham Young, settled here in 1847. They developed a large-scale irrigation system in this arid region. More than 60 percent of Utah's population is Mormon.

Human Geography The earliest inhabitants of the Intermountain West were the Native Americans who arrived in the region more than 12,000 years ago. Agricultural Native American cultures were established by A.D. 700. When the Spanish arrived in the sixteenth century, there were many different Native American groups living in the region.

A large Native American population lives on the Colorado Plateau of Arizona and New Mexico. The Navajo and the Hopi are the largest of these groups. The Navajo reservation, which is located mainly in northeastern Arizona but also extends into Colorado, New Mexico, and Utah, is the largest reservation in the United States. For many years, the Navajo lived as nomadic herders, moving over the plateau with their sheep, goats, and horses in search of pasture. Today, some Navajo are farmers as well as herders. Though deposits of coal, uranium, and oil have been found on the Navajo reservation, poverty here persists.

The Hopi reservation, also in northeastern Arizona, is surrounded completely by Navajo land. The Hopi are mainly farmers and craftsworkers who live in small farming villages on high plateaus. Some Hopi also graze sheep and goats.

The impact of Spanish settlement is illustrated in the many Spanish **toponyms** that can be found on maps of the southwest United States. Toponyms are names of places and often reflect the history and culture of a region. Examples are Las Vegas (the meadows), Casa Grande (grand house), and Las Cruces (the crossroads).

Spanish architecture and language in the region reflect this cultural influence. Today, the majority of the population of New Mexico speaks Spanish. Many of the farms in the southwestern United States are dependent on migrant labor from Mexico. The continuing migration from south of the border ensures that Hispanic (Latino) culture will remain important in this region.

3 SECTION REVIEW

1. Given the region's landforms and climates, how is agriculture possible?
2. What is the cultural background of the region?
3. **Critical Thinking** **Imagine** that aqueduct systems and air conditioning had not been invented. Would the Intermountain West region still be undergoing rapid growth? Explain your answer.

CHAPTER 14 CHECK

Hoover Dam on the Colorado River between **Nevada** and **Arizona**

Reviewing the Main Ideas

1. Flat plains and a steppe climate dominate the Great Plains. The land has been used for grazing, but now wheat farming and strip mining are the major economic activities. Various political, economic, and environmental challenges continue to face the region.
2. The Rocky Mountains, a series of mountain ranges, form the Continental Divide. Climates and vegetation vary according to elevation. Mining activities have been important in the Rocky Mountains region but have decreased with the depletion of some minerals, and tourism is now one of the region's most important industries.
3. Despite dry climates in the Intermountain West region, irrigation has made agriculture possible. Mining and tourism are also important parts of the region's economy. Cities of the region are growing rapidly. Retirees, Mormons, Native Americans, and Spanish-speakers make up a large part of the region's population.

Building a Vocabulary

1. What are badlands?
2. What term is used to describe a strong, dry mountain wind?
3. Define *shelterbelt*.
4. Why do parts of the Great Plains appear to be a series of green circles when viewed from above?
5. What technique is used to mine coal? What does land reclamation accomplish?

Recalling and Reviewing

1. How is the Great Plains region internationally important?
2. Compare and contrast the economic activities found in the Rocky Mountains with those of the Intermountain West.
3. What environmental concerns do the Great Plains and Intermountain West regions share?

Thinking Critically

1. In what respects do the Great Plains resemble the Midwest? How do they differ?
2. Propose a way to reduce the poverty found on Native American reservations without radically altering Native American cultures.

Learning About Your Local Geography

Individual Project

Conduct library research to find out the origin and/or meaning of your community's toponym. Does it reflect the history or culture of your community? Was it the name of an individual or family? Does it describe the physical setting of your community? Write a paragraph explaining the toponym. You may want to include your paragraph in your individual portfolio.

Using the Five Themes of Geography

Illustrate how the themes of **region, place,** and **human-environment interaction** apply to the Interior West by drawing at least one picture of each region discussed in the chapter. Each illustration should show physical and human place characteristics and evidence of human-environment interaction in one of the regions. Do not label your illustrations. Have other class members identify the region shown in each illustration.

CHAPTER 15

The Pacific States

> "The last of the Coast Range foothills were in near view. . . . Their union with the valley is by curves and slopes of inimitable beauty. They were robed with the greenest grass and richest light I ever beheld, and were colored and shaded with myriads of flowers of every hue, chiefly of purple and golden yellow."
>
> *John Muir*

GEOGRAPHY DICTIONARY

agribusiness
multicultural region
caldera
panhandle
hot spot

At first glance, you might wonder why the Pacific states of California, Oregon, Washington, Alaska, and Hawaii are grouped together as a region. After all, Alaska boasts the record low temperature for the entire United States, while Hawaii is tropical and the only island state. California is home to world-famous tourist attractions such as Hollywood and the Golden Gate Bridge. Washington and Oregon delight residents and visitors with rugged mountains, wild and scenic rivers, craggy coastlines, and green valleys.

Upon closer look, however, the similarities among the five states become more apparent. Physical geographers would note that these states share an active geological environment characterized by mountains, volcanoes, and earthquakes. Cultural, political, and economic geographers might point out the increasing trade, investment, and immigration that connect these states with the Pacific Rim countries.

Furthermore, the Pacific states face a common challenge. They must preserve and protect the fragile wilderness areas, fertile agricultural lands, and valuable natural resources that first attracted people to the region.

Wheel from a wagon that carried settlers to **California**

The Pacific Coast Highway north of **Half Moon Bay, California**

PACIFIC STATES: PHYSICAL–POLITICAL

LEGEND

ELEVATION

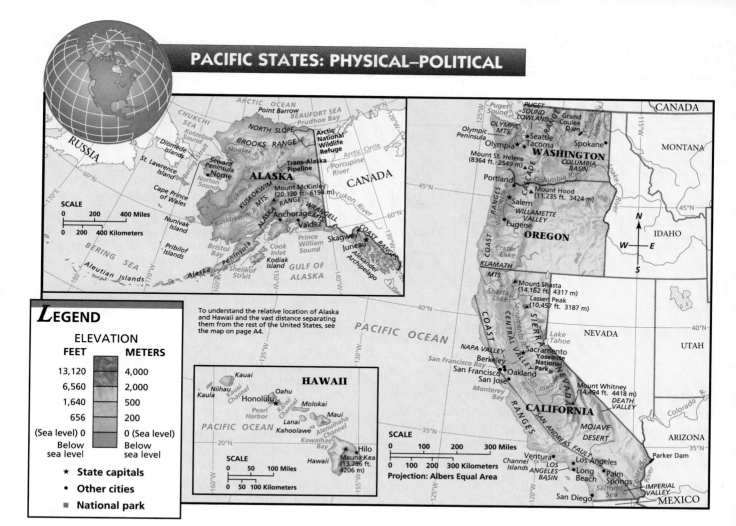

FEET	METERS
13,120	4,000
6,560	2,000
1,640	500
656	200
(Sea level) 0	0 (Sea level)
Below sea level	Below sea level

★ State capitals
• Other cities
■ National park

To understand the relative location of Alaska and Hawaii and the vast distance separating them from the rest of the United States, see the map on page A4.

SCALE 0 100 200 300 Miles; 0 100 200 300 Kilometers
Projection: Albers Equal Area

CALIFORNIA

FOCUS

- *What are the major landform and climate regions of California?*
- *Where are the three major urban areas of California located?*

REGION Landforms California can be divided into four major landform areas: the Coast Ranges, the Sierra Nevada, the Central Valley, and the desert basins and ranges. The Coast Ranges form a rugged coastline along the Pacific. They are made up of parallel mountain ranges and are separated by earthquake-fault systems. There are a few lowland areas between the mountains and the Pacific, the largest of which is the Los Angeles Basin.

The Sierra Nevada range lies inland and east of the Coast Ranges and is one of the longest and highest mountain ranges in the United States. Mount Whitney, reaching an elevation of 14,494 feet (4,418 m), is the highest peak in the 48 contiguous states.

Between the Sierra Nevada and the Coast Ranges is a narrow plain known as the Central Valley, one of the richest agricultural regions in the United States. Stretching more than 400 miles (644 km), the Central Valley is irrigated by the many rivers that flow from the Sierra Nevada.

To the east of the Sierra Nevada are desert basins and mountain ranges separated by fault zones. At 282 feet (86 m) below sea level, Death Valley, California, is the lowest point in all of North America.

Earthquake activity is common in California due to the geologic structure of the Pacific States region. The San Andreas Fault system, formed where the Pacific and North American plates meet, stretches from offshore north of San Francisco to the southeast through southern California. The Pacific plate is slowly moving northward along the San Andreas Fault, past the

FACTS IN BRIEF The Pacific States of the United States

STATE POPULATION (1993)	POP. RANK (1993)	AREA (1990)	AREA RANK (1990)	PER CAPITA INCOME (1993)
ALASKA 599,000	48	615,230 sq. mi. 1,593,446 sq. km	1	$22,846
CALIFORNIA 31,211,000	1	158,869 sq. mi. 411,471 sq. km	3	$21,821
HAWAII 1,172,000	40	6,459 sq. mi. 16,729 sq. km	47	$23,354
OREGON 3,032,000	29	97,093 sq. mi. 251,471 sq. km	10	$19,443
WASHINGTON 5,255,000	15	70,637 sq. mi. 182,950 sq. km	20	$21,887

Rankings are for all 50 states.
Source: *Statistical Abstract of the United States 1994*

Which of the Pacific states has the largest area of all 50 states? Which state has the largest population of all 50 states?

North American plate. The shock waves caused by moving rock along the fault line can create severe earthquakes, such as the 1994 Los Angeles earthquake. Scientists warn that future severe earthquakes threaten California.

Climate A marine-west-coast climate covers the northern coast of the state, while a Mediterranean climate dominates most of coastal southern and central California. Temperatures are warm all year, even in winter. Dry, hot summer winds heighten the chance of dangerous brushfires.

The western slopes of the Sierra Nevada act as the main watershed for California. The mountains store snow and rainwater in snowfields, in lakes, and behind dams in reservoirs. The water is then distributed to farmlands and cities below.

The Sierra Nevada range blocks Pacific moisture from reaching the basins of eastern California.

The San Andreas Fault scars the **California** landscape between Los Angeles and San Francisco.

Desert and steppe climates exist here. The hottest temperature and lowest rainfall records in North America have been recorded in Death Valley, where summer temperatures frequently soar to more than 120°F (49°C) and annual rainfall averages less than two inches (5 cm).

Agriculture California is the nation's leading agricultural producer. The southern and central parts of the state have a year-round growing season. This enables California to be the major supplier of fresh vegetables and fruits to the rest of the nation throughout the winter. Most farming in California takes place on large modern farms, many of them owned by big corporations. This is called **agribusiness**.

Because of the dry climate, California's agriculture can be productive only with the help of irrigation, which uses 85 percent of California's water supply. The greatest variety of crops in North America, including nut trees and fruit trees, is grown in the flat, irrigated Central Valley. The southern portion of the Central Valley—the San Joaquin Valley—is one of the world's major cotton-growing areas.

The valleys of coastal California grow specialty crops such as artichokes and garlic, while the Napa Valley near San Francisco is world famous for its vineyards. The valleys and basins of southern California supply many citrus fruits, such as oranges, lemons, and limes.

INTERACTION Resources and Industry

California is rich in mineral resources. When gold was discovered in the Sacramento Valley in 1849, a gold rush brought thousands of fortune seekers to California. Today, however, California's most valuable mines are those that provide the sand and gravel needed in construction.

California's oil deposits lie offshore on the state's continental shelf. The future development of these resources is a source of controversy. An oil spill could be disastrous to the area's fishing industry, tourism, and rich ocean life. On the other hand, future oil production would benefit the state's economy.

California's most precious natural resource is water. Aqueducts transport water hundreds of miles from the Colorado River and the rivers of northern California and the Sierra Nevada to meet the needs of dry southern and central California. Overuse of underground water for irrigation from

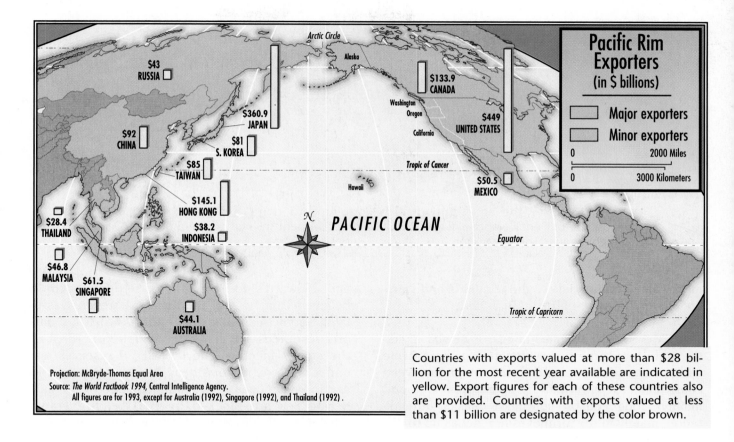

Pacific Rim Exporters
(in $ billions)

Major exporters
Minor exporters
0 2000 Miles
0 3000 Kilometers

Arctic Circle
Alaska
$43 RUSSIA
$133.9 CANADA
Washington
Oregon
California
$449 UNITED STATES
$360.9 JAPAN
$92 CHINA
$81 S. KOREA
$85 TAIWAN
Tropic of Cancer
Hawaii
$50.5 MEXICO
$145.1 HONG KONG
$28.4 THAILAND
PACIFIC OCEAN
$38.2 INDONESIA
Equator
$46.8 MALAYSIA
$61.5 SINGAPORE
$44.1 AUSTRALIA
Tropic of Capricorn

Projection: McBryde-Thomas Equal Area
Source: *The World Factbook 1994*, Central Intelligence Agency.
All figures are for 1993, except for Australia (1992), Singapore (1992), and Thailand (1992).

Countries with exports valued at more than $28 billion for the most recent year available are indicated in yellow. Export figures for each of these countries also are provided. Countries with exports valued at less than $11 billion are designated by the color brown.

agricultural produce and other goods to the United States. Political and economic problems within Middle and South America, however, have kept these nations from sharing many of the benefits of Pacific Rim trade.

On the Horizon

The exchange of goods and cultures taking place in the Pacific Rim may increase as China, Mexico, and other countries begin to play a more important economic role in the region. Ties among the Pacific Rim states and nations could weaken, however, if Asian nations join together to increase trade mainly among themselves. The Pacific states, too, may concentrate on expanding trade with other nations in the Americas. Already, the United States, Mexico, and Canada have sought to lower trade barriers and join together to form a vast free-trade zone.

Regardless of what nations decide to do, the Pacific Rim will remain an important economic and cultural region. The links of trade, investment, immigration, and culture that have been forged will not be broken easily.

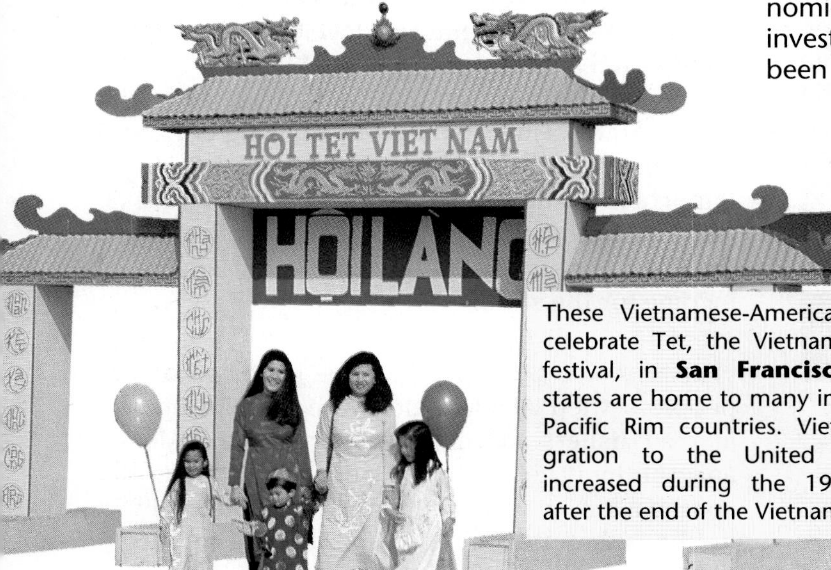
HOI TET VIET NAM

These Vietnamese-Americans prepare to celebrate Tet, the Vietnamese New Year festival, in **San Francisco**. The Pacific states are home to many immigrants from Pacific Rim countries. Vietnamese immigration to the United States greatly increased during the 1970s, especially after the end of the Vietnam War in 1975.

Your Turn

Imagine you live in one of the states or nations of the Pacific Rim. **Write** a letter to a friend living in the eastern United States. **Explain** how location and movement tie the Pacific Rim states and nations together as a region. **Describe** a few daily activities in your region that your friend might find interesting.

Mount McKinley, the highest mountain in North America, dominates the landscape of Denali National Park in **Alaska**. Native Americans named the mountain *Denali,* which means "The Great One."

more than 3 million lakes, and ice fields and glaciers cover 4 percent of the state.

Climate The mountain ranges of Alaska divide the state into three major climate regions: marine west coast, subarctic, and tundra. A marine-west-coast climate with mild temperatures is found mainly along coastal southeast Alaska. Precipitation, mainly in the form of rain, reaches up to 200 inches (508 cm) per year here, supporting a dense temperate forest. The interior of Alaska has a subarctic climate. Summers here are short, while winters are long and severe. Most precipitation is in the form of snow and supports a boreal forest. The North Slope area along the Arctic Ocean coast has a tundra climate.

Resources In 1867, U.S. Secretary of State William H. Seward purchased Alaska from Russia for $7.2 million. Alaska was called "Seward's Folly" and "Seward's Ice Box" because many people believed that he had purchased a frozen wasteland. Today, Alaska is known for its wealth of natural resources. Oil generates about 85 percent of the state's income. The largest oil reserves are under the North Slope near Prudhoe Bay on the Arctic Ocean coast. Because the Arctic Ocean and Bering Sea are blocked by sea ice, the Trans-Alaska Pipeline System (TAPS) was built to transport the oil to the ice-free port of Valdez, 800 miles (1,287 km) south.

The Trans-Alaska Pipeline snakes across mountain ranges and tundra. Completed in 1977, the pipeline has become a human addition to the environment of **Alaska**.

As Alaska's oil resources begin to decline, however, there is pressure to seek new sources in the Arctic National Wildlife Refuge (ANWR). Proposals to drill in the refuge have sparked debate between people who want increased resource development and those who see such development as a threat to the area's environment. In 1989, the nation's largest oil spill occurred when an oil tanker ran aground in Alaska's Prince William Sound. The spill brought into question the future development of the state's oil resources.

Alaska is the leading fishing state in the United States. Huge catches of cod, king crab, and salmon are made each year. Forestry also is important to the economy of southeast Alaska. Alaska's scenic forests, mountains, and glaciers have made tourism another important source of income for the state.

Human Geography Three major groups of people settled Alaska before the first Europeans arrived: fishers and nomadic herders, the Aleuts, and the Inuit, or Eskimo. Fishers lived along the southeastern coast, and nomadic hunters lived in the subarctic interior of Alaska. The Aleuts occupied the Alaskan Peninsula and the Aleutian Islands and were engaged mainly in seal hunting and fishing. The largest native Alaskan group was the Inuit (Eskimo). They lived along the tundra coasts of the Arctic Ocean and Bering Sea. Although this region has one of the most severe climates of any inhabited area in the world, the Inuit (Eskimo) were able to meet all their needs by hunting seals, walrus, and whales.

Today, native Alaskans account for about 15 percent of the state's population. During the settlement of Alaska, many native Alaskans lost their homelands. The Alaska Native Land Claims Settlement Act of 1971 helped them regain much of what they had lost. Native-owned corporations now earn income from oil, mining, and forestry on native lands.

The first Europeans to reach Alaska were Russian explorers. Russia ran a fur-trading colony in Alaska from 1743 until 1867, when Russia sold Alaska to the United States. Alaska's population grew after gold was discovered in the late 1800s. Since statehood in 1959, the oil industry has spurred more population growth. Today, more than a half million people live in the state, mostly in urban areas.

Issues The Alaskan economy suffers from a high cost of living and seasonal unemployment in forestry and fishing. With little agriculture and industry, most food and goods must be shipped in from other states.

The federal government owns more than 72 percent of the land, which is protected in national parks, wildlife refuges, national forests, and wilderness areas. Concern for the Alaskan wilderness is a consideration in any future exploitation of the state's rich resources.

3 SECTION REVIEW

1. Why are Alaska's oil reserves a source of controversy?
2. Identify the three native Alaskan groups. Where did each group live? How did each group survive?
3. **Critical Thinking** **Decide** whether William H. Seward was wise or foolish to purchase Alaska for the United States in 1867. Explain your answer.

The setting sun casts a golden glow on the cliffs of Kalalau Valley on **Kauai**. Rugged cliffs such as these make road building impossible in some parts of this Hawaiian island.

4 HAWAII

FOCUS

- *How does Hawaii's physical environment affect its economy?*
- *What major economic and environmental issues does Hawaii face?*

LOCATION **Island Environment** Hawaii lies in the Pacific Ocean, about 2,500 miles (4,023 km) south of Alaska. The Hawaiian Islands are a chain of eight major islands and more than 100 smaller islands that stretch for over 1,500 miles (2,414 km). The largest and southernmost island is Hawaii. The other large islands include Maui, Kauai, Molokai, Lanai, and Oahu, the most populated island. (See the map on page 159.)

The islands are volcanic mountains rising above the Pacific, though the only island with active volcanoes is Hawaii. Hawaii sits on a volcanic **hot spot** in the Pacific plate. A hot spot is an area where molten material from the earth's interior rises through the crust. The larger islands are mountainous with flat land only along narrow coastal plains and in a few mountain valleys. The coast is rugged and scenic, with many coral reefs that erode into fine, white beach sand.

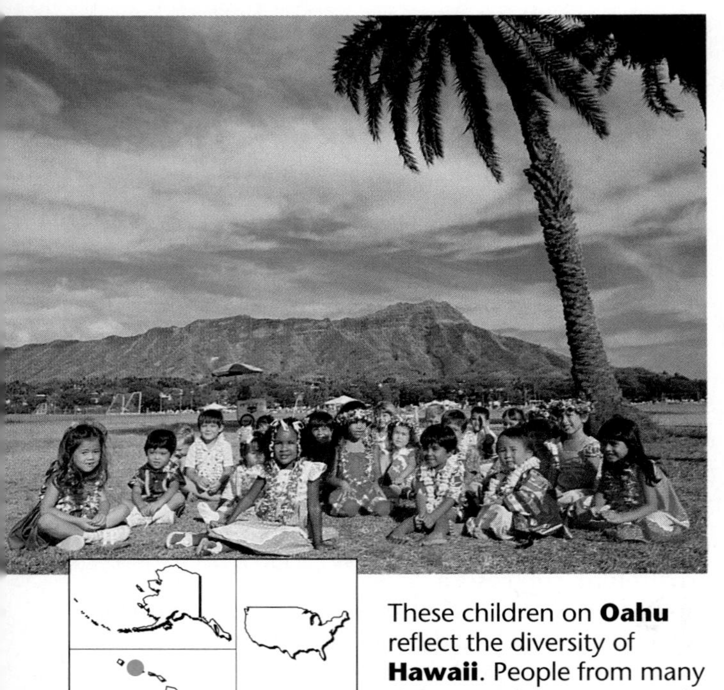

These children on **Oahu** reflect the diversity of **Hawaii**. People from many different ethnic groups call Hawaii home.

Climate Hawaii has little daily or seasonal temperature change. Honolulu's coldest month is January, which averages 72°F (22°C). August, the warmest month, averages 81°F (27°C). Only the high volcanic peaks have cool temperatures.

The northeast trade winds dominate the weather and cause heavy rainfall on the eastern, windward slopes of the islands. This produces a humid-tropical climate. Part of Kauai is considered the wettest place in the world. The western, leeward slopes of the islands generally receive less annual rainfall.

Economic Geography Hawaii's tropical climate and fertile volcanic soils provide excellent growing conditions for sugarcane, pineapples, and other tropical crops. Although less than 3 percent of the population is engaged in agriculture, it is important to the local economy. Tourism now dominates Hawaii's economy. Several million tourists, mostly from other states and Japan, visit the islands each year. More than 30 percent of all jobs are in the tourist industry.

Cultural Geography The original settlers of Hawaii were Polynesians from the South Pacific islands. They arrived as early as the fourth century, forming chiefdoms on each of the major islands. The early European settlers on the islands were missionaries and whalers. They drastically changed life on the islands. Settlers and traders brought farm animals, plants, and manufactured goods. They also brought diseases that were fatal to many native Hawaiians.

During the nineteenth century, European settlers developed sugarcane and pineapple plantations. Laborers emigrated from China, Japan, the Philippines, Samoa, and Portugal to work these fields. Hawaii became a U.S. territory in 1900, and gained statehood in 1959. Since 1900, there has been a continuous flow of workers and settlers from the mainland United States and the Pacific Rim nations, especially Japan.

Today, Hawaii has more than 1 million people representing a great mixture of cultures. Asian Americans form more than 50 percent of the population. Of these, the greatest percentage is Japanese American, followed by Filipino American and Chinese American. European Americans are the second largest group, accounting for one-third of the total population. Native Hawaiians make up about 13 percent of the population. More than 80 percent of the population lives in and around the capital city of Honolulu on Oahu.

Issues The major concern for Hawaii is its dependence on materials and income from outside sources. Because the islands lack industry and a diversified agriculture, more than 80 percent of all materials, energy, and food must be imported. This economic situation keeps the cost of living in Hawaii high. Like the other states of the Pacific States region, Hawaii must find a balance between development and preservation of its natural environment.

SECTION REVIEW

1. How do Hawaii's landforms and climate affect its economy?
2. What problem results from Hawaii's dependence on outside resources for most of its income and materials?
3. **Critical Thinking** **Analyze** how severe hurricane damage might affect Hawaii's economy.

CHAPTER 15 CHECK

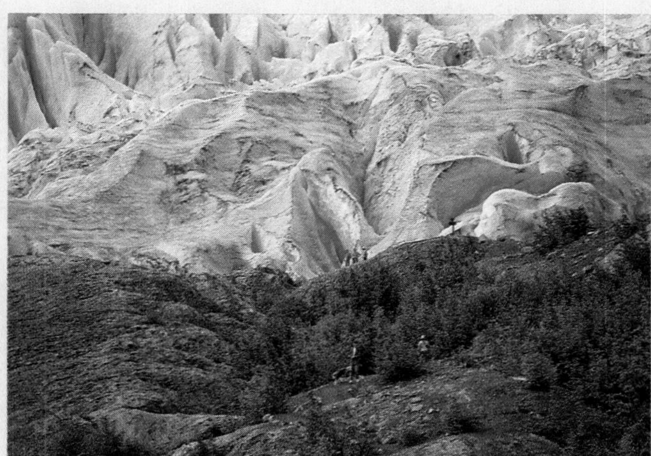
Exit Glacier in Kenai Fjords National Park, **Alaska**

Reviewing the Main Ideas

1. The Pacific States region includes California, Oregon, Washington, Alaska, and Hawaii. Although landforms and climate throughout the states vary, the whole region experiences volcanic activity and earthquakes.
2. California is the most urbanized state, with three major urban areas—the Los Angeles Basin, the San Francisco Bay Area, and San Diego County. Oregon and Washington also have urbanized areas, particularly in the Portland and Puget Sound metropolitan areas.
3. Alaska is the largest state, and oil is its most important resource. Hawaii is tropical and is the only island state. Hawaii depends mostly on tourism for income.
4. The Pacific states must balance future development with environmental considerations.

Building a Vocabulary

1. What is a multicultural region? Give an example of a multicultural region from the chapter.
2. Define *caldera.* What lake discussed in this chapter fills the caldera of an extinct volcano?
3. Name two states other than Alaska that have panhandles.
4. What causes a hot spot? Where in the Pacific States region would you find a hot spot?

Recalling and Reviewing

1. What are the major climate regions of California?
2. What are the major resources in Oregon and Washington?
3. What are Alaska's important resources?
4. Why does Hawaii have a high cost of living?

Thinking Critically

1. Imagine that you live in California in the year 2010, when there will be no ethnic majority group. What new issues might this population shift create for California?
2. What do you think everyday life is like for many of the people who live in Alaska? Hawaii? Are there any similarities between the two states?

Learning About Your Local Geography

Cooperative Project

With your group, use the library and other community resources to find out if there is a conflict in or near your area between resource development and environmental protection. Present your information to the rest of the class in an interview, debate, or news report.

Using the Five Themes of Geography

Sketch a map of the Pacific states. On your map, place some arrows to show **movement** into the various parts of the region. From what directions have people arrived into the region? Also, use symbols and/or shading to show some of the challenges to the region brought about by **human-environment interaction**. Make sure your map has a legend.

Harvesting grapes in the San Joaquin Valley, **California**

CHAPTER 16

Canada

GEOGRAPHY DICTIONARY

- *newsprint*
- *portage*
- *deport*
- *dominion*
- *parliament*
- *hinterland*
- *potash*
- *muskeg*
- *regionalism*

"In northern Canada, the roads are civilization, owned by the collective human *we*. Off the road is *other*. Try walking in it, and you'll soon find out why all the early traffic here was by water. "Impenetrable wilderness" is not just verbal."

Margaret Atwood

A Native American totem pole in **British Columbia, Canada**

Canada is the world's second largest country, after Russia. It covers about two-fifths of the North American continent, stretching east to west for over 3,400 miles (5,471 km) and north to south for over 2,000 miles (3,218 km). The population of this large nation numbers only about 27 million—less than that of California. Canada is divided into 10 provinces and three northern territories. The harsh climates of the Canadian north have kept Canada from becoming evenly populated. More than 75 percent of Canadians live along the nation's southern border within 100 miles (161 km) of the United States.

Today, the economies and cultures of Canada and the United States are closely linked. The Canadian–United States border is the longest unguarded boundary in the world. Each country is the other's most important trading partner, and many companies do business in both countries. Cultural exchange between the two countries has helped shape both societies.

Kluane National Park, **Yukon, Canada**

CANADA: PHYSICAL–POLITICAL

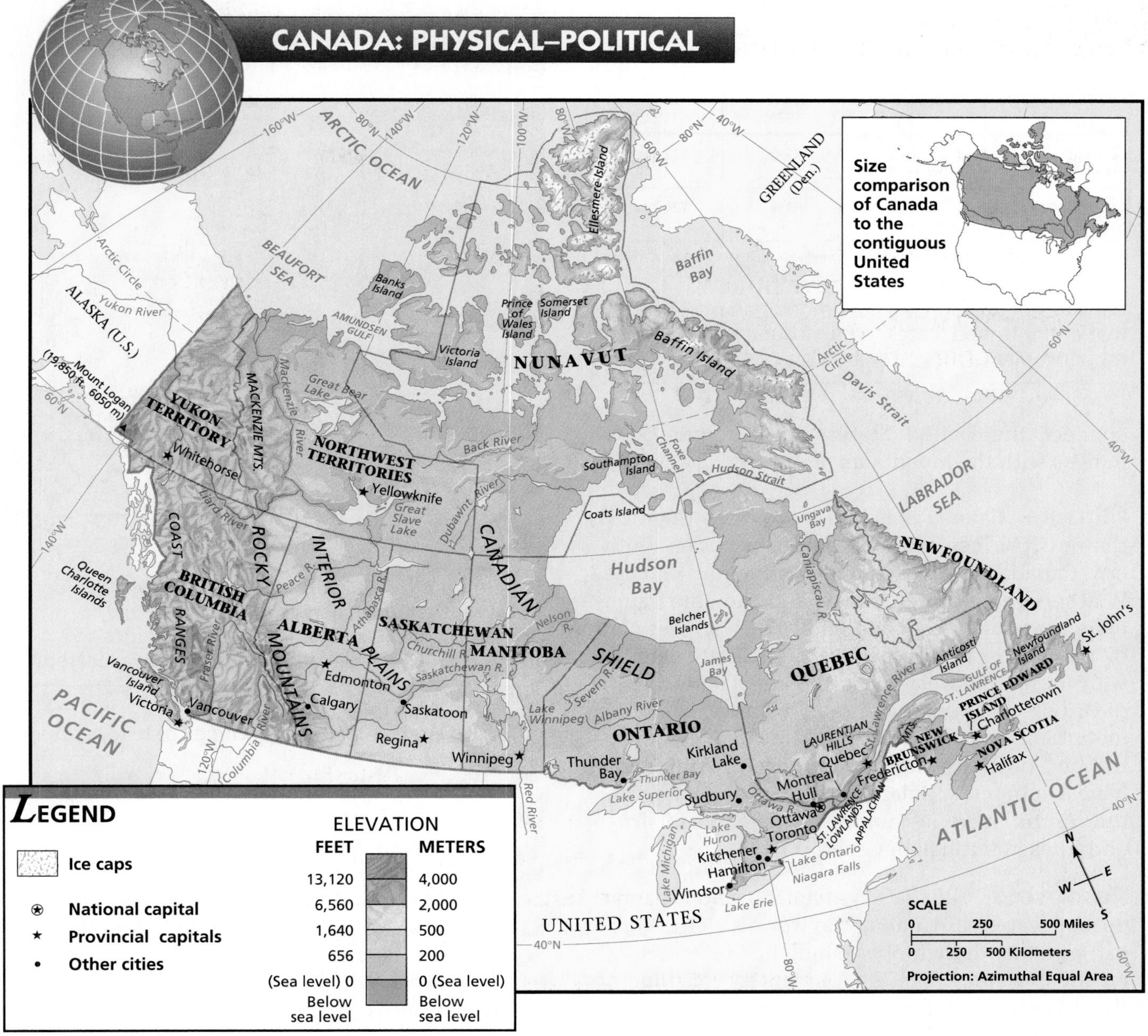

Size comparison of Canada to the contiguous United States

LEGEND

Ice caps

National capital

Provincial capitals

Other cities

ELEVATION

FEET	METERS
13,120	4,000
6,560	2,000
1,640	500
656	200
(Sea level) 0	0 (Sea level)
Below sea level	Below sea level

SCALE

0 250 500 Miles

0 250 500 Kilometers

Projection: Azimuthal Equal Area

PHYSICAL GEOGRAPHY

FOCUS

- *What landforms, climates, and vegetation does Canada have?*
- *What are Canada's primary resources?*

REGION Landforms Canada can be divided into six major landform regions: the Appalachian Mountains, the St. Lawrence Lowlands and Great Lakes, the Canadian Interior Plains, the Canadian Shield, the mountains of the Canadian west, and the mountains of the eastern Arctic.

Canada's Appalachian Mountains are in southeastern Quebec and in the Maritime Provinces. The St. Lawrence River valley and the Great Lakes were major routes of waters from melting glaciers. Today, these parts of Ontario and Quebec contain some of Canada's most fertile soils. The Canadian Interior Plains stretch east to west across Manitoba, Saskatchewan, and eastern Alberta. The Canadian Shield is a semicircular band of rocky uplands and plateaus surrounding Hudson Bay. The mountains of the Canadian west

FACTS IN BRIEF Canada

COUNTRY POPULATION (1994)	LIFE EXPECTANCY (1994)	LITERACY RATE (1991)	PER CAPITA GDP (1993)
CANADA 28,113,997	75, male 82, female	99%	$22,200
UNITED STATES 260,713,585	73, male 79, female	97%	$24,700

Sources: *The World Factbook 1994*, Central Intelligence Agency; *The World Almanac and Book of Facts 1995*

Canada's population is a little more than one-tenth that of the United States. Which country has the highest per capita GDP?

connect the United States Rockies and Coast Ranges with the mountains of Alaska.

Climate Canada is frozen more often than it is thawed. The central and eastern parts of southern Canada have a humid-continental climate. Winters are cold and snowy, and the short summers are usually cool and mild. The warmest part of Canada is in southwestern British Columbia with its marine-west-coast climate. The lowlands of British Columbia have rainy winters, while the mountains have heavy snow. The climates of central and northern Canada are severe. Wide areas have a subarctic climate. The coast and islands of the Arctic have a tundra climate. Permafrost underlies about half of Canada.

Resources Canada's Atlantic and Pacific coastal waters are among the world's richest fishing grounds, though overfishing has taken its toll on these waters. The long, indented coastline of the Atlantic seaboard provides many good harbors. On the Pacific coast, salmon fishing is of major importance. The many inland lakes and streams of Canada provide a plentiful supply of freshwater fish and attract many tourists.

A belt of needle-leaf forests stretches all the way from Labrador to the Pacific coast. These softwood trees provide good lumber and pulp for paper. The United States, the United Kingdom, and Japan look to Canada for many of their lumber, **newsprint** (an inexpensive paper used mainly for newspapers), and pulpwood products.

Minerals are the most valuable of Canada's natural resources. The Canadian Shield, once thought to be a wasteland, has proven to be a source of many mineral deposits. Canada is a leading source of the world's nickel, zinc, and uranium. It also is a leading producer of lead, copper, gold, and silver.

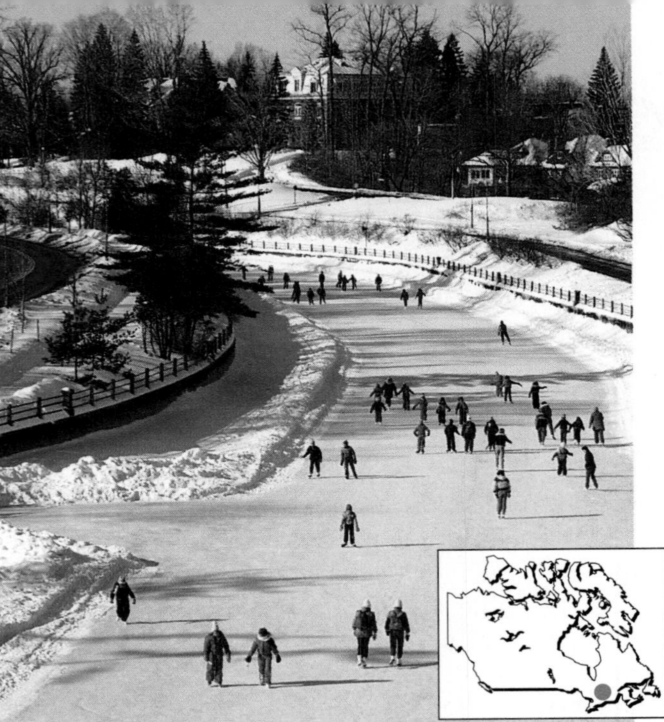

Skating on frozen canals is a popular activity in **Ottawa**, the Canadian capital. Compare the physical–political and climate maps on pages 107 and 108. In which climate region is Ottawa located?

Experts believe that the coal fields of Canada may be among the largest in the world. The coal deposits are generally very deep and are located in unpopulated areas. Soft coal is mined in Nova Scotia and in the western provinces of Saskatchewan, Alberta, and British Columbia.

Canada is a large-scale producer of petroleum. About two-thirds of Canada's petroleum and about four-fifths of its natural gas come from Alberta. Major deposits have been discovered off the east coast and off the Arctic coast as well. Canada also is one of the world's largest producers of hydroelectricity. Hydroelectric power plants generate about two-thirds of Canada's electricity.

SECTION REVIEW

1. List four landform regions and two climates found in Canada.
2. What mineral resources does Canada have, and where is each one found?
3. **Critical Thinking** **Discuss** the importance of water and forest resources to Canada.

HISTORICAL GEOGRAPHY

FOCUS

- *How have other countries played a role in Canada's history?*
- *How were various parts of Canada settled, and by whom?*

MOVEMENT **New France** The first Europeans to land on Canada's eastern shores were Viking adventurers, who arrived between A.D. 1000 and 1200. They left no permanent settlements, however. The formal exploration of Canada by Europeans began in 1497 when John Cabot landed on Newfoundland and other islands. During the first part of the 1500s, Canada's eastern coast was visited by people from many European nations. Trade between Canada's Native Americans and Europeans developed quickly.

The first great explorer of the St. Lawrence River was Jacques Cartier of France. In 1535, he traveled upstream as far as the site of Montreal. Cartier and the French explorers who followed him had three goals: to search for a northwest passage to China and India, to develop a fur trade, and to convert the Native Americans to Roman Catholicism. The territories claimed by the French during this time were known as New France.

The French realized that the St. Lawrence River and the Great Lakes were an avenue into the interior of North America. The lakes and rivers were connected by easy **portages**. A portage is a low land area across which boats and their cargoes can be carried. French settlers came to farm on the fertile lowlands along the St. Lawrence and in Nova Scotia. Fur trappers and trading posts spread into central North America.

British North America In 1713, England acquired Nova Scotia for its fishing ports and farmlands. The French who were living in Nova Scotia were **deported** to what was then French Louisiana. To deport means to send out of a country. In the early 1700s, England's expanding American colonies and France's interests in North America came into conflict, which, along with other conflicts elsewhere, caused France and England to go to war in 1756. In North America, this war drew to a close in 1759 when the British captured the city

The Canadian Parliament buildings in **Ottawa, Ontario**, reflect the European influence of the country's colonial past. Canada's government is a combination of U.S. and British government systems.

of Quebec. To promote peace, the British recognized French traditions and laws. Both French and English became official languages.

British settlement of Canada increased dramatically during the American Revolution. To remain under British rule, many colonists who remained loyal to Britain left the colonies to live in southern Ontario. Their descendants have been economically and politically powerful throughout Canada's history.

In 1791, Canada was divided into Upper Canada (present-day Ontario) and Lower Canada (present-day Quebec). At this time, Canada's Maritime Provinces along the eastern coast were separate British colonies. In the early 1800s, Canada grew steadily. The English and French communities began arguing over language and immigration policy. The cities of Quebec, Montreal, and Toronto grew into major commercial centers. Farther west, the Hudson's Bay Company and the North West Company extended their fur-trading posts to the Pacific coast and to the subarctic region.

Dominion of Canada In 1867, the British Parliament passed the British North America Act, which created the Dominion of Canada. A **dominion** is a territory or sphere of influence. The dominion joined together the provinces of Ontario, Quebec, Nova Scotia, and New Brunswick to be governed by a **parliament** (a lawmaking body) and a prime minister. The city of Ottawa, in Ontario but on the Quebec border, became the capital. The other present-day provinces joined the dominion later. Each Canadian province has its own parliament, headed by a provincial prime minister. Provincial governments have control over taxes, education, and civil rights.

During the 1800s, many African Americans from the United States moved to eastern Canada to escape slavery. Most of Canada's west was settled after the country's transcontinental railroad was finished in 1885. Today, many of the nation's citizens are immigrants from Europe, East and Southeast Asia, and the Caribbean.

Ties between Canada and Britain have remained close. Canada's national government has a parliament like Britain's, and Britain's queen also is Canada's queen.

2 SECTION REVIEW

1. What countries have ruled Canada? What was the British North America Act?
2. Over time, what groups of people have settled in Canada?
3. **Critical Thinking** **Compare and contrast** historical reasons for French and British settlement in Canada.

Fish dry in the afternoon sun of **Newfoundland**. As in Canada's other eastern provinces, most residents live near or along the Atlantic coast, where fishing is a major occupation.

3 CANADA'S PROVINCES AND ARCTIC FRONTIER

FOCUS

- *What are Canada's provinces and territories?*
- *What economic characteristics does each province or territory have?*
- *What cultures are found in Canada, and how do they interact?*

Maritime Provinces and Newfoundland

New Brunswick, Nova Scotia, and Prince Edward Island are called the Maritime Provinces. Newfoundland includes the island of Newfoundland as well as Labrador on the nearby mainland. These provinces are Canada's eastern **hinterland**. A hinterland is a region beyond the center of a country or city.

Life often is hard in eastern Canada. Economic development has depended on the resources of the sea and forests. The short growing season and thin, rocky soils make farming difficult, although small farms that produce a variety of crops are typical.

Commercial fishing is an important economic activity. The Grand Banks, an area of shallow waters off the eastern shore of Newfoundland, forms one of the world's great fishing grounds. The region's thick fog and winter storms can make fishing dangerous, however.

Many people in the Maritime Provinces are descended from families that originally emigrated from the British Isles. They are separated from the rest of English-speaking Canada by French-speaking Quebec. Many families from Quebec have moved to New Brunswick, however, and New Brunswick claims to be Canada's only truly bilingual province.

Most people in Canada's eastern provinces reside in coastal cities. These cities are manufacturing centers, as well as fishing and freight ports. The region's other population centers are in the few inland areas best suited for farming. Halifax, Nova Scotia, is the region's leading city.

Quebec The chain of cities that extends from Windsor, Ontario, to the city of Quebec is Canada's greatest urban region. The province of Quebec is distinctive in North America because of its strong French-Canadian culture. More than one-fourth of all Canadians live in the province. For most of them, French is their first language. With nearly four centuries of experience in North America, these French Canadians are proud to call themselves Quebecois (kay-beh-KWAH), rather than French. Many people speak only French, official papers of the provincial government are prepared only in French, and signs along roads and on businesses appear only in French.

Throughout Quebec history, many French Canadians have argued that the province should become an independent country. The issue of Quebec's independence has threatened to break up Canada many times. An independent Quebec would separate the eastern provinces from Ontario. The western provinces might then demand their independence as well.

Quebec, Canada's first city and the provincial capital, is located where the shores of the St. Lawrence estuary pinch together. Many of the city's older sections, which feature narrow streets and stone walls, have been preserved. Quebec is a symbol of Quebecois culture and is a popular tourist destination.

Montreal is the financial and industrial center of Quebec. The city began on an island in the St. Lawrence River. Today, Montreal serves as an important port on the St. Lawrence Seaway. To combat the winter cold, many of the buildings in the city center are connected by underground arcades and overhead glass tunnels.

Montreal was once an island of British-Canadian culture in French Quebec. Now, important jobs are being filled by the Quebecois. After Paris, Montreal is the second largest French-speaking city in the world.

The French colonial past is clearly seen in the architecture of the city of **Quebec**. Where in the United States is French influence most noticeable?

Ontario As in Quebec, most of the people of Ontario live in the far south. The cities of southern Ontario are the fastest growing in Canada. Southeastern Ontario is the chief manufacturing district of the country. Toronto, Hamilton, Kitchener, Windsor, and several smaller cities are grouped together in this region.

Toronto is the provincial capital and the industrial, financial, cultural, and educational center of Canada. It houses Canada's largest stock exchange, and most large Canadian corporations are headquartered here. The nation's finest museums, universities, symphony orchestras, and other cultural institutions are concentrated in the Toronto region. Metropolitan Toronto has a population of about 3 million.

Toronto reflects Canada's British heritage much like the city of Quebec symbolizes French Canada. British customs are still widespread, and French is seldom heard or seen on signs in Toronto. Instead, recent immigrants from Italy, the Caribbean, and China give the city its international flavor.

Ottawa, Canada's capital city, lies in the province of Ontario. Its twin city is Hull, just across the Ottawa River in Quebec. The capital region is fully bilingual. Ottawa has grand government buildings, more than 1,500 acres (607 ha) of parks, and several universities. Manufacturing activities concentrate on forest products.

Prairie Provinces Manitoba, Saskatchewan, and Alberta are called the Prairie Provinces. Much of the southern parts of these provinces was once covered with prairie grasslands. Today, the region is the center of a rich wheat belt. Highly mechanized farms require few people, and the Prairie Provinces remain only thinly settled. Still fewer people live to the north, where the grasslands give way to forests.

Canada's population is quite small in proportion to its production of wheat. Consequently, in most years, Canada has more wheat available for export than any other country. Changes in the foreign wheat market and uncertain weather conditions cause difficulties for the wheat farmers, however. As in the United States, the uncertainty of prices and growing conditions has led Canadian farmers to request aid from the government.

Winnipeg, the capital of Manitoba, is the chief city of the eastern prairies. All east-west rail

NEWS FROM AROUND THE WORLD Canada

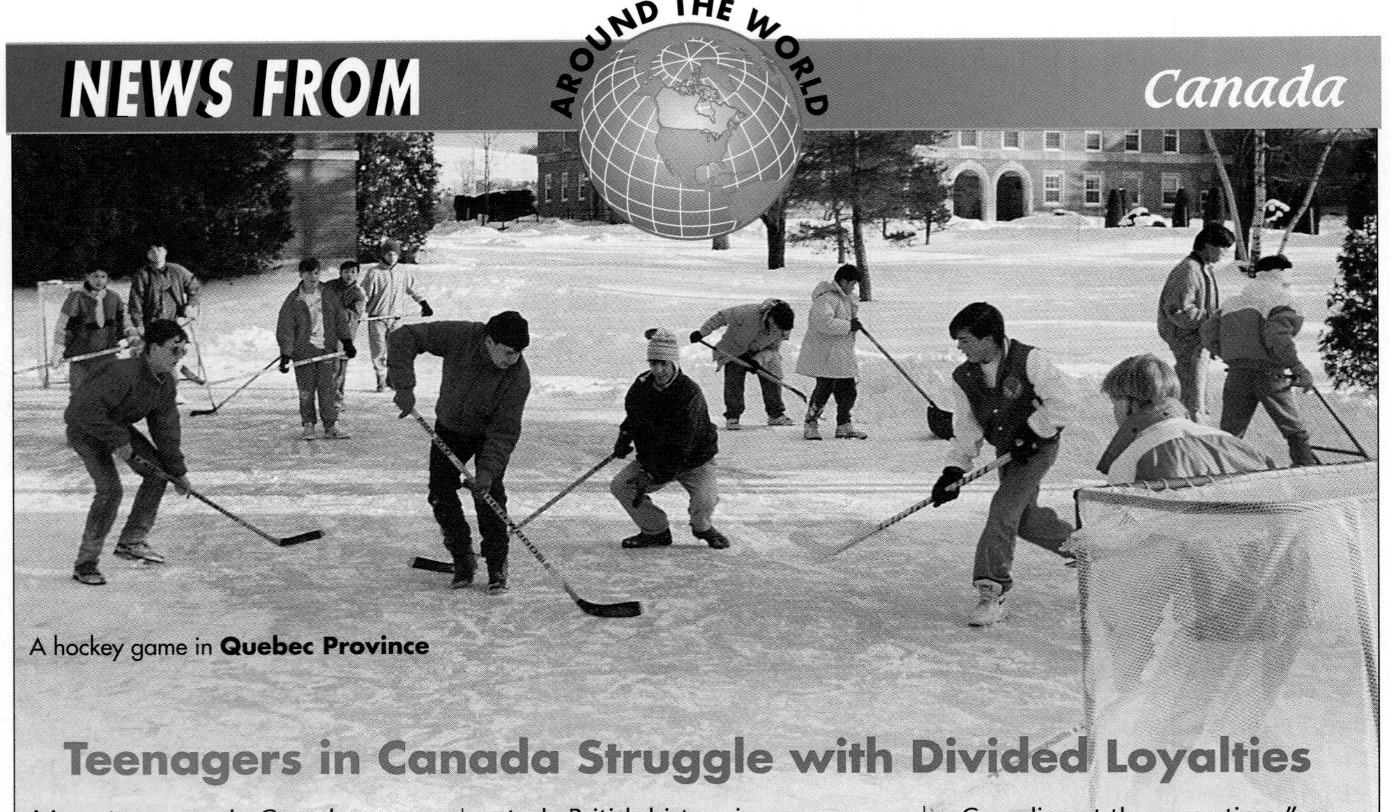

A hockey game in **Quebec Province**

Teenagers in Canada Struggle with Divided Loyalties

Many teenagers in Canada are accustomed to waking before dawn on chilly Saturday mornings. That may be the only time their hockey team can practice at one of the busy 24-hour rinks. When rinks are full, teens put up homemade nets and play hockey on frozen driveways and streets.

Winter sports are a way of life in much of Canada. The winters are cold, snowy, and long. Still, the snow and ice cannot cover up some basic differences that have not only divided Canadians but have threatened to split the country as well.

Imagine that you are a typical student in Ontario, Canada. English is your first language at home and in school, and you study British history in your classes. Perhaps you come home to tea in the afternoon.

In contrast, imagine that you live in Quebec, Canada's only predominantly French-speaking province. Your classes are probably taught in French. You might eat croissants for breakfast and tourtiere, or pork pie, for lunch. Indeed, as a Quebecois, your life might be quite different from that of a student living only a few hundred miles away in the province of Ontario.

Teenagers are keenly aware of the differences that exist between Quebec and other regions of Canada. "Sometimes there seems to be a feeling that you can't be Quebecois and Canadian at the same time," a young man comments.

Still, young Canadians have a great deal in common with each other and with young people around the world. Many Canadian high school students work at part-time jobs to earn money for college. Many also are concerned about their future careers and job opportunities.

Many of them worry, too, about the possible breakup of their homeland. They hope that the Canadian provinces can overcome their differences and remain unified. One young woman says, "I often wonder how I'd feel if Quebec actually separated. I love Montreal, but I'm Canadian."

traffic passes through Winnipeg. As an important collection and shipping point for the region's products, Winnipeg serves Canada much like Chicago serves the United States.

Saskatchewan is predominantly agricultural, although it contains one of the world's largest deposits of **potash**. This mineral is an important raw material for manufacturing fertilizers. Regina, Saskatchewan's capital, is a center of the potash industry as well as of the province's food-processing industries.

Alberta's wealth is based on fossil fuels, particularly oil. The province's oil and natural gas travel in pipelines to eastern and western Canada and to the United States. The spectacular Rocky Mountains of western Alberta are popular with tourists from around the world. Alberta has two rapidly growing cities, Calgary and Edmonton. Each is an important oil and agricultural center.

British Columbia The province of British Columbia lies on the Pacific coast of Canada. It is a land of rugged mountains, intermountain plateaus, and fertile river valleys. Large mountain ranges extend the length and width of the province. These mountains have long isolated the province from the rest of Canada.

British Columbia is rich in natural resources. Much of the province is covered with forests of fir, spruce, and cedar trees. Salmon fishing and mining industries also are important to the province's economy. Because of its location on the west coast, British Columbia can trade easily with other ports around the Pacific Rim. Water resources play a large role in British Columbia's economy, and mountain streams provide inexpensive electric power.

Vancouver is Canada's third largest urban region. It grew to become western Canada's greatest city after the first transcontinental railway built its end point at the mouth of the Fraser River. Vancouver has Canada's major ice-free harbor and is Canada's outlet to the Pacific Ocean. Victoria, the capital of British Columbia, is located at the southern tip of Vancouver Island. This government center is the home port for a large fishing fleet.

The Canadian North Canada's vast northern lands are divided into the Yukon, the Northwest Territories, and Nunavut. Nunavut is a new territory that was created for and is administered by the native Inuit (Eskimo) who live there. These three regions occupy more than one-third of Canada's land surface but are home to only about 80,000 people. Dense forests and tundra, mostly underlain by permafrost, separate the isolated towns and villages. Thousands of rivers and millions of lakes characterize the Canadian North. **Muskegs** are other common landforms. A muskeg is a forested swamp that melts only during the mosquito-breeding season in the summer.

Airplanes and satellite communications recently have brought northern Canada into the national sphere. In spite of its severe climate, the region holds important promise for Canada's future. Rich deposits of metals and fossil fuels have been discovered here, and the supplies of fresh water and potential for hydroelectricity are enormous.

The people of the Canadian north are a mixture of hardy frontier people. The residents include the Inuit (Eskimo) and Native Americans whose ancestors fished and hunted in the region hundreds of years ago, as well as people who have migrated from Canada's cities in the south.

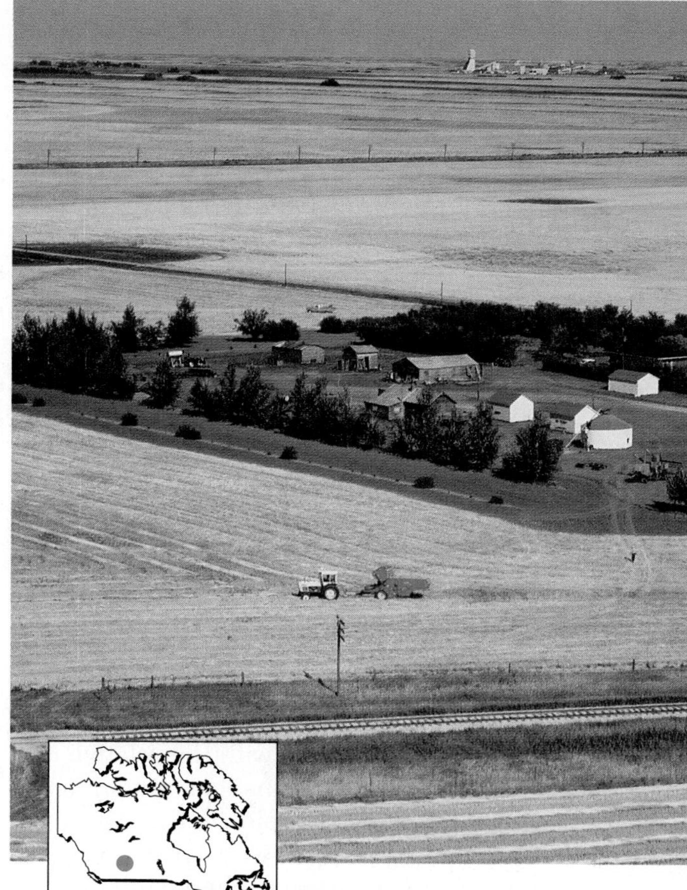

A farmer works his land in the Canadian wheat belt near **Saskatoon, Saskatchewan**. Wheat is the primary crop throughout the Great Plains from the Canadian prairie provinces to Texas.

3 SECTION REVIEW

1. List Canada's provinces and territories.
2. Which Canadian provinces are primarily industrial? agricultural?
3. **Critical Thinking** **Analyze** Canada's cultural geography. How have cultural differences been a challenge at times?

CANADIAN ISSUES

FOCUS

- *What are the origins of regionalism in Canada? What effects has it had?*
- *How have economic ties been an issue for Canada and its regions?*

REGION **Regionalism** When Canadians deal with each other, they often show considerable **regionalism**. Regionalism is the political and emotional support for one's region before support for one's country. The threat to divide Canada into several countries has surfaced repeatedly in Canadian history. Although Canada's national government has international duties similar to those of the U.S. federal government, the Canadian provinces are much more powerful than the states in the United States.

This situation, combined with the relative isolation and separation of Canadian settlements, keeps regionalism alive. People in the Maritime Provinces and in the Canadian west have long been suspicious of events in Ottawa, where Ontario and Quebec seem to dominate. People in Quebec and Ontario have been haggling over language and culture for more than two centuries. Alberta would rather not share its petroleum industry revenues with Ottawa. British Columbia, separated by high mountains and long distances, has often voiced detachment from the rest of Canada.

Even so, most Canadians support a united Canada rather than several small countries. Canadians continue to find strength in their diversity. In doing so, they set a positive example for many other countries where ethnic diversity has led to unrest and war.

Economic Ties The many economic ties between the various provinces are another factor holding Canada together. Poor regions, such as Newfoundland and rural Manitoba, depend on the bankers and industrialists of Toronto and Montreal. Development of resources in the Northwest Territories relies on trading at the Vancouver Stock Exchange. All Canadians depend on the oil resources of Alberta.

Internationally, Canada's main economic ties are with its southern neighbor. Often, the United States buys Canada's natural resources in exchange for manufactured goods. Both Toronto and Montreal do more business with U.S. firms than they do with each other. Many U.S. companies are giants in the Canadian economy.

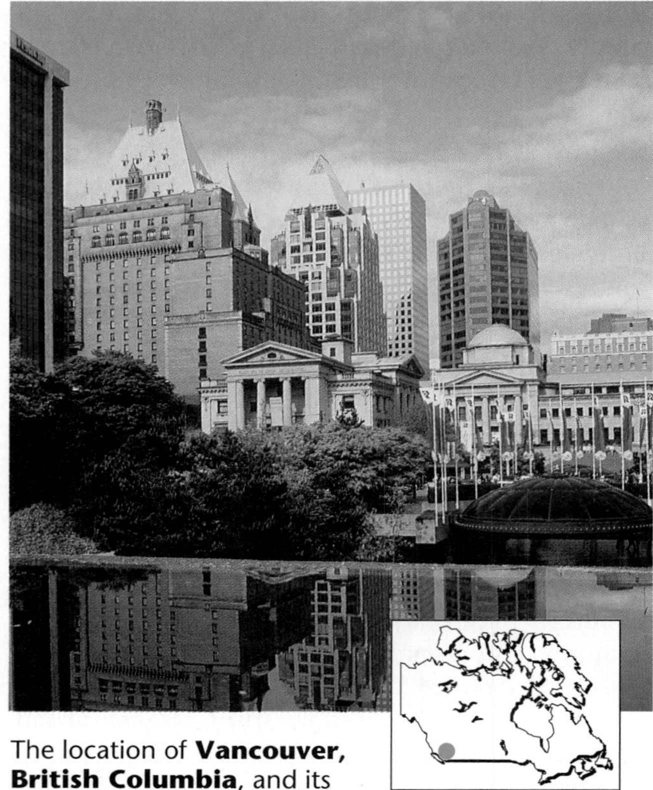

The location of **Vancouver, British Columbia**, and its busy port make the city Canada's main trade link to other Pacific Rim countries.

The United States has nearly 10 times the population of Canada, and U.S. mass culture frequently seems to overwhelm Canadians. Canadians, however, continue the struggle to maintain their own distinct culture. They provide another voice in the varied culture of North America.

4 SECTION REVIEW

1. Which Canadian provinces have been most affected by regionalism, and in what ways?
2. How are Canada's regions dependent on each other? What economic ties exist between Canada and the United States?
3. **Critical Thinking** **Discuss** the relationship between regionalism and interregional economic ties.

CHAPTER 16 CHECK

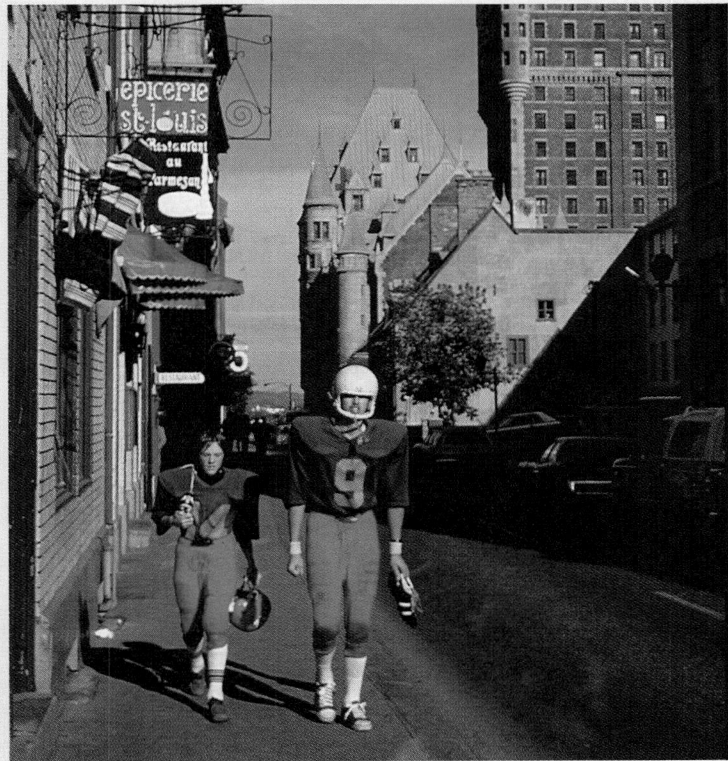
epicerie
st-louis
Old city section of **Quebec**

Reviewing the Main Ideas

1. Canada has diverse landforms and mostly cold climates. Resources include fish, forest products, minerals, and hydroelectricity.
2. The French formed the first permanent European settlements in Canada. Over time, Britain gradually gained control over most of Canada. Canada has maintained ties to Britain. Today, immigrants to Canada come from Europe, East and Southeast Asia, and the Caribbean.
3. Canada has 10 mostly self-governing provinces and three northern territories. The provinces and territories have a wide range of economic and cultural characteristics.
4. One major Canadian issue has to do with regionalism, the support for one's region before one's country. Another Canadian issue involves economic ties, on both national and international levels.

Building a Vocabulary

1. How did portages help the French explore and settle the Canadian interior?
2. Define *hinterland.* Which provinces make up Canada's eastern hinterland?
3. What is potash used in?
4. What term describes a cold forested swamp that melts only in the summer?

Recalling and Reviewing

1. Why does a large portion of Canada's population live close to the border with the United States? What effects has this closeness had on Canada?
2. How is Canada's history reflected in the country today?
3. What challenges has Canada's cultural diversity presented? How has this diversity also been a source of strength to the country?

Thinking Critically

1. How might going to school in a country that has two official languages be different from going to school in a country where only one language dominates?
2. Imagine that several Canadian provinces became independent countries. What effects might Canada's breaking up have on the United States?

Learning About Your Local Geography

Individual Project

Research your community's cultural background. From what countries or areas did the first settlers of your community originate? What ethnic backgrounds do current residents have? What languages are spoken in your community? Make a list of the various cultural and ethnic groups that have lived in your community over time. You may want to include your list in your individual portfolio.

Using the Five Themes of Geography

Write a paragraph about the themes of **place** and **location** in Canada. How have physical place characteristics influenced the location of Canada's people and cities? How has location influenced Canada's human place characteristics?

SKILLS

Using Models in Geography

When geographers present ideas, they sometimes find it helpful to use **models**. Here, model does not mean a small structure made of cardboard and glue, like a "model" airplane. Rather, the models used by geographers are more like the plans or patterns of different types of airplanes. You might look at these plans to construct an actual airplane (miniature or otherwise), or you might just study the plans to gain a greater understanding of the parts of different types of airplanes.

Models are particularly useful for understanding how cities are arranged. For example, why are skyscrapers and houses located in different areas? Over the past decades, geographers have suggested several models to answer such questions. The goal of these models is not to explain the layout of an existing city, such as New York or Los Angeles, but to make generalizations that will apply to many cities. Models change over time as geographers gain new information and as cities themselves change.

Most models of cities show three major categories of land use. The first category is the central business district, or CBD—the area people often call "downtown." It contains banks, office buildings, hotels, and many kinds of stores and shops. Because these businesses need to be reached by people from all over the city, most models place the CBD in the middle of the city. The second category of urban land use is the industrial and manufacturing area. The third category consists of the residential areas, where most people live.

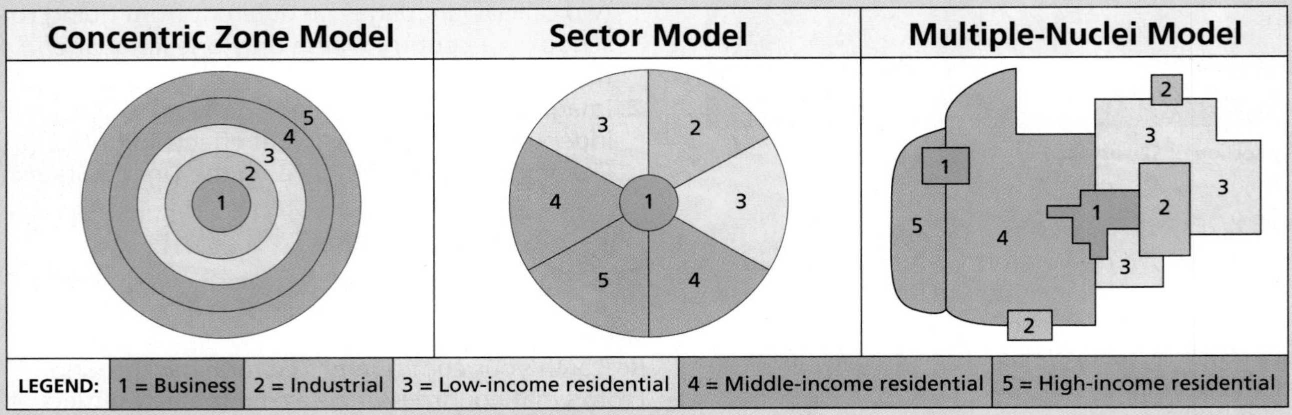
Concentric Zone Model
Sector Model
Multiple-Nuclei Model
LEGEND: 1 = Business
2 = Industrial
3 = Low-income residential
4 = Middle-income residential
5 = High-income residential

In this early model, the city looks like a bull's-eye. The center circle is the CBD. The next ring is the manufacturing area. The three outer rings are the residential areas, with low-income areas toward the inside and high-income suburban areas as the outermost ring.

The Sector Model also places the CBD in the middle of a circle, but the rest of the circle looks like a pie rather than a bull's-eye. Transportation routes, such as freeways, rail lines, and rivers, run between the CBD and the different sectors, or pieces of the pie.

In this non-circular model, developed in the 1940s, there are several small business and industrial districts, or nuclei, in some of the outer areas in addition to the more central CBD and industrial areas. This model attempts to show the increased complexity of cities.

PRACTICING THE SKILL

1. Why do you suppose low-income residential areas tend to be next to industrial areas?
2. Recent models of cities attempt to combine elements of the previous models with the idea that, in modern times, a city's surrounding areas tend to overlap with those of other cities and towns. What might such a model look like?
3. Identify the central business district, industrial area, and residential areas of the nearest city. Which model does the city resemble most?

BUILDING YOUR PORTFOLIO

Individually or in a group, complete one of the following projects to show your understanding of the geography concepts involved.

A Economic and Demographic Shifts

In this unit, you have been reading about some of the major changes the United States and Canada have been experiencing recently: the shift from primary and secondary industries to service industries, the displacement of traditional industries by high-tech industries, the increase of corporate-owned farms as compared to family-owned farms, and the population shifts to the Sunbelt and to suburbs. Research how these changes have affected your local area, and prepare a report on your findings. To complete your report, you will need to do the following.

1. Make bar or pie graphs showing how the economic activities of your area have changed over the past several years.
2. Write a paragraph discussing how manufacturing industries in your area have changed, and how the changes have affected local workers.
3. In a chart, show the extent to which patterns of farm ownership have been changing in your area.
4. Construct a line graph showing how your area's population has changed in size over time as a result of the national population shift to the Sunbelt.
5. With maps, show whether the demography of a local town or city indicates a widespread move into suburbs.
6. In several illustrations and captions, predict how these changes will affect the future of your area and your own future, as an individual.

Organize your materials, and present your report to the rest of the class.

B Urban Renewal

You are a city planner. For a conference called "The Renewal of American Cities," prepare a proposal for how cities might go about revitalizing themselves. For your proposal, you will need to do the following.

1. Draw a series of illustrations, with captions, that show problems faced by urban areas. What characteristics might make some areas unpleasant for residents and unattractive to visitors?
2. Construct a table identifying three U.S. and/or Canadian cities that have areas in need of revitalization and showing some of the causes of urban decline in each city.
3. Make a list of three cities that have already gone through renewal. Next to each city, discuss when the renewal was finished, what forms it took, and what problems the city faced during the renewal.
4. Graph before-and-after crime rates, income levels, and other information about the revitalized areas of the three cities that have already gone through renewal. With your graphs, include captions discussing what the renewal accomplished and how it changed residents' daily lives.
5. Write a paragraph about how to avoid the problems during revitalization that other cities have faced. Are some problems unavoidable?
6. Create a graphic organizer showing your plan for urban renewal. What problems does your plan address, and how does it address them? Does your plan create any new problems?

Organize your materials, and present your proposal at the conference (in class).

Plan for
Urban Renewal

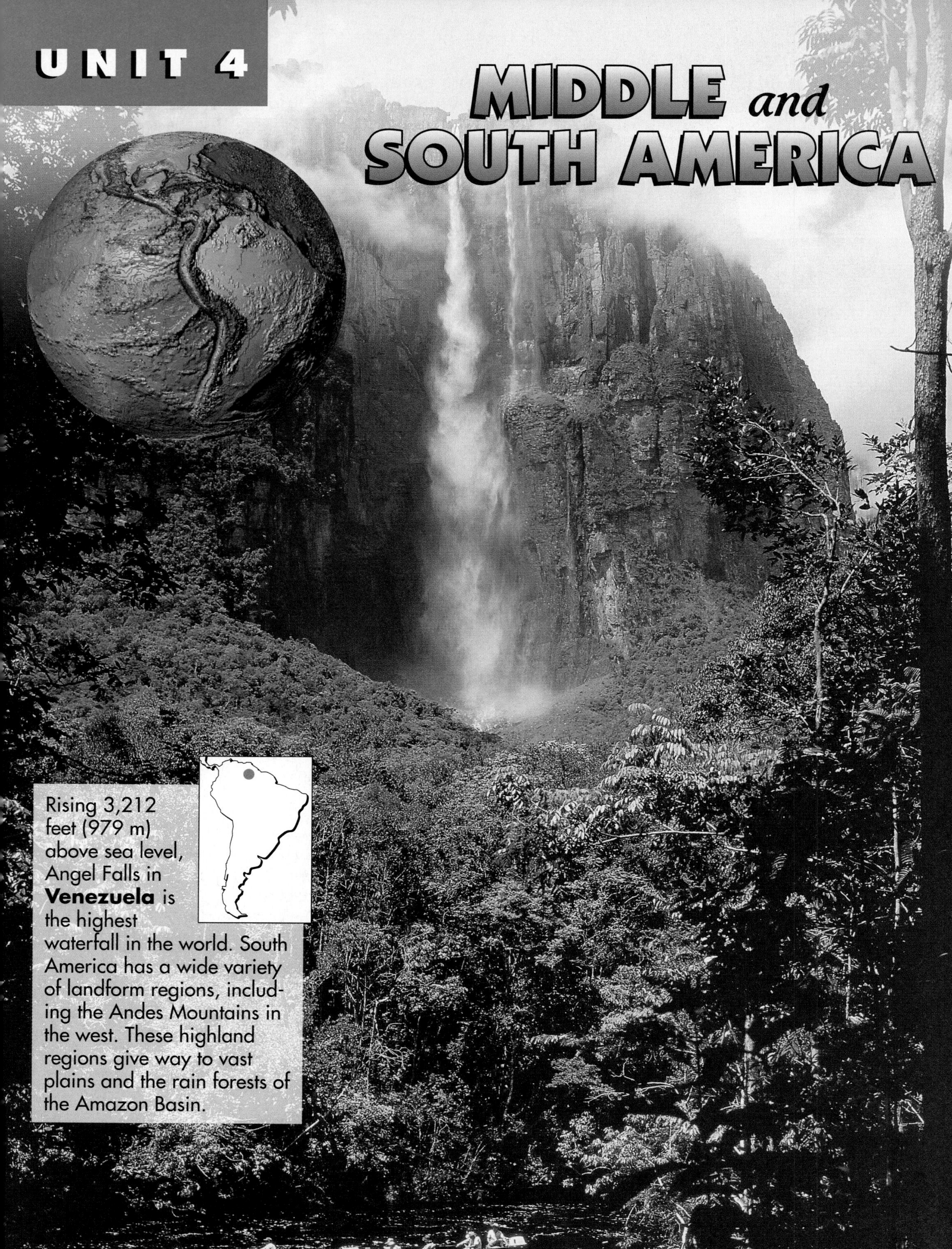
UNIT 4
MIDDLE and SOUTH AMERICA

Rising 3,212 feet (979 m) above sea level, Angel Falls in **Venezuela** is the highest waterfall in the world. South America has a wide variety of landform regions, including the Andes Mountains in the west. These highland regions give way to vast plains and the rain forests of the Amazon Basin.

UNIT 4 CHAPTERS

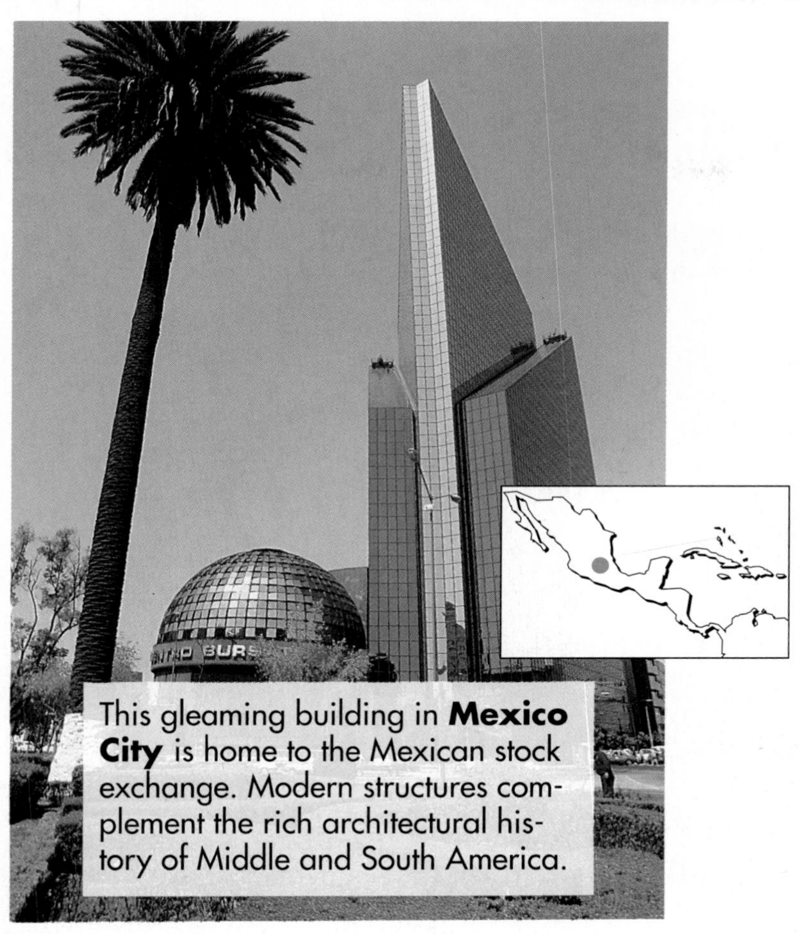

This gleaming building in **Mexico City** is home to the Mexican stock exchange. Modern structures complement the rich architectural history of Middle and South America.

Gold Incan medallion from **Bolivia**

Clouds shroud the ancient Incan city of **Machu Picchu** high in the Andes Mountains of **Peru**. Middle and South America was home to many thriving civilizations before the arrival of the first Europeans in the late fifteenth century. Many people in the region can trace their ancestry to the original inhabitants of the region.

Painted decoration from **Puerto Rico**

LITERATURE

**"1911: Machu Picchu
The Last Sanctuary of the Incas**

isn't dead; it only sleeps. For centuries the Urubamba River, foaming and roaring, has exhaled its potent breath against these sacred stones, covering them with a blanket of dense jungle to guard their sleep. Thus has the last bastion of the Incas, the last foothold of the Indian kings of Peru, been kept secret.

Among snow mountains which appear on no maps, a North American archeologist, Hiram Bingham, stumbles upon Machu Picchu. A child of the region leads him by the hand over precipices to the lofty throne veiled by clouds and greenery. There, Bingham finds the white stones still alive beneath the verdure [green vegetation], and reveals them, awakened, to the world."

Translated by Cedric Belfrage

© MARCELO ISARRUALDE

Eduardo Galeano (1940–) is an Uruguayan writer who is the author of *OPEN VEINS OF LATIN AMERICA, THE BOOK OF EMBRACES, WE SAY NO,* and the three-volume history of North and South America, *MEMORY OF FIRE*. He has worked as a journalist and political cartoonist.

In this excerpt from *MEMORY OF FIRE,* Galeano writes about the discovery of a great Incan city located near the peak of Machu Picchu. Until 1911, only Native Americans living high in the Andes Mountains were aware of its location or knew what remained of the ancient city.

Interpreting Literature

1. **What elements of physical geography kept the Incan city hidden for centuries?**
2. **Examine the photographs of Machu Picchu on this page and on page 185. What purposes do you think the "white stones" mentioned in this passage served in this city?**

FOR THE RECORD

HIRAM BINGHAM COLLECTION, COURTESY OF THE NATIONAL GEOGRAPHIC SOCIETY AND YALE UNIVERSITY, PEABODY MUSEUM OF NATURAL HISTORY.

Hiram Bingham (1875–1956) was an archaeologist and historian. Bingham led an expedition to find the ancient Incan capital city. He chanced upon the site after crossing the rugged Andes Mountains and following the course of the Urubamba River through dense jungle terrain. The excerpt below describes his amazement at the physical beauty of the Urubamba canyon.

"It will be remembered that it was in July 1911, that I began the search for the last Inca capital. . . . I had entered the marvelous canyon of the Urubamba below the Inca fortress of Salapunco near Torontoy.

Here the river escapes from the cold plateau by tearing its way through gigantic mountains of granite. The road runs through a land of matchless charm. . . . In the variety of its charms and the power of its spell, I know of no place in the world which can compare with it. Not only has it great snow peaks looming above the clouds more than two miles overhead; gigantic precipices of many-colored granite rising sheer for thousands of feet above the foaming, glistening, roaring rapids, it has also, in striking contrast, orchids and tree ferns, the delectable [pleasing] beauty of luxurious vegetation, and the mysterious witchery of the jungle. One is drawn irresistibly onward by ever-recurring surprises through a deep, winding gorge, turning and twisting past overhanging cliffs of incredible height."

Analyzing Primary Sources

1. What words does Bingham use to show that the Urubamba was not a calm river along the course that he followed?
2. What aspects of the physical geography were most surprising to Bingham?

CHAPTER 17

Introducing Middle and South America

GEOGRAPHY DICTIONARY

isthmus
indigenous
pampas
llanos
indigo
mestizo
mulatto
cacao
hacienda
labor intensive
shantytown
land reform

Fifteenth-century plaster carving from a monastery in **Oaxaca, Mexico**

Geographers have used cultural and historical geography to define a region called Middle America. Middle America includes Mexico, the seven countries of Central America, and the many islands of the Caribbean Sea. The region also includes the Isthmus of Panama, which links North America to South America. An **isthmus** is a narrow neck of land that acts as a bridge to connect two larger bodies of land.

Middle America is one of the most culturally diverse regions in the world. Though the Caribbean islands were originally inhabited by **indigenous** (native) peoples and later colonized by European countries, millions of African slaves were transported to the Caribbean to work its sugar plantations. As a result, the major cultural influence is African.

South America extends from the warm and sunny beaches of Point Gallinas in Colombia to the cold and stormy seas around Cape Horn, about 4,500 miles (7,241 km) to the south. Its peoples are as diverse as its physical geography. The continent's many ethnic groups are separated by geographical obstacles, including the Andes Mountains and the forests of the Amazon. The region is united, however, by the Roman Catholic religion and the Spanish language (except for Brazil, where Portuguese is spoken). The region's architecture, laws, language, and religion are lasting reminders of Europe's colonial rule of South America.

Ancient Mayan pyramid in **Yucatán, Mexico**

Mapping the Region

MIDDLE AND SOUTH AMERICA

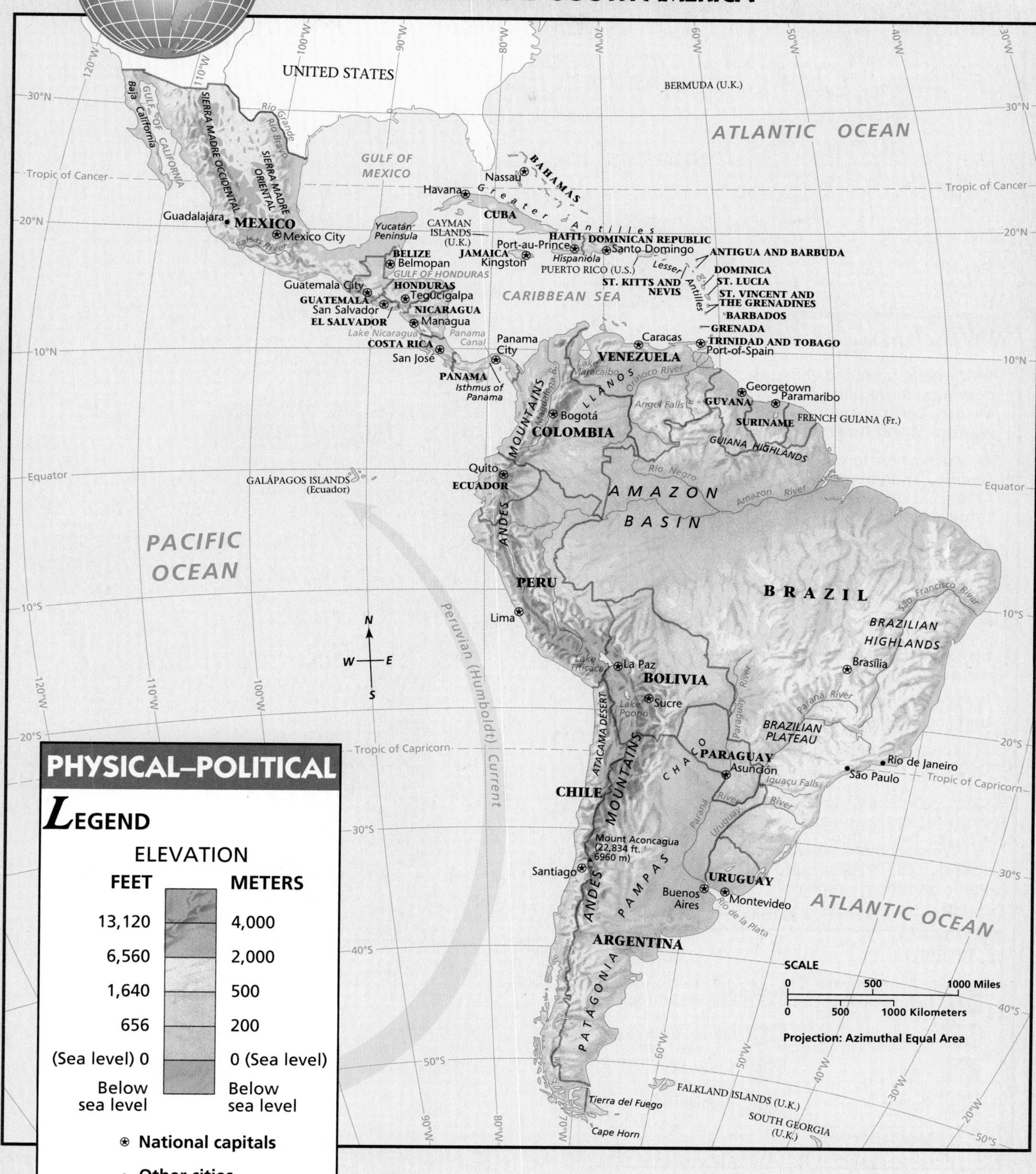
UNITED STATES
BERMUDA (U.K.)
ATLANTIC OCEAN
GULF OF MEXICO
Baja California
GULF OF CALIFORNIA
SIERRA MADRE OCCIDENTAL
SIERRA MADRE ORIENTAL
Río Grande (Río Bravo)
Balsas River
Guadalajara
MEXICO
Mexico City
Tropic of Cancer
Yucatán Peninsula
Nassau
BAHAMAS
Havana
Greater Antilles
CUBA
CAYMAN ISLANDS (U.K.)
HAITI
DOMINICAN REPUBLIC
Port-au-Prince
Santo Domingo
Hispaniola
JAMAICA
Kingston
PUERTO RICO (U.S.)
ANTIGUA AND BARBUDA
DOMINICA
ST. LUCIA
ST. VINCENT AND THE GRENADINES
BARBADOS
GRENADA
TRINIDAD AND TOBAGO
Port-of-Spain
ST. KITTS AND NEVIS
Lesser Antilles
BELIZE
Belmopan
GULF OF HONDURAS
Guatemala City
GUATEMALA
HONDURAS
Tegucigalpa
San Salvador
EL SALVADOR
NICARAGUA
Managua
Lake Nicaragua
CARIBBEAN SEA
COSTA RICA
San José
Panama Canal
Panama City
PANAMA
Isthmus of Panama
Caracas
VENEZUELA
Lake Maracaibo
Orinoco River
LLANOS
Magdalena R.
Georgetown
Paramaribo
GUYANA
SURINAME
FRENCH GUIANA (Fr.)
Angel Falls
GUIANA HIGHLANDS
Bogotá
COLOMBIA
Quito
ECUADOR
Equator
GALÁPAGOS ISLANDS (Ecuador)
Río Negro
AMAZON BASIN
Amazon River
ANDES MOUNTAINS
PACIFIC OCEAN
PERU
Lima
BRAZIL
São Francisco River
BRAZILIAN HIGHLANDS
Brasília
Peruvian (Humboldt) Current
N
W
E
S
Lake Titicaca
La Paz
BOLIVIA
Sucre
Lake Poopó
Paraguay River
Paraná River
BRAZILIAN PLATEAU
ATACAMA DESERT
Tropic of Capricorn
CHACO
PARAGUAY
Asunción
Rio de Janeiro
São Paulo
Iguaçu Falls
CHILE
Uruguay River
Mount Aconcagua (22,834 ft. 6960 m)
Santiago
PAMPAS
URUGUAY
Buenos Aires
Montevideo
Río de la Plata
ATLANTIC OCEAN
ARGENTINA
PATAGONIA
SCALE
0
500
1000 Miles
0
500
1000 Kilometers
Projection: Azimuthal Equal Area
FALKLAND ISLANDS (U.K.)
Tierra del Fuego
Cape Horn
SOUTH GEORGIA (U.K.)
120°W
110°W
100°W
90°W
80°W
70°W
60°W
50°W
40°W
30°W
20°W
30°N
20°N
10°N
10°S
20°S
30°S
40°S
50°S
PHYSICAL–POLITICAL
LEGEND
ELEVATION
FEET
METERS
13,120
4,000
6,560
2,000
1,640
500
656
200
(Sea level) 0
0 (Sea level)
Below sea level
Below sea level
National capitals
Other cities

Mapping the Region

MIDDLE AND SOUTH AMERICA

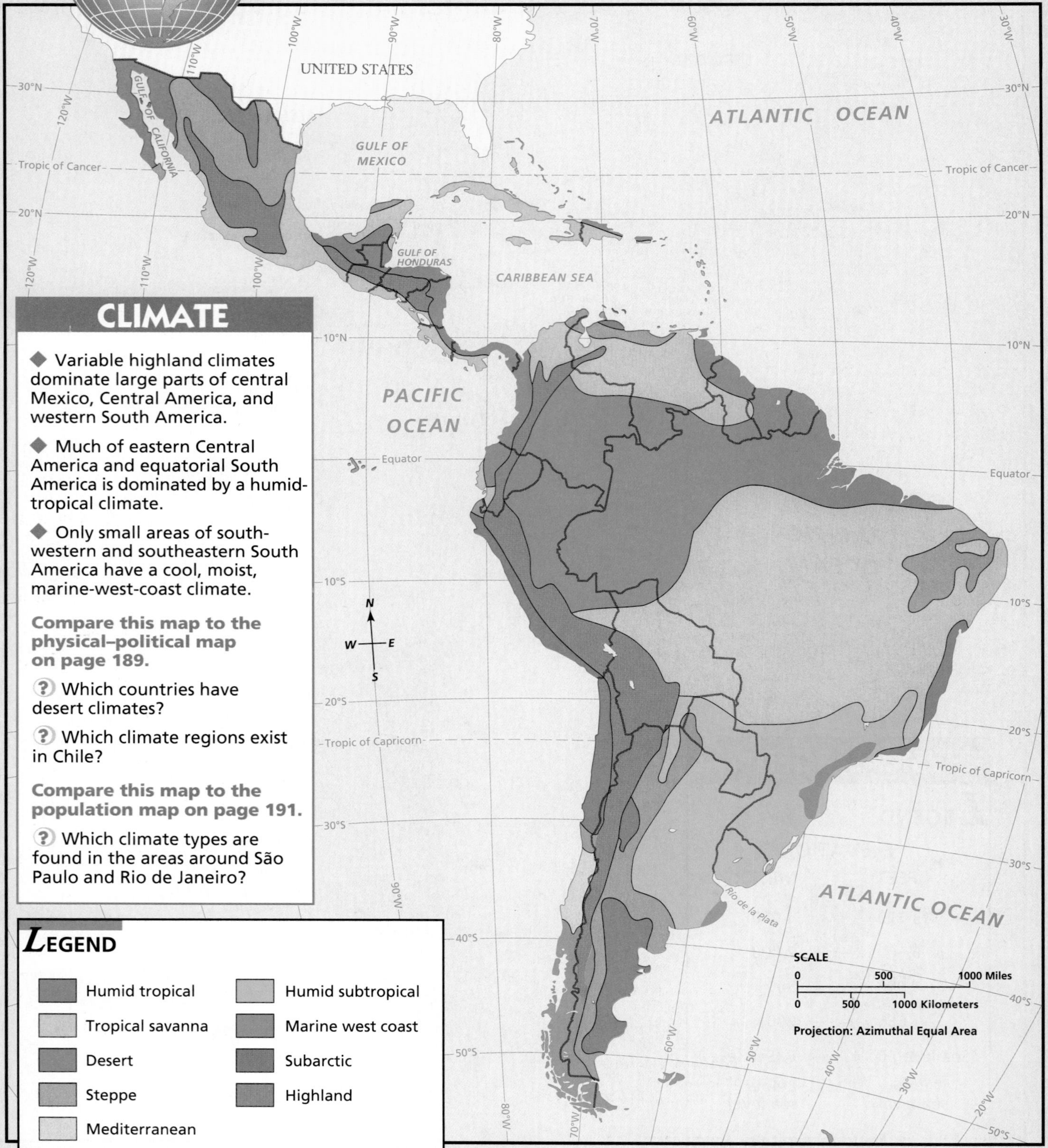

CLIMATE

◆ Variable highland climates dominate large parts of central Mexico, Central America, and western South America.

◆ Much of eastern Central America and equatorial South America is dominated by a humid-tropical climate.

◆ Only small areas of southwestern and southeastern South America have a cool, moist, marine-west-coast climate.

Compare this map to the physical–political map on page 189.

? Which countries have desert climates?

? Which climate regions exist in Chile?

Compare this map to the population map on page 191.

? Which climate types are found in the areas around São Paulo and Rio de Janeiro?

Legend

Humid tropical	Humid subtropical
Tropical savanna	Marine west coast
Desert	Subarctic
Steppe	Highland
Mediterranean	

SCALE: 0 – 500 – 1000 Miles; 0 – 500 – 1000 Kilometers

Projection: Azimuthal Equal Area

MAPPING THE REGION

MIDDLE AND SOUTH AMERICA

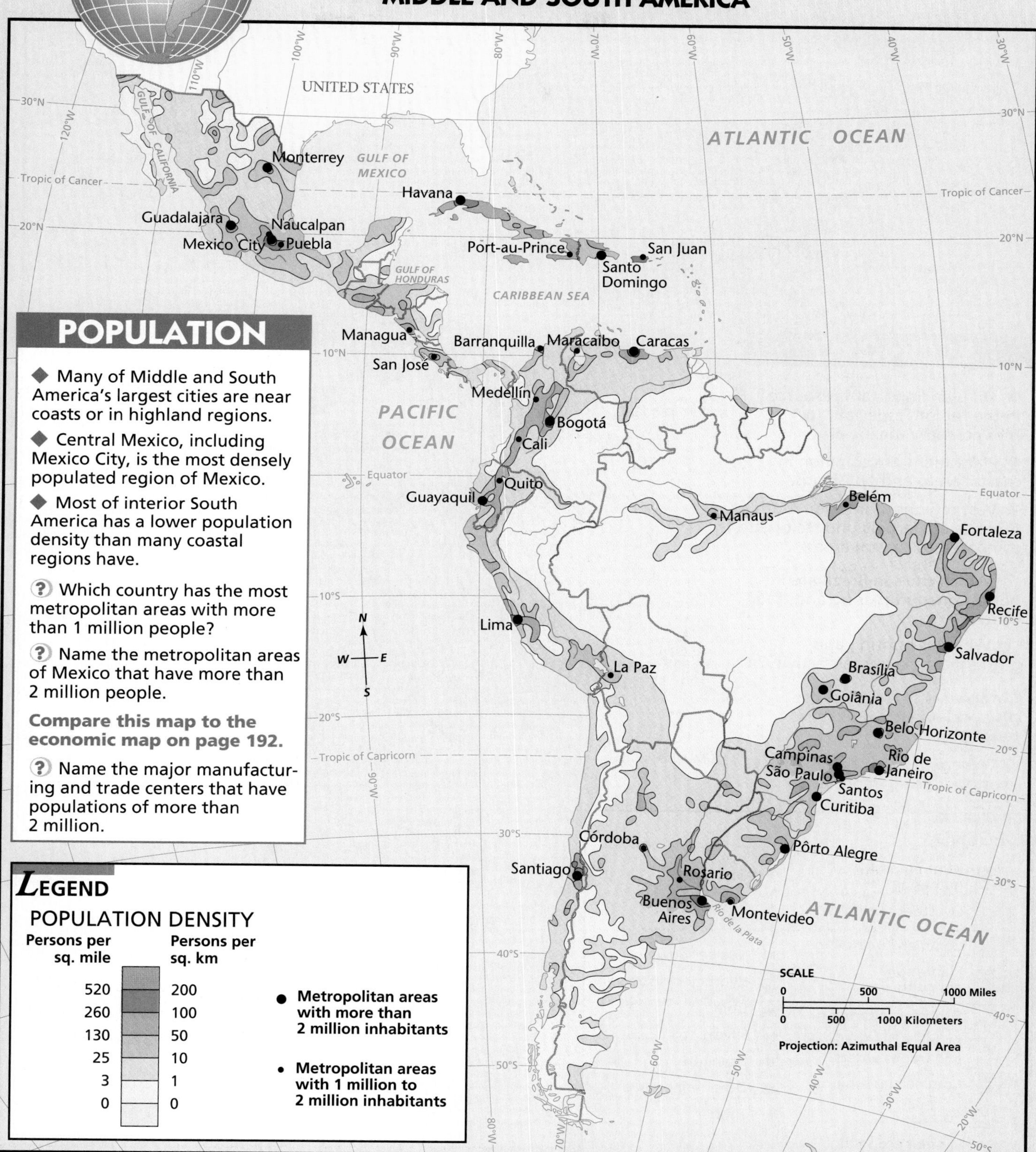

UNITED STATES
ATLANTIC OCEAN
GULF OF MEXICO
GULF OF CALIFORNIA
PACIFIC OCEAN
CARIBBEAN SEA
GULF OF HONDURAS
Tropic of Cancer
Equator
Tropic of Capricorn
30°N
20°N
10°N
10°S
20°S
30°S
40°S
50°S
120°W
110°W
100°W
90°W
80°W
70°W
60°W
50°W
40°W
30°W
20°W
Monterrey
Guadalajara
Naucalpan
Mexico City
Puebla
Havana
Port-au-Prince
Santo Domingo
San Juan
Managua
San José
Barranquilla
Maracaibo
Caracas
Medellín
Bogotá
Cali
Quito
Guayaquil
Manaus
Belém
Fortaleza
Recife
Salvador
Lima
La Paz
Brasília
Goiânia
Belo Horizonte
Campinas
São Paulo
Rio de Janeiro
Santos
Curitiba
Córdoba
Pôrto Alegre
Santiago
Rosario
Buenos Aires
Montevideo
Río de la Plata
N
S
E
W
SCALE
0
500
1000 Miles
0
500
1000 Kilometers
Projection: Azimuthal Equal Area

POPULATION

- Many of Middle and South America's largest cities are near coasts or in highland regions.
- Central Mexico, including Mexico City, is the most densely populated region of Mexico.
- Most of interior South America has a lower population density than many coastal regions have.

? Which country has the most metropolitan areas with more than 1 million people?

? Name the metropolitan areas of Mexico that have more than 2 million people.

Compare this map to the economic map on page 192.

? Name the major manufacturing and trade centers that have populations of more than 2 million.

LEGEND

POPULATION DENSITY

Persons per sq. mile	Persons per sq. km
520	200
260	100
130	50
25	10
3	1
0	0

● Metropolitan areas with more than 2 million inhabitants

• Metropolitan areas with 1 million to 2 million inhabitants

MAPPING THE REGION

MIDDLE AND SOUTH AMERICA

ECONOMY

◆ Oil is an important resource in the region, especially in Mexico and Venezuela.

◆ Plantations are common in coastal areas and on islands.

◆ Vast regions of interior South America are largely undeveloped compared to coastal areas.

? Which cities in Brazil are important manufacturing and trade centers?

? Where is hunting and gathering an economic activity?

Compare this map to the physical–political map on page 189.

? Where in Argentina is livestock raising important?

LEGEND

- Hunting and gathering
- Livestock raising
- Commercial farming
- Plantation agriculture
- Subsistence farming
- Major manufacturing and trade centers
- Commercial fishing
- Forests
- Limited economic activity
- Coal
- Natural gas
- Oil
- Nuclear power
- Hydroelectric power
- Au Gold
- Ag Silver
- U Uranium
- Other minerals
- Timber

Tijuana
Ciudad Juárez
UNITED STATES
Monterrey
Guadalajara
Mexico City
Puebla
GULF OF CALIFORNIA
GULF OF MEXICO
Freeport
Havana
Port-au-Prince
San Juan
GULF OF HONDURAS
CARIBBEAN SEA
Willemstad
Caracas
Colón
Panama City
Bogotá
PACIFIC OCEAN
ATLANTIC OCEAN
Quito
Guayaquil
Equator
AMAZON BASIN
Manaus
Belém
Callao
Lima
La Paz
Arica
Belo Horizonte
São Paulo
Rio de Janeiro
Tropic of Cancer
Tropic of Capricorn
Valparaíso
Santiago
Buenos Aires
Montevideo
Río de la Plata
Cape Horn
SCALE
0 500 1000 Miles
0 500 1000 Kilometers
Projection: Azimuthal Equal Area

1 PHYSICAL GEOGRAPHY

FOCUS

- *How were the various landforms of Middle and South America created?*
- *What climate types and what rivers are found in Middle and South America?*

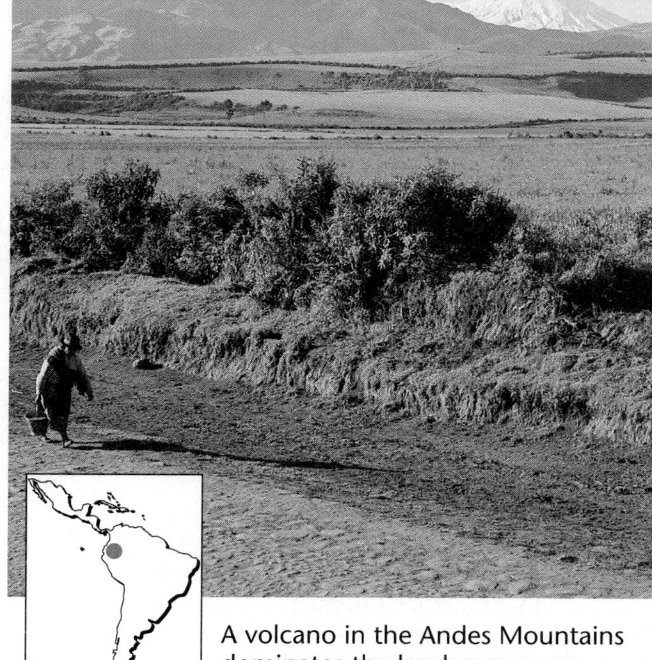

A volcano in the Andes Mountains dominates the landscape near **Quito, Ecuador**. Despite the area's tropical location, many people in the region, such as the woman in the foreground, dress for the cooler highland climates of the Andes Mountains.

REGION **Landforms** According to the theory of plate tectonics, South America and Africa broke apart from Gondwanaland more than 100 million years ago. As South America moved westward, other continental pieces from the Pacific region collided with the west coast. These collisions may have formed the first Andes Mountains. A few million years ago, South America began to ride up over the Nazca plate under the southeastern Pacific Ocean. Since this first collision, the Andes have continued to build as the result of folding, faulting, and volcanic activity. The other mountains of South America are the Guiana (gee-AN-uh) Highlands and the highlands of eastern Brazil. These eroded highlands were formed much earlier than the Andes.

Plains cover most of South America. The largest plains are in the Amazon River basin. The other large plains are the **pampas** of Argentina and the **llanos** (YAHN-ohz) of Colombia and Venezuela. These plains are the result of erosion steadily smoothing the land's surface over millions of years.

The theory of plate tectonics also explains the creation of islands around the Caribbean Sea. As the South American plate drifted westward from Africa, the smaller Caribbean plate cruised eastward. This Caribbean plate carried only a few continental pieces. These pieces became the mountainous islands of the Greater Antilles—Cuba, Jamaica, Hispaniola, and Puerto Rico. At the eastern edge of the Caribbean plate was a chain of volcanic islands—the Lesser Antilles.

Central America and Mexico are believed to have been formed from a jumble of small plates attached to the rim of the North American and Caribbean plates. Mexico consists mostly of a high central plateau with mountains along each side. Central America has many short mountain ranges and huge volcanoes. These mountains and an ocean trench off the Pacific coast of southern Mexico and Central America indicate a continuing plate collision. Scientists believe that Central America and South America became connected only a few million years ago. Earthquakes and volcanoes are a constant threat throughout Middle America and the western side of South America.

Climate Because the region of Middle and South America extends across nearly 90 degrees of latitude, it includes a full range of middle-latitude as well as tropical climate regions. South America contains the largest humid-tropical region in the world—the huge forested area centered on the Amazon River. Rain falls here almost every day. The region's other areas of humid-tropical climate are along the eastern coasts where the moist trade winds move on shore.

Many areas of Middle and South America experience wet summers and dry winters typical of the tropical-savanna climate. The many areas of dry forests on the Caribbean islands and the Pacific coast of Central America have a similar climate.

Southern South America has a variety of middle-latitude climates. Far southern Chile is influenced by the westerlies. The climate found here is cool and wet, similar to that of the United States' Pacific Northwest. Snow and rainfall are particularly heavy in the southern Andes. East of the Andes in southern Argentina, however, the mountains block moisture from reaching the area. As a result, semiarid steppe and desert climates

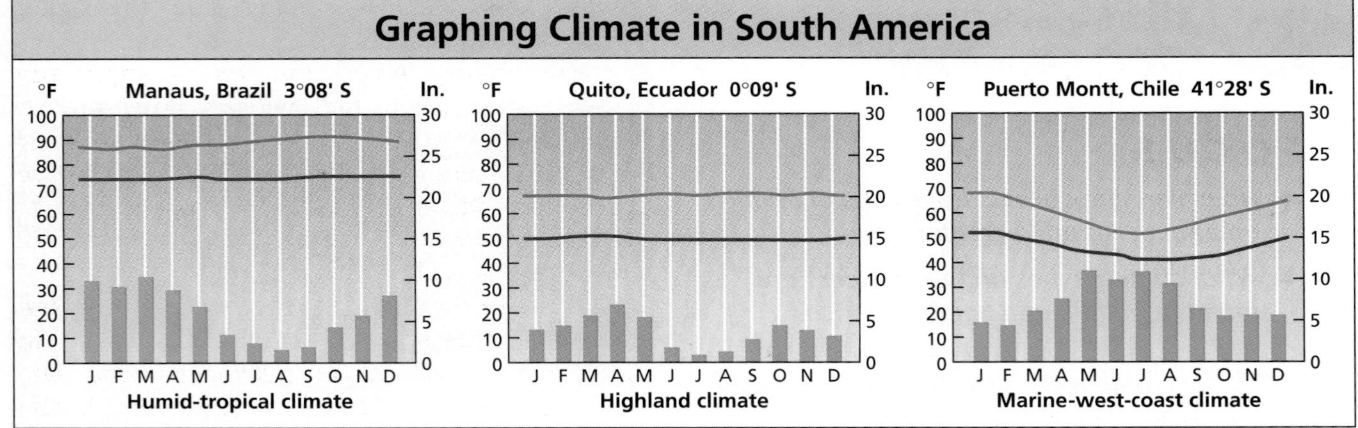
Graphing Climate in South America
°F
Manaus, Brazil 3°08' S
In.
100
90
80
70
60
50
40
30
20
10
0
30
25
20
15
10
5
0
J F M A M J J A S O N D
Humid-tropical climate
Quito, Ecuador 0°09' S
Highland climate
Puerto Montt, Chile 41°28' S
Marine-west-coast climate

SKILL STUDY The first two climographs above show the effects of the Andes Mountains on climates in South America. To learn how to read climographs, see page S11 of the Skills Handbook. Although both are located near the equator, the humid-tropical climate of Manaus, Brazil, is much warmer and moister than the highland climate of Quito, Ecuador. The Ecuadorean capital is located in the higher elevations of the Andes Mountains. Other than having cooler temperatures, what is the weather of Puerto Montt, Chile, like during the winter?

are found. Farther north, the central valley of Chile has a Mediterranean climate, with winter rains and summer droughts. A large region with humid-subtropical climates is found in northeastern Argentina and southern Brazil.

The driest region of Middle and South America includes the Atacama Desert of northern Chile and the nearby coastal deserts of Peru and southern Ecuador. Subtropical high pressure brings dry weather to these areas all year. Because the high Andes block moisture from the east, the western slopes of the mountains are very dry.

The mild climates on mountain slopes between the elevations of 3,000 and 6,000 feet (914 and 1,829 m) are particularly important in tropical Middle and South America. The mountain valleys have fertile soils and water from mountain streams. These pleasant environments are some of the densest areas of settlement in the region.

The pattern of the highland climates of the region is important in understanding population distribution in the region. (See the maps on pages 190 and 191.) This climate pattern can be seen from the mountains of Mexico and Central America through the Andes of South America. You know that temperatures become cooler at higher elevations. For this reason, the region's highland climates have been divided into elevation zones along mountain slopes. Geographers are able to study how population distributions and ways of life relate to these climate patterns. You will learn more about these elevation zones in Chapter 20.

River Systems Three major river systems drain the eastern side of South America: the Amazon, the Paraná, and the Orinoco. Beginning on the eastern side of the Andes, the Amazon River stretches more than 4,000 miles (6,436 km) before it reaches the Atlantic Ocean. Curiously, the Orinoco and Amazon river basins are linked. (See the map on page 189.) The Río Negro, which flows mainly south into the Amazon, has an upper tributary that flows north into the Orinoco during part of the year. Farther south, the Paraná River system empties into the Río de la Plata estuary, located between Argentina and Uruguay. The headwaters of the Paraná River system are located in the highlands of eastern Brazil and in western Argentina along the eastern slopes of the Andes.

1 SECTION REVIEW

1. What forces created the different mountain ranges of Middle and South America?
2. Why does the region of Middle and South America have several climate types, and what are they? What are the region's major river systems?
3. **Critical Thinking** **Imagine** a country in Middle America experiences a severe earthquake. How might the physical geography of the region affect that country's ability to respond?

HISTORICAL GEOGRAPHY

FOCUS

- *What Native American civilizations existed in Middle and South America before the arrival of Europeans in the region?*
- *How did European colonization affect the region's land and peoples? What effects did independence have?*

Settlement Patterns By the time Christopher Columbus arrived in the Americas, Native Americans had lived in the region for thousands of years. Columbus thought he had reached the islands off the coast of India. He called the indigenous peoples he found living there "Indios," or Indians. Columbus actually had landed on an island in the Bahamas, southeast of Florida. He claimed the island for Spain and named it San Salvador (Holy Savior). Emigration from Spain was encouraged, and by the mid-1500s, there were thousands of Spanish colonists on the larger Caribbean islands.

Most Native Americans were farmers and hunters who lived in small villages. Some groups traded salt, shells, and other goods with their neighbors. There were also, however, great Native American civilizations that built impressive monuments and large cities housing tens of thousands of people.

The ancient Olmecs thrived in eastern and southeastern Mexico from about 1200 B.C. to about 400 B.C. They developed extensive trade and were the first people to build religious pyramids, greatly influencing later Middle American cultures.

The Mayas, considered one of the most advanced Native American cultures, flourished between about A.D. 250 and 900. Ruins of once brilliant Mayan cities now lie in the forests of Guatemala, Honduras, and the Yucatán Peninsula of Mexico and Belize. Sophisticated Native American societies also flourished elsewhere on the mainland of Middle America and in northern South America.

At the time the Spanish arrived on the continents, there existed two monument-building civilizations. The empires of the Aztecs in central Mexico and of the Incas in western South America developed about A.D. 1300 and continued the achievements of earlier civilizations. The Aztec Empire was centered at Tenochtitlán (tay-nawch-tee-TLAHN) on the present site of Mexico City. Tenochtitlán and the Aztec Empire were conquered in 1521 by the Spanish explorer Hernán Cortés (kawr-TEHZ). The Incan Empire extended along the Andes Mountains from Ecuador to Chile.

Many factors contributed to the decline of Native American civilizations. The Europeans, with guns and armor, had a technological advantage in warfare. Furthermore, diseases that the Europeans brought with them, such as smallpox and measles, were devastating to the Native

Early Civilizations of Middle and South America

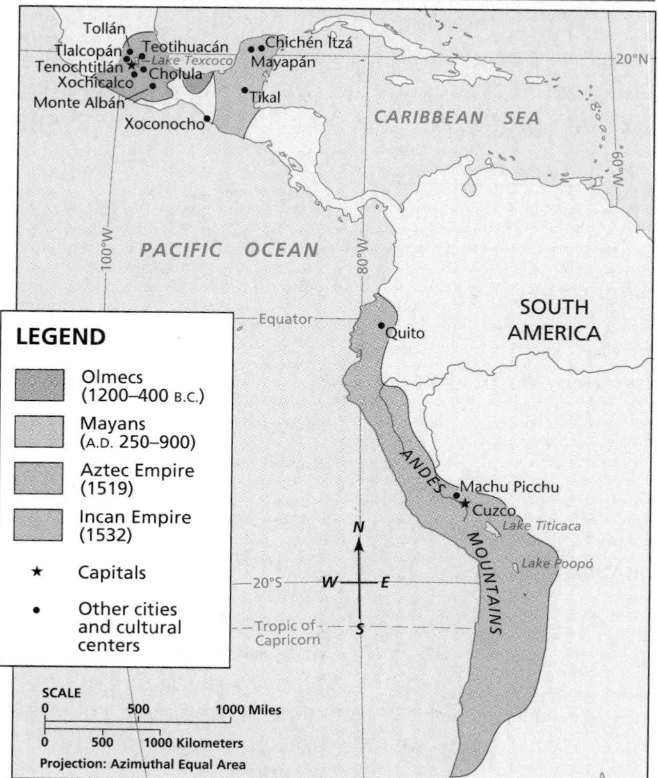
Tollán
Tlalcopán
Tenochtitlán
Xochicalco
Teotihuacán
Lake Texcoco
Cholula
Monte Albán
Xoconocho
Chichén Itzá
Mayapán
Tikal
20°N
CARIBBEAN SEA
60°W
PACIFIC OCEAN
100°W
80°W
Equator
Quito
SOUTH AMERICA
ANDES MOUNTAINS
Machu Picchu
Cuzco
Lake Titicaca
Lake Poopó
N
W
E
S
20°S
Tropic of Capricorn
LEGEND
Olmecs (1200–400 B.C.)
Mayans (A.D. 250–900)
Aztec Empire (1519)
Incan Empire (1532)
Capitals
Other cities and cultural centers
SCALE
0 500 1000 Miles
0 500 1000 Kilometers
Projection: Azimuthal Equal Area

MAP STUDY Historical maps are important tools for understanding earlier times. Some historical maps identify earlier civilizations, major cities, and important events. This map shows the location and dates of civilizations in Middle and South America before European colonization. Which civilization dominated western South America at the time the Spanish reached the region in the first decades of the sixteenth century?

Americans. When Cortés arrived in the Valley of Mexico, perhaps three-quarters of the Aztec people were ill. Diseases from Europe eventually killed millions of Native Americans.

European Colonization By the late fifteenth century, Spain and Portugal were rich and powerful. Both countries were eager for the profits possible by having overseas empires. A treaty in 1494 divided the Western Hemisphere into Spanish and Portuguese territories. Portugal received a large triangle of eastern South America, including present-day Brazil.

Spanish and Portuguese colonists quickly settled the region. They extended the Native Americans' gold and silver mines and opened new ones. Ranching and farming were the key economic activities, however. The colonists brought with them European crops, livestock, and farming methods. They also adapted Native American crops, such as maize (corn), beans, and potatoes, to their needs. Spanish colonists were awarded large land grants by the Spanish king. Native Americans living on this land were forced to work the farms and ranches. The idea that land could be owned was new to the indigenous peoples.

By the mid-1600s, the French, British, and Dutch began to acquire lands that the Spanish had not colonized. These included the smaller islands of the Caribbean, as well as Jamaica and Haiti. Their only mainland settlements were British Honduras (now Belize), British Guiana (now Guyana), French Guiana, and Suriname (soohr-uh-NAMH-uh). These colonies were all based on plantation agriculture. Major plantation crops were sugarcane and **indigo**. Indigo is a plant used to make a blue dye. Later, bananas became a key plantation crop. For more than two centuries, African slaves provided most of the labor for the plantations. When slavery ended, the British imported field workers from India, while the Dutch brought in Indonesians.

Because there were few women among the colonists, European settlers often married Native American or African women. Many Middle and South Americans are **mestizos**, or people with both Native American and European ancestors. The descendants of European plantation colonists and Africans are known as **mulattos**. Today, the geography of the peoples, languages, and cultures in the region is mostly a result of the process of colonization.

Independence By 1800, ideas of independence introduced by the American and French revolutions were well known throughout Middle and South America. Between 1808 and 1825, many colonies gained independence. Yet, the economies and societies of the various countries usually changed little. Rich landowners, merchants, and military generals merely replaced the colonial government. The lives of rural peoples remained much the same.

With independence, Middle and South America began trading more with Great Britain and the United States rather than with Spain. Great Britain protected the new Middle and South

Languages of Middle and South America

Language		Where spoken
Spanish ►		► Mexico, Argentina, most other countries of Middle and South America
Portuguese ►		► Brazil
French ►		► Haiti, French Guiana, Guadeloupe, Martinique
Dutch ►		► Suriname, Aruba, Netherlands Antilles
English ►		► Belize, Guyana, some Caribbean countries and territories
Native American ►	*Nahuatl*	► Mexico, El Salvador
	Quiché	► Guatemala
	Quechua	► Peru, Bolivia, Ecuador, Argentina, Colombia
	Aymara	► Bolivia, Peru
	Guaraní	► Paraguay, Argentina, Brazil

NOTE: Nahuatl, Quiché, Quechua, Aymara, and Guaraní are only five of the many Native American languages and dialects spoken in Middle and South America, especially in southern Mexico, Central America, and Andean South America.

Spanish and Portuguese are the most widely spoken languages in Middle and South America. Many Native American languages, however, still are spoken throughout the region.

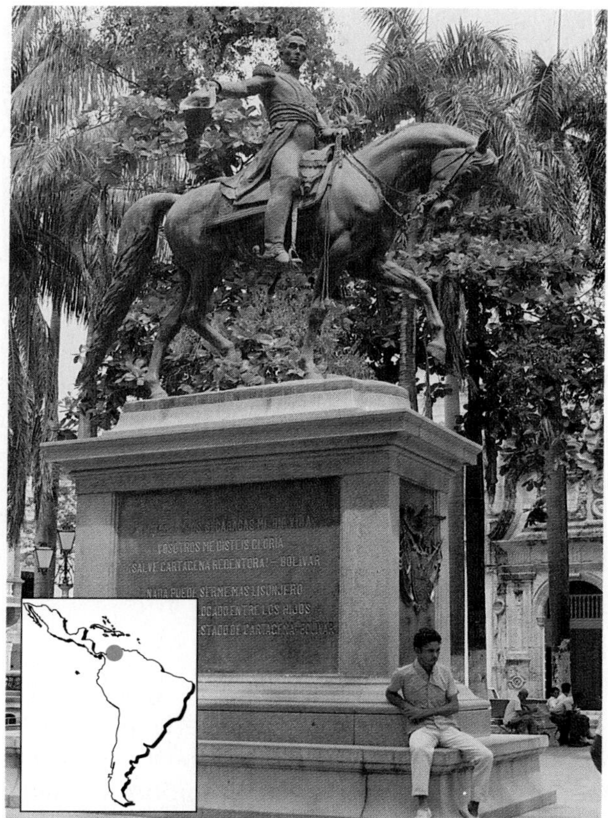

This statue in **Cartagena, Colombia**, honors Simón Bolivar for his role in leading South America's fight for independence. Statues of Bolivar, who is called *El Libertador* (The Liberator) by some South Americans, can be found in cities throughout the continent.

American states with its navy. The United States offered diplomatic protection under the Monroe Doctrine, which declared the countries of Middle and South America to be off-limits to new European colonization. The influence of the United States has remained strong throughout the region.

SECTION REVIEW

1. What two Native American empires were encountered by Spanish explorers in Middle and South America? Where was each empire located?
2. How did the region's agriculture change as a result of European colonization?
3. **Critical Thinking** **Agree or disagree** with the following statement: "The colonization of Middle and South America by Spain and Portugal had both positive and negative effects." Support your answer.

3 ECONOMIC GEOGRAPHY

FOCUS

- *What natural resources and industries are important to the economies of the region's countries?*
- *What forms of agriculture are common in Middle and South America today?*

Natural Resources The countries of Middle and South America have rich mineral deposits, fertile soils, and climates suitable for growing many different crops. Dams on many rivers provide irrigation water and generate electricity. Several countries also have forest and fish resources.

The mineral wealth that attracted the Spanish to the region centuries ago is still being developed. Gold and silver deposits have been found in Brazil, Mexico, Colombia, and Peru. Metals such as copper are being mined in parts of the Andes, in central Brazil, and in Mexico's mountains. Brazil has enormous iron-ore reserves. Bauxite, the ore of aluminum, is widely available in northeastern South America and in Jamaica.

Petroleum is found in many countries of Middle and South America as well. The greatest oil reserves are in Mexico and Venezuela. Oil deposits also have been developed in the upper Amazon Basin. More recent oil discoveries have been made off the coasts of Brazil, Argentina, and Chile.

Agriculture Several types of agriculture are practiced in Middle and South America today. Subsistence agriculture remains common in many remote areas, such as in inland forests and in the mountains. Subsistence farmers often live in small villages. Each day, they walk to their nearby fields.

In contrast, modern commercial agriculture, or agribusiness, uses the latest farming methods and mechanized equipment. Farm products are processed and exported to international markets. Areas with agribusiness are generally prosperous agricultural regions. Commercial farming is well developed in southern Brazil and southward to the pampas of Argentina, in the central valley of Chile, and in northern Mexico. Fruits and vegetables from each of these regions can be bought in stores throughout the United States.

Manufacturing jobs, such as those provided by this steel plant in **Monterrey, Mexico**, attract many people to urban areas throughout Middle and South America. The pace of economic growth must increase to catch up with the region's soaring population growth.

Today, plantation agriculture is a form of modern commercial agriculture. It is practiced on many Caribbean islands and along the region's humid coasts. Most plantation products, including sugar, bananas, and **cacao**, are exported to developed countries. Cacao is a small tree on which the cocoa bean grows. The cocoa bean is used in making chocolate. Modern-day plantations usually are part of large, foreign-owned corporations.

Another form of agriculture common in Middle and South America is the **hacienda** system. Introduced by the Spanish, haciendas are large family-owned estates. Peasants live on these estates and have small plots on which to grow their own crops. In return for these plots and services granted by the landowner, the peasants also must work the owner's land as a form of payment. More symbols of the landowners' social status than profit-making operations, haciendas are often unproductive. Even so, they remain a basic form of land organization in much of the region.

Industry As manufacturing increases in Middle and South America, the number of jobs in urban areas also increases. The largest industrial region is located in an arc of cities that extends from Rio de Janeiro in southeastern Brazil to Buenos Aires (bway-nuhs AR-eez) in eastern Argentina. Most capital cities of Middle and South America have developed into industrial regions.

Most Middle and South American factories produce food items, consumer goods, or building materials for local markets. Automobiles, trucks, and railway cars are produced in the larger countries. Industries that manufacture export products are often **labor intensive**. That is, they require a large work force. Often assembled by hand, export products include clothing, furniture, and small appliances. While these labor-intensive industries provide jobs and the training that may lead to better jobs, factory workers often face a life of hard work with little reward.

Tourism is a growing industry in the region, particularly in the islands of the Caribbean. The income generated from tourism is important to many countries' economies. The benefits of the tourist industry to local island communities are sometimes less than expected, however, because hotel and resort workers generally receive low pay.

SECTION REVIEW

1. List four natural resources and four industries found in Middle and South America.
2. Compare and contrast plantations and haciendas.
3. **Critical Thinking** **Determine** which forms of agriculture are labor intensive, and explain your answer.

Economic progress, especially in the developing nations of the world, often has proceeded at the expense of the environment. Large parts of the Amazon rain forests, such as here in **Brazil**, have been burned and cleared. Developers want to use the cleared land for farming and mining.

REGIONAL ISSUES

FOCUS

- *How are economics and politics linked in the region?*
- *How do countries outside the region affect countries in the region? How do countries in the region affect countries outside the region?*

REGION **Population Issues** Middle and South America are experiencing rapid population growth. If current growth rates continue, many countries will double their populations in 20 to 30 years. Building schools and roads as well as creating jobs for this increasing number of people will be difficult.

Rapid population growth often leads to increased migration from poor rural areas to cities. Rural migrants come to the cities seeking jobs and a better life. Most find housing in the **shantytowns** that form the large slums surrounding many Middle and South American cities. A shantytown is a poor settlement of small makeshift shelters. City services such as water and waste disposal are not available, and most roads are unpaved.

Even in the richest countries of Middle and South America, there are many poor people. Middle and South America contain sharp economic and social contrasts. Lack of economic development has contributed to political instability in the region.

Political Issues While much of Middle and South America today is democratic, most countries have histories of repeated changes in government. Typically, different groups within the wealthy families and the military have battled for control of a country's government. Changes in government usually have been minor and have done little for most of the country's people. Wealth and land often remain in the hands of the fortunate few.

Political instability may be lessened in Middle and South America by **land reform**. Land reform means breaking up the large landholdings and allowing small farmers to own their own land. For generations, farmers have worked for shares of the crops they produced or for low wages on lands that were part of plantations and haciendas. These large landholdings usually have been owned by powerful people. Though land reform has been slow in many countries, it became widespread in Mexico after the 1910 Mexican Revolution.

Economic Issues Sometimes, citizens become angry at the policies of their national government. Often, money that was supposed to aid development of a poor region goes to meet less important needs in a wealthier capital city. Also, many people in Middle and South America believe that their countries are not truly independent. These people think that there is too much foreign investment and control.

Foreign governments, trying to encourage development, lent a great deal of money to various countries in the region during the economic boom years of the 1970s. These loans, however, have caused almost as many problems as they have solved. The borrowing countries now have difficulty paying back the loans. As a result, many Middle and South American countries have large debts, which make further economic development even more difficult.

Although multinational corporations from the United States, Europe, and Japan provide investments and technology, these companies have been accused of interfering in local politics and of exploiting local resources. Today, most countries in the region have a cautious attitude toward foreign corporations. Tax laws encourage multinational corporations to share ownership and profits with the local communities.

Many of the region's economic troubles echo throughout the world. For example, cutting the rain forests for lumber, or clearing them for farmland, provides an important source of income for many people in Central America and Brazil. This widespread deforestation, however, may contribute to global warming and the extinction of some species. In addition, deforestation may be leaving homeless some indigenous peoples who live in the rain forests.

Another regional issue with global consequences is the growing and trafficking of illegal drugs. Illegal drugs carry a high price. Many Middle and South American farmers find that they can earn more money by growing crops used for drugs than they can from growing other kinds of crops. Likewise, people who are willing to trade these drugs among countries often can make large profits. The large illegal-drug markets in the United States and Europe often are the final destinations for these dangerous products. Political and military conflicts have broken out over the illegal-drug issue, such as the conflict between the United States and Panama in 1989.

REGION Regional Cooperation In an effort to promote increased regional cooperation, almost all of the countries of Middle and South America, with the United States, formed the Organization of American States (OAS) in 1948. The goals of the OAS include keeping peace and increasing security in the Western Hemisphere, settling disputes among member nations, and promoting economic, social, and cultural cooperation.

Several of the countries of Middle and South America also have made efforts at regional economic cooperation on a smaller scale by forming organizations such as the Latin-American Integration Association, the Central American Common Market, and the Andean Group. Disagreements among the member countries and their differing goals and levels of development, however, have hindered these organizations' efforts and progress. For example, Ecuador and Peru fought a brief border skirmish in 1995.

4 SECTION REVIEW

1. What factors have caused political instability in some of the countries of Middle and South America? How might greater stability be achieved?
2. Using examples, discuss the international forces at work within the region.
3. **Critical Thinking** **Explain** why many countries in Middle and South America have large debts. Do you think that there is any solution to these "debt crises"? Why or why not?

CHAPTER 17 CHECK

Harvesting sugarcane in **Jamaica**

Reviewing the Main Ideas

1. The theory of plate tectonics helps explain the origins of landforms found in Middle and South America. The region has a wide range of tropical and middle-latitude climates, and three major river systems.
2. Ancient Native American civilizations flourished in Middle and South America. When Europeans colonized the region, many Native Americans died or were forced to work for the Europeans. Cultures in the region today are a mix of Native American, European, and African heritages.
3. The region of Middle and South America has many natural resources. Subsistence and commercial agriculture both are common in the region. Industry is increasing.
4. Many of the countries of Middle and South America have rapidly growing populations and much poverty, which have contributed to political instability. Other regional issues involve relationships with countries outside the region.

Building a Vocabulary

1. What term describes a strip of land that acts as a bridge between two larger bodies of land?
2. Define *indigenous*, and list some of the indigenous groups of Middle and South America.
3. What are pampas and llanos? Where are they found, and how were they created?
4. For what is indigo used?
5. What is the purpose of land reform?

Recalling and Reviewing

1. How have the landforms, climates, and rivers of Middle and South America influenced the region's economic activities?
2. Other than Native Americans and the Spanish, what groups of people have resided in Middle and South America?
3. How did European ideas of land ownership create problems in Middle and South America? How are these ideas still visible in the region today?
4. What interests has the United States had in the countries of Middle and South America, historically and currently?

Thinking Critically

1. The term *revolution* means a complete and total change in the way of life of a majority of a country's people. Given this definition, discuss the following statement: "There have been very few true revolutions in Middle and South America."
2. Which of the following factors do you think is most responsible for the challenges facing Middle and South America today: past colonialism, present foreign involvement, or rapid population growth? Support your answer.

Learning About Your Local Geography

Individual Project

Many of the countries of Middle and South America have experienced much political instability. Investigate politics in your local area. Are elections held regularly? For what purposes are they held? Use library resources to make a chart showing this information. You may want to include your chart in your individual portfolio.

Using the Five Themes of Geography

Create and write the clues for a crossword puzzle about the themes of **region** and **place** in Middle and South America. The words in the puzzle should include subregions and human and place characteristics of the region, as discussed in the chapter.

CHAPTER 18

Mexico

GEOGRAPHY DICTIONARY

- *cash crop*
- ejido
- maquiladora
- *inflation*
- *monopoly*
- *privatization*

"The geography of Mexico is abrupt. . . . To ascend to Mexico City from the burning sandy beaches of Veracruz is to travel through every landscape, from the suffocating vegetation of the tropics to the temperate lands of the meseta [plateau]. In the lowlands the air is warm and humid; on the high tableland it is dry and, by night, quite cool."

Octavio Paz

Mexico is officially called "The United Mexican States." Its natural environments vary from deserts in the north to tropical forests in the south. Economic contrasts between wealth and poverty and between modern cities and ancient villages exist throughout the country.

Mineral resources mined beneath Mexico's large land area support a growing industrial society. With more than 90 million people, Mexico has more than twice the population of Spain and is the largest Spanish-speaking country in the world. Political and social changes within Mexico are creating a more open society. These and other developments in Mexico have an impact on the political, economic, and human geography of all of North America.

The relationship between Mexico and the United States continues to strengthen. Mexico's oil resources are important to the U.S. economy, and Mexico is the United States' third largest trading partner. In turn, the United States is Mexico's most important foreign customer. These economic connections affect thousands of workers in both countries.

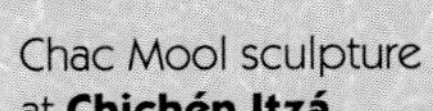

Chac Mool sculpture at **Chichén Itzá**

Ruins of a Mayan observatory at Chichén Itzá in **Yucatán, Mexico**

MEXICO: PHYSICAL–POLITICAL

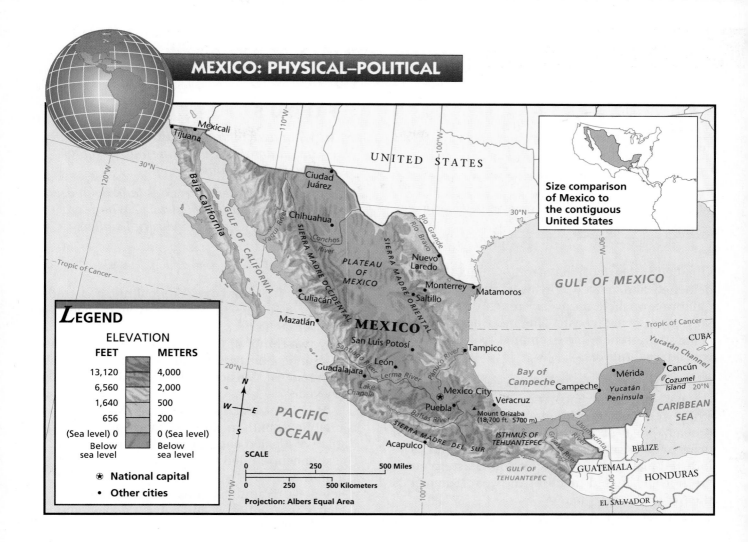

Size comparison of Mexico to the contiguous United States

LEGEND

ELEVATION

FEET	METERS
13,120	4,000
6,560	2,000
1,640	500
656	200
(Sea level) 0	0 (Sea level)
Below sea level	Below sea level

⊛ National capital
• Other cities

SCALE: 0 – 250 – 500 Miles; 0 – 250 – 500 Kilometers

Projection: Albers Equal Area

1 PHYSICAL GEOGRAPHY

FOCUS

- *What are Mexico's major landforms?*
- *What climates and vegetation does Mexico have?*

REGION Landforms The landforms of Mexico are more diverse than those of some entire continents. (See the map above.) Most of the country is made up of a rugged central plateau, called the Plateau of Mexico. Its wide plains average more than 6,000 feet (1,829 m) above sea level. Isolated mountain ridges rise up to 3,000 feet (914 m) higher. Two mountain ranges—the Sierra Madre Oriental to the east and the Sierra Madre Occidental to the west—border the Plateau of Mexico.

At the south end of the Plateau of Mexico is the Valley of Mexico, where Mexico City, the capital, is located. The mountains south of Mexico City include towering snowcapped volcanoes. The highest, Mount Orizaba, rises to 18,700 feet (5,700 m). Smaller mountain ranges, volcanoes, and fertile valleys continue southward into Central America.

The outline of Mexico narrows like a funnel in the south. This is the Isthmus of Tehuantepec (tuh-WAHNT-uh-pehk), where the Pacific Ocean and Gulf of Mexico are separated by only about 137 miles (220 km). Before the Panama Canal was built, this served as one of the major routes between the Atlantic and Pacific oceans. The Yucatán Peninsula of southeastern Mexico is the country's flattest region.

The theory of plate tectonics is important in understanding Mexico's landforms. Geographers believe that millions of years ago the peninsula of Baja California in the northwest was

FACTS IN BRIEF Mexico

COUNTRY POPULATION (1994)	LIFE EXPECTANCY (1994)	LITERACY RATE	PER CAPITA GDP (1993)
MEXICO 92,202,199	69, male 77, female	90% (1993)	$8,200
UNITED STATES 260,713,585	73, male 79, female	97% (1991)	$24,700

Sources: *The World Factbook 1994,* Central Intelligence Agency; *The World Almanac and Book of Facts 1995*

What is the life expectancy of men and women in Mexico?

attached to Mexico's mainland. As Baja California detached and moved northwest and the Gulf of California opened, an ocean trench developed off Mexico's Pacific coast. The heat and pressure along this trench have created volcanoes and caused earthquakes.

Climate and Vegetation Mexico extends through the subtropical and tropical latitudes, where the savanna, steppe, and desert climates prevail. Most of northern Mexico is arid with desert scrub vegetation and dry grasslands. The forested plains along Mexico's southeastern coast are hot and humid much of the year. Summer is the rainy season.

Elevation is a key climate variable throughout much of Mexico. Away from the dry north, climates vary quickly with elevation. Many people have settled in the mild environments of the mountain valleys. The valleys along Mexico's southern coast also have pleasant, subtropical climates. The high elevations on the plains of the Plateau of Mexico, however, cause cooler temperatures than would be expected at these latitudes. Sometimes during winter, cold polar air flows southward across the Plateau of Mexico, bringing freezing temperatures as far south as Mexico City.

1 SECTION REVIEW

1. What is Mexico's central plateau called? What mountain ranges border it?
2. Why does Mexico City sometimes experience freezing temperatures?
3. **Critical Thinking** **Compare and contrast** the climates and vegetation in northern Mexico and southern Mexico.

2 ECONOMIC GEOGRAPHY

FOCUS

- *What agricultural products, industries, and natural resources are important to Mexico's economy?*
- *In what different ways has land been used in Mexico during its history? Where are the country's major industrial centers located?*

REGION **Agriculture** Agriculture is a traditional focus of life in Mexico, even though the country has limited amounts of good farmland. Nearly two-thirds of the country is mountainous, and farming here is difficult. In fact, only about 12 percent of Mexico's lands produce a crop every year.

Most of Mexico's agriculture that relies on rainfall is found in the southern part of the Plateau of Mexico and in the valleys of the south. The sediment eroded from the mountains and the ash and lava from volcanoes have created fertile valley soils. Traditional Mexican food crops such as maize (corn), beans, and squash remain important for subsistence farmers. Although slightly more than 20 percent of all Mexicans work in agriculture, farming only provides about 9 percent of the country's gross national product.

Commercial agriculture in Mexico is expanding. Yields are high on these modern farms. Nearby markets in the United States have encouraged major investments in **cash crops**. A cash crop is a crop produced primarily for direct sale in a market, rather than for other purposes. The most extensively cultivated lands produce fruits and vegetables, including tomatoes and melons. Cotton, wheat, and alfalfa for cattle feed occupy much of the land on large estates in central Mexico. The drylands of the north are important for large-scale livestock ranching.

Plantation agriculture is found in the coastal lowlands and on mountain slopes along the Gulf of Mexico. Cash crops include coffee, sugarcane, cacao, bananas, and pineapples. Despite improvements in food supply, Mexico must import basic food items in order to feed its rapidly growing population.

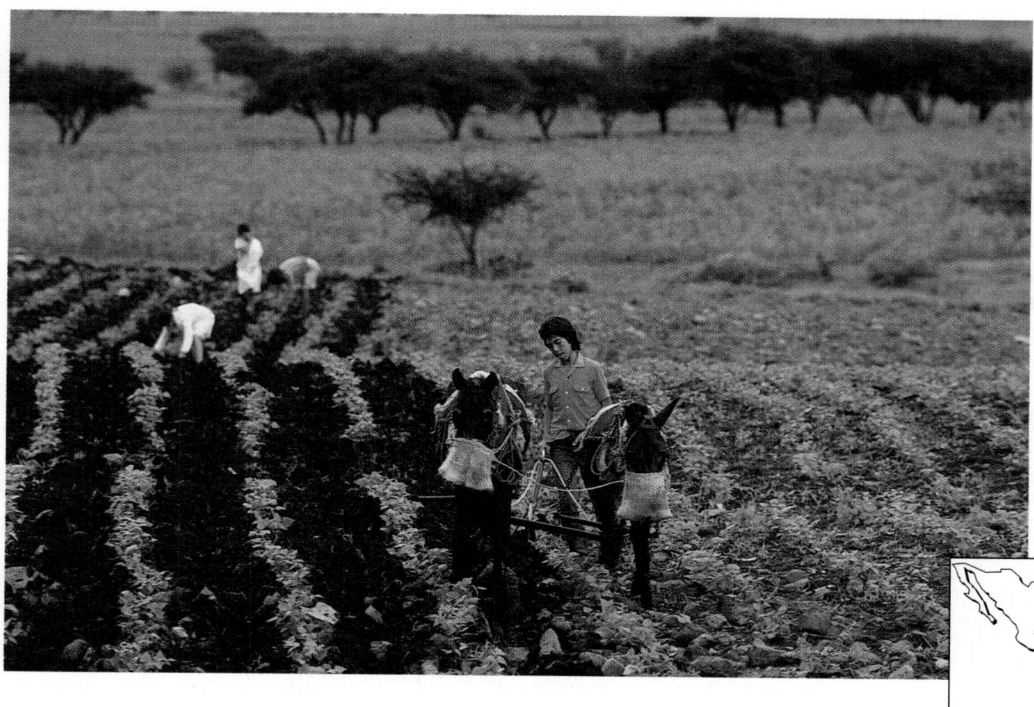

This Mexican teenager and his sisters cultivate a beanfield in **Guanajuato**, a state in central **Mexico**. Mexicans are debating reforms aimed at increasing farm production in their country.

Land Reform The system of land use established by the Spanish divided most of Mexico into haciendas. Workers and their families lived on the haciendas. The workers owned no land and received low wages. Less than a century ago, 96 percent of Mexico's farm families owned almost no land.

Land ownership changed after the Mexican Revolution of 1910. A major goal of this revolution was to increase the number of farmworkers who owned land. Many of the large haciendas were broken up into communal farms, called ***ejidos*** (eh-HEE-thos). Under this system, farmers either worked the land together or gained rights to farm individual plots of land.

Agricultural development proceeded very slowly, however. The *ejido* farmers did not own the land, and there was little money for new equipment, seeds, and fertilizers. Often, workers grew discouraged and moved to the cities seeking industrial jobs. Today, the *ejido* system is changing. Some farmers are acquiring full property rights and larger farms. Agricultural corporations also are buying up the land to build large, modern farms. The government hopes these changes will lessen rural poverty. Yet, people will continue to move to the cities in search of a better way of life and greater opportunities.

Industrial Growth Despite the traditional importance of agriculture in Mexico, more than three-quarters of Mexico's workers are employed in industrial and service jobs. Petroleum has become Mexico's major mineral resource. The country's petroleum reserves, among the largest in the world, are located primarily under the plains and offshore along the Gulf of Mexico coast.

Mining remains important in Mexico as well. Many of the gold and silver mines begun by the Aztecs have developed into major mining districts. In Mexico's northern and southern mountains, new mines have been developed that employ many workers in those regions.

Mexico City, Puebla, Guadalajara (gwahd-uh-luh-HAHR-uh), and Monterrey (mahnt-uh-RAY) are Mexico's major industrial centers. Mexico's factories produce food products, consumer goods, building materials, steel, and chemicals. In rural areas, handicrafts provide cash income.

Mexico's most rapidly growing industrial centers are the cities along the United States border, such as Tijuana (tee-uh-WAHN-uh) and Ciudad Juárez (see-u-thad HWAHR-ehz). Taking advantage of Mexico's lower wages, many U.S. and other foreign companies have established

Text continues on page 208.

Mexico's Northern Border Region

Bordering the United States, northern Mexico lies hundreds of miles north of Mexico's heartland.

Northern Mexico is a vast territory that stretches for almost 2,000 miles (3,218 km) from the Pacific Ocean to the Gulf of Mexico. It is up to 300 miles (483 km) deep and includes more than one-third of Mexico's territory and about 20 percent of its population.

Most of northern Mexico is a semiarid plateau whose eastern and western edges rise into mountain ranges. Its vegetation is mainly desert scrub or short steppe grasslands. A strip of coastline along the Gulf of Mexico in the east and part of the western mountains receive more than 20 inches (51 cm) of rain per year.

MOVEMENT Only the shallow waters of the Rio Grande separate Ciudad Juárez, Mexico, from El Paso, Texas. Each day, many people and products cross the bridge linking the two cities. This constant two-way traffic flow creates the impression that although they are in different countries, these two cities are one metropolitan area. **Name other cities along the U.S.–Mexico border whose relationship is similar to that of Ciudad Juárez and El Paso. What advantages result when this relationship exists between two countries?**

Farther west, the Gulf of California separates the main part of northern Mexico from Baja California, a long, slender peninsula.

To tie the border region more closely to the rest of the country, the Mexican government has tried to increase the region's population and improve its economic conditions. Mexico's most ambitious and successful program to develop the northern border region began in the mid-1960s. Under the Border Industrialization Program, the government encouraged foreign companies to build factories in Mexico to manufacture goods for export.

Today, there are more than 2,000 of these foreign-owned factories, or *maquiladoras*. Most are located within 20 miles (32 km) of the United States. At these plants, workers assemble foreign-made parts into automobiles, computers, sunglasses, and other products for sale outside Mexico. About two-thirds of the *maquiladoras* are owned by U.S. companies, and by the early 1990s, about 70 percent of Mexico's exports went to the United States.

Several of northern Mexico's most important cities are located close to large U.S. cities. Tijuana, one of Mexico's largest border cities, lies less than 20 miles (32 km) from San Diego, California. These two cities have nearly grown into one large metropolitan area. Other Mexican cities and their U.S. neighbors in California, Arizona, New Mexico, and Texas are separated only by the international border.

Although northern Mexico's economic growth and its growing ties with the United States have benefited both countries, these developments have caused problems as well. Because Mexico's environmental regulations often

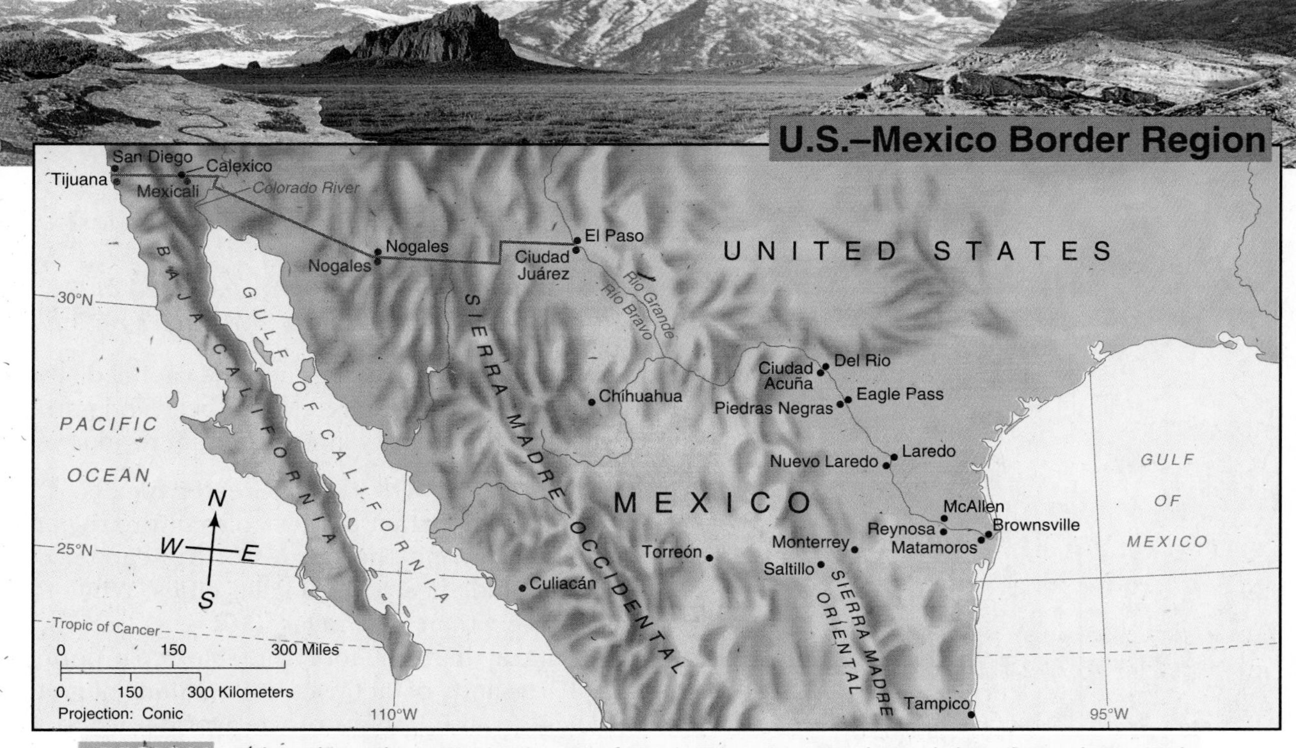
U.S.–Mexico Border Region
San Diego
Calexico
Tijuana
Mexicali
Colorado River
Nogales
Nogales
El Paso
Ciudad Juárez
UNITED STATES
30°N
Rio Grande
Río Bravo
BAJA CALIFORNIA
GULF OF CALIFORNIA
SIERRA MADRE OCCIDENTAL
Ciudad Acuña
Del Rio
Chihuahua
Eagle Pass
Piedras Negras
PACIFIC OCEAN
Laredo
Nuevo Laredo
GULF OF MEXICO
MEXICO
McAllen
Reynosa
Brownsville
Matamoros
Monterrey
25°N
Torreón
Saltillo
Culiacán
SIERRA MADRE ORIENTAL
Tropic of Cancer
0 150 300 Miles
0 150 300 Kilometers
Projection: Conic
Tampico
110°W
95°W

REGION Although northern Mexico is part of Mexico, its economy depends heavily on the United States, and its people are influenced by U.S. culture. Northern Mexicans in turn influence Americans across the border. **How does a border area such as northern Mexico differ from regions in the central part of a country?**

are not enforced, the *maquiladoras* along the U.S.–Mexico border pour industrial wastes into the air and water of both countries. In addition, population and industrial growth along the border have led to a demand for water that exceeds the available natural supply. As a result, groundwater reserves are being exhausted or destroyed from overuse.

The rapid industrialization of northern Mexico has led to additional concerns. Many Mexicans fear that the United States is "Americanizing" northern Mexico. North of the border meanwhile, some U.S. workers worry that northern Mexico's economic growth is taking place at their expense. They have seen U.S. companies close plants in the United States and relocate in Mexico to take advantage of lower wages and costs there. Both countries hope that the North American Free-Trade Agreement (NAFTA), approved by the U.S. Congress in 1993, will spur economic growth on both sides of the border.

Northern Mexico is a region where two countries and cultures meet and blend. Today and in the future, Mexico and the United States must continue to work together on trade agreements and environmental concerns to resolve issues between the two countries.

INTERACTION Many companies, such as the one that runs this plant on the Texas–Mexico border, have sought ways to limit pollution. Businesses and citizens realize that continued population growth and industrialization will cause more pollution on both sides of the border. **How can the United States and Mexico work together to solve problems caused by industrial pollution?**

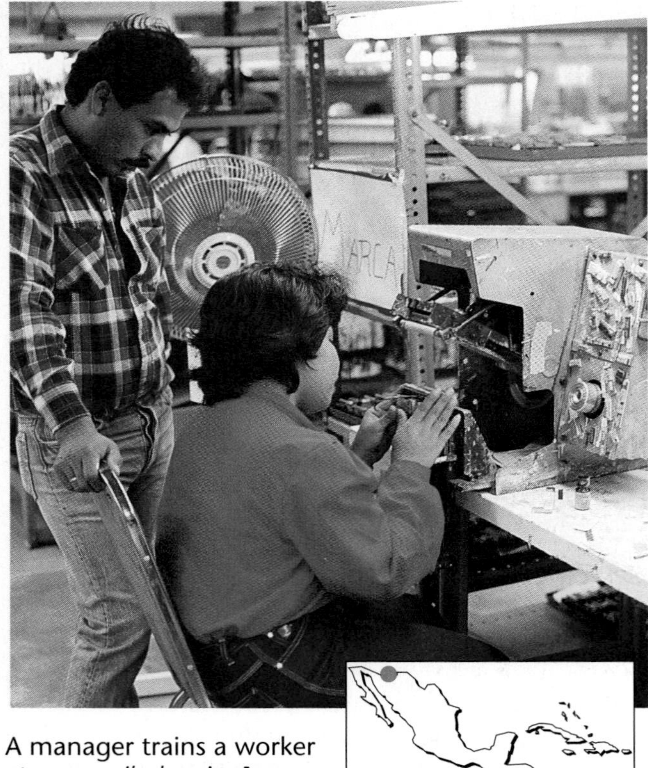
MARCA

A manager trains a worker at a *maquiladora* in **Agua Prieta**, across the border from Arizona in **Mexico**. The *maquiladoras* have provided jobs and training for hundreds of thousands of Mexicans.

factories called ***maquiladoras*** (mah-KEEL-ah-dor-ahs) in the border cities. In these plants, products are assembled for export to the United States and other nations. (See "Themes in Geography" on pages 206–207.)

Mexico also has a large tourist industry. Tourists, particularly from the United States, travel to the old colonial cities and ancient monuments of the Plateau of Mexico. Coastal cities such as Acapulco, Mazatlán, and Cancún draw tourists from around the world.

2 SECTION REVIEW

1. What are Mexico's major sources of income?
2. How has land been used in Mexico over time?
3. **Critical Thinking** **Discuss** the role played by the United States in Mexico's agricultural and industrial development.

3 MEXICO'S CULTURE REGIONS

FOCUS

- *Into what six culture regions can Mexico be divided?*
- *What are the economic, political, and social characteristics of Mexico's regions?*

Mexico consists of 31 states and the capital district. The country's geography can be best understood by dividing the country into six culture regions.

Greater Mexico City Greater Mexico City is the most important of Mexico's culture regions. It extends beyond the border of the capital district to include nearly 50 smaller cities. While this area is less than 100 miles (161 km) across, it holds more than one-fourth of Mexico's population. Thousands of factories and millions of motor vehicles generate some of the world's worst air pollution.

Mexico City is one of the most populous cities in the world. (See "Cities of the World" on pages 210–211.) Rural migrants are drawn to Mexico City from throughout Mexico, looking for new opportunities. Many do not find jobs, and the slums grow larger every day. At the same time, there are many modern highways, office buildings, high-rise apartments, fine museums, and major universities. Wealth and poverty often exist side by side. Economic growth, however, will likely continue as wealth from Mexico's industries, natural resources, and tax revenues flows into the Greater Mexico City region.

As Mexico's capital, Mexico City is the country's center of political power. Since the 1910 revolution, Mexico's government has been dominated by one political party. Officials of the Institutional Revolutionary Party, or PRI, control all of the government ministries and most of the labor unions.

Central Interior Northward from Mexico City and extending toward both coasts is Mexico's historic core of Spanish settlement. Many cities began here as mining and ranching centers during the colonial period. This region has many fertile valleys and small family farms. Many old-style haciendas still exist in this region, and small towns with a central square and colonial-style church are common.

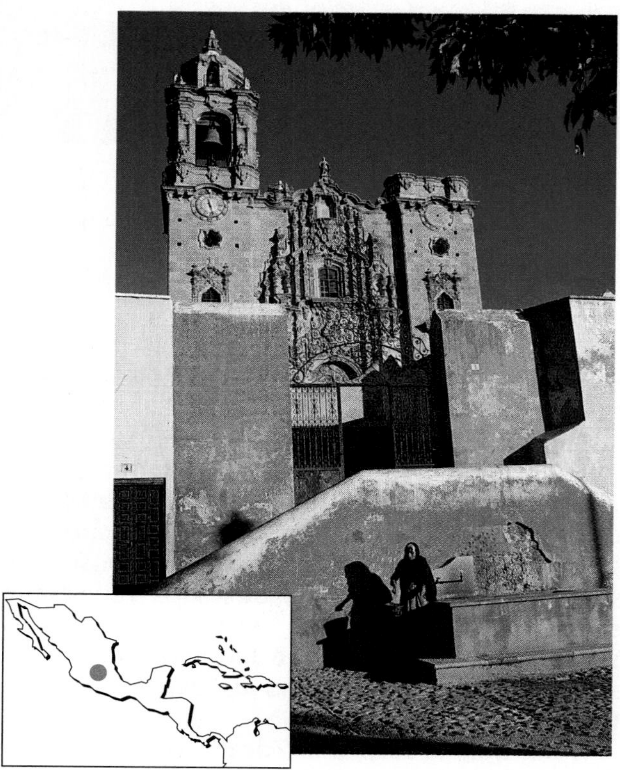

This woman gets water from a faucet at a colonial-era church in the state of **Guanajuato**, northwest of Mexico City.

The Central Interior region is losing many people to Mexico City and the northern region because the funds for improving agriculture and modernizing industry here are limited. As Mexico City becomes increasingly crowded, however, more industries are moving new factories and businesses to this region.

Oil Coast Oil, Mexico's greatest natural resource, is found along the shore of the Gulf of Mexico between Tampico and Campeche. Historically, these coastal plains were lightly settled, and they still have extensive forests. Oil production first began here in the early twentieth century. During the 1970s, new discoveries dramatically increased production and brought oil refineries and petrochemical complexes to the region.

Now-booming cities along this coast were founded during the colonial period and for centuries acted as service centers for plantations. Large tracts of forest have been cleared for commercial farming and ranching.

Southern Mexico Southern Mexico, the mountainous land south of Mexico City, is Mexico's poorest region. It has few large cities and little industry. Basic infrastructure, including roads, telephone services, schools, and electric power, is poorly developed. It is a region of traditional villages, many of which are connected to the outside world only by dirt roads. Mayan and other indigenous languages are still spoken by about half the region's population. Subsistence agriculture is typical. Increasingly, migrants head north for new opportunities.

Northern Mexico The dry region of northern Mexico has become one of the country's most prosperous and modern. The merchants, miners, and ranchers of the region generally have cared little about events in Mexico City. Instead, they look to the United States for investment funds. Many firms here have business connections north of the border.

Monterrey is the north's industrial city, and it is developing closer economic relations with large Texas cities such as San Antonio and Houston. In the far northwest, Tijuana, Mexico, and San Diego, California, are beginning to merge into one metropolitan area. Northern Mexico's industries, such as the *maquiladoras*, and commercial farms are drawing migrants from throughout Mexico.

The Yucatán Few people live in the Yucatán, a peninsula in the southeast of Mexico. In ancient times, the Yucatán's population was probably greater than it is today. The Mayas here developed an effective agricultural system. They placed their fields on raised platforms built between drainage canals. Modern farmers are using some of the Mayas' techniques, and crop yields have improved.

Mérida is the Yucatán's major city. Its wealth came from rope made from agave fiber, which was a key product when sailing ships rigged with rope connected the world's ports. Other important centers include Cancún and Cozumel, resort cities on the Caribbean coast.

3 SECTION REVIEW

1. Where is each of Mexico's six culture regions located?
2. Which regions of Mexico are industrial?
3. **Critical Thinking** **Decide** what region of Mexico you might visit if you wanted to learn about the country's history. Support your answer.

Text continues on page 212.

Mexico City

Mexico City is a rich fabric of peoples, cultures, history, and technology woven together.

Mexico City is one of the oldest cities in the Western Hemisphere. It lies high in the mountains of Mexico on a plain that was once a lake. Originally, it was the site of the Aztec city of Tenochtitlán, founded in about 1325. At that time, the city was built on a series of islands in the lake. Spaniards led by Hernán Cortés invaded the city about 200 years later. They destroyed the original city, then built a new one, Mexico City, on the site of the old. They drained the water from the lake and filled in the land so the new capital city could expand.

Building on the Past

Today, the main square of the city stands on the site of the old Aztec palace and temple. The square is named the Plaza of the Constitution, commonly called the Zócalo. A cathedral, the national palace, government buildings, museums, and commercial buildings surround the Zócalo.

To the west of the Zócalo is the wide, tree-lined boulevard Paseo de la Reforma, often called simply Reforma. At first, grand mansions lined Reforma. Now, the boulevard has become an extension of the central business district, and tall office buildings tower over both sides. Reforma passes through Bosque de Chapultepec, the largest park in Mexico City. Aztec emperors used the park hundreds of years ago.

Until the 1900s, the Zócalo contained almost the whole city. Since that time, however, migration and industrialization have led to a dramatic increase in Mexico City's population. In the 1970s, the number of people in the city rose more than 70 percent. The 1990 Census of Mexico set Mexico City's population at 15 million. By 2010, the city may contain as many as 19 million people.

The Challenges Ahead

Due to its rapid growth, Mexico City faces a number of problems, including a severe shortage of housing. Land is scarce, and construction has not been able to keep up with the need for homes. The government has provided some housing, but these efforts have helped only a small percentage of the people. As a result, large squatter settlements have grown on the outskirts of the city. Migrants from rural areas of the country pour into these settlements daily, searching for work and a better life in the city.

Mexico City's rapid population increase also has placed a great strain on the transportation system. Traffic congestion clogs the city's freeways and streets, and spews exhaust fumes into the air. The mountains surrounding the

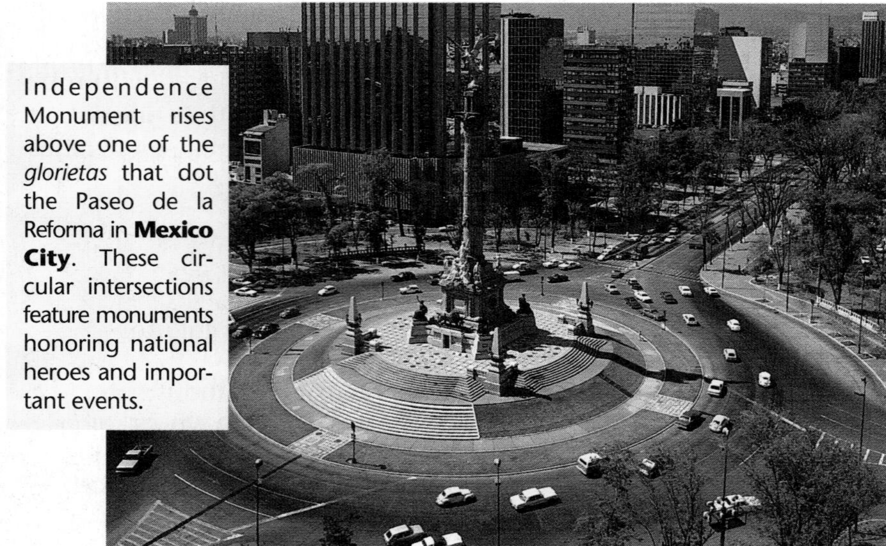

Independence Monument rises above one of the *glorietas* that dot the Paseo de la Reforma in **Mexico City**. These circular intersections feature monuments honoring national heroes and important events.

GINZA
HOLLYWOOD

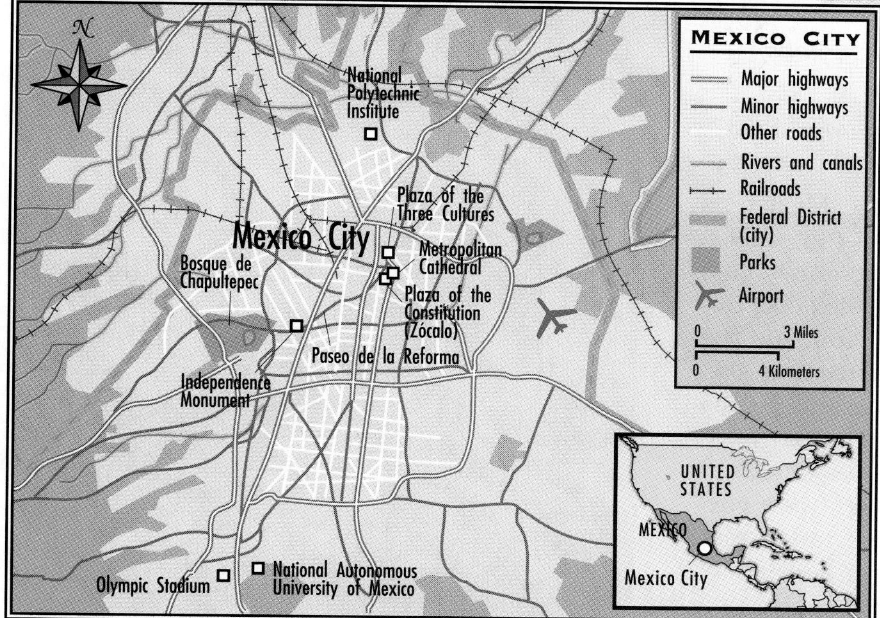
N
National Polytechnic Institute
Plaza of the Three Cultures
Mexico City
Metropolitan Cathedral
Bosque de Chapultepec
Plaza of the Constitution (Zócalo)
Paseo de la Reforma
Independence Monument
Olympic Stadium
National Autonomous University of Mexico
MEXICO CITY
Major highways
Minor highways
Other roads
Rivers and canals
Railroads
Federal District (city)
Parks
Airport
0
3 Miles
0
4 Kilometers
UNITED STATES
MEXICO
Mexico City

As the national capital, Mexico City is a transportation center and a home to numerous government offices, universities, and cultural sites. Name two buildings that are located near the Zócalo.

city trap the fumes as well as pollution from factories. These mountains act as a wall, blocking winds from blowing into the valley. As a result, thick smog often blankets the city.

Earthquakes present another threat to Mexico City. Because the city is built on the soft, spongy soil that was a lake bed in Aztec times, shock waves from earthquakes are magnified. When an earthquake strikes, the soil quivers almost like jelly. Tall buildings are particularly at risk. In 1985, an earthquake struck the city, resulting in some 7,200 deaths and destroying hundreds of buildings.

Another problem is that the buildings have been sinking as much as one foot (30 cm) per year into the soft ground. To combat this problem, the government has regulated the height of buildings in the city, and new buildings have special foundations that prevent them from sinking. The foundations also reduce damage from earthquakes.

Although faced with these and other challenges, Mexico City remains one of the world's major cultural centers. The city's natural beauty, rich history, and mild climate continue to attract people from around the globe.

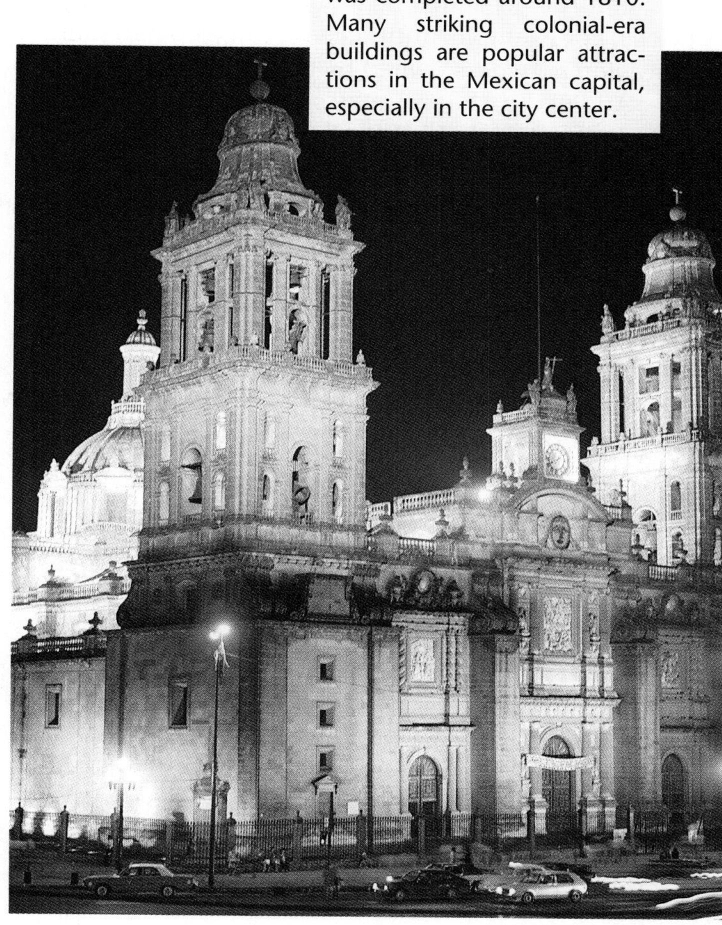

The Metropolitan Cathedral anchors one side of the Zócalo in **Mexico City**. Construction of the massive cathedral began in 1570 and was completed around 1810. Many striking colonial-era buildings are popular attractions in the Mexican capital, especially in the city center.

MEXICO'S FUTURE

FOCUS

- *What economic and political challenges has Mexico faced recently?*
- *How are conditions in Mexico improving?*

Social and Economic Issues Mexico is undergoing rapid change. During the 1980s, Mexico's economy grew little. The country had invested unwisely in previous years. Mexico's foreign debt had risen to nearly $100 billion, and the country was nearly out of money. **Inflation** skyrocketed. Inflation is the rise in prices caused by a decrease in value of a nation's currency. In 1987, prices in Mexico more than doubled in one year.

Mexico's industries, many of which were government owned, were inefficient. High tariffs protected most from foreign competition, and many companies were monopolies. A **monopoly** is a business that has no competitors. Such companies can control an industry's supply of goods and prices. For example, until recently, Mexico had four television networks owned by the same company.

Adding to the country's economic troubles was decay within the PRI. Instead of working to improve Mexico's economy, some PRI leaders and associates gained wealth through corruption. Several elections were won by fraud.

In the late 1980s, Mexico's economy began to show signs of improvement. The government sold many of its large firms to private individuals or companies. This selling of government-owned businesses or lands to private owners is called **privatization**. The *ejido* system also was reformed to increase agricultural productivity. These changes led to Mexico's developing a better relationship with the United States and other countries. Foreign competition forced many of Mexico's industries to become more efficient and productive. Mexico's government encouraged foreign trade, and many U.S. and Japanese firms began to compete in Mexico. Many Mexican firms became active in the United States.

Mexico now hopes to be considered a developed country. Vast differences, however, exist between the wealthy and the poor, and between the modern and the traditional. Ernesto Zedillo Ponce de Leon, elected president in 1994, has had to deal with new economic and political crises, including a rebellion in the poor, southern state of Chiapas. Yet, the country's infrastructure is steadily improving, industries are modernizing, and agricultural production continues to increase. Politically, the nation is moving toward true democracy.

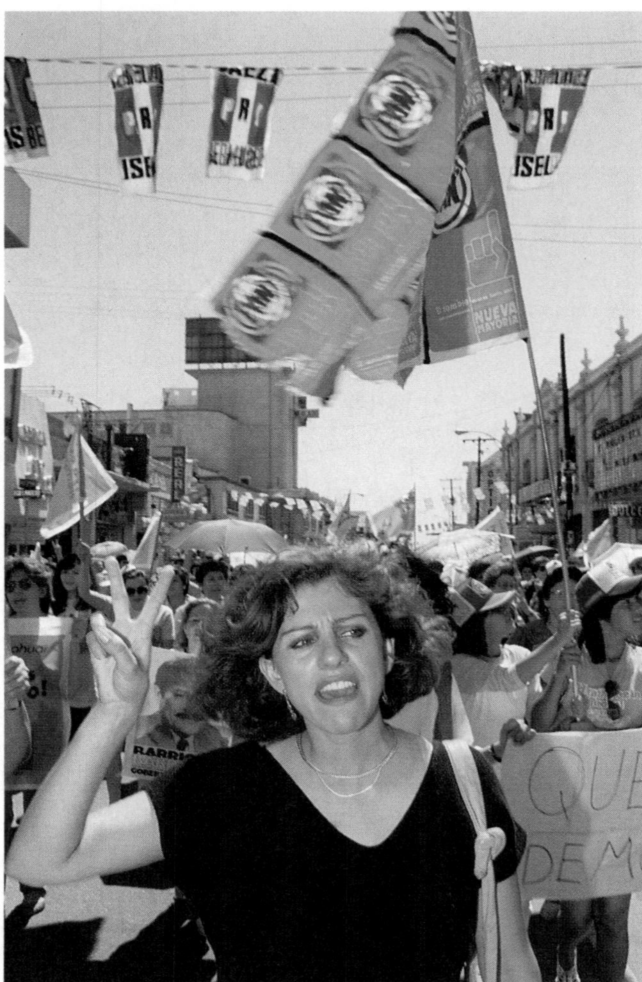

Mexican politics are more competitive after decades of one-party rule by the PRI. Opposition parties, such as the National Action Party (PAN), appeal to more Mexican voters than before.

SECTION REVIEW

1. Why were many of Mexico's industries inefficient in the late 1980s?
2. What reforms were begun in the late 1980s?
3. **Critical Thinking** **Evaluate** the following statement: "Monopolies do not benefit consumers."

CHAPTER 18 CHECK

Library of the National Autonomous University of Mexico in **Mexico City** with a close-up view of the building's mosaic murals

Reviewing the Main Ideas

1. Most of Mexico consists of a large plateau bordered by two mountain ranges. Other landforms include volcanoes, plains, and valleys. The climates of Mexico range from humid tropical and subtropical to desert and highland.
2. Agriculture is widely practiced in Mexico, though good farmland is scarce. Commercial agriculture is becoming more widespread. Even so, food must be imported because of Mexico's rapidly expanding population. More than 75 percent of Mexico's workers are employed in industrial and service jobs.
3. Mexico can be divided into six culture regions. Much of Mexico's population lives in and around Mexico City, the nation's capital. Wealth and poverty often exist side by side in the urban regions of the country.
4. In the 1980s and early 1990s, Mexico experienced economic hardships. With recent political and economic reforms, officials have worked to aid development and to improve conditions.

Building a Vocabulary

1. What is a cash crop?
2. How does an *ejido* differ from a hacienda?
3. What are *maquiladoras*, and where are they often located?
4. If a country suddenly introduced a large amount of currency into its economy without changing the amount of goods available for purchase, what might occur?

Recalling and Reviewing

1. Why does Mexico City have severe air pollution? Consider the city's physical location as well as its economic activities.
2. For what reasons do many Mexicans move from rural areas to urban areas? What are some consequences of this migration?
3. Although the term *monopoly* usually refers to a business, how might it also apply to a political party? Explain, using information from the chapter.
4. Compare and contrast daily life in southern and northern Mexico.

Thinking Critically

1. Do you think an *ejido* system could ever be effective? If so, how? If not, why not?
2. Do tourist brochure pictures of Mexico's beaches, palm trees, and hotels represent an accurate portrayal of the country? Explain.

Learning About Your Local Geography

Individual Project

Contact your state or local air quality board, a local environmental organization, or the weather department of a local television station. Request information on the air quality of your area. If possible, collect data going back several years so that you can see how the air quality of your area has been changing. Graph the data. What geographic factors account for the patterns you see?

Using the Five Themes of Geography

Create a graphic organizer showing how the themes of **location**, **place**, and **movement** interact in Mexico. How do relative location and place characteristics influence movement within Mexico and between Mexico and other countries?

CHAPTER 19

Central America and the Caribbean Islands

"It is true that the early morning is the most beautiful time of day on the island [Antigua]. The sun has just come up and is immediately big and bright, the way the sun always is on an island, but the air is still cool from the night; the sky is a deep, cool blue; . . . the red in the hibiscus and the flamboyant flowers seems redder; the green of the trees and the grass seems greener."

Jamaica Kincaid

GEOGRAPHY DICTIONARY

- ***cardamom***
- ***calypso***
- ***reggae***
- ***merengue***
- ***cooperative***
- ***plantain***
- ***voodoo***
- ***refugee***
- ***commonwealth***

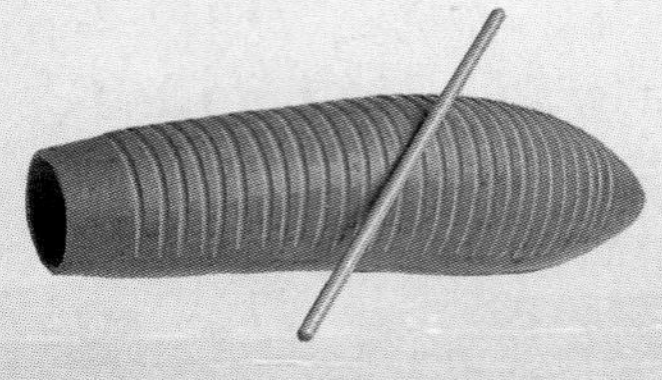

A güiro musical instrument from the **Lesser Antilles.**

Central America and the Caribbean include 20 small countries and more than a dozen territories. The countries of Central America lie on a natural land bridge between North and South America. The Caribbean islands lie in an arc from Cuba to Trinidad and Tobago.

Cultural history helps explain the development of so many countries and territories. The Spanish, English, French, and Dutch introduced different languages, technologies, and laws to their territories in the region. These European traditions mixed with the cultures of indigenous peoples and of African slaves.

In recent decades, the region has been strongly influenced by the United States, which has offered democratic ideas, modern technology, tourist dollars, and a market for agricultural products. Most Central American and Caribbean countries now have better transport links with the United States than with each other.

Ocho Rios, **Jamaica**

CENTRAL AMERICA AND THE CARIBBEAN: PHYSICAL—POLITICAL

CENTRAL AMERICA

FOCUS

- *What landforms and climate types are found in the countries of Central America?*
- *What activities support the economies of the Central American countries? How are economics and politics linked in the region?*

LOCATION **Isthmus Region** Central America, the isthmus between North and South America, also separates the Caribbean Sea and the Pacific Ocean. (See the map above.) Earthquakes and volcanic eruptions are frequent threats.

Along the Caribbean coast are humid-tropical plains with dense forests. Inland from the coasts, the mountains rise into mild, highland climates, where most of the people live. Coffee is the main cash crop on the humid mountain slopes. The Pacific coast of Central America has warm, sunny weather. In many areas, the savanna vegetation has been cleared for plantations and ranches.

Because the seven countries of Central America have similar environments, they grow similar export crops. Competition rather than cooperation between the countries has been typical. Free regional trade is a goal of most of the Central American countries.

Guatemala Guatemala is the most populous country in Central America. The heartland of Guatemala is found in the fertile valleys below the volcanic mountains in the southern part of the country. The capital, Guatemala City, is located here. Coffee is Guatemala's primary export. Another major export is **cardamom**, a spice grown in the central highlands. Cardamom is popular in many of the foods of Southwest Asia.

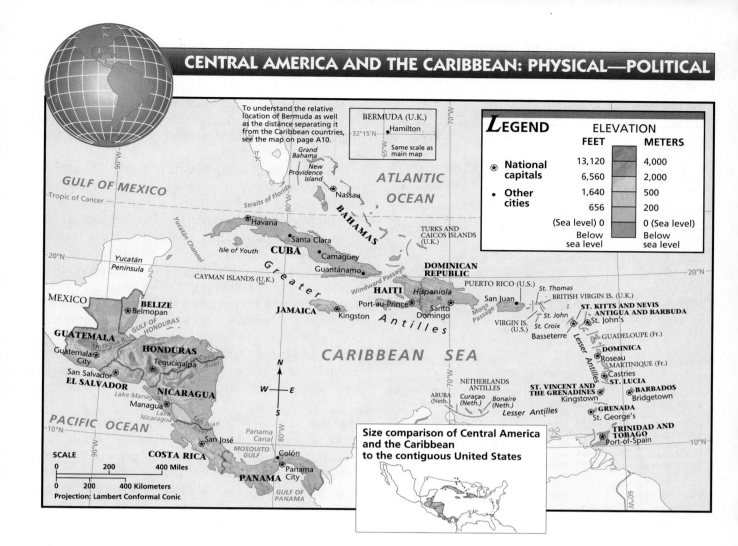

FACTS IN BRIEF Central America and the Caribbean Islands

COUNTRY POPULATION (1994)	LIFE EXPECTANCY (1994)	LITERACY RATE	PER CAPITA GDP (1993)
ANTIGUA AND BARBUDA 64,762	71, male 75, female	90% (1990)	$5,800
BAHAMAS 273,055	68, male 75, female	95% (1992)	$16,500
BARBADOS 255,827	71, male 77, female	99% (1992)	$8,700 (1990)
BELIZE 208,949	66, male 70, female	93% (1991)	$2,700
COSTA RICA 3,342,154	76, male 80, female	93% (1992)	$5,900
CUBA 11,064,344	75, male 79, female	99% (1992)	$1,250 (GNP)
DOMINICA 87,696	74, male 80, female	94% (1992)	$2,100 (1992)
DOMINICAN REPUBLIC 7,826,075	66, male 71, female	83% (1991)	$3,000
EL SALVADOR 5,752,511	64, male 70, female	75% (1991)	$2,400
GRENADA 94,109	68, male 73, female	95% (1991)	$3,000 (1992)
GUATEMALA 10,721,387	62, male 67, female	55% (1991)	$3,000
HAITI 6,491,450	43, male 47, female	53% (1991)	$800
HONDURAS 5,314,794	65, male 70, female	73% (1991)	$1,950
JAMAICA 2,555,064	72, male 77, female	98% (1990)	$3,200 (1992)
NICARAGUA 4,096,689	61, male 67, female	57% (1991)	$1,600
PANAMA 2,630,000	71, male 78, female	87% (1991)	$4,500
ST. KITTS AND NEVIS 40,671	63, male 69, female	98% (1991)	$4,000 (1992)
ST. LUCIA 145,090	67, male 72, female	78% (1989)	$3,000
ST. VINCENT AND THE GRENADINES 115,437	71, male 74, female	85% (1989)	$2,000 (1992)
TRINIDAD AND TOBAGO 1,328,282	68, male 73, female	97% (1988)	$8,000
UNITED STATES 260,713,585	73, male 79, female	97% (1991)	$24,700

Sources: *The World Factbook 1994*, Central Intelligence Agency; *1995 Britannica Book of the Year; The World Almanac and Book of Facts 1995*

Which Central American or Caribbean country has the highest per capita GDP? Which countries of the region have life expectancy rates of more than 70 years for both men and women?

Land reform is a continuing issue in Guatemala. Much of the land is owned by a small percentage of farmers. Some traditional haciendas and coffee estates remain. With modern technology, sugar and cotton cultivation have increased, particularly on the Pacific slopes of the mountains.

About half of Guatemala's nearly 11 million inhabitants are indigenous peoples, most of whom live in small, isolated villages throughout the highlands. Subsistence farming is typical. Like their Mayan ancestors, farmers grow maize (corn), beans, squash, and chili peppers.

Economic inequality and discrimination against the indigenous population have led to violent politics in recent decades. Guatemalan leaders now are realizing the importance of political and economic freedom. Land hunger, or the desire of farmers to own their own land, is being reduced through settlement of the once nearly empty plains of Petén in the north.

Belize Formerly British Honduras, Belize (buh-LEEZ) is located along the Caribbean coast of the Yucatán Peninsula. Its humid-tropical, dense forests limited settlement for many years.

The country's population is made up of many kinds of people. Blacks and mulattos who speak English live along the coast. Indigenous peoples and peoples of African ancestry live in the inland forests, and settlers from Mexico and Guatemala live along the inland borders. Few crops can grow in the poor soils. Business leaders are concentrating on the fishing and tourist industries. Visitors find beautiful coral reefs and coastal resorts.

Honduras South of Belize is Honduras, a land of rugged mountains and valleys. Coffee grown in the highlands supplies most of the country's export income. Yet, small subsistence farms and large colonial-style estates dominate locally. Less than 20 percent of the land can be farmed.

About 90 percent of Honduras's people are mestizos. Most of the population lives in small mountain valleys in the highlands and in river valleys along the north coast. The rugged terrain makes transportation difficult.

The key development of modern Honduras is in the humid-tropical lowlands along the north coast. Many people have migrated out of the highlands to work on the large banana and sugarcane plantations there.

El Salvador El Salvador is a small country of volcanoes and plains on the Pacific side of Central America. Although volcanic ash has made El Salvador's soils the most fertile in the region, earthquakes and volcanic eruptions are continuing threats. El Salvador's climate is wet in the summer and dry in the winter.

As elsewhere in Central America, subsistence agriculture on small plots and commercial farming on large estates dominate rural areas. Coffee and cotton from large estates provide most of the export income. The majority of people, however, live a life of poverty.

El Salvador faces an increasing population and the need for effective land reform. Much of the best land is held by a few powerful families. The country's population is increasing faster than the government can provide schools and jobs.

These Maya women are taking part in a religious celebration in **Guatemala**. Native American communities in the region are distinguished from one another by their distinct, bright styles of clothing.

During much of the 1980s, civil war raged in El Salvador, hindering economic development. The United States backed the Salvadoran government in the conflict. Finally, in 1992, a peace treaty ended the civil war.

Nicaragua The physical geography of Nicaragua is much like that of neighboring Honduras. There is a hot, humid plain along the Caribbean coast, low rugged mountains inland, and a volcanic plain along the Pacific coast. Nicaragua, however, also has a broad valley with large lakes inland from the Pacific coast. The valley is tectonically active with earthquake faults and volcanoes. Its fertile volcanic soils and those of the nearby mountain slopes make this area the most densely settled part of the country. The capital city, Managua, is located here.

Throughout Nicaragua's history, economic progress has been slow and political unrest typical. Most of the people continue to farm. Primary exports are coffee, beef, cotton, and bananas. The highlands contain deposits of gold and silver.

Severe problems of social inequality and land reform have always plagued Nicaragua. In 1979, the country began a period of dramatic political change. Its longtime dictator was overthrown and a revolutionary government took power. Many rich Nicaraguans fled the country. The government called itself Sandinista, named after César Sandino, a Nicaraguan patriot of the 1930s.

The United States put economic pressure on the Sandinistas and supported Nicaraguan rebels, called Contras. The economy of Nicaragua nearly collapsed. In 1990, free elections were held and the Sandinistas were defeated, although they remained the largest political party. Today, Nicaragua is a democracy with many political voices.

Costa Rica Costa Rica's central highlands with fertile soils and mild climates are the nation's heartland. Located here is San José, the capital city. Many coffee farms, which provide the country's main exports, also are found in this area.

When the European colonists first arrived in Costa Rica, most of the local peoples quickly died of diseases or migrated elsewhere. As a result, the colonists did not have local labor to develop large estates. Coffee cultivation and favorable government policies encouraged small farms. This development eventually created a large middle class

Teenagers in Costa Rica Think Life Is *PURA VIDA!*

Costa Ricans live in a democratic republic where education is valued highly. Over 90 percent of Costa Ricans are literate, one of the highest rates in Middle and South America. Education is providing increased opportunities to young Costa Ricans, like those pictured above.

For most teens, weekdays are filled with a rigorous study schedule. After school, there is always work to be done around the house. Whether they are taking care of younger siblings, helping to prepare the evening meal, or doing other chores, most teens have family obligations to fulfill. Family ties here are very important, and children learn respect, discipline, and pride at an early age.

In general, Costa Ricans have a strong interest in the United States, and many students study English. Many Costa Ricans also watch U.S. television programs that have been dubbed in Spanish. The latest U.S. fashions, music, food, and movies are all hot topics of interest among Costa Rican teenagers.

Weekends are dedicated to spending time with family and friends. There is always something to do—parties, dances, weddings, church, and more. Most of these events are family oriented. On weekends, many teens watch soccer games and soap operas on television, listen to music, talk on the phone, spend time with their friends, and dream about their futures.

If you ask Costa Rican teens what they think about life in their country, you will most likely hear *"PURA VIDA!"* Life is great!

that has supported stable democratic governments. Costa Rica has the highest standard of living and literacy rate in Central America.

Like the other countries of the region, Costa Rica has a high population growth rate. In the past, the opening of new lands closer to the coasts often relieved population pressure. Only recently have landless people and suburban slums become a serious concern. The government has encouraged industrial growth and expanded tourism, and has set aside land for national parks.

Panama Made up mostly of low hills and mountains, Panama has only narrow coastal plains. High volcanic peaks tower over the western part of the country. Most Panamanians live along the southern (Pacific) side of the nation.

To some, modern Panama seems like three separate countries. In the east is a densely forested region that is nearly uninhabited. In the middle of the country is a prosperous area surrounding the Panama Canal, with large cities. In the west are more rural areas, where subsistence farming and haciendas are common.

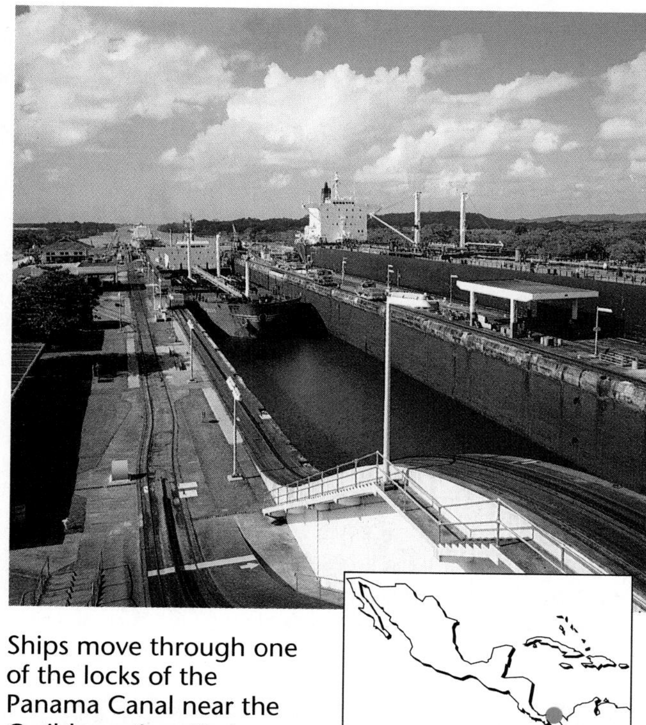

Ships move through one of the locks of the Panama Canal near the Caribbean Sea. To learn how a ship moves through a lock, see the illustration on page 142.

Panama gained its independence from Colombia in 1903. Eager to construct a canal across Panama linking the Atlantic and Pacific oceans, the United States aided Panama in its separation from Colombia. Completed in 1914, the Panama Canal allows ships and trade to travel more quickly and safely between the two oceans.

Canal fees and canal-side industries provide nearly half of Panama's gross national product. The cities at both ends of the canal—Colón on the Caribbean and Panama City on the Pacific—are major industrial centers. The United States administers the Canal Zone, a ten-mile-wide strip of land straddling the canal. In 1979, a treaty between the United States and Panama began a transition that will return control of the canal to Panama in the year 2000.

SECTION REVIEW

1. What landforms are shared by most of the countries of Central America?
2. What crops are grown in Central America?
3. **Critical Thinking** **Suggest** a way that Central American countries might reduce political instability.

THE ISLANDS

FOCUS

- *What are the physical characteristics of the Caribbean islands?*
- *What cultural, economic, and political patterns are found on the islands?*

LOCATION **Physical Geography** To the east of Central America lie the islands Columbus called the West Indies, now called the Caribbean islands. They extend in a wide arc from south of Florida to the coast of Venezuela. Separate island groups in the Caribbean are the Greater Antilles, the Lesser Antilles, and the Bahamas. (See the map on page 215.)

The islands of the Greater Antilles are Cuba, Hispaniola, Puerto Rico, and Jamaica. The Lesser Antilles, which are smaller than the Greater Antilles, include the Virgin Islands, Barbados, Trinidad and Tobago, and many other small islands. The Bahamas, north of the Greater Antilles, include more than 700 islands.

The islands of the Caribbean have pleasant humid-tropical climates, and winters usually are drier than summers. The islands receive a yearly average of 40 to 60 inches (102 to 152 cm) of rainfall. Crops, particularly fruits and vegetables, are grown throughout the year because of the region's favorable climate. Fertile soils, however, are found only on some of the islands. On other islands, including Jamaica and Puerto Rico, there are many limestone rocks through which water rapidly drains. As a result, drought conditions are common, even though it may rain often.

Human Geography Nearly every island in the Caribbean still shows the effects of colonialism and slavery. Languages reflect the region's cultural diversity. English, Spanish, French, Dutch, Hindi, and Chinese are spoken. Blendings of these languages with African languages also can be heard. The islands' musical styles are popular inside and outside the Caribbean. Trinidad and Tobago is the home of steel-drum and **calypso** music. Jamaica is famous as the birthplace of **reggae**. The **merengue** is the national music and dance of the Dominican Republic.

Text continues on page 222.

Case Study: Saving the Rain Forests

Along Panama's Caribbean coast, a great rain forest reaches down to the sea. Giant anteaters, nocturnal tapirs, and armadillos move along the forest floor, while eagles and migrating birds perch high in the forest's canopy. Thousands of species of plants thrive in this humid-tropical environment.

The Kuna have practiced conservation of their rain forest environment for generations.

This rain forest also is home to some 30,000 Kuna, whose villages occupy the small offshore islands. The Kuna reservation, which encompasses more than 1,100 square miles (2,849 sq. km), stretches some 100 miles (161 km) along the coast of Panama. The coastal waters provide the Kuna with fish, while the fertile mainland plains allow them to grow bananas, sweet manioc, yams, and maize (corn).

Caretakers of the Land

In 1980, a Panama official met with the Kuna. "Why do you Kuna need so much land?" he demanded. "You don't do anything with it. You don't use it. And if anyone else cuts down so much as a single tree, you shout and scream." Rafael Harris, a Kuna leader, responded. "We Kuna need the forest, and we use it and take much from it. But we take only what we need, without having to destroy everything as your people do."

In this area secluded by the rugged San Blas Mountains, the Kuna have managed to head off development and clearing of their precious rain forest. For the Kuna, the plants of the rain forest are a source of sacred medicine. The rain forest also protects their farmlands and fishing areas. Without the trees and plants that absorb rain and hold the soil in place, the land would erode and eventually destroy the fishing grounds on which the Kuna depend.

To protect the rain forest, the Kuna have set up a 230-square-mile (596-sq.-km) protected reserve on the edge of the ridge that separates the Kuna reservation from the rest of Panama. To help finance their plans, the Kuna called on international environmental groups. In addition, after building a dormitory near the rain forest, they welcomed paying visitors, including scientists, bird watchers, and the Panamanian Boy Scouts. The Kuna collect fees from tourists and sell research rights to scientists visiting the reserve. The Kuna have worked to have their reservation declared a Biosphere Reserve and a World Heritage Site. These United

A Kuna woman stitches a mola, a colorful, needlework garment. The Kuna, who live on coral islands off the northern coast of **Panama**, have maintained control over their own cultural and political ways of life.

Nations classifications can help the Kuna in their struggle for international recognition as an independent people.

The Kuna have not relied on outside sources alone, using $100,000 of their own funds. "The main thing about the project is the incredible initiative that has come from the Kuna," said R. Michael Wright of the World Wildlife Fund. "They led us, we didn't lead them. Conservationists come and go, but the Kuna are going to stay." Geographers hope that the Kuna's experiment will provide a model for other peoples living in the world's rain forests.

Panama and the Kuna

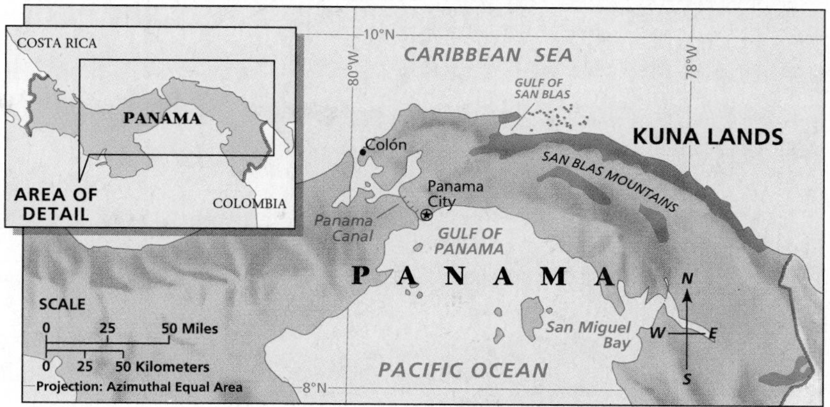

COSTA RICA
PANAMA
AREA OF DETAIL
COLOMBIA
10°N
80°W
78°W
CARIBBEAN SEA
GULF OF SAN BLAS
KUNA LANDS
SAN BLAS MOUNTAINS
Colón
Panama City
Panama Canal
GULF OF PANAMA
PANAMA
San Miguel Bay
PACIFIC OCEAN
SCALE
0 25 50 Miles
0 25 50 Kilometers
Projection: Azimuthal Equal Area
8°N
N
S
W
E

Panama City and **Colón** offer job opportunities for many Kuna, a well-educated people. What is the distance from the Kuna reservation to Colón?

The Rain Forests

The wettest and hottest places on Earth, tropical rain forests teem with plant and animal life. Of the 30 million or so species of plants and animals living on Earth, more than half live in rain forests. Rain forest trees and plants use carbon dioxide, helping to keep the air clean. Rain forest vegetation also holds rain-drenched soil in place, preventing it from flowing into valleys or clogging rivers. In addition, plants in rain forests provide one-fourth of all medicines found in drugstores.

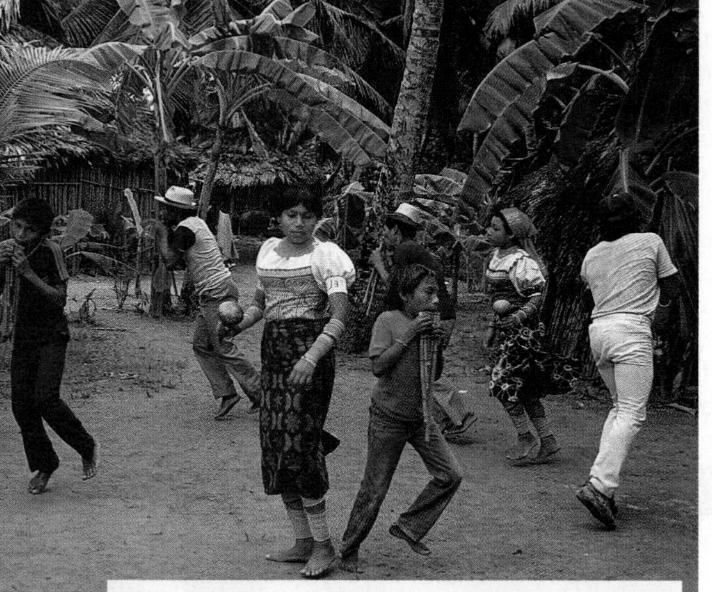

These Kuna are performing traditional dance and music in their village on an island in the **Gulf of San Blas**. Most Kuna villages depend on the fish in the coastal waters around the islands.

Today, the world's tropical rain forests are endangered. Thirty to fifty acres (12–20 ha) of rain forest disappear each minute, which means the world is losing some 10–12 million acres (4–5 million ha) of rain forest a year. The destruction of trees and other vegetation is accompanied by the loss of animals that depend on them for habitat and food.

Several environmental groups and nations around the world are working to stop the destruction of the rain forests. They are training people in developing countries to use farming methods that do not deplete the soil. They also are educating rain forest peoples about conservation. Environmental groups are also pressuring wealthy nations to decrease their demand for the valuable hardwoods of the rain forests. As the Kuna have shown, the goal of saving the rain forests is in reach but must involve all the world's peoples.

YOUR TURN

Many geographers wish to stop the destruction of the world's rain forests. How might you enlist the help of the people who live in, or earn a living from, the rain forests in this effort? **Plan** a conservation training workshop for a rain forest community. **Outline** the topics you will cover in your workshop, and provide a brief description of the information you will discuss for each topic.

Havana, Cuba, was established by Spanish colonizers in the early sixteenth century. Havana's location just across the Straits of Florida from the United States helped make the Cuban capital a popular tourist destination before the 1959 revolution.

Much of the best land is part of large plantations or estates, many of which are controlled by foreign firms. When people abandon farming, most travel to the islands' capital cities seeking jobs in manufacturing, government, or tourism.

Beautiful beaches and sunny weather have made tourism a rapidly growing industry. The social impacts of tourism are mixed, however. Many of the jobs in tourism pay low wages. Many resorts are hidden behind large gates and high fences, isolating them from most island residents.

Even though most of the islands are politically independent, they remain economically dependent on foreign countries. The islands export raw materials and import most of their manufactured goods. In this environment of poverty, inequality of wealth, and foreign dependence, political unrest is common. The countries of the Caribbean are discovering that they have many common issues. Regional cooperation is a hope for the future.

Cuba Cuba, the largest island of the Greater Antilles, lies just 90 miles (145 km) off the southern coast of Florida. Mountains and hills cover about one-fourth of the island. The remainder of the land consists of gently rolling hills and fertile valleys.

The United States was once Cuba's most important market for exports. The Cuban economy, however, was controlled by a few wealthy people and suffered from corruption. A revolution in 1959 brought Fidel Castro to power. By 1961, Castro's government had taken over most privately owned property. Cuban society remains organized along Communist lines and has a command economy. Most farmlands have been put into **cooperatives** and large state sugarcane plantations. A cooperative is an organization owned by and operated for the mutual benefit of its members.

Much of the farmland is used to grow sugarcane, Cuba's most important crop. Cuba is one of the world's largest sugar exporters. Its success as a sugar producer is the result of several factors. First, most of Cuba's soil is fertile and holds moisture well. Second, the climate is wet for the six-month growing season but dry during the winter harvest. Third, because the agricultural land has been leveled, machinery operates easily on the sugarcane plantations. Since the 1959 revolution, the economy has depended increasingly on sugarcane. As a result, lands that once produced food now grow sugarcane. Food production cannot meet the island's needs.

Most of Cuba's industry is based on the processing of agricultural products. The country is famous for its cigars. Cuba also produces rice, coffee, vegetables, and tropical fruits.

Havana, the capital of Cuba, is the largest city in the Caribbean. Once the major port of the Spanish colonies, Havana is the country's most important trade, transportation, and manufacturing center. The city's more than 450-year history is evident in the old forts and churches and in the narrow streets. Havana also is a city of wide avenues, large parks, and modern buildings.

After Castro's revolution, Cuba became dependent on trade and aid from the Soviet Union and Eastern Europe. When the Soviet Union collapsed in the early 1990s, aid to Cuba decreased dramatically. Since then, the Cuban economy has deteriorated. Large numbers of people opposed to Castro's government have fled Cuba. Many have made their new homes in the United States.

Haiti Haiti occupies the mountainous western third of the island of Hispaniola. It is the poorest and most densely populated country in the Americas. Haiti was the major French sugar-supplying colony until a slave uprising in 1791 brought independence. Since then, the country has experienced a series of corrupt governments and financial disorder. The United States sent troops to Haiti in 1994 to restore to office a democratically elected president.

Most Haitians farm small plots, which produce yams and **plantains**, a type of banana used in cooking. Coffee, cotton, and sugarcane provide some export income.

The official languages of Haiti are French and Creole. Creole is mostly French, with some Spanish, English, and African words. While Roman Catholicism is the recognized religion, **voodoo** is important throughout Haitian culture. Followers of voodoo believe that spirits of good and evil play an important part in daily life.

The capital city of Haiti, Port-au-Prince (pohrt-oh-PRINS), is the country's manufacturing center. Industries often depend on hand labor and simple machines. Because of few economic opportunities and many political problems, a large number of Haitians have tried to flee their country. Many have become **refugees** in the United States. A refugee is someone who flees his or her own country to another, often for economic or political reasons.

Dominican Republic The Dominican Republic occupies the eastern part of Hispaniola. The country was Spain's earliest colony, and the traditions of Spanish culture remain strong. Santo Domingo, the capital, was the first permanent European settlement in the Western Hemisphere.

The Dominican Republic is a land of steep mountains and fertile valleys, and its forests contain valuable tropical woods. Sugarcane and coffee are the key commercial crops, while tourism has become an important industry. The country's economy, as well as its education, health care, and housing, is more developed than that of nearby Haiti.

Puerto Rico Puerto Rico is the easternmost island of the Greater Antilles. Most of the island is mountainous, with narrow coastal plains. Once a Spanish colony, Puerto Rico is a **commonwealth**, a self-governing political unit, associated with the United States. Puerto Ricans are U.S. citizens, although they cannot vote in presidential elections.

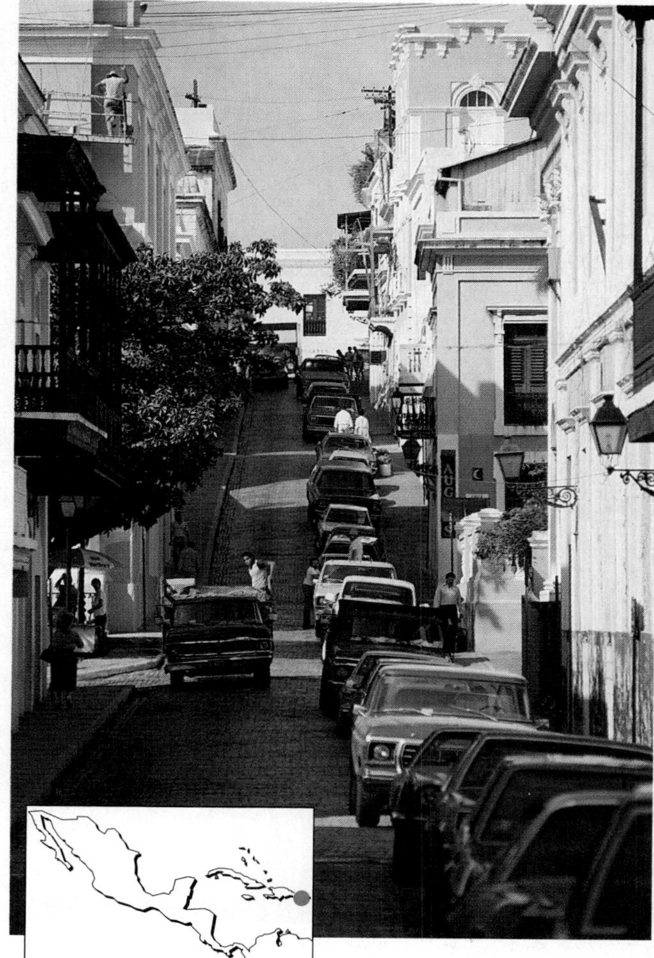

Spanish-style architecture lines the narrow streets of the older parts of **San Juan, Puerto Rico**.

Sugarcane is the commonwealth's most valuable crop, though coffee, tropical fruits, and tobacco also are grown. However, the most important feature of Puerto Rico's economy is its industrialization. Puerto Rico's low wages, inexpensive manufacturing, and easy access to U.S. markets have attracted U.S. firms.

Unemployment is high and incomes are low in Puerto Rico when compared to the United States. Yet, Puerto Rico's economy is much more developed than that of any other island in the Antilles. Puerto Ricans continue to debate whether the island should become a U.S. state or an independent country, or remain a commonwealth.

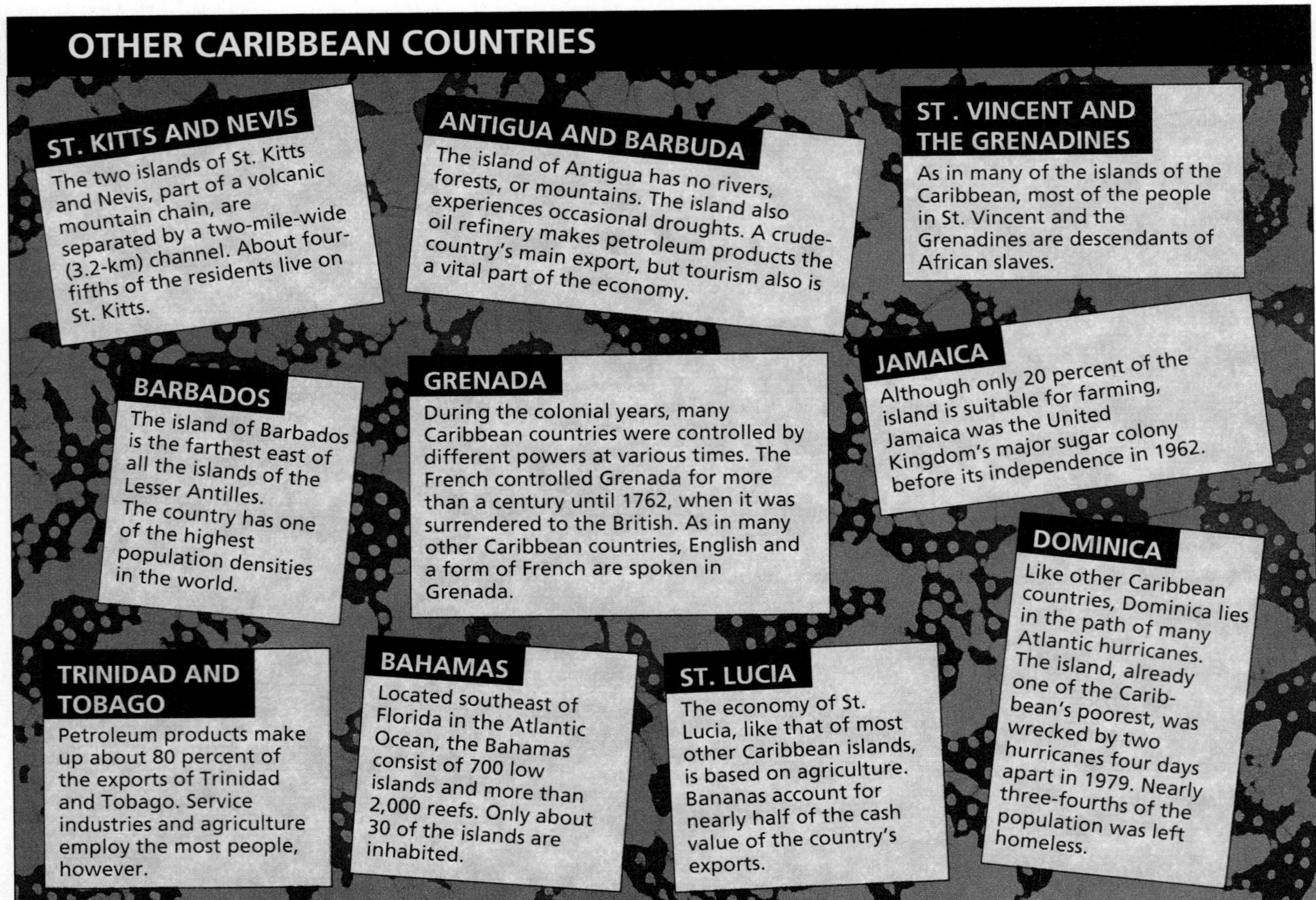
OTHER CARIBBEAN COUNTRIES
ST. KITTS AND NEVIS
The two islands of St. Kitts and Nevis, part of a volcanic mountain chain, are separated by a two-mile-wide (3.2-km) channel. About four-fifths of the residents live on St. Kitts.
ANTIGUA AND BARBUDA
The island of Antigua has no rivers, forests, or mountains. The island also experiences occasional droughts. A crude-oil refinery makes petroleum products the country's main export, but tourism also is a vital part of the economy.
ST . VINCENT AND THE GRENADINES
As in many of the islands of the Caribbean, most of the people in St. Vincent and the Grenadines are descendants of African slaves.
BARBADOS
The island of Barbados is the farthest east of all the islands of the Lesser Antilles. The country has one of the highest population densities in the world.
GRENADA
During the colonial years, many Caribbean countries were controlled by different powers at various times. The French controlled Grenada for more than a century until 1762, when it was surrendered to the British. As in many other Caribbean countries, English and a form of French are spoken in Grenada.
JAMAICA
Although only 20 percent of the island is suitable for farming, Jamaica was the United Kingdom's major sugar colony before its independence in 1962.
DOMINICA
Like other Caribbean countries, Dominica lies in the path of many Atlantic hurricanes. The island, already one of the Caribbean's poorest, was wrecked by two hurricanes four days apart in 1979. Nearly three-fourths of the population was left homeless.
TRINIDAD AND TOBAGO
Petroleum products make up about 80 percent of the exports of Trinidad and Tobago. Service industries and agriculture employ the most people, however.
BAHAMAS
Located southeast of Florida in the Atlantic Ocean, the Bahamas consist of 700 low islands and more than 2,000 reefs. Only about 30 of the islands are inhabited.
ST. LUCIA
The economy of St. Lucia, like that of most other Caribbean islands, is based on agriculture. Bananas account for nearly half of the cash value of the country's exports.

Virgin Islands The Virgin Islands, which lie just east of Puerto Rico, total only 195 square miles (505 sq. km)—about the same amount of area covered by the U.S. city of New Orleans. The Virgin Islands are politically divided between the United States and the United Kingdom. The United States territory includes the three largest islands: St. Thomas, St. John, and St. Croix (CROI). Once Denmark's sugar colony, the islands were purchased by the United States in 1917 as a base to protect the Panama Canal. Today, the United States Virgin Islands include some of the Caribbean's most popular tourist centers. Farmlands are disappearing as vacation homes cover more land. Remaining agricultural lands produce fruits, vegetables, and meat for local markets.

The British Virgin Islands include Tortola, Anegada, Virgin Gorda, Jost Van Dyke, and 32 smaller islands. The area of these islands is just 59 square miles (153 sq. km). As with the U.S. Virgin Islands, tourism is a major part of the economy.

Other Islands Between the Virgin Islands and Trinidad are more than a dozen important, though small, island territories. French possessions in the Lesser Antilles include Martinique (mahrt-uhn-EEK) and Guadeloupe (GWAHD-uhl-oop). The Dutch territories include three small tourist islands in the northeast Caribbean and the "ABC" islands—Aruba, Bonaire, and Curaçao (kyur-uh-SOE)—off the coast of Venezuela. Bermuda, a favorite tourist spot, is a prosperous possession of the United Kingdom.

Many other islands lie in the Caribbean. To learn some "quick facts" about the other independent islands, study the chart on this page.

2 SECTION REVIEW

1. How do the Greater Antilles differ from the Lesser Antilles? What economic activities are most important to the islands?
2. What is unique about Antigua's physical geography?
3. **Critical Thinking** **Compare and contrast** political and economic characteristics of Cuba and Puerto Rico.

CHAPTER 19 CHECK

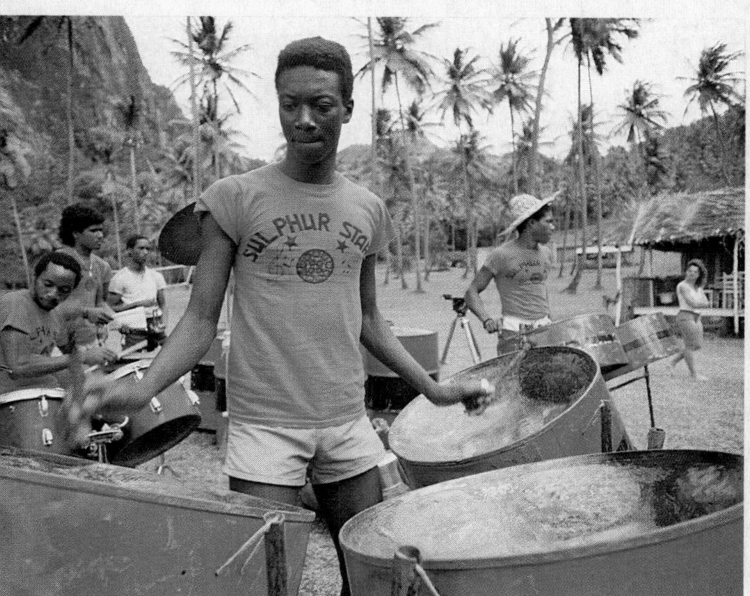

A steel-drum band in **St. Lucia**

Reviewing the Main Ideas

1. The Central American and Caribbean region includes 20 small countries and more than 12 territories. Physical and cultural characteristics have divided the region.
2. The countries of Central America have coastal plains with humid-tropical climates, inland valleys, and mountains with highland climates. Coffee, sugarcane, cotton, and bananas are major exports. Subsistence farming also is common. Population growth and the need for land reform have created economic and political challenges.
3. The Caribbean islands include the Greater Antilles, the Lesser Antilles, and the Bahamas. The climate and beauty of these islands attract many tourists. Nearly every island in the Caribbean shows the effects of colonialism and slavery. Dependence on foreign countries is a major issue, and political unrest is common.

Building a Vocabulary

1. What is cardamom? What are plantains?
2. What term refers to a person who flees a country for economic or political reasons?
3. Define *commonwealth*. Which Caribbean island is a commonwealth?
4. What are calypso, reggae, and merengue? Where did each originate?

Recalling and Reviewing

1. Compare and contrast the major economic activities of the Central American countries with those of the Caribbean islands. What factors account for the differences?
2. What political characteristics are shared by most Central American and Caribbean nations?

Thinking Critically

1. The United States purchased the Virgin Islands in 1917 as a base to protect the Panama Canal. Do you think the United States' interests in protecting the canal were primarily economic, or do you think they were primarily military?
2. Many of the countries of the Caribbean are heavily dependent upon foreign aid and imports. How might these countries become more self-sufficient?

Learning About Your Local Geography

Individual Project

You have read about calypso, reggae, and merengue. What types of music and dance are popular in your community? Interview community members about their favorite forms of music and dance. Is your community "known" for a particular form of entertainment, as Trinidad, Jamaica, and the Dominican Republic are? Design a poster advertising music and dance in your community.

Using the Five Themes of Geography

Write a paragraph about different forms of **human-environment interaction** that occur when tourists visit **regions** composed of tropical islands, such as the Caribbean. What attracts tourists to tropical island regions? How do tourists both adapt to and change the islands' environments?

CHAPTER 20

Caribbean South America

GEOGRAPHY DICTIONARY

tierra caliente
tierra templada
tierra fría
páramo
tierra helada
guerrilla
slash-and-burn farming

"You think we're going to make it?" I asked the conductor.

"You will be in Bogotá tonight," he said. . . .

Soon after, the mountains appeared, the cordillera of the Andean chain; and with them the brown Magdalena River, on which men paddled dugout canoes or fished from the shore with contraptions that looked like butterfly nets. The mountains were at first scattered buttes and solitary peaks, and some were like citadels, squarish with fortresslike buildings planted around the summits."

Paul Theroux

Gold artifact from an early culture in present-day **Colombia**

The northern rim of South America contains Colombia, Venezuela, Guyana (gie-AN-uh), Suriname, and French Guiana. In the west, the Andes thrust skyward over 17,000 feet (5,182 m). In the east, the Guiana Highlands still reach above 7,000 feet (2,134 m) even though scientists believe the highlands have been eroding for millions of years. Between these uplands are the vast plains of the Orinoco River basin. Most of the region's cities and agricultural areas are found near the Caribbean shore and in mountain valleys.

Fertile valleys in the Andes and coastal ranches made Colombia and Venezuela key parts of Spain's colonial empire. The other former territories were colonized by European powers who arrived during the sugar boom of the 1700s. Throughout the region's history, rugged land and dense forests often have separated peoples and cultures here.

Andes Mountains of western **Colombia**

CARIBBEAN SOUTH AMERICA: PHYSICAL–POLITICAL

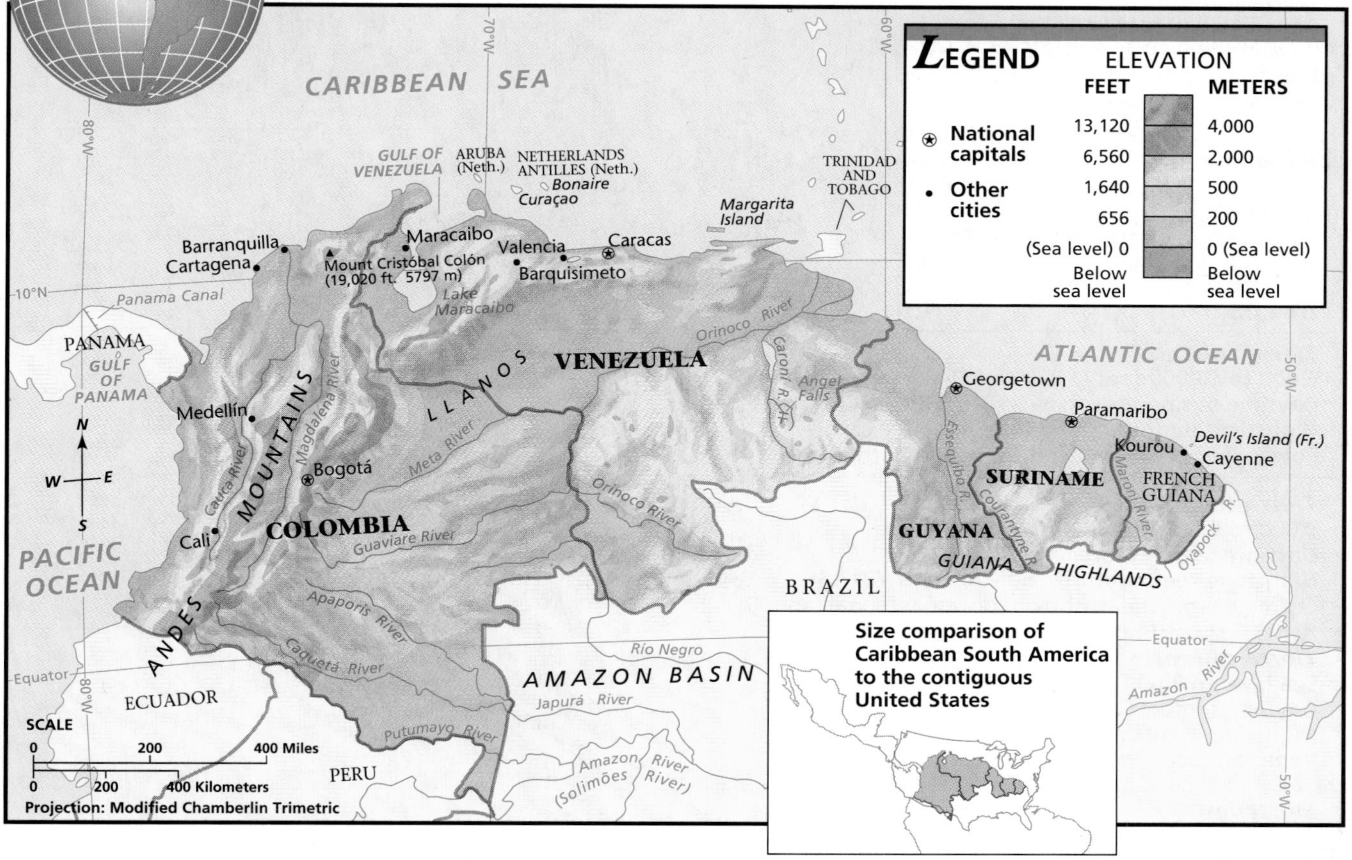

LEGEND

- National capitals
- Other cities

ELEVATION FEET	METERS
13,120	4,000
6,560	2,000
1,640	500
656	200
(Sea level) 0	0 (Sea level)
Below sea level	Below sea level

SCALE 0 200 400 Miles; 0 200 400 Kilometers

Projection: Modified Chamberlin Trimetric

Size comparison of Caribbean South America to the contiguous United States

COLOMBIA

FOCUS

- *What are Colombia's landform regions? What are the five elevation zones of the Andes?*
- *What economic activities are important to Colombia? What are the country's human and political characteristics?*

Physical Geography Colombia is the only South American country with coastlines on both the Caribbean Sea and the Pacific Ocean. (See the map above.) The Andes, however, prevent easy travel between the two coasts. With the opening of the Panama Canal, road construction, and the development of air transportation, communications have improved.

Colombia has two major landform regions: the Andes Mountains and valleys in the west, and the plains of the llanos and the Amazon Basin to the east. In the north, the plains are covered with savannas, but southward, the vegetation thickens to become dense tropical forests. Colombia's hot and humid coastal lowlands extend only a short distance inland.

LOCATION **Andes Environments** The highland climates of the Andes slopes are divided by elevation into five zones. (See the illustration on page 228.) The ***tierra caliente***, or "hot country," refers to the hot and humid lower elevations between sea level and 3,000 feet (914 m). Daytime temperatures here range from 85°F to 90°F (29°–32°C). During the nights, temperatures usually range from 70°F to 75°F (21°–24°C). There is little difference between summer and winter temperatures. The *tierra caliente* produces tropical products, including bananas and cacao.

Elevation Zones in the Andes

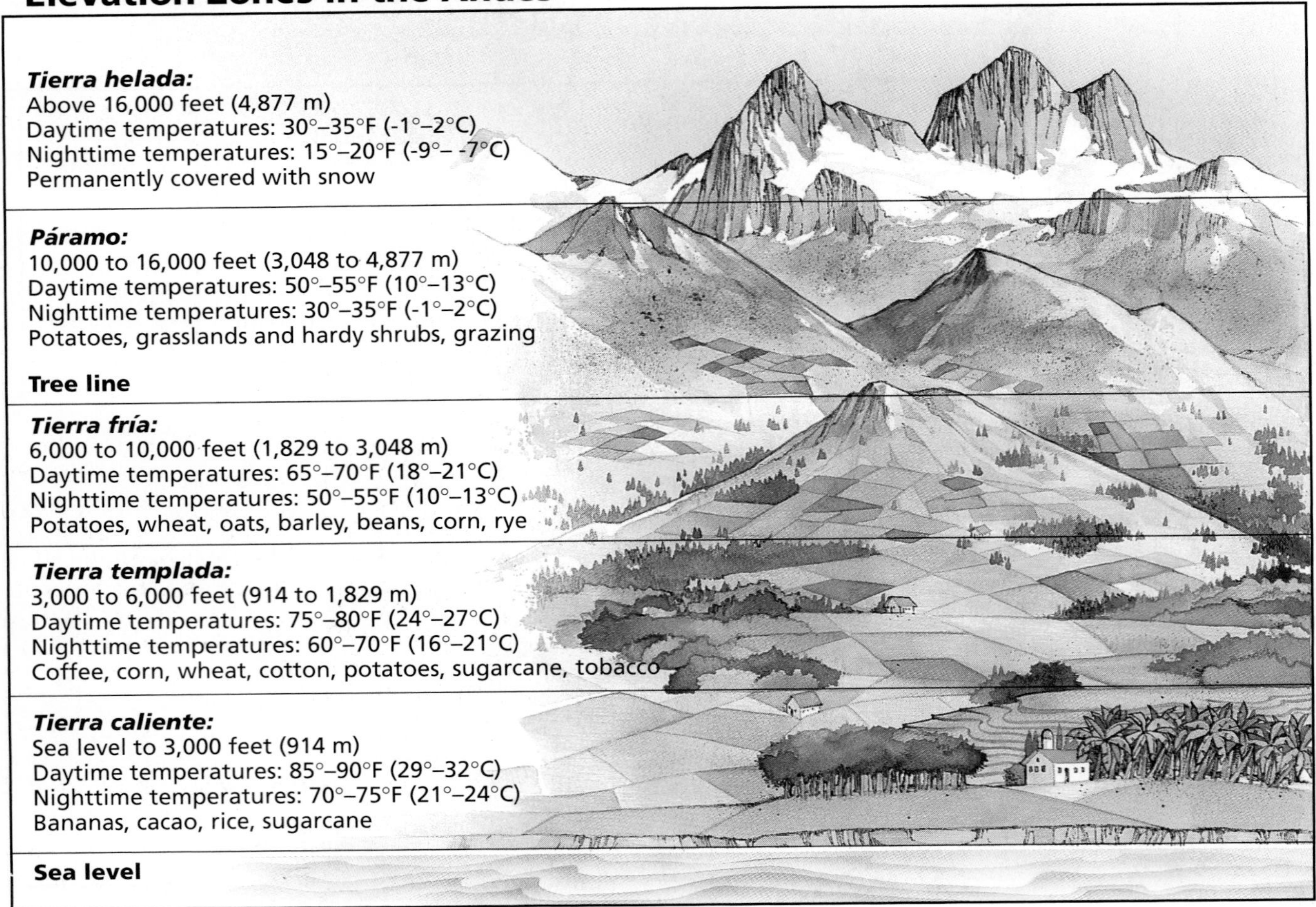

Tierra helada:
Above 16,000 feet (4,877 m)
Daytime temperatures: 30°–35°F (-1°–2°C)
Nighttime temperatures: 15°–20°F (-9°– -7°C)
Permanently covered with snow
Páramo:
10,000 to 16,000 feet (3,048 to 4,877 m)
Daytime temperatures: 50°–55°F (10°–13°C)
Nighttime temperatures: 30°–35°F (-1°–2°C)
Potatoes, grasslands and hardy shrubs, grazing
Tree line
Tierra fría:
6,000 to 10,000 feet (1,829 to 3,048 m)
Daytime temperatures: 65°–70°F (18°–21°C)
Nighttime temperatures: 50°–55°F (10°–13°C)
Potatoes, wheat, oats, barley, beans, corn, rye
Tierra templada:
3,000 to 6,000 feet (914 to 1,829 m)
Daytime temperatures: 75°–80°F (24°–27°C)
Nighttime temperatures: 60°–70°F (16°–21°C)
Coffee, corn, wheat, cotton, potatoes, sugarcane, tobacco
Tierra caliente:
Sea level to 3,000 feet (914 m)
Daytime temperatures: 85°–90°F (29°–32°C)
Nighttime temperatures: 70°–75°F (21°–24°C)
Bananas, cacao, rice, sugarcane
Sea level

The names for Andean elevation zones vary somewhat from country to country. The names in this illustration are commonly used in Colombia. In which elevation zone is growing crops not possible?

Farther up the mountains, the air becomes cooler, and moist climates with mountain forests are typical. This zone of pleasant climates is called ***tierra templada***, or "temperate country." Here, temperatures during the day range from 75°F to 80°F (24°–27°C). Nighttime temperatures range from 60°F to 70°F (16°–21°C). Many people live in this zone, where both tropical and middle-latitude types of crops are grown.

The next higher zone is the ***tierra fría***, or "cold country." The *tierra fría* has a natural vegetation of cool forests and grasslands. It extends from about 6,000 feet (1,829 m) to 10,000 feet (3,048 m). Daytime temperatures here range from 65°F to 70°F (18°–21°C). During the night, the temperature range is from 50°F to 55°F (10°–13°C). The *tierra fría* is important for growing potatoes and wheat.

Above the tree line is a zone called the ***páramo***, which extends from about 10,000 feet (3,048 m) to as high as 16,000 feet (4,877 m). Temperatures range from 50°F to 55°F (10°–13°C) during the day and from 30°F to 35°F (-1°–2°C) during the night. Grasslands and hardy shrubs are the usual vegetation. Llamas, alpacas, and sheep graze on these lands, while potatoes grow in the lower elevations of the *páramo*. Frost may occur on any night of the year.

The ***tierra helada***, or "frozen country," is the zone of highest elevation. It is permanently covered with snow. Temperatures during the day range from 30°F to 35°F (-1°–2°C) and from 15°F to 20°F (-9° to -7°C) during the night.

Economic Geography Colombian coffee, produced mostly on family farms, is world famous.

The main coffee-producing area is found along the Andes ridges between the Magdalena and Cauca rivers. The country is second only to Brazil in coffee production. One of Colombia's newest and most profitable industries is flower farming. Colombia exports large amounts of cut flowers. Other commercial crops grown in Colombia include rice, maize (corn), sugarcane, tobacco, bananas, and cotton. Cattle grazing is an important occupation in the llanos.

The mountains of Colombia are rich in minerals. Ninety percent or more of the world's emeralds come from Colombia. Colombia's iron-ore deposits, however, are of more value to the country's economy. Oil is found in the eastern part of Colombia along the Venezuelan border. Recent discoveries have made oil a larger export product than coffee.

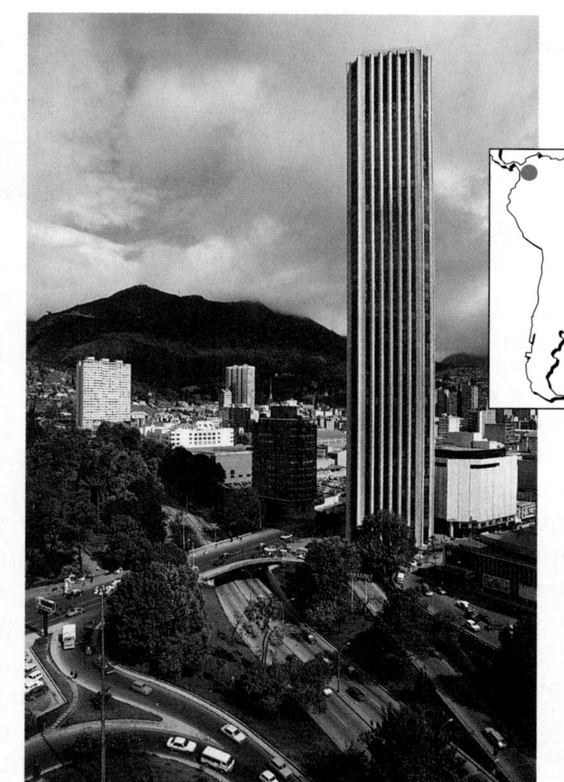

Modern high rises and highways of the banking and international district stand out in **Bogotá**. The capital and largest city of **Colombia** sits on a plateau of the Andes Mountains that is 8,563 feet (2,610 m) above sea level.

As in other Middle and South American countries, urban poverty, rapid population growth, and land reform are pressing problems. A good trade location, plentiful natural resources, and improved interior transportation, however, give Colombia the potential to prosper.

Human Geography Colombia's capital, Bogotá, is a modern city located on a plateau in the eastern Andes. Cartagena on the Caribbean coast has preserved colonial architecture among its modern buildings.

Colombia has a history of civil war and corrupt governments. Much of Colombia is ruled more by violence and intimidation than by laws. **Guerrilla** groups often have frightened people away from elections and have controlled large regions of the country. A guerrilla is one who fights as a member of an armed band and takes part in irregular warfare, such as harassment and raids.

Much of the country's recent violence is related to a large illegal drug trade. Drugs are smuggled out of the country in great quantities. Smugglers have used their illegal profits to support private armies and corrupt judges and politicians. Income from the illegal drug trade accounts for a significant portion of Colombia's true GNP.

FACTS IN BRIEF Caribbean South America

COUNTRY POPULATION (1994)	LIFE EXPECTANCY (1994)	LITERACY RATE (1990)	PER CAPITA GDP (1993)
COLOMBIA 35,577,556	69, male 75, female	87%	$5,500
GUYANA 729,425	62, male 68, female	95%	$1,900
SURINAME 422,840	67, male 72, female	95%	$2,800
VENEZUELA 20,562,405	70, male 76, female	88%	$8,000
UNITED STATES 260,713,585	73, male 79, female	97% (1991)	$24,700

Source: *The World Factbook 1994,* Central Intelligence Agency

What is the most populous country of Caribbean South America?

SECTION REVIEW

1. Describe each elevation zone of the Andes.
2. What are Colombia's most important exports?
3. **Critical Thinking** **Discuss** how Colombia's physical geography relates to the country's political unrest.

VENEZUELA

FOCUS

- *What landforms, climates, and vegetation are found in Venezuela?*
- *What are Venezuela's economic, cultural, and social characteristics?*

LOCATION **Landforms** Venezuela is located on South America's northern Caribbean coast. The country is divided into four landform regions: (1) the northern mountains and valleys of the Andes; (2) a narrow tropical-lowland coast that is surprisingly arid in parts; (3) the llanos, the savanna plains between the Andes and the Orinoco River; and (4) the Guiana Highlands in the southeast. (See the map on page 227.)

The Orinoco River and its tributaries drain the llanos as well as the Guiana Highlands. When streams plunge over the steep cliffs in the highlands, they create many high waterfalls. Angel Falls, with a drop of 3,212 feet (979 m), is the world's highest. In contrast, parts of the llanos are so flat that the water from summer floods stands on the plains for months.

Climate and Vegetation Venezuelans experience climates that are typical of the lower latitudes. The temperatures are warm to hot year-round. A dry winter season gives way to a rainy summer season. The highlands of Venezuela, however, have many climate variations. Mountains along the north coast may receive rain throughout the year as the trade winds blow moisture inland from the Caribbean.

The coastal region has humid air, but in some locations there is very little rain. Scrub savannas with cacti are common. Thick forests were once widespread in the Andean valley, but 400 years of clearing the land has opened many areas to grasses. Farther south are the llanos, one of the world's largest short-tree and grass savannas. Huge expanses of grassland are flooded during rainy seasons. Like in nearby Brazil, the plains in the far south of Venezuela are covered with humid-tropical forests. The vegetation of the Guiana Highlands includes tall-tree savannas as well as dense forests.

Economic Geography Venezuela's mineral wealth provides the country's chief source of income. Until 1920, Venezuela was a poor country with few opportunities. Then, the vast oil deposits surrounding Lake Maracaibo (mar-uh-KIE-boh) were developed. Today, the country is a leading oil exporter. Although Venezuela has several oil refineries, much of the production is refined outside the country on offshore islands such as Curaçao and Trinidad. A large amount is also shipped to refineries in the United States.

The Venezuelan government has used oil profits to build schools, improve living conditions, and expand medical care. Venezuela's economy has become dependent on oil revenues, however. The country suffers when oil prices drop. For this reason, the government of Venezuela is encouraging new industries that process food and minerals.

The Guiana Highlands are rich in minerals, including large deposits of iron ore. Large dams have been built on the Orinoco River. Steel mills using oil and hydroelectric power are producing a steel surplus.

As for agriculture, small family farms and large commercial farms are spread throughout the

Ranchers herd cattle on the rich grazing lands of **Venezuela**.

The Teresa Carrena Theater is one of the many modern and beautiful structures in **Caracas**, the capital of **Venezuela**. Money from oil sales has helped finance many public works in the country.

northern part of the country. The main crops are sugarcane and coffee. Maize (corn) and wheat are grown on the cooler mountain slopes. The llanos is a land of large cattle ranches. Until recently, few people have lived in the region. With the development of irrigation projects, more farmers are moving there. Except for large settlements near mines, the Guiana Highlands are nearly empty of people. Small communities of indigenous peoples practice traditional **slash-and-burn farming**, a type of agriculture in which forests are cut and burned to clear land for planting.

Human Geography Most of the people of Venezuela live along the Caribbean coast and in the valleys of the nearby mountains. These areas were the focus of early settlement. Most Venezuelans are mestizos. Their national pride is expressed in a popular small statue, which displays a European man, an African man, and a Native American woman standing shoulder to shoulder.

Caracas (kuh-RAK-uhs) is the capital of Venezuela and the center of Venezuelan culture. It has a new subway system and a ring of modern expressways. There are modern office buildings and many job opportunities.

As in other Middle and South American cities, however, shantytowns and slums encircle Caracas. Rural poverty is widespread. Oil wealth and new lands to develop in the south suggest that Venezuela can continue as an economic and political leader in the region.

SECTION REVIEW

1. What are Venezuela's four landform regions? Describe the climate of each region.
2. What resources are found in Venezuela?
3. **Critical Thinking** **Explain** how slash-and-burn farming might affect the environment.

THE GUIANAS

FOCUS

- *What are the physical and economic features of the Guianas?*
- *What are the political and cultural characteristics of the Guianas?*

Guyana Guyana and its two neighbors on South America's northeast coast have similar humid-tropical environments. Nearly all agricultural lands are found on a narrow coastal plain. Guyana's main crops are rice, usually grown for food, and sugarcane, usually exported as sugar. Guyana's chief mineral resource is bauxite. Hydroelectric power enables the country to process its bauxite into aluminum for export.

Guyana is the former British Guiana. Since independence in 1966, the country's political situation has been unstable, and the government has had close relations with Cuba and other Communist countries. Guyana's economy has declined as government-run industries and businesses have collapsed.

More than half of Guyana's population is of South Asian descent. Most of these people farm small plots and run small businesses. Most of the remaining population is made up of Africans and mulattos who are descendants of African slaves. This sector of the population controls most of the large businesses and holds most of the government positions. Economic differences between the two ethnic groups have created tension.

Because Venezuela claims more than half of Guyana's territory, Guyana faces an uncertain future. In Venezuela, this area is shown on maps as "occupied territory." If Guyana, already one of the poorest countries in the world, were to lose this territory, its future prosperity would be threatened.

Suriname Suriname is a former Dutch colony that gained its independence in 1975. As in Guyana, Suriname's coastal area produces sugarcane, rice, and tropical fruits. Lumber from the interior forests is exported in the form of wood and wood products. Suriname's economy depends on the country's bauxite and aluminum exports.

Suriname's population, which totals less than 500,000, is an ethnic mixture. There are Africans, mulattos, Chinese, Indonesians, and South Asians. Muslim, Hindu, Roman Catholic, and Protestant houses of worship line the streets of Paramaribo (par-uh-MAR-uh-boe), the country's capital.

Violent ethnic political disputes are common in Suriname. As various political parties representing different ethnic groups have come to power, other ethnic groups have worked behind the scenes to support guerrilla groups.

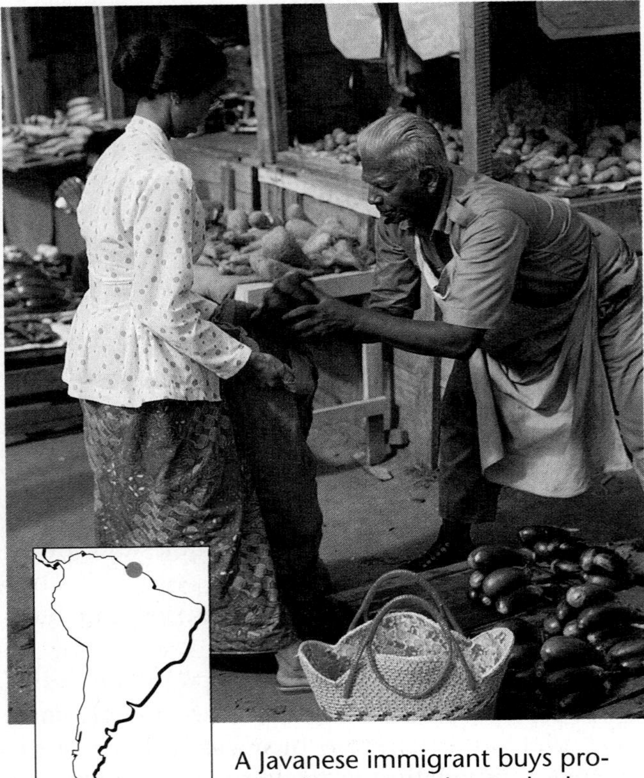

A Javanese immigrant buys produce from a merchant who has immigrated to **Suriname** from India. Just over half of Suriname's population is made up of ethnic groups from South Asia and island Southeast Asia.

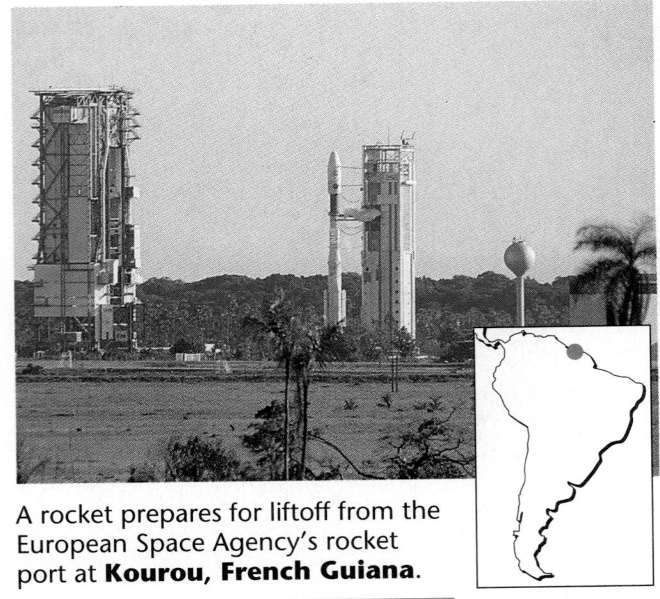

A rocket prepares for liftoff from the European Space Agency's rocket port at **Kourou, French Guiana**.

French Guiana French Guiana is the largest of France's overseas territories. Its status as a French territory is much the same as that of a state in the United States. French Guiana sends three representatives to the National Assembly in Paris, France.

The population of French Guiana is only about 139,000, and large parts of the territory are uninhabited. Cacao, rice, sugarcane, and bananas are grown on the coastal plain. Chief exports are gold, timber, and shrimp. France used the famous offshore Devil's Island as a prison. Today, Kourou on the coast is the site of the European Space Agency's rocket port.

3 SECTION REVIEW

1. What physical features are shared by Guyana, Suriname, and French Guiana? What resources do these countries have?
2. How does French Guiana's political situation differ from that of Guyana and Suriname?
3. **Critical Thinking** **Agree or disagree** with the following statement: "The political conflicts within Guyana are as much related to economic factors as to ethnic factors." Support your answer.

CHAPTER 20 CHECK

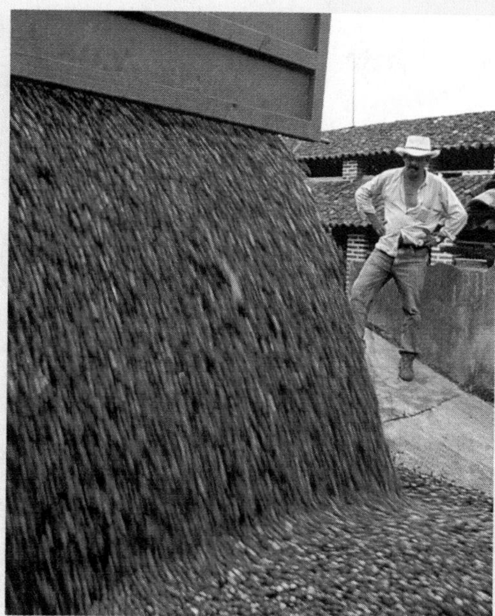

Unloading coffee berries in **Colombia**

Reviewing the Main Ideas

1. The countries of Colombia, Venezuela, Guyana, Suriname, and French Guiana make up the northern rim of South America.
2. Unlike other South American countries, Colombia has coastlines on both the Caribbean Sea and the Pacific Ocean. Coffee is Colombia's leading crop and a major export.
3. Venezuela's chief source of income is its mineral wealth. It is one of the world's leading oil-exporting countries and is an economic and political leader in South America.
4. Guyana, Suriname, and French Guiana are situated on South America's northeast coast. Since gaining independence from Britain in 1966, Guyana has endured political instability. Suriname, a former Dutch colony, depends upon bauxite and aluminum exports. French Guiana is the largest of France's overseas territories.

Building a Vocabulary

1. What is the *páramo*?
2. Define *guerrilla*. In which countries discussed in this chapter do guerrillas play a major role?
3. What is the term used to describe the type of agriculture in which forests are cut and burned to clear land for planting?

Recalling and Reviewing

1. Name three major cities in Colombia and Venezuela.
2. What political relationship exists between Venezuela and Guyana?
3. List five ethnic groups that live in the countries discussed in this chapter. Which country's people are mostly mestizos?

Thinking Critically

1. If French Guiana became an independent country, do you think it would experience increased political violence? Why or why not? Do you think independence is worth the troubles it sometimes brings?
2. What might happen to Venezuela's economy if scientists discover an inexpensive and convenient way to produce large quantities of solar energy?

Learning About Your Local Geography

Cooperative Project

Your local area is probably too small to have different climate zones like the Andes do, but how else might you divide your area into zones? Each member of your group should select one characteristic upon which to base zoning and then should sketch a map of the area showing the different zones. As a group, compare all the maps. Do the maps look similar in terms of the zones each one shows? Or do the locations of zones on each map vary widely?

Using the Five Themes of Geography

Make two illustrations linking **location** and **movement** in Colombia. One of your illustrations should show how the location of the country helps the movement of goods in and out of the country. In the other illustration, show how the location of the Andes hinders movement within Colombia. Write captions for your illustrations.

CHAPTER 21

Atlantic South America

GEOGRAPHY DICTIONARY

buffer state
soil exhaustion
favela
growth pole
tannin
landlocked

Carnival mask from **Brazil**

"Despite the immense size of Brazil and the diverse ethnic backgrounds of her peoples, there is one festive occasion that unites the entire country. Each year, in the final days before Lent, Carnival erupts in an explosion of dance, music, and color. From Pôrto Alegre in the south to Macapa . . . to the north, all Brazilians join together to share in the magic and mystery of Carnival."

Robert Fried

If the world's continents were pieces of a puzzle, the east coast of South America and the west coast of Africa would fit together almost perfectly. This realization was the first major clue in the development of the theory of plate tectonics. As South America drifted westward from Africa, erosion gradually wore down the eastern side of the South American continent. As a result, the landforms of this region are mostly wide plains.

Brazil and Argentina, the two largest countries of eastern, or Atlantic, South America, are among the most prosperous of the world's developing countries. Sandwiched between Brazil and Argentina are Uruguay and Paraguay, two **buffer states.** A buffer state is a small country that separates two larger competing countries. Brazil and Argentina regularly assert influence over these small countries, and Uruguay and Paraguay generally are eager to accept favors from their giant neighbors.

Carnival celebration in **Rio de Janeiro, Brazil**

ATLANTIC SOUTH AMERICA: PHYSICAL–POLITICAL

Size comparison of Atlantic South America to the contiguous United States

BRAZIL

FOCUS

- *What are Brazil's physical characteristics? How has the country changed through history?*
- *Into what four economic and cultural regions can Brazil be divided? What challenges face the country?*

REGION Physical Geography Brazil is the largest country in South America and covers more than half of the continent's land area. It has borders with every other South American country except Ecuador and Chile. (See the map above.) Brazil is about 2,700 miles (4,344 km) from north to south and about the same distance from east to west. This huge country is made up mostly of lowland plains and level plateaus. Forests and savanna woodlands once covered the entire country.

Brazil can be divided into three major landform and climate regions. In the north is the Amazon region of rolling plains. This is the broad lowland of the Amazon River basin where a humid-tropical climate supports the world's largest tropical rain forests.

Along the east coast are the Brazilian Highlands. Most of these old, eroded mountains are less than 9,000 feet (2,743 m) high. In the north, the coastal region is covered mostly with rain forests and savannas. A little inland from the coast, the highlands are semiarid with dry grasslands. The southeastern highlands have humid-subtropical and marine-west-coast climates. These moist southeastern environments have become Brazil's major agricultural regions.

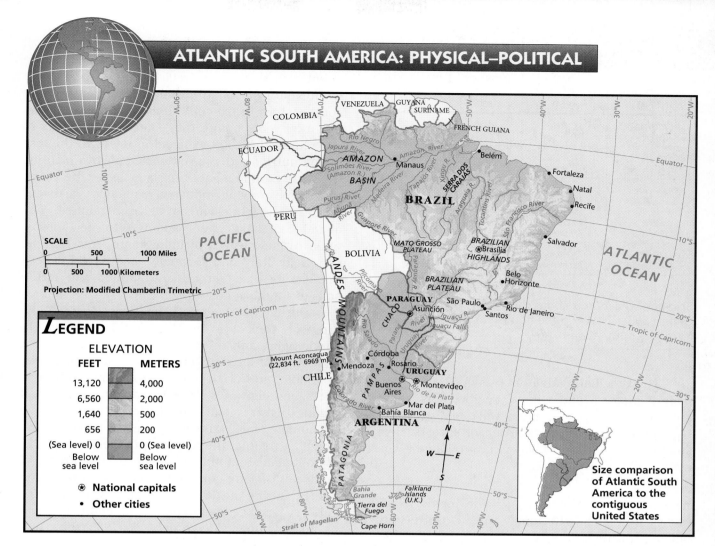

FACTS IN BRIEF Atlantic South America

COUNTRY POPULATION (1994)	LIFE EXPECTANCY (1994)	LITERACY RATE (1990)	PER CAPITA GDP (1993)
ARGENTINA 33,912,994	68, male 75, female	95%	$5,500
BRAZIL 158,739,257	57, male 67, female	81%	$5,000
PARAGUAY 5,213,772	72, male 75, female	90%	$3,000
URUGUAY 3,198,910	71, male 77, female	96%	$6,000
UNITED STATES 260,713,585	73, male 79, female	97% (1991)	$24,700

Source: *The World Factbook 1994,* Central Intelligence Agency

Brazil's population is about three-fifths that of the United States. How do the figures for per capita GDP for each country compare?

The Brazilian Plateau makes up Brazil's third major landform and climate region. It is a vast area of upland plains found west of the coastal highlands and south of the Amazon forests. Here, summer rains support savanna woodlands that extend all the way to Bolivia.

Historical Geography The name *Brazil* comes from the "pau-brasil" tree. This tree provides a purple-red dye that attracted the first Portuguese colonists in the early 1500s. Brazil became the Americas' first large sugar colony. Slaves, imported from Portugal's African territories, joined Brazil's indigenous peoples in the hard work of cutting cane. Eventually, sugar plantations spread along much of the Atlantic coast, steadily replacing the tropical forests. In the northeast, Salvador and Natal grew to become Brazil's first cities. Farther inland, cattle ranches of Portuguese settlers provided hides and dried beef for world markets.

Beginning in the 1690s, gold and precious gems were discovered inland from Rio de Janeiro. A mining boom brought adventurers from around the world. São Paulo, founded in the 1500s as a mission outpost, became the economic center of southeastern Brazil. In 1763, Rio de Janeiro became the colonial capital. By the time of independence in 1822, Brazil's population of about 4 million people was twice that of Portugal. Brazil remains the largest Portuguese-speaking country in the world.

During the late nineteenth century, Brazil was well known for the coffee produced in the country's southeastern region, and it became the world's largest coffee producer. But the farms of southeastern Brazil could only support coffee for about 30 years before they showed the effects of **soil exhaustion**. Soil exhaustion refers to the loss in soil nutrients that results from always planting the same crop in a particular area. Over the years, the coffee-growing industry expanded westward as the older coffee areas were abandoned. The abandoned lands were adapted to new crops, often by new waves of immigrants from Europe, especially Italy and Germany, and Japan. These immigrants also settled and farmed the fertile soils of southernmost Brazil.

During the twentieth century, Brazil's government took the lead in South America in encouraging and redirecting economic growth. A collection of huge government-owned companies assumed authority over expanding railways, ports, and highways. These companies developed mining districts and built factories. Though the economic plan was designed to benefit the Brazilian people, the already prosperous families of politicians and military leaders gained most of the wealth. Many companies have been managed poorly. During the 1970s and 1980s, when Brazil's imports far exceeded exports, the country began to amass a huge foreign debt. By 1993, the country's foreign debt was more than $100 billion. Today, much of Brazil's export earnings go to pay just the interest on these loans. This debt burden continues to weaken the economy.

The Northeast Modern Brazil can be divided into four cultural and economic regions: the Northeast, the Southeast, the Campo Cerrado, and the Amazon.

The Northeast refers to the coastal region of Brazil that includes the colonial cities of Fortaleza, Natal, Recife, and Salvador. Plantation agriculture made this region the center of the Portuguese Empire during the 1600s. Decline set in, however, and the capital was moved from Salvador to Rio de Janeiro in 1763. After centuries of intensive cultivation, this region's land is worn out. Many farmers move every few years, seeking more fertile farmland.

Today, the Northeast is Brazil's poorest region. Literacy and health standards are low. Attempts to move industry to this region have been unsuccessful. Droughts in the northeast interior have caused many farms to fail, and people have moved to the cities seeking jobs. The northeastern cities are filled with huge slum areas called ***favelas*** (fah-VEH-lahs). With so much poverty, social unrest has become common in the region.

The Southeast The Southeast region provides a sharp contrast to the Northeast. It includes the modern cities of São Paulo and Rio de Janeiro and extends southward to the border with Uruguay. More than 120 million people, nearly 80 percent of all Brazilians, live here. It is the nation's most prosperous region, with rich natural resources, productive agriculture, and most of the country's industries.

Though the Southeast is highly urban, it is Brazil's agricultural heartland. Hot, rainy summers and dry, mild winters make the Southeast the greatest coffee-growing region in the world. Brazil's economy is no longer dependent on this single crop, however. The Southeast region also is a leader in two of Brazil's newest agricultural exports—soybeans and orange juice. The Southeast also produces a variety of middle-latitude crops, such as cotton and maize (corn).

The modern mines of the Southeast cluster around Belo Horizonte, Brazil's iron and steel center. Huge iron-ore deposits have been developed here, and limited coal resources are found nearby. Oil deposits have been found off the coast. Belo Horizonte's steel industry has supported the development of heavy-machinery, automobile, and shipbuilding industries. Other industries in the region produce chemicals, cement, plastics, processed food, textiles, and computers. There are growing aircraft and defense industries here as well.

Rio de Janeiro is a major tourist area and an important shipping, manufacturing, and trade center. The Rio area has a population of more than 11 million. São Paulo is one of the world's largest cities. It is an awesome industrial area with a population above 20 million. São Paulo and its port, Santos, form the largest and wealthiest industrial area in South America. Although the Southeast is prosperous, many *favelas* are found here, too.

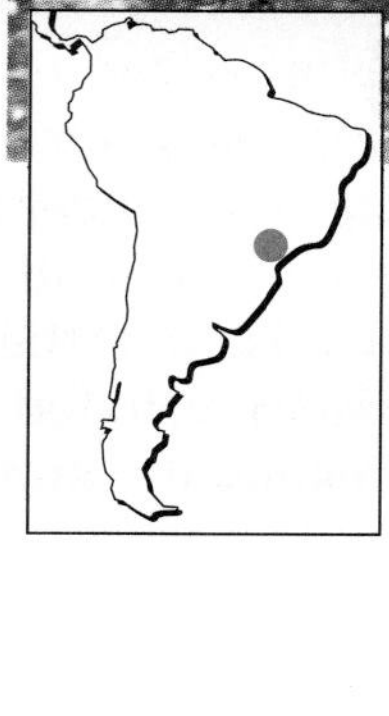

One of the world's largest urban areas, **São Paulo** is haunted by poverty despite its position as the wealthiest industrial area in South America.

Campo Cerrado The Campo Cerrado region is a frontier land of savannas and dry woodlands that extends east to west across central and southern Brazil. The region begins in the upper São Francisco River basin and continues westward past Brasília to the Mato Grosso Plateau. Until Brazil's capital was moved to Brasília in 1960, this region was nearly empty.

Brasília, located about 600 miles (965 km) from the coast, was built in a deliberate attempt to develop the interior region of the country. Brasília features modern architecture and highways. The city was designed to accommodate 500,000 people. Today, its population exceeds 3 million. As with Brazil's other cities, Brasília has growing *favelas*.

The soils of the Campo Cerrado were once thought infertile because of high levels of acid and aluminum. After agricultural scientists discovered how to make the soils fertile, however, commercial agriculture became possible. Wealthy investors from the Southeast have established large farms, and new towns are appearing. If farming spreads throughout the region, the Campo Cerrado could become the world's greatest agricultural region within the next century.

This Jivaro man in **Brazil** is playing a traditional flute. Economic development threatens the ways of life for many Native Americans in the Amazon Basin.

The Amazon The Amazon region covers the northern and western parts of Brazil. Its humid-tropical climate and dense rain forests have made settlement and transportation difficult. Native American villages are scattered through the forests. Some of the indigenous peoples living here have had little contact with the outside world. They grow sugarcane, yams, bananas, and other crops for their own use. A few still hunt with bows and arrows or blowguns and darts.

The Amazon and its tributaries form a network of waterways thousands of miles long. The Amazon River carries more water than any other river in the world. Before the government built highways through the rain forests, riverboats were the only way to travel across the Amazon region.

Brazil's newest and largest mining district is in the Amazon. This is the Serra dos Carajás (sehr-uh dus car-uh-ZHAHS), south of the port of Belém (buh-LEHM). Here, major deposits of iron ore, copper, and tin are found, as well as major reserves of gold, bauxite, and manganese. Brazil could become the world's largest producer of these minerals.

The new roads and mining projects in the Amazon have encouraged new settlers and treasure seekers to spread into the region. Conflicts among the settlers, miners, and Native Americans have increased. Many Native Americans have lost their forest homes and have become farm workers in the developing areas.

Natural rubber, nuts, medicinal plants, and fine woods used in furniture are traditional products of the Amazon. The destruction of the Amazon forests, which is leading to the extinction of many plants and animals, has become a global issue. Some people predict that the rain forest will disappear within the next 50 years. This would be a tragic loss. The biological resources of the Amazon region are just beginning to be investigated.

Manaus, the Amazon region's largest city, is the historic center of rubber collection and is now a growing industrial area. The government of Brazil has designated Manaus as a **growth pole**. A growth pole is a metropolitan area given official help to strengthen its economic development. In Manaus, special laws encourage industrial development by foreign companies. There are few business taxes, and foreign goods enter the city with few import fees.

Brazil's Future Today, Brazil faces several social, economic, and environmental challenges. Its population exceeds that of the rest of South America. The Southeast is prosperous, but the remainder of the country is underdeveloped. While Brazil has rich natural resources and

Population Growth in the Americas

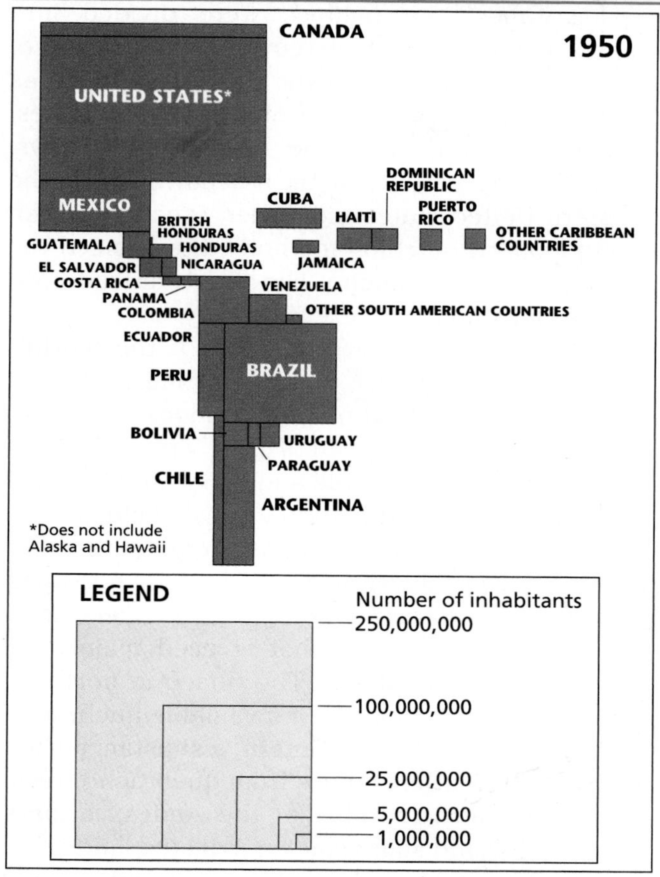

1950
CANADA
UNITED STATES*
MEXICO
CUBA
HAITI
DOMINICAN REPUBLIC
PUERTO RICO
OTHER CARIBBEAN COUNTRIES
BRITISH HONDURAS
HONDURAS
GUATEMALA
EL SALVADOR
NICARAGUA
JAMAICA
COSTA RICA
PANAMA
VENEZUELA
COLOMBIA
OTHER SOUTH AMERICAN COUNTRIES
ECUADOR
PERU
BRAZIL
BOLIVIA
URUGUAY
PARAGUAY
CHILE
ARGENTINA
*Does not include Alaska and Hawaii
LEGEND
Number of inhabitants
250,000,000
100,000,000
25,000,000
5,000,000
1,000,000

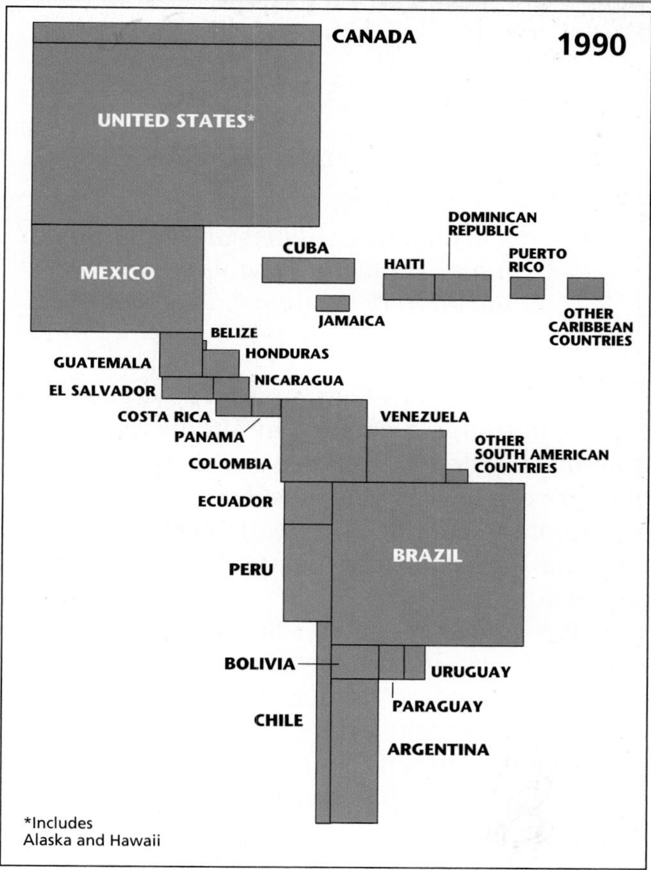

1990
CANADA
UNITED STATES*
MEXICO
CUBA
HAITI
DOMINICAN REPUBLIC
PUERTO RICO
JAMAICA
OTHER CARIBBEAN COUNTRIES
BELIZE
HONDURAS
GUATEMALA
EL SALVADOR
NICARAGUA
COSTA RICA
PANAMA
VENEZUELA
OTHER SOUTH AMERICAN COUNTRIES
COLOMBIA
ECUADOR
PERU
BRAZIL
BOLIVIA
URUGUAY
PARAGUAY
CHILE
ARGENTINA
*Includes Alaska and Hawaii

SKILL STUDY You can use cartograms to compare characteristics of countries. The size of each country in these cartograms is determined by its population and the scale in the legend. The more populous a country is, the larger it will appear in the cartogram. Cartograms also can be based on other data, such as economic or health information.

The cartogram on the left shows that the combined populations of the United States and Canada were about the same as the combined populations of Middle and South America in 1950. By 1990, however, the combined populations of Middle and South America were far larger. Which Middle or South American country had the largest population in 1950?

agricultural potential, unemployment is high. And as elsewhere on the continent, enormous differences still exist between the wealthy and the poor. Many of the country's people do not have enough to eat, even though Brazil is a major world food exporter.

Based on its natural and human resources, Brazil is a country of the future. The blending of cultures in Brazil has produced beautiful and vibrant art, architecture, and music. Products from Brazil are becoming common in the developed countries. In addition, Brazil has enormous mineral resources to be developed, and it has increasing supplies of fossil fuels. Hydroelectric plants on Brazil's rivers also provide energy for the future.

1

SECTION REVIEW

1. What different products attracted people to Brazil from the 1500s through the 1800s?
2. Name a city found in each of Brazil's economic and cultural regions. What challenges do these cities face?
3. **Critical Thinking** **Discuss** the social and environmental consequences of the development of Brazil's interior regions.

ARGENTINA

FOCUS

- *What environmental regions does Argentina have?*
- *What economic activities are important to Argentina? How are economics and politics linked in the country?*

REGION **Physical Geography** Argentina is the second largest country in Middle and South America. It extends from just north of the Tropic of Capricorn to cold Tierra del Fuego at the southern tip of the continent. Argentina can be divided into four environmental regions. (See the map on page 235.)

The Andes Mountains form a long narrow region of grassy and forested highlands along Argentina's western border. These are the country's only mountains. Here, there are ski resorts and beautiful lakes. From north to south, the elevations in the Andes steadily decrease. In the north, the peaks are more than 22,000 feet (6,706 m) above sea level. In the far south, they rise less than 10,000 feet (3,048 m).

East of the Andes in northern Argentina is the Chaco (CHAHK-oh), a region of low plains covered with dry forests and savannas. These plains are so flat that water sometimes stands for months after the summer rainy season ends.

Argentina's third natural region is made up of the pampas. The pampas are wide plains with grasslands located southwest of the estuary of the Río de la Plata. The pampas stretch for almost 400 miles (644 km) and are nearly flat. With their rich soils and humid middle-latitude climates, these plains have become one of the world's major agricultural regions. They are the heartland of Argentine life and are by far the country's most densely settled region.

Argentina's fourth region, Patagonia, is found south of the pampas. Patagonia's landforms are dry plains and windswept plateaus. Because the Andes Mountains block the region from the Pacific's rain-bearing storms, the climates are arid and semiarid. Few people live in Patagonia. Oil deposits off the coast of Patagonia promise new energy resources.

INTERACTION **Economic Geography** In earlier times, the pampas were divided into ranches, some of which covered thousands of acres. Livestock was allowed to wander, much as it did in the Great Plains of the United States. Gauchos (GAU-chohz), the Argentine cowboys, managed the cattle just as the cowboys of the western United States did. When ranchers began using modern ranching methods, they fenced the livestock in. The gaucho, like the American cowboy, is vanishing.

Today, the pampas are one of the world's largest sources of beef, wheat, and vegetable oils. Maize (corn) is raised north of Buenos Aires and, like wheat, is mainly an export crop. Farmers have developed vegetable and dairy farms to feed the populations of Argentina's growing cities.

Compared to the pampas, Argentina's other regions are less developed. North of Buenos Aires between the Paraná and Uruguay rivers is a region of grassy plains that is used mainly for grazing cattle and sheep. The Chaco in northern Argentina is important for its valuable quebracho (kay-BRA-choh) forests. **Tannin**, a substance used in preparing leather, comes from quebracho trees. In the northern foothills of the Andes, farmers grow grapes and sugarcane on irrigated lands.

In Patagonia, sheep grazing on huge ranches is the chief industry. Though many crops could be grown here with irrigation, food is plentiful in Argentina, and there has been little need for developing this potential.

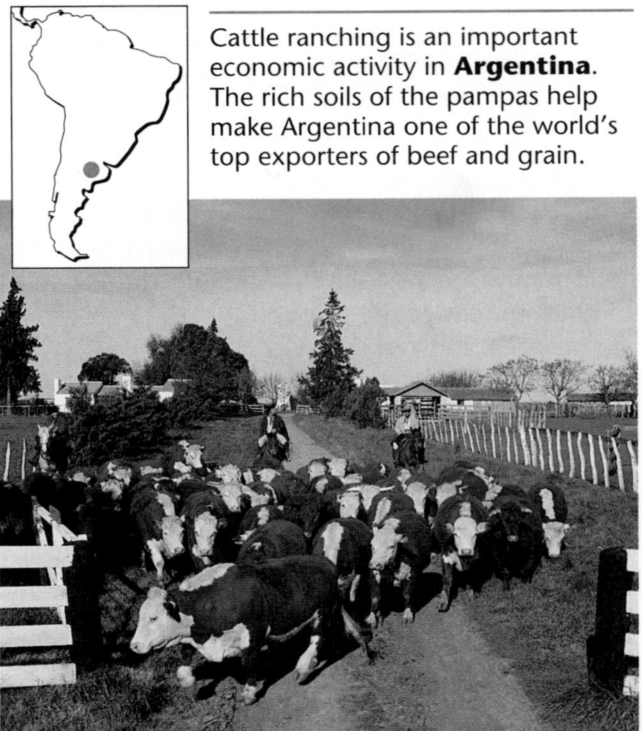

Cattle ranching is an important economic activity in **Argentina**. The rich soils of the pampas help make Argentina one of the world's top exporters of beef and grain.

Buenos Aires, the capital of **Argentina,** is South America's third largest metropolitan area. High-rise apartment buildings in the Retiro District, a middle-class part of the city shown here, stretch for many blocks.

Argentina competes with the United States, Canada, and Australia in selling beef and wheat on the world market. Argentina often can sell these products at lower cost because of lower production costs. The country has lower land values and pays lower wages than its competitors. Argentina also exports large amounts of wool and other products to the United States.

Even though agriculture is the country's main source of wealth, only about 10 percent of Argentina's people farm. Many more are employed in manufacturing. Most of Argentina's industry provides consumer goods, such as textiles and beverages, for local markets. Food processing is the primary export industry. Argentina has few mineral resources.

Argentina has several large cities, including Rosario (roh-ZAHR-ee-oh) and Córdoba. Buenos Aires, however, dominates the country. Its urban region has more than one-third of the nation's people and an even larger share of the industries.

Political Geography Because Argentina has abundant natural resources and great human potential, many believe that Argentina should be the richest country in South America. Unfortunately, the politics and economy of Argentina are nearly inseparable. Ineffective governments run by dictatorships and the military have been typical. During a particularly violent period of unrest in the 1970s, many Argentines were killed by the military government. In 1982, a brief war between Argentina and the United Kingdom broke out when Argentina invaded the Falkland Islands, a British territory in the South Atlantic. When Argentina was defeated, the military returned power to a civilian democratic government. Even so, economic progress has been slow, and the country has accumulated a large foreign debt.

2 SECTION REVIEW

1. Name Argentina's four environmental regions.
2. What are Argentina's major resources?
3. **Critical Thinking** **Suggest** an explanation for why a country with rich natural resources might need to borrow money.

3 URUGUAY AND PARAGUAY

FOCUS

- *What are Uruguay's physical, cultural, and economic characteristics?*
- *What are Paraguay's physical, cultural, and economic characteristics?*

Uruguay As a buffer state, Uruguay often has been dominated by its giant neighbors. (See the map on page 235.) As in Argentina, Spanish is the national language, but many of Uruguay's people also speak Brazilian Portuguese. The beach resorts of Uruguay are popular with both Argentines and Brazilians. Tourism is a rapidly growing industry.

The Río de la Plata estuary, along Uruguay's southern edge, is the major waterway of southern South America. On the shore of the estuary is the port of Montevideo (MAHNT-uh-vuh-DAY-oh), the business and government center of Uruguay. In the southern part of the country are Uruguay's rolling plains, a continuation of the pampas. Farther north, the country becomes more hilly like neighboring parts of Brazil.

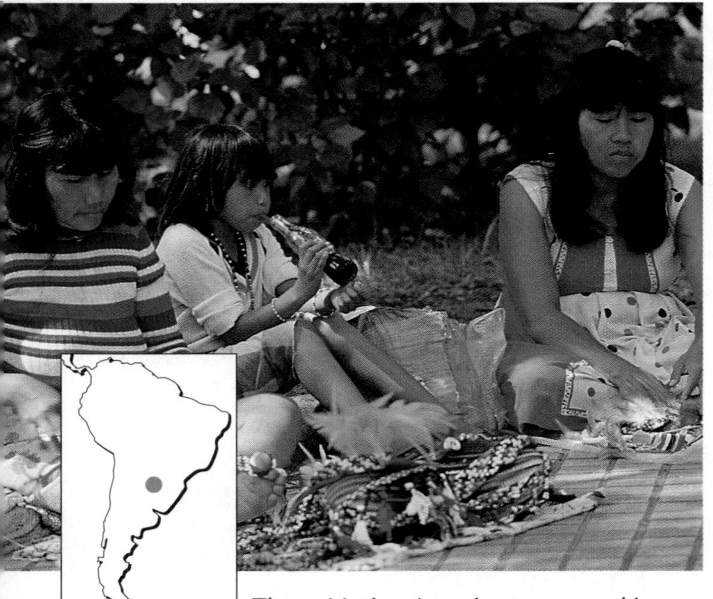

These Native Americans are making various crafts in **Asunción, Paraguay.** Traditional handicrafts are still practiced widely.

Uruguay is similar to the Argentine pampas in several ways. It has a humid-subtropical climate, and most of the country is a natural grassland. The soil is rich and well watered. Ranchers and farmers on the inland plains graze livestock and grow grains. In the coastal region, farmers grow vegetables and fruits. Rice has replaced wheat as the most important export crop. Uruguay has few mineral resources. Hydroelectric projects, however, provide an important energy source.

Uruguay has a high literacy rate and a large middle class. Despite occasional military dictatorships, Uruguay has strong democratic traditions and offers many economic opportunities. Uruguay's main challenge is to bring the prosperity of Montevideo into the less-developed rural areas. Many parts of the interior remain in large, unproductive haciendas. As in other countries in South America, people are abandoning rural areas. They hope the Montevideo region, where more than one-third of the country's people already live, will offer a better life.

Paraguay Landlocked Paraguay shares borders with Bolivia, Brazil, and Argentina. (See the map on page 235.) A landlocked country is one that is completely surrounded by land and does not have access to the ocean. The Paraguay River divides the country into two separate regions. East of the river, humid-subtropical lowlands rise to join the rolling highlands of southern Brazil. This is Paraguay's most productive agricultural region. The chief commercial crop is soybeans, though farmers grow fruits, corn, and other vegetables for local use. Rice, tobacco, sugarcane, and a tea called yerba mate (YEHR-buh mah-TAY) also are grown here.

West of the Paraguay River is the less humid region of forests. It is a continuation of the Chaco area, which begins in northern Argentina, crosses Paraguay, and extends into southeastern Bolivia. Paraguay's Chaco is used mostly for grazing. The dry forests of this region also contain tannin-producing quebracho trees. Because quebracho trees grow slowly, and because they are cut down for their tannin, the production of tannin is depleting the forests.

Paraguay has productive soil, adequate rainfall, and subtropical temperatures. Yet, the country has long been poorer than nearby Argentina and Brazil. Most of Paraguay has few people, and much of the land is underused. Because there are few industries, there are few jobs. As a result, many Paraguayans work as migrant laborers in neighboring countries.

Paraguay's people primarily are mestizos, but the indigenous influence remains strong. About 95 percent of the people speak Guaraní (gwah-ruh-NEE), an indigenous language. As in most developing countries, much of Paraguay's wealth is in the hands of a few wealthy families and large companies, which have a great deal of influence over the government.

Paraguay's future development is promising. Hydroelectric projects such as the Itaipu Dam on the Paraná River are providing Paraguay with an enormous surplus of power—far more than it can use. As a result, Paraguay is exporting power to Brazil and Argentina at a profit.

SECTION REVIEW

1. Name one thing Uruguay has in common with Brazil, and one thing Uruguay has in common with Argentina.
2. What economic activities occur east of the Paraguay River? west of the river?
3. **Critical Thinking** **Compare and contrast** cultural and political aspects of Uruguay and Paraguay. What factors account for the differences?

CHAPTER 21 CHECK

Rio de Janeiro, Brazil

Reviewing the Main Ideas

1. Brazil, the largest country in South America, is made up mostly of plains and plateaus. Originally a Portuguese colony, the country today has four major economic and cultural regions: the Northeast, the Southeast, Campo Cerrado, and the Amazon. Brazil faces several social, economic, and environmental challenges.
2. Argentina is the second largest country in South America. It is divided into four natural regions: the western highlands, the northern Chaco, the pampas, and Patagonia. Argentina has rich agricultural resources, but political unrest has limited the country's economic progress.
3. Uruguay is a small country with strong democratic traditions. Agriculture and tourism are important industries.
4. Paraguay, a landlocked country, is sparsely populated and has much underused land and little industry. Hydroelectricity is exported for profit.

Building a Vocabulary

1. What is a buffer state? Give an example of a buffer state from this chapter.
2. How does soil exhaustion occur? What are its consequences?
3. What are *favelas*? Where are they found?
4. How might a landlocked country be at a disadvantage compared to other countries?

Recalling and Reviewing

1. How is wealth distributed among people within Brazil and Paraguay?
2. To which countries discussed in this chapter are quebracho trees important? Why?
3. In which country is the Amazon region located? How is the region changing, and what consequences might these changes have?
4. What languages are spoken in each country of Atlantic South America?

Thinking Critically

1. Do you think it is more to a country's advantage or disadvantage to be a buffer state? Use examples from the chapter to support your answer.
2. What conditions in a country might cause the military to take over the government? Do you think military takeovers should be and/or can be prevented? Explain.

Learning About Your Local Geography

Individual Project

Both Brazil and Argentina have difficulty paying the interest on their debts. Interview an economics student or check an economics textbook for an explanation of interest. Then, contact a local bank or credit union and ask for the interest rates on various types of loans, such as car loans, home improvement loans, and personal loans. Make a bar graph showing these different rates. You may want to include your graph in your individual portfolio.

Using the Five Themes of Geography

Show how the themes of **region** and **human-environment interaction** are related in the countries of Atlantic South America. Select a geographic region discussed in the chapter. Construct a chart with three columns. In the first column, list natural physical characteristics of the region you have selected. In the second column, list economic activities for which the characteristics in column 1 are suited. In the third column, list ways in which humans have changed the region's environment as a result of the economic activities in column 2.

CHAPTER 22

Pacific South America

GEOGRAPHY DICTIONARY

selvas
altiplano
El Niño
terrorism
coup
cartel
latifundia
minifundio

"Anyone approaching the narrow, elongated country of Chile will find in the uppermost portion the Great North, the desert-like regions of salt and copper mines: inclement [harsh] weather, silence, and struggle. In the extreme South, the great frozen expanses, ranging from the silence of Patagonia to Cape Horn, crossed so many thousands of times by the flight of the albatross, and then, the glittering Antarctic."

Pablo Neruda

Incan vase from near **Cuzco, Peru**

The Andes Mountains dominate the Pacific, or western, side of South America. Some mountain ridges and volcanic peaks here rise more than 20,000 feet (6,096 m) above sea level. The Andes continue to rise steadily upward due to the subduction of a crustal Pacific plate under the continent. The region's frequent earthquakes and volcanic eruptions have triggered landslides and ashfalls, changed the courses of rivers, and engulfed villages under flowing mud.

The countries of Pacific South America are Ecuador, Peru, Bolivia, and Chile. Traditionally, these countries have relied on North American and European markets for trade. Today, the prosperity of the Pacific Rim has captured economic interests. Goods from East Asia are found throughout South America. South American factories are turning out products bearing Japanese labels. Meanwhile, the peoples of the region continue their struggles to reduce poverty and to reform their political systems.

Andes Mountains in **Chile**

PACIFIC SOUTH AMERICA: PHYSICAL–POLITICAL

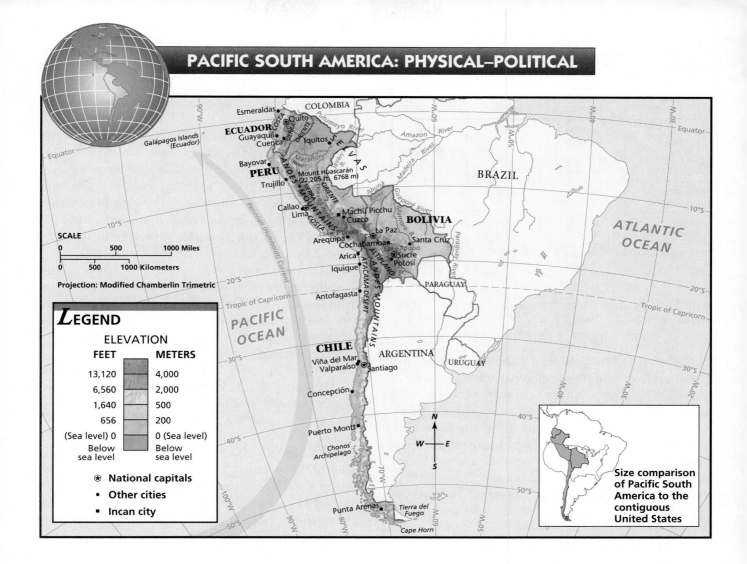

SCALE
0 500 1000 Miles
0 500 1000 Kilometers
Projection: Modified Chamberlin Trimetric
LEGEND
ELEVATION
FEET METERS
13,120 4,000
6,560 2,000
1,640 500
656 200
(Sea level) 0 0 (Sea level)
Below sea level Below sea level
National capitals
Other cities
Incan city
Size comparison of Pacific South America to the contiguous United States

ECUADOR

FOCUS

- *What are the physical and economic characteristics of Ecuador's regions?*
- *Where do Ecuador's peoples live? What challenges face the country?*

LOCATION Physical Geography The equator passes directly through Ecuador—in fact, the name *Ecuador* is Spanish for "equator." Ecuador can be divided into three landform and climate regions. (See the map above.) In the east are humid-tropical lowlands of the Amazon Basin. In Ecuador, this forested region is known as the Oriente, which means "east." South Americans call the thick rain forest vegetation found in the eastern region the ***selvas***, or "forest."

The Andes form two high mountain ranges through the middle of Ecuador. The valleys in the ranges lie at elevations up to 10,000 feet (3,048 m), with peaks rising up to 10,000 feet higher. More than two dozen high volcanic peaks have permanent snowcaps, even at the equator. The high valleys have mild climates all year. About 45 percent of Ecuador's population lives in the mountain region, known as the Sierra.

Ecuador's third region is the lowlands along the Pacific coast, known as the Costa. The climates in the Costa region vary from humid tropical in the north to a narrow strip of desert in the south. A savanna region is found midway along the Costa. This region has rich agricultural potential and the fastest growing population and economy in Ecuador. About half of Ecuador's people live in the Costa.

FACTS IN BRIEF Pacific South America

COUNTRY POPULATION (1994)	LIFE EXPECTANCY (1994)	LITERACY RATE (1990)	PER CAPITA GDP (1993)
BOLIVIA 7,719,445	61, male 66, female	78%	$2,100
CHILE 13,950,557	72, male 78, female	93%	$7,000
ECUADOR 10,677,067	67, male 73, female	88%	$4,000
PERU 23,650,671	63, male 68, female	85%	$3,000
UNITED STATES 260,713,585	73, male 79, female	97% (1991)	$24,700

Source: *The World Factbook 1994,* Central Intelligence Agency

Which Pacific South American country has the highest literacy rate?

Economic Geography Ecuador's economy closely reflects its physical geography. Farmers in the Sierra grow maize (corn), potatoes, and barley. The fields are often on steep slopes where soil erosion is severe, and many of the farms are too small to support a family. Landless farmers from the Sierra are moving down the Andes slopes to clear small farms from the dense forests in the Oriente. The Oriente's greatest resource is oil, which accounts for more than 50 percent of Ecuador's export income.

The Costa has emerged during the twentieth century as Ecuador's most important agricultural region. The key export crops are bananas, coffee, and cacao. Many of the Costa's farms are large enough for prosperous family farming. Offshore from the Costa are rich fishing grounds. Natural-gas deposits have been found beneath the Costa's lowlands.

Human Geography The Sierra is a historic region of dense Native American settlement. Many people still speak Quechua (KECH-oo-ah), the language of the ancient Incas. Descendants of African slaves and of European colonists have settled primarily in the Costa region.

Guayaquil (gwie-uh-KEEL), on the Pacific coast, is Ecuador's largest city. It also is Ecuador's most important port and manufacturing center. Guayaquil's modern buildings, banks, and factories are symbols of a modern Ecuador.

Ecuador's capital city is Quito (KEE-toh), in the Sierra. A northern kingdom of the Incan Empire, it became the center for Spanish colonial government. Despite repeated earthquakes over the centuries, Quito has survived as a classic example of a Spanish colonial city. Old churches, monuments, and open-air markets are tourist favorites.

The oil wealth of the Oriente and the agricultural progress of the Costa have moved Ecuador to world prominence. The country faces several challenges, however. Providing jobs, schools, and health care for a still rapidly growing population is an ongoing concern. Large numbers of people are moving to the Oriente, threatening the survival of the rain forest.

Another major challenge is transportation. Despite many roads and a railroad stretching from the coast to Quito, the country's regions are not well linked. The Andes make construction and maintenance of transportation systems both difficult and expensive.

After decades of corrupt governments, Ecuadoreans are having more political success. Democratically elected governments have ruled the country in recent years.

Galápagos Islands The arid Galápagos (guh-LAHP-uh-guhs) Islands, located in the Pacific Ocean about 650 miles (1,046 km) off the coast, are part of Ecuador. The animals and plants that are native to the islands are specially adapted to their habitats. Some scientists believe that the Galápagos' famous wildlife demonstrates the theory of biological evolution. The islands are home

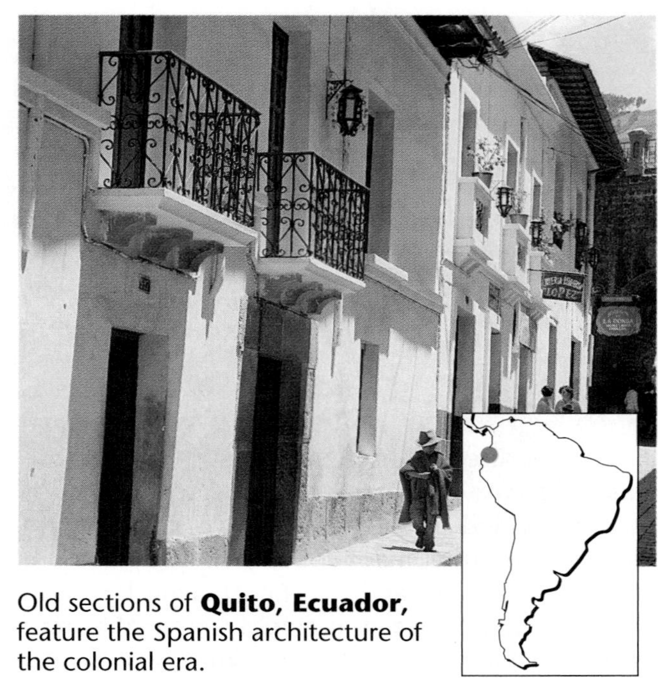
LOPEZ

Old sections of **Quito, Ecuador,** feature the Spanish architecture of the colonial era.

to about 4,000 people, most of whom are employed in the growing tourist trade. Though profitable, the tourist trade disrupts the natural environments on the islands and is threatening the existence of the unique plant and animal species.

SECTION REVIEW

1. What economic activities are important in Ecuador's three regions and on the Galápagos Islands?
2. How is Ecuador's population distributed? Why is transportation a challenge in Ecuador?
3. **Critical Thinking** **Compare and contrast** Guayaquil and Quito.

PERU

FOCUS

- *What are the characteristics of Peru's three major regions?*
- *What challenges face Peru?*

Like Ecuador, Peru can be divided into three regions: the Oriente, the Andes highlands, and the coastal lands. (See the map on page 245.)

The Oriente The Oriente of Peru begins on the eastern side of the Andes and continues onto the Amazon lowlands. The Oriente has dense rain forests that provide lumber and tree crops such as rubber. New settlers are clearing away the forests for subsistence farming. Preserving the rain forests on this modern frontier is an international concern.

The coca plant is native to the eastern slopes of the Andes in Peru, Bolivia, Ecuador, and Colombia. The extract of coca leaves is the source of the powerful drug cocaine. Illegal drug smugglers encourage farmers here to grow coca, which has become their most profitable crop. The United States is working with the governments of the Andean countries to find substitute crops for the farmers.

The major port city of the Peruvian Amazon is Iquitos. Ships travel more than 2,000 miles (3,218 km) on the Amazon from the Atlantic Ocean to Iquitos.

The Andes Highlands The Andes highlands make up Peru's second region. Traditional agriculture is practiced by farmers in the Andes highlands. Crops include potatoes, maize (corn), and barley. Local livestock includes the llama, alpaca, and vicuña. The easily tamed llama can be used for transport, and its coarse wool is used for making blankets. Alpacas are herded like sheep, and their wool makes fine cloth. The vicuña is a wild animal whose wooly fur is highly prized.

In southern Peru, the eastern and western ridges of the Andes separate, leaving a broad highland plain in between. This plain is called the ***altiplano*** and extends well into Bolivia. At an elevation of about 12,000 feet (about 3,658 m), the vegetation of the *altiplano* is a high-elevation grassland with few trees.

The *altiplano* and nearby valleys were the heartland of the Incan Empire. The buildings at Machu Picchu, an ancient Incan city, are impressive remains of the empire. This region continues to be one of the densest areas of Native American settlement in the Americas. Quechua is spoken by perhaps as many as 3 to 4 million Peruvians.

This gold handle from a ceremonial knife honors a legendary leader of the Chimú civilization. The Chimú were conquered by the Incan Empire in the fifteenth century.

The Coastal Region West of the Andes is Peru's dry desert coast. Here, the rugged lower sides of the Andes often extend down to the sea. Settlement is concentrated in small river valleys that have a water supply from melted Andean snow.

A cool ocean current flowing northward off the coast of Peru contributes to the desert conditions. In South America, this is the Peruvian (Humboldt) Current. Air passing over the current becomes cool and very moist. Because the air masses are under high pressure, the moist air cannot travel upward. The result is little, if any, rain, although coastal fogs are common. The coastal desert is also in the rain shadow of the Andes. The Andes' slopes far to the east prevent rain by blocking the moist winds from the Amazon Basin.

Every five to ten years, the dry coasts of Peru and Ecuador are visited by unusual weather patterns called **El Niño**. The ocean near the coast warms, and most fish flee from what is normally a rich fishing area. People along the coast may suffer severe flooding from heavy rains. Scientists believe El Niño conditions are caused by the buildup of unusually warm waters in the mid-Pacific Ocean. These warm waters then slosh eastward, turning aside the cool Peruvian (Humboldt) Current. El Niño affects ocean and weather events around the world.

The coastal region of Peru is the most modern and developed. The coastal rivers have many sites suitable for development of hydroelectric power, and the region holds Peru's major mineral deposits. Fishing has become a major industry in the coastal ports. Anchovies are the primary catch.

The people who live in the coastal region are predominantly mestizo. Lima, the capital of Peru, and its port city of Callao (kuh-YAH-oh) are located in the coastal region. Both are increasingly important to the country's economy. The population of the Lima and Callao region now numbers nearly one-third of the country's people. Lima's many government jobs and manufacturing industries draw people from throughout the country, particularly from the highlands.

Political Geography Peru's problems are related to underdevelopment, poor governing, and the contrasts between the mountain and coastal regions. Large amounts of food must be imported. To manage a large foreign debt, the government printed more money to pay for government services. The result was inflation.

Peru has made progress in recent years. The government has eased inflation problems, and a violent revolutionary movement has weakened. The government also has worked to end **terrorism,** or the use of violence as a means of political force. Poverty remains a major concern, however. Some critics also have accused the government of using undemocratic methods to solve the country's problems.

SECTION REVIEW

1. List Peru's three major regions, and give one characteristic of each.
2. Why has Peru experienced inflation?
3. **Critical Thinking** **Assess** whether the rain-shadow effect experienced by Peru's coastal region is the same effect experienced by Oregon and Washington in the United States.

BOLIVIA AND CHILE

FOCUS

- *Into what regions can Bolivia and Chile be divided? What are the features of each region?*
- *How are political and economic conditions in Bolivia similar to those in Chile?*

LOCATION **Bolivia** Bolivia is a landlocked country south of Peru. (See the map on page 245.) The country once extended west to the Pacific but lost its coastal province to Chile in 1884, following the War of the Pacific. Bolivia has the highest percentage of indigenous peoples of any South American country.

Bolivia's government has changed through revolution or military **coup** (koo) many times since the country became independent in 1825. From the French phrase *coup d'etat*, a coup is an overthrow of an existing government by another political or military group. Bolivia has had a democratically elected government, however, for more than 10 years.

Most of Bolivia can be divided into two major physical regions: the western highlands

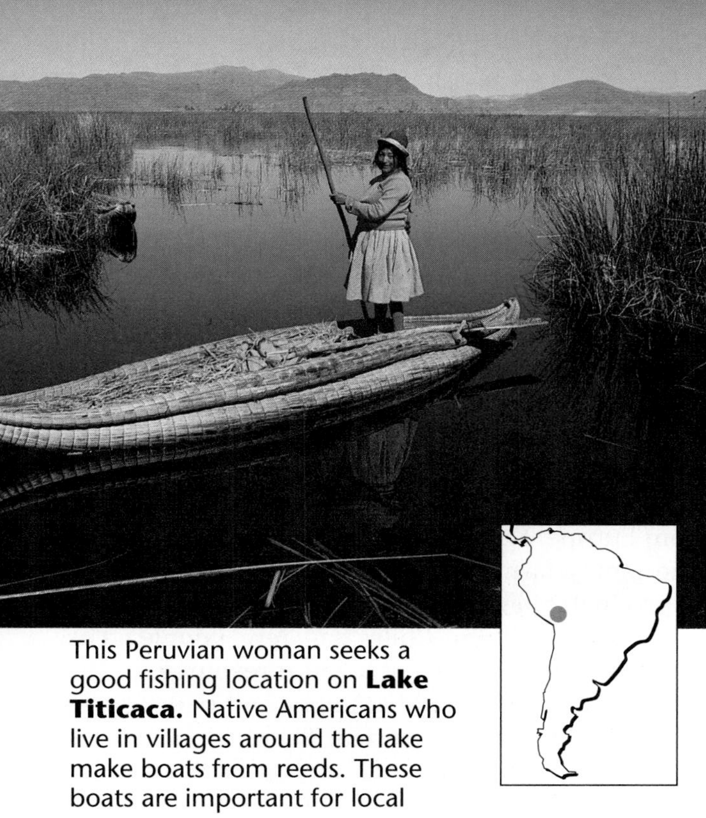

This Peruvian woman seeks a good fishing location on **Lake Titicaca.** Native Americans who live in villages around the lake make boats from reeds. These boats are important for local commerce.

and the eastern lowlands. Most Bolivians live in the highlands, despite the cold temperatures and barren lands. Western Bolivia is a continuation of the Andes highlands of Peru. More than 500 miles (805 km) long, the *altiplano* contains two great lakes. These are freshwater Lake Titicaca, which Bolivia shares with Peru, and salty Lake Poopó (poh-uh-POH). Lake Titicaca covers about 3,200 square miles (8,288 sq. km). At an elevation of 12,500 feet (3,810 m), it is the highest lake in the world on which ships travel.

The Bolivian Andes are famous for their natural resources. Spanish colonial governments mined great amounts of silver near the city of Potosí. Later, tin became the key mineral export. Today, valuable mineral resources include copper, lead, and zinc. Oil and natural gas also provide income for the country.

East of Lake Titicaca, in a valley below the *altiplano*, is La Paz (luh PAHZ), the country's administrative capital. At 12,001 feet (3,658 m), it is the highest capital in the world. Despite the semiarid environment and the frequently freezing weather, La Paz is the principal industrial city of Bolivia. It is also the country's center for imports and exports.

Much of eastern Bolivia is a series of low plateaus covered with dry forests and grasslands. The plains have fertile soils, adequate rain, and good grazing lands. But there are few roads, few people, and little money for investment. Development of this region would offer the nation economic improvement.

Bolivia's greatest growth region surrounds the city of Santa Cruz in the lowlands. Santa Cruz was connected by road with La Paz in the 1950s, making it possible to provide many agricultural products to the highlands. Recent oil discoveries in eastern Bolivia may make this region more important in the future.

Chile Chile's name comes from a Native American word meaning "land's end." Chile's shape makes it one of the most unusual countries in the world. The country is 2,650 miles (4,264 km) long but only about 110 miles (177 km) wide. It lies between the Andes Mountains and the Pacific Ocean. (See the map on page 245.) From north to south, the country can be divided into three landform regions: northern, central, and southern Chile.

The northern region of Chile is the Atacama Desert, one of the driest regions in the world. South of the desert, the region becomes semiarid with just enough rain for grazing. Croplands along the valleys are irrigated from streams draining the Andes. For more than a century, the minerals of northern Chile provided most of the country's export income.

The first products to be mined in the north were guano (the droppings of seabirds) and nitrate, which were processed into fertilizer and exported to Europe and North America. While nitrate remains important, synthetic fertilizers are replacing it in most of the market. Nitrate also is an important raw material for explosives.

Northern Chile is one of the world's largest producers of copper, its key export mineral since the 1920s. Most often, the copper is exported as nearly pure bars that can be manufactured into wire and sheets. Chile's copper reserves remain very large, and its copper production dominates the world market. Even so, the country's economy rises and falls with world copper prices. Chile often has considered starting a **cartel** to control prices. A cartel is a group of business organizations that agree to limit supplies of a product and thereby keep prices high.

The main cities of the Atacama region are Arica, Iquique, and Antofagasta. Each of these cities is a port with small fishing fleets and some

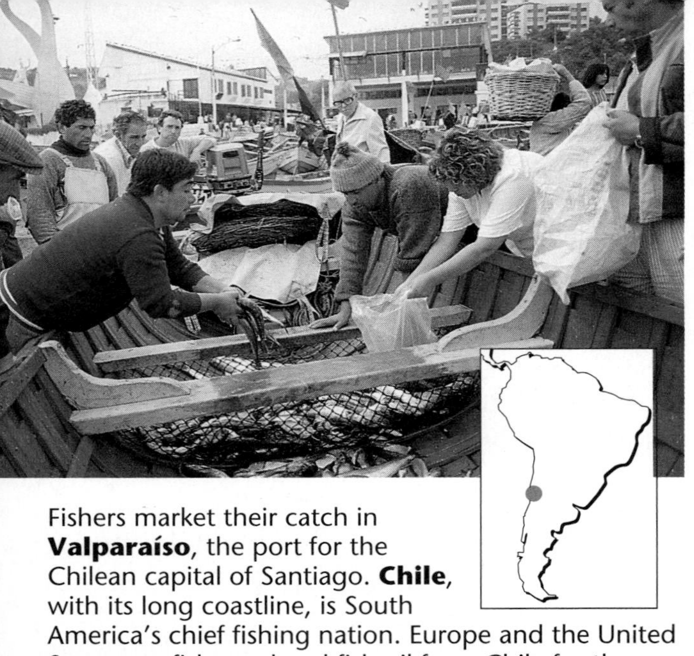

Fishers market their catch in **Valparaíso**, the port for the Chilean capital of Santiago. **Chile**, with its long coastline, is South America's chief fishing nation. Europe and the United States use fish meal and fish oil from Chile for the production of animal feed and industrial oil.

industries. To encourage industrial development, earlier Chilean governments gave special import rules to these port cities. These rules no longer apply, however. The Atacama is far from the markets in central Chile, and manufacturing may decline without these special rules. Landlocked Bolivia uses Arica and Antofagasta as its main ports.

Farther south is central Chile, the second environmental region and the country's heartland. Here, between the coastal mountains and the Andes, is the broad central valley. It has a Mediterranean climate and fertile soils. This region extends for almost 700 miles (1,126 km).

Central Chile was the center of Spanish colonial settlement. The land remains organized by the ***latifundia*** system, which was brought to the continent by European colonists. In this system, tenant farmers work part of the land for their own food in return for their labor. The term *latifundia* refers to the land tenure system as well as to the large estates on which tenants work. Today, productivity on the *latifundia* is low. Now, the trend is toward commercial farming of fruits such as grapes, peaches, and plums for export.

In the past, the government divided some large estates into small farms for families. Called ***minifundios***, many of these small farms are not large enough to be profitable. In addition, social unrest often has interfered with food production, forcing Chile to import food in some years.

About 90 percent of Chile's people live in the central region. In fact, the urban population around the capital city, Santiago, and its port, Valparaíso, now totals more than one-third of all of Chile's people. As the center of the country's trade and government, the Santiago region dominates Chile's economy.

Chile's third environmental region is southern Chile, with its thick forests and towering mountains. The marine-west-coast climate brings heavy autumn and winter rains. South of the town of Puerto Montt, the region is wild, with cold, stormy weather and rugged, snowcapped mountains.

Southern Chile was settled by immigrants from Europe. Here, there are prosperous medium-size family farms. These farms provide a variety of products, including wheat, dairy products, and beef. Far southern Chile has few people. The region's forests represent a major resource for the future and could provide lumber and paper products for all of South America. There are also large deposits of low-grade coal and oil in this region. With further development, oil deposits should provide enough oil for the country as well as some for export.

Like most Middle and South American countries, Chile has been in turmoil throughout its history. After the failures of an elected Communist government in the early 1970s and a dictatorship throughout the 1980s, Chile's economic and political life has been renewed. The country's large foreign debt is under control, and prosperity has increased. Chile now has a stable democratic government, and optimism is widespread.

Although urban slums remain, Chile's healthier economy seems to be improving the status of the urban poor. After decades of little economic growth, Chile's small businesses and factories are growing rapidly. Unemployment is decreasing and wages are rising. Chile's resources are once again being developed, and the country is working to join the United States, Mexico, and Canada in NAFTA.

3 SECTION REVIEW

1. In which regions of Bolivia and Chile is mining important? What products have been mined in these regions?
2. How are political conditions in Bolivia similar to political conditions in Chile?
3. **Critical Thinking Evaluate** advantages and disadvantages of cartels. How does a cartel resemble a monopoly?

CHAPTER 22 CHECK

Southern **Chile**

Reviewing the Main Ideas

1. Ecuador has forest, mountain, and coastal regions. The country's major challenges are population growth and the lack of a good transportation system.
2. Like Ecuador, Peru has forest, mountain, and coastal regions. The coastal region is the most developed. Political unrest continues to be an issue in Peru.
3. Bolivia is a landlocked country. La Paz is the administrative capital. Bolivia's western highlands hold most of the country's people. The eastern lowlands have development potential.
4. Chile is a long, narrow country situated between the Andes Mountains and the Pacific Ocean. The minerals of northern Chile provide much of the country's export income. Chile has a stable government, and the country's economy is improving.

Building a Vocabulary

1. What does *selvas* mean?
2. What term is used to describe the plain located between the eastern and western ridges of the Andes?
3. What is El Niño? What countries does it affect?
4. Define *terrorism.*
5. Is a coup the same as a revolution? Explain.

Recalling and Reviewing

1. What environmental concerns do Ecuador and Peru have?
2. Both Ecuador and Peru have an inland forest region, a highland region, and a coastal region. Why are Bolivia and Chile divided differently?
3. Where are most of the major industrial cities of Pacific South America located?

Thinking Critically

1. The ancient Incan city of Machu Picchu was discovered not by Spanish explorers but by Hiram Bingham in 1911. The stone structures of the city, although empty of people, were still in very good condition. What do these facts suggest about the city and its inhabitants?
2. Chile has agreed to grant Bolivia a strip of land that would give Bolivia access to the coast. The two countries, however, have not been able to agree on exact details. How do you think conditions in Bolivia would change if the country were no longer landlocked?

Learning About Your Local Geography

Individual Project

One of the major challenges facing the countries of Pacific South America is a lack of good transportation. Contact your state or local department of transportation to find out about the last major road or highway built in your area. Request information about why this road was built, when it was built, and how it has affected the area as a whole. Synthesize this information into a brief written report. You may want to include your report in your individual portfolio.

Using the Five Themes of Geography

How is the Pacific coast **location** of the countries discussed in this chapter likely to affect future **movement** of people and goods into and out of these countries? What global **region** are these countries a part of because of their location? Sketch a map of this region and draw arrows to indicate patterns of movement.

SKILLS

Using the Geographic Method

You may have learned in a science class about the "scientific method." Similarly, geographers use the "geographic method" to collect and analyze geographic information and to make sense of that information.

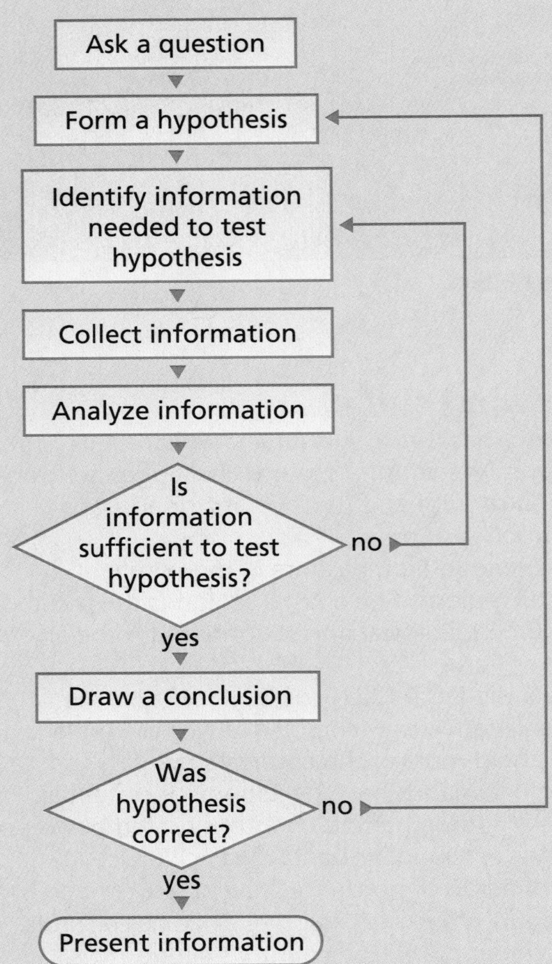
Ask a question
Form a hypothesis
Identify information needed to test hypothesis
Collect information
Analyze information
Is information sufficient to test hypothesis?
no
yes
Draw a conclusion
Was hypothesis correct?
no
yes
Present information

1. *Ask a question.* What would you like to know more about? Most questions start with one of the following words: who, what, where, when, why, or how. For example, you might ask the question "Why does Peru import food?"
2. *Form a hypothesis.* An educated guess is called a hypothesis. A hypothesis gives you something to focus on in answering your question. There may be several reasons that a country imports food, and your hypothesis should identify one or more of these reasons. Your hypothesis might be "Peru imports food because its agricultural methods are not efficient enough to feed the country's population."
3. *Identify what information you need in order to determine whether your hypothesis is correct.* For example, to test the hypothesis that Peru must import food because of its agricultural methods, you would want to find out what those methods are. You might also want to find out the size of Peru's population.
4. *Collect the information necessary to test your hypothesis.* Once you have identified what information you need, you must decide how to collect this information. Collecting information can take many forms. Ideally, you would visit a place and observe it firsthand. This is called field work. Unfortunately, field work is not always possible. Instead, you can gather information at the library. To find out about Peru's agricultural methods and its population, you might look in an atlas, an almanac, or in a book about Peru. Your information might be in the form of maps, graphs, charts, or photographs.
5. *Analyze the information you have collected.* Look for patterns or relationships in your information, particularly if your source is a map, graph, chart, or photograph. Is the information you have collected sufficient to test your hypothesis? If not, you may need to return to Step 3.
6. *Draw a conclusion.* Based on the information you have analyzed, was your hypothesis correct? If not, you might want to go back to Step 2 and form a new hypothesis. If your hypothesis was correct, you may want to present your information to others.

PRACTICING THE SKILL

Identify a topic in this unit about which you would like to learn more, such as political instability, deforestation, or the growth of cities in Middle and South America. Then, follow the steps of the geographic method listed above to come to a conclusion. Write a report summarizing your findings and the process you followed.

BUILDING YOUR PORTFOLIO

Individually or in a group, complete one of the following projects to show your understanding of the geography concepts involved.

A Middle and South America: A Regional Approach

As you know, ***region*** *is one of the Five Themes of Geography. In what ways can the concept of region be applied to the countries of Middle and South America? For this project, you will sketch a series of maps, each map showing a different aspect of region with regard to Middle and South America.*

1. Some geographers consider Mexico to be part of the North American region, rather than part of the Middle and South American region. Decide whether you agree or disagree, and sketch a map to defend your opinion. Use symbols and shading to show regional features.
2. Your next map should show the regional divisions within any country of Middle or South America. With symbols and a legend, show how the country's regions differ in physical, economic, or human characteristics.
3. Because people move among different regions, the geographic theme of movement is closely related to the theme of region. For your third map, select any country in Middle or South America, and with arrows, show methods and routes of transportation within that country.
4. One type of movement that occurs in many Middle and South American countries is migration from rural regions to urban regions. Select a country in which this movement is happening and, on your fourth map, indicate migration to the country's cities. Next to the map, write several sentences explaining the causes and effects of this type of movement.
5. Another type of migration that occurs in some Middle and South American countries is from crowded developed regions into regions that have few people. Select a country where this type of migration has taken place. For your fifth map, sketch the country, showing the migration. Include several sentences explaining the causes and effects of this type of migration.

Present and explain your maps to the rest of the class.

B Land Use and Reform

A theme common to many of the countries of Middle and South America has to do with the ways in which land is used, distributed, and reformed. Construct a series of models showing these different patterns, including one model showing what you think is the best use for land. Your models may be physical representations, or they may be diagrams on paper as explained in the Unit 3 Skill.

1. Your first model should show subsistence agriculture in Middle and South America. Include with the model a list of the crops grown in subsistence agriculture, where subsistence agriculture occurs, and who practices subsistence agriculture.
2. Another model should show agribusiness and plantation agriculture in Middle and South America. Again, include with the model a list of the crops grown in agribusiness and plantation agriculture, where agribusiness and plantation agriculture occurs, and who practices agribusiness and plantation agriculture. Who owns the land, and who provides the labor?
3. A third model should show the hacienda system in Middle and South America. Include with the model a description of the hacienda system. Also include a statement comparing the hacienda system to the *latifundia* system. Conduct research to find out what similar systems are called in Brazil and in Argentina.
4. Your next set of models should show how any of the systems you have modeled above have been reformed as *ejidos* (Mexico), cooperatives (Cuba), or *minifundios* (Chile). Include with these models statements of how, when, where, and why the land has been reformed, and how successful the reforms were or have been.
5. Finally, decide what form of land use you think is best. You may wish to consider several factors, including how productive the land is, how the land is divided among various people, and environmental effects. Your model may be one you have already made from the steps above, or it may be a new one altogether.

Present and explain your models to the rest of the class.

UNIT 5

EUROPE

The Arc de Triomphe (ARK duh tree-OMPH) commands the western end of the Champs-Élysées (SHAHN-say-lee-ZAY), the most famous avenue in **Paris**. Paris, like many European cities, is a showcase of architecture, history, and beauty.

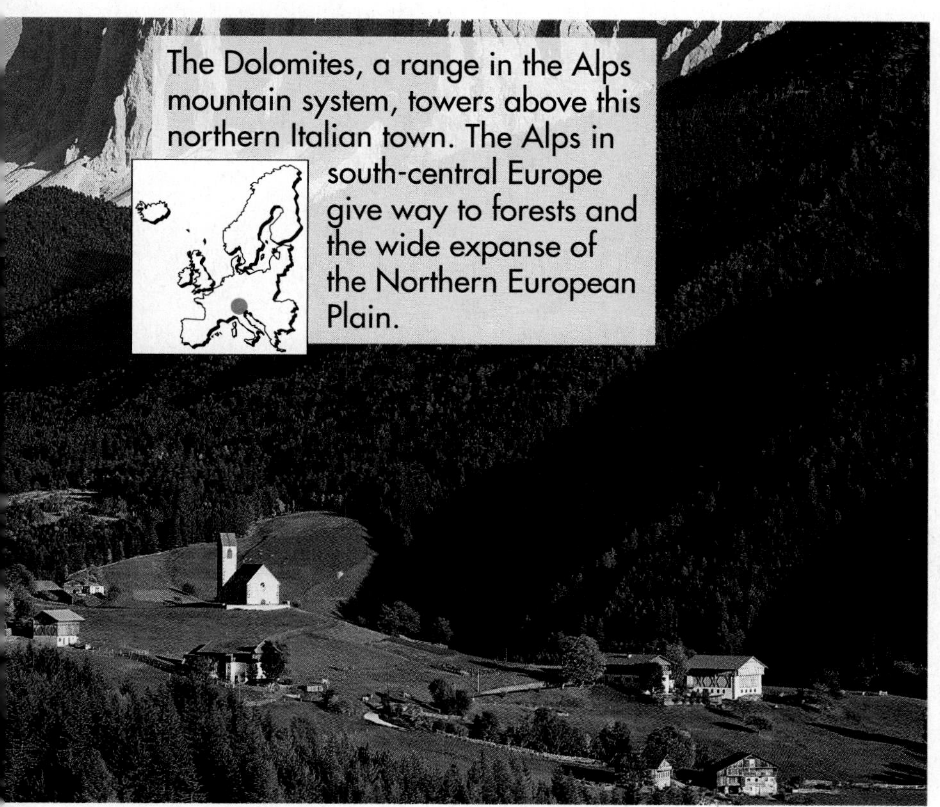

The Dolomites, a range in the Alps mountain system, towers above this northern Italian town. The Alps in south-central Europe give way to forests and the wide expanse of the Northern European Plain.

UNIT 5 CHAPTERS

Gold Spanish coin

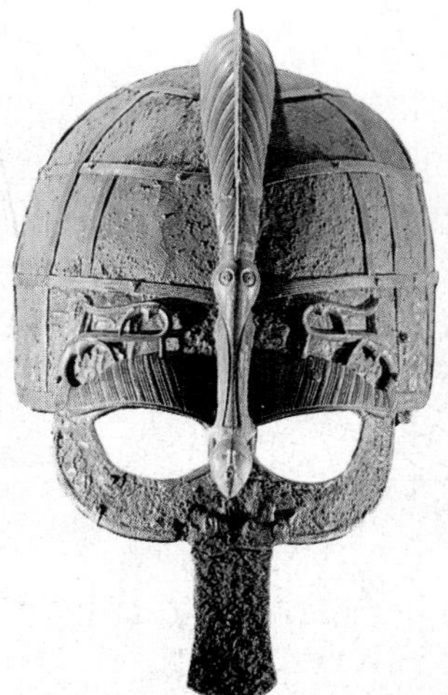

Early Viking mask

Dawn breaks over the Danube River in **Budapest, Hungary**. The Danube River, which is important to European commerce, links nine countries and four national capitals from southwestern Germany to the Black Sea.

One World, Many Voices

LITERATURE

"At this point they caught sight of thirty or forty windmills which were standing on the plain there, and no sooner had Don Quixote laid eyes upon them than he turned to his squire [attendant] and said, 'Fortune is guiding our affairs better than we could have wished; for you see there before you, friend Sancho Panza, some thirty or forty lawless giants with whom I mean to do battle. I shall deprive them of their lives, and with the spoils from this encounter we shall begin to enrich ourselves; for this is righteous warfare, and it is a great service to God to remove so accursed a breed from the face of the earth.'

'What giants?' said Sancho Panza.

'Those that you see there,' replied his master, 'those with the long arms some of which are as much as two leagues in length.'

'But look, your Grace, those are not giants but windmills, and what appear to be arms are their wings which, when whirled in the breeze, cause the millstone to go.'

'It is plain to be seen,' said Don Quixote, 'that you have had little experience in this matter of adventures. If you are afraid, go off to one side and say your prayers while I am engaging them in fierce, unequal combat.'

Saying this, he gave spurs to his steed [horse] Rocinante, without paying any heed to Sancho's warning that these were truly windmills and not giants that he was riding forth to attack. Nor even when he was close upon them did he perceive what they really were, but shouted at the top of his lungs, 'Do not seek to flee, cowards and vile creatures that you are, for it is but a single knight with whom you have to deal!'"

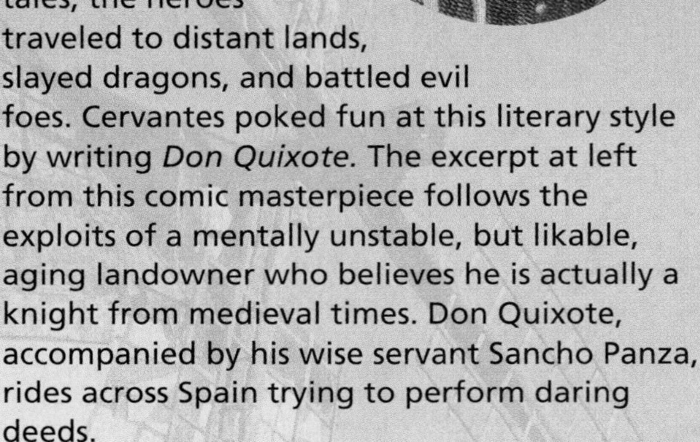

Miguel de Cervantes (1547–1616) was a Spanish writer from Madrid. Many people of his day enjoyed reading romantic adventure stories about medieval knights. In these tales, the heroes traveled to distant lands, slayed dragons, and battled evil foes. Cervantes poked fun at this literary style by writing *Don Quixote.* The excerpt at left from this comic masterpiece follows the exploits of a mentally unstable, but likable, aging landowner who believes he is actually a knight from medieval times. Don Quixote, accompanied by his wise servant Sancho Panza, rides across Spain trying to perform daring deeds.

Interpreting Literature

1. **What does Don Quixote see on the Spanish plain that he mistakes for "giants"?**
2. **How does Sancho Panza attempt to reason with Don Quixote?**

FOR THE RECORD

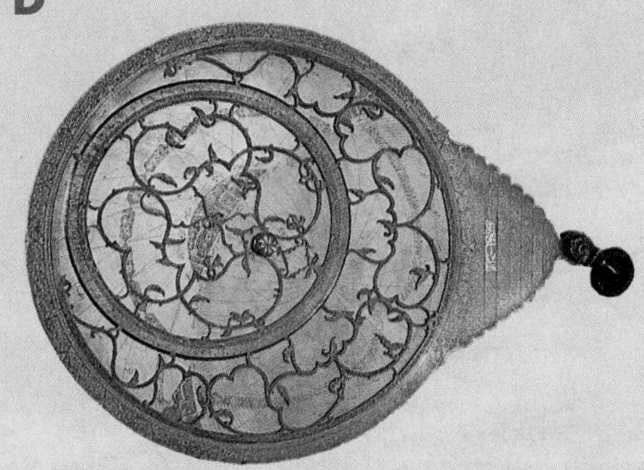
Eighteenth-century brass astrolabe

Abu Abdallah Mohammed Idrisi (1100–1165) was a famous Arab geographer from Morocco. His ancestors were among the Muslim princes and caliphs that ruled Spain after the Muslim conquest of the Iberian Peninsula in 711. Idrisi's interest in geography took him to many corners of western Europe, including Spain, Portugal, England, Italy, and the Atlantic coast of France. Below is a passage from Idrisi's greatest geographical work, which described what Arabs and Europeans knew about the world by the twelfth century. The text combined the findings of Greek and Arab geographers, eyewitness accounts, and Idrisi's own firsthand observations about foreign lands.

"Norway is a large island, but most of it is a desert. This island has two capes, of which one in the west is opposite the island of Denmark, a half-day's sail away, while the other is opposite the coast of Finmark.

It is said that there are three flourishing cities in Norway, two are opposite Finmark and the third opposite Denmark. These cities look all the same, they are not very populous, and rather poor, for it rains a great deal in this country and fog prevails a good deal of the time. The Norwegians, having sown their grain, harvest it while it is still green and take it to their dwellings to dry it by the fire, for there is but little sunshine. There are a great many trees in this country, with immense trunks, that are very heavy. They also state that there lives there a race of savages, whose heads are next to their shoulders so that they do not have any neck at all. They live in the midst of dark forests where they build their homes and live entirely on nuts and chestnuts."

Coastal **Norway**

Analyzing Primary Sources

1. Read about the physical geography and climate of Norway on pages 277–278 in your text. In what ways is Idrisi's description of Norway incorrect?
2. Study the map on page 259. Why might early geographers have assumed Norway was an island?

CHAPTER 23

Introducing Europe

Fifteenth-century illuminated French manuscript

Geographers define Europe as the region stretching from the Atlantic Ocean to the Ural Mountains and from the Arctic Ocean to the Mediterranean Sea. Europe is actually the western end of the world's largest landmass, known as **Eurasia**. Eurasia is made up of the continents of Europe and Asia. According to the theory of plate tectonics, about 300 million years ago, the two continents collided and formed the Ural Mountains at their joining.

The first geography textbooks referred to Europe as the "peninsula of peninsulas." This observation is still true. While Europe contains many peninsulas of its own, it also can be viewed as a giant peninsula of Eurasia. Europe consists of more than 35 countries, some of which are among the most economically developed in the world. Several are tiny **microstates**, or very small countries.

In this textbook, the geography of Europe is divided into two parts. The western part is discussed in the five chapters that follow. The eastern part is discussed in Unit 6, which covers Russia and the other countries that once were part of the Soviet Union.

GEOGRAPHY DICTIONARY

Eurasia
microstate
natural boundary
navigable
icebreaker
imperialism
alliance
balance of power
reunification
famine
multilingual
economic association
European Union (EU)

Clifden Castle, **County Galway, Ireland**

Mapping the Region

EUROPE

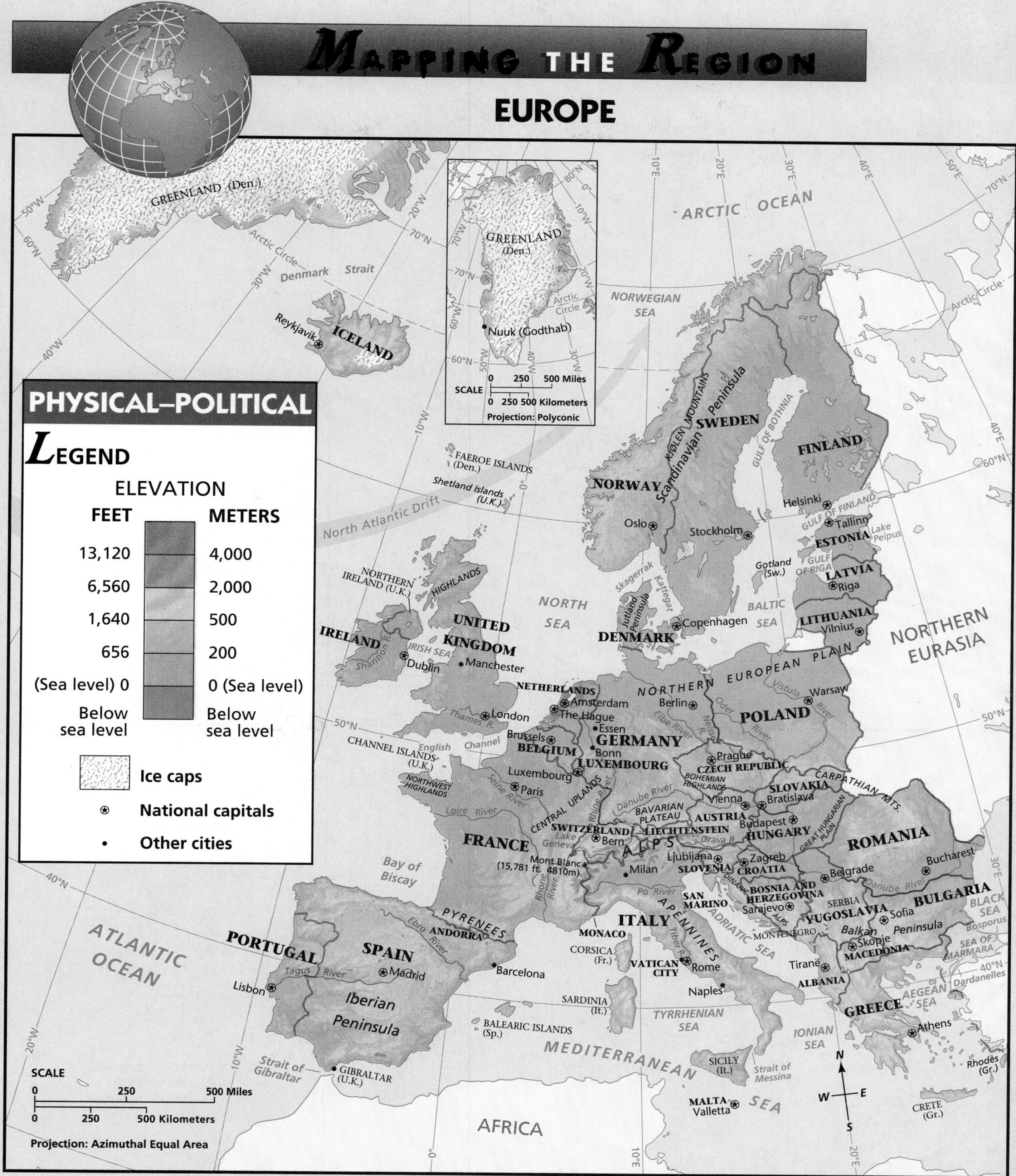

PHYSICAL–POLITICAL
LEGEND
ELEVATION
FEET
METERS
13,120
4,000
6,560
2,000
1,640
500
656
200
(Sea level) 0
0 (Sea level)
Below sea level
Below sea level
Ice caps
National capitals
Other cities
GREENLAND (Den.)
Nuuk (Godthab)
SCALE
0 250 500 Miles
0 250 500 Kilometers
Projection: Polyconic
ARCTIC OCEAN
NORWEGIAN SEA
Denmark Strait
Arctic Circle
Reykjavik
ICELAND
FAEROE ISLANDS (Den.)
Shetland Islands (U.K.)
North Atlantic Drift
NORWAY
SWEDEN
FINLAND
KØLEN MOUNTAINS
Scandinavian Peninsula
GULF OF BOTHNIA
Oslo
Stockholm
Helsinki
GULF OF FINLAND
Tallinn
Lake Peipus
ESTONIA
Gotland (Sw.)
GULF OF RIGA
LATVIA
Riga
LITHUANIA
Vilnius
BALTIC SEA
NORTHERN EURASIA
Skagerrak
Kattegat
Jutland Peninsula
DENMARK
Copenhagen
NORTH SEA
NORTHERN IRELAND (U.K.)
HIGHLANDS
UNITED KINGDOM
IRELAND
IRISH SEA
Shannon R.
Dublin
Manchester
Thames R.
London
English Channel
CHANNEL ISLANDS (U.K.)
NETHERLANDS
Amsterdam
The Hague
Essen
Bonn
Brussels
BELGIUM
GERMANY
Berlin
Elbe River
Oder River
Neisse R.
NORTHERN EUROPEAN PLAIN
POLAND
Warsaw
Vistula River
LUXEMBOURG
Luxembourg
Paris
Seine River
NORTHWEST HIGHLANDS
Loire River
CENTRAL UPLANDS
Rhine River
Danube River
BAVARIAN PLATEAU
Prague
CZECH REPUBLIC
BOHEMIAN HIGHLANDS
SLOVAKIA
Bratislava
CARPATHIAN MTS.
Vienna
AUSTRIA
LIECHTENSTEIN
Budapest
HUNGARY
GREAT HUNGARIAN PLAIN
ROMANIA
Bucharest
FRANCE
SWITZERLAND
Bern
Lake Geneva
ALPS
Mont Blanc (15,781 ft. 4810m)
Drava R.
Ljubljana
SLOVENIA
Zagreb
CROATIA
Belgrade
Danube River
Milan
Bay of Biscay
Rhone River
Po River
SAN MARINO
BOSNIA AND HERZEGOVINA
Sarajevo
DINARIC ALPS
SERBIA
YUGOSLAVIA
MONTENEGRO
Sofia
BULGARIA
BLACK SEA
Bosporus
PYRENEES
ANDORRA
Ebro River
ITALY
APENNINES
ADRIATIC SEA
Balkan Peninsula
Skopje
MACEDONIA
SEA OF MARMARA
Dardanelles
ATLANTIC OCEAN
PORTUGAL
SPAIN
Madrid
Tagus River
Lisbon
Iberian Peninsula
Barcelona
MONACO
CORSICA (Fr.)
VATICAN CITY
Rome
Tiber R.
Tiranë
ALBANIA
Naples
SARDINIA (It.)
TYRRHENIAN SEA
GREECE
AEGEAN SEA
Athens
IONIAN SEA
BALEARIC ISLANDS (Sp.)
MEDITERRANEAN SEA
Strait of Gibraltar
GIBRALTAR (U.K.)
SICILY (It.)
Strait of Messina
MALTA
Valletta
Rhodes (Gr.)
CRETE (Gr.)
AFRICA
SCALE
0 250 500 Miles
0 250 500 Kilometers
Projection: Azimuthal Equal Area
N
S
E
W

MAPPING THE REGION

EUROPE

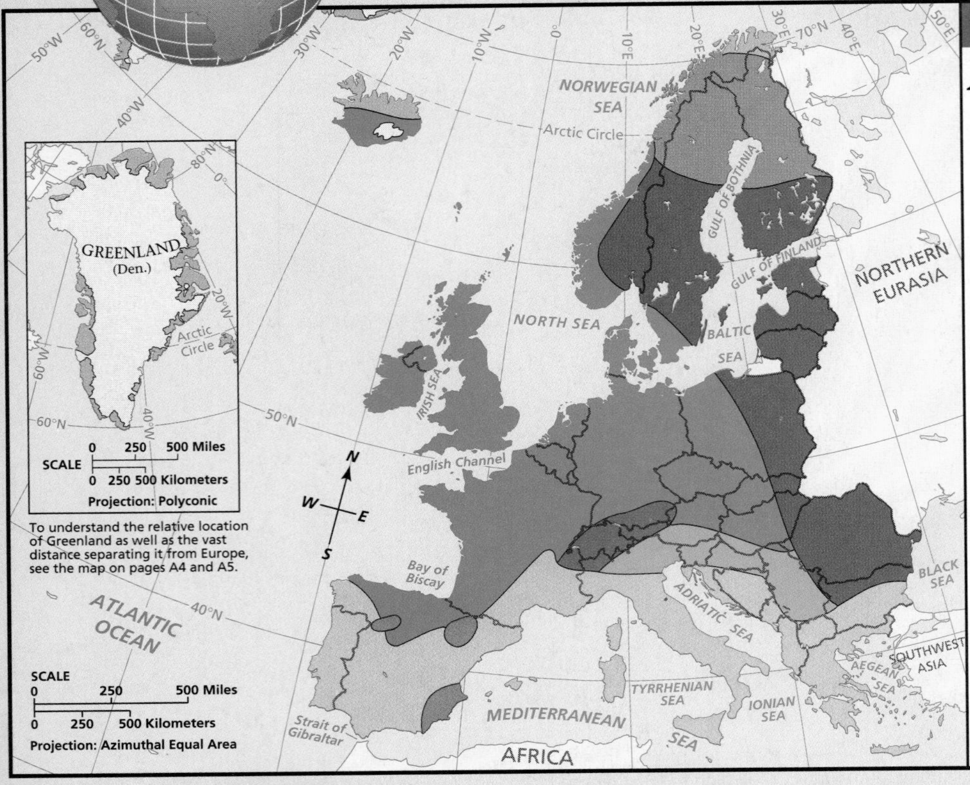

CLIMATE

LEGEND

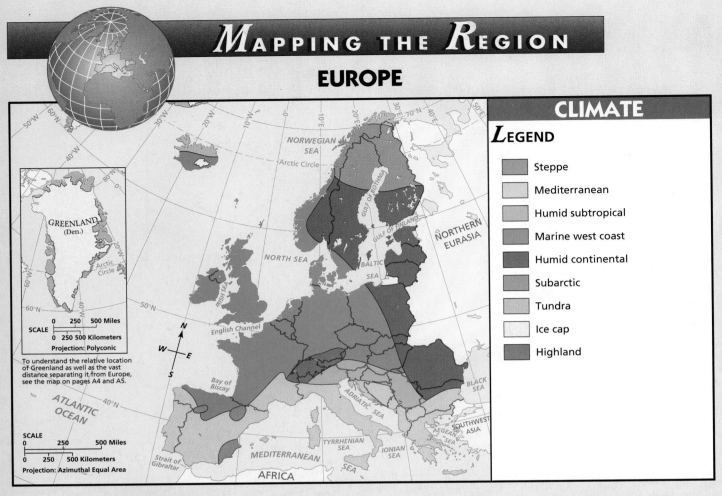

- Steppe
- Mediterranean
- Humid subtropical
- Marine west coast
- Humid continental
- Subarctic
- Tundra
- Ice cap
- Highland

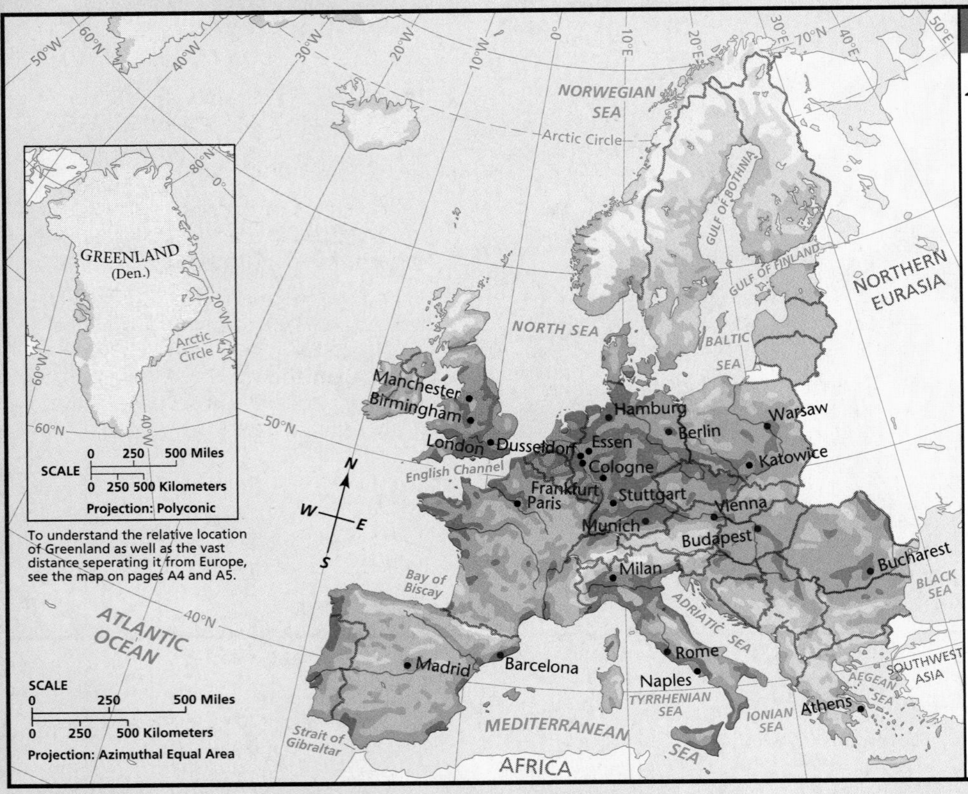

SCALE: 0–250–500 Miles; 0–250–500 Kilometers
Projection: Polyconic

To understand the relative location of Greenland as well as the vast distance separating it from Europe, see the map on pages A4 and A5.

SCALE: 0–250–500 Miles; 0–250–500 Kilometers
Projection: Azimuthal Equal Area

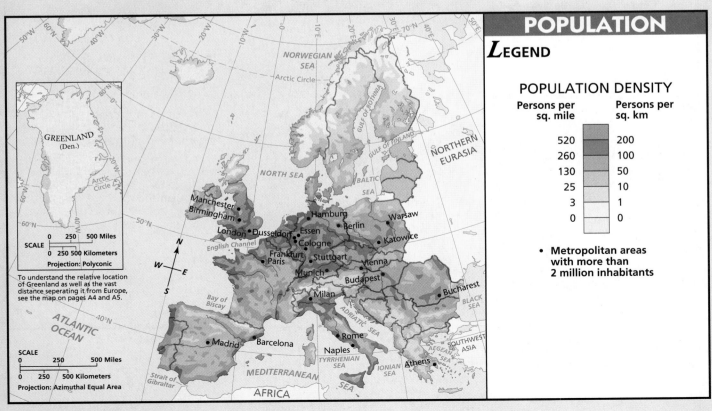

POPULATION

LEGEND

POPULATION DENSITY

Persons per sq. mile	Persons per sq. km
520	200
260	100
130	50
25	10
3	1
0	0

- **Metropolitan areas with more than 2 million inhabitants**

SCALE: 0–250–500 Miles; 0–250–500 Kilometers
Projection: Polyconic

To understand the relative location of Greenland as well as the vast distance seperating it from Europe, see the map on pages A4 and A5.

SCALE: 0–250–500 Miles; 0–250–500 Kilometers
Projection: Azimuthal Equal Area

MAPPING THE REGION

EUROPE

ECONOMY
LEGEND
Herding
Livestock raising
Commercial farming
Dairying
Commercial fishing
Forests
Manufacturing and trade
Limited economic activity
Coal
Natural gas
Oil
Nuclear power
Hydroelectric power
Geothermal power
Uranium
Other minerals
Timber
GREENLAND (Den.)
Arctic Circle
SCALE
0 250 500 Miles
0 250 500 Kilometers
Projection: Polyconic
To understand the relative location of Greenland as well as the vast distance separating it from Europe, see the map on pages A4 and A5.
NORWEGIAN SEA
Arctic Circle
NORTH SEA
BALTIC SEA
GULF OF BOTHNIA
NORTHERN EURASIA
ATLANTIC OCEAN
Bay of Biscay
Strait of Gibraltar
MEDITERRANEAN SEA
ADRIATIC SEA
AEGEAN SEA
BLACK SEA
SOUTHWEST ASIA
AFRICA
RUHRSTADT
Narvik
Helsinki
Oslo
Stockholm
Copenhagen
Dublin
Rostock
Hamburg
Bremen
Berlin
Warsaw
London
Rotterdam
Brussels
Frankfurt
Prague
Paris
Munich
Vienna
Bratislava
Budapest
Zurich
Lyons
Ljubljana
Milan
Turin
Genoa
Marseilles
Bucharest
Sofia
Oporto
Lisbon
Madrid
Barcelona
Rome
Naples
Salonika
Piraeus
Athens
Palermo
Catania
N
S
W
E
SCALE
0 250 500 Miles
0 250 500 Kilometers
Projection: Azimuthal Equal Area

CLIMATE

◆ A marine-west-coast climate dominates most of Northern and West Central Europe.

◆ Southern Europe, including most of Italy and all of Greece, has a Mediterranean climate.

◆ Away from the oceanic influence of the mild North Atlantic Drift, much of Eastern Europe has a humid-continental climate.

Compare this map to the physical–political map on page 259.

? Which countries have an entirely marine-west-coast climate?

? Name the major mountain ranges where highland climates are shown.

Compare this map to the economy map above.

? What economic activities are common in the subarctic and tundra climates of northern Scandinavia?

POPULATION

◆ One of Europe's most crowded areas is in a belt stretching through Belgium, the Netherlands, and central Germany.

◆ Except along coastal areas, the northern Scandinavian countries of Norway, Sweden, Finland, and Iceland are sparsely populated.

◆ Italy has many of the most crowded areas in Southern Europe.

? Which metropolitan areas west of the prime meridian (0°) have populations of more than 2 million?

Compare this map to the physical–political map on page 259.

? Which country has the most metropolitan areas with populations of more than 2 million?

Compare this map to the climate map on page 260.

? Which climate regions of Scandinavia are most sparsely populated?

ECONOMY

◆ Europe's irregular coastline has created many natural harbors that have become important manufacturing and trade centers.

◆ The rivers of Europe are of great importance for trade, water resources, and hydroelectricity.

◆ Petroleum and natural gas have been found beneath the waters of the North Sea claimed by Norway and the United Kingdom. Most other European countries must import all of their oil and gas.

Compare this map to the physical–political map on page 259.

? Name the manufacturing and trade centers shown in Spain and Portugal.

? What area of western Germany is an important manufacturing region?

? Which country of Eastern Europe is a producer of petroleum?

1 PHYSICAL GEOGRAPHY

FOCUS

- *What are Europe's four major landform regions?*
- *What climate types are found in Europe?*

Landforms Europe west of the former Soviet Union can be divided into four major landform regions: the Northwest Highlands, the Northern European Plain, the Central Uplands, and the Alpine mountain system. (See the map on page 259.) Landforms in Europe vary over very small distances.

The Northwest Highlands is an ancient, eroded region of rugged hills and low mountains. The region includes the hills of Ireland and England, the Scottish Highlands, Brittany in northwestern France, most of the Iberian Peninsula, and the Kjølen (CHUHL-uhn) Mountains in Norway and Sweden. Much of the region was affected by the continental ice sheets during the last ice age. When the ice melted, it left thousands of lakes, poor river drainage, and thin soils.

The largest landform region in area is the Northern European Plain. This broad coastal plain extends from the Atlantic coast of France into Russia and borders the North Atlantic Ocean, the North Sea, and the Baltic Sea. Most of the region is less than 500 feet (152 m) above sea level.

The third landform region is the Central Uplands. This ancient, hilly area of Europe consists of the Massif Central (ma-SEEF sehn-TRAHL) of France and the Jura Mountains on the French-Swiss border. It also includes the Black Forest and the Bavarian Plateau of Germany, the Bohemian Highlands of Germany and the Czech (CHEHK) Republic, and the Ardennes (ahr-DEHN), in the Benelux countries.

The Alps are western and central Europe's highest mountain range. They stretch from the Mediterranean coast of France through Switzerland, Austria, and northern Italy to the Balkan Peninsula. Many peaks in the Alps reach more than 14,000 feet (4,268 m). The highest peak, France's Mont Blanc (mohn BLAHN), reaches to 15,771 feet (4,807 m). Because of their high elevations, the Alps have large glaciers and frequent avalanches due to heavy winter snowfall.

Other ranges in the Alpine system include the Carpathian Mountains in Eastern Europe and the Apennines, which run the length of the Italian peninsula. The Pyrenees (PIR-uh-neez) form a **natural boundary** between Spain and France. When physical features such as mountain ranges and rivers determine political borders, they are called natural boundaries.

The Rhine River cuts a path through western Europe toward the North Sea. Ships can sail the Rhine all the way from Switzerland to the sea.

REGION **Climate** Compared to world regions of similar latitude, much of Europe has mild conditions throughout the year. The moderating influence of the North Atlantic Ocean helps prevent extreme temperatures. (See the map on page 260.) The high latitudes mean long summer days and long winter nights.

Europe has three major climate types: marine west coast, humid continental, and Mediterranean. The marine-west-coast climate is found throughout most of Northern and West Central Europe. The prevailing westerly winds blowing over the warm ocean current called the North Atlantic Drift bring mild temperatures and rain. Frequent Atlantic storms bring clouds and even more rain to the region. Rainfall averages between 20 and 80 inches (51 and 203 cm) a year. The windward western slopes of the coastal mountains of Scotland and Norway receive the region's highest rainfall totals due to the orographic effect. Occasional snow and frosts occur in winter.

To the east and inland of the marine-west-coast climate is a humid-continental climate region. This region is located away from the direct influence of the mild Atlantic currents. It stretches across central Sweden and Finland and southward across most of Eastern Europe as far south as northern Bulgaria. The humid-continental climate has four distinct seasons, including a cold, snowy winter and a humid summer. Winters are severe from Finland to Poland. Periodic summer droughts affect Hungary and Romania.

High mountains, particularly the Alps, separate the marine-west-coast and humid-continental climates from Europe's third major climate region, the Mediterranean climate. This climate region includes most of Portugal, Spain, Italy, Albania, Macedonia, and Greece, and parts of France, Croatia, and Bulgaria. (See the map on page 260.) The Mediterranean region receives between 10 and 30 inches (25 and 76 cm) of rainfall a year. Most of the rainfall comes during the mild winter when occasional North Atlantic storms pushed by the westerly winds bring rain into the Mediterranean basin. A long sunny, dry summer is typical.

Pockets of minor climate types cover parts of Southern and Northern Europe. A small area of humid-subtropical climate is located south and southeast of the Alps, mainly in the Po Valley of northern Italy and extending into the Balkan region between the Adriatic and Black seas. In some areas of Spain, where high mountains block moist ocean air from reaching farther inland, a dry steppe climate prevails. A subarctic climate, meanwhile, stretches across northern Norway, Sweden, and Finland. The very northern portions of these countries and the island nation of Iceland have a tundra climate.

1 SECTION REVIEW

1. To which landform regions does France belong?
2. What are Europe's three major climate types? Name one characteristic of each climate type.
3. **Critical Thinking** **Determine** which areas of Europe are likely to be agriculturally productive, based on landforms and climates. Explain your reasoning.

2 NATURAL RESOURCES

FOCUS

- *How are various bodies of water important to Europe?*
- *What are Europe's major resources?*

MOVEMENT **Water** Europe's long, irregular coastline is indented with hundreds of excellent ice-free natural harbors. Because these harbors generally are located near the mouths of **navigable** rivers, Europe is ideally situated for trade by sea. A navigable river is one that is deep and wide enough for ships.

Of the seas bordering Europe, the Mediterranean is the largest. Although it suffers from pollution, the Mediterranean supports some fishing. The Mediterranean is connected to the Black Sea by the strait of Bosporus (BAHS-puh-ruhs) in Turkey. This strait gives Bulgaria and Romania access to world sea routes. Geographers consider the Bosporus to be a boundary between Europe and Asia.

Other major European seas are the North Sea and the Baltic Sea. The North Sea is vital for trade and fishing, and beneath its waters lie Europe's major oil reserves. The Baltic Sea ices over during the winter months, however. Ships

The docks of **Hamburg, Germany**, are crowded with products ready for export to world markets.

called **icebreakers** are needed to keep sea lanes open to Sweden, Finland, and the Baltic countries of Estonia, Latvia, and Lithuania. The warm North Atlantic Drift keeps Europe's other important seaways free of ice during winter.

The rivers of Europe are of great importance for trade, water resources, and hydroelectricity. Most are connected by a series of canals. Even cities far from the coast have access to the sea through this extensive river and canal system. The Rhine and Danube are Europe's most developed rivers. Unfortunately, massive amounts of pollution enter the sea via these and other rivers. Cleaning up and controlling pollution in Europe's rivers is a serious environmental challenge.

Some scientists believe acid rain killed much of the dense tree growth that once surrounded this home in the Black Forest of **Germany**.

INTERACTION Forests and Soil

Most of Europe's original forests were cut for timber or cleared for farming centuries ago. In recent years, air pollution and acid rain have killed many of the remaining trees. Only in Sweden and Finland do large areas of timber-producing forest still exist. Clearing and overgrazing have destroyed nearly all of the Mediterranean's original oak forests, leaving only a scrub-plant community covering the hillsides. Europe must now import wood to satisfy its timber needs. Most European nations, however, have begun reforestation programs.

Europeans have made good agricultural use of their soil, and farmers grow a wide variety of crops. Intensive farming techniques and modern technology have made Europe's crop yields among the highest in the world. Much of Eastern Europe, however, lags in agricultural production, because its farming technology is not as modern as that in other European countries.

Minerals and Energy To meet its current industrial and energy needs, Europe relies heavily on imports. The United Kingdom, Germany, the Benelux countries, and the Czech Republic, however, all have deposits of iron and coal, which were important in the growth of these industrial countries. Sweden and France also have large iron deposits.

The oil reserves and natural-gas deposits beneath the North Sea and the natural-gas deposits in the Netherlands do not satisfy Europe's industrial demands. Imported oil and gas from Southwest Asia, Africa, and Russia supply more than 80 percent of the energy needs of most European countries.

Europe has other energy sources, however. Nuclear power fills some energy needs, particularly in France, Belgium, Great Britain, and Hungary. Hydroelectricity, meanwhile, is plentiful in the mountainous nations such as Norway and Switzerland. France has been successful in producing ocean tidal power and in developing and using solar energy.

2 SECTION REVIEW

1. Name three seas and two rivers found in Europe. Why are seas and rivers important to Europe?
2. What mineral and energy resources does Europe have? Why must Europe import oil and gas?
3. **Critical Thinking** **Compare and contrast** deforestation in Europe with the deforestation that is taking place in Middle and South America.

HISTORICAL GEOGRAPHY

FOCUS

- *What were the major historical developments in Europe prior to World War II?*
- *How has Europe changed since World War II?*

Rise of Nations Throughout history, different people have fought for the control of Europe. From about 900 to 300 B.C., while the Greek civilization was flourishing in the Mediterranean area, warring tribes ruled the rest of Europe. Wars and conquests continued throughout the expansion of the Roman Empire during the second century B.C. The Roman Empire at its peak included northern Africa, southwestern Asia, and most of western and central Europe.

The fall of the Roman Empire marked the beginning of a period in Western European history, from about A.D. 500 to 1500, called the Middle Ages. During the first part of this period, various Germanic groups established a number of new kingdoms. The most important of these was the Frankish kingdom. It reached its height in about 800, under the reign of the emperor Charlemagne (SHAHR-luh-mayn).

During the latter part of the Middle Ages, Europeans increasingly ventured out onto the oceans, dominating world trade for centuries. From 1000 to 1500, the Mediterranean ports of Venice and Genoa were the main centers of trade.

By 1500, after centuries of bitter fighting among kingdoms, France, Spain, Portugal, and England had become nations. France rose as a strong power in Europe in the mid-1600s, and by the early 1800s, it controlled most of western Europe. Then in 1815, the British defeated Napoléon's French army at Waterloo, a town in Belgium. For nearly a century after Waterloo, Britain was Europe's leading political and economic power.

Imperialism During the sixteenth and seventeenth centuries, Spain and Portugal became the first European nations to establish large foreign colonies. By the second half of the nineteenth century, the British, French, Dutch, and Belgians also had established colonies in the Americas, Africa, Asia, and the Pacific. The policy of gaining control over territory outside a nation is called **imperialism**.

Britain's colonial empire became the largest in the world. Its trade connections, skilled population, and control of the seas made Britain the world's leading power during the nineteenth century. At its height, the British Empire included one-fourth of the world's population and one-fifth of the world's land area. It was true that "the sun never set on the British Empire" because, with the earth's rotation, some part of the empire was always in daylight.

The World Wars Germany, which did not become a unified nation until 1871, was late to enter the colonial race. But by the end of the nineteenth century, Germany, too, had obtained colonies in Africa and the Pacific. By then, Germany was one of the strongest military and industrial powers in Europe. Its military buildup frightened many other European nations.

In this atmosphere of fear and mistrust, the European nations formed military **alliances**. An alliance is an agreement between countries to support one another against enemies. Countries joined in an alliance are called allies.

These European alliances helped keep a **balance of power** in the region for several years. A balance of power exists when members of a group have equal levels of strength. This balance in Europe, however, was lost in the early 1900s. World War I broke out between the two main alliances of Germany and Austria-Hungary, and Great Britain, France, and Russia.

Germany lost World War I and faced harsh peace terms from Great Britain and its allies. Bitterness over the war combined with severe economic hardship led to World War II only 21 years later. Germany and its ally Italy conquered much of Europe before being defeated by the allied forces of Great Britain, the United States, the Soviet Union, and several other nations.

Post–World War II After the war, Germany was divided into two nations, East Germany and West Germany. East Germany and the rest of Eastern Europe came under Communist rule and the control of the Soviet Union. The end of World War II marked the beginning of the period known as the Cold War.

The military alliance of the North Atlantic Treaty Organization (NATO) formed in 1949 to

War has helped shape the history and landscape of Europe. The cross-covered beaches of **Normandy, France,** are evidence of the sacrifices and the costs of these wars.

provide for the common defense of western Europe. Its goal was to protect its members from possible invasion by the Soviet Union and its Eastern European allies, who in 1955 formed their own alliance called the Warsaw Pact. Members of NATO include 13 European nations plus Turkey, Canada, and the United States.

During the early 1990s, Europe again experienced historic changes. As the Soviet Union began to break up, the Cold War thawed and East and West Germany underwent **reunification**. Reunification is the process of reuniting, or rejoining, into one unit. Czechoslovakia (chehk-uh-sloh-VAHK-ee-uh) peacefully split into two nations, while the three major ethnic groups in the new nations of former Yugoslavia engaged in a terrible civil war. With the end of the Cold War, the Warsaw Pact dissolved. The role of NATO likely will change as well.

SECTION REVIEW

1. What countries dominated Europe from the 1600s through the 1800s?
2. Why was NATO formed? Why will its role likely change?
3. **Critical Thinking** **Agree or disagree** with the following statement: "The history of Europe is a history of wars and conflict."

HUMAN GEOGRAPHY

FOCUS

- *What are the demographic and cultural characteristics of Europe?*
- *How are the economies of Europe's country's linked?*
- *What challenges does Europe face?*

Population Changes The population of Europe grew rapidly in the years between 1000 and 1300. Between 1300 and 1450, however, population growth in Europe experienced major declines due to diseases, **famines**, and wars. A famine is an extreme shortage of food.

Then during the seventeenth and eighteenth centuries, agricultural advances led to another increase in population. As the Industrial Revolution took hold, trade increased and living standards rose. Workers took factory jobs and moved from farms and villages to the cities. Today, Europe's main centers of population reflect the urbanization that took place during the Industrial Revolution. London, Paris, Milan, Amsterdam, and Berlin all grew dramatically during the 1800s.

Over the past several centuries, Europe's surge in population led in part to increasing emigration. Millions of Europeans have immigrated to the United States, Canada, and Australia. Others have moved to overseas colonies or settled in South American nations such as Brazil, Argentina, and Chile. Many of these emigrants were seeking new opportunities or were escaping religious persecution, wars, famine, and poverty.

Today, mainly due to economic progress, much of western Europe is attracting immigrants to its shores. Foreign laborers, called guest workers, have moved to western European countries from poorer Eastern and Southern European countries and from Turkey, North Africa, and Asian nations such as India and Pakistan. In some urban areas, overcrowding has led to slum conditions and increasing ethnic tension.

Even with immigration, however, European population growth rates are the lowest in the world. In fact, some European countries, such as Hungary and Germany, are experiencing a population decline. In addition, life expectancy rates that are among the highest in the world coupled with low birthrates means that the region is becoming a population of older people.

Language Most of the languages spoken in Europe are of the Indo-European family. This language family includes the Germanic, Celtic, Romance (Latin or Italic), Slavic, and Greek language groups among others. (See the chart below.)

Although English is the most widely spoken language in Europe, German and French are also widespread. Ninety percent of all Europeans between the ages of 15 and 24 speak a second language. Some countries, including Luxembourg, Switzerland, the Netherlands, and Belgium, are considered **multilingual**. That is, many people speak three or more languages.

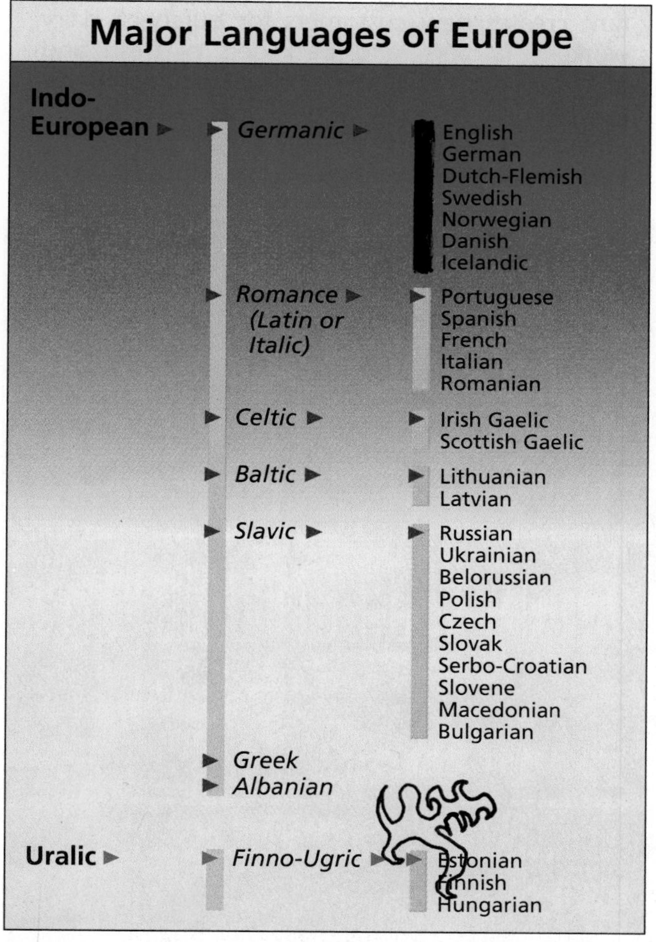
Major Languages of Europe
Indo-European
Germanic
English
German
Dutch-Flemish
Swedish
Norwegian
Danish
Icelandic
Romance (Latin or Italic)
Portuguese
Spanish
French
Italian
Romanian
Celtic
Irish Gaelic
Scottish Gaelic
Baltic
Lithuanian
Latvian
Slavic
Russian
Ukrainian
Belorussian
Polish
Czech
Slovak
Serbo-Croatian
Slovene
Macedonian
Bulgarian
Greek
Albanian
Uralic
Finno-Ugric
Estonian
Finnish
Hungarian

Languages from two major language families are spoken across much of Europe. Languages that are related are grouped together within these language families. Find the language group to which English belongs. What other languages are part of that language group?

Religion Many different religions are practiced in Europe. Roman Catholicism is the dominant religion in the Southern European countries such as Spain, Italy, and Portugal. Catholics also are in the majority in Poland, France, Austria, Ireland, and Belgium. Most of Northern and Central Europe is mainly Protestant, including the United Kingdom, the Scandinavian countries, and Germany. Small numbers of Jews live in many parts of western Europe. Significant numbers of Muslims live in southeastern Europe and among the new immigrants and guest workers from North Africa, Turkey, and Southwest Asia.

REGION **Economic Cooperation** World War II put severe stress on the economies of the European nations. Moreover, after the war, several European colonies in Asia and Africa gained their independence. The colonies had been major sources of inexpensive raw materials that supported industrial growth in some European countries. To strengthen their economies, the nations of western Europe began forming **economic associations**. An economic association is an organization formed to break down trade barriers among member nations.

The most important of these cooperative economic associations has been the **European Union**, commonly known as the **EU**. The EU, once called the European Community, was formed in 1957 by Belgium, the Netherlands, Luxembourg, France, West Germany, and Italy. Since then, nine other countries have joined: the United Kingdom, Ireland, Denmark, Greece, Spain, Portugal, Austria, Finland, and Sweden. Most other European nations, as well as Turkey, are seeking membership.

Combining the natural resources and industries of its members, the EU has become the world's largest exporter. The present 15-nation organization has a population far exceeding that of the United States and Japan, its two main trade rivals. The EU seeks a common European currency, and its ultimate goal is a unified region with a free flow of people and goods. In 1993, several European nations passed the Maastricht Treaty, which outlines steps toward further economic and political integration in the EU.

Switzerland, Norway, Liechtenstein, and Iceland belong to the European Free Trade Association (EFTA). The EFTA holds trade agreements with the EU.

Text continues on page 270.

Global Connections

Tourism

Tourism is growing so rapidly that it is now the world's largest industry. In 1993, 500 million people traveled the world. By 2000, the number will have exploded to 650 million.

Responsible tourism requires a balance between economic concerns and cultural and environmental concerns.

With an estimated 8 million jobs depending on tourism, the European Union (EU) has pushed specific plans to attract visitors to member countries. Nearly 300 million tourists visited Europe in 1993, spending more than $160 billion. Despite the importance of tourism to the region's economy, some people question whether European nations are wise to encourage more of it. For years, France's Lascaux caves have been closed to tourists after the discovery that such traffic endangered the prehistoric cave paintings. Restrictions may also be placed on entrance to Paris's Cathedral of Notre Dame. More than 11 million people shuffled through the cathedral in 1990.

Throngs of tourists are flocking to the rest of Europe and the globe as well. More than 19 million visitors poured into Britain in 1993. Venice, Italy, has played host to as many as 100,000 tourists on peak summer days. In East Asia, meanwhile, tourism in Thailand, Malaysia, Singapore, and Japan has made the region the world leader in growth in international tourism.

The Tourist Industry

Despite tourism-related problems such as pollution and overcrowding, most nations continue to encourage tourism. In some countries, tourism brings in billions of dollars. Italy, for example, took in more than $21 billion in 1992.

Money spent by tourists creates jobs. Hotels and restaurants employ a variety of workers, including bellhops, waitpersons, and cooks. Tourism also supports other industries. The building of a hotel provides jobs for construction workers, plumbers, electricians, and the makers of brick, glass, and steel. The opening of a restaurant creates new customers for butchers, dairy workers, farmers, and packagers of fresh and frozen foods. For some developing nations, tourism may mean the difference between poverty and a standard of living that provides citizens with basic education and health care.

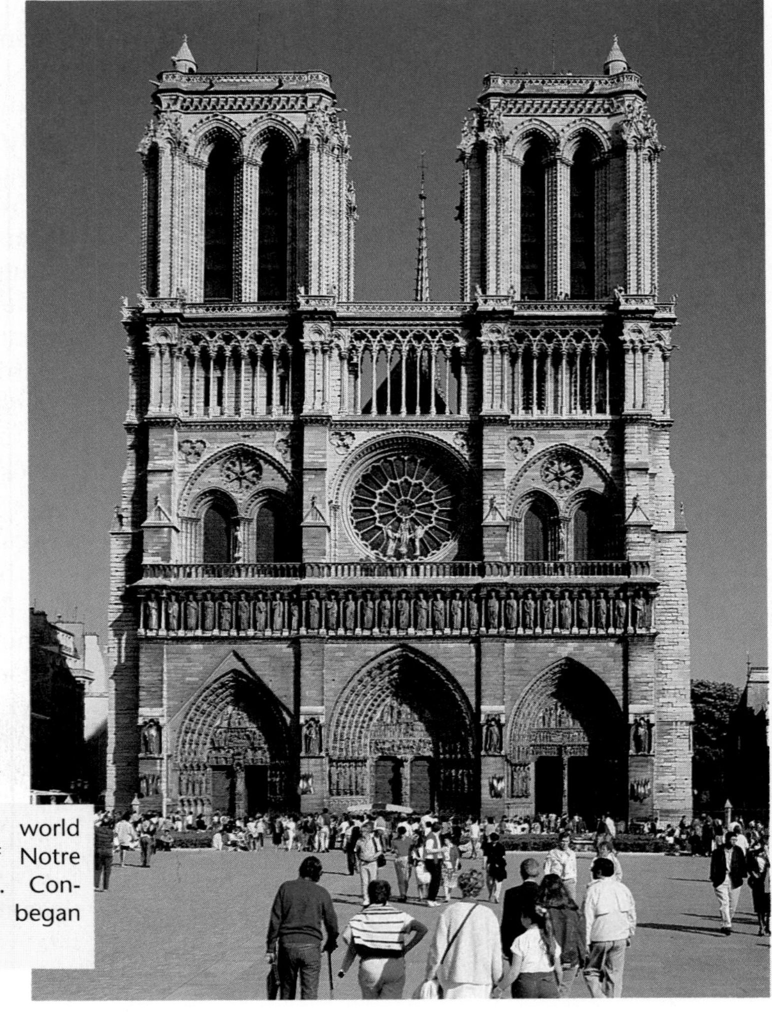

Tourists from around the world flock to the Cathedral of Notre Dame in **Paris, France**. Construction of the cathedral began in A.D. 1163.

Sometimes, however, nations reap few rewards from tourism. When foreign developers build hotels and restaurants that serve tourists, much of the money from tourism flows out of the country. Tourism also can disrupt cultures. For example, hula skirts and shell necklaces are alien to a number of Polynesian cultures. Yet because tourists expect to see these items throughout the Pacific, many Polynesians have adopted the skirts and jewelry to satisfy visitors. Although some areas do not go so far as to adopt new customs, they often mass-produce souvenirs such as machine-made baskets that cheapen the fine handicrafts their cultures have developed over centuries.

Tourists look over crafts at a "shell market" on a beach in the **Yasawa Islands**, part of the South Pacific country of **Fiji**. Natural beauty and mild climates draw many tourists to South Pacific islands.

Unrestrained tourism also has left its mark on much of the world's natural environment. Desperate for photographs, some tourists disobey the rules of African wildlife preserves, persuading guides to drive as close as possible to rare and endangered animals. Tourism has been blamed for spoiling the French Riviera, a region along the Mediterranean coastline. Overconstruction has led to overcrowding, and the dumping of wastes has polluted the Mediterranean Sea. Now, the Alps and the Himalayas also are threatened by tourism.

Responsible Tourism

Tourists as well as host nations gradually are learning important lessons. Both realize that a balance must be reached between economic concerns and a responsibility to a region's culture and environment. Nations know that tourists will eventually turn away from places where indigenous cultures have disappeared and natural environments have been destroyed. The French Riviera, for example, experienced a 30 percent decline in tourism in recent years.

Many tourists have become more sensitive travelers. Some are signing up for "ecotours" that promote tourism to wilderness areas and wildlife reserves while encouraging conservation of the environment. On most ecotours, tourists must be willing to do without the conveniences of fancy hotels and luxury rental cars.

Responsible tourists approach other nations and cultures with respect for languages, customs, and ways of life different from their own. Responsible host nations recognize that, in the long run, the economic rewards of tourism will multiply when cultures are preserved and natural environments are left unspoiled.

This climber nears the top of the world on one of the mighty peaks of the Himalayas near **Mount Everest** in South Asia. The Himalayas, which have some of the highest mountains in the world, are popular destinations for mountain climbers and hikers.

Your Turn

Plan a trip in which you will visit three different areas of the world. **Write** a brief description of each place, and **explain** why you have selected each location. Also explain how you will plan your travels so that you show respect for cultures and do not harm natural environments. With your plans, include an outline map of the world with each of your chosen locations labeled.

Historic Cologne Cathedral rises above modern buildings in this city along the Rhine River in **Germany**. Many European cities feature historic buildings nestled among modern structures in areas once leveled by fighting in World War II.

European Ways of Life Today, European city skylines include shopping centers, fast-food restaurants, and high-rise buildings next to their centuries-old architecture. Europe is a society of consumers, much like the United States. Teenagers on both sides of the North Atlantic Ocean share similar tastes in clothing, music, movies, and sports.

Advanced transportation and communication networks crisscross much of Europe. The region's railroads, highways, seaports, river-canal systems, and airline connections are generally modern and efficient. Europeans are also leaders in the development of high-speed rail and aerospace technology. In some European countries, however, particularly in Eastern and Southern Europe, transportation is not as well developed.

Most European nations provide many social services to their citizens. Social programs often provide for the health care, education, and welfare of citizens throughout their lives. All these programs are supported by taxes that are high by U.S. standards. Taxes collected by governments also provide public transportation and support education, sports, and the arts.

Issues Americans and Europeans face some similar challenges, such as unemployment, crime, traffic congestion, and limited energy resources. Environmental pollution is another concern among Europeans. Antinuclear movements and environmental political parties such as the "Greens" influence the politics of some European nations. Overcrowding, loss of wilderness areas, and a long history of environmental abuse have brought environmental issues to the top of Europe's list of concerns.

Some minority groups in Europe, such as the Basques in Spain and various ethnic groups in the former Yugoslavia, are seeking independence. In some instances, these movements have led to violence. In addition, many European cities have large foreign populations of guest workers. Violence toward these immigrants has broken out, especially in former East Germany, where unemployment is high. In addition, terrorist groups have carried out bombings, assassinations, hijackings, and kidnappings.

Other European concerns are at the international level. The political and economic changes and ethnic tensions in the former Soviet Union and in the Balkan nations are affecting the entire region. Europe also faces continued economic competition from Japan and the United States.

To face these challenges, Europe is seeking to strengthen economic and political unity among its many countries. While political unity remains a distant vision, greater economic cooperation promises increased prosperity for all of Europe.

SECTION REVIEW

1. Name five language groups found in Europe.
2. What is the European Union?
3. **Critical Thinking** **Explain** the causes of modern-day violence in Europe.

CHAPTER 23 CHECK

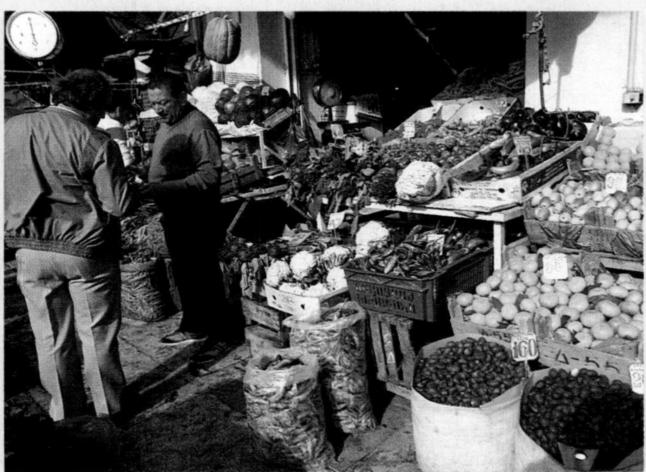
Produce market in **Athens, Greece**

Reviewing the Main Ideas

1. Europe has four major landform regions and three major climate types. The largest landform region is the Northern European Plain. Most of Europe has mild climate conditions.
2. Europe's navigable rivers and long coastline make it ideally situated for trade. Europe also has agricultural, mineral, and energy resources, although some imports are necessary.
3. Throughout history, various groups have fought for control of Europe and of the world's oceans and trade. Europe's control of overseas territories, or imperialism, reached its peak during the late nineteenth century. Two world wars were fought during the first half of the twentieth century. From World War II until the 1990s, much of Eastern Europe was under Communist rule.
4. Europeans speak many different languages and practice several religions. The European Union is one of the world's most important economic associations. Most of Europe is prosperous and has a high level of social services, but Europeans still face many economic, cultural, political, and environmental challenges.

Building a Vocabulary

1. To what does the term *Eurasia* refer?
2. What is a microstate?
3. What type of boundary do the Pyrenees form?
4. Define *alliance*. What are the members of an alliance called?
5. What happened to East and West Germany in 1990?
6. Why might a famine cause a population decline?

Recalling and Reviewing

1. Which sea in Europe requires the use of icebreakers during the winter? What countries border this sea, and what climates do they have?
2. What environmental issues surround the use of Europe's forests? its rivers?
3. How did the loss of overseas colonies affect some of Europe's economies?
4. Identify four European countries that are multilingual.

Thinking Critically

1. Would you say that Europe is a region of increasing unity, or do you think it is a region of increasing disunity? Support your answer.
2. Some people believe that one must understand Europe's history in order to understand the rest of the world. Do you agree? Why or why not?

Learning About Your Local Geography

Individual Project

You read that Europe's social programs are supported by taxes that are high by U.S. standards. Make a list of the different taxes that people in your area must pay. For what purposes is each tax used? Conduct a brief survey of your community members asking their feelings about the taxes they pay. Do they think their taxes are low or high? How do they feel about the way their tax money is spent? You may want to include your survey in your individual portfolio.

Using the Five Themes of Geography

Sketch a map of the world. With arrows, symbols, and shading, show how **location** and **movement** interacted to make Europe an economic leader. Be sure to include a legend for any symbols you use. Write a caption explaining the map's contents.

CHAPTER 24

Northern Europe

"The English love . . . the pleasant tranquillity of the countryside. But deeper down, they feel the sea as their native land, their own domain. . . . Nothing is more enticing for the English than strange tales of the sea, pirate adventures, sea heroes. From the time they are small children, they dream of distant travels and sea-washed glory. The sea satisfies the romantic, anxious necessities of the English spirit."

Nikos Kazantzakis

GEOGRAPHY DICTIONARY

constitutional monarchy

loch

nationalize

bog

peat

fjord

socialism

neutral

uninhabitable

geyser

Gold Celtic jewelry

Northern Europe is a land of many islands and narrow peninsulas where fingers of the sea cut into the long coastlines. The nations of Northern Europe have been heavily influenced by their locations. Their nearness to the sea and, for some, their far north location, have greatly affected their climate, cultural heritage, and economic development.

Northern Europe can be divided into two regions: the British Isles and Scandinavia. The British Isles lie on the continental shelf off the European coast and consist of Great Britain and Ireland. These islands are separated by the Irish Sea and are divided politically into two nations, the United Kingdom and the Republic of Ireland.

Scandinavia, or the Nordic countries, covers a broad area stretching from the western Russian border to the North Atlantic Ocean. This region includes the nations of Norway, Sweden, Denmark, Iceland, and Finland.

Southern coast of the **United Kingdom**

NORTHERN EUROPE: PHYSICAL–POLITICAL

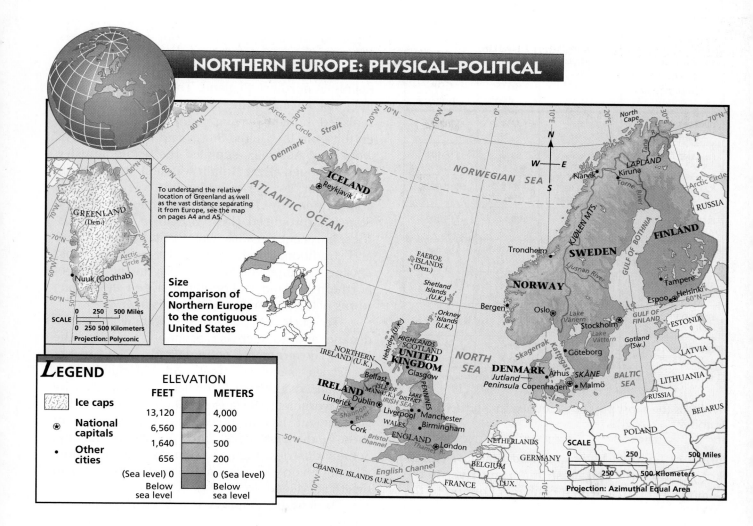

To understand the relative location of Greenland as well as the vast distance separating it from Europe, see the map on pages A4 and A5.

Size comparison of Northern Europe to the contiguous United States

1 THE UNITED KINGDOM

FOCUS

- *What are the United Kingdom's physical and economic characteristics?*
- *What role does the United Kingdom play in the global community?*

REGION The United Kingdom (UK) includes England, Scotland, and Wales (all on the island of Great Britain) and Northern Ireland. Because the people of the United Kingdom live mainly on the island of Great Britain, they are often referred to as the British. The United Kingdom is one of the most populous nations in Europe. It also has one of Europe's highest population densities.

The United Kingdom is a **constitutional monarchy**. That is, it has a reigning king or queen, but a parliament serves as the lawmaking branch of government. The royal family is a symbol of history and patriotism to the British.

Physical Geography Great Britain can be divided into two landform regions, lowland and highland Britain. (See the map above.) The southeastern half of the island is lowland Britain. This region is part of the Northern European Plain. Only the narrow English Channel separates the British portion from the rest of the Northern European Plain. Highland Britain is in the north and west. The major highland areas are made up of ancient eroded mountains. Highland Britain includes the beautiful English Lake District and the large Scottish lakes, or **lochs**. Lochs are long, deep lakes carved by glaciers.

For centuries, the rivers of Britain have been important for trade. They connect the major

FACTS IN BRIEF Northern Europe

COUNTRY POPULATION (1994)	LIFE EXPECTANCY (1994)	LITERACY RATE (1991)	PER CAPITA GDP (1993)
DENMARK 5,187,821	73, male 79, female	99%	$18,500
FINLAND 5,068,931	72, male 80, female	99%	$16,100
ICELAND 263,599	77, male 81, female	99% (1992)	$16,000
IRELAND 3,539,296	73, male 79, female	99%	$13,100
NORWAY 4,314,604	74, male 81, female	99%	$20,080
SWEDEN 8,778,461	75, male 81, female	99%	$17,600
UNITED KINGDOM 58,135,110	74, male 80, female	99%	$16,900
UNITED STATES 260,713,585	73, male 79, female	97%	$24,700

Sources: *The World Factbook 1994,* Central Intelligence Agency; *The World Almanac and Book of Facts 1995*

Which is the most populous country in Northern Europe?

ports of London, Liverpool, and Glasgow (GLAS-koh) to the sea. The most important river is the Thames (TEHMZ), which connects London to the North Sea.

Even though Great Britain is relatively far north (50° to 60° north latitude), it has a marine-west-coast climate. The prevailing westerly winds blowing off the North Atlantic Drift help keep its climate mild. Britain is known for its rainy, cloudy, and foggy weather. Britain's west coast, especially in Scotland and the Hebrides Islands, averages more than 60 inches (152 cm) of rainfall per year.

INTERACTION **Economic Geography** In the 1700s, Great Britain had abundant coal and iron deposits, a large labor force, and an excellent transportation network of rivers and canals. By the early 1800s, the British had built the first railroads. All these resources helped Britain lead the Industrial Revolution.

Three of Great Britain's early industries were iron and steel, shipbuilding, and textiles. Deposits of iron ore and coal found in the central part of England fueled the iron and steel industry. During the middle of the nineteenth century, with Great Britain's steel industry already established, steel replaced wood as a material for building ships. The island's coastal cities were a perfect home for this industry. The British textile industry began mainly in northern England. The invention of spinning machines and the harnessing of steam power replaced home weaving. Local wool and, later, cotton from the United States and India, as well as wool from Australia, supplied the textile industry.

Although Great Britain was first to industrialize, it lost its world leadership role to foreign competition. By 1900, Germany and the United States outproduced Great Britain in many industries. In recent decades, Great Britain's coal mines and traditional industries have suffered rapid decline. Unemployment, especially in the cities of northern England and Scotland and in the mining towns of the Midlands and Wales, is high.

To curb its post–World War II industrial decline and to increase economic equality, the British government **nationalized** many of its industries, such as steel and shipbuilding. A nationalized industry is owned and operated by the government. Because of limited success, the government has in recent years sold some of these industries back into private ownership. Some industries, such as railroads, remain nationalized, as do services such as the postal and health services.

Agriculture employs less than 2 percent of the British work force yet produces more than half of the nation's food. The southern part of Great Britain has prosperous, modernized farms. In contrast, most of highland Britain has limited farming because of the rocky, glaciated terrain. Grazing of cattle and sheep is common on the island.

The United Kingdom has more fossil fuels than any other European nation. In 1969, oil was discovered under the North Sea, and the United Kingdom became Europe's major oil producer and exporter. The oil deposits will last only a few more decades, however. The United Kingdom also depends on coal, offshore natural-gas deposits, and hydropower to generate electricity. In addition, it was one of the first countries to use nuclear power; about one-fifth of its electricity is nuclear generated.

Today, most British are employed in service industries, especially banking, insurance, communications, and government. In education, British universities are considered some of the best in the world. The tourist industry is also important to the economy.

Kings
SAMPAN

A British police officer provides security during a celebration by immigrants from South Asia in **London**. Many immigrants to the **United Kingdom** are from former colonies of the British Empire.

PLACE **London** London is the United Kingdom's capital city and largest urban area. It is home to the Houses of Parliament and Buckingham Palace, the residence of the British monarch, and many cathedrals and museums. London is also a center of trade, industry, and services, such as banking and insurance. Its early economic importance was due to its location on the Thames River. London is a major theater, retail, and tourist center as well.

Like most other large cities, London suffers from urban decay, pollution, and traffic. The government has made effective efforts to clean up London's air and the waters of the Thames. A vast rail network helps ease some traffic congestion. The government also maintains a "greenbelt" of parks and woodlands that circle the city.

Northern Ireland The six northern counties of Ireland are called Northern Ireland or sometimes Ulster. Part of the United Kingdom, Northern Ireland sends representatives to Parliament in London. The majority of Northern Ireland's population is descended from Scottish and English Protestants who arrived during the seventeenth century. Irish Roman Catholics form almost one-third of the population.

There has been bitter fighting between the Protestant majority and the Roman Catholic minority of Northern Ireland. The Roman Catholics are seeking more social, political, and economic opportunities. Many would like Northern Ireland to join the Republic of Ireland, which is mostly Roman Catholic. The Protestants, however, fear becoming the minority on the island and want to remain a part of the UK.

Due to the violence, British troops have been stationed in Northern Ireland since 1969. Several thousand people have been killed by fighting and terrorist activities. Open talks to end the conflict began after the Irish Republican Army (IRA), one of the most active of the terrorist groups, agreed to stop terrorist activities in 1994.

A Global Role At its height during the nineteenth and early twentieth centuries, the British Empire ruled the seas. The empire's colonies provided raw materials for British industries and served as markets for finished goods. When World War II ended in 1945, many colonies sought independence. New nations appeared on the map, including India, Pakistan, Jamaica, Kenya, and Nigeria. The United Kingdom still holds some territories scattered around the world.

The once-powerful British Empire has been replaced by a commonwealth. In this context, a commonwealth is an association of self-governing states with similar backgrounds and united by a common loyalty. The Commonwealth of Nations is a voluntary association of 50 former British colonies. Member nations meet to discuss economic, scientific, and business matters.

The United Kingdom joined the European Union in 1973. The British now look to other European nations for trade rather than to their distant former colonies. For instance, sugar may come from France instead of Jamaica, and dairy products come from Denmark rather than New Zealand. The UK remains a major international trading nation, exporting more than 30 percent of its goods.

SECTION REVIEW

1. Into what two landform regions can Great Britain be divided?
2. What is the Commonwealth of Nations?
3. **Critical Thinking** **Explain** how Britain's location, resources, and industry combined to make Britain the leading world power during the nineteenth century.

THE REPUBLIC OF IRELAND

FOCUS

- *What are Ireland's physical features?*
- *What human, political, and economic characteristics does the Republic of Ireland have?*

Physical Geography Across the Irish Sea from Great Britain is Ireland, sometimes called the "Emerald Isle" because of its green countryside. (See the map on page 273.) The entire island has a marine-west-coast climate. Rainfall varies from 30 to 100 inches (76 to 254 cm) per year.

Rolling green fields and pastures cover the island's central part. The Shannon River, the longest river in the British Isles, flows through the middle of the island. Low, rocky mountains form a rugged coastline. Much of Ireland is either rocky or boggy. **Bog** is soft ground that is soaked with water. For centuries, **peat** deposits from the bog areas have been used for fuel. Peat is composed of decayed vegetable matter, usually mosses. Less than 5 percent of the island is forested.

Human Geography The Irish, like the Welsh and Scots, are descendants of the Celts who settled in Ireland around 400 B.C. Irish Gaelic, a Celtic language, is still considered an official language but is spoken by only a small percentage of the Irish. Today, English is the main language.

In the thirteenth century, the Norman rulers from Britain conquered Ireland. Eventually, the Protestant British controlled most of the land, reducing the Roman Catholic Irish to landless farmers. Lack of economic opportunity coupled with a potato famine during the nineteenth century caused many Irish to emigrate, particularly to the United States.

For several hundred years, the Irish rebelled against British control. Then in 1921, Ireland became a dominion, similar to Canada, called the Irish Free State. In the 1930s, the Irish government cut most ties with Britain, and a 1937 constitution renamed Ireland as Eire (EHR-uh). The completely independent Republic of Ireland, recognized by Britain, was formed in 1949. Today, the Irish Republic is made up of 26 counties and covers about 80 percent of the island.

Economic Geography Although Ireland's farming population has declined in recent decades, farm income has risen. Since Ireland joined the European Union in 1973, Irish farmers have modernized agriculture and now grow a wider variety of crops. The country produces dairy products, beef, poultry, potatoes, and grains both for export and home consumption. More than half of Ireland's farm products are exported to the EU nations.

Most Irish workers are employed in manufacturing. Some of Ireland's industries are based on food products, including dairies and breweries. Other factories manufacture glassware and fine textiles. The greatest industrial development has

Traffic along O'Connell Street crosses the River Liffey in **Dublin, Ireland**. O'Connell Street is the Irish capital's main road and is part of the central shopping district. Statues of famous Irish people line the center of the street.

occurred around the seaport city of Dublin. Nearly one-third of the nation's population now lives in Dublin, the capital and center for education, banking, and shipping. The country's other, smaller cities lie mainly along the coast.

Except for peat deposits used to generate some electricity, Ireland is an energy-poor island, with little oil, natural gas, or coal. Its high taxes and high unemployment rates are depressing the economy. People continue to emigrate from Ireland to seek jobs elsewhere, especially in the United Kingdom, the United States, Canada, and Australia. In addition, by borrowing money to help its unemployed, Ireland has acquired a huge debt. The country's future economic development depends on more foreign investment, new industries, and increased tourism.

2 SECTION REVIEW

1. Why is Ireland sometimes called the "Emerald Isle"?
2. What economic activities are important to Ireland?
3. **Critical Thinking** **Suggest** three reasons that the Irish rebelled against British control in the nineteenth century.

3 THE NORDIC COUNTRIES

- *What are the environments of the Nordic countries?*
- *What are the economic and human characteristics of the region?*

Once the home of fierce, warlike Vikings, today the Nordic countries are peaceful, prosperous, and industrialized nations. In general, the people are urbanized and well educated and have some of the highest standards of living in the world. Because of their high health standards, long lifespans, and low birthrates, they also have some of the oldest populations in the world. Except for Finnish, all the Scandinavian languages are closely related, and all the Nordic nations are Protestant (Lutheran) in religion.

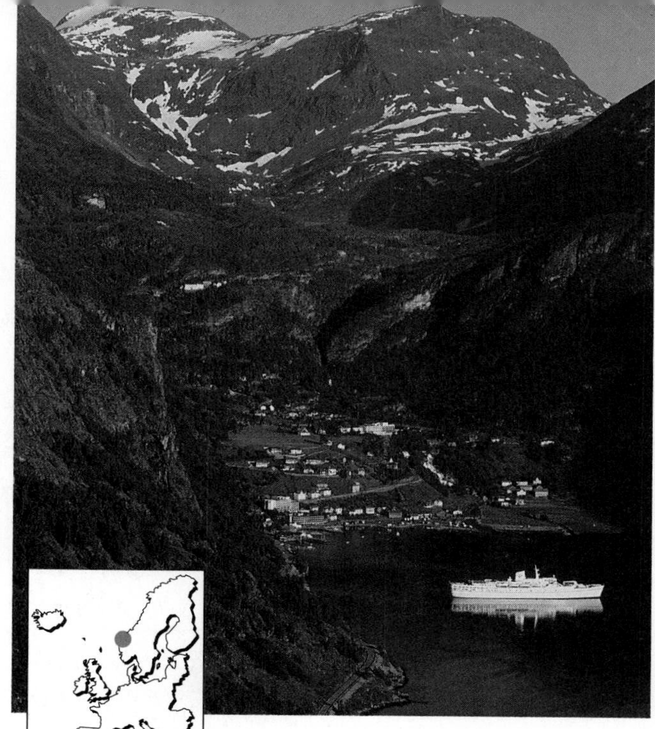

This fjord in southern **Norway** provides a scenic setting for ships and tourists. How were fjords formed?

Norway Norway is Europe's northernmost nation. It is a long, narrow country bounded by the North Atlantic Ocean and the North Sea on the west. It has a long eastern border with Sweden, which occupies the other half of the Scandinavian Peninsula. In the far northeast, Norway borders Finland and Russia. The country also includes thousands of islands. (See the map on page 273.)

Norway's climate is influenced by the westerly winds and the warm North Atlantic Drift. Most of coastal Norway has a marine-west-coast climate, and the ocean waters remain ice-free all year. The far north is Arctic tundra and barren mountains. In winter, snow blankets northern Norway under almost continuous darkness. In contrast, there is almost continuous daylight during the summer. This is why the area north of the Arctic Circle is called the "land of the midnight sun."

Norway is one of Europe's most mountainous nations. In fact, transportation by sea or over mountains on skis was once the only way to reach many Norwegian towns. The Kjølen Mountains tower over most of the country. These steep mountains and the country's wet climate make Norway a major producer of hydroelectricity. Its inexpensive electricity has attracted many industries.

Most Norwegians live on narrow coastal plains or on the shores of the **fjords** (fee-AWRDS).

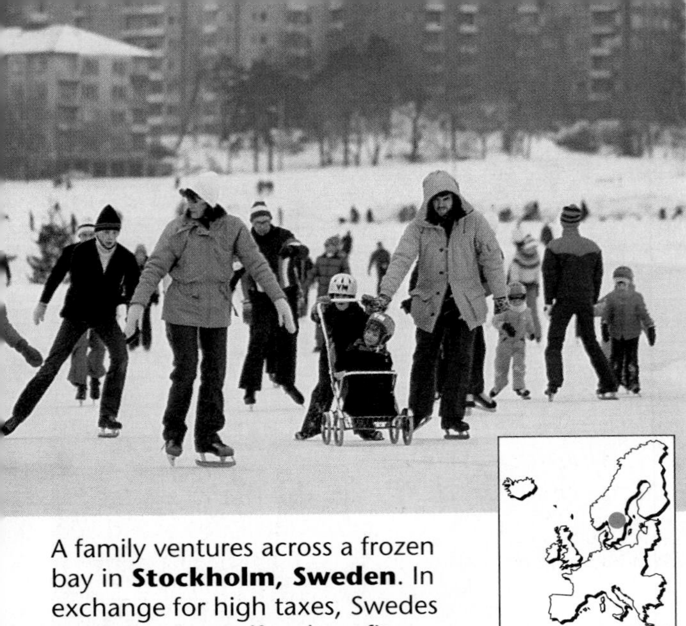

A family ventures across a frozen bay in **Stockholm, Sweden**. In exchange for high taxes, Swedes receive many welfare benefits, including incentives for raising children.

Fjords are narrow, deep inlets of the sea between high, rocky cliffs. Ice-age glaciers that stretched down to the sea carved the fjords out of the mountains. Spectacular waterfalls dropping from vertical cliffs along the coast attract tourists to this area each summer.

The long, rugged coastline, sheltered harbors, and the closeness of fishing grounds make Norway Europe's leading fishing nation. Fishing is one of the oldest industries in Norway and provides major export products. Shipping also is important to the Norwegian economy. Even Norway's most valuable resource, oil, comes from the sea. Norway's North Sea oil fields, developed during the 1970s, are expected to produce for only a few more decades, however. Today, the country exports most of its oil to other European nations.

More than half of Norway's population lives in the very southeast corner of the nation. Oslo, the capital, cultural center, and leading seaport and industrial center, is located here. This modern city lies at the end of a wide fjord. Well-planned parks, wide avenues, and houses built along the hillsides are typical. Government buildings and the royal palace also grace the city. Oslo is close to hiking and skiing trails. Business people are often seen taking winter lunch breaks on cross-country skis through the city parks.

Sweden Sweden is the largest Scandinavian country, both in area and in population. Except for a marine-west-coast climate along its southern tip, Sweden is divided into two cold climates. The southern-central portion has a humid-continental climate, while the northern portion is subarctic. Unlike ice-free Norway, Sweden's east coast on the Gulf of Bothnia usually freezes during winter. The Kjølen Mountains line Sweden's western border with Norway. The rest of the country is generally a rolling, glaciated plain with thin soils and thousands of lakes. (See the map on page 273.)

Sweden's four main sources of wealth are forestry, farming, mining, and manufacturing. Forests cover more than half of Sweden and provide large quantities of wood pulp for paper and building materials. Sweden is the world's third largest exporter of wood products. To help sustain the forests, the Swedish government limits cutting and encourages replanting.

The richest farmland in Sweden is in the southern region, called Skåne. Here, the soil is fertile, and the climate is mild. Wheat, rye, potatoes, sugar beets, cattle, and dairy products are sources of income. Less than 5 percent of the Swedes are farmers, yet they produce nearly all of the country's food.

Northern and central Sweden have rich iron-ore deposits. Kiruna, in the north, is the major iron-mining center. From here the ore is exported to other European countries from the port of Narvik in northern Norway. Hydroelectric power from Sweden's rivers provides more than half of the nation's electricity.

Sweden's industries are highly diversified, technologically advanced, and well managed. The nation produces a variety of products for export, such as its high-quality steel. The country also produces automobiles, aircraft, household appliances, chemicals, furniture, glassware, and a variety of electrical products. It is a world leader in the production and use of industrial robots.

Sweden is one of the most prosperous and democratic nations in the world. The country has a mixed economy with some elements of **socialism**, an economic system in which the government owns and controls the means of producing goods. Sweden's government controls about 10 percent of the nation's industries and services, including telephones, railroads, airlines, and some electric utilities. The remaining 90 percent of the country's economy is in the private sector. A number of Swedish businesses are run by cooperatives owned by their customers.

Sweden's socialist system includes an extensive social-welfare program. Government programs take care of almost all the educational and medical needs of Swedish citizens. Such programs are expensive and take up about one-third of the national budget. The Swedes pay for these services with some of the world's highest taxes.

About 85 percent of Sweden's population lives in cities and towns, mainly in the country's southern part. Stockholm, the capital, is the nation's main commercial center. It is a beautiful city of islands and forests.

Sweden has a **neutral** foreign policy and is not a member of NATO. A country is neutral when it chooses not to take sides in international conflicts. Sweden ended its membership in the EFTA and joined the EU in 1995.

Denmark Denmark is the smallest of the Nordic countries. Most of the country's land area lies on the low Jutland Peninsula, which borders Germany on the south. More than 400 islands make up the rest of the country, including the large island of Greenland. Denmark faces the North Sea on the west and the Baltic Sea on the east. Narrow straits separate Denmark from Norway and Sweden. The country has benefited economically from this central location between land and water routes. (See the map on page 273.)

Denmark is the most densely populated Nordic country. One-fourth of its population lives in the area surrounding Copenhagen, the capital city. Copenhagen lies on an island along a narrow strait leading to the Baltic Sea. The city is a center of trade, shipping, and manufacturing.

Part of the Northern European Plain, Denmark's flat lowlands are excellent for farming. In fact, about 60 percent of the country's land is used for agriculture. The nation's mild marine-west-coast climate and rich pastureland support widespread dairy farming and hog raising. The Danes are famous for their fine butter, cheese, and ham. Together, dairy and meat products account for most of the country's farm income.

Although Denmark is a largely agricultural nation, it also is modern and industrialized. The Danes have succeeded in producing quality goods for world markets, such as machinery, furniture, medicines, packaged foods, and electronics. All of Denmark's industry is privately owned.

Denmark has a large fishing fleet to take advantage of its access to the sea. Danish North Sea waters also hold natural-gas and oil deposits, which may decrease Denmark's dependence on imported fuels.

Denmark is one of the most prosperous of the EU nations, and the Danes have one of the world's highest per capita incomes. The country's social-welfare system includes medical care, child care, and care for the aged. High taxes help fund these and other welfare programs.

Greenland Greenland lies between the North Atlantic and Arctic oceans. Athough located nearer to North America, it is a self-governing province of Denmark. Greenland is not a descriptive name for this island as thousands of feet of ice cap cover nearly 85 percent of it. The entire icy interior of the island is **uninhabitable**. An uninhabitable region is one that cannot support human life and settlements. Only along the coasts are there areas of tundra climate where people can live. Even much of the coastal area is glaciated barren rock.

Most of the people of the island are Eskimo, also called Inuit, or are of European-Eskimo descent. Most Greenlanders live along the island's southwestern coast. Fishing is the chief occupation, but some Eskimos still practice traditional hunting of seals and small whales.

Iceland The island nation of Iceland lies in the middle of the North Atlantic Ocean just south of the Arctic Circle. (See the map on page 273.) Situated on the active Mid-Atlantic Ridge where the North American and Eurasian plates are separating, Iceland is Europe's westernmost nation.

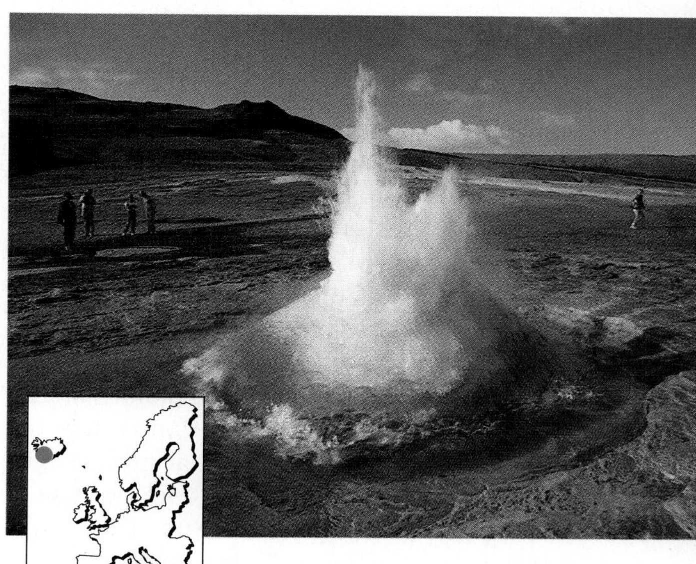

An erupting geyser in **Iceland** scatters some wary visitors. Iceland uses its underground heat for geothermal energy.

This is a land of lava rock and more than 100 volcanoes. Three-quarters of Iceland is mountainous and uninhabitable. More than 100 glaciers cover the high mountain interior, while much of the coast is extremely rugged and battered by waves. The name Iceland is misleading, since coastal temperatures are mild and remain above freezing. Due to its mid-ocean location, Iceland has mainly a marine-west-coast climate, which is wet and cloudy much of the year.

Iceland's most important natural resource is the rich North Atlantic fishing waters that surround the island. More than 70 percent of the nation's export income comes from fishing. Iceland's other major resource is energy. Because of its mountains and wet climate, Iceland produces large amounts of hydroelectricity.

Iceland also harnesses geothermal energy. As a result of volcanic activity, underground water rises all over the island as steam, forming **geysers**. The word *geyser* is an Icelandic term for hot springs that shoot hot water and steam into the air. This hot water heats most Icelandic homes, as well as many greenhouses, where crops such as tomatoes, grapes, and cucumbers grow year-round.

The world's northernmost capital, Reykjavik (RAYK-yuh-veek) contains more than half of Iceland's population. The country has little crime or unemployment.

Finland Most of Finland lies between two arms of the Baltic Sea: the Gulf of Bothnia to the west and the Gulf of Finland to the south. Finland borders Russia on the east and Sweden and Norway in the far north. (See the map on page 273.) Finland's landforms show the effects of the last ice age, when a thick sheet of ice covered the whole country. Finland is mainly a glaciated plain with about 50,000 lakes, extensive areas of bog and wetlands, and thousands of small offshore islands.

Because the Gulf of Bothnia freezes during most winters, Finland has more severe climates than the other Nordic countries, which are near ice-free seawater. Finland has three major climate regions. In the far south is a humid-continental climate with summer temperatures that permit agriculture. The central part of the country has a subarctic climate of evergreen forests. The northern tip of the country has a tundra climate.

The severe climates and glaciated terrain make only 10 percent of Finland suitable for farming. Hay and oats are grown to feed cattle, and potatoes and grains are important food crops. Forests, which cover about 60 percent of the country, are Finland's greatest natural resource and make the nation a world leader in paper and other forest products. Finland also produces electronics and is a major shipbuilder. Finnish exports include televisions, mobile telephones, icebreakers, and cruise ships.

Finland's major trade partners are other Nordic nations and the nations of the European Union. The Soviets used to buy manufactured goods from Finland in exchange for oil and gas. In an effort to adjust to the market loss resulting from the collapse of the Soviet Union, Finland joined the EU in 1995.

About one-fifth of the population of Finland lives in or around Helsinki, which lies along the Gulf of Finland. Helsinki is the Finnish capital, leading industrial area, and major seaport.

The Finns came from northern Asia. Their language belongs to the same language family as Hungarian and Estonian and is unrelated to other languages of Europe. Finland's second official language is Swedish, which almost all Finns learn in school.

Lapland Across northern Finland, Sweden, and Norway is a region known as Lapland. This tundra region is populated by the Lapps, or Sami as they call themselves. These reindeer herders probably originated from hunters of northern Asia.

The Lapps have managed to maintain some of their culture, even with a modern Nordic way of life in their midst. Although only about 5 percent of their population is involved in traditional reindeer herding, the herders are important to the Lapps as the keepers of their culture. Most Lapps now earn a living from tourism.

3 SECTION REVIEW

1. What two countries do the Kjølen Mountains border?
2. What are Iceland's major resources? How has Finland's economic situation changed in recent years?
3. **Critical Thinking** **Specify** one advantage and one disadvantage of the social-welfare systems in Sweden and Denmark.

CHAPTER 24 CHECK

Big Ben and the Houses of Parliament along the Thames River, **London**

Reviewing the Main Ideas

1. Great Britain can be divided into two landform regions and has a marine-west-coast climate. Although Great Britain was the world's first industrialized nation, its traditional industries have declined. Northern Ireland is part of the United Kingdom. The British Empire has been replaced by the Commonwealth of Nations, which includes former British colonies.
2. Ireland is covered with rolling green fields. Once under British control, the southern 80 percent of the island of Ireland is now the independent Republic of Ireland. Most Irish workers are employed in manufacturing.
3. The Nordic countries, including Norway, Sweden, Denmark, Iceland, and Finland, have some of the highest standards of living in the world. Landforms range from mountains to plains, and climates are cold. In general, the economies of the Nordic countries are industrial.

Building a Vocabulary

1. What term describes Britain's government?
2. What is a loch? Where are lochs found?
3. Define *bog*. What fuel product comes from bogs?
4. What is a fjord? Which country has many fjords?
5. Define *uninhabitable*. Which countries discussed in the chapter have large uninhabitable regions?

Recalling and Reviewing

1. Compare the economic systems of the United Kingdom and Sweden.
2. How is the population of each Northern European country distributed, particularly with regard to cities? Name a major city in each country.
3. What energy resources are found in the Northern European countries?

Thinking Critically

1. Why do you think Northern Ireland remains part of the United Kingdom when the rest of Ireland is an independent republic? What are the causes of conflict in Northern Ireland?
2. Many sports that people enjoy today began as activities essential to everyday life. How do you think skiing has been essential to daily life in the Nordic countries? What other sports might have originated from activities that people performed to meet their basic needs?

Learning About Your Local Geography

Cooperative Project

In Sweden, government programs take care of the educational and medical needs of Swedish citizens. In your own area, who is responsible for paying the costs of people's educational and medical needs? With your group, research how each set of needs is addressed. Present your information in a chart.

Using the Five Themes of Geography

Using Iceland as an example, write a paragraph discussing the effects that **location** and **place** characteristics can have on **human-environment interaction**. How have Iceland's location and place characteristics affected the ways in which the Icelandic people interact with their environment?

CHAPTER 25

West Central Europe

"Meanwhile, a thousand miles to the north, the wind that had started in Siberia was picking up speed for the final part of its journey. We had heard stories about the Mistral. . . . Typical Gallic [French] exaggeration, we thought. . . . And so we were poorly prepared when the first Mistral of the year came howling down the Rhone valley, turned left, and smacked into the west side of the house with enough force to skim roof tiles into the swimming pool and rip a window that had carelessly been left open off its hinges. The temperature dropped twenty degrees in twenty-four hours."

Peter Mayle

GEOGRAPHY DICTIONARY

mistral
primate city
city-state
loess
cosmopolitan
dike
polder
confederation
canton
foehn
arable

Shield and helmet of King Charles IX of **France** (1550–1574)

The history of West Central Europe is rich and varied, and the region's influence has extended over distant parts of the world. Today, West Central Europe is the continent's industrial heartland and center of population and agricultural production. This core region includes France; Germany; the Benelux countries of Belgium, the Netherlands, and Luxembourg; and the Alpine countries of Switzerland and Austria.

France is a leading industrial and agricultural nation, whose people share a great cultural pride in their country. Germany is one of the world's largest industrial powers. West of Germany, along the North Sea, lie the modern and industrialized Benelux countries. The landlocked, rugged Alpine countries, meanwhile, have prospered by taking advantage of their central European location.

Geneva, Switzerland

WEST CENTRAL EUROPE: PHYSICAL–POLITICAL

LEGEND

ELEVATION

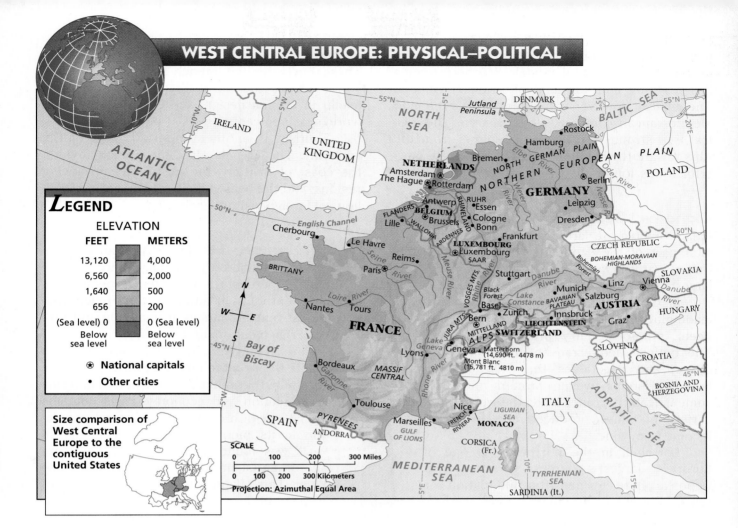

FEET	METERS
13,120	4,000
6,560	2,000
1,640	500
656	200
(Sea level) 0	0 (Sea level)
Below sea level	Below sea level

⊛ National capitals

• Other cities

Size comparison of West Central Europe to the contiguous United States

SCALE 0 100 200 300 Miles; 0 100 200 300 Kilometers

Projection: Azimuthal Equal Area

FRANCE

FOCUS

- *What is France's physical geography? economic geography?*
- *What human and urban characteristics does France have? What role does France play in the global community?*

REGION Physical Geography France's landforms vary from high mountains to hills and flat plains. Its three coastlines face the North Atlantic Ocean to the west, the English Channel to the northwest, and the Mediterranean Sea to the southeast. Numerous rivers and two major climate regions add to the country's diverse geography. (See the map above.)

Except for Brittany, which is part of Europe's Northwest Highlands landform region, most of northern France is part of the Northern European Plain. This lowland area is the largest landform region of France.

The mostly hilly Central Uplands lie mainly in central and eastern France and include the Massif Central, a volcanic area in south central France. The Ardennes, Vosges, and folded Jura mountain ranges form France's eastern borders with Germany and Switzerland. The rocky and rugged Mediterranean island of Corsica also falls in this landform region.

The mountainous region of southern France includes the Pyrenees and the Alps. The beautiful French Riviera is located where the Alps meet the Mediterranean coast.

France has four important rivers: the Seine, the Loire, the Garonne, and the Rhone. The canals that connect these rivers are being modernized to provide a better network for transporting goods to neighboring nations.

Except for the high mountain areas, almost all of France has a mild climate. Northern France has a marine-west-coast climate with mild temperatures and plentiful rainfall. Southern France has a sunny and warm Mediterranean climate. The **mistral**, however, sometimes breaks the mild Mediterranean winter weather pattern. The mistral is a strong, cool wind that blows from the Alps toward the Mediterranean coast. These powerful winds follow the valley of the Rhone River.

Economic Geography Although few of the French people are farmers, agriculture remains a vital part of the economy. Second only to the United States in farm exports, France is by far the largest farm exporter in the European Union.

In the north and interior of France, the leading agricultural products are wheat, sugar beets, potatoes, dairy products, and meat. The typical Mediterranean crops of grapes, olives, and vegetables thrive in southern France. Here, however, farmers must use some irrigation. France leads the world in the variety of wines produced and in export income from wine.

Lorraine, in the northeast, is one of Europe's largest sources of iron ore. Deposits here formed the foundation of France's early steel industry and were a major export. The traditional steel, coal, and textile industries are being replaced by new industries, however.

Some French industries, particularly those using high technology, have expanded rapidly. France produces satellites, jet aircraft, and helicopters. The nation's aerospace industry, which requires aluminum, is centered in southern France near the region's bauxite mines.

The country's main economic concerns lie in the area of energy resources. The French use some hydroelectricity and solar energy, and operate a tidal power plant on the north coast. Still, the nation relies on nuclear energy for more than half of its electricity and must import oil as well.

The French are famous for the production and export of perfumes, jewelry, glassware, and furniture. These finely crafted French goods are sold in stores around the world. France also remains a leader in fashion design, and Paris is one of the world's fashion centers.

Tourism is another important part of the economy. The French Alps, the French Riviera, the coast of Brittany, the many wine-producing regions, and the beautiful city of Paris attract millions of foreign visitors each year.

PLACE **Human Geography** France has a rich culture unified by language and religion. About 75 percent of the French people are Roman Catholic. In addition, the French government budgets money to maintain and promote French culture and language.

Paris, France's capital and largest urban area, is the **primate city** of France. A primate city is one that ranks first in a nation in terms of population and economy. Paris contains almost one-fifth of the nation's total population in its metropolitan area. The city of Paris was founded more than 2,000 years ago on an island in the middle of the Seine River. This island was at the crossroads of water and land routes that were important to France and the rest of Europe. As Paris grew, it spread from the island onto both banks of the Seine.

Paris, a nodal region, is the nation's center for education, government, communications, banking, business, and tourism. As the transportation center of France, Paris is the hub of all major rail, canal, air, and highway systems in the country. The city is served by a modern subway system called the *métro,* two major airports, and the nearby seaport cities of Le Havre (luh HAHVR) and Cherbourg (SHEHR-bur). Like other major cities, Paris suffers from air pollution, traffic, noise, and urban sprawl.

The busy seaport of Marseilles (mahr-SAY) on the Mediterranean is France's oldest and second largest city. Marseilles also has a major French

Eiffel Tower, **Paris**

naval base, and is an industrial center. Lyons (lee-OHN), on the Rhone River in eastern France, is the nation's third largest city. Lyons also is an important manufacturing center.

Issues Like most industrialized nations, France faces problems such as unemployment and international competition in industry and agriculture. Dependence on imported oil is a continuing concern for the nation.

France enjoys close ties with most former colonies through an organization called the French Community. Immigration from these former colonies and elsewhere is a heated issue in France. Politicians have debated the effects of immigration on the economy and French culture.

The country still maintains overseas territories such as Guadeloupe in the Caribbean and Réunion in the Indian Ocean, which send representatives to France's parliament. The people of the Mediterranean island of Corsica have a degree of self-government, yet some want complete independence from France. Meanwhile, independence movements on the French island of New Caledonia in the Pacific have led to violence.

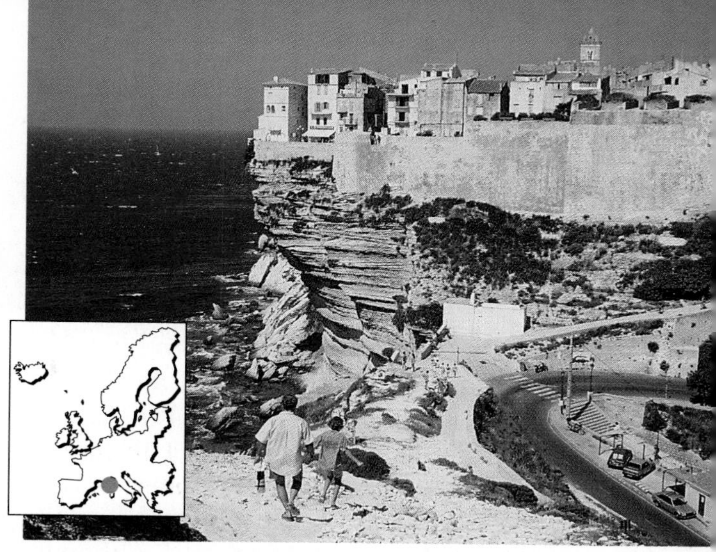

The historic town of **Bonifacio** overlooks the Mediterranean Sea at the southern tip of **Corsica**. The French island's coastline consists of high cliffs with few natural harbors.

France has had a long relationship with the United States. The French helped the American colonies fight for their independence from Britain. Over a century and a half later, during World War II, the United States helped drive the Germans out of France. Signs of the countries' special relationship mark both cultures. For example, the United States' Declaration of Independence and the French Declaration of the Rights of Man express similar ideas. Perhaps the most obvious symbol is the Statue of Liberty in New York harbor, a gift from France to the United States.

Militarily, France is strong, with its own nuclear-defense system. Although France withdrew its armed forces from NATO in the 1960s, the country has kept some ties with the alliance for the defense of Europe. France is one of the most forceful influences in the European Union.

FACTS IN BRIEF West Central Europe

COUNTRY POPULATION (1994)	LIFE EXPECTANCY (1994)	LITERACY RATE	PER CAPITA GDP (1993)
AUSTRIA 7,954,974	73, male 80, female	99% (1992)	$17,000
BELGIUM 10,062,836	74, male 80, female	98% (1992)	$17,700
FRANCE 57,840,445	74, male 82, female	99% (1991)	$18,200
GERMANY 81,087,506	73, male 80, female	99% (1991)	$16,500
LIECHTENSTEIN 30,281	74, male 81, female	100% (1992)	$22,300 (1990)
LUXEMBOURG 401,900	73, male 81, female	100% (1989)	$20,600
MONACO 31,278	74, male 82, female	99% (1989)	$16,000 (1991)
NETHERLANDS 15,367,928	75, male 81, female	99% (1991)	$17,200
SWITZERLAND 7,040,119	75, male 82, female	99% (1991)	$21,300
UNITED STATES 260,713,585	73, male 79, female	97% (1991)	$24,700

Sources: *The World Factbook 1994*, Central Intelligence Agency; *The World Almanac and Book of Facts 1995*

The German economy is recovering from the high costs of reunification in the early 1990s.

1 SECTION REVIEW

1. What landform regions and climates does France have? What economic activities are found in different areas of France?
2. How are Paris, Marseilles, and Lyons important?
3. **Critical Thinking** **Compare** France's relationships with its former and present colonies to those of the United Kingdom with its former and present colonies.

GERMANY

FOCUS

- *What are Germany's physical and economic features?*
- *What are Germany's five major regions? What challenges does Germany face?*

The German federal system consists of 16 *Lander* (states), of which three are **city-states**: Berlin, Bremen, and Hamburg. A city-state is a self-governing city and its surrounding area. Although the seat of government presently remains in Bonn, Berlin is the capital of Germany.

REGION **Physical Geography** Germany is divided into three main landform regions: the North German Plain, the Central Uplands, and the Alps. (See the map on page 283.) The elevation of Germany increases from north to south. The lowest region is the North German Plain, which has coastlines on the North and Baltic seas and is part of the Northern European Plain. The German portion of the Central Uplands contains the hilly Black Forest and Bohemian Forest. The Alps rise along the southwestern border of Germany.

Germany's major rivers include the Rhine, the Danube, the Elbe, and the Weser. An excellent canal system links the rivers to form a network of waterways. The Rhine is particularly vital to Germany's transport system and industry.

Most of Germany has a marine-west-coast climate. Landforms and distance from the sea cause the temperature and rainfall to vary. The northern part of the country, which is affected by moist and mild ocean winds, has mild winters and cool summers. The region to the south and east, toward the interior of Germany, experiences a more continental influence. Here, summers are warmer, and winters are colder. Most of Germany receives enough rainfall for productive farming.

INTERACTION **Economic Geography** Although Germany is one of the world's giant industrial nations, it does not have a wealth of natural resources. Coal deposits boosted its industrialization in the last century and continue to supply about one-third of the nation's energy. Germany is Europe's leading steel producer, yet it must import iron ore. Germany also imports most of its oil and natural gas.

Germany's forests are mostly in the Central Uplands. Conservation and careful harvesting of timber is strictly practiced. The nation's recent forest losses have resulted mainly from air pollution rather than from logging or land development. Tree loss from pollution and acid rain is of national concern to Germans.

Germany's industrious population is still its most important economic resource. Its workers, among the world's best paid and best educated, are considered world leaders in scientific, technological, and organizational skills.

German industry is extremely diverse. The country is a world leader in the production and

Historic castles are nestled throughout the Bavarian Plateau of southern **Germany**. Look at the physical–political map on page 283. What landform region dominates northern Germany?

export of automobiles, electronics, chemicals, and optical and surgical instruments. Its transportation network, with modern railroads, airports, and *autobahns* (freeways), is admired around the world.

Reunification has allowed the nation to combine many valuable resources. Reunified Germany is a large country with varied geographical characteristics. The country can be divided into five major regions.

Northwestern Germany Northwestern Gemany lies along the windy and damp coast of the North and Baltic seas. Soils are poor here, and sand dunes cover most of the offshore islands. Farther south and inland are some of Europe's richest soils. This productive agricultural belt is created by thick deposits of **loess** (LEHS), or dust-sized soil particles deposited by the wind. Grains, potatoes, and sugar beets are important crops. Hamburg, on the Elbe River, is Germany's major seaport and second largest city.

The Ruhr The Ruhr in western Germany is the industrial heart of Europe. The region developed around Europe's largest coal deposits and became Germany's most densely populated and most polluted area. The Ruhr has become almost one continuous belt of cities and industry, called the Ruhrstadt (Ruhr City). This "smokestack region" manufactures chemicals, pharmaceuticals, plastics, and automobiles. To the south, along the Rhine River, are the cities of Cologne and Bonn, the seat of government.

The Rhineland South of the Ruhr is the fertile Middle Rhine Valley, also called the Rhineland. This is both an agricultural and an industrial area. The Rhine Valley is famous for its castles, forests, vineyards, and farms lining the steep river gorge. This region's main city, Frankfurt, is Germany's transportation and financial center. West of the Rhine River, along the French border, is an old industrial area called the Saar. The region's coal deposits and steel industry helped it prosper. Also an agricultural area, the Saar is famous for its wine and cheese.

Southern Germany Southern Germany includes Bavaria, the largest German state, and the major city and manufacturing center of Munich (MYOO-nik). Munich produces electrical and scientific equipment, automobiles, textiles, and toys. Munich and nearby Stuttgart have been growing rapidly due to new high-tech industries. Southern Germany also has rich farmland and mountains, forests, and lakes.

East Germany The former East Germany lies mainly on the Northern European Plain. Its soils are fertile, with sugar beets, grains, and potatoes the most important crops. Inefficient government-run farms in the region are being returned to private landowners. In the southeast, coal deposits lie beneath the hilly region of the Central Uplands. The city of Rostock in the north is the region's major Baltic port, while Berlin is the region's largest city and the country's capital. (See "Cities of the World" on pages 288–289.)

One of the most serious challenges facing eastern Germany is cleaning up pollution. The former East German Communist government used low-grade coal for its energy needs and dumped chemical wastes into the region's rivers.

Because East Germany lagged behind West Germany in transportation, housing, and industry, a "wealth gap" exists between eastern Germany and the rest of the country. Eastern Germany's services and its inefficient and outdated industries are being modernized at great cost.

Issues Although unification was a goal of Germany for more than 40 years, the financial and emotional realities of reunification have been challenging. Many Germans did not realize the costs of elevating eastern Germany to western standards. The nation must deal with eastern Germany's serious environmental problems, crumbling infrastructure, and high unemployment rate. Many eastern Germans have moved to western Germany to seek employment.

Germany is a key member of the EU and NATO. As an economic powerhouse of Europe, Germany will continue to play an important role in Europe's future.

2 SECTION REVIEW

1. What are Germany's major rivers, and how are they economically important? What products does Germany export?
2. Which of Germany's regions are agricultural? Which are industrial? Which have coal resources?
3. **Critical Thinking** **Explain** both the positive and negative aspects of Germany's reunification.

Text continues on page 290.

CITIES OF THE WORLD
HOLLYWOOD
GINZA

Berlin

Berlin, once again the capital of a united Germany, is home to many symbols of historic German power and influence.

Historically, Berlin has symbolized Germany's greatest triumphs and achievements as well as its worst tragedies and crimes. Today, Berlin is again Germany's capital. For many, the city represents the end of the Cold War that divided both Germany and Europe for nearly 45 years.

The *Quadriga of Victory,* a statue of a horse-drawn chariot, commands the top of the Brandenburg Gate at the western end of Unter den Linden. The Brandenburg Gate was completed in 1791.

The Heart of Germany

Present-day Berlin officially became a city in 1307 when two towns on opposite banks of the Spree River merged. At this time, Germany was divided into hundreds of kingdoms, or states. Munich, Cologne, and Dresden, all well-established cultural centers, were among the major cities of the German states. Smaller Berlin, a part of Prussia, grew slowly during the next 300 years.

Rapid development characterized Berlin in the second half of the seventeenth century. By the 1680s, the city had several new cultural institutions, including its first public library. In 1701, the King of Prussia officially made the city his royal capital.

In the decades that followed, the Prussian kings continued to build, rebuild, and expand their capital. Berlin soon became an important political, cultural, and economic center. The Brandenburg Gate and a wide, tree-lined avenue called Unter den Linden are among the famous Berlin landmarks built by the Prussians.

By 1815, Berlin had a population of about 200,000. Growth intensified as the Industrial Revolution lured hundreds of thousands more people to the city to work in the new factories. In 1871, Prussia completed its task of creating a unified Germany. Berlin became the capital of the German domain and Europe's strongest economic and military power. By the early 1900s, Berlin was a large metropolis of more than 2 million people.

Berlin remained the capital of Germany through two world wars. From 1933 to 1945, when Adolph Hitler and the Nazi party ruled the country, Berlin represented in the eyes of the free world the horrors that Nazi Germany had unleashed upon the world.

After Germany's defeat in World War II, Berlin, like the country itself, was divided. East Berlin became the capital of East Germany, a Communist state controlled by the Soviet Union. West Berlin became part of democratic West Germany, an ally of the United States. Bonn, a small city on the Rhine River, became West Germany's capital. West Berlin was entirely surrounded by East German territory.

Divided Berlin immediately arose as a crisis spot in the Cold War between the United States

GINZA
HOLLYWOOD

and the Soviet Union. Between 1945 and 1961, almost 3 million East Germans, fleeing Soviet control, escaped to the West. Many crossed from East Berlin into West Berlin.

To stop this mass flight, Soviet-backed East Germany built the Berlin Wall in 1961. This barrier of concrete, electric fences, barbed wire, floodlights, trenches, armed soldiers, and attack dogs eventually surrounded all of West Berlin. In time, the massive wall came to symbolize the division between East and West.

For 28 years, democratic West Berlin stood out as a prosperous industrial center highlighted by many cultural attractions. In contrast, Communist East Berlin, though a showplace city of Eastern Europe, experienced slower economic growth and a deteriorating infrastructure. This division existed until 1989 when the collapsing East German government opened the Berlin Wall. News reports throughout the world showed Berliners tearing down the barrier. What was once a symbol of division now symbolized freedom and unity.

BERLIN
Major highways
Other roads
Rivers and canals
City limits
Railroads
Parks and forests
Airports and airfields
Former location of Berlin Wall
0 3 Miles
0 4 Kilometers
Berlin
GERMANY
N
Falkensee
Spandau Citadel
Olympic Stadium
Brandenburg Gate
Berlin
Kaiser Wilhelm Memorial Church
Havel River
Spree River
Potsdam

The historic Brandenburg Gate is in central **Berlin** just east of where the Berlin Wall once stood. The wall cut across many of Berlin's busiest thoroughfares.

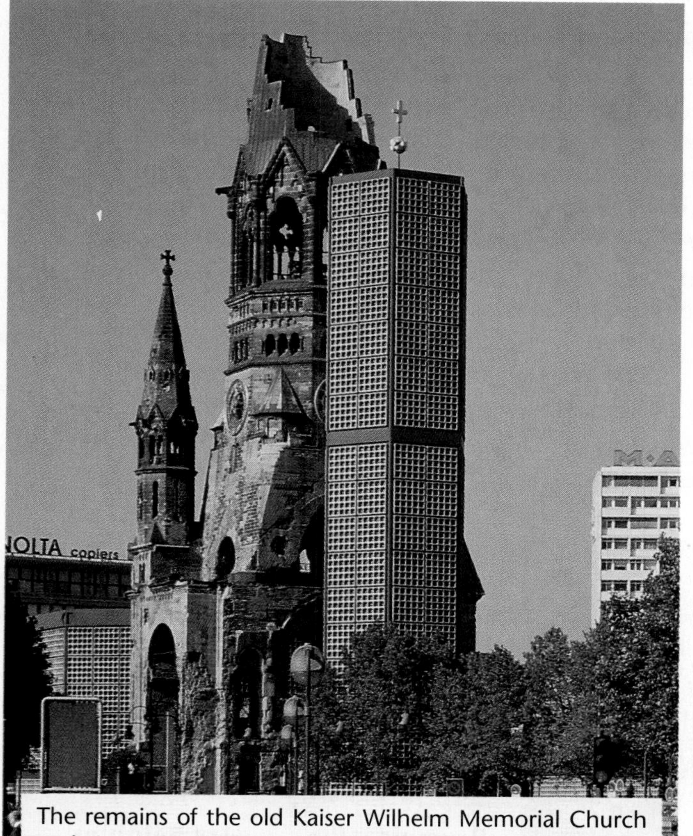

The remains of the old Kaiser Wilhelm Memorial Church and its towering, modern replacement stand as a reminder of the devastation of World War II and the postwar rebirth of **Berlin**.

The Reunited City

Today, reunified Berlin is home to more than 3 million people. Except for a few fragments, the Berlin Wall is gone, and landmarks in what was once East Berlin are being restored. Berlin holds many museums, universities, and theaters. Parks, woodlands, and waterways occupy a large part of its land area.

Berlin, however, faces many challenges. Rebuilding the eastern part of the city is very expensive. In addition, many former East Berliners have been slow to adjust to a capitalist economy. Still, Berlin's location and well-educated population promise to attract many new businesses to the city. With its difficulties related to reunification, combined with high hopes for the future, Berlin continues as a symbol for all of Germany.

3 THE BENELUX COUNTRIES

FOCUS

- *What are the physical and economic characteristics of the Benelux countries?*
- *What are the urban and cultural characteristics of the Benelux countries?*

Belgium Belgium can be divided into three landform regions: a coastal region, a central plain, and the Ardennes. Much of the coastal plains, which are used mainly for farming, is reclaimed from the sea and from wetlands. The central plain between the Schelde and Meuse rivers contains Belgium's most productive soils. In the southeast, the forested hills of the Ardennes reach to 2,283 feet (696 m). (See the map on page 283.)

Belgium has two language regions. In the northern coastal part of the country, people speak Flemish, a language related to Dutch. This region is called Flanders and contains about 70 percent of the country's population. In the southern part of Belgium, the French-speaking people, called Walloons, make up most of the region's population. This language region is called Wallonia. Most of the nation's citizens view themselves as either Flemish or Walloon rather than as Belgians. The government now has assemblies for each language region.

Belgium's main resource is its skilled and productive labor force, which produces high-quality machinery and textiles. Belgium is known for quality carpets, cut diamonds, and fine chocolate as well. With its small population, Belgium depends on international trade to maintain its high standard of living. Its small but efficient farms provide most of the nation's food and produce some exports.

Brussels, Belgium's capital and largest city, is one of Europe's most **cosmopolitan** cities. A cosmopolitan city is one that is characterized by its many foreign influences. Brussels' central location in Europe and excellent transportation connections have made the city a center for finance. Numerous organizations have their headquarters there, including the EU and NATO.

INTERACTION **The Netherlands** *Netherlands* means "low lands," and indeed, nearly one-half of the Netherlands lies below sea level. (See the map on page 283.) These flat, low-lying lands were once covered by wetlands or the North Sea. First, the country built **dikes**, or walls to keep out water. Then it constructed a system of canals to remove the water. Finally, the land was drained, or pumped dry, by windmills, which became a symbol of the country. Lowland areas that have been drained are called **polders**. Today, polders are the sites of farms, cities, industries, and airports. (See "Themes in Geography" on pages 292–293.) The name Holland is sometimes used to refer to the Netherlands. Its people are known as the Dutch.

Although the country is well known for its agriculture, the Netherlands is a highly industrialized nation. The diversified Dutch industry is geared toward the export market, and the EU nations are its major customers. One of the world's largest oil refineries and the headquarters for the world's largest oil company are located in the Netherlands. Also important are food processing, dairy products, and breweries. Important crops include potatoes, sugar beets, and flowers, especially tulips.

The Netherlands must import almost all of its industrial raw materials. Though there are coal deposits in the south, it is cheaper for the Netherlands to import coal than to mine it. The

Brussels is the capital of **Belgium** and headquarters for many international organizations, including NATO.

The Dutch have little breathing room in this crowded city in the southwestern part of the country. The **Netherlands** is one of Europe's most densely populated countries.

country has large natural-gas deposits. About half of this gas is exported by pipeline to other European nations. The country imports almost all of its oil.

The old port city of Amsterdam is the Netherlands' official capital. Famous for its canals, museums, and art, Amsterdam also is home to international banking and modern industries. The North Sea Canal connects Amsterdam to the North Sea.

The city of Rotterdam to the southwest serves the trade traffic that enters and exits Europe on the Rhine River. Rotterdam is the world's busiest seaport and the site of oil refineries, chemical plants, and shipbuilding. The Hague is the seat of government, where parliament, the royal family, and the International Court of Justice are all located.

The Netherlands is a modern democratic society. Dutch is the national language, but most people speak English, German, or French as well. The country has Europe's most expensive social security system. Government benefits for education, unemployment, and medical care are among the highest in the world. Dutch workers receive high wages and are considered to be some of the world's most productive and skilled.

Luxembourg Tiny Luxembourg is a prosperous country bordered by France, Germany, and Belgium. (See the map on page 283.) Luxembourg's population is almost all Roman Catholic, and the people speak a Germanic language with many French influences. Most citizens are multilingual and well educated.

Luxembourg's iron and steel industry is located in the south near the iron-ore mines. This was once the nation's major industry. In recent decades, however, the nation has attracted international banking and rubber and chemical industries.

3 SECTION REVIEW

1. What is Belgium's major resource? What is significant about the Netherlands' elevation?
2. What are Belgium's two language regions?
3. **Critical Thinking** **Suggest** three reasons for Luxembourg's prosperity.

4 THE ALPINE COUNTRIES

FOCUS

- *How are Switzerland's political, cultural, physical, economic, and urban characteristics linked?*
- *What are Austria's main characteristics?*

Switzerland Switzerland is a **confederation** of 26 **cantons**, or states. A confederation is a group of states joined together for a common purpose. Each canton governs itself, except for military defense and international relations, which are under federal control.

Although Switzerland appears to be a culturally divided land, it is one of the world's most stable countries. Most Swiss speak German, but many also know French, Italian, and Romansh. (See the map on page 294.) The nation is peacefully and equally divided into two major religious groups, Protestant and Roman Catholic.

Switzerland contains three landform regions: the forested Jura Mountains, the Alps, and the Swiss Plateau. (See the map on page 283.) Most of Switzerland is mountainous. Glaciers, long narrow lakes, and deep winter snows dominate the spectacular Alpine scenery. Many of Europe's major rivers, including the Rhine and the Rhone, begin in the Swiss Alps.

The Swiss Plateau, or Mittelland, separates the Jura Mountains and the Alps. The Mittelland

Text continues on page 294.

Polders of the Netherlands

The building of polders is an example of people adapting the environment to suit their needs.

Much of the Netherlands was once covered by shallow sea, inland lakes, and swamps. As the demand for agricultural and industrial land increased, the Dutch found that they could reclaim, or take back, large areas of valuable land from the sea if they drained the water from low-lying areas. They believed that reclamation was necessary because most of the Netherlands consists of low-lying and flooded land, with few areas of higher elevation.

At first, the Dutch built walls, called dikes, around the area to be drained. Then they pumped the water from this area into canals that drained into the North Sea. In the early days, the Dutch used windmills to generate the power to operate the pumps. The windmill became a symbol of the Netherlands. Later, they used steam power and, eventually, electric power. The areas of land that were once submerged beneath the sea or inland waters are called polders. Because the polders lie below sea level, they have no natural drainage. All excess water must be pumped from the polders into the canals.

As a result of the creation of the polders, and because of the poor natural drainage of the land, the Dutch landscape is covered by a network of canals. These are important for drainage, for irrigating farmlands, and as waterways for transporting goods. The polders have provided very fertile farmland.

The Netherlands is one of Europe's most densely populated countries. Because it is also a very small country, there is much competition among farming, industry, and cities for use of the reclaimed land. Many of the cities, including Amsterdam, the capital and largest city, have been built on polders.

LOCATION The city of **Amsterdam**, the capital of the Netherlands, is built on land that was once below water. Using the map on page 293, identify other Dutch cities that are built on or near reclaimed land.

Since the first construction of the dikes and creation of polders, the Dutch have struggled constantly to keep the sea from flooding their land. Storm tides from the North Sea have broken through the dikes many times in the country's history. If it were not for the dikes and sand dunes that line the shores, half the country would be flooded at high tide. Major projects are under way to improve the dikes along the country's coastline.

The cost of repairs is high, however. The Netherlands spends several hundred million dollars each year to repair dikes and drain water.

This high cost combined with sinking lowlands and other environmental damage has prompted the Netherlands to reconsider its use of the land. The Dutch have made plans to restore a portion of the land to wetlands, lakes, and forests.

In addition, the Dutch have recently been reminded that flooding can come from inland areas as well as from the North Sea. Heavy rainfall in early 1995 caused flooding that forced more than 300,000 people, mostly Dutch, from their homes in parts of West Central Europe. Northeastern France, western Germany, and the Benelux countries received some of the heaviest rainfall. Development along rivers, such as the Rhine River in western Germany and the Netherlands, kept much of the rainwater from being absorbed in marshes and floodplains instead of spilling into populated areas. Europeans, and especially the Dutch and their neighbors, are searching for ways to prevent similar disasters in the future.

REGION Large areas of land in the Netherlands lie below sea level. This means that excess water must be drained artificially by the use of an extensive system of canals, dikes, and pumping stations. The photograph above shows land after the polders were created. **To where is water from polders drained?**

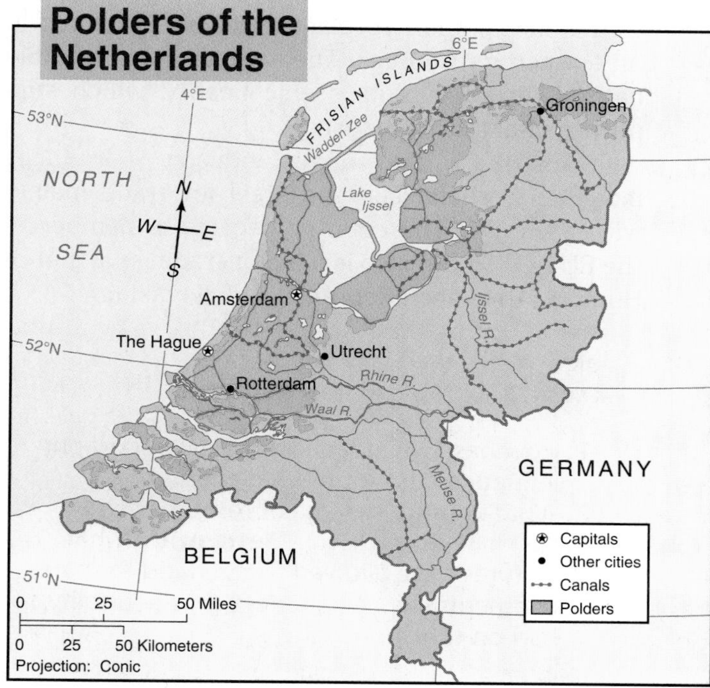
Polders of the Netherlands
6°E
4°E
53°N
52°N
51°N
FRISIAN ISLANDS
Wadden Zee
Groningen
NORTH
SEA
N
W
E
S
Lake Ijssel
Amsterdam
The Hague
Utrecht
Rotterdam
Rhine R.
Waal R.
Ijssel R.
Meuse R.
GERMANY
BELGIUM
Capitals
Other cities
Canals
Polders
0 25 50 Miles
0 25 50 Kilometers
Projection: Conic

INTERACTION Since the first polders were built, the Dutch people have fought a constant battle with the waters of the North Sea. A barrier dike protects Lake Ijssel and surrounding polders from the Wadden Zee and the North Sea. Look at the map of the Netherlands to the left. **What areas would be flooded if the coastal dunes and dikes were removed?**

Languages of Switzerland

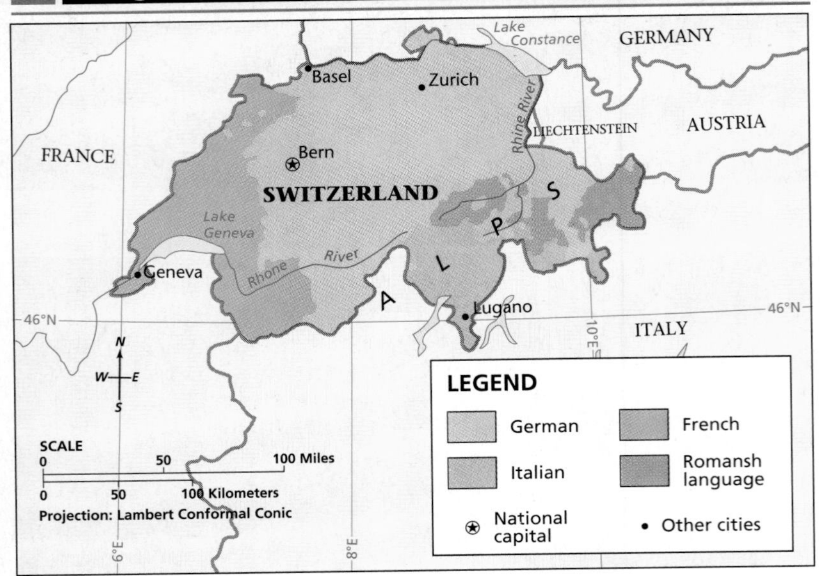

Lake Constance
GERMANY
Basel
Zurich
Rhine River
LIECHTENSTEIN
AUSTRIA
FRANCE
Bern
SWITZERLAND
Lake Geneva
Rhone River
A L P S
Geneva
Lugano
46°N
10°E
ITALY
N
W
E
S
SCALE
0 50 100 Miles
0 50 100 Kilometers
Projection: Lambert Conformal Conic
6°E
8°E
LEGEND
German
French
Italian
Romansh language
National capital
Other cities

MAP STUDY

Maps can be used to show areas where specific languages are spoken. Most people in Switzerland are multilingual. The colors on this map show where various languages are dominant. In which language region is the Swiss capital located?

stretches from Lake Geneva on the western border with France to Lake Constance on the northern border with Germany.

Switzerland's mainly highland climate has variable temperatures and precipitation within short distances. The mountains surrounding the country protect much of it from Northern Europe's cold winter air. In the Alpine valleys and the Mittelland, the warm southern **foehn** (FUHRN) keeps winter temperatures mild. A foehn is a warm, dry wind that blows down the slopes of the Alps.

Except for hydroelectric power, Switzerland is resource poor. The country's mountainous terrain and poor soils mean there is little **arable** land, or land that is suitable for growing crops. Dairy farming and grape growing in the Mittelland are the country's main agricultural activities.

Switzerland's economy depends largely on foreign trade. Most Swiss goods are exported, mainly to EU nations. Swiss industries include machinery, chemicals, and watches. The Swiss also are famous for their chocolate and cheese. The snow-covered Alps support a year-round tourist industry.

International banking, however, brings the most foreign funds into Switzerland. The country's peaceful history has made it a stable world finance center. To remain neutral, Switzerland has not joined the United Nations, the EU, or NATO.

All of Switzerland's major cities are in the Mittelland. Zurich, the largest, is a leading world-banking center. Bern, the nation's capital, is centrally located between the country's German-speaking and French-speaking populations. The city of Geneva is known for diplomacy. It is home to many international organizations and conferences.

Austria Austria is about twice as large as neighboring Switzerland. (See the map on page 283.) Almost all of the population is German-speaking and Roman Catholic. The Alps loom over the western two-thirds of the nation. Vienna, Austria's capital city and heart of commerce and culture, lies along the banks of the Danube River.

Austria is mainly an industrial country, but agriculture also is important. Austrian farmers supply almost all of the nation's food. Austria has forest resources and hydroelectricity but is limited in other energy resources. The nation's most valuable natural resource is its scenic beauty, which supports a vital tourist industry.

Austria's diversified industry produces machinery, chemicals, and steel for trade mainly within the EU. The country became a member of the EU in 1995 but also is a major trading partner of Hungary, the Czech Republic, and Slovakia.

4

SECTION REVIEW

1. How has Switzerland's physical geography influenced its culture and economy?
2. What are Austria's major resources?
3. **Critical Thinking** **Determine** what features have allowed Switzerland to maintain neutrality. What effects has neutrality had on the country?

CHAPTER 25 CHECK

FLORE
SALLE AU 1ER
Sidewalk cafe, **Paris**

Reviewing the Main Ideas

1. France's landforms vary from high mountains to hills and flat plains. Paris is France's capital and largest urban area. The country is one of the most forceful influences in Europe.
2. Germany is a major industrial power, although it lacks many natural resources. The country can be divided into five major regions: northwestern Germany, the Ruhr, the Rhineland, southern Germany, and East Germany. Reunification has posed some serious challenges for the country.
3. The Benelux countries are Belgium, the Netherlands, and Luxembourg. These three industrialized countries share many geographic similarities.
4. The Alpine countries include Switzerland and Austria. Tourism, banking, manufacturing, and agriculture are major economic activities.

Building a Vocabulary

1. What is a primate city? Give an example of a primate city.
2. Define *loess*. Where in Germany is it found?
3. How were dikes used in the development of polders?
4. What does it mean to say that Switzerland is a confederation of cantons?

Recalling and Reviewing

1. In which countries discussed in this chapter are the Alps located?
2. Compare and contrast the economic geography of France and Germany.
3. How are Belgium and Switzerland similar?
4. Name four West Central European cities that are located along rivers. Name the rivers.

Thinking Critically

1. How do you think the French people would react to the idea of France joining with several other European countries to become one large, unified country?
2. Can a nation with few resources or with challenging physical conditions become prosperous? Explain, using two examples from the chapter.

Learning About Your Local Geography

Individual Project

France's mistral and Switzerland's foehn influence weather and climate patterns in West Central Europe. How is the climate of your local community influenced by various winds? Contact the weather bureau or the weather department of a local television station, and request information on your area's winds over the past week—where the winds came from, how strong they were, how they affected the area. Graph your information, and include captions explaining the information shown. Present your graph to the class. You may want to include your graph in your individual portfolio.

Using the Five Themes of Geography

Sketch a map showing the different language **regions** of West Central Europe. With shading or symbols, indicate how a region's **location** determines which languages are spoken in that region. In particular, consider the countries that border any given region or country. How are the languages spoken in that region or country influenced by the bordering countries?

CHAPTER 26

Southern Europe

GEOGRAPHY DICTIONARY

- *Meseta*
- *plaza*
- *cork*
- *dialect*
- *autonomy*
- *Renaissance*
- *coalition government*
- *balance of trade*

"This is still a country of local specialties: wines vary widely from town to town, and so do foods. In San Sebastian they make small pastries designed to look like ham and eggs. In Toledo they make marzipan. Oviedo is famous for its stews, richly compounded of vegetable-broth and black pudding, and Vigo for its eel pies. . . . Uniformity has not yet fallen upon Spain . . . this one [town] makes paper, this one lace, this one swords, this one cars, this one wooden figures of Don Quizote, and the little town of Jijona, near Alicante, makes nothing else but nougat."

Jan Morris

Floor mosaic from a former Roman villa in **Spain**

The countries of Southern Europe, also known as Mediterranean Europe, are Spain, Portugal, Italy, Greece, and several microstates. The four larger nations occupy peninsulas on Europe's southern edge.

The Mediterranean Sea stretches some 2,320 miles (3,733 km) from east to west. *Mediterranean* comes from a Latin word meaning "middle of the land." Indeed, in ancient times, the Mediterranean Sea was considered the center of the Western world, as it is surrounded by Europe, Africa, and Asia. The Strait of Gibraltar (juh-BRAWL-tuhr) links the Mediterranean to the North Atlantic Ocean.

Small waves, weak currents, and low tides characterize the Mediterranean. As a result, its poorly circulated waters are easily polluted. A growing population, coastal industrialization and a flood of tourists continue to stress the Mediterranean environment.

Mijas, southern **Spain**

SOUTHERN EUROPE: PHYSICAL–POLITICAL

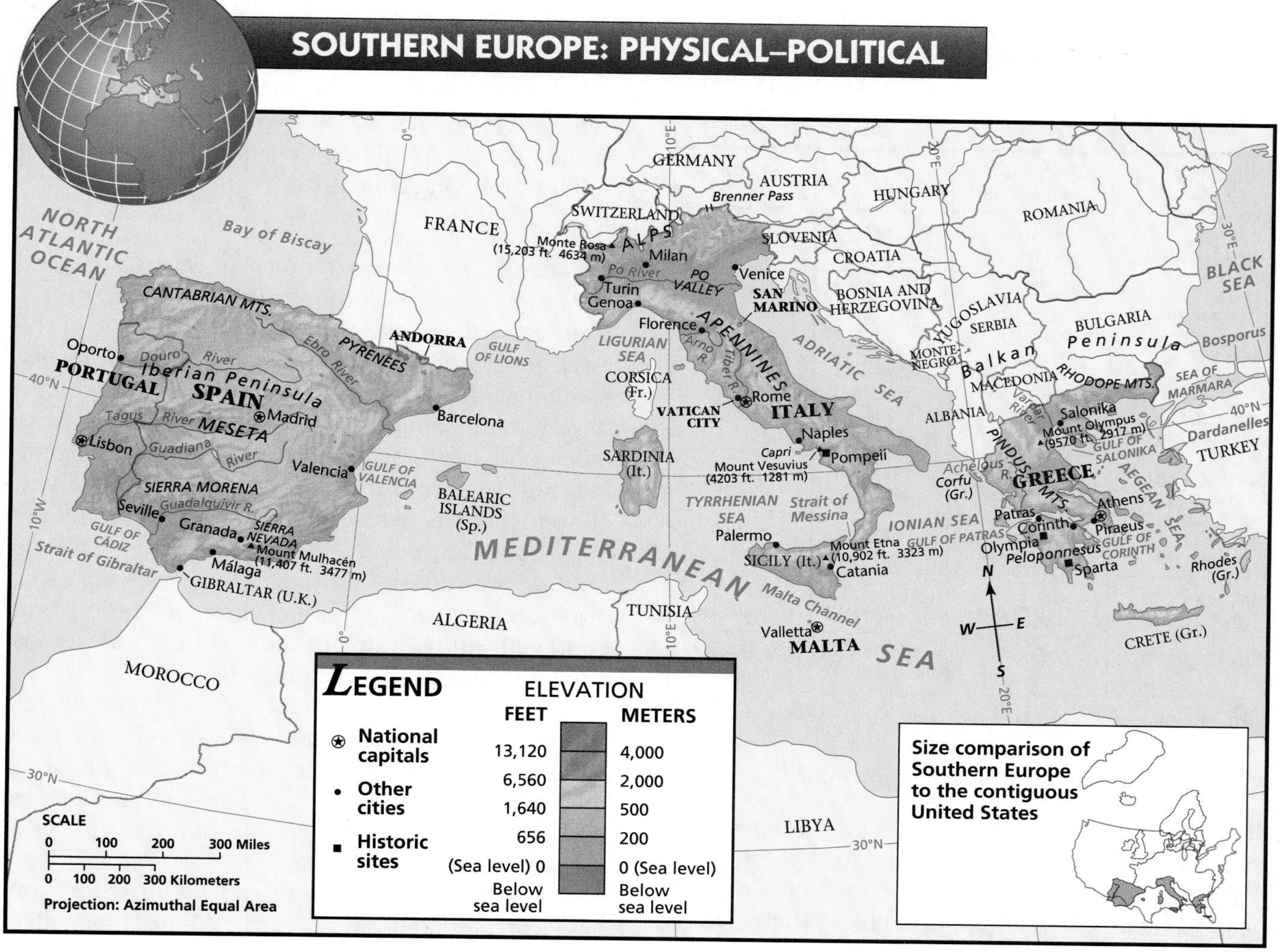

LEGEND

	ELEVATION FEET		METERS
⊛ National capitals	13,120		4,000
• Other cities	6,560		2,000
■ Historic sites	1,640		500
	656		200
	(Sea level) 0		0 (Sea level)
	Below sea level		Below sea level

SCALE
0 100 200 300 Miles
0 100 200 300 Kilometers
Projection: Azimuthal Equal Area

Size comparison of Southern Europe to the contiguous United States

SPAIN AND PORTUGAL

FOCUS

- *What landforms and climates are found in Spain? in Portugal?*
- *What economic activities are important in Spain and Portugal? What are each country's major cities?*
- *What are the historical, political, and cultural characteristics of Spain and Portugal?*

REGION **Physical Geography** Spain and Portugal lie on the Iberian (ie-BIR-ee-uhn) Peninsula. Spain covers over 80 percent of the peninsula, while Portugal occupies the remaining western portion. The peninsula is bounded on the north and west by the North Atlantic Ocean and on the south and east by the Mediterranean Sea. (See the map above.) Spain also includes the hilly Balearic (bal-ee-AR-ik) Islands in the Mediterranean and the volcanic Canary Islands in the North Atlantic Ocean.

Plateaus dominate the interiors of both countries. A high plateau called the **Meseta** (muh-SAYT-uh) covers much of Spain. This rocky and treeless plateau averages 2,100 feet (640 m) in elevation and is almost surrounded by mountains. It is the driest and least populated part of Spain. Portugal's interior plateau is cut by rivers, which have formed steep valleys that slope westward toward narrow coastal lowlands.

Spain is one of Europe's most mountainous nations. The Cantabrian (kan-TAY-bree-uhn) Mountains rise up in the northwest while the

FACTS IN BRIEF Southern Europe

COUNTRY POPULATION (1994)	LIFE EXPECTANCY (1994)	LITERACY RATE (1990)	PER CAPITA GDP (1993)
ANDORRA 63,930	75, male 81, female	99% (1992)	$14,000 (1992)
GREECE 10,564,630	75, male 80, female	93%	$8,900
ITALY 58,138,394	74, male 81, female	97%	$16,700
MALTA 366,767	75, male 79, female	90% (1988)	$6,600 (1992)
PORTUGAL 10,524,210	71, male 78, female	85%	$8,700
SAN MARINO 24,091	77, male 85, female	97% (1991)	$16,000 (1992)
SPAIN 39,302,665	74, male 81, female	95%	$12,400
VATICAN CITY 821	Not available	Not available	Not available
UNITED STATES 260,713,585	73, male 79, female	97% (1991)	$24,700

Sources: *The World Factbook 1994,* Central Intelligence Agency; *The World Almanac and Book of Facts 1995*

Which country has the lowest literacy rate?

Sierra Morena and Sierra Nevada ranges are located in the south. The Pyrenees stretch from the Bay of Biscay to the Mediterranean with peaks over 10,000 feet (3,048 m) high. Historically, these mountain ranges have isolated the peoples of Spain from each other. Even today, regional differences exist, partly because of this geographical separation.

The most populated lands are the coastal plains and river valleys. The Ebro, the Douro, the Tagus, the Guadiana (gwahd-ee-AHN-uh), and the Guadalquivir (gwahdl-KWIV-uhr) are the most important rivers on the Iberian Peninsula. The Guadalquivir in southern Spain is one of the country's few navigable rivers.

Spain's climate types include marine west coast, Mediterranean, and steppe. The marine-west-coast climate is found in the north between the Cantabrian Mountains and the Atlantic coast. This area of lush forest and pasture is the country's coolest and wettest region.

A Mediterranean climate bathes southern and central Spain and all of Portugal. Here, the summers are sunny, hot, and dry, while the winters are mild and wet. Portugal's northern region is at higher elevations and is somewhat cooler, wetter, and forested.

The Meseta of central Spain has a semiarid steppe climate with very hot summers and very cold winters. Because the surrounding mountains block moist ocean air from reaching inland, the Meseta receives little rainfall. Some of the plateau suffers from desertification.

Agriculture Because much of Spain is rugged and dry, there is little arable land. Even so, Spanish farmers make up about 10 percent of the work force and produce a wide variety of crops. Spain has three agricultural regions, which are based on its climates. In the north, ranchers raise dairy and beef cattle, and farmers grow vegetables and fruit. In the Mediterranean region, olives, almonds, citrus fruits, grapes, and vegetables are important products. The high and dry Meseta is not as productive, although farmers can grow grains, olives, and grapes where irrigation and good soils allow. Much of the rest of the Meseta is used for grazing sheep and goats.

Many Spanish farmers live in small villages and travel each day to the fields. Activity in small agricultural villages centers around a **plaza**, a public square in the center of town. Often the site of a marketplace, the plaza usually serves as a gathering place for people.

Portugal is a leading exporter of wine. The dry, warm climate south of the Tagus River also supports citrus fruits and olives. North of the Tagus River, farmers grow cereal grains, especially wheat and corn.

Economic Geography Spanish agriculture supports a large food-processing and export industry and is a major part of the national economy. The country also has a good natural resource base for industrialization. The iron-ore mines in northern Spain support the country's steel industry.

Some industries, such as automobiles, machine tools, and processed food, are supported by foreign investment and have been very successful. Spain is a world leader in automobile manufacturing, and automobiles account for a large share of the nation's exports. Other industries, however, especially steel and coal mining, have suffered labor cutbacks. Spain also must import almost all of its oil.

Spain has one of the world's largest tourist industries. Its sunny beaches and rich cultures have made the country Europe's leading tourist destination. As a result, construction is booming along the country's Mediterranean coast and in the Canary Islands. Such growth has caused pollution and overcrowding, however. The many high-rise buildings block out some of Spain's most scenic coasts. The tourist industry as a whole also competes with agriculture for the region's limited water supplies.

Portugal imports more goods than it exports. The country's income from its tourist industry, however, helps offset its trade deficit. A mineral-poor nation, Portugal must import almost all of its energy resources. Much of the nation's industry and exports are based on food processing, textiles, shoes, and electronics. The Portuguese also have a long tradition of fishing the North Atlantic, and seafood remains an important economic resource to the nation.

The Iberian Peninsula is a leading producer of **cork**, which is a bark stripped from the trunk of the Mediterranean oak. Cork is used to make a variety of products, including insulation and flooring. Its main use is for bottle stoppers.

PLACE **Urban Geography** Madrid, Spain's capital, is located on the Meseta. Although it is separated from other populated areas, Madrid serves as Spain's transportation, financial, and administrative center. Spain is investing large sums of money in high-speed rails and highways to better connect Madrid to other Spanish cities and to the rest of Europe.

Barcelona (bahr-suh-LOH-nuh) is Spain's second largest city. Located on the Mediterranean coast with a good harbor, it is a leading seaport and industrialized urban area. This ancient Roman port is a popular tourist destination as well.

Southwest of Barcelona on the Mediterranean coast lies Valencia, Spain's third largest city. Valencia is the center of a rich agricultural area and is known for its citrus fruits, especially oranges. The city is eastern Spain's railroad center and has a booming beach resort and tourist industry.

Portugal's two largest cities are Lisbon and Oporto (oh-POHRT-oh). Lisbon, the nation's capital and largest city, is located at the mouth of the Tagus River. It serves as a major transportation, industrial, and cultural center. Oporto lies in northern Portugal on the Douro River. It is a major seaport and industrial area and is the center for bottling and exporting wine.

Historical Geography From 200 B.C. to A.D. 400, Iberia was a province of Rome. In 711, the Moors, a Muslim people from North Africa, invaded the peninsula. They ruled much of the Iberian Peninsula for more than 700 years. During those years, their art, architecture, agriculture, and science influenced much of the region. Grenada, the last city held by the Moors, was reconquered by Spain in 1492.

From about the fifteenth to the seventeenth century, Spain and Portugal were the most powerful European nations and controlled much of the ocean trade. Their explorations for new trade routes brought them to the Americas, Asia, and Africa. Both nations built large colonial empires, bringing back gold, spices, and other valuable materials from these lands. To their colonies,

After stripping the Mediterranean oak of its bark, workers in western **Spain** load trucks to take their harvest to the factory. Workers at the factory steam the bark to produce cork stoppers for bottles.

Spain and Portugal brought their languages, religion, and culture traits.

Political Geography Today, Spain is a prosperous democracy with a king. About 75 percent of the population speaks Castilian, the official national language. Castilian is a Spanish **dialect** from the Castile region around Madrid. A dialect is a regional variety of a major language. Many dialects are spoken throughout Europe.

Although the Roman Catholic religion and the Spanish language are unifying forces, many divisions exist in Spanish society. In fact, the Spanish government has granted **autonomy** to 17 regions within Spain. Autonomy is the right to self-government. The central government, however, controls national matters such as defense.

The status of the Basques, an ethnic minority, is a major concern for the future of Spain. The Basques live primarily along the Bay of Biscay and the Pyrenees in northeastern Spain and an adjoining area of France. They lived in the area long before the French or the Spanish settled near them. Their language, which they call Euskara (ehws-kuh-RAH), deepens the mystery surrounding them. Euskara is not related to any other known language. Although the government has granted the Basques autonomy, some Basques want complete independence.

In the past few decades, social, economic, and political conditions have improved in Spain and Portugal. Yet the nations still suffer from an unequal distribution of wealth. Spain and Portugal joined the European Union in 1986. While membership in the EU is improving economic conditions, it also is changing the way of life of many rural peoples by pressuring them to abandon traditional farming methods.

1 SECTION REVIEW

1. What landform features do Spain and Portugal share? Which climate type is found in both countries?
2. What resources must Spain and Portugal import? What is each country's capital?
3. **Critical Thinking** **Suggest** two reasons that the Basques speak a language unrelated to other languages. Consider both physical and historical factors.

2 ITALY

FOCUS

- *What is Italy's physical geography? What role has Italy played in Europe's history?*
- *How are Italy's regions similar and different?*

Physical Geography Italy is a boot-shaped peninsula stretching southward from Europe some 760 miles (1,223 km) into the Mediterranean Sea. It is bounded on the west by the Tyrrhenian Sea and on the east by the Adriatic Sea. Italy also includes two large islands, Sicily and Sardinia. (See the map on page 297.)

The Alps form Italy's northern borders with France, Switzerland, Austria, and Slovenia. The Apennines extend from the Alps on the northwest coast to the southern tip of Italy, creating a rugged spine down the center of the peninsula. Italy's largest plain is the Po Valley, situated in northern Italy between the Alps and the Apennines. The Po River flows eastward through the center of the valley to the Adriatic Sea. The Arno and the Tiber in central Italy flow to the west coast.

La Scala, in **Milan, Italy**, is one of Europe's leading opera houses. The origins of modern opera, a drama in which characters sing most of their lines, lie in Italy in the late 1500s.

The great power and wealth of the ancient Roman Empire can be seen in the ruins of the Colosseum of **Rome**. The Colosseum, which seated about 50,000 spectators, is one of many Roman ruins scattered throughout Europe.

Most of Italy has a Mediterranean climate. The Alps and the Apennines prevent the cold winds of Northern and Eastern Europe from reaching the country. Italy's surrounding seas also help prevent temperature extremes in winter and summer. The Po Valley has a humid-subtropical climate with warm, humid summers and cool, rainy winters. Rainfall and winter snows are heavy in the Alps and northern Apennines. In contrast, overgrazing and deforestation have eroded the drier southern mountains. When it does rain, floods and mud slides may occur.

Historical Geography Italy's rich culture dates back as far as 500 B.C., when Rome was a thriving city. At its peak in about A.D. 117, the Roman Empire stretched from Britain to Mesopotamia. Unable to protect such a huge area from invaders, Rome collapsed in the fifth century. Italy was divided into many small kingdoms, city-states, and papal (church) states. San Marino and Vatican City still exist as separate independent countries.

Almost 1,000 years after the fall of the Roman Empire, Italy became the center of the **Renaissance** (REHN-uh-sahnts), which means "rebirth." During this period, a renewed interest in learning spread throughout Europe. It was during this period that powerful Italian city-states, such as Florence, Genoa, and Venice, developed. Rivalry among these city-states, and invasions by Spain and Austria, prevented Italy from becoming a unified nation until the late nineteenth century. Although Italy has been unified for over 100 years, differences between its north and south regions remain strong.

Northern Italy The Po Valley in northern Italy is the country's economic heartland. Warm summers, year-round water resources, and rich soils make the valley the "breadbasket" of Italy.

Industry is also important in the Po Valley. The large cities of Milan, Turin, and Genoa form an industrial triangle that benefits from its nearness to the rest of Europe. This industrialization, however, has caused severe air pollution, which blankets much of the area.

Italy's Alps provide water, hydroelectric power, and some timber to the industrial areas. The Alps are also a major tourist region, with their glacial lakes, ski resorts, and vineyards. The eastern Po Valley is less industrialized. Venice, near the mouth of the Po River, is the main city here. (See "Planet Watch" on pages 302–303.)

Central Italy Central Italy is the country's political and cultural center. Rome, Italy's capital and largest city, lies in this region. The city has spread out over both banks of the Tiber River. Ancient Rome was the center of the once-vast Roman Empire. The Forum, the Colosseum, and other Roman ruins still stand in the city today. Rome handles Italy's political, educational, scientific, and international affairs.

The city of Florence, on the Arno River, is an important art and cultural center. A beautiful and historic city, Florence is one of Europe's leading tourist destinations. The Apennines' summer and winter resorts and the region's volcanic lakes also attract thousands of tourists.

Southern Italy Southern Italy, the country's poorest region, is characterized by rural poverty and small farms. Because the area has very dry summers, water is scarce. Throughout history, southern Italy has suffered from natural disasters, including earthquakes, volcanic eruptions, floods, and droughts. Human destruction of valuable forest and soil cover has contributed to desertification in the region.

Unlike northern Italy, southern Italy has few industrial centers. With mixed results, the

Text continues on page 304.

PLANET WATCH

Saving Venice

> *Venice is famous for its historic architecture—and the efforts now being waged to save these treasures.*

The city of Venice is built on a group of about 120 islands in the middle of a lagoon. For hundreds of years, the water separating the islands from the Italian mainland protected Venetians from invasion and wars that rocked the rest of Europe. By the early fifteenth century, the city was Europe's center of trade with Asia.

Although Venetians benefited from the sea, they knew it could destroy all they had built. Venice is located along the Adriatic Sea, only a few feet above sea level. Floods caused by high tides are a constant danger. As a result, for much of Venice's early history, a watchful government kept the canals free of silt. They also prevented people from building anything in the lagoon that might interfere with the natural flow of the tides.

After a terrible flood in the late 1800s, the sea caused few problems. Then, in 1966, Venice faced disaster. On November 4, strong winds prevented the morning tide from flowing out of the lagoon. As the afternoon tide flowed in and heavy rains pounded the city, the high waters burst the city's fuel oil storage tanks, releasing thick black oil into the lagoon. Oily black water threatened historic churches and palaces in and around St. Mark's Square.

The devastating flood of 1966 almost drowned the city, along with some of the world's most famous art and architecture. Suddenly, people began to look closely at the city, and many did not like what they saw. Venice was decaying. Paintings, sculptures, and ancient buildings had been neglected, and the city was sinking. Industrial air pollution made these problems worse.

A Victim of Growth

The rapid growth of industry in post–World War II Italy brought wealth to the nation but was harmful to Venice. In the 1940s, industry began to expand facilities at the Marghera site, an industrial complex built on the mainland near Venice. Soon, Marghera became one of Italy's largest industrial centers.

Developers bought up mainland farms so they could build cheap apartment complexes for industrial workers. To make room for large ships and oil tankers, new channels were dug in the lagoon. Two artificial islands also were created in the lagoon to house more industry.

By the 1960s, Venice was showing the effects of this development. Smoke belched from factories, polluting the air and damaging

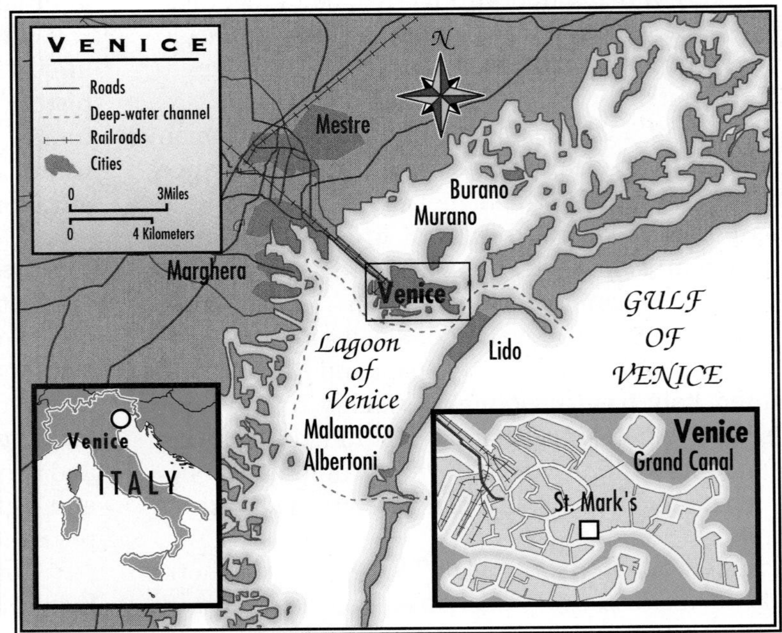
VENICE
Roads
Deep-water channel
Railroads
Cities
0
3Miles
0
4 Kilometers
N
Mestre
Burano
Murano
Marghera
Venice
Lagoon of Venice
Lido
GULF OF VENICE
Malamocco
Albertoni
Venice
ITALY
Venice
Grand Canal
St. Mark's

Deep-water channels provide access to the open sea for shipping traffic from **Venice** and other cities around the Lagoon of Venice. The inset map offers a close-up view of Venice and the canals that cut through the city.

historic palaces and churches. Wells that had been dug beneath the lagoon to provide industry with water had caused the city to sink nearly a foot (31 cm). Chemicals polluted the waters, spurring the growth of algae. As the algae rotted, the lagoon's murky waters gave off a horrible odor. In addition, pollutants had severely damaged the foundations of buildings along the waterways.

Boats and canals replace cars and streets as the primary means of transportation in **Venice**. About 400 permanent bridges span the city's canals.

Plans for Saving Venice

Critics claim that the city and national governments have been slow to respond to Venice's problems. In an effort to repair the damage caused by years of neglect, the city has dug new channels to replace those destroyed, and the mayor has promised to clean the silt from the canals. During this process, water will be pumped from waterways, and owners of buildings along them will be forced to restore foundations. A ban on automobiles in the city's center remains.

The Italian government, which is short of funds, has acted even more slowly than the city. In 1984, the government organized a group of companies to develop plans and carry out the work of saving Venice. One of the group's suggestions is that sea barriers be built at openings to the lagoon. Acting as gates, the barriers will lie flat on the bottom of the lagoon, allowing ships in and out and tides to flow freely. When high tides threaten flooding, the barriers will be raised to hold back the water. The group also hopes to focus attention on the need to protect the region's fragile ecosystem.

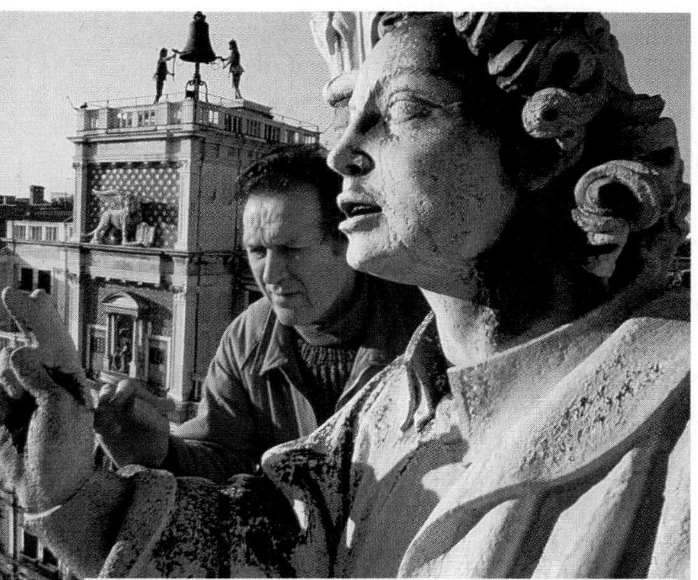

This man is cleaning a statue damaged by decades of air pollution in **Venice**. Time, pollution, and the heavy flow of tourists all take their toll on the city's famous architecture and artwork.

Plans like these depend on money. A $1.8 billion loan approved by the national government will pay for part of the work. The fund-raising efforts of Save Venice, Inc., an international organization formed shortly after the 1966 flood, continue to focus on restoring the city's historic structures. Venetians are hopeful that Italy is serious about saving and protecting their city. They would like to use their experience to inform other people about environmental issues. Some hope that a restored Venice can one day become a center of environmental research.

Your Turn

As a visiting environmentalist from Venice, **prepare** a speech that you will present to high school students around the United States. Include the problems that Venice faces, the attempts to solve them, and reasons that students should look to Venice as an environmental case study.

Italian government has attempted to develop southern Italy's economy. It has introduced nationalized industries, such as steel mills and refineries. Naples, the largest city in southern Italy, is the region's leading manufacturing center and major port. The nearby island of Capri, Mount Vesuvius, and the ancient city of Pompeii attract many tourists.

The narrow Strait of Messina separates the island of Sicily from the southern tip of the Italian peninsula. Mount Etna, an active volcano, rises 10,902 feet (3,323 m) above the eastern part of the island. Because Sicily's interior is mountainous, most of the island's population lives along the coast. The rocky island of Sardinia is located across the Tyrrhenian Sea southwest of the Italian peninsula. Tourism, farming, and sheep grazing are the main economic activities.

Political Geography Historical, social, and economic differences exist between northern and southern Italy. At the same time, the Italian language, strong family loyalties, and the Roman Catholic Church bind Italians together.

Italy's politics are fragmented, and many political parties exist. No single political party has been strong enough to gain control or form a stable **coalition government**. A coalition government is one in which several political parties join together to run the country. Italy has had more than 50 governments since World War II.

Today, the government must deal with poverty and unemployment in the nation's south, serious pollution problems, and government corruption. Also, nearly all of Italy's energy resources must be imported, which has a major effect on the country's **balance of trade**. Balance of trade is the relationship between the value of a country's exports and its imports. A favorable balance of trade exists when the value of a country's exports is higher than the value of its imports.

2 SECTION REVIEW

1. What mountain ranges are found in Italy? Where is the Po Valley?
2. Which of Italy's regions is most agriculturally productive? Which regions attract tourists?
3. **Critical Thinking** **Explain** the historical background of Italy's regional differences.

GREECE

FOCUS

- *How have Greece's physical features influenced the country's economy and history?*
- *What are Europe's microstates?*

LOCATION **Physical Geography** Greece is located at the southern end of the Balkan Peninsula, which extends into the Mediterranean Sea. The Ionian (ie-OH-nee-uhn) Sea lies to the west and the Aegean (i-JEE-uhn) Sea to the east. This strategic Mediterranean location is close to major trade routes and the entrances to the Black and Red seas. (See the map on page 297.)

Greece consists of many peninsulas, islands, and rugged mountains. Hundreds of narrow bays and inlets form the country's jagged coastline. About one-fifth of the area of Greece is made up of more than 2,000 islands. The largest of these is Crete. Mountains divide mainland Greece into many isolated valleys. The highest peak in Greece is Mount Olympus, with an elevation of 9,570 feet (2,917 m).

Economic Geography The dry climate and mountainous terrain limit agriculture to less than 30 percent of the land. Even so, agriculture supports about 20 percent of the Greek work force. The food-processing and food-export industries are based on agricultural products such as olive oil, tobacco, grapes, and citrus fruits. Oil refining is a major industry; however, Greece is heavily dependent on imported oil.

Shipping is a vital part of the Greek economy, and its fleet includes oil tankers, cargo ships, cruise ships, and fishing boats. Ancient ruins, beautiful beaches, and the sunny climate make the tourist industry an important part of the Greek economy as well. While financial aid from the European Union has helped improve its economic situation, Greece remains one of the poorest nations in the EU.

Historical Geography Ancient Greece was vital to the development of Western culture. More than 2,500 years ago, the Greeks developed ideas that today are the roots of our democratic system

NEWS FROM AROUND THE WORLD — Europe

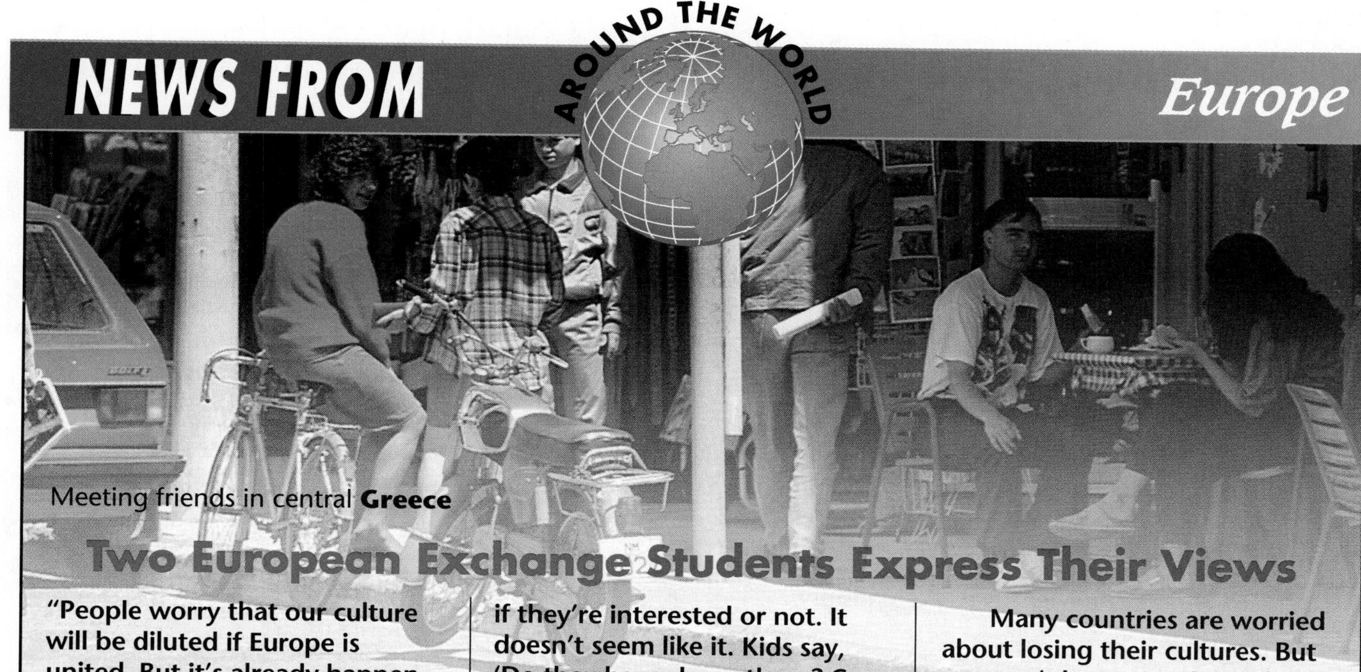

Meeting friends in central **Greece**

Two European Exchange Students Express Their Views

"People worry that our culture will be diluted if Europe is united. But it's already happening. Holland is trying to be like America.

There are some things I like about America. Everyone is so polite. . . . But I think America is too big. You can travel 1,000 miles this way and that way and you're still in America speaking English. In Europe you'd go through five countries.

It's more important for Americans to go to Europe than for Europeans to go to America. In school here [the United States], nobody knows anything about my country. I don't know if they're interested or not. It doesn't seem like it. Kids say, 'Do they have drugs there? Can you drink there? Can you say something in your language?' And that's the end of their questions. That's all they want to know."

Alice van Exel,
the Netherlands

"European political alliances have been disastrous in the past. But maybe we can learn from the past. Every country is very willing. We want to see a united Europe.

Many countries are worried about losing their cultures. But we won't lose ours. Our culture is centuries old, it would be very difficult to lose that. Even in America, Greek people still have their customs, even if they've lived here for generations.

After unification, it will be more difficult to find a job, much harder because you'll be competing with people from all over Europe. You will have to know more languages, have a better education, have skills that will separate you from everyone else."

Yolanda Grammaticaki,
Greece

of government. Greek art, philosophy, and science formed much of the basis of Western civilization.

Although it was the birthplace of democracy, Greece has had little democratic government in its long history. Its location at the crossroads of the Mediterranean exposed the country to invasions. Through the centuries, Persians, Romans, and Turks have all ruled Greece. Finally, in 1829, Greece became a free country after 400 years of Turkish rule. During World War II, the country was occupied by Germany. Then, immediately after the occupation ended, a bloody civil war broke out. After several governments, including a military dictatorship, the Greek government became a parliamentary republic in 1974.

Southern Greece Joined to the mainland by the Isthmus of Corinth, the Peloponnesus (pehl-uh-puh-NEE-suhs) Peninsula forms southern Greece. The peninsula's Mediterranean climate makes the region ideal for growing a variety of crops. Along with agriculture, fishing and tourism are important to the economy.

The ancient cities of Olympia and Sparta are in southern Greece. Olympia was the site of the first Olympic Games, which were held in 776 B.C. Sparta was famous for its athletes and warriors.

Crete lies directly south of the Peloponnesus Peninsula. Greece's largest and most populated island, Crete is primarily agricultural. To the west, in the Ionian Sea, are a number of scenic islands,

OTHER COUNTRIES IN EUROPE

ANDORRA
Andorra has no railway system, and its location in the Pyrenees does not permit air transportation. Roads link Andorra to neighboring Spain and France.

LIECHTENSTEIN
A huge wooden cross marks the spot where visitors to Naafkopf Peak in southern Liechtenstein can sit in three different countries at one time: Liechtenstein, Austria, and Switzerland.

MALTA
Malta's location in the Mediterranean Sea between Sicily and North Africa has made the islands important in struggles between Mediterranean powers throughout history.

VATICAN CITY
Tiny Vatican City is an independent state surrounded by the city of Rome. The pope, the head of the Roman Catholic Church, has absolute political authority within the city.

GIBRALTAR
Gibraltar, a British possession since 1713, sits at the entrance to the Mediterranean Sea from the Atlantic Ocean. The narrow peninsula at the southern tip of Spain is dominated by a limestone and shale ridge known as the Rock.

MONACO
Monaco draws people from around the world to its Mediterranean beaches, port, casino, and cultural attractions. Two well-known auto races, the Monte Carlo Rally and the Grand Prix de Monaco, take place in this tiny country.

SAN MARINO
Tradition has it that San Marino was founded early in the fourth century by Christians escaping persecution. It is one of the world's oldest and smallest republics.

Andorra, Liechtenstein, Monaco, San Marino, and Vatican City are microstates. Gibraltar is a dependent territory of the United Kingdom.

such as Corfu. The islands to the east, in the Aegean Sea, extend Greece's borders nearly to the mainland of Turkey. Most of these Greek islands are rocky and dry and depend on tourism for income. Greece and Turkey, rivals for centuries, both claim the offshore waters of the Aegean Sea, where there may be gas and oil deposits. Tension over the island of Cyprus remains a major issue. Although both nations are members of NATO, their relations with each other remain poor.

Central Greece Central Greece is the country's most populated region. Athens, the nation's capital and largest city, is located here. Athens was chosen as the capital in 1834 because of its historical importance. Most of the ancient city is located in the Acropolis, on a high hilltop overlooking the modern part of the city. The ancient marble temple, called the Parthenon, is one of the world's most photographed and famous buildings. After surviving almost 2,500 years of warfare and weather, it is now suffering from the effects of air pollution. Athens and its seaport of Piraeus (pie-REE-uhs) have become the country's largest center of urban and industrial growth.

Northern Greece The Pindus Mountains form a spine dividing northern Greece. To the west of the mountains, there are forests and pasture to graze sheep and cattle but almost no flat land to grow crops. This area is the wettest part of the nation.

In contrast, the northeast has the country's flattest land. Here, the plain of Thessaly and the Vardar River valley are the most productive agricultural areas. Salonika, located in this agricultural area, is the nation's second largest city and the major seaport for northern Greece.

3 SECTION REVIEW

1. How does Greece's physical geography affect economic activities? What effect has location had on Greece's history?
2. Which microstates border two other countries?
3. **Critical Thinking** **Propose** a way to protect the Parthenon from further damage caused by pollution.

CHAPTER 26 CHECK

The ancient Acropolis above **Athens, Greece**

Reviewing the Main Ideas

1. Spain covers over 80 percent of the Iberian Peninsula. It is a mountainous nation with three main climate types. Spain has a good natural resource base for industrialization and is one of Europe's leading tourist destinations.
2. Portugal occupies the western part of the Iberian Peninsula. Its climate is primarily Mediterranean. Portugal must import many goods and almost all of its energy resources.
3. Italy is a long, boot-shaped peninsula in the Mediterranean region of Europe. It was the center of the Roman Empire and the Renaissance. Northern Italy is industrialized and prosperous, while southern Italy is dry and poor.
4. Greece is a mountainous country with many islands and peninsulas. Agriculture, shipping, and tourism are important to the economy.
5. Europe's microstates are Andorra, Liechtenstein, Monaco, San Marino, and Vatican City. Gibraltar is a British possession at the southern tip of Spain. Malta is an independent island country.

Building a Vocabulary

1. What is Spain's high interior plateau called?
2. How is the plaza significant to agricultural villages in Spain?
3. What is cork?
4. Is Euskara a dialect? Explain.
5. Define *autonomy*. Which countries discussed in this chapter have autonomy?
6. What is a coalition government?

Recalling and Reviewing

1. What one word would you use to describe the landforms of Spain, Italy, and Greece?
2. For what crops is a Mediterranean climate suited? What natural resource must be imported by most of the countries in Southern Europe?
3. What cultural and political influences have the Southern European countries had on the rest of Europe and on the world?
4. What are two major port cities of the Iberian Peninsula?

Thinking Critically

1. Using what you learned in this chapter, write two sentences about the colonization of Middle and South America from the point of view of a Spanish or Portuguese colonist. Then, use what you learned in Unit 4 to write two sentences about the same experience from the point of view of one of the indigenous peoples.
2. The first Olympic Games were held in Greece in 776 B.C. What purposes do you think these games served? What purposes do today's Olympic Games serve?

Learning About Your Local Geography

Cooperative Project

Many dialects are spoken throughout Europe. In the United States, different regions often have their own accents. With your group, brainstorm and make a list of words that are pronounced differently in your area than they are in other areas. Using the pronunciation guide in this book, write out the local pronunciations for these words. What factors contribute to the growth of dialects and accents?

Using the Five Themes of Geography

Create a graphic organizer linking **location** and **human-environment interaction** in the countries of Southern Europe. How does the Mediterranean location influence human activities in these countries? How do these activities affect the physical environment?

CHAPTER 27

Eastern Europe

> **"W**e [in the former Czechoslovakia] are what history has made us. We live in the very centre of Central Europe, in a place that from the beginning of time has been the main European crossroads of every possible interest, invasion, and influence of a political, military, ethnic, religious, or cultural nature."
>
> *Vaclav Havel*

GEOGRAPHY DICTIONARY

- *regional specialization*
- *legacy*
- *complementary region*

Eastern Europe stretches from the often cold, stormy shores of the Baltic Sea south to the warmer and sunnier beaches along the Adriatic and Black seas. Eastern Europe is actually located in the middle of the European continent. Parts of the former Soviet Union occupy the true eastern part of the continent.

Historically, the countries of Eastern Europe have found themselves caught between larger and more powerful neighbors, including the Turks, Austrians, Germans, and Russians. The region's many ethnic groups often survived by becoming allies of nearby empires.

Conflicts among the peoples of Eastern Europe have also surfaced throughout the region's history. Ethnic and religious differences and extreme nationalism have become hallmarks of the region. Such conflicts continue in the region today.

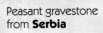

Peasant gravestone from **Serbia**

Charles Bridge, **Prague, Czech Republic**

EASTERN EUROPE: PHYSICAL–POLITICAL

LEGEND

ELEVATION

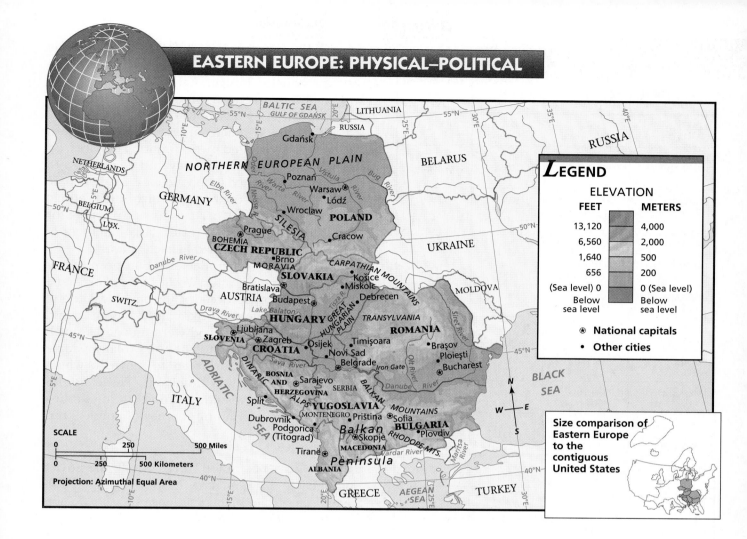

FEET	METERS
13,120	4,000
6,560	2,000
1,640	500
656	200
(Sea level) 0	0 (Sea level)
Below sea level	Below sea level

⊛ National capitals

• Other cities

SCALE 0 250 500 Miles; 0 250 500 Kilometers

Projection: Azimuthal Equal Area

Size comparison of Eastern Europe to the contiguous United States

AN OVERVIEW OF EASTERN EUROPE

FOCUS

- *What is Eastern Europe's physical geography? cultural background?*
- *How did Soviet control affect Eastern Europe politically and economically? What effects has the end of Soviet control had?*

REGION **Physical Geography** Eastern Europe is a region of varied landforms, including mountain ranges, plains, and plateaus. The plains of Poland are part of the vast Northern European Plain. To the south, the curving Carpathian Mountains stretch from the Czech Republic eastward across southern Poland to Ukraine, before heading south and west into Romania. Large plains lie between the mountain ranges. One of the largest of these is the Great Hungarian Plain. Another large plain area lies along the lower Danube River in southern Romania.

Low, rugged mountains and valleys characterize the Balkan Peninsula. The Dinaric Alps extend inland from the eastern shore of the Adriatic Sea, while the Rhodope and Balkan mountains rise up in Bulgaria. Communication among the peoples of the mountains has often been difficult. This isolation has contributed to the survival of many different ethnic groups in the region.

Two major climate types are found in Eastern Europe. In the north is the humid-continental climate. Winters are cold and snowy, and summers are warm and rainy. Farther south, winters are mild and summers become drier. The Mediterranean climate found along the Adriatic coast draws vacationers for much of the year.

FACTS IN BRIEF Eastern Europe

COUNTRY POPULATION (1994)	LIFE EXPECTANCY (1994)	LITERACY RATE	PER CAPITA GDP (1993)
ALBANIA 3,374,085	70, male 77, female	92% (1989)	$1,100
BOSNIA AND HERZEGOVINA 4,651,485	72, male 78, female	90% (1991)	Not available
BULGARIA 8,799,986	70, male 77, female	98% (1990)	$3,800
CROATIA 4,697,614	70, male 77, female	96% (1991)	$4,500
CZECH REPUBLIC 10,408,280	69, male 77, female	99% (1992)	$7,200
HUNGARY 10,319,113	67, male 76, female	98% (1992)	$5,500
MACEDONIA 2,213,785	72, male 76, female	90% (1990)	$1,000
POLAND 38,654,561	69, male 77, female	98% (1991)	$4,680
ROMANIA 23,181,415	69, male 75, female	96% (1991)	$2,700
SERBIA AND MONTENEGRO 10,759,897	71, M; 76, F 77, M; 82, F	90% (1991)	$1,000
SLOVAKIA 5,403,505	69, male 77, female	99% (1993)	$5,800
SLOVENIA 1,972,227	70, male 78, female	90% (1991)	$7,600
UNITED STATES 260,713,585	73, male 79, female	97% (1991)	$24,700

Sources: *The World Factbook 1994,* Central Intelligence Agency; *The World Almanac and Book of Facts 1995*

Considerable fighting makes statistics for Bosnia and Herzegovina unreliable.

MOVEMENT **A Cultural Crossroads** For centuries, waves of migrating peoples have swept into Eastern Europe, making the region a cultural crossroads. The new residents brought their own languages, religions, and other culture traits. By the start of the twentieth century, the human geography of Eastern Europe was extremely diverse.

The languages of Eastern Europe reveal the history of migrations and the many different people who settled in the region. The most widespread languages in Eastern Europe belong to the Slavic language group. Other languages spoken in the region include Romanian, Albanian, and Hungarian.

Religion also has played a part in separating the many ethnic groups. People in the eastern parts of the region primarily follow the traditions of the Eastern Orthodox church. Northern peoples are mostly Roman Catholic. Islam spread into southern sections of the region with the Turkish (Ottoman) invasions of the 1300s.

Throughout much of its history, Eastern Europe was controlled by neighboring empires. By the end of World War I, however, the countries of the region had gained independence. Many people migrated to the country in which their ethnic group was the majority. This was not possible for everyone, however. As a result, large numbers of ethnic minorities remained in each country.

Borders between Eastern European countries changed again after World War II. For example, the eastern portions of Czechoslovakia and Romania were joined to Ukraine. The eastern and western borders of Poland were moved as much as 125 miles (201 km) to the west. Many people again moved or were forced to move to where their ethnic group was dominant. These border changes and the resulting problems for groups left behind are continuing sources of political turmoil in the region.

Soviet Control After World War II, the Soviet Union came to dominate Eastern Europe. The Soviets controlled the nations of the region through military treaties and the placement of Soviet military bases in most of the countries. Communist governments took power and controlled almost all aspects of people's lives. The western boundary of Eastern Europe became known as the "Iron Curtain."

Beginning with Poland in 1989, the countries of Eastern Europe were successful in establishing independent governments. As Communist rule collapsed, personal freedoms increased and market economies emerged. The countries of Eastern Europe are now seeking closer ties to democratic countries around the world. The new political order in the region, however, has allowed old ethnic conflicts to resurface.

Economic Geography During their years under Communist rule, the countries of Eastern Europe had command economies. In most countries, farmers were forced to merge their lands to create large farms controlled by the government. The Council for Mutual Economic Assistance, or COMECON, oversaw economic ties between the Soviet Union and Eastern Europe.

Different religious cultures meet in **Hungary**, where this former Muslim mosque now is a Christian church. On top of the dome, note the cross, a symbol of Christianity, that has been added to the crescent, a symbol of Islam. Most Hungarians are Christians.

Under COMECON, **regional specialization** was the main plan. That is, each country specialized in products that it could make or grow best. COMECON gave the Eastern European countries access to the minerals and raw materials of the Soviet Union at discount prices. Many consumer goods then were shipped to the Soviet Union. Roads, railroads, and pipelines still connect the countries of Eastern Europe to Russia and Ukraine.

With the breakup of the Soviet Union, economic reorganization among Eastern European countries has varied. Poland, Hungary, and the Czech Republic have made rapid progress toward building free-market economies. Other countries, such as Bulgaria and Romania, have been slower to change. For all former Soviet countries, the end of Communist-controlled economies has meant some degree of financial distress. Many factories have closed, and unemployment has increased. Today, privately owned businesses are beginning to appear, offering workers new opportunities.

Another problem facing Eastern Europe is environmental decay, a **legacy** of the Communist era. A legacy is something received from a previous era. During the Communist era, production was more important than environmental concerns. As a result, air, soil, and water pollution, deforestation, and the destruction of natural resources were widespread. Many countries have begun the long and expensive task of cleaning up their environments.

Many Eastern European countries are middle-income nations, somewhere between the world's developed and developing countries. The region has good education and health-care facilities, highways, and railroads. The peoples of Eastern Europe look forward to greater economic progress and political freedom.

SECTION REVIEW

1. How has Eastern Europe's physical geography influenced the region's cultural background? What factors have caused ethnic conflicts in Eastern Europe?
2. How has the breakup of the Soviet Union affected daily life in Eastern Europe?
3. **Critical Thinking** **Suggest** why the words *iron* and *curtain* were used to describe the western boundary of Eastern Europe.

COUNTRIES OF EASTERN EUROPE

FOCUS

- *What are the economic characteristics of each country in Eastern Europe?*
- *What is the historical and cultural background of each country?*

Poland Poland is Eastern Europe's largest and most populous country. Despite some industrial progress, Poland is well known as an agricultural region.

Warsaw, the country's capital, is located on the plains. (See the map on page 309.) With about

Visitors look over the art and crafts for sale in the Old Town Market Square in **Warsaw**. The homes on the square have been rebuilt to recapture their original fifteenth-century style.

1.6 million people, it is the cultural, political, and historical center of Polish life. Its location on the Vistula River makes Warsaw the center of the national transportation and communications networks as well.

In 1980, an independent organization of workers, called Solidarity, formed in Poland. Solidarity threatened the authority of the Communist party and became a popular focus for Polish nationalism and anti-Soviet activities. In 1989, the organization succeeded in bringing democratic elections to the country. Today, Poland is trying to model its economic and political system after the developed countries of Europe.

Czech Republic and Slovakia In 1993, Czechoslovakia peacefully divided into the Czech Republic and Slovakia. Czechoslovakia had formed in 1918 from the combination of Czech and Slovak territories. Czech lands held mineral resources important for industrialization, while Slovakia was primarily agricultural. This is an example of what geographers call **complementary regions**. The merging of the two regions benefited both.

Czechoslovakia gained its independence from Soviet control in 1989. This split took place so peacefully that it has been called the "velvet revolution." Cultural differences between the Czechs and Slovaks remained, however. The Slovaks, who had never had their own independent country, soon voted for separation.

Prague is the capital and industrial heartland of the Czech Republic. This prosperous city features historic buildings and one of Europe's oldest universities. Prague is the center of Czech life and is undergoing a cultural rebirth.

Slovakia is more rural than the Czech lands, and incomes are lower than in the Czech Republic. Bratislava, the capital of Slovakia, is on the Danube River. As the cultural, economic, and industrial center of Slovakia, Bratislava draws rural Slovaks seeking better-paying jobs. The Slovaks have been less enthusiastic than the Czechs in embracing a free-market economy.

Hungary Hungary is a major agricultural producer. Hungary's past Communist government put most of the nation's farmland into state-controlled, cooperative farms. Local farm managers sometimes were allowed to make key business decisions, however. They kept technology current, chose their crops, and marketed their products. This experience is helping Hungary improve its economy today.

Budapest is Hungary's capital city as well as the country's political, cultural, and economic center. About 20 percent of Hungary's people live in the Budapest region, where manufacturing is most developed.

Hungary's future looks promising. The country is making cultural and economic links with the developed countries of Europe and with the United States. Russia and Ukraine, meanwhile, remain a ready market for Hungarian products and an important source of raw materials.

Romania Although the Communist government expanded industry after World War II, Romania remains a developing country. With the exception of oil, most of Romania's potential mineral resources are undeveloped. The famous Ploiesti (plaw-YEHSHT) oil fields are some of the richest in Eastern Europe. Even so, Romania's oil resources do not meet the country's needs.

Romania's cultural, industrial, and commercial center is Bucharest (BOO-kuh-rehst), the capital city. About 70 percent of all Romanians belong to the Romanian branch of Eastern Orthodox Christianity. A large number of ethnic Hungarians live in northern Romania.

For over 20 years, Romania had a particularly repressive Communist dictatorship. A violent uprising in 1989 overthrew the government. The new Romanian government promises a parliamentary democracy and a market economy.

Bulgaria Bulgaria occupies the eastern side of the Balkan Peninsula. During the Communist era, the Bulgarians were the most loyal followers of the Soviet Union. With extensive Soviet economic aid, Bulgaria became more prosperous than some other countries of Eastern Europe.

Important industries are concentrated near Sofia, the capital and largest city. Sofia is also the historic center of Bulgaria's cultural and political life. Today, it is a bustling, modern city.

The new government contains many leaders from the Communist era, and life in Bulgaria is changing only gradually. The people do enjoy more freedoms, and a market economy is growing slowly.

Albania Physically and culturally isolated from the rest of Eastern Europe, Albania is Europe's least developed and poorest country. Albania's capital city, Tirane (ti-RAHN-uh), the site of the country's few factories, has fewer than 300,000 people.

After World War II, a Communist government took power, but the Albanian Communists feuded with the Soviet Union and received no economic aid. A command economy coupled with isolation meant little economic progress for the country.

In 1990, the Communist government collapsed, and Albanians began to welcome foreign contact. However, economic change and movement toward democracy are coming only slowly. Economic and political progress could slow even more if Albania becomes involved in ethnic conflicts with neighboring countries.

2 SECTION REVIEW

1. What economic activities support the countries discussed in this section?
2. What roles are played by Eastern Europe's capital cities?
3. **Critical Thinking** **Compare and contrast** Poland's and Bulgaria's relationships with the Soviet Union during the Communist era.

3 A REGION IN CONFLICT

FOCUS

- *How did Yugoslavia form, and why did it break up? How is the territory divided now?*
- *What role have ethnic conflicts played in the region?*

The Breakup of Yugoslavia Along the western half of the Balkan Peninsula lie the new nations of the former country of Yugoslavia. Yugoslavia was created after World War I mainly by U.S., British, and French policy makers. *Yugoslavia* means "land of the southern Slavs." Lands populated by several Slavic groups were combined from pieces of the Turkish (Ottoman) Empire and the Austro-Hungarian Empire. The plan was that the region's many Slavic groups would share common goals that would benefit the new country. The main Slavic groups in the former Yugoslavia were the Serbs, Croats, Bosnians, Slovenians, and Macedonians.

Although Yugoslavia became a Communist country after World War II, its leader, Josef Broz Tito steered a course independent from the Soviet Union. Tito kept a tight grip to keep the country unified, and indeed, Yugoslavia survived as a single nation until 1991. The strong central government had been able to force the ethnic groups to live together. The Serbs and Croats, however, have thought of themselves as separate peoples since the seventh century. With Tito's death and, later, the end of communism in the Soviet Union, the country soon began to unravel violently.

Serbia and Montenegro together have claimed the name Yugoslavia. Bosnia and Herzegovina, Croatia, Slovenia, and Macedonia have all declared their independence. Due to ethnic and religious differences, civil war broke out in Bosnia between Serbs, Croats, and Bosnians. Most Croats and Slovenians are Roman Catholic, the Serbs and Macedonians are Eastern Orthodox, and many Bosnians are Muslims. Because of the mixture of ethnic groups in many regions, additional border changes are possible. As these ethnic struggles continue, the economy of the region is in steady decline, along with the quality of people's lives.

Serbia and Montenegro Serbia is in the eastern plains of former Yugoslavia. In the southwest, Serbia borders Montenegro, a mountainous coastal territory north of Albania. The capital of Serbia is Belgrade, also the capital of former Yugoslavia. Located on the Danube River, Belgrade is one of the major cities of Eastern Europe.

The Serbs have been demonstrating extreme nationalism since the late 1980s. Radical Serbs maintain that no Serb should be governed by someone from another ethnic group. Of particular concern is the future status of the Vojvodina and Kosovo regions. Vojvodina is part of the Great Hungarian Plain north of Belgrade. Its people are mostly Hungarians who fear Serbian nationalism. Many have fled Vojvodina into Hungary. Kosovo is a southern Serbian region along the border with Albania. Most of the people here are Muslim Albanians. Friction between Kosovo's Albanians and Serbs has escalated.

Bosnia and Herzegovina Bosnia and Herzegovina generally are referred to as Bosnia. This is a mountainous territory located between the plains of Serbia and the coastal mountains of Croatia. Within the mountains of Bosnia lie isolated valleys where rural life has changed little in the past century. The once-beautiful capital of Sarajevo (SAIR-uh-yay-voe) has been destroyed in the civil war. No single ethnic group is a majority.

Croatia Croatia occupies the northern part of the Dinaric Alps and has a long, narrow coastal region along the Adriatic Sea. Zagreb is the capital and cultural center. About 12 percent of the people in Croatia are ethnic Serbs who would like to see Yugoslavia reestablished. Tensions between Croatia and Serbia will likely continue.

Slovenia Slovenia is a former Austrian territory situated in the mountains northwest of the Balkan Peninsula. Slovenians have long thought of themselves as part of Western rather than Eastern Europe. Few minorities live in the region, and independence came with little bloodshed. The territory remains largely agricultural, and the farms have always been privately owned. Ljubljana (lee-yoo-blee-AHN-uh), the capital city and principal industrial center, lies on the country's major plain.

Macedonia Macedonia refers to both a province of northern Greece and the southern region of former Yugoslavia. The capital of the latter is the city of Skopje (SKAWP-yeh). About 65 percent of the people in Macedonia are Macedonians. The largest minority, about 20 percent, are Muslim Albanians. Many of the farms in this agricultural land provide little more than subsistence incomes. With few jobs available, many young Macedonians seek work in other European countries. There is a history of cooperation between Macedonians and Serbs in Yugoslavian affairs. As a result, this region has been spared much of the impact of the civil war. The region's economic future remains uncertain, however.

Ethnic Albanians worship at a Muslim mosque In **Priština**, the capital of the Kosovo province in **Serbia**. Although a minority in Serbia, ethnic Albanians make up most of the population of Kosovo.

Future Changes The history of the Balkan region is being rewritten every day. It is difficult to understand the turmoil and bitterness that has plagued the Balkans for centuries. Many attempts have been made by the United States, the EU, and the United Nations to negotiate peace, but there is no clear or easy solution to the conflict.

3 SECTION REVIEW

1. When and why was the country of Yugoslavia formed? What held the country together until 1991?
2. What ethnic groups live in the region? How do these groups relate to each other? Where have ethnic conflicts been most extreme?
3. **Critical Thinking** **Compare and contrast** the breakups of Yugoslavia and Czechoslovakia.

CHAPTER 27 CHECK

Black Sea vacationers in **Bulgaria**

Reviewing the Main Ideas

1. Eastern Europe is a region of varied landforms and cultures. The borders of Eastern European countries have changed many times. From World War II to the late 1980s, the Soviet Union controlled Eastern Europe and introduced communism to the region. Today, the countries are becoming democratic and are building free-market economies.
2. Some of the countries of Eastern Europe are Poland, the Czech Republic, Slovakia, Hungary, Romania, Bulgaria, and Albania. The economic activities in these countries vary, as do the ways in which the countries have responded to independence from Soviet control.
3. The country formerly called Yugoslavia began to unravel violently when communism in the Soviet Union ended and the central government could no longer control ethnic tensions. Rivalries between ethnic groups wrack Serbia and Montenegro, Bosnia and Herzegovina, and Croatia.

Building a Vocabulary

1. What term describes the main plan of COMECON for the countries of Eastern Europe? Describe the plan.
2. Define *legacy*, and give an example of a legacy.
3. What are complementary regions? How were the Czech and Slovak territories complementary regions?

Recalling and Reviewing

1. What climates are found in Eastern Europe?
2. What languages do Eastern Europeans speak? What religions do they practice?
3. What do Romania and Albania have in common economically?

Thinking Critically

1. How might regional specialization be a disadvantage for a country?
2. The ethnic tensions in the countries that were formerly called Yugoslavia are centuries-old. Why do you think that even after 40 years of repression by a Communist government, ethnic identities were still alive?

Learning About Your Local Geography

Cooperative Project

As the countries of Eastern Europe become more open to the other countries of the world, Eastern Europeans are eager to learn about the ways of life of people in these other countries. As a group, write a letter to a teenager in Eastern Europe about your daily lives. You might want to include what you do for fun, what your favorite foods are, and what you study in school.

Using the Five Themes of Geography

Make a flowchart showing how physical and human **place** characteristics have influenced peoples' **movement** in the countries of Eastern Europe. You might want to make your flowchart from the point of view of an Eastern European trying to decide whether to move to a different area within the region. What factors might keep you from moving? What factors might encourage you to move?

SKILLS

Linking Geography and Fine Art

Throughout the centuries, artists have been interested in the physical world around them. By studying works of art, we can learn much about the physical and human geography of a particular region.

Peter Graham, a Scottish artist who lived from 1836 to 1921, often painted scenes from the Scottish highlands. As you know from Chapter 24, these highland areas contain ancient, eroded rocky hills, as shown in the painting to the right. Also shown clearly is the cloudy weather for which Great Britain is known. The painting even portrays the economic activities for which the region is suited: grazing and fishing. If you were to visit the area shown in the painting today, you would probably find a similar scene.

Wandering Shadows, Peter Graham

The Gleaners, Jean-Francois Millet

Jean-Francois Millet was a famous French painter who lived from 1814 to 1875. Most of his paintings, such as this one at left, show scenes from French rural life.

PRACTICING THE SKILL

Study the painting *The Gleaners* above, and then answer the following questions.

1. What economic activity is shown in the painting? Use a dictionary to identify the meaning of the word *glean*.
2. Based on the landform and on the activity shown in the painting, what part of France does this painting probably show?
3. The main characters of the painting are women. How is this significant?
4. Millet completed this painting in 1857. If you were to visit the place portrayed in the painting today, would you find a similar scene? Why or why not?

BUILDING YOUR PORTFOLIO

Individually or in a group, complete one of the following projects to show your understanding of the geography concepts involved.

A A Unified Europe?

In this unit you have read much about disunity in Europe. Some countries, such as Czechoslovakia and Yugoslavia, have broken up into smaller countries. In addition, minority groups in some countries, such as the Roman Catholics in Northern Ireland and the Basques in Spain, want independence. Yet, as you have also read, many Europeans hold an ideal of a united Europe. Write a draft of what a constitution of the new united Europe might say. Your constitution will consist of four written articles, each outlining a different aspect of the new united Europe.

1. Your first article will address political aspects of the new united Europe. What form of government will it have? How will leaders be chosen? Will individual regions or countries have any political powers?
2. The second article of your constitution will be about the new united Europe's military. How will the region defend itself? Will it have a national military force, and if so, who will serve in it?
3. Your third article will relate to the new united Europe's economic system. Will industries be publicly or privately owned? What currency will be used? How will taxes be collected, and what will they be used for? How will trade among the different regions be regulated?
4. The constitution's fourth article will address cultural matters. What will be the new united Europe's official language or languages? Will all cultural groups have the same rights and privileges? How would any ethnic conflicts be handled?

Present your constitution to the rest of the class. In doing so, address the following questions. Do you think it is desirable for the countries of Europe to unite? Do you think such unification is possible?

B Water Pollution in Europe

As you know, water is one of Europe's most important resources. Rivers and harbors are used for trade and transportation, for fishing, and for hydroelectricity. Yet many of Europe's waters suffer from pollution. What are the causes of this pollution? Which bodies of water are particularly polluted, and why? How might the problem of pollution be solved? Investigate the answers to these questions, and write a script for a brief documentary film presenting your findings. Include with your script illustrations, maps, and other visual materials that will help your audience understand your points. Your documentary will be organized according to Europe's regions.

1. First, consider water pollution in Northern Europe. You might want to focus on Great Britain's rivers, or the Irish, North, or Baltic sea.
2. Second, discuss the pollution of rivers in West Central Europe, such as the Rhine or the Seine.
3. Next, address the pollution of the Mediterranean Sea. Include a discussion of the environmental effects of tourism.
4. Finally, your documentary should discuss water pollution in Eastern Europe, such as along the Danube.

Present your script and supporting visual materials to the rest of your class.

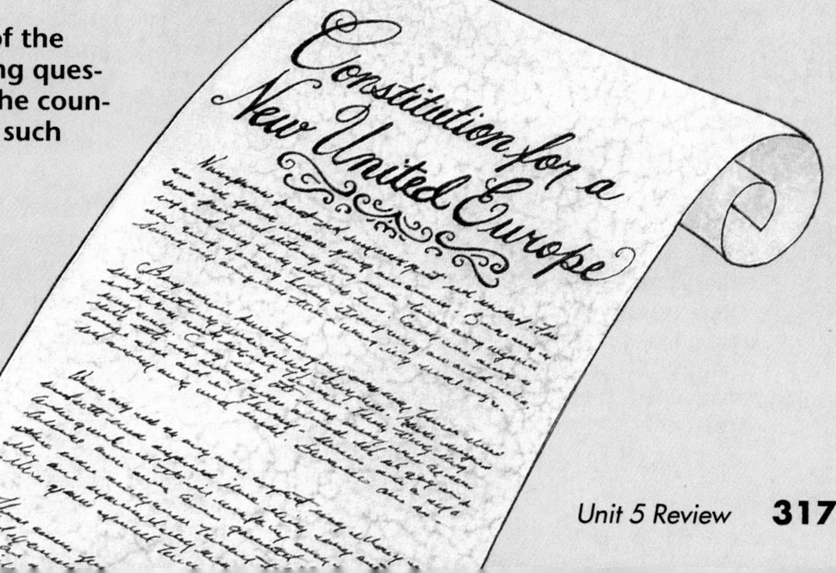
Constitution for a
New United Europe

Redrawing the European Map

The collapse of the Soviet Union marks one of the most significant events of the twentieth century. Geographers are defining new regions in Europe that include countries that were once part of the Soviet Union. Among these are Moldova, sandwiched between Romania and Ukraine, and the three small Baltic countries of Lithuania, Latvia, and Estonia.

The Baltic Countries

Lithuania, Latvia, and Estonia were once part of the Russian Empire. They enjoyed brief independence after the Communist Revolution in 1917, but in 1940, they were forced into the Soviet Union. The Baltic states regained their political independence in 1991. Achieving economic and cultural independence, however, could take years for several reasons. First, Estonia and Latvia include large numbers of ethnic Russians. Second, Russian army and navy bases in the region have long been important to the Baltic states' local economies. And third, transportation and business links tie the Baltics to the former Soviet Union.

Each of the Baltic countries is particularly dependent on minerals and fossil fuels from Russia and Ukraine. Similarly, the traditional markets for Baltic goods are in Russia and Ukraine. For these reasons, continued cooperation with the other regions of the former Soviet Union is necessary. Establishing strong economic links to the rest of the world will come gradually.

These Latvian children in local dress are playing a traditional stringed musical instrument from their country. People in the Baltic countries proudly maintained their cultural traditions during the Soviet era.

FACTS IN BRIEF The Baltic Countries and Moldova

COUNTRY POPULATION (1994)	LIFE EXPECTANCY (1994)	LITERACY RATE	PER CAPITA GDP (1993)
ESTONIA 1,616,882	65, male 75, female	100% (1989)	$5,480
LATVIA 2,749,211	64, male 75, female	99% (1992)	$4,810
LITHUANIA 3,848,389	67, male 76, female	98% (1989)	$3,240
MOLDOVA 4,473,033	65, male 72, female	96% (1992)	$3,650
UNITED STATES 260,713,585	73, male 79, female	97% (1991)	$24,700

Sources: *The World Factbook 1994,* Central Intelligence Agency; *1995 Britannica Book of the Year*

Which country has life expectancy rates closest to that of the United States?

As the Baltic countries reexamine their cultural makeup and economies, geographers ponder how best to include these countries in the subregions of Europe. Perhaps Estonia and Latvia will be identified as Scandinavian countries of Northern Europe. The Estonian language is related to Finnish, and ties between Estonia and Finland are growing stronger. Many Estonians watch Finnish television broadcasts, and ferries link the capitals of Tallinn, Estonia, and Helsinki, Finland. Latvia, like Estonia, was heavily influenced by Sweden in the past. Lutheranism is the traditional religion of Latvia, Estonia, and Scandinavia.

In contrast to Estonia and Latvia, Lithuania shares a border with Poland and has ancient ties to that country. About 7 percent of the people in Lithuania are ethnic Poles. In both countries, Roman Catholicism is the dominant religion. Geographers speculate that many of these cultural ties will grow, and that Lithuania, like Poland, will be considered part of Eastern Europe.

THE BALTIC COUNTRIES AND MOLDOVA: PHYSICAL–POLITICAL

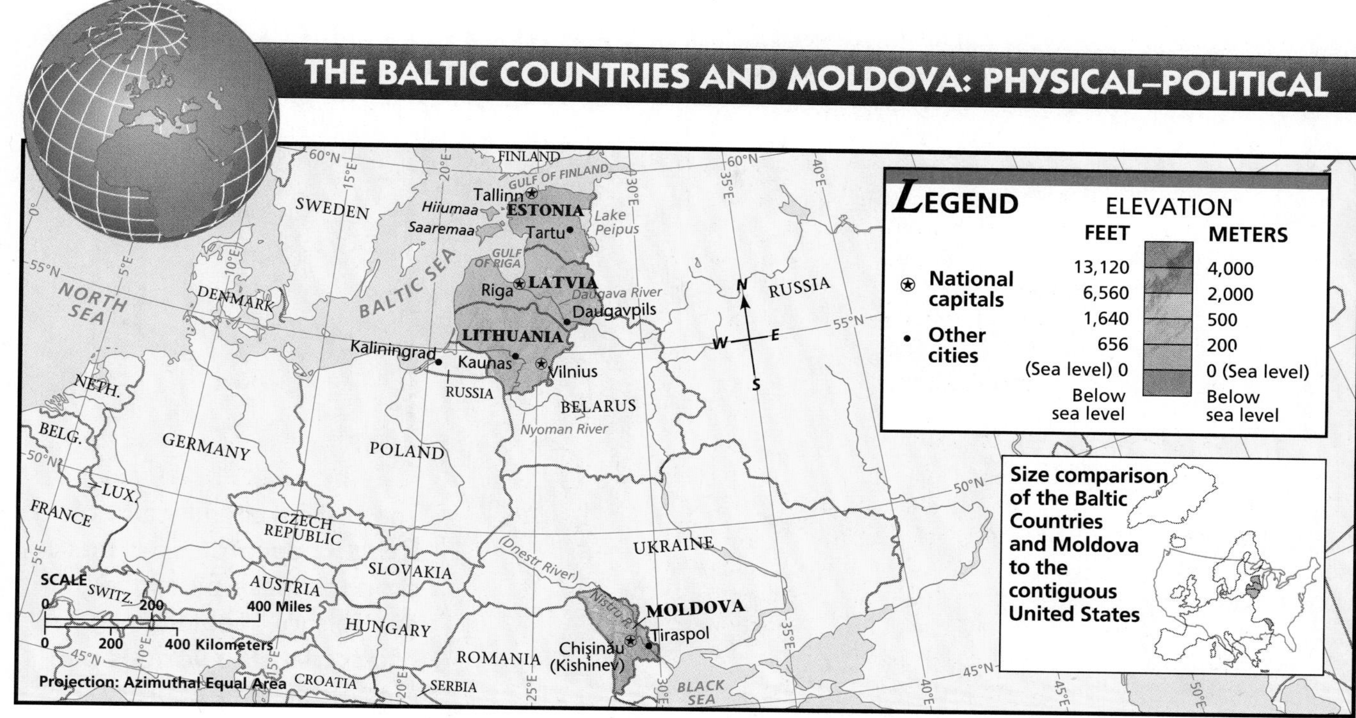

LEGEND
ELEVATION
FEET
METERS
13,120
6,560
1,640
656
(Sea level) 0
Below sea level
4,000
2,000
500
200
0 (Sea level)
Below sea level
National capitals
Other cities
Size comparison of the Baltic Countries and Moldova to the contiguous United States
SCALE
0
200
400 Miles
0
200
400 Kilometers
Projection: Azimuthal Equal Area
FINLAND
GULF OF FINLAND
Tallinn
ESTONIA
Hiiumaa
Saaremaa
Tartu
Lake Peipus
GULF OF RIGA
Riga
LATVIA
Daugava River
Daugavpils
LITHUANIA
Kaliningrad
Kaunas
Vilnius
RUSSIA
BELARUS
Nyoman River
SWEDEN
BALTIC SEA
DENMARK
NORTH SEA
NETH.
BELG.
LUX.
FRANCE
GERMANY
POLAND
CZECH REPUBLIC
SLOVAKIA
AUSTRIA
SWITZ.
HUNGARY
CROATIA
SERBIA
ROMANIA
Dnestr River
Nistru R.
MOLDOVA
Chişinău (Kishinev)
Tiraspol
UKRAINE
BLACK SEA
N
S
E
W
60°N
55°N
50°N
45°N
0°
5°E
10°E
15°E
20°E
25°E
30°E
35°E
40°E
45°E
50°E

Kaliningrad

The future status of the Kaliningrad region along the Baltic Sea also is of interest to geographers. This territory remains part of Russia, even though it is now separated from Russia by independent Lithuania. Before World War II, this region was part of Germany. After the war, many of the region's ethnic Germans were forced to move to Siberia and Central Asia. Ethnic Russians replaced the German population. Kaliningrad could become valuable to Russia as a window on Europe. Some geographers believe, however, that it will be next to demand its independence.

Moldova

Throughout history, control of Moldova has shifted many times. It has been dominated by the Turks, Polish princes, Austria, Hungary, Russia, and Romania. The Soviet Union acquired Moldova by force from Romania during World War II. Like the Baltic states, Moldova became independent of the Soviet Union in 1991.

Moldova's ethnic composition is diverse. Ethnic Romanians make up about two-thirds of the population, but the country also has many Russians and Ukrainians. Ethnic conflicts among these groups have emerged and remain to be settled. Gradually, geographers may include Moldova in the same region as Romania. It is even possible that the two countries will eventually merge.

Mapping the New Europe

As countries realign themselves culturally and politically, borders may shift again and again. The new Europe may stretch across the geographical boundaries of the European continent, though ties to the former Soviet Union will remain. Geographers are hopeful that these changes can be made peacefully and with respect for the rights of all ethnic groups.

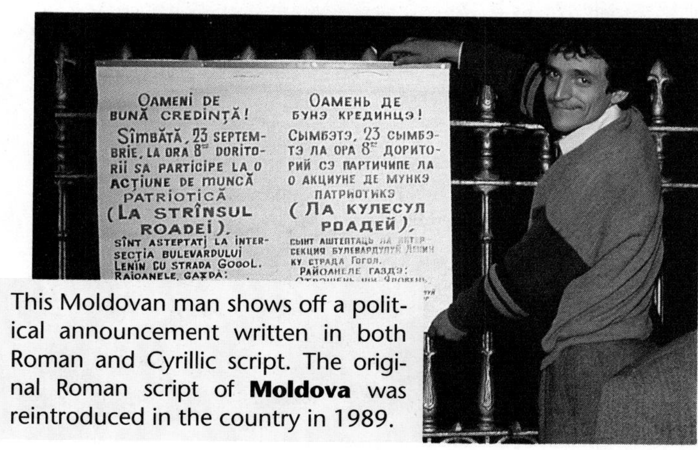

OAMENI DE BUNĂ CREDINŢĂ!
Sîmbătă, 23 septembrie, la ora 8 doritorii să participe la o acţiune de muncă patriotică
(La strînsul roadei).
ОАМЕНЬ ДЕ БУНЭ КРЕДИНЦЭ!
Сымбэтэ, 23 сымбэтэ ла ора 8 доритоpий сэ партичипе ла о акциуне де мункэ патриотикэ
(Ла кулесул роадей).

This Moldovan man shows off a political announcement written in both Roman and Cyrillic script. The original Roman script of **Moldova** was reintroduced in the country in 1989.

UNIT 6

RUSSIA *and* NORTHERN EURASIA

The distinct architecture of **Russia** decorates the backdrop for these performers at the Bolshoi Theater in **Moscow**. The various peoples of Northern Eurasia have a rich cultural heritage. Many famous writers, composers, and performers throughout history have been from Russia and its neighboring republics.

UNIT 6 CHAPTERS

The rolling plains of the steppe sweep toward the distant Ural Mountains. Stretching from **Ukraine** to **Central Asia**, the vast steppe is the most productive agricultural region of Northern Eurasia.

A gold comb from the early Scythian kingdom

Russian *matryoshka* nesting doll

Gingerbread trim decorates this *dacha* near **Kursk, Russia**. Country homes such as this are popular vacation locations for many Russians. Some *dachas* are small, one-room houses or even rented rooms in a larger house.

ONE WORLD, MANY VOICES

LITERATURE

Russian Railroad

"The Donetz Railroad. A cheerless station, quiet and lonely, gleaming white on the steppe, with walls hot from the sun, with not a speck of shade and, it appears, with not a single human being. The train which brought you here has left; the sound of it is scarcely audible and at last dies away. The neighborhood of the station is deserted, and there are no carriages but your own. You get into it—this is so pleasant after the train—and you roll along the road through the steppe, and by degrees, a landscape unfolds such as one does not see near Moscow—immense, endless, fascinating in its monotony. The steppe, the steppe, and nothing else; in the distance an ancient grave-mound or a windmill; oxcarts laden with coal file by. Birds fly singly low over the plain and the monotonous beat of their wings induces a drowsiness. It is hot. An hour or two passes, and still the steppe, the steppe, and still in the distance the grave-mound. The driver rambles on telling you some long-drawn-out, irrelevant tale, frequently pointing at something with his whip, and tranquillity takes possession of your soul, you are loath to think of the past."

Anton Chekhov (1860–1904) is one of Russia's most important literary artists. Ill health from tuberculosis, along with a love of writing, caused him to give up a career in medicine. He wrote many plays and hundreds of short stories. Most of his work concerns the everyday lives of people from many classes of pre-revolutionary Russian society in the late 1800s. Chekhov's tales reveal his understanding of human nature and also express his affection for the Russian physical landscape. The selection on this page is from a story entitled "At Home."

Interpreting Literature

1. What type of climate and vegetation region does Chekhov describe?
2. What words does the narrator use to indicate that the scenery changes very little throughout the carriage ride?

FOR THE RECORD

Laurens van der Post (1906–) is a noted travel writer and novelist. He has said that "one never really knows another country" unless one observes it firsthand through the lives of its citizens. Many of his books describe Africa, where he was born, and his travels to England, France, and Japan. Van der Post was curious about Russia because he was not satisfied with the images of the country he received through newspapers, magazines, and television. In 1961, he began the first of four lengthy trips across the Soviet Union. This excerpt is from *Journey into Russia,* published three years later.

"This Russian now told me that it was right to begin with a feeling of the immensity of the land. I must build on that as perhaps the single greatest fact for understanding his people and his native country. But he feared that it would take time before a stranger could realize how immense the land was. . . .

So wide apart are the two extremities of land that it is day in one when it is still night in another. He himself had once flown in a fast jet aircraft from Baikal [Baykal] to Moscow, made four landings on the way and each time as they landed the sun had set and each time they took to the air the sun had risen again. . . . Already in Turkmenia [Turkmenistan], ploughing and sowing would be done and the spring nearly over, but in Uzbekistan where we would land in a few hours we would find the earth aglow with purple heliotrope, scarlet poppies and wild red tulips. Yet the apple trees around Moscow would not blossom until late in May and in the far north of Siberia the tundra rivers would not be free of ice until the end of June. There by August the ice and frost would start their march in the south. In the extreme south the bamboo shot up at a rate of a yard a day. On the northern Siberian shore it took the straggling larches a century to grow the thickness of a man's middle finger. In the south they would already be harvesting the grain while in Kamchatka the snow was still falling. That, measured in the seasons, was the physical frame of the Soviet Union."

Analyzing Primary Sources

1. What does this selection suggest is most significant about Russia's physical geography?
2. How does Van der Post illustrate Russia's vast size?

CHAPTER 28

Introducing Russia and Northern Eurasia

GEOGRAPHY DICTIONARY

- ***czar***
- ***abdicate***
- ***soviet***
- ***autarky***
- ***superpower***
- ***client state***
- **glasnost**
- **perestroika**

Nineteenth-century jeweled Fabergé egg

For most of the twentieth century, much of the vast region of northern Eurasia was dominated by the Union of Soviet Socialist Republics (USSR), commonly known as the Soviet Union. The Soviet Union's beginnings reach back to 1917 when the Russian Empire ended in violent revolution.

The main goal of the Soviet Communist government was to achieve economic equality for all. In trying to reach this goal, however, the government used oppressive methods of control on its citizens. It limited economic and social freedoms and suspended basic political rights. This command economy system could not feed or house all of the country's people and fell behind the Western free-market economies.

After the collapse of the Soviet Union in 1991, the political geography of northern Eurasia began to change dramatically. The land that was once the Soviet Union is now organized into 15 independent countries, including Russia, the world's largest country. There are new borders, new political systems, new place-names, and new goals for the future. And perhaps most important, the many peoples of northern Eurasia are rediscovering their ethnic heritages.

St. Basil's Church in **Moscow**

MAPPING THE REGION

RUSSIA AND NORTHERN EURASIA

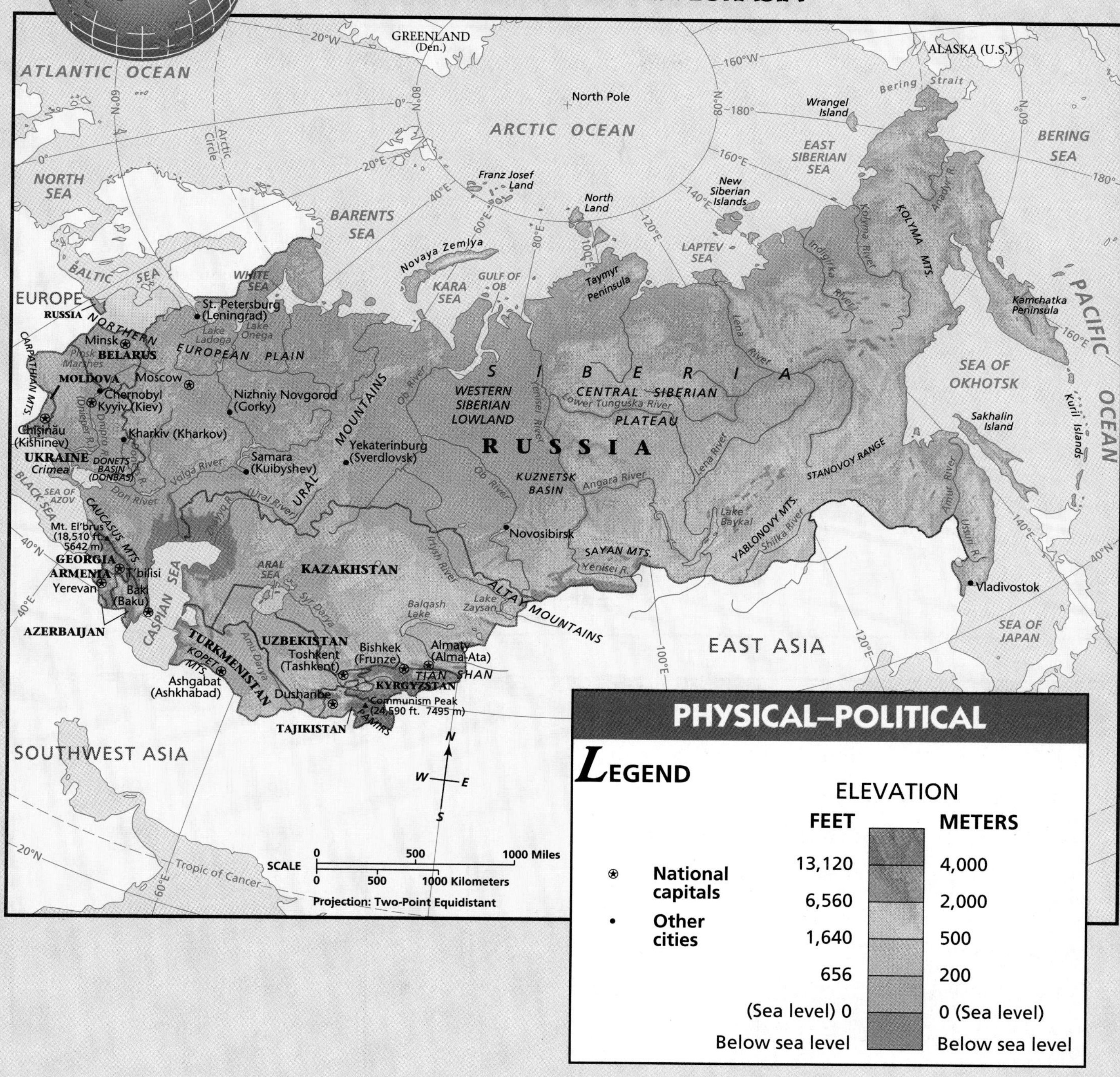

GREENLAND (Den.)
ALASKA (U.S.)
ATLANTIC OCEAN
ARCTIC OCEAN
North Pole
Arctic Circle
Wrangel Island
Bering Strait
BERING SEA
EAST SIBERIAN SEA
NORTH SEA
Franz Josef Land
North Land
New Siberian Islands
BARENTS SEA
Novaya Zemlya
KARA SEA
GULF OF OB
LAPTEV SEA
Taymyr Peninsula
KOLYMA MTS.
Kolyma River
Indigirka River
Anadyr R.
Kamchatka Peninsula
PACIFIC OCEAN
BALTIC SEA
WHITE SEA
EUROPE
RUSSIA
NORTHERN EUROPEAN PLAIN
St. Petersburg (Leningrad)
Lake Ladoga
Lake Onega
Minsk
BELARUS
Pinsk Marshes
CARPATHIAN MTS.
MOLDOVA
Moscow
Chernobyl
Kyyiv (Kiev)
Nizhniy Novgorod (Gorky)
Dnieper R.
Dnipro R.
Chișinău (Kishinev)
Kharkiv (Kharkov)
UKRAINE
Crimea
DONETS BASIN (DONBAS)
Don R.
Volga River
Samara (Kuibyshev)
URAL MOUNTAINS
Ural River
Ob River
Yekaterinburg (Sverdlovsk)
WESTERN SIBERIAN LOWLAND
Yenisei River
SIBERIA
CENTRAL SIBERIAN PLATEAU
Lower Tunguska River
Lena River
RUSSIA
SEA OF OKHOTSK
Sakhalin Island
Kuril Islands
STANOVOY RANGE
Amur River
Ussuri R.
KUZNETSK BASIN
Angara River
Lake Baykal
YABLONOVY MTS.
Shilka River
BLACK SEA
SEA OF AZOV
Don River
CAUCASUS MTS.
Mt. El'brus (18,510 ft. 5642 m)
Zhayyq R.
Novosibirsk
SAYAN MTS.
Yenisei R.
GEORGIA
T'bilisi
ARMENIA
Yerevan
Baki (Baku)
AZERBAIJAN
CASPIAN SEA
ARAL SEA
KAZAKHSTAN
Irtysh River
Syr Darya
Balqash Lake
Lake Zaysan
ALTAY MOUNTAINS
Vladivostok
SEA OF JAPAN
EAST ASIA
TURKMENISTAN
KOPET MTS.
Amu Darya
UZBEKISTAN
Toshkent (Tashkent)
Bishkek (Frunze)
Almaty (Alma-Ata)
TIAN SHAN
KYRGYZSTAN
Ashgabat (Ashkhabad)
Dushanbe
Communism Peak (24,590 ft. 7495 m)
PAMIRS
TAJIKISTAN
SOUTHWEST ASIA
Tropic of Cancer
N
W
E
S
SCALE
0 500 1000 Miles
0 500 1000 Kilometers
Projection: Two-Point Equidistant
PHYSICAL–POLITICAL
LEGEND
National capitals
Other cities
ELEVATION
FEET
METERS
13,120
4,000
6,560
2,000
1,640
500
656
200
(Sea level) 0
0 (Sea level)
Below sea level
Below sea level

MAPPING THE REGION

RUSSIA AND NORTHERN EURASIA

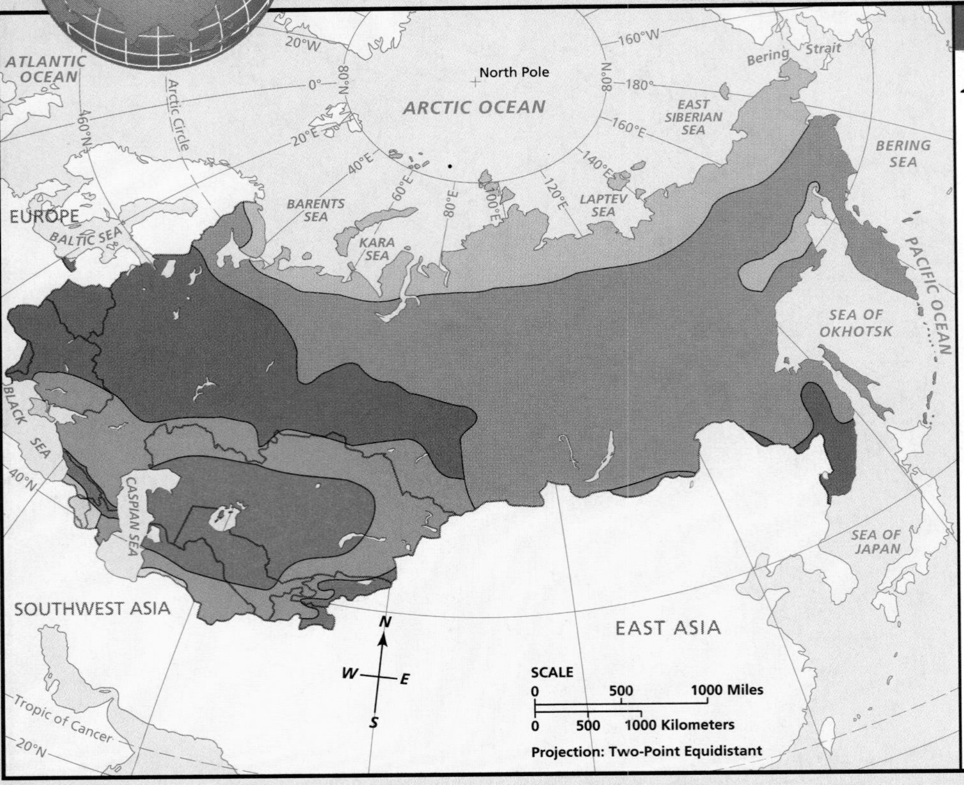

CLIMATE

LEGEND

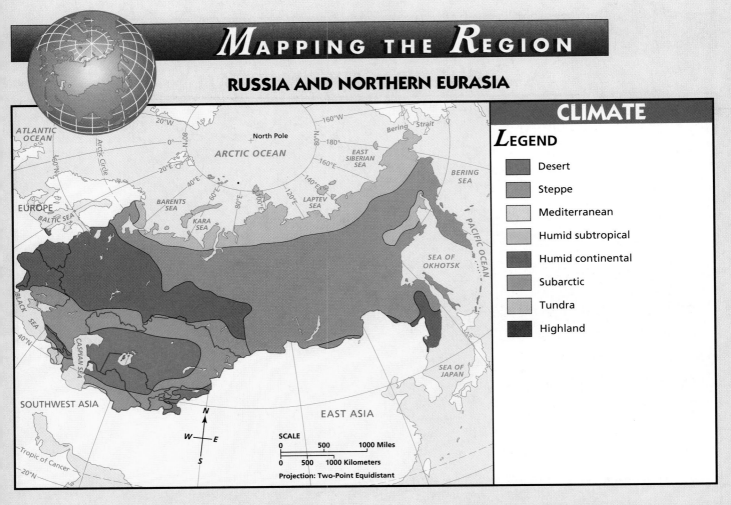

- Desert
- Steppe
- Mediterranean
- Humid subtropical
- Humid continental
- Subarctic
- Tundra
- Highland

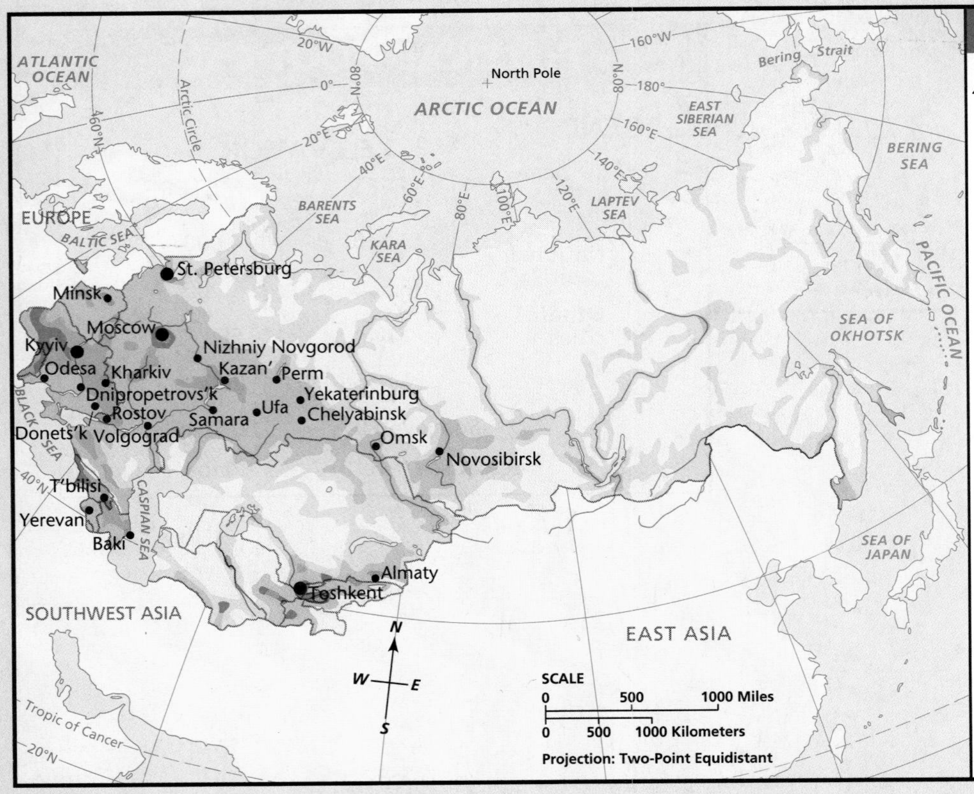

ATLANTIC OCEAN
ARCTIC OCEAN
North Pole
EUROPE
BALTIC SEA
BARENTS SEA
KARA SEA
LAPTEV SEA
EAST SIBERIAN SEA
Bering Strait
BERING SEA
SEA OF OKHOTSK
PACIFIC OCEAN
SEA OF JAPAN
BLACK SEA
CASPIAN SEA
SOUTHWEST ASIA
EAST ASIA
Arctic Circle
Tropic of Cancer

SCALE
0 500 1000 Miles
0 500 1000 Kilometers
Projection: Two-Point Equidistant

POPULATION

LEGEND

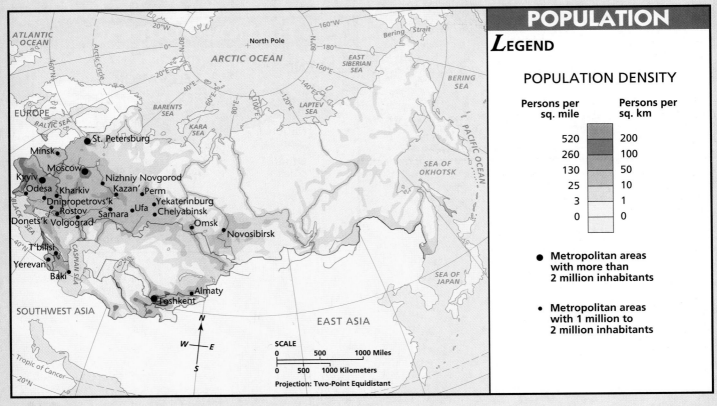

POPULATION DENSITY

Persons per sq. mile	Persons per sq. km
520	200
260	100
130	50
25	10
3	1
0	0

- ● Metropolitan areas with more than 2 million inhabitants
- • Metropolitan areas with 1 million to 2 million inhabitants

St. Petersburg
Minsk
Moscow
Kyyiv
Nizhniy Novgorod
Odesa
Kharkiv
Kazan'
Perm
Dnipropetrovs'k
Yekaterinburg
Rostov
Ufa
Samara
Chelyabinsk
Donets'k
Volgograd
Omsk
Novosibirsk
T'bilisi
Yerevan
Baki
Almaty
Toshkent

SCALE
0 500 1000 Miles
0 500 1000 Kilometers
Projection: Two-Point Equidistant

MAPPING THE REGION

RUSSIA AND NORTHERN EURASIA

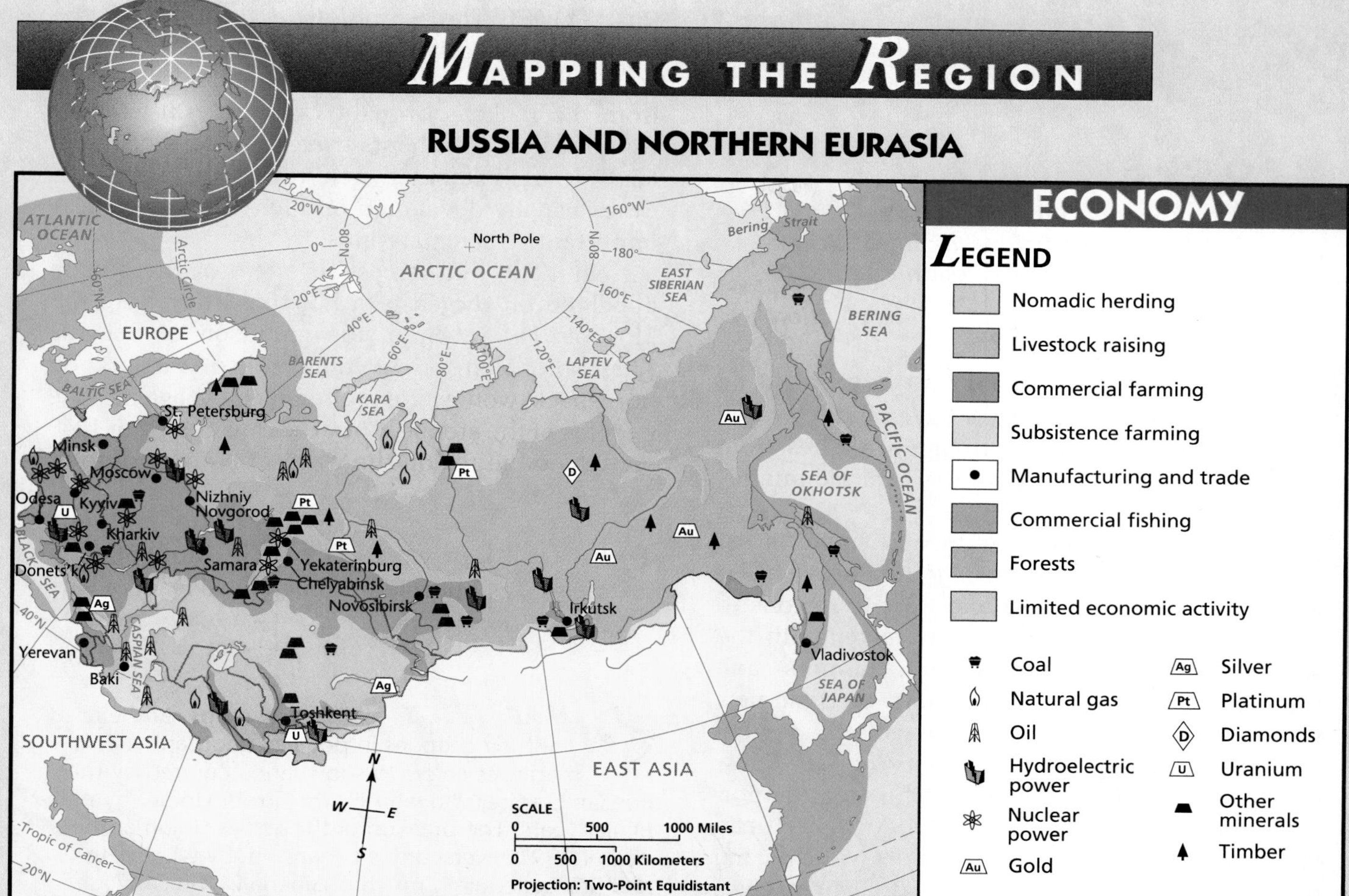

ECONOMY
LEGEND
Nomadic herding
Livestock raising
Commercial farming
Subsistence farming
Manufacturing and trade
Commercial fishing
Forests
Limited economic activity
Coal
Natural gas
Oil
Hydroelectric power
Nuclear power
Gold
Silver
Platinum
Diamonds
Uranium
Other minerals
Timber
ATLANTIC OCEAN
ARCTIC OCEAN
North Pole
EUROPE
BALTIC SEA
BARENTS SEA
KARA SEA
LAPTEV SEA
EAST SIBERIAN SEA
Bering Strait
BERING SEA
PACIFIC OCEAN
SEA OF OKHOTSK
SEA OF JAPAN
BLACK SEA
CASPIAN SEA
SOUTHWEST ASIA
EAST ASIA
Arctic Circle
Tropic of Cancer
St. Petersburg
Minsk
Moscow
Odesa
Kyyiv
Kharkiv
Nizhniy Novgorod
Samara
Yekaterinburg
Chelyabinsk
Novosibirsk
Irkutsk
Vladivostok
Donets'k
Yerevan
Baki
Toshkent
SCALE
0 500 1000 Miles
0 500 1000 Kilometers
Projection: Two-Point Equidistant

CLIMATE

◆ Most of Russia, the largest of the republics of the former Soviet Union, lies in the cold subarctic and tundra climate regions.

◆ The dry steppe and desert climates, with hot summers and cold winters, dominate much of the Central Asian republics.

◆ Ukraine, a large agricultural producer, has a humid-continental climate in the northwest and a warmer and drier steppe climate in the southeast.

? Regions with tundra climates lie no farther south than which latitude?

Compare this map to the physical–political map on page 325.

? Where are highland climates located?

? Which climate type dominates the region surrounding the Aral Sea?

POPULATION

◆ Moscow and St. Petersburg are the two largest of many metropolitan areas in the region.

◆ The largest cities and many of the more densely populated regions of eastern Russia lie along the route of the Trans-Siberian Railroad.

◆ Large areas of the Central Asian republics and Siberia are sparsely populated.

? Which metropolitan areas east of 60°E longitude have populations of more than 1 million?

Compare this map to the physical–political map on page 325.

? Which republics have metropolitan areas with populations of more than 2 million?

Compare this map to the climate map on page 326.

? Which climate region in Russia is the most sparsely populated?

ECONOMY

◆ Many of the manufacturing centers are located near large deposits of natural resources.

◆ The most productive farming lands are in the steppe regions of Ukraine, Russia, and Kazakhstan.

◆ Although the climate and terrain discourage large-scale farming in much of Siberia, the region has a rich supply of other resources.

Compare this map to the physical–political map on page 325.

? What energy resource is found in the Donets and Kuznetsk basins?

? Name the manufacturing and trade centers in the three republics of the Caucasus region.

Compare this map to the climate map on page 326.

? What do you think is necessary to support farming in the desert region that surrounds the Aral Sea?

A DIVERSE HERITAGE

FOCUS

- *What peoples inhabited and ruled the Russian steppe before the fifteenth century? What was the Russian Empire?*
- *What led to the Communist Revolution?*

MOVEMENT **Invaders of the Steppe** The roots of the Russian Empire lie deep in the grassy plains of the steppe. The steppe stretches from what is now Ukraine west to the Ural Mountains. (See the map on page 325.)

For thousands of years, people moved west across the steppe. These migrants drove their herds away from droughts and wars in regions to the east. In time, the immigrants mixed with the settled population and became farmers and traders. Each wave of immigrants brought new ways of life, languages, and religions to the region.

In the ninth century, the city of Kiev (now called Kyyiv) emerged in the region as an important trading center between the Mediterranean and Baltic seas. Kiev's first leaders were Viking traders from Scandinavia who called themselves Rus (ROOS). The word *Russia* comes from the name of these early traders.

During the following centuries, missionaries from the Balkans brought Orthodox Christianity to the region. Merchants traveling into the forests farther north founded new towns there as well. Cities usually developed on high riverbanks near important stream junctions.

In 1223, the fiercest invaders of the steppe appeared on the eastern horizon. They were the Mongols led by Batu, the grandson of Genghis Khan, the Mongol chieftain. These warriors on horseback won victory after victory against the peoples of the steppe. They destroyed Kiev in 1240 and made the region the western outpost of their growing empire.

The Russian Empire The Mongols demanded taxes but ruled through local leaders who

MAP STUDY Historical maps can be used to show changes in political geography. This map uses colors to show land acquired by the Russian Empire, and later by the Soviet Union, by particular years. The breakup of the Soviet Union at the end of 1991 reversed this expansion. By what date had Russia acquired much of Siberia in the East?

History of Russian Expansion

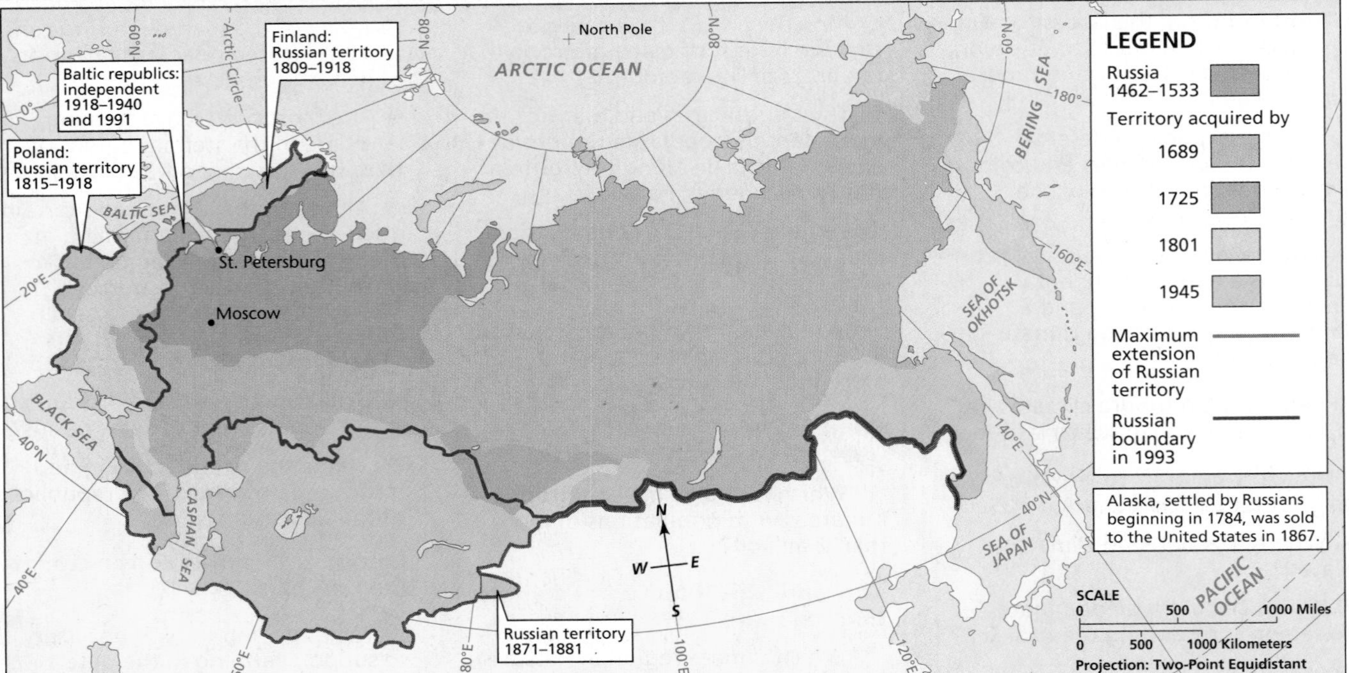

Finland: Russian territory 1809–1918
Baltic republics: independent 1918–1940 and 1991
Poland: Russian territory 1815–1918
North Pole
ARCTIC OCEAN
Arctic Circle
BALTIC SEA
St. Petersburg
Moscow
BLACK SEA
CASPIAN SEA
BERING SEA
SEA OF OKHOTSK
SEA OF JAPAN
PACIFIC OCEAN
Russian territory 1871–1881
LEGEND
Russia 1462–1533
Territory acquired by
1689
1725
1801
1945
Maximum extension of Russian territory
Russian boundary in 1993
Alaska, settled by Russians beginning in 1784, was sold to the United States in 1867.
SCALE
0 500 1000 Miles
0 500 1000 Kilometers
Projection: Two-Point Equidistant
N
S
E
W
60°N
80°N
40°N
20°E
40°E
60°E
80°E
100°E
120°E
140°E
160°E
180°

controlled different areas. As a result, several states emerged, the strongest of which was Muscovy, north of Kiev. Its chief city was Moscow. In the late 1400s, Ivan III, the prince of Moscow, won control over parts of Russia from the Mongols.

Then in 1547 at age 16, Ivan IV, who became known as "Ivan the Terrible," crowned himself **czar** (ZAR) of all Russia. The word *czar*, or *tsar*, comes from the Latin *Caesar* and means "emperor." The Russian Empire under Ivan IV stretched from north of Kiev to the Arctic Ocean and east to the Urals.

Ivan's victories over neighboring rivals allowed Russian fur trappers, hunters, and pioneers to migrate eastward into Siberia. By 1637, these explorers reached the Russian Pacific coast. The Russian czars then built a land-based empire larger than any in the world. In the 1860s, the Russians spread into Central Asia, where much of the population was Muslim. Russian territory even spread eastward to what is now Alaska and California.

Several factors, however, hindered Russia's expansion into Europe. The Russian tradition of political control conflicted with western European ideals of government limited by laws and customs. Religions of the two regions also clashed as Russia was mainly Eastern Orthodox and most of Europe was either Protestant or Roman Catholic. In addition, Russia faced the powerful kingdoms of Sweden and Poland to the west and the Ottoman Empire to the south.

Despite these factors, Russia was able to gain some European territory. Czar Peter the Great, who ruled from 1682 to 1725, acquired lands along the Baltic Sea. Catherine the Great, who ruled from 1762 to 1796, took the northern side of the Black Sea. By this time, many non-Russian peoples found themselves within Russian borders.

The Empire Falls Although industrialization had begun in Russian cities by the late 1800s, Russia remained largely a country of poor farmers. Only a small percentage of the people were royalty, landowners, or wealthy merchants. A slightly larger number were factory workers and craftspeople. Most, about 85 percent, were peasants. The poor economic condition of these people was Russia's greatest problem.

In the 1890s, life for many Russians worsened. Droughts led to food shortages, and a global economic depression slowed trade and reduced exports. In 1905 and 1906, riots broke out in several large cities. Workers destroyed factories, and peasants burned crops and seized many estates.

Many people were forced to move to Siberia; thousands more went there willingly to escape the violence. The government began new social and economic programs, but they were too little and too late. By the start of World War I in 1914, the foundations of Russian society were on shaky ground.

Russia was unprepared to fight in this war that spread across Europe. The loss of life and the high costs of World War I, combined with Russia's social and economic problems, caused the czar to **abdicate**, or resign, in early 1917. A temporary republic was set up but met with little success. In the fall of 1917, the small Bolshevik party overthrew the government in what became known as the Communist Revolution.

SECTION REVIEW

1. What factors hindered the westward expansion of the Russian Empire?
2. What was the Communist Revolution? What were its causes?
3. **Critical Thinking** **Assess** the extent to which physical geography played a role in the region's early history.

THE SOVIET UNION

FOCUS

- *On what principles was the Soviet Union based? What were the country's political and economic realities?*
- *What was Soviet daily life like?*
- *How was the Soviet Union internationally significant?*

The Soviet Union The Bolsheviks, led by V.I. Lenin, wanted to remake Russia following the ideas of Karl Marx, a nineteenth-century German philosopher. Marx believed that the people of the working classes were victims of capitalism. He argued that capitalist business leaders kept their workers poor to maintain their own power and wealth. Lenin believed the solution was for Russia's government to own and control all farms and factories. Many disagreed with Lenin's ideas, however, and a civil war raged until 1921, when the Bolsheviks finally gained control.

The country was renamed the Union of Soviet Socialist Republics (USSR), commonly known as the Soviet Union. Each republic had its own local governing council, or **soviet**. The republics were founded on the idea that, where possible, each of the large ethnic groups should have its own territory. In reality, however, the republics had little power. Communist party leaders in Moscow made all economic and political decisions. The Soviet Union became a one-party, totalitarian state in which there were few freedoms.

After Lenin's death in 1924, Joseph Stalin took power. His brutal dictatorship lasted until 1953. Both Lenin and Stalin tried to promote a single culture. Names of cities and streets were changed to honor communism's heroes, and Russian became the official language. Many other languages and cultures survived, however.

Soviet Economic Geography

The enormous land area of the Soviet Union stretched from the Arctic Sea south to the mountains of Iran and east from the Baltic Sea to the Pacific Ocean. To develop this huge country, Soviet leaders set goals based on detailed five-year economic plans. Government officials decided what crops would be grown, where industries would be built, and how much workers would be paid.

After bitter struggles with independent farmers, particularly in Ukraine, almost all the farmland in the Soviet Union came under government control. The government, however, invested little in agriculture. As a result, food production was poor on the government-run farms. By the late 1950s, the nation had to import large amounts of grain. Private plots, which farm families were allowed to work in their spare time, produced about one-fourth of the country's food.

Dramatic industrial growth, however, did take place under Stalin's plans. The government developed mines and built dams. Huge factories specializing in heavy industries, such as steel and military goods, sprang up in many cities.

Soviet planners followed a policy of **autarky** (aw-tahr-kee), under which a country strives for self-sufficiency in economic production. That is, the Soviet Union would try to produce all the goods that the country needed. Planners decided that only a few huge factories were needed to accomplish this goal. Indeed, the Soviet Union became a world leader in the production of many minerals and industrial products. Without competition, however, efficiency and product quality were often low. The production of consumer goods and service industries fell well behind that of the United States and western Europe.

Life in the Soviet Union

By U.S. standards, life in the Soviet Union was dismal. Working conditions were often poor, and food supplies were limited in quality and quantity. The Soviet people often waited in line for hours to buy everyday products. For many families, limited housing meant crowding into cramped two-room apartments. Yet many believed that such sacrifices were necessary to build the Communist ideal. And for most people, life in the Soviet Union by the mid-1950s was as good as it had ever been.

This Russian woman and 9 of her 11 children sit down to eat in their crowded six-room apartment. As **Russia** lurches toward a free-market economy, consumer prices, especially for food, have risen dramatically. Those pushing for a free-market economy believe that reforms will eventually improve living conditions over those under communism.

The government did achieve some successes in education and health care. Under the czars, only about 25 percent of the population could read and write. By the 1980s, well over 90 percent could. Much of Soviet education stressed science and technology, because training in these fields could help increase Soviet production. Many people also became doctors, and basic health care was widely available.

Personal freedoms, however, were strictly limited in the Soviet Union. Citizens who disagreed with government leaders could be jailed. Under Stalin, millions were sent to dreaded prison camps in the wilds of Siberia and the Arctic. Soviet leaders also set out to eliminate religious worship so that it could not compete with people's devotion to communism. Many Christian, Jewish, and Muslim houses of worship were closed or destroyed.

The Cold War Although they were World War II allies, the United States and the Soviet Union became bitter rivals after that war. These two huge, powerful nations, or **superpowers**, competed in the production of military arms and in the development of space programs. This intense political struggle became known as the Cold War and for more than 40 years created the dangerous potential for nuclear war.

The United States and the Soviet Union also competed for the support of other countries around the world. Both gained allies by supplying military goods and economic aid to these countries. Many of these allies became **client states** of the superpowers. A client state is a country that is politically, economically, or militarily dependent on a more powerful country. Violent struggles among these client states nearly drew the United States and the Soviet Union into direct conflict several times.

2 SECTION REVIEW

1. What economic policies did the Soviet Union follow? What economic problems did the country have?
2. List five characteristics of daily life in the Soviet Union.
3. **Critical Thinking** **Explain** why the rivalry between the Soviet Union and the United States was called the Cold War.

A NEW GEOGRAPHY OF NORTHERN EURASIA

FOCUS

- *Why did the Soviet Union break apart?*
- *What challenges does the region face?*

Collapse of the Soviet Union During the 1980s, the Soviet economy continued to decline. In 1985, Mikhail Gorbachev became what turned out to be the last leader of the Soviet Union. Gorbachev saw the need for major reforms in his country. He promoted a policy of ***glasnost*** (openness), which allowed more open discussion of the country's problems. He also initiated major economic reforms called ***perestroika*** (restructuring).

Under Gorbachev, the Communist party began to lose much of its authority. The Soviet people grew tired of sacrifice without gain, and as their demand for consumer goods rose, their frustrations increased. With the central Soviet government losing its authority, its client states in Eastern Europe collapsed in 1989.

Then in 1991, the Soviet Union itself broke apart. Officially, the Soviet Union was replaced by the Commonwealth of Independent States (CIS). This organization does not have a strong central government. Instead, it provides a way for the various regions of the former Soviet Union to address shared problems. The first democratically elected president of Russia, Boris Yeltsin, has called for more democracy and a market economy. Politics in the former Soviet Union, however, has been unstable. Tensions between supporters and opponents of reform and between ethnic groups have made the political future of the former Soviet republics uncertain.

Human Geography Life has changed rapidly for the peoples of the former USSR. News from around the world now flows more freely. Religious and democratic rights are expanding.

Within the 15 independent countries of the former Soviet Union, however, ethnic disputes continue. The Soviet Union contained more than 100 ethnic groups, and the region's division into separate countries still leaves many ethnic minorities. Civil wars are now raging in some areas.

Economic Geography Despite these dividing forces, economic geography may create unity. Soviet planners made the various regions dependent on each other. For example, the Baltic republics relied on the oil and minerals of Russia. Cotton grown in Central Asia was made into cloth in Belarus and Ukraine. And the grains from Central Asia were made into bread in Moscow and St. Petersburg (formerly Leningrad). Breaking such well-established links may make little sense.

The economic geography of the region is changing, however. Although little money is available for investment, individually owned farms and small businesses are appearing. Some factories are now corporations in which the workers own some stock. Some former Soviet republics, especially Russia, have made important progress in privatizing formerly government-owned businesses.

The country's economic upheaval has caused some problems. Unemployment and crime have increased, troubles that some blame on capitalism. For much of the population, the Soviet system represented order, stability, and security. Many remain unsure about the future.

The new-found freedom of local communities and regions to make their own decisions has created some conflict as well. Many resource-rich regions believe that the income from selling their products should remain in their own territories. Resource-poor regions, on the other hand, fear that their economies will suffer greatly under such a system. In all regions, finding people who can become effective local leaders is a continuing problem after so many years of the central government in Moscow making all decisions.

In the past, officials in Moscow also filtered the republics' contact with foreign countries. Today, however, communities all over the former Soviet Union are beginning to develop international ties of their own. For example, the city of Novosibirsk (noh-voh-suh-BIRSK) has built an international airport, hoping to profit as the midpoint on the many flights between western Europe and East Asia.

These **Moscow** demonstrators reflect the political confusion that has followed the collapse of the Soviet Union. Images of Jesus Christ, Lenin, and Czar Nicholas II all compete for attention in this 1993 anti-government demonstration.

INTERACTION **Environmental Issues** A disturbing legacy of the Soviet Union is its record of environmental pollution and destruction. Soviet industrial leaders often felt pressured to produce immediate results at the lowest cost. As a consequence, factories and industries seldom bothered with controlling pollution or preserving the environment.

Throughout the region, there is evidence of deforestation, river and lake pollution, and the improper disposal of toxic wastes. For example, the Soviet Union dumped several nuclear reactors and large amounts of radioactive waste in the Arctic Ocean. Many sewers still dump directly into rivers and lakes as well.

Even though geographers and environmentalists are drawing worldwide attention to these problems in the former Soviet Union, solutions still lie far in the future. The various republics have little money to invest in the expensive and overwhelming job of cleaning up their environment.

3 SECTION REVIEW

1. What role did the reforms of Mikhail Gorbachev play in the collapse of the Soviet Union?
2. What economic and environmental challenges face the region?
3. **Critical Thinking** **Predict** which is more likely to characterize the region over the next decade: unity or disunity. Support your answer.

CHAPTER 28 CHECK

Old country church in **Ukraine**

Reviewing the Main Ideas

1. Northern Eurasia has been settled by many groups of people over time. The Mongol Empire covered the area for most of the thirteenth, fourteenth, and fifteenth centuries. The czars of the Russian Empire then ruled until 1917, when the Communist Revolution took place.
2. Following the Communist Revolution, the country was renamed the Union of Soviet Socialist Republics (USSR), commonly known as the Soviet Union. The government had total economic and political control. After World War II, the United States and the Soviet Union became bitter rivals.
3. The Soviet Union collapsed in 1991. Freedom, democracy, and capitalism are spreading. As it adjusts to change, however, the region faces numerous challenges.

Building a Vocabulary

1. What does *czar* mean? Does Russia still have a czar?
2. Define *abdicate*. Why did the Russian ruler abdicate in 1917?
3. What were the governing councils of the Soviet Union's republics called?
4. Why might a country want to follow a policy of autarky?
5. What is a client state?

Recalling and Reviewing

1. Where did the roots of the Russian Empire lie?
2. What is the Commonwealth of Independent States?
3. Why did the Soviet Union have little concern for the environment? What environmental problems face the region today as a result of policies followed by the Soviet Union?

Thinking Critically

1. Agree or disagree with the following statement: "Modern-day ethnic conflicts in the former Soviet Union have their roots in the region's early history."
2. Despite the many hardships of daily life in the Soviet Union, the country survived for nearly 70 years. In your opinion, what factors were responsible for keeping the Soviet Union together?

Learning About Your Local Geography

Cooperative Project

Largely because of their physical geography, the plains of the Russian steppe have been subject to many invasions over the centuries. Organize your group into two teams. One team should design a "plan of invasion" of your local area. The other team should design a "plan of defense." Both teams should consider the area's physical geography. Teams should share their completed plans.

Using the Five Themes of Geography

Over time, northern Eurasia has been divided into several different **regions**. What roles did **location** and **movement** play in the changes to regional boundaries before 1917? Sketch one or more maps of the area to illustrate your answer to the question.

CHAPTER 29

Russia

“Traveling along country roads in central Russia, you begin to understand why the Russian countryside has such a soothing effect. It is because of its churches. They rise over ridge and hillside, descending towards wide rivers like red and white princesses, towering above the thatch and wooden huts of everyday life with their slender, carved, and fretted belfries. From far away they greet each other; from distant, unseen villages they rise towards the same sky.”

Aleksandr Solzhenitsyn

GEOGRAPHY DICTIONARY

habitation fog
taiga
serf
light industry
heavy industry
smelter

Russia, officially “the Russian Federation,” is a giant among the nations of the world. It stretches 6,000 miles (9,654 km) from the center of Europe and the shores of the Baltic Sea to the eastern tip of Asia and the coast of the Bering Sea. Russia’s northern border is on the icy shores of the Arctic seas, while its southernmost border, 2,000 miles (3,218 km) away, climbs the rocky peaks of the Caucasus Mountains. The country’s 150 million people live in a territory more than 6.5 million square miles (16.8 million sq. km) in area, one-ninth of the entire land surface of the planet. Russia is nearly twice the size of Canada, the world’s second largest country. Because Russia covers so much of Europe and Asia, for the past several centuries it has played a major role in the affairs of both continents.

Snowy taiga forest of **Russia**

The painting on this late-eighteenth-century lacquer box, titled “Princess of the North Wind,” is based on a Russian fairy tale.

RUSSIA: PHYSICAL–POLITICAL

PHYSICAL GEOGRAPHY

FOCUS

- *What landform regions are found in Russia?*
- *What types of climates and vegetation does Russia have?*

REGION **Landforms** Plains are the dominant landform in Russia. Extending west of the Urals is the Russian Plain, a part of the Northern European Plain that connects with plains in Poland and northern Germany. (See the map above.) This region is where most of the Russian people live. The Ural Mountains, which divide Europe from Asia, are high rolling hills. Because the Urals run north to south, they provide no protection from the cold winds that sweep out of the Arctic each winter. Nor did the Urals ever stop the invaders who came from inner Asia over the centuries.

The lands east of the Urals are called Siberia. Directly east of the Urals is the Western Siberian Lowland, a large, flat area with widespread marshes. The huge Ob River system drains the area. In the spring, the Ob and other rivers of Siberia that drain toward the Arctic Ocean thaw first in the south. Downstream to the north, however, the rivers remain frozen much longer. Ice jams here block the passage of the melting snow and ice upstream. As a result, annual floods can be expected.

Situated between the Yenisei (yehn-uh-SAY) and Lena rivers is the Central Siberian Plateau.

FACTS IN BRIEF The Russian Federation

COUNTRY POPULATION (1994)	LIFE EXPECTANCY (1994)	LITERACY RATE	PER CAPITA GDP (1993)
RUSSIA 149,608,953	64, male 74, female	99% (1993)	$5,190
UNITED STATES 260,713,585	73, male 79, female	97% (1991)	$24,700

Sources: *The World Factbook 1994,* Central Intelligence Agency; *The World Almanac and Book of Facts 1995*

Russia's population is only about three-fifths that of the United States, although Russia's land area is nearly twice as large. Russia is the largest of the republics of the former Soviet Union.

This land of upland plains and valleys holds many undeveloped mineral deposits of lead, iron ore, and coal, which should provide income for centuries.

While flat lowlands make up Russia's heartland, soaring mountains stand guard along much of its eastern rim. Mountains stretch all the way along the border with Mongolia and then to the Pacific coast. These mountain peaks are comparable in size to those of the Rocky Mountains in the United States. The mountains north of Mongolia also contain Lake Baykal (bie-KAWL), the world's deepest lake. Known as the "Jewel of Siberia," Lake Baykal is home to many endangered species, including the world's only freshwater seal. (See "Planet Watch" on pages 338–339.)

Along Russia's Pacific coast is the peninsula of Kamchatka. This is a land of mountains much like the volcanic Cascade Mountains of the U.S. Pacific Northwest. The mineral riches that might exist in the east could be important to future economic development in Russia and East Asia.

Russia's many rivers are important resources, forming a network of valuable waterways for transportation and the production of hydroelectric power. The Volga, Europe's longest river, covers 2,293 miles (3,689 km) on its gently sloping journey from the plains north of Moscow to its mouth on the Caspian Sea. Canals between rivers connect the Volga with the Baltic Sea in the northwest, while the Volga-Don Canal allows barges on the southern Volga to reach the Black Sea. In Siberia, the dam at Bratsk on the Angara River downstream from Lake Baykal is one of the world's largest.

REGION Climate

Russians sometimes joke that winter lasts for 12 months and then summer begins. Nearly all of Russia is located at high latitudes that correspond to those of Canada and Alaska. Most of the country lies in the humid-continental and subarctic climate regions. (See the map on page 326.) Winters are cold and long, and winter days are very short. During the five coldest months of the year, it is possible to ice-skate on frozen rivers and canals from St. Petersburg (formerly Leningrad) on the Baltic Sea south to Astrakhan on the Caspian Sea.

Russian winters are surprisingly dry. The snow remains on the ground for many months, continuously blowing into new drifts. Snow and rain in Russia come mostly from moisture that has evaporated from the Atlantic Ocean far to the west. Only the coastal areas of eastern Russia receive rain-bearing winds from the Pacific Ocean.

Winters are particularly severe in Siberia, where temperatures often drop below -40°F (-40°C). During winter, **habitation fogs** hang over cities. A habitation fog is a fog that is caused by fumes and smoke from cities. The Siberian summer often lasts for less than three months. Surprisingly, temperatures can rise as high as 100°F (38°C) during this short season. Russians take advantage of the warmer weather to tend gardens, attend sporting events, and stroll through parks.

The long winter isolates Russia from the rest of the world. Travel between cities is difficult, and

The Shilka River meanders between mountain ranges in southeastern **Siberia** near the Mongolian border.

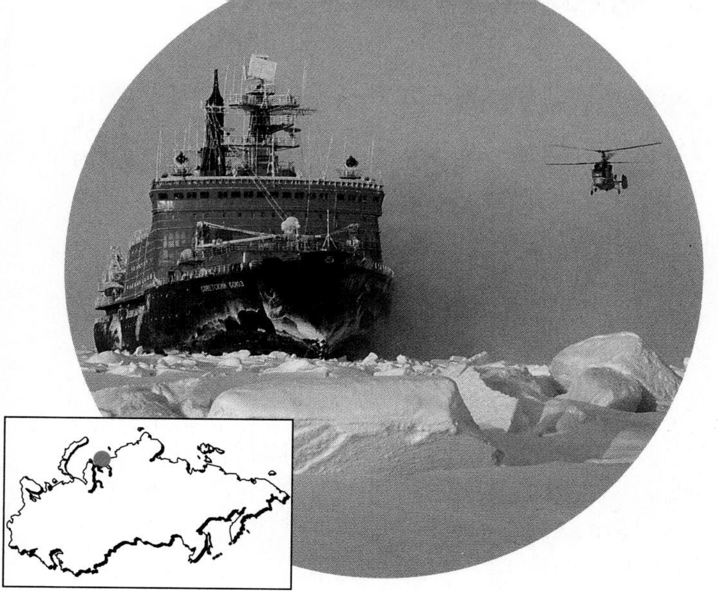

This nuclear-powered icebreaker is clearing a path through the frozen **Kara Sea**. Icebreakers allow ships to travel along the frozen northern Russian coast throughout the winter.

even friends inside the same city rarely visit each other. The country has few ports that are free of ice year-round. Except in the summer, the Arctic Ocean freezes all the way to Russia's northern shores. Barge traffic here requires the use of icebreakers. Similarly, the Pacific shore and most of the Baltic coast are iced over for more than four months each year. Russia's only large, ice-free port is Murmansk, located in the far northwest where the warm waters of the North Atlantic Drift reach around northern Norway.

The cold climates of Siberia and the Arctic create permanently frozen ground beneath the surface of the earth. This permafrost causes many problems. Tree growth is stunted because the roots cannot penetrate the frozen soil. When the surface layer melts in summer, buildings tilt, highways buckle, and railroad tracks slip sideways.

Vegetation Vegetation in Russia varies from north to south as the climates vary. Along Russia's northern coast, temperatures rise above freezing for only a few weeks each year. Tundra vegetation of mosses, wildflowers, and stunted trees grows here.

To the south is the **taiga**, a huge forest of tall, evergreen needle-leaf trees such as spruce, fir, larch, and pine. The dark taiga provides the setting in much of Russian literature. Historically, the taiga has served as protection against foreign invaders.

The taiga provides a large supply of softwood for building products and paper pulp. Steady logging of the taiga west of the Ural Mountains has left many cleared areas, but in Siberia, most of the taiga remains intact. With good management, it will provide forest resources for a long time. Although farming is possible in the taiga, yields are low because of the short growing season.

Farther south in European Russia is an area of forests with deciduous, hardwood trees. This type of forest also is found along the Sea of Japan in eastern Russia. Still farther south are the drier grasslands known as the steppe. The steppe stretches west into Ukraine, east into Kazakhstan, and south to the Caucasus Mountains. Many of the forests here have been cut for their lumber or cleared for farms. The grasslands now are plowed under for crops or used for grazing.

SECTION REVIEW

1. Where are Russia's plains located? its mountains? What are some of the country's major rivers?
2. What is the dominant characteristic of Russia's climates? How do the climates influence Russia's vegetation?
3. **Critical Thinking** **Explain** how landforms, climates, and vegetation have influenced Russia's relationships with people outside the country.

RESOURCES AND AGRICULTURE

FOCUS

- *What resources are found in Russia?*
- *What forces have shaped Russian agriculture?*

Resources Russia has more forest, energy, and mineral resources than any other country in the world. Unfortunately, the former Soviet government wasted many of these resources. For example, the taiga west of the Urals is badly depleted. Because the trees of the taiga grow slowly, it will be many years before these forests grow back. As a result, forest products must now be transported

Text continues on page 340.

Lake Baykal

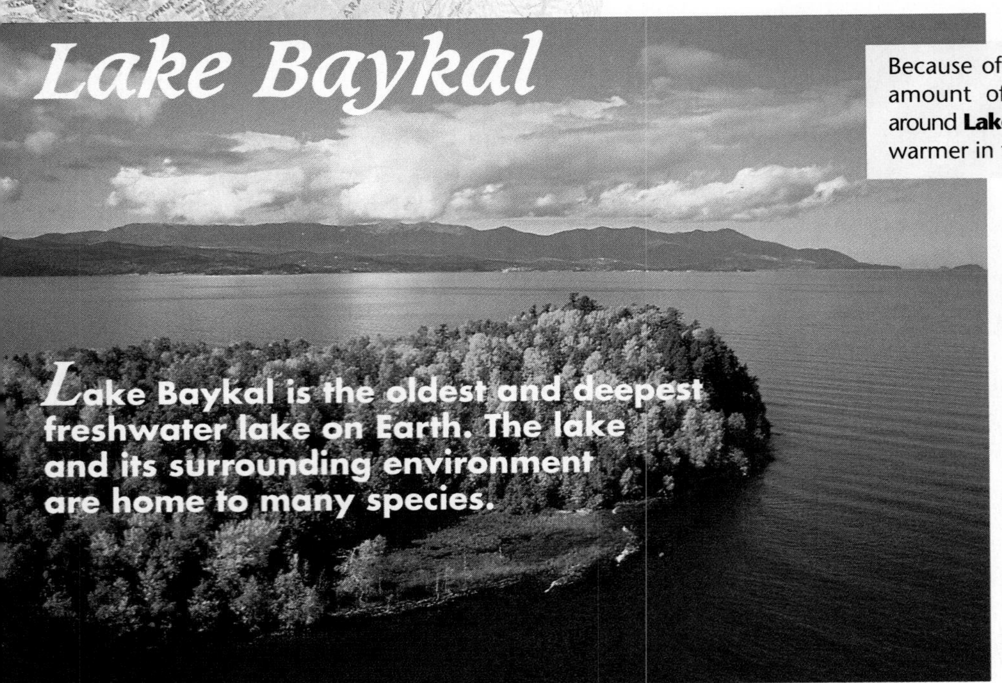

Lake Baykal is the oldest and deepest freshwater lake on Earth. The lake and its surrounding environment are home to many species.

Because of the moderating effect of the large amount of water in the lake, temperatures around **Lake Baykal** are cooler in the summer and warmer in the winter than areas farther inland.

North of the Gobi lies crescent-shaped Lake Baykal. Grayling, pike, perch, and many other species of fish swim through its cold waters as eagles, falcons, and osprey fly overhead. Moose, wolves, and brown bears roam the surrounding forest. The lake's islands serve as a resting place for silver-furred nerpas, the world's only known freshwater seals.

Set in a basin seven times as deep as the Grand Canyon, Lake Baykal holds as much water as all of North America's Great Lakes. Formed 25 million years ago by the movement of plates along the earth's crust, Lake Baykal sits on active faults along which small earthquakes occur daily.

Baykal is like no other lake. In parts of the lake floor, warm-water springs sustain a variety of plant and animal life. Oxygen traveling down to the lake's deepest parts allows flatworms and freshwater shrimp, snails, and sponges to survive on the lake bottom. One species of freshwater shrimp helps keep the water pure by filtering algae and bacteria.

The Human Presence

The Buryat, a Mongol people, have lived along Baykal's eastern and northern shores for more than 1,700 years. Conquered by Russia in the late 1600s, the Buryat were part of that nation until the 1900s. In 1923, their lands became a separate Soviet republic. With Soviet rule came new attitudes toward Baykal. Soviet planners believed that Baykal's water would be ideal for producing cellulose, a material used in jet airplane tires. In 1957, the government announced plans for construction of the Baykalsk Cellulose-Paper Plant at one end of Lake Baykal.

Despite the risks of criticizing the government, scientists, environmentalists, and local residents spoke out. Industrialization, they claimed, would spoil the clean waters of Baykal. Even so, the government forged ahead, completing Baykalsk in 1966. A Buryat woman contrasted Buryat with Soviet attitudes. "In ancient times all life was considered sacred. Now those times are gone, nobody thinks about it anymore."

Critics' fears have been realized. Chemicals dumped into the lake by the Baykalsk plant have damaged Baykal's southern part. Pollutants from its smokestacks have killed trees in hundreds of square miles of forest nearby.

The Baykalsk Cellulose-Paper Plant spews pollution into the environment around **Lake Baykal.** In Russia, as elsewhere around the world, people must balance environmental protection with economic development.

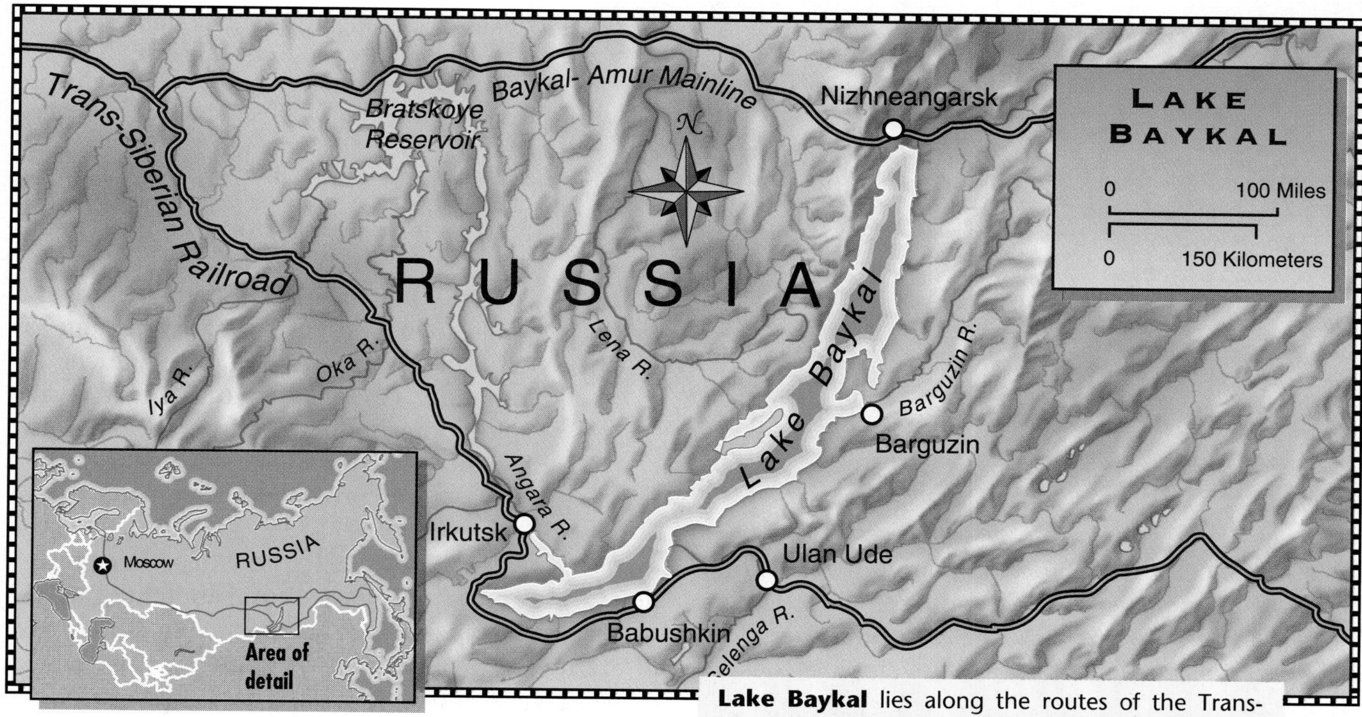

Trans-Siberian Railroad
Baykal- Amur Mainline
Bratskoye Reservoir
Nizhneangarsk
N
LAKE BAYKAL
0
100 Miles
0
150 Kilometers
RUSSIA
Lake Baykal
Lena R.
Barguzin R.
Oka R.
Iya R.
Barguzin
Angara R.
Irkutsk
Ulan Ude
Babushkin
Selenga R.
Moscow
RUSSIA
Area of detail

Lake Baykal lies along the routes of the Trans-Siberian Railroad and the Baykal-Amur Mainline. These two railroads link western Russia with the Russian Far East.

After building Baykalsk, the Soviet government was eager to develop industry throughout Siberia. New coal-burning factories gave off pollutants that produced acid rain in the region. Factories dumped chemicals and waste into the Selenga River, which flows into Lake Baykal.

With industrialization came the need for more transportation and fuel sources. Construction of the Baykal-Amur Mainline, a railroad linking Siberia with the Pacific Ocean, caused soil erosion along Lake Baykal's northern shore. Oil tankers began to appear on the lake despite the threat of oil spills.

These events have endangered the plants and animals that live in and around Baykal, many of which are found nowhere else on Earth. The number of shrimp is decreasing, as is the population of *omul,* a whitefish important to Baykal's fishing industry. To the dismay of scientists around the world, the lake's more than 60,000 nerpa also face an uncertain future due to the buildup of toxic chemicals in the food chain.

Protecting Lake Baykal

As industry's effects on the lake became apparent and the Soviet Union changed policies under Mikhail Gorbachev, environmentalists began to win small but important victories. They stopped a proposed Soviet-Italian project to mine limestone near Baykal. They also defeated a government plan to transport Baykalsk's toxic waste through pipes and dump it into one of Siberia's large rivers. In 1987, their actions finally encouraged the government to issue a decree protecting Lake Baykal.

Dramatic events have shaken the world since 1987, however. The Soviet Union has broken up, and Russia's economy is struggling. It is feared that Baykal's fate is no longer a priority among those who have the power to save Russia's "Sacred Sea."

Used to overcoming obstacles, environmentalists continue to work to protect Baykal, and concern about the lake's future is spreading among the Russian population. An American writer visiting Lake Baykal saw Russian citizens demonstrating against the destruction of the lake, a sight that would have been rare just a decade or so ago. "We're here for our children," said one Russian woman. "We must save Baykal for our children."

YOUR TURN

Disputes often arise when environmental protection conflicts with economic development. Think of examples of such conflict that have taken place in the United States. Work with a partner to **write** a debate that might take place between an environmentalist and an executive of a company that is about to construct an industrial plant in an area of environmental concern. In your debate, **explain** how the conflict will be resolved.

Workers lay a pipeline that will transport natural gas westward from **Siberia**. Although it has many economic problems, Russia is rich in natural resources.

long distances from Siberia. Many of Russia's other remaining natural resources also are located in remote areas. Large, new investments will be necessary to continue developing these resources.

Russia has been a major oil producer since the early twentieth century. At first, most of the oil came from the shore of the Caspian Sea that today belongs to Azerbaijan. After World War II, however, oil fields were developed between the Volga River and the Ural Mountains. As these reserves were depleted, huge gas and oil discoveries were made in the lower Ob River region in northwest Siberia. About 70 percent of Russia's oil and natural gas now comes from this region. Pipelines carry the oil and gas to Moscow and St. Petersburg. Natural-gas pipelines run even farther west to Germany and France. Smaller oil and gas fields are located on Sakhalin Island and on the Kamchatka Peninsula. Soon, new fields will be developed in Siberia, in the Arctic, and north of the Caucasus.

Russia also has large coal reserves. Low-grade coal lies beneath the region south of Moscow, but the most developed coal fields are in the Kuznetsk (kuz-NEHTSK) Basin in Siberia. Even more coal lies undeveloped in the Tunguska and Lena river basins. Reserves here could last for centuries. Unfortunately, these deposits are far from Russian cities and markets. They also are distant from potential export customers around the Pacific Rim. Even so, companies from Japan and South Korea have plans to help develop mines in these regions.

Russia also has the potential to become a major supplier of more than a dozen metals. Among these metals are iron ore, copper, manganese, gold, platinum, and nickel. Russia produces specialized, high-technology metals such as tungsten and vanadium as well. Russia also is a major producer of diamonds. Large areas of Siberia await exploration for even more minerals. Efficient development of natural resources could provide future prosperity.

INTERACTION **Agriculture** Russia's climate limits the country's agricultural potential. The growing season is short, and the soils under much of the taiga are poor. Russia's finest soil, the black earth of the steppe, often does not receive enough rainfall. In addition, a long history of political problems related to agriculture slowed Russia's agricultural success. Before the middle of the nineteenth century, much of the farming in Russia was done by **serfs**, peasants who were bound to the land. Even after the serfs were freed in 1861 and agricultural improvements were made, food production continued to fall behind needs. In 1891, a crop failure led to one of the worst famines in Russia's history.

The Communist government, which took power in 1917, wanted to modernize the country's agriculture. During the late 1920s and early 1930s, the government seized farmers' lands to form collective and state farms. On collective farms, workers were to share any profits the farm earned. In reality, however, the Soviet government controlled the farms and took most of what was produced. Farmers were paid such low prices for agricultural goods that they could barely survive. State farms were larger and were run like factories. The farmers worked the land for a salary. Because farmers were not given any reward for increased production, agriculture under the Soviets was unproductive and inefficient.

Returning land ownership to the farmers is a primary goal of the Russian government today. Officials now believe that the farmers will produce the highest yields when they own the land themselves.

Russia's most productive agricultural region is found south of a line drawn from St. Petersburg

Village children watch this Russian farmer work his small plot of land south of **St. Petersburg**. Where were the most productive farmlands of the former Soviet Union?

to Kazan' to Novosibirsk. It is an area of mixed farming, focusing mainly on growing grains and raising livestock. Near cities, small gardens provide fresh fruits and vegetables for summer markets. Cattle ranching is important in the dry lands north of the Caucasus.

Much of the food that Russia produces still comes from the small plots that were established during the Soviet era. The farmers use these one- or two-acre plots as their private land and sell the products at competitive prices on the open market. This intensive cultivation produces a variety of summer fruits and vegetables as well as high-grade pork, beef, and eggs. The Russian government hopes to increase food production with the privatization of the larger farms. When the Soviet Union broke up, however, Russia lost the productive farmlands of Ukraine and northern Kazakhstan. For now, Russia must continue to rely on imports to feed its people.

SECTION REVIEW

1. What are Russia's major resources? What resources are likely to be important in the future?
2. Name three agricultural products raised in Russia. Where is each one produced?
3. **Critical Thinking** **Construct** a time line showing how agriculture has changed in Russia over the past two centuries.

ECONOMIC AND INDUSTRIAL REGIONS

FOCUS

- *What economic changes is Russia undergoing?*
- *How are each of Russia's economic regions similar and different?*

The Soviet Union was an industrialized country able to produce all of the products a modern country needs. Generally, however, Soviet industrialization concentrated on military goods. As a result, consumer goods usually were expensive and in short supply. Because Soviet industrial firms were monopolies and thus had little competition, quality was often poor.

Today, Russian industry is rebuilding and restructuring so that the country can compete in the world market. Former defense-related industries are beginning to produce consumer goods, and new businesses are developing. The companies that have focused on one product now are diversifying. While the country's move to a free-market economy is a very long and difficult process, the Russian people are eager for a better selection of consumer goods.

The Moscow Region Moscow, with its massive Kremlin and onion-shaped church domes, has symbolized Russia for centuries. The city became the home of the Russian Orthodox Church in the fourteenth century and Russia's capital when the country was unified in the fifteenth century. Even during the 200 years that St. Petersburg was the country's capital, most Russians still looked to Moscow as their country's heart and soul.

Today, with more than 9 million people, Moscow is the country's largest city and the heart of Russia's most important industrial region. As the national capital and a cultural center, Moscow's economic advantages are many. Moscow also is the center of communications and transportation for the nation. Roads, railroads, and air routes radiate from the capital to all points in Russia.

The Moscow industrial region stretches 250 miles (402 km) to the east and includes the city of

The colorful, onion-shaped domes of St. Basil's Church and the walled Kremlin draw tourists to Red Square. The Kremlin is the political heart of **Moscow**, the national capital. Where was the Russian capital before it was moved to Moscow?

Nizhniy Novgorod (Gorky) as well as smaller cities surrounding Moscow. The region has more than 50 million people, about one-third of the country's total population. Many people here work as government officials.

To try to ease industrial pollution problems in Moscow, planners decided that **light industry**, rather than **heavy industry**, would dominate. Light industry focuses on producing lightweight goods, such as clothing. Heavy industry usually involves manufacturing based on metals such as steel. Advanced-technology and electronics industries also have been developed here, and mineral resources are found nearby.

The St. Petersburg Region While Moscow symbolizes Russia's values and traditions, St. Petersburg represents the country's desire to modernize and adapt ideas and practices from the West. Located on the Gulf of Finland about 420 miles (676 km) northwest of Moscow, St. Petersburg has been called the "Venice of the North." The city is built on the delta of the Neva River and has many canals instead of streets. With more than 5 million people, St. Petersburg is the second largest city in Russia. (See "Cities of the World" on pages 344–345.)

St. Petersburg is situated in a region of low, poorly drained land. As a result, the city is compact and does not sprawl outward like Moscow. The surrounding area has few natural resources, so fuel and raw materials must be brought in from other regions.

The city's harbor, canals, and rail connections provide good facilities for trade. The harbor and Gulf of Finland are kept open during winter by icebreakers. The city's chief industrial products are textiles, ships, chemicals, and machinery. St. Petersburg benefits from its location close to the innovative industrial centers of Europe. The city hopes the number of high-technology industries will grow.

The Volga Region Russia's third major industrial region stretches along the middle part of the Volga River. Easy transportation on the Volga, energy from hydroelectric plants, and fossil fuel resources benefit the region. During World War II, many factories were moved to the Volga region to keep them safe from German invaders. Through much of its course, beginning near the city of Kazan', flowing past Samara (Kuibyshev) and Saratov (suh-RAHT-tuhf) and on to Volgograd (once known as Stalingrad), the Volga River is more like a chain of lakes.

The Volga region is famous for its heavy-machine industries and giant factories that produce such goods as motor vehicles, chemicals, and food products. Famous Russian caviar, a delicacy, comes from the fishery based at the ancient city of Astrakhan on the Caspian Sea.

The Urals Region Mining in the southern Urals began more than two centuries ago, and the rich mineral deposits laid the base for the region's industrial development. Nearly every important mineral except coal and oil has been discovered here. Copper and iron **smelters**, factories that process the ores, are still important in the region.

The Soviet government built many industries in the Urals region during its drive to industrialize the country in the 1930s. As with the Volga region during World War II, whole factories were transported to the Urals ahead of the advancing Germans. The great cities of the Urals, such as Yekaterinburg (formerly Sverdlovsk), Chelyabinsk (chehl-YAH-buhnsk), and Magnitogorsk (mag-NEET-uh-gawrsk), started as commercial centers for mining districts. Today, these cities manufacture machinery and metal goods. As coal from Siberia and Kazakhstan was brought in to process iron ore into steel, heavy industry grew.

Siberia Siberia stretches from the Urals to the Pacific Ocean. The popular image of Siberia is that of a vast, frozen wasteland. In many ways, this image is accurate. Large areas of Siberia are almost uninhabited and have little industry. This enormous region has been a rich frontier luring Russian adventurers for more than 400 years. It continues to do so today.

Settlement in Siberia mostly follows a line along the route of the Trans-Siberian Railroad. Construction of this railway started in 1891, and by 1916, it connected Moscow to Vladivostok on the Sea of Japan. More than 5,700 miles (9,171 km) long, the Trans-Siberian Railroad is the longest single rail line in the world. For many towns in the region, the train provides the only transportation link to the outside world.

Many natural resources have been developed in Siberia during the twentieth century. So that these resources could be more easily transported, a more direct railway, called the Baykal-Amur Mainline (BAM), was completed across eastern Siberia in 1989. The building of BAM was an impressive construction effort. Built by youth organizations, BAM crosses five mountain ranges and 17 major rivers. It is connected by four tunnels and more than 3,000 bridges. Most of the construction of BAM took place in winter so that trucks could be driven over frozen soils.

A plant worker at a fishery in **Astrakhan** on the Caspian Sea prepares caviar for market. Russia and Iran produce nearly all of the world's true caviar from fish in the Black and Caspian seas.

Less than 10 percent of Russia's industry is in Siberia. Government plans for settling and developing Siberia have not gone smoothly, partly because of the harsh climate and difficult terrain. Many people would rather live in the cities of European Russia even though wages are higher in Siberia.

Lumbering and mining are the region's most important industries. Most of the lumbering and mining towns are connected by rail or river routes to the Trans-Siberian Railroad or to BAM. Siberia's mineral ores and forest products are needed by consumers and industries in western Russia.

Rich coal deposits have made the Kuznetsk Basin, or the Kuzbas, Siberia's most important industrial region. *Kuznetsk* means "blacksmith" in Russian. The Kuzbas is located in southwestern Siberia between the Ob and Yenisei rivers. Coal from the Kuzbas is sent to the Urals region and even farther west. Novosibirsk is the manufacturing and transportation center of Siberia. Its population of almost 1.5 million makes it the largest city in Siberia. As the resources of Siberia are further developed, Novosibirsk and nearby cities will gain in importance.

The Russian Far East Russia has more coastline on the Pacific Ocean than any other country. This area bordering the Sea of Okhotsk and the Sea of Japan is known as the Russian Far East. The climate of the Russian Far East is less severe than that of the rest of Siberia. The summer weather is mild enough for some successful farming, particularly in the Amur River valley. Yet, the region cannot produce enough food for itself. Much of the land remains heavily forested, and the region's minerals are only beginning to be developed.

Khabarovsk (kuh-BAHR-uhfsk), on the Amur River, is the principal city of the Russian Far East. Its central location enables it to process forest and mineral resources from throughout the region. Vladivostok, which means "lord of the east" in Russian, is the chief seaport of the Russian Far East. It is a major naval base and the home port for a large fishing fleet. The city of Komsomol'sk-na-Amure produces iron and steel. Russia hopes that the Russian Far East will become part of the Pacific Rim trading network and find new markets for its natural resources and products.

Sakhalin Island, with its petroleum and mineral resources, lies off the eastern coast of Siberia

Text continues on page 346.

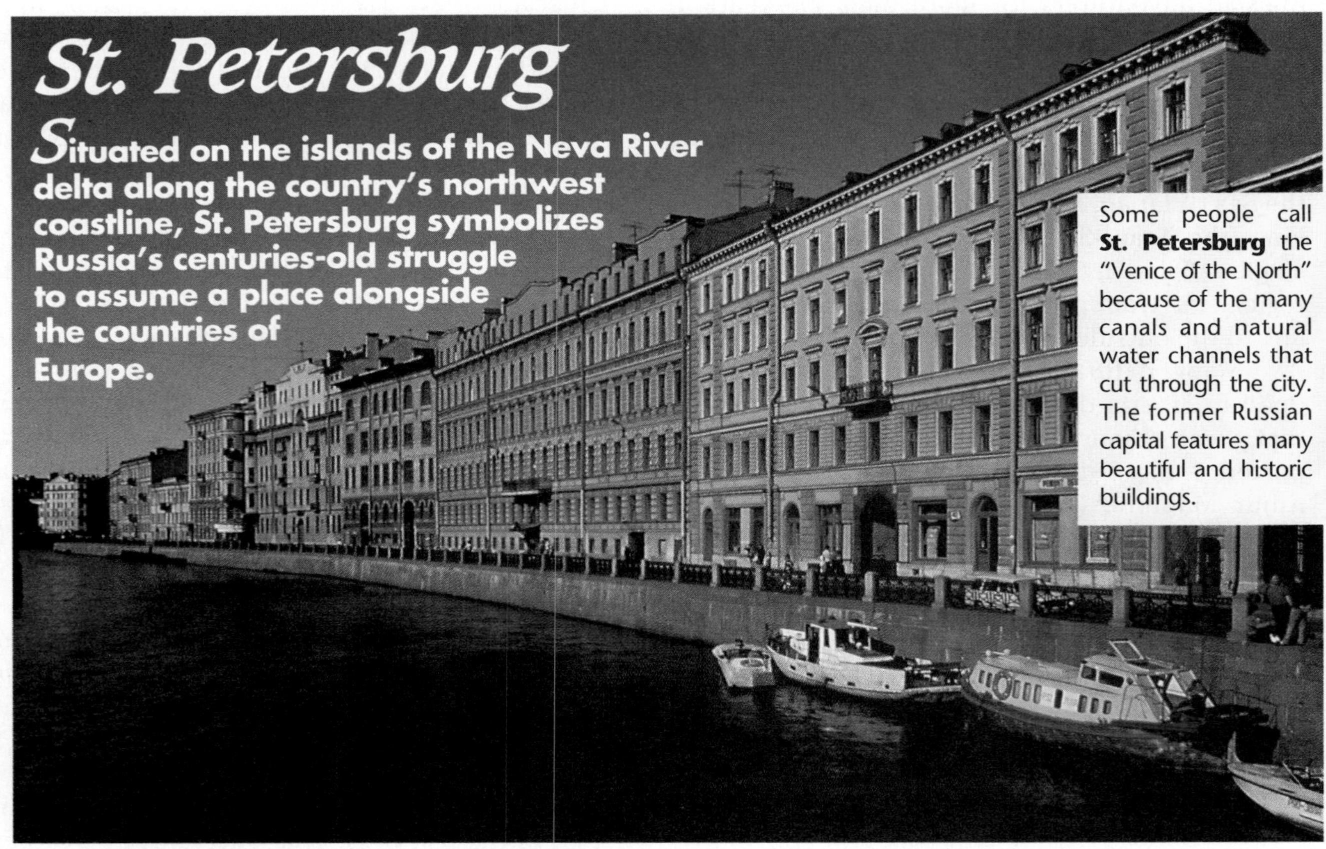
GINZA
HOLLYWOOD

St. Petersburg

Situated on the islands of the Neva River delta along the country's northwest coastline, St. Petersburg symbolizes Russia's centuries-old struggle to assume a place alongside the countries of Europe.

Some people call **St. Petersburg** the "Venice of the North" because of the many canals and natural water channels that cut through the city. The former Russian capital features many beautiful and historic buildings.

St. Petersburg is Russia's second largest city, a major transportation center, and one of the most beautiful urban centers in Europe. The city was founded in 1703 by Peter the Great, one of Russia's most brilliant but brutal rulers. The site that Peter chose for the city was poor. The land was swampy and the climate unhealthy. Though thousands of forced laborers died from malaria and other diseases while building the new city, Peter the Great was determined to build a city and port on the Baltic coast. This city would serve as Russia's "window on Europe." Through that window, Russia could adopt the scientific and technical skills of the countries of Europe and begin to modernize.

To emphasize this plan, Peter gave his city the German-sounding name St. Petersburg. In 1712, St. Petersburg replaced Moscow as the capital of the Russian Empire. Remembering the thousands who died building the city, Russian people called St. Petersburg "the city built on bones." But as Peter imported some of Europe's finest craftsworkers and architects to St. Petersburg, a well-planned city soon took shape. Peter's ambitious plan was continued by his successors. By the late eighteenth century, St. Petersburg was a large and impressive city in which Russians took great pride.

During the nineteenth and early twentieth centuries, St. Petersburg rose to become the center of modern Russian culture. Great nineteenth-century writers such as Alexander Pushkin and Fyodor Dostoyevsky set some of their most important works here. Peter Tchaikovsky, a famous Russian composer, wrote his music in St. Petersburg. The city has also been home to some of Russia's most famous ballet dancers.

Many of Russia's most significant historical events have taken place in the heart of St. Petersburg. In March 1917, the uprising that ended the Russian monarchy began here. Eight

CHAPTER 29 CHECK

Summer Palace in **St. Petersburg**

Reviewing the Main Ideas

1. Russia is a huge country dominated by plains, although it also has hills and mountains. Winters are cold and long, making travel difficult. Vegetation varies from forest to steppe.
2. Russia has rich forest, energy, and mineral resources. Its climate, however, limits the country's agriculture, which has undergone much change over the past two centuries. Today, much of the food that Russia produces comes from farmers' private plots. Fruits, vegetables, and meats are among the products raised.
3. Russian industry is in the process of rebuilding and restructuring as the country makes the transition to a free-market economy. Economically, Russia can be divided into Siberia, the Far East, and the regions around Moscow, St. Petersburg, the Volga River, and the Urals.
4. The Russian Federation's government is divided over reform. About 15 percent of its territory is divided into 21 largely self-governing republics. Major national issues include building stable political relationships and bringing freedom and prosperity to the country's peoples.

Building a Vocabulary

1. Where are habitation fogs found, and why?
2. What is the taiga?
3. What are serfs? Do they still exist in Russia?
4. What is the difference between light industry and heavy industry?
5. Define *smelter*.

Recalling and Reviewing

1. What large lake is found in Siberia? Why is this lake significant?
2. How has Russia's size influenced development of the country's resources?
3. Where do most of Russia's people live? What is the least populated area?
4. How have agricultural and industrial activities in Russia changed as a result of the breakup of the Soviet Union? What effect do these changes have on the country's people?

Thinking Critically

1. Do you think Russia is best considered a country of Europe or a country of Asia? Keep in mind demographic, cultural, and physical factors.
2. As you know from the chapter, the republics of the Russian Federation are based to some extent on the location of ethnic groups. What are some advantages and disadvantages of forming political units, such as republics or countries, in this way?

Learning About Your Local Geography

Individual Project

Siberia's Baykal-Amur Mainline (BAM) was built largely by youth organizations. Using school or community resources, investigate youth organizations in your own area and the projects with which they have been involved. Make a list of each organization, and next to each item, write down some of its projects. If possible, volunteer to spend an afternoon working with one of the local youth organizations. You may want to include your list in your individual portfolio.

Using the Five Themes of Geography

Imagine you are taking a week-long tour of Russia in the winter. Keep a journal of your **movement**, of the physical **place** characteristics you encounter, and of evidence of **human-environment interaction**. How do place characteristics affect what places you visit and how you move between them? What forms of human-environment interaction have made your movement possible?

CHAPTER 30

Ukraine, Belarus, and the Caucasus

GEOGRAPHY DICTIONARY

Cossack

homogeneous

> "The farther they [the Cossacks] penetrated the steppe, the more beautiful it became. . . . Nothing in nature could be finer. The whole surface resembled a golden-green ocean, upon which were sprinkled millions of different flowers. . . . Oh, steppes, how beautiful you are!"
>
> *Nikolay Gogol*

Door woodcarving from **Georgia**

The political geography of Eurasia has changed dramatically since the collapse of the Soviet Union. The former Soviet republics have become independent countries, making previous internal Soviet boundaries international frontiers. Border posts and customs stations dot once-deserted roads. Gradually, a new system of political and economic relationships is developing as local political leaders take on the authority that was once centered in Moscow. The new geography is being tested and adjusted, again and again.

The countries of Ukraine, Belarus, Georgia, Armenia, and Azerbaijan form the western rim of the Russian Federation. The latter three countries are part of a region called the Caucasus. All of these countries are working to build new societies. In many cases, they have combined ideas and values from their own ancient histories with those from the former Soviet system and from the democratic countries of western Europe.

Ukraine field

UKRAINE, BELARUS, AND THE CAUCASUS: PHYSICAL–POLITICAL

LEGEND

National capitals
Other cities

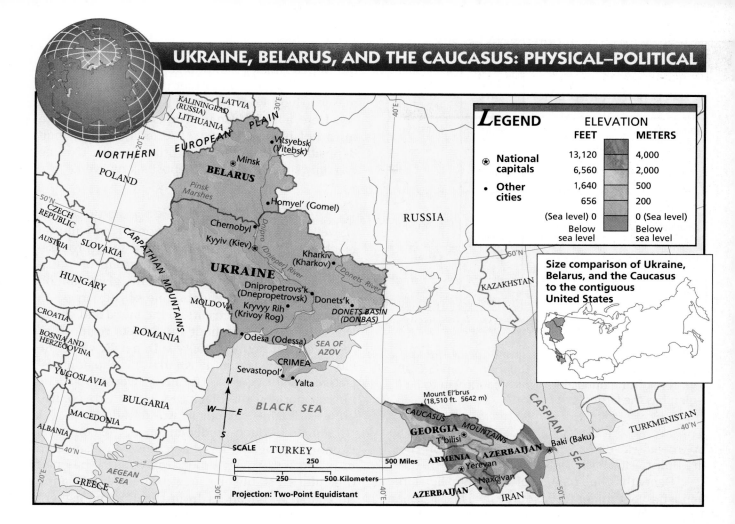

ELEVATION FEET	METERS
13,120	4,000
6,560	2,000
1,640	500
656	200
(Sea level) 0	0 (Sea level)
Below sea level	Below sea level

Size comparison of Ukraine, Belarus, and the Caucasus to the contiguous United States

SCALE 0 250 500 Miles; 0 250 500 Kilometers

Projection: Two-Point Equidistant

UKRAINE

FOCUS

- *What physical features, resources, and economic activities are found in Ukraine?*
- *What is the human geography of Ukraine?*

REGION **Physical Geography** A portion of the Carpathian Mountains runs through western Ukraine, but most of the country lies in a region of plains and low hills that stretch north from the Black Sea. (See the map above.) In plains regions such as this, climate and vegetation are linked in broad, east-to-west bands. The climates here become warmer and drier from north to south, though winters throughout the country are cold. Ukraine's latitudes are the same as those of southern Canada.

Deciduous forests were once widespread in northern Ukraine. In the south, the forests opened onto the grasslands of the steppe. The capital city of Kyyiv grew near the former forest-steppe boundary, where the products of the steppe and forests could be traded. Today, little of Ukraine's original vegetation remains. Farm fields now cover much of the land.

Resources Since ancient times, rich farmlands have been Ukraine's greatest natural resource. The region's fertile black soils and summer rains produce high yields of grains and vegetables. Indeed, Ukraine was the breadbasket for the Russian Empire and, later, for the Soviet Union. Even though farming under the Soviet system was inefficient, Ukraine's farmlands helped make the Soviet Union the world's largest producer of wheat, oats, barley, and sugar beets.

Ukraine's level plains also permit commercial farming using modern machinery. Irrigation water for large-scale agriculture comes from

FACTS IN BRIEF Ukraine, Belarus, and and the Caucasus

COUNTRY POPULATION (1994)	LIFE EXPECTANCY (1994)	LITERACY RATE	PER CAPITA GDP (1993)
ARMENIA 3,521,517	69, male 76, female	99% (1992)	$2,040
AZERBAIJAN 7,684,456	67, male 75, female	97% (1992)	$2,040
BELARUS 10,404,862	66, male 76, female	98% (1992)	$5,890
GEORGIA 5,681,025	69, male 77, female	99% (1990)	$1,390
UKRAINE 51,846,958	65, male 75, female	98% (1993)	$3,960
UNITED STATES 260,713,585	73, male 79, female	97% (1991)	$24,700

Sources: *The World Factbook 1994,* Central Intelligence Agency; *1995 Britannica Book of the Year*

Because of major economic changes since 1990 in the republics of the former Soviet Union, per capita GDP figures might be unreliable. Which country in the region has the largest population?

hydroelectric dams on the region's major river, the Dnipro (formerly Dnieper).

Beneath the plains of Ukraine lie large deposits of mineral resources. The Russian Empire's first important industrial region developed near the still-productive coal fields of the Donets Basin (also called the Donbas) in eastern Ukraine. Iron and steel industries were established here in the late 1800s.

During the Soviet period, the government developed many additional mining areas in Ukraine. The most important is at Kryvyy Rih (formerly Krivoy Rog) between Kyyiv and the Black Sea. Here, a huge open-pit iron-ore mine supports a region of heavy industry. The area also has rich deposits of manganese and other metals. Ukraine has only small oil and gas deposits, however.

INTERACTION **Economic Geography** Ukraine is well known for its production of agricultural machinery, transport vehicles, and food processing equipment. South of Kyyiv, a group of industrial cities has developed around the giant dam on the Dnipro River. This dam, one of the largest in Europe, supplies electricity to the metal and chemical industries throughout the Donets region. The region's most important city is Donets'k, while the city of Kharkiv (formerly Kharkov) is the administrative center. Odesa (formerly Odessa) is the busiest port on the Black Sea and is expected to grow as Ukraine expands its world trade.

Since Ukraine's independence, the country's economic progress has been slow. As elsewhere in the former Soviet Union, the country has had difficulties establishing new trade links. In addition, holdover leaders from the Communist era delayed reforms for moving the economy away from central planning. As a result, the transition to a free-market economy has been slow.

A major concern for Ukraine is making its farmland more productive. Agriculture is still

Wheat: From Field to Consumer

- The head of the wheat plant contains the wheat kernels, wrapped in husks. The kernel includes the bran or seed coat, the endosperm, and the germ from which new wheat plants grow.

Bran (seed coat)
Endosperm
Germ
Harvesting

organized in the inefficient state farms and collectives of the Soviet period. Ukraine's farm population also is in decline as younger Ukrainians move to the cities.

Human Geography In area and population, Ukraine is about the same size as France. Kyyiv serves as the country's political capital as well as its cultural and historical center. The city's churches, museums, and opera houses are stately symbols of Ukrainian life.

A large number of ethnic Russians live in Ukraine, mostly in the Donets region and along the Russian border. Ukrainians and Russians share similar cultures. As a result, many Russians call Ukrainians—to their displeasure—"little Russians" and talk of Ukraine as if it were part of Russia.

Through centuries of Russian domination, **Ukraine** has held onto its traditions. Ukrainian folk musicians, such as these, sing ballads and historical songs.

Ukrainians, however, believe there are important differences between themselves and Russians. Nationalism is on the rise among many Ukrainians, who have long cherished a separate identity. Some Ukrainians claim that they are descended from the independent, often warlike **Cossacks**. The Cossacks controlled the steppe region beginning in the sixteenth century and fiercely fought the southward expansion of the Russian Empire.

Major religions practiced in Ukraine include Eastern Orthodox Christianity and Roman Catholicism. Many Ukrainians also belong to the Uniate Church, whose members share the beliefs of Eastern Orthodox Christianity but recognize the authority of the Roman Catholic Pope. This church arose in part because the lands of western Ukraine were once part of Poland.

Crimea Debate in Ukraine continues over the status of Crimea, a small peninsula that juts southward into the Black Sea. Although Crimea was part of the Ukrainian republic in the Soviet Union, it has never thought of itself as part of Ukraine. Instead, Crimea's ties have been to Russia. In fact, about

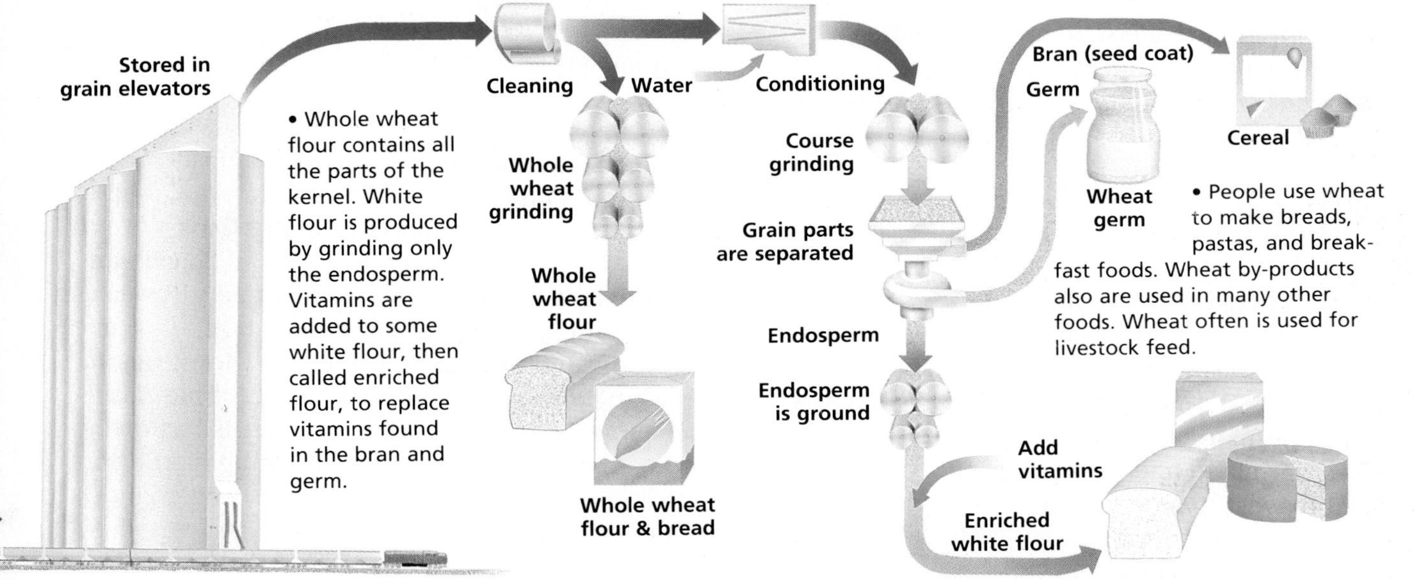

Stored in grain elevators
• Whole wheat flour contains all the parts of the kernel. White flour is produced by grinding only the endosperm. Vitamins are added to some white flour, then called enriched flour, to replace vitamins found in the bran and germ.
Cleaning
Water
Whole wheat grinding
Whole wheat flour
Whole wheat flour & bread
Conditioning
Course grinding
Grain parts are separated
Endosperm
Endosperm is ground
Add vitamins
Enriched white flour
Bran (seed coat)
Germ
Wheat germ
Cereal
• People use wheat to make breads, pastas, and breakfast foods. Wheat by-products also are used in many other foods. Wheat often is used for livestock feed.

70 percent of the population is ethnic Russian. The Russian royal family built summer palaces in Crimea, and during the Soviet era, the area was a favorite vacation spot of Russian officials.

Vineyards and fruit orchards thrive in Crimea's Mediterranean climate, and the region's beaches are still popular resorts. Valuable iron-ore deposits in the area have led to the development of an iron-and-steel industry at the city of Kerch, and the port of Sevastopol' (suh-VAS-tuh-pohl) was the site of an important Soviet naval base. Many Russians would like to see Crimea become part of Russia.

SECTION REVIEW

1. Why is Ukraine agriculturally productive? What resources have helped industrialization in the region?
2. Why do many Ukrainians dislike being called "little Russians"? What is significant about Crimea?
3. **Critical Thinking** **Explain** how Ukraine was significant to the Soviet Union. What legacies are left in Ukraine from the Soviet era?

BELARUS

FOCUS

- *What are the physical and economic characteristics of Belarus?*
- *What are the cultural characteristics of Belarus?*

INTERACTION **Physical Geography** Belarus, formerly the Soviet republic of Byelorussia (bee-ehl-oh-RUSH-uh), is a region of plains located north of Ukraine. (See the map on page 351.) The country has a cool, humid-continental climate similar to that of neighboring Poland and Russia.

Although farming is important to Belarus, swamps and marshes cover much of the country. The best farmland and highest ground is found in an east-to-west band across the middle of the country. This hilly countryside, made up of sandy, glacial moraine, formed during the last ice age.

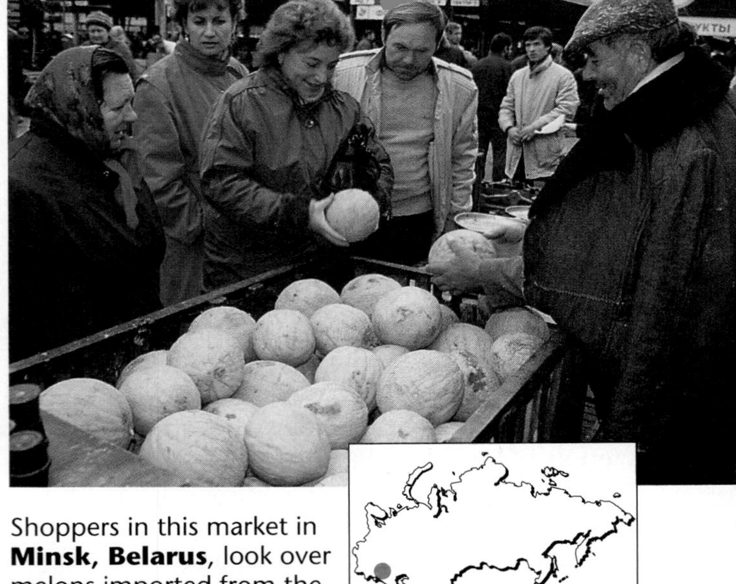

Shoppers in this market in **Minsk, Belarus,** look over melons imported from the Central Asian republic of Uzbekistan.

The railroad from Moscow to Warsaw, Poland, runs along this landform. It is part of the route used to invade Russia by Napoléon in 1812 and by Hitler in 1940.

Economic Geography With few mineral resources and generally poor soil, Belarus has relied on its traditions of education and technical training to build its economy. Enough of its forests remain to support some wood products industries. Peat, however, is still used as a fuel, even though its burning creates severe air pollution. The country's industries, many of which produce consumer goods, are concentrated in Minsk, the capital. Minsk is also an important transportation center. During the Soviet era, most of Belarus's products went to other Soviet republics. Replacing these lost trade links is proving difficult.

Human Geography Throughout history, the lands of Belarus have been culturally connected to Poland as well as to Russia. The people of Belarus, often called White Russians (*byelo* means "white" in Russian), speak a Slavic language that is similar to both Polish and Russian. And the Belarus people practice both Eastern Orthodox and Roman Catholic religions. As in other regions of the former Soviet Union, the once-large Jewish population of Belarus has dwindled to very few, in part due to recent emigration.

Before 1991, Belarus had never been an independent country. In 1995, Belarussians voted to strengthen economic ties with Russia. Belarus has not moved forward with important economic reforms, and many former communists remain in power in the country.

SECTION REVIEW

1. Why is agriculture limited in much of Belarus?
2. How has Belarus's location influenced its cultural geography?
3. **Critical Thinking** **Suggest** a reason that Belarus has not broken with many traditions from the Soviet era.

THE CAUCASUS

FOCUS

- *What are the physical features of the Caucasus?*
- *What conflicts are taking place within and among Georgia, Armenia, and Azerbaijan? What are the sources of these conflicts?*

REGION Physical and Human Geography

The Caucasus region is the broad isthmus of mountains and isolated valleys that separate the Black Sea and the Caspian Sea. (See the map on page 351.) The Caucasus can be divided into three landform regions. In the north along the border with Russia is a wide mountain range. The region's highest peak is Mount El'brus at 18,510 feet (5,642 m)—the highest elevation in Europe. Lowland regions, meanwhile, lie along the Black and Caspian seas. In the south of the Caucasus is a rugged, volcanic plateau with mountains. Both earthquakes and volcanic eruptions threaten this tectonically active region.

Ethnically, the Caucasus is one of the most complex regions in the world. Dozens of different languages are spoken. The Caucasus is also rich in mineral resources, including manganese, lead, zinc, and copper.

Since ancient times, the armies of invading powers swirled around the Caucasus. The region's people often retreated into the mountain valleys to defend themselves against the invaders. As a result, the region has traditions of fierce independence and ethnic pride.

Today, the Caucasus is organized into three independent countries: Georgia, Armenia, and Azerbaijan. Future divisions based on ethnicity are possible, however. Georgia recognizes some additional ethnic subregions. Nonetheless, ethnic violence has spread throughout the Caucasus. Permanent peace in the region may take years to achieve.

Georgia Georgia stretches inland from the lowlands along the eastern side of the Black Sea. The mild climate of the Georgia coast, which is much like that of the Carolinas in the United States, has made it home to several busy resorts. Citrus fruits and tea are grown in the region.

The Georgians have long been famous for their enterprise and business skills. Vegetables from Georgia still appear for sale in Moscow's street markets. An independent Georgia must extend its trading links, however. The country has little oil or other fossil fuels, and its hydroelectric power does not meet its needs.

Georgia's capital of T'bilisi, located at the crucial pass between the Black and Caspian seas, has been the center of Georgian culture for nearly 2,000 years. During the Soviet era, industrialization around T'bilisi and other cities in Georgia expanded, resulting in serious air pollution. Georgia's most immediate concern since independence, however, is violent unrest among its ethnic groups. In some areas, Islamic groups have longstanding ethnic quarrels with the Georgians, who are traditionally Eastern Orthodox. Russian troops have tried to intervene in the fighting, which has strained Russia's relations with Georgia.

Armenia Landlocked Armenia lies in the volcanic plateaus and mountains of the southern Caucasus, where the average elevation is more than a mile (1.6 km) above sea level. Severe earthquakes have rocked the country.

Compared to the rest of the Caucasus, the people of Armenia are unusually **homogeneous** (hoe-muh-JEE-nee-uhs), or of the same kind. More than 90 percent of the population is ethnically Armenian, and most people are Christian as well. In fact, Armenians claim to be the first nationality that adopted Christianity. The Islamic countries bordering Armenia—Turkey, Azerbaijan, and Iran—historically are its enemies.

This Christian monastery is nestled in the Caucasus Mountains of **Armenia**. Nearly all the residents of the country are Armenian Orthodox Christians.

Armenia's roots go back to ancient times. Written Armenian literature can be traced to the fifth century, and Armenian states once existed in what is now eastern Turkey. Armenian communities could be found throughout Southwest Asia as well. As the Ottoman Empire began to collapse, however, a civil war raged in Turkey, and about 1.5 million Armenians were killed. Some fled to the United States and other Western countries, but most retreated to their ancient homeland in the south Caucasus. Here they found protection in the windswept, grassy countryside and the ruins of ancient Armenian churches.

Armenia industrialized during the Soviet era. The heavy industries of the capital of Yerevan produced some of the Soviet Union's worst air pollution. Today, Armenia is developing its economy and democratic government. Traditionally, Armenians have valued education and community life.

The most pressing issue in Armenia is the status of Nagorno-Karabakh, a region that was part of ancient Armenia. This rural territory, which is populated by Armenians, now lies entirely within the neighboring country of Azerbaijan. Armenia and Azerbaijan have fought over this region. Until the conflict is resolved, Armenia's economy is unlikely to move forward much.

Azerbaijan Azerbaijan stretches west from the southern end of the Caspian Sea. Because much of the country is in the rain shadow of the mountains of Georgia and Armenia, steppe and semiarid environments are typical. The land is mostly low mountains, foothills, and plains. About three-fourths of the Azerbaijan people are Azeri. They speak a Turkic language and first arrived in the region during the eleventh century. Most are farmers and herders.

Baku, Azerbaijan's capital city, is the center of a large oil-refining industry. Development here began during the late 1800s, and the oil reserves should last well into the future. Indeed, new deposits under the shallow Caspian Sea will provide income for many years. In the past, Baku's oil products were shipped north into Russia. Today, they are also exported to world markets.

Azerbaijan has strong ties to parts of Iran to the south. In fact, more than twice as many Azeri people live in Iran as in Azerbaijan. In addition, the Azeri are rediscovering Islam. Many Azeri students are studying Islam in Iran and other Islamic countries.

SECTION REVIEW

1. What landforms make up the Caucasus?
2. What is the main source of the violence in Georgia? Why have Armenia and Azerbaijan fought each other?
3. **Critical Thinking** **Explain** how Armenia's location has presented problems for the country.

CHAPTER 30 CHECK

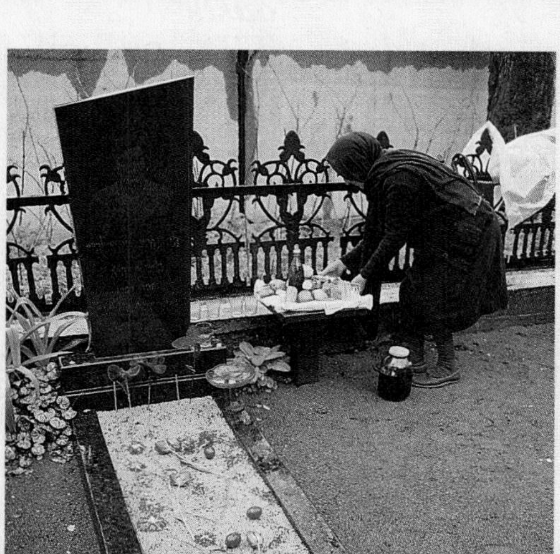

Leaving flowers and Easter eggs at the grave of a loved one, a tradition in **Georgia**

Reviewing the Main Ideas

1. Ukraine has rich agricultural and mineral resources. Although industry is well developed in its cities, economic progress has been slow since the country became independent. Nationalism is increasing among Ukrainians, but some parts of Ukraine, such as Crimea, still identify with Russia.
2. Belarus, north of Ukraine, has poor soils and few mineral resources. The people of Belarus have been culturally connected to Poland and to Russia. Now independent for the first time, Belarus has made little progress with economic and political reform.
3. The Caucasus is the broad isthmus of mountains and valleys that separate the Black and Caspian seas. Ethnically, the Caucasus is one of the most complex regions in the world. The Caucasus countries are Georgia, Armenia, and Azerbaijan.

Building a Vocabulary

1. From what group of people do some Ukrainians claim to be descended?
2. Define *homogeneous*. Which Caucasus country has the most ethnically homogeneous population?

Recalling and Reviewing

1. How do Belarus's resources differ from Ukraine's?
2. Why might the countries of the Caucasus be redivided or subdivided in the future?
3. Which of the countries discussed in this chapter have large Christian populations? Which have large Islamic populations?

Thinking Critically

1. The word *ukraine* comes from a Slavic word meaning "borderland." How might the word also be applied to other countries discussed in this chapter?
2. The United States has many immigrants from the former western Soviet Union republics, including Belarus and Armenia. Why do you think these people immigrated to the United States?

Learning About Your Local Geography

Individual Project

Investigate the dam nearest to your community, and write a paragraph about its origins and uses. When and why was it built? How big is the dam? What purposes and areas does it now serve? You may want to include your paragraph in your individual portfolio.

Using the Five Themes of Geography

Sketch a map of the countries discussed in this chapter. With symbols or shading, show how the **location** of each country has influenced its human **place** characteristics. Consider each country's location relative to other countries and regions. What effects does location have on the religions practiced in these countries? on the languages spoken by their peoples? What other cultural characteristics of these countries have been determined or influenced by location? Do any political issues in these countries concern their locations?

CHAPTER 31

Central Asia

GEOGRAPHY DICTIONARY

- *oasis*
- *mosque*
- *dryland agriculture*
- *clan*

"For twelve days the course is along this elevated plain, which is named Pamer [Pamir Plateau] . . . . So great is the height of the mountains, that no birds are to be seen near their summits; and however extraordinary it may be thought, it was affirmed, that from the keenness of the air, fires when lighted do not give the same heat as in lower situations, nor produce the same effect in cooking food."

Marco Polo

Handwoven textile from **Turkmenistan**

The dry lands of Central Asia lie east of the Caspian Sea and north of the world's highest mountains. The nearest ocean to Central Asia is the Indian Ocean, but the towering mountains block the humid ocean air blowing inland that might bring needed rains.

Throughout its history, Central Asia often has been a buffer zone for neighboring empires. In ancient times, the region marked the eastern edge of Greek civilization under Alexander the Great. A few centuries later, it became the western outpost of China's empire. In the nineteenth century, the British colonial empire in India and the expanding Russian Empire fought for control of the area. The Soviets took control of the region after the Communist Revolution.

With the collapse of the Soviet Union, Central Asia's five former Soviet republics became independent countries. These are Kazakhstan (kuh-ZAK-stan), Kyrgyzstan (kir-JEEZ-stan), Uzbekistan (uz-BEHK-is-tan), Turkmenistan (turk-MUHN-is-tan), and Tajikistan (tah-JIK-is-tan).

Spring tulips in the Pamirs of **Tajikistan**

CENTRAL ASIA: PHYSICAL–POLITICAL

RUSSIA
Petropavl
Tobol River
Irtysh River
Zhayyq (Ural) River
Semey (Semipalatinsk)
Qaraghandy (Karaganda)
Lake Zaysan
KAZAKHSTAN
Aral'sk
Baikonur Cosmodrome
ARAL SEA
Syr Darya
Balqash Lake
Ili River
Almaty (Alma-Ata)
Bishkek (Frunze)
CASPIAN SEA
AZERBAIJAN
UZBEKISTAN
Shymkent
Amu Darya
Toshkent (Tashkent)
KYRGYZSTAN
TIAN SHAN
CHINA
MONGOLIA
TURKMENISTAN
Bukhoro (Bukhara)
Samarqand
Khujand (Leninabad)
Osh
TAJIKISTAN
Dushanbe
Communism Peak (24,590 ft.; 7495 m)
PAMIRS
Ashgabat (Ashkhabad)
Chärjew
Mary
KOPET MOUNTAINS
Kara Kum Canal
IRAN
AFGHANISTAN
PAKISTAN
INDIA
50°N
40°N
50°E
60°E
70°E
80°E
90°E
100°E
SCALE
0 250 500 Miles
0 250 500 Kilometers
Projection: Two-Point Equidistant

Size comparison of Central Asia to the contiguous United States

LEGEND

ELEVATION

FEET	METERS
13,120	4,000
6,560	2,000
1,640	500
656	200
(Sea level) 0	0 (Sea level)
Below sea level	Below sea level

National capitals
Other cities

1 PHYSICAL GEOGRAPHY

FOCUS

- *What are the landforms and climates of Central Asia?*
- *What effects have various irrigation methods had on the region?*

REGION Landforms and Climate Plains and low plateaus cover much of Central Asia. To the south and east, the spectacular mountain peaks of the Kopet Mountains, the Pamirs, and the Tian Shan rim the region. Summer rains and winter snows in the mountains provide water for irrigation far out on the plains.

Central Asia is a region of mainly steppe, desert, and highland climates. Its latitudes compare to those of Nevada and Idaho, and hot summers and cold winters are typical. Climates become drier from north to south, as reflected in the region's vegetation. North of the Aral Sea, dry steppe vegetation is usual, whereas south of the sea, desert shrubs and shifting sands are common. Grasslands reappear on the lower mountain slopes and give way to forests at higher elevations.

INTERACTION Water Resources In dry regions such as Central Asia, water resources must be managed carefully. The main sources of water in the region are the narrow streams draining north from the mountains. These streams have been used for irrigation for thousands of years. The Syr Darya (sir DAHR-yuh) and the Amu Darya (ahm-oo DAHR-yuh) drain the snowcapped Pamirs and carry their water all the way to the deserts near the Aral Sea. Irrigation projects and natural **oases** are nearly continuous along both channels. An oasis is a site in the desert where water can be found in springs or wells or under dry riverbeds.

The Soviet government, seeking increased farm production, invested heavily in water developments in the region. The more than 500-mile-long (805-km) Kara Kum Canal, one of the world's largest and longest canals, carries water from the Amu Darya to the deserts near the Caspian Sea. Unfortunately, poor irrigation methods and overwatering wasted large amounts of water. In addition, the Soviets directed farmers to plant cotton, which uses a great deal of water. As the irrigation water evaporated, salts built up in the fields, decreasing crop growth.

The Aral Sea in particular has suffered from these policies. Once the world's fourth largest inland body of water, the Aral Sea has lost more than 60 percent of its water since 1960. Its level has dropped by 50 feet (15 m) and continues to fall. Aral'sk, once the sea's major fishing port, was by 1993 more than 60 miles (97 km) from the shore. The deltas of the Amu Darya and Syr Darya, once rich in wildlife, are now dry and desolate. Windstorms sweep the dry seafloor, blowing salty dust hundreds of miles downwind.

To restore the waters of the Aral Sea, all irrigation in the region would have to cease for several decades. Such a regional solution is difficult to achieve. Indeed, the countries of the region have not been able to agree on reducing their use of water headed toward the sea in order to stabilize it, let alone restore it.

A cemetery of rusted ships lies on the dried bottom of what once was part of the **Aral Sea** in **Central Asia**.

1 SECTION REVIEW

1. What landforms and climates are found in Central Asia?
2. What has happened to the Aral Sea? Why did it happen?
3. **Critical Thinking** **Suggest** why the Central Asian countries might decide to try to stabilize the Aral Sea's level rather than restore it to earlier levels.

2 HUMAN GEOGRAPHY

FOCUS

- *What is the cultural background of Central Asia? the economic background?*
- *What challenges does the region face?*

Most of Central Asia's major groups speak Turkic languages. One exception is the Tajiks, whose language is related to Iranian. As for religion, Islam has been the dominant faith for nearly 1,000 years and is once again central to people's lives. In 1989, the region had only 160 functioning **mosques**; by 1993, it had more than 10,000. A mosque is an Islamic place of worship. Each of the region's largest ethnic groups has its own country now; however, there are many minority ethnic groups.

INTERACTION **Economic Geography** For thousands of years, life in Central Asia was sustained by fields of grains and vegetables, orchards of apricots and apples, and herds of sheep, goats, and horses. Some communities practiced subsistence farming using simple irrigation methods; others were involved in herding.

Many of the cities of southern Central Asia began as trading centers along the Silk Road, an overland trade route from Southwest Asia to China. (See "Global Connections" on pages 362–363.) Caravans along the route stopped to rest and trade in towns in the fertile, irrigated valleys at the base of the mountains. These towns developed into great cultural and economic centers.

During the Soviet era, government planners attempted to modernize the region's economy.

почта

This knife grinder practices his trade in **Bukhoro, Uzbekistan**. Traditional as well as modern ways of life are found throughout the republics of Central Asia, the least developed region of the former Soviet Union.

Herders were forced onto collectives and state farms, while massive irrigation projects attempted to expand the production of cotton. Cities grew dramatically, as wide avenues and rows of high-rise apartments surrounded the old mud-brick buildings and narrow streets. The government built factories to process nearby minerals.

Issues Central Asia is the least-developed region of the former Soviet Union. Incomes are low, and birthrates and illiteracy rates are high. Although Central Asians welcome their political independence from Moscow, the break has caused an economic decline in the region. Central Asia receives foreign aid from the world's developed countries as well as from oil-rich Islamic countries such as Libya and those in Southwest Asia.

Economic growth will depend on increased exports of cash crops and minerals. Development will take place only slowly, however. The combination of decreasing water resources, increasing populations, economic decline, and traditional ethnic conflicts will heighten and prolong conflict in the region.

2 SECTION REVIEW

1. How did people's economic activities change during the Soviet era?
2. What social and economic challenges does the region face?
3. **Critical Thinking** **Decide** whether Central Asia is a culturally homogeneous region. Explain your answer.

3 COUNTRIES OF CENTRAL ASIA

FOCUS

- *How has each Central Asian country reacted to independence and the collapse of the Soviet Union?*
- *What are each country's economic characteristics? cultural characteristics?*

Kazakhstan Kazakhstan is the second largest territory of the former Soviet Union, second only to the Russian Federation. The country is gradually dismantling the Soviet system by privatizing collectives and state farms and by selling the government's assets.

The country's major industrial region is found around Qaraghandy (formerly Karaganda) in the north. Here lies a huge coal field, with copper, lead, and zinc mines nearby as well. Kazakhstan also has major undeveloped oil deposits along the Caspian Sea.

During the Soviet era, huge areas in northwest Kazakhstan were plowed for **dryland agriculture**, which depends on rainfall rather than irrigation to water crops. The Soviets' goal was to use modern methods and machinery to increase grain production on huge state farms. But dust storms and frequent periods of drought hindered this goal.

Ethnic Kazakhs make up only about 40 percent of the population. Russians make up the country's second largest ethnic group and are particularly numerous in the districts along the Russian border, in the Qaraghandy region, and in the capital city of Almaty (formerly Alma-Ata). The former Soviet Union's great space port, the Baikonur Cosmodrome, is near the lower Syr

Text continues on page 364.

The Silk Road

For centuries, goods, ideas, technology, and religions traveled back and forth from China to the Mediterranean Sea over an ancient route that came to be called the Silk Road. The 5,000-mile (8,045-km) trade route linked China and the Western world.

Many of Central Asia's cities began as trading centers where caravans traveling the Silk Road stopped to trade and rest in the region's fertile valleys.

The Silk Road's beginnings reach far into the past, to a time when groups of nomads roamed Central Asia. These nomads eventually began trading goods produced in Iran, Mesopotamia, China, and India. While carrying gold, precious stones, and jade between countries, the nomads also spread ideas and technology. They may have introduced metalworking and the wheel to eastern Asia.

The nomads were warriors as well as traders. Around 200 B.C., one fierce group, the Huns, moved from the Asian steppes to threaten China. In 138 B.C., China sent an expedition westward to form alliances with other nomads against the Huns. Soon, other groups traveled westward to gather allies and obtain horses for the Chinese army. The expeditions included diplomats and brought livestock, gold, and silk to trade. The Chinese finally defeated the Huns and established important trade links with their neighbors.

These merchants show off beautiful silk fabrics manufactured in **Samarqand, Uzbekistan**. Silk is a popular natural fiber often used in the production of clothing, upholstery, and curtain materials.

The Golden Age of the Silk Road

By about 100 B.C., Chinese silk had traveled as far west as the Mediterranean, and the Silk Road had entered its golden age. Trade flourished as great caravans traveled across Asia. Gold, ivory, amber, and linen flowed eastward to China on the backs of camels and in carts pulled by oxen or mules. Westward, the caravans carried pottery, iron weapons, and, of course, silk. The various rulers of the region controlled the caravans that traveled through their territories.

Along with traded goods came ideas and religions. In the first century, Buddhist missionaries from India introduced their religion to China. Buddhism later spread to Korea and Japan. In the seventh century, Christian missionaries carried their form of Christianity to eastern Asia.

Perils of the Silk Road

Travelers and traders on the Silk Road encountered natural geographic obstacles. From China, caravans wound westward around the Great Wall and then from oasis to oasis across the desert. From there, they continued through icy mountain passes and over a great plateau. Caravans then descended onto the plains of Mesopotamia, to arrive at trading cities near the Tigris and Euphrates rivers. The Silk Road usually ended in Antakya (Antioch), a Mediterranean seaport. From there ships carried goods to Europe. Other land routes branched off from the Silk Road, reaching ports along the Arabian Sea and the Persian Gulf.

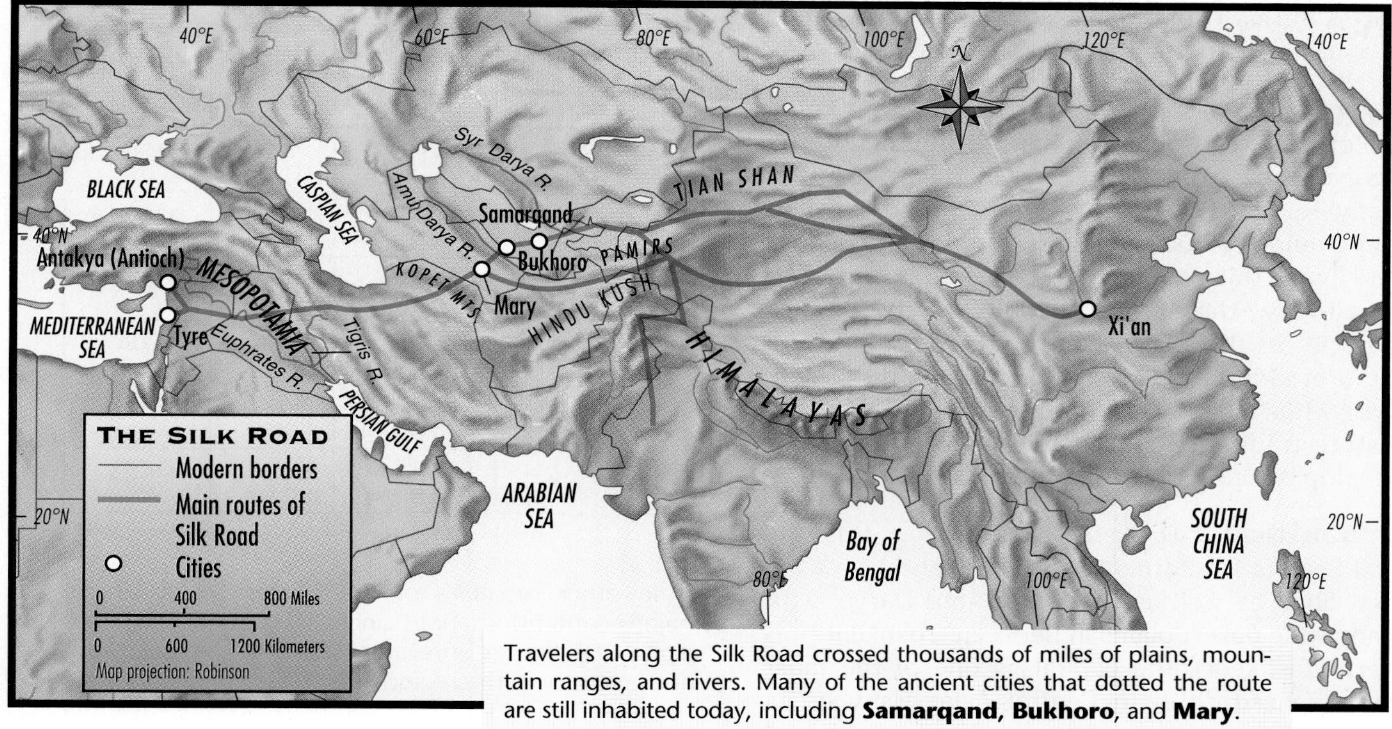
40°E
60°E
80°E
100°E
120°E
140°E
N
BLACK SEA
CASPIAN SEA
Syr Darya R.
Amu Darya R.
TIAN SHAN
Samarqand
Bukhoro
PAMIRS
40°N
Antakya (Antioch)
MESOPOTAMIA
KOPET MTS
Mary
HINDU KUSH
Xi'an
MEDITERRANEAN SEA
Tyre
Euphrates R.
Tigris R.
HIMALAYAS
PERSIAN GULF
THE SILK ROAD
Modern borders
Main routes of Silk Road
Cities
0 400 800 Miles
0 600 1200 Kilometers
Map projection: Robinson
20°N
ARABIAN SEA
Bay of Bengal
80°E
100°E
SOUTH CHINA SEA
120°E

Travelers along the Silk Road crossed thousands of miles of plains, mountain ranges, and rivers. Many of the ancient cities that dotted the route are still inhabited today, including **Samarqand, Bukhoro**, and **Mary**.

Although few people traveled over the entire route of the Silk Road, each part was hazardous. Bandits often raided the caravans. Some travelers lost their way in the desert and died of thirst in the blistering heat. Others perished in sandstorms, and some were buried by avalanches, never to emerge from the mountains.

The Decline of the Silk Road

From the seventh to the mid-thirteenth century, religious and political turmoil often blocked the Silk Road. Muslims, followers of the growing religion Islam, fought with the peoples of Persia and China. From Chinese prisoners, Muslims learned paper-making techniques that had long been kept secret. They also successfully smuggled silkworms out of China. Once these techniques spread to Europe, the need for trade with Asia decreased. Besides, the collapse of the Roman Empire left western Europe too weak and too poor to trade much with eastern Asia.

Thus, by the time Marco Polo journeyed to China over the Silk Road in the late thirteenth century, the route had experienced a decline. Upon his return to Europe, Polo described great trading cities and resting areas that had deteriorated into rubble.

Although the road remained open through the middle of the fourteenth century, in 1368 it again fell victim to political forces. Determined to isolate China from the rest of the world, China's Ming Dynasty closed the eastern parts of the route. A centuries-old tradition of trade between China and the Western world came to an end.

The deep, rich appearance of dyed silk makes beautiful clothing. This woman sells fabrics made from silk and other materials in **Bukhoro, Uzbekistan.**

Your Turn

Imagine you are a trader during the early days of the Silk Road. **Write** an entry for a daily journal. **Explain** what goods you are trading, and tell from where they came and to where you are taking them. Then **describe** the natural environment you encounter on your journey.

Darya. Russia's continuing space program requires cooperation with the Kazakh government.

Kyrgyzstan Kyrgyzstan is a landlocked country of the western Tian Shan. Traditionally, the Kyrgyz have lived as herders and farmers, but many Russians came to Kyrgyzstan to develop mines and hydroelectric projects. Many Uzbeks also live in this country.

Today, the Kyrgyz people barely form a majority in their own country. Newspapers and radio broadcasts are in both Kyrgyz and Russian. Kyrgyzstan's new government is working to restore traditional land management and further develop the country's mineral resources.

Uzbekistan Uzbekistan stretches east from the Aral Sea. Its landforms include the foothills of the Tian Shan as well as the lower Amu Darya, with many wide desert plains in between. Toshkent (formerly Tashkent), the capital, is one of the most industrialized cities in Central Asia. Many of its industries, however, were run by Russians, most of whom have left the region. Large numbers of Uzbeks, Tajiks, and Kazakhs remain.

Few changes have taken place in the country since independence, and the economy has declined steadily under the strict central government. The government hopes that expansion of the mining and oil industries can bring new income.

Turkmenistan During the Soviet era, Turkmenistan became a country of two desert waterways: the channel of the Amu Darya and the Kara Kum Canal, which passes the capital city of Ashgabat (formerly Ashkabad). Much of the rest of the country is desert with large areas of shifting sands. Turkmens make up a little more than 70 percent of the population.

Turkmenistan's Soviet system has changed little since independence. The government still owns nearly everything, and the collectives and state farms are still operating. Turkmenistan has little industry. Its economic future lies in the large oil deposits that have been discovered along the eastern shore of the Caspian Sea.

Tajikistan Within Tajikistan's borders lie the mineral-rich Pamirs and the great valleys of the upper Amu Darya system. More than half of the country is about 10,000 feet (3,048 m) above sea level. The mountain passes across Tajikistan form the major route connecting Central Asia to India.

FACTS IN BRIEF Central Asia

COUNTRY POPULATION (1994)	LIFE EXPECTANCY (1994)	LITERACY RATE	PER CAPITA GDP (1993)
KAZAKHSTAN 17,267,554	63, male 73, female	97% (1992)	$3,510
KYRGYZSTAN 4,698,108	64, male 72, female	97% (1993)	$2,440
TAJIKISTAN 5,995,469	66, male 72, female	98% (1992)	$1,180
TURKMENISTAN 3,995,122	62, male 69, female	98% (1992)	$3,330
UZBEKISTAN 22,608,866	65, male 72, female	97% (1992)	$2,430
UNITED STATES 260,713,585	73, male 79, female	97% (1991)	$24,700

Sources: *The World Factbook 1994,* Central Intelligence Agency; *1995 Britannica Book of the Year*

As in other republics of the former Soviet Union, major economic changes since 1990 might make per capita GDP figures unreliable. Which is the most populous country in the region?

Traditionally, Tajiks are a mountain people who are divided into many **clans**. A clan is a group of people descended from a common ancestral family. Tajiks and Uzbeks make up most of the population. Tajiks' historic ties are to the south in Afghanistan.

The 10-year Soviet occupation of Afghanistan disrupted Tajik life. As the Soviet armies retreated, they left behind unsettled conditions and many weapons. Since the collapse of the Soviet Union, Tajikistan has known only turmoil. Traditional and modern Tajiks have been fighting each other, and the country's limited industry has nearly stopped.

3 SECTION REVIEW

1. In which Central Asian countries has there been little change since the Soviet Union's collapse?
2. What ethnic groups live in each Central Asian country? What resources hold the most economic potential for these countries?
3. **Critical Thinking** **Imagine** that you are a Russian astronaut. Why might you have had some reservations about the breakup of the Soviet Union?

CHAPTER 31 CHECK

Mosque in **Kazakhstan**

Reviewing the Main Ideas

1. Plains and low plateaus cover much of Central Asia, and climates become drier from north to south. Poor irrigation methods, combined with a growing population, have threatened water resources and shrunk the Aral Sea.
2. Many different ethnic groups live in Central Asia. Most of the major groups speak Turkic languages, while Islam is the dominant religion. Traditional economic activities include herding and subsistence farming. Soviet planners tried to modernize the economy by introducing state-run farms and industry. Even so, Central Asia is the least-developed region of the former Soviet Union and faces many challenges.
3. The countries of Central Asia are Kazakhstan, Kyrgyzstan, Uzbekistan, Turkmenistan, and Tajikistan. With development, mineral resources and oil could provide income to these countries.

Building a Vocabulary

1. Define *oasis*. Why are oases important?
2. What is a mosque?
3. What is dryland agriculture's major source of water?
4. What term describes a group of people descended from a common ancestral family?

Recalling and Reviewing

1. How did irrigation methods in Central Asia change when the Soviets took control of the region?
2. What was the Silk Road? How did it affect Central Asia's towns?
3. What has happened in Tajikistan since it became independent?

Thinking Critically

1. Imagine you are an official in one of the Central Asian countries. Design a plan for the development of oil deposits in your country. Consider the extent to which you would invite the involvement of other countries and multinational oil companies.
2. Although cotton has been the major cash crop of much of Central Asia, the region's landforms and climates might be better suited to growing grains such as wheat. Why do you think the Soviets focused on cotton instead?

Learning About Your Local Geography

Cooperative Project

With your group, investigate a major scientific facility, one comparable to Kazakhstan's Baikonur Cosmodrome, in your area or state. Make a list of some of the projects being completed at the facility. Below the list, write a few sentences explaining the national or international significance of the projects.

Using the Five Themes of Geography

Sketch a series of diagrams of the Aral Sea shrinking over time to show the effects that **human-environment interaction** can have on a **region**. You may wish to draw cross-sectional diagrams that indicate the drop in the Aral Sea's level in addition to its decrease in area. Below your diagrams, write several sentences explaining what factors have been responsible for the Aral Sea's loss of water.

SKILLS

Linking Geography and Economics

As you know, Russia and several other former Soviet republics are now making the transition from a command economy to a free-market one. No handbooks exist to smoothly guide these new countries through the process. In addition, the region's unique geography and history make the transition a step into unknown territory.

Gaining a clear perspective on the current changes may take years, even decades. Already experts disagree on exactly what effects the move to a system of free enterprise is having on the region. Some geographers and economists point to the region's high inflation, rising crime rates, and monthly wages equivalent to less than 50 U.S. dollars. These experts claim that economic conditions in the region have in fact worsened since the Soviet Union's collapse.

For example, a textile factory in the Russian city of Ivanovo once produced one-quarter of the entire Soviet Union's fabric needs. Since the Soviet Union's collapse, however, the factory has had a shortage of raw cotton, as its links to the cotton-growing Central Asian republics have broken. Also, the Soviet government bought much of the factory's cloth. When the breakup took place, the factory lost its major customer. Of the 6,500 workers employed by the factory at one time, only 2,000 workers were needed by the beginning of 1993. At that time, each earned about $12 a month.

Yet other geographers and economists point out that while wages are low by Western standards, basic goods and services, such as heat and public transportation, are much cheaper in Russia than they are elsewhere. These experts also claim that while wages now may be lower than they were several years ago, there was little to buy before the breakup, so higher wages were meaningless. In addition, many Russians have private gardens in which they can raise some of their own foods. This reduces the amount of money they must spend on groceries. While admitting that the former Soviet people are experiencing economic hardships during the transition, many experts believe that in the end the free market will be of much greater benefit than harm.

There are already some people for whom the economic changes have been undeniably positive. One "success story" has to do with housing. During the Soviet era, most people lived in cramped apartments, and the government declared that private homes could be no larger than 20 feet by 20 feet (6 m by 6 m).

Today, in places such as Desna, a village not far from Moscow, more land is becoming available for private ownership. People are starting to build their own—larger—homes. Construction is often expensive and usually must start from scratch, including the building of access roads and the installation of plumbing and utilities. Yet many people believe that home ownership is well worth the costs and the troubles. And as more and more building takes place, it in turn sparks economic growth for many other people and businesses—suppliers of lumber and other building materials, equipment manufacturers, construction workers, real estate agencies, and so on.

Sorting out all of the different issues regarding the economic transition to a free-market economy is not easy. Yet one thing is clear. There are few world regions in which geography and economics are changing as much as, or are more closely linked than, they are in northern Eurasia.

PRACTICING THE SKILL

The economic success of the former Soviet Union is linked not only to the region's own geography but also to the world's geography. Several countries, including the United States, want to show their support for the former Soviet Union's economic and political shifts.

Research the issue of international economic support to the former Soviet Union. What forms has the support taken? What are the major debates among experts regarding economic assistance to the region? Summarize your findings in a brief written report.

BUILDING YOUR PORTFOLIO

Individually or in a group, complete one of the following projects to show your understanding of the geography concepts involved.

A *A World of Change*

While the geographic changes we read about in magazines, newspapers, and textbooks may seem far away from our own lives, in reality these changes affect people all around the world. Using the breakup of the Soviet Union as an example, prepare a report on the consequences, both personal and global, and both far and near, of geographic change. To complete your report, consider the following questions.

1. How has the collapse of the Soviet Union affected the daily lives of people who live in the region? Imagine that you live in one of the countries of the former Soviet Union. Write, in the form of a journal entry, about a typical day in your life before the Soviet Union collapsed. Then write another journal entry about a typical day in your life now.
2. How did the collapse of the Soviet Union and the end of the Cold War affect people in other parts of the world? Interview some members of your community about how they personally have been affected by the changes in northern Eurasia. How have their views of the former Soviet Union changed? Did their economic livelihoods change as a result of military base closings or the decline of military industries? Write a summary of your findings from your interviews.
3. How did the collapse of the Soviet Union and the end of the Cold War have an impact on global political and economic relationships? Did political relationships between countries change? How was international trade affected? Present this information using charts or graphs.
4. What areas of the world are likely to experience geographic change within the near future? Select one of the areas, and write one page explaining why the area is likely to change, what forms the change will take, and what consequences the change will have.

Organize your materials and present your report to the rest of your class.

B *Economic Systems: A Closer Look*

The collapse of the Soviet Union and the transition of some of its former republics from communism to a free-market economy has raised many questions about the world's different economic systems—not only communism but also capitalism and socialism. You already know a little bit about these systems but now research them in greater depth. Make a poster-sized chart showing the information you collect on the systems. Your chart will be divided into three rows: one for capitalism, one for communism, and one for socialism. The chart will have four columns, as explained below.

1. The first column should be called "Origins." For each economic system, investigate how it arose. Where did each system originate? What groups or individuals played a major role in the origins and development of each system, or in the ideas behind it? What social and economic conditions set the stage for the formation of each system and its ideas?
2. The chart's second column should discuss the "Key Ideas" of the theory behind each economic system. On what assumptions is each system based? What does each system seek to do, and what are its methods? What other ideas are important in each system?
3. The third column, "Practice and Reality," should discuss examples of how the system has worked in practice. In which countries has each system been tried? How closely do the systems in practice resemble the systems in theory (as described in the chart's second column)? What challenges has each system faced? In places where any of the systems have failed, what were the reasons for the failure?
4. The fourth column should be called "Interaction with Politics." Although each system is technically an economic system, economics and politics are often closely related. What roles do governments play in each system? How does each system involve political processes?

Present your chart to the rest of your class.

Facing the Past and Present

Patterns of trade and culture can change quickly in our modern world. For example, until just a few decades ago, the United States looked primarily to Europe for trade. Many innovations in culture and technology and most of America's immigrants came from Europe. Today, the American focus on Europe has faded. U.S. trade with Japan and other Pacific Rim nations now far exceeds U.S. trade with Europe. New ideas, technology, and immigrants to the United States now come from all around the globe.

Central Asia

With the collapse of the Soviet Union, similar changes are taking place in Central Asia. In the past, many ties linked Central Asia to the Soviet Union. For instance, the economies of the two regions were integrated. Central Asia exported cotton and oil to Russia and the countries of Eastern Europe. In exchange, Central Asia received a variety of manufactured goods from these areas. The Soviet Union also heavily influenced the culture of Central Asia, and many Central Asians learned to speak Russian.

Looking South

Today, Central Asia's links to the former Soviet Union have weakened. At the same time, its ancient ties to Southwest Asia have grown stronger. The Silk Road, and the travelers that came before it, formed valuable trade links between the oasis cities of Central Asia and the busy Southwest Asian ports on the Mediterranean. Now the peoples of Central Asia are looking southward once again. For example, links are forming between the region's new republics and Turkey. Many people in Central Asia are traditionally Turkic in culture and language. As a result, Turkey's business leaders are working to expand their industries in Central Asia. Regular air travel from Turkey to the capitals of Central Asia now is possible as well.

Religion provides another firm link between Central Asia and Southwest Asia. Islam first arrived in Central Asia in the eighth century, but its role diminished during the Soviet era. Missionaries from the Arab countries and Iran are now committed to reestablishing this connection. In fact, Iran is spending millions of dollars to build roads and rail lines to Central Asia.

Central Asia and Southwest Asia also share a similar climate, environment, and way of life. In both of these dry regions, water conservation and irrigation are basic to daily life. Many of both regions' people grow cotton and herd animals, though this way of life may soon change. Both Central Asia and Southwest Asia are experiencing an invasion of modern Western culture and are struggling to deal with its impact. Many people believe that the ideas and trends that arrive daily through compact discs, videotapes, and satellite television are a threat to their traditional beliefs and ways of life. Such shared fears of cultural loss are bringing Central Asia and Southwest Asia even closer together.

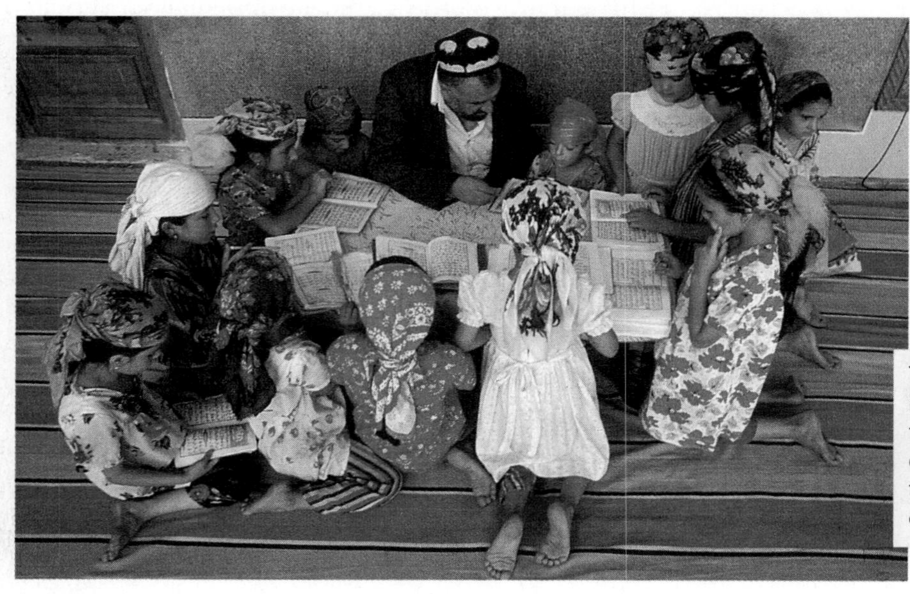

These children are learning about Islam in **Dushanbe, Tajikistan**. Although the former Communist government discouraged the practice of religion, today Islam flourishes in the independent Central Asian republics.

Language Groups of Southwest and Central Asia

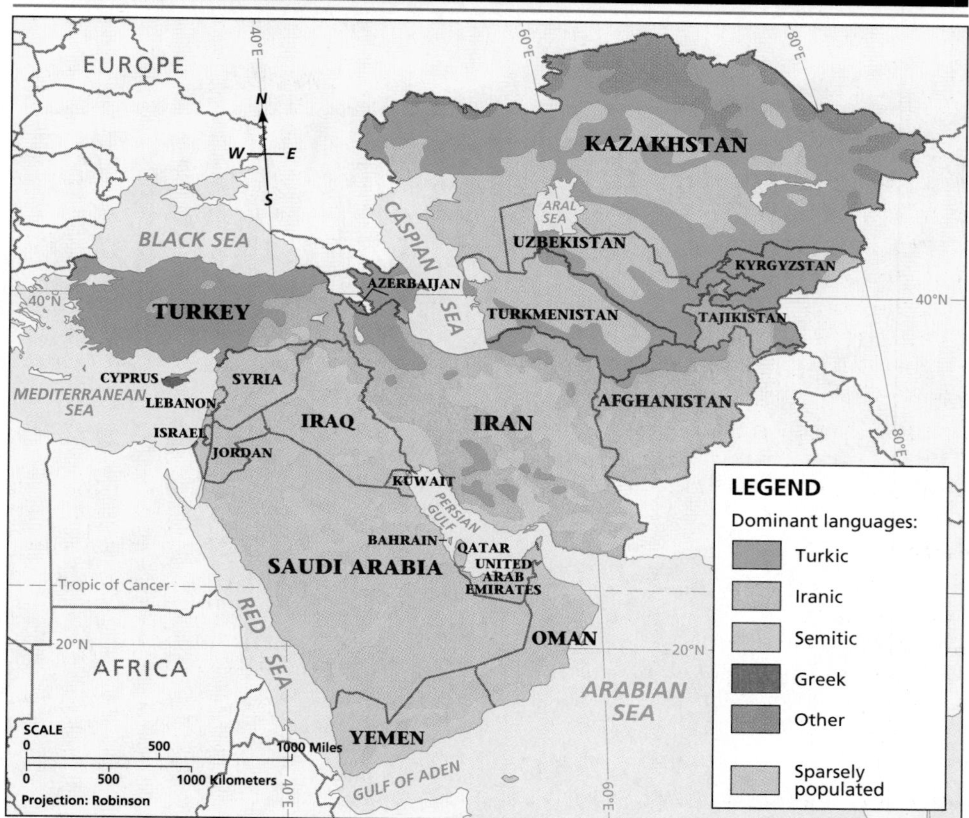
EUROPE
KAZAKHSTAN
BLACK SEA
CASPIAN SEA
ARAL SEA
UZBEKISTAN
KYRGYZSTAN
AZERBAIJAN
TURKEY
TURKMENISTAN
TAJIKISTAN
CYPRUS
SYRIA
MEDITERRANEAN SEA
LEBANON
ISRAEL
JORDAN
IRAQ
IRAN
AFGHANISTAN
KUWAIT
PERSIAN GULF
BAHRAIN
QATAR
UNITED ARAB EMIRATES
SAUDI ARABIA
Tropic of Cancer
RED SEA
OMAN
AFRICA
ARABIAN SEA
YEMEN
GULF OF ADEN
40°E
60°E
80°E
40°N
20°N
SCALE
0 500 1000 Miles
0 500 1000 Kilometers
Projection: Robinson
LEGEND
Dominant languages:
Turkic
Iranic
Semitic
Greek
Other
Sparsely populated

MAP STUDY This map uses colors to show the major language groups that link ethnic peoples throughout Southwest and Central Asia. Turkic languages, such as Azerbaijani and Uzbek, and Iranic languages, such as Kurdish and Persian, are spoken by ethnic peoples throughout Central Asia and northern Southwest Asia. Jews and Arabs also are linked by language: both Hebrew and Arabic are Semitic languages. Very often, however, the links between ethnic groups are overshadowed by the differences in culture and history.

Defining the Region

For all these reasons, geographers are constantly reexamining this and other regions of the world. Will they decide to include the new countries of Central Asia in the region of Southwest Asia? Or will Russia regain control of Central Asia and once again push aside the region's ancient economic and cultural links to Southwest Asia? The geographers are watching and waiting.

In the tradition of many Turkic communities, these community elders are meeting to discuss local schools in **Tajikistan**. The Central Asian republics share many traditions with both northern Eurasia and Southwest Asia.

UNIT 7

SOUTHWEST ASIA

The desert dunes of the Rub' al-Khali (Empty Quarter) roll toward the horizon in **Saudi Arabia**. Dry desert and steppe climates dominate much of Southwest Asia.

UNIT 7 CHAPTERS

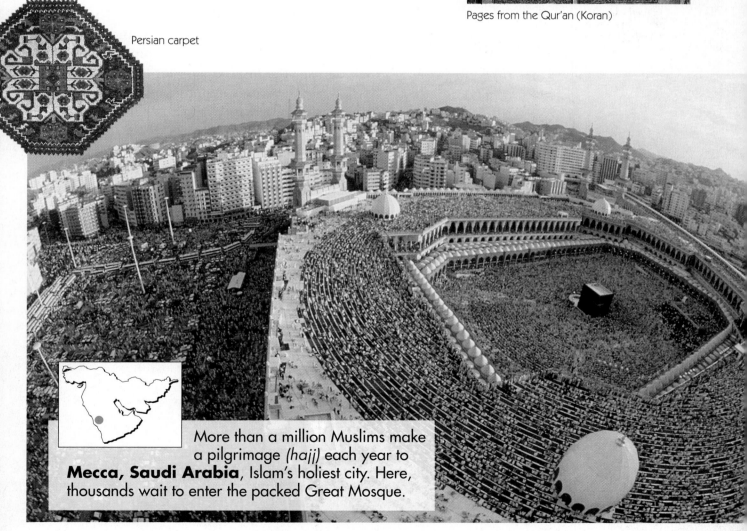

The sun sets over **Izmir**, along the Aegean coast of **Turkey**. The towers of a mosque rise at left-center. Although Turkey has historical ties to Christianity, nearly all Turks today are Muslim.

Pages from the Qur'an (Koran)

Persian carpet

More than a million Muslims make a pilgrimage *(hajj)* each year to **Mecca, Saudi Arabia**, Islam's holiest city. Here, thousands wait to enter the packed Great Mosque.

One World, Many Voices

LITERATURE

Khalil al-Fuzay (1944–) is a journalist and short story writer from Saudi Arabia. The selection below is from a story entitled "Scattered Voices." It reflects aspects of Saudi Arabian culture and traditional ways of life in the coastal villages on the Persian Gulf. Decades before this country's economy was transformed by the discovery of vast oil reserves, the waters around southern Saudi Arabia, Kuwait, and the island of Bahrain were noted for their many pearl-producing oyster beds.

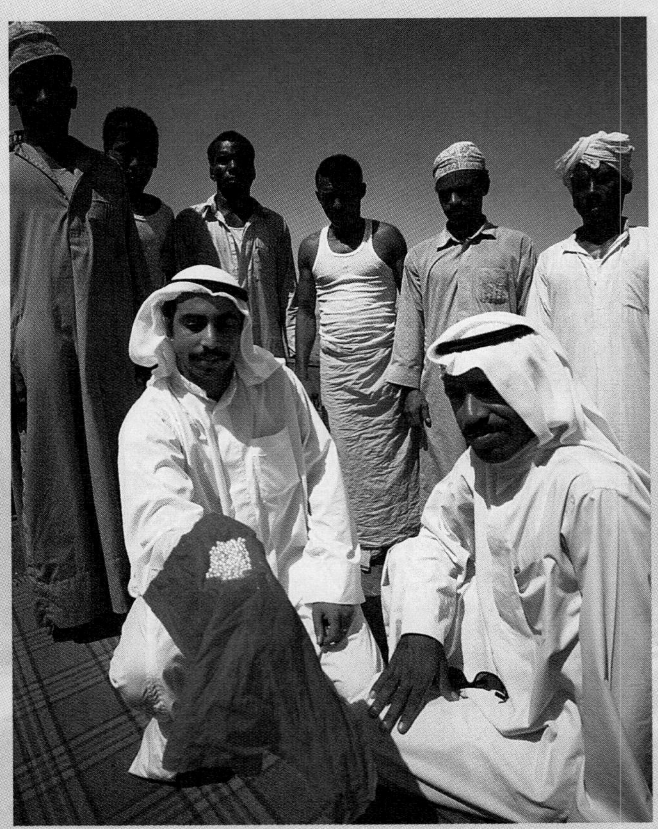
Pearls from offshore **Bahrain**

"The peace-loving village keeps a hold on hope at all times. When the men set off for the pearl-hunt, the women send them off with prayers; and they welcome them back with sky-piercing ululations [wailing]. The traditional diving-songs place no value on composure: at the return of his son, the old man dances; the mother sings at the return of hers, while the wife's happiness is such that it brings tears to her eyes. This happens when they come back with plenty of pearls, but also when they come back with nothing. Reunion was always a joyous occasion.

After sunset the men gather in a semicircle. Their dark faces flicker in the light of the fire at which they warm their drums and tambourines. The communal singing begins. It is exquisite [beautiful] in both its joy and its sorrow. The dancers dance themselves into a frenzy. In the pauses between songs, you can hear the sounds of the Gulf waves as they enfold the golden sands of the shore."

Translated by Ahdaf Soueif and Thomas G. Ezzy

Interpreting Literature

1. How do the people in this story make their living?
2. What details in this selection indicate the village's location near the Persian Gulf?

FOR THE RECORD

190 MILES To CHINA
To RUSSIA 600 MILES
200 MILES TO TIBET

William O. Douglas (1898–1980) was a justice on the United States Supreme Court for more than 35 years. He was also a noted conservationist, outdoorsman, and world traveler. The passage below is from *Strange Lands and Friendly People*, a book about his travels throughout Southwest Asia. The beautiful cedars of Lebanon he writes about have been described by many writers since biblical times. Although most of Lebanon's forests were cut centuries ago, some of the cedars can still be found in the northern Lebanon Mountains.

"We had walked four days under a sweltering sun. There was no cover of any kind—for man, birds, or beasts. The Mediterranean sky was cruel. When at last I reached the cedars at Hadeth, . . . I lay on my back under the first cedar and thanked God for trees. I felt as if I had reached home, as if protective arms were around me. I had escaped an enemy that pursued me with a hot breath. I, a refugee from the sun who was burned to a frazzle, was at last safe. . . .

These cedars rise a hundred feet or more. While the trunks are usually four or five feet thick, like our tamarack [a North American tree], some are over thirty feet in diameter. These are the monarchs of the forest, perhaps a thousand years old.

The branches are long and swooping and cover the ground for fifty feet or more around. A forest of them is so closely woven with branches that it is difficult to get a picture of a single tree. At Bsharreh there are only about four hundred cedars left, packed together in a small ten-acre tract, the remnants of a mighty race."

Analyzing Primary Sources

1. What connection is shown in this selection between the region's climate and admiration for the cedars of Lebanon?
2. Why does the author believe these trees are important?

CHAPTER 32

Introducing Southwest Asia

Jewish menorah

Southwest Asia is the crossroads where Asia, Europe, and Africa meet. The region stretches from Turkey and the eastern shore of the Mediterranean to the high plateaus of Iran (i-RAN), and south to the Indian Ocean coast of the Arabian Peninsula. Southwest Asia is often called the Middle East, a term that originated with the European colonial powers. The Europeans viewed the region as midway to the Pacific shores of eastern Asia, which was called the Far East.

The ancient foundations of both farming and city life were laid in Southwest Asia. This is where people first learned to read and write, and where societies first built great monuments. Three of the world's major religions—Judaism, Christianity, and Islam—first appeared here as well. Today, the discovery of oil, modernization, and the growing influence of Western cultures are changing traditional ways of life in Southwest Asia.

GEOGRAPHY DICTIONARY

- *exotic river*
- *furrow*
- *fossil water*
- *Judaism*
- *Christianity*
- *Islam*
- *Muslim*
- *Qur'an*
- *monotheism*
- *secularism*
- *drip irrigation*
- *Bedouin*

Irrigation canal in **Jordan**

MAPPING THE REGION

SOUTHWEST ASIA

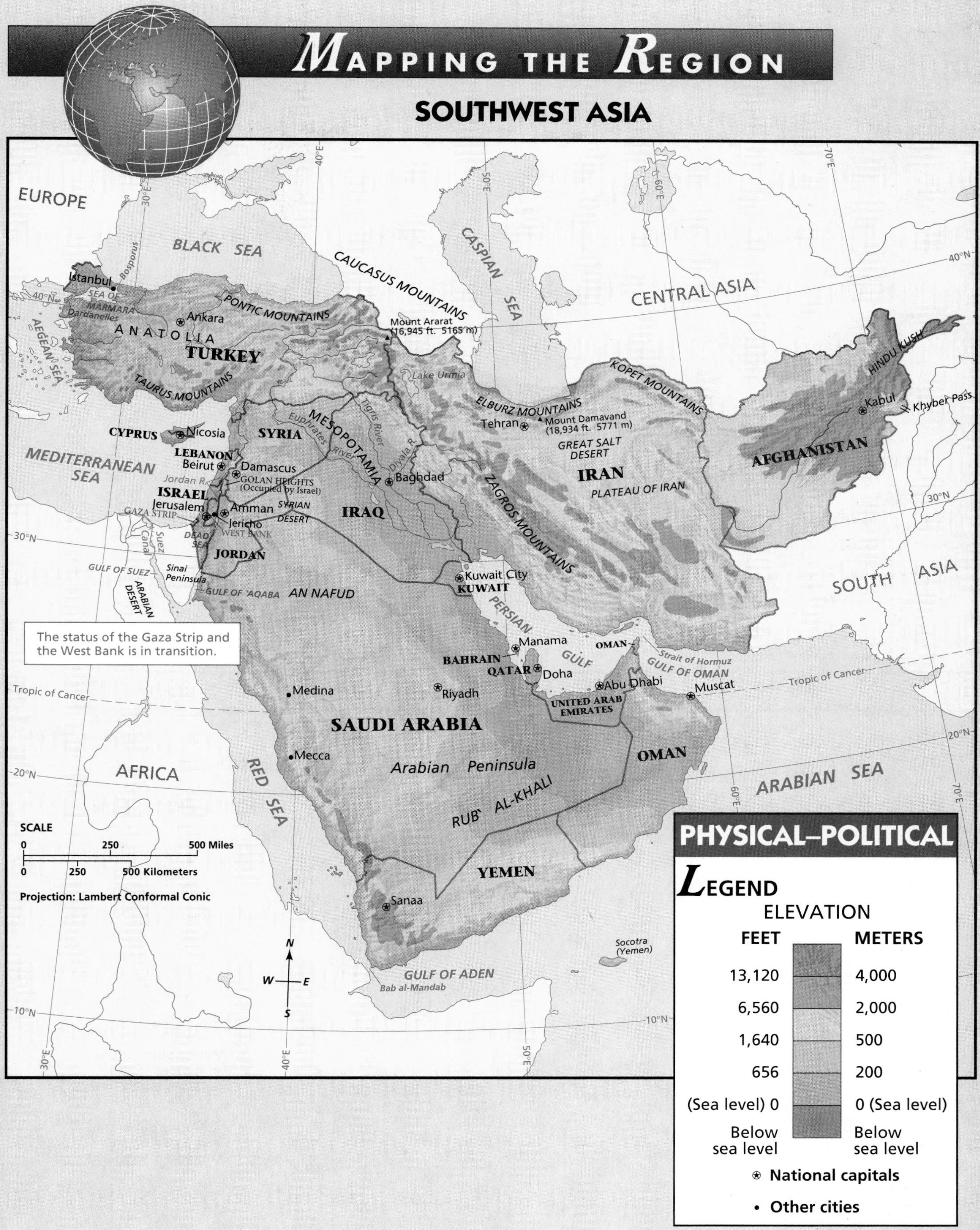

EUROPE
BLACK SEA
CASPIAN SEA
CAUCASUS MOUNTAINS
CENTRAL ASIA
Istanbul
Bosporus
SEA OF MARMARA
Dardanelles
AEGEAN SEA
PONTIC MOUNTAINS
Ankara
ANATOLIA
TURKEY
TAURUS MOUNTAINS
Mount Ararat (16,945 ft. 5165 m)
Lake Urmia
KOPET MOUNTAINS
HINDU KUSH
ELBURZ MOUNTAINS
Tehran
Mount Damavand (18,934 ft. 5771 m)
GREAT SALT DESERT
Kabul
Khyber Pass
AFGHANISTAN
CYPRUS
Nicosia
SYRIA
MESOPOTAMIA
Euphrates River
Tigris River
Diyala R.
MEDITERRANEAN SEA
LEBANON
Beirut
Damascus
Jordan R.
GOLAN HEIGHTS (Occupied by Israel)
Baghdad
IRAN
PLATEAU OF IRAN
ZAGROS MOUNTAINS
ISRAEL
Jerusalem
GAZA STRIP
Amman
Jericho
WEST BANK
SYRIAN DESERT
IRAQ
DEAD SEA
Suez Canal
JORDAN
GULF OF SUEZ
Sinai Peninsula
ARABIAN DESERT
GULF OF 'AQABA
AN NAFUD
Kuwait City
KUWAIT
SOUTH ASIA
PERSIAN GULF
Manama
BAHRAIN
OMAN
QATAR
Doha
Strait of Hormuz
GULF OF OMAN
Abu Dhabi
UNITED ARAB EMIRATES
Muscat
Tropic of Cancer
Medina
Riyadh
SAUDI ARABIA
Mecca
AFRICA
RED SEA
Arabian Peninsula
OMAN
ARABIAN SEA
RUB' AL-KHALI
YEMEN
Sanaa
Socotra (Yemen)
GULF OF ADEN
Bab al-Mandab
N W E S
40°N 30°N 20°N 10°N 30°E 40°E 50°E 60°E 70°E

The status of the Gaza Strip and the West Bank is in transition.

SCALE

0 — 250 — 500 Miles

0 — 250 — 500 Kilometers

Projection: Lambert Conformal Conic

PHYSICAL–POLITICAL

LEGEND

ELEVATION

FEET	METERS
13,120	4,000
6,560	2,000
1,640	500
656	200
(Sea level) 0	0 (Sea level)
Below sea level	Below sea level

⊛ **National capitals**

• **Other cities**

MAPPING THE REGION

SOUTHWEST ASIA

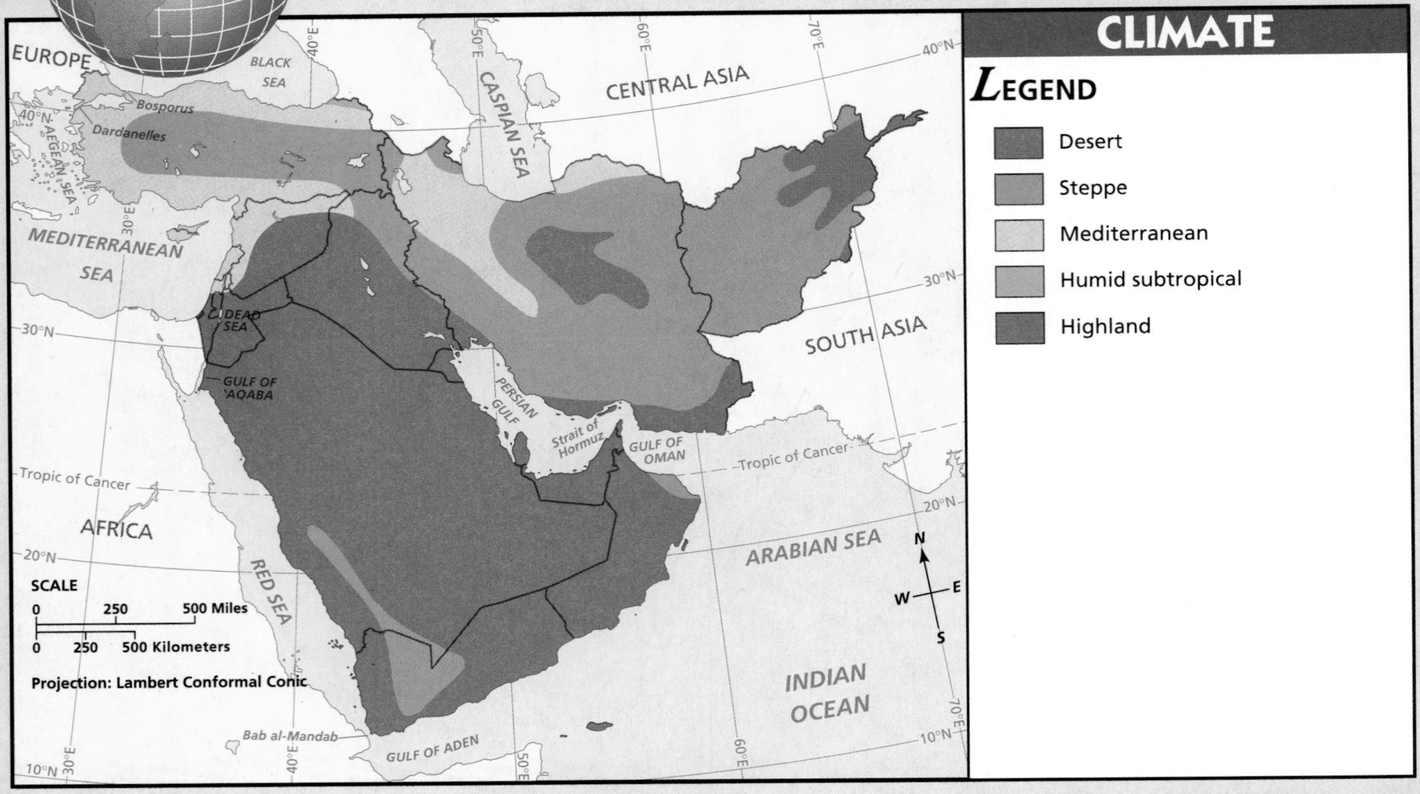

CLIMATE

LEGEND

- Desert
- Steppe
- Mediterranean
- Humid subtropical
- Highland

EUROPE
BLACK SEA
Bosporus
Dardanelles
AEGEAN SEA
MEDITERRANEAN SEA
CASPIAN SEA
CENTRAL ASIA
SOUTH ASIA
DEAD SEA
GULF OF AQABA
PERSIAN GULF
Strait of Hormuz
GULF OF OMAN
Tropic of Cancer
AFRICA
RED SEA
ARABIAN SEA
INDIAN OCEAN
Bab al-Mandab
GULF OF ADEN
40°N
30°N
20°N
10°N
30°E
40°E
50°E
60°E
70°E

SCALE
0 250 500 Miles
0 250 500 Kilometers
Projection: Lambert Conformal Conic

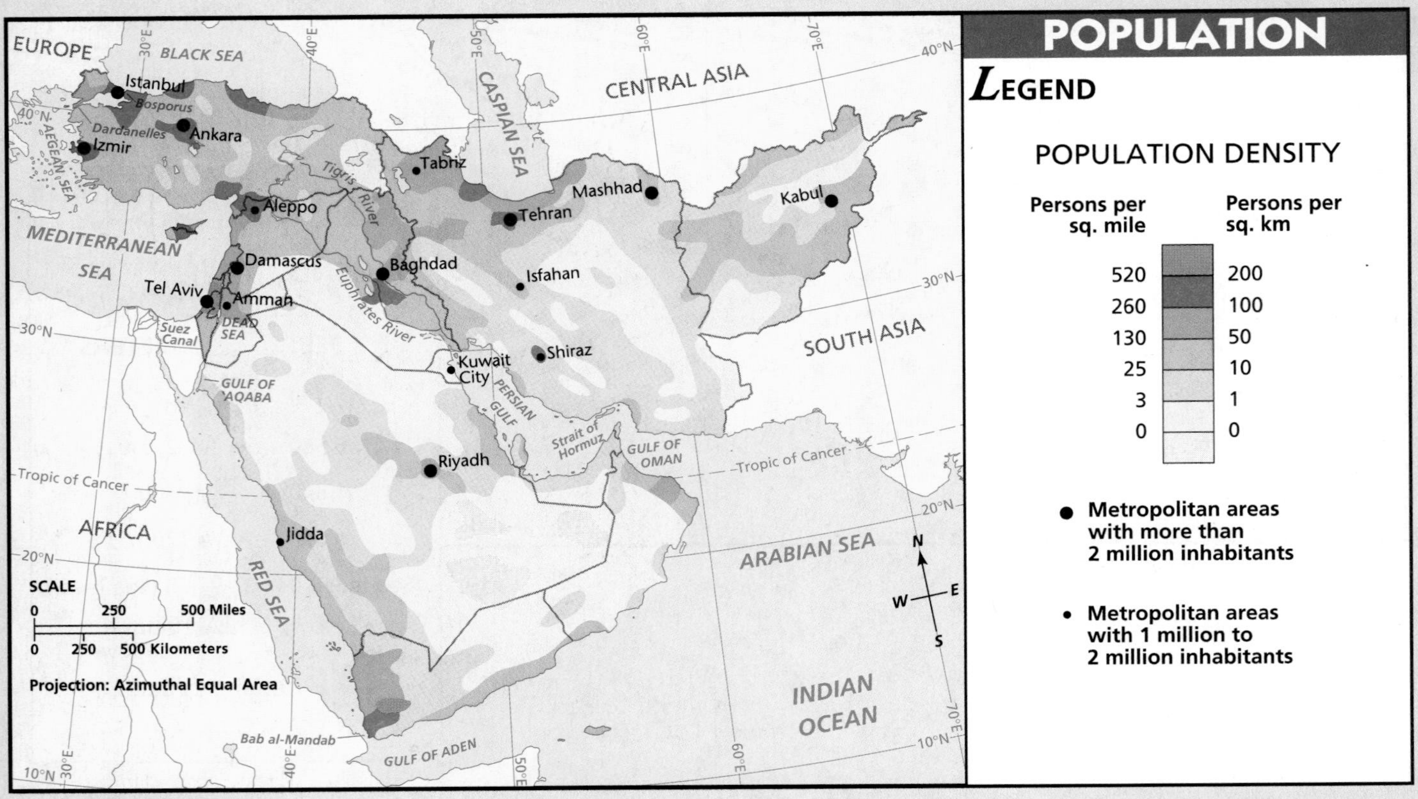

POPULATION

LEGEND

POPULATION DENSITY

Persons per sq. mile	Persons per sq. km
520	200
260	100
130	50
25	10
3	1
0	0

- ● Metropolitan areas with more than 2 million inhabitants
- • Metropolitan areas with 1 million to 2 million inhabitants

EUROPE
BLACK SEA
Istanbul
Bosporus
Dardanelles
Ankara
Izmir
AEGEAN SEA
Aleppo
MEDITERRANEAN SEA
Damascus
Tel Aviv
Amman
Suez Canal
DEAD SEA
Tigris River
Euphrates River
Baghdad
Tabriz
Tehran
Mashhad
Isfahan
Shiraz
Kabul
Kuwait City
CASPIAN SEA
CENTRAL ASIA
SOUTH ASIA
GULF OF AQABA
PERSIAN GULF
Strait of Hormuz
GULF OF OMAN
Riyadh
Jidda
Tropic of Cancer
AFRICA
RED SEA
ARABIAN SEA
INDIAN OCEAN
Bab al-Mandab
GULF OF ADEN
40°N
30°N
20°N
10°N
30°E
40°E
50°E
60°E
70°E

SCALE
0 250 500 Miles
0 250 500 Kilometers
Projection: Azimuthal Equal Area

MAPPING THE REGION

SOUTHWEST ASIA

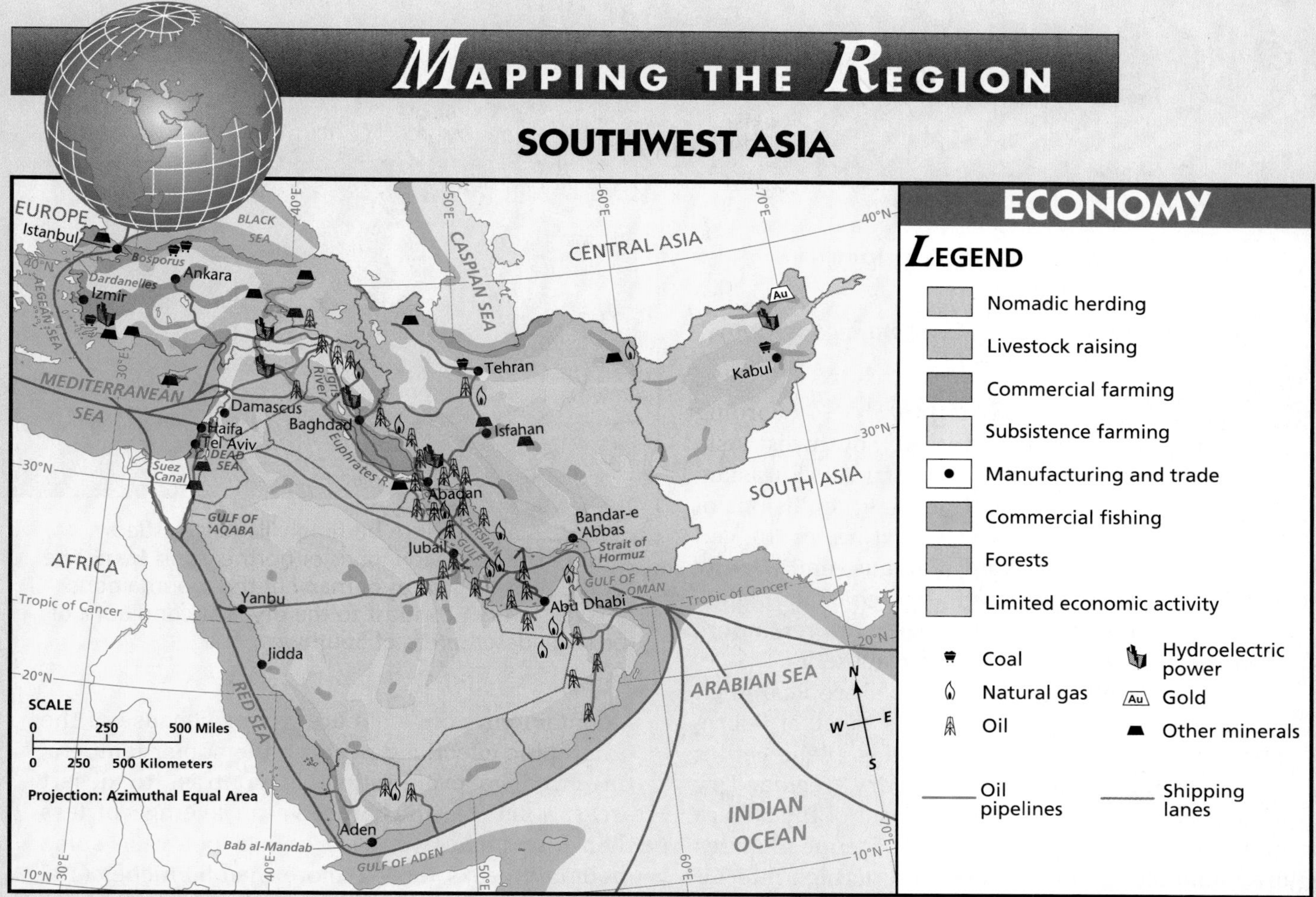
ECONOMY
LEGEND
Nomadic herding
Livestock raising
Commercial farming
Subsistence farming
Manufacturing and trade
Commercial fishing
Forests
Limited economic activity
Coal
Natural gas
Oil
Hydroelectric power
Gold
Other minerals
Oil pipelines
Shipping lanes
EUROPE
Istanbul
BLACK SEA
Bosporus
Dardanelles
Ankara
Izmir
AEGEAN SEA
MEDITERRANEAN SEA
CASPIAN SEA
CENTRAL ASIA
Tehran
Kabul
Au
Damascus
Haifa
Tel Aviv
DEAD SEA
Baghdad
Tigris River
Euphrates R.
Isfahan
Suez Canal
Abadan
SOUTH ASIA
GULF OF AQABA
Bandar-e 'Abbas
Strait of Hormuz
Jubail
PERSIAN GULF
AFRICA
GULF OF OMAN
Abu Dhabi
Yanbu
Tropic of Cancer
Jidda
ARABIAN SEA
RED SEA
SCALE
0 250 500 Miles
0 250 500 Kilometers
Projection: Azimuthal Equal Area
N
S
E
W
Aden
Bab al-Mandab
GULF OF ADEN
INDIAN OCEAN
40°N
30°N
20°N
10°N
30°E
40°E
50°E
60°E
70°E

CLIMATE

◆ Desert and steppe climates are found throughout most of Southwest Asia.

◆ A Mediterranean climate dominates the coastal areas along the Mediterranean Sea and parts of northern Iran.

◆ Afghanistan lies at the edge of a large highland climate region that spreads east across Asia.

? Which climate types are found in Southwest Asia?

Compare this map to the physical–political map on page 375.

? Which climate type dominates nearly all of the Arabian Peninsula?

? Which climate types are found in Israel?

POPULATION

◆ Some of the most densely populated areas of Southwest Asia are along the Mediterranean coast and along the Tigris and Euphrates rivers.

◆ The Arabian Peninsula, despite large sparsely populated areas, has several cities with more than 1 million inhabitants.

◆ Most of the largest cities of Southwest Asia are in the northern and western parts of the region.

? Which Saudi Arabian cities have populations of more than 1 million?

Compare this map to the physical–political map on page 375.

? Name the physical region where the population of the Arabian Peninsula is the least densely populated.

Compare this map to the climate map on page 376.

? Which climate types dominate the more densely populated northern and western regions of Southwest Asia?

ECONOMY

◆ Oil production is the major economic activity for many Southwest Asian countries, particularly those along the Persian Gulf and in Mesopotamia.

◆ Pipelines carry oil across large areas of Southwest Asia to ports on the Persian Gulf and the Mediterranean Sea.

◆ The irrigated lands along the Tigris and Euphrates rivers form an important agricultural region.

? Why do you think many manufacturing and trade centers are located along the Persian Gulf?

? Name the major manufacturing and trade centers shown along the Mediterranean coast.

Compare this map to the climate map on page 376.

? How do you suppose most agriculture can be successful in some of the dry regions of Southwest Asia?

PHYSICAL AND ECONOMIC GEOGRAPHY

FOCUS

- *What is the physical geography of Southwest Asia?*
- *What is the region's economic geography?*

REGION **Physical Geography** According to the theory of plate tectonics, Southwest Asia lies at the intersection of the African, Eurasian, and Arabian plates. The grinding collision of these plates has created a mixture of rugged mountains, valleys, upland plateaus, and narrow gulfs and seas bordered by coastal plains. Frequent earthquakes in the region are reminders of the earth's continuing tectonic activity.

As a result of prevailing subtropical high pressure, a desert climate covers most of Southwest Asia. In fact, Southwest Asia is at the center of the world's largest arid, or dry, region. Dry lands stretch from the Atlantic coast of the Sahara in North Africa to western China. Steppe climates are found along desert edges and on high interior plateaus. Winter rains are common in these areas. A Mediterranean climate dominates the coastal areas along the Mediterranean Sea.

In many of the desert areas, rain falls only a few times each year, and surface water is rare. People can live in the desert only where there is water. Springs, where water rises naturally from cracks in the surrounding rock, are an important water resource, often forming oases. Over the years, many springs in the region have been made into productive wells. Water can also be found beneath dry riverbeds. The arid lands between these sources of water are usually uninhabited, visited only by migrating herders.

The desert traveler can suffer from both extreme heat and cold. The sun is able to heat the ground quickly because the skies are clear and there is little vegetation for shade. Afternoon summer temperatures can range from 100°F to 120°F (38°–49°C). During the night, however, the temperature can drop more than 30°F (17°C). Frosts occur often in winter.

Mountains receive most of the region's precipitation, both as summer rain and winter snow.

A herding village is nestled within the Kopet Mountains of northeastern **Iran**. The cooler, wetter climates of many of the mountainous regions stand in contrast to the dry, hot conditions of the large desert areas of Southwest Asia.

Mountain environments are wetter because of the orographic effect and because the cooler temperatures reduce evaporation. In contrast to desert areas, some of which receive an average of less than four inches (10 cm) of rain per year, some mountain peaks receive more than 50 inches (127 cm) of rain each year. The streams that drain these highlands provide water to the dry valleys below. Streams that flow from humid environments across deserts into seas are called **exotic rivers**.

INTERACTION **Agriculture** Despite arid climates and rugged landforms that make farming difficult, agriculture is a major occupation in much of Southwest Asia. Fertile soils are found only in the river valleys, on high mountain plateaus, and in a few oases. In many of the wetter parts of the region, overgrazing has led to increased soil erosion. The few mountain forests have been cut to near destruction as well. The countries of the region have begun attempts at conserving their soil and plant resources.

Most farmers raise only enough crops and animals to support their own families. Any extra food and animal products are sold for cash at city markets. Barley and wheat are the important grains in the area, while citrus fruits, olives, figs, nuts, and grapes are grown in the Mediterranean coastal areas. Livestock, mostly sheep, goats, and some cattle, is raised on some farms.

Commercial agriculture is practiced throughout Southwest Asia but is widespread only in Israel (IZ-ree-uhl), where citrus fruits and other crops are grown for sale to European markets. The other countries of the region are developing commercial agriculture, particularly to provide fruits and vegetables for local markets.

Irrigation first developed in Southwest Asia, and it has steadily expanded over the centuries. In the mountains, most irrigation developments are small in scale. Often, water is simply diverted into a small canal or **furrow** (narrow trench) that leads directly to a field. Large-scale irrigation usually depends on an exotic river such as the Tigris (TIE-gruhs) or Euphrates (yoo-FRAY-teez), both of which begin in the mountains of Turkey. Wide, deep canals lead away from the exotic rivers. Smaller and smaller canals then lead into individual fields. These large irrigation systems are much larger than one community can maintain.

In many places, deep wells can provide irrigation. Such wells require modern drilling technology and powerful pumps. The groundwater in these wells is usually **fossil water**, or water that is not being replaced by rainfall. Eventually, overuse will deplete this water source.

Resources and Industry Many people think of oil when they think of Southwest Asia. Indeed, oil is the region's richest mineral resource. The main oil deposits lie along the shores of the Persian Gulf (also called the Arabian Gulf) and in Iraq (i-RAHK).

Since many of the oil-rich countries have small populations and little industry, most of their oil is available for export. As a result, the economies of the oil-exporting countries rise and fall with the price of oil. When oil prices fall, unemployment and business failures increase in these countries. To have more control over world oil prices, the oil-rich countries of the region joined with other oil-rich countries around the world to form the Organization of Petroleum Exporting Countries (OPEC). OPEC was quite successful in the 1970s, but its effectiveness has since declined. Nevertheless, world demand for oil has continued to rise. Large profits should flow into the oil-rich countries as long as their enormous oil reserves last.

Other than oil, the countries of Southwest Asia have few resources that are important for developing industry. Turkey does have large deposits of coal, iron ore, and chromium, a metal used to coat and protect other metals. Iran has a copper industry, and the Dead Sea on the border between Israel and Jordan contains salts that are used in the chemical industry.

Of the countries of Southwest Asia, only Israel is considered a developed country. The oil-rich countries have high incomes, but their industry is limited mainly to oil refining and related chemicals. Most other industries in the region center on food processing and consumer goods.

SECTION REVIEW

1. What climate type dominates most of Southwest Asia?
2. What crops are grown in Southwest Asia? What is the region's most valuable resource?
3. **Critical Thinking** **Imagine** that you will be traveling through a desert in Southwest Asia. Given the desert environment, what items might you want to bring with you?

HISTORICAL AND HUMAN GEOGRAPHY

FOCUS

- *What were some of Southwest Asia's ancient civilizations, and how were they significant? What powers have controlled the region in its history?*
- *What religious and ethnic groups live in Southwest Asia?*
- *What challenges face the region?*

Historical Geography The world's first civilizations developed in the area known as the Fertile Crescent. The Fertile Crescent is a wide arc of productive land that runs along the eastern shore of the Mediterranean Sea, through the plains along the Tigris and Euphrates rivers, to the Persian Gulf. The ancient name for these plains is Mesopotamia. Many of the plants and animals found on farms throughout the world today were first domesticated in the Fertile Crescent.

The period from about 3500 B.C. to 600 B.C. saw the rise and fall of many important civilizations in Southwest Asia. Each culture left a part of its knowledge and way of life for later civilizations. In Mesopotamia, the Sumerians developed writing. Later, the Hittites developed iron weapons. Along the Mediterranean shore, Phoenician merchants formed a large trading network.

After 600 B.C., most of the region was controlled by one empire after another. These included the Persian, Macedonian (Greek), and Roman empires. By A.D. 800, Muslim armies had conquered most of Southwest Asia and had ventured far to the east and west. Following these conquests, Southwest Asia divided into many small empires. A political and religious prince known as either a caliph, emir, or sultan ruled each territory.

In Turkey, an influential leader named Osman founded the Ottoman Empire in about 1300. In 1453, the Ottomans captured the city of Constantinople, now called Istanbul, and made it the empire's capital. At its peak in the 1600s, the Ottoman Empire stretched in the east from the Persian Gulf to the Caucasus, and in the south along the eastern Mediterranean, around the Red Sea, and across much of North Africa. In the north, the empire included Europe's Balkan Peninsula and Crimea.

As Ottoman power declined in the 1700s and 1800s, the powers of Europe expanded their control over the region. The Ottoman Empire finally collapsed at the end of World War I in 1918. Some parts of the empire then became colonies of France, Britain, and Italy. The Europeans were interested in the natural resources and transportation routes of Southwest Asia. They guarded these important shipping routes through their military bases in the region.

Most of the present-day countries of Southwest Asia achieved their independence during the past 60 years. Today, the leaders of these countries vary from traditional kings, sheiks, and emirs to dictators. There also are a few democracies. Since independence, nationalism and modernization have been the main forces shaping the cultures of the region.

This tablet honors the founding of the Babylonian Sun Temple in the ninth century B.C. near the present-day city of **Baghdad, Iraq**.

Religion Three of the world's major religions originated in Southwest Asia. **Judaism**, the religion of the Jews, appeared during the period between 3500 B.C. and 600 B.C. in a region called Palestine. Judaism was the first major religion to center around the belief in a single god.

Christianity developed out of Judaism and spread during the Roman era. It is a religion founded on the teachings of Jesus Christ, whom Christians believe was the son of God. Although the Romans persecuted Christians for many years, Christianity eventually became the main religion of the Roman Empire. After the Roman Empire divided into the Eastern Roman Empire and the Western Roman Empire in the fourth century, Christianity split into the Eastern Orthodox Church, based in Constantinople, and the Roman Catholic Church, based in Rome. In the 1500s, another group of European Christians, called Protestants, broke away from the Roman Catholic Church. Today, Christianity in its many variations is the dominant religion of the Western Hemisphere.

Islam was founded by the prophet Muhammad (moe-HAM-uhd), who lived from about A.D. 570 to 632. Muhammad worked as a merchant in Mecca, a trading city in western Saudi Arabia. When he was about 40 years of age, he often went to nearby mountains to think about the social injustices in Mecca. Muhammad said that during one of his trips, he was told by the angel Gabriel, a messenger of God, to preach religion to the Arabs.

After its founding, Islam spread rapidly. In 10 years, Muhammad was able to unite the peoples of western Saudi Arabia, and within a century of his death, Islam had spread much farther. A person who follows Islam is called a **Muslim**.

The words that Muhammad received from God (whom Muslims call Allah) were recorded by his followers in the **Qur'an** (Koran), Islam's sacred book of writings. Islam remains a unifying culture trait of Southwest Asia.

Islam shares many of the traditions of Judaism and Christianity. One of these shared traditions is **monotheism**, or the belief in one god. Muslims believe in one god, Allah. The Qur'an also has several passages similar to the Old Testament of the Bible. Like Christianity, however, Islam has split over the years into different groups. This division has resulted from varying interpretations of the Qur'an and from questions of leadership throughout the Islamic world. There are two major branches of Islam, the Sunni and the Shi'a (SHEE-ah). The Sunni emphasize the basic Islam of the Qur'an and personal prayer. The Shi'a follow imams (i-MAHMS), who are religious as well as political leaders.

Today, about 90 percent of Muslims are Sunni, and 10 percent are Shi'a. The Shi'a are concentrated in Iran, southern Iraq, Yemen, and Lebanon (LEHB-uh-nuhn). The Shi'a are particularly powerful in Iran and hope to spread their political ideas throughout Southwest Asia.

In recent years, Islamic fundamentalism has grown. Muslim fundamentalists stress family unity and the importance of the oldest male family member. They also wish to reduce foreign influences on the region. Equal opportunity for women and the independence of children are seen as a reflection of Western culture and thus are considered wrong. Similarly, **secularism**, an indifference to or rejection of religion, is considered evil.

Some Muslim leaders teach that religion is more important than laws and customs created by governments. This belief, however, conflicts with the growing nationalism throughout Southwest Asia. Sorting out these cultural issues is difficult because Islam has no single powerful leader. The Muslim leaders within each country often act independently.

Culture and Conflict As a result of the region's long history of many empires and religions, there are several ethnic groups in Southwest Asia. Arabs are the dominant group in all countries of the region except Turkey, Cyprus, Israel, Iran, and Afghanistan. Nearly all Arabs are Muslims. Therefore, the many Arab countries share several common cultural and religious views. Although both the Turks and the Iranians are Muslim, their histories and cultures are distinctive and non-Arab.

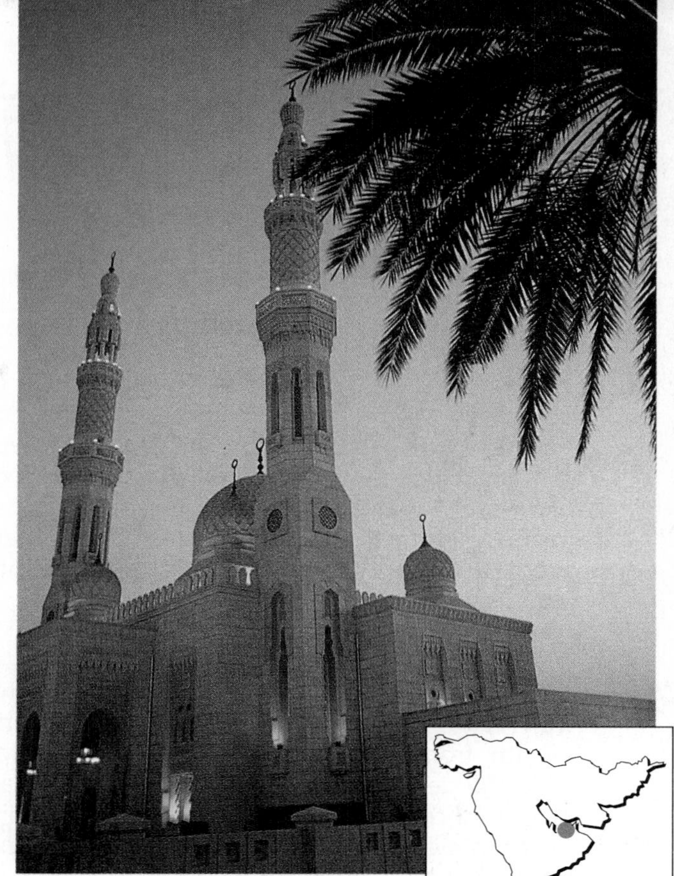

A *muezzin* (moo-EH-zuhn) will call fellow Muslims to prayer from the tower of this mosque in **Dubai, United Arab Emirates**.

The Jews in Israel are another important group in the region. After the spread of Islam in the seventh century, only small communities of Jews remained in Southwest Asia. Then in the late 1800s, Jews from Europe began moving to the ancient homeland in Palestine. In 1948, the Jewish state of Israel formed in Palestine, and Jews from Europe and elsewhere began moving to Israel in large numbers. The people of Israel view themselves as descendants of the ancient Hebrews, whose kingdom thrived here some 3,000 years ago.

Before 1948, the lands of Israel had been the home of the Arab Palestinian people. This Arab-Israeli conflict over homelands has led to several wars and has dominated political issues in Southwest Asia. In 1993 and 1995, Israel and the Palestinians reached major agreements providing for gradual and limited Palestinian self-rule in certain areas. Israel and Jordan signed a peace treaty in 1994, and efforts toward peace in the region have continued.

Text continues on page 384.

The Politics of Water

Did the ancient civilization of Mesopotamia in Southwest Asia disappear because it did not have enough water to survive? Will competition for water rock this already tense region of the world today? Control of the region's water resources is a key political issue. One expert predicts, "In the year 2000, water will be more expensive than oil."

Water Resources

The countries of Southwest Asia obtain water from different sources. In Saudi Arabia, the wealth from the nation's oil resources helps fund the desalinization plants along the Persian Gulf. Without desalinization, there would be little fresh water available for cities or irrigation farming. Many other countries in the region, however, do not have the money to build and run such plants.

Israel, with a vast expanse of desert, depends on several methods to obtain fresh water. Since 1964, much of Israel's water has been supplied by a pipeline called the National Water Carrier, which carries water from the Sea of Galilee to Tel Aviv and down the coastal plain toward the Negev Desert. Israel also relies on desalinization plants and conserves water in agriculture through use of **drip irrigation**. Drip irrigation is a process by which pipes placed near plant roots drip only as much water as the plants need to grow. Even so, droughts and a population explosion, resulting from a large number of refugees immigrating to Israel from the former Soviet Union and elsewhere, have strained the country's water resources.

Even though they border many seas, the countries of Southwest Asia face a severe water shortage.

Droughts and a growing population have strained water supplies to the west in Jordan as well. Even though Jordan and Israel share some water resources, including the Jordan River, political conflicts in the past kept the two countries from reaching agreements to share water effectively. Leaders in Israel and Jordan hope their 1994 peace treaty will help them find joint solutions. In addition, Jordan has begun using drip irrigation methods and is searching for deeper groundwater sources.

Qanat Irrigation System

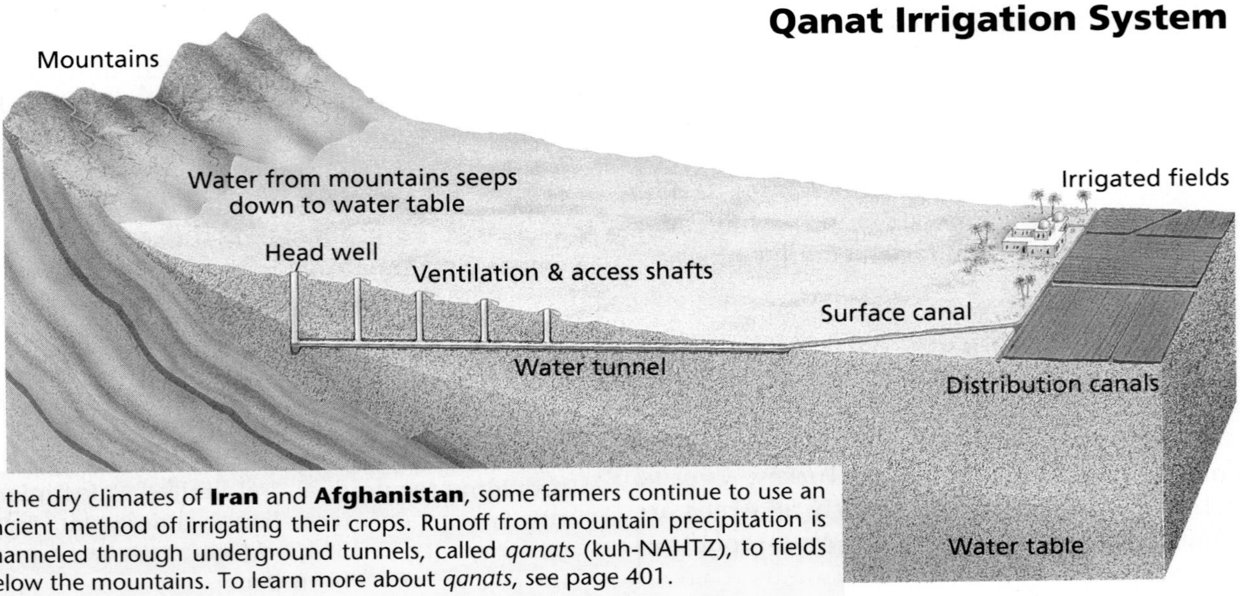

Mountains
Water from mountains seeps down to water table
Head well
Ventilation & access shafts
Surface canal
Irrigated fields
Water tunnel
Distribution canals
Water table

In the dry climates of **Iran** and **Afghanistan**, some farmers continue to use an ancient method of irrigating their crops. Runoff from mountain precipitation is channeled through underground tunnels, called *qanats* (kuh-NAHTZ), to fields below the mountains. To learn more about *qanats*, see page 401.

Syria, with a large amount of desert land, gets most of its water from the Euphrates River. Droughts, rapid population growth, and wasteful water use, however, have led to shortages. In addition, harmful pesticides and chemicals have been dumped into the Euphrates, making it a dangerous source of drinking water. Syria also has neglected its irrigation and sewage systems, allowing pipes to crumble and leak.

Conservation Efforts

Faced with growing populations, periodic droughts, and limited resources, the nations of Southwest Asia are looking closer at cutting their water use. For some nations, however, water conservation involves difficult economic decisions. For example, some crops grown in the region, such as bananas and cotton, need large amounts of water. Nations must weigh the cost of using the water to grow these crops against the cost of saving the water and importing the crops from other countries.

Effective conservation methods must be accompanied by policies that protect the environment. Countries are realizing that pesticides and chemical fertilizers can make their way into rivers, aquifers, and wells. Also, unrestricted digging of wells can lower the water table and allow seawater to enter aquifers.

Cooperation Among Nations

Because most rivers and aquifers do not begin or end at a nation's borders, Southwest Asia's water problems cannot be solved without regional cooperation. If countries upstream take too much water from or pollute a river, countries downstream suffer. For example, because the headwaters of the Tigris and Euphrates rivers are in Turkey, Turkey's use of the rivers affects Syria and Iraq downstream.

Before the countries of Southwest Asia can solve their water problems, however, they must solve or set aside their political differences. As long as conflicts remain, the threat of war hovers over the region and makes nations unwilling to invest in pipelines and desalinization plants that could be attacked. Future peace negotiations in the region will likely be linked to water agreements. In turn, peace could set the stage for easing the growing water crisis.

Leaders of the United States, Israel, and Jordan gathered in 1994 for the signing of a peace treaty between the two Southwest Asian countries. Israelis and Jordanians hope the peace treaty, signed along the border between their two countries, will aid them in finding common ground on important water issues.

Vegetation clings to life along this irrigation canal in a desert region of southern **Israel**. Israel, as well as other countries of Southwest Asia, uses well-developed irrigation systems for agriculture.

YOUR TURN

Draw a map of Southwest Asia and label all the countries, bodies of water, and major rivers of the region. Then **assess** each country's natural water resources, noting if any of these resources are shared among countries. **Write** a paragraph outlining what might be each country's concerns about use of the region's water resources. **Describe** how a regional water plan might address these concerns.

This modern home in **Saudi Arabia** is furnished with many items of Arab culture. Many Arab men, such as the one pictured here, wear a *kaffiyeh* (kuh-FEE-yuh), a cloth headdress that offers protection against the sun and wind.

Many other ethnic and religious minorities live in Southwest Asia. In Lebanon, there is a large community of Christians. In Iran, there are Zoroastrians (zohr-uh-WAS-tree-uhns), who follow the religion of ancient Persia. The Kurds, who are also Muslim but non-Arab, are a large ethnic group living in the bordering regions of Iraq, Iran, Syria, Armenia, and Turkey. The Kurds do not have their own country.

Migrating herders, known as nomads or **Bedouins**, still roam the outer parts of the deserts. Many years ago, caravans of camels carried traders and their goods from oasis to oasis. Camels, long a symbol of life in Southwest Asia, are now used for transport in isolated areas only. Some people still use donkeys as well, but for the most part, trucks have taken the place of camels and donkeys. Some wealthy families in the region keep camels in much the same way as some Americans keep horses.

Issues Although the Arab-Israeli conflict has dominated the political geography of Southwest Asia, many other challenges face the region. Modern health care practices together with a tradition of large families have led to rising populations throughout the region. These growing populations need food and want more basic consumer goods and opportunities.

Also, as populations in the region continue to grow, so will conflicts over water resources. In most of Southwest Asia, farms and cities depend on careful management of rivers, springs, and wells for water. Many communities are miles away from the rivers that provide their water. To make sure that everyone receives water, governments must maintain cooperation among communities. In ancient times, wars often led to the collapse of irrigation systems, and starvation resulted. Today, all of the important exotic rivers of Southwest Asia cross international borders, further complicating the issue. (See "Planet Watch" on pages 382–383.)

Another issue of international significance concerns the region's oil resources. The importance of oil in world trade has given the oil-rich countries of Southwest Asia more political power. For example, Arab countries used their oil supplies to influence international attitudes toward the Arab-Israeli conflict during a war in 1973. Arab countries would not sell oil directly to countries that supported Israel in the war.

In the long run, the oil wealth of the region, as well as foreign aid, will change the region more than any issue or event since the spread of Islam. Whole industrial complexes are being built, including chemical and oil-related manufacturing plants. New schools, housing, and Western influences are appearing throughout the region. The enormous changes taking place in the countries of Southwest Asia make predicting the region's future difficult.

2 SECTION REVIEW

1. What were three major innovations developed by the ancient civilizations of Southwest Asia? What empires have existed in the region?
2. What three major religions originated in Southwest Asia? What ethnic and religious groups live in the region today?
3. **Critical Thinking** **Discuss** the roles that natural resources can play in a region's conflicts. Use two examples from Southwest Asia.

CHAPTER 32 CHECK

Mounds of potash, taken from the **Dead Sea**, will be used to make fertilizer.

Reviewing the Main Ideas

1. Southwest Asia is at the center of the world's largest arid region. Agriculture is a major occupation in the region, but oil provides the most income.
2. The world's first civilizations developed in Southwest Asia. After 600 B.C., a series of empires controlled most of the region. When the last empire fell in 1918, parts of the region came under European control. Most countries here achieved independence during the past 60 years.
3. Several religious and ethnic groups live in Southwest Asia. Tensions between these groups, rising populations, scarce water resources, and oil wealth all are factors that will shape the region's future.

Building a Vocabulary

1. What term describes a stream that flows from a humid environment, such as mountains, across a desert into a sea? Why are mountain environments wetter?
2. Define *monotheism.* In what three major religions is monotheism a central belief?
3. What is Islam's sacred book of writings?
4. Why is secularism considered evil by Muslim fundamentalists?

Recalling and Reviewing

1. What types of irrigation are used in Southwest Asia?
2. Where is the Fertile Crescent? How is it historically significant?
3. Why were Europeans interested in Southwest Asia following the collapse of the Ottoman Empire?
4. What are Islam's two major branches?
5. What conflict has dominated political issues in Southwest Asia?

Thinking Critically

1. Both Christianity and Islam are divided into different branches; certain divisions exist among the followers of Judaism as well. What factors might cause divisions within a major religion?
2. During the Cold War, some countries in Southwest Asia sought support from the United States, while others turned to the Soviet Union. How do you think the end of the Cold War affects political events in the region as a whole?

Learning About Your Local Geography

Individual Project

Gasoline is a major oil product. Interview a few community members to find out how much they drive each week and how often they buy gasoline for their cars. Summarize your findings in a written report. Conclude with a few statements discussing how people in your community might be affected by a conflict in Southwest Asia that reduced or cut off oil supplies to the United States. What actions might community members take to reduce the local impact of such a conflict?

Using the Five Themes of Geography

Write a paragraph discussing how Southwest Asia's **location** near Europe, Africa, and the rest of Asia has influenced the region's physical and human **place** characteristics. Also consider the role that **movement** has played in determining these characteristics.

CHAPTER 33

The Eastern Mediterranean

GEOGRAPHY DICTIONARY

kibbutz
Zionism
Holocaust
suq
minaret

"The kibbutz to which we came in the autumn of 1921 consisted of a few houses and a cluster of trees. . . . There were no orchards, no meadows, no flowers, nothing, in fact, except wind, rocks and some sun-scorched fields. . . . Today it is quite an impressive forest, and whenever I pass it, I remember how we dug endless holes in the soil between the rocks and then carefully planted each sapling, wondering whether it would ever grow to maturity there and thinking how lovely the roadside, the whole country in fact, would be if only those trees of ours managed to survive."

Golda Meir

Pieces of the Dead Sea Scrolls

Each day members of three different religions go to the Cave of Machpelah (mak-PEE-luh) in the Israeli-occupied West Bank. To Muslims, the cave is a mosque. To Christians, it is a cathedral. To Jews, it is a holy temple. Many believe that Abraham and Sarah, biblical ancestors honored by Jews, Muslims, and Christians, are buried there.

Such complex interactions of history, religion, and geography are typical in the Eastern Mediterranean. The peoples of this region have built their lives among the ruins of past civilizations. Like their ancestors, they remain committed to strong beliefs that shape their lives and the future of the region.

A grove of pine saplings in **Israel**

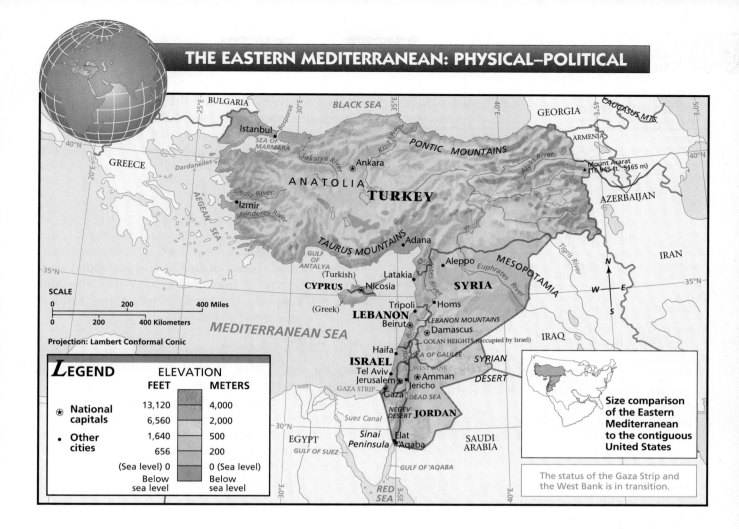

THE EASTERN MEDITERRANEAN: PHYSICAL–POLITICAL

SCALE
0 200 400 Miles
0 200 400 Kilometers
Projection: Lambert Conformal Conic

LEGEND

⊛ National capitals
• Other cities

ELEVATION FEET	METERS
13,120	4,000
6,560	2,000
1,640	500
656	200
(Sea level) 0	0 (Sea level)
Below sea level	Below sea level

Size comparison of the Eastern Mediterranean to the contiguous United States

The status of the Gaza Strip and the West Bank is in transition.

TURKEY

FOCUS

- *What are the physical and economic characteristics of Turkey? What are the country's major cities?*
- *How did Turkey change during the 1920s? What issues face Turkey today? What is the status of Cyprus?*

REGION Physical Geography Turkey is often described as a country of two peninsulas. A small part of Turkey lies on a corner of Europe's Balkan Peninsula. The larger part of Turkey occupies the Asian peninsula of Anatolia. Two narrow straits, the Bosporus (BAHS-puh-ruhs) and the Dardanelles (dahrd-uhn-EHLZ), divide the two parts of Turkey, thereby dividing Europe from Asia. These straits, along with the Sea of Marmara, also connect the Black Sea with the Mediterranean Sea. (See the map above.)

The landforms of European Turkey are mainly rolling plains and hills. Anatolia, meanwhile, has rugged coastlines, broad interior plateaus, and mountains that rise steadily in the east. Central Turkey has two main mountain ranges: the Taurus (TAWR-uhs) Mountains in the south and the Pontic Mountains in the north along the Black Sea coast. Farther east, rows of rugged mountains and narrow valleys cross the country. These mountains merge with the Caucasus Mountains. Along Turkey's border with Iran and Armenia, the higher of the two peaks of Mount Ararat rises up to 16,945 feet (5,165 m).

Turkey has two climate types. The coasts of the Mediterranean and Black seas have winter rains and summer droughts typical of the Mediterranean climate. Western Turkey has a similar climate, but its good rains make it the country's most productive agricultural region and one of the wettest areas of Southwest Asia. The

FACTS IN BRIEF The Eastern Mediterranean

COUNTRY POPULATION (1994)	LIFE EXPECTANCY (1994)	LITERACY RATE	PER CAPITA GDP (1993)
CYPRUS 730,084	74, male 79, female	95% (1991)	$11,390 (Greek area) $3,130 (Turkish area) (1992)
ISRAEL 5,050,850	76, male 80, female	92% (1991)	$13,350
JORDAN 3,961,194	70, male 74, female	80% (1990)	$3,000
LEBANON 3,620,395	67, male 72, female	80% (1990)	$1,720
SYRIA 14,886,672	65, male 68, female	64% (1990)	$5,700
TURKEY 62,153,898	69, male 73, female	81% (1990)	$5,100
UNITED STATES 260,713,585	73, male 79, female	97% (1991)	$24,700

Sources: *The World Factbook 1994,* Central Intelligence Agency; *The World Almanac and Book of Facts 1995*

Which country in the region is the most populous?

interior of Anatolia has a drier steppe climate. To the east, the climate becomes even drier and the land more rugged. Rain and snow in the eastern mountains form the sources of the Tigris and Euphrates rivers.

Economic Geography With its rugged landforms, less than half of Turkey is suitable for farming. The most densely settled parts of the country are around the Sea of Marmara and along the Aegean coast. Here, farmers grow typical Mediterranean crops. Wheat and barley are the important grains, while tobacco, hazelnuts, cotton, and citrus fruits are grown for export. The drier and cooler climate of the high interior plateau limits farming to grains and raising livestock.

The eastern mountains are the least developed part of Turkey. Most people here live in small, isolated villages. The region also has many ethnic Kurds, whose relations with the Turks have always been cautious. The central government hopes that the construction of large dams will bring progress to eastern Turkey. Deposits of oil and mineral ores are being developed here as well. In the meantime, mohair, the wool of Angora goats, brings in some cash income.

Because Turkey has a better network of roads and railroads than most other countries in Southwest Asia, industrial development has steadily increased since the collapse of the Ottoman Empire after World War I. Turkey hopes eventually to become a member of the European Union, and many Turks are already guest workers in EU nations.

Urban Geography Combining the old with the new, Istanbul and Ankara are the leading cities of Turkey. As the Ottoman Empire's capital city, Istanbul was a center of political and economic power. Today, it lies in the European part of Turkey and is the country's largest city and leading seaport. Its harbor, called the Golden Horn, opens onto the Bosporus.

When the Turkish republic formed in 1923, the more centrally located Ankara became the capital city. Today, Ankara is home to some of Turkey's most important industries. It is also a center for railroad transportation and for the development of the eastern part of the country.

Turkish cities were among the first in the Muslim world to modernize. Industrial development, education, and better health care and housing are improving the standard of living in Turkey. Contact with people from the cities is slowly changing the way of life among people in rural areas as well.

Modern Turkey The roots of modern Turkey reach back to the early 1920s, when a revolution reorganized Turkish society and opened it to Western ideas and modernization. Although most people remained Muslims, Islam lost its status as the state religion. The government took over the many schools previously controlled by religious groups. The Roman alphabet also replaced the Arabic alphabet, and laws changed to reflect European codes. Even the wearing of European-style clothing was required. In addition, Turkey became the first Muslim country to begin recognizing women's rights. Since the revolution, Turkey has distanced itself from Arab political affairs and from other Muslim countries. After several periods of military dictatorship, democratic traditions are on the rise.

Other than conflict with ethnic Kurds, which included large-scale military operations in 1995, Turkish foreign affairs focus on two main issues. The first is the bitter, centuries-old conflict between

Istanbul, Turkey sits along the Bosporus, an important route for shipping between the Black Sea and the Aegean and Mediterranean seas beyond. Turkey's largest city has been known by several names and ruled by many empires over a period of more than 2,500 years.

the Turks and the Greeks over the island of Cyprus and the potential oil resources in the Aegean Sea. The sea's resources are a matter of pride as well as economic gain to both countries. The second issue is cooperation with the independent countries of Central Asia. During the periods of the Russian Empire and the Soviet Union, Turkey had little contact with Central Asia. Today, Turkey is leading development efforts there. In many ways, the country is still a physical and cultural bridge between Asia and Europe.

Cyprus Cyprus is a rocky, mountainous island located in the eastern Mediterranean Sea. (See the map on page 387.) A wide, fertile plain separates two major mountain ranges on the island.

Cyprus gained its independence from Britain in 1960. The island, however, has experienced cultural conflict for decades. About 78 percent of the people of Cyprus consider themselves to be Greeks. About 18 percent think of themselves as Turks. Each group is suspicious and resentful of the other. As a result of the struggle, the Greeks now live in the south part of the island, and the Turks live in the north. Their dividing line passes through the capital city of Nicosia (NIK-uh-SEE-uh). Each community has declared its own republic and has its own government. The intense dispute alternates between periods of violent bitterness and peaceful discussion. United Nations forces help keep peace on the island.

The Greek part of Cyprus, which has more industry and tourist facilities, has three times the per capita income of the more rural Turkish part. Yet all of the people of Cyprus know that without peace, the island will attract few tourists and its economy cannot grow. Many hope stable political relationships soon will emerge in Cyprus.

SECTION REVIEW

1. Into what two parts is Turkey physically divided, and what are the landforms of each part? Which region of Turkey is the least economically developed? What are the country's two major cities?
2. List four social or political changes that have taken place in Turkey since the early 1920s.
3. **Critical Thinking** **Explain** why Cyprus might be considered part of Europe rather than part of Asia.

ISRAEL

FOCUS

- *What are Israel's major physical and economic features?*
- *When and how did Israel become a country? What challenges have faced Israel since its formation?*

LOCATION **Physical Geography** Israel lies along the eastern shore of the Mediterranean Sea on what once was the ancient land of Palestine. (See the map on page 387.) This small country has many environments. A strip of fertile land with a Mediterranean climate lines the country's coast. Inland is the hilly, semiarid region of Galilee. Beyond this region is the hot, arid valley of the Jordan River and the Dead Sea. The Dead Sea is nearly 1,302 feet (397 m) below sea level, the lowest point on any continent. The Dead Sea, which has no outlet to another body of water, is so salty that swimmers cannot sink in it. Southwest of the Dead Sea is the rocky Negev Desert, which stretches to the Gulf of 'Aqaba.

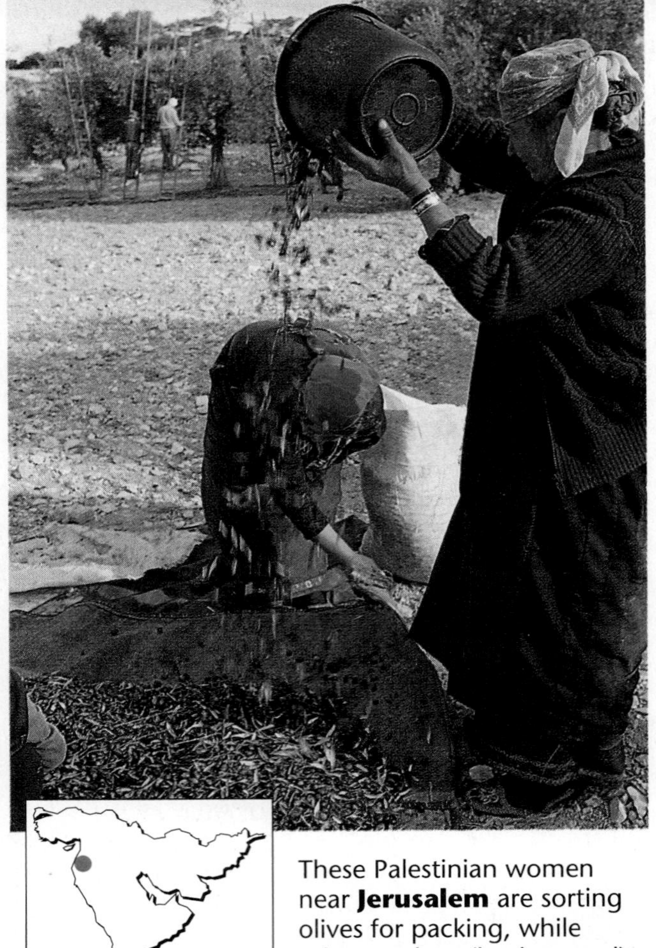

These Palestinian women near **Jerusalem** are sorting olives for packing, while other workers (background) pick the olives from trees. Olive cultivation is common in countries along the Mediterranean Sea.

Agriculture Israel is an economically developed country. Unlike in other countries of Southwest Asia, only about 5 percent of Israel's people work in agriculture. Modern commercial agriculture is typical, and Israel's farms are some of the most productive in the world. The best farmland lies along the coastal plain and on irrigated farms farther inland. Israel exports fruits, particularly oranges, as well as vegetables and cut flowers to European markets.

The **kibbutz**, or Israeli collective farm, remains an important element of Israeli identity. These farms were important in the development of the country. Today, many people from foreign countries, including the United States, travel to Israel to work with Israelis on a kibbutz.

Israel has one of the world's most sophisticated water management systems and uses nearly every drop of its water resources. The main system is the National Water Carrier, whose pipelines carry water from the Sea of Galilee throughout the country, even to the Negev Desert's northern borders. Here, irrigation agriculture has replaced dry farming and herding. In the coastal region, modern wells provide additional water.

Resources and Industry Israel has almost no oil or coal, little hydroelectricity, and relatively few minerals. The Dead Sea does contain potash, magnesium, bromide, and other salts, which are used for fertilizers and chemical products. One Israeli company has found another use for the minerals. Sales of its Dead Sea mud cosmetics are booming. The firm was not the first to use the mineral-rich mud, however. The Egyptian queen Cleopatra came to the Dead Sea for her facials centuries ago. For the most part, though, Israel's economic growth relies on the skills of its well-educated population.

Israel has developed high-technology industries, such as the manufacturing of computers, complex military weapons, and precision engineering equipment. Rough diamonds are imported, cut, polished, and then exported around the world.

Tourism is another major industry. Many people come to visit the holy places and the sunny beaches along the Mediterranean Sea and the Gulf of ‘Aqaba. In addition, Israel's economy is boosted by aid from the United States and from Jews worldwide.

MOVEMENT **Historical Geography** The Jews established their first kingdom, with its capital at Jerusalem, in the eastern Mediterranean more than 3,000 years ago. Over time, however, a series of conflicts forced most of the Jews to leave the region, and it became populated largely by Arabs.

In the nineteenth century, a movement known as **Zionism** began to grow among Jews outside Palestine who longed for a homeland. The roots of Zionism lay in Europe, where Jews suffered persecution for many centuries. The Zionists believed that a Jewish state should be established in Palestine. They also believed that Jews from around the world should move there. As a result of this movement, more Jews began settling in Palestine in the late 1800s and early 1900s.

After World War II, many nations throughout the world supported the idea of creating the modern nation of Israel as a Jewish homeland. These nations were sympathetic to the persecution, known as the **Holocaust**, that Jews had suffered

NEWS FROM AROUND THE WORLD Israel

Israeli Teens Live with the Ancient and the Modern

Israel covers less than 8,000 square miles (20,720 sq. km), roughly the size of New Jersey. Yet nearly 5 million people crowd into the country. Jewish peoples have immigrated to Israel from almost every part of the world, bringing their languages and traditions with them. As a result, Israeli cities are a mixture of many cultures. American, Chinese, French, Greek, Hungarian, Indian, Iraqi, Italian, Moroccan, Turkish, and Vietnamese restaurants can be found even in small cities.

Most Israelis grow up in the cities, where the population density is highest. Because space is scarce, many city dwellers live in crowded high-rise apartment buildings. Families can find these apartments particularly cramped. Bunk beds are popular with teenagers, who often must share bedrooms with younger brothers or sisters.

For Jewish teens in Israel, schools are a unifying force, bringing together students of many different backgrounds. Arab teens attend their own schools, however, where subjects are taught in Arabic rather than Hebrew. This division is a daily reminder of the religious, cultural, and political beliefs that separate the peoples in this troubled part of the world.

In Jerusalem, Israeli teenagers can look down from the rooftops of their apartment buildings on the holy places of three religions—Christianity, Judaism, and Islam. The rooftops and the walls surrounding and dividing the city's quarters are a kind of transportation network for young Israelis. They race across the buildings and explore the rooms and stairways hidden within the walls.

Although daily life in Israel is often overshadowed by political issues and turmoil, Israeli teens, like those pictured above, are much like most young people in the United States. Television, rock music, and other symbols of American popular culture are widespread, and many young Israelis hope someday to travel or study in the United States.

during the war. The word *holocaust* means "widespread destruction." Nazi Germany persecuted and killed millions of Jews during the Holocaust. As a step toward creating a Jewish homeland, Britain, which then controlled Palestine, gave the region its independence after the war. How the area was to be divided was unclear, however.

Finally, in 1948, a United Nations plan divided the lands west of the Jordan River between Jews and Arabs. The Arab lands west of the Jordan River, called the West Bank, were joined to the country of Jordan. At this time, Jews occupied little land in the region and totaled only about one-third of the population.

Neighboring Arab countries and the Arab people of Palestine, called Palestinians, rejected the establishment of Israel. The Palestinians and the Arab countries believed that Israel's existence denied Arabs their essential rights to their homeland. As a result, war broke out between Arabs and Israelis in 1948. The Israeli forces defeated the Palestinians and their Arab allies.

Many Palestinians fled to refugee camps in nearby Arab states. From these camps, some Palestinians continued to attack Israeli settlements in hopes of regaining the land they felt was theirs. Wars again broke out between the Arab countries and Israel in 1956, 1967, and 1973.

Text continues on page 394.

Jerusalem

The capital of ancient Israel 3,000 years ago, Jerusalem is the capital of modern Israel today. The city's past and present merge in its skyline of ancient ruins, medieval towers, and modern high-rise buildings.

Jerusalem is of central importance to people of the Jewish faith. For centuries, Jews living outside the city have turned toward Jerusalem in prayer three times each day. Hundreds of millions of non-Jews also care deeply about the city. Christians cherish it as the place where Jesus lived, preached, and was crucified. Muslims honor the site in Jerusalem where they believe the prophet Muhammad rose to heaven.

Perhaps more than in any other city, the threads of human history wind through Jerusalem's streets.

An Ancient History

Because it is sacred to so many, several groups have fought for control of Jerusalem in its 5,000-year history. It became the center of Jewish life in 1003 B.C., when King David made it his capital. Thirty-three years later, David's son Solomon built his magnificent Temple there. Although the Babylonians conquered Jerusalem and destroyed the Temple in 586 B.C., the Israelites recovered the city and built a second Temple.

Jerusalem was under Roman control when Jesus lived and preached there. A generation later, in A.D. 66, the Jews revolted against Roman rule. The Romans crushed the revolt, however, and in A.D. 70 destroyed the second Temple. The Temple's only remnant, known today as the Western Wall, remains Judaism's most sacred religious shrine.

After the Romans were finally driven out, Jerusalem came under the control of a series of groups, including the Byzantines, Arabs, Christian Crusaders, Egyptians, and Turks. In the late seventh century, the Arabs built the Dome of the Rock and the al-Aqsa Mosque where the Jewish Temple once stood. Later, during the sixteenth century, the Turks constructed the high stone walls that still surround what today is called the Old City.

Visitors view the Children's Memorial, part of the Holocaust Memorial, or Yad Vashem, in **Jerusalem**. Israel was founded in 1948, three years after the end of World War II and the defeat of Nazi Germany.

Modern Jerusalem

The modern history of Jerusalem begins in the 1860s, when the city's growing Jewish population built the first neighborhoods outside the city's walls. Jerusalem continued to spread outward, especially after the British wrestled the city from the Turks in 1917.

When modern Israel formed in 1948, another round in the struggle for Jerusalem began. Neighboring Arab states attacked Israel, and Jordan occupied the Old City. Israel recovered the western part of the city in 1949 and made it the country's capital. The eastern part, including the Old City, remained under Jordanian control.

Jerusalem was reunified when Israel defeated Jordan, Egypt, and Syria in the Six-Day War in 1967. Today, the Old City is part of a metropolis of more than 500,000 people. Within

GINZA
HOLLYWOOD

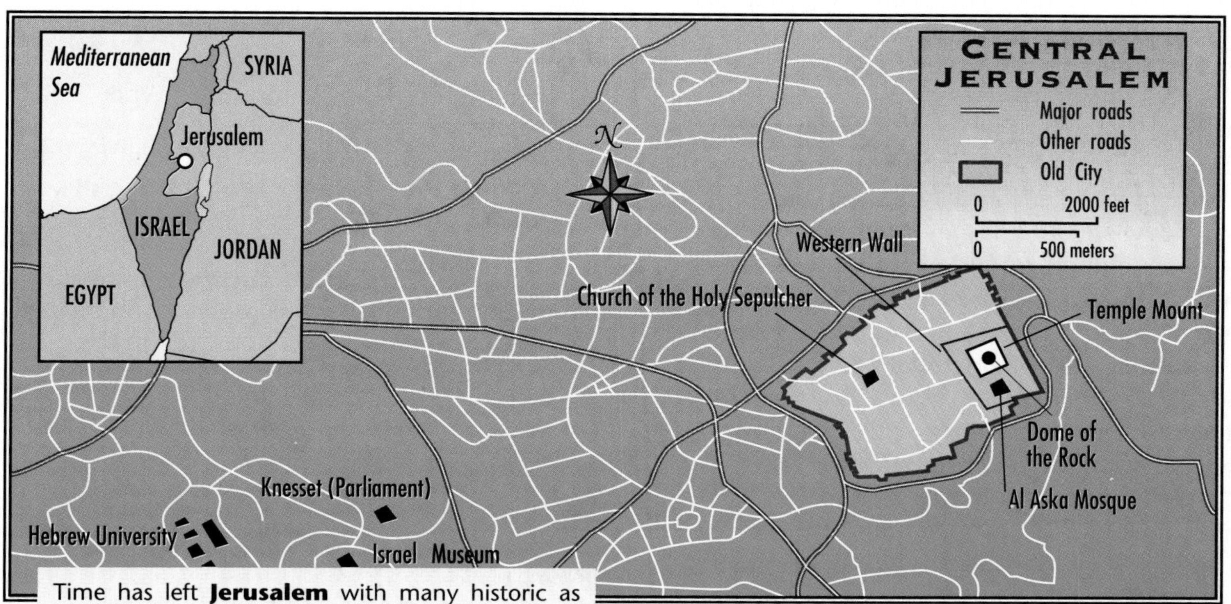
Mediterranean Sea
SYRIA
Jerusalem
ISRAEL
JORDAN
EGYPT
CENTRAL JERUSALEM
Major roads
Other roads
Old City
0 2000 feet
0 500 meters
N
Western Wall
Church of the Holy Sepulcher
Temple Mount
Dome of the Rock
Al Aska Mosque
Knesset (Parliament)
Hebrew University
Israel Museum

Time has left **Jerusalem** with many historic as well as modern structures scattered throughout the city. In which part of Jerusalem would a visitor find holy sites for three major world religions?

its walls are the major Jewish, Christian, and Muslim religious sites, access to which is guaranteed to all by Israeli law. Jewish, Christian, and Muslim authorities control their respective sacred sites.

The focal point of the Old City remains the Temple Mount, a huge 30-acre (12 ha) platform on which the Temple once stood. Muslims pray in the mosques located on the Temple Mount, while Jews gather at the Western Wall that forms one of the Temple Mount's edges. Nearby, Christians retrace the "Way of the Cross," which leads to the Church of the Holy Sepulcher.

Jerusalem's western part is home to more modern attractions such as the Israel Museum, the Hebrew University, Israel's parliament building, and popular shops and cafes. Near the city's western edge is Yad Vashem, Israel's memorial to the 6 million Jews killed during the Holocaust. Even in its modern sections, Jerusalem remains a city cloaked in history.

Worshippers welcome the Jewish Sabbath on Friday evening at the Western Wall. The golden Dome of the Rock, sacred to Muslims, rises to the left.

As a result of the war in 1967, called the Six-Day War, Israel occupied land from Egypt, Syria, and Jordan. This land included all of the Sinai (SIE-nie) Peninsula west to the Suez Canal, the Gaza (GAHZ-uh) Strip, the West Bank of the Jordan River, East Jerusalem, and the Golan (goh-lahn) Heights. (See the map on page 387.) The Sinai was returned to Egypt in 1982. In 1993 and 1995, Israel and the Palestinians agreed to limited Palestinian self-rule in Gaza and parts of the West Bank. Negotiations on expanding this agreement and on the status of the Golan Heights have continued.

Population and Urban Geography Today, Israel is a modern, urban nation with a democratic government. Nearly 85 percent of its people are Jewish, and large numbers of Jews from both developed and developing countries continue to immigrate to Israel. Integrating Jews from many different cultures has not always been easy. In addition, many Palestinians live in Israel. Even though the Palestinian population in Israel is growing rapidly, the country will remain a primarily Jewish state for years to come.

As Israel's population has grown, so have the country's cities. Tel Aviv (tehl uh-VEEV) is the nation's largest city. Haifa is the country's major port and an important center of oil-refining and chemical industries. Jerusalem, the capital of Israel, is the nation's cultural and political center. (See "Cities of the World" on pages 392–393.)

Issues Israel's most serious problem is the ongoing conflict with its Arab neighbors. A crucial question is the fate of the Palestinian people and the territories Israel has occupied since 1967. Many Israelis have settled in these lands, further complicating the situation. Although Israel and the Palestinians have reached some agreements, the road to a stable and lasting peace is a long one.

Palestinians living in Israel continue to stage demonstrations seeking their independence. Palestinians living outside Israel also long to return to the land they view as their home. The Palestinian Liberation Organization (PLO) is the leader of the Palestinian cause and is pursuing further negotiations for Palestinian self-rule with Israel. The PLO has many internal divisions, however, which may threaten its stability and effectiveness.

Because of the threat of war and frequent border clashes, Israel maintains a large army. In fact, all young Israeli men and women must serve in the military. The cost of supporting this military force places a burden on the country's economy, even though Israel receives a large amount of foreign aid from the United States and other countries. Despite agricultural and industrial successes, Israel's future remains troubled until permanent peace comes to countries of the Eastern Mediterranean.

SECTION REVIEW

1. How is the Dead Sea unusual? What resources does it contain?
2. What role did World War II play in the creation of the state of Israel? How are the effects of the 1967 Six-Day War felt today?
3. **Critical Thinking** **Suggest** an explanation for why Israel is more economically developed than other countries in the Eastern Mediterranean.

SYRIA, LEBANON, AND JORDAN

FOCUS

- *What are the physical and economic characteristics of Syria, Lebanon, and Jordan?*
- *How are culture and politics related in these three countries?*

Syria Syria stretches east from the mountainous Mediterranean coast to the plains of Mesopotamia. (See the map on page 387.) The Syrian Desert lies in the center and southeast of the country. Accordingly, most Syrians live in the western one-fourth of the country, which has a Mediterranean climate. Settlement farther east is dependent on the water resources of oases and of the Euphrates River.

Syria is trying to develop modern agriculture, but despite programs to encourage high-technology farming, most of the agricultural land is made up of small subsistence farms and large inefficient estates. In the mountain valleys of western Syria, farmers grow citrus fruits and other Mediterranean crops. In the drier environments east of the coastal region, cotton, wheat, and barley are key crops, and livestock is important. In the Syrian Desert, a few Bedouin herders still move their flocks with the seasons. Even though about one-third of Syria's people work in agriculture, the country must import food.

Agricultural progress depends on developing water resources. Syria's irrigation water comes mostly from the Euphrates River in the northeast and the Orontes (aw-RAHNT-eez) River in the northwest. A huge dam on the Euphrates River has brought thousands of acres of land into production. The dam also provides electric power for industry in the northern part of the country. The government hopes this area will become a center for commercial agriculture.

Although traditional livestock raising, farming, textiles, and crafts are still the main occupations in Syria, the country has some industry and is producing more basic consumer goods. In addition, Syria has a small amount of oil available for export, and new oil deposits are being developed in eastern Syria. Public education has spread widely in recent years, and all large cities have universities.

Damascus, Syria's capital, was one of the great oasis cities of the ancient world as well as a center of Islamic civilization. The city lies on the plains east of the mountains of Lebanon. The water from the rains and snows in the mountains have supplied Damascus and the surrounding farmlands for centuries.

Modern Damascus is a city of contrasts. Wide boulevards remain from French colonial times. These are mixed with the narrow streets of the old city and marketplaces, called ***suqs*** (sooks). Traditional textiles and metalwork from the Damascus *suqs* are famous throughout the world for their beauty and design. Ancient Roman ruins also still stand in various parts of the city. And rising over the city's mosques are dozens of **minarets**. A minaret is the tower of a mosque from which a *muezzin* (crier) calls Muslims to pray five times a day.

Syria has been a leader in the Arab opposition to Israel, and its relations with Turkey and Iraq are strained as well. Syria claims some Turkish territory along the Mediterranean, while Iraq fears that Syria will divert too much water from the Euphrates River.

Once aided by the Soviet Union, Syria has changed its foreign policy since the end of the Cold War and has proposed peace negotiations with Israel. It now seeks aid from Europe and the United States. Syria's leading role in Southwest Asian affairs is likely to last for many years.

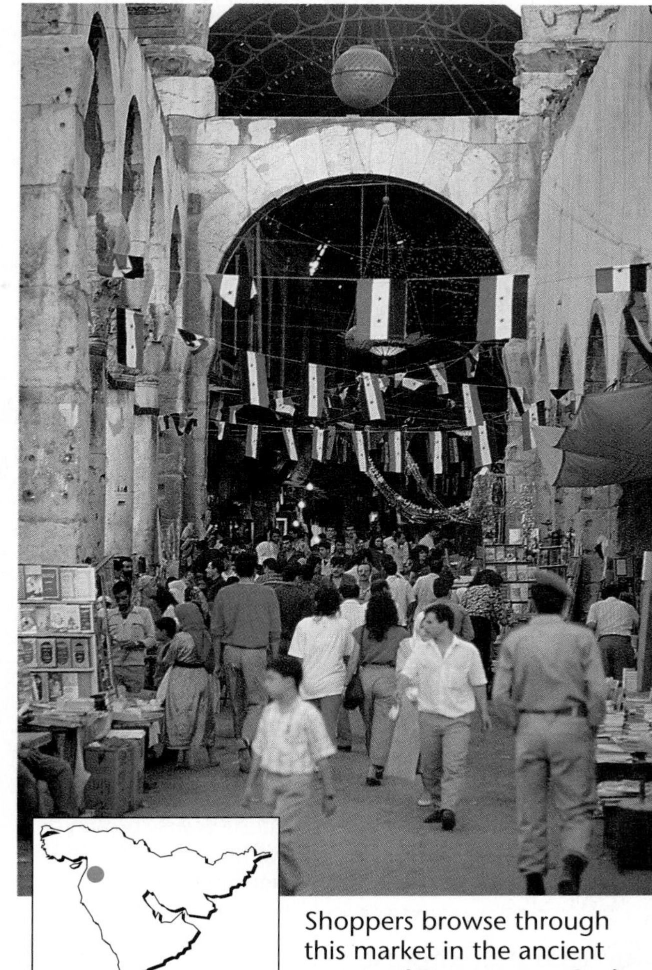

Shoppers browse through this market in the ancient streets of **Damascus, Syria**. The Syrian capital is one of the world's oldest cities.

Lebanon Lebanon lies between Israel and Syria, wedged between the Mediterranean coast and the Lebanon Mountains. (See the map on page 387.) Its shore was the ancient home of the Phoenician traders. Later, it was an important province of the Roman and Ottoman empires. Unlike in any other country in the region, about one-fourth of the people of Lebanon are Christian. The beautiful mountains and coast of Lebanon have attracted many other groups as well. These include people representing more than four different varieties of Islam. Traditionally, the Lebanese have one of the most modern and Westernized cultures in Southwest Asia.

Beirut (bay-ROOT), the capital, used to be called the "Paris of the Middle East." Its schools and universities once attracted students from

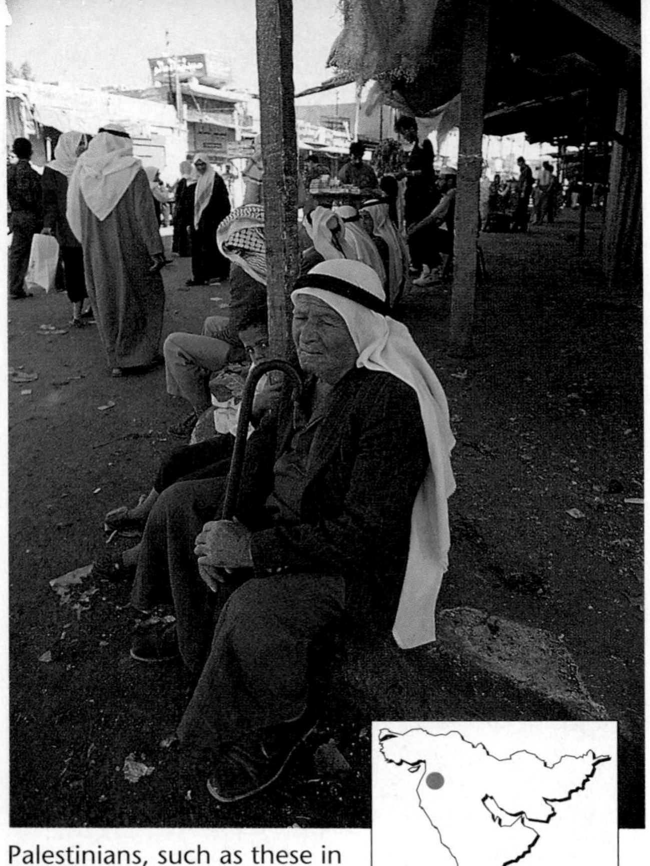

Palestinians, such as these in **Jordan**, are a people without a country of their own. In recent years, however, Palestinians have won the right to limited self-rule in some territories occupied by Israel since 1967.

throughout Southwest Asia, and tourists flocked to its beaches and historic sites. Unfortunately, periodic civil wars that began in the late 1950s and intensified between 1975 and 1990 have wracked Lebanon. Christians and the various Muslim groups living in Lebanon are the main opponents. Fighting has also broken out among factions within these religious groups.

The Arab-Israeli conflict has further complicated the situation in Lebanon, as several hundred thousand Palestinians have settled in refugee camps there. Foreign countries and the United Nations have intervened many times in Lebanon to resolve the conflict. The most heated intervention has come from Syria and Israel. Both countries have invaded and occupied areas of Lebanon, mainly to keep the civil war from spilling over into their countries and because of the Palestinian refugee issue.

Lebanon's economy lies largely in ruins, and most of the international business community has left the country. Today, after years of fighting, peace may be on the horizon. A stable government, however, has yet to come to power.

Jordan Jordan stretches east from the Dead Sea and the Jordan River into the rocky Arabian Desert. Only the northwest corner of the country is not arid. (See the map on page 387.)

When Jordan became an independent kingdom in 1946, fewer than 400,000 people lived east of the Jordan River. They were mostly migrating herders and subsistence farmers. When Israel formed in 1948, nearly a million Palestinians took refuge in Jordan. Even more refugees came after the 1967 Arab-Israeli war. As a result, Palestinians in Jordan outnumber the country's original inhabitants. Relations between these groups often have been tense. The large refugee population has strained the country's resources and placed Jordan on the front line of the Arab-Israeli conflict.

Jordan has few minerals or other natural resources, and the dry environment makes farming difficult. In fact, most of Jordan's desert lands can never be cultivated. Large-scale irrigation is only possible in the northwestern area. As a result, much of the country's food must be imported. Aid from many countries, including the United States and the oil-rich Arab states, has helped the country develop. Negotiations between Jordan and Israel over water rights and other issues are progressing.

Modern Jordan centers around the capital city of Amman (a-MAHN). Jordan's port city of 'Aqaba on the Gulf of 'Aqaba also is an important trade center and the nation's only seaport. Goods headed for Damascus and northern Iraq travel through the port daily. The country is trying to build more industries. Food products, simple consumer goods, and building materials are produced for local markets. Jordanians hope that peace in the region will help the country's economic development.

3 SECTION REVIEW

1. Which country discussed in this section does not border the Mediterranean Sea? To which country does the Euphrates River provide irrigation water and hydroelectric power?
2. How have cultural factors played a role in Lebanon's civil wars?
3. **Critical Thinking** **Discuss** the ways in which Syria, Lebanon, and Jordan have been involved in the Arab-Israeli conflict.

CHAPTER 33 CHECK

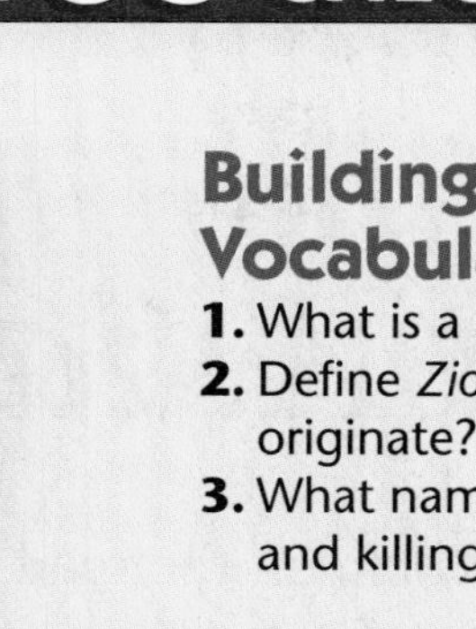

Hagia Sophia in **Istanbul, Turkey**

Reviewing the Main Ideas

1. Turkey lies on two peninsulas and has two climate types. Less than half of the country is suitable for farming. Ankara became the capital of Turkey in 1923 when a revolution reorganized Turkish society.
2. Cyprus, a Mediterranean island, is divided between Greeks and Turks. Each community has declared its own republic.
3. Israel lies along the eastern shore of the Mediterranean Sea. Israel has few resources but is economically developed, with commercial agriculture and high-tech industries. The country was formed in 1948 as a Jewish homeland. Arab countries and people rejected Israel's establishment. Israelis and Arabs have reached some agreements on Palestinian self-rule, but conflicts continue.
4. Syria, which stretches from the Mediterranean coast to Mesopotamia, is trying to modernize its agriculture. Lebanon lies between Israel and Syria and has been wracked by civil wars since the late 1950s. Jordan, stretching east of Israel, hosts many Palestinian refugees.

Building a Vocabulary

1. What is a kibbutz?
2. Define *Zionism*. Where did the movement originate?
3. What name has been given to the persecution and killing of Jews during World War II?

Recalling and Reviewing

1. What are the most densely settled areas of Turkey? What crops are grown in these areas?
2. What is the Palestinian Liberation Organization? What challenges does it face?
3. Why do most Syrians live in Syria's western region? On what does Syria's agricultural progress depend?
4. How is Lebanon culturally distinct in the region?
5. How have Jordan's Palestinian refugees presented a challenge for the country?

Thinking Critically

1. Would you describe the Arab-Israeli conflict as a religious conflict? Why or why not?
2. Using three examples from the chapter, discuss the effects that political conflicts can have on a country's economy.

Learning About Your Local Geography

Cooperative Project

On a kibbutz, members share equally all the responsibilities for its operation. To what extent does your classroom or school function in the same way? With your group, make a list of all of the maintenance chores required to keep your classroom or school a pleasant environment. Then, make a chart showing who is responsible for completing each chore. If certain chores are not already assigned, complete these chores with your group (if possible).

Using the Five Themes of Geography

Sketch a map of Israel, Lebanon, and Jordan. With arrows, show how the relative **locations** and borders of these countries have affected peoples' **movement** within the **region**.

CHAPTER 34

Interior Southwest Asia

GEOGRAPHY DICTIONARY

fallow
theocracy
pilgrimage
perimeter
sultanate

"We meet with many of them on the road to Mecca. From Java and Sumatra, from India and Afghanistan, from Morocco . . . they seek the fountain of eternal life and bliss. Colour and race and rank, like the distance of land and sea, yield to the supreme purpose, make the world of the pilgrim move in a common faith toward a goal unique, a shrine in the desert."

Ameen Rihani

Clay cuneiform tablet from ancient **Mesopotamia**

According to the theory of plate tectonics, a new ocean is being born under the Red Sea as Arabia and Africa slowly separate. While the mountains of western Saudi Arabia push upward, the earth's crust is buckling beneath the plains of Mesopotamia and the Persian Gulf.

The human geography of Interior Southwest Asia, however, is changing much more quickly than the landforms. Oil is the key reason. Beneath Mesopotamia and the Persian Gulf lie the world's largest deposits of petroleum. The discovery of this "black gold" in the region has brought wealth and increased Western contact. The inheritors of this vast treasure are now struggling to deal with the resulting changes in their societies.

Muslims at prayer in **Mecca, Saudi Arabia**

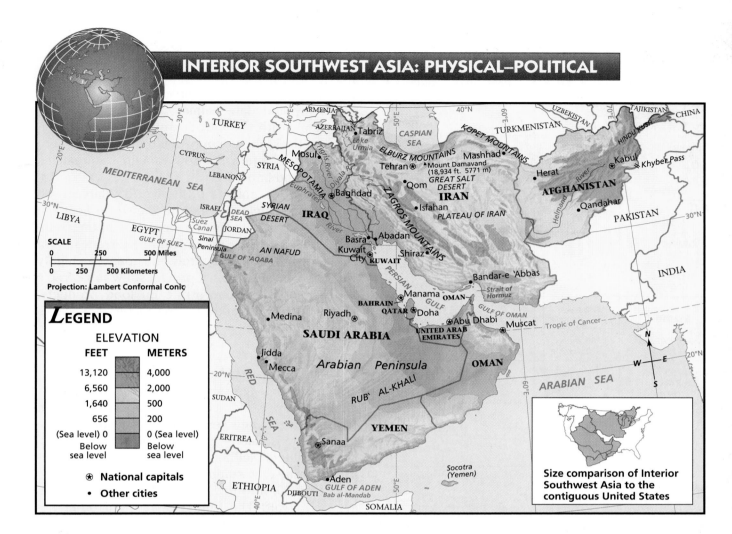

INTERIOR SOUTHWEST ASIA: PHYSICAL–POLITICAL

SCALE
0 250 500 Miles
0 250 500 Kilometers
Projection: Lambert Conformal Conic

LEGEND
ELEVATION

FEET	METERS
13,120	4,000
6,560	2,000
1,640	500
656	200
(Sea level) 0	0 (Sea level)
Below sea level	Below sea level

⊛ National capitals
• Other cities

Size comparison of Interior Southwest Asia to the contiguous United States

IRAQ

FOCUS

- *How has Iraq's physical geography affected the country's settlement patterns?*
- *What is the economic geography of Iraq? the political geography?*

LOCATION **Physical Geography** Iraq, the ancient land of Mesopotamia, lies at the southeastern end of the Fertile Crescent. The country's rich plains along the Tigris and Euphrates rivers are bounded by the Syrian and the An Nafud deserts in the west and by the Zagros Mountains of Iran in the east. In the north is a mountainous area, similar to nearby parts of Iran and Turkey. Most of Iraq's rain falls in these northern mountains. The plains, deserts, and mountains form Iraq's three landform regions.

Population In the northern part of Iraq, most of the people live in the mountain valleys and grow grains such as wheat and barley. Sheep, goats, and cattle graze on the mountain slopes. While some valleys have small irrigation projects, rain is the sole source of water for most crops. Many people in the north are Kurds, who continue to seek independence.

In the western deserts and dry grasslands, a few Bedouins still graze their flocks over wide areas. In the dry seasons, the nomadic Bedouins stop at oases and river settlements to trade their animal products for grain and manufactured goods. The traditional life of the nomad is ending, however. More people are settling in villages and on permanent pastures, as well as in cities.

The most populated areas of Iraq are the regions of farming villages located on the plains along rivers and irrigation canals. The capital city of Baghdad on the Tigris River is the country's largest city. It is a center of manufacturing,

FACTS IN BRIEF Interior Southwest Asia

COUNTRY POPULATION (1994)	LIFE EXPECTANCY (1994)	LITERACY RATE (1990)	PER CAPITA GDP (1993)
AFGHANISTAN 16,903,400	46, male 44, female	29%	$200 (1989)
BAHRAIN 585,683	71, male 76, female	77%	$12,000
IRAN 65,615,474	65, male 67, female	54%	$4,780
IRAQ 19,889,666	65, male 67, female	60%	$2,000 (GNP)
KUWAIT 1,819,322	73, male 77, female	73%	$15,100
OMAN 1,701,470	66, male 70, female	41% (1992)	$10,000
QATAR 512,779	70, male 75, female	76% (1992)	$17,500
SAUDI ARABIA 18,196,783	66, male 70, female	62%	$11,000
UNITED ARAB EMIRATES 2,791,141	70, male 74, female	73% (1992)	$24,000
YEMEN 11,105,202	50, male 53, female	38%	$800
UNITED STATES 260,713,585	73, male 79, female	97% (1991)	$24,700

NOTE: Recent GDP figures for Afghanistan are unavailable because of continuing conflict and instability.

Sources: *The World Factbook 1994,* Central Intelligence Agency; *1995 Britannica Book of the Year*

Why do you suppose the per capita GDP figures for some of these countries are as high as they are?

government services, and commerce. Baghdad is also an ancient city of Islamic culture and learning. The government is attempting to make Baghdad the grandest city in the Arab world. It has built impressive public buildings and freeways that contrast against the narrow, crowded streets of the old parts of the city.

Economic Geography Even though less than 15 percent of Iraq's land is arable, about one-third of the country's people are involved in agriculture. To conserve moisture and nutrients, croplands must be left **fallow**, or unused, every two or three years.

Irrigation has always been vital for agriculture in the region, and irrigation maintenance—clearing mud from canals and repairing flood damage—is a constant task for Iraq. Soil salinization presents another challenge to the region's agriculture. As irrigation water evaporates, salts build up in the soils. Over the centuries, a large part of Iraq's irrigated fields has become too salty for most crop growth. Some crops, however, such as barley and date palms, can grow in slightly salty soil. With the introduction of modern farming practices, the country could increase its agricultural production. For now, though, Iraq remains a major food importer.

Iraq has some of the world's largest oil deposits. Oil fields lie both in the south near the Persian Gulf and in the northern part of the country. Pipelines carry the oil to Mediterranean and Red Sea ports. Unfortunately, the country's highly mechanized oil industry provides few jobs. Iraq's other industries concentrate on food products, consumer goods, and building materials such as bricks and cement.

Political Geography In recent years, Iraq has been an aggressive state. After agriculture, the military is the country's largest employer. Hoping to gain new territory, Iraq invaded Iran in 1980. A decade-long war between the two countries resulted. Neither side gained significant territory in the devastating conflict.

In 1990, Iraq invaded the tiny, oil-rich country of Kuwait. The United States and other nations sent troops to keep Iraq from invading neighboring Saudi Arabia. Although the United Nations demanded that Iraq withdraw from

Although most of the population of **Iraq** lives in urban areas, these Arabs live in the marshlands of the southeastern part of the country.

Kuwait, Iraq refused. Finally, after a brief air and ground conflict, known as the Persian Gulf War, Iraq suffered a crushing defeat, and Kuwait was freed. Iraq's totalitarian government still clings to its ambition to become the leader of all of the Arab nations.

SECTION REVIEW

1. Which of Iraq's three landform regions is the most populated? Why?
2. What agricultural challenges does Iraq face? Where are the country's oil fields?
3. **Critical Thinking** **Suggest** why the United States became involved in the prevention of Iraq's expansion in 1990.

IRAN

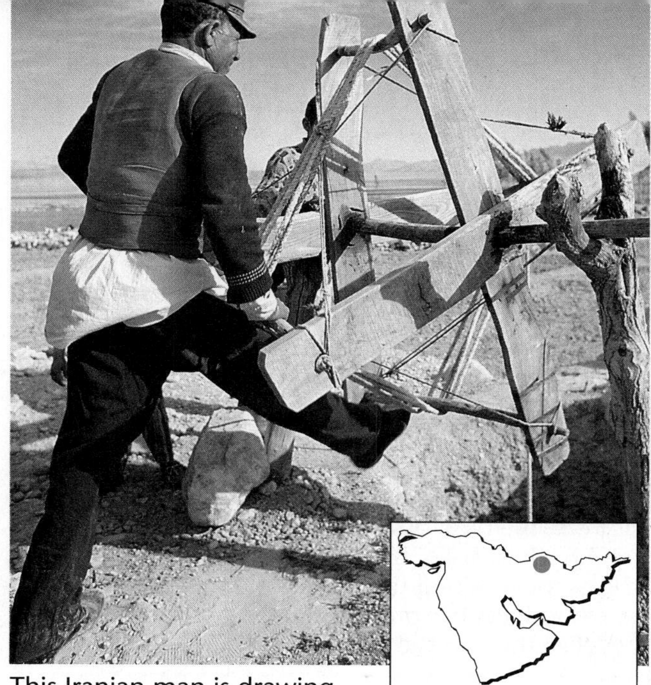

This Iranian man is drawing water from a well that is part of a *qanat* in northeastern **Iran**. Runoff from the mountains in the background provides water for the surrounding farmlands and communities.

FOCUS

- *What are the physical features of Iran? economic activities? towns and cities?*
- *What cultural characteristics does Iran have? How has the country changed politically over the past several decades?*
- *What are Afghanistan's key features?*

LOCATION **Physical Geography** Iran, formerly called Persia, is a country of arid highland plains and rugged mountains. (See the map on page 399.) The heart of the country is the high Plateau of Iran. The Zagros Mountains rise to the southwest, while to the north are the high Elburz Mountains and the Kopet Mountains. Iran's many mountains are steadily rising as a result of the region's tectonic activity, which also causes frequent earthquakes.

High pressure dominates the atmosphere over Iran most of the year, resulting in mainly dry, hot weather. Steppe and desert climates are typical. Central and southeastern Iran are particularly desolate. Winds blowing over the many dry lake basins create huge dust storms here. Rains come mostly during the winter and spring when middle-latitude storms sweep across the country. The wettest part of Iran lies along the Caspian Sea.

INTERACTION **Economic Geography** About one-third of Iran's people battle rugged land and difficult climates to make their living from farming. Barley and wheat are the major grains produced. Tree crops, particularly pistachio (puh-STASH-ee-oe) nuts and almonds, are grown in the southeast. In fact, Iran is the world's largest producer of pistachios. The lands along the shore of the Caspian Sea also are agriculturally important, producing fruits, tea, cotton, rice, and mulberry trees. The mulberry trees feed silkworms, which support Iran's carpet and clothing industries.

Rain-fed agriculture is possible only in the high plateaus and mountains of northern Iran. Elsewhere in the country, the waters of the mountains must be diverted to the fertile valley soils. The Iranian government has built many large-scale irrigation projects over the years. Individual communities manage thousands of additional small-scale irrigation projects. Iran also is famous for its *qanats* (kuh-NAHTZ). These long tunnels carry water from springs at the foot of the mountains to the plains. *Qanats*, which need a great deal of maintenance, are very deep near the mountains but rise to the surface near the farms. Water shortages and dry *qanats* have caused many people to migrate to towns from villages. (See "Planet Watch" on pages 382–383.)

This petrochemical plant near **Abadan, Iran**, is one of many similar industrial facilities found along the shores of the Persian Gulf. The oil business has provided some jobs for people at plants like these in the region.

Land reform programs have led to widespread individual farm ownership. Yet many farming communities still use the methods and tools of past centuries. The government has provided few funds for farmers and has kept food prices low. As a result, farmers cannot afford to invest in technology, and yields remain low by world standards. Like Iraq, Iran is a major food importer.

Other farming traditions remain as well. For example, a few peasants take part in a seasonal migration of people and livestock. Hoping for good winter rains, they plant grain on the plateaus during the fall. After the spring harvest, they move to mountain pastures, where their herds graze during the summer.

Iran was the first Persian Gulf country to develop its oil deposits, and in 1951, it nationalized its oil industry. Today, refineries, petrochemical plants, and busy ports provide some jobs. Iran's mountain regions also hold deposits of metallic ores, many of which were discovered recently through the analysis of satellite images. Iran once had an important fishery in the Caspian Sea, but pollution has caused its decline.

In many of Iran's smaller towns, local wool, silk, and metals are used in handmade products sold around the world. Persian rugs in particular have been prized for centuries. The wool for making rugs is cleaned, dyed, combed, and sent to rural homes, where it is spun into yarn. The yarn is then sent to skilled artisans, who knot designs that are hundreds of years old. Many Iranians buy rugs as investments.

PLACE **Urban Geography** As in other developing countries, rural-to-urban migration is typical in Iran. In the older sections of the towns and cities, life goes on today much as it did centuries ago. As in the Arab countries, a central marketplace, shielded from the sun and rain, displays goods such as rugs, cloth, household utensils, and spices. A mosque is always nearby. In contrast, the newer sections of Iranian cities have modern high-rise buildings and shopping centers clustered on wide avenues.

Tehran (tay-uh-RAN), Iran's capital city, lies on the northern edge of the Plateau of Iran at the foot of the Elburz Mountains. The city is more than 5,000 feet (1,524 m) above sea level. Tehran is the site of many government agencies and is Iran's most industrialized city. The city's population is over 7 million.

Abadan was founded along the Persian Gulf to market and transport Iran's oil exports. The once-empty desert land around the city became a busy industrial center with modern housing, hospitals, and schools. Abadan, however, was nearly destroyed in the Iran-Iraq war and will take years to rebuild. Iran's other important cities include Tabriz (tuh-BREEZ), Isfahan (is-fuh-HAHN), and Qom (KOHM). These cities are most famous as historic sites and religious centers, with beautiful mosques and religious schools and shrines.

Human Geography The Shi'a branch of Islam is the dominant religion in Iran and provides a key source of national unity. The population of Iran is ethnically diverse, however. Ethnic Persians make up about half the country's people. They speak Farsi, an Indo-European language. Many smaller ethnic groups live in the country's mountain valleys, where the surrounding mountains have protected them from repeated waves of invaders during past centuries.

Stormy politics have long characterized Iran. In ancient times, city-states and empires battled for control of the countryside. Feuds between the many ethnic groups raged almost continuously. More recently, in the decades after World War II, Iran's ruler, titled the Shah, used the country's oil income to launch a program of modernization and industrialization. The Shah was popular among developed countries, who benefited from Iran's industrial development.

Opposition to the Shah, however, grew steadily inside Iran. The Shah's economic changes widened the gap between the country's rich and poor. A small number of Iranians became wealthy and began enjoying Western ways of life. The growing number of poor Iranians had little to show for the country's new wealth. Also, the Shah's Westernization of Iran angered Iran's powerful Shi'a leaders, who believed the Shah's policies were not in line with traditional Islamic views.

After increasing unrest and violence, the Shah fled Iran in 1979. A fundamentalist Islamic government then came to power under the Ayatollah Khomeini. Among the Shi'a, an ayatollah is a religious leader of the highest rank and greatest respect.

Peace eventually returned to Iran through the use of religious courts. People who disobeyed religious law were imprisoned; some were executed. Many skilled and educated Iranians left the country. Khomeini died in 1989, but Iran remains a **theocracy**, a country governed by religious law.

The Islamic government of Iran is struggling to redirect the country's future. Its population is rising much faster than its economic growth, and poverty is increasing. Many industries also run irregularly. In addition, Iran's relations with many foreign countries, including some of its Arab neighbors and the United States, are strained.

LOCATION **Afghanistan** Afghanistan is a dry, mountainous country landlocked between Iran, Pakistan, and the Central Asian republics. The towering Hindu Kush range, which reaches up to 25,000 feet (7,620 m), cuts across northern Afghanistan. The country's location on the route between India and Central Asia has attracted nearby empires throughout its history. Merchants, warriors, and missionaries have all used the Khyber Pass on Afghanistan's border with Pakistan for thousands of years. (See the map on page 399.)

Afghanistan is one of the world's least developed countries. Farming and grazing livestock are the main occupations, though the country's steppe and desert climates limit agriculture to isolated, irrigated regions. Some Afghan farmers also raise opium poppies as a cash crop. This flower is the source of the powerful drugs heroin and morphine.

Kabul is the country's capital and largest city. Afghan cities employ people mainly in food-processing industries and government. Traditional crafts, though, remain an important source of cash income. Although Afghanistan's mountains likely hold important minerals, only the country's natural-gas deposits are developed.

The Pathans make up the largest ethnic group in Afghanistan. They are closely related to the Tajiks to the north. There are dozens of other ethnic groups as well. As for religion, almost all Afghans are Muslim. Disputes between and among Shi'a and Sunni are common.

Building a modern state in this culturally diverse region has been difficult. A civil war, which included intervention by the Soviet Union, raged during much of the 1980s. Although the Soviets withdrew from Afghanistan in 1989, fighting between rival political and religious groups has continued. An end to the fighting and an easing of political tensions are required for rebuilding the war-torn country.

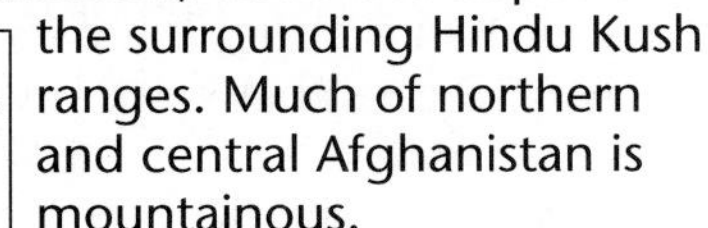

Residential neighborhoods in **Kabul**, the capital and largest city of **Afghanistan**, climb the slopes of the surrounding Hindu Kush ranges. Much of northern and central Afghanistan is mountainous.

2 SECTION REVIEW

1. What landforms dominate in Iran? In what part of the country is rain-fed agriculture possible? What is Iran's capital?
2. What branch of Islam is widespread in Iran? Why were the leaders of this branch offended by the Shah's policies?
3. **Critical Thinking** **Agree or disagree** with the following statement: "Afghanistan has more in common with its neighbors to the north than with its neighbors to the west." Explain your answer.

SAUDI ARABIA

FOCUS

- *How is Saudi Arabia's physical geography related to settlement in the country?*
- *How has Saudi Arabia's oil affected the country, internally and globally?*

REGION **Physical Geography** Saudi Arabia occupies much of the Arabian Peninsula. (See the map on page 399.) Most of the country is desert plains, which stretch from the Persian Gulf to the mountains along the Red Sea.

In the southeastern corner of Saudi Arabia lies one of the world's largest areas of windswept sand. It is the Rub' al-Khali, also called the Empty Quarter. This area is uninhabited, visited only occasionally by Bedouin herders. Beyond the sands rise the western mountains, the most prominent landforms in Saudi Arabia. These mountains line the Red Sea coast for more than 1,000 miles (1,609 km). Peaks in the southern end of the range reach nearly 10,000 feet (3,048 m).

Population The people of Saudi Arabia live where there is either water or oil. Mecca, built on an ancient oasis spring site in the western mountains, is the most important city of Islam. It was in Mecca that Muhammad lived and began teaching Islam. Muslims are required to make a **pilgrimage** (journey) to Mecca at least once in their lifetimes. Medina, also in the western mountains, is another holy city. Today, more than 2 million pilgrims from all parts of the world visit the holy cities each year. A more recent area of settlement in Saudi Arabia lies near the oil fields along the Persian Gulf. These underground fields may contain as much as one-fourth of the world's remaining oil.

Between the western mountains and the shores of the Persian Gulf, only isolated oases can support human settlement. The largest oasis is in the area surrounding Riyadh (ree-YAHD), the capital. The oases of the Saudi Arabian interior have been settled for thousands of years by nomadic Bedouins.

Economic Geography The Saudi economy is based on petroleum. Oil sales provide nearly all the government's income, and the total is almost

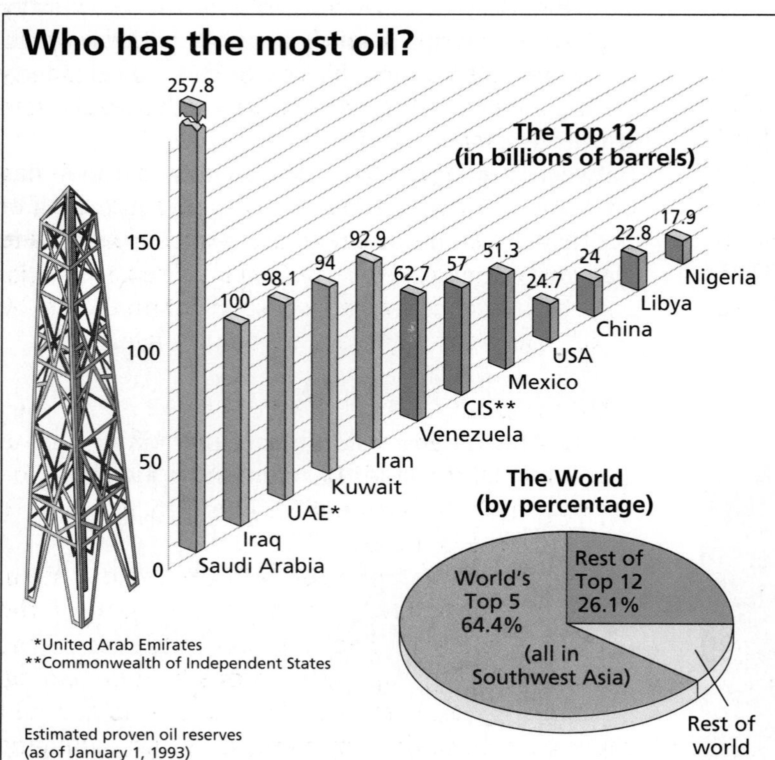
Who has the most oil?
The Top 12
(in billions of barrels)
257.8
150
100
50
0
100
98.1
94
92.9
62.7
57
51.3
24.7
24
22.8
17.9
Saudi Arabia
Iraq
UAE*
Kuwait
Iran
Venezuela
CIS**
Mexico
USA
China
Libya
Nigeria
The World
(by percentage)
World's Top 5 64.4%
(all in Southwest Asia)
Rest of Top 12 26.1%
Rest of world 9.5%
*United Arab Emirates
**Commonwealth of Independent States
Estimated proven oil reserves
(as of January 1, 1993)
Source: Oil and Gas Journal

SKILL STUDY The bar graph shows the 12 countries in the world with the largest estimated proven reserves of oil. The country with the most oil, Saudi Arabia, has more than two-and-a-half times the oil of Iraq, a distant second. The bars for the Top 5 countries, all of which are located along the Persian Gulf in Southwest Asia, are shown in orange. The bars for the remaining Top 12 countries are shown in green.

The pie graph shows that the Top 5 countries, shown in orange, have nearly two-thirds of the world's estimated proven reserves of oil. In all, the Top 12 countries control more than 90 percent of the world's oil. Some of the world's oil-producing countries cooperate with each other in setting prices and production goals for their oil. What effect would such an organization have if it decided to increase oil prices or lower the amount of oil produced, or both?

more than the country can spend. Government-paid social services, including education, are available to all.

Saudi Arabia's modern oil fields and related chemical industries, however, provide few jobs. To increase employment, the country has encouraged industry and the modernization of agriculture. Despite the country's desert environment, Saudi Arabia now boasts modern dairies and ranches. Dates, wheat, and barley are the main crops. Construction of seawater desalinization plants has been particularly important to the country's economy. Saudi Arabia produces more desalinized seawater than any other country in the world.

Young Saudis can be found behind desks as office workers, businesspeople, and bankers. Meanwhile, foreign workers from Egypt, Yemen, and South Asia perform much of the manual labor in the country. Before the development of the Saudi petroleum industry, the government received most of its money by taxing visitors to the holy cities. Pilgrims still provide a large amount of money for the government.

A Global Role As the world's largest oil exporter, Saudi Arabia has a powerful role both in Southwest Asia and in the world. By increasing or decreasing their oil exports, the Saudis have the power to influence oil prices worldwide.

The Saud family, which has ruled the country since its founding in 1932, is devoutly Muslim. In fact, Saudi society is organized under Islamic law. As a rich Muslim state, Saudi Arabia gives economic and military aid to the poorer states of the region. Although Saudi Arabia strongly defends the rights of the Palestinians, it has been patient in working for a peaceful settlement of the Arab-Israeli issue.

3 SECTION REVIEW

1. What parts of Saudi Arabia support human settlement?
2. How has the government of Saudi Arabia used its oil wealth? What has the oil industry not provided to most Saudis?
3. **Critical Thinking** **Predict** what might happen to world oil prices if Saudi Arabia decided to drastically reduce its oil exports. Explain your answer.

4 THE ARABIAN PERIMETER

FOCUS

- *What are the countries of the Arabian perimeter?*
- *What are the physical, economic, and political characteristics of these countries?*

Along the southern and eastern **perimeter** (outer boundary) of the Arabian Peninsula is a string of small, traditional Arab states. (See the map on page 399.) Most consist of a strip of coastline, one or two port cities, and an isolated area of dry, undeveloped countryside. All of these small countries have significant locations in relation to world shipping. They also hold important oil deposits.

Kuwait Kuwait is an oil-rich slice of desert land on the northern Persian Gulf between Iraq and Saudi Arabia. The country's population has grown rapidly with the development of its oil deposits. The government also receives substantial income from foreign investments.

Wealthy Kuwaiti families have ruled the country for many years, and foreign laborers perform much of the country's work. Before the 1990–1991 Persian Gulf War, many of these workers were Palestinians. Now, workers come mostly from poorer Arab countries and from South and Southeast Asia.

Soon after the war, Kuwait began rebuilding from the devastation caused by the Iraqi army. The country again has one of the highest per capita incomes in the world, as well as an extensive social welfare program. Seawater desalinization plants, oil fields, and highways are functioning again.

Gulf States Along the Persian Gulf between Saudi Arabia and Oman are the United Arab Emirates, Qatar, and Bahrain. All of these Gulf states are oil rich. Except for a few oases, most of the land here is desert.

Until oil was discovered in the region, life in the thinly populated states of this part of the Arabian Peninsula had changed little over the centuries. Traditional occupations include fishing, pearl diving, and trading. Oil wealth has brought rapid economic development, however, changing people's lives. Seaports, roads, modern housing,

A minaret stands out amid the distinct architecture of the old section of **Sanaa, Yemen**. Sanaa was founded before the first century A.D.

airports, and irrigation systems are springing up. The United Arab Emirates and Bahrain in particular have become Westernized. They are popular destinations for many tourists from Kuwait and Saudi Arabia.

Oman Oman, formerly known as Muscat and Oman, is a **sultanate**. A sultanate is a country that is ruled by a Muslim monarch (sultan). Oman has long been isolated from the modern world, and most of the country's people live by fishing and traditional agriculture. The government, however, has begun to use the income from the country's large oil deposits to improve its citizens' standard of living.

Yemen Yemen lies in the mountainous southwestern corner of the Arabian Peninsula. North Yemen and South Yemen merged in 1990 to form one country. South Yemenis sought independence again in 1994 but were defeated in a short war. Yemen has a strategic location next to the Bab al-Mandab, the narrow strait that connects the Red Sea with the Indian Ocean. Many ships pass through the strait on their way to the Suez Canal and the Mediterranean Sea.

Yemen is not an industrialized country. Instead, traditional agriculture, raising sheep and goats, and small-village life are typical. Crops grown on the coastal drylands include dates and grains. Humid mountain slopes support cultivation of fruit trees and vegetables. Coffee, though, is Yemen's most famous crop. In fact, the word *coffee* has its origin in Yemeni Arabic, and some believe that coffee may have been first cultivated here. Today, however, Yemen's coffee is not particularly successful on the world market. It is considered of low quality in comparison to the coffee from countries with better growing conditions.

Yemen has two principal cities. Its capital, Sanaa, lies on the high plateau east of the mountains. This ancient, walled city has a few factories that provide basic goods for the population. On the south coast of Yemen is the port city of Aden. Aden has an oil refinery and ancient trading links to the Persian Gulf and East Africa.

Many of Yemen's people exhibit more loyalty to their families and clans than to the national government. Conflicts among these groups take place frequently, hindering Yemen's development. The country's per capita income ranks as one of the world's lowest, and about 60 percent of the people are illiterate. Many young people seek unskilled jobs in Saudi Arabia and the small Gulf states. The money they send home is crucial to the country's economy.

In the early 1980s, oil was discovered in Yemen. Money gained from oil exports is providing Yemen with increasing financial independence. Yemen's foreign policy now stresses an independent course and seeks a more active role among the Arab nations.

4 SECTION REVIEW

1. Which Arabian perimeter countries border the Persian Gulf? Which country is next to the Bab al-Mandab?
2. What is the major source of income in the region? What kinds of political systems exist here?
3. **Critical Thinking** **Determine** what effects newly discovered oil deposits may have on the daily lives of people in Oman and Yemen. Explain your answer.

CHAPTER 34 CHECK

Khyber Pass between **Afghanistan** and **Pakistan**

Reviewing the Main Ideas

1. Iraq lies at the southeastern end of the Fertile Crescent. Baghdad is the capital and largest city. Iraq has large oil reserves, but many of the country's people are involved in agriculture. In recent years, Iraq has been politically aggressive.
2. East of Iraq lies Iran. Iran was the first Persian Gulf country to develop its oil deposits. Rural-to-urban migration is common. Iran's population is ethnically diverse, and turbulent politics have long characterized the country.
3. Afghanistan is a dry, mountainous country located east of Iran and south of the Central Asian republics.
4. Saudi Arabia occupies much of the Arabian Peninsula and holds one of the world's largest areas of windswept sand as well as the most important city of Islam, Mecca. The Saudi economy is based on petroleum.
5. The countries of the eastern and southern Arabian perimeter include Kuwait, the Gulf states, Oman, and Yemen. All of these countries have valuable oil deposits.

Building a Vocabulary

1. Define *fallow*. Why must Iraq's croplands be left fallow occasionally?
2. What is a pilgrimage? To what city must Muslims make a pilgrimage at least once in their lifetimes?
3. Define *sultanate*. Which Arabian perimeter country is a sultanate?

Recalling and Reviewing

1. What people of northern Iraq seek their independence?
2. What purpose do *qanats* serve? Which country is famous for its *qanats*?
3. Which two countries discussed in the chapter engaged in a decade-long war during the 1980s?
4. Which country discussed in this chapter is not oil rich?
5. What is Saudi Arabia's capital? Why is the city located where it is?
6. How has the recent discovery of oil in Yemen affected the country as a whole?

Thinking Critically

1. What factors do you think have prevented Iraq and Iran from using the profits from their oil to modernize agriculture?
2. Is Saudi Arabia a theocracy? Support your answer.

Learning About Your Local Geography

Individual Project

Persian rugs are world famous. Conduct a survey of community members to find out what craft industries are important in your community. Is your community "famous" for any crafts? What crafts do community members produce? Do they sell them, and if so, where? Does the community have many crafts shops? Draw some illustrations of different crafts produced by your community members. You may want to include your illustrations in your individual portfolio.

Using the Five Themes of Geography

Write a paragraph about the theme of **human-environment interaction** in the countries discussed in this chapter. How have people changed the natural environment? How have they adapted to it? What forms of interaction do you think will take place in the future?

SKILLS

Linking Geography and History

No understanding of the present can be complete without an understanding of the past. Only by examining the history of a region can we make sense out of its modern geography. Similarly, only by studying the history of a group of people—where they originated and how and where they spread—can we understand the group's human geography today.

For example, to explain the current human geography of the Muslim world, we must study how Islam and Arab culture spread in earlier centuries. At the start of the seventh century, Arabs were a small, divided group of peoples scattered across the arid Arabian Peninsula. By 750, however, Arabs controlled an area stretching from Spain, across North Africa and western Asia, to India, and from Central Asia to the Arabian Peninsula. In just over a century, the Arabs had built the largest empire the world had ever seen.

What brought the Arabs together and allowed them to expand over such a vast area during those 150 years? The unifying force was the new religion of Islam. Muhummad, Islam's founder, began preaching in Mecca in 610. Muhummad left Mecca in 622 and settled in the town of Medina 200 miles (322 km) away. He returned to Mecca several years later, now at the head of an army of devoted followers. The simple message of Islam—that there was only one God, Allah, whose prophet was Muhammad—had great appeal in the troubled world of the seventh century. Islam was not just a system of religious beliefs, but a way of life.

Inspired by their new religion, Arabs swept out of the Arabian Peninsula to spread Islam. Partly as a result of internal divisions among the peoples they encountered, the Arabs met with great success. In some places, peoples who had been the victims of religious persecution welcomed the often more tolerant Muslims. During the seventh and eighth centuries, Islam became a major world religion and Christianity's main rival.

The spread of Islam and the Arabic language gave unity to a vast area and created a vibrant Islamic culture. The center of the Islamic world was the newly built city of Baghdad in Mesopotamia. Baghdad also became a center of education, art, and trade. Tigers, rubies, and ebony came to the city from India. Porcelain, paper, and ink came from China. Ivory, gold, and slaves came from Africa.

Islamic culture drew on a number of ancient civilizations. The philosophy and science of the Greeks, the artistic and administrative skills of Persia, and the mathematical discoveries of India mingled in the Islamic world. Scholars built on the knowledge of these older civilizations to make important progress in astronomy, geography, physics, chemistry, and mathematics.

Although Islamic power began to decline in the thirteenth century, the spread of Islam continued. Muslim traders and missionaries traveled east to Malaysia, Indonesia, and the Philippines and south farther into Africa. The Mongols, who built a Eurasian empire in the thirteenth century, converted to Islam and carried it deeper into Asia. The Turkish Ottoman Empire's capture of Constantinople in the fifteenth century and other Ottoman conquests spread Islam into Eastern Europe's Balkan Peninsula. In the sixteenth century, Muslim rulers created the powerful Mogul Empire in India that lasted for 300 years.

Islam continues to thrive. Today, it is the religion of the majority of the population throughout North Africa, in most of western and Central Asia, and in much of southern and southeastern Asia. Islam also has a large following in parts of Africa south of the Sahara and keeps a foothold in Europe's Balkan Peninsula. The United States has a large and growing Muslim population as well.

PRACTICING THE SKILL

Select a local, regional, or national historical event in which you are interested. Research the event. How was geography involved? Did land or water play a role? What types of movement took place to cause the event or as an effect of it? Did the event change the physical or human geography of any areas? Use maps and illustrations, with captions, to show how the event demonstrates the link between geography and history.

BUILDING YOUR PORTFOLIO

Individually or in a group, complete one of the following projects to show your understanding of the geography concepts involved.

A *Three Religions*

You have read that Southwest Asia was the birthplace of three major world religions—Judaism, Christianity, and Islam. Find out more about each of these three religions, beyond what you read in the unit. Make a poster to display the information you collect. To make your poster, you will need to do the following.

1. Investigate the origins and histories of the three religions. Exactly where in Southwest Asia did each one form? How did each of the three religions spread to other parts of the world? What people were involved in the formation of each one, and what roles did they play? For your poster, sketch a map showing the places of origin and the spreading of each religion. Under the map, include a caption discussing other aspects of the religions' origins and histories.
2. Find out more about particular beliefs and practices of each religion. What are the central ideas of each religion? What customs do the followers of each religion have? What forms of worship do they participate in? How do their beliefs and customs affect their daily lives? Sketch drawings showing some of the different beliefs or practices of each religion. Next to each drawing, write several statements explaining the belief or practice. Put your drawings and statements on your poster.
3. Explore the different groups and branches within each of the three religions. How is each religion divided internally? How did these divisions arise? How do each religion's branches differ in beliefs or practices from the other branches of the same religion? Make a chart showing this information and include it on your poster.

Present your poster to the rest of your class.

B *Oil Makes the World Go Round*

Southwest Asia contains some of the world's largest oil deposits. The region's oil resources are of great economic value and political significance, not only in the region but around the world. Write the script for a brief documentary film about the region's oil. Include with your script illustrations, maps, and other visual materials that will help your audience understand your points. To prepare your script, think about the following questions.

1. Where are Southwest Asia's major oil resources located? How do people remove the oil from the earth, and how do they process it afterwards? To where is the oil exported, and how is it transported from Southwest Asia to its destinations? Your documentary film might use a map to show this information.
2. How is oil used, in the region and in the world? How do the countries of Southwest Asia use the money they make from oil? To what extent has oil wealth changed the daily lives of most people in these countries? Make rough sketches of the images your documentary film might use to show oil-based products and the effects of oil wealth in Southwest Asia.
3. How does Southwest Asia's oil affect political events, both in the region and around the world? What role has oil played in recent conflicts within the region? Your documentary might include a made-up interview with or dialogue between experts discussing these questions.

Present your script and supporting materials to the rest of the class.

Differences and Connections

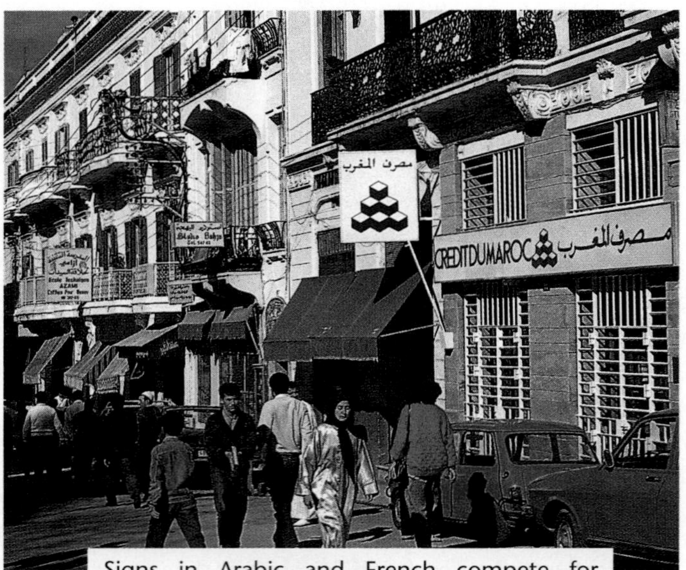
مصرف المغرب
CREDITDUMAROC
مصرف المغرب

Signs in Arabic and French compete for attention in the North African country of **Morocco**. Arabic is spoken throughout North Africa and many parts of Southwest Asia.

In this textbook the countries of Africa are grouped together to emphasize the region's connections. Yet in any region covering an area as large as Africa, there are important differences within the region. Thus, Southern Africa is different from West Africa, and West Africa is different from North Africa. This is no less true of other large regions of the world. Venezuela, for example, is different from Argentina in many important ways.

North Africa and Southwest Asia

One subregion of Africa that geographers often include in another region is North Africa. These geographers see greater connections between North Africa and Southwest Asia than they see between North Africa and the rest of the continent. For example, in both areas Arabic is the main language, and Islam is the major religion. Political issues also tie North Africa to its eastern neighbors, as does the physical geography of the region. The countries of North Africa and Southwest Asia are all a part of a vast desert region. They face common issues such as water conservation and management.

The African Continent

There are just as many reasons for placing North Africa in a region with the rest of the African continent. People in this part of the world share important historical, cultural, and economic ties. For example, the ancient Egyptians maintained close contact with peoples farther up the Nile River, in the interior of Africa. In turn, cultures south of the Sahara contributed to the development of the great Nile Valley civilizations of Egypt. Farther west, Mediterranean peoples have a long history of trading with West African kingdoms south of the Sahara. The Islamic religion, meanwhile, spread southward through contact between the peoples of North Africa and the rest of the continent. Today, mosques can be found as far south as Nigeria and Tanzania.

North Africa also has political connections to the rest of Africa. Many African countries face similar political and economic issues that

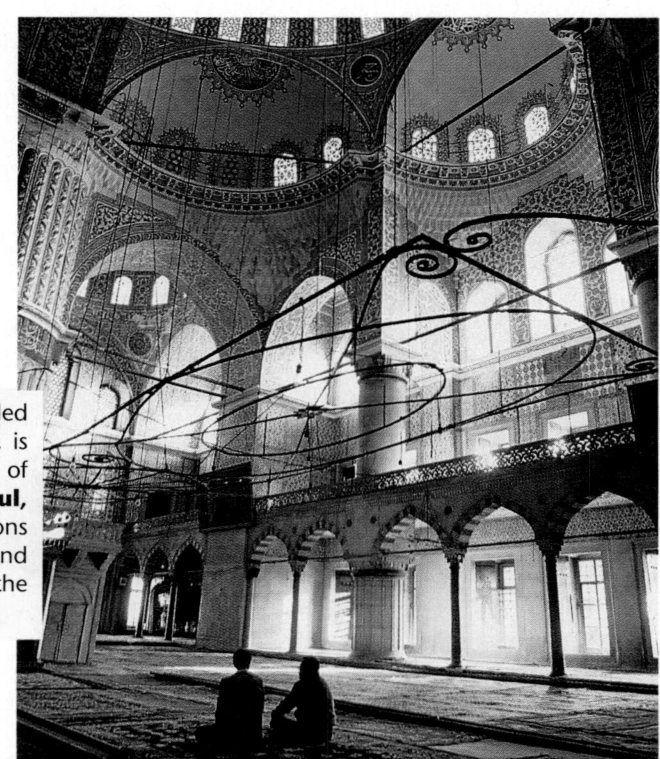

The Blue Mosque, so called because of its interior blue tiles, is one of many beautiful places of worship for Muslims in **Istanbul, Turkey**. Although various religions are practiced in North Africa and Southwest Asia, most people in the two regions are Muslims.

Regional Links

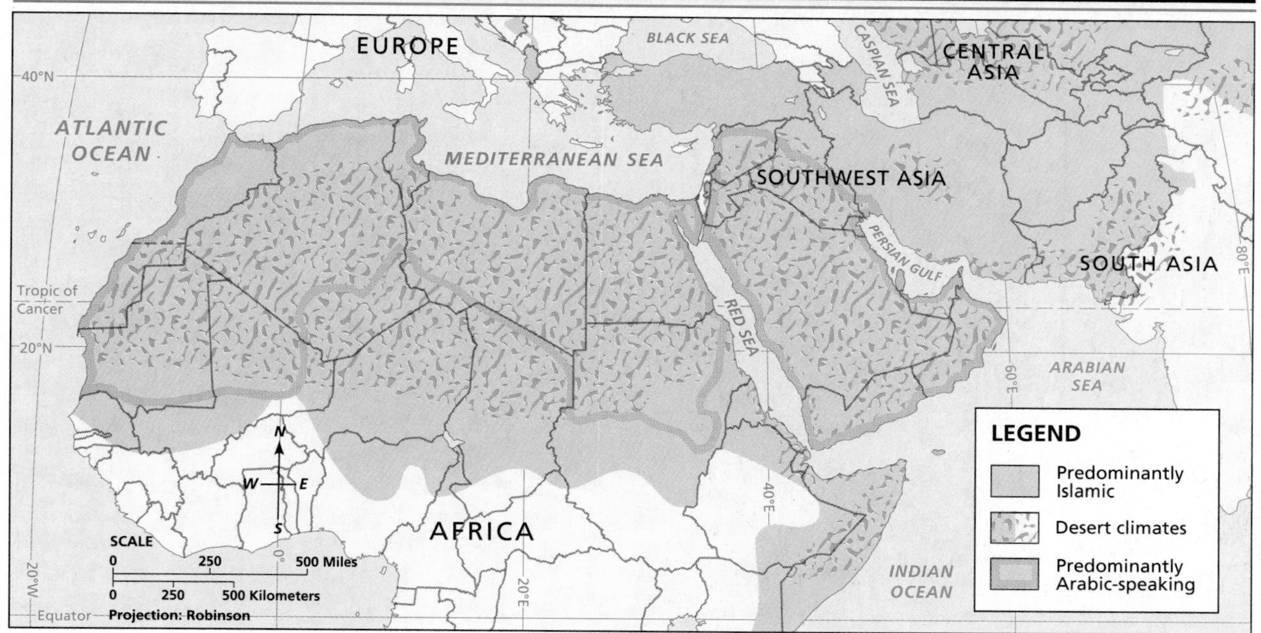
EUROPE
BLACK SEA
CASPIAN SEA
CENTRAL ASIA
40°N
ATLANTIC OCEAN
MEDITERRANEAN SEA
SOUTHWEST ASIA
PERSIAN GULF
SOUTH ASIA
80°E
Tropic of Cancer
RED SEA
20°N
60°E
ARABIAN SEA
LEGEND
Predominantly Islamic
Desert climates
Predominantly Arabic-speaking
N
W
E
S
40°E
AFRICA
SCALE
0 250 500 Miles
0 250 500 Kilometers
Projection: Robinson
20°W
0°
20°E
INDIAN OCEAN
Equator

Linguistic, religious, and climatic characteristics are only some of the links between the regions of North Africa and Southwest Asia. This map shows Arabic speakers, Muslims, and desert climates are dominant throughout much of the two regions and in parts of the surrounding transition zones.

stem in part from their shared colonial history. To resolve some of these issues, the countries of North Africa are working with other African nations through associations such as the Organization of African Unity.

Transition Zones

The differences and connections between North Africa and the rest of the continent are perhaps most evident in a band of countries that lie just south of the Sahara. These countries, from Mauritania in the west to Ethiopia and Sudan in the east, form a transition zone because they are as similar to North Africa as to Africa south of the Sahara. In Chad and Sudan, for example, there are strong political and cultural differences between north and south. The northern areas of these countries are tied closely to North Africa, while the southern parts of these countries are tied closely to African nations to the south.

In effect, North Africa and the countries in the transition zone have important connections to two major world regions. How they are organized depends on the geographer's point of view. Understanding the differences among countries and their connections to other areas of the world is more important than determining where on a map a region begins or ends.

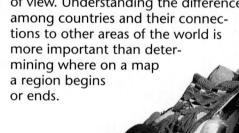

This Arab woman and her child trek through one of the desert regions that cover much of North Africa and Southwest Asia.

UNIT 8

AFRICA

GIVE WAY
GIVE WAY

Purple-blossomed jacaranda trees line many streets of springtime **Harare, Zimbabwe**. In addition to being the national capital, Harare is the commercial and banking center of the country.

Early African mask from present-day **Côte d'Ivoire**

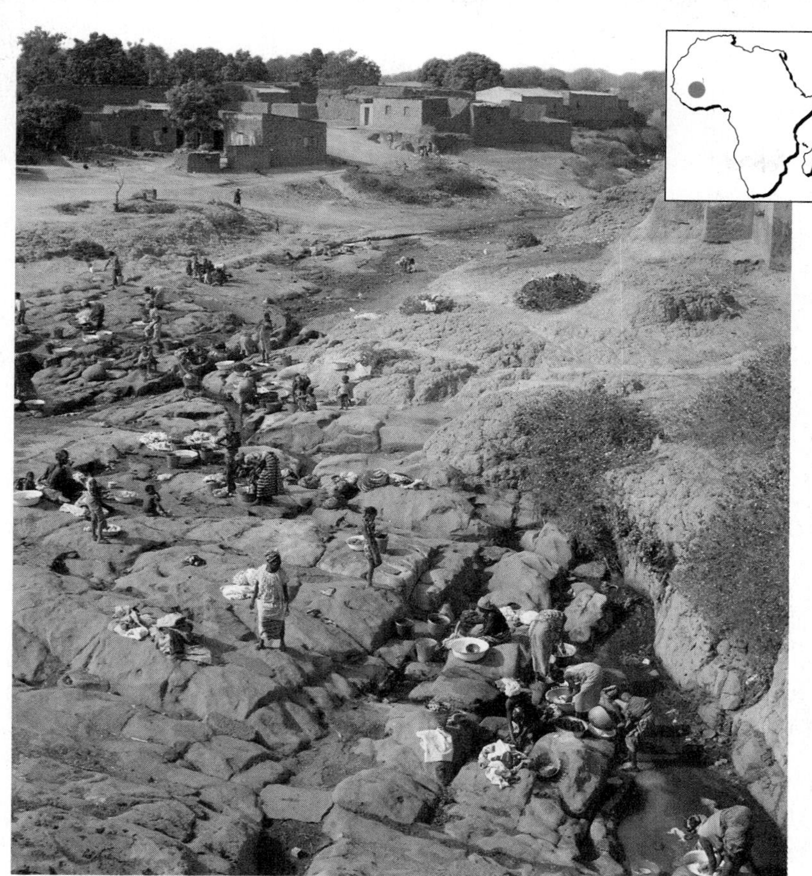

UNIT 8 CHAPTERS

Water, a valuable resource everywhere, is scarce in some parts of Africa. This nearly dry riverbed is near **Bobo-Dioulasso** in southwestern **Burkina Faso** in the Sahel. Burkina Faso is a landlocked country that was plagued by severe drought throughout the 1970s and 1980s.

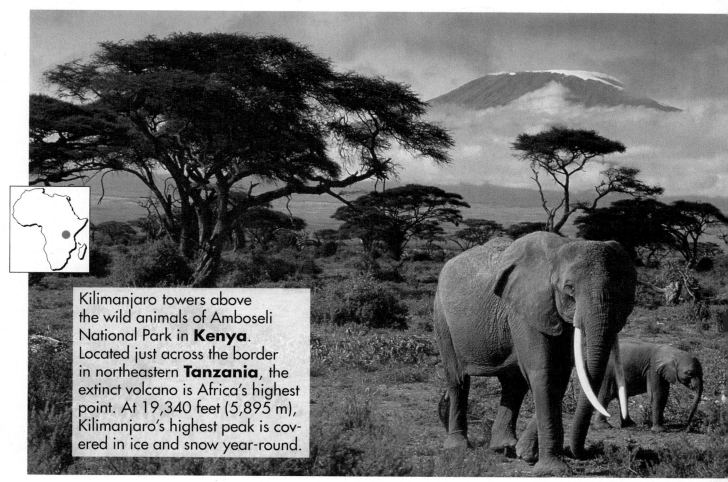

Kilimanjaro towers above the wild animals of Amboseli National Park in **Kenya**. Located just across the border in northeastern **Tanzania**, the extinct volcano is Africa's highest point. At 19,340 feet (5,895 m), Kilimanjaro's highest peak is covered in ice and snow year-round.

LITERATURE

SOUFFLES
(IT IS THE BREATH OF THE ANCESTORS)

LISTEN MORE OFTEN TO THINGS THAN TO BEINGS
HEAR THE FIRE'S VOICE,
HEAR THE VOICE OF WATER.
HEAR, IN THE WIND, THE SOBBING OF THE TREES.
IT IS THE BREATH OF THE ANCESTORS.

THE DEAD ARE NOT GONE FOREVER
THEY ARE IN THE PALING SHADOWS,
THEY ARE IN THE DARKENING SHADOWS.
THE DEAD ARE NOT BENEATH THE GROUND,
THEY ARE IN THE RUSTLING TREE,

IN THE MURMURING WOOD,
THE FLOWING WATER,
THE STILL WATER,
IN THE LONELY PLACE, IN THE CROWD;
THE DEAD ARE NEVER DEAD.

LISTEN MORE OFTEN TO THINGS THAN TO BEINGS.
HEAR THE FIRE'S VOICE.
HEAR THE VOICE OF WATER.
IN THE WIND HEAR THE SOBBING OF THE TREES.
IT IS THE BREATH OF THE ANCESTORS.
THEY ARE NOT GONE
THEY ARE NOT BENEATH THE GROUND
THEY ARE NOT DEAD.

THE DEAD ARE NOT GONE FOREVER.
THEY ARE IN A WOMAN'S BREAST,
A CHILD'S CRY, A GLOWING EMBER.
THE DEAD ARE NOT BENEATH THE EARTH,
THEY ARE IN THE FLICKERING FIRE,
IN THE WEEPING PLACE, THE HOME.
THE DEAD ARE NEVER DEAD.

LISTEN MORE OFTEN TO THINGS THAN TO BEINGS
HEAR THE FIRE'S VOICE,
HEAR THE VOICE OF WATER.
HEAR, IN THE WIND, THE SOBBING OF THE TREES.
IT IS THE BREATH OF THE ANCESTORS.

Translated by Samuel Allen
From Poems from Africa

Birago Diop (1906–1989) is one of Africa's best-known writers. He was born in Senegal, a former French colony. While completing his university studies in Paris, Diop joined other French-speaking African writers in a literary movement that emphasized African traditions. Diop feared that European colonialism threatened the existence of many unique African cultures. When he returned to his homeland, he developed his literary talents while pursuing a career as a veterinary surgeon. The poem at left reminds modern Africans to remember the teachings and culture of their ancestors. It also reflects important themes in contemporary African literature—respect for nature and the connection between the physical world and the spiritual realm.

Interpreting Literature

1. **How do the ancestors mentioned in the poem pass on their teachings and culture to the living?**
2. **How might the message in this poem apply to cultures outside of Africa?**

FOR THE RECORD

Captain Harry Dean (1864–1935) was an African American sea captain. Before he was 15 years old, Dean had sailed around the world with his uncle. On his first voyage to Africa, the young man formed an emotional bond with Africa that lasted his lifetime. Dean had studied the accomplishments and glories of powerful ancient African trading kingdoms. He longed to restore "Africa for the Africans" and undo the effects of European colonialism. As a grown man, Dean sailed his own ship, *Pedro Gorino*, to Africa. He believed the continent's development depended on establishing a great African merchant fleet. The excerpt below is from his autobiography, *Umbala*. Dean recorded his thoughts during an overland journey across southern Africa in the early 1900s.

As we were travelling less than thirty miles a day we had plenty of time for thought. In three days of dreaming I rebuilt the Ethiopian Empire. In three days of dreaming I recaptured Africa for the Africans. . . . The ruins of Zimbabwe were no longer ruins, but stately masonry. The sons of the ancient race who raised those piles of stone to forgotten gods once more were proud possessors of all they surveyed. And those dark descendants of the Phoenicians, still worshipping the crane and the ram, reattained the genius of their ancestors, sailing their ships to every country, bearing the wealth of Africa. As in the ancient days, precious stones and metals poured from Sheba northward through all Arabia, and westward down the wide rivers of the jungle. Nowhere were there slaves, or poverty, or ignorance. In three days of dreaming I dammed the rivers to water the karroo [a dry region of southern Africa] until the desert bloomed like the rose. I built cities in trackless thickets, and from the forests of Africa constructed such a fleet of graceful ships as the world has never seen.

A fisher casts a net on the **Zaire River**.

Analyzing Primary Sources

1. What natural resources and physical geography features of Africa does Captain Harry Dean mention?
2. What dreams does Dean have for Africa?

CHAPTER 35

Introducing Africa

GEOGRAPHY DICTIONARY

escarpment
rift
Sahel
animism

Cave paintings, **South Africa**

Africa contains the world's oldest archaeological sites. They tell us about the places where humans first gathered around campfires to share food and tell stories. The African cultures descended from these early societies have given the world a wealth of art, architecture, music, and folktales. Africa also is rich in natural resources that are a key part of the world economy.

Africa stretches some 5,000 miles (8,045 km) from north to south. The shifting sands and gravel plains of the Sahara separate Africa's Mediterranean coast from the greater part of Africa lying farther south. Current political boundaries divide the region into more than 50 countries. Modern Africa also includes several hundred ethnic groups with different languages and religions. This diversity adds a richness to African cultural life but has also led to conflicts among and within countries.

Africa has modern cities with skyscrapers and superhighways. These symbols of economic development, however, are often overshadowed by more common signs of poverty. Many Africans have too little to eat, poor health services, and few opportunities to go to school. Populations are rising rapidly. Creating jobs for the increasing number of people is a struggle all African governments share.

A young girl near **Timbuktu, Mali**

MAPPING THE REGION

AFRICA

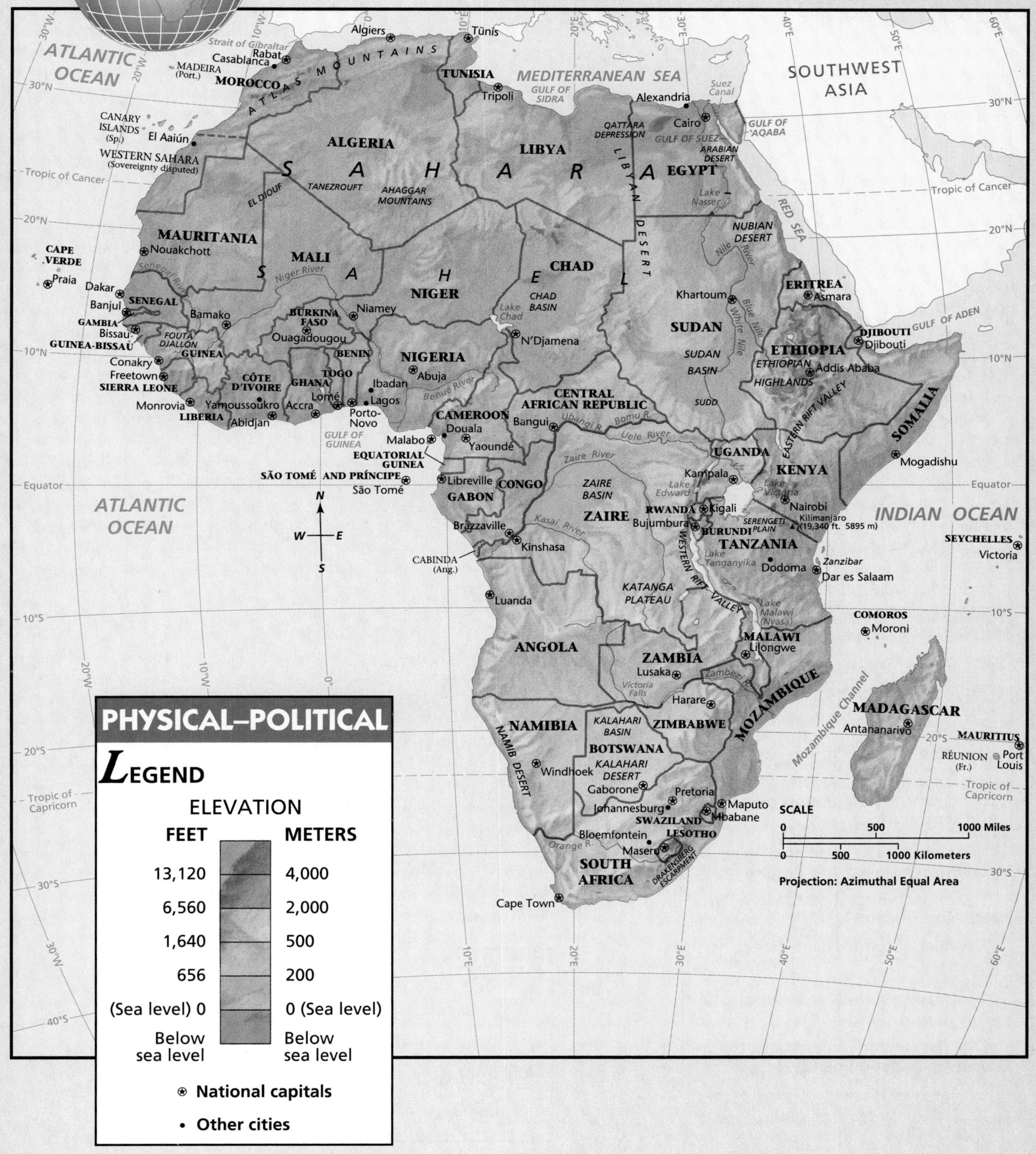

ATLANTIC OCEAN
MADEIRA (Port.)
Strait of Gibraltar
Rabat
Casablanca
MOROCCO
ATLAS MOUNTAINS
Algiers
Tūnis
TUNISIA
Tripoli
MEDITERRANEAN SEA
GULF OF SIDRA
Alexandria
Suez Canal
SOUTHWEST ASIA
CANARY ISLANDS (Sp.)
El Aaiún
WESTERN SAHARA (Sovereignty disputed)
ALGERIA
LIBYA
QATTARA DEPRESSION
Cairo
GULF OF SUEZ
GULF OF AQABA
ARABIAN DESERT
EGYPT
SAHARA
LIBYAN DESERT
TANEZROUFT
AHAGGAR MOUNTAINS
EL DJOUF
Tropic of Cancer
Lake Nasser
RED SEA
NUBIAN DESERT
MAURITANIA
Nouakchott
CAPE VERDE
Praia
MALI
SAHEL
Niger River
Senegal R.
CHAD
CHAD BASIN
NIGER
Nile River
Dakar
SENEGAL
Banjul
GAMBIA
Bissau
GUINEA-BISSAU
Bamako
BURKINA FASO
Niamey
Lake Chad
N'Djamena
Khartoum
ERITREA
Asmara
GULF OF ADEN
SUDAN
SUDAN BASIN
Blue Nile
White Nile
DJIBOUTI
Djibouti
FOUTA DJALLON
GUINEA
Ouagadougou
BENIN
NIGERIA
ETHIOPIA
ETHIOPIAN HIGHLANDS
Addis Ababa
Conakry
Freetown
SIERRA LEONE
CÔTE D'IVOIRE
GHANA
TOGO
Abuja
Ibadan
Lagos
Benue River
Lomé
Accra
Porto-Novo
Yamoussoukro
Monrovia
LIBERIA
Abidjan
CENTRAL AFRICAN REPUBLIC
SUDD
EASTERN RIFT VALLEY
SOMALIA
CAMEROON
Douala
Bangui
Ubangi R.
Bomu R.
Uele River
GULF OF GUINEA
Malabo
Yaoundé
EQUATORIAL GUINEA
Zaire River
UGANDA
KENYA
Mogadishu
SÃO TOMÉ AND PRÍNCIPE
São Tomé
Libreville
GABON
CONGO
ZAIRE BASIN
Kampala
Lake Edward
Lake Victoria
Equator
Nairobi
ZAIRE
RWANDA
Kigali
Bujumbura
BURUNDI
SERENGETI PLAIN
Kilimanjaro (19,340 ft. 5895 m)
INDIAN OCEAN
Brazzaville
Kasai River
Kinshasa
WESTERN RIFT VALLEY
TANZANIA
Lake Tanganyika
Dodoma
Zanzibar
Dar es Salaam
SEYCHELLES
Victoria
CABINDA (Ang.)
Luanda
KATANGA PLATEAU
Lake Malawi (Nyasa)
COMOROS
Moroni
MALAWI
Lilongwe
ANGOLA
ZAMBIA
Lusaka
Zambezi R.
Victoria Falls
MOZAMBIQUE
Mozambique Channel
Harare
MADAGASCAR
Antananarivo
NAMIBIA
KALAHARI BASIN
ZIMBABWE
MAURITIUS
Port Louis
RÉUNION (Fr.)
NAMIB DESERT
BOTSWANA
Windhoek
KALAHARI DESERT
Gaborone
Pretoria
Tropic of Capricorn
Johannesburg
Maputo
Mbabane
SWAZILAND
Bloemfontein
LESOTHO
Maseru
Orange R.
SOUTH AFRICA
DRAKENSBERG ESCARPMENT
Cape Town
N
W
E
S
SCALE
0
500
1000 Miles
0
500
1000 Kilometers
Projection: Azimuthal Equal Area
30°N
20°N
10°N
10°S
20°S
30°S
40°S
30°W
20°W
10°W
0°
10°E
20°E
30°E
40°E
50°E
60°E
PHYSICAL–POLITICAL
LEGEND
ELEVATION
FEET
METERS
13,120
4,000
6,560
2,000
1,640
500
656
200
(Sea level) 0
0 (Sea level)
Below sea level
Below sea level
National capitals
Other cities

MAPPING THE REGION

AFRICA

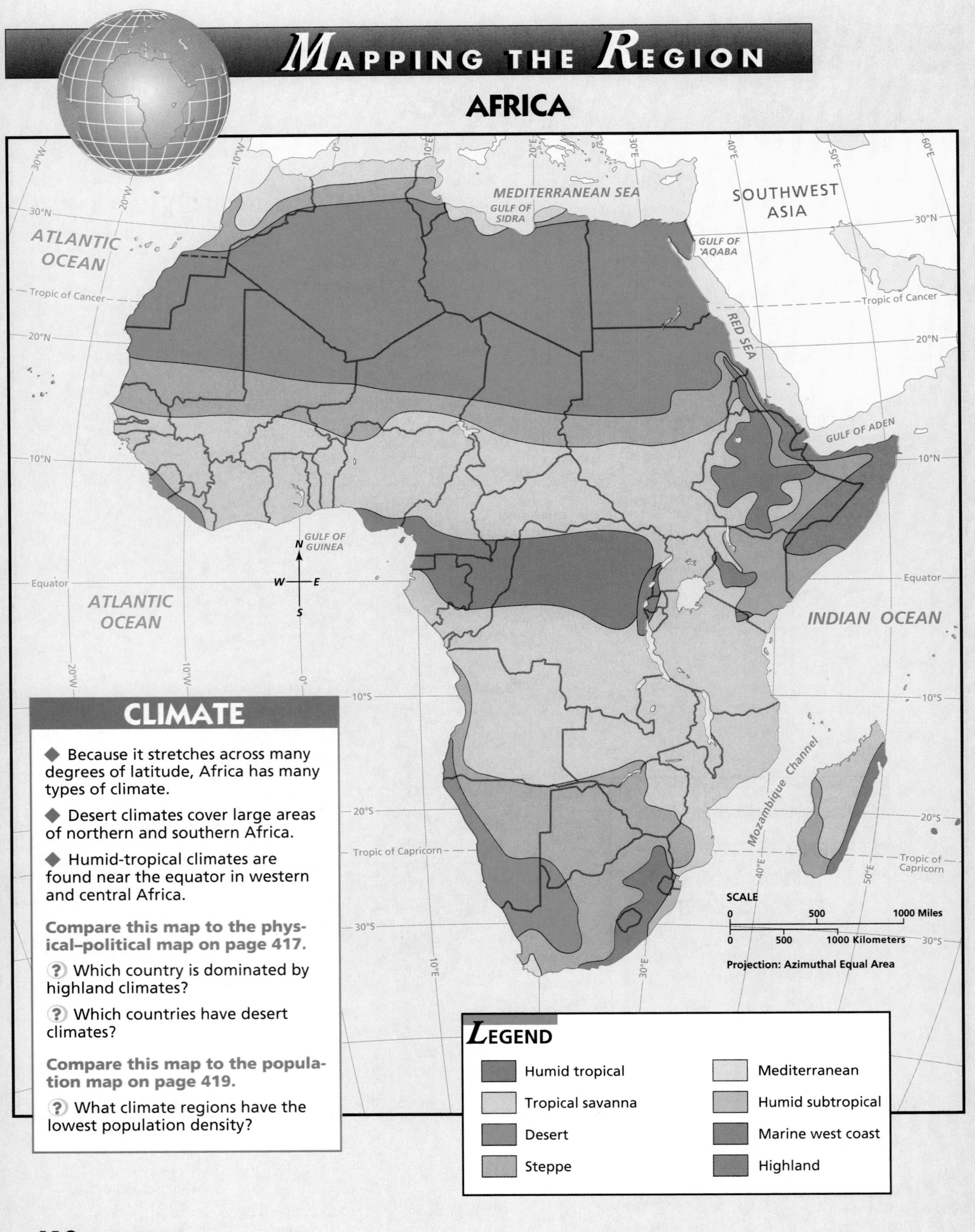
MEDITERRANEAN SEA
GULF OF SIDRA
SOUTHWEST ASIA
GULF OF AQABA
ATLANTIC OCEAN
RED SEA
GULF OF ADEN
GULF OF GUINEA
ATLANTIC OCEAN
INDIAN OCEAN
Mozambique Channel
Tropic of Cancer
Equator
Tropic of Capricorn
30°N
20°N
10°N
10°S
20°S
30°S
30°W
20°W
10°W
0°
10°E
20°E
30°E
40°E
50°E
60°E
N
S
E
W
SCALE
0
500
1000 Miles
0
500
1000 Kilometers
Projection: Azimuthal Equal Area
LEGEND
Humid tropical
Tropical savanna
Desert
Steppe
Mediterranean
Humid subtropical
Marine west coast
Highland

CLIMATE

◆ Because it stretches across many degrees of latitude, Africa has many types of climate.

◆ Desert climates cover large areas of northern and southern Africa.

◆ Humid-tropical climates are found near the equator in western and central Africa.

Compare this map to the physical–political map on page 417.

? Which country is dominated by highland climates?

? Which countries have desert climates?

Compare this map to the population map on page 419.

? What climate regions have the lowest population density?

MAPPING THE REGION

AFRICA

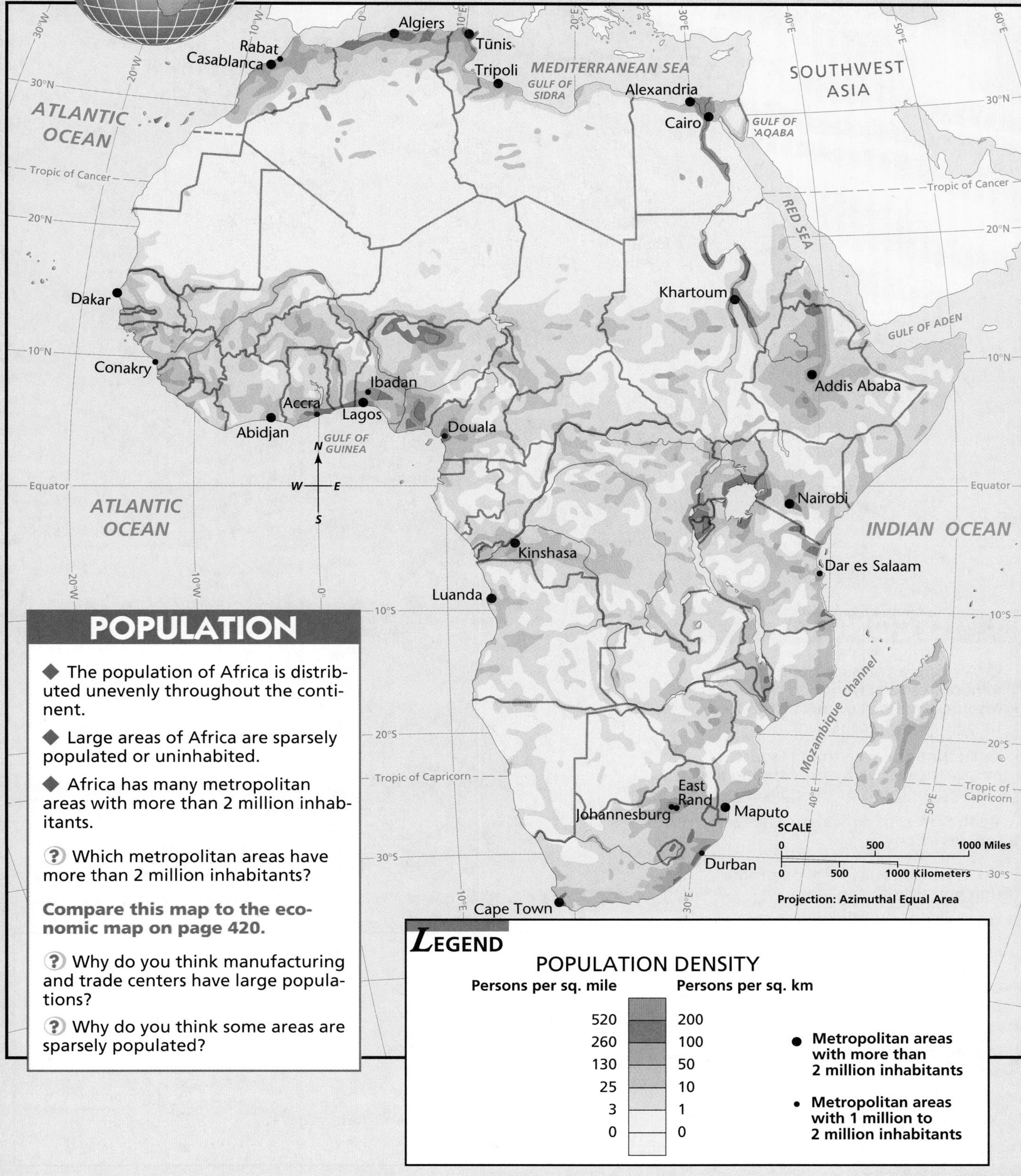

Algiers
Rabat
Casablanca
Tūnis
Tripoli
MEDITERRANEAN SEA
GULF OF SIDRA
Alexandria
Cairo
SOUTHWEST ASIA
GULF OF 'AQABA
ATLANTIC OCEAN
Tropic of Cancer
RED SEA
Khartoum
GULF OF ADEN
Dakar
Conakry
Addis Ababa
Ibadan
Accra
Lagos
Abidjan
GULF OF GUINEA
Douala
Equator
ATLANTIC OCEAN
Nairobi
INDIAN OCEAN
Kinshasa
Dar es Salaam
Luanda
Mozambique Channel
Tropic of Capricorn
East Rand
Johannesburg
Maputo
Durban
Cape Town
SCALE
0 500 1000 Miles
0 500 1000 Kilometers
Projection: Azimuthal Equal Area
LEGEND
POPULATION DENSITY
Persons per sq. mile
Persons per sq. km
520 200
260 100
130 50
25 10
3 1
0 0
Metropolitan areas with more than 2 million inhabitants
Metropolitan areas with 1 million to 2 million inhabitants

POPULATION

◆ The population of Africa is distributed unevenly throughout the continent.

◆ Large areas of Africa are sparsely populated or uninhabited.

◆ Africa has many metropolitan areas with more than 2 million inhabitants.

? Which metropolitan areas have more than 2 million inhabitants?

Compare this map to the economic map on page 420.

? Why do you think manufacturing and trade centers have large populations?

? Why do you think some areas are sparsely populated?

Mapping the Region

AFRICA

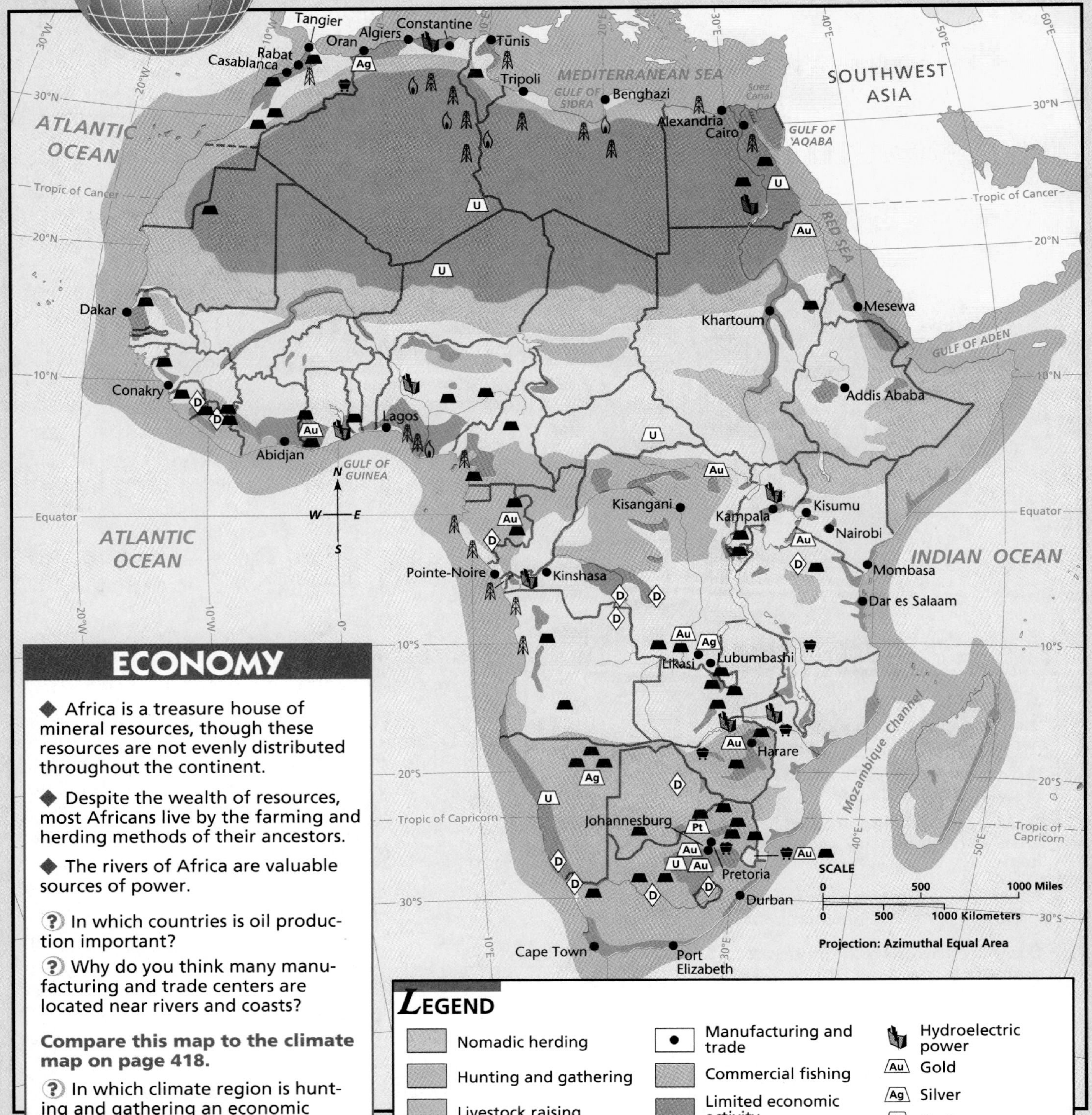

ECONOMY

◆ Africa is a treasure house of mineral resources, though these resources are not evenly distributed throughout the continent.

◆ Despite the wealth of resources, most Africans live by the farming and herding methods of their ancestors.

◆ The rivers of Africa are valuable sources of power.

? In which countries is oil production important?

? Why do you think many manufacturing and trade centers are located near rivers and coasts?

Compare this map to the climate map on page 418.

? In which climate region is hunting and gathering an economic activity?

Legend

Nomadic herding
Hunting and gathering
Livestock raising
Commercial farming
Subsistence farming
Manufacturing and trade
Commercial fishing
Limited economic activity
Coal
Natural gas
Oil
Hydroelectric power
Au Gold
Ag Silver
Pt Platinum
D Diamonds
U Uranium
Other minerals

SCALE
0 500 1000 Miles
0 500 1000 Kilometers
Projection: Azimuthal Equal Area

PHYSICAL GEOGRAPHY

FOCUS

- *What are the major landforms of Africa?*
- *What types of climate and vegetation are found in Africa?*

LOCATION **Landforms** Africa is a continent mostly of high plateaus and wide plains. Erosion has shaped Africa's landforms for millions of years. There are few major mountain ranges. Only the Atlas Mountains north of the Sahara and the mountains of Ethiopia (ee-thee-OE-pee-uh) compare to the mighty ranges on other continents.

The most famous river in Africa is the Nile, an exotic river that has its sources in the highlands of East Africa, where there is abundant rainfall. More than 95 percent of Egypt's (EE-juhpt) people depend directly on the Nile.

Africa has several other great rivers with curious courses. (See the map on page 417.) The Niger (NIE-jer) begins in the Fouta Djallon (FOOT-uh juh-LONE) highlands along Africa's west coast. Instead of flowing toward the coast, it flows northeast to its inland delta in the southern Sahara before turning south toward the Atlantic. The Zambezi (zam-BEE-zee) River, meanwhile, heads south toward the Kalahari (KAL-uh-HAHR-ee) but then turns east over the Victoria Falls to the Indian Ocean. In Southern Africa, the Orange River begins in a plateau edge called the Drakensberg Escarpment near the east coast. An **escarpment** is a steep slope capped by a nearly flat plateau. The river then flows west all the way to the Atlantic. Africa's rivers typically have waterfalls or rapids near the coast. As a result, traveling inland by river can be difficult. The Zaire (zah-IHR) (Congo) River is navigable only inland above the falls. Geographers think these stream patterns can be traced to when Africa was the center of Gondwanaland millions of years ago.

Victoria Falls are on the Zambezi River in **Southern Africa**. Over a mile (1.6 km) wide, the falls produce a deafening roar and mist that can be seen for nearly 40 miles (64 km). Local inhabitants called the mighty falls "Smoke That Thunders."

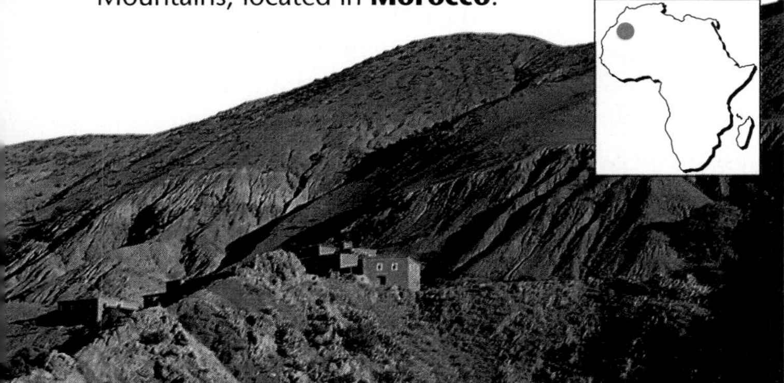

The Atlas Mountains, which stretch for about 1,200 miles (1,931 km) across northwestern Africa, are divided into several ranges. Shown here is part of one of the southern chains, the High Atlas Mountains, located in **Morocco**.

Africa has five huge depressions on its surface. These depressions, or basins, are each more than 625 miles (1,006 km) across and drop up to 5,000 feet (1,524 m) below the surrounding highlands. Three of these basins lie along the southern edge of the Sahara. In the west is El Djouf (ehl JOOF), which contains the inland delta of the Niger River. In the center is the Chad Basin, which holds shallow Lake Chad. In the east is the Sudan Basin, with the Sudd swamps at its center. The two remaining basins are in Central and Southern Africa. They are the forested Zaire Basin in Zaire and the arid Kalahari Basin in Botswana (baht-SWAHN-uh).

Over long periods of time, these basins have filled with sediment eroded from the surrounding highlands. Before Gondwanaland broke up, these basins were the ending points for major rivers. Since then, however, Africa's four greatest rivers—the Nile, the Niger, the Zambezi, and the Zaire—have cut channels to the sea.

The eastern third of Africa is the continent's highland region. The highlands are cut by two deep troughs called **rift** valleys. The Eastern Rift Valley and the Western Rift Valley begin near Lake Malawi (Lake Nyasa) and continue north into the Red Sea and then Syria. In many areas, the rifts have narrow valley floors with mountainous slopes on each side. In other areas, such as the Serengeti Plain in Tanzania (tan-zuh-NEE-uh), the rifts are wide and not as deep. Volcanoes have been active along both rifts. In fact, the highlands of Ethiopia are made up of layers of hardened lava.

REGION **Climate** Africa is the second largest continent in area. Because it stretches across many degrees of latitude, it has many climate types. (See the map on page 418.) Humid-tropical climates are found near the equator in the Zaire River basin and along the coast of the Gulf of Guinea (GIN-ee). These regions receive rain more than 200 days each year, and temperatures are the same day after day. Rain forests here have tall hardwood trees with leaves that stay green all year. The soils of the rain forests are poor, however, and farming is difficult. Few people live in the central parts of the rain forests.

Much larger regions with tropical-savanna climates surround the rain forests. Instead of rain year-round, these areas have dry weather during the winter. In regions with good rains, the savanna trees and grasses are tall. These regions are called tall-tree savannas. Farther away from the equator, seasonal drought is common. The trees become shorter and are better adapted to water shortages. These drier bush savannas have low, thorny trees and short, drought-resistant grasses. Clearing the savannas has created some of Africa's prime agricultural land.

Along the southern edge of the Sahara is a region of bush savannas and grasslands called the **Sahel** (suh-HAYL). The Sahel sometimes experiences years of drought. In the past, the people of the Sahel coped with droughts by moving south with their cattle, goats, and sheep. Today, however, moving is difficult because governments stop people from crossing their borders. As a result, a growing population is crowding the available grazing lands in the Sahel. The desert is expanding due to overgrazing, and many people and farm animals have died. Many more people barely survive in refugee camps with little hope for the future.

Africa's deserts are found in both the north and the south. The Sahara is the world's greatest desert, both in area and in climatic extremes. The Namib (NAHM-ib) and Kalahari deserts are in Southern Africa. Like the driest parts of the Sahara, the Namib is a severe coastal desert with sparse plant and animal life. In contrast, the Kalahari has low grasses and shrubs. Parts of it receive up to 10 times as much rain as the Namib.

These vineyards thrive in the Mediterranean climate region of **South Africa**. Africa has a great variety of climates. Some provide ideal conditions for agriculture, but others discourage farming. Produce from mild African climates is found in European markets when crops there are out of season.

Much of Southern Africa lies about the same distance from the equator as the southern United States. For example, Durban (DUHR-buhn), South Africa, is about as close to the equator as San Antonio, Texas. Southern Africa has many different climate regions. The Kalahari's eastern edge is a bush savanna that receives about 10 inches (25 cm) of rain each year. Rains increase toward the east coast, and tall-tree savannas gradually become dominant. In summer, rainstorms move into Southern Africa from the Indian Ocean, producing regions of humid-subtropical and marine-west-coast climate in the southeast. The southern tip of Africa has a Mediterranean climate, making this region a center for fruit growing.

Africa's highlands stretch from the Cape of Good Hope to Ethiopia. These high plateaus and mountains have cooler climates than are usually found at these latitudes. The plateau region of South Africa, for example, is not tropical like the lowlands along the east coast. Instead, subhumid to semiarid grasslands are present. Even though the equator is near, the highlands of Kenya (KEN-yuh) have mild weather all year.

The northern part of Africa lying along the Mediterranean Sea has a Mediterranean climate with good winter rains. The long, dry summer means that productive agriculture requires irrigation and wise conservation of winter rainwater. Vegetation ranges from desert plants in the northeast to grasslands and Mediterranean scrub forest in the northwest. The Sahara begins only a short distance south from the Mediterranean shore.

Natural Challenges Life in Africa can be difficult. Almost every year, some regions suffer from drought while others are ruined by floods. Earthquakes and volcanic eruptions occur along the rift valleys. These news-making events, however, cannot compare with the natural disadvantages and hazards Africans must face each day.

Africa has little fertile land compared with areas of the world of similar size. Much of the soil in dry regions has too much salt or lime to be agriculturally productive. In humid regions, soils often are not fertile enough for good crop yields. Good soils are common only in the highlands of eastern and southern Africa and along river valleys. Swarms of insects sometimes wipe out crops.

Diseases are a constant threat in Africa. Malaria and tuberculosis are widespread. In areas with intense irrigation, such as Egypt, certain parasites cause disease. In the humid tropics, sleeping sickness, carried by the tsetse (TSEHT-see) fly, attacks people and cattle. And in many areas, poor diets encourage a host of related diseases. AIDS (Acquired Immune Deficiency Syndrome) also is destroying a large part of the African population. By the beginning of 1995, the World Health Organization (WHO) estimated that as much as 60 percent of the global adult population infected with the AIDS virus lived in Africa.

1 SECTION REVIEW

1. What are the major landforms in Africa?
2. What type of vegetation would you find in a tropical-savanna climate with good rains?
3. **Critical Thinking** **Compare and contrast** the climates of North Africa and Southern Africa.

Early African Kingdoms

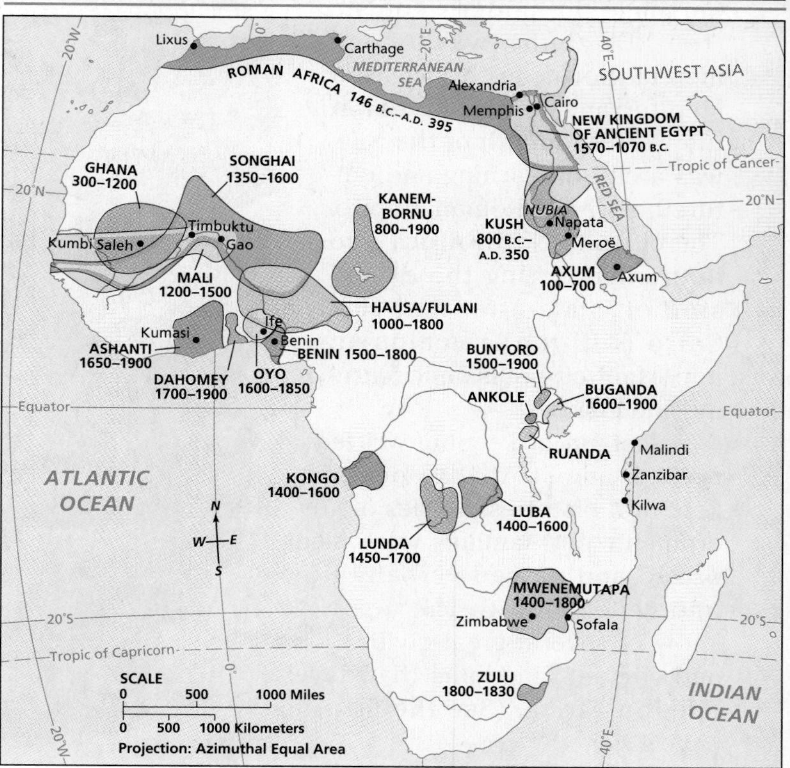

Lixus
Carthage
MEDITERRANEAN SEA
ROMAN AFRICA 146 B.C.–A.D. 395
Alexandria
Cairo
Memphis
SOUTHWEST ASIA
NEW KINGDOM OF ANCIENT EGYPT 1570–1070 B.C.
Tropic of Cancer
20°N
20°W
20°E
40°E
GHANA 300–1200
SONGHAI 1350–1600
KANEM-BORNU 800–1900
NUBIA
Napata
Meroë
RED SEA
KUSH 800 B.C.–A.D. 350
Timbuktu
Gao
Kumbi Saleh
MALI 1200–1500
AXUM 100–700
Axum
HAUSA/FULANI 1000–1800
Kumasi
Ife
Benin
ASHANTI 1650–1900
BENIN 1500–1800
BUNYORO 1500–1900
DAHOMEY 1700–1900
OYO 1600–1850
ANKOLE
BUGANDA 1600–1900
Equator
RUANDA
Malindi
Zanzibar
Kilwa
ATLANTIC OCEAN
KONGO 1400–1600
LUBA 1400–1600
LUNDA 1450–1700
N
S
E
W
MWENEMUTAPA 1400–1800
Zimbabwe
Sofala
20°S
Tropic of Capricorn
SCALE
0 500 1000 Miles
0 500 1000 Kilometers
Projection: Azimuthal Equal Area
ZULU 1800–1830
INDIAN OCEAN
0°

Numerous African civilizations flourished long before Europeans set out to build colonial empires. This map shows many early African kingdoms. Which West African kingdom existed from about 1350 to 1600?

HISTORICAL GEOGRAPHY

FOCUS

- *What economic activity supported most early African civilizations?*
- *What effect did European colonization have on the land and peoples of Africa?*

MOVEMENT **Early African Civilizations** Archaeologists have discovered more evidence of early humanity in Africa than on any other continent. More than 8,000 years ago, residents of Africa began planting barley and herding animals along the Nile River. Early in their history, Egyptians learned to use water from the Nile for irrigation. The people of the Nile Valley soon created one of the world's first great civilizations.

Later, cities along the African Mediterranean traded with the ancient Greeks and Phoenicians.

Carthage, in modern Tunisia (too-NEE-zhuh), was a commercial power in the western Mediterranean. Alexandria, at the western mouth of the Nile, was a city of learning and culture before the Roman Empire. The cities of North Africa also flourished during the expansion of Islam after A.D. 700. Cairo (KIE-roe) remains an important city of Islamic learning and culture.

Instead of keeping written records, most African peoples kept oral histories. Stories of governments and families were memorized and passed verbally from one generation to the next. We know of several great civilizations and ancient kingdoms that developed in Africa. (See the map on page 423.)

The kingdom of Kush in present-day Sudan had close connections to ancient Egypt and controlled much of the middle Nile River valley. The capital of Kush was Meroë (MEHR-uh-wee). After a period of decline, Kush was conquered by rulers of powerful Axum. Axum, located south of Kush in the highlands of present-day Ethiopia, was founded as a city of traders and merchants. Today, its ruins reveal the innovation of dry-stone construction, by which walls were built without cement or mortar.

Other empires arose at different times in other parts of Africa. Perhaps the most famous ones were founded in the dry savannas of West Africa. These early empires grew powerful by controlling trade across the Sahara between tropical Africa and the Mediterranean coast. Caravans brought salt from the Sahara and goods from Mediterranean cities. They also brought gold, ivory, and copper from the coastal forests to the south. Probably the most famous city of this region was Timbuktu, a major trading center.

While empires were developing in the east and west, civilizations were growing in Central and Southern Africa as well. These kingdoms also were involved in establishing trade routes across Africa. More than 1,000 years ago, Central African kingdoms were trading gold, spices, precious woods, and slaves with Arabia, Persia, India, and China.

This mask was crafted from ivory in the first half of the sixteenth century in the early African forest kingdom of **Benin**.

In Southern Africa, there are ruins of ancient cities. One of these cities was Great Zimbabwe (zim-BAHB-way), which began as a village of ironworkers during the third century A.D. Impressive stone structures built on a hilltop around the thirteenth century still stand. Civil wars within the region and new trading patterns brought about by contact with Europe led to the empire's decline.

Contact with Europe A dramatic change in the history of Africa began with the arrival of the European explorers in the fifteenth century. Most Europeans first dealt with African ethnic groups from trading posts along the African coast.

After Europeans began settling in the Americas, workers were needed for the plantations the colonists were building there. Africans, sold as slaves, became a major source of this labor. The main slave-trading ports stretched from the southern edge of Mauritania (mahr-uh-TAY-nee-uh) to the northern edge of the Namib Desert. The trading posts were located at sites where there was a harbor along with a river or trail leading inland.

Often, European merchants bought the slaves from coastal African leaders. These leaders were paid with European manufactured goods, particularly guns. Many of the slaves had been captured inland. As the demand for slaves grew, slave raids to the interior increased. The lives of millions of Africans were destroyed by the horrific practice of enslavement. The slave trade left a lasting imprint on Africa.

European exploration of the African interior did not begin until a few hundred years after the slave trade began. Much of this exploration centered on the search for the sources of Africa's rivers. Among the early explorers was Mungo Park. In 1795 and 1805, he traveled to West Africa,

Colonialism and Independence in Africa

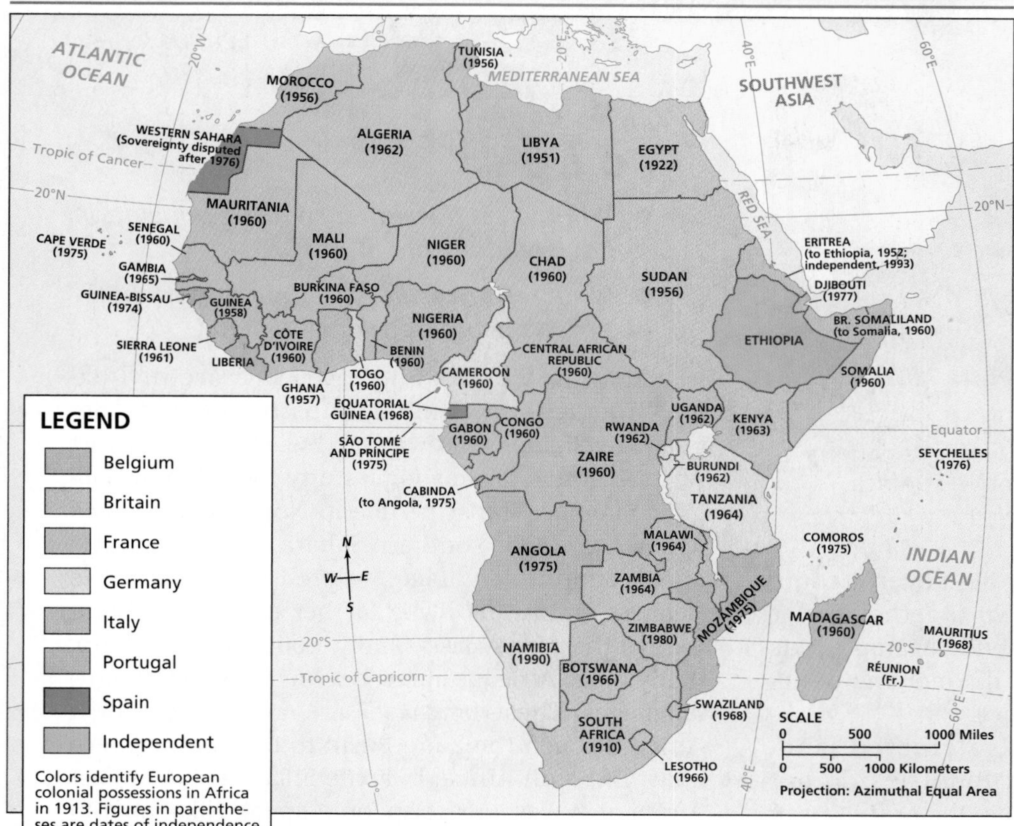

ATLANTIC OCEAN
MEDITERRANEAN SEA
SOUTHWEST ASIA
RED SEA
INDIAN OCEAN
Tropic of Cancer
Tropic of Capricorn
Equator
20°N
20°S
20°W
0°
20°E
40°E
60°E
MOROCCO (1956)
TUNISIA (1956)
ALGERIA (1962)
LIBYA (1951)
EGYPT (1922)
WESTERN SAHARA (Sovereignty disputed after 1976)
MAURITANIA (1960)
MALI (1960)
NIGER (1960)
CHAD (1960)
SUDAN (1956)
ERITREA (to Ethiopia, 1952; independent, 1993)
DJIBOUTI (1977)
BR. SOMALILAND (to Somalia, 1960)
ETHIOPIA
SOMALIA (1960)
SENEGAL (1960)
CAPE VERDE (1975)
GAMBIA (1965)
GUINEA-BISSAU (1974)
GUINEA (1958)
SIERRA LEONE (1961)
LIBERIA
CÔTE D'IVOIRE (1960)
BURKINA FASO (1960)
GHANA (1957)
TOGO (1960)
BENIN (1960)
NIGERIA (1960)
CAMEROON (1960)
CENTRAL AFRICAN REPUBLIC (1960)
EQUATORIAL GUINEA (1968)
SÃO TOMÉ AND PRÍNCIPE (1975)
GABON (1960)
CONGO (1960)
CABINDA (to Angola, 1975)
ZAIRE (1960)
UGANDA (1962)
KENYA (1963)
RWANDA (1962)
BURUNDI (1962)
TANZANIA (1964)
SEYCHELLES (1976)
ANGOLA (1975)
ZAMBIA (1964)
MALAWI (1964)
COMOROS (1975)
MOZAMBIQUE (1975)
MADAGASCAR (1960)
MAURITIUS (1968)
RÉUNION (Fr.)
ZIMBABWE (1980)
NAMIBIA (1990)
BOTSWANA (1966)
SOUTH AFRICA (1910)
SWAZILAND (1968)
LESOTHO (1966)
N
S
E
W
SCALE
0 500 1000 Miles
0 500 1000 Kilometers
Projection: Azimuthal Equal Area
LEGEND
Belgium
Britain
France
Germany
Italy
Portugal
Spain
Independent

Colors identify European colonial possessions in Africa in 1913. Figures in parentheses are dates of independence.

MAP STUDY Often, historical maps use colors to identify important features and dates to highlight events. In this map, colors identify European colonial possessions in Africa in 1913. The dates tell when colonies became independent. Note that South Africa became a self-governing country within the British Empire in 1910. It achieved complete independence in 1931. Which African countries had not been colonized by 1913?

exploring the upper parts of the Niger River. In 1857, Richard Burton and John Speke went to Africa in search of the source of the Nile River in the African interior. Four years later, Speke returned to the interior with J.A. Grant and discovered the White Nile's source, which he named Lake Victoria in honor of Queen Victoria.

European Colonization Tales of Africa's beauty and treasures quickly reached Europe, and mapmakers began charting the continent. With little consideration for the Africans living there, European powers scrambled to claim territories throughout Africa. To settle conflicts among themselves, the Europeans divided Africa at a conference in Berlin in 1884–1885. They took a map of Africa and drew boundaries without regard for Africa's landforms, climate regions, or cultures. European boundaries divided ethnic groups and cut through paths of migration and agricultural lands.

The Africans had little say about either these boundaries or the colonization. Except for Liberia (lie-BIR-ee-uh) and Ethiopia, all of Africa was under the rule of one European country or another by 1900. The Europeans imposed their culture, technology, and economic policies on Africa. The consequences of this process still exist today.

When the Europeans settled in Africa, they set up plantations to grow cash crops such as coffee, cotton, cacao, palm products, and peanuts. The Europeans also mined gold, iron ore, and copper. In contrast, most Africans still practiced traditional subsistence agriculture.

The Europeans also built roads, schools, ports, and hospitals in the areas where there were exportable resources. All of Africa's modern capital cities developed during this colonial era. The colonizers also paid for the European education of a small number of Africans. Many of these Africans later became leaders of independence movements.

Independence Africans began struggling for their independence as soon as the colonial powers appeared. After World War II, pressure for independence grew. Beginning with Libya (LIB-ee-uh) in 1951, country after country became independent. Today, all African countries are independent. Ties remain, however, to Europe and other developed countries.

For the most part, Africa's political boundaries today are those set by the Europeans. These boundaries have created many problems. Some African people, such as the Somalis of Kenya, Ethiopia, and Somalia (soe-MAHL-ee-uh), have

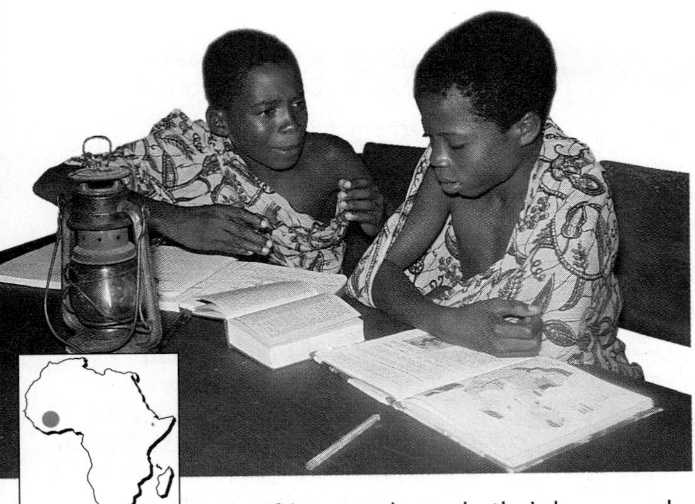

Two African students do their homework in a village in **Côte d'Ivoire**, a country on the Atlantic coast of West Africa.

found themselves divided among several countries. Other groups were thrown together under one rule. Nigeria (nie-JIR-ee-uh), for example, has more than 200 ethnic groups. This merging of different cultures has caused conflict.

European boundaries also created landlocked states. Because these countries do not have direct access to any oceans, they often have economic and political problems importing and exporting goods across the boundaries of other countries. Finally, some countries were carved out of only one climate region, limiting the kinds of crops that can be grown.

African nations are trying to squeeze into a short period of time progress that has taken developed countries many years to achieve. Most Africans appreciate the value of Western democracy, and democracy has spread in recent years. Yet some African leaders feel that a one-party government can move faster to achieve national goals. Other governments are dictatorships. Often, unrest—and sometimes revolution—threaten the stability of many African countries.

2 SECTION REVIEW

1. How did early African empires gain power in a region?
2. What were the effects of European colonization?
3. **Critical Thinking** **Compare and contrast** the permanence of oral histories and written records.

3 HUMAN AND ECONOMIC GEOGRAPHY

FOCUS

- *What effect does an increasing population have on daily life in Africa?*
- *What resources in Africa are important to Africa's economic development?*

Human Geography Africans are divided into hundreds of ethnic groups by different languages, religions, and ways of life. One way to sort Africa's ethnic groups is to study the different language families. Most people in North Africa speak Arabic. Along the southern Sahara are peoples who speak Semitic languages related to those of the Berbers and Arabs living farther north. These are part of the Afro-Asiatic family. South of the Sahara, many black Africans speak one of the Niger (Bantu) languages. These languages are spoken from West Africa through the Zaire Basin to the southeastern coast of South Africa. In some inland areas of East Africa and West Africa, people speak Nilo-Saharan languages.

Another language family is the ancient Khoisan, which is still spoken by isolated peoples in Southern Africa, such as the San (Bushmen). These people, now few in number, are not black Africans. Their yellow-brown skin, short stature, and unique blood type make them unlike any other people. Madagascar (mad-uh-GAS-kuhr), off the east coast of Africa, meanwhile, was settled by migrants from southeast Asia more than a thousand years ago. The language here is related to those of the southwestern Pacific region.

The colonial period brought European languages to Africa. In South Africa, the Cape Dutch language developed into Afrikaans. Many African countries have adopted French or English as an official language. In the countries of North Africa, however, Arabic is the official language.

Religion and family traditions are particularly important for modern Africans. Christian churches and Islamic mosques can be found in most African towns. Africans often combine these organized religions with more ancient forms of traditional worship. Many people believe in the spirits of the waters, animals, trees, and mountains. In most cultures, the memories of ancestors are honored, and

LANGUAGE FAMILIES OF AFRICA

Language Family:		Afro-Asiatic	Niger-Congo	Khoisan	Nilo-Saharan	Malayo-Polynesian
Language:	*English*	*Arabic*	*Swahili*	*Nama*	*Kanuri*	*Malagasy*
	mother	umm	mama mzazi	//gûs	yâ	reny
	child	walad	mtoto	/gôaï	táda	zaza
	head	rā`s	kichwa	tanas	kəlâ	loha
	water	mā`	maji	/gami	njî	rano
	tree	šajarah	mti	heis	kəská	hazo
	house	bayt	nyumba	omi	fáto	trano
	red	aḫmar	-ekundu	/awa	cimê	mena
	eat	akala	la	‡û	búkin	homana
	go/walk	ḏahaba/mašā	enda	!gû	lengîn	mamìndra

Africans speak a wide variety of languages. These languages originated from different language families. The chart above shows examples from five African language families. Nine English words are translated into five languages. For the Nama language, /, //, !, and ‡ represent different clicks, or sounds made by the tongue.

their spirits are believed to be powerful forces in daily life. This traditional belief in nature and spiritual beings is called **animism**. (See "One World, Many Voices" on pages 414–415.)

During the past 25 years, populations have increased in Africa. The use of modern health practices is largely the reason for this growth. Families have become larger and people are living longer. Many farming communities are now overcrowded, however. Overgrazing by cattle and goats has caused severe soil erosion. The traditional rural cultures of Africa are undergoing dramatic change as people move to cities to seek jobs.

Unfortunately, the number of people moving to the cities is usually greater than the number of jobs. The unemployed and those with temporary work live in the slums that surround African cities. Providing jobs and better housing is the most pressing issue for many African governments. Governing under these conditions is difficult.

REGION Resources Africa is a treasure house of mineral resources. Gold, copper, chromium, manganese, uranium, and cobalt are some of the important minerals. Africa's mineral resources are not evenly distributed, however, and some countries are much richer than others. Most of Africa's mineral wealth is exported to industrial countries of the Northern Hemisphere. As a result, the resource-rich countries of Africa are often targets of economic competition among more powerful countries.

Africa's oil deposits also are distributed unevenly. The greatest oil and natural gas deposits lie in Libya, Nigeria, and Algeria (al-JIR-ee-uh). Oil also is found offshore between Cameroon (kam-uh-ROON) and Angola (an-GOE-luh). Because oil is so important to modern industrial nations, all African countries have active oil-exploration programs. More oil deposits will be developed along the west coast of Africa and in the Sudan.

The rivers of Africa are a valuable source of power. Hydroelectric power stations have been built or are planned at almost every rapids or waterfall. Perhaps 30 percent of the world's future hydroelectric potential exists in Africa.

Despite its resources, Africa remains the least developed continent. Many Africans live by the farming and herding methods of their ancestors. South of the Sahara, cattle and goats are the most important animals. In fact, in many African cultures, cattle are a measure of wealth. Teenagers discuss cows in much the same way that teens in wealthy countries talk about cars.

Manufacturing often is limited to producing simple consumer goods and processed food to meet the needs of local markets. Few products manufactured in Africa travel to world markets. Economic progress is limited mainly by the region's poor educational systems, a lack of trained business managers, and little money for investment.

Some African countries are small in population and land area, which can create an economic

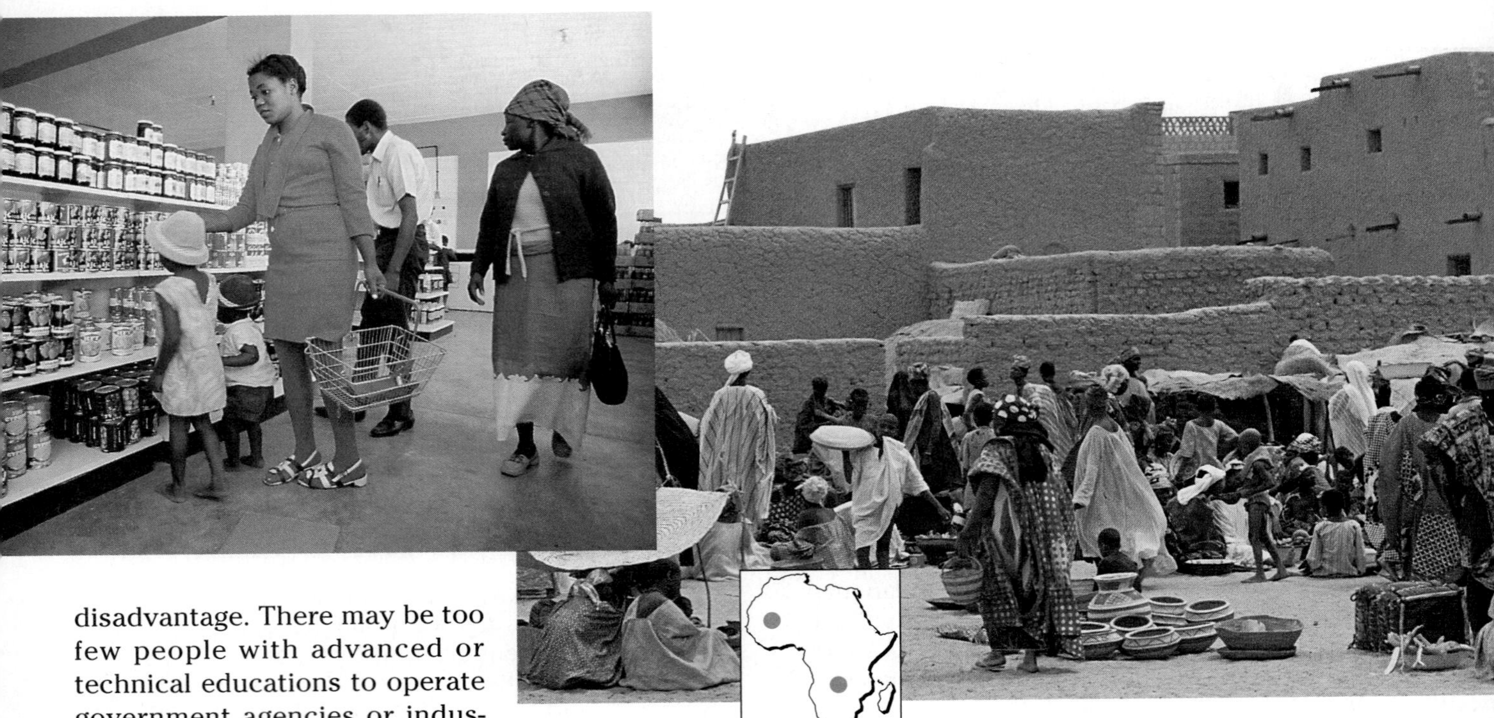

disadvantage. There may be too few people with advanced or technical educations to operate government agencies or industries. More important, there may be too few people to buy the goods produced by factories. A solution is for neighboring countries to cooperate economically. Goods can then be sold in several countries. Workers can seek jobs in other nations as well. Recently, organizations such as the Economic Community of West African States have developed in each of Africa's regions. These kinds of economic ties represent a break from the trading patterns of the colonial past.

Africans shop in modern supermarkets, such as in **Zambia** (left), and in open-air markets, such as this one in **West Africa** (right). Are there any open-air markets in your hometown?

REGION Issues Most countries of Africa face several challenges. First, struggling against often illogical colonial boundaries, African leaders must build stable countries from territories that contain peoples of different languages, religions, and cultures.

A second challenge is slow economic progress. African economists explain that African agricultural exports are not always treated fairly on the global market. African countries do not receive the same trade protection as other countries. Another cause of slow economic development is poor government economic policies. For example, governments often want their people to grow valuable export crops such as cotton and coffee. This discourages food production in rural areas. As a result, earnings from exports are spent on food imports instead of on development. Money for investment is in short supply. Furthermore, foreign aid from the developed countries often helps the city dwellers rather than the rural people.

A third challenge is Africa's rapidly increasing population. As a result of its current rate of population growth, Africa is becoming a continent of children who need to be cared for and educated. Also, the quality of Africa's natural environments is declining as more areas are cleared for farming, and overgrazing increases.

Improving people's lives in Africa will not be easy. Foreign aid is needed, mainly to improve agriculture and to develop appropriate manufacturing industries. Income from mining and export crops is necessary to improve infrastructure and transportation and to provide better schools. Cooperation among African countries will help in finding answers to these shared problems.

3 SECTION REVIEW

1. Why are many Africans moving from rural to urban areas?
2. How are Africa's rivers important economic resources?
3. **Critical Thinking** **Explain** how a country might be disadvantaged by either a large or a small population.

CHAPTER 35 CHECK

Senegal village

Reviewing the Main Ideas

1. Africa is a continent primarily of high plateaus and wide plains. Soils are generally poor for agriculture. The climate varies from humid tropical to desert.
2. Before European colonization in the late nineteenth century, Africa was home to great empires. Europeans divided Africa with little regard to ethnic, economic, or climate regions. Most African countries did not receive their independence from colonial rule until the 1960s.
3. The countries of Africa have many characteristics in common, but they have many differences as well. Several hundred ethnic groups inhabit the region, each with its own language, religion, and customs.

Building a Vocabulary

1. Define *escarpment*. Give an example from this chapter.
2. What are the two deep troughs called that are found in the highlands of East Africa?
3. Describe the Sahel. Where in Africa is it found? How has the region changed since the late 1960s?
4. What is animism?

Recalling and Reviewing

1. Name the five huge basins in Africa. Where are they located?
2. What were the major effects of European colonization on the lands and peoples of Africa?
3. What economic resources are important to development in Africa?

Thinking Critically

1. Describe what the development of Africa might have been like if European colonization had not taken place. Consider the formation of the countries and their economies.
2. How would you explain to an African village leader the economic advantages for his village if the farmers produced more food crops instead of cash crops for export?

Learning About Your Local Geography

Cooperative Project

With your group, research and report on the original boundary of your town or city. Who drew the boundary? Does the original boundary make sense geographically? Why or why not? Draw a map of the original boundary to include in the report.

Using the Five Themes of Geography

Carefully examine the photographs on this page. For each photograph, write two or three sentences that explain how one or more of the Five Themes of Geography are evident in that photograph.

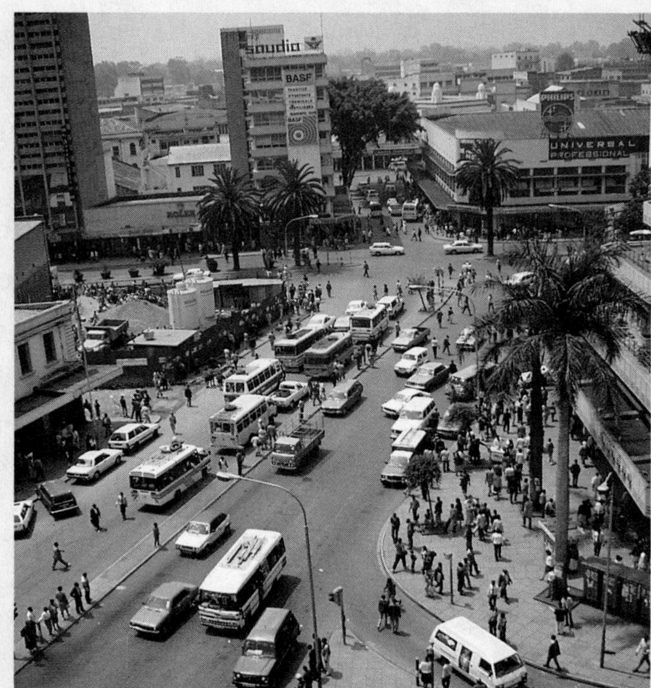
BASF
PHILIPS
UNIVERSAL
PROFESSIONAL

Nairobi, Kenya

CHAPTER 36

North Africa

> "A dune is a curiously dynamic creature. It can form behind any impediment—a bush or rock—or simply as the child of wind currents. Once formed, a dune can grow. It can change shape and move with the wind. It can even breed new dunes. Some of these offspring may be carried on the back of the mother dune. Others are born and race down wind, outpacing their parents. Some dunes in the Western Desert [of the Sahara in Egypt] move as much as 100 meters a year."
>
> *Farouk El-Baz*

GEOGRAPHY DICTIONARY

erg
reg
basin irrigation
perennial irrigation
silt
capital
free port

The Suez Canal cuts through the arid isthmus separating the Red Sea and the Mediterranean Sea. Some geographers mark this human construction as the boundary between Africa and Asia. Completed in 1869, the canal was the great engineering work of its time. It is a 100-mile-long (161-km), sea-level ditch that saves ships thousands of miles on their route from the Mediterranean Sea to the Indian Ocean. The Suez Canal passes through the North African country of Egypt and is one of the world's busiest waterways.

Along with Egypt, the countries of Africa's northern rim include Libya, Tunisia, Algeria, and Morocco. The countries west of Libya often are called the *Maghreb,* which in Arabic means "west." Culturally, North Africa is predominantly Arabic and Islamic. The region is a key part of the Arab world.

Alabaster image of King Tutankhamen of **Egypt** from the fourteenth century B.C.

The Suez Canal, **Egypt**

NORTH AFRICA: PHYSICAL–POLITICAL

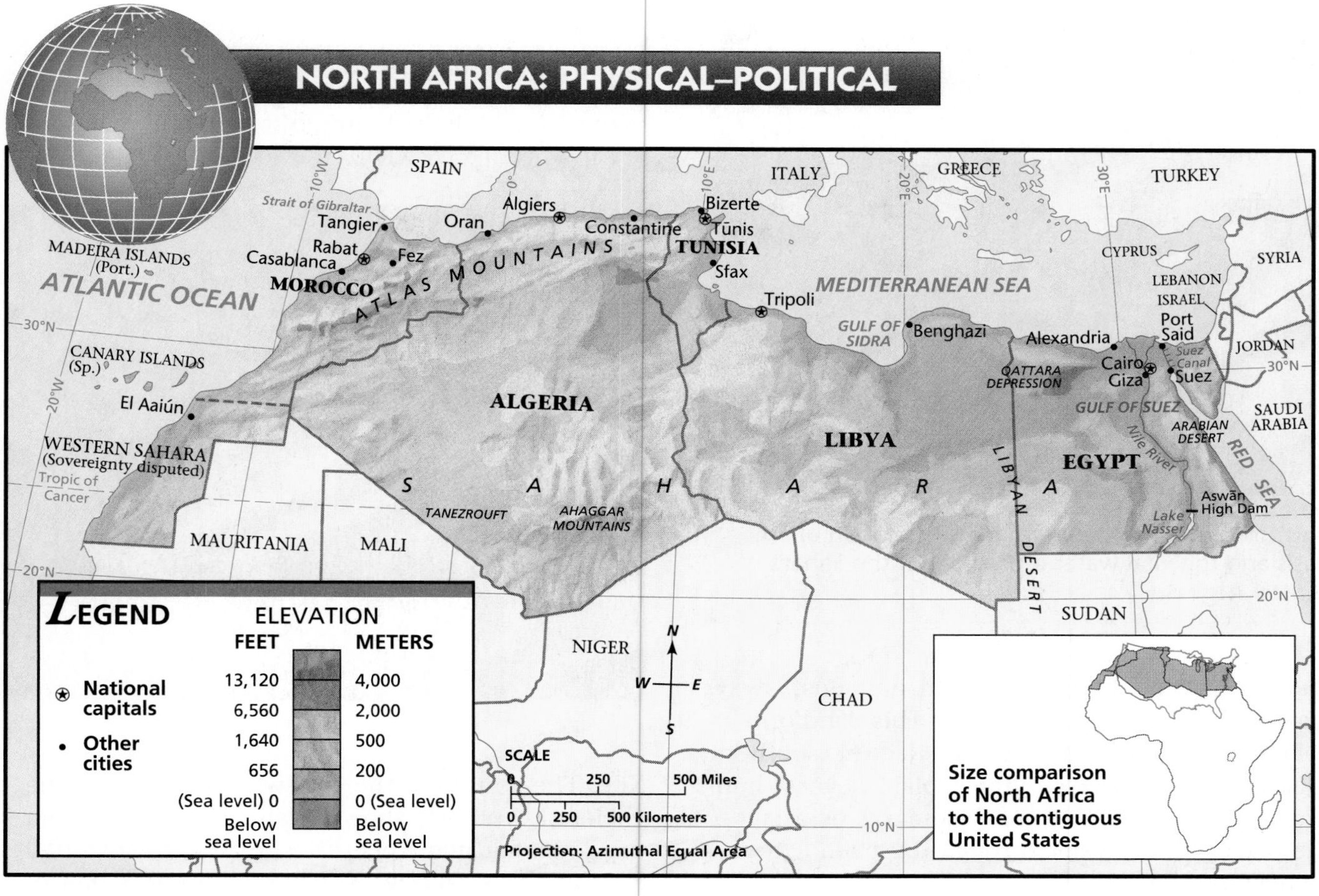

SPAIN
ITALY
GREECE
TURKEY
CYPRUS
SYRIA
LEBANON
ISRAEL
JORDAN
SAUDI ARABIA
Strait of Gibraltar
Tangier
Algiers
Oran
Constantine
Bizerte
Tunis
TUNISIA
Sfax
Rabat
Casablanca
Fez
MOROCCO
ATLAS MOUNTAINS
MADEIRA ISLANDS (Port.)
ATLANTIC OCEAN
MEDITERRANEAN SEA
Tripoli
GULF OF SIDRA
Benghazi
Alexandria
Port Said
Suez Canal
Suez
Cairo
Giza
QATTARA DEPRESSION
GULF OF SUEZ
ARABIAN DESERT
Nile River
RED SEA
Aswān High Dam
Lake Nasser
CANARY ISLANDS (Sp.)
El Aaiún
WESTERN SAHARA (Sovereignty disputed)
Tropic of Cancer
ALGERIA
LIBYA
EGYPT
S A H A R A
TANEZROUFT
AHAGGAR MOUNTAINS
LIBYAN DESERT
MAURITANIA
MALI
NIGER
CHAD
SUDAN
30°N
20°N
10°N
10°W
0°
10°E
20°E
30°E

LEGEND

National capitals
Other cities

ELEVATION

FEET	METERS
13,120	4,000
6,560	2,000
1,640	500
656	200
(Sea level) 0	0 (Sea level)
Below sea level	Below sea level

SCALE
0 250 500 Miles
0 250 500 Kilometers
Projection: Azimuthal Equal Area

Size comparison of North Africa to the contiguous United States

1 PHYSICAL GEOGRAPHY

FOCUS

- *How do the Sahara's physical characteristics influence settlement?*
- *Why is the Nile River important, historically and agriculturally?*

The countries of North Africa share similar environments. Along their Mediterranean coasts, summers are dry, and winter rains are welcomed. Oak trees and grasses cover much of the plains and plateaus here, and irrigation is crucial for agriculture. From Tunisia westward, the Atlas Mountains parallel the coast. Within the mountains are pleasant valleys as well as snow-covered peaks. North Africa holds two of the world's geographic wonders—the immense Sahara and the Nile River.

LOCATION **The Sahara** The word *sahara* comes from the Arabic word for "desert." The Sahara, combined with the deserts of the Arabian Peninsula, Iran, and Central Asia, forms the world's largest arid region. Deserts may be created by two separate factors. The first is a weather pattern dominated by regional high pressure. In high-pressure cells, the air spirals downward toward the earth's surface. Even if the descending air is damp, rain does not form. The second factor is the absence of moist sea air. Most of the world's rain starts as evaporation from an ocean. A desert results if an ocean is far away, or if the ocean air is blocked by mountains. In the Sahara, both of these factors are at work. In parts of central Libya and Algeria, summer temperatures are regularly above 120°F (49°C).

Because it has little water, the Sahara has few plants. Therefore, wind and a rare rain can be very effective in eroding the land. Bare rock surfaces are common in the Sahara, particularly in highland regions such as the Ahaggar (uh-HAHG-uhr) Mountains in southern Algeria. The basins below the rocky ridges fill with the eroded sediment. Some basins are covered with high, shifting sand dunes. Such a sea of sand is called an **erg**. In other

This town and vegetation mark the location of an oasis and the vital water it provides in the Sahara of central **Algeria**.

Much of the fertile **Nile River valley** south of Cairo is bordered by steep, dry cliffs. The valley's width narrows to a few hundred yards in some places, but is as much as 14 miles (23 km) wide in other areas.

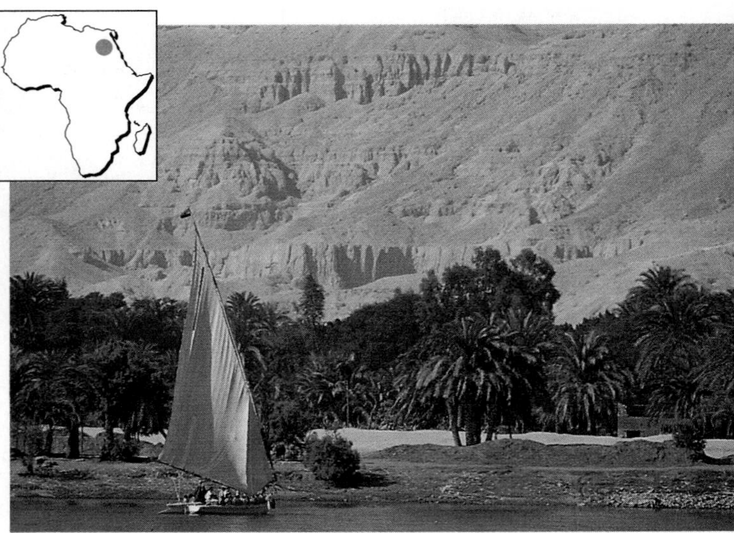

areas, the wind blows the sand and dust away, leaving a gravel-covered plain. This landform is called a **reg**. The Tanezrouft Reg in southern Algeria is a featureless gravel plain several hundred miles across. With no landmarks as a guide, the occasional traveler through the plain can easily become lost.

The Sahara also has broad depressions. In western Egypt, the Qattara Depression, 436 feet (133 m) below sea level, forms a vast wilderness of quicksand and salt marsh. Other basins have large, dry lake beds where rainwaters briefly accumulate. In Arabic, a dry lake bed is called a *sebka. Sebka* surfaces are severely eroded by winds. Dust storms seem like fog and can even blot out the sun. Sometimes the dust travels very far. Saharan dust occasionally falls with rain in the Caribbean.

The Sahara is almost uninhabited. Human settlement is limited to sites where there is water. Ancient oasis towns are found at springs and where wells can reach water below dry riverbeds. The few roads crossing the Sahara follow old caravan routes from oasis to oasis.

The Nile River In striking contrast to the Sahara, the valley and delta of the Nile contain thousands of acres of rich, fertile land. On the banks of this famous river, ancient Egyptians developed their science and art and built pyramids, temples, and towns.

At 4,187 miles (6,737 km) long, the Nile is the longest river in the world. It is formed by the union of two rivers: the White Nile and the Blue Nile. The source of the Nile was unknown to the ancient Egyptians. They explained it as a gift from the gods. For many centuries, explorers searched for the Nile's source. Not until the latter half of the eighteenth century, however, was the source of the Blue Nile discovered in the highlands of Ethiopia. Nearly a century later, the source of the White Nile was found in the tributaries of Lake Victoria in East Africa. The two rivers meet at Khartoum, Sudan, to form the Nile River.

The Nile River valley is like a long oasis in the middle of a desert. The fertile lands along the river are a product of the Nile itself. Before dams were built upstream, the Nile regularly overflowed its banks in late summer, restoring richness to the soil century after century. In addition to water, these floods brought rich, fertile **silt** to the farm fields. Silt is sediment that is coarser than clay yet finer than sand. It feels similar to flour.

SECTION REVIEW

1. Why is settlement sparse in the Sahara?
2. How has the Nile River been important to people from ancient times to the present?
3. **Critical Thinking** **Suggest** a reason why the Nile carries less water than the Amazon, even though the Nile is longer.

EGYPT

FOCUS

- *What forms of irrigation and agriculture have been used in Egypt, and what challenges have they presented for the growing population?*
- *What two major political forces are influencing Egypt today?*

Almost 2,500 years ago, a Greek geographer wrote that "Egypt is an acquired country, the gift of the river." This statement is still true today. The Nile River remains Egypt's main resource. In Egypt, the Nile flows north through 1,000 miles (1,609 km) of desert and then empties into the Mediterranean Sea. (See the map on page 431.)

Little rain falls in Egypt. During some years, there is no rainfall at all in the southern part of the country. Cairo receives about 1.5 inches (3.8 cm) a year. Coastal areas around the city of Alexandria receive from 4 to 8 inches (10 to 20 cm) a year. For this reason, irrigation from the Nile River is of utmost importance.

INTERACTION Agriculture Nearly 97 percent of Egypt's cultivated land is irrigated by the Nile. For many centuries, the Egyptians used **basin irrigation**. Farmers divided their land into shallow basins by building walls of earth up to about six feet (2 m) high. When the Nile overflowed each year, water and silt were deposited on the fields. The earth walls helped trap the water on the land. Seeds were then sown in the rich mud. Under the strong sun, the moist, fertile soil yielded good crops. By using basin irrigation, fields up to about nine miles (14 km) from the Nile were irrigated. Only crops that would ripen during the short time that the land was moist, however, could be grown. Beans, wheat, barley, and onions were the traditional food crops grown with basin irrigation, but only one crop could be grown each year.

In most cases, basin irrigation has given way to **perennial** (continual) **irrigation**. This type of irrigation was first practiced by pumping water directly from the Nile. Today, huge dams stop the Nile's annual flood by storing the floodwater. Then the water is released steadily throughout the year and carried to the fields by networks of canals.

FACTS IN BRIEF North Africa

COUNTRY POPULATION (1994)	LIFE EXPECTANCY (1994)	LITERACY RATE (1990)	PER CAPITA GDP (1993)
ALGERIA 27,895,068	67, male 69, female	57%	$3,300 (1992)
EGYPT 60,765,028	59, male 63, female	48%	$2,400
LIBYA 5,057,392	62, male 66, female	64%	$6,600
MOROCCO 28,558,635	66, male 70, female	50%	$2,500
TUNISIA 8,726,562	71, male 75, female	65%	$4,000
UNITED STATES 260,713,585	73, male 79, female	97% (1991)	$24,700

Source: *The World Factbook 1994,* Central Intelligence Agency

Which North African country has the highest GDP per person? Why do you think this is so?

Thus, two or three crops can be grown instead of only one. As a result of perennial irrigation, rice and cotton have become important crops in Egypt.

The Nile's largest dam is located about eight miles (13 km) south of Aswān (a-SWAHN). The Aswān High Dam is 364 feet (111 m) high and about two miles (3.2 km) long. It can irrigate more than 2 million acres (809,400 ha) of farmland. The dam provides about one-third of Egypt's electricity.

Although the Aswān High Dam provides electric power and water throughout the year, the dam has created problems. The Nile no longer deposits the fertile silt needed to enrich the land, because the sediment is trapped behind the dam. Expensive artificial fertilizers are now used. The fishing industry in the Mediterranean also has suffered because the trapped sediment no longer provides nutrients for the fish. The fishing industry has been helped, however, by fishing in the new lake created by the dam. Evaporation from this lake, though, reduces the amount of water available to farmers.

A serious problem associated with perennial irrigation is an increase in a disease produced by small organisms that invade the liver and blood of people. These organisms are carried by snails that live in the irrigation canals. Because people are in close contact with irrigation water, it is difficult to keep the disease from spreading.

INTERACTION Population Growth Population density in Egypt's rural areas is among the highest in the world. Egypt's cities are growing rapidly, too. As a result, the amount of productive land per person is steadily declining. Nearly all of Egypt's agricultural land is found along the Nile, particularly in the Nile Delta. New farmland is increasing slowly as irrigation canals are built farther from the river and as salt marshes in the delta are reclaimed.

Still, it is difficult to put new land and technology into use fast enough to keep pace with Egypt's rapid population growth. With such intensive land use, there is danger that the soil will become unproductive. More and more fertilizer is needed. In addition, overwatering has brought to the surface salts that are harmful to crops. Once self-sufficient in food, Egypt now must import more than half of its food.

Cotton is Egypt's main cash crop and agricultural export. Egyptian cotton, prized around the world, once was grown mainly on large estates. Today, it is produced primarily on small farms, some no bigger than gardens. Many farmers now proudly work the land they own. By cooperating in seed purchases, marketing, and fertilizer applications, small farmers can use modern technology with their hand labor.

PLACE **Urban Geography** Cairo, Egypt's capital, has more people than any other city in Africa. In fact, it is one of the world's fastest growing cities. About 10 million people crowd into the city and its surrounding region, which sprawl along the banks of the Nile River. Cairo was founded more than 1,000 years ago on high ground east of the river at the head of the Nile Delta. From this location, Cairo's political leaders commanded both the delta and the river valley. In the late 1800s, railways connected the city to ports along the Mediterranean and Gulf of Suez. The city spread south along the river.

Modern Cairo is a mixture of building styles, reflecting the cultural diversity of the city. High-rise offices, department stores, hotels, and government buildings tower over historic Islamic mosques and universities. Much of the city, however, consists of tiny, mud-brick houses situated on maze-like streets. Overcrowding, traffic jams, and inefficient services plague the city.

Despite limited housing available in Cairo and other crowded Egyptian cities, migration from rural to urban areas continues. Such population pressure has resulted in the growth of satellite cities, which are connected to Cairo by commuter rail lines. The once-sleepy village of Giza, across the Nile near the pyramids, is now an industrial suburb with more than 1 million people.

Egypt's second largest city is Alexandria, located just west of the delta along the Mediterranean. Founded by Alexander the Great in 332 B.C., Alexandria has about 3 million people and handles much of the country's trade.

Industry and Trade Although Egypt is one of Africa's most industrialized countries, it must expand and modernize its industries to meet the economic needs of its growing population. Textile and yarn factories have been developed, and Egypt now exports these products. Other industrial growth has been only partly successful.

Egypt has limited natural resources. It has no forests and only small coal deposits. Oil, Egypt's main export, is found on both sides of the Gulf of Suez. As Egypt develops, more of this oil will be needed inside the country, and less will be available for export. Egypt has manganese and iron ore, which form the basis for a growing steel industry. Gypsum and other minerals are used for making cement, and phosphate and nitrates are used in fertilizers.

A lack of skilled workers also has contributed to Egypt's slow industrial development. Skilled,

The Nile Delta in **Egypt** stands out sharply from the surrounding desert in this satellite photograph. Cultivated areas, shown in red, follow a narrow path south along the Nile River.

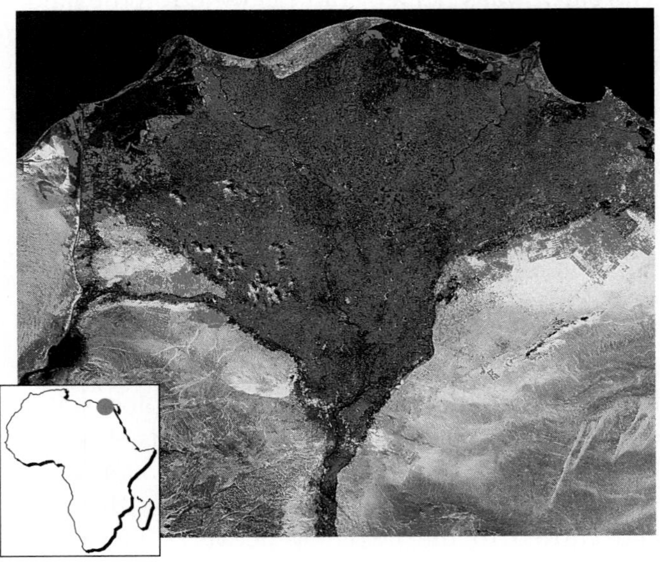

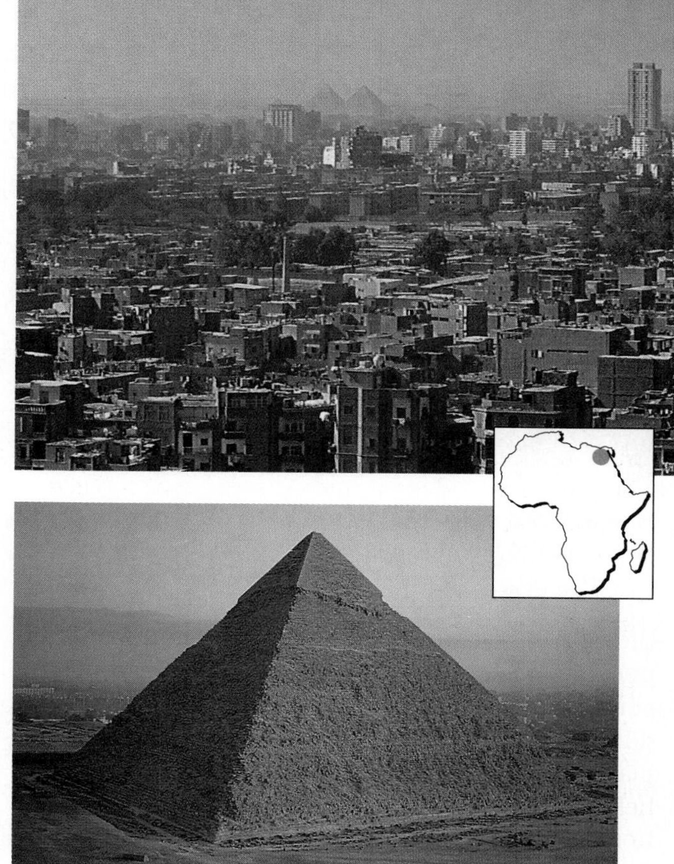

The building styles of **Cairo** (right) are a mix of old and new. West of the city, across the Nile, is the suburb **Giza**, where ancient Egyptians built many great pyramids, including the one shown below right.

educated Egyptians often choose to work on development projects in the oil-rich Arab states. In addition, like other developing countries, Egypt lacks a good home market. Most people can afford to buy only the most necessary items.

Another problem is Egypt's balance of trade. The total value of Egypt's imports is greater than the value of its exports. As a result, little money is available for investment, and foreign aid from the developed countries and from the rich Arab nations makes up an important part of the government budget. Workers living out of the country often send money to their families in Egypt.

Issues Two political forces are pulling at modern Egypt. The first is the struggle between Arab regionalism and Egyptian nationalism. Egypt has significant influence in the Arab countries. Regional as well as international issues of concern to Arabs often are settled in Egypt. At odds with these international and regional concerns are the local issues emphasized by Egyptian nationalists. Most Egyptians prefer that their government focus on economic issues that will assist Egyptians. For example, they would like to see improvements made in the many inefficient government-owned industries. Egyptian farmers want more aid from their government.

The second political force tugging at Egypt is Islamic fundamentalism. Egyptians are divided on whether their government should change to closely follow Islamic rules. If the movement toward Islamic fundamentalism strengthens, traditional religious laws would make social life more conservative and, for many, less tolerable.

SECTION REVIEW

1. How have different forms of irrigation and agriculture presented problems for Egypt's rapidly growing population?
2. Over what two political issues are Egyptians divided, and why?
3. **Critical Thinking** **Imagine** that you are an Egyptian farmer. Why might you migrate to Cairo?

LIBYA AND TUNISIA

FOCUS

- *How did the discovery of oil in Libya affect the country?*
- *Why has Tunisia been a relatively successful country in North Africa?*

Libya Situated along the most desolate shore of the Mediterranean, Libya was ignored for much of its history. (See the map on page 431.) Then, sudden oil wealth and a radical government dramatically changed the country and its image. Major oil deposits were discovered in the southern desert in 1959, and Libya is now a major oil exporter. Libyan crude oil is particularly valuable because it contains little sulphur and, therefore, causes less pollution. Libya also exports natural gas to the nearby markets of Europe.

Unlike Egypt, Libya has no great river and is sparsely populated. More than 75 percent of the nation's people live within 10 miles (16 km) of

Humans and machines carve a path through the desert to create the Great Manmade River Project in **Libya**.

the Mediterranean Sea because the country is 93 percent desert. Agriculture is possible, but the summer heat and desert winds make farming difficult. Along the northeastern coast are the plateaus of Cyrenaica (sir-uh-NAY-uh-kuh), where herding sheep and goats has long been the traditional occupation. The major city in this region is Benghazi (ben-GAHZ-ee). To the west is the region of Tripolitania, which receives slightly more rain. Tripoli, Libya's capital and major city, is located in this region. Typical Mediterranean crops, including wheat, citrus fruits, and olives, are grown here.

Libya's oil wealth is large in relation to its small population. The government has used oil revenues to establish generous programs for public health, housing, and transportation. Educational opportunities and health conditions continue to improve. Oil profits also have been spent on irrigation projects, the most spectacular of which is called the "Great Manmade River Project." During oil exploration in the country's southwestern desert, a large aquifer in the Sahara was discovered. Huge pipes now transport this water northward to the coast. Because this precious water from beneath the Sahara is not replaced by rains, however, it will be depleted some day.

Oil wealth has given Libya political power. Libya's leader, Muammar al-Qaddafi, is an avid Arab nationalist who has supported radical causes around the world. He has tried to overthrow governments in neighboring countries and has engaged in border disputes with Egypt. People opposed to his policies have been killed. For these reasons, many Western governments view Libya as a potential threat and pay close attention to events there.

LOCATION **Tunisia** Tunisia has a strategic location between the eastern and western basins of the Mediterranean Sea. (See the map on page 431.) For thousands of years, this location made Tunisia a center of Mediterranean culture. In ancient times, the city of Carthage, now in ruins, controlled trade throughout the Mediterranean. Today, European tourists flock to these sunny shores.

Most Tunisians live on coastal lands, which have a mild Mediterranean climate. In these lowlands, farmers grow wheat and olives, and olive oil is an important export. More than 25 percent of Tunisia's land is suitable for agriculture. This is a large amount compared with that of other North African countries whose territories extend much farther into the Sahara. Plateaus in western Tunisia mark the beginning of the Atlas Mountains.

While agriculture in Tunisia is important, most of its people live in cities. Tunisia's transportation system is good, and the government has encouraged industrial expansion. Tunisia has a variety of minerals to export, including some oil. Many industries rely on hand labor, which provides more jobs.

Tūnis, the capital, and the other major cities of Sfax and Bizerte, are very European. In fact, as a former French colony, Tunisia maintains ties to France. Many Tunisians work in France and send part of their income home. Influences of French culture are common, and the ideals of European democracy are widely appreciated. For example, Tunisian women have the right to vote, and the national school system is envied by neighboring countries. Literacy is relatively high. These ideals earn Tunisia large foreign aid contributions.

Tunisians visit a marketplace in **Tūnis**. About 1 million people live in Tūnis.

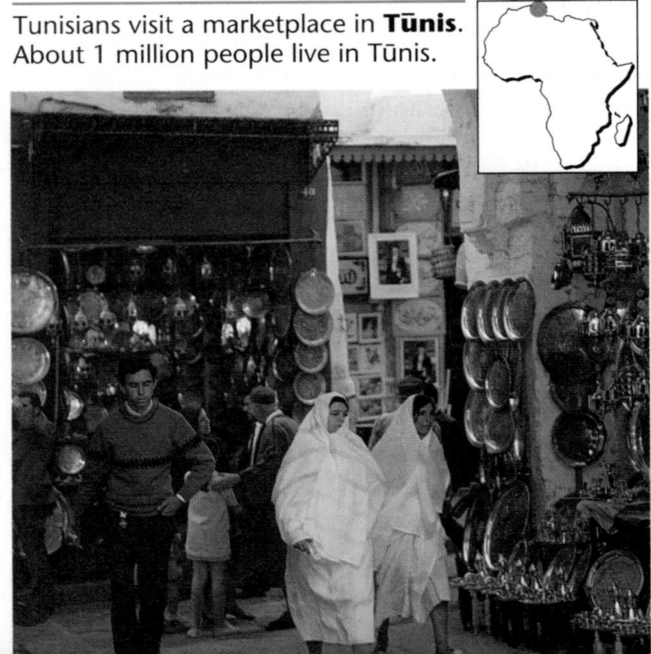

As in other areas of North Africa, the population in Tunisia is increasing faster than the number of jobs. Migration from rural areas is bringing people into the cities. Islamic fundamentalism is gaining in popularity, although it is repressed by the government.

SECTION REVIEW

1. Why was the discovery of oil in Libya important?
2. What advantages does Tunisia have over some of its neighbors?
3. **Critical Thinking** **Explain** why Libya and Tunisia are different politically even though they are both well off economically.

ALGERIA AND MOROCCO

FOCUS

- *What are the physical, cultural, and economic characteristics of Algeria? What problems does Algeria face?*
- *Why can Morocco be considered a nation of contrasts?*

Algeria Like Egypt and Libya, Algeria is a country both of the Mediterranean rim and of the Sahara. (See the map on page 431.) More than 80 percent of all Algerians live in the north along the Mediterranean coast, where lands are fertile. To the south is the desolate Ahaggar, the highlands of the Sahara.

Modern Algerians are a blend of Arab, Berber, and European cultures. The Berbers were Algeria's original residents, and their language still is spoken along the Atlas Mountains. A few Berbers can be seen leading their flocks of sheep and goats on annual migrations in and out of the Atlas valleys.

During 132 years of French colonization, large investments and more than a million French settlers made Algeria one of the most prosperous regions in Africa. After a bitter and bloody war, the Algerians ousted the French, and Algeria gained its independence in 1962. France left behind the infrastructure that is the basis of modern Algeria. The United States and the European Community are now Algeria's most important trading partners.

Most of Algeria's income comes from oil and natural gas fields scattered across the Sahara. Pipelines carry these products to the coast where they are shipped to markets in Europe and eastern North America. A new pipeline under the Mediterranean Sea will increase Italy's dependence on Algeria's oil and gas resources.

Algerians can proudly claim that much of the profits from petroleum are invested in education and industrial developments. Today, Algeria has a higher proportion of factory workers than any other Arab country. A wide range of consumer goods are produced for the local market. Industries are concentrated in three main cities: Algiers, the capital and major port; Oran, another major port; and Constantine, located farther inland.

Algeria shares many problems with its neighboring countries, however. Many of these problems stem from its rapid population growth. More than half of the country's food is imported, and the schools are producing twice as many graduates as there are jobs. Some people believe that the government should encourage more free enterprise, which could lead to rapid job growth. Also, Islamic fundamentalism is gaining popularity. Since 1992, violence in the conflict between fundamentalists and their opponents, especially the government, has claimed thousands of lives.

LOCATION **Morocco** Morocco is an Islamic kingdom on both the Mediterranean and the Atlantic coasts. (See the map on page 431.) In the

Women at this plant in **Algeria** work on parts for televisions and computers. Although the Algerian economy is dependent on oil and natural-gas production, manufacturing continues to grow in importance.

Irrigated farms, such as these outside the city of **Safi** on the Atlantic coast of **Morocco**, produce citrus fruits, grapes, and vegetables.

north is the Strait of Gibraltar, which separates Europe and Africa by only eight miles (13 km). Inland, a series of plateaus rises toward the Atlas Mountains. The soils and rocks of the mountains act as a reservoir, storing water for the streams that flow south. In sharp contrast to Morocco's mountain region, which has the wettest climate of the Arab nations, the Saharan coast in southern Morocco is extremely dry.

While Morocco has only limited deposits of oil and metal ores, the country contains three-fourths of the world's total reserves of phosphate, the raw material in many fertilizers. This valuable mineral is recovered from limestone found along the lower slopes of the Atlas Mountains.

Despite its rich natural resources, the country has a shortage of skilled labor and **capital** for investment. Capital is any source of wealth used to produce more wealth. For example, the money people invest in a savings account is capital because it produces more money through the interest it earns. The machinery used in an industry also is considered capital because it is used to produce goods that will, in turn, create new wealth. A shortage of capital is common in developing countries.

Morocco's location has encouraged communication with Europe and the Mediterranean region throughout the country's history. Rabat, the capital, and Casablanca are busy centers for Atlantic trade. The port city of Tangier is located on the Strait of Gibraltar. This beautiful Muslim city is a brief ferry ride for tourists traveling from Spain. Tangier is a **free port**, which means that almost no taxes are placed on goods unloaded there.

Sharp contrasts of Moroccan life are evident in the cities. In the ancient sections, sometimes called "old town," narrow streets are lined with rectangular, mud-brick houses that seem stacked against each other. In the center of town is a mosque and a covered market. Also in the old town is the *Kasba*, the palace of the local sultan or lord. Nearby, the modern "new town" features wide streets and large European-style houses. The new and old towns were originally separated by open land or by cemeteries. Gradually, the two towns grew together.

The population of Morocco includes Arabs and Berbers. Jews, people of European descent, and some black Africans also live here. Despite this diversity, the Arabic language and Islamic faith dominate the culture. The Moroccan way of life can be described as traditional versus modern. Traditional Morocco is evident in rural areas where about half of the country's people are subsistence farmers. They till small plots of barley, wheat, and vegetables on the plateaus below the Atlas Mountains. The best example of modern Morocco is found in Casablanca, the country's largest city. Its population is about 3 million. The wide city streets are lined with office towers, shopping districts, and apartments. Slums on the city's edge are filled with recent rural migrants.

Morocco's king has personal influence throughout the Islamic world because he is a direct descendant of the prophet Muhammad. Morocco hopes to become a powerful Arab state. This goal was evident in Morocco's annexation of the former Spanish colony of Western Sahara, which holds large phosphate reserves. The king of Morocco also has allied the country with the United States and western Europe. As a result of this alliance, Morocco has received large amounts of foreign aid. The government also encourages trade and cultural exchange with developed countries.

4 SECTION REVIEW

1. What natural resources does Algeria have? How has Algeria used the profits from these resources?
2. In what respects is Morocco a nation of contrasts?
3. **Critical Thinking** **Assess** the extent to which, in your opinion, a country's government should be involved in religious matters.

CHAPTER 36 CHECK

Cropland near **Aswān, Egypt**

Reviewing the Main Ideas

1. The Sahara, a desert, and the Nile River are two of North Africa's most important physical features. They have had a major impact on the settlement and economic patterns of people in the region.
2. Most of Egypt's agriculture depends on water from the Nile River. Egypt is currently undergoing rapid population growth and urbanization. Regional, national, and religious issues influence Egypt's politics.
3. The discovery of oil in Libya brought the country wealth and power. Tunisia's location and appreciation of democracy have helped the country. Like its neighbors, Tunisia faces the challenges of rapid population growth and religious extremism.
4. Algeria has substantial natural resources, on which some European countries are becoming increasingly dependent. Algeria is more industrialized than other countries in North Africa. Morocco is a nation of physical, economic, and social contrasts.

Building a Vocabulary

1. Define *erg* and *reg*. Where would one find them?
2. What is the term for sediment that is coarser than clay yet finer than sand? From where does this sediment come, and why is it significant?
3. What is capital, and why can a shortage of capital be a problem?
4. How is a free port different from other ports?

Recalling and Reviewing

1. How has the Nile River been important to Egypt?
2. Why is Tunisia's relative location with regard to the Sahara significant?
3. What do Libya and Algeria have in common?
4. Name three different cultures represented by inhabitants of the countries of North Africa.

Thinking Critically

1. Do you think that there should be a free trade agreement between the countries of Europe and the countries of North Africa? What advantages would such an agreement have, and what obstacles might prevent it from occurring?
2. Make a list of strategies the governments of North African countries might use to boost their economies.

Learning About Your Local Geography

Individual Project

Use an almanac to find your state's five most populous cities. Make a table listing these cities and their populations. Do you live in one of these cities? If so, how often do you leave it, and for what purposes? If not, how often do you visit one of them, and for what purposes? List some ways in which living in or out of a city affects your life. You may want to include your table in your individual portfolio.

Using the Five Themes of Geography

Using **place** characteristics as a guide, divide North Africa into nonpolitical **regions**. You might base your regions on landforms, climates, natural resources, historical background, cultures, or anything else you think relevant. Sketch a map showing your regional divisions.

CHAPTER 37

West Africa

"The Niger is the cradle of West Africa. It is a moving path into the heart of the continent, a long, liquid magic wand, that makes fertile the soil it touches. Cattle drink its water and graze on its green banks. Its fish and game birds provide food. Its trees give wood for dugouts. The river means an end to hunger, thirst, and isolation."

Sanche de Gramont

GEOGRAPHY DICTIONARY

zonal
pastoralism
millet
sorghum
staple
cassava
lingua franca

The rich history and remarkable culture of West Africa are two of the major characteristics that define this area as a region. The great kingdoms of Mali, Ghana, Songhai, and Benin (buh-NIN), among others, were sophisticated and highly developed. Today, West Africa is famous for the art that came out of these early civilizations.

The interior kingdoms of Mali, Ghana, and Songhai were well known for their earthenware. The Sapi, ancestors of ethnic groups in present-day Sierra Leone (see-ER-uh lee-ONE), were expert ivory and stone sculptors. The quality of Sapi artwork led to trade with the Portuguese in the early sixteenth century. The city of Ife (EE-fay), in present-day western Nigeria, was famous for its metal and clay sculptures. In the forest kingdom of Benin, craftsworkers and artists lived in special neighborhoods and created art only for the Oba, or king. Benin was especially well known for its beautiful brass sculptures.

The vitality of West Africa was interrupted first by the slave trade and then by European colonization. Today, that vitality is being renewed by talented artists and writers.

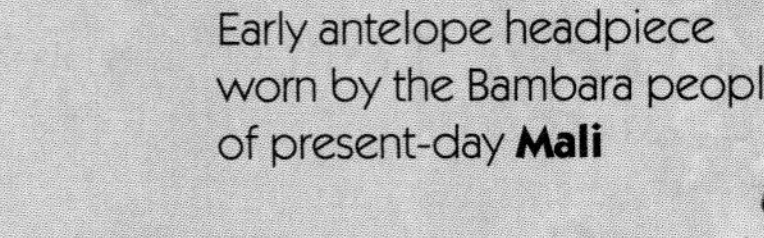

Early antelope headpiece worn by the Bambara people of present-day **Mali**

The Niger River, **Niger**

WEST AFRICA: PHYSICAL–POLITICAL

LEGEND

ELEVATION

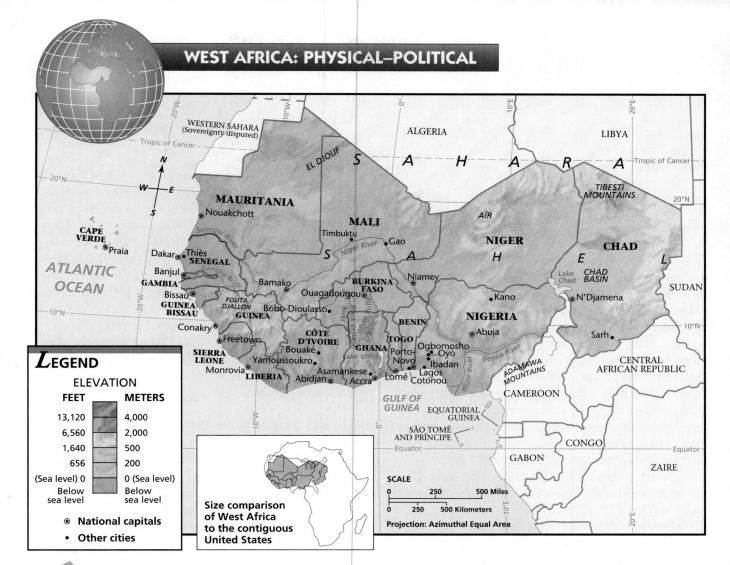

FEET	METERS
13,120	4,000
6,560	2,000
1,640	500
656	200
(Sea level) 0	0 (Sea level)
Below sea level	Below sea level

⊛ National capitals
• Other cities

Size comparison of West Africa to the contiguous United States

SCALE
0 250 500 Miles
0 250 500 Kilometers
Projection: Azimuthal Equal Area

PHYSICAL GEOGRAPHY

FOCUS

- *What are the three environmental zones of West Africa?*
- *What are the physical features of West Africa's environmental zones?*

REGION Climate West Africa is lower in elevation than eastern and southern Africa. It is mostly wide, rolling plains, with a few highland areas. There are no major mountain ranges here to break up the region's climate pattern. Thus, geographers refer to the climate regions of West Africa as **zonal**. This means that, when mapped, the climate pattern runs in bands from east to west. (See the map on page 418.) West Africa contains three main zones, each with distinct climates and vegetation. In the north are the drylands of the southern Sahara, the areas dominated by desert and steppe climates. In the middle is the tropical-savanna zone, with winter drought and summer rains. In the south is the forest zone, with humid-tropical climates and dense forests of the West African coastal belt.

Dryland Zone The dryland zone lies along the southern edge of the Sahara. Here, rainfall varies greatly from year to year. Climatic variations are typical in desert margins. In some years, there is no rain. In others, rain fills the dry lake beds, and rivers flow across the desert. During the 1950s and 1960s, the Sahel had many years of good rains. Grazing was good and crops grew well. Health standards improved and populations increased dramatically.

A long period of drought began in the Sahel during the 1970s, however. Herds of cattle, sheep,

People make their way through a dust storm in **Mali**, in West Africa's desert region. Some of the dust will be carried hundreds of miles out over the Atlantic.

and goats quickly overgrazed the savanna and desert grasslands, and many animals died. The Sahel's inhabitants cut trees for firewood. As the vegetation was stripped, the Sahara spread southward. During this period of desertification, crops failed for several years. People often ate their seed grain instead of saving it to grow crops the next year. With little or no food, many people died. Many more became refugees, migrating to the large cities in search of food from the government and relief agencies. Today, refugee camps, or shantytowns, surround the cities.

Some years in the past decade have been rainy. It is doubtful, however, that the droughts have ended for good. Geographers know that droughts in the Sahel have taken place repeatedly in past centuries. Fewer people and fewer animals existed then, however. Herds could be moved to the better lands in the south. This movement is not possible now. The lands to the south are blocked by settled farmers and the political boundaries of new countries. Still, the people of the Sahel who depend on farming and grazing for their livelihood have adapted to this unpredictable climate. They have discovered new methods of conserving water and growing crops. They have also improved trading relationships with their southern neighbors.

Savanna Zone In general, the rains in Africa increase toward the equator. The thorny, bush savannas of the southern Sahel gradually give way to tall trees and lush grasses. When the summer rains are good, the farmers of the savanna zone prosper.

Despite good rains and fertile soils, however, this region has few people. In the past, raids for the slave trade devastated the population. Today, the region is home to the dreaded tsetse fly and black fly. The tsetse fly carries sleeping sickness, a disease that is often fatal to cattle as well as people. The black fly carries a tiny parasitic worm from rivers to humans, often causing blindness. The black fly is common along rivers; therefore, people in the savanna zone often avoid settling in the river valleys. Instead, they live in the drier upland plains, where poor soil makes farming difficult. Although the tsetse fly and black fly can be controlled using powerful insecticides, the chemicals are expensive and harmful to the environment, animals, and humans.

Forest Zone Along the coast of West Africa are forested plains. Here, it rains almost daily for up to 10 months a year. This is West Africa's most densely populated region. As populations have increased during the last decades, much of the forest has been cut. The effects of commercial lumbering, especially in Côte d'Ivoire (kote deev-WAHR) (Ivory Coast), Nigeria, and Ghana, have been particularly damaging.

West Africa's coastal plains have few natural harbors. The coastline in many places is muddy and lined with mangrove trees. These trees form dense thickets that make transportation difficult. Elsewhere along the coast, sandy beaches with coconut palms and fishing villages are typical. The sunny beaches attract vacationers to newly built coastal resorts.

SECTION REVIEW

1. What do geographers call the pattern of West Africa's climate and vegetation regions? Why?
2. Why do more people live in the forest zone than in the dryland zone?
3. **Critical Thinking** **Predict** how West African settlement patterns might change if the diseases carried by the tsetse fly and black fly could be easily prevented.

2 HUMAN GEOGRAPHY

FOCUS

- *How do the ways of life of people in West Africa relate to the area in which they live?*
- *What economic challenges face West Africans?*

REGION **Cultural Patterns** West Africa is one of the most densely settled parts of Africa. People are concentrated in the coastal region and in the dry savannas of the southern Sahel. (See the map on page 419.) There are more than 500 African ethnic groups living in the region. Most countries contain at least four different ethnic groups. Some countries have as many as 50. Nigeria is home to more than 200 ethnic groups. Most of these ethnic groups were forced to live together in the new countries formed by European political boundaries. Today, many of the ethnic groups get along. Sometimes, however, there are religious, cultural, and land use conflicts.

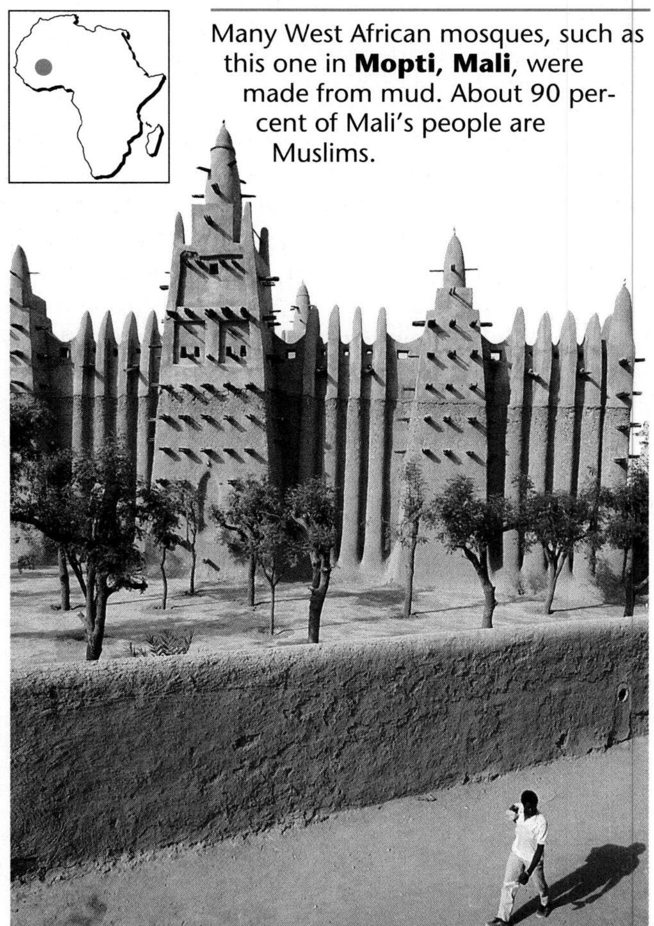

Many West African mosques, such as this one in **Mopti, Mali**, were made from mud. About 90 percent of Mali's people are Muslims.

West Africa is a region of religious diversity as well as ethnic diversity. Generally, the peoples of the dry northern lands are Muslims. Islam came with trade across the Sahara. In the south, people are primarily Christians. A smaller number of animists are found throughout West Africa. In many cases, animism has blended with Christianity and Islam.

To a large extent, cultural and religious diversity in West Africa is related to location and place. For example, the northern drylands are wide open, and travel is easy. Resources necessary for daily life are often far apart. Thus, in the past, a few ethnic groups could control large territories. In contrast, in the forested south travel was more difficult and less necessary. Many ethnic groups controlled much smaller areas. The coastal peoples, meanwhile, had more contact with European traders and explorers, and European missionaries spread Christianity inland. Africans often adapted European Christianity into Christian organizations of their own.

Economic and Settlement Patterns

Economic patterns that developed in West Africa during the colonial period continue. The region exports its raw materials and agricultural products mainly to developed countries. Most manufactured goods, in turn, are imported from these nations. All West African countries receive some form of foreign aid from developed countries and from international agencies.

West Africa is primarily an agricultural region, although only a small percentage of the land is arable. In the northern areas, pastoral nomads move their herds of cattle, sheep, and goats, following the rains. **Pastoralism** is an agricultural way of life that involves raising livestock and usually includes some farming. Often, pastoralists will grow crops to tide them over between the seasonal rains. **Millet** and **sorghum** are grain crops most often grown in the north because they are resistant to drought. The terrible drought in the 1970s forced many pastoralists to give up their way of life and move to cities in search of jobs. This migration of people, especially of young men, to the cities has lowered agricultural productivity. With fewer workers, traditional agriculture has declined.

In the savanna and forest zones of West Africa, small, compact farming villages are common. Most farmers practice traditional agriculture. The **staple**, or main, foods include root crops such as yams and **cassava**. Cassava has fleshy roots that provide a nutritious carbohydrate.

NEWS FROM AROUND THE WORLD *Niger*

Tuareg tending their livestock in **Niger**.

Tuareg Teenagers on the Move

Imagine growing up in a traditional nomadic society, moving constantly. The Tuareg (TWAH-reg), a Berber group who live mainly in Niger, Mali, and Algeria, spend their lives moving from place to place in search of water and grazing lands. At a young age, Tuareg boys and girls learn to ride dromedaries, Arabian one-humped camels. Children also learn to care for the goats, sheep, and cattle of the Tuareg caravans.

The history of the Tuareg is passed down by the women, who share stories of the Arab invasion of West Africa centuries ago and of French colonizers who later left their imprint on the region. Religious education is generally left to the men, who recite from the Qur'an, the holy book of Islam.

Although most Tuareg are Muslims, their daily lives reflect a freedom and independence from traditional Muslim societies. Tuareg women may own and inherit property, and many women are wealthier than their husbands. Also, the faces of Tuareg women are not veiled, while those of Tuareg men are. Forbidden to show their faces in public, Tuareg males wear a long scarf wound around their heads, covering the mouth, nose, and chin.

Tuareg society is divided into social classes. A girl is encouraged to wed a Tuareg of her family's social position or of a higher rank. Men often are judged by their dromedaries—the animals' appearance and the fineness of the saddles they bear.

The Tuareg come into contact with other peoples when they travel to villages to trade cattle for rice, sugar, tea, and other supplies. In these settled communities, Tuareg teenagers might participate in a soccer game and mingle with French-speaking teens. Such visits, though, are only a pause in the lives of most Tuareg. To support their nomadic way of life, they must move with the changing seasons.

West Africa has large-scale commercial farms, but crops are grown for export, not for feeding the people. To increase income from exports, some African leaders have sought to increase the export of cash crops and minerals. This economic strategy, however, forces African countries to continue to buy more goods from the developed countries. It also forces the economies of the African countries to be dependent on the global market. Even though a West African country might have a good crop yield, the price that other countries are willing to pay for that crop might be extremely low that year.

One result of this practice is that progress has been slow for the people in rural areas who produce cash crops for export rather than grow food for the people. Many West African countries must now import basic food supplies. Thus, there is even less money to invest in the education of the growing population and in infrastructure.

Economic growth remains the most challenging issue facing the countries of West Africa. Some of the countries have rich resources. Given more education and training programs, their chance for economic improvement increases. For the countries without adequate resources, more foreign aid and regional cooperation are the keys to growth. Indeed, at economic summits, African leaders meet to discuss the economic challenges facing their countries. The leaders have concluded that export-led development is hurting Africa's chances to provide a better life for its people. They hope to change this economic pattern.

SECTION REVIEW

1. What challenges do the pastoral nomads face in the Sahel?
2. What happens when a poor country focuses on growing cash crops for export?
3. **Critical Thinking** **Compare and contrast** northern West Africa and southern West Africa.

SAHEL COUNTRIES

FOCUS

- *What effects did the drought of the 1970s and 1980s have on the Sahel countries?*
- *Why is the Niger River important to the countries of the Sahel?*

REGION The Sahel countries of West Africa are Mauritania, Mali, Niger, Burkina Faso (boor-KEE-nuh FAH-soe), and Chad. (See the map on page 441.) These countries share many common features. Most of the people live in the savanna region to the south, where herding and farming are the traditional occupations. In contrast, the stony plains and shifting sands of the Sahara are nearly empty of people. Struggles between the desert people and the settled farmers are frequent.

In the developing countries of the Sahel, there are few railroads or good roads. French is the ***lingua franca*** of these former French colonies. A *lingua franca* is a common language spoken and understood by the majority of the people, although other languages may be spoken at home. Arabic is often heard, a sign of the importance of Islam in the region. Many people speak one of the many languages of the diverse ethnic groups in the region.

The drought of the 1970s and early 1980s was devastating in the Sahel. In Niger, for example, nearly 90 percent of the herd animals died. Desertification forced one-fourth of the population to move to nearby countries. Recovery has been slow, and refugee settlements surround all of the capital cities in the Sahel countries.

Mauritania, Mali, and Niger Mauritania stretches east from the Atlantic coast into the Sahara. About two-thirds of the population are Moors, an Arab and Berber mixture. Almost all of the people are Muslims. Arabic and French both are official languages. The capital of Mauritania is the coastal city Nouakchott (nu-AHK-shaht). Nouakchott had 5,000 people when it was founded in 1960. Migration from rural and drought-stricken areas has increased the population to more than 500,000. Most of the people live in refugee camps. International aid,

FACTS IN BRIEF West Africa

COUNTRY POPULATION (1994)	LIFE EXPECTANCY (1994)	LITERACY RATE (1990)	PER CAPITA GDP (1993)
BENIN 5,341,710	50, male 54, female	23%	$1,200
BURKINA FASO 10,134,661	46, male 48, female	18%	$700
CAPE VERDE 423,120	61, male 65, female	66% (1989)	$1,070 (1991)
CHAD 5,466,771	40, male 42, female	30%	$500
CÔTE D'IVOIRE 14,295,501	47, male 51, female	54%	$1,500
GAMBIA 959,300	48, male 52, female	27%	$800
GHANA 17,225,185	54, male 58, female	60%	$1,500
GUINEA 6,391,536	42, male 46, female	24%	$500
GUINEA-BISSAU 1,098,231	46, male 49, female	36%	$800
LIBERIA 2,972,766	55, male 60, female	40%	$800
MALI 9,112,950	44, male 48, female	17%	$650
MAURITANIA 2,192,777	45, male 51, female	34%	$1,050 (1992)
NIGER 8,971,605	43, male 46, female	28%	$650
NIGERIA 98,091,097	54, male 57, female	51%	$1,000
SENEGAL 8,730,508	55, male 58, female	38%	$1,400
SIERRA LEONE 4,630,037	44, male 49, female	21%	$1,000
TOGO 4,255,090	55, male 59, female	43%	$800
UNITED STATES 260,713,585	73, male 79, female	97% (1991)	$24,700

Source: *The World Factbook 1994,* Central Intelligence Agency

In which West African countries is the literacy rate at least 50 percent?

Motorbikes (mopeds) are assembled in **Ouagadougou, Burkina Faso**. Two-wheel commuters choke the city's streets during rush hour.

particularly from the oil-rich Islamic countries, provides much of the country's food.

Because its northern half is in the Sahara, Mali sometimes seems to be two countries. Bamako, the capital, is in the savanna lands along the Niger River. Nearly all of the people of Mali live in this southern region. The government of Mali hopes to use the Niger's waters to expand irrigation. Potential also exists to increase tourism to Mali's ancient trading cities of Timbuktu and Gao.

Landlocked Niger consists of a small savanna region along the Niger River valley and a large expanse of the Sahara and Sahel to the northeast. The capital is Niamey (nee-AHM-ay), situated on the Niger River. Only 2 percent of the country's land is arable, all of it located along the river. Ethnic groups include the Hausa, Djerma, Fulani, and Tuareg. As in the other northern countries of West Africa, Islam is the predominant religion. Democratic reforms are beginning to take place.

Burkina Faso and Chad Burkina Faso, formerly Upper Volta, is a landlocked country of dry savannas situated between Mali and Niger. Ethnic groups and subgroups include (among many others) the Mossi, Gurunsi, Senufo, Lobi, Mande, and Fulani. Most of the people are animists, and cattle raising is traditional here. Before the severe drought of the 1970s and 1980s, great cattle drives moved herds to countries to the south. Today, many workers migrate to the coastal countries in search of jobs. Near the center of the country is the capital city of Ouagadougou (wahg-uh-DOO-goo). Great slums have grown around the city since the drought, and the search for firewood has removed nearly every tree within 25 miles (40 km). The government is supported mainly by aid from developed countries.

Chad stretches from the central Sahara to the tall-tree savanna region in the south. Chad has more than 200 ethnic groups, and more than 100 local languages are spoken. The few people in the north are mostly Islamic herders with Arabic customs. The numerous people in the south are Christian and animist farmers. As a result, the politics of Chad are complex. Many groups have fought for control of the government in N'Djamena (ehn-juh-MAY-nuh), the capital. Meanwhile, the drought of the 1970s and 1980s reduced Lake Chad to one-third the size it was in 1950. An important fishery there has nearly collapsed. With few resources and internal conflict, Chad has little chance for economic prosperity in the near future.

SECTION REVIEW

1. How did the drought of the 1970s and 1980s affect the Sahel countries?
2. To which countries is the Niger River an important resource?
3. **Critical Thinking** **Determine** how a *lingua franca* might unite the people of a country.

ATLANTIC COAST COUNTRIES

FOCUS

- *How do people in the Atlantic coast countries earn a living?*
- *How are the countries of the Atlantic coast alike? How are they different?*

REGION The Atlantic coast countries of West Africa are Senegal, Gambia (GAM-bee-uh), Guinea, Guinea-Bissau (GIN-ee bis-OW), Liberia, and Sierra Leone. (See the map on page 441.) The dry savannas of Senegal give way to forests in Sierra Leone and Liberia. Much of the original forest has been cleared by farmers, however. To the south, the wetter climates permit the growing of bananas, coffee, and cassava. Farmers grow rice in river valleys throughout the region. In the wettest areas, farmers tend oil palms, rubber trees, and vegetable gardens.

Rains are heavy on the mountain slopes. Supposedly, Sierra Leone (which means "mountain lion" in Portuguese) was named by Portuguese navigators who thought that the rainstorms coming out of the mountains sounded like lions. Mountain areas in Sierra Leone have received more than 240 inches (600 cm) of rain in one year.

Gambia and Senegal The boundary between Gambia and Senegal is one of the most unusual in the world. Gambia, a former British colony, is composed of the land 15–30 miles (24–48 km) on both sides of the Gambia River. The country is 295 miles (475 km) long, and apart from a tiny coastline, entirely surrounded by Senegal. Senegal and Gambia have proposed to merge, but they have many issues to settle. In Gambia, the largest ethnic group is the Mandinka, and Islam is the predominant religion. Gambia is primarily an agricultural country, and peanuts are the major crop.

Senegal, a former French colony, is more developed with better commercial traditions. For centuries, peanuts have been the dominant cash crop in Senegal. Cotton and corn also are important. Senegal's capital, Dakar (duh-KAHR), was a major administrative center for the nearby French colonies. There, manufacturing income is growing. Dakar is a modern city, with an international airport, sports stadiums, cinemas, and museums. Factories process peanuts, cotton, rice, fish, and phosphate ores. French is the official language, and Wolof (also the name of the largest ethnic group) is widely spoken. Most of the population is Muslim. Senegal's political stability encourages tourism.

Guinea After Guinea gained independence from France, the new country's first leader was a cruel dictator. Initially promising prosperity, he came to control both the lives of the people and the economy. More than 2 million people fled the country; many others died in prisons. Poverty increased in the rural areas, and hardships continued for 25 years until the dictator died.

Today, the government of Guinea is making progress. Conakry (KAHN-uh-kree), the capital of Guinea, is the center of education for the country, while most industrial activity takes place northeast of the capital. Particularly important to the economy are Guinea's large reserves of bauxite, the ore of aluminum. These reserves total about one-third of the world's supply. The products of the oil palm

African and French cultures have influenced the development of **Dakar**. Once the capital of French West Africa, Dakar is one of Africa's major cities and the capital of independent **Senegal**.

also are important to Guinea's economy. Palm oil is used for making cosmetics and soap and in cooking, palm kernels are processed for starch and food, and palm fiber is used for making floor mats and rope.

The major ethnic groups in Guinea include the Fulani, Malinke, and the Susa. Islam is practiced by more than 75 percent of the population; about 5 percent are animists.

Guinea-Bissau and Cape Verde Guinea-Bissau and Cape Verde were once Portuguese colonies. In fact, the city of Bissau was the major port for the Portuguese West African slave trade. The colonies were often administered together, and they shared common paths to establish Communist governments after independence. Having abandoned communism, both countries are now attempting to establish democratic governments. The ethnic makeup in Guinea-Bissau is diverse. More than 50 percent of the population is animist. The people survive on subsistence agriculture, growing rice and cassava. Poverty in Guinea-Bissau is widespread, however,

Cape Verde is a group of islands located 400 miles (640 km) off the coast of Senegal. When the Portuguese arrived in 1456, Cape Verde was uninhabited. The 10 islands are volcanic in origin and appear stark and barren. There is little farmland.

The Portuguese brought African slaves to the islands in the 1460s. Intermarriage between the Africans' descendants and Europeans produced most of the present population. Many islanders have emigrated to the United States.

Liberia and Sierra Leone Liberia and Sierra Leone are unusual in that both countries were founded as settlements for freed slaves. The name *Liberia* comes from the Latin word *liber*, which means "free." Liberia's settlers came from the United States beginning in 1822. The country became a republic in 1847 and has continued to maintain close relations with the United States. Sierra Leone was founded in 1787 for freed slaves from the British Empire. It remained a British colony until 1961. In Liberia, the descendants of African American slaves are the most prosperous people in the country, though they total only about 2 percent of the country's population. In Sierra Leone, the descendants of freed slaves, who call themselves Creoles, make up about 2 percent of the population. Recent politics in both countries have been violent. The descendants of the freed slaves are often resented by the majority of the people.

Liberia and Sierra Leone are poor countries. In its long history, Liberia has received little economic support from foreign countries. Sierra Leone, meanwhile, was probably the least developed of Britain's African colonies. In Liberia, rubber plantations, most of which were owned by a single U.S. company, once dominated the country's economy. Today, manufacturing in Sierra Leone and Liberia is centered around processed food, textiles, and building supplies. With large reserves of iron ore, diamonds, and bauxite, both countries hope expanded mining will improve their economies.

SECTION REVIEW

1. What are the major cash crops in the Atlantic coast countries? Where are the major manufacturing centers?
2. What is unusual about Liberia and Sierra Leone?
3. **Critical Thinking** **Compare and contrast** Senegal and Guinea-Bissau.

GUINEA COAST COUNTRIES

FOCUS

- *Why do the Guinea coast countries look to Nigeria for leadership?*
- *What is the economic situation in the various Guinea coast countries?*

REGION The coastline of West Africa that faces south is called the Guinea coast. The countries here are Nigeria, Ghana, Côte d'Ivoire, Togo (TOE-goe), and Benin. These humid countries are densely settled near the coasts. Climates become drier inland.

Nigeria Compared to the other countries of West Africa, Nigeria is enormous, and it is by far Africa's most populous country. Nigeria stretches

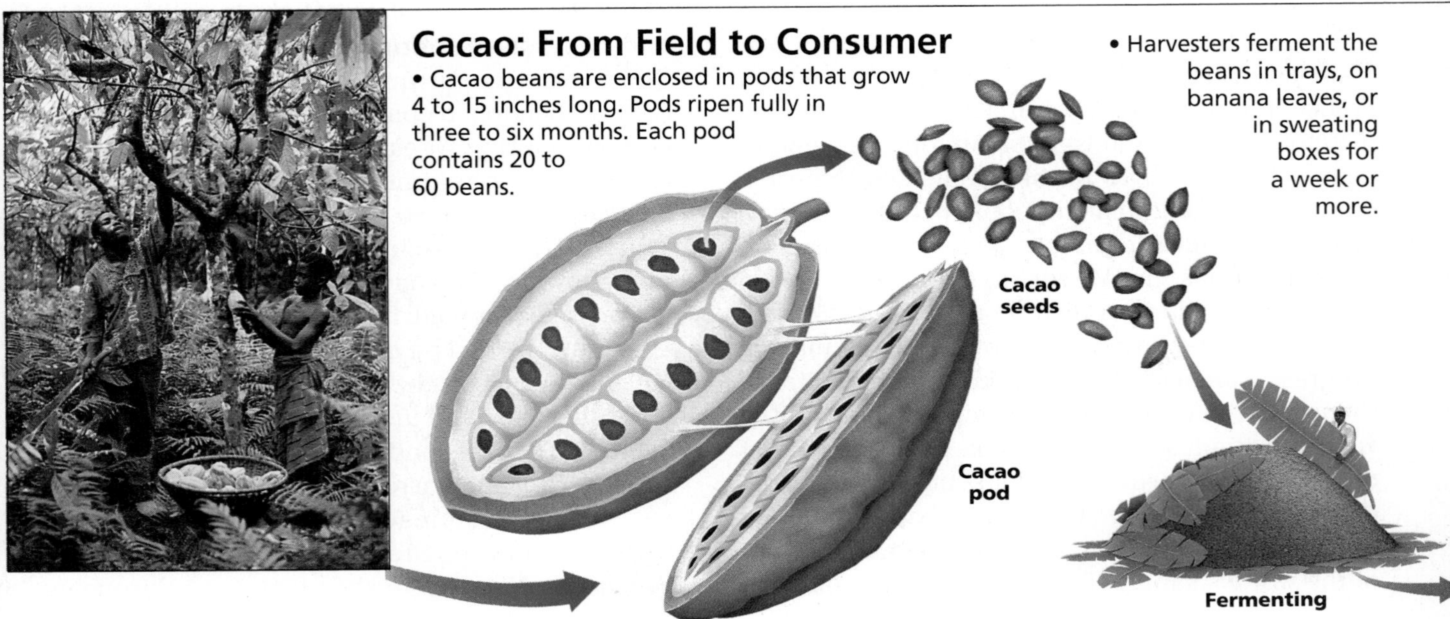
Cacao: From Field to Consumer
• Cacao beans are enclosed in pods that grow 4 to 15 inches long. Pods ripen fully in three to six months. Each pod contains 20 to 60 beans.
• Harvesters ferment the beans in trays, on banana leaves, or in sweating boxes for a week or more.
Cacao seeds
Cacao pod
Fermenting

Oil production is an important economic activity in the Niger River delta of **Nigeria**. Nigeria has some of the largest oil reserves in the world.

from the humid climate of the Gulf of Guinea to the dry savannas along Lake Chad. In the tropical south, oil-palm products, cacao, and rubber are important. In the north, peanuts and cotton are grown.

Nigeria's greatest mineral resources are the large oil deposits of the Niger River delta. Nigeria's economy depends so heavily on oil, however, that it can be severely hurt when world oil prices are low. Coal deposits also are located in the delta. Inland, there are large iron-ore and manganese mines. Nigeria holds about half of the world's supply of columbite, a metal used to harden stainless steel. The Niger River, meanwhile, is a major hydroelectricity and fishing resource.

When Nigeria became independent from Britain in 1960, a network of roads and railroads was already in place. Lagos (LAY-gahs), Nigeria's former capital city, is an industrial and commercial center. Its bustle and congestion are known worldwide. The government has moved the capital to the new city of Abuja (ah-BOO-juh), in the center of Nigeria. New factories also are being built throughout the country. Nigeria has truck and automobile assembly plants in addition to other factories that export goods throughout the region.

Cities have been important in Nigeria for centuries. In the dry north, walled cities such as Kano (KAH-noe) were once centers for prosperous Muslim kingdoms. Kano now has an international airport and is the largest city in the northern region. Ibadan (ee-BAHD-uhn) in the south, whose origins remain a mystery, was an important town before colonization. Today, this hilly city has a prominent university and is second in size only to Lagos.

There are four main ethnic groups in Nigeria. The two large groups in the south are the Yoruba of the Lagos-Ibadan region and the Ibo of the lower Niger River region. These and other southern groups were those most influenced by Christian missionaries and European ideas. The two large groups in the north are the Fulani and the Hausa. These and most other people in the northern regions are Muslims who follow traditional ways of life.

Nigeria has been the stage for ethnic and religious conflict. The Ibo in the southeastern province of Biafra (bee-AF-ruh) tried to secede from the rest of the country in the late 1960s. A terrible civil war resulted. Each of the four large ethnic groups, with its own customs, religion, and language, is suspicious of the others. English has become the *lingua*

• Cacao beans are then cleaned, roasted, and hulled. Shells can be used for fertilizer, organic mulch, and the production of cocoa butter. Roasted beans are crushed, producing a thick, dark liquid.

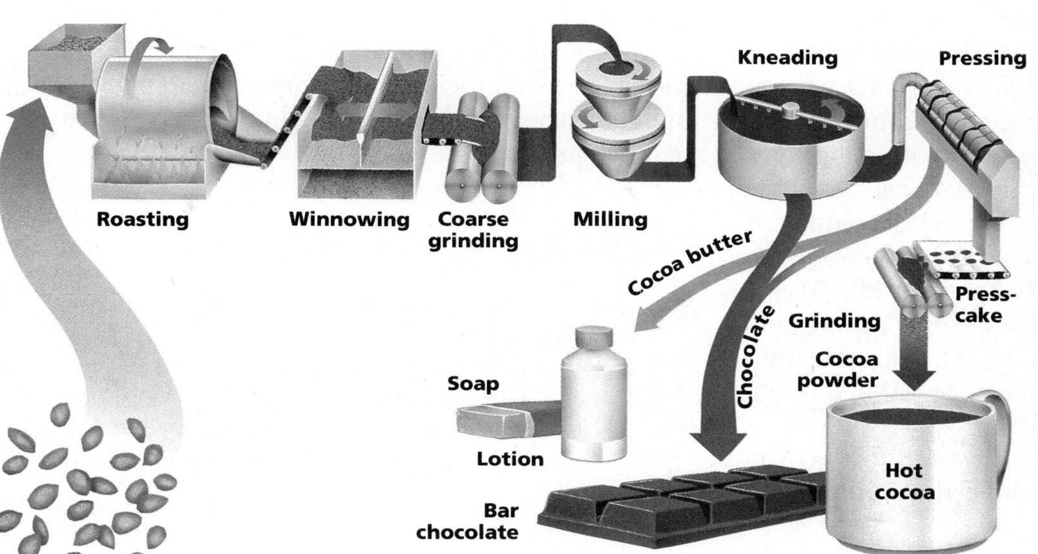
Roasting
Winnowing
Coarse grinding
Milling
Kneading
Pressing
Cocoa butter
Chocolate
Grinding
Press-cake
Cocoa powder
Soap
Lotion
Bar chocolate
Hot cocoa

• The thick liquid can be combined with sugar and vanilla to make sweet chocolate. Cocoa butter can be used in the preparation of medicines and other consumer products, such as soaps and cosmetics.

• Pressing the liquid to remove most of the cocoa butter, or cacao fat, produces powdered cocoa. Chocolate is mostly cacao fat, while cocoa can contain as little as 10 percent fat.

franca because it is not the language of any single group. Despite these problems, the idea of a united Nigeria is becoming stronger. Moving the capital from Lagos to Abuja, where no ethnic groups have land claims, may further unite the Nigerian people.

Nigeria influences West Africa militarily, politically, and economically. Because of its large population and its oil wealth, Nigeria has often assumed a leadership role in Africa. However, conflict between supporters and opponents of democratic reforms has added to Nigeria's political instability.

Ghana and Côte d'Ivoire Côte d'Ivoire and Ghana have similar climates, soils, and natural resources. Ghana took its name from the former African empire. (The word *ghana* means "war chief.") When Ghana gained independence from Britain in 1957, the country had prosperous farms, good roads, industrial opportunities, and a promising economy. Since then, its economy has had ups and downs. Côte d'Ivoire became independent from France in 1960 with fewer economic opportunities. Nonetheless, it is one of Africa's most prosperous countries today.

During the colonial period, Ghana was called the "Gold Coast." European traders built forts at the country's natural harbors and began trading inland for gold, ivory, and slaves. In 1919, the British began building roads, harbors, and agricultural processing centers. Ghana soon became the world's largest cacao producer. After independence, the poor economic policies of several dictatorships and elected governments caused the economy to weaken. The governments of Ghana failed to invest in agriculture. Farmers were not given fair prices for their products and, in turn, could not reinvest in their farms. As a result, yields declined. Cacao production dropped by 80 percent.

The current government in the capital of Accra (uh-KRAH) has realized these past errors, and recent economic reforms have improved the economic outlook. Ghana has rich gold mines and the ability to produce high-quality cacao. Another hope for Ghana's future is continued development of the Volta River Project. This great dam on the Volta River has created one of the world's largest human-made lakes. The hydroelectric power generated here, easy shipping access, and irrigation opportunities could help restore Ghana's economic position.

The healthy economy of Côte d'Ivoire is related to its history of political stability. The French developed prosperous farms and large plantations here. After independence, Félix Houphouet-Boigny, a rich African planter and Boule chief, was elected president and stayed in power for more than 30 years. The stable government invested in agriculture, and Côte d'Ivoire became the largest producer of cacao. The country maintained good relations with France and managed its finances well. Côte d'Ivoire has relied on free enterprise and foreign investment to encourage economic growth.

The economic heartland of Côte d'Ivoire is the southern forested zone, although much deforestation has occurred. Lumber, coffee, and cacao are grown here for export. The coffee crop is the largest in Africa and an important part of the world market. Abidjan (ab-i-JAHN), the capital, is a modern city and major port. With many small and large industrial firms, it has the largest manufacturing output of all the former French African cities. Abidjan is crowded, however. In order to encourage development in the central and northern parts of the country, the government has considered moving the capital to the inland town of Yamoussoukro (yahm-i-SOO-kro).

Togo and Benin Togo and Benin are small, poor countries situated between Ghana and Nigeria. Both have savanna environments with tropical forests along the coast and on the uplands. As in the other countries along the Guinea coast, coffee, cacao, and palm oil are important products.

Togo was a German colony before World War I. After that war, it came under French control until gaining independence in 1960. Benin, the former French colony of Dahomey (duh-HOE-mee), is named after the kingdom of Benin. Togo and Benin both have numerous ethnic groups, and most of the people are animists or practice African religions. Both are countries of small farmers who provide little tax revenue for their governments. Manufacturing industries are few and small, and there are not enough jobs.

5 SECTION REVIEW

1. Why is Nigeria such an important country in the region?
2. What economic activity is important to every Guinea coast country?
3. **Critical Thinking** **Suggest** two reasons that countries move their capitals.

CHAPTER 37 CHECK

Construction of Notre Dame de la Paix Cathedral (Our Lady of Peace Cathedral) in **Yamoussoukro, Côte d'Ivoire**, was completed in 1990. The cathedral is one of the largest Christian churches in the world.

Reviewing the Main Ideas

1. The climate and vegetation pattern in West Africa is divided into three zones. Most people live in the forest zone and in the dry savannas of the southern Sahel.
2. West Africa is a region of religious and ethnic diversity. More than 500 ethnic groups live here. West Africa has been influenced by the spread of Islam and by Christianity, though many people still retain their traditional beliefs.
3. Agriculture is the predominant economic activity in West Africa. In the north, pastoral nomads raise livestock. Villages of small farmers are common in the savanna and forest zones.
4. The most serious issue facing West Africa today is economic development. Development of resources, education and training programs, foreign aid, and regional cooperation are the keys to economic improvement.

Building a Vocabulary

1. Why do geographers use the word *zonal* to describe West Africa's climate and vegetation pattern?
2. What activities does pastoralism involve?
3. What is a staple food? Give two examples of staple foods in West Africa.
4. What are millet and sorghum? Why are they suitable for West Africa?

Recalling and Reviewing

1. Describe West Africa's three environmental zones.
2. What effects has the drought of the 1970s and 1980s in the Sahel had on West Africa?
3. What two Atlantic coast countries were founded as settlements for freed slaves?
4. Why are the Guinea coast countries of Ghana, Côte d'Ivoire, and Nigeria more prosperous than Togo and Benin?

Thinking Critically

1. One problem in West Africa is that countries often compete rather than cooperate with one another. What specific steps do you think the leaders of West Africa could take to bring about regional cooperation?
2. Using three examples from the chapter, discuss the various ways by which countries get their names. Also, suggest a reason that the people or government of a country might want to change that country's name.

Learning About Your Local Geography

Cooperative Project

With your group, conduct research to determine how the population density in your area has changed over the last 30 years. What factors have contributed to this change? Present your information using charts, graphs, or illustrations.

Using the Five Themes of Geography

Create a graphic organizer that illustrates the relationships among the geographic themes of **location**, **place**, **movement**, and **human-environment interaction** in the Sahel.

CHAPTER 38

East Africa

GEOGRAPHY DICTIONARY

Swahili

gum arabic

sisal

"On our left rose a long, dark-crested mountain range from which sprang rivers that watered a great part of the Kikuyuland. These rivers, no larger than streams, had dug down through soil red as a fox and rich as chocolate to form steep valleys whose sides were now green with young millet and maize."

Elspeth Huxley

Ceramic mural in **Burundi**

Deep beneath the scenic landscapes of East Africa, the earth's mantle is churning, its intense heat struggling to escape to the surface. Active volcanoes and hot springs here tell us that the processes of plate tectonics are trying to break Africa apart. So far, the continent has arched up and split along the rift valleys.

As the rifts opened, the geography of the region was dramatically altered. People settled near newly formed lakes in the rift valley. They brought cattle and goats to the region and began farming there. These settlers found greatest agricultural success in the fertile highlands. Later, the highland peoples discovered that the ivory tusks of elephants were prized by merchants of the Mediterranean and India. Ivory was traded for goods from as far away as China.

Today, tourists travel from throughout the world to admire the beauty of East Africa's national parks. Important to Africa's economy, tourism creates jobs in the region's hotels, restaurants, and transportation facilities. Best of all, the tourists take pictures, not ivory.

Agricultural plots in **Kenya**

EAST AFRICA: PHYSICAL–POLITICAL

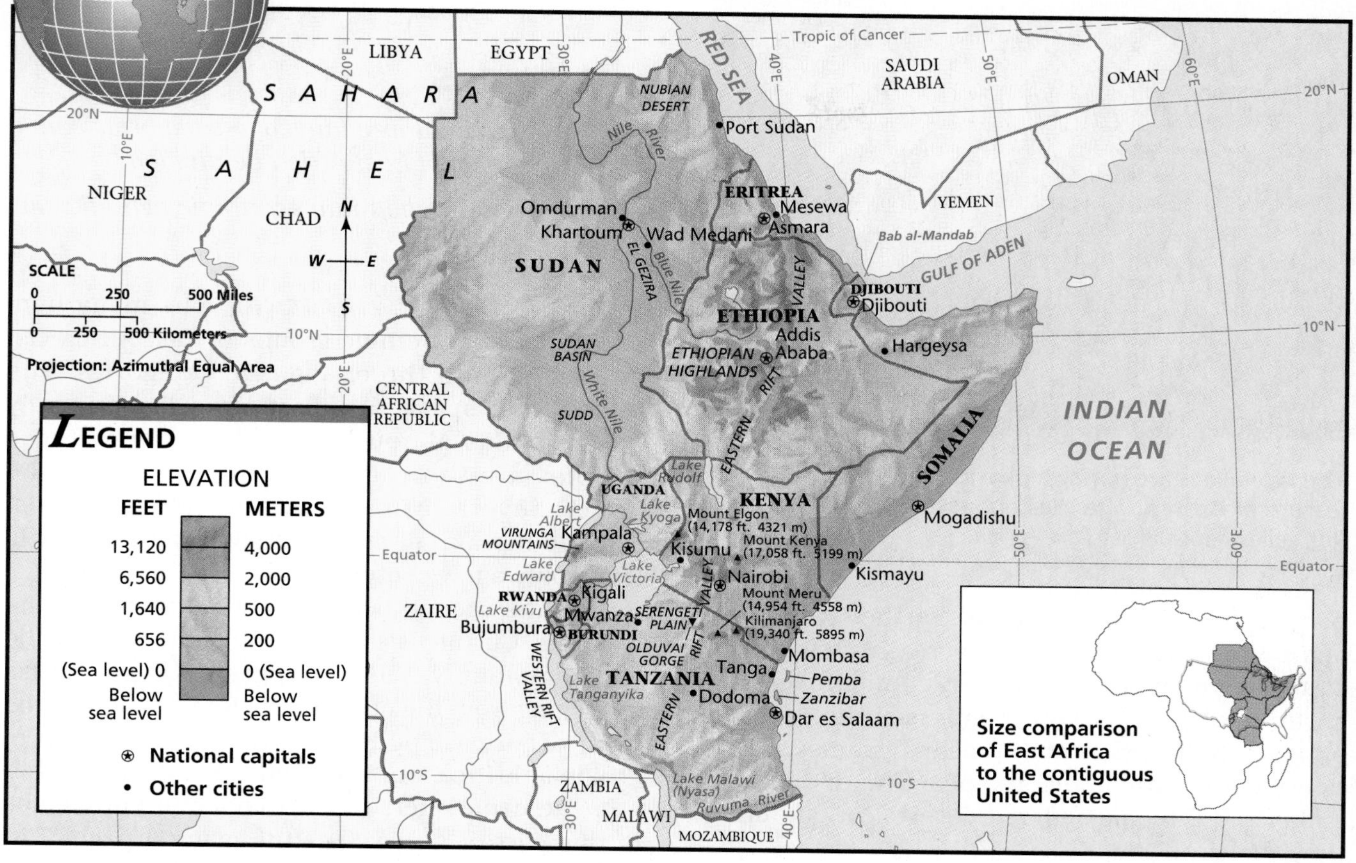

SCALE: 0 – 250 – 500 Miles; 0 – 250 – 500 Kilometers

Projection: Azimuthal Equal Area

LEGEND

ELEVATION

FEET	METERS
13,120	4,000
6,560	2,000
1,640	500
656	200
(Sea level) 0	0 (Sea level)
Below sea level	Below sea level

⊛ National capitals

• Other cities

Size comparison of East Africa to the contiguous United States

PHYSICAL GEOGRAPHY

FOCUS

- *What major landforms are found in East Africa?*
- *What is the climate like in East Africa?*

REGION **Landforms and Climate** East Africa is a region mostly of high plains and plateaus. The plateaus are topped by snow-capped volcanoes and cut by the Eastern and Western Rift valleys. Much of East Africa's scenery is spectacular. Mount Kenya in central Kenya is more than 17,058 feet (5,199 m) high. Mount Elgon (EHL-gahn) on the Kenya-Uganda (yoo-GAN-duh) border is 14,178 feet (4,321 m) high. The highest point, Kilimanjaro, rises to 19,340 feet (5,895 m). Even though Kilimanjaro is near the equator, it is so high that its twin peaks are always covered with snow.

From above, the rift valleys appear as giant scars on Africa's surface. The rift walls are usually steep cliffs, with the flat valley floors sometimes more than 10,000 feet (3,048 m) below. The Eastern Rift begins in southern Tanzania and continues northward to Ethiopia and onto the floor of the Red Sea. (See the map on page 417.) The dry plains along the Eastern Rift in Tanzania and Kenya are sites of famous game parks. The more humid Western Rift is filled with a chain of lakes, which are often located between high volcanoes. Lake Victoria, the largest lake in Africa, occupies a shallow, saucer-shaped basin on the plateau between the Eastern and Western rifts. To the east, the land slopes downward to the Indian Ocean. The sandy beaches and beautiful lagoons of the East African coastlines are beginning to attract tourists.

East Africa is not only a region of magnificent landforms, but also one of great climate contrasts. The region contains some of Africa's driest deserts and wettest forests. The deserts of northern Sudan are like those of neighboring Egypt. The Nile River forms a long oasis within the surrounding desert of

The rift valleys are a major physical feature of **Kenya.** The Western and Eastern Rift valleys cut across East Africa.

bare rocks and shifting sands. Farther south, the lands along the Nile become increasingly wet. The desert gives way to the more common thorny-shrub desert. Eventually, a grassy-shrub savanna appears. Still farther south, near Uganda, vegetation changes to tall-tree savanna. The drylands of Sudan and Ethiopia suffered desertification during the Sahel droughts.

The moist climates of the Ethiopian plateau stand high above the deserts of Somalia and northern Kenya. Along the coast of the Indian Ocean is a tropical-savanna climate with low grasslands and woodlands. Inland, dry savannas with a steppe climate are typical up to the forested highlands. The greatest climate changes occur along the sides of the rift valleys. The floors of the rifts are usually semiarid, with thorn scrub and dry grasslands. In contrast, the surrounding mountains have a humid highland climate with dense forests. The mountains enjoy rains caused by the orographic effect, while the rift valleys are in rain shadows.

1 SECTION REVIEW

1. Why can it be said that East Africa has some of the most spectacular scenery in all of Africa?
2. What are the various climate types found in East Africa?
3. **Critical Thinking** **Explain** why plains make good game parks.

2 HUMAN GEOGRAPHY

FOCUS

- *Into what three language categories can peoples of East Africa be divided?*
- *What are the main economic activities in East Africa?*

REGION **Ethnic Heritage** The many hundreds of modern ethnic groups in East Africa can be divided into three language categories. The Nilo-Saharan speakers of East Africa live primarily on the plains of southern Sudan. Among these peoples are the cattle-keeping Dinka and the Nauer. Several Nilo-Saharan speaking groups, such as the Masai, migrated south into the highlands only a few centuries ago.

In the second language category are Cushite speakers. Cushite is a subdivison of the Afro-Asiatic language family. Cushite speakers live in areas stretching from the coast of the Red Sea across the Horn of Africa. The Horn of Africa is made up of Ethiopia, Eritrea, Somalia, and Djibouti (juh-BAHT-ee). The name refers to the rhinoceros-horn shape of this part of Africa. (See the map on page 453.) Within this language group are the Amhara of the Ethiopian highlands and the Somali of the coastal region.

The third language group is the Niger-Congo (Bantu) language family of East Africa. These Bantu languages are similar to those of West and Southern Africa. Included in this group are the Kikuyu of Kenya and the Sukuma of Tanzania.

In part, the wide ethnic diversity in East Africa is due to its location. East Africa has always attracted visitors and settlers. The ancient Egyptians regularly sent trading and military expeditions up the Nile. Later, Arab traders established trading cities along the east coast. Trading ships carried gold, ivory, and slaves from the interior to the ports of the Persian Gulf and India.

One of the legacies of this period is **Swahili** (swah-HEE-lee). Swahili developed as the key trade language along the coast. The language contains many Arabic words within a distinctly African grammar. Through the centuries, Swahili has developed a rich literature and has become the common language of the region. Swahili is the official language of Kenya and Tanzania.

During the colonial period, immigrants from India and Europe settled in the cities as merchants and skilled workers. East Africa's many ethnic groups make the cities of East Africa particularly interesting. Every sort of clothing can be seen, and in the evenings, the spicy aromas of many different foods float through the streets.

Economic Geography Although East Africa has several modern cities, most of the people live in rural areas as traditional farmers. There, economic opportunities vary with the natural environment. In the drylands, herding is important. Cattle are highly valued and often are a measure of wealth. Major food crops include maize, sorghum, and millet. **Gum arabic**, which comes from the sap of acacia trees, is an important cash crop. It is used in some medicines and also gives many candies their chewy texture. Cotton grown on irrigated lands is the most valuable crop of the drylands.

In the humid highlands, peas, beans, and cassava are chief crops. Bananas and tomatoes also grow year-round. The most valued cash crop in the highlands is coffee. Though coffee is grown on large plantations, a dozen coffee trees can provide a small cash income for a family.

East Africa's cities have factories that produce consumer goods, processed foods, and building materials. As in West Africa, however, many manufactured goods are imported from developed countries.

Graduating students at the University of Nairobi in **Kenya** reflect the ethnic diversity of East Africa.

SECTION REVIEW

1. What are the three major language groups of East Africa?
2. How do land use patterns in the drylands of East Africa differ from those of the humid highlands?
3. **Critical Thinking** **Assess** the extent to which peoples' movements shape their languages. Give an example from the section.

COUNTRIES OF EAST AFRICA

FOCUS

- *What are the physical and human environments of the countries of East Africa?*
- *What economic and political challenges face the countries of East Africa?*

Kenya Kenya's first cities were founded along the coast of the Indian Ocean by Arab merchants from the Persian Gulf. Beginning in the 1500s, Portugal controlled this coast for about 200 years until Arab forces recaptured it. British merchants began trading on the coast during the 1800s. In the 1890s, Britain extended its sphere of influence inland as far as the Western Rift valley. This territory included modern Kenya and modern Uganda. The British soon built a railway from Mombasa to Lake Victoria.

In 1900, the railway reached the highlands. There, near some railroad repair shops, the city of Nairobi (nie-ROE-bee) emerged. The local people, the Kikuyu, had recently moved from these rich farmlands because of a terrible smallpox epidemic in 1898 and 1899. British administrators encouraged settlers from England to move to the "empty" area. Soon, the fertile highlands were covered by large, productive farms whose products were shipped back to Europe. Along with the British settlers came workers and businesspeople from India and Pakistan. The remaining Kikuyu and similar groups had little choice but to move out of their traditional areas or to become farm workers. Others found menial jobs in the cities.

After World War II, black Africans protested British colonial rule. There were peaceful demonstra-

Text continues on page 458.

Women Tree Planters of Kenya

Avocado, mango, and acacia trees thrive in the Kenyan highlands, where not long ago farm workers could not even find shade.

Many of the trees in the highlands of **Kenya** have been cleared for crops.

Ten million young trees scattered throughout the country offer new hope to Kenyan farmers and families. The trees were planted not by foresters or government workers but by the many women who farm the lands. "We are planting trees to ensure our own survival," says Wangari Maathai, founder of Kenya's Green Belt Movement. The organization encourages tree planting in Kenya's rural communities as a way of fighting deforestation and desertification.

Deforestation

Like many other developing nations, Kenya is a country mainly of farmers and herders. The most productive farmlands are in the highlands, an area once green with trees and lush vegetation. Over the past century, however, much of the land has been stripped. Today, only about 10 percent of the original forests remain. Many trees were cut for firewood, which farm families use in open-hearth cooking.

One by one, the trees began to disappear from the highlands. Without tree roots to hold the soil in place, the land eroded. Erosion of banks along streams and rivers then threatened the water supplies.

The land was losing its ability to sustain the region's growing population as well as its fertility. Farmers moved down to the savannas in search of better land. These plains are home to nomadic peoples and much of Kenya's wildlife.

Wangari Maathai recognized what was happening to her country. The daughter of a farm worker, she is a biologist and the first woman in Kenya to receive a PhD. "When I would visit the village where I was born," she says, "I saw whole forests had been cleared for cultivation and timber. People were moving onto hilly slopes and riverbeds and marginal areas that were only bush when I was a child. Springs were drying up." Maathai was shocked to find children suffering from malnutrition. "[M]y community was supposed to be a rich, coffee-growing area." Instead of eating nutritious, traditional foods, such as beans and corn, she explains, people were relying on refined foods such as rice because they need less cooking—and thus less firewood.

The Green Belt Movement

On June 5, 1977, in honor of World Environment Day, Maathai and a few supporters planted seven trees in Nairobi, Kenya's capital. This small tree planting began the Green Belt

Movement, which soon swept through the Kenyan highlands and captured the attention of people around the world.

From the beginning, Maathai knew that the success of her efforts depended mainly on women. In Kenya, as in much of Africa, men tend cash crops such as coffee and cotton, while women collect firewood, tend livestock, and grow corn, beans, and other food crops. It is the women, then, who provide food for their families and who were the first to see the connection between poor soil and famine.

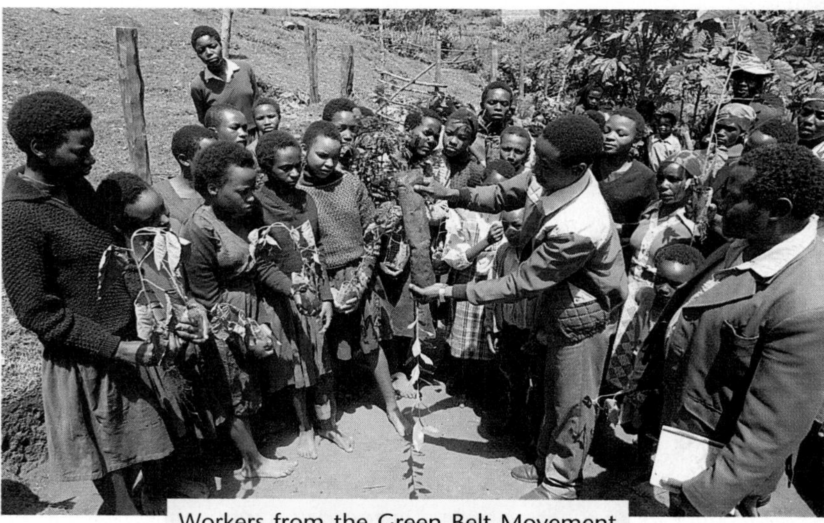

Workers from the Green Belt Movement teach schoolchildren in **Kenya** how to care for tree seedlings.

The movement's organizers encouraged women to plant trees by pointing out that they would no longer have to walk miles to collect firewood. They would have their own wood available for fires, fences, and buildings. If the tree seedlings survived, the women also would be rewarded with a small sum of money.

First, a few small nurseries were established to distribute free seedlings. The nurseries are staffed by local women, who are paid for their work. Nurseries also train and pay local people called Green Belt Rangers to visit farms, check on seedlings, and offer advice.

Soon, nurseries were appearing in communities throughout the highlands. Kenyan women talked to friends and neighbors, sharing the benefits of planting seedlings. Five years after neighbors encouraged Esther Wairimu to plant seedlings, her fields were surrounded by mango, blue gum, and other trees. "I have learned that a tree, in another way altogether, is life," she says. Today, the advantages of tree planting are apparent. Once-barren farms are ringed by greenery. Farmers now have fuel and shade trees. Even more important, the soil is being protected from erosion.

The number of Green Belt nurseries has grown to 1,500, and nearly all of them are run by women. "My greatest satisfaction is to look back and see how far we have come," Wangari Maathai says. "Something so simple but meaning so much, something nobody can take away from the people, something that is changing the landscape."

Expanding the Movement

Maathai believes that local people must work together to protect the environment. She stresses that the Green Belt Movement did not rely on outside experts but on Kenyan farmers. Furthermore, the movement receives little support from the government. Most of its funds come from small donations of money from people around the world and from gifts from groups like the United Nations Development Fund for Women.

Now Maathai dreams of spreading her movement to other African nations. Recognizing the obstacles she and other environmentalists face, she says, "[W]e must never lose hope. When any of us feels she has an idea or an opportunity, she should go ahead and do it. . . . One person *can* make a difference."

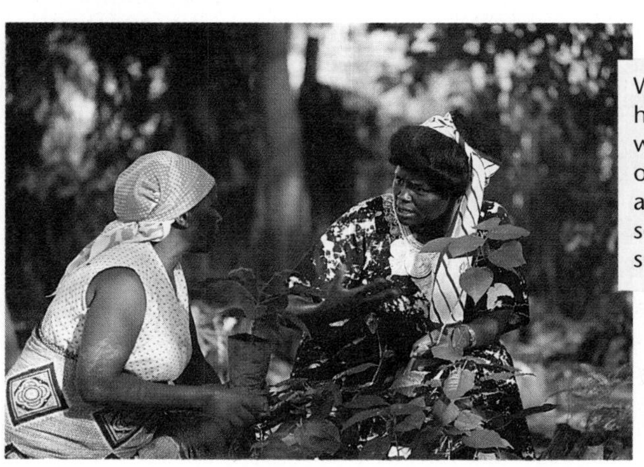

Wangari Maathai (right) has encouraged Kenyan women to plant millions of trees. Women are paid, although very little, for seedlings they plant that survive.

YOUR TURN

Working in groups of three or four, think of an environmental issue facing your community or state. **Develop** a plan that demonstrates how "one person can make a difference." **Present** your group's plan to the rest of the class.

tions as well as an outbreak of violence by the Mau Mau. The Mau Mau was a secret guerrilla organization made up of Kikuyu. At the heart of this conflict were different attitudes toward land. The British viewed land as personal wealth, power, and property. In contrast, the Kikuyu valued land for the amount of food it could produce. For them, land could not be bought or sold.

Since independence from Britain in 1963, Kenya's leaders have provided a stable, though not a fully democratic, government. Encouraging tourism, controlling population growth, and improving Kenya's small-farm agriculture have been the government's chief goals.

Despite the productive natural environments in the highlands, most of Kenya is too dry for farming. Only about 13 percent of the country is suitable for agriculture. Agriculture in the highlands is increasingly concentrated on small farms. Many large farms that were developed during the colonial period have been divided into small holdings. The key cash crops are tea, coffee, cotton, and **sisal**. Sisal is a strong fiber used to make rope and twine.

Kenya's industry is concentrated in small factories located primarily in Nairobi and Mombasa, the country's two largest cities. Kenyan canned foods and some manufactured goods are sold throughout the region. Kenya is Africa's favorite tourist attraction. It features cool, green highlands with rich farmlands, forests, and popular game parks on its dry savanna plains. There are also wide vistas of the Eastern Rift and snowcapped mountains. Tourism is a major source of income.

Kenya's greatest challenge is its rapidly increasing population. Due to this population growth, there are no empty farmlands in most of the highlands. Many people migrate to the towns to look for work. Kenya's capital city, Nairobi, is filled with young people, many of whom have only farming skills, little education, and high hopes. Very few will find a good job. Some may eventually return to the family land. Most will remain in the city, however, living on the income of occasional work. This increasing number of people has also put pressure on the game parks. Many people would like to farm these lands. If the game parks were converted to farmland, however, the tourist industry would suffer, along with the animals that the game parks protect.

A modern city, **Nairobi** plays an important political role in East Africa. The city also hosts some United Nations offices.

Tanzania Along Kenya's southern border is the United Republic of Tanzania. (See the map on page 453.) It was created in 1964 when Tanganyika (tan-guhn-YEE-kuh), which had been independent since 1961, merged with the small island country of Zanzibar (ZAN-suh-bahr). Zanzibar had been independent since 1963. Tanzania is mostly upland plateaus covered by bush and tall-tree savannas. Humid forests are found where elevations rise above 6,000 feet (1,829 m). The greatest landform in Tanzania is Kilimanjaro on the border with Kenya. The mountain's southern slopes are a rich agricultural region much like the Kenya highlands.

The rift valleys also pass through Tanzania. In the Western Rift is deep Lake Tanganyika. Between the Eastern Rift and Lake Victoria is the large Serengeti Plain, where herds of antelope and zebras still roam free. Nearby is the famous archaeological site of Olduvai (old-duh-way) Gorge, where evidence of some of the earliest humanlike settlements has been discovered.

Most of the people of the Tanzanian mainland belong to one of many Bantu-speaking groups. Along the coast and on the islands of Zanzibar and Pemba are many Arabs. These traditional Islamic societies maintain a separation from Tanzanian life. Zanzibar has the world's major source of cloves.

Tanzania is a country of subsistence farmers. Infertile soils and limited technical support hinder farm productivity. Tanzania has many minerals, but the deposits are neither large nor rich. Most of Tanzania's factories make goods such as foods and cement for the local market. Dar es Salaam (dahr ehs suh-LAHM) is the country's major city and chief port. A modern railroad connects the city with the copper fields in the neighboring country of Zambia (ZAM-bee-uh). Tanzania is in the process of moving its capital from Dar es Salaam to Dodoma (DOHD-uh-mah), a new town inland.

After independence, Tanzania launched an ambitious program of village modernization to make the country more self-sufficient. The government pressured farmers to work in cooperative villages in which farmers would share the farmlands, field work, processing of products, and profits. For numerous reasons, this plan failed. Some farmers complained of unequal work loads or resented losing their private land. Some people began to rely on government assistance. As a result, production dropped and food imports steadily increased. Most of the cooperative villages have closed, and the lands have returned to private control.

Olduvai Gorge, in northern **Tanzania**, is a canyon about 30 miles (48 km) long and 295 feet (90 m) deep. Many human fossils, some of which are more than 1 million years old, have been found there.

FACTS IN BRIEF East Africa

COUNTRY POPULATION (1994)	LIFE EXPECTANCY (1994)	LITERACY RATE (1990)	PER CAPITA GDP (1993)
BURUNDI 6,124,747	38, male 42, female	50%	$700
DJIBOUTI 412,599	47, male 51, female	48%	$1,200
ERITREA 3,782,543	46 (all) (1993)	20% (1993)	$500
ETHIOPIA 54,927,108	51, male 54, female	24% (1991)	$400
KENYA 28,240,658	51, male 55, female	69%	$1,200
RWANDA 8,373,963	39, male 41, female	50%	$800
SOMALIA 6,666,873	54, male 55, female	24%	$500
SUDAN 29,419,798	53, male 55, female	27%	$750
TANZANIA 27,985,660	42, male 45, female	94% (1991)	$600
UGANDA 19,121,934	37, male 38, female	48%	$1,200
UNITED STATES 260,713,585	73, male 79, female	97% (1991)	$24,700

Sources: *The World Factbook 1994,* Central Intelligence Agency; *The World Almanac and Book of Facts 1995*

Large-scale ethnic conflict makes statistics for Rwanda unreliable.

Rwanda and Burundi In the pleasant highlands east of the Western Rift are the small landlocked countries of Rwanda (ru-AHN-duh) and Burundi (bu-ROON-dee). (See the map on page 453.) Once a German colony, these territories were administered by the Belgians after World War I. At independence in 1962 they were divided into two countries. Rwanda and Burundi have the densest rural settlement in Africa. Foreign aid has led to improvements in farming technology and better public health.

Coffee and tea, grown on small farms on the fertile mountain slopes, are the main exports. Rwanda also has some tin mines, and Burundi has nickel deposits. Rwanda is popular with tourists, who come to see the mountain gorillas.

Politics in these two countries are complex. Most of the people are Hutu, a Bantu-speaking group. When the Germans arrived, the Hutu were controlled by the Tutsi. The Tutsi are related to the Masai and the peoples along the Nile. The Hutu are farmers; the Tutsi are cattle keepers. The Hutu staged an uprising

Hydroelectric power provided by Owen Falls Dam has enabled **Jinja** to become **Uganda's** main industrial center. The town is located where the Nile River flows out of Lake Victoria.

in 1959, and after independence in 1962, violence between the Tutsi and Hutu in both countries claimed more than a half million lives. In 1994, hundreds of thousands of Rwandans, mostly Tutsi, were killed in renewed ethnic violence. Millions of Rwandans fled to refugee camps outside the country. Ethnic tensions and periodic violence have continued in both Rwanda and Burundi.

Uganda The railroad from Mombasa on the Indian Ocean goes beyond Nairobi to Kampala (kahm-PAHL-uh), the capital of landlocked Uganda. (See the map on page 453.) Most of Uganda is a high plateau, and the country's highlands are some of Africa's most beautiful farmlands. Inland swamps and lakes cover about 20 percent of the country. The lands along the northern shore of Lake Victoria were part of the ancient kingdom of Buganda. Dams across the upper Nile provide hydroelectricity.

Unfortunately, the promise of prosperity for Uganda at independence in 1962 was destroyed. The country's decline began in 1971 when Idi Amin seized power after a military takeover. People who disagreed with Amin often were murdered. Successful farmers as well as British and Asian businesspeople fled. Foreign investment stopped, and the country's economy collapsed. Though Amin was driven out in 1979 by an invasion of Ugandan and Tanzanian forces, civil unrest continued until 1988 when peace was restored.

Rebuilding Uganda has not been easy, but the return of foreign aid, investment, and peace offers Ugandans new hope. The country relies heavily on its agricultural potential. Adding to Uganda's woes is the country's battle against AIDS, which has hit Uganda particularly hard. International agencies continue to offer support.

Sudan Sudan is Africa's largest country in area. (See the map on page 453.) It also is one of Africa's least developed countries. The word *sudan* refers to the dry savannas that extend across the middle of the country. Northern Sudan is in the dryland zone. Here, summer dust storms make life miserable. In the south, Sudan has moist savannas. The Nile runs the length of the country, providing a crucial water resource.

For the most part, modern agricultural development in Sudan has been the result of irrigation projects. The largest of these projects is the Gezira (juh-ZIR-uh) Scheme near the capital city of Khartoum. El Gezira is the area between the White Nile and the Blue Nile near their junction. Former grazing land, this area has been transformed into a major producer of cotton and food crops. Although Sudan is mainly an agricultural country, it has developed some of its mineral resources. Oil has been discovered in the south but is undeveloped.

Like other countries along the southern edge of the Sahara, Sudan includes strong influences from Arab as well as black African cultures. The Nubians and Arabs of the north total about 50 percent of the people. They are Islamic herders who have a long history of contact with nearby Southwest Asia. The people in southern Sudan are primarily Christians

A worker surveys part of an irrigation system in a sugarcane field in **Sudan**. Most of Sudan's commercial crops are grown in the irrigated northern and eastern areas of the country.

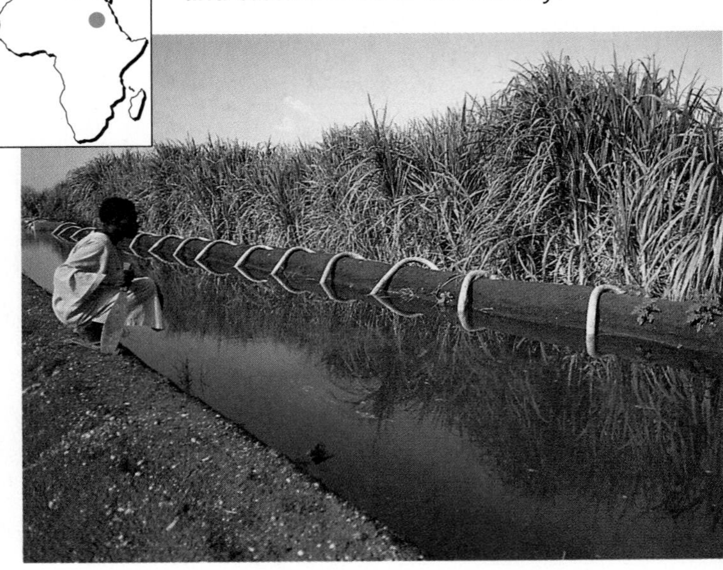

who were influenced by European missionaries. Farming is their agricultural tradition.

Uniting these peoples is the most difficult political problem for Sudan. Since independence in 1956, Sudan has experienced many periods of unrest. The government in Khartoum has tried to force Islam, as well as Arabic ways of life, on the southern region. The southern people have resisted and have been in rebellion against the central government since 1980. Development has stopped and will not continue until this conflict is settled. Meanwhile, life for everyone in Sudan deteriorates as desertification increases and civil war continues.

SECTION REVIEW

1. What physical features in Kenya and Tanzania might attract tourists?
2. Compare and contrast economic and political issues in Uganda and Sudan.
3. **Critical Thinking** **Categorize** the causes of conflicts experienced over the past 50 years by the countries discussed in this section.

THE HORN OF AFRICA

FOCUS

- *What is the physical environment of the Horn of Africa?*
- *What is the economic and political situation in the Horn of Africa?*

Ethiopia Dry lowlands of savannas and deserts surround Ethiopia's humid mountain region. (See the map on page 453.) The densely populated highlands are the core of Ethiopian life. Yet while the rich soils of the mountain slopes and upland plateaus offer agricultural promise, there has been little progress. Life for the herders in the drylands below has been harsh. In fact, with more than 50 million people, Ethiopia is one of the world's poorest countries. The capital city of Addis Ababa (ad-uh-SAB-uh-buh) holds more than 1 million people.

Typical of mountain regions, Ethiopia's climates and agricultural products vary with elevation. The hot, semiarid lowlands reach up to about 5,000 feet (1,524 m). With irrigation, bananas, dates, and other fruits can be grown. Cattle and goats are grazed on the dry savannas. From about 5,000 feet (1,524 m) up to about 9,000 feet (2,744 m), the land was once forested. Wild coffee, grains, oranges, and other Mediterranean-type crops now grow here. Farther up, wheat and barley are grown.

In the last 30 years, droughts have hit Ethiopia hard. Despite aid from around the world, several million people starved during the 1980s. The country's mountainous terrain makes communication difficult, and the few roads are often impassable. In most of the lowlands, there are no roads.

Except for the period between 1935 and 1942, the Ethiopian highlands have never been under foreign rule. The mountainous slopes have protected the interior from foreign invaders. Missionaries brought a form of Christianity to the highlands between the fourth and sixth centuries, but the Ethiopian Orthodox Church survived several attempted invasions by Islamic armies. Today, the highland people practice Christianity, while most of the lowland people are Muslims.

Cooperation among Ethiopia's many ethnic groups is limited. While most of the people are Gallas and Amharas, there are other groups, including

Christians visit a religious shrine in **Ethiopia**. More than one-third of Ethiopians are Christian.

some Somalis. Nearly half of the people speak Amharic, once the official language. English is taught as a second language. Tigre (TEE-gray) is spoken in the northern, Islamic region of Ethiopia.

Since the days of the powerful ancient kingdom of Axum, Ethiopia's history often was one of quarreling small states. The majority of people were landless peasants. A ruling class, along with the royal family and the church, held most of the economic and political power. They controlled most of the good farmland, industry, and the government.

This system collapsed in 1974 when the combination of drought and ethnic struggles led to a major change of government. A Communist government took over, seizing most lands, industries, and businesses. Many rural people were forced to move to new villages where the government could control them. Eventually, a multisided civil war spread across the country. In 1991, the Communist government was overthrown. Ethiopians hope a general election held in 1995 will aid in the rebuilding of their country.

Despite its political tensions, Ethiopia is the home of the Organization of African Unity (OAU). The members of the OAU believe that African states can solve their problems through cooperative efforts. Some of these problems involve disputes among member nations, threats to independence, lack of foreign investment, and economic and social issues associated with modernization.

Eritrea Eritrea (ehr-uh-TREE-uh), on the Red Sea coast north of Ethiopia, was once an Italian colony. In the 1960s, the Eritreans began a long fight to separate from Ethiopia. During the struggle, Eritrea's infrastructure was destroyed, and drought and famine took their toll on the province. Eritreans gained control of their country in 1991, but full independence was not granted until 1993. Agricultural output has increased, and roads, bridges, hotels, and factories are being repaired. There may be oil and natural gas deposits off the coast. Although the country is still poor, Eritreans are enjoying peace for the first time in many years.

Somalia Somalia is a land of deserts and dry savannas. Only along the coastal edge and in the far southern part of the country, near the capital of Mogadishu (mahg-uh-DISH-ooh), is there enough rainfall for farming. Most Somalis are nomadic herders. Livestock and hides are the main exports. The trees of the limited coastal woodlands produce frankincense and myrrh, which are marketed to Christian churches around the world.

Somalia differs from most other African countries in that it contains mostly one group of people, the Somali. Most Somali have the same culture, religion (Islam), language (Somali), and way of life (herding). Many geographers falsely believed that this cultural uniformity would save Somalia from ethnic conflict. But during the early 1990s, Somalia was devastated by civil war.

Longstanding disputes between various clans over water rights and political authority led to fighting. The violence came during a period of severe drought. Livestock died and crops failed. Nearly a whole generation of children died, with their suffering documented on television and in newspapers around the world. The United Nations sent troops to Somalia to help stabilize the country and to provide aid. Somalis have worked to stop fighting that resumed after UN troops left the country in 1995.

Djibouti Although Djibouti is a small desert country, it is strategically located on the Bab al-Mandab (bab uhl-MAN-duhb), the strait that connects the Red Sea and the Indian Ocean. All the ships that travel between the Mediterranean Sea and the Indian Ocean must pass nearby. The city of Djibouti was built primarily as a French colonial port. Though independence came in 1977, the French government still contributes economic support to the country. The city of Djibouti also is the main port for the railway that runs into Addis Ababa. This railway carries about half of Ethiopia's trade. Foreign aid and trade have made Djibouti more prosperous than its neighboring countries.

The people of Djibouti are almost equally divided between the Issa, who are closely tied to the people of Somalia, and the Afars, who are linked to leading families in Ethiopia. Somalia and Ethiopia have vied for control of Djibouti, but so far the country has maintained self-government.

4 SECTION REVIEW

1. How does the physical environment of Somalia differ from that of Ethiopia?
2. Why has economic development in Ethiopia been slow despite its rich agricultural potential?
3. **Critical Thinking** **Compare and contrast** Eritrea and Djibouti.

CHAPTER 38 CHECK

KLB
976
Tourists take in the sights of a **Kenya** wildlife preserve.

Reviewing the Main Ideas

1. Magnificent landscapes and climate contrasts characterize East Africa.
2. East Africa is a region of great ethnic diversity. Most of the ethnic groups can be divided into three main language categories.
3. Most East Africans are traditional farmers, though industrialization has begun in the region.
4. In order for the pace of economic development to increase in East Africa, ethnic groups and nations must cooperate.

Building a Vocabulary

1. What is the official language of Kenya and Tanzania?
2. What is gum arabic? For what is it used?
3. Is sisal a food crop? Where is it grown?

Recalling and Reviewing

1. What physical features make East Africa the most scenic region in Africa?
2. Into what three language groups can East Africans be divided?
3. What techniques for economic development have been used in Kenya and Tanzania?
4. How has Ethiopia's mountainous terrain both helped and hindered the country?

Thinking Critically

1. What evidence is there of cultural diffusion in East Africa? Has cultural diffusion been helpful or harmful? Explain.
2. What effects might civil war have on a country's infrastructure and economy?

Learning About Your Local Geography

Individual Project

Create a travel brochure that highlights the physical features of your local environment that would attract tourists. With your classmates, create a bulletin board display using the brochures. You may wish to include your brochure in your individual portfolio.

Using the Five Themes of Geography

Identify three different patterns or types of **human-environment interaction** mentioned in the chapter. Make an illustration of each. Write a caption for each illustration, explaining how the illustration shows the relationship between human-environment interaction and the **movement** of people and goods.

Farmers harvest wheat in **Ethiopia**.

CHAPTER 39

Central Africa

GEOGRAPHY DICTIONARY

periodic market
trust territory
exclave

"Westerners always ask me about the jungles, as if there were no other kind of home for Africans. On the contrary, more than half of Africa is covered with steppe, savanna, or grassland; one-third is desert and lakes; while but one-sixth is forest or bush. These forest and bush lands sometimes grow into dense jungles along the equator where heaviest rainfall makes the soil most fertile."

Mbonu Ojike

The languages of the Niger-Congo family, often called the Bantu languages, are found throughout Africa south of the Sahara. Over a period of 2,000 years, Bantu speakers spread into Central and Southern Africa from the border of present-day Nigeria and Cameroon. This expansion is often referred to as the Great Bantu Migration.

Evidence indicates that the Bantu speakers were ironworkers as well as farmers and cattle keepers. As they moved into the lands of hunters and gatherers, they either absorbed these peoples, teaching them a Bantu language and culture, or displaced them. Great Bantu speaking states emerged in Central Africa, including the Kongo kingdom, and the Kuba, Luba, Lunda, and Lozi states.

From study of this great expansion, we learn that although the ethnic groups of Central Africa are extremely diverse, common threads tie their histories together. Linguists have found that the Bantu languages, which number around 300, are no more different from each other than are the Germanic languages of northern Europe. Today, Bantu languages are spoken throughout much of the African continent.

Wood-carving and beadwork image honoring an early Bamileke leader in present-day **Cameroon**

Rice fields near the coast of **Angola**

CENTRAL AFRICA: PHYSICAL–POLITICAL

SCALE 0 250 500 Miles; 0 250 500 Kilometers

Projection: Azimuthal Equal Area

LEGEND

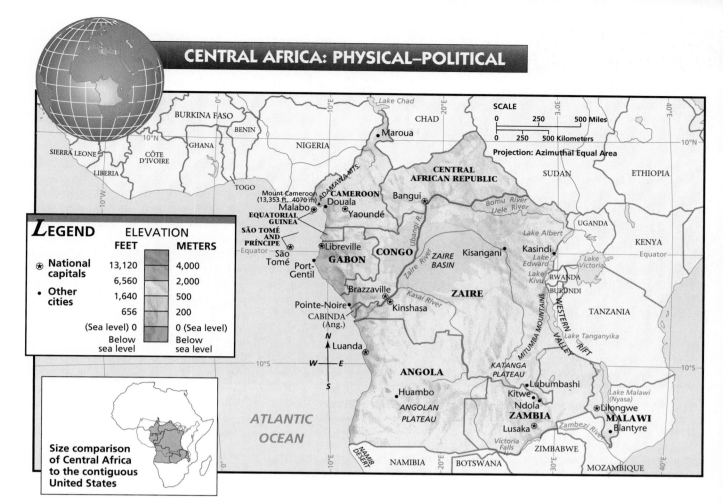

	ELEVATION	
FEET		METERS
13,120		4,000
6,560		2,000
1,640		500
656		200
(Sea level) 0		0 (Sea level)
Below sea level		Below sea level

⊛ National capitals

• Other cities

Size comparison of Central Africa to the contiguous United States

PHYSICAL AND HUMAN GEOGRAPHY

FOCUS

- *What is the physical geography of Central Africa?*
- *What is the economic geography of Central Africa?*

REGION **Landforms and Climate** The heart of Africa has dense tropical forests, wide plains covered with savanna woodlands, and pleasant forested highlands. Humid-tropical climates stretch from the Atlantic coast inland to the mountains along the Western Rift. Much of Central Africa has a tropical-savanna climate with wet summers and dry winters that favor savanna vegetation. Small desert areas lie along the northern edge of Cameroon near Lake Chad and in southwestern Angola along the edge of the Namib Desert.

Central Africa is mostly plateaus and rolling plains. Along the coast are narrow lowlands that connect to several escarpments inland. The greatest highlands in Central Africa are the Adamawa Mountains in Cameroon and Mitumba Mountains along the western side of the rift-valley lakes. Similar mountains cross Malawi (muh-LAH-wee) and continue south of the Zambezi River. (See the map above.)

Central Africa has two important river systems—the Zaire and the Zambezi. The Zaire River, also called the Congo, leads out of the huge interior Zaire Basin into the Atlantic. From Kinshasa (kin-SHAHS-uh) upstream, riverboats can navigate the Zaire for about 1,085 miles (1,736 km) before reaching the falls near Kisangani (kee-suhn-GAY-nee). The Zambezi River drains Zambia and northern Zimbabwe before heading across Mozambique to its mouth on the Indian Ocean.

Ways of Life Population density is low across most of Central Africa. Although most Central Africans are subsistence farmers, the economy is shifting to commercial agriculture. Cash crops are

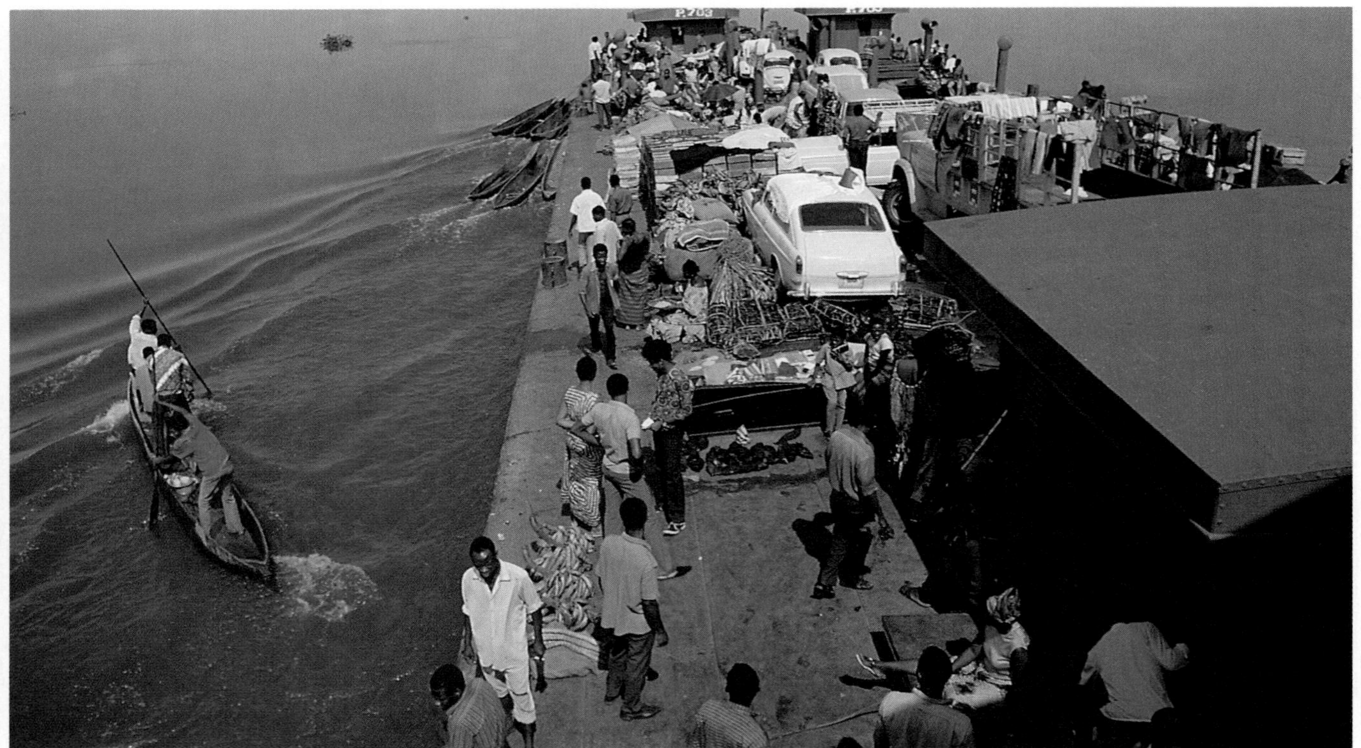

A boat transports passengers and freight along the **Zaire River**.

produced on large commercial farms and on small plots. Depending on local customs, farmers either live in small villages or on their farms. Many farmers have fields for cash crops as well as food crops. As in many African countries, women are the primary farmers, and they perform most of the household duties as well.

In rural areas, people trade their products at **periodic markets**. A periodic market is an open-air trading market held regularly at a crossroads or in a town. It also is a social event where people visit with their friends and share news. (See "Themes in Geography" on pages 468–469.)

As in other countries in Africa, many people in Central Africa are migrating away from rural areas to seek jobs with cash wages. These migrants are usually men, and their absence strains African family life. Some find jobs in the many mines, while others look for work in the tourist industry and at game parks. Most of the new jobs are found in factories and businesses in the cities and towns, where the government may offer jobs. Central Africa's capital cities are overflowing with new residents. City governments struggle to provide basic services as slums sprawl ever outward.

The countries of Central Africa import most manufactured goods from developed countries. International aid provides some of the funds with which to purchase these foreign goods. The export of cash crops and minerals remains the main source of income, however. In some cases, governments have encouraged cash crops to the extent that production of food crops has decreased. At times, world and government-controlled prices are so low that farmers earn little for their food crops. When this happens, the farmers usually raise only enough food for their own communities and have nothing extra to sell to the towns.

Food production often has not kept up with rapidly increasing populations. All Central African countries must now import food despite the region's many fertile farmlands. Investment in agriculture must become a priority in order to improve food production throughout the region.

SECTION REVIEW

1. Describe the vegetation, climate, and landforms of Central Africa.
2. What various types of jobs are available in Central Africa?
3. **Critical Thinking** **Suggest** possible reasons that farmers in Central Africa often are women while migrants to cities are usually men.

ZAIRE

FOCUS

- *What problems hinder economic development in Zaire?*
- *What resources does Zaire have that can be used to its advantage?*

Zaire is the largest and most populous country in Central Africa. Its physical geography is dominated by the plains along the Zaire River. The river's many navigable stream channels allow transportation by riverboats and canoes. The surrounding uplands are the Angolan and Katanga (kuh-TAHNG-uh) plateaus to the south and the Mitumba Mountains to the east. The Zaire River tumbles down steep falls near the coast before it reaches its mouth at the Atlantic Ocean. Zaire's Atlantic coastline is a short 25 miles (40 km) long. (See the map on page 465.)

Zaire is home to some 250 different ethnic groups, whose staple crops and economic activities vary with the local climate and vegetation. People in the open lands of the savannas primarily are grain farmers and cattle keepers. In the humid forests, coffee and cassava are harvested. Many large and small communities in Zaire are often isolated because of inadequate transportation and communication facilities. The country's poor roads mean that a 60-mile (96-km) journey by truck can take more than a day. Even on a curvy paved road in the United States, a trip of the same length would take less than two hours. Telephone systems are not linked nationwide, and the postal system is inefficient. Throughout Zaire, electricity service is unreliable. In many rural areas, kerosene lamps provide the only light after sunset.

Historical Geography Zaire once was called the Belgian Congo. During the past century, its history has been one of Africa's most turbulent. The people of Zaire have suffered alternating periods of poor government or almost no government.

The impact of European control was severe. The Berlin Conference of 1884–1885 awarded the Belgian Congo to King Leopold II of Belgium. The king considered the area his personal empire. The peoples of the Congo were forced to labor under threat of injury and often were treated as slaves.

Public outrage forced King Leopold to turn over the territory to the Belgian government in 1908. Under the government, workers received better treatment and had better living conditions. Belgian businesses invested heavily in the territory, and prosperous copper mines opened. Many Belgians came to work in the Congo and soon held nearly all of the jobs in government and business. Many Africans, however, had little contact with the European customs and economic system. Few efforts were made to train Africans for the new jobs.

After World War II, running the Congo began to cost the Belgian government more and more money. Unemployment increased dramatically when copper prices fell in the late 1950s. Adding to the discontent, the people in neighboring French and British colonies were gaining their independence. Riots broke out in the capital of Leopoldville (now called Kinshasa), and unrest increased. The colony finally won its independence from Belgium, and in June 1960, with only six months' notice, the Democratic Republic of the Congo formed.

Text continues on page 470.

A woman carries her child through fields in the Virunga Mountains of eastern **Zaire**. Although about 75 percent of Zaire's people work in agriculture, most of them are subsistence farmers who use traditional methods.

THEMES IN GEOGRAPHY

Periodic Markets

Periodic markets are a regular part of rural life in much of Africa. They are held one or more days each week in a village, at a crossroad, or in a convenient open field. These markets give people the opportunity to shop, sell their products, and share news about families and friends. They are events that break up the steady pace of village life.

Periodic markets shape the daily economic, social, and cultural life of many African communities.

Because markets usually start early in the morning, people often begin their walk to market before sunrise. Most of their marketing is completed before the heat of the afternoon. The markets are well organized by a village leader or a local government official.

Sellers can rent stalls and display tables for a small fee. Some spread their wares on the ground or on makeshift tables. People selling the same products usually set up next to each other so shoppers can compare the quality and prices of their offerings. Buyers bargain over prices, always hoping to make a better deal. Children may have jobs wrapping parcels and delivering them to customers.

Traveling merchants also come to the periodic markets. They usually have trucks so they can transport their goods to a different market each day. Merchants have an assortment of manufactured goods to sell. One may offer drugstore items such as cosmetics, soap, detergents, brushes, and music cassettes. Another merchant may sell household goods such as cloth, pails, lanterns, batteries, and blankets.

As populations in Africa have increased in the last decade, some periodic markets have become daily markets with permanent residents. Some have become towns. Kasindi (kuh-SIN-dee), in eastern Zaire, is an example of a small village that has developed as the result of periodic markets. It began as a border outpost on the main road between Zaire and Uganda. This location made it accessible to sellers and to customers.

Kasindi lies in the Western Rift Valley, just south of the Ruwenzori (roo-wuhn-ZOHR-ee)

MOVEMENT Before sunrise, people can be seen walking along the roadside on their way to the **Kasindi** market. Merchants from near and far bring their goods to sell. Buyers walk from table to table, inspecting goods and bargaining over prices. How is a periodic market different from a U.S. supermarket or shopping mall? How is it similar?

mountain range and north of Lake Edward. The area can supply varied products for its periodic markets. The immediate environment is a tall-tree savanna; however, environments vary widely over short distances along the rift valley, permitting many products to be grown.

Kasindi's location on an important transportation route makes it accessible to merchants who bring manufactured products such as gasoline. Traffic across the border also affects Kasindi. In recent decades, neighboring Uganda often has experienced serious unrest. At these times, products from Zaire have helped the Ugandans survive. In turn, many of the manufactured goods needed by Zaire come through Uganda from Nairobi, Kenya, an international trade center.

Kasindi
UGANDA
Kasindi
ZAIRE
30°E
Semliki River
Fort Portal
RUWENZORI
Mount Margherita (16,763 ft. 5109 m)
ZAIRE
UGANDA
Butembo
Equator
Kasindi
Lake George
Ishango
Katwe
Lake Edward
N
W
E
S
SCALE
0
25 Miles
0
25 Kilometers
Major roads

LOCATION **Kasindi's** absolute location is 0°5′ north latitude, 29°45′ east longitude. The market's relative location is on the road between **Zaire** and **Uganda**, north of Lake Edward and south of Mount Margherita. Can you describe the absolute and relative locations of your favorite places to shop?

PLACE **Kasindi** is located near the Semliki River, shown here, and an international border. Officials from **Zaire** and **Uganda** often stop in Kasindi on government business. It is a place where goods from both countries can be traded. What might be the eventual result of this interchange?

Although the Belgians left behind school buildings and many hospitals, they did not leave behind education or medical systems. The Belgians had trained few African educators or medical personnel. When the Belgians fled the country after independence, Zaire was left with few teachers or doctors. Nor did the Belgians train Africans in government jobs, and most Africans lacked the experience to manage the country effectively.

After independence, several political parties battled to control the central government. The idea of a united Congo was not widely supported. People often thought first of their own ethnic groups and region. Lawlessness and rebellion were widespread. To make matters worse, the mineral-rich province of Shaba (then called Katanga) tried to withdraw from the new nation. Civil war resulted as several armies stormed across the country. Economic conditions worsened, and famine threatened.

After several changes of government, the commander of the central army, Mobutu Sese Seko, took control. Many industries owned by Europeans were given to Mobutu's friends and relatives. Such corruption hurt the economy. The Mobutu government also changed the country's name to Zaire in 1971, and in 1975, it changed the names of cities and provinces in an effort to remove signs of the colonial past. The capital Leopoldville, for example, was changed to Kinshasa. Mobutu succeeded in keeping the country together into the 1990s, but the country's economic troubles have continued.

FACTS IN BRIEF Central Africa

COUNTRY POPULATION (1994)	LIFE EXPECTANCY (1994)	LITERACY RATE (1990)	PER CAPITA GDP (1993)
ANGOLA 9,803,576	44, male 48, female	42%	$600
CAMEROON 13,132,191	55, male 59, female	55%	$1,500
CENTRAL AFRICAN REPUBLIC 3,142,182	41, male 44, female	27%	$800
CONGO 2,446,902	46, male 49, female	57%	$2,900
EQUATORIAL GUINEA 409,550	50, male 54, female	50%	$700
GABON 1,139,006	52, male 58, female	61%	$4,800
MALAWI 9,732,409	39, male 41, female	25% (1989)	$600
SÃO TOMÉ AND PRÍNCIPE 136,780	61, male 65, female	50% (1988)	$450 (1990)
ZAIRE 42,684,091	46, male 49, female	72%	$500
ZAMBIA 9,188,190	44, male 45, female	73%	$800
UNITED STATES 260,713,585	73, male 79, female	97% (1991)	$24,700

Sources: *The World Factbook 1994,* Central Intelligence Agency; *The World Almanac and Book of Facts 1995*

From what you have read in the chapter, why does Gabon have the highest per capita GDP?

Resources If successfully developed, Zaire's rich resources could bring about a better economic life for the people. There are vast areas that could be developed as game parks for tourists. Commercial crops also can be grown successfully.

Zaire's mineral resources are even more substantial than its agricultural ones. The copper deposits in the Shaba province in the southeast are among the largest in the world. Deposits of cobalt, iron ore, manganese, gold, and tin are nearby. Combined with adjacent areas in Zambia, the copper belt of Shaba is the second largest mining region in Africa. Zaire also has enormous reserves of industrial diamonds. The rivers of Zaire have the greatest hydroelectric potential in Africa, and oil has been found off Zaire's narrow Atlantic coast.

Life in Zaire Zaire has three major cities. The first is the capital, Kinshasa, located on the Zaire River near the Atlantic coast. The river docks in Kinshasa receive goods from throughout the Zaire Basin. Kinshasa is a congested city where traditional markets and modern stores exist side by side. Colonial buildings, high-rise apartments, and shantytowns often share the same neighborhood. Kinshasa is the center of political power in Zaire.

Another important city is Lubumbashi (loo-boom-BAHSH-ee), the industrial center of the Shaba copper belt. The railroad from Lubumbashi carries copper to ports in South Africa. Many manufactured goods arrive on return trips. A third important city is Kisangani, located on the great falls of the Zaire River. Industries use the electricity from the power plant there to process products from surrounding forests and mines.

Zaire's only experience with public service was that imposed by colonial powers. Today, people take jobs in the government to gain personal income. Indeed, most government ministers become wealthy. Meanwhile, teachers and government workers in distant areas sometimes wait months for their paychecks. Skilled workers often keep a small farm to ensure that food is available for their families.

There were some 80,000 miles (128,720 km) of good roads at the time of independence. Today, paved roads total less than 12,000 miles (19,308 km) and are in poor repair. The ruins of colonial factories and farms stand silent as much land returns to subsistence agriculture.

SECTION REVIEW

1. How has Zaire's history affected its economic development?
2. What economic potential does Zaire have in its natural resources?
3. **Critical Thinking** **Construct** a time line showing political and economic changes in Zaire.

NORTHERN CENTRAL AFRICA

FOCUS

- *What are the major economic activities in the northern part of Central Africa?*
- *What are the similarities and differences among the economies of the countries in northern Central Africa?*

Congo Situated in the forested lands west of northern Zaire is the People's Republic of the Congo, usually called "the Congo." (See the map on page 465.) Brazzaville (BRAZ-uh-vill), its capital, is located across the Zaire River from Kinshasa. In the Congo, the Zaire River is referred to as the Congo River.

As the capital of French Equatorial Africa during the colonial period, Brazzaville grew into a cosmopolitan city. After independence in 1960, the city became less important, and the Congo has lost its leadership role in the region. Economic development has been uneven. Foreign aid has supported agricultural projects, though with little success.

Two-thirds of the people live in the southern part of the country between Brazzaville and the coast. Here, most villages are within a day's walk of the Congo-Ocean Railway, which connects the port city of Pointe-Noire (pwant-nuh-WAHR) with Brazzaville. This transportation route is essential to the economic development of the Congo. Farmers can get their crops to market with less spoilage, and imports and exports move easily. The transportation route also allows the government to provide better schools, farmer education, and health services.

Away from the Congo-Ocean Railway, the Congo seems desolate. Beautiful hardwoods were exported from the country's forests for many years. Once-forested areas are now bare. On the coastal plain, several small oil fields have been developed, and an oil refinery at Pointe-Noire serves the region. Improvements in transportation services along the Congo-Ubangi (oo-BANG-ee) river system would benefit the interior of the Congo and the Central African Republic to the north.

Central African Republic North of the Congo and Zaire is the landlocked Central African Republic. (See the map on page 465.) The northern half of the country is savanna, while dense forest covers the southern half. More than 80 percent of the people are subsistence farmers with little government contact, although only about 2 percent of the land is farmed. During the colonial period, the country was exploited by foreign companies. These companies had complete authority over the region. Programs of forced labor did not end until after 1950, and the local Africans came to view European business ventures with distrust.

The country's traditional exports are coffee and cotton. Industrial diamonds are the only commercial minerals. Most overseas exports travel by river to Brazzaville and then by rail to the coast. The economy, however, is barely functioning because of poor governments, and the country is dependent on foreign aid.

Cameroon Much like neighboring Nigeria, Cameroon stretches all the way from the humid-tropical forests along the Atlantic coast into the dry savanna regions. (See the map on page 465.) The country's farmers grow a wide range of forest and savanna products. Mount Cameroon, a

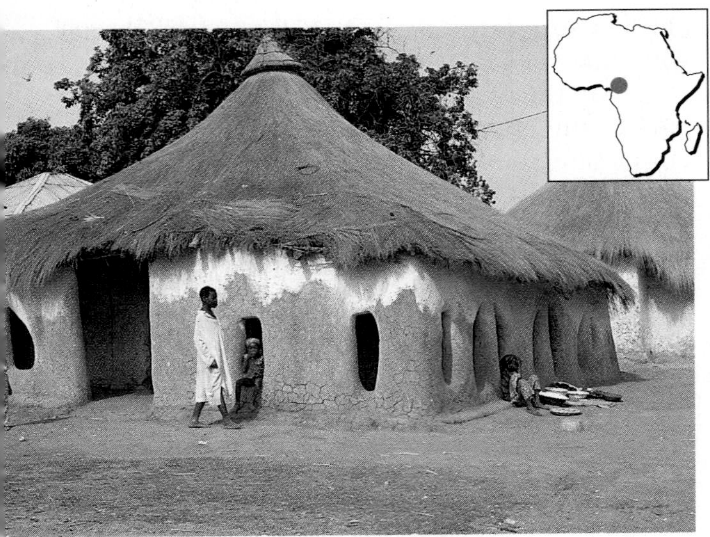

Traditional forms of architecture are common outside the modern cities of Africa. In rural areas and villages of northern **Cameroon,** many homes, such as these, are made from clay.

volcanic peak, rises 13,353 feet (4,070 m) and can be seen from the Atlantic coast.

Cameroon was a German colony that was divided between France and Great Britain after World War I. It later became part of a trust territory. A **trust territory** is a region that is placed under the control of another nation until it can govern itself. In 1960, the larger, French part of the trust territory received its independence and became Cameroon. The British part was divided between Nigeria and Cameroon. Both French and English are used by the government, and schoolchildren study both languages.

Cameroon has dozens of ethnic groups, yet the country enjoys relatively stable political leadership. Although dictatorships often have governed the country, the economy has prospered. There are many skilled business managers and prosperous towns. Coastal oil deposits have been developed and now provide almost half of the country's foreign income. Even so, the government has always maintained that agricultural development is most important to the country's welfare. Most of the people have small farms and receive government loans and up-to-date information on farming technology. Food crops are given such importance that Cameroon is nearly self-sufficient in food. Manufacturing and mining also are important to economic development.

The Island Countries The Central African region includes two small island countries near the Atlantic coast. One of these, São Tomé (sow tuh-MAY) and Príncipe (PREEN-sip-uh), is a tiny country made up of two volcanic islands. Although the country has dense forests and fertile land, the economy is struggling. The country completed its move toward democracy in the early 1990s.

Equatorial Guinea is a former Spanish colony made up of a mainland province and a group of offshore islands. Equatorial Guinea is underdeveloped, with low health standards. A dictator led the country to financial disaster during the 1970s. He had many people killed as others fled the country or into the forests. Today, the people live as subsistence farmers and hope for a better future. Spain and France provide some aid and technical support.

INTERACTION **Gabon** Rich natural resources have made Gabon (guh-BONE) the most prosperous country in Central Africa, although the country's wealth is not evenly distributed. Three-fourths of the people of Gabon are subsistence farmers.

Gabon is nearly covered with dense rain forests, and few people have settled in these areas. Only 1 percent of the country's land is farmed. Major cash crops include cacao and coffee. In the past, Gabon's main resource was hardwoods that were exported to Europe and used to manufacture furniture. Deforestation has taken its toll, however, and today, oil provides most of the government income. Gabon also has active gold, diamond, manganese, and iron mines. New roads and railroads are being built to carry these resources to the coast. The wealth in the capital of Libreville (LEE-bruh-vill) and the city of Port-Gentil (pawr-zhahn-TEE) continues to grow. The government provides educational opportunities and public-health facilities.

3 SECTION REVIEW

1. How is the Congo-Ocean Railway important to the Congo?
2. What are the reasons for Cameroon's relative economic prosperity?
3. **Critical Thinking** **Compare and contrast** the economic geography of the Central African Republic with that of Gabon.

4 SOUTHERN CENTRAL AFRICA

FOCUS

- *What are some of the challenges facing Angola, Zambia, and Malawi?*
- *What are the economic prospects for Angola, Zambia, and Malawi?*

LOCATION **Angola** Angola lies along Africa's west coast between Zaire and Namibia. (See the map on page 465.) Luanda (looh-ANN-duh), Angola's capital, is a harbor city founded by Portuguese traders in 1575. From the sea, Luanda's high-rise buildings and industrial facilities are impressive. But a closer look reveals decades of strife and neglect. From the 1960s through the early 1990s, Angola was a war zone.

During the colonial period, Portugal had little money to invest in the colony. Many poor Portuguese people came to Angola as settlers, and they competed directly with local Africans for jobs and land. Relations between the European settlers and rural Angolans remained tense and sometimes violent. When Angola became independent of Portugal in 1975, the new leaders chose a Communist form of government. Groups from the southern part of the country rebelled against the Communist government, and Angola became a center of the rivalry between the world powers. South Africa and the United States supported the rebels, and Cuba and the Soviet Union backed Angola's Communist government. Although a peace settlement was reached in 1994, chances for peace were unclear.

Copper mines near **Kitwe, Zambia**, are some of the country's richest. Factories near the mines process the copper.

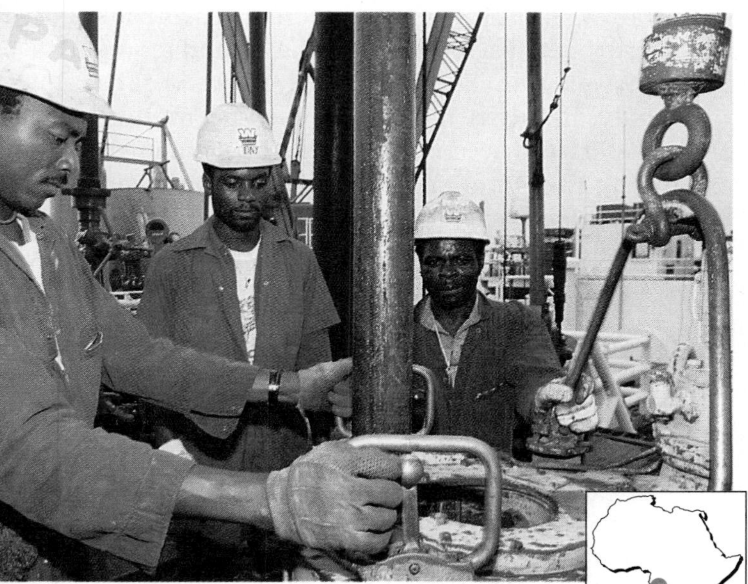

These Angolans work for a multinational oil company in **Cabinda**.

Angola has substantial natural resources that could aid its development. One of Africa's largest petroleum reserves lies along the coast north of Luanda and in Cabinda (kuh-BIN-duh), an **exclave** north of the Zaire River mouth. An exclave is a part of a country that is separated from the rest of the country and surrounded by foreign territory. Angola also has rich deposits of minerals and fertile agricultural lands. The problems of poverty, unemployment, and poor health services need to be addressed.

Zambia Zambia lies on a savanna-covered high plateau. (See the map on page 465.) The country's copper and iron have been mined for more than 2,000 years. In fact, Zambia's mineral deposits in the copper belt are even richer than those in Zaire, and the country is a leading copper producer. Zinc, lead, manganese, cobalt, gold, and vanadium are mined as well. Vanadium is a gray mineral that can be combined with iron to make very strong steel. Mines provide needed jobs for Zambians and supply much of the government's income.

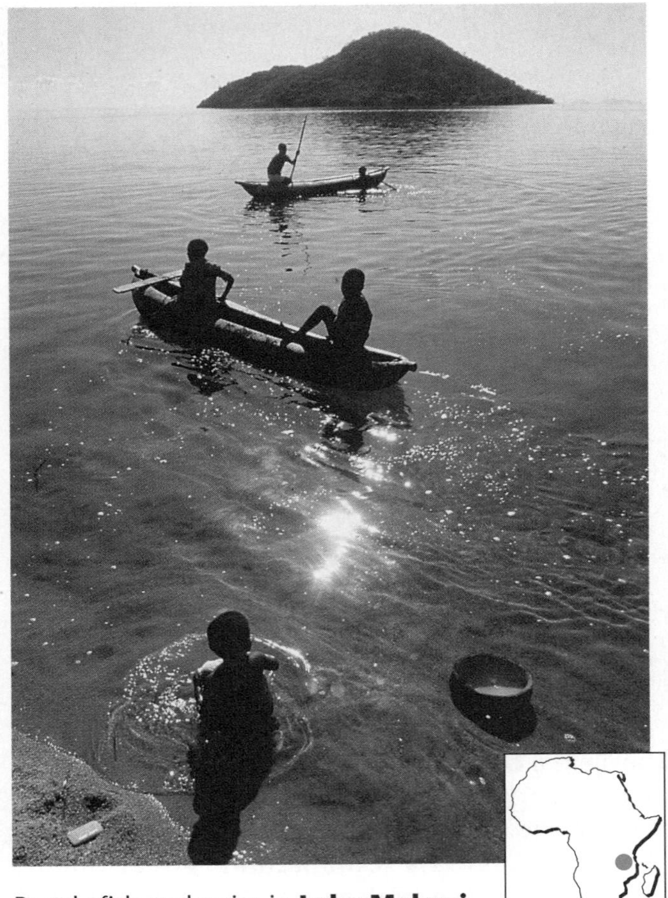

People fish and swim in **Lake Malawi (Lake Nyasa)**. The lake, which drains into the Zambezi River, is about 360 miles (579 km) long but averages only 25 miles (40 km) in width.

Much of Zambia's land is suitable for agriculture and grazing. Farming on large commercial fields as well as on small plots remains important. About half of Zambians live in cities, however, and the country must import food.

Zambia, formerly the British colony of Northern Rhodesia (roe-DEE-zhuh), made the change to independence with ease. Large revenues from copper mines helped the country quickly achieve one of the highest per capita GNPs in Central Africa. In recent years, however, Zambia has experienced economic decline.

As a landlocked country, Zambia's economic stability depends on its ability to move imports and exports. Zambia is most dependent on South Africa but also relies on other neighboring countries. During the civil war in Angola, the railway leading to the Atlantic Ocean often was blocked. Minerals could not be easily exported, and imports became difficult to get. Zambia also depended on Zimbabwe for many manufactured goods whose production was disrupted by civil war in that country.

Several periods of severe drought and fluctuating copper prices in the world market also have affected life in Zambia. That the country's copper reserves are steadily declining adds to the discontent. Today, however, Zambians are hopeful because democracy has returned to the government. Neighboring countries are at peace and are rebuilding Zambia's vital transportation links to the ocean. A railway and oil pipeline now go to Tanzania as well.

Malawi In contrast to Zambia, Malawi is densely populated. With fewer natural resources, Malawi must base its economic development on its agricultural potential. Most of the country's foreign income comes from tobacco, its most important export. Some new factories have been built to produce items such as textiles and farm tools, but industrialization is taking place slowly.

Malawi's major challenge is its landlocked location. (See the map on page 465.) If important minerals are discovered in the future, it would be difficult and costly to move ores to the coastal ports. For this reason, Malawi's cautious government has worked hard at getting along with its neighbors.

Meanwhile, the government continues to invest in farm improvements. Both commercial farms and small farms are supported with road and railroad construction. The government sells fertilizers and seeds to farmers at low prices.

Malawi's major undeveloped resources include its beautiful forested highlands and mountains, as well as Lake Malawi (Lake Nyasa). The rift valley, lined with grass-covered plateaus, crosses part of the country. The government is expanding the national parks and forest areas. With new roads and hotels, Malawi could profit from tourism.

4 SECTION REVIEW

1. How does being landlocked affect the economic situation of Zambia and of Malawi?
2. What natural resources can be developed in Angola, Zambia, and Malawi?
3. **Critical Thinking** **Discuss** the conflicts experienced by Angola since 1975. Why did many other countries become involved?

CHAPTER 39 CHECK

Cameroon rock band

Reviewing the Main Ideas

1. Central Africa is a plains region covered with tropical forests and savannas. Most of the region has a tropical-savanna climate.
2. Central Africa imports many manufactured goods from developed countries. International aid and exports of minerals and cash crops provide income to pay for these goods. Most of the countries in the region must import food.
3. Many people in Central Africa are migrating to the cities in search of work. Cities are overcrowded, and many people live in slums surrounding the cities.
4. Most Central African countries have rich natural resources. Tourist industries can be developed around the beautiful lands and lakes. The development of these resources and the move toward democracy provide Central Africans with hope for the future.

Building a Vocabulary

1. Define *periodic market*. How do periodic markets serve a social as well as an economic function?
2. What term is used to describe a region that is under control of another nation until it can govern itself?
3. What is an exclave? Give an example of an exclave from the chapter.

Recalling and Reviewing

1. What are the two important river systems of Central Africa?
2. What political problems have caused Zaire's economy to suffer?
3. Why is Cameroon's economic situation better than that of most of the other countries in Central Africa?
4. What challenges face Angola?

Thinking Critically

1. Using examples from the chapter, explain how the quality and availability of transportation is related to the economic development of an area.
2. Predict which country in Central Africa will enjoy the most economic success by the year 2000. Give reasons for your prediction.

Learning About Your Local Geography

Cooperative Project

Plan a periodic market. In your group, make a list of items grown or produced locally that could be sold or bartered at the market. Your group should decide the best time and location in your community for the market and provide reasons for your choices. Share your group's list of items for sale or barter and the choice for time and location for the market with the whole class. Work with other groups to make a final decision for the time and place to hold the market.

Using the Five Themes of Geography

Make a list of the landlocked countries discussed in this chapter. Describe the relative **location** of each in terms of neighboring countries. Make a list of common problems faced by these countries due to their relative location. Explain how **movement** is affected by each country's relative location.

CHAPTER 40

Southern Africa

“I was born in that country known as the Union of South Africa. The heart of it is a great interior plateau that falls on all sides to the sea. But when one thinks of it and remembers it, one is aware not only of mountains and valleys, not only of the wide rolling stretches of the veld, but of solemn and deep undertones that have nothing to do with any mountain or any valley, but have to do with men [people].”

Alan Paton

GEOGRAPHY DICTIONARY

- *veld*
- *apartheid*
- *embargo*
- *sanction*
- *exile*
- *copra*

Southern Africa has been the scene of clashes between Africans and Europeans since the 1600s. The struggle between cultures has dominated the history of South Africa, the most powerful and influential country in this part of the world. Because of South Africa's importance, events there have affected all other nations in Southern Africa.

Beneath the region's surface are some of the world's greatest mineral deposits. The wealth that comes from these rich natural resources has long been controlled by people of European ancestry. Blacks have struggled to share in control of the land here and the wealth it has generated. The struggle is most clear in the cities of South Africa, where wealth and poverty stand side by side.

Domination of the governments and economies of Southern Africa by whites has been breaking down in recent years. The great changes taking place throughout the region are being watched around the world.

Wood sculpture from **Madagascar**

Southern coast of **South Africa**

SOUTHERN AFRICA: PHYSICAL–POLITICAL

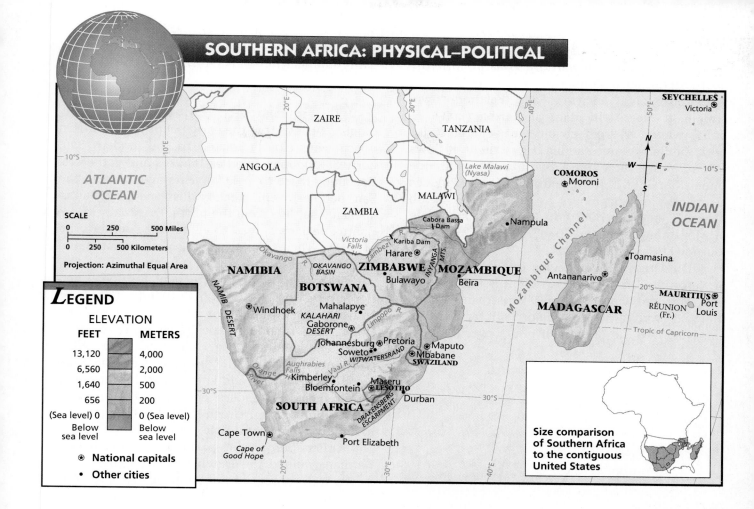

PHYSICAL AND HISTORICAL GEOGRAPHY

FOCUS

- *How does the Drakensberg Escarpment affect Southern Africa's climate?*
- *What various groups of peoples settled in Southern Africa?*

LOCATION **Landforms and Climate** Southern Africa stretches from the subtropical low latitudes south to the edge of the middle latitudes. Climates range from tropical savanna in the north to Mediterranean in the south. Except for a rim of lowlands around the coast, Southern Africa is mostly a region of high plains, with plateaus rising above 4,000 feet (1,219 m).

The Drakensberg Escarpment rises near the coast of the Indian Ocean and reaches up to 11,425 feet (3,482 m). (See the map above.) From the east, the Drakensberg looks like a mountain, but its western side actually forms the high edge of a wide region of plains. Moisture flows into Southern Africa primarily from the Indian Ocean, where it is forced upward by the Drakensberg Escarpment and Inyanga Mountains. Beyond these rainy eastern slopes, climates become drier farther inland.

The term used for South Africa's various natural regions is ***veld*** (vehlt). The grassy, high plains in central South Africa north and west of Lesotho make up the highveld. The bushveld is a savanna region of short trees and bushes that is found at lower elevations. Still another of South Africa's regions is the lowveld, which includes the dry, tall-tree savannas as well as the moist forests at the base of the Drakensberg.

The major river systems of Southern Africa are the Orange and the Zambezi. Water resource development along these rivers is important to the region. Dams, irrigation canals, and tunnels deliver water and hydroelectric power to nearby farms and distant cities.

The Zambezi winds southward from Central Africa before turning toward the coast to form the northern border of Zimbabwe. The Zambezi's first great drop is at Victoria Falls. Farther downstream, it meets the dam at Kariba (kuh-REE-buh), where hydroelectricity is generated for both Zimbabwe and Zambia. From there, it flows toward rapids at Cabora Bassa in Mozambique (moe-zihm-BEEK). Here, another dam produces electricity for consumers as far away as South Africa. Finally, the river crosses the coastal plain of Mozambique to end its 1,700-mile (2,735-km) journey at the Indian Ocean.

The Orange River begins its westward journey atop the Drakensberg where it is fed by summer rains and winter snows. Downstream, it joins the Vaal (vahl) River, its major tributary. Before emptying into the Atlantic, it pours over the high Aughrabies (aw-GRAHB-eez) Falls near the border with Namibia (nuh-MIB-ee-uh).

The Drakensberg stretches from northeastern **South Africa** to the southern part of the country.

MOVEMENT **Historical Geography** Archaeologists believe that the early peoples of Southern Africa were closely related to the modern San. Early San peoples lived as hunters and gatherers. Beginning more than 4,000 years ago, most of the San peoples were steadily displaced by Bantu-speaking settlers from the north. Eventually, these Bantu speakers spread to all but the southern rim of Africa. Their farming societies raised cattle and sorghum, and they were expert ironworkers.

The region's first permanent European settlers were from the Dutch East India Company. Drawn to a natural harbor near the Cape of Good Hope, they founded Cape Town in 1652. Here, they encountered the San and the Khoi (Hottentots). Unlike the foraging San, the Khoi were cattle herders and knew about farming. The Dutch business leaders encouraged farming in the region to supply fresh food for ships passing to and from Southeast Asia.

Other Europeans, particularly French and Germans, joined the Dutch settlers at the Cape. Slaves were brought from Southeast Asia as well. Far removed from Europe, the European settlers soon called Africa home and called themselves Afrikaners. Over the years, many of the Europeans developed into frontier farmers, or "Boers." They spoke a mixture of Dutch, words learned from Malayan slaves, and Bantu languages. This language eventually became modern Afrikaans. Today, about 70 percent of white South Africans are descendants of these first European settlers.

As a result of political shifts in Europe, the region of the Cape of Good Hope at the tip of Southern Africa became a British colony in 1814. The British administrators made the English language, English laws, and English education the rule. English traders and missionaries spread across the region. Many Boers were unhappy with the British takeover. In the 1830s, Boers began moving farther into the interior, a migration sometimes called the Great Trek.

Before the Boers began their northward trek, the Zulus and other groups of Bantu speakers were migrating southward. By the early 1800s, the Zulu Empire extended into Southern Africa. The Zulus were led by Shaka, a great warrior who forged his armies into powerful forces. As the

Boers moved northward, they clashed with the Zulus over control of the land. The Zulus resented the loss of their land to people they saw as invaders. Eventually, the Zulu Empire was defeated by the British and the Boers.

The Boers established two independent countries in the region—the Orange Free State and the Transvaal. Meanwhile, the British were establishing the colony of Natal (nuh-TAHL) on the Indian Ocean coast. Many people from India came there to work as laborers in the growing sugar industry.

At first, few Europeans penetrated what they thought was the barren interior of Southern Africa. Then, in 1868, diamonds were found near Kimberley. About 20 years later, the world's largest gold deposits were discovered near present-day Johannesburg (joe-HAHN-uhs-buhrg). The vast mineral wealth brought many British settlers to the Southern African interior. Black Africans also began moving to the cities and mines to look for work. Soon Maputo (muh-POO-toe), the tiny capital of Portuguese Mozambique, became a booming port for shipments from Johannesburg.

Competition for mineral wealth and the Boers' distrust of British expansion led to the Anglo-Boer War (1899–1902). The Boers lost this bitter struggle. In 1910, the Transvaal and the Orange Free State were joined with the Cape and Natal colonies to form the Union of South Africa. The government, however, was nearly all white. Pretoria (prih-TAWR-ee-uh), in the Transvaal, became the administrative capital. Bloemfontein (BLOOM-fahn-tayn), in the Orange Free State, became the site of the judicial capital, where the Supreme Court is located. Cape Town, in the Cape Province, became the legislative capital, where the parliament meets. The four original provinces were divided into nine in 1994.

1 SECTION REVIEW

1. What is Southern Africa's single greatest landform, and how does it affect the region's climate?
2. Beginning with the San, list the various groups of people who have settled in Southern Africa.
3. **Critical Thinking** **Explain** why a country might want to have three separate capitals.

2 REPUBLIC OF SOUTH AFRICA

FOCUS

- *What was the system of apartheid, and what were its effects?*
- *How did people protest apartheid?*
- *Why is South Africa the continent's most economically developed country?*

Political Geography Despite the country's beauty and rich natural resources, political issues dominate the geography of South Africa. After the 1948 elections, South Africa's white-run government formally established the separation of the country's many ethnic groups. This policy of "separate development" became known by its Afrikaans name, **apartheid** (uh-PAHR-tayt). The idea behind apartheid was that each ethnic group would progress best under its own culture and traditions. Ideally, each group would be equal. In practice, however, separate could not be equal.

Apartheid was a series of laws that separated the races where they lived, worked, and went to school. It affected the daily lives of all people. Status, power, and wealth belonged to white South Africans, who controlled the national government and the economy. Coloureds—South Africans of mixed race—occupied a middle status under apartheid. Coloureds included descendants of the San and Khoi as well as people of Asian ancestry. At the bottom of the racial divisions were black South Africans, who make up about 75 percent of the population. South Africa became divided into black and white lands.

The black population was further divided into territories based upon African ethnic groups such as Zulu, Xhosa, Tswana, Venda, Sotho, and others. These territories, called homelands, extended in a broad horseshoe of lands along the east coast and in northern South Africa. Under the homelands policy, blacks were given citizenship in the homeland of their ethnic group, whether they lived there or not. The homelands policy created less-developed regions with limited medical care, educational opportunities, and progress. Many families were separated. Often, men found work in the cities, leaving women and children behind in the impoverished rural areas.

The national government gave some of the homelands status as independent countries. The government then claimed that many black South Africans living outside their designated homelands were not citizens of South Africa. They could be treated like foreigners without civil rights in cities where they may have been born. The independence of the homelands was never recognized by other countries.

Apartheid made South Africa an outcast among other nations. In 1961, foreign opposition to apartheid forced the Union of South Africa to leave the British Commonwealth. The country became the Republic of South Africa. South African sports teams and products were refused in other countries. Some countries placed **embargoes** on South African products. An embargo is a legal restriction against the movement of freight. Such economic measures are called **sanctions**, which are put in place by other nations in the hopes of forcing a nation to cease an illegal or immoral act. Countries that placed sanctions argued that economic pressure would cause South Africa to find better and quicker solutions to apartheid. While sanctions affected the white community, the economic pressure also increased unemployment among black workers and created hardships for the country's poorest people.

Protests Even before apartheid was established, organizations in South Africa were fighting against racial discrimination. The most well known of these organizations is the African National Congress (ANC), founded in 1912. In the 1950s, the ANC openly protested apartheid by disobeying apartheid laws. The South African government responded quickly and brutally. The ANC leaders, including the well-known black lawyer Nelson Mandela, believed that, in self-defense, blacks would have to confront violence with violence. In response, the government banned the ANC, charged its leaders with treason, and gave Mandela and other ANC leaders sentences of life in jail. Other black South Africans fled to nearby countries, becoming **exiles**, people who have been forced to leave their country.

Ending Apartheid Apartheid did not work for many reasons. Not only did it discriminate against the majority of the people, it failed economically. No country can prosper by holding back most of its people. Apartheid began to disappear as official government policy in 1989. At that time, the government opened relations with the ANC. This generally peaceful dialogue resulted in the repeal of South Africa's apartheid laws. In 1990, Nelson Mandela and others who had been jailed for disagreeing with apartheid were released. Finally, in 1994, Mandela and the ANC won the country's first all-race elections.

Nelson Mandela, an important leader in the long struggle against apartheid, was elected president of **South Africa** in 1994.

Today, public facilities are being integrated, and people of different races meet each other socially more often. Although there has been resistance to school integration among some white South Africans, public schools must now accept students of all races. Memories of the injustices of apartheid live on, however, and race relations remain tense. Also, cultural and political differences between various black ethnic groups, such as the Zulus and the Xhosas, are more apparent now that they are no longer overshadowed by the black-white divisions of the nation. The tensions between these ethnic groups have influenced the negotiations involved in drafting a new constitution for South Africa by 1999.

Economic Geography South Africa faces numerous challenges, many of which have roots in the old system of apartheid. For example, as apartheid ended, about three-fourths of black South Africans were farmers. Many of them were subsistence farmers in the former homelands. In contrast, one-fifth of white South Africans were farmers and ranchers. Agriculture among white South Africans has been as modern and productive as any other in the world. The new government has had to battle problems caused by severe soil erosion in the former homelands and high unemployment among black South Africans. To improve the economic status of black South Africans, government plans have included jobs programs and other employment reforms.

South Africa is the continent's most economically developed country. Although the country has only 6 percent of Africa's population, it has more than half of Africa's industrial firms. South Africa's economic activity also accounts for about half of Africa's total output. The basis for South Africa's wealth is its minerals, which are exported throughout the world. South African mines use the most advanced technology. Mineral resources include gold, uranium, copper, platinum, iron ore, coal, chromium, and diamonds. South Africa has nearly every mineral necessary for a modern, industrial society except bauxite and petroleum. South African factories manufacture almost all the industrial products the country needs.

PLACE Urban Geography The largest industrial region on the African continent is the Witwatersrand (WIT-wawt-uhrz-rahnd). Johannesburg, the capital, is located almost at its center. (See "Cities of the World" on pages 482–483.) Although most industries are South African owned, many large U.S., European, and Japanese companies are represented as well. Many of South Africa's automobiles are built near Port Elizabeth, on the southeast coast. Cape Town and Durban in Natal are other active ports.

Now that apartheid has ended, cities are open to all races, and the integration of neighborhoods has begun. There is, however, a continuing common pattern of settlement in South Africa. Many white South Africans live close to city centers and in suburbs away from industrial districts. White neighborhoods are similar to those in American suburbs, with modern-style houses and shopping centers.

Black areas are usually on the outer edges of cities, often across industrial districts and away from the white neighborhoods. Often, blacks must commute long distances to work. Coloured and Indian neighborhoods are usually situated between the black and white districts. Black neighborhoods vary in style from small, government-built modern houses to slum settlements built from scrap lumber and sheet metal. The new government has worked to improve utility service to existing homes and to help black South Africans build hundreds of thousands of new homes.

South Africa's Future The effects of apartheid will remain for years. Black South Africans have had little access to their country's riches and economic opportunities, while white South Africans have the best educations and job skills. They enjoy higher incomes, hold the best jobs, own most of the good land, and live in the best neighborhoods. Blacks only recently have moved into positions of authority.

South Africa's black population is growing at a very fast rate, yet its white population is barely increasing. There are not enough schools in black neighborhoods, while schools in white neighborhoods are sometimes nearly empty. South Africa's economy will prosper only when equal educational, economic, and political opportunities exist for all the country's people.

SECTION REVIEW

1. What was apartheid, and how did it affect the daily lives of South Africans?
2. Why is South Africa the continent's most economically developed nation?
3. **Critical Thinking** **Evaluate** the different ways that apartheid was protested, both in and out of South Africa.

OTHER COUNTRIES OF SOUTHERN AFRICA

FOCUS

- *How are the histories of the countries of Southern Africa alike and different?*
- *What are the major economic activities in the countries of Southern Africa?*

The conflicts in South Africa often have overshadowed other countries of the region. Although these countries are strongly affected by South Africa, they are culturally diverse, independent nations. Mineral resources give these countries potential for industrial development.

Text continues on page 484.

Johannesburg

"Gold! Gold! Gold!"

The cry of "gold!" marked the beginning of Johannesburg.

Unlike many other major cities around the world, Johannesburg, South Africa, owes its existence and growth largely to gold. In 1886, a wandering prospector came across gold-bearing rocks on a farm near a range of low hills known as the Witwatersrand, or "ridge of white waters." The news of the gold strike spread far and wide. Thousands of fortune seekers from around the world made their way to the area. People came mainly from Europe, America, Australia, and other parts of southern Africa. A dusty mining camp soon sprang up, its streets filled with a jumble of languages. As more gold was discovered and more people arrived, the small town of Johannesburg began to expand and take shape.

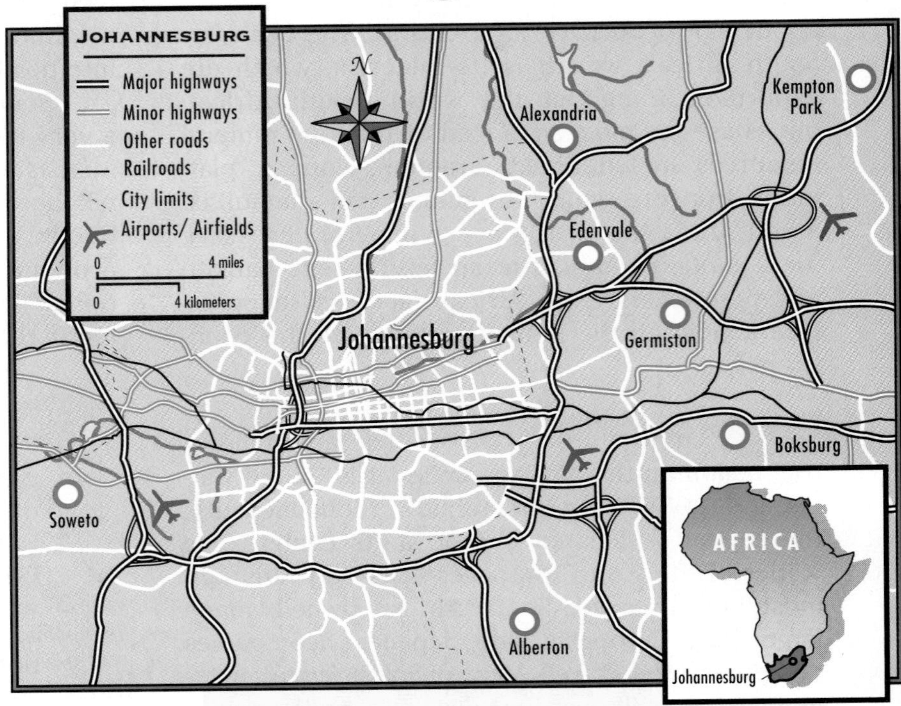
JOHANNESBURG
Major highways
Minor highways
Other roads
Railroads
City limits
Airports/ Airfields
0
4 miles
0
4 kilometers
N
Alexandria
Kempton Park
Edenvale
Germiston
Johannesburg
Boksburg
Soweto
Alberton
AFRICA
Johannesburg

Johannesburg is at the heart of a bustling urban area that includes other modern cities and black townships. Many residents from these cities and townships, such as Soweto, commute to Johannesburg for work.

The Mining Town

In Johannesburg's earliest years, gold mining was by far the most important part of its economy. But as the town grew, so did its need for other goods and services. Factories supplied machinery for the mines. New businesses provided food and clothing for the region's growing population. Banks, schools, and hospitals were built. Today, a century after gold was first discovered, Johannesburg forms the heart of a bustling, modern metropolitan area.

These South African workers are mining for gold more than one mile (1.6 km) underground. Many gold-mining companies have headquarters in **Johannesburg**, the country's leading business center.

Gold and businesses related to gold are still important to the city's economy, and the Witwatersrand region is a major source of gold. In fact, nearly half of all the world's gold mined during the past 100 years has come from this region. Although most of the mines have closed, miners continue to work as much as two miles underground to haul out gold-bearing ore east and west of the city in the 62-mile-long (100-km) Witwatersrand.

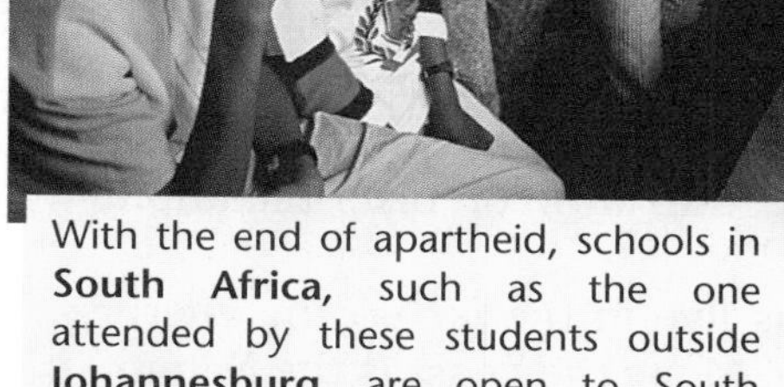

With the end of apartheid, schools in **South Africa,** such as the one attended by these students outside **Johannesburg,** are open to South Africans of all races.

A Changing City

Today, Johannesburg is the financial and commercial capital of South Africa. It is a city of modern skyscrapers, busy industrial districts, and sprawling suburbs. Most of Johannesburg's residents now work in jobs that have little to do with mining.

Although most blacks still live in vast townships, housing projects, and slums on the outskirts of the city, many have moved to the city center. In contrast, most whites live in comfortable homes in pleasant suburbs of the city. But some blacks, South Africans of mixed race (called coloureds), and those of South Asian descent also can now afford to live in suburbs that once were all white. In addition, schools have become more integrated.

As more blacks move into positions of authority in government and employment, the townships also have begun to change. The more than 2 million residents of Soweto, a large township southwest of downtown Johannesburg, once had to travel to the main city to do even basic shopping. Today, Soweto's homes have better utility services, such as electricity, and many stores and other businesses have opened in the township.

Downtown Johannesburg, once exclusively white, was declared open to all races when apartheid ended in the early 1990s. South Africans of all races can now live in the same apartment buildings, eat in the same restaurants, and work and go to school together. Divisions remain, but Johannesburg is leading South Africa into a future free of apartheid.

Vans from this minibus station carry many **Soweto** residents to their jobs in Johannesburg. Road and rail lines link Soweto to the capital city.

Namibia Namibia lies on the Atlantic coast north and west of South Africa. (See the map on page 477.) Its road to independence was slow and violent. Once a German colony, it was called German Southwest Africa. During World War I, South African forces drove out the Germans. It then became South West Africa. South Africa's administration of the region was challenged repeatedly by the United Nations. In 1990, South Africa finally granted the territory its independence as Namibia.

The new government is democratic and is represented by several political parties in the parliament. At one time, apartheid was introduced into Namibia. However, lands once reserved for whites and blacks, as well as for Khoisan peoples, have been opened to settlement by all groups. With independence, the laws of discrimination have disappeared.

Most Namibians live in the far north, where there is dry savanna. Along the coast is the Namib Desert; inland is the Kalahari Desert. People also live in the central highlands, where the climates are cooler and not as arid. Ranching is the only commercial agriculture. The fishing fleet of this territory also provides important income. The economy is most dependent, however, on mining. Namibia has enormous mineral reserves, including literally mountains of copper, lead, zinc, and uranium. Along the southern coast are large diamond deposits.

Botswana Like Gabon in Central Africa, Botswana is becoming one of Africa's success stories. Formerly the British territory of Bechuanaland (bech-uh-WAHN-uh-land), Botswana is an arid, landlocked country occupying much of the Kalahari Desert. (See the map on page 477.) Britain took control of the area in 1886 to connect the Cape Colony with Rhodesia. Britain invested little in Botswana, however. At the time of independence in 1966, there were no paved roads, no electricity, and only one important factory, a meat-canning plant.

Today, the country earns much income from its enormous diamond and copper deposits and has one of the fastest-growing economies in Africa. Cattle herding and mining are the principal economic activities. Botswana also is one of the more democratic countries in the region and has opened its doors to refugees from neighboring countries. A new capital stands at Gaborone.

FACTS IN BRIEF Southern Africa

COUNTRY POPULATION (1994)	LIFE EXPECTANCY (1994)	LITERACY RATE (1990)	PER CAPITA GDP (1993)
BOTSWANA 1,359,352	60, male 66, female	23%	$4,500
COMOROS 530,136	56, male 60, female	15% (1989)	$700
LESOTHO 1,944,493	60, male 64, female	59%	$1,500
MADAGASCAR 13,427,758	52, male 56, female	80%	$800
MAURITIUS 1,116,923	67, male 75, female	80%	$7,800
MOZAMBIQUE 17,346,280	47, male 50, female	33%	$600
NAMIBIA 1,595,567	59, male 64, female	72%	$2,500
SEYCHELLES 72,113	66, male 73, female	85% (1991)	$5,900 (1992)
SOUTH AFRICA 43,930,631	62, male 68, female	76% (1992)	$4,000
SWAZILAND 936,369	52, male 60, female	65%	$2,500
ZIMBABWE 10,975,078	40, male 44, female	67%	$1,400 (1991)
UNITED STATES 260,713,585	73, male 79, female	97% (1991)	$24,700

Sources: *The World Factbook 1994*, Central Intelligence Agency; *The World Almanac and Book of Facts 1995*

Which country in Southern Africa has the lowest life expectancy rates?

The Tswana (from which Botswana gets its name) is the largest ethnic group, and many of these people are cattle herders and corn farmers. Most people live along the wetter eastern edge of the country, where dry savannas are the typical vegetation. In the Kalahari Desert are the remaining groups of hunting-and-gathering San.

Botswana's major river is the Okavango (oe-kuh-VANG-oe), which starts in the wet highlands of Angola and flows into the Okavango Basin. There, it forms the rich wetland forests of the inland Okavango Delta. Large game reserves are located in the region. These wetlands in the desert could become a major tourist attraction.

Zimbabwe Zimbabwe takes its name from the ancient walled ruins in the central part of the country. There, a city of stone-walled buildings and ceremonial structures was built between the twelfth and sixteenth centuries. These ruins are a symbol of the achievements of African cultures before the arrival of Europeans.

Southern Africa: Transportation Network

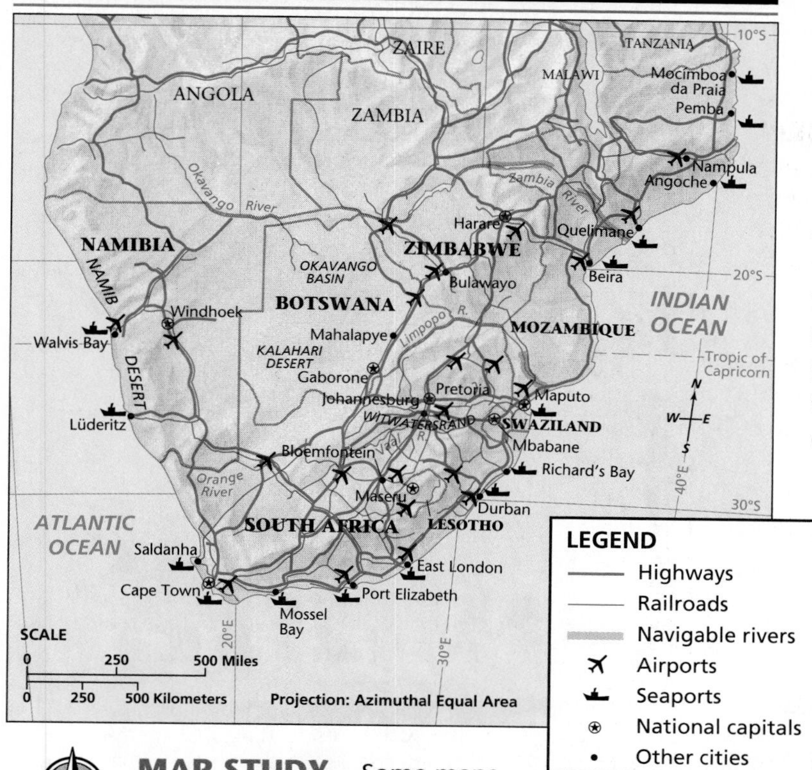
ZAIRE
TANZANIA
MALAWI
ANGOLA
ZAMBIA
Mocimboa da Praia
Pemba
Nampula
Angoche
Okavango River
Zambia River
Harare
Quelimane
NAMIBIA
ZIMBABWE
Beira
OKAVANGO BASIN
Bulawayo
INDIAN OCEAN
NAMIB DESERT
BOTSWANA
Limpopo R.
Windhoek
Walvis Bay
Mahalapye
MOZAMBIQUE
KALAHARI DESERT
Gaborone
Tropic of Capricorn
Pretoria
Johannesburg
Maputo
Lüderitz
WITWATERSRAND
SWAZILAND
Mbabane
Bloemfontein
Vaal
Richard's Bay
Orange River
Maseru
Durban
ATLANTIC OCEAN
SOUTH AFRICA
LESOTHO
Saldanha
East London
Cape Town
Port Elizabeth
Mossel Bay
10°S
20°S
30°S
20°E
30°E
40°E
SCALE
0 250 500 Miles
0 250 500 Kilometers
Projection: Azimuthal Equal Area
LEGEND
Highways
Railroads
Navigable rivers
Airports
Seaports
National capitals
Other cities

MAP STUDY Some maps, such as the one above, provide information that shows how people and goods are transported between locations. The legend shows symbols for major roads, railroads, airports, and seaports that are vital to trade and travel. In some places, rivers also are important parts of transportation networks.

During the colonial period, Zimbabwe, Zambia, and Malawi were British territories. The British had combined them into the Federation of Rhodesia and Nyasaland. The members were to profit economically, because each region would offer something important to the federation. Malawi, formerly Nyasaland, was to provide a labor force. Zambia, formerly Northern Rhodesia, was to provide mineral wealth. Zimbabwe, formerly Southern Rhodesia, would provide industrial strength. The few Europeans, mainly in Southern Rhodesia, controlled most of the economic and political affairs of the federation, however.

Northern Rhodesia and Nyasaland objected to the influence of white-dominated Southern Rhodesia. In 1964, Nyasaland and Northern Rhodesia became independent as the separate countries of Malawi and Zambia. Southern Rhodesia, calling itself Rhodesia, remained a self-governing British colony. In 1965, the white-dominated parliament of Rhodesia declared itself independent, an illegal action under British law. Furthermore, most of the people living in Rhodesia were black Africans who were not being equally represented by the white government. In response, the United Nations ordered a trade embargo against Rhodesia. Only white-dominated South Africa openly traded with the country.

In 1979, this economic pressure, combined with successes of guerrilla fighters, brought black majority rule to Rhodesia. Soon after, Rhodesia became Zimbabwe, but the long period of conflict weakened the once prosperous economy.

In the years since the civil war, Zimbabwe's leaders have focused on rebuilding the country. The many factories started during the embargo are exporting goods to neighboring countries. The country has food for export when the region is not suffering from drought, and the commercial agriculture on farms owned by whites is prosperous again. Although the majority of commercial farms in Zimbabwe are owned by whites, the government recently passed legislation that would equalize the distribution of land among black and white farmers. Other economic and political reforms are taking place as well.

Mozambique Mozambique, located on the east coast of Africa between the Republic of South Africa and Tanzania, has a good base of economic resources. The country produces tea, sugarcane, cotton, cashew nuts, sisal, and **copra**. Copra is the dried meat of the coconut, from which coconut oil is obtained. Because coal in Africa is uncommon, the country's coal reserves are valuable. Other sources of income are tax collections at the ports of Maputo, the capital, and Beira (BAY-ruh). Beira's port handles many products from interior countries, particularly Malawi, Zimbabwe, and Zambia.

Until this century, there was little economic development in this former Portuguese colony. Today, the government has several development plans under way. One is an irrigation project in the Limpopo River valley. Plans call for growth of an extensive network of irrigation canals to support small farms. Also important is the dam in the Zambezi River valley at Cabora Bassa.

Other Countries of Southern Africa

Swaziland

Economic Resources

Timber is an important resource in Swaziland, which also mines gold, coal, asbestos, and iron ore. The country also has valuable grazing lands, and it exports sugarcane and citrus fruits.

At a Glance

Although the vast majority of Swaziland's population is Swazi, Europeans own nearly half the land. Income from European-owned mining operations is vital to the landlocked country.

Seychelles
Comoros
Mauritius
Swaziland
Lesotho

Seychelles

Economic Resources

Seychelles' economy is based on agriculture. Its main products are coconuts, bananas, and cinnamon. Tourists also provide important income for the islands.

At a Glance

Seychelles, which is made up of 92 islands in the Indian Ocean, is a popular destination for tourists.

Lesotho

Economic Resources

Lesotho's limited resources and poor land for farming produce small exports of wheat, diamonds, and wool. The building of large dams will provide hydroelectricity and water for export in the future.

At a Glance

Lesotho, which is completely surrounded by South Africa, is located on the high plateau of the Drakensberg Escarpment.

Comoros

Economic Resources

The economy of Comoros revolves around tourism, fishing and farming, and the export of perfumes, spices, and copra.

At a Glance

The poor, rocky soil of Comoros forces the islands, located in the Mozambique Channel, to import most of their food.

Mauritius

Economic Resources

The economy of Mauritius is heavily dependent on sugar production. The cultivation of sugarcane was begun during the period of European control, first by the Dutch, then the French, and finally the British.

At a Glance

The volcanic island of Mauritius, once a home of the now-extinct dodo bird, is almost completely surrounded in the Indian Ocean by reefs.

Mozambique has suffered continuing violence and rebellion. It began with the war for independence in the 1960s and has continued since independence in 1975. Rebels trying to oust the Communist government cut roads and bombed rail lines, nearly collapsing the economy. In the early 1990s, pressure from neighboring countries forced an uneasy peace treaty between the government and rebel forces.

Madagascar The island countries off the east coast of Africa share connections with Africa as well as with Asia. The largest of these islands is Madagascar. (See the map on page 477.)

Madagascar is rich in natural wonders. Along Madagascar's east coast is a narrow, humid, coastal plain and a steep escarpment. Atop the escarpment are a mild highland climate and rich farmlands. In the west, the island becomes more arid as it slopes down to the coast on the Mozambique Channel. The Southeast Asians who settled Madagascar more than 1,000 years ago brought Asian farming systems and crops. Later, Arabs traded between Madagascar and Africa's east coast. The Malagasy language has a rich literature and is spoken throughout Madagascar.

When the French took control of Madagascar in the 1890s, they often neglected this isolated island. Schools and roads were built, but little else. Since independence in 1960, industrial development has been slow. The rapidly increasing population has meant that more and more available farmland is being used. Soil erosion has been severe, and deforestation is increasing in the highlands. The government has nationalized many businesses, and the economy has steadily declined. The country is dependent on foreign aid.

SECTION REVIEW

1. What European countries colonized each country discussed in this section?
2. Which South African countries have economies based on mining? on agriculture? on both mining and agriculture?
3. **Critical Thinking** **Assess** whether the conflicts of the 1960s–1980s in Zimbabwe and Mozambique had similar causes.

CHAPTER 40 CHECK

Early African ruins, **Zimbabwe**

Reviewing the Main Ideas

1. The region of Southern Africa has a long history of settlement by various groups of people. San (Bushmen), Bantu speakers, Dutch Boers, Indians, the British, and Southeast Asians were some of the peoples who settled the area.
2. Under apartheid, South Africa was divided into black and white regions, and political and economic power rested with the white minority. With the end of apartheid, blacks increasingly hold positions of authority in politics and employment.
3. Most of the countries of Southern Africa are rich in mineral resources, although some countries have not been able to develop these resources to their advantage.

Building a Vocabulary

1. Name and describe the three natural regions of Southern Africa.
2. What was the system of apartheid?
3. What are embargoes? For what reasons do countries impose sanctions on other countries?
4. What is the term for a person who has been forced to leave his or her country?
5. Define *copra*. Where in Southern Africa is it produced?

Recalling and Reviewing

1. Describe the relationship between landforms and climate in Southern Africa.
2. How did apartheid affect daily life in South Africa?
3. Why is South Africa the wealthiest country on the African continent?
4. Compare the economic situation in Botswana immediately following independence to its economic situation today.
5. What is the major economic activity in Southern Africa?

Thinking Critically

1. In your opinion, what are the positive and negative aspects of industrialization? If you were an economic adviser to Namibia, for example, would you advise the country to develop its mineral resources? Why or why not?
2. Imagine you have been living on Madagascar and are moving to Lesotho. Write a paragraph comparing the physical and economic geography of the two countries. In which country would you prefer to live? Why?

Learning About Your Local Geography

Cooperative Project

As a class, make a list of the mineral resources found in or near your area. Then, organize into groups of four. Each group should select a different mineral resource and research where it is found, how it is mined, to where it is transported, and for what it is used. Groups should present their information to the rest of the class.

Using the Five Themes of Geography

Write a paragraph about how the geographic themes of **region** and **place** apply to the countries of Southern Africa. Into what kinds of political and social regions have these countries been divided? How do the various regions differ in place? Discuss how certain ways of defining regions can put some groups of people at a disadvantage.

SKILLS

Interpreting Photographs

Imagine this textbook without any photographs. Not only would it be less interesting, but you also would not have any visual images of the places you are studying. Photographs are important tools for studying geography. They provide information about a location's physical features, economy, people, and climate.

There are several different kinds of photographs, all serving different purposes. Aerial photographs, taken from above, are useful for showing land use and settlement patterns, and they aid cartographers in making maps. Landsat photographs, taken from satellites, provide much the same information, but they can show more land area because they are taken from farther away. In contrast, photographs taken closer to Earth reveal a location's vegetation, landforms, and weather, as well as culture traits. Cultural geographers study photographs to discover how people live—their housing, their dress, and ways in which they have adapted to or changed their environment.

Photographs record things as they happen—effects of floods or severe drought, crops at harvest time, a highway being built. Photographs also record changes over time, such as the effects of deforestation or urbanization. Historical photographs can be compared to modern-day photographs to find out how a place has changed.

How can we analyze photographs to obtain useful geographic information? One way to examine a photograph is to seek its main idea or theme. Ask, "What does this photograph tell me about this location?" How a photographer arranges a subject in the camera's view is a good clue. If you were asked to take only one photograph of the city or town in which you live, you might choose a wide view from far away to show its size, or perhaps you might show people at work to reveal something about the local economy.

When you study a photograph, note what is in the foreground, the middle ground, and the background. What part of the photograph provides the clearest details? Try to estimate the sizes of various features by comparing them to other objects in the same part of the photograph. Remember, however, that photographs have limitations. The distance between the camera and its subject often distorts what you see—the actual distance between the foreground and the background can be difficult to judge. Because of this, geographers do not use photographs to determine exact distance or size.

Identifying geographic characteristics from photographs takes practice, because the camera captures insignificant details as well as important ones. Learning to interpret photographs effectively will enhance your study of geography.

PRACTICING THE SKILL

The photograph above was taken on an animal reserve in Southern Africa. Using what you have learned, examine the photograph for geographic information. Then answer the following questions.

1. What biome is depicted in the photograph?
2. What is in the foreground? the middle ground? the background?
3. If the photograph had been taken from the top of the mountains, how would the view change?

BUILDING YOUR PORTFOLIO

Individually or in a group, complete one of the following projects to show your understanding of the geography concepts involved.

A Redrawing the Map of Africa

The year is 1900, 16 years after the Berlin Conference. You are an African who has been asked to attend the Africa Conference. (Reminder: No such conference took place, and Africans were not invited to participate in the Berlin Conference.) The goal of this conference is to create a new map of Africa, one that will better serve the needs of Africans.

Your report to the Africa Conference will consist of a map of Africa with borders and country names that you think best, along with written and visual materials that support your decisions. Be prepared to explain any changes you have made to the European map of Africa. To prepare your report to the Africa Conference, you will need to do the following.

1. Identify the countries that were landlocked after European boundaries were set. What resources do these countries have? How do the countries transport these resources to market? You might want to present this information in a chart.
2. Identify the minerals and other natural resources under European control. How are these resources used? How would you use these resources differently? Present your information using flowcharts.
3. Locate maps of Africa that show language and cultural groups before European colonization. Compare to a post-colonization map. Make your own map that illustrates how cultures and/or language groups were divided or thrown together.
4. Create a new map of Africa, including boundaries and names of the African nations. Write two or three paragraphs explaining any changes you have made to the European map of Africa.

Organize the information you have gathered, and present your information to the other groups at the Africa Conference (the rest of the class).

B Investigating Desertification

The United Nations has estimated that 40 percent of Africa's nondesert land is in danger of desertification. Imagine that you are on the Planet Watch committee. Your subcommittee has been assigned to document the causes and effects of desertification and to suggest a solution to the problem. Although your report will focus on Africa, you will need to conduct research to find out more about desertification and its effects worldwide. Complete the following steps to prepare for your presentation to Planet Watch.

1. Prepare a poster that illustrates the definition of *desertification*. Include on the poster a written definition of *desertification*.
2. Draw a cross-sectional diagram or flowchart illustrating how desertification occurs. Use labels to explain what is happening. Be sure to include characteristics of the physical environment that contribute to desertification.
3. Create two maps. One should show countries in Africa most affected by desertification. The other map should be a world map illustrating the worldwide extent of desertification.
4. Consider the following questions: What will happen if desertification continues at its present rate? What will have to change in Africa to control desertification? Work the answers to these questions into a proposed solution to desertification.

Present your research to the Planet Watch committee (the rest of the class). Your proposed solution to desertification should be the conclusion of your presentation.

UNIT 9

EAST and SOUTHEAST ASIA

Visitors to the Great Wall near **Beijing, China** can see only part of the structure as it snakes across northern China. The wall, which includes a series of fortified towers, was used for defense and for communication. It was built in the 200s B.C.

UNIT 9 CHAPTERS

Dancers perform the *legong,* a popular folk dance in **Bali, Indonesia**. *Legong* dancers wear elaborate costumes and use their dance to tell stories about such things as adventures, battles, and love.

Japanese lacquer bowl and tray

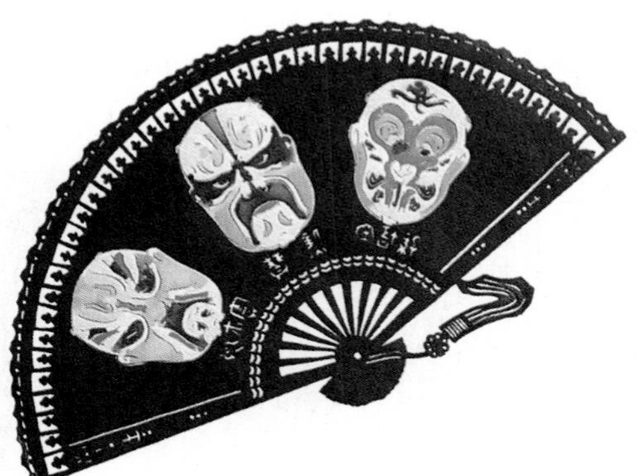

Chinese paper-cut fan

A Chinese junk sails past the skyscrapers of **Hong Kong**, one of East Asia's major commercial centers. China is to take possession of the prosperous British territory in 1997.

One World, Many Voices

LITERATURE

Yuan-tsung Chen (1932–) was born in Shanghai, China. When the People's Republic of China formed in 1950, Chen and many other young people were enlisted by the Communist government to work toward land reform. She spent 20 years helping peasants in agricultural cooperatives. The passage below is from a novel she wrote about her experiences in Gansu, a province in northwestern China. It describes the physical geography of the remote region.

"We had traveled for a whole day over this ancient land. . . . At first, we had marveled at the strangeness of the landscape. It was a plain riven [cut] by deep gullies so that the dirt road either meandered wildly to avoid the slits in the earth or plunged zigzag down and up the sides of the ravines that couldn't be avoided. The earth had been ravaged and made desolate. I knew from my history books that these eroded lands were once pastures and forested plateaus. Then the pastures had been ploughed up to grow crops and the forests cleared for farmland. The natural rhythm of nature had been disturbed. The animals had disappeared and so had their dung. War had devastated the farms. Marauding [plundering] warlord armies had chopped down the remaining trees for firewood. Without vegetation to hinder them, the rain and run-off rivulets [small streams] of centuries had eaten into the fields and carried them away. The green clothing of the earth had been filched [stolen] and the naked earth was dying of cold.

I sighed disconsolately [unhappily]."

Interpreting Literature

1. What human activities have changed or damaged the land in Gansu?
2. Why is the narrator saddened by what she observes there?

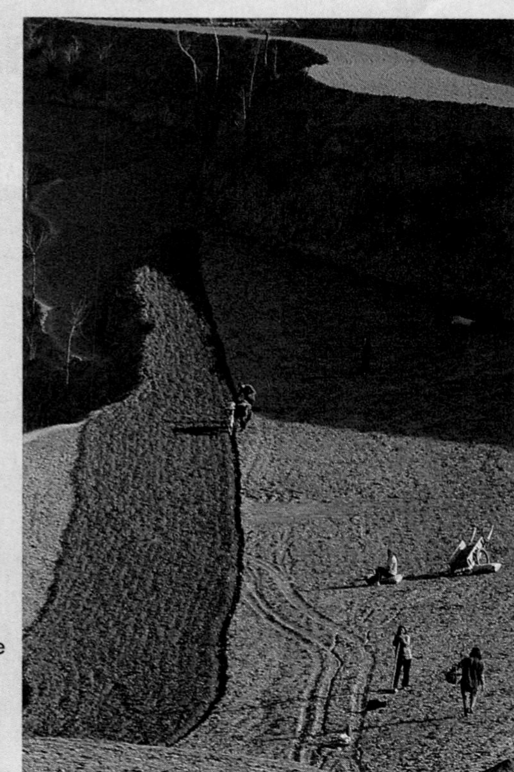

Farming in Gansu province

FOR THE RECORD

Ma Huan (c. 1380–c. 1460) of China was the official interpreter for Cheng Ho, a diplomat and explorer who led several naval expeditions to many places in Asia and Africa during the fifteenth century. The expeditions were undertaken to expand China's influence as a trading power throughout the lands bordering the "Western Ocean"—all the bodies of water between China and the east coast of Africa. Ma Huan recorded his impressions of the people and places they encountered. The passage below describes an empire in central Vietnam called Champa.

Cheng Ho's ship

"You come to the city where the king resides; its foreign name is Chan city. The city has a city-wall of stone, with openings at four gates, which men are ordered to guard. The king of the country is a So-li man, [and] a firm believer in the Buddhist religion. . . . The house in which the king resides is tall and large. It has a roof of small oblong tiles on it. The four surrounding walls are ornately constructed of bricks and mortar, [and look] very neat. The doors are made of hard wood, decorated with engraved figures of wild beasts and domestic animals.

The houses in which the people live have a covering made of thatch; the height of the eaves cannot exceed three *ch'ih* (36.7 inches), [people] go in and out with bent bodies and lowered heads; a greater height is an offense.

As to the colour of their clothing; white clothes are forbidden, and only the king can wear them; for the populace, black, yellow, and purple coloured [clothes] are all allowed to be worn; to wear white clothing is a capital offense."

Translated by Feng Ch'eng-Chün

Analyzing Primary Sources

1. What is the first thing Ma Huan observes about the city of Chan?
2. How does Ma Huan note the differences between royalty and common people in Champa?

CHAPTER 41

Introducing East and Southeast Asia

GEOGRAPHY DICTIONARY

paddy
aquaculture
newly industrialized country (NIC)
pictogram
Buddhism
Taoism
Confucianism
Shinto

East and Southeast Asia is home to some 2 billion people, roughly one-third of Earth's human population. With its forbidding mountains, blazing deserts, fertile river valleys and deltas, and volcanic islands, this vast region is one of great geographic as well as cultural diversity. Here, traditional farming villages, huge modern cities, and some of the planet's last large wilderness areas all compete for space.

Korean temple wall painting

The diverse resources and peoples of East and Southeast Asia are connected to the world community. Each day, automobiles, televisions, cameras, computers, bicycles, clothing, and canned foods from Japan, South Korea, Indonesia, China, and Taiwan (TIE-WAHN) pour into the United States, Europe, and other countries around the globe. At the same time, however, some of the world's poorest nations struggle to survive and develop their economies in this region.

Celebrating the Chinese New Year in **Singapore**

MAPPING THE REGION

EAST AND SOUTHEAST ASIA

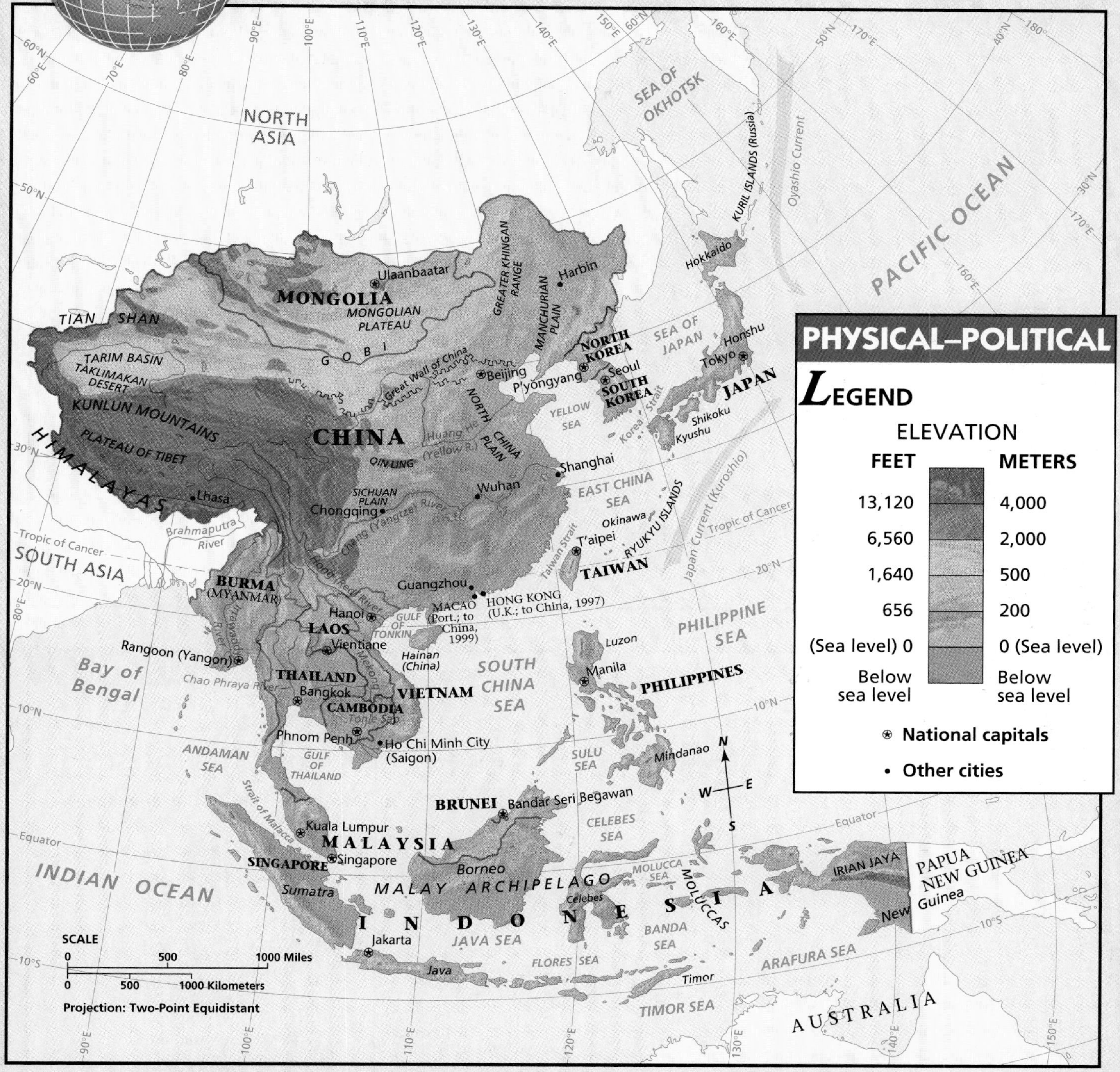

NORTH ASIA
SEA OF OKHOTSK
KURIL ISLANDS (Russia)
Oyashio Current
PACIFIC OCEAN
Hokkaido
Ulaanbaatar
MONGOLIA
MONGOLIAN PLATEAU
GREATER KHINGAN RANGE
MANCHURIAN PLAIN
Harbin
TIAN SHAN
TARIM BASIN
TAKLIMAKAN DESERT
G O B I
Great Wall of China
Beijing
P'yongyang
NORTH KOREA
Seoul
SOUTH KOREA
SEA OF JAPAN
Honshu
Tokyo
JAPAN
KUNLUN MOUNTAINS
PLATEAU OF TIBET
HIMALAYAS
CHINA
Huang He (Yellow R.)
QIN LING
NORTH CHINA PLAIN
YELLOW SEA
Korea Strait
Shikoku
Kyushu
Shanghai
SICHUAN PLAIN
Chongqing
Wuhan
Chang (Yangtze) River
EAST CHINA SEA
Okinawa
RYUKYU ISLANDS
Japan Current (Kuroshio)
Lhasa
Brahmaputra River
Tropic of Cancer
SOUTH ASIA
Taiwan Strait
T'aipei
TAIWAN
BURMA (MYANMAR)
Irrawaddy River
Hong (Red) River
Guangzhou
MACAO (Port.; to China, 1999)
HONG KONG (U.K.; to China, 1997)
Hanoi
GULF OF TONKIN
LAOS
Vientiane
Mekong
Hainan (China)
Rangoon (Yangon)
Bay of Bengal
Chao Phraya River
THAILAND
Bangkok
CAMBODIA
Tonle Sap
VIETNAM
SOUTH CHINA SEA
PHILIPPINE SEA
Luzon
Manila
PHILIPPINES
Phnom Penh
Ho Chi Minh City (Saigon)
ANDAMAN SEA
GULF OF THAILAND
SULU SEA
Mindanao
Strait of Malacca
BRUNEI
Bandar Seri Begawan
CELEBES SEA
Kuala Lumpur
MALAYSIA
SINGAPORE
Singapore
Equator
INDIAN OCEAN
Borneo
Sumatra
MALAY ARCHIPELAGO
Celebes
MOLUCCA SEA
MOLUCCAS
INDONESIA
IRIAN JAYA
PAPUA NEW GUINEA
New Guinea
Jakarta
JAVA SEA
BANDA SEA
Java
FLORES SEA
Timor
ARAFURA SEA
TIMOR SEA
AUSTRALIA

SCALE
0 500 1000 Miles
0 500 1000 Kilometers
Projection: Two-Point Equidistant

N W E S

PHYSICAL–POLITICAL

LEGEND

ELEVATION

FEET	METERS
13,120	4,000
6,560	2,000
1,640	500
656	200
(Sea level) 0	0 (Sea level)
Below sea level	Below sea level

⊛ National capitals
• Other cities

MAPPING THE REGION

EAST AND SOUTHEAST ASIA

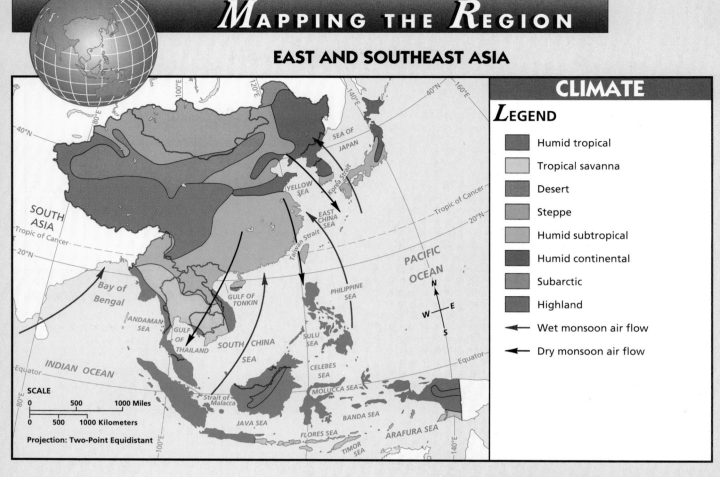

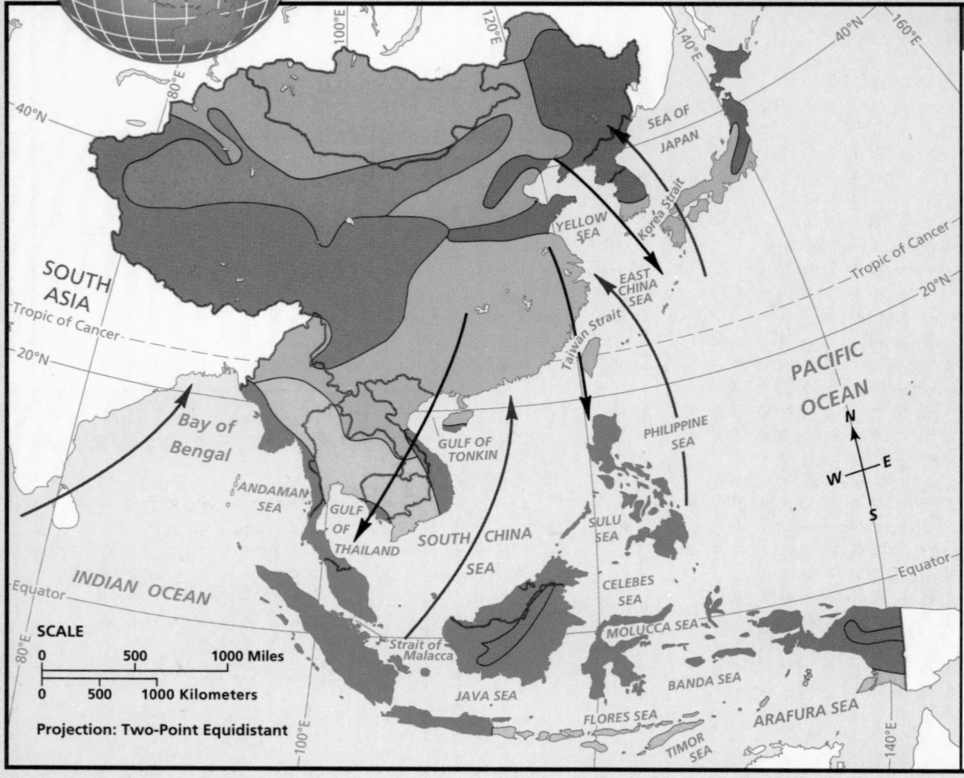

CLIMATE

LEGEND

- Humid tropical
- Tropical savanna
- Desert
- Steppe
- Humid subtropical
- Humid continental
- Subarctic
- Highland
- Wet monsoon air flow
- Dry monsoon air flow

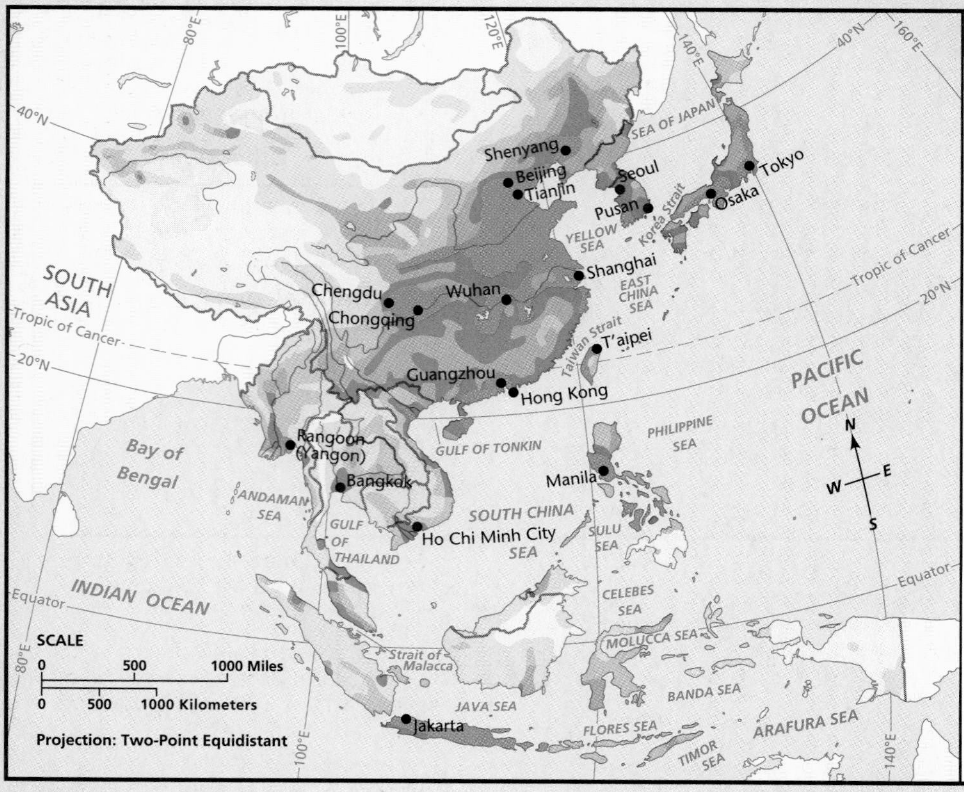

SOUTH ASIA, Bay of Bengal, ANDAMAN SEA, GULF OF THAILAND, INDIAN OCEAN, Strait of Malacca, JAVA SEA, SOUTH CHINA SEA, GULF OF TONKIN, YELLOW SEA, EAST CHINA SEA, SEA OF JAPAN, Korea Strait, Taiwan Strait, PHILIPPINE SEA, PACIFIC OCEAN, SULU SEA, CELEBES SEA, MOLUCCA SEA, BANDA SEA, FLORES SEA, TIMOR SEA, ARAFURA SEA

Tropic of Cancer, Equator, 40°N, 20°N, 80°E, 100°E, 120°E, 140°E, 160°E

SCALE: 0–500–1000 Miles; 0–500–1000 Kilometers

Projection: Two-Point Equidistant

POPULATION

LEGEND

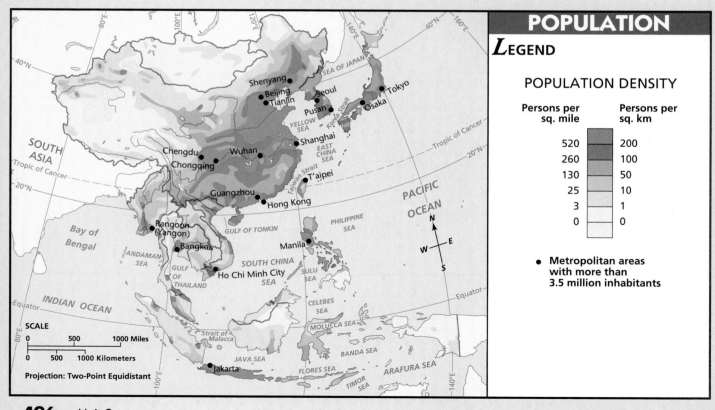

POPULATION DENSITY

Persons per sq. mile	Persons per sq. km
520	200
260	100
130	50
25	10
3	1
0	0

● Metropolitan areas with more than 3.5 million inhabitants

Shenyang, Beijing, Tianjin, Seoul, Pusan, Tokyo, Osaka, Shanghai, Chengdu, Chongqing, Wuhan, T'aipei, Guangzhou, Hong Kong, Rangoon (Yangon), Bangkok, Ho Chi Minh City, Manila, Jakarta

SOUTH ASIA, Bay of Bengal, ANDAMAN SEA, GULF OF THAILAND, INDIAN OCEAN, Strait of Malacca, JAVA SEA, SOUTH CHINA SEA, GULF OF TONKIN, YELLOW SEA, EAST CHINA SEA, SEA OF JAPAN, Korea Strait, Taiwan Strait, PHILIPPINE SEA, PACIFIC OCEAN, SULU SEA, CELEBES SEA, MOLUCCA SEA, BANDA SEA, FLORES SEA, TIMOR SEA, ARAFURA SEA

Tropic of Cancer, Equator, 40°N, 20°N, 80°E, 100°E, 120°E, 140°E, 160°E

SCALE: 0–500–1000 Miles; 0–500–1000 Kilometers

Projection: Two-Point Equidistant

MAPPING THE REGION

EAST AND SOUTHEAST ASIA

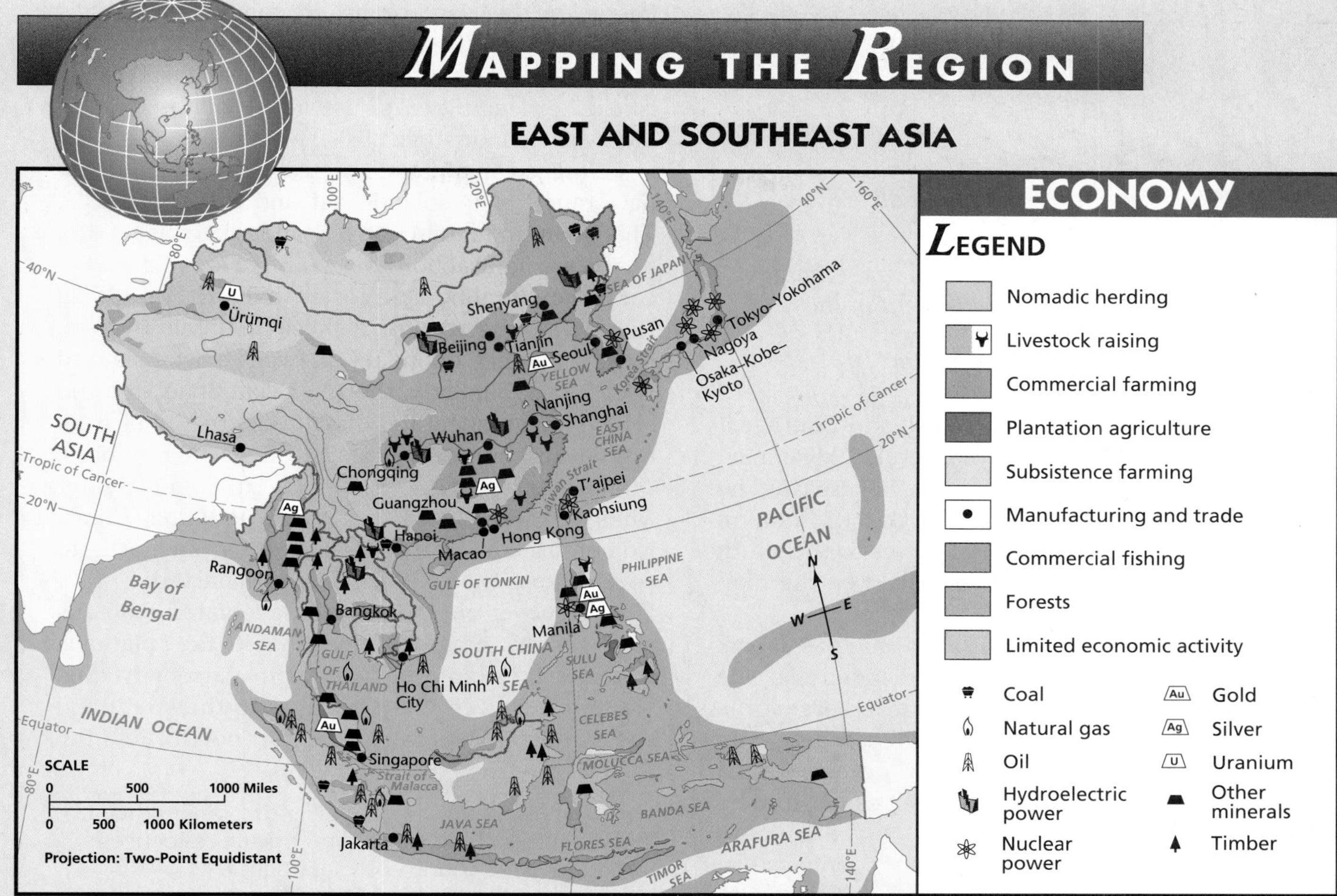

ECONOMY
LEGEND
Nomadic herding
Livestock raising
Commercial farming
Plantation agriculture
Subsistence farming
Manufacturing and trade
Commercial fishing
Forests
Limited economic activity
Coal
Natural gas
Oil
Hydroelectric power
Nuclear power
Gold
Silver
Uranium
Other minerals
Timber
Au
Ag
U
Ürümqi
Lhasa
Beijing
Tianjin
Shenyang
Seoul
Pusan
Tokyo-Yokohama
Nagoya
Osaka-Kobe-Kyoto
Nanjing
Shanghai
Wuhan
Chongqing
Guangzhou
T'aipei
Kaohsiung
Hong Kong
Macao
Hanoi
Rangoon
Bangkok
Manila
Ho Chi Minh City
Singapore
Jakarta
SOUTH ASIA
Bay of Bengal
ANDAMAN SEA
GULF OF THAILAND
GULF OF TONKIN
SOUTH CHINA SEA
PHILIPPINE SEA
PACIFIC OCEAN
SULU SEA
CELEBES SEA
MOLUCCA SEA
BANDA SEA
JAVA SEA
FLORES SEA
TIMOR SEA
ARAFURA SEA
INDIAN OCEAN
SEA OF JAPAN
YELLOW SEA
EAST CHINA SEA
Korea Strait
Taiwan Strait
Strait of Malacca
Tropic of Cancer
Equator
40°N
20°N
80°E
100°E
120°E
140°E
160°E
N
S
E
W
SCALE
0 500 1000 Miles
0 500 1000 Kilometers
Projection: Two-Point Equidistant

CLIMATE

◆ Dry desert and variable highland climates dominate large regions of northern and western China.

◆ The coasts of Southeast Asia and much of Indonesia have a humid-tropical climate, with warm, rainy weather all year.

◆ A humid-continental climate, with cold winters and warm, moist summers, dominates northern Japan. Southern Japan, with warm, moist summers and mild winters, has a humid-subtropical climate.

Compare this map to the physical–political map on page 495.

? What blocks the southern flow of moisture to the desert areas of western and northern China?

? Which two large islands in the Malay Archipelago have highland climates?

? Which countries have only a humid-tropical climate?

POPULATION

◆ The huge population of East and Southeast Asia is unevenly distributed, with large sparsely populated areas as well as densely populated regions and islands.

◆ The most crowded part of China, the world's most populous country, is its eastern half.

◆ Many metropolitan areas with populations of more than 3.5 million each are located throughout East and Southeast Asia.

? Which Japanese metropolitan areas have populations of more than 3.5 million?

Compare this map to the climate map on page 496.

? Why do you think western areas of China are less populated than eastern regions of the country?

Compare this map to the physical–political map on page 495.

? Which Indonesian island is most densely populated?

ECONOMY

◆ Large deposits of coal and minerals such as tungsten, tin, and iron ore can be found in East and Southeast Asia.

◆ Tropical-forest lumber, especially valuable hardwoods, are major exports of Southeast Asia.

◆ In the desert and high-plateau areas, herders raise sheep and other livestock.

? Name two major manufacturing and trade centers of the Malay Archipelago.

Compare this map to the climate map on page 496.

? Which climate types dominate the regions used for commercial farming in China?

? Why do you think rice farming is an important economic activity in many parts of East and Southeast Asia?

PHYSICAL GEOGRAPHY

FOCUS

- *What landforms and climates are found in East and Southeast Asia?*
- *How have these landforms and climates influenced human settlement in the region?*

LOCATION **Landforms** East and Southeast Asia stretches from Burma (also called Myanmar) in the southwest to Japan in the northeast. The region borders Russia on the north and India on the southwest. The Indian Ocean lies to the south and the Pacific Ocean to the east. (See the map on page 495.)

Mountainous volcanic islands, such as Japan, the Philippines (FIL-uh-peenz), and Indonesia, line East and Southeast Asia to the south and east, forming part of the Pacific Ring of Fire. Tectonic activity caused by the subduction of the oceanic plates under the continental plates in this area has formed the world's largest concentration of active volcanoes. Eruptions from these volcanic mountains can be deadly, but they are also the source of rich soils. Paralleling the island chains are some of the world's deepest marine trenches. The differences in elevation between these trenches and the summits of the islands' volcanic peaks just a few hundred miles away can be more than 35,000 feet (10,668 m).

Rugged mountains, plateaus, and hills dominate much of mainland East and Southeast Asia. The great mountain ranges and plateaus of the region are natural borders and often serve as political and cultural boundaries as well. The forbidding Himalayas, the world's highest mountain system, form a distinct boundary between East Asia and South Asia. Similarly, the high Tian Shan (tee-ahn SHAN) range separates northwestern China from Russia and some of the Central Asian republics. The Kunlun (KOON-LOON), Qin Ling, and Greater Khingan (SHING-AHN) ranges all lie within China. Numerous smaller ranges fan out from north to south across the Southeast Asian mainland.

Between the mountains lie isolated plateaus and basins. Few people inhabit the plateaus because they are very dry with extremely cold winters. Between the Himalayas to the south and the Kunlun Mountains to the north lies the

The Japanese mold the natural landscape into beautiful gardens in public places and private residences throughout the country. This garden is located on the island of **Shikoku**.

Plateau of Tibet, the world's highest plateau. Its elevation averages 16,000 feet (4,877 m) above sea level. In contrast, part of the arid Tarim Basin located between the Kunlun and the Tian Shan ranges drops below sea level. To the east of the Tarim Basin is the Gobi, the desert that makes up part of the plateau of Mongolia.

The region's high mountains are the source of all of East and Southeast Asia's major rivers, including two of the world's longest, the Huang He (Yellow River) in northern China and the Chang (Yangtze) in central China. The region's river valleys and deltas are the most densely populated places on Earth. When the rivers overflow their banks during floods, rich alluvial soil is spread over the surrounding land. Some of this fertile land has been under cultivation for more than 5,000 years and continues to feed the region's huge population. The river waters also are important for transportation and hydroelectricity.

The region's limited lowlands hold the other major centers of population. People crowd onto the narrow coastal plains and the interior areas of the North China Plain, the Manchurian Plain, and the Sichuan Basin.

INTERACTION **Climate** Climate factors play a part in the high population density of the river valleys and plains of East and Southeast Asia. The climates outside these lowlands, especially those of the desert areas of the interior, cannot support large human populations. For example, the Gobi of Mongolia and northern China and the Taklimakan Desert of northwestern China have hot summers and bitterly cold winters. The rain-shadow effect of the surrounding mountains prevents moist ocean air from bringing rain to these desolate areas. Likewise, areas in the highland climates of interior East Asia and in the severe subarctic climate along the borders of Mongolia, China, and Russia are too cold for much human settlement.

Other climates of the region include a steppe climate in areas of China and a humid-continental climate in parts of northeastern China, North Korea, and northern Japan. The cold, snowy winters and hot summers of the humid-continental regions limit agriculture. Southern Japan, South Korea, Taiwan, and southern China have a humid-subtropical climate, with warm, moist summers and mild winters. This climate region lies in the path of severe tropical storms, or typhoons. These storms form in the tropical North Pacific Ocean and flow to the northwest, following the warm waters of the Japan Current (Kuroshio). They usually strike the Philippines, Japan, South Korea, southern China, or Taiwan.

The western islands of Indonesia and the coasts of Southeast Asia have a humid-tropical climate. Prior to recent deforestation, these areas were completely covered by dense tropical rain forests. Temperatures are warm, and rainfall is heavy all year. Some of the eastern islands of Indonesia and the interior mainland of Southeast Asia have a tropical-savanna climate with a distinct dry season.

Rainfall Extremes of rainfall characterize much of East and Southeast Asia. In some coastal and island locations of Southeast Asia, yearly rainfall totals average more than 80 inches (203 cm). Around 250 inches (635 cm) of rain fall in some mountain areas of Irian Jaya in Indonesia. In contrast, in the far continental interior, such as in the Taklimakan and Gobi deserts, annual rainfall totals less than 10 inches (25 cm). Interior cities, such as Beijing (BAY-JING), China, receive from 10 to 20 inches (25 to 51 cm) of rain per year.

Much of the rainfall in East and Southeast Asia is seasonal, due to the monsoon system. The winter months are dry, as winds blow from the north and west off the Asian continent. The summer months are humid and rainy, with winds bringing in moisture from the tropical waters of the Indian and Pacific oceans.

Unlike East Asia's mainland areas, island nations of the region, such as Japan and the Philippines, receive rain even during the winter. Dry winds flowing out of Asia must cross over the Sea of Japan or the South China Sea. Here they pick up moisture before reaching the islands.

SECTION REVIEW

1. Name five mountain ranges found in East and Southeast Asia. Between which ranges is the Tarim Basin located?
2. List five climate types found in East and Southeast Asia.
3. **Critical Thinking** **Explain** how the mountains of East and Southeast Asia both limit and support human settlement in the region.

ECONOMIC GEOGRAPHY

FOCUS

- *What are the major resources of East and Southeast Asia? How have these resources been used?*
- *What effects has rapid industrialization had on the region?*

INTERACTION **Agriculture** East and Southeast Asia's land area is huge. Because of terrain and climate factors, however, only a small percentage of the land is suitable for farming. Yet most of the people of East and Southeast Asia are involved in agriculture. This combination of many farmers and relatively little arable land means that most of the region's farms are tiny. For example, in Japan the average farm size is about two and one-half acres (1 ha). In contrast, the average size of a farm in the U.S. state of Iowa is 328 acres (133 ha).

The type of farming practiced varies with the climate. In humid-tropical and humid-subtropical climates, rice is the major food crop. In general, rice is grown in water-covered fields known as **paddies** (also spelled *padis*). These flooded fields produce most of the food consumed by the region's large population.

In tropical climates, commercial planters produce export crops such as rubber, tea, coffee, coconuts, sugarcane, and various spices. In the cooler and drier humid-continental and steppe climates of China, wheat is the main food crop. In the desert and high-plateau areas, herders keep flocks of sheep or other animals.

Harvesters have left rice to dry along this road in southern **Vietnam**. The wet, warm climates of East and Southeast Asia are ideally suited to growing rice.

Fishing Farming in seas and ponds, called **aquaculture**, is an important local activity throughout much of the region. Seafood such as shrimp and oysters are commercially farmed in protected bays and river mouths. For a local food supply, many nations in East and Southeast Asia depend upon freshwater fish from lakes, rivers, and flooded rice paddies.

Surrounded partially or entirely by water, several nations in the region, especially Japan, Taiwan, and South Korea, have large fishing fleets. In supplying the great demand for seafood, these fleets have exhausted local waters and now fish across the entire Pacific Ocean. Although vast, even the Pacific Ocean has limited resources. Parts of its fish population are now threatened.

Forests The forests of the region are generally divided into two types, middle-latitude and tropical. The middle-latitude forests lie mainly in Japan, China, and North and South Korea. The tropical rain forests of Thailand (TIE-land), Indonesia, Burma, and the Philippines once stretched from border to border in these countries.

Logging of these forests has increased at an amazingly rapid rate because of the demand in Japan and other countries for tropical-forest lumber, especially valuable hardwoods. In Thailand, for example, only one-fifth of the country's original forests remain. If this rate and the destructive methods of removal continue, most of the region's tropical-hardwood forests will not survive beyond the next few decades. In addition, deforestation leads to more flooding and soil erosion. It also threatens the habitats of the region's endangered animals, such as the panda, Asian rhino, and orangutan.

Mineral and Energy Resources Some of East and Southeast Asia's most abundant mineral resources are tin, tungsten, coal, and iron ore. More

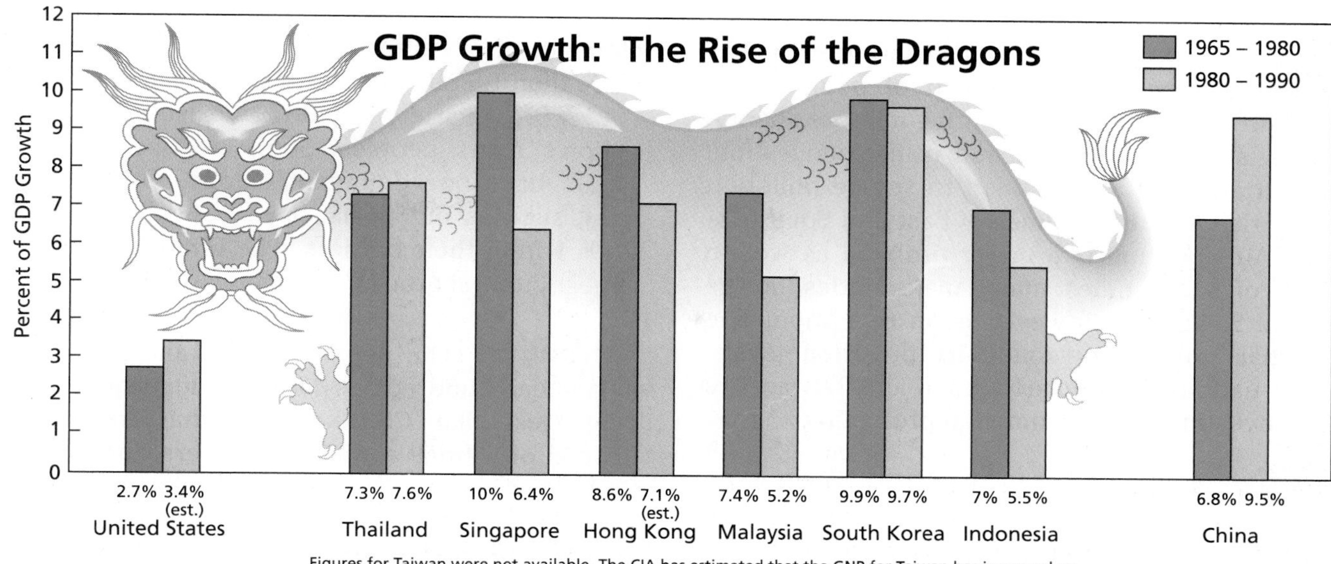
GDP Growth: The Rise of the Dragons
1965 – 1980
1980 – 1990
Percent of GDP Growth
0
1
2
3
4
5
6
7
8
9
10
11
12
2.7% 3.4% (est.)
United States
7.3% 7.6%
Thailand
10% 6.4%
Singapore
8.6% 7.1% (est.)
Hong Kong
7.4% 5.2%
Malaysia
9.9% 9.7%
South Korea
7% 5.5%
Indonesia
6.8% 9.5%
China

Figures for Taiwan were not available. The CIA has estimated that the GNP for Taiwan has increased an average of nine percent annually over the past three decades. For a discussion of GNP and GDP, see page 80.

Source: *World Development Report 1992: Development and the Environment,* The World Bank

SKILL STUDY These bar graphs show how the economic growth of the "Four Dragons" and "Little Dragons" of Asia outpaced that of the United States in the 25 years after 1965. Although the United States still had the largest economy in the world, the economies of these East and Southeast Asian countries and China grew much faster. Which two countries experienced the fastest rate of growth during the 1980s?

than half of the world's supply of tin is found in the region, mainly in Malaysia, Thailand, and Indonesia. The region also holds more than one-third of the world's supply of tungsten, which is used in the production of electronics and lighting materials.

Parts of the region hold large oil deposits. Indonesia, Malaysia, and Brunei (broo-NIE) are the major producers and exporters of this resource. Although there are natural-gas and oil deposits in China, Burma, and Thailand, none of these countries appears to have enough to support major exporting. China has developed several medium-size petroleum fields and may become a more important oil exporter as more reserves are discovered and developed.

Melting snow from the region's high mountains and large amounts of rainfall provide dependable sources of hydroelectric power. Japan has effectively developed its hydroelectric resources. China and other countries in the region have the potential to do the same.

Industrialization Japan is the region's major economic power and the world's second leading industrial nation after the United States. It is also a leader in foreign investment and international banking, and is a major producer and exporter of automobiles and high-tech products such as electronics. Although Japan's exports are far greater than its imports, the nation is a major importer of raw materials, such as coal and oil.

Rapid industrialization now characterizes several other countries of East and Southeast Asia. These nations are known as **newly industrialized countries (NICs)**. Collectively called the "Four Dragons" or "Four Tigers," the NICs of Taiwan, Singapore, Hong Kong, and South Korea have experienced explosive economic growth and are now very competitive, even against Japan. Ironically, much of their growth has been fueled by Japanese investors and industries seeking lower labor and land costs.

As industrialization and trade expand, other NICs have appeared. In recent years, Malaysia, Thailand, and Indonesia, called the "Little Dragons" or "Little Tigers," have shown impressive economic growth. Some nations of East and Southeast Asia, however, remain isolated and poor. Vietnam, Laos, Cambodia, Burma, and Mongolia all have struggling economies.

The region's rapid industrial growth has caused a large-scale movement of people from rural areas to the cities, resulting in more shantytowns, sewage problems, and traffic congestion. Industrialization also has led to severe pollution of the rivers and coastlines of East and Southeast Asia. Air pollution is a major problem as well; a cloak of smog hides many major cities in the region. Singapore, however, is an exception. The city has greatly controlled its pollution levels. Other nations, particularly Japan and Taiwan, are now investing in environmental protection.

2 SECTION REVIEW

1. What major crops are raised in East and Southeast Asia? What mineral and energy resources are important to the region?
2. What challenges face the region as a result of its industrial growth?
3. **Critical Thinking** **Compare** ways in which people of East and Southeast Asia have used fish resources to the ways they have used forest resources.

HUMAN GEOGRAPHY

FOCUS

- *In what ways is the population of East and Southeast Asia diverse?*
- *How have foreign countries played a role in the region's history? What different political systems are found in the region?*

Population and Culture Roughly one-third of the world's people live in East and Southeast Asia. The population is not distributed evenly among the region's countries, however. China has more than 1 billion people, while tiny Brunei has fewer than a million.

Population growth in much of the region continues at a rapid rate. Birthrates have increased and death rates have decreased because of improved medical care. Some nations now have programs to control their population growth, and Japan and Singapore have extremely low birthrates. Even China's population growth rate has slowed somewhat, though some officials estimate the country's population may reach more than 1.5 billion in the next 30 years.

A great variety of cultures exists among the region's many peoples. Some nations, such as Japan, have one dominant culture. Other countries, such as Indonesia, have a diversity of cultures within their borders. The Chinese form the largest cultural group.

Language The people of East and Southeast Asia speak hundreds of different languages. Some languages, like Chinese, are character-based. Instead of a small alphabet of letters, the written Chinese language has nearly 50,000 characters. Many scholars believe some of the characters are symbols or **pictograms**, simple pictures of the objects and ideas they represent. For example, many of the Chinese characters, such as the characters for mountain and tree, reflect the physi-

山
Mountain
木
Tree

Scholars have studied how the Chinese written language reflects the country's geography. Here, you can see that mountain peaks are suggested in the character for *mountain*. Similarly, the character for *tree* resembles the trunk and bare limbs of a tree.

cal geography of China. (See the illustration on page 502.) Characters can also be combined. Two tree characters placed side by side become the character for *forest.* Three tree characters grouped together may mean "thick forest," "many trees," or "dark." The characters and their meanings have gradually changed over time. Tracing their origins and development reveal part of the cultural history of China and its people.

Within a single country in East and Southeast Asia, dozens of languages may be spoken. As a result, some countries have adopted one official language to unify the people, as in China and Indonesia. Even in these countries, however, many regional dialects exist. In contrast, Japan and Korea each are relatively uniform in terms of languages. The Japanese language dominates in Japan, and Korean is the major language used in North and South Korea. South Korea even has a public holiday to celebrate its written alphabet, called *han'gul*, which was developed in the fifteenth century.

English and French are still used in many East and Southeast Asian countries with colonial histories. English also has become the business language of several nations in the region.

Religion Most of the world's major religious systems are represented in East and Southeast Asia. Islam is the main religion in Indonesia, Malaysia, and Brunei. In fact, Indonesia has more Muslims than any other country in the world. Christianity is practiced primarily in the areas of the region, such as the Philippines, that were once controlled by European countries. South Korea also has a fairly large Christian population.

Buddhism (BOO-diz-uhm) is the major religion in Thailand, Burma, Tibet (in China), Laos, Cambodia, Vietnam, and Mongolia. Buddhism was founded in India around 500 B.C. by a teacher called the Buddha. *Buddha* means "Enlightened One," and Buddhists seek enlightenment through meditation (deep thought). Buddhism is divided into a few major branches, which are practiced in different regions. The two main branches are Theravada (ther-uh-VAHD-uh) Buddhism and Mahayana (mah-huh-YAHN-uh) Buddhism. Theravada Buddhism, which is practiced mainly in Southeast Asia, is the earlier branch. It believes that humans must look only to themselves for salvation. In contrast, Mahayana Buddhism, practiced in China, Korea, and Japan, holds that humans may also seek the aid of spiritual beings. There are also many local varieties of Buddhism.

China's religions are a mixture of Buddhism, Taoism, and Confucianism. **Taoism** (DOW-iz-uhm) was founded in China in the sixth century B.C. by a man named Lao-Tzu. He taught that there is a natural order to the universe, called the *Tao*, or "way." The basic idea of Taoism is to live a simple life close to nature in order to be in perfect harmony with the *Tao*. **Confucianism**, which is based on the teachings of Confucius, a Chinese philosopher who lived from 551 to 479 B.C., spread throughout China and later to Korea and Japan. It is more a code of ethics than an organized religion. The Confucian code centers around family, social relationships, and duty.

Unlike Judaism, Christianity, and Islam, Buddhism, Taoism, and Confucianism do not claim to have been revealed by a god. Instead, they stem from the teachings of individual

This Japanese bride prepares for her Shinto wedding. The people in East and Southeast Asia follow a variety of religions and philosophies, including Shinto, Buddhism, Taoism, and Confucianism.

human beings. Meanwhile **Shinto** (SHIN-toe), a religion practiced in Japan, does not claim to have any founder at all. Shintoists believe that gods, called *kami*, inhabit natural objects such as rivers, trees, rocks, and mountains. The worship of these gods involves a loosely structured set of rituals focused around community activities at local shrines. An important symbol of Shinto is the *torii*, a gateway marking the entrance to these shrines. The daily lives of most Japanese people mix Shinto with Buddhism. For example, many Japanese weddings are Shintoist, whereas funerals might be Buddhist.

MOVEMENT **Colonial History** In the eighteenth and nineteenth centuries, many parts of East and Southeast Asia came under the control of foreign nations. Great Britain controlled the area that now contains the countries of Burma, Malaysia, Singapore, and Brunei. It also gained the small colony of Hong Kong on the southeast coast of China. The Netherlands oversaw the thousands of islands of the Dutch East Indies, known today as Indonesia. France controlled an area known as French Indochina, which later became Cambodia, Laos, and Vietnam. Portugal acquired small possessions on the island of Timor, in Southeast Asia, and Macao (muh-KAU), on the coast of China. And for over 45 years, the United States controlled the Philippines, which it had won from Spain in 1898 as a result of the Spanish-American War.

China also was a target of foreign control. Europeans dominated its coastal ports in the nineteenth century, and the Japanese invaded China during the 1930s and 1940s. Japan had also colonized Taiwan in 1895 and Korea in 1910. During World War II, Japan expanded its empire to control much of East and Southeast Asia.

Post–World War II After World War II, the countries of East and Southeast Asia sought independence and an end to colonialism. Maps of the region were redrawn to include the new nations of the Philippines, Indonesia, Burma, Vietnam, Cambodia, and Laos. In China in 1949, after a bitter civil war, the Communists established the People's Republic of China. The Nationalists, who opposed the Communists, escaped complete defeat by forming a government-in-exile on the island of Taiwan. The country's official name is the Republic of China.

The region continued to be a site of international interest and conflict into the second half of the twentieth century. In the early 1950s, the Korean War pitted South Korean, U.S., and United Nations forces against North Korea and China. After more than three years of fighting, the two sides agreed to a truce and Korea remained divided into two countries.

In the late 1960s and early 1970s, the Vietnam War again focused world attention on the region. This war, in which South Vietnamese and U.S. troops battled North Vietnamese forces, ended in the unification of North and South Vietnam under a Communist government.

Political Geography As a result of the countries' varied histories and experiences, political systems vary widely among the nations of East and Southeast Asia. While Japan has had a democracy for decades, other nations, such as the Philippines and South Korea, have only recently gained democracy. China and North Korea are Communist countries, while others, such as Burma, remain under the control of strict military dictatorships. Hong Kong and Macao are colonies of the United Kingdom and Portugal, respectively, but both will become part of China in the late 1990s.

Today, tensions among and within the countries of the region still dominate the political landscape. Indeed, thousands of refugees have had to leave their homelands as a result of conflicts in these countries. Border disputes and claims to various islands continue to affect relations between China and Vietnam, and between the Philippines and Malaysia. Vietnam has occupied parts of Cambodia and Laos.

3 SECTION REVIEW

1. What religions are practiced in East and Southeast Asia? In what countries is each one practiced by a large part of the population?
2. What European countries have been involved in East and Southeast Asia? How has the United States played a role in the region?
3. **Critical Thinking** **Assess** the extent to which East and Southeast Asia was influenced culturally by colonialism.

CHAPTER 41 CHECK

CHOI MENG

Portuguese **Macao**, to be turned over to China in 1999

Reviewing the Main Ideas

1. Mountains, plateaus, and hills dominate much of East and Southeast Asia. A wide range of climates exists in the region, from desert and subarctic to humid tropical. Landforms and climates strongly affect patterns of human settlement in East and Southeast Asia.
2. Most of the people of East and Southeast Asia are involved in agriculture. The region also has important fishing, forest, mineral, and energy resources. Rapid industrialization has recently occurred in many of the region's countries.
3. Roughly one-third of the world's people live in East and Southeast Asia. Among these people is a diversity of cultures, languages, and religions. Many of the region's countries have colonial histories, although most became independent after World War II. Political systems now vary widely.

Building a Vocabulary

1. What crop is usually grown in paddies? Do you think the crop could be grown in a very dry climate? Why or why not?
2. What is aquaculture?
3. Which nations in East and Southeast Asia are considered to be newly industrialized countries?
4. State the central emphasis of each of the following religions: Buddhism, Taoism, Confucianism, and Shinto.

Recalling and Reviewing

1. What areas of East and Southeast Asia receive the most rainfall? How are water resources important throughout the entire region?
2. Has recent economic success marked every country in East and Southeast Asia? Explain.
3. What political tensions have the nations of East and Southeast Asia experienced since World War II?

Thinking Critically

1. Use two examples from the chapter to show how economic goals sometimes seem to be at odds with the environment. Do you think economic success must always come at the expense of the environment? Explain.
2. As you read, Confucianism is more a code of ethics than an organized religion. How do you think a "code of ethics" differs from a "religion"?

Learning About Your Local Geography

Individual Project

Is fishing important to many people in your community, either as a source of income or as recreation? What rivers, lakes, bays, or other bodies of water in your area support fishing? Where do the fish in your local grocery store or market originate? Draw illustrations of three different types of fish that can be caught near or bought in your community. Under each illustration, write the name of the type of fish and where the fish can be caught. You may want to include your illustrations in your individual portfolio.

Using the Five Themes of Geography

Write a paragraph discussing how the theme of **movement** applies to East and Southeast Asia. Consider the movement of resources as well as the movement of people.

CHAPTER 42

China, Taiwan, and Mongolia

> "As we climb into the Trans-Himalayan range just before Lhasa we enter a spectacular landscape of gorge and cliff and turbulent river. . . . When I next look out we are already in the broad valley of the Lhasa River—with fields of wheat and barley, tall trees, buildings of cement, and, from far away, the dominating vertical plane of the Potala palace, monolithic [huge] and of immense grandeur, white and pale pink and red and gold."
>
> *Vikram Seth*

GEOGRAPHY DICTIONARY

dynasty

puppet government

commune

double cropping

Special Economic Zone (SEZ)

martial law

With more than 1 billion people, China is the world's most populous nation. In fact, one out of every five people on Earth lives in China. It is also the world's third largest country after Russia and Canada. Some of the world's highest mountains, driest deserts, and longest rivers make China a land of great contrasts.

Ming vase from **China**

China has one of the world's oldest cultures. For centuries, Chinese art, architecture, science, and philosophy flourished when most other nations had not yet formed. Today, China is a land of rapid change. Although it remains a Communist nation, the country is opening its doors to some free enterprise and foreign trade.

This chapter also examines the nearby bustling island nation of Taiwan and the isolated country of Mongolia to the north of China.

Central **Tibet**

CHINA, TAIWAN, AND MONGOLIA: PHYSICAL–POLITICAL

SCALE
0 250 500 Miles
0 250 500 Kilometers
Projection: Two-Point Equidistant

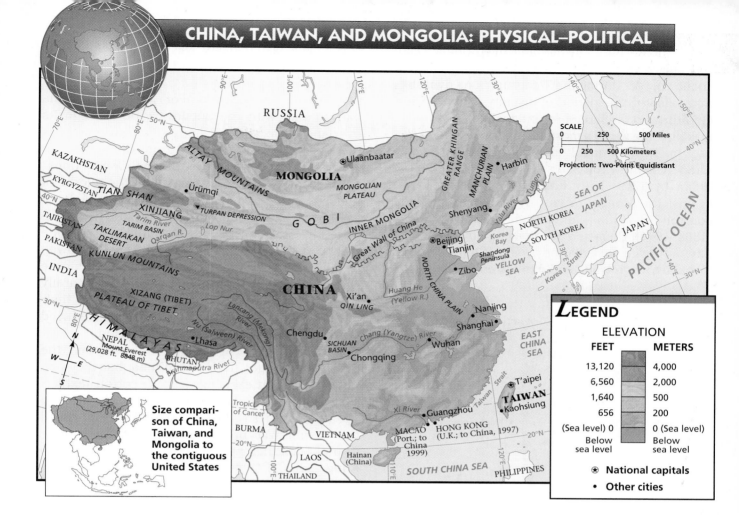

Size comparison of China, Taiwan, and Mongolia to the contiguous United States

LEGEND

ELEVATION

FEET	METERS
13,120	4,000
6,560	2,000
1,640	500
656	200
(Sea level) 0	0 (Sea level)
Below sea level	Below sea level

⊛ National capitals
• Other cities

HISTORICAL GEOGRAPHY

FOCUS

- *What were the various Chinese dynasties? How did China react to contact with foreign countries? What groups of people controlled China in the first half of the twentieth century?*
- *Who was Mao Zedong? How did he change China, and how has China changed since his death?*

MOVEMENT **Cultural Development** The Chinese culture has the longest continuous history of any on Earth. By 5000 B.C., the first rice farmers worked the land near the Chang River. One thousand years later, to the north along the Huang He, an organized society of people lived in small agricultural villages. These Mongoloid people occupied much of northern China for thousands of years. Culture traits such as the use of chopsticks and the Chinese style of writing developed here. The Mongoloid people also used irrigation agriculture along the Huang He.

During the third century B.C., China's first emperor conquered much of eastern China and ordered the building of the Great Wall as protection against invaders from the north. The Qin (CHIN) **dynasty** was one of the earliest of many Chinese dynasties to follow. A dynasty is a government ruled by a family whose power lasts for a long period of history. The term *Chinese* comes from the Qin dynasty.

During the Han dynasty, which ruled from 206 B.C. to A.D. 220, the military might of China grew stronger and Chinese culture expanded into southern China. Cities grew rapidly, and property rights were recognized. Chinese art and architecture flourished as well. Today, the Chinese call themselves Han after this great era.

After the decline of the Han dynasty due to invasion, corruption, and much internal fighting, numerous warlords divided and ruled China. Unity was not restored until some 400 years later, in about A.D. 626. During the T'ang and Sung dynasties that followed this period of disunity, China's cities, literature, art, architecture, and technology thrived, making it the world's most advanced nation.

In 1215, Mongol invaders led by Genghis Khan overthrew the Sung dynasty and ruled all of China by 1279. The Chinese eventually rebelled against the Mongol (Yuan) dynasty and regained control of China under the Ming dynasty (1368–1644). Chinese culture again advanced rapidly, although in isolation as the Chinese rulers forbade foreign trade and minimized outside influences. The last dynasty was the Manchu (Qing [CHING]), which lasted from 1644 to 1912.

Outside Influences For centuries, China had contact with the outside world through its overland trade routes. The best-known route was the Silk Road, which wound from China through Central Asia to Europe. (See "Global Connections" on pages 362–363.) It was not until the middle of the sixteenth century, however, that outside influences reached China by sea. The first to arrive were the Portuguese, who established a trade colony at Macao in 1557. They were soon followed by the British, French, and Germans. Chinese leaders, believing that their society was culturally superior, largely ignored these visitors and their ideas. Even so, European missionaries introduced Christianity into China during this period.

Wanting Chinese silk, tea, and porcelain, the Europeans sought trade with China, but they had little to offer in exchange except cotton textiles, silver, and opium. When the Chinese declined to open more ports to British trade, a conflict called the Opium War broke out in 1839. After three years of fighting, the British won the port of Hong Kong. Soon, France, Germany, Russia, and the United States claimed other Chinese port cities. Later, in 1895, the Japanese captured the island of Taiwan. The Chinese were unprepared for these attacks on their country by foreign interests.

During this period, some Chinese emigrated to other countries seeking jobs. Many went to the United States and Canada to build railroads and work in gold and silver mines. Others emigrated to European colonies in Southeast Asia to work

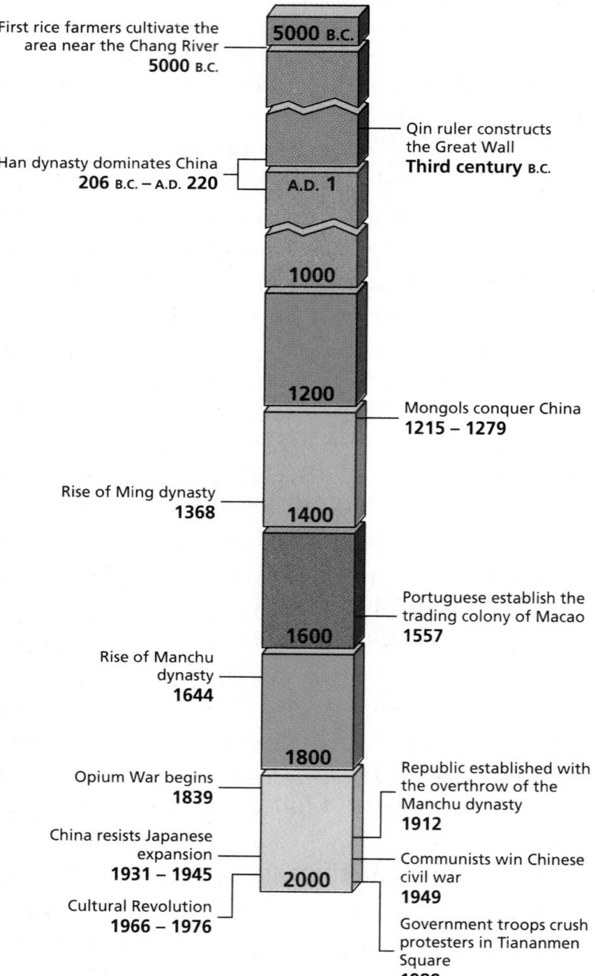
Historic China: A Time Line
First rice farmers cultivate the area near the Chang River 5000 B.C.
5000 B.C.
Qin ruler constructs the Great Wall Third century B.C.
Han dynasty dominates China 206 B.C. – A.D. 220
A.D. 1
1000
1200
Mongols conquer China 1215 – 1279
Rise of Ming dynasty 1368
1400
Portuguese establish the trading colony of Macao 1557
1600
Rise of Manchu dynasty 1644
1800
Opium War begins 1839
Republic established with the overthrow of the Manchu dynasty 1912
China resists Japanese expansion 1931 – 1945
Communists win Chinese civil war 1949
2000
Cultural Revolution 1966 – 1976
Government troops crush protesters in Tiananmen Square 1989

SKILL STUDY This time line reviews major events in China's rich history. The last Chinese dynasty was overthrown in 1912. Why is history important to the study of geography?

Archaeologists discovered armies of clay figures near **Xi'an, China**, in 1974. The life-size clay figures are located near the tomb of an early Chinese emperor and date from the 200s B.C.

on plantations or in mines there. Soon, distinct Chinese communities, called Chinatowns, sprang up in major cities around the world.

Political Turmoil In 1912, rebels under the leadership of Sun Yat-sen overthrew the last of the Chinese dynasties and formed the Republic of China. Sun Yat-sen tried to unify the country and make economic improvements.

After Sun Yat-sen's death in 1925, Chiang Kai-shek became the country's leader. Like Sun Yat-sen, Chiang Kai-shek was a member of the Nationalist party. To establish a strong central government, he set out to defeat the warlords who then controlled parts of China. Chiang Kai-shek faced opposition from another source as well, the Chinese Communist party. Formed in 1921, the party wanted to put more power into the hands of the working peasants. Chiang Kai-shek's Nationalist forces attempted to destroy the Communist opposition.

Sensing a weak and divided China, Japan took control of China's resource-rich northern region of Manchuria in the early 1930s. Here, Japan established a **puppet government**. A puppet government is one that is controlled by outside forces. Then in 1937, Japan invaded China's east coast and controlled much of eastern China until the end of World War II.

After Japan's defeat in 1945, a civil war broke out between the Chinese Nationalists and the Communists. In 1949, the Communists won control of mainland China and established the People's Republic of China (PRC) under Mao Zedong. Mao changed China to fit his own vision of communism, which became known as Maoism. The Nationalists retreated to the island of Taiwan and set up a government-in-exile called the Republic of China (ROC).

China Under Mao The Communists' first challenge was to feed the country's people. In the past, as the population increased, fields were divided into smaller and smaller plots to be worked by farmers and their families. The Communist leaders believed that larger, government-controlled plots could be worked more efficiently and would produce the food necessary to feed China's growing population. The government organized farmers into collectives and seized all private land. Many families were separated or forcibly relocated, their personal property taken and their individual freedoms lost.

FACTS IN BRIEF **China, Taiwan, and Mongolia**

COUNTRY POPULATION (1994)	LIFE EXPECTANCY (1994)	LITERACY RATE (1990)	PER CAPITA GDP (1993)
CHINA 1,190,431,106	67, male 69, female	78%	$2,200
MONGOLIA 2,429,762	64, male 69, female	89%	$1,200
TAIWAN 21,298,930	72, male 79, female	90% (1991)	$10,600
UNITED STATES 260,713,585	73, male 79, female	97% (1991)	$24,700

Sources: *The World Factbook 1994,* Central Intelligence Agency; *The World Almanac and Book of Facts 1995*

Find the population of China in this chart. Look at the population map on page 496. What part of China is the most densely populated?

Communism also changed the traditional family structure in China. In the past, the oldest male had been the dominant family member. Under communism, women had equal status and were sent to the fields to work along with men. The government also tried to slow population growth by making a rule that families could have only one child. Despite much opposition, this rule still holds today.

To improve the economy and develop small-scale industries nearer to natural resources in the country's interior, Mao launched the Great Leap Forward in 1958. Under this program, workers dug new mines and built small blast furnaces to process iron ore into steel.

At the same time, work groups of thousands of people built dams and dikes and labored to improve the transportation system. In an attempt to increase agricultural output, many people were forced to leave the cities to work in large cooperative groups called **communes**. Rather than look to the market needs of the people, the government set fixed agricultural and industrial goals.

The Great Leap Forward was a huge failure. It forced people to work much harder without increasing the personal rewards of their labor. It also delayed the economic development of China and caused great hardship and even starvation among much of the Chinese population. The program was an environmental disaster as well. Entire regions of forest were destroyed as trees were cut for charcoal to burn in the many blast furnaces. No attempts at reforestation were made, and massive soil erosion resulted.

Terraced vegetable gardens rise above this farming village near **Guangzhou**, in southeastern **China** (left). Despite greater economic development in recent years, most Chinese peasants, such as the one watering his fields (right), still must rely on traditional farming methods.

The Great Leap Forward was followed by the Cultural Revolution, which lasted from 1966 to 1976. The Cultural Revolution's goal was for everyone to follow the peasant way of life. Followers of Mao, especially youths known as the Red Guards, banded together to rid China of Mao's enemies and critics. Anyone with an education was suspect, especially intellectuals, scientists, and business leaders. Schools and universities were closed. In response to the government's policies, riots and street violence broke out across China. Millions suffered as they were taken from their jobs and homes and sent to communes to work in the fields. Millions more were imprisoned in harsh labor camps or executed.

A New China After Mao's death in 1976, a new government under the leadership of Deng Xiaoping (DUHNG SHAU-PING) came to power. Admitting that the Great Leap Forward had been an economic disaster and that the Cultural Revolution had been a mistake, Deng proposed new policies to modernize the country's agriculture, industry, and technology.

Today, many farmers grow and market their own crops and build their own private homes. They do not own the land but lease it from the government for long terms. Agricultural productivity has increased, and China is nearly self-sufficient in food. A greater variety of fruits, vegetables, and meat has improved the diet of many Chinese people.

Industry and cities have also changed. Although state-owned industry still employs many workers, private enterprise and self-employment are expanding. Instead of only large state-run factories and businesses, there are now more than 19 million small companies in rural and village areas. These industries, known as town village enterprises (TVEs), are the fastest growing sector of the economy.

Although China remains a poor country by world standards, in general, daily life has improved. For example, more consumer products are now available, including televisions, refrigerators, bicycles, cars, and clothing. The Communist government still has control over the Chinese people, however. While the government seems to be in favor of increasing economic freedoms, it does not want to make political reforms. This resistance to political reform became clear in 1989 when students and workers held pro-democracy demonstrations in Beijing, China's capital. Chinese army troops and tanks crushed the protests, killing and injuring hundreds of demonstrators in the city's Tiananmen Square.

SECTION REVIEW

1. Name, in order, China's major dynasties. How and when did the last dynasty end? What enabled Japan to take control of parts of China in the 1930s?
2. What goals did Mao Zedong have for China? What were the Great Leap Forward and the Cultural Revolution? How has China changed since Mao's death, and how has it remained the same?
3. **Critical Thinking** **Suggest** an explanation for why the Chinese were not prepared for attacks by foreign powers in the 1800s, even though China had earlier been the most advanced nation in the world.

CHINA'S REGIONS

FOCUS

- *What is the physical geography of each of China's regions? the economic geography?*
- *What are the major cities of southern and northern China?*

REGION To study a country of China's huge size and vast regional differences, geographers divide the country into four geographic regions. These four regions are southern China, northern China, northeastern China, and western China. (See the map on page 507.)

Southern China Southern China is bordered on the north by the Qin Ling range, on the west by the Plateau of Tibet, and on the east by the East China and South China seas. Vietnam, Laos, and Burma lie to the south. All of southern China has a humid-subtropical climate except for the island of Hainan, which has a humid-tropical climate.

Southern China is the nation's most productive region and contains more than one-fourth of its population. The most densely populated areas are the river basins along the southeastern coast, especially near the mouths of the large Chang and Xi rivers.

Despite continuous cultivation for more than 4,000 years, the soil of southern China remains fertile. The alluvial deposits left by flooding rivers and careful management renew the soils. Southern China is often called "China's Rice Bowl." Indeed, the production of rice in this region has helped make China the world's leading rice producer and a rice exporter.

The Chang Delta is one of China's major rice-growing regions. The humid-subtropical climate here provides a long growing season that allows **double cropping** on the same land. This practice raises two crops in one year. In the far south, three crops may be grown, usually two crops of rice and one of vegetables.

Rice is usually grown on small terraced paddies. In hilly terrain, each paddy is built at a different level, or terrace, from the next, increasing the amount of area available for cultivation. The paddy is plowed, fertilized, and flooded to make it ready for the seedlings. Traditionally, water buffalo have supplied the power for plowing, but small mechanical tillers are now replacing buffalo power.

Southern China's other important crops include tea and cotton. The mountainous region between the Xi and Chang rivers is the center of tea growing. Production here has helped make China the world's second largest producer of tea. Because most Chinese tea is used in China, however, the country ranks only third as a tea exporter, after India and Sri Lanka. Cotton is the main raw material for China's textile industry. The major cotton-growing area is the Chang River valley, where farmers often practice double cropping, alternating cotton with a food crop.

In addition to agriculture, much of China's industry and trade lies along the Chang River. Shanghai, the country's largest city and major seaport, is also located on the river's delta. With more than 50 universities and colleges, Shanghai is an important educational as well as industrial center.

Upstream lie the industrial cities of Nanjing and Wuhan. Their early industrialization was based on local iron-ore and coal deposits. The city of Chongqing, farther inland along the upper Chang River, lies in the Sichuan (Red) Basin, which has rich soils as well as coal and minerals.

Rice, sugarcane, and soybeans are grown here. The region is also a major silk producer.

Farther south, at the mouth of the Xi River, is the famous trading center of Guangzhou (Canton). Guangzhou is now the largest city south of Shanghai and has attracted much industrial and commercial development. In addition, the region's warm, wet climate allows a variety of crops to be grown in the fertile delta area surrounding the city. More resources may be found near Guangzhou's front door—large, untapped oil fields may lie just offshore in the shallow waters of the South China Sea.

Located on the southeast coast of China just south of Guangzhou is the British crown colony of Hong Kong. It is one of the world's busiest ports and largest financial centers. Much of China's trade and foreign investment passes through this port. Major industries include textiles, electronics, plastics, food processing, and tourism. All of this trade and industry has helped make Hong Kong one of the world's most densely populated places. The colony faces an ongoing struggle to house its more than 5 million people. Great Britain and China have agreed that Hong Kong will return to Chinese control in 1997. A capitalist system will continue, and the former colony will have some autonomy.

Near Hong Kong lies the Portuguese province of Macao. Its half million people are crowded onto only six square miles (15.5 sq. km) near the mouth of the Xi River. Textiles, fireworks, fish exports, and tourism provide the province's main income. After more than 400 years of Portuguese rule, Macao will be returned to China in 1999. Like Hong Kong, it will be allowed to keep a capitalist system.

Along the coast of southern China are **Special Economic Zones (SEZs)** designed to attract foreign companies and investment to China. The SEZs nearest to Hong Kong have experienced rapid economic growth. Areas that were rice fields a few years ago now bustle with modern factories, high-rise buildings, traffic jams, and new freeways. In these "open-door" ports, foreign goods and money circulate freely. Most foreign investment pours in from Taiwan, South Korea, and Japan, although investments from the United States are increasing.

Northern China Northern China lies north of the city of Nanjing and the Qin Ling range. It includes the Huang He valley and the fertile North China Plain. Chinese culture first developed here, and the region remains the nation's center of culture and political power. About one-half of China's population lives in this region. In fact, the North China Plain, which is about the size of Texas, has a population larger than that of the entire United States.

Unlike humid southern China, the drier and cooler north sometimes experiences droughts and dust storms. The climate is mainly steppe, with

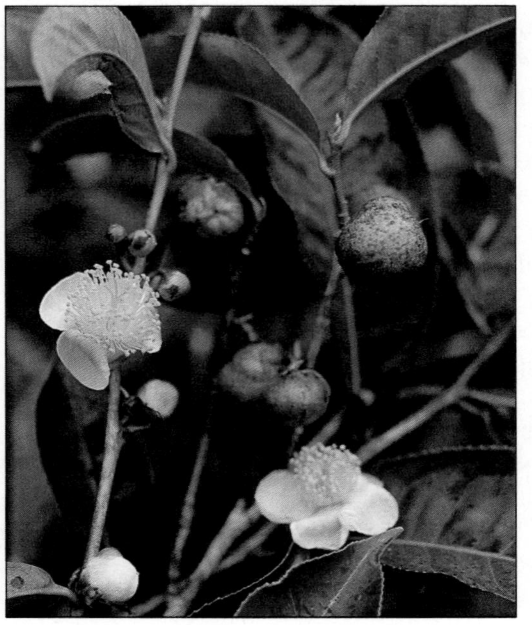

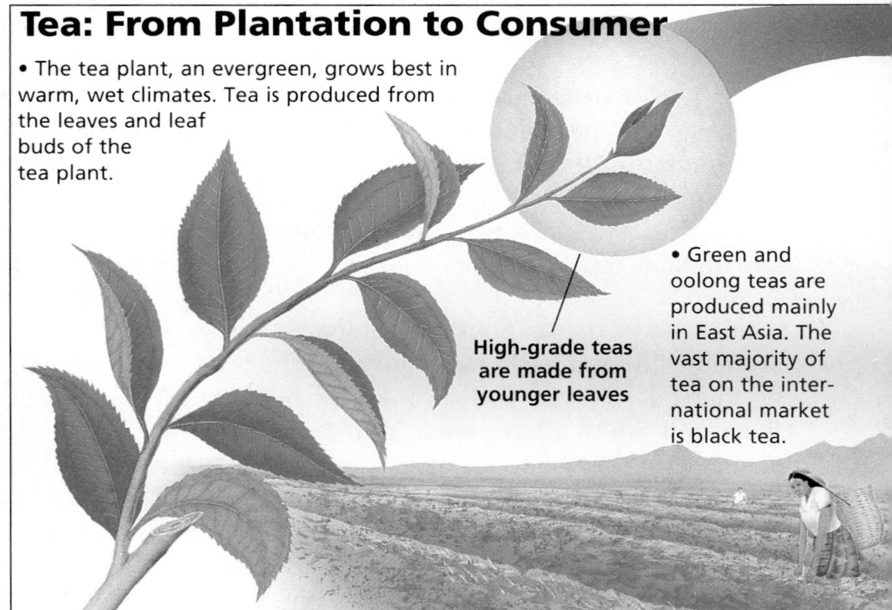
Tea: From Plantation to Consumer
• The tea plant, an evergreen, grows best in warm, wet climates. Tea is produced from the leaves and leaf buds of the tea plant.
High-grade teas are made from younger leaves
• Green and oolong teas are produced mainly in East Asia. The vast majority of tea on the international market is black tea.

very cold winters, especially in the interior. Along the coastal areas and on the Shandong Peninsula, the climate is humid continental. Here, winters are severe, but summer rainfall is reliable.

The Huang He begins in Tibet, flows across northern China, and empties into the Yellow Sea. It has changed its course many times and overflowed its banks, causing serious flood damage. Thousands have drowned over the centuries in the river's flood waters. For this reason, the Huang He has been called "China's sorrow." The Chinese have worked to tame the river's waters for irrigation and hydroelectric power.

Wind-blown deposits of fine yellowish-brown soil, called loess, cover much of northern China, making its lands yellowish. In fact, the dust turns houses, clothing, and water in the region all the same shade. The Yellow Sea and the Huang He (Yellow River) take their names from this yellow-colored loess.

The steppe climate of northern China has made this region the nation's wheat belt and made the country one of the world's largest wheat growers. Yet, because of its huge population, China imports wheat, especially from the United States, Canada, and Australia. Other grains grown in this region include millet and corn.

Beijing is northern China's largest city as well as China's cultural center and national capital. *Beijing* means "north capital." The city, which dates back more than 3,000 years, has served as China's capital almost continuously since the thirteenth century. The ancient, walled part of Beijing is divided into two sections, the Outer City and the Inner City. Within the Inner City lies the Imperial City, from which the government of the emperors ruled. Nested in the Imperial City itself was the Forbidden City, where the emperors lived, secluded from the world.

Tiananmen Square lies near the heart of **Beijing** within the Inner City. Chinese troops violently ended weeks of massive demonstrations for political reform here in 1989.

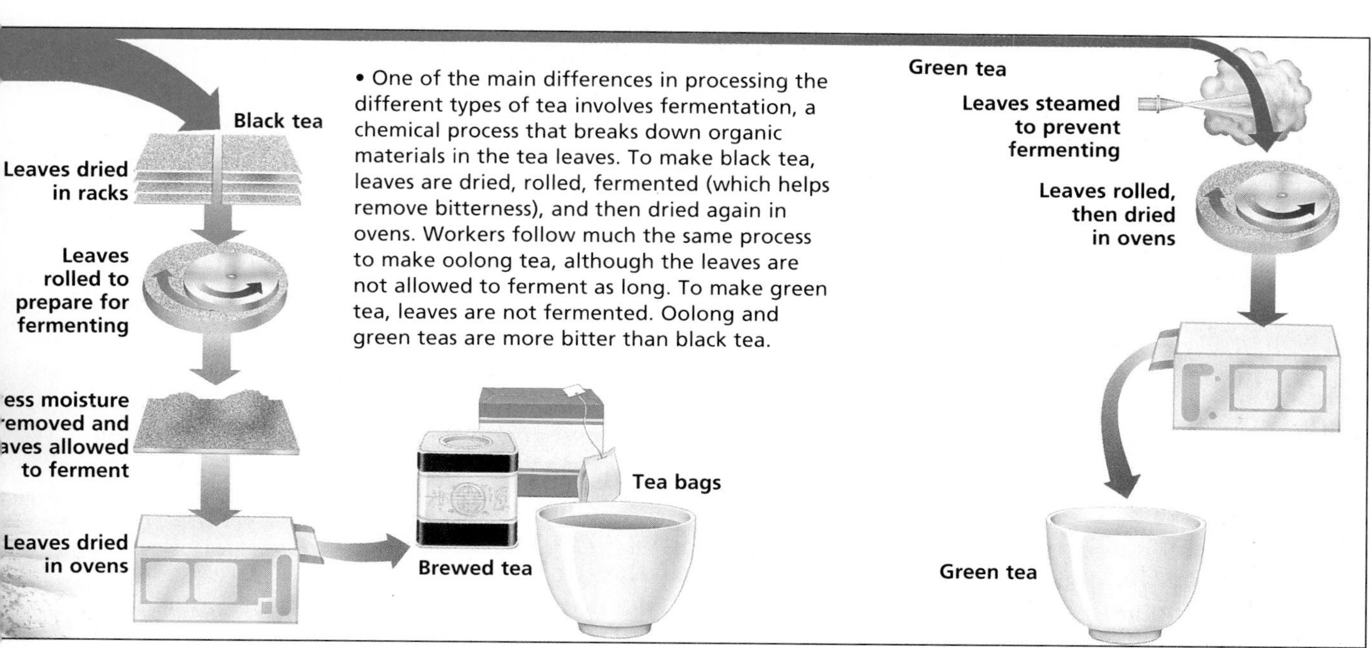
Black tea
Leaves dried in racks
Leaves rolled to prepare for fermenting
ess moisture removed and aves allowed to ferment
Leaves dried in ovens
Brewed tea
Tea bags
Green tea
Leaves steamed to prevent fermenting
Leaves rolled, then dried in ovens
Green tea

• One of the main differences in processing the different types of tea involves fermentation, a chemical process that breaks down organic materials in the tea leaves. To make black tea, leaves are dried, rolled, fermented (which helps remove bitterness), and then dried again in ovens. Workers follow much the same process to make oolong tea, although the leaves are not allowed to ferment as long. To make green tea, leaves are not fermented. Oolong and green teas are more bitter than black tea.

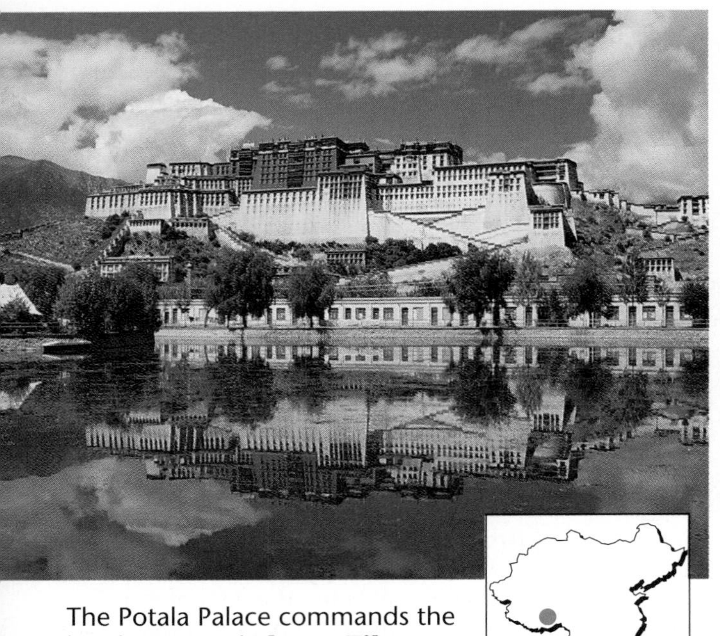

The Potala Palace commands the heights around **Lhasa, Tibet**. The palace, with more than 1,000 rooms, is the former home of the Dalai Lama.

Beijing grew beyond its original walls and spread out across the plains. Today, it is a modern city of industry, subways, department stores, hotels, and wide boulevards. Parliament, housed in the Great Hall of the People, and the Museum of the Chinese Revolution surround Beijing's Tiananmen Square.

Northeastern China Northeastern China consists of the three provinces that were once known as Manchuria, and the autonomous region of Inner Mongolia. The main landforms of the northeastern provinces are the Manchurian Plain and a surrounding rim of mountains. The region has a humid-continental climate with severe winters. Forested areas of subarctic climate cover the very far northern sections. The climate here is similar to that of neighboring Siberia (in Russia). Northeastern China's short summer limits agriculture to mostly wheat, corn, and soybeans.

Oil, coal, iron ore, zinc, lead, and manganese are plentiful in the region. In particular, the oil fields here have helped make China self-sufficient in energy. They produce more than 40 percent of China's oil, some of which is exported to Japan. The region also contains China's remaining forest resources. Other important industries include iron and steel, chemicals, paper, textiles, food processing, and shipbuilding. The largest city and the center of the iron-and-steel industry is Shenyang.

Just to the west, behind the Greater Khingan Mountains, is Inner Mongolia, an autonomous region of China. The area was originally populated by Mongolians, but today the Han Chinese outnumber the Mongolians many times over. This dry region is at the southern fringes of the Gobi. Irrigation farming is the major activity.

Western China The two large, autonomous regions of Tibet (officially called Xizang) and Xinjiang (SHIN-jee-AHNG) make up most of Western China. These regions have some local government and retain their own culture and languages. Both Tibet and Xinjiang were originally populated by people who were not Han Chinese. Han Chinese colonists, though, may soon outnumber the Tibetans and other minority groups.

Western China covers more than one-third of the nation's territory. The land, however, is too high, too dry, or too cold to support a large population. Most of the people are either nomadic herders or irrigation farmers.

Tibet has been occupied by China since 1950. Taking up most of the Plateau of Tibet, it is one of the highest and most barren regions in the world. It averages over 14,500 feet (4,420 m) in elevation. In the south, the towering Himalayas separate Tibet from India and Nepal. Mt. Everest, the world's highest peak, looms over the border between China and Nepal.

The huge Buddhist palace of the Dalai Lama towers above Lhasa, the capital of Tibet. The Dalai Lama is the spiritual leader of the Tibetans, who practice Lamaism, a form of Buddhism. The Dalai Lama fled to India in 1959 after an unsuccessful revolt against Chinese control.

The Chinese are allowing Tibetan Buddhism to be practiced again after decades of banning it. Even so, Tibetans continue to call for independence. The Tibetans fear that they will become an oppressed minority in their own land because economic development has brought thousands of Han Chinese into the region. Although Tibet remains one of the poorest and most isolated parts of China, it has unexplored mineral wealth.

Xinjiang, north of Tibet, is populated mostly by Muslim Turkic people, mainly the Uygurs and Kazakhs. Mountains ring the dry desert basins of Xinjiang on every side except the east, where they meet the Gobi. The largest and driest basin of Xinjiang is the Tarim. Its lowest point, the Turpan Depression, drops to 505 feet (154 m) below sea level. The Tarim Basin also holds the Taklimakan Desert and Lop Nur, a huge, dry lake bed and salt flat.

Xinjiang has minerals, including oil, coal, and iron ore. Ürümqi, its capital city, produces textiles, steel, and other products. Most Uygurs and Kazakhs, however, live as they have for centuries, herding animals.

2 SECTION REVIEW

1. What climates dominate southern and northern China, and what crops are grown in these two regions? Which of China's regions have major mineral resources?
2. What major cities in southern China are industrial? What is the capital of China, and where is it located?
3. **Critical Thinking** **Predict** whether the return of Hong Kong and Macao will present any challenges for China. Explain your answer.

3 TAIWAN AND MONGOLIA

FOCUS

- *What is the physical geography of Taiwan? the human geography?*
- *What challenges does Mongolia face?*

LOCATION **Taiwan** Early Portuguese explorers called the island of Taiwan *Formosa*, meaning "beautiful island." Taiwan lies 100 miles (161 km) across the Taiwan Strait from the southeast coast of China. The island and 86 smaller islands nearby make up the Republic of China. (See the map on page 507.)

Taiwan has a humid-subtropical climate with year-round rainfall and occasional typhoons. The eastern half of the island is mountainous, with peaks soaring 13,113 feet (3,997 m). These mountains form a rugged coastline. The western half of the island consists of fertile plains.

Because of its warm, wet climate and fertile soil, Taiwan is self-sufficient in rice and has a large food-processing industry. Major crops are rice, fruits, and vegetables. Taiwan also has a large fishing fleet.

Today, the population of Taiwan is about 85 percent Taiwanese. The ancestors of these people migrated to the island from China over hundreds of years. A small percentage of the population is descended from the island's original Malay settlers. The remainder is made up of "mainlanders," or Nationalist Chinese, who escaped from mainland China after the Communists took over in 1949. The Nationalist Chinese party controlled Taiwan under **martial law** (military rule) for 38 years. Only in the past few years have democratic rights expanded.

Taiwanese near **T'aipei** celebrate at a festival honoring the dead. Elements of Taoism and Buddhism are part of the celebration.

Much of the population of Taiwan is urbanized. T'aipei, on the north end of the island, is the largest city, a financial center, and the nation's capital. Millions of motorbikes clog the streets of this modern city each day. On the south end of the island is Kaohsiung, the country's second largest city, the major seaport, and a center of heavy industry.

Taiwan is one of Asia's most prosperous and industrialized nations. Export oriented, the country is a world leader in the production of computers, calculators, scientific instruments, and sports equipment. Taiwan's major trading partners are the United States, Japan, and China (through Hong Kong).

The average per capita income of Taiwan is more than 10 times that of China. Taiwan's rapid industrial success has led to environmental problems, however. The government has initiated an expensive clean-up program. In addition, the building of new highways, rapid transit, sewage plants, and recreational facilities promises to improve the quality of life for Taiwan's people.

Both the People's Republic of China and the Republic of China, or Taiwan, claim to be the true government of China. Until 1971, Taiwan represented China in the United Nations. Its membership was then replaced by the People's Republic of China. This loss, along with the loss of other international memberships, has not affected Taiwan's trade and successful industrialization.

The People's Republic of China seeks reunification and, should it occur, is willing to grant some political autonomy to Taiwan. Presently, Taiwan is not considering immediate reunification. The Taiwanese, however, are major investors in the rapidly industrializing regions of coastal China. This increasing economic interdependence between the two countries is drawing them closer together.

LOCATION **Mongolia** Mongolia is a large landlocked nation surrounded by China and Russia. It includes the Gobi, vast rocky plateaus covered with steppe grasses, and several mountain ranges. (See the map on page 507.) Due to its continental location, Mongolia experiences extreme temperatures. In the winter, temperatures may drop as low as -54°F (-48°C), and severe blizzards occasionally hit the country. Blistering temperatures during the dry summers can lead to raging brushfires.

During the thirteenth century, Genghis Khan led his fierce horse-soldiers out of Mongolia to conquer much of Asia, including China. Their Mongol Empire also reached into Europe and India. A few centuries later, Mongolia fell under the rule of China. In 1911, Outer Mongolia, with Russian support, broke away from China and declared its independence. Inner Mongolia remains part of China.

In 1924, Outer Mongolia installed a Communist government and became the Mongolian People's Republic. At this time, the country became economically dependent on the Soviet Union. Soviet aid to Mongolia, however, ended with the collapse of the Soviet Union in the early 1990s. Mongolia now seeks foreign aid and investment from other countries, especially Japan and South Korea.

Mongolia's natural resources include coal, copper, and perhaps oil. About half the nation's exports come from copper mining. Industrialization, meanwhile, is limited to textiles and carpets. There is little farming, and the country faces food shortages. The capital and only city is isolated Ulaanbaatar (oo-lahn-BAH-tawr).

Mongolia is three times larger than France but has a population of only 2 million. Today, about half of the population still depends on herds of sheep, goats, cattle, camels, horses, and yak (oxen) to earn a living. In fact, livestock outnumbers people many times over. The country's major religion is Buddhism. Much of the Mongolian culture, however, vanished under Soviet influence.

In 1990, free elections were held in Mongolia, and the government began allowing some free enterprise. Although Mongolia is one of the world's poorest and most isolated countries, there is hope for development of its natural resources and increased trade.

3 SECTION REVIEW

1. How does Taiwan's physical geography influence its economic geography?
2. What challenges does Mongolia face as a result of its physical geography?
3. **Critical Thinking** **Suggest** a reason why the People's Republic of China seeks reunification with Taiwan but Taiwan is not presently considering reunification.

CHAPTER 42 CHECK

Inside the Forbidden City in **Beijing**

Reviewing the Main Ideas

1. China has the world's oldest continuous civilization. For more than 2,000 years, a series of dynasties ruled China. In 1912, the last of these dynasties was overthrown. In 1949, China became a Communist country. Today's leaders are promoting more economic freedoms but are resisting political reforms.
2. China's four geographic regions include the fertile valleys of southern China; northern China, the most populated part of the country; northeastern China, rich in mineral resources; and dry, rugged western China.
3. Taiwan, an island off the southeast coast of China, is one of Asia's most prosperous and industrialized countries. Many of Taiwan's people are Chinese who left China since the Communists took over in 1949.
4. Mongolia is a large landlocked nation surrounded by China and Russia. It is one of the world's poorest and most isolated countries.

Building a Vocabulary

1. What is a puppet government? In what region of China did Japan establish a puppet government in the early 1930s?
2. Define *double cropping*. Why is southern China able to practice double cropping?
3. Along the coast of southern China are five areas designed to attract foreign investment. What are these areas called?
4. What is martial law? For how long was Taiwan under martial law?

Recalling and Reviewing

1. In what region did Chinese culture first develop?
2. Explain the difference between the People's Republic of China and the Republic of China.
3. What two major rivers run through southern China? What major river runs through northern China, and how did it get its name?
4. What mountains separate Tibet from India and Nepal?
5. How was Mongolia affected by the collapse of the Soviet Union?

Thinking Critically

1. Do you think that Maoism had any positive aspects? Explain your answer.
2. In what ways is the political situation in Tibet similar to the situation in Northern Ireland?

Learning About Your Local Geography

Cooperative Project

Under China's communism, the role of women in the work force changed. What economic opportunities are open to women in your area? First, identify the meaning of the term "equal opportunity employment." Then, identify several different companies or businesses in your area. Each member of your group should select one of these companies and research its hiring policies. To do this, contact the personnel department of the company. You may also ask to interview a female employee of the company about her career. As a group, construct a chart that shows the information you have gathered.

Using the Five Themes of Geography

Construct a chart with four columns, one for each major **region** of China. In each column, explain how the region's **location** and **place** characteristics have influenced forms of **human-environment interaction** in the region.

CHAPTER 43

Japan and Korea

"Because I was born in Mishima in Izu, which is connected to the southern base of Mt. Fuji, Fuji was for me part of the home that I was used to seeing every day. . . . The Fuji that is the most familiar and nostalgic to me is the water from the melted snow on the mountain that passes through the rock beds deep in the earth and gushes forth in the western and northern parts of the town of Mishima. The water is plentiful, and it is in this clear river water that I have my Fuji. Once when asked, 'What does Mount Fuji mean to you?' I answered, 'Mt. Fuji is water.' "

Makoto Ōoka

GEOGRAPHY DICTIONARY

tsunami
annex
subsidy
work ethic
trade surplus
urban agglomeration
export economy
demilitarized zone (DMZ)

Less than 150 miles (241 km) of water separate the Korean peninsula, on the eastern edge of the Asian mainland, from the islands of Japan. As a result, the two nations of the Korean Peninsula share with Japan several geographic and cultural characteristics. In other ways, however, the three countries are very different.

Japan's small land area is very mountainous. The country is subject to floods, typhoons, and earthquakes, and has few mineral and energy resources. Yet despite these disadvantages, Japan has become one of the world's leading industrial and economic powers.

In contrast to Japan, much of the Korean peninsula contains valuable mineral deposits. Political geography, however, has split the peninsula into two very different countries, Communist North Korea and industrialized South Korea.

Japanese sandals

Mount Fuji, **Japan**

JAPAN AND KOREA: PHYSICAL–POLITICAL

JAPAN

FOCUS

- *What are the physical features of Japan? What relationships has Japan had with other countries throughout its history?*
- *In what ways has Japan been economically successful?*
- *What are the cultural characteristics of Japan's people? What are their daily lives like?*

REGION **Landforms** Japan is a country about the size of California. The Sea of Japan separates the island nation from mainland Asia. (See the map above.) The country is made up of four large "home" islands and more than 3,000 smaller ones. The home islands from north to south are Hokkaido (hah-KIDE-oh), Honshu (HAHN-shoo), Shikoku (shi-KOH-koo), and Kyushu (kee-YOO-shoo). Honshu is the largest and most populated island. The Ryukyu (ree-OO-kyoo) Islands to the south are also part of Japan. Okinawa is the largest of these. Japan claims the southern Kuril Islands located northeast of Hokkaido as well. Russia has held the Kurils since the end of World War II.

More than 70 percent of Japan is mountainous. The longest mountain range is the Japanese Alps, which form a volcanic spine through the island of Honshu. About one-fourth of Japan's 200 volcanoes are active. Mount Fuji (FOO-jee), Japan's highest peak and national symbol, is an inactive volcano. It rises 12,388 feet (3,776 m).

Japan lies along a subduction zone, where the Pacific plate dives beneath the Eurasian and Philippine plates. Earthquakes and volcanic eruptions rock the region and have killed thousands of people. The 1995 earthquake in Kobe killed more than 5,000 people, for example. Large

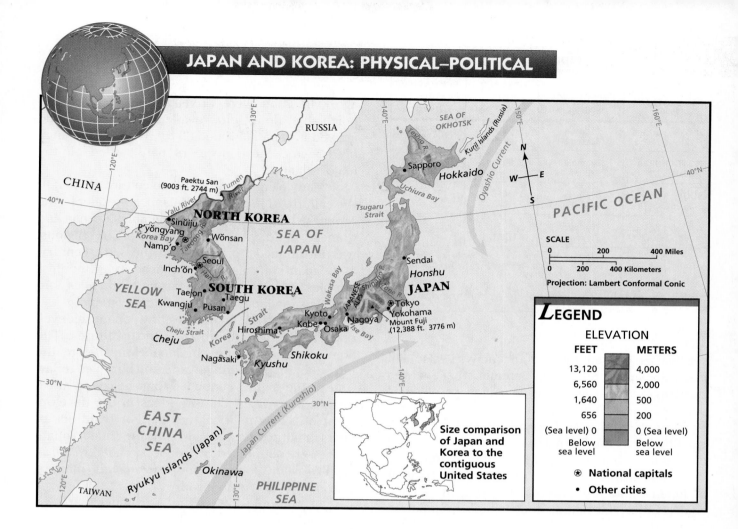

FACTS IN BRIEF Japan and Korea

COUNTRY POPULATION (1994)	LIFE EXPECTANCY (1994)	LITERACY RATE	PER CAPITA GDP (1993)
JAPAN 125,106,937	76, male 82, female	99% (1991)	$20,400
NORTH KOREA 23,066,573	67, male 73, female	99% (1990)	$1,000 (1992)
SOUTH KOREA 45,082,880	67, male 74, female	96% (1990)	$9,500
UNITED STATES 260,713,585	73, male 79, female	97% (1991)	$24,700

Sources: *The World Factbook 1994,* Central Intelligence Agency; *The World Almanac and Book of Facts 1995*

Which country has the highest life expectancy rates?

sea waves caused by tectonic activity also kill many Japanese. Known as **tsunami,** these waves can travel across an ocean at hundreds of miles per hour and rise up to a height of 100 feet (30 m) or more by the time they reach shore. A 1993 earthquake in northern Japan created waves and caused fires that destroyed hundreds of homes.

Plains, which make up only about 20 percent of Japan's land area, lie mainly along the Pacific coast. Cities, farmland, and Japan's population compete for space on these small coastal plains. Here, a population four times that of California crowds into a land area about one-fifth the size of that state.

Climate Japan lies at about the same latitudes as the east coast of the United States, and its climates are similar to that region. Hokkaido and northern Honshu have a humid-continental climate with summers cooled by the Oyashio Current from the north. Icy winds blowing off the Asian continent make winters in northern Japan cold, and the windward Sea of Japan coast experiences heavy snowfall.

Southern Japan, which includes Kyushu, Shikoku, and the southern part of Honshu, has a humid-subtropical climate. Here, winters are mild, and summers are warm and humid. The Japan Current, which flows from the tropical Pacific, warms southern Japan. The area receives rain year-round and has a long growing season. Some areas receive more than 80 inches (203 cm) each year. In late summer and early fall, typhoons bring destructive winds but necessary rainfall.

MOVEMENT Historical Geography Japan's first inhabitants were the Ainu, who probably arrived from Central Asia several thousand years ago. The Ainu were driven farther and farther north by Mongoloid invaders from Asia who arrived in Japan around 300 B.C. Within a few hundred years, these invaders had settled all the major islands.

Many elements of Japanese culture are tied to the cultures of the Asian mainland. For example, the Japanese language strongly resembles Korean and Chinese. Buddhism, one of Japan's major religions, was introduced from China in the sixth century, and Japan still has many Chinese-style Buddhist temples. Also, Chinese ideas and practices laid the basis for Japan's early political systems.

Despite these influences, the Japanese developed a distinct culture over the centuries. The Shinto religion existed in Japan long before the introduction of Buddhism and still survives today. And in the eighth century, Japan began to move further away from Chinese ideas in the development of a new political system. For hundreds of years, military leaders, or *shoguns*, ruled Japan. They were supported by warriors, called *samurai*. Japan's island location and the *shoguns'* military power kept out Asian invaders.

In the mid-1500s, the Portuguese arrived in Japan seeking trade. The Spanish and Dutch soon followed, resulting in the introduction of Christianity. Soon after, however, European traders and missionaries were either driven out of Japan or restricted to the port of Nagasaki. Japanese leaders feared that foreign ideas might cause instability in Japanese society.

Japan remained cut off from the world from the 1600s to the mid-1860s, when merchants and other groups in Japanese society pressed for greater contact with the outside world. As a result, the United States and many European nations soon entered into treaties with Japan.

By this time, Japan had begun its modernization period, called the Meiji (MAY-jee) Restoration *Meiji* means "enlightened rule." The *shoguns* were overthrown, and an emperor took their place. Japan's modernization took many forms, including rapid industrialization and changes in education, law, and government. By 1890, Japan had a constitution and parliamentary system of government.

Japan soon became recognized as a world power. In 1895, it defeated China in a war over control of Korea. In 1905, it defeated Russia in the Russo-Japanese War. Japan **annexed**, or formally joined, Korea to its empire in 1910. By 1920, Japan had further expanded its borders to include many North Pacific islands.

Then in 1940, Japan signed an alliance with Germany and Italy and, in 1941, entered World War II by attacking the U.S. naval base at Pearl Harbor, Hawaii. Japan conquered much of Southeast Asia and numerous Pacific islands before being pushed back by U.S. and allied forces.

The end of World War II marked the end of Japan's military empire. Japanese cities had suffered massive bombing during the war. Two major cities, Hiroshima and Nagasaki, were almost completely destroyed by atomic bombs dropped by the United States.

After World War II, Japan established a democratic government. With U.S. aid, Japan also began to rebuild into a major world industrial power. Today, Japan is a constitutional monarchy with several political parties. The government is made up of an elected law-making body, called the Diet (DEE-it), and a prime minister. Japan's emperor remains a symbol of the nation but has no real political power.

INTERACTION **Agriculture** Less than one-fifth of Japan's land is arable. Yet Japanese crop yields are some of the highest in the world, and the nation produces about 40 percent of its food needs. With a population half that of the United States, Japan has twice as many farmers.

Japanese agricultural success is due to terraced cultivation and modern farming methods. The farmers own their land and live in small villages. The size of the average Japanese farm is only two and one-half acres (1 ha). Many Japanese farms are under constant pressure from expanding urban areas.

Sixteenth-century *samurai* armor

Almost half of Japan's farmland, mostly in southern Japan, is used to grow rice. Tea, mulberry trees (for silkworms), soybeans, and a variety of fruits and vegetables also grow well in this warmer climate. Northern Japan's shorter growing season produces wheat, barley, potatoes, and vegetables. The cool island of Hokkaido is home to a successful dairy industry.

Japan imports about half of its food, especially grains and meat. Food imports have increased because many Japanese have added more Western foods, such as meat, bread, and potatoes, to their usual diet of rice, fish, and vegetables. Packaged foods and fast-food restaurants also are on the rise. French fries are now so popular that the nation imports $85 million in frozen fries each year, about half of which come from the U.S. state of Washington. Other foreign products are imported under a quota or completely banned.

The Japanese farmer is protected by high government **subsidies**, especially for rice. A subsidy is financial support given by a government. Subsidies give Japanese farmers an economic advantage over foreign competitors in the Japanese market.

Resources Japan's location near seas rich in marine life, as well as its long coastline, has helped create the largest fishing industry in the world. In fact, Japan makes over 10 percent of the world's fish catch. The country has a local fishing fleet of thousands of small boats as well as a large oceangoing fleet. Aquaculture programs also supply Japanese markets with fish, shellfish, seaweed, and pearls (from oysters). The Japanese have continued to practice commercial whaling despite international protests.

Forests cover more than two-thirds of Japan, making it one of the world's most forested nations. The cutting of Japan's timber is carefully controlled to prevent the loss of forests, limit soil erosion, and provide national parks. As a result, Japan imports timber from the United States, Canada, and Southeast Asia.

Japan has few energy resources and industrial raw materials such as iron ore. Oil is the country's major import. Japan also imports coal from Australia and the United States. Hydroelectric power plants provide about 12 percent of Japan's electricity needs. Nuclear power plants provide another 25 percent. Conservation efforts and new nuclear power plants have helped lower the country's dependence on imported oil.

Industry Despite its dependence on imported natural resources, Japan produces goods that are considered some of the best in the world. The country is the second largest industrial power in the world, after the United States. It is a world leader in the production of cameras, televisions, videocassette recorders, radios, watches, and motorcycles. The country ranks high in the production of steel, computers, telephones, and many household products. The Japanese are also leaders in such high-tech fields as electronic miniaturization and industrial robotics.

Many nations are looking at Japan to find the reasons for its economic success. Some reasons lie in Japan's cultural geography. Nearly everyone in Japan is of Japanese origin. This makes communication and the setting of common goals much easier. Also, Japan's people have a strong **work ethic**, a belief that work is good in and of itself. Many Japanese work on weekends and until late at night.

Other reasons for Japan's success lie in the relationship between employers and employees. Workers often participate in decision making in their companies. In general, Japanese workers are loyal to their employers, and some keep the same jobs their whole lives. Strikes are rare by Western standards. In addition, the Japanese are familiar with the markets and nations that buy their exported products and are generally well educated. In fact, Japan has one of the world's highest rates of literacy and of high school graduation.

The Japanese government, through the Ministry of International Trade and Industry (MITI), works closely with businesses and offers financial aid and import protection. Also, Japan's domestic market is one of the most competitive in the world. A Japanese product that survives in the home market will probably compete successfully in international markets as well.

Workers feed mulberry leaves to silkworms on the island of **Honshu, Japan**. Silk comes from the cocoons these worms will spin when they begin the process of changing into moths.

International Japan The Japanese have succeeded in moving whole industries overseas. For example, there are several Japanese-owned factories in the United States producing automobiles and televisions. The Japanese have also invested in foreign industries, such as entertainment and real estate. Many of the buildings and hotels in cities such as Los Angeles and Honolulu are Japanese owned. Most of the world's largest banks are Japanese-owned as well.

At the same time, however, Japan is feeling competition from other newly industrialized Asian countries, especially China, the "Four Dragons," and the "Little Dragons." Many of Japan's older industries, such as shipbuilding, steel, and textiles, are suffering from this competition. Also, the country has built up a huge **trade surplus**. A trade surplus exists when a nation exports more than it imports. For instance, Japanese exports to the

United States are two times greater than its imports from the United States. This trade imbalance could cause some nations to set up trade barriers to Japanese goods.

Urban Heartland The core of Japan's cities and industry lies on Honshu's eastern coastal plains. The three nodal areas are the Tokyo Bay area, the Kansai region, and the Nagoya area. A dense network of roads, high-speed rails, seaports, and air routes connects the entire region.

With more than 25 million people, the Tokyo Bay area is the largest **urban agglomeration** in the world. An urban agglomeration is a densely inhabited and contiguous region surrounding a central city. Tokyo is Japan's capital and center of communications, government, banking, education, and trade. It is the site of the Imperial Palace and the *Ginza*, the world's largest and busiest shopping district. Because little land is available, Tokyo's real estate prices are among the highest in the world. (See "Cities of the World" on pages 524–525.) Yokohama (yoh-kuh-HAHM-uh), located just south of Tokyo, is one of the world's busiest seaports.

The Kansai region has three major industrial centers: Osaka (oh-SAHK-uh), Japan's second largest city; the nearby port city of Kobe (KOH-bee); and Kyoto (kee-OHT-oh), the ancient capital. Kyoto's many temples and traditional handicrafts attract large numbers of visitors. The Nagoya region, meanwhile, lies on Ise (EE-say) Bay. Nagoya is its major city and industrial center.

All of Japan's major urban areas are faced with unaffordable housing, air pollution, and traffic congestion. Unlike in large U.S. cities, however, crime rates in Japan are low.

Japanese Way of Life Japan has many millionaires as well as urban poor, but most Japanese belong to the middle class. Most families live in suburbs and spend several hours each day commuting to and from jobs in the cities.

A typical middle-class home in Japan is small—usually about the size of a small apartment in the United States—and expensive. Sliding wooden screens separate sparsely furnished rooms. People generally sit on cushions or mats placed on the floor. Families are used to living in crowded spaces, and most children must share their bedrooms. Western-style homes, recently introduced to Japan, have proved popular, though they are very expensive.

Japanese society has undergone many changes over the past several decades. In urban and suburban areas, Western clothing, such as blue jeans, has largely replaced the traditional kimono. Rock music and Western soft drinks are popular, and baseball, golf, and skiing are favorite forms of recreation.

The birthrate in Japan is as low as that of Europe, and the Japanese now boast the world's highest life expectancy—82 years for women and 76 for men. As the average age of the population becomes higher, many worry about the increased demands on social services. The roles of Japanese women also are changing as women become more educated and urbanized. More than half of Japanese women are in the work force and are seeking a larger role in Japanese society.

Japan's traditional culture, however, remains an important part of Japanese society. Japanese artists practice traditional forms of poetry, music,

Text continues on page 526.

4097
4099
4101

These traveling businessmen in **Osaka, Japan**, are spending the night in one of the country's "capsule hotels." Many businesspeople find these hotels more convenient than commuting long distances to and from crowded Japanese cities.

Tokyo

Tokyo is Japan's largest industrial and urban area. The city is the center of the country's government and finance.

In the fifteenth century, a Japanese feudal lord built his castle overlooking a natural harbor on the island of Honshu. In time, the city of Edo grew up around the castle. In 1868, a new emperor changed the city's name to Tokyo, meaning "eastern capital," and made it the capital of Japan.

The City Rebuilt

The growth of Tokyo followed the original city plan, with the castle in the center. The town expanded outward in a circle, similar to the growth patterns of Moscow and Paris. Most of present-day Tokyo, however, has been rebuilt since the end of World War II, when bombing destroyed about half of the city.

Earthquakes also have disrupted the Tokyo skyline, as in 1923 when a huge earthquake devastated large areas of the city. On average, three tremors shake the foundations of Tokyo each day! The city's modern skyscrapers have been specially designed to endure these frequent quakes.

Tokyo and its neighboring cities have become so large that their boundaries overlap. Tokyo, Nagoya, and Osaka form one huge, continuous urban area that stretches along the narrow plain of Japan's east coast and includes

ITOYA
MITSUKOSHI
Illuminated signs, in English and Japanese, scream at traffic and shoppers crowding into the Ginza district. The district is famous for its stores and entertainment.

Metropolitan **Tokyo** sprawls outward from the Imperial Palace. Stone walls and a moat insulate the palace from the bustling city.

the port cities of Yokohama and Kawasaki.

Looking for Space

The chief concern in Tokyo, as in the rest of Japan, is space. The city has expanded to fill all the available land on the narrow coast, in addition to landfill areas. The severe shortage of housing leads some to look upward rather than outward to find space.

Indeed, Tokyo is already crowded with skyscrapers, and more are springing up all the time. Tall buildings often tower over smaller neighbors, blocking out the natural light. As a result, a law called *nisshoken*, or "the sunshine law," requires developers to pay money to those who live in the shadows of their towering buildings.

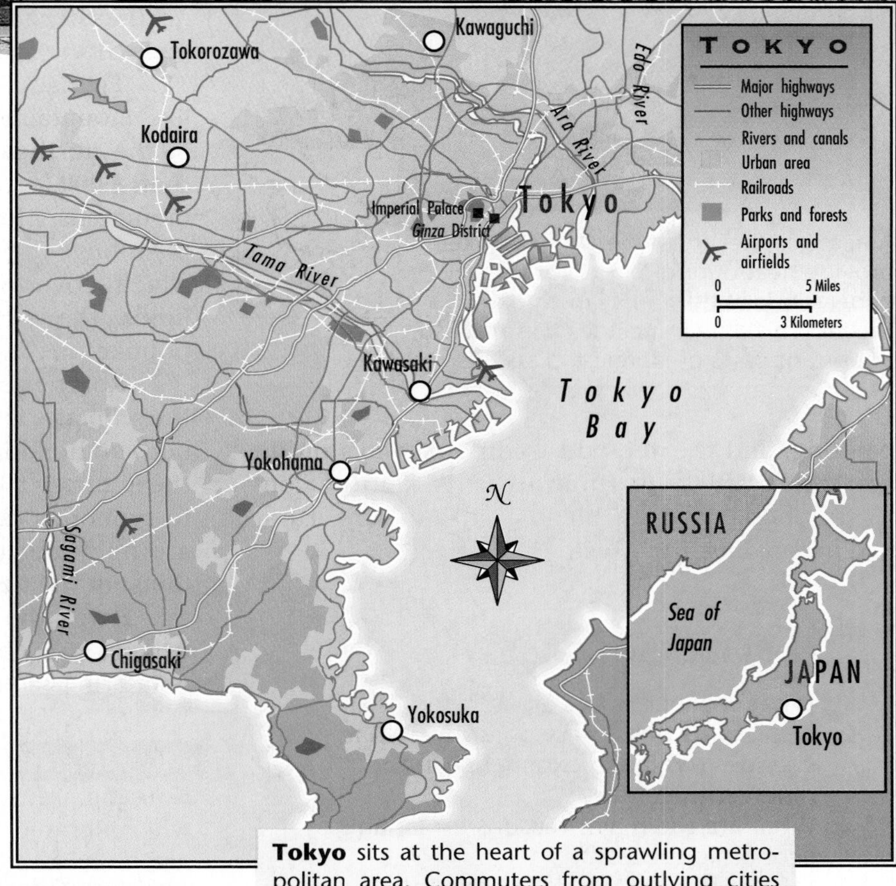

TOKYO
Major highways
Other highways
Rivers and canals
Urban area
Railroads
Parks and forests
Airports and airfields
0 5 Miles
0 3 Kilometers
Tokorozawa
Kawaguchi
Edo River
Ara River
Kodaira
Tokyo
Imperial Palace
Ginza District
Tama River
Kawasaki
Tokyo Bay
Yokohama
N
Sagami River
Chigasaki
Yokosuka
RUSSIA
Sea of Japan
JAPAN
Tokyo

Tokyo sits at the heart of a sprawling metropolitan area. Commuters from outlying cities use railway transportation to travel to and from their jobs in Tokyo.

A weekend getaway to the country might be on the minds of these Japanese commuters being crammed onto trains in crowded **Tokyo**. Workers are employed to shove as many passengers onto commuter trains as possible.

Only some 10 percent of the land area in Tokyo is open space, much of which is devoted to beautiful public gardens and parks featuring Shinto shrines and Buddhist temples. The Japanese take great pride in their parks and strive to keep them clean and safe. The crime rate in Tokyo is much lower than that of other large cities of the world.

As in most other major cities, however, traffic congestion is a problem in Tokyo. Efficient transportation networks and high-speed commuter trains help ease traffic problems and cut down pollution. The city used to be badly polluted from automobiles and factories. Since 1965, though, the government has worked to cut the level of air pollution and keep the city clean. As a large industrial city, Tokyo still faces many challenges, but its officials and citizens have taken important steps toward managing several pressing urban issues.

A heater under the table and a blanket provide heat during dinner for three generations of this family in **Kyoto, Japan**. This dinner includes rice and fish, important parts of Japanese diets.

painting, and theater, and traditional ways of life are still followed in rural areas. A land of contrasts, modern Japan blends old and new, rural and urban, East and West.

SECTION REVIEW

1. What challenges has Japan faced as a result of its physical geography? What changes did Japan experience in the nineteenth century?
2. What are the major reasons for Japan's economic success?
3. **Critical Thinking** **Explain** how Japanese ways of life are changing today. What do you think accounts for the changes?

NORTH AND SOUTH KOREA

FOCUS

- *What landforms and climate types are found on the Korean peninsula?*
- *How is the peninsula politically divided? How do North and South Korea differ?*

REGION **Physical Geography** Korea is a 600-mile-long (965-km) peninsula, about the size of Florida. The narrow Korea Strait and the Sea of Japan separate the peninsula from Japan. The peninsula is divided into two nations, North Korea and South Korea. The Yalu River on the north and the Yellow Sea on the west separate North Korea from China. North Korea shares its northeastern border with Russia. (See the map on page 519.)

The landforms of Korea are mostly hills and low mountains. The most mountainous region is in the northeast, where several peaks rise more than 8,000 feet (2,439 m). On the east coast, steep mountains plunge into the sea, while a coastal plain hugs the western coast. This crowded plain, which covers about one-fifth of the peninsula, holds the richest farmland and most of the population.

Korea has two climate regions. In the south, the climate is humid subtropical. Summers are hot and humid, and heavy rain usually falls during the summer monsoon season. The area is subject to typhoons during the summer and fall. Winters are mild along the coast but cold in the mountains. In the north, the climate is humid continental. Here, the growing season is shorter, and winters are colder. The continental influence of Asia brings cold air and snow during winter.

MOVEMENT **Historical Geography** The Koreans are descendants of a number of peoples who migrated to the peninsula from Central Asia and northern China. Separated from other lands by sea and mountains, the Korean culture evolved in isolation for centuries, even though China and Japan invaded Korea many times.

The last such major invasion occurred in 1895, when Japanese forces defeated the Chinese and annexed Korea. The Japanese removed the Korean monarchy in 1910 and took complete control, making Korea an economic colony of Japan. The Japanese controlled Korea until the end of World War II.

After World War II, the victorious powers agreed to temporarily divide the Korean peninsula along the 38th parallel. The Soviet Union occupied the northern part, and the United States occupied the southern part. Political differences between the two regions, however, prevented the expected reunification. In the north, the Soviets set up a Communist government, and North Korea became the Democratic People's Republic of Korea. In the south, the United Nations supervised a government election, and South Korea became the Republic of Korea. Soviet and U.S. occupation forces withdrew in 1949.

In 1950, North Korea invaded South Korea, sparking the Korean War. The United Nations sent troops to defend South Korea, while Communist China sent forces to North Korea. Eventually, a truce line was established near the original 38th-parallel boundary. The war devastated much of Korea's land and left more than 1 million people dead. The Koreans remained a divided people.

North Korea North Korea is a strict Communist state with few political and religious freedoms. Kim Il Sung, who ruled for more than 40 years, died in 1994. It is unclear whether his son can hold onto power. North Korea has a large military, and the United States has worked to prevent the country from obtaining atomic weapons.

The country has several hydroelectric power sources and large deposits of minerals, including coal, iron, and copper. Major industries include steel and textiles, and all industry is government owned. As for agriculture, about 40 percent of North Koreans work on state farms; private citizens can own only small garden plots. Rice, potatoes, and wheat are the main crops. The capital and largest city is P'yŏngyang (pee-AWNG-yahng).

South Korea South Korea is slightly smaller than North Korea but has more than twice as many people. The flatter land and warmer climate of South Korea allow for greater agricultural development and a wider variety of crops. Only about 15 percent of South Koreans farm, yet the nation is self-sufficient in most foods. Crops include rice, cabbage, sweet potatoes, and many fruits and vegetables. Fishing is an important economic activity as well.

With the help of the United States, South Korea rebuilt its industry after the Korean War. The nation progressed rapidly from an agricultural nation to an industrial one. South Korea has an **export economy**, which means most of its manufactured products are made for export to other nations. Indeed, trade accounts for a large part of the South Korean economy. The country's major industries are electronics, steel, automobiles, shipbuilding, and textiles. It is also a major producer of televisions, microwave ovens, fax machines, and computers.

South Korea's industrial base is heavily diversified and competitive, and the work force is literate and well trained. Lower wages in South Korea have helped the country to produce goods of high quality but with a lower price than those from Japan. South Korea is trying to expand its nuclear power and mining to help make up for large imports of oil and raw materials.

The country's recent economic progress has led to much social change. Industrialization and prosperity have created a large urban middle class interested in consumer goods. Seoul (SOLE), the capital and largest city, is a huge, rapidly growing industrial, commercial, and cultural center. Small agricultural villages are disappearing as young people move to the cities and as industrial suburbs and factories expand outward.

South Koreans have many political and religious freedoms. About 20 percent of the population are Buddhist, and another 20 percent are Christian. Although government leaders have enjoyed nearly unlimited power, demands by the people have recently led to the expansion of democratic rights.

Korean Reunification Koreans have long wished for peace and unity. In 1972 and again in 1991, North and South Korea agreed to seek

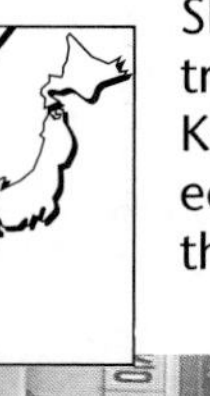

Shoppers throng this shopping district in **Seoul, South Korea**. South Koreans have experienced dramatic economic growth since the end of the Korean War.

가을 신상품 대매출
대도 마케트
그릇도매상가 C동 3층
혼수품일체
남문교자

NEWS FROM AROUND THE WORLD — South Korea

Teenagers in South Korea Prepare for the Future

Today, as in the past, teenagers in South Korea are taught to honor tradition, respect their elders, and place family loyalty above all other ties. Now, however, they must also adjust to great changes taking place in their society.

Instead of parents, grandparents, children, uncles, and aunts living under one roof, most youngsters live with only their parents. The size of families has decreased, and many young South Korean adults live in cities, where family bonds are weakening.

Education remains highly valued in South Korean society, and the opportunities for education are expanding. Schools reflect a mixture of traditional and modern elements. Students must study English as well as the Korean language. They participate in centuries-old sports, such as *ssirum*, a form of wrestling, and the martial art of *tae kwon do*. (See the photograph above.) Students also play sports introduced into the country only in the last century or so, such as soccer, volleyball, and basketball.

Korean teenagers are serious students, and about ninety percent of South Koreans attend high school. Those planning to attend a university must take an entrance examination after graduation from high school. Study for this examination begins as early as seventh grade. By the time students are high school seniors, they usually attend school six days a week. Most of their waking hours are spent in class or studying. About 40 percent of all students go on to college.

Sending a child to college is a priority for South Korean parents. "Our ancestors would sell their cows, their oxen and even their paddies, if necessary, to pay for an education for their children. And it was the best investment they ever made," claims one father whose 18-year-old son is studying day and night for the college examination.

peaceful reunification. However, many obstacles lie in the way of this goal, and the process has not moved forward significantly.

There is almost no trade or communication between North and South Korea. Their border is a **demilitarized zone (DMZ)**, a buffer zone just under two and one-half miles (4 km) wide into which military forces may not enter. Yet just a few miles to the north stands North Korea's large army, and to the south are South Korean troops and thousands of U.S. soldiers. In many respects, the two countries remain worlds apart.

SECTION REVIEW

1. Where are Korea's mountainous regions? Where is its plains region? What are the peninsula's two climate regions?
2. When was the peninsula divided into North and South Korea? What differences do these two countries have?
3. **Critical Thinking** **Compare and contrast** post–World War II political developments in Korea to those in Germany.

CHAPTER 43 CHECK

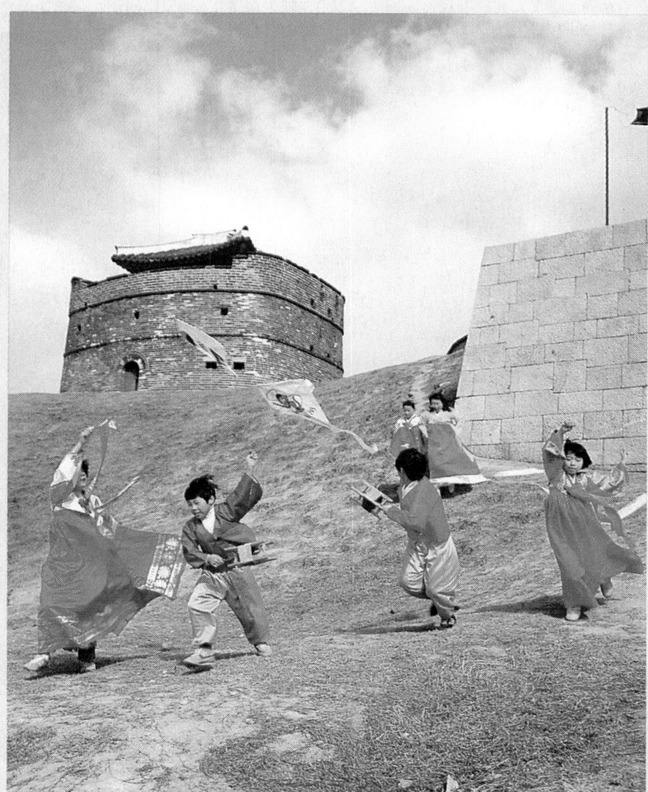

Children playing in traditional dress in **South Korea**

Reviewing the Main Ideas

1. Mountainous Japan is made up of four large islands and more than 3,000 smaller ones. Most of Japan's people live in the few coastal plains areas. The country is a major world economic power. Despite rapid modernization during this century, Japan has held on to much of its traditional culture.
2. The Korean peninsula is divided into two countries, Communist North Korea and democratic and industrial South Korea. Although the two countries have discussed reunification, they have not reached any agreements.

Building a Vocabulary

1. What are tsunami? Why are they dangerous?
2. When one country annexes a territory, what is it doing?
3. Define *subsidy*.
4. What is an urban agglomeration? Give an example of an urban agglomeration discussed in the chapter.
5. What term describes the border between North and South Korea? What does the term mean?

Recalling and Reviewing

1. What are Japan's major resources? What resources must the country import?
2. What role did foreign countries play in the division of the Korean peninsula?
3. Which country does Japan resemble more—North Korea or South Korea? Why?

Thinking Critically

1. In your opinion, should the United States adopt some of Japan's ideas about management and economic development? Why or why not?
2. What obstacles do you think lie in the way of Korean reunification?

Learning About Your Local Geography

Cooperative Project

Western foods, such as french fries, are more and more common in Japan. What "foreign" foods are popular in your area? With your group, make a list of the foods that you like to eat whose ingredients or recipes probably originated in another country. If your group has access to and permission to use kitchen facilities and supplies, prepare one of the items on your list. Share it with the rest of the class and explain its origins.

Using the Five Themes of Geography

Using symbols or shading, sketch a map showing how **location** has influenced human **place** characteristics in Japan, South Korea, and North Korea. Consider economic, cultural, and political characteristics of the three nations. Include a legend of the symbols and shading you use, and write captions explaining the relationship between location and place in these countries.

CHAPTER 44

Mainland Southeast Asia

GEOGRAPHY DICTIONARY

shifting agriculture

intensive agriculture

defoliant

"I glance out of the window. . . . As we bank toward the south, climbing over the sandy coastline, I see the upturned thimbles of Marble Mountain and the lazy snakes of the Song Han delta. . . . I spot a loamy patch of ground covered with thatchroofed houses rising among the paddies like a mother's knee. There is a scraggly jungle on two sides—fighting . . . to come back after a decade of fire and poison to become the lush paradise I once knew."

Le Ly Hayslip

Southeast Asia lies between the Indian and Pacific oceans. Key sea routes joining these oceans cross the region, including the narrow Strait of Malacca and the South China Sea. These important waters, as well as the region's valuable resources, have attracted Chinese, Indians, Arabs, and Europeans for centuries. In recent decades, much of Southeast Asia has undergone rapid industrialization. Some of the world's fastest growing economies are found here.

Southeast Asia has a diverse cultural geography. Because it borders both China and India, the region shows influences of these two large neighbors. Nonetheless, mountains, peninsulas, river valleys, and islands have isolated the region to an extent and nurtured several distinct cultures. This chapter examines the region's physical, cultural, and economic geography and then explores the nations of mainland Southeast Asia. The region's island countries are explored further in Chapter 45.

Thai mask

Farming village in northern **Vietnam**

MAINLAND SOUTHEAST ASIA: PHYSICAL–POLITICAL

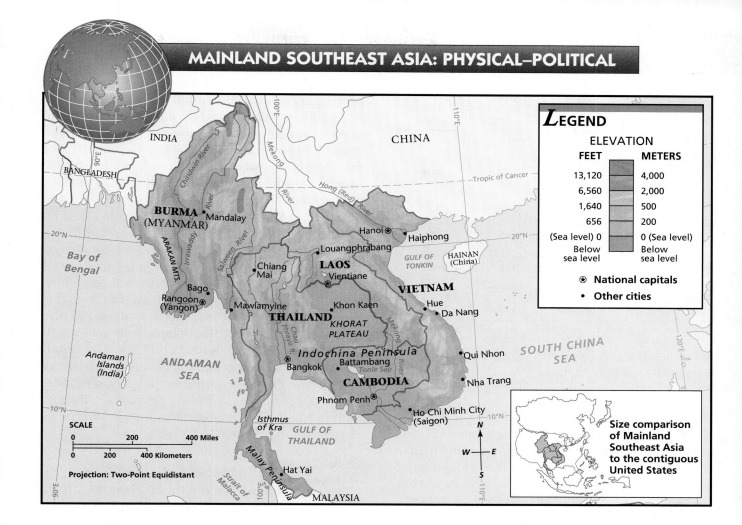

1 PHYSICAL AND HUMAN GEOGRAPHY

FOCUS

- *What landform regions, rivers, and climates are found in Southeast Asia?*
- *What groups of people have settled in Southeast Asia over time? What is the region's economic geography?*

REGION **Landforms** Southeast Asia contains three landform regions. (See the map above.) In the north, mountain ranges fan out from the high Himalayas and the Plateau of Tibet into Burma, Thailand, Laos, and Vietnam. A central region of flat plains and low plateaus reaches between the mountain ranges. It then stretches across southern Thailand, Cambodia, and Borneo. The third landform region consists of thousands of islands. Some islands formed from single volcanoes that grew from the sea bottom. Larger islands, such as Java, have hundreds of volcanoes, many of them active. Most of Indonesia and all of the Philippines are volcanic and part of the Pacific Ring of Fire. Earthquakes and volcanic eruptions shake this region frequently.

Mainland Southeast Asia is also known as the "land of great rivers." Four important rivers flow from Asia's mountainous interior through the region. They are (1) the Irrawaddy in Burma; (2) the Chao Phraya (chow PRIE-uh) in Thailand; (3) the Mekong (MAY-KAWNG), which borders Thailand and flows through Laos, Cambodia, and Vietnam; and (4) the Hong (Red) River in northern Vietnam. The valleys and deltas of these rivers support rich agriculture.

Climate All of Southeast Asia has tropical climates. The Philippines, the western islands of Indonesia, the Malay Peninsula, and most of the

FACTS IN BRIEF Mainland Southeast Asia

COUNTRY POPULATION (1994)	LIFE EXPECTANCY (1994)	LITERACY RATE (1990)	PER CAPITA GDP (1993)
BURMA 44,277,014	58, male 62, female	81%	$950
CAMBODIA 10,264,628	48, male 51, female	35%	$600
LAOS 4,701,654	50, male 53, female	64% (1993)	$900
THAILAND 59,510,471	65, male 72, female	93%	$5,500
VIETNAM 73,103,898	63, male 68, female	88%	$1,000 (GNP)
UNITED STATES 260,713,585	73, male 79, female	97% (1991)	$24,700

Sources: *The World Factbook 1994,* Central Intelligence Agency; *The World Almanac and Book of Facts 1995*

Which country ranks last in figures for life expectancy, literacy rate, and per capita GDP?

coasts of the mainland are humid tropical. These areas have hot, year-long growing seasons, and heavy rainfall supports dense tropical rain forests.

Most of the interior of the mainland countries and the eastern islands of Indonesia have a tropical-savanna climate and are affected by the seasonal monsoon rains and dry seasons. Deadly typhoons, which bring high seas, strong winds, and torrential rainfall, are common in the entire region, especially in the Philippines and Vietnam.

MOVEMENT **Settlement** Fossils found on the island of Java indicate early human settlement in Southeast Asia. Some tribal groups still live on the rugged mountainous islands, including Mindanao, Borneo, and New Guinea. These dark-skinned peoples were driven out of most of the region or mixed with the Caucasians and Mongoloids who migrated to Southeast Asia thousands of years ago.

The Burmese migrated to Southeast Asia from China around 200 B.C. Soon, the Annamese people arrived from northern China and settled in Vietnam. The largest and most advanced culture was that of the Khmer (kuh-MEHR), who dominated present-day Cambodia beginning in the ninth century. By the twelfth century, they had spread their empire to include most of mainland Southeast Asia. In the 1200s, the Thais migrated from southern China to areas settled by the Khmers.

Peoples from beyond Southeast Asia's borders continued to pour into the region for the next several centuries. Indians set up trading cities and brought a Buddhist-Hindu influence to the region. Arab traders, although they did not settle in the region, brought a Muslim influence. European colonization began in the early 1500s with the arrival of the Portuguese, who were followed by the British, Dutch, French, and Spanish. Only Siam (sie-AM), now called Thailand, was not colonized by the Europeans. Many Chinese also migrated to the region during the colonial period.

During World War II, the Japanese conquered all of Southeast Asia. After the war, several Asian nationalist groups sought independence. Soon, new countries appeared on world maps. The Philippines became independent of the United States. French Indochina split into three nations: Vietnam, Cambodia, and Laos. The Dutch East Indies became Indonesia, and the many British colonies and protectorates became independent countries. These included Burma, Malaya and North Borneo (now Malaysia), Singapore, and Brunei.

INTERACTION **Agriculture** Agriculture provides many nations of Southeast Asia with their most valuable exports. Three major types of farming are common in the region: slash-and-burn agriculture, intensive paddy agriculture, and plantation agriculture.

Farming is difficult in the poor soils of the region's rugged, forested mountains. In addition, the slash-and-burn techniques used in many areas force farmers to abandon the fields after only two or three years due to soil depletion. They must move to a new area to start over. Because of the movement from area to area, this form of farming is called **shifting agriculture**. Most farmers in these forested areas are subsistence farmers who grow a variety of crops, including cassava, yams, rice, bananas, pineapples, beans, and sugarcane.

In the lowlands, farmers grow rice on fertile slopes and along river banks. Rice accounts for more than half of the region's agricultural land use, and the tropical climates allow two or even three crops of rice per year. People move the seedlings by hand from seedbeds to the flooded paddy fields. Harvesting also is done by hand. Because paddy agriculture uses much human labor, it is called **intensive agriculture**. This farming method is especially successful in areas such as India, China, and Japan, where there is little arable land. Rice farmers also raise animals in the paddy fields, particularly ducks, water buffalo, pigs, and fish.

During the colonial period, a plentiful supply of low-cost labor, tropical climates, and nearness to sea routes made Southeast Asia an ideal site for large European-owned plantations. Today, most plantations are controlled by the local population. Crops grown for export include rice, coconuts, oil palms, spices, sugarcane, and rubber, the most valuable commercial crop in Southeast Asia. In fact, Malaysia, Indonesia, and Thailand produce most of the world's supply of natural rubber. Coffee also is a valuable export crop, especially that produced in the rich soils of the mountainous islands of Indonesia.

Resources Southeast Asia has large areas of tropical forest, which hold valuable hardwoods such as mahogany, ebony, and teak (an evergreen tree valued for furniture making). As a result, large areas of these forests are being cut for their woods. In most cases, there is little concern about replanting the forests.

This continued clearing of the region's tropical rain forests has already led to serious environmental problems, including loss of habitat for wildlife, soil erosion, and water shortages. It could also affect climates. Some countries in the region now recognize the effects of this deforestation. These countries have set aside national park areas and have instituted logging export bans. For some countries' forests, however, it is almost too late.

Southeast Asia has valuable mineral deposits, particularly tin and oil. The world's major tin deposits are found in Malaysia, Thailand, and Indonesia. Other important minerals include tungsten, iron ore, and manganese. In addition, Thailand has natural-gas fields, Vietnam and Indonesia have coal deposits, and Indonesia, Malaysia, and Brunei export petroleum.

Regional Cooperation Although the Southeast Asian countries are far from a unified group, several are members of the Association of Southeast Asian Nations (ASEAN). These countries—Thailand, Malaysia, Singapore, Brunei, Indonesia, the Philippines, and Vietnam—have treaties of cooperation with each other. Except for the Philippines, these nations are experiencing rapid economic growth. The region's poorer nonmember nations now also are considering joining. The ASEAN countries, meanwhile, are exploring the formation of a free-trade area which will remove tariffs on some products moving among member nations.

SECTION REVIEW

1. What are Southeast Asia's three landform regions? What are its two major climates?
2. What are the region's most important exports?
3. **Critical Thinking** **Explain** how Southeast Asia's physical geography and history may have hindered unity in the region today.

Elephants haul timber in the tropical rain forest of northern **Thailand**. Elephants are used in the timber industry in areas where the terrain is too difficult to use machines.

This street vendor carts his products through the streets of **Ho Chi Minh City, Vietnam**. Vietnamese officials hope that some free-enterprise reforms will help improve the country's economy.

THE COUNTRIES OF MAINLAND SOUTHEAST ASIA

FOCUS

- *What are the physical, economic, and cultural characteristics of each country in mainland Southeast Asia?*
- *What are the political backgrounds of these countries? What are their current political situations?*

From east to west, the countries of mainland Southeast Asia are Vietnam, Laos and Cambodia (with Laos to the north and Cambodia to the south), Thailand, and Burma. (See the map on page 531.)

Vietnam Vietnam is a long, narrow country that occupies the eastern portion of the Indochina Peninsula. The country's east coast faces the South China Sea. Laos and Cambodia form its rugged western border, and China is to the north. Mountains dominate Vietnam's interior.

Vietnam's population is heavily concentrated in the deltas formed by the Hong River in the north and the Mekong River in the south, and on the coastal plain between the two deltas. The Vietnamese, who arrived from China more than 2,000 years ago, make up more than 85 percent of the country's population. The largest minority groups are the Chinese, who live mainly in the cities, and the Montagnards, who live in the mountains.

Vietnam, along with Laos and Cambodia, was part of French Indochina. When the French returned to Southeast Asia after World War II, they found that Communist forces (the Vietminh) had gained power in the north. The Vietminh, under the leadership of Ho Chi Minh, demanded independence.

After eight years of fighting, the Vietnamese defeated the French in the north. As a result, in 1954 Vietnam was divided along the 17th parallel. North Vietnam established a Communist government, with its capital at Hanoi (ha-NOI). South Vietnam established a government friendly to the United States, with its capital at Saigon (sie-GAHN). Both governments wanted reunification but could not agree on the terms. In 1963, North Vietnamese troops invaded South Vietnam. The U.S. military supported the South Vietnamese forces.

During the Vietnam War, more than 55,000 Americans and more than 2 million Vietnamese were killed. Due largely to the high human costs of the war, the United States withdrew its troops in 1973. Soon after, the South Vietnamese government collapsed. In 1975, North Vietnamese forces occupied Saigon and renamed it Ho Chi Minh City. The country was officially united in 1976 as the Socialist Republic of Vietnam. More than 1 million South Vietnamese tried to escape the Communist takeover. Seeking a new homeland, many of these refugees found their way to the United States, Australia, and other Western nations.

The long war devastated Vietnam's economy. Forests and other vegetation over about one-fifth of South Vietnam's area still show the effects of **defoliants** sprayed by U.S. forces. A defoliant is a chemical that makes plants' leaves drop off, usually killing the plants. In addition, much of Vietnam's infrastructure, especially roads and airports, still must be rebuilt. To speed economic development, the Communist government recently opened the nation to a free-market economy and has welcomed outside investment. The United States government ended a 19-year ban on trade with Vietnam in 1994 and restored diplomatic relations with the country the following year.

Vietnam has the resources necessary to build a strong economy. In the north, there are coal, oil, and mineral resources, while the delta of the Hong River is a fertile agricultural region. The south has the productive rice fields of the Mekong Delta as well as rubber plantations. Vietnam is now one of the world's largest rice exporters after Thailand and the United States. Other major exports include oil, rubber, and seafood. Although Vietnam remains one of Asia's poorest nations, it has the potential to become strong and self-sufficient.

Laos Landlocked and mountainous, Laos is one of the poorest countries in the world. It was first unified in the fourteenth century under a Lao prince. From the 1890s until 1954, Laos was under French rule.

Although the country was officially neutral during the Vietnam War, major supply routes for the North Vietnamese crossed Laos. The United States bombed parts of the country in an attempt to block these routes. In 1975, Communist forces took over the monarchy and formed the Lao People's Democratic Republic.

Laos's most important crop is rice, but the country also grows tobacco, corn, cotton, and coffee. Hydroelectric power exported to neighboring Thailand is the nation's major source of income. Lumber is the other leading export. In 1990, in an effort to conserve its tropical forests, Laos banned the export of logs. Illegal logging by corrupt military leaders for export to Japan and other nations continues, however. The country has almost no industry and survives largely on foreign aid.

More than half of the population is made up of Lao, who are closely related to the Thais. The Lao are Buddhists and live mainly in rural valleys, especially along the Mekong River. The remainder of the population consists of highlanders and isolated mountain groups, such as the Hmong people. In total, Laos has more than 50 ethnic groups. The capital and only large city is Vientiane in the Mekong River valley.

The main problem facing Laos is poverty. More than 80 percent of the people are subsistence farmers who live in small, isolated villages. Most of these villages are not served by roads. There is little government spending on health care, and life expectancy is nearly 20 years less than that in neighboring Thailand.

The government, though Communist, allows privately owned businesses and religious freedom. Since the Vietnam War, more than 10 percent of the Laotian population has fled the country. Many now live in the United States and Australia. Thousands more remain in refugee camps in Thailand.

Cambodia Cambodia, the successor of the Khmer Empire, lies between Vietnam and Thailand. Most of its population are Khmer and Buddhist. The people are concentrated in Cambodia's fertile rice-growing region along the Mekong River. Rice and fish are the main food sources. The capital is Phnom Penh, located in the rice-growing region of the south.

Located in the northwest is Tonle Sap, Southeast Asia's largest lake. Tonle Sap supports a fishery and acts as a reservoir during the monsoon season. It helps reduce flooding downstream by filling with overflow water from the Mekong River. After the monsoon, water drains back into the Mekong.

Since independence from France in 1953, Cambodia has experienced much political unrest. During the Vietnam War, the United States bombed North Vietnamese supply lines in Cambodia. In 1975, Communist forces called the Khmer Rouge seized power. The Khmer Rouge

These fishers check the haul in this fish trap at Tonle Sap, a lake in western **Cambodia**. The freshwater lake supports an important fishing industry in the country.

government isolated Cambodia and launched a program to establish a "Khmer peasant nation."

The goals of the Khmer Rouge were similar to those of the Cultural Revolution in China. The methods of achieving a simple, pure society were much more brutal in Cambodia, however. Cities were emptied, the educated were executed, families were separated, and citizens were forced into field labor. More than 1 million Cambodians and others, such as Vietnamese, were killed by their own government. At this time, the nation was called Kampuchea.

Vietnamese forces invaded Cambodia in 1978. They overthrew the Khmer Rouge and set up a government friendly to Vietnam. Today, despite periodic violence, the nation is trying to rebuild after years of warfare. Thousands of Cambodians remain in refugee camps in neighboring Thailand. United Nations troops are stationed in Cambodia to oversee the country's peaceful transition to democracy. Cambodia has sought economic aid from Japan and other countries.

Thailand *Thailand* means "land of the free," and it is the only Southeast Asian country that was not a European colony. The Kingdom of Thailand occupies the central part of the Southeast Asian mainland and extends south into the Malay Peninsula. Rugged, forested mountains form Thailand's border with Burma in the north and west. The Mekong River forms its eastern boundary with Laos. To the southeast is Cambodia.

The plains along the Chao Phraya River in central Thailand are the most populated and productive part of the nation. The climate is mostly tropical savanna and is driest on the Khorat Plateau in the northeast portion of the country. In the south along the narrow Malay Peninsula, the climate is humid tropical. All of Thailand is subject to seasonal monsoonal rainfall.

Thailand leads the world in rice exports and is a major fish exporter. Sugarcane, corn, and rubber are other export crops. Many of the country's food exports go to Japan, Thailand's major trade partner. Forests provide valuable tropical hardwoods, but extensive forest loss has affected the water supply and has caused other serious environmental problems. Thailand's minerals include tin, tungsten, and lead. The country also has oil resources and large offshore natural-gas deposits in the Gulf of Thailand.

The economy of Thailand has grown rapidly with the help of foreign investment, especially from Japan. The textiles, electronics, and automobile assembly industries have experienced rapid growth, while tourism provides one of the nation's largest sources of foreign income. Major imports include automobiles and machinery from Japan.

Bangkok, Thailand's largest city and capital, is built on the Chao Phraya. Much of Bangkok is connected by canals, called *klongs*, for which the city is famous. Bangkok also is well known for its palaces and Buddhist monasteries. In recent years, it has become known for its traffic jams and air pollution. As more and more water is drawn out of wells, the city is slowly sinking.

Thailand is a constitutional monarchy with a king, and popular support for the royal family is strong. Through much of Thailand's history, however, political power has been in the hands of military leaders.

More than three-quarters of the population is made up of Thais, who are Buddhist. The remaining population includes Chinese, Lao, and many mountain peoples. Along Thailand's eastern border areas are thousands of refugees from Cambodia, Laos, and Vietnam. This border region has also been an area of tension between Thailand and neighboring Cambodia and Laos.

Burma Burma occupies the mountainous northern and western part of the Southeast Asian mainland. It borders China on the north, India and Bangladesh on the west, and Thailand and Laos on the east. The country's southernmost tip extends to the narrow Isthmus of Kra along the Andaman Sea.

All of Burma is tropical and strongly influenced by the monsoon. The southern part of the country has a humid-tropical climate and receives up to 110 inches (279 cm) of rainfall annually, mostly during monsoonal downpours between June and September. Burma's northern part, which falls in a mountain rain shadow, is drier and has a tropical-savanna climate. Because mountains block most of the summer monsoon rains from the Indian Ocean, annual rainfall here can be as little as 32 inches (81 cm).

The broad north-south Irrawaddy River valley, much of which is cultivated, dominates the central part of Burma. A series of mountain ranges runs north to south over most of the rest of the country. Tropical forests cover much of the nation and contain the world's largest reserves of teak.

Burmans form more than half of the population. They are Buddhists and have a culture rich in art, architecture, and literature. Many ethnic minorities live in the mountains, but most of the population is crowded into the Irrawaddy River valley. The country's capital and largest city is Rangoon (also called Yangon).

This *klong* (canal) in **Bangkok, Thailand**, features a floating market. Bangkok's system of canals is important to the local transportation of people and goods.

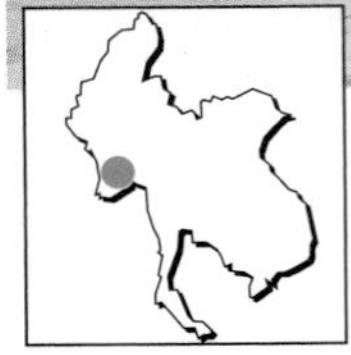

The Shwe Dagon pagoda, a Buddhist place of worship, is one of the most famous buildings of **Rangoon, Burma**. Nearly nine of every ten Burmese are Buddhists.

Rice, Burma's chief crop, is grown on the river floodplains. The country was once Asia's largest rice exporter, but today rice accounts for only a small part of the country's exports. Instead, lumber and seafood are the major exports. Burma also supports a large illegal opium trade. Smugglers distribute the drug as heroin in the United States, Europe, and many other parts of East and Southeast Asia.

Oil production in the lower Irrawaddy River valley is limited and goes to meet local needs. Mines in the northern mountains contain jade, and the central and southern parts of the country are rich in other minerals, including silver, lead, zinc, copper, tin, and tungsten. Offshore gas deposits have been discovered but not yet developed.

Burma gained its independence from Britain in 1948. In 1962, a military government took over all business and trade and created a socialist economy cut off from foreign influence. Indians and Chinese who lived in the country were driven out to neighboring nations.

In 1989, the military government changed the country's name from Burma to Myanmar, but the U.S. government still recognizes the country as Burma. Although pro-democracy protests have taken place in Burma, the military government remains in power. Government troops continue to battle minority and rebel groups in the northern mountains, and thousands have fled into Bangladesh to the west. To strengthen its position and the economy, the government now seeks foreign investment.

SECTION REVIEW

1. What ethnic groups live in each country of mainland Southeast Asia? To which countries is the production of rice important, and how?
2. How would you characterize political events in the region over the past few decades? What type of government does each country have today?
3. **Critical Thinking** **Compare and contrast** post–World War II political developments in Vietnam to those in Korea.

CHAPTER 44 CHECK

Angkor Wat temple ruins in northwestern **Cambodia**

Reviewing the Main Ideas

1. Southeast Asia contains three landform regions and has tropical climates. Many different groups of people have settled in Southeast Asia over time. The region has valuable forest and mineral resources. Economic cooperation among the countries of the region is growing.
2. The countries of mainland Southeast Asia are Vietnam, Laos, Cambodia, Thailand, and Burma. These countries have experienced much political unrest over the past few decades.

Building a Vocabulary

1. What term describes the form of farming in which farmers periodically move to new areas? Why is this movement necessary, given the techniques the farmers use?
2. What is intensive agriculture? Give an example of a crop whose cultivation often involves intensive agriculture.

Recalling and Reviewing

1. What are the four major rivers of Southeast Asia? How are these rivers important to the region?
2. Why are large forest areas of Southeast Asia being logged? What effects might continued deforestation have?
3. How were Laos and Cambodia involved in and affected by the Vietnam War?
4. Name one city from each country of mainland Southeast Asia. List a characteristic of each city.

Thinking Critically

1. Thailand has numerous refugees from other countries in Southeast Asia. Why have these refugees left their own countries? Why do you think have they gone to Thailand? How might Thailand's history be related to the reasons refugees go there?
2. For what reasons might a country use a defoliant during war?

Learning About Your Local Geography

Individual Project

Interview community members and read old newspapers (available at the public library) to find out how your local community was affected by the Vietnam War. How did community residents feel about the war at the time it was happening? Did many people from the community fight in the war? If so, how did the experience affect them? Put your findings together into a brief written report. You may want to include your report in your individual portfolio.

Using the Five Themes of Geography

Draw two illustrations of different forms of **human-environment interaction** in Southeast Asia. How has the **region's** absolute **location** influenced human-environment interaction? In particular, you might want to focus on different forms of agriculture that are made possible by the region's location, or on the use of certain natural resources.

CHAPTER 45

Island Southeast Asia

GEOGRAPHY DICTIONARY

archipelago
entrepot
kampong
sovereignty

"The frill of shrubby ferns and grass at the edge quickly became swallowed in tangles of vines, bamboo, rattan, palms, acacias, and tree ferns over which towering mahoganies and oaks cast a ragged, sun-shielding canopy. The moist jungle sprang up sawtooth slopes and down into ravines, across limestone cliffs, and broad plateaus, climbing 5,000-foot peaks before it slanted onto the sandy shores of the Celebes Sea."

John Nance

Part of a dance costume from **Bali, Indonesia**

In 1991, Mount Pinatubo, on the Philippine island of Luzon, began to erupt for the first time since 1380. The volcano shot ash over 9 miles (14 km) into the air and sent 1,500°F (816°C) lava flowing down the mountain at more than 60 miles per hour (97 kph).

Luzon is just one of the some 20,000 tropical islands of Southeast Asia. Physical geographers call the region the Malay Archipelago (ahr-kuh-PEHL-uh-goh). An archipelago is a large group of islands. These islands first attracted the seafaring Malay people centuries ago. The area's riches also have attracted Chinese, Indians, and various European colonial powers, who dubbed the region the East Indies. They also called part of the region the Spice Islands because of its valuable spices, such as cinnamon and pepper. Today, island Southeast Asia has a wealth of cultures, languages, foods, music, and art.

Bamboo forest on **Sumatra, Indonesia**

ISLAND SOUTHEAST ASIA: PHYSICAL–POLITICAL

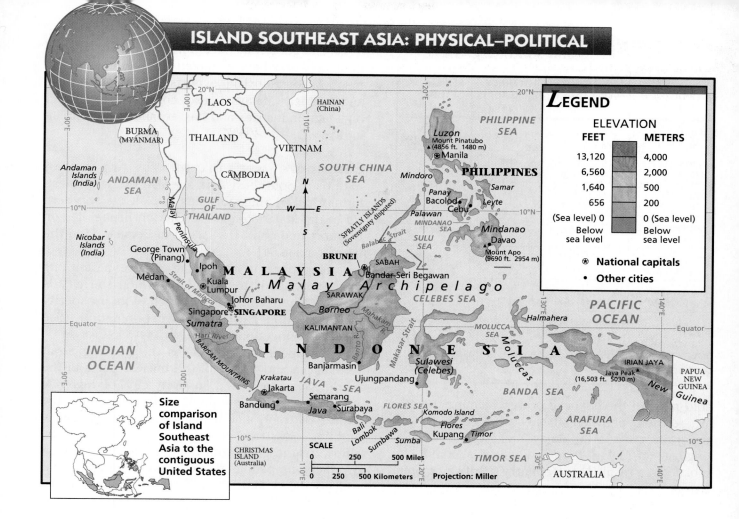

Size comparison of Island Southeast Asia to the contiguous United States

1 MALAYSIA, SINGAPORE, AND BRUNEI

FOCUS

- *What are the physical, cultural, and economic characteristics of Malaysia?*
- *What factors account for the prosperity of Singapore and Brunei? How are the countries' citizens affected by the prosperity?*

Malaysia The South China Sea divides the country of Malaysia into two parts. Western Malaysia occupies part of the narrow Malay Peninsula and is connected to mainland Southeast Asia. Eastern Malaysia lies on the northern part of the island of Borneo. (See the map above.)

The British governed Malaya, on the Malay Peninsula, until 1957. Then in 1963, Malaya joined with three other former British colonies—Sarawak and Sabah, both on north Borneo, and Singapore—to form a new country called the Federation of Malaysia. Singapore left the Federation in 1965 and became an independent country.

All of Malaysia lies near the equator and has a humid-tropical climate. The interior of the country is forested and mountainous. Its population is concentrated along the Malay Peninsula's western coastal plain, where the capital city of Kuala Lumpur (kwahl-uh LUM-pur) lies.

Slightly more than half of the country's people are Muslim Malays. The largest minority groups are the Chinese, who make up one-third of the population, and Indians, who make up one-tenth. Other minority groups include the Dayaks, a tribal people of Eastern Malaysia. The Malays control the nation's political power, while the Chinese, who live mainly in urban areas,

FACTS IN BRIEF Island Southeast Asia

COUNTRY POPULATION (1994)	LIFE EXPECTANCY (1994)	LITERACY RATE (1990)	PER CAPITA GDP (1993)
BRUNEI 284,653	69, male 73, female	95% (1987)	$9,000 (1991)
INDONESIA 200,409,741	59, male 63, female	77%	$2,900
MALAYSIA 19,283,157	66, male 72, female	78%	$7,500
PHILIPPINES 69,808,930	63, male 68, female	90%	$2,500
SINGAPORE 2,859,142	73, male 79, female	88%	$15,000
UNITED STATES 260,713,585	73, male 79, female	97% (1991)	$24,700

Sources: *The World Factbook 1994,* Central Intelligence Agency; *The World Almanac and Book of Facts 1995*

Which country in island Southeast Asia has the highest per capita GDP?

hold most of the financial power. Calming the ethnic tension between these two groups is an important concern in Malay society. The national language is Malay, though English is commonly spoken as a result of the country's British colonial history.

Malaysia is rich in natural resources and is a major exporter of rubber, palm oil, timber, tin, and oil. Rubber plantations and tin mines are located mainly on the Malay Peninsula, while offshore oil and natural-gas deposits line the Strait of Malacca. The economy of Eastern Malaysia depends heavily on forestry and petroleum.

Industrialization in Malaysia is taking place at a rapid pace. The country's stable government, educated work force, and good transportation and telecommunications networks have attracted much foreign investment. Major industrial exports include electronics and textiles. Malaysia also continues to attract high-tech industries and has expanded its tourism industry.

Singapore and Brunei The two smallest nations in Southeast Asia are also the wealthiest. Industrialized Singapore lies just off the southern tip of the Malay Peninsula. Oil-rich Brunei lies on the north coast of the island of Borneo. (See the map on page 541.)

Singapore is the name of the nation, its major city, and the island on which the city is located. The country owes much of its prosperity to its location and productive population. (See "Global Connections" on pages 544–545.)

Singapore's harbor faces the bustling Strait of Malacca and is one of the busiest ports in the world. Singapore is a modern city of high-rise hotels, banks, and shopping malls. The city is the center of trade and banking for Southeast Asia, and its industries produce refined-oil products, ships, textiles, electronics, and many high-tech products. Its people are generally prosperous and well educated. More than 75 percent of the population are Chinese. The remainder is Malay and Indian.

Brunei is an independent, oil-rich sultanate. The government provides its citizens with basic education, health care, and some housing. About two-thirds of the population is Malay, while about

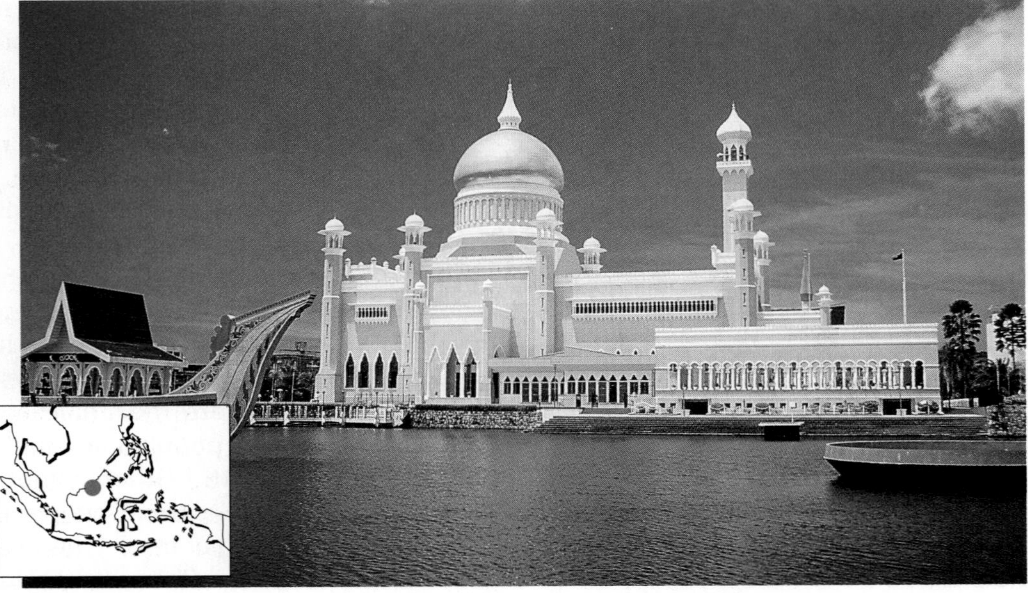

This mosque in **Bandar Seri Begawan, Brunei,** is evidence of the wealth provided by the small country's oil industry. Revenues from the oil industry account for more than half of the country's GDP.

one-fifth is Chinese. Thousands of guest workers from other Southeast Asian nations also live in the country.

Brunei's economy is based primarily on petroleum exports. Knowing that its oil supplies will run out one day, the government has invested for the future. It has plans for agricultural expansion and, to attract new industry, it has a tax policy favorable to businesses.

SECTION REVIEW

1. On what two major land features does Malaysia lie? Why is there tension between the country's two largest ethnic groups?
2. How does the overall prosperity of Singapore and Brunei benefit the countries' citizens?
3. **Critical Thinking** **Decide** which country's economy has a stronger foundation for future growth: that of Singapore or of Brunei. Explain your reasoning.

INDONESIA

FOCUS

- *What is the physical geography of Indonesia? What are the country's major islands?*
- *What economic, cultural, and political characteristics does Indonesia have?*

For more than 300 years, Indonesia was ruled by the Dutch. The colony was known as the Netherlands East Indies (also called the Dutch East Indies). Indonesia became independent in 1950. Today, the country has one of the world's most rapidly increasing populations.

REGION **Physical Geography** Indonesia is the largest country in Southeast Asia. Its 13,660 islands stretch some 3,000 miles (4,827 km) between the Indian and Pacific oceans, nearly the same distance as that between Los Angeles and New York. (See the map on page 541.)

All of Indonesia lies along the equator, giving the entire nation a humid-tropical climate. The islands to the west receive rain year-round, while the islands to the east experience more seasonal monsoon rainfall. Most of the islands, especially Kalimantan and Irian Jaya, are covered by vast tropical rain forests. A wide variety of wildlife lives on the islands, including the Komodo Island dragon, a lizard that weighs up to 300 pounds (136 kg) and grows up to 12 feet (3.6 m) in length. Most coastal areas have dense mangrove swamps and many coral reefs.

Its location at the junction of several major tectonic plates makes Indonesia the most active volcanic region on Earth. The nation has more than 150 volcanoes, about 50 of which are active. In 1883, the volcanic island of Krakatau erupted, killing thousands of people on nearby Java and Sumatra. Although the volcanoes present a danger, they also provide the lava and ash that enrich the islands' soils.

Major Islands From west to east, the largest islands are Sumatra, Java, Kalimantan (part of Borneo), Sulawesi (also called Celebes), Timor, and Irian Jaya (the western part of New Guinea). Holding most of the nation's industry and political power, Java is the core island of Indonesia. Although Java makes up only seven percent of the country's land area, it is home to two-thirds of the population. It is one of the world's most densely populated areas and is home to the huge, rapidly growing capital city of Jakarta (juh-KART-uh). Vast slums are expanding around the city.

The larger island of Sumatra lies at the far western edge of Indonesia. The island has rich natural resources, including volcanic soils, dense tropical forests, and large oil and natural-gas deposits. Many different cultures exist on the island.

Bali, Java's island neighbor to the east, has a rich cultural background. The Balinese practice a Hindu-Buddhist religion, and thousands of elaborate temples and shrines dot the island. The islanders are known in particular for their art, festivals, and dances. Bali supports one of Indonesia's largest tourist industries.

The island of Timor was once split between the Dutch, who controlled west Timor, and the Portuguese, who controlled east Timor. Indonesia took over the western half of the island when it gained its independence from the Netherlands. East Timor, however, was seized by Indonesia

Text continues on page 546.

Global Connections

Singapore

Travelers visiting Singapore should take care to read the rules posted all over the city. Laws prohibit spitting, jaywalking, littering, smoking in restaurants, and eating or drinking on the subways. Fines over the equivalent of 300 U.S. dollars can be the price for breaking one of these rules. These laws help make Singapore one of the cleanest nations in the world. It is also one of the most efficient, and there is little unemployment, poverty, or crime. Life in Singapore was not always so ordered and prosperous, however.

In just a few decades, Singapore has risen from a poor country to become an economic force on the Pacific Rim.

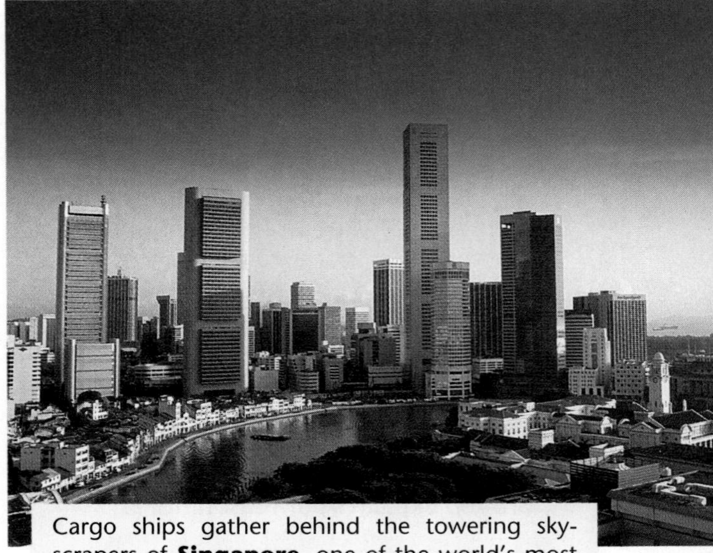

Cargo ships gather behind the towering skyscrapers of **Singapore**, one of the world's most important shipping and trade centers.

The Importance of Location

A tiny island-nation, Singapore has little land and few natural resources. Its one economic advantage is its location. Situated between the Indian Ocean and the South China Sea, Singapore is at a crossroads of Southeast Asian trade. Its location attracted the British to Singapore in 1819, and Singapore soon became a busy Asian port and a major British naval base.

When Singapore gained its independence in 1959, many people were still poor farmers. The nation lacked adequate housing, and about half the population had never attended school. For a few years, Singapore was united with Malaysia, but the two separated in 1965. Then Britain decided to close down its naval base in Singapore. With little industry, Singapore's economy seemed too weak to support its people.

Singapore was still able to profit from its location, however. The island served as an **entrepot** (AHN-truh-poe), or intermediate port, for nations sending goods between the Pacific and Indian oceans. These nations used Singapore to temporarily store their goods before shipping them to other countries. Southwest Asian nations such as Indonesia also began shipping their oil to Singapore, where it was refined before being exported to countries around the world.

Warehousing and oil refining were not enough to support a strong economy, however. Singapore's then Prime Minister Lee Kuan Yew realized that sustained economic growth depended on the development of industry. The government set out to create new economic resources where there were none before.

The government launched a long-term industrialization program. Lacking the money to build large factories, Lee sought to lure foreign investment. To encourage multinational companies to locate in Singapore, the government improved transportation facilities, educated workers, and offered low tax rates on businesses. Construction began on roads, railroads, power stations, housing for workers, a new port, and an airport.

By the early 1990s, 3,000 foreign companies had operations in Singapore. Of these, 800 were American, making the United States the largest of Singapore's foreign investors. Today, Singapore is one of the world's busiest ports. It is southern Asia's banking center, the world's third largest oil refiner, and a leader in communications and information technology.

Major World Shipping Routes

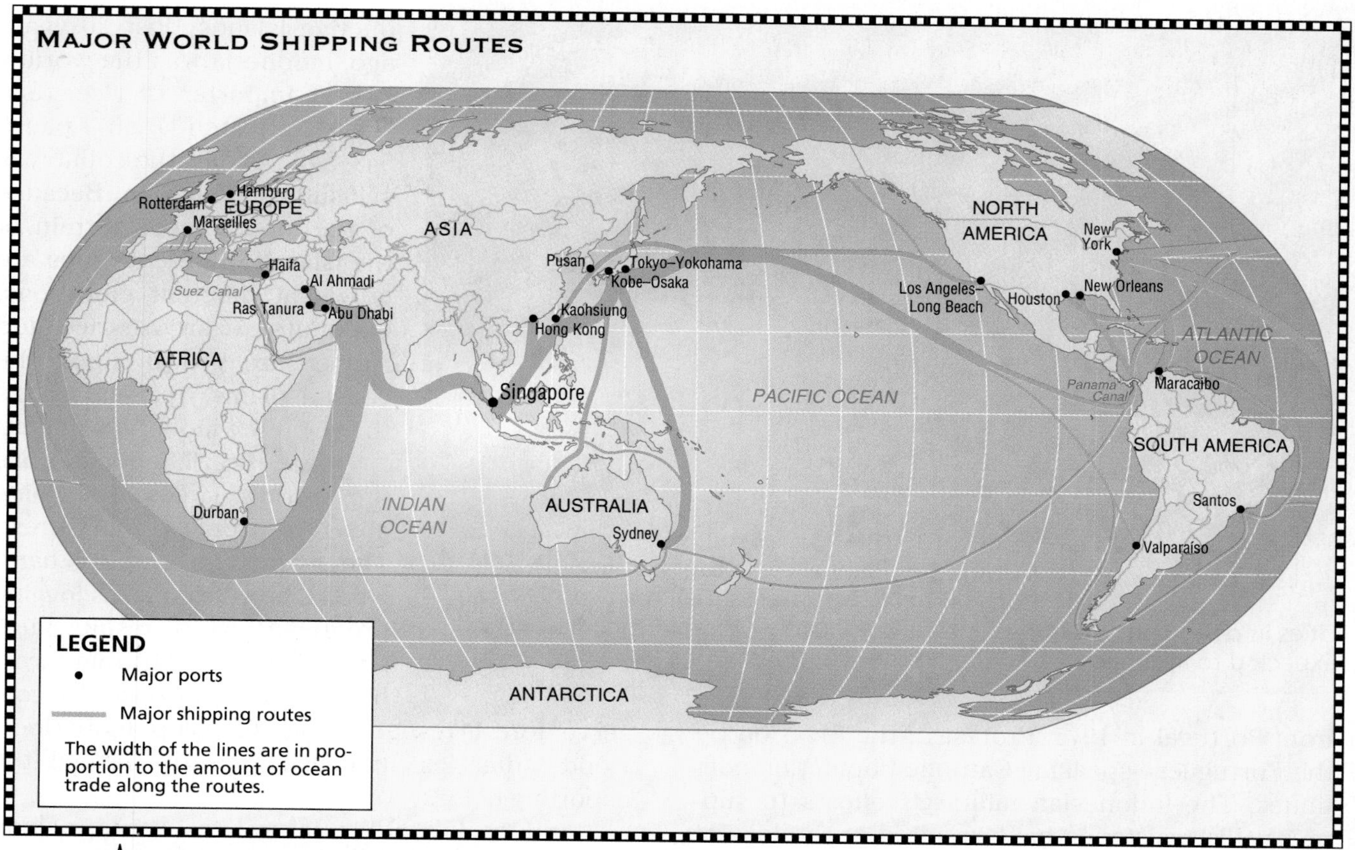

LEGEND

- Major ports
- Major shipping routes

The width of the lines are in proportion to the amount of ocean trade along the routes.

MAP STUDY Maps can provide useful information to businesses trying to get products to markets. The thick bands on this map show major shipping routes throughout the world. The thicker the width of a band, the greater the amount of goods that are shipped along the route. A prime location helps make Singapore a hub of international shipping. Which two canals on this map are important links along shipping routes?

Another Side to Success

Singapore is the product of a focused government with significant economic and political control. In fact, the government controls nearly every aspect of life in Singapore. Elections mean little because one political party is so strong that it almost always wins. The government owns many of the nation's companies and controls the press, making sure that newspapers do not criticize government decisions. It also bans certain movies and music.

According to one human-rights organization, "Singapore is not a country in which individual rights have significant meaning." This may be the price of Singapore's economic success. But for now, most of Singapore's people are content to enjoy the rewards of a government "dedicated to the pursuit of capitalism."

Singapore residents enjoy their beautiful city as they relax at this festival market. Strict antilittering laws help keep Singapore one of the world's cleanest cities.

Your Turn

Choose a country in Southeast Asia and **investigate** its economic resources, political system, and history. Imagine you are a government official of this country, and **write** a report suggesting policies for developing the economy. Discuss the policies' positive and negative consequences.

Crowded residential areas continue to push outward from **Jakarta, Indonesia**. Like many other large cities in developing countries, Jakarta's population is rapidly increasing and is expected to approach 14 million by the year 2000.

from Portugal in 1975. Protest of the invasion by the Portuguese-speaking Catholic population continues. The Indonesian military's efforts to suppress this resistance have resulted in the loss of thousands of lives.

The three-fourths of the large island of Borneo that is part of Indonesia is called Kalimantan. The population of this island is made up of numerous groups with different languages and cultures. To the east of Kalimantan, across the Makasar Strait, is the island of Sulawesi. Two peninsulas jutting out to the east and southeast from a north-south backbone make Sulawesi look a little bit like a capital K. With 3,404 miles (5,477 km) of coastline, no place on the island is more than 70 miles (113 km) from the sea.

Irian Jaya, the former Dutch colony on the western half of the huge island of New Guinea, became part of Indonesia in 1969. Irian Jaya is extremely rugged and forested. Its snow-capped mountain spine reaches an elevation of 16,500 feet (5,029 m). Various peoples live in the highlands, forests, and swamps of this undeveloped land. New Guinea's eastern half is the independent country of Papua New Guinea. (See Unit 11: The Pacific World and Antarctica.)

Economic Geography Indonesia remains a poor agricultural country, but it has taken steps toward increased industrialization and prosperity. Agriculturally, Java and Sumatra are the most productive islands. Just 20 years ago, Indonesia was the world's largest importer of rice. (See "Planet Watch" on pages 548–549.) Today, the country is self-sufficient in rice. Because of the mountainous terrain of the islands, different crops are grown at different elevations. Coconuts, sugarcane, rice, and rubber are grown along the coastal plains. Tea and coffee thrive on the mountain slopes.

After Brazil, Indonesia has the world's largest remaining areas of tropical rain forest. The market for tropical hardwoods, however, makes logging profitable. As a result, huge tracts of Indonesia's rain forest have been destroyed. To conserve forest resources and develop more local wood industries, Indonesia recently banned the exporting of logs.

Indonesia's other natural resources include large deposits of tin, coal, and oil. Petroleum and natural gas are major exports, with petroleum being the country's main source of income. Most of the oil deposits are on Sumatra, Java, Kalimantan, and Irian Jaya. Petroleum deposits also lie under the Java Sea.

Industrialization in Indonesia is in an early stage of development. The goal is to decrease the country's dependence on oil exports and to further develop industry. Most of Indonesia's present factories produce textiles, food and wood products, and chemicals related to petroleum. Indonesia seeks to attract more high-tech industries from the developed nations. Its supply of labor and low wages—some Indonesian factory workers earn less than the equivalent of one U.S. dollar a day—have already drawn many multinational corporations. In addition, Indonesia hopes to modernize the economy by developing a reliable communications system for all the islands.

Most industry is centered in a few cities on Sumatra and Java, especially in Jakarta. Millions of people have crowded into Jakarta, many of whom live in the ***kampongs*** that ring the city. A *kampong* is a traditional village, but it has also come to be the term for the slums around Jakarta.

Human Geography Almost 90 percent of Indonesians are Muslims, giving this area the largest Muslim population in the world. Hindus, Buddhists, Christians, and animists also live in Indonesia.

Indonesia's complex cultural geography is evident in the more than 300 ethnic groups and more than 350 languages and dialects spoken on the various islands. To help unify the nation's population, the government has started a program to persuade all citizens to use a national language called Bahasa Indonesia. Basic education for all Indonesians is another major goal of the government. The national motto is "Unity in Diversity."

Like in many other Southeast Asian countries, however, a serious "wealth gap" exists among the different groups in Indonesian society. Chinese Indonesians are among the wealthier groups, as they own most of the nation's businesses. The economic status and Christian religion of the Chinese in Indonesia have led to resentment and even violent attacks against them by Muslim Indonesians.

Since independence in 1950, Indonesia has had only two political leaders, both of whom have supported strong military control of the nation. Java dominates the country politically and economically. The Indonesian government's major goal is economic development.

This man crafts a guitar at a small factory in the **Philippines**. Manufacturing, even in small industries, is an important source of jobs in the country.

SECTION REVIEW

1. Name the major islands of Indonesia and list one main characteristic of each.
2. What is Indonesia's main source of income? Why have multinational corporations been attracted to the country?
3. **Critical Thinking** **Explain** the meaning of the motto "Unity in Diversity." Do you think the motto applies to the United States? Why or why not?

THE PHILIPPINES

FOCUS

- *What are the major characteristics of the Philippines' islands, population, and economy?*
- *What challenges do the Philippines face?*

LOCATION **Physical Geography** The Republic of the Philippines lies north of Indonesia and east of mainland Southeast Asia across the South China Sea. (See the map on page 541.) The Philippines include two main islands, Luzon in the north and Mindanao in the south. Between these two large islands are more than 7,000 smaller ones. Only about 700 are inhabited.

The Philippine islands lie on the Pacific Ring of Fire, and most of the islands are volcanic. Most of the Philippines' rugged terrain is covered by dense tropical rain forests. All the islands have a humid-tropical climate and lie in the path of dangerous Pacific typhoons.

Population The central plain of Luzon holds the greatest concentration of people. The capital, Manila, is located here and is a major seaport and the largest industrial area. This already crowded urban area of Luzon continues to grow, drawing workers from the rural areas of all the islands.

The population of the Philippines, called Filipino, is of Malay origin, with strong Chinese

Text continues on page 550.

PLANET WATCH

Rice Growers Take a Stand

In 1986, 1,000 agriculture experts covered much of Indonesia on motorcycles to bring new techniques to farmers and save the nation's rice crops. In that one year alone, the brown planthopper, an insect that causes rice to dry out and rot, had wiped out enough food to feed 3 million people. Part of an ambitious and highly successful program called Integrated Pest Management (IPM), these field trainers helped bring the planthopper plague under control.

Fighting Pests with Chemicals

In the late 1960s, Indonesia was one of the world's largest importers of rice. To become self-sufficient in rice, the government set out to help farmers improve crop yields. To do this, the government built modern irrigation systems for rice paddies and introduced new strains of rice. To protect crops from insects and disease, the government spent as much as $150 million a year on pesticides. Often, farmers had to agree to use these chemicals before the government would loan them money.

The belief that increased rice production depended on increased use of pesticides outweighed the environmental and health concerns about pesticide use.

At first, the plan seemed to work. By 1983, the nation no longer had to import rice. Success proved to be short-lived, however. By 1986, about 600,000 tons of rice, ravaged by planthoppers, lay rotting in farmers' fields. The pests seemed unstoppable, even though farmers were using more and more pesticides. The Indonesian government then learned of research conducted by the United Nations Food and Agriculture Organization (FAO). FAO research showed that pesticides were not the solution to the planthopper plague. Instead, the chemicals were a large part of the problem.

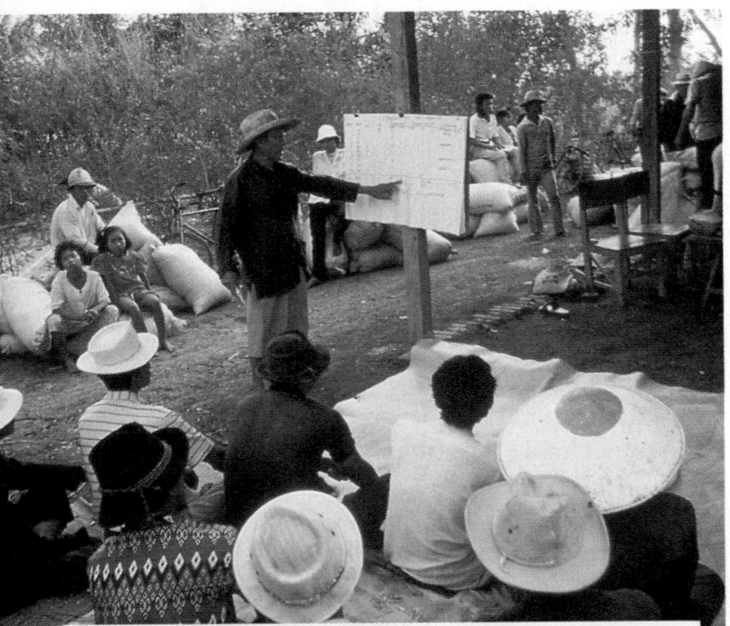

Farmers on **Java, Indonesia**, learn about efficient pest-control methods at this class sponsored by the United Nations Food and Agriculture Organization. These classes help prevent the overuse of chemical pesticides.

Managing Pests a New Way

Peter Kenmore, the FAO manager and researcher whose work brought new rice-growing methods to Indonesia, says, "Trying to control population outbreaks with insecticides is like pouring kerosene on a house fire." Kenmore found that uncontrolled spraying of pesticides—some farmers sprayed their fields up to 15 times a month—was in fact helping the planthoppers to thrive. The chemicals were killing insects that could control the population of planthoppers naturally, such as wolf spiders, ladybugs, and wasps. Furthermore, pesticides had brought about changes in the planthoppers, allowing them to reproduce in greater numbers.

The FAO offered Indonesia a long-term solution to its problem through Integrated Pest Management. Farm experts toured the country, identifying places with planthopper problems and training farmers to farm without pesticides. They visited the small theaters that are at the center of life in most Indonesian villages. There,

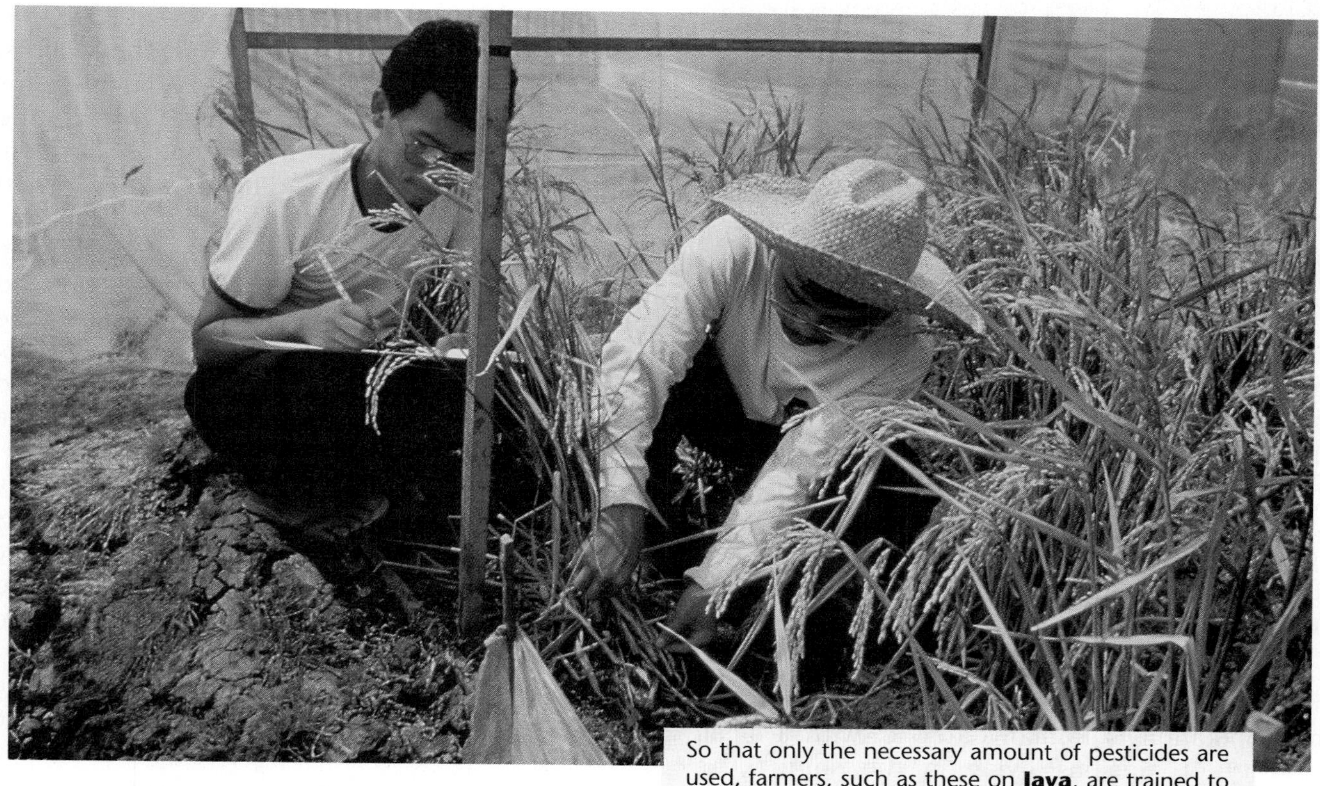

So that only the necessary amount of pesticides are used, farmers, such as these on **Java**, are trained to identify and count insects. Pesticides also kill beneficial insects that otherwise would help keep the numbers of harmful pests in check.

villagers staged plays, acting out the roles of farmers and pests. In "Farmer Field Schools," small groups met weekly to learn IPM methods. "By going out into the rice paddies," Kenmore says, "the trainers showed farmers how to diagnose problems, calculate the ratio of good bugs to bad and decide how much damage the crop could take without harming the yield." Farmers also learned to identify whether crop damage was caused by pests or by other factors. Armed with this knowledge, farmers could determine whether pesticide use would be helpful or harmful.

The Benefits of IPM

Support from farmers and the government has made the Integrated Pest Management program a huge success in Indonesia. Today, Indonesian farmers use pesticides only when absolutely necessary and in small amounts.

A lady beetle, a beneficial insect, feasts on an aphid that was attacking an alfalfa.

Limited spraying has allowed beneficial insects like the wolf spider to survive, and careful planning of crops has denied the planthoppers a continuous source of food. In the years since the government banned the use of most pesticides, the planthopper plague has disappeared and rice yields have increased. Government dollars once spent on pesticides are now used for IPM programs that teach farmers new ways of managing crops.

Now farmers want IPM techniques for crops other than rice. "They are demanding integrated pest management for their other crops and asking why we don't have a system for beans or cabbage," Kenmore points out. "I tell them it will just take more research."

Your Turn

Review the definition of *ecosystem* on page 54. **Write** a paragraph explaining how Integrated Pest Management preserves a rice paddy ecosystem.

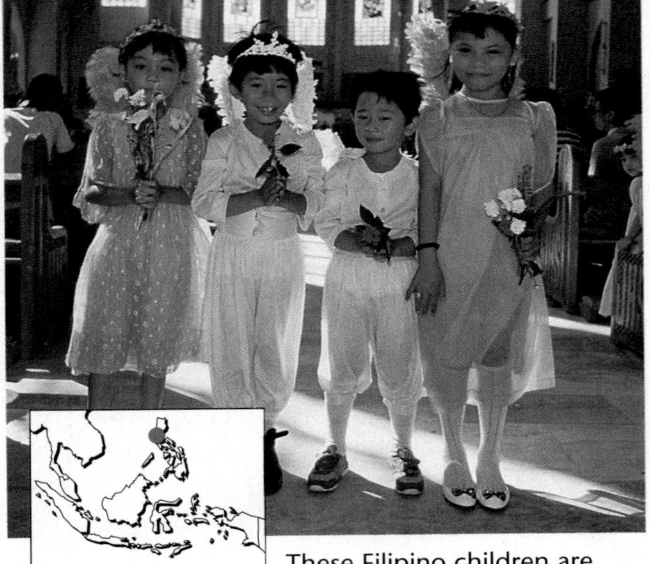

These Filipino children are dressed in their church clothes for Mass at a Roman Catholic church. The Spanish colonial past of the **Philippines** has resulted in a population that is mainly Roman Catholic.

and Spanish influences. About 85 percent of the population is Roman Catholic, with a small minority of Muslims living in the south. The national language is Pilipino, though English is used widely in education and business. Many Filipinos also speak an island language called Tagalog (tuh-GAHL-uhg).

Economic Geography The Philippines is a primarily agricultural country. Many Filipino farmers, however, are poor and own no land. A small group of rich landlords controls most of the best land. Huge sugarcane and coconut plantations cover the islands, and commercial farming is widespread. Sugar and coconut oil are the major agricultural exports, but small farms also produce tobacco, tropical fruits, and corn.

The forested mountains of the Philippines support a tropical hardwood lumber industry. Deforestation along steep slopes on some of the islands, however, has led to serious environmental problems. Minerals mined include iron ore, copper, gold, and silver. Unlike in neighboring Indonesia, little petroleum has been discovered in the Philippines. Rather, the nation's economy depends on its exports of electrical products and textiles. Its major trade partners are the United States and Japan.

Historical and Political Geography The Philippines have a long colonial history. Spain controlled the islands for 300 years, until the Spanish-American War ended in 1898. At that time, the United States took over the islands. Japanese forces occupied the Philippines for three years during World War II, before the islands became an independent republic in 1946. The new republic's government was modeled after that of the United States.

Corruption, however, has troubled the Philippine government. Despite attempts at land reform, most of the land and industry remain in the hands of a few powerful families. In addition, violent Muslim rebel movements in the south and attacks by Communist rebel guerrilla forces in the north are ongoing problems.

In 1986, a popularly supported military revolt led to the fall of the 20-year government of Ferdinand Marcos, which had set aside basic political rights. A democratic government formed under the nation's first female president, Corazon Aquino.

Although democracy has returned to the Philippines, prosperity has not. The nation's huge foreign debt and high unemployment have weakened the economy. In fact, the Philippines is one of the poorest nations in Southeast Asia. The gap between the rich and poor of this island nation is steadily widening.

The Spratly Islands The Spratly Islands, a collection of more than 100 islands and reefs in the South China Sea, are one of the most disputed territories in the world. The Philippines, China, Vietnam, Malaysia, and Taiwan all have claimed **sovereignty**, or authority to rule, over the Spratly Islands. The intense interest in these tiny islands stems from the oil and natural-gas deposits lying off the islands' coasts.

3 SECTION REVIEW

1. On which Philippine island do most Filipinos live? What are the country's major resources and exports?
2. What physical, political, and economic challenges have the Philippines faced recently?
3. **Critical Thinking** **Suggest** an explanation for why the Philippines have a much higher percentage of Roman Catholics than do the other countries of East and Southeast Asia.

CHAPTER 45 CHECK

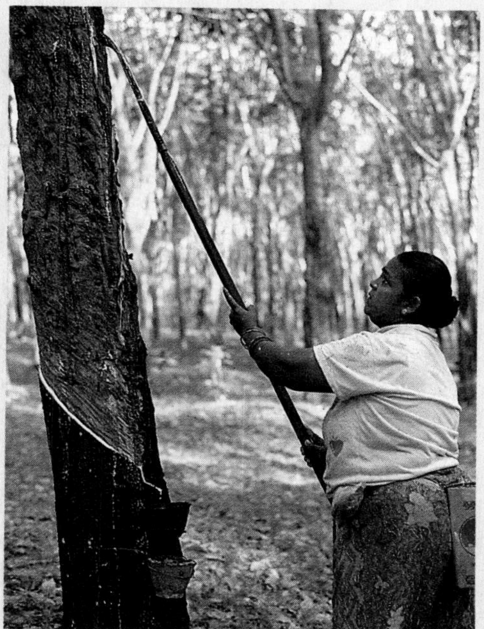

Collecting latex from a rubber tree in **Malaysia**

Reviewing the Main Ideas

1. The South China Sea divides Malaysia into two parts. More than half of Malaysia's people are Muslim Malays. The country is rich in natural resources, and industrialization is occurring at a rapid pace.
2. Singapore and Brunei are the smallest and wealthiest nations in Southeast Asia.
3. Indonesia is Southeast Asia's largest country and the world's most active volcanic region. Java is its core island. Indonesia remains mostly agricultural but is becoming more developed. The country has more than 300 ethnic groups.
4. The two main islands of the Philippines are Luzon and Mindanao. Like Indonesia, the Philippines is primarily agricultural. It has a long colonial history. The country faces several political and economic challenges.

Building a Vocabulary

1. What is an archipelago? Give an example of an archipelago.
2. What are *kampongs*? In what city are many *kampongs* found?
3. Define *sovereignty*.

Recalling and Reviewing

1. Describe the locations of Singapore and Brunei in relation to the two parts of Malaysia.
2. What physical characteristics do Indonesia and the Philippines have in common?
3. What languages are spoken in Malaysia, in Indonesia, and in the Philippines?

Thinking Critically

1. As you know, European colonists called the islands of the Malay Archipelago the East Indies. You also know, from Unit 4, that Europeans called the Caribbean islands the West Indies. Why do you think these two terms are no longer commonly used?
2. Why do you think the Spratly Islands are claimed both by Malaysia and the Philippines but not by Indonesia?

Learning About Your Local Geography

Individual Project

Despite many years of political problems, democracy has recently returned to the Philippines. Democracy, however, carries responsibilities. People must make informed choices. Contact local political organizations to find out when your area's next election will be held, and who the candidates for any election race are. Contact the campaign headquarters of each candidate to collect information on their positions. Decide which candidate best represents your own point of view on the issues. If possible, volunteer to work for the campaign of that candidate. Give a report to your class outlining your candidate's platform. You may want to include your report in your individual portfolio.

Using the Five Themes of Geography

Write a paragraph discussing how the theme of **human-environment interaction** in the islands of Southeast Asia is related to their **location** on the Pacific Ring of Fire. How have volcanoes both hurt and helped the people of the islands?

Detecting Cultural Bias

Recognizing cultural bias is an important skill in the study of human geography. Cultural bias is the influence that a person's own culture has on his or her view toward other cultures and peoples. Cultural bias can be expressed in many ways.

Sometimes, a person's actions reflect a cultural bias. At its most extreme, cultural bias can be expressed in violent actions against members of a particular culture group. Other culturally biased actions might include attempts to change another people's cultural beliefs. For example, many of the earliest Europeans to visit East and Southeast Asia believed that Asian beliefs and practices were inferior to their own. As a result, they tried to force some elements of their own cultures on the region's peoples. Similarly, many early Asians, such as the Chinese, believed that foreigners were culturally inferior. But the Chinese chose a different set of actions to express this cultural bias. Wanting to keep "barbarians" out of their society, Chinese leaders tried to close China off to foreigners by limiting trade with them.

Cultural bias, however, is often expressed not in actions, but in words. Ma Huan, the Chinese explorer about whom you read in "For the Record" on page 493, made a number of judgments about the peoples he encountered on his travels in the fifteenth century. About some of the peoples on the island of Java he said, "they have very ugly and strange faces" and "the food which these people eat is very dirty and bad." Both of these remarks, which may seem like simple observations, are actually judgments, as indicated by the words "ugly," "strange," and "bad." Even if a judgment seems positive, using words such as "beautiful," "normal," or "good," it may still express cultural bias.

A less obvious form of cultural bias is a **stereotype**, or a generalization about a group of people. Stereotypes tend to ignore differences within groups, instead treating everyone in a group as being alike. For example, after visiting the island of Sumatra, Ma Huan said "The people are very rich and prosperous." While it may have been true that some of the people Ma Huan met in Sumatra were "rich and prosperous," it is unlikely that all Sumatrans were. In fact, Ma Huan probably did not meet all of the people of Sumatra, and he probably did not stay long enough to gain a good understanding of the people he did meet. Thus his statements were based on limited, and possibly inaccurate, information.

Probably the hardest form of cultural bias to detect has to do with perspective, or point of view. When we use our own culture and experiences as a point of reference from which to make statements about others, we are showing a form of cultural bias called **ethnocentrism**. A good geographical example of ethnocentrism is in how people have named places. Europeans called East and Southeast Asia "the Far East," which is a culturally biased term because it describes East and Southeast Asia in terms of its location relative to Europe. Similarly, in Ma Huan's accounts of his travels abroad, he refers to his homeland, China, as "the Central Country." This reference reflects his belief that China was the geographical center of the world.

Detecting cultural bias, with its many forms, is a great challenge. In our study of geography and of the world's peoples, we must constantly be on the lookout for such bias. Too often it can cloud our view of the world and hide its rich diversity of peoples and cultures.

PRACTICING THE SKILL

Using books, magazines, newspapers, movies, television shows, news reports, and observations in your own school and community, identify one example of each of the following:

(1) an action that shows cultural bias
(2) a culturally biased judgment
(3) a stereotype
(4) ethnocentrism

List each example and write a few sentences explaining how it shows cultural bias.

BUILDING YOUR PORTFOLIO

Individually or in a group, complete one of the following projects to show your understanding of the geography concepts involved.

A *Following the Dragons*

You are a government official in one of the world's poorer developing countries. Your job is to assess your country's ability to successfully and quickly industrialize, and to decide whether such industrialization is desirable. To make these assessments, it will be helpful to look at Japan and the newly industrialized countries of East and Southeast Asia—Taiwan, Singapore, Hong Kong, South Korea, Malaysia, Thailand, and Indonesia. Your study of these countries will require you to do the following.

1. Research the population characteristics of each country. Make a graph showing each country's population size. Under your graph, write a caption identifying other population characteristics of each country, including quality and availability of education, general wage level, and degree of cultural uniformity.
2. Make a list of each country's resources. Consider location and infrastructure as well as agricultural, mineral, and energy resources.
3. Make a table comparing the historical and political characteristics of each country. What historical background does each country have? What type of government and political system are in place in each country now, and how stable have each country's political affairs been in recent decades? To what extent have foreign countries and companies been involved in each country?
4. Write a paragraph discussing the effects of industrialization in each country. How has industrialization affected the environment? Has the process benefited all groups within each country equally?
5. Summarize the information you have gathered by constructing a large chart comparing the population characteristics, resources, historical and political characteristics, and the effects of industrialization in each country you have studied. Examine the chart. Are there any characteristics that are shared by every country, without which industrialization would not have been possible? Does industrialization have any unavoidable negative effects?

Present your chart to the leaders of your country (the rest of the class). Discuss the implications of your findings for the industrial growth of your developing country.

B *Speak, Write, Read, Listen*

As you know, communication plays an important role in geography. It is one of the major links between peoples across and within regions. Create a "Communications Almanac" to discuss the spoken languages, writing systems, literacy rates, and telecommunications of East and Southeast Asia. To prepare your almanac, you will need to do the following.

1. Research and make a list of the region's major language families and language groups. Write a short paragraph describing the origins of each language group, and how each one relates to other languages of the world or region.
2. Make a language map of the region. Sketch where language groups are, and note on the map the official languages of each country.
3. Investigate different types of writing systems. Identify and write down definitions of the following words: ideography, logography, phonetization, syllabary, alphabetic writing. Research Chinese, Japanese, and Korean writing systems. Use the terms you have defined to compare and contrast the three systems in a paragraph.
4. Make a bar graph showing the literacy rates of each country of East and Southeast Asia.
5. Find out how widespread telecommunications systems are in each country. You may wish to focus on the numbers of televisions, radios, and/or telephones per person in each country. Present this information using charts or graphs.

Organize your materials into your almanac, and present it to the rest of the class.

ALMANAC

UNIT 10

SOUTH ASIA

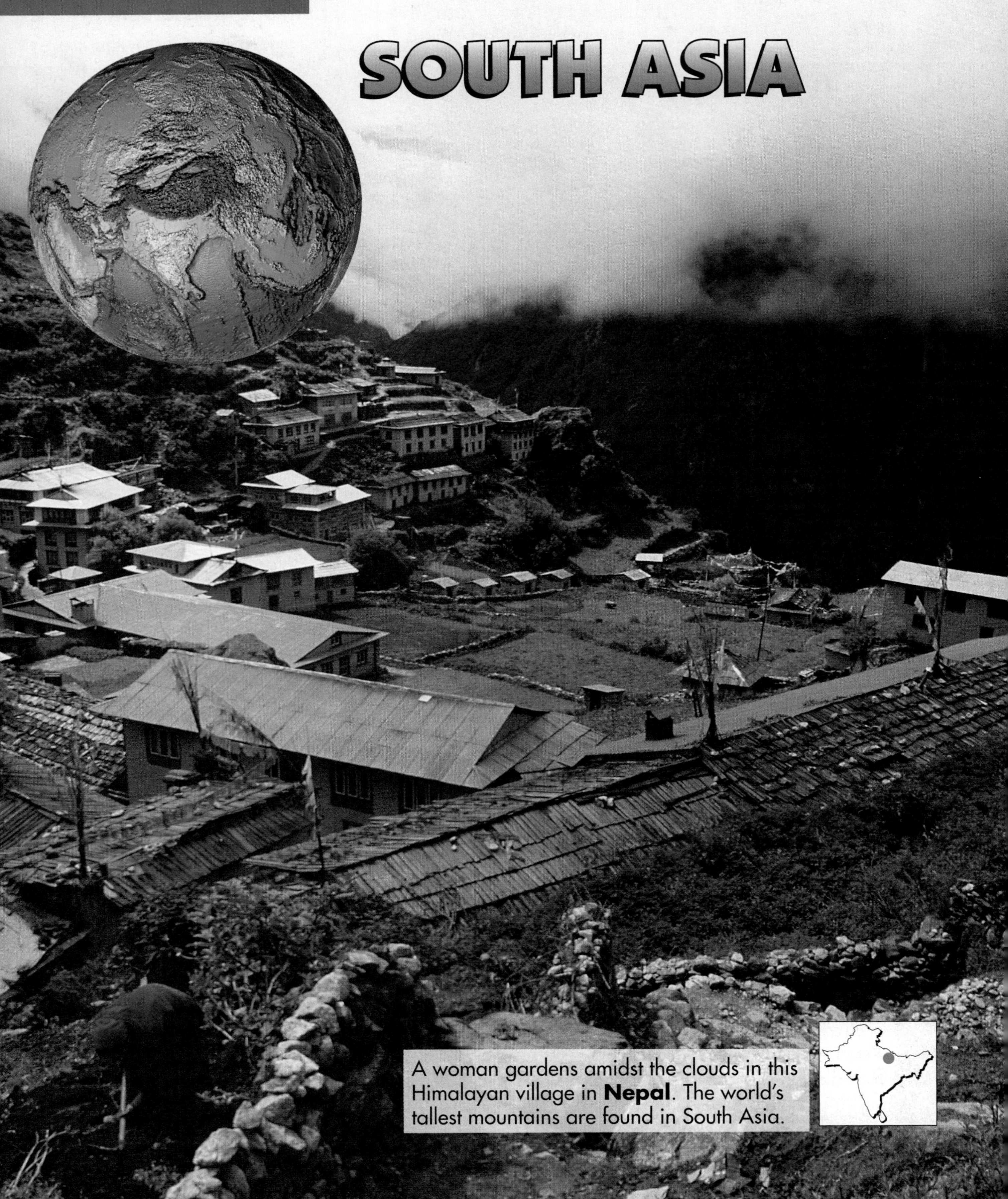

A woman gardens amidst the clouds in this Himalayan village in **Nepal**. The world's tallest mountains are found in South Asia.

Completed in 1643, the Taj Mahal in **Agra, India**, is one of India's most famous landmarks.

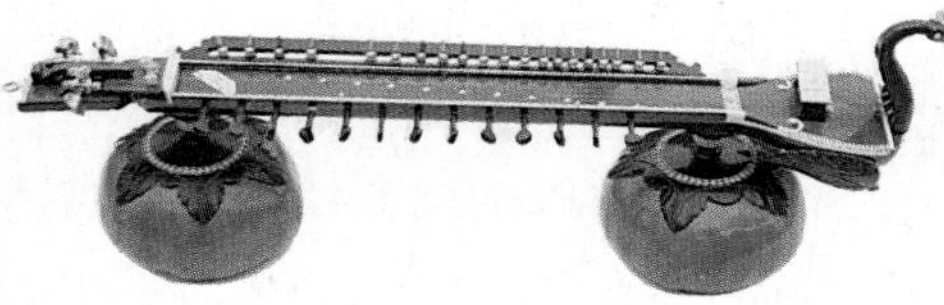

A *vina* (VEE-nuh), a stringed musical instrument from **South Asia**

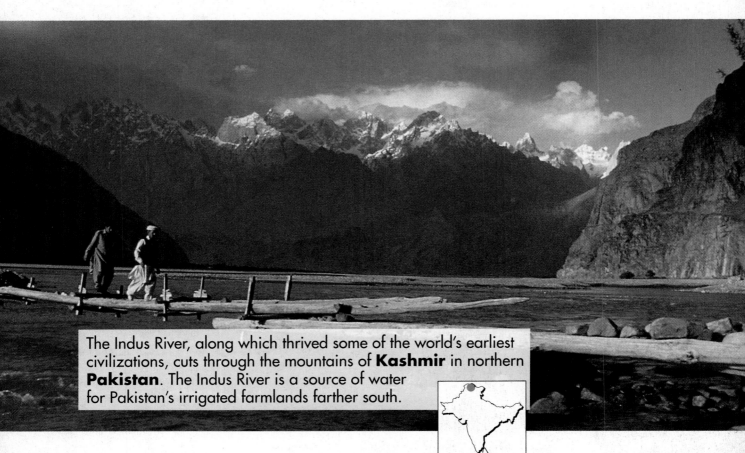

The Indus River, along which thrived some of the world's earliest civilizations, cuts through the mountains of **Kashmir** in northern **Pakistan**. The Indus River is a source of water for Pakistan's irrigated farmlands farther south.

One World, Many Voices

Literature

"Kavadi was a wonderful place for one like me from the village—a street full of all sorts of shops, sewing machines rattling away, coloured ribbons streaming down from shop-fronts. My father had his favourite shop. . . . He would buy us each a toy—a ball, a monkey dangling at the end of a rubber-piece, and a doll, and invariably [always] an exercise book and a pencil for my elder brother, declaring that he was past the age of having toys, a reminder which made him smart every time he heard it. The road would be ankle deep in bleached dust and the numerous cattle and country carts passing along stirred it up so much that a cloud always hung over the road, imparting an enchanting haze to the whole place, though, by the time we started back, so much of this dust settled on our skins and hair that our mother had to give us a bath as soon as we reached home."

R. K. Narayan (1906–) has written many novels and short stories about life in India. The passage at left is from his novel *The English Teacher,* published in 1945. It describes a man's pleasant boyhood memories of visits his family made to a town near their rural village.

Interpreting Literature

1. What does the boy like about this town that is different from his rural village?
2. What type of roads does the town have? What details does the author use to show that many people from rural areas visit this town?

FOR THE RECORD

V. S. Naipaul (1932–) is a well-known writer of fiction and travel books. Naipaul's family moved from India to the Caribbean island of Trinidad before he was born. He did not visit India until he was a grown man. The selection below from Naipaul's travel book *India* describes his impressions of Chandigarh, a city in northern India.

"The traffic was of all sorts: buses, trucks with towering loads, packed three-wheeler taxi-buses with about 20 people each (I counted), mule carts, tractors with trailers, some of the trailers carrying very wide loads of straw in sacking, or carrying logs placed crosswise, so that they occupied a good deal more of the width of the road than you thought from a distance. There seemed to be no limit to a load. . . . Many bicycles carried two or three people each: the cyclist proper, someone on the cross bar, someone sitting sideways on the carrier at the back. A motor-scooter could carry a family of five: father on the main saddle, one child between his arms, another behind him holding on to his waist, mother on the carrier at the back, sitting sideways, with the baby.

Always in India this feeling of a crowd, of vehicles and services stretched to their limit: the trains and the aeroplanes never frequent enough, the roads never wide enough, always needing two or three or four more lanes. . . . Hooters and horns, from scooters and cars and trucks, sounded all the time, seldom angrily. The effect was more that of celebration, as with a wedding procession."

Analyzing Primary Sources

1. What examples does Naipaul use to show the variety of transportation methods in Chandigarh?
2. What type of human geography does this excerpt concern?

CHAPTER 46

Introducing South Asia

GEOGRAPHY DICTIONARY

subcontinent
tropical cyclone
Hinduism
Sikhism
jute
Green Revolution
calico

A representation of the Hindu god Vishnu

South Asia, sometimes called the Indian **subcontinent**, is a large triangular peninsula that juts into the Indian Ocean. A subcontinent is a large landmass that is smaller than a continent. The region is one of physical extremes. Barren deserts contrast with flooded river deltas and tropical islands. The world's highest mountains rise up in the north, earning this area the nickname "the roof of the world."

South Asia is a region of great cultural contrasts as well. More than one-fifth of the world's people live on the South Asian peninsula, making it one of the most populated regions on Earth. The region's long cultural history has resulted in a mixture of religions, languages, and other culture traits.

In recent decades, South Asia has experienced rapid population growth but limited economic development. Some of the world's poorest nations struggle here with serious economic and environmental challenges.

Historic Brhadisvara temple in southern **India**

MAPPING THE REGION

SOUTH ASIA

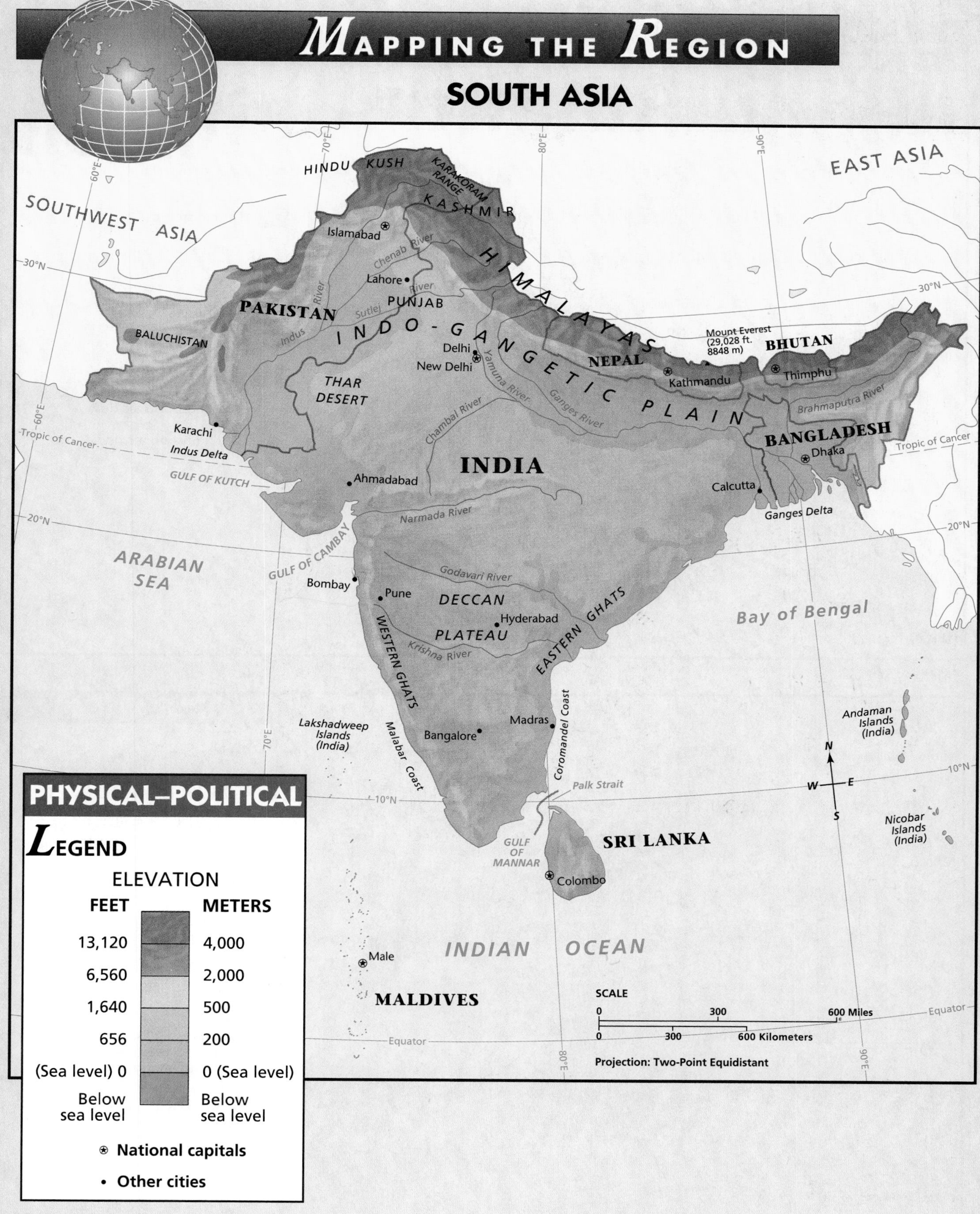

HINDU KUSH
KARAKORAM RANGE
KASHMIR
EAST ASIA
SOUTHWEST ASIA
Islamabad
Chenab River
HIMALAYAS
Lahore
30°N
PAKISTAN
Indus River
Sutlej River
PUNJAB
BALUCHISTAN
INDO-GANGETIC PLAIN
Delhi
New Delhi
Yamuna River
NEPAL
Mount Everest (29,028 ft. 8848 m)
BHUTAN
Kathmandu
Thimphu
THAR DESERT
Ganges River
Brahmaputra River
Chambal River
Karachi
Tropic of Cancer
Indus Delta
BANGLADESH
Dhaka
INDIA
GULF OF KUTCH
Ahmadabad
Calcutta
Ganges Delta
Narmada River
20°N
ARABIAN SEA
GULF OF CAMBAY
Godavari River
Bombay
Pune
DECCAN PLATEAU
Hyderabad
EASTERN GHATS
Bay of Bengal
Krishna River
WESTERN GHATS
Lakshadweep Islands (India)
Malabar Coast
Bangalore
Madras
Coromandel Coast
Andaman Islands (India)
Palk Strait
10°N
GULF OF MANNAR
SRI LANKA
Colombo
Nicobar Islands (India)
INDIAN OCEAN
Male
MALDIVES
SCALE
0 300 600 Miles
0 300 600 Kilometers
Projection: Two-Point Equidistant
Equator
60°E
70°E
80°E
90°E
N
S
E
W
PHYSICAL–POLITICAL
LEGEND
ELEVATION
FEET
METERS
13,120
4,000
6,560
2,000
1,640
500
656
200
(Sea level) 0
0 (Sea level)
Below sea level
Below sea level
National capitals
Other cities

MAPPING THE REGION

SOUTH ASIA

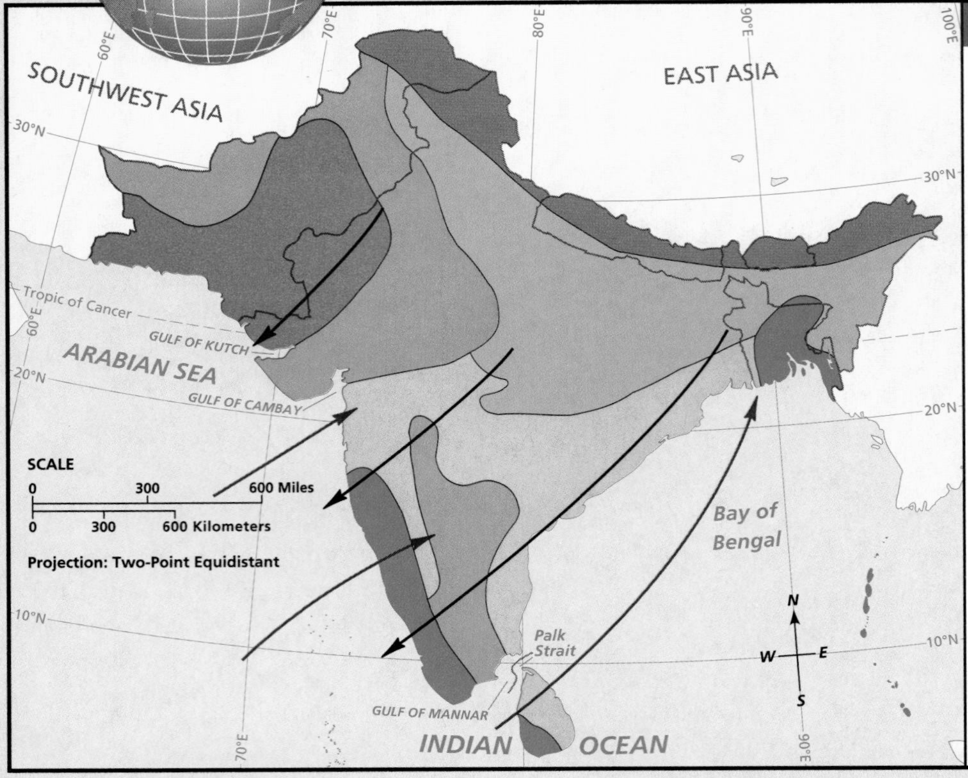

CLIMATE

LEGEND

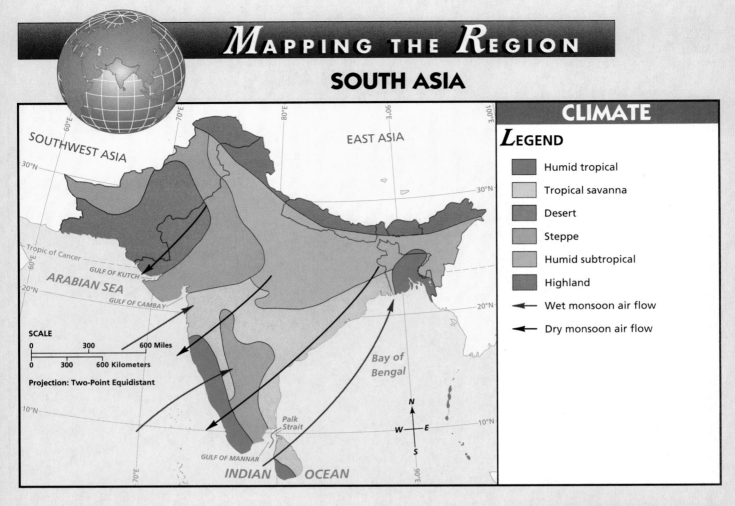

- Humid tropical
- Tropical savanna
- Desert
- Steppe
- Humid subtropical
- Highland
- Wet monsoon air flow
- Dry monsoon air flow

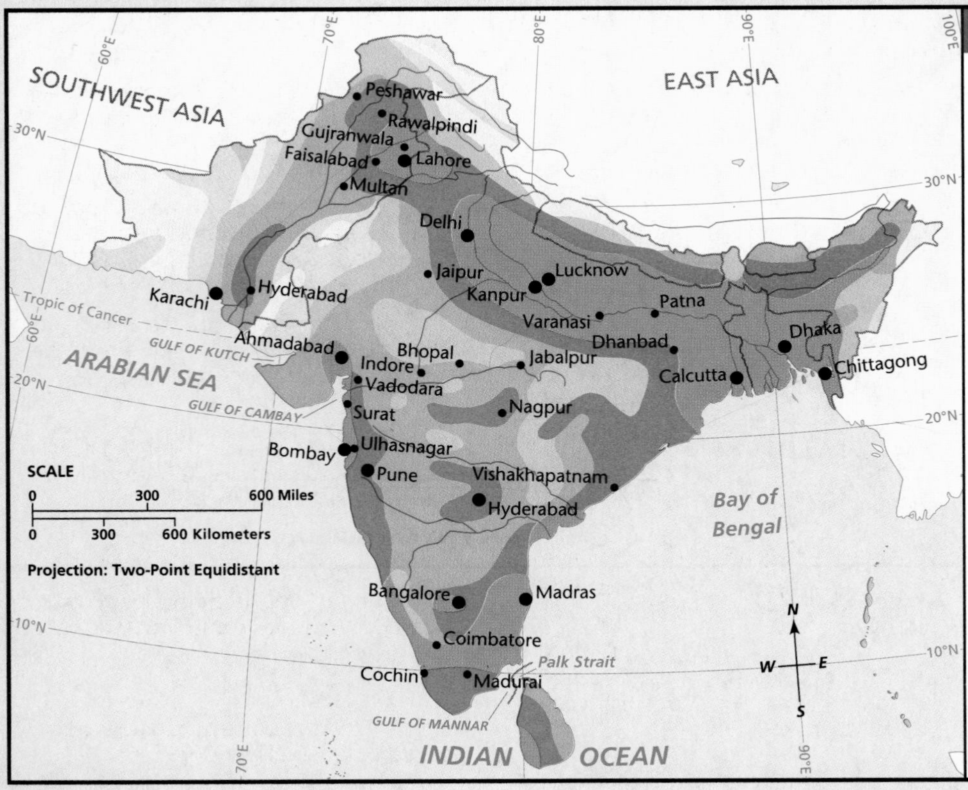

SCALE
0 300 600 Miles
0 300 600 Kilometers
Projection: Two-Point Equidistant

POPULATION

LEGEND

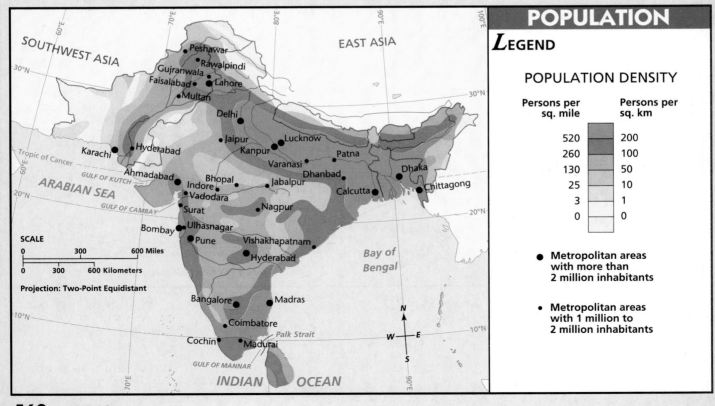

POPULATION DENSITY

Persons per sq. mile	Persons per sq. km
520	200
260	100
130	50
25	10
3	1
0	0

- ● Metropolitan areas with more than 2 million inhabitants
- • Metropolitan areas with 1 million to 2 million inhabitants

SCALE
0 300 600 Miles
0 300 600 Kilometers
Projection: Two-Point Equidistant

MAPPING THE REGION

SOUTH ASIA

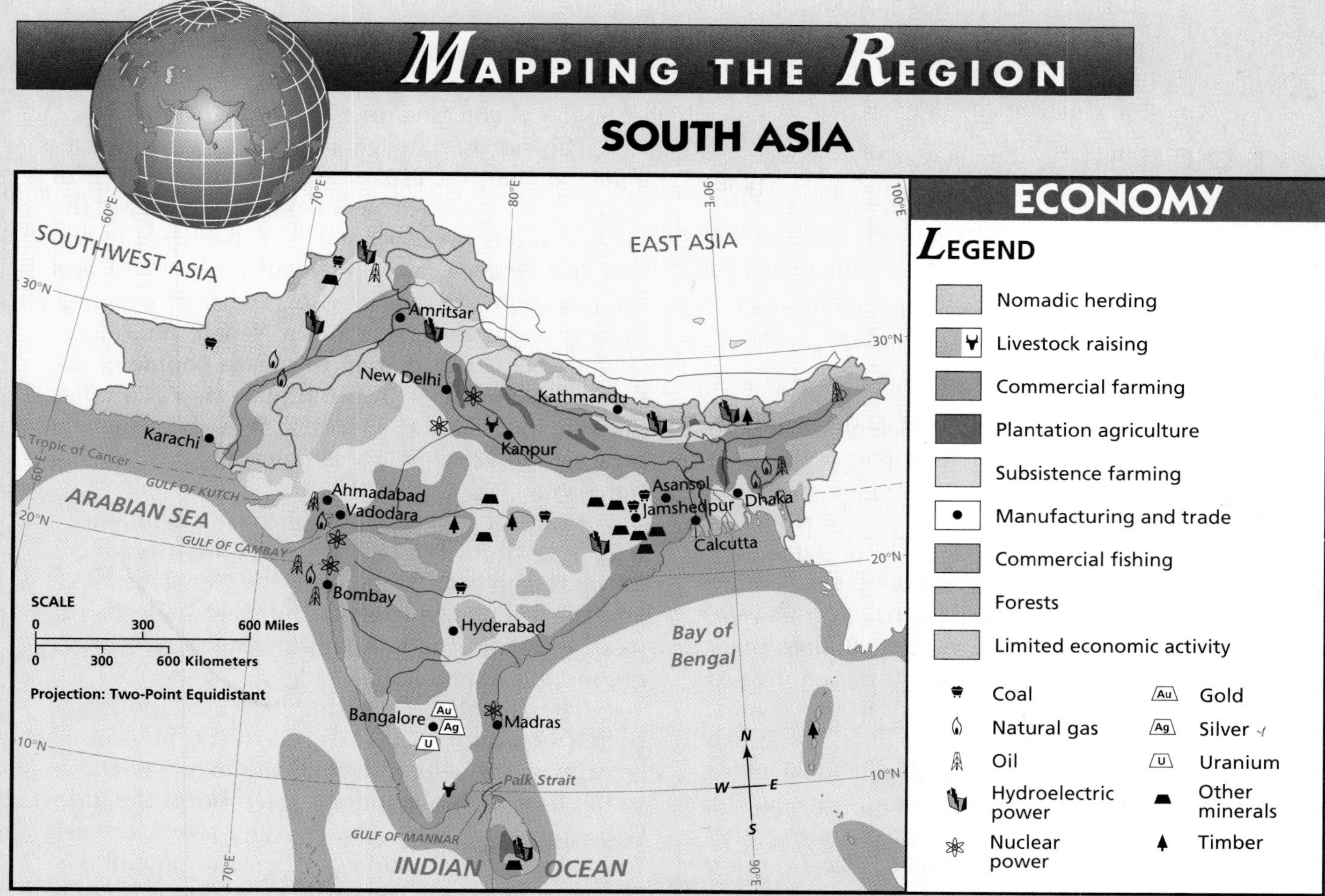

ECONOMY
LEGEND
Nomadic herding
Livestock raising
Commercial farming
Plantation agriculture
Subsistence farming
Manufacturing and trade
Commercial fishing
Forests
Limited economic activity
Coal
Natural gas
Oil
Hydroelectric power
Nuclear power
Gold
Silver
Uranium
Other minerals
Timber
Au
Ag
U
SOUTHWEST ASIA
EAST ASIA
Amritsar
New Delhi
Kathmandu
Kanpur
Karachi
Tropic of Cancer
GULF OF KUTCH
ARABIAN SEA
Ahmadabad
Vadodara
GULF OF CAMBAY
Bombay
Asansol
Jamshedpur
Dhaka
Calcutta
Hyderabad
Bay of Bengal
Bangalore
Madras
Palk Strait
GULF OF MANNAR
INDIAN OCEAN
SCALE
0
300
600 Miles
0
300
600 Kilometers
Projection: Two-Point Equidistant
N
S
E
W
60°E
70°E
80°E
90°E
100°E
30°N
20°N
10°N

CLIMATE

◆ Climate patterns in South Asia are greatly influenced by the monsoons that alternately bring wet and dry seasons to most of the region.

◆ Much of the northernmost areas of South Asia are dominated by variable highland climates.

◆ The driest climates of South Asia are found in the northwestern portions of the region.

? What climate types are found in South Asia?

Compare this map to the physical–political map on page 559.

? Other than those dominated by highland climates, which country's climate is least influenced by the monsoons?

? What physical landform region in central India is dominated by a steppe climate?

POPULATION

◆ Much of northern India is densely populated.

◆ Many densely populated areas and large metropolitan areas are located in coastal regions and along major rivers.

◆ The most sparsely populated areas of South Asia are in the western and northern parts of the region.

Compare this map to the climate map on this page 560.

? Why do you think the western part of Pakistan is not densely populated?

Compare this map to the physical–political map on page 559.

? What is the name of the landform region in densely populated northern South Asia?

? Which two major rivers flow through densely populated areas in the northwestern and northeastern parts of the region?

ECONOMY

◆ South Asia's most heavily cultivated regions are located along major rivers, especially the Indus and Ganges rivers.

◆ India's supply of uranium ore supports a number of nuclear power plants and has enabled the country to develop an atomic bomb.

◆ Many major manufacturing and trade centers are located along the coasts and in regions along the Indus and Ganges rivers.

? Name the manufacturing and trade centers shown along the coast of South Asia.

Compare this map to the climate map on page 560.

? One of Bangladesh's major crops is rice. Why do you suppose this is so?

Compare this map to the population map on page 560.

? Why do you suppose northern India is so densely populated?

PHYSICAL GEOGRAPHY

FOCUS

- *What are the three landform regions of the Indian subcontinent? How was each region formed?*
- *What are South Asia's climates? How do the monsoons affect the region?*

REGION **Tectonic Origins** Scientists believe that India was once part of the supercontinent of Gondwanaland. Over 65 million years ago, the Indian plate broke away from other portions of the supercontinent and became an island. It drifted north toward the Eurasian plate at a rate of about three to four inches (8–10 cm) per year. About 40 million years ago, the Indian plate slammed into Eurasia. This collision compressed and lifted the sea floor between the two plates, forming mountains. India continued to push under the Eurasian plate for about 1,200 miles (1,931 km), thickening the crust and pushing the mountains even higher. Earthquakes along this mountain rim indicate continued plate motion and mountain building.

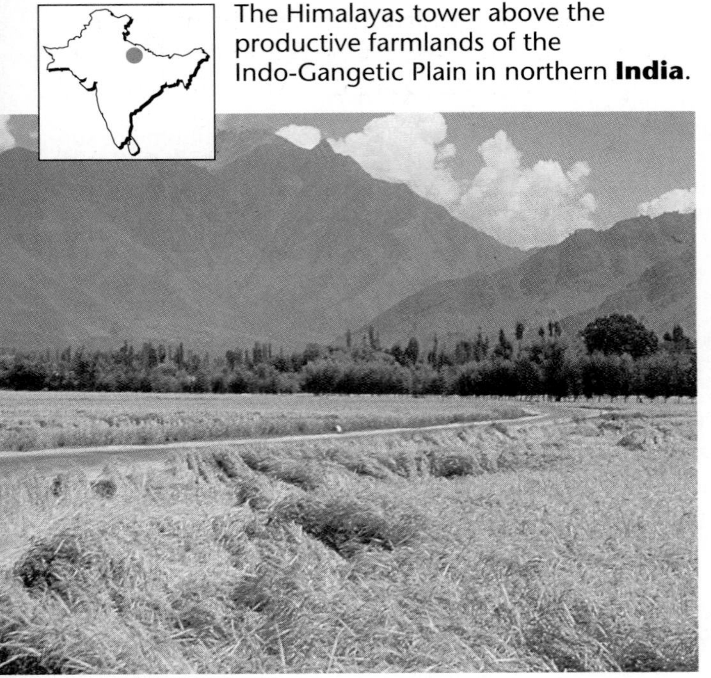

The Himalayas tower above the productive farmlands of the Indo-Gangetic Plain in northern **India**.

Landforms South Asia can be divided into three major landform regions: the Himalayan Mountain rim, the Indo-Gangetic Plain, and the Deccan Plateau. (See the map on page 559.)

The northern edge of the Indian subcontinent contains the world's highest mountains, of which the Himalayas (him-uh-LAY-uhz) are the longest and highest system. This northern mountain rim crosses most of Nepal (nuh-PAWL) and Bhutan (boo-TAN), as well as the northern portions of India and Pakistan. In Nepali, the official language of Nepal, *Himalaya* means "home of the snows." One glacier in the Himalayas is 20 miles (32 km) long. Mount Everest, the world's highest peak at 29,028 feet (8,848 m), is part of the Himalayas.

Most of the people of this mountainous region live in the valleys that separate the ranges of the Himalayas. To the east, the Himalayas split into the various ranges of Southeast Asia. To the west, they join the Karakoram range, the world's second highest mountain range.

The Himalayas are the source of the major rivers of South Asia. These rivers and other streams carry sediment from the high Himalayas to the base of the mountain rim to form the Indo-Gangetic Plain. Here, the flooding rivers deposit rich alluvial soil eroded from the mountains. These productive river plains are densely populated, even though severe flooding often has destroyed crops and killed many people here.

Three major rivers—the Indus, the Ganges (gan-JEEZ), and the Brahmaputra—cut through the region. Their vast alluvial plains stretch south from the Himalayas to the Arabian Sea in the west and to the Bay of Bengal in the east. The Indus River flows southward through Pakistan and is the lifeblood of that desert and mountain country. The Ganges flows eastward across northern India, where half of India's huge population lives. The Brahmaputra originates north of the Himalayas in Tibet and flows eastward. It then makes a sharp turn to the south, entering India in the country's far northeastern corner. Where the Ganges and Brahmaputra meet, they form Bangladesh's low delta plain.

The ancient Deccan Plateau, the third landform region, covers much of the southern two-thirds of the Indian subcontinent. The plateau tilts downward toward the east. In the west are the Western Ghats (GAWTS), the steep, rugged hills that face the Arabian Sea. In the east are the more gentle slopes of the Eastern Ghats. These hills face the Bay of Bengal.

Part of the Deccan Plateau is a shield of granite formed more than 600 million years ago. The younger portion of the Deccan is made up of volcanic layers of lava, which are more than 10,000 feet (3,048 m) thick. Weathering of the volcanic rocks has produced fertile soils in the region. With irrigation, these areas of the Deccan support productive grain fields. The plateau also holds most of India's mineral resources.

Between the Ghats and the coastlines of the Arabian Sea and the Bay of Bengal are narrow coastal plains. These fertile plains are productive agricultural areas and are heavily populated. Offshore to the southeast is the large island of Sri Lanka (sree LAHNG-kuh). To the southwest is a chain of low coral islands called the Maldives.

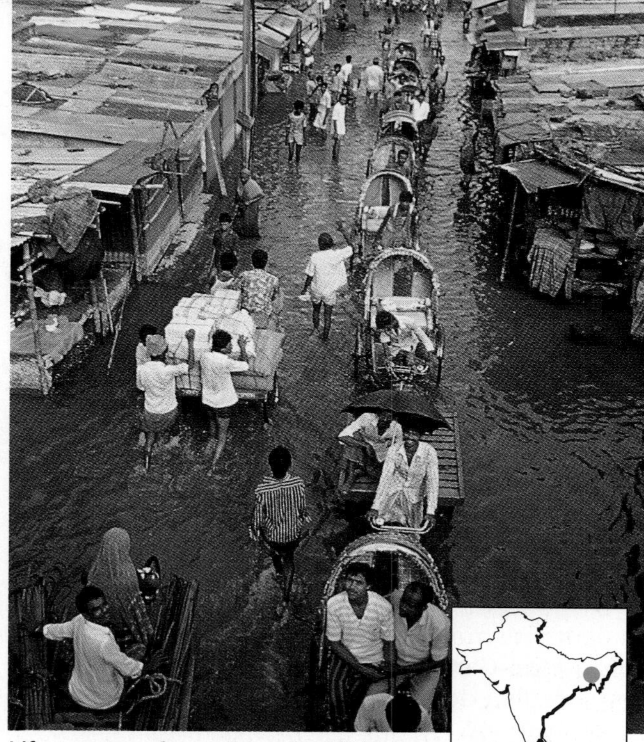

Life goes on for residents of **Dhaka, Bangladesh**, a regular flooding victim of the heavy rains brought by the annual wet monsoon.

Climate South Asia has several climate types. A humid-tropical climate dominates in Bangladesh and along the southwest coasts of India and Sri Lanka. Most of interior India, which includes most of the Deccan Plateau, has a tropical-savanna climate. A humid-subtropical climate covers the foothills along the southern edges of the mountains of northern India and northern Pakistan. Highland climates are found in the mountains of the northern rim.

To the northwest lie the drier climates. A desert climate stretches from the Thar (TAHR) Desert, or Great Indian Desert, in northwest India and eastern Pakistan across the irrigated Indus Valley of Pakistan to the country's borders with Afghanistan and Iran. Part of the central Deccan Plateau, meanwhile, has a steppe climate. The Western Ghats block moist winds off the Arabian Sea from reaching the area.

REGION The Monsoon The monsoons strongly influence the climates of South Asia. These seasonal winds blow from the oceans in summer and from the interior of the Asian continent in winter.

The wet summer monsoon usually begins around mid-June. The high summer sun rapidly heats the interior of the Asian continent, causing the air above the land to warm and rise. This creates a large low-pressure area that pulls moist air, the summer monsoon, inland from the Indian and Pacific oceans. As the summer monsoon flows inland, it brings high humidity, heavy rains, and thunderstorms. The heaviest rains fall where the warm, moist winds meet the Western Ghats and the foothills of the Himalayas. Some areas receive as much as 400 inches (1,016 cm) of rain, and flooding threatens many villages. In the fall, a period of hot, still weather brings some relief from the rains.

During the dry winter monsoon, the interior of Asia becomes bitterly cold. This heavy, cold air causes a huge high-pressure center to form over the interior of Asia. Cold, dry winds flow toward the surrounding oceans. As the winds descend down the mountains toward India, they warm and become even drier, making South Asia's winter dry and warm as well. In the spring, the air again becomes hot, humid, and still. Schools and some businesses close for vacation at this time.

The timing of the monsoon is critical to the farmers and to the economy of South Asia. Seeds are planted in the spring so that they have time to develop and take hold in the soil. If the monsoon rains come too soon, the seeds will wash away. If the monsoon rains come late or are in short supply, the seedlings will die from a lack of moisture.

The Thar Desert and the lower Indus Valley of Pakistan are not in the path of the wet monsoon winds. Thus, the amount of annual rainfall in these areas is low, and irrigation is needed to grow crops. The Thar Desert receives as little as five inches (13 cm) of rain per year. Here, dust storms are more common than rain storms.

The coastal regions of the Bay of Bengal are occasionally struck by severe **tropical cyclones** during the spring and fall. These hurricane-like storms produce winds above 100 miles per hour (161 kph) and dangerous high seas. The storms have taken hundreds of thousands of lives, especially in the low delta area of Bangladesh.

SECTION REVIEW

1. How were the mountains of the Indian subcontinent formed, and what role have these mountains played in forming the Indo-Gangetic Plain? To which landform region do the Western and Eastern Ghats belong?
2. Where are the subcontinent's drier climates found? How is the summer monsoon important to South Asia's agriculture?
3. **Critical Thinking** **Compare and contrast** the origins of the summer and winter monsoons. How is the earth's position relative to the sun a factor?

HISTORICAL AND HUMAN GEOGRAPHY

FOCUS

- *What cultural groups lived in or controlled South Asia from ancient times through the colonial era? How did the region achieve its independence?*
- *What effects did the 1947 division of the subcontinent have on its peoples?*

MOVEMENT **Cultural Influences** The Indus civilization, one of the world's first advanced civilizations, developed in South Asia's Indus River valley around 2500 B.C. This early society was based on irrigation agriculture and had a written language with an alphabet of 250 to 500 characters. It also produced stone and bronze sculptures, metal tools, and pottery. The Indus civilization declined around 1750 B.C. for reasons not completely clear. Soil salinization and overpopulation may have been part of the cause.

In about 1500 B.C., nomads from Central Asia, called Aryans, conquered the Indus area. By 1000 B.C., the Aryans controlled northern India as well. The local farming peoples, the Dravidians, were driven to the south. Some of them remained, however, and mixed with the Aryan population. This new Indo-Aryan culture was centered near the upper Ganges River.

The Indo-Aryan culture produced many distinctive culture traits of South Asia, including the Hindu religion. **Hinduism**, whose roots date to prehistoric times in South Asia, is the major religion of India and Nepal. Hindus worship many gods and believe that salvation is offered through *dharma*, or good conduct. They also believe in reincarnation, or rebirth after death. Sanskrit, an Indo-Aryan language which flourished from about 500 B.C. to A.D. 1000, is the classical literary language of Hinduism. Hindu scholars still use Sanskrit today.

Hindu pilgrims crowd into the Ganges River in **Varanasi, India**. Bathing in the sacred Ganges is an important spiritual activity for Hindus.

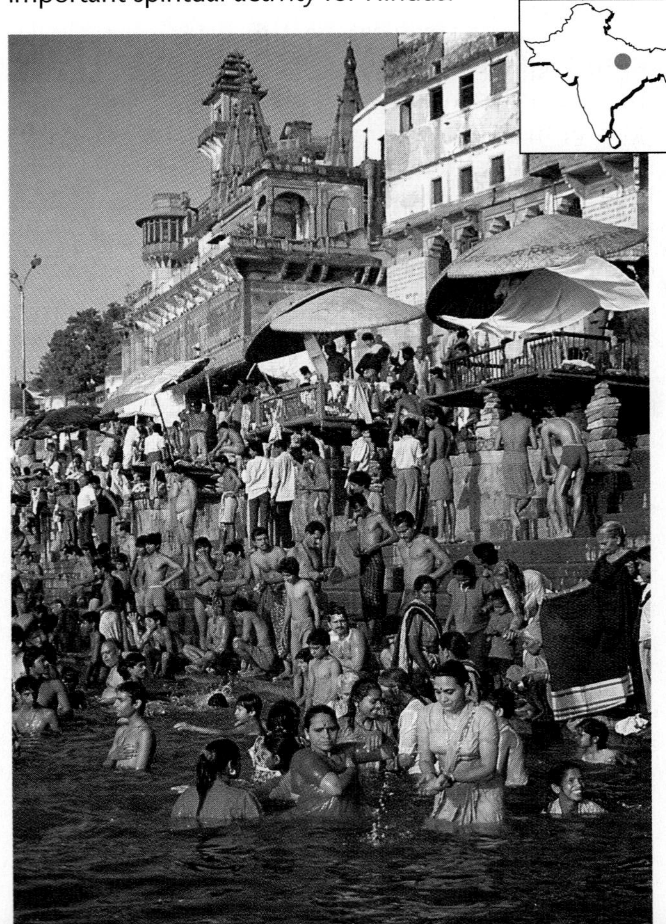

Another religion, Buddhism, had begun to form in northern India by about 500 B.C. At about the same time, a series of Indian empires and kingdoms rose to power in the region and ruled for the next 1,000 years. The Mauryan Empire, which ruled during the third century B.C., helped spread Buddhism throughout much of Asia. After A.D. 500, the Indian empires declined as peoples from Central Asia again invaded northern India.

Beginning around A.D. 1000, several Muslim invasions in the northwest brought Islam into the region. During the late fifteenth century, Muslim teachings blended with Hinduism to form yet another religion, **Sikhism** (SEEK-iz-uhm), in the region. Today, Sikhs make up a small minority. They are concentrated in the state of Punjab in northwestern India.

English architectural styles are common in **Simla,** located in the Himalayan foothills of northwestern **India**. The cool climate and scenic setting made the city the summer capital of British colonial India.

The last Muslim invasion took place in 1526 when the Mogul Empire established Muslim control over most of India. Today, Muslims are the majority in Pakistan and Bangladesh and form a large minority in India.

Colonialism Europeans arrived in India during the sixteenth century. The Portuguese came first, seeking to form new trade links, expand their empire, and spread Christianity. Today, millions of Indians are Christians.

The Dutch, the French, and the British soon followed the Portuguese. For over a century, large trading companies from these countries sought to earn profits in India. At first, the products available in India were not directly valuable to the Europeans back in their home countries. Instead, the products were valued as goods to be traded for yet other goods elsewhere. For example, the Europeans longed for spices from the East Indies (present-day Indonesia). In exchange for the spices, the Europeans needed goods such as cotton to trade, and India was a ready source of that raw material. The European companies developed many such trading networks.

In time, however, the Dutch came to monopolize the East Indies spice trade, and the French and British began to take more of an interest in India for its own resources. And as Europe industrialized, Great Britain developed a textile industry that could make direct use of Indian cotton.

For most of this period, the Europeans had competed peacefully for the region's trade. A series of conflicts, however, began in the 1740s. British victories over other European powers as well as over native Indian powers steadily increased British political control of the region. As a result, in 1858 almost all of the Indian subcontinent officially became part of the British Empire, controlled by the British government rather than by trading companies. British India included present-day India, Pakistan, Bangladesh, and Sri Lanka.

For the next several decades, the Indian subcontinent was considered the "jewel in the crown" of the British Empire. The region's many products, especially cotton, traveled to Great Britain's factories. Other important products shipped to Great Britain included tea, indigo, coffee, and **jute**. Jute is a plant fiber used to produce coarse fabrics, such as burlap, and twine.

India also served as a market for goods manufactured in Great Britain. To ensure an efficient flow of goods, the British built railroads, roads, and ports in the region. In addition, they brought the English language, English law and education, and the English political system.

Independence Demands for independence increased in the early twentieth century, particularly after the Massacre at Amritsar in 1919. British troops fired on a large crowd of people protesting British policies. Some 400 Indians were killed and 1,200 wounded.

The main independence movement in British India was led by Mohandas Gandhi. Gandhi believed that all life is sacred and thus should never be harmed. Related to this philosophy, called *ahimsa*, was Gandhi's belief that a policy of "nonviolent noncooperation" was the best way to bring about political and social change, including independence. Thus Gandhi's methods often took the forms of boycotts, marches, and fasts, but never violence.

Gandhi's efforts were effective. Britain granted independence to the colony in 1947, ending the colonial era. First, however, the British divided the subcontinent into the countries of India and Pakistan. Pakistan itself had two parts: West Pakistan and East Pakistan (present-day Bangladesh). The two parts were separated by 1,000 miles (1,609 km) of Indian land.

The British divided the region because many feared that war might break out between Hindus and Muslims if they were forced to live in the same country. Hindus were in the majority in India, while Muslims formed a majority in Pakistan. A mountainous region called Kashmir was split between the two nations, even though the local population wanted independence and was mainly Muslim. The island of Ceylon (now called Sri Lanka) gained its independence in 1948, just one year after the region broke away from Great Britain. Sri Lanka is a Buddhist country but has a large Hindu minority.

The political division of the subcontinent at independence was traumatic for millions of people in South Asia. Minorities found themselves surrounded by hostile majority populations, and violence broke out. Millions of India's Muslims fled to Pakistan, just as millions of Hindus and Sikhs fled from Pakistan to India. Hundreds of thousands of people died in riots and massacres. Even Gandhi, who had millions of followers, was not spared by the violence he so detested. He was shot in 1948 by a Hindu extremist.

In 1971, East Pakistan, with the help of India's military and with much bloodshed, broke away from West Pakistan and formed the independent nation of Bangladesh. In 1973, a new constitution made Pakistan, formerly West Pakistan, an Islamic republic.

Today, religious and ethnic tension is still one of the most pressing issues facing South Asia. Independence movements among different groups have led to continuing violence.

SECTION REVIEW

1. List three different groups of people that lived in South Asia before the colonial era. What European empire did the subcontinent become part of in the 1800s? Who was Mohandas Gandhi?
2. Why was the subcontinent divided into India and Pakistan? What problems did this division create?
3. **Critical Thinking** **Construct** a time line showing the cultural and political history of the region.

ECONOMIC GEOGRAPHY

FOCUS

- *What agricultural challenges does South Asia face?*
- *What has hindered industrialization in much of South Asia?*

INTERACTION **Agriculture** Most of the people of South Asia are farmers who live in small villages. The small, inefficient farms, the high cost of fertilizers, and a lack of mechanization keep most South Asian countries from achieving consistent food production. Also, crop disease and natural disasters such as monsoon floods, droughts, tropical cyclones, dust storms, and locust swarms damage crops.

To improve agricultural output, new types of rice and wheat that produce higher yields and require shorter growing seasons were developed. At first, these new types of grain produced such high yields that their introduction was known as the **Green Revolution**. Natural disasters and population increases, however, reduced the gains that had been made. Also, these new improved grains needed larger amounts of water and costly fertilizers, which many farmers could not afford.

Today, there is little land left to develop in South Asia, and soil erosion and salinization threaten existing farmlands. As the region's population grows, supplying the demand for more farmland and for more wood is increasingly difficult. Often, the nearby forests are destroyed in the process. This deforestation, in turn, causes a loss of forest resources, destroys wildlife habitat, and creates increased flooding and soil loss. The cycle is hard to break.

REGION **Resources and Industry** Much of South Asia has experienced a slow rate of industrialization. Part of the explanation lies in the region's colonial history. During the colonial period, the British did not promote industry in South Asia because they did not want the region's industries to compete with British industries. To the British colonists, then, the region was mainly a source of raw materials. As a result, some Indian industries declined during the colonial era.

A good example of the British policy and its devastating effects on Indian industry involved **calicoes**. Calico, a fine fabric woven from cotton and printed with colorful patterns, originated in the subcontinent's southwestern town of Calicut (now called Kozhikode) in the eleventh century. By the 1600s, the handcrafted calicoes were popular throughout Europe, and India was the major producer and exporter. By 1700, however, British cloth manufacturers had successfully fought for a ban on imports of India's calicoes. Even though the British cloths were not of as high quality, the ban kept the Indian calicoes from competing and forced their decline. The ban not only boosted the British industries but also gave them an even greater supply of raw cotton from South Asia because the region was using less cotton itself.

Then during the mid-nineteenth century, events halfway around the world affected India's cotton market. When cotton exports from the southern United States dropped dramatically during the U.S. Civil War, Great Britain's textile factories ran short of raw cotton. To supply their cotton needs, the British relied more heavily on, and paid more for, India's cotton. When the U.S. Civil War ended, though, Great Britain turned back to U.S. supplies. This sudden loss of income devastated India's economy, to which cotton exports had become key. Farmers who had stopped growing food in order to grow cotton found themselves with neither food nor money with which to buy food. For the next 35 years, the people of India suffered a series of severe famines.

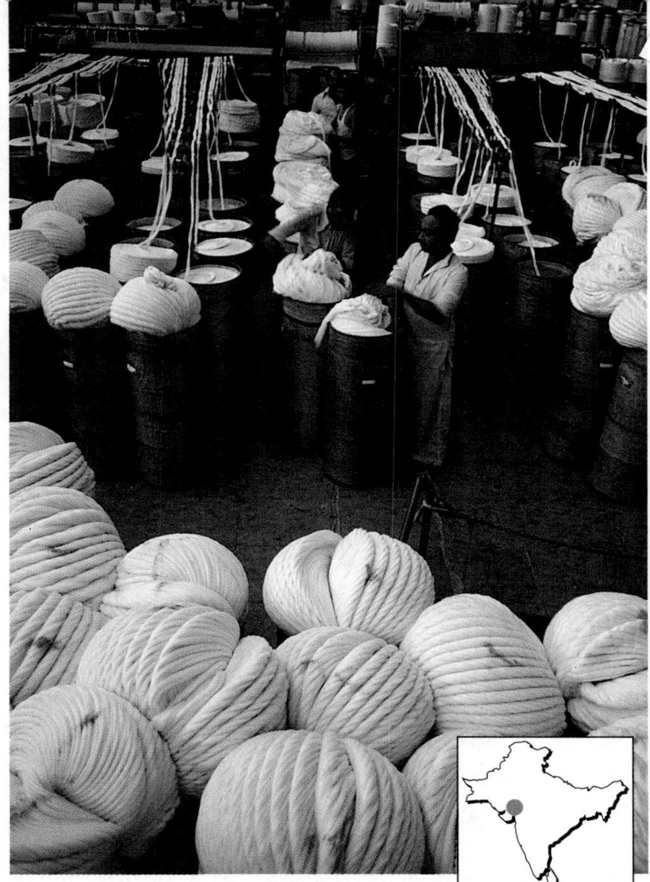
Cotton is prepared for market and for use in various industries at this mill in **Ahmadabad, India**. India is one of the world's major producers of cotton.

After independence, the new countries of South Asia focused on feeding their growing populations. As a result, few resources could be devoted to developing industry. The region's leaders sought foreign investment, but political turmoil, social unrest, and natural disasters interfered.

Even so, some industry did develop, particularly in India. Emphasis was placed on state-owned heavy industries, such as iron and steel, supported by the region's coal and iron-ore deposits. Cotton clothing and textile industries once again sprang up and became the region's most successful industries. Most of this industrialization took place in the cities.

A lack of industrial resources, especially petroleum, may hinder future development in the region. Although India has increased petroleum production and Pakistan has expanded development of its natural-gas supplies since the 1980s, petroleum imports are still a major economic burden to South Asian countries.

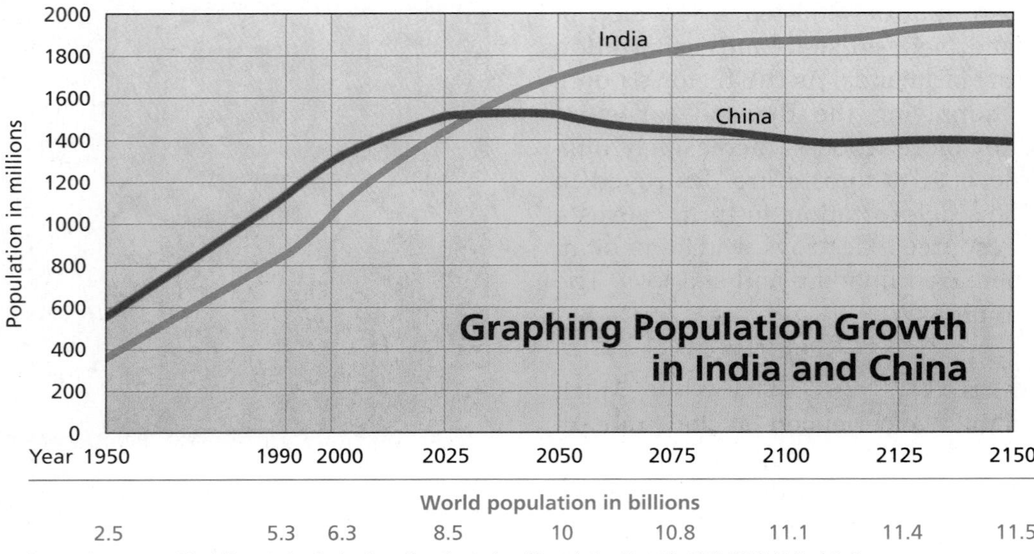

Graphing Population Growth in India and China
India
China
Population in millions
2000
1800
1600
1400
1200
1000
800
600
400
200
0
Year 1950 1990 2000 2025 2050 2075 2100 2125 2150
World population in billions
2.5 5.3 6.3 8.5 10 10.8 11.1 11.4 11.5

Source: *Long-range World Population Projections: Two Centuries of Population Growth, 1950–2150,* United Nations

SKILL STUDY This line graph shows past and projected population increases of India and China over time. Despite government efforts to hold down birthrates, the combined populations of China and India are projected to climb to more than 3.3 billion by 2150. When is India projected to replace China as the world's most populous country?

Despite these challenges, the potential for growth is strong. The region has a huge low-wage labor force, which could draw labor-intensive manufacturing industries. These industries make small consumer goods, clothing, chemicals, and some high-technology items, such as computers and telecommunication systems. With industrialization, a South Asian middle class could grow. In turn, this growth would increase the demand for more locally manufactured goods. This economic growth may help cut the huge foreign debts that most of the region's countries have acquired over the years.

Population Growth All of the nations of South Asia are faced with rapid population growth. The population of India alone is more than 900 million. Its growth rate of about 2 percent means that the country gains more than 15 million people each year. Most other South Asian nations will double their populations in 20 to 40 years. Poverty in the region continues to grow along with the population.

Although the region's population is mainly rural, millions of people have migrated to the larger cities, seeking jobs. Several South Asian cities have some of the largest populations in the world. Unfortunately, these large cities are now surrounded by huge shantytowns with few city services. Unemployment is high, and many people struggle to survive by working at temporary jobs. Urban poverty has led to increased crime.

Part of the answer to South Asia's problem of balancing people and resources might be to control population growth. This solution faces many obstacles, however. For example, poor communication systems keep government family-planning programs from reaching many of the rural areas, where more than 70 percent of the population live. In addition, most rural farmers want large families to help work the fields. The region's low literacy rates add to the problem. Education must be expanded to reach the majority of the population before meaningful change can take place.

3 SECTION REVIEW

1. What agricultural obstacles must South Asia overcome?
2. What factors have hindered South Asia's industrialization?
3. **Critical Thinking** **Agree or disagree** with the following statement: "The Green Revolution proves that technology can provide complete answers to many of the problems of developing countries."

CHAPTER 46 CHECK

Calcutta, India

Reviewing the Main Ideas

1. South Asia, once part of the supercontinent of Gondwanaland, can be divided into three major landform regions and has several climate types. The monsoons strongly influence life in the countries of South Asia.
2. South Asia has had many cultural influences, from ancient times to the present. Europeans arrived in South Asia during the sixteenth century. Britain later came to dominate much of the region. Partly through the efforts of Mohandas Gandhi, the region gained its independence in 1947. The division of the subcontinent into India and Pakistan was traumatic for millions of people.
3. Many South Asians are farmers who live in small villages. The region's agriculture faces several challenges. South Asia also lags behind much of the world in industrialization, though the potential for growth is strong. All of the countries of South Asia are experiencing rapid population growth, which strains their resources.

Building a Vocabulary

1. Define *subcontinent*.
2. What are tropical cyclones? What regions of South Asia do tropical cyclones tend to hit?
3. State three beliefs of Hinduism. How is Sikhism related to Hinduism?
4. What is the name of the plant fiber used to produce coarse fabrics and twine?

Recalling and Reviewing

1. What are the three major rivers of South Asia? Through which landform regions do they flow? Near which river was the Indo-Aryan culture centered?
2. What interests did European colonists have in South Asia?
3. How is the timing of the monsoon critical to South Asia's farmers?
4. What resources supported the region's early heavy industries? Where are most of India's mineral resources located?

Thinking Critically

1. Mohandas Gandhi believed that nonviolent noncooperation was the best way to bring about political and social change. Do you agree? Why or why not?
2. Compare and contrast the twentieth-century British division of the Indian subcontinent with the nineteenth-century European division of Africa. Did the divisions of both regions take place for the same reasons? Did they have similar effects? Explain your answer.

Learning About Your Local Geography

Individual Project

To many South Asians, the change of the seasons is of great significance to daily life. Write a paragraph describing your area's seasons in terms of average temperatures, precipitation, and humidity. Also discuss the extent to which your daily activities change with the seasons over the course of one year. You may want to include your paragraph in your individual portfolio.

Using the Five Themes of Geography

Sketch a map of South Asia. With arrows, show how the region's **location** has influenced the **movement** of peoples, ideas, and goods in and out of the region over time. Label each arrow and include a caption explaining what exchanges and events are taking place on the map.

CHAPTER 47

India

GEOGRAPHY DICTIONARY

sanctuary
cottage industry
caste system

"In India the monsoon is predictably unpredictable. . . . Rains can skip entire regions. It can rain for days, even weeks, until the earth is squirming with life and walls are slippery with mildew. For another week the skies may be clear, but then rain will come again, relentless, finally tedious, until memories of cracked, dry earth and fierce sun fade into wishful thinking."

Priit J. Vesilind

One out of every six people in the world lives in India, the largest country in South Asia. In fact, India is second only to China in population. By the middle of the twenty-first century, India could become the most populous country in the world. Today, India's more than 900 million people crowd into an area about one-third the size of the United States.

India is one of the world's most complex culture regions. Its peoples practice several major world religions and speak languages from more than 50 language groups. The 25 states of this democratic republic are for the most part divided along ethnic lines. Economic contrasts between rural and urban, rich and poor, and industrial and agricultural also define India's cultural landscape.

Inlaid stonework from **Agra, India**

Farming during the wet monsoon in southwestern **India**

INDIA: PHYSICAL–POLITICAL

SCALE
0 300 600 Miles
0 300 600 Kilometers
Projection: Two-Point Equidistant

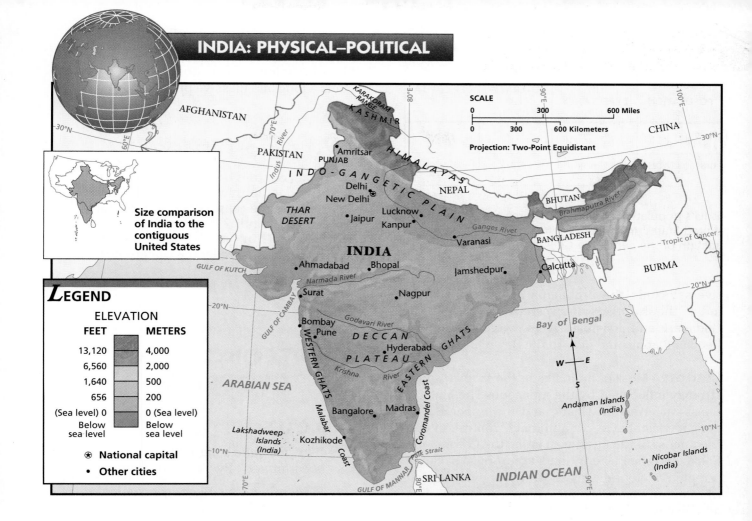

Size comparison of India to the contiguous United States

LEGEND

ELEVATION

FEET	METERS
13,120	4,000
6,560	2,000
1,640	500
656	200
(Sea level) 0	0 (Sea level)
Below sea level	Below sea level

⊛ National capital
• Other cities

PHYSICAL GEOGRAPHY

FOCUS

- *Where is India located relative to other countries and to seas and oceans? What are India's major rivers and climate types?*
- *What environmental challenges does India face?*

LOCATION **Landforms** The shape of India resembles an arrowhead pointing south. The towering Himalayas in the north form the base of the arrowhead and mark a natural boundary between India and China. The Himalayas also cross the small kingdoms of Nepal and Bhutan in the northeast. These countries have served as buffer states between India and China's autonomous region of Tibet. To the east, India borders Burma in the hill country and almost surrounds Bangladesh. To the west of India lies Pakistan, while in the far northwest is the disputed Kashmir region, a territory claimed by both India and Pakistan.

Water surrounds much of India. The southern tip of the country plunges into the Indian Ocean. The eastern Coromandel Coast faces the Bay of Bengal, and the western Malabar Coast borders the Arabian Sea. India also includes the Andaman and Nicobar islands in the Bay of Bengal and the Lakshadweep Islands in the Arabian Sea. (See the map above.)

Water is vital to the interior of the country as well. Here, India's major rivers act as lifelines. The Ganges, which flows from the Himalayas southeastward to the Bay of Bengal, is called the "mother river" and is sacred to Hindus. Over one-third of the country's population lives in the Ganges' fertile plains region. Other important

FACTS IN BRIEF India

COUNTRY POPULATION (1994)	LIFE EXPECTANCY (1994)	LITERACY RATE (1991)	PER CAPITA GDP
INDIA 919,903,056	58, male 59, female	52%	$1,300 (1994)
UNITED STATES 260,713,585	73, male 79, female	97%	$24,700 (1993)

Sources: *The World Factbook 1994,* Central Intelligence Agency

India's population is more than three times that of the United States. India's per capita GDP, however, is a fraction of the per capita GDP of the United States.

rivers include the Brahmaputra in northeastern India, the Narmada in western India, and the Godavari and Krishna rivers, which drain the Deccan Plateau and flow to the Bay of Bengal.

Climate Most of India lies in the tropics and is strongly influenced by the monsoon. The Malabar Coast, the Western Ghats, and the upper Bay of Bengal coastal area are all humid tropical. In contrast, most of the interior of India, the Coromandel Coast, and the Eastern Ghats are tropical savanna. A region of steppe climate stretches over the central Deccan Plateau, while a humid-subtropical climate covers the northern edges of India along the base of the mountains. A desert climate extends along the western border with Pakistan and includes portions of the Thar Desert.

INTERACTION **Environmental Issues** A major challenge for India is to clean up and protect its environment. As much as 70 percent of India's surface water may be polluted. Even the sacred Ganges has been polluted by industrial wastes. Industry is poorly regulated in India, and pollution controls often are not enforced. This lack of regulation has had disastrous results. For example, in 1984 in the city of Bhopal (boh-PAHL), several thousand people were killed when a chemical plant accidentally released a cloud of poisonous gas.

Deforestation, overgrazing, and soil erosion have caused India's most serious environmental damage. Indeed, much of the country's natural habitat has disappeared. In some places, the government has created **sanctuaries**, areas in which forests and endangered animals are protected. (See "Planet Watch" on pages 574–575.) An environmental movement called "Embrace the Trees" is helping to slow the pace of deforestation in India. Protecting and restoring India's forests remains a long and massive project, however.

SECTION REVIEW

1. What countries border India? What climate does most of India's interior have?
2. What are the causes of environmental damage in India?
3. **Critical Thinking** **Suggest** a reason why Hindus consider the Ganges River to be sacred.

ECONOMIC GEOGRAPHY

FOCUS

- *Why has India's agricultural production been limited? Why is it now improving?*
- *What mineral and energy resources does India have? How has India's industry changed?*

INTERACTION **Agriculture** About three-fourths of India's people are involved in agriculture, mostly as subsistence farmers. The country does, however, produce many cash crops for export, including tea, cashews, tobacco, and coffee. India's cotton fields support the country's large textile industry.

Even though much of India's soil is fertile, crop yields are very low. The country's poor yields stem from a lack of widespread irrigation and modern farming methods. Most modern farm equipment and fertilizers cost too much for many farmers to buy. Cow dung, which is readily available, could be used for fertilizer. Farmers often burn it, however, for heating and cooking, because firewood and charcoal are in short supply.

Cattle are used as work animals and as a source of dung and milk. Cows are sacred in India, however, and many are allowed to wander freely. Farmers also use water buffalo to work the land. In addition, they raise sheep for wool and meat, and goats for meat and milk.

This cow threatens to stop traffic in **Jaipur, India**. Cows, considered sacred by Hindus, can be found roaming freely in many urban and rural areas.

JCT FABRICS
TATA

One of India's main goals is to make better use of its agricultural resources. To this end the Indian government and foreign countries have aided several rural agricultural development projects. New irrigation systems are increasing farmland, and flood-control and soil-erosion programs are saving precious soil. Better grain-storage facilities reduce damage by pests, and the use of high-yield crops allows higher production and shorter growing seasons. Mechanization also makes the farms more efficient, though it sometimes adds to rural unemployment because fewer workers are needed. All of these improved agricultural methods have helped make India self-sufficient in food production in most years. Although poverty persists, the people of India no longer suffer major famines.

Mineral and Energy Resources India has many mineral resources, including large deposits of iron ore, bauxite, and coal. Much of the iron ore and bauxite is exported. India has some petroleum deposits but must import oil to meet its consumption needs. The country's supply of uranium ore supports a number of nuclear power plants and has enabled India to develop an atomic bomb.

India's great river systems are a source of hydroelectric power, though only a small part of this resource is developed. A large hydroelectric-irrigation project planned for the Narmada River will be among the world's largest, with 30 major dams and hundreds of smaller ones.

Industry Even though India is largely agricultural, it ranks about tenth among the world's industrial countries. Manufacturing generates around one-fourth of India's income.

In recent decades, the government wanted to make India self-sufficient in manufactured goods. Many large, state-owned industries and some private ones formed to produce chemical fertilizers, steel, railroad and farm machinery, automobiles, and many other items. The iron and steel mills in the Jamshedpur region, west of Calcutta, grew to become some of the world's largest. The cotton textile mills in Bombay and the jute mills in Calcutta also became major production centers. Although the government's approach did decrease India's dependence on imported goods, the nation still acquired a huge trade deficit. In addition, inefficiency and corruption marked the state-owned industries.

The government has since changed its policies. Today, small privately owned industries are the fastest-growing sector of the economy. India has also opened its doors to foreign capital and industry, which are attracted by India's huge pool of low-cost labor. The government has been particularly successful at promoting high-tech industry in the country. For instance, Bangalore, a city in southern India, has attracted so many foreign computer companies it is called India's "Silicon Valley."

To draw more foreign industry, India must strengthen its infrastructure. The country's roads, railroads, airlines, electricity, and telecommunications systems, little changed from colonial times, are inadequate and unreliable. In addition,

Text continues on page 576.

PLANET WATCH

Case Study: Saving the Tiger

The tiger is a solitary animal that usually stalks its prey at night, deep in the forests. Tigers hunt deer and pigs and sometimes young elephants and rhinoceroses. The tiger is prey, though, as well as hunter. Humans have killed tigers both for sport and profit. Rugs made from tigers' skins were once popular throughout much of the world, as were clothes trimmed with tiger fur. Even today, laws designed to protect the tiger are ignored by poachers, who kill the animals for the thousands of dollars their skins will bring.

Once roaming over much of Asia, the tiger has become one of Earth's most endangered species.

The Tiger in India

According to one source, tigers in India in the early 1800s were "so numerous it seemed to be a question as to whether man or the tiger would survive." Exactly how many tigers existed in India is not clear, but by 1900, they numbered only 40,000 due to hunting. Among India's British rulers as well as the country's maharajas, or princes, a fallen tiger was a symbol of status.

To slow the decline of the Indian tiger population, the British government limited hunting to certain areas. They also set up seasons, or time periods, during which hunting tigers was illegal. Laws against poaching were enforced so harshly that few people dared to hunt illegally.

As a result of these policies, India's tiger population thrived for decades. Then, in 1947, India gained its independence from Britain. Free of foreign rule, people began to ignore many of the British hunting laws, and poaching increased.

At this same time, India's tigers faced an added threat. Between 1947 and 1990, India's population swelled from less than 500 million to more than 800 million. As the number of people soared, the forests began to disappear rapidly. The amount of India's forested lands decreased by almost 40 percent from 1900 to the early 1990s. Farmers began to plant crops and graze buffalo and other farm animals on lands once covered by trees and roamed by tigers.

With their habitat shrinking, large numbers of tigers fell prey to hunters, starved, or attacked humans and were killed in return. By 1972, fewer than 2,000 tigers remained in India. The Caspian, Javan, and Balinese tigers are already extinct.

Project Tiger

In 1973, the Indian government took steps to prevent the extinction of the tiger. With support from the World Wildlife Fund, nine tiger reserves were created as part of Project Tiger. Since then, six more reserves have been added. The core

A tiger makes its way through a river in one of India's Project Tiger reserves. The total core area of the reserves, where people are not allowed, is more than 3,000 square miles (7,770 sq. km).

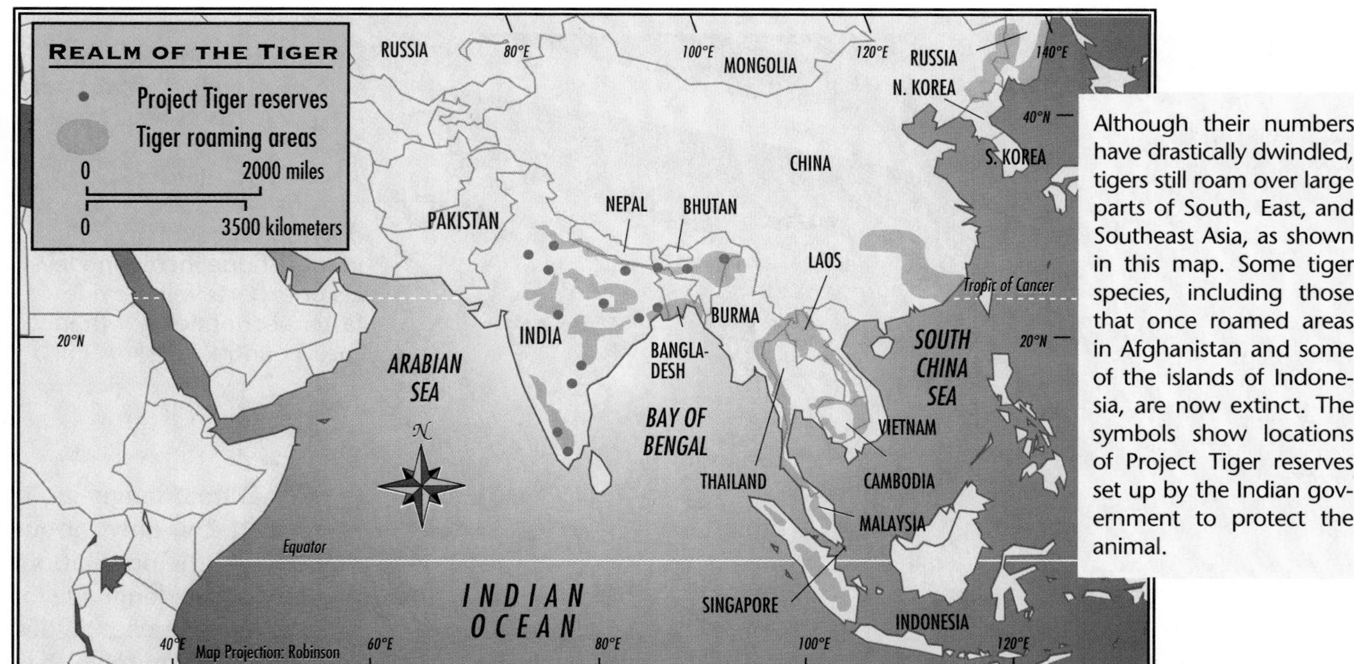
REALM OF THE TIGER
Project Tiger reserves
Tiger roaming areas
0
2000 miles
0
3500 kilometers
RUSSIA
80°E
100°E
MONGOLIA
120°E
RUSSIA
140°E
N. KOREA
40°N
S. KOREA
CHINA
PAKISTAN
NEPAL
BHUTAN
LAOS
Tropic of Cancer
20°N
INDIA
BURMA
BANGLA-
DESH
ARABIAN
SEA
SOUTH
CHINA
SEA
20°N
BAY OF
BENGAL
VIETNAM
N
THAILAND
CAMBODIA
MALAYSIA
Equator
SINGAPORE
INDIAN
OCEAN
INDONESIA
40°E
Map Projection: Robinson
60°E
80°E
100°E
120°E

Although their numbers have drastically dwindled, tigers still roam over large parts of South, East, and Southeast Asia, as shown in this map. Some tiger species, including those that once roamed areas in Afghanistan and some of the islands of Indonesia, are now extinct. The symbols show locations of Project Tiger reserves set up by the Indian government to protect the animal.

areas of the reserves, which are surrounded by less-restrictive buffer zones, are closed to people. No one is allowed to enter these areas unless accompanied by a park ranger. In addition, tens of thousands of acres have been set aside as national parks and wildlife preserves.

Many poor, rural Indians have reacted negatively to the reserves. Scarcely able to feed their families, villagers have little sympathy for the tiger. They argue the reserve lands are desperately needed for grazing livestock, collecting firewood, and gathering wild honey. Many believe that tigers are being protected at the expense of people.

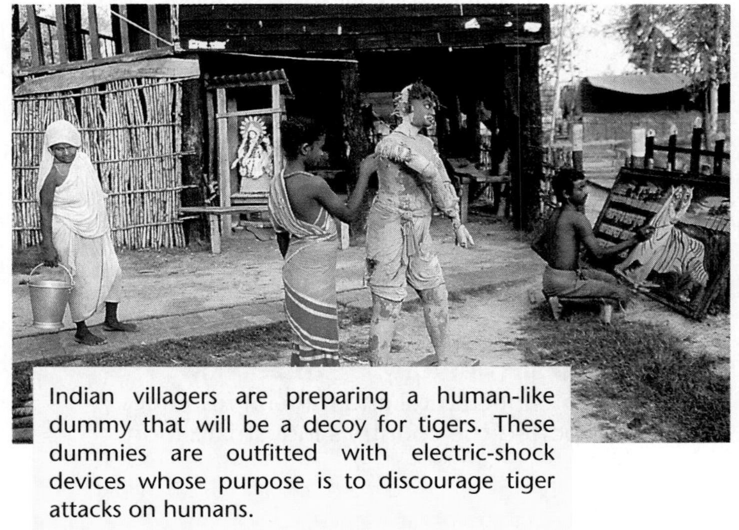

Indian villagers are preparing a human-like dummy that will be a decoy for tigers. These dummies are outfitted with electric-shock devices whose purpose is to discourage tiger attacks on humans.

A wildlife expert from India's Ranthambhore National Park understands the concerns of India's rural poor. "They're right that the park does not yet benefit them. It's our job to change that, to help them see that their survival and that of the park are linked, that if the forest is destroyed, their lives and all our lives will be destroyed."

He and other environmentalists belong to a foundation that reaches out to villagers to show them that they and the reserves can coexist. In an effort to reverse deforestation, the foundation distributes more than 40,000 tree seedlings a year. To deal with the problem of too much livestock and too little grazing land, it has introduced new types of buffalo that need little land. It has tried to aid the local economy by forming a crafts cooperative, in which women make scarves and other handicrafts for sale in urban areas.

Since 1973, the number of India's tigers has doubled to about 4,000. This number is still dangerously low, however, and poaching remains a problem. As many as 20 tigers were killed by poachers in Ranthambhore in 1992. Environmentalists, however, are determined to succeed in their efforts to save the tiger and its habitat.

YOUR TURN

Imagine you are a worker on a tiger reserve in India. **Write** a brief talk explaining to villagers why it is important to all Indians to save the tiger and its habitat. **Address** the specific concerns of the farmers.

EXPORT
AUSTRALIA

A worker at a factory in central **India** checks heavy machinery manufactured for export to Australia. As in other developing countries, officials in India hope increasing levels of exports will provide faster economic growth in their country.

manufacturing employs only a small percentage of the working population. Nonetheless, India's overall economy is expected to grow at a rapid pace.

SECTION REVIEW

1. What effects have recent agricultural development projects had on India?
2. What mineral and energy resources does India have? How did India's government change its policies regarding industry, and why did it do so?
3. **Critical Thinking** **Explain** how India's agricultural and mineral resources supported the country's industrialization.

HUMAN GEOGRAPHY

FOCUS

- *What are India's cultural and political characteristics?*
- *How does rural India differ from urban India?*

REGION **Language and Religion** India is one of the most culturally complex countries in the world. Eighteen major languages and many others are spoken in the country. Although Hindi is the national language, it is spoken by only about one-third of the population. English is the language of India's government, higher education system, and business community. It is spoken mainly in urban areas. Most people of the Indo-Aryan north speak Hindi, while most people in the nation's south speak Dravidian. Both languages have many dialects.

India is also complex in terms of religion. The country is about 80 percent Hindu, 14 percent Muslim, 3 percent Christian, and 2 percent Sikh. Millions of Buddhists, animists, and followers of other religions also live in India.

For hundreds of years, religious tensions have plagued India. Ancient hostility still exists between the country's Hindu and Muslim populations. Hindu nationalism has grown stronger in recent years, leading to violent riots between the two groups. In two months alone, from December 1992 to January 1993, violence among members of various sects killed more than 1,700 people, after militant Hindus destroyed a Muslim mosque in northern India. The Hindus charged that the mosque was built on the site where a Hindu god was born and where a Hindu temple had stood.

Independence movements have also led to violence. In the Punjab province, hundreds of people have died in conflicts between Hindus and Sikh separatists. Meanwhile, separatists in Kashmir have been fighting for their independence.

Political Geography Under the leadership of Jawaharlal Nehru following independence, India adopted a constitution making the country a democratic republic. The nation is made up of a union of states held together by a strong central

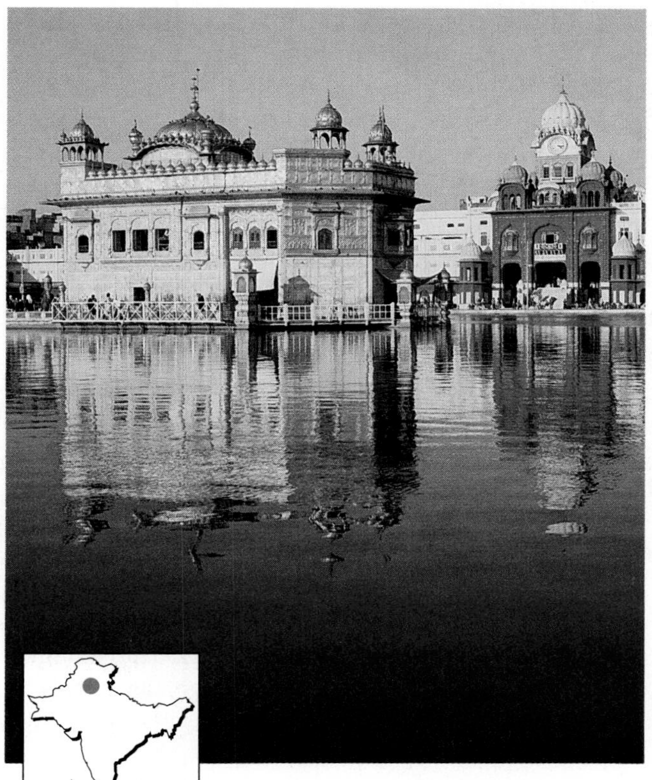

The Golden Temple in **Amritsar, India**, is the chief house of worship for Indian Sikhs. Sikh demands for an independent state in Punjab, and opposition to such demands by the Indian government, has at times brought violence to the region.

government and headed by a prime minister and a parliament.

Nehru died in 1964 and was succeeded as prime minister two years later by his daughter, Indira Gandhi. Indira Gandhi, India's first female prime minister, remained in office until 1984, when she was assassinated by Sikhs who disliked her resistance to Sikh independence. Indira Gandhi's son, Rajiv Gandhi, then became prime minister. In 1991, he too was assassinated, bringing an end to the Gandhi family leadership and throwing India's politics into even greater turmoil. Although India's government has tried to keep politics and religion separate, a movement in favor of a Hindu nation has been growing stronger.

Rural Life The differences between rural and urban life are striking. About three-quarters of India's people are poor farmers who live in rural villages. (See "Themes in Geography" on pages 580–581.) Many have seen little progress in their lifetimes, and some 60 percent cannot read or write.

Most villages are not connected to paved roads. Electricity, telephones, and running water are rare, and sanitation is poor. Often, one doctor must try to serve thousands of people. Rural houses are built of local materials. In the alluvial river valleys, straw-roofed homes are made of mud. In other areas, the houses might be made of dried cattle-dung plaster, brick, or bamboo.

Families tend to be large. In fact, one of every five children born in the world today is born in India. Most of India's rapid population growth occurs in rural areas, where children are needed to help in the fields. In addition, both the Hindu and Muslim cultures value large families and have resisted family planning.

Despite government attempts at land reform, landless field workers make up more than 30 percent of India's rural population. Those who do own small plots of land are often forced into debt to meet farm and living expenses. Village moneylenders may charge as much as 300 percent interest on loans. Other farmers work the land of large landowners and must give a large share of the crops they grow to the owners.

Rice is the staple food in the south, wheat in the north, and millet in the Deccan Plateau region. The diets of most Indians contain little meat. The Hindu religion forbids the eating of beef, and the Muslim religion does not allow the eating of pork. Some religions in India call for a completely vegetarian diet.

Village industries are limited mostly to **cottage industries**. In cottage industries, workers make small consumer items in homes and small workshops. These products include handwoven carpets, metal tools, locks, jewelry, and shoes and other leather goods.

Most Indian men in rural areas wear a *dhoti* (DOHT-ee), a simple white cloth wrapped around the hips and between the legs. Hindu women wear a *sari*, a wide piece of cloth draped to cover the body from shoulder to ankle. Indians usually can tell what part of the country a Hindu woman is from by the way she drapes her *sari*.

Rural India is a male-dominated society. When a woman marries, she is no longer part of her own family, but rather becomes a member of her husband's family. A rural wife shares field work with her husband and must gather drinking water and firewood for the family. She may also care for the cattle, as well as raise the children. Literacy rates among rural Indian women are very low.

This Indian woman is wearing a *sari*, a light, loose-fitting garment made of a brightly colored cloth, such as silk. Many Indians wear light, cool clothing because of the warm climates that dominate much of the country.

The Caste System Most rural villages are still based on the **caste system**. The Hindu Aryans developed this social system thousands of years ago. It divided India's society into four major classes, or castes, by occupation. The Brahmins were the highest caste. They were priests and intellectuals and the only people who could read and write. Below the Brahmins in order of rank were the Kshatriyas (kuh-SHA-tree-uhz), or warriors; Vaisyas (VISHE-yuhz), the traders and merchants; and Sudras, the farmers and laborers.

Below these four classes were the untouchables, who were considered so lowly as not to be part of any caste. They were not allowed to enter a temple and did only the most unpleasant tasks. They were forbidden from having contact with Indians of any caste.

Under the caste system, a person is born into a caste and cannot move into another. Some Indians believe that taking up the occupation of another caste or marrying into another caste is a violation of moral code, punishable by rebirth in another life as an untouchable or a non-human form of life.

Since independence from British rule, the government has worked hard to abolish the caste system. It declared that the poor treatment of untouchables was illegal. Mohandas Gandhi, who helped free India from British control, saw the untouchables as human beings who deserved to be treated with respect and dignity, just as people of other classes were. Gandhi called the untouchables *harijans*, or "children of God."

Yet, regardless of the law, the caste system remains a part of Hindu life, and violence against the untouchables still takes place. As improved education and transportation reach the villages, the caste system will likely change, as it has in many cities.

City Life In general, Indian city dwellers wear European-style clothing, read English-language newspapers, and work in factories and office buildings. Smog, traffic, and noise are part of their daily lives. Education in universities and technical schools is available in the cities, and urban women work with men in professional and factory jobs. Some are members of trade unions and seek pay equal to that of men.

The cities are home to a small number of wealthy industrialists and landowners. These prosperous few live in large walled-in homes or guarded high-rise apartments. The working middle class, however, makes up a larger, and growing, sector of urban society. Its members have become part of India's growing consumer economy. Most live in small apartments. Some can afford televisions, motor scooters, and electric kitchen appliances. They can also afford entertainment in the form of movies, soccer games, shopping, and restaurants.

Despite some economic progress, however, most of the urban population remains poor. Many are homeless and live in the streets. Many others live in slums in makeshift homes built of scraps of wood, cardboard, and cloth. They do not have clean water, sewage systems, or garbage collection. They look for work each day for whatever pay they can get.

NEWS FROM AROUND THE WORLD — India

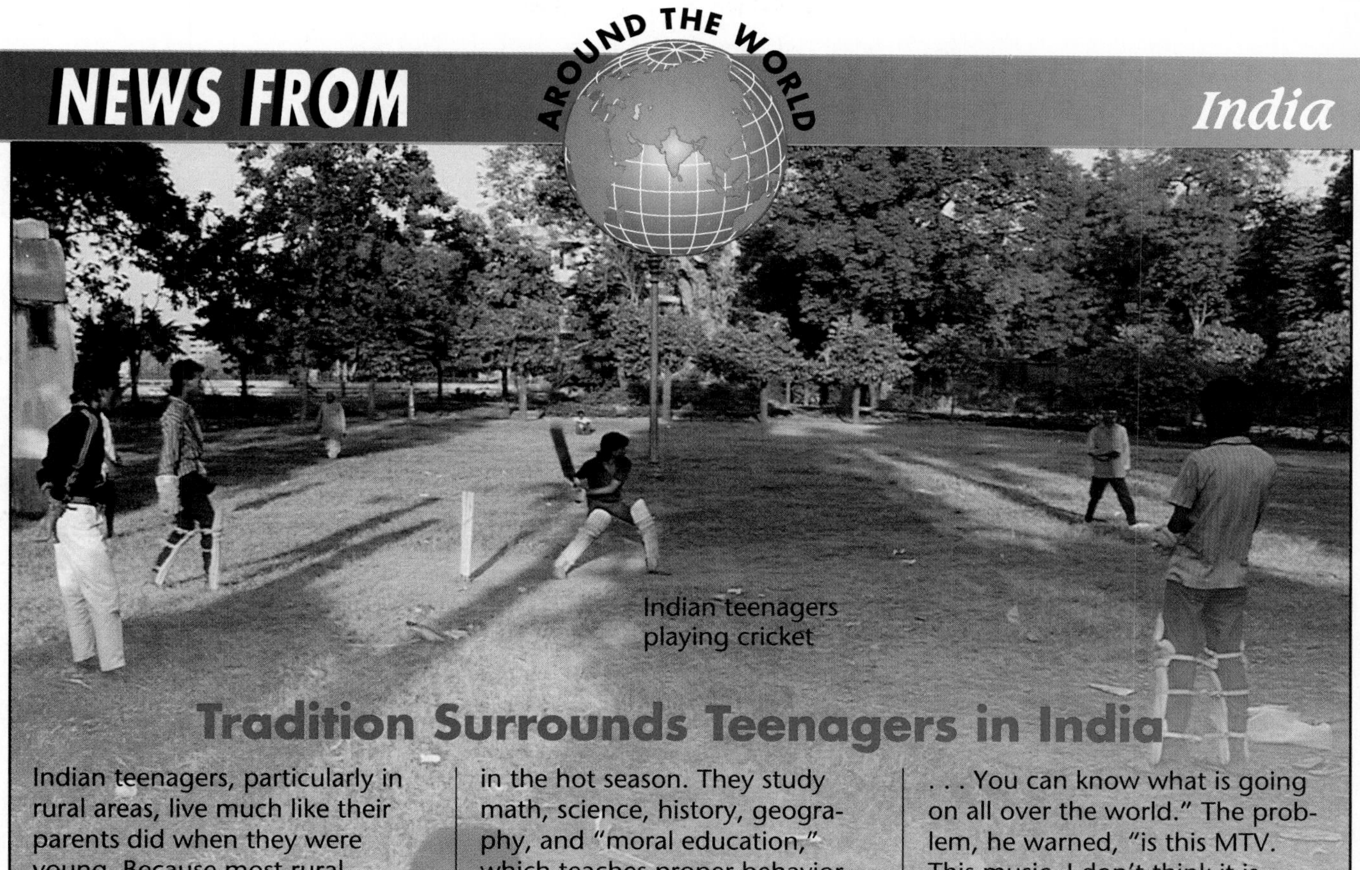

Indian teenagers playing cricket

Tradition Surrounds Teenagers in India

Indian teenagers, particularly in rural areas, live much like their parents did when they were young. Because most rural Indians are farmers, teens and children help their families work the land and raise crops. As a result, many Indian children do not attend elementary school. In fact, half the population over age 15 is unable to read and write.

For those who are able to attend school in rural areas, conditions are poor. There are too few teachers, and materials and equipment are outdated. School often takes place under makeshift circumstances—on a porch, under a tree, in a village square. Children go to school from 10:00 A.M. to 4:00 P.M. in the cool season and from 7:00 A.M. to noon in the hot season. They study math, science, history, geography, and "moral education," which teaches proper behavior.

Life for many teenagers in Indian cities is much different. Technology and communications expose them to a larger world. MTV® music television now reaches an estimated 400,000 homes in India. Eager Indian teenagers tune in to the channel's rap and rock music. "Most of my friends, they watch," says one teen. "In every house, they have their MTV."

Parents, however, often view such western influences as threats to traditional values. For example, one father says, "Satellite TV is very good, no doubt. There are so many programs you can get. . . . You can know what is going on all over the world." The problem, he warned, "is this MTV. This music, I don't think it is good for children."

One tradition most Indian families, both rural and urban, still observe is arranged marriages. The process of arranging a marriage begins when a girl and boy are still in their early teens. The process can continue for years while parents exchange and analyze their children's horoscopes and arrange property settlements. Although many teenagers accept their parents' decisions, others growing up in the cities are beginning to choose their own spouses. Traditional ways of life in India, however, are slow to change.

Government family planning programs have been more successful in urban areas than in rural areas. Migration to the cities, however, has raised the urban population to hold about one-fourth of the country's people. Millions of people from rural areas seek a better income in the cities and an escape from the caste system. Unfortunately, many move from rural poverty to urban poverty.

Urban Centers New Delhi (DEHL-ee), in northern India, is the capital of the country. Like Washington, D.C., this city was built to house the national government. Along with the neighboring old city of Delhi, New Delhi is the largest industrial area of northern India.

Bombay, located on the western coast, is India's largest city and a commercial and industrial

Text continues on page 582.

Indian Villages

Although India is gradually industrializing, the majority of its people are still peasant farmers. These farmers live in the hundreds of thousands of small villages scattered throughout the country.

Surrounded by farmland, each village consists of a number of small huts with one or two rooms. These huts cluster around a well or a stream, which forms the central point of the village. Each day, the women gather at the well with large clay pots for carrying water. Most do not have luxuries such as running water or electricity. And even in villages that do have electricity, many of the huts are not connected to the electricity supply. There are few shops in the villages.

Small farming villages set the stage of daily life in rural India.

Some villagers own their own plots of land, but they usually manage to grow only enough food to feed their own families. Most villagers, however, do not own land. Because

INTERACTION Most of the farming methods practiced in Indian villages today have been used for generations. Because the peasant farmers are very poor, they cannot afford modern farm equipment. They obtain their water from wells, using cattle to drive the pumps. Most crops, however, must depend on rainfall rather than irrigation. The great differences in rainfall throughout the country mean people must live off the crop that grows best in their area. **Why are there great differences in rainfall in India?**

PLACE Family ties are very important in Indian communities. Marriages often are arranged by parents. After a couple marries, they usually live in the home of the husband's parents. In fact, grandparents, uncles, aunts, cousins, and other relatives may all live in one house. They combine their earnings, and all share in the work on the farm and in the house. **Can you name other countries in which marriages are still arranged?**

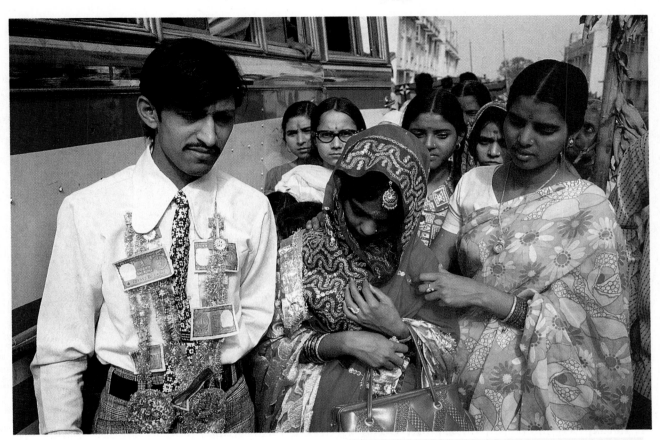

India has such a large population, there is not enough land for everyone. Instead, many farmers must rent plots from landowners. It may take a village farmer a year to earn as much as an American farm laborer could earn in two weeks. Most Indian farmers make their money at markets in nearby towns, where they sell any extra produce they grow.

People living in neighboring villages may not be able to communicate with each other because India is divided into 15 states, each with its own language. Each of these different languages has more than 500 dialects. As a result, many villagers spend their entire lives without ever leaving their own village.

Most peasant farmers in India are illiterate. Their lives revolve around their land, and they have little time for learning to read and write. The government is trying to change this situation, however. Some villages now have schools, which also serve as meeting places for the community. Adults and children alike can attend these schools.

The community schools play an important part in teaching villagers about new and better ways of farming. Many government officials know that the only way India can feed its rapidly increasing population is by using modern farming methods. They also are recognizing that education can help the people of India's villages improve their lives.

Movie advertisements cover this small theater in **Madras**. India's large moviemaking industry is supported by millions of movie fans.

center. Actually, Bombay lies on a small island just off India's western coast. Several bridges link the island and the city to the mainland. Bombay's wide variety of industries includes India's large textile and movie industries. In fact, India is the number-one moviemaking nation in the world, and Bombay is sometimes called the "Hollywood of India." Its constant stream of movies offering romance and adventure delights large Indian audiences. Bombay is also a busy seaport, well situated for world trade.

Calcutta, located in eastern India on a branch of the Ganges River, is the other giant industrial and seaport city of India. Calcutta produces a wide range of goods, from tea and hides to steel and electronics. As thousands of people live in poverty in the streets, high-rise buildings tower above them, and a modern subway system runs beneath them. Most of the city's people are Hindu, though Muslims, Sikhs, and Christians make up about one-fifth of the population. Even today, these groups tend to live in separate sections of the city.

Madras is southern India's leading industrial city and port. This busy shipping center on the country's southeastern coast also has automobile assembly plants, cotton mills, and iron, cement, and glass works.

SECTION REVIEW

1. List three languages spoken and four religions practiced by the people of India.
2. What factors are responsible for population growth in rural India? in urban India? What social class divisions exist in each of the two areas?
3. **Critical Thinking** **Discuss** the influences of religion on life in India in rural areas, urban areas, and in the country as a whole.

CHAPTER 47 CHECK

Palace of the Winds in **Jaipur**

Reviewing the Main Ideas

1. India borders several countries to the north and is surrounded by water to the west, east, and south. Most of India lies in the tropics and is strongly influenced by the monsoon. A major challenge for India is to clean up and protect its varied environment.
2. About three-fourths of India's people are involved in agriculture. Agricultural development programs are increasing crop yields. The country has a variety of mineral resources and industries. Today, small privately owned industries are the fastest-growing sector of the economy, and foreign investment is increasing.
3. India, a democratic republic, is one of the most culturally complex nations in the world, with many languages spoken and religions practiced. Religious tensions have plagued India for centuries. The differences between rural and urban life in India are striking.

Building a Vocabulary

1. What are cottage industries? What are some of the products they produce?
2. What is the caste system? What are Hinduism's four castes? What group is considered outside the caste system?

Recalling and Reviewing

1. What bodies of water does India border?
2. How is farmers' use of cow dung as a source of fuel related to India's deforestation?
3. What attracts foreign capital and industry to India? What might India do to further attract them?
4. What are four major cities in India?

Thinking Critically

1. India became a democratic republic after it became independent. In what way, however, did the country's government resemble a dynasty in the decades following independence? What do you think was responsible for this resemblance?
2. As you read in the chapter, many farmers in India work the land of large landowners and must give a share of the crops they grow to the owners. What Middle and South American system of agriculture does this resemble?

Learning About Your Local Geography

Individual Project

In India, rural houses are built of local materials such as straw, mud, and bamboo. From what materials are houses or other buildings made in your community? How do these materials reflect local geography in terms of resources found in or near your area? Draw some illustrations of buildings in your community. Write captions for your illustrations, describing the materials used in the buildings and where these materials might have originated.

Using the Five Themes of Geography

Write a paragraph discussing the theme of **human-environment interaction** in India. Focus on India's rivers. How have India's rivers affected human settlement, influenced religious beliefs, and served as resources? How has human activity affected the rivers?

CHAPTER 48

The Indian Perimeter

GEOGRAPHY DICTIONARY

partition
cease-fire
storm surge

"We took a long wheeling look from the highest point in the world. There was Makalu and Lhotse, Nuptse, and Kanchenjunga looming on the horizon and a host of other giants, now all far below us, in a maze of rock and snow, fluted ice-walls plunging to glaciers and glaciers plunging to the valleys below. We gazed north, towards the Tibetan plateau, and south, towards the plains of India. . . . The view was unforgettable."

Hari Pal Singh Ahluwalia

Decorative buckle from **Pakistan**

India borders Pakistan in the west and the Himalayan kingdoms of Nepal and Bhutan in the north. To the east is Bangladesh, and to the south are the island nations of Sri Lanka and the Maldives. These countries make up the Indian perimeter.

A majority religious group dominates each of the region's countries. Pakistan, Bangladesh, and the Maldives all have a Muslim majority. Sri Lanka and Bhutan are Buddhist, and Nepal is Hindu. Today, ethnic tension between majority and minority groups, along with economic development, challenge most of these countries.

Mount Everest

THE INDIAN PERIMETER: PHYSICAL–POLITICAL

SCALE
0 250 500 Miles
0 250 500 Kilometers
Projection: Two-Point Equidistant

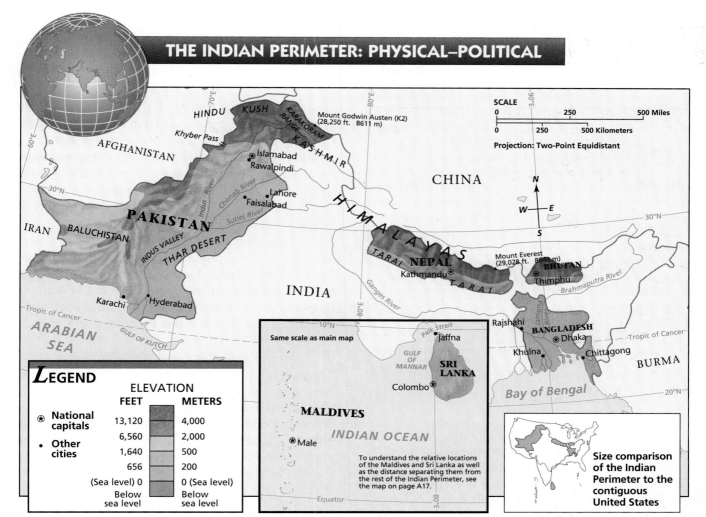

HINDU KUSH
KARAKORAM RANGE
Mount Godwin Austen (K2) (28,250 ft. 8611 m)
Khyber Pass
AFGHANISTAN
KASHMIR
Islamabad
Rawalpindi
CHINA
Chenab River
Indus River
Lahore
Faisalabad
Sutlej River
HIMALAYAS
PAKISTAN
IRAN
BALUCHISTAN
INDUS VALLEY
THAR DESERT
TARAI
NEPAL
Kathmandu
Mount Everest (29,028 ft. 8848 m)
BHUTAN
Thimphu
Brahmaputra River
INDIA
Ganges River
Karachi
Hyderabad
Tropic of Cancer
ARABIAN SEA
GULF OF KUTCH
Same scale as main map
Palk Strait
Jaffna
GULF OF MANNAR
SRI LANKA
Colombo
MALDIVES
Male
INDIAN OCEAN
Equator
To understand the relative locations of the Maldives and Sri Lanka as well as the distance separating them from the rest of the Indian Perimeter, see the map on page A17.
Rajshahi
Jamuna River
BANGLADESH
Dhaka
Khulna
Chittagong
BURMA
Bay of Bengal
Size comparison of the Indian Perimeter to the contiguous United States

LEGEND

National capitals
Other cities

ELEVATION

FEET	METERS
13,120	4,000
6,560	2,000
1,640	500
656	200
(Sea level) 0	0 (Sea level)
Below sea level	Below sea level

PAKISTAN

FOCUS

- *What are Pakistan's landform regions? What are the characteristics of each?*
- *What are the economic, political, and cultural characteristics of Pakistan?*

Pakistan is the second largest country in South Asia. It formed in 1947 when the British government **partitioned**, or divided, British India into Muslim and Hindu areas. Pakistan, with its Muslim majority, originally was divided into West Pakistan and East Pakistan. Then in 1971, East Pakistan became the independent country of Bangladesh. Pakistan today consists of the former West Pakistan.

Physical Geography Pakistan can be divided into three landform regions: the mountains in the north, the deserts in the west and east, and the Indus River valley. (See the map above.)

The northern mountain region includes the glacier-covered Karakoram Range, which forms Pakistan's northern border with China. The mountains of this range are some of the highest and most rugged in the world. Mt. Godwin Austen is the world's second highest peak, soaring 28,250 feet (8,611 m). The mountain is also called K2 because it was the second peak measured in the Karakoram Range. This mountain region is the source of the Indus River system. It also holds most of the nation's remaining forests. Less than five percent of Pakistan is forested.

Deserts lie in the east and west of Pakistan. In the east, the Thar Desert separates Pakistan and India. In the west, a small strip of desert runs to the border with Iran. Rugged hills surround this arid land.

Pakistan's third landform region follows the Indus River, which flows from the Himalaya and Karakoram ranges south to the Arabian Sea. The

FACTS IN BRIEF The Indian Perimeter

COUNTRY POPULATION (1994)	LIFE EXPECTANCY (1994)	LITERACY RATE (1990)	PER CAPITA GDP (1993)
BANGLADESH 125,149,469	55, male 55, female	35%	$1,000
BHUTAN 716,380	51, male 50, female	15% (1989)	$700
MALDIVES 252,077	63, male 66, female	93% (1989)	$620 (1991)
NEPAL 21,041,527	52, male 53, female	26%	$1,000
PAKISTAN 128,855,965	57, male 58, female	35%	$1,900 (GNP)
SRI LANKA 18,129,850	69, male 75, female	88%	$3,000
UNITED STATES 260,713,585	73, male 79, female	97% (1991)	$24,700

Sources: *The World Factbook 1994*, Central Intelligence Agency; *The World Almanac and Book of Facts 1995*

Which two countries of the Indian perimeter have the lowest per capita GDP?

river's valley, a wide and flat alluvial plain, dominates the central and eastern central parts of the country. The valley was the birthplace of the Indus civilization and has been the site of irrigation farming for several thousand years.

Today, the Indus River valley is one of the world's largest irrigated regions. In fact, virtually all of Pakistan's agriculture and population are located along the river and its tributaries. Deforestation and overgrazing in the Indus River's upper drainage basin has led to severe flooding in the Indus Valley in recent years.

Economic Geography Pakistan is a poor country in which agriculture produces about half of the nation's exports. Cotton is the major export crop, while wheat and rice are the major food crops. Most of Pakistan's industries produce cotton textiles and food products.

Hydroelectric power from the Indus River is an important source of energy. Even so, the country must import oil. To reduce this dependence, Pakistan is developing its natural-gas deposits.

The Pakistan government is looking for other ways to improve the country's economy. For example, the government now promotes the privatization of inefficient state-owned industries. It also is trying to attract foreign investment to help increase export earnings. The nation's poor infrastructure, particularly its roads and communications systems, is improving as well.

Human Geography Pakistan is officially an Islamic republic based on Islamic law. In reality, though, power has often been in the hands of military governments since the country's independence. There have been a few brief periods under elected civilian governments, but military coups have usually put an end to these periods. Political corruption is a serious problem.

More than 95 percent of the country's people are Muslim. Punjabis form the largest ethnic group. The official language is Urdu, but English is spoken in the cities. The Baluchi of Baluchistan, a western desert region, speak their own language. Many other languages are spoken in the isolated northwest.

Like neighboring India, Pakistan faces rapid population growth. Its more than 120 million people may double in number by about 2025. Pakistan also has a large number of refugees. (See "Global Connections" on pages 588–589.) The country is struggling to provide schools, health care, and housing for this exploding population.

More than one-quarter of all Pakistanis live in urban areas. The country's major cities include the capital city of Islamabad (is-LAHM-uh-bahd) in the far north and Karachi, the nation's largest city and major seaport. Karachi lies near the Indus River delta on the Arabian Sea coast.

Merchants display fabrics for sale in their shop in **Lahore, Pakistan**. Cotton and textiles are major exports for Pakistan.

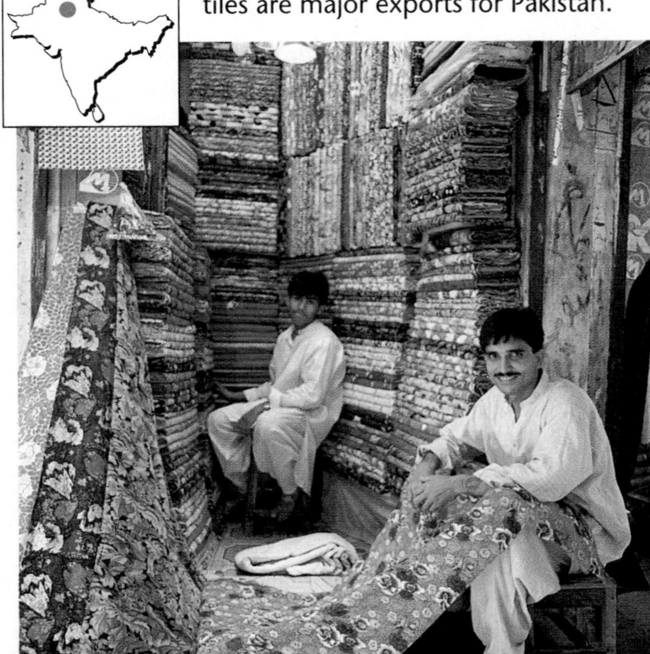

Lahore, near the border with India, is Pakistan's cultural center and second largest city.

LOCATION **Kashmir** For nearly 50 years, Pakistan and India have fought over the mountainous Muslim region of Kashmir. Pakistan tried to take the region after the partition of British India in 1947, sparking a war with India. In 1949, the United Nations initiated a **cease-fire**, an agreement to halt active hostility, between Pakistan and India. Pakistan was given control over the northern portion of Kashmir, and India had control over the southern part. More fighting, however, broke out in 1965, 1971, and 1990. Kashmir remains a major source of conflict between the two countries.

SECTION REVIEW

1. What are Pakistan's three landform regions? Which region holds most of the country's people and agriculture?
2. How is Pakistan's government trying to improve the nation's economy? What types of governments has Pakistan had since independence? To what religion do most Pakistanis belong?
3. **Critical Thinking** **Agree or disagree** with the following: "Pakistan should be considered part of Southwest Asia rather than South Asia." Support your answer.

THE HIMALAYAN KINGDOMS

FOCUS

- *What are the physical, economic, cultural, and environmental characteristics of Nepal?*
- *How is Bhutan similar to Nepal? How is it different?*

REGION Sandwiched between India and China are the small landlocked kingdoms of Nepal and Bhutan. (See the map on page 585.) Both are extremely mountainous and depend on India for most of their trade and foreign aid.

Nepal Nepal rises steadily from south to north. In the south, near the border with India, lies a low, tropical plain called the Tarai. Just north of the Tarai is a region of hills. Farther north tower the ice-capped Himalayas. Mount Everest, the world's highest mountain, rises above Nepal's border with China.

Over 90 percent of Nepal's population is involved in agriculture, most of it subsistence. The Tarai is one of Nepal's most agriculturally productive regions. It supports typical tropical crops, such as jute, rice, and sugarcane.

In the hills north of the Tarai, farmers grow corn, millet, and wheat. This central region includes the Kathmandu (kat-man-DOO) Valley, where most Nepalese live. The capital city of Kathmandu lies here as well. At an elevation of 4,344 feet (1,324 m), the city's back door opens on the Himalayas. *Kathmandu* means "wooden temple" and refers to a Buddhist temple that was built in the area in the late sixteenth century. A temple in the city's central square is believed to be the original structure. Kathmandu is Nepal's only major city, and it is the center of the country's business and transportation.

For the most part, the Himalayas are not agriculturally productive. Some of the lower slopes can be used for grazing, and some herbs grown here are used in medicines in demand around the world. In addition, some wheat grows in the highlands. Farmers raise the grain along narrow terraces that conserve soil moisture and prevent erosion. Most of the higher mountain areas, however, are uninhabitable. Indeed, snow covers more than 15 percent of Nepal's total land area year-round.

Although many of the people in northern Nepal are Buddhists, most Nepalese live farther south and are Hindus. Nepali is the official language, but there are many local languages. The government is a constitutional monarchy in which the king has limited power. Democracy has been increasing, particularly since 1990 when the government legalized political parties.

With few mineral resources and little access to world trade routes, Nepal's economic development has been limited. Tourism is important to the country's economy. Visitors browsing in the streets of Kathmandu will find shops so small that shopkeepers can sit on the floors of their stores and reach all items for sale simply by moving their arms. Most tourists in Nepal,

Text continues on page 590.

Global Connections

The Movement of People

In the 1500s, Ferdinand Magellan's ship took three years to sail around the world. Today, airplanes fly across entire continents in a matter of hours, and communication technology brings distant lands into contact. As news circles the earth, the world's people are moving from place to place in greater numbers than ever before.

Some agencies estimate that at least 100 million people live outside the countries in which they were born.

Immigration

Many people move to find employment or political freedom. Others seek a community of people who share their particular religion or culture. Fathers often send back money to their families to ease the economic hardship in their home countries, but women and children are moving in larger numbers as well.

In the 1960s, for the first time in its history, the United States received more immigrants from Middle and South America than from Europe. Most arrived in search of jobs, though some came to escape civil wars and political persecution. These immigrants brought distinct cultures with them and helped make cities such as Los Angeles and Miami rich in diversity.

Many western European nations also have received large numbers of immigrants in recent decades. Arabs have moved to France, Turks to Germany, and Pakistanis to Britain. These newcomers have introduced new food, languages, religions, and customs to western Europe. In the opposite direction, more than 1 million Jews have emigrated to Israel in recent years from the former Soviet Union. They seek not jobs but the opportunity to live without discrimination among other Jews.

The largest migrations, however, are taking place within and between developing countries. In 1993, the United Nations Population Fund reported that each year 20 to 30 million people move from rural to urban areas in developing countries. By the year 2000, 18 of the world's largest urban areas will be in non-Western countries such as India. Mexico City, Mexico, and São Paulo, Brazil, will lead the list. London and Paris will not make the top 20.

Refugees

Many people migrate because they are in danger of losing their lives. They are refugees, fleeing war-torn nations and governments that persecute them because of their political beliefs, religion, or ethnic background. In 1995, the world's refugees numbered as many as 23 million.

Recent refugees include the 5 million who fled from the 1990–1991 Persian Gulf War. Kuwaitis, along with foreign-born workers from Yemen, Egypt, Pakistan, Bangladesh, Sri Lanka, and the Philippines, escaped to Jordan. Then at the war's end, Palestinian workers in Kuwait fled or were deported. About 2 million Rwandans

Hundreds of thousands of refugees from Rwanda fled in 1994 to camps such as this one in **Zaire** to escape ethnic violence in their country. War and ethnic violence have been only two of the reasons for the mass movement of people throughout history.

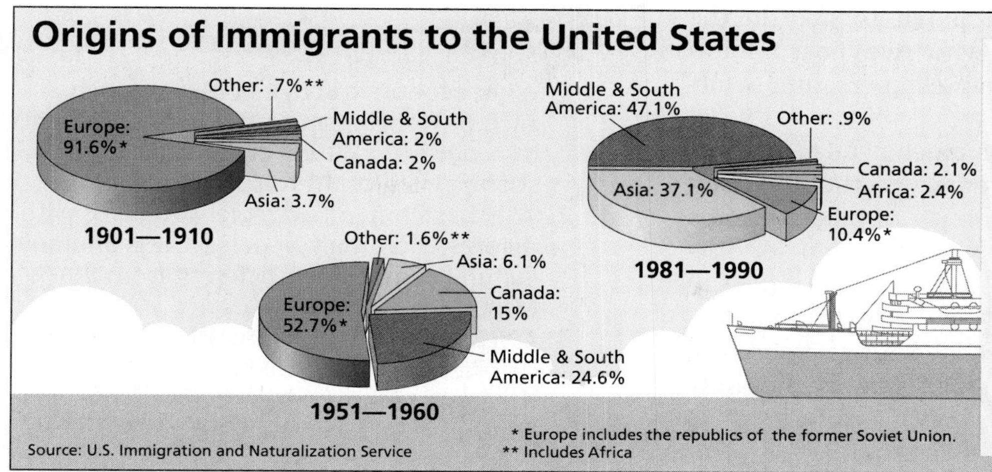
Origins of Immigrants to the United States
Other: .7%**
Europe: 91.6%*
Middle & South America: 2%
Canada: 2%
Asia: 3.7%
1901—1910
Other: 1.6%**
Asia: 6.1%
Europe: 52.7%*
Canada: 15%
Middle & South America: 24.6%
1951—1960
Middle & South America: 47.1%
Other: .9%
Asia: 37.1%
Canada: 2.1%
Africa: 2.4%
Europe: 10.4%*
1981—1990
Source: U.S. Immigration and Naturalization Service
* Europe includes the republics of the former Soviet Union.
** Includes Africa

These pie graphs break down by percentage the origins of immigrants to the United States during three periods of this century. Today, immigrants to the United States are more likely to be from Asia or Middle or South America than from anywhere else. In the decade after the turn of the century, 92 percent of immigrants were from Europe. Most Americans are immigrants or are decendants of immigrants.

fled ethnic fighting in their country in 1994. Many refugees, fearing for their lives, refuse to return home.

Some regions of the world are filled with refugees who have left their homes to escape a devastated environment. This is particularly true in parts of Africa, where severe drought and overgrazing have forced many to migrate in search of fertile soil. Often, such migration leads to a cycle of environmental damage as people crowd onto neighboring lands.

Immigration Policies

Because migration can influence a nation's economy, culture, and physical environment, migrants are not always made to feel welcome in their new country. People worry that newcomers will take jobs or introduce customs that threaten accepted ways of life. Such attitudes often produce social conflict.

For example, in the early 1990s, more than 1 million Germans from Eastern Europe and the former Soviet Union returned to Germany. At the same time, another 1 million non-Germans were seeking to immigrate to the country. Germans, however, were struggling with their own economic problems brought about by the unification of East Germany and West Germany. Some Germans resented immigrants, who were provided with food and housing by the government. Anti-foreigner groups attacked not only recent immigrants but also many Turks who had lived in Germany for decades.

Most nations have policies dealing with the admission of immigrants and refugees. The U.S. Congress sets the annual limit for immigrants and refugees allowed into the United States. Americans remain divided on immigration. Some believe that the United States should have an open-door policy, extending a welcome to all who knock. Others, however, believe that the country should close its doors to all but a few immigrants.

Regardless of nations' policies, people continue to move, seeking better lives. As a result of this migration, cultures mix, languages and religions spread, areas gain and lose population, and the natural environment is sometimes altered.

The large number of Jewish immigrants from the former Soviet Union has forced the government of **Israel** to speed up the pace of home building.

YOUR TURN

Imagine you are an immigrant who has just arrived in a new country. **Write** a few journal entries describing your experiences in your new home. **Explain** why you have moved and what you are hoping to find.

however, come not to shop but to hike in the mountains. The Sherpa, who live near Mount Everest, are skilled mountain climbers, and sometimes serve as guides.

Tourism has brought much-needed income to Nepal, but it has also damaged the country's environment. A constant stream of hikers tramps over the hillsides. These visitors' need for firewood has contributed to deforestation and soil erosion already begun by Nepal's growing population with its own fuel needs. Serious flood and landslide hazards now exist. To combat the problem, the government has restricted access to some of the more popular hiking routes.

Bhutan The tiny Buddhist kingdom of Bhutan lies east of Nepal. Until 1974, the Bhutan government promoted a policy of almost total isolation. Even today, the country remains somewhat cut off from the rest of the world. Bhutan has worked to preserve its forests, particularly on the steepest slopes. These policies and Bhutan's smaller population have led to fewer of the deforestation problems that plague Nepal.

Bhutan's culture is closely related to that of nearby Tibet. About 70 percent of the people are Buddhists. The rest are mainly Hindus from Nepal. Most of the people are subsistence farmers who grow rice and wheat in the few fertile valleys. Fruit and cardamom are cash crops, and the country sells lumber and hydroelectricity to India.

Even though Bhutan may appear poor by the standards of developed countries, most of its people are self-sufficient in food. The king believes that people's ability to provide food for themselves is more important than the development of an export economy.

This golden statue of Buddha is draped in colorful clothing and jewelry for a festival in **Kathmandu, Nepal**. Many Hindu and Buddhist temples are found throughout Nepal, although the vast majority of the people are Hindus.

SECTION REVIEW

1. What economic activities take place in each of Nepal's landform regions? How do the country's cultural characteristics change from north to south?
2. Why has Bhutan had fewer problems with deforestation than Nepal? How is Bhutan's cultural composition different from that of Nepal?
3. **Critical Thinking** **Decide** whether deforestation in Nepal has occurred for the same reasons it has occurred in Southeast Asia. Explain your answer.

BANGLADESH

FOCUS

- *What is the physical geography of Bangladesh?*
- *What challenges does the country face?*

LOCATION **Physical Geography** Bangladesh is surrounded by India on the north, west, and east. Its only other neighbor is Burma, with whom it shares a short border at its southeastern tip. To the south, Bangladesh has a low, marshy coastline on the Bay of Bengal. (See the map on page 585.)

The low flood plain and delta of the Ganges and Brahmaputra (called the Jamuna in Bangladesh) rivers make up most of the country. A complex network of more than 5,000 miles (8,045 km) of waterways make river transportation the most common form of travel. Except for a few hilly areas along the eastern border, most of Bangladesh has suffered deforestation.

The entire country has a humid-tropical climate and depends on the monsoon's seasonal rainfall. Temperatures average about 80°F (27°C) year-round, and rice and vegetables thrive throughout the year in the alluvial soils.

Natural disasters are common in Bangladesh. During the wet monsoon, between June and September, heavy monsoon rains flood the lower river areas and delta islands. Often, more than half the nation is flooded for several weeks each year.

Bangladeshis crowd onto these ferry boats, which are an important form of transportation on the many rivers that cross the low-lying country.

The greatest danger, however, is from the high **storm surges** that accompany tropical cyclones in the Bay of Bengal. A storm surge is a rise in tide level that occurs as a storm approaches the coast, washing huge waves of seawater over the land. In 1991, more than 130,000 people in Bangladesh drowned when a 20-foot-high (6-m) storm surge from a tropical cyclone hit the low-lying coast. Outbreaks of deadly diseases, such as cholera, often follow such disasters. Carried by contaminated water supplies, the diseases are hard to control.

Human Geography The Bangladeshis are closely related to the West Bengalis of neighboring India. They speak the same language but differ in religion. Most Bangladeshis are Muslims, while many West Bengalis are Hindus.

The country's capital and largest city is Dhaka (DAK-uh). Most of the city's several million people are poor. Many live in the streets or in surrounding shantytowns without electricity, waste disposal, or a safe water supply. Most industry is located in polluted factory towns near Dhaka.

Rice and jute are the country's most important crops. Jute is the major export, though its value is declining because of competition from synthetic fibers. Bangladesh also has a textile industry that is growing in importance.

The major challenge facing Bangladesh is how to feed, educate, and employ its more than 125 million people. Over 80,000 children are born each week in Bangladesh, and the country has the eighth largest population in the world. There is little room for all these people. The country already has a population density of about 2,200 people per square mile (868 per sq. km). In contrast, the U.S. state of Iowa is about the same size as Bangladesh and has a population density of only about 50 people per square mile (19 per sq. km). For Iowa to have the same population density as Bangladesh, almost half of the entire U.S. population would have to crowd into the state. At its present rate of growth, Bangladesh's population will double by 2025.

3 SECTION REVIEW

1. What countries border Bangladesh? What climate does Bangladesh have?
2. What economic challenges must Bangladesh overcome?
3. **Critical Thinking** **Explain** how storm surges and other natural disasters can be deadly in more than one way. How might Bangladesh reduce the effects of these natural disasters?

4 ISLAND COUNTRIES

FOCUS

- *What are the physical and economic characteristics of Sri Lanka? of the Maldives?*
- *What challenges does each country face?*

LOCATION **Sri Lanka** Sri Lanka, formerly Ceylon, is a pear-shaped tropical island located just off the southern tip of India. It sits on the continental shelf of the Indian subcontinent. (See the map on page 585.) Sri Lanka has three major landform regions: the central mountains, the coastal plain, and the northern limestone plain.

The low mountains of the island's interior reach an elevation of 8,280 feet (2,524 m) and are

These women are picking tea leaves on a plantation in southern **Sri Lanka**. To learn how tea is harvested and processed for consumers, see the illustration on pages 512–513.

an important tea-growing region. In fact, Sri Lanka is the world's largest exporter of tea, and tea is the nation's leading source of income. The mountain area also supplies the nation with forest products and hydroelectricity. In addition, mines tap into deposits of precious gems and graphite. Graphite is a mineral used in a wide variety of products, including pencils.

The coastal plain along the south and west is Sri Lanka's most densely populated area. Coconuts and rubber are the major crops here. Most of the nation's rice and other food crops also are produced in this area. Colombo, the capital and largest city, lies on the western coastal plain.

Sri Lanka's third landform region is the northern limestone plain, which is wide but drier and less arable than the coastal plain. Irrigation projects have helped turn some of this land into a food-growing area.

The region's monsoons and Sri Lanka's location just north of the equator strongly influence the island's climate. The central mountains are the coolest and wettest part of the island, while the southwest portion of the island has a wet humid-tropical climate. The drier northeast part has a tropical-savanna climate.

The population of Sri Lanka is divided into two major culture groups. About three-quarters of the people are Sinhalese, who are Buddhist and of Aryan background. They arrived in Sri Lanka from northern India more than 2,000 years ago. About one-fifth of the population is Tamil, who are Hindu. The Tamils have migrated from southern India over many centuries and live mostly in the northeast portion of the island.

During the British colonial era, the two groups lived together peacefully. Since independence in 1948, however, the Sinhalese have discriminated against the Tamils. Some Tamils initiated terrorist activities against the government in the 1980s. The democratic government granted some autonomy to the Tamils, but the Tamils demanded separation. Over the past decade, the violence has intensified. Since 1983 more than 35,000 people have died as a result of civil war. The war has drained the country economically and has prevented Sri Lanka from developing its industry and tourism.

LOCATION **The Maldives** The Maldives are a group of tropical islands in the Indian Ocean. They lie just north of the equator, about 400 miles (644 km) southwest of Sri Lanka. (See the map on page 585.) The nation's 1,200 tiny coral islands sit on top of an ancient submerged volcanic plateau. About 200 of the islands are inhabited, mainly by Muslim Sinhalese. The highest point on the islands is less than 10 feet (3 m) above sea level. Tropical cyclones threaten all of the islands.

Fish products account for almost half of the country's income. Coconuts and yams are its main food crops. Most other food must be imported. Tourism and a small clothing industry are a growing source of income.

SECTION REVIEW

1. What economic activity provides Sri Lanka's major source of income? What products provide much of the Maldives' income?
2. What is the major challenge facing Sri Lanka? What type of storm threatens the Maldives?
3. **Critical Thinking** **Predict** how the Maldives might be affected by global warming.

CHAPTER 48 CHECK

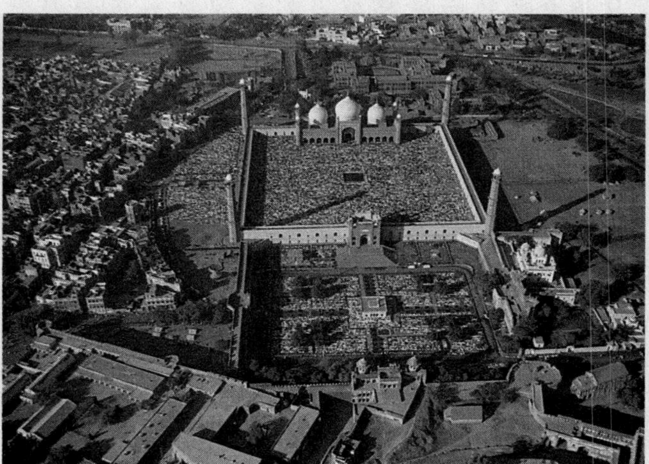
Muslims worship at this mosque in **Lahore, Pakistan**.

Reviewing the Main Ideas

1. Pakistan, officially an Islamic republic, is South Asia's second largest country. Agriculture produces about half of the nation's exports. The region of Kashmir remains a source of controversy between Pakistan and India.
2. Nepal and Bhutan are small landlocked kingdoms located between India and China. Over 90 percent of Nepal's population is involved in agriculture, but tourism is the country's largest industry. Bhutan is somewhat isolated from the rest of the world.
3. Bangladesh is one of the world's poorest and most densely populated countries. Most of Bangladesh has suffered deforestation, and natural disasters are common. The major challenge facing Bangladesh is how to feed, educate, and employ its many people.
4. Sri Lanka is a tropical island located off the southern tip of India. Its population is divided into two major culture groups, the Sinhalese and the Tamils. The Maldives are a group of tropical islands in the Indian Ocean.

Building a Vocabulary

1. Define *partition*. Into what two countries did the British partition British India in 1947?
2. What is a cease-fire? Was the 1949 cease-fire between Pakistan and India effective?
3. What term describes a rise in tide level that occurs as a storm approaches the coast?

Recalling and Reviewing

1. Which countries discussed in this chapter do not have a mountain region?
2. What is Pakistan's most important river? What two rivers are important to Bangladesh?
3. Which two countries of South Asia are monarchies? Which of these has had more of a problem with deforestation?
4. What are Sri Lanka's major agricultural products? natural resources?

Thinking Critically

1. Why do you think Pakistan was originally divided into West Pakistan and East Pakistan, when the two parts were separated by 1,000 miles (1,609 km) of Indian territory? Why do you think East Pakistan eventually separated from West Pakistan to become Bangladesh?
2. Three of the countries discussed in this chapter have borders with China. Do you think China takes a great strategic or economic interest in these countries? Why or why not?

Learning About Your Local Geography

Cooperative Project

With your group, research the population densities of various communities within your town or city, of towns and cities within your county, or of counties within your state. Make a bar graph showing the different population densities. Below the graph, suggest a few possible reasons for why the different areas have different population densities.

Using the Five Themes of Geography

Construct a table with six columns, one for each country discussed in this chapter. In each column, list some of that country's major **place** characteristics. When you have finished filling in the table, decide whether place characteristics alone support the grouping of the six countries together as one **region**. Next to your table, write a few sentences explaining your decision.

Linking Physical and Human Geography

In the summer of 1993, parts of the United States and South Asia were both devastated by rains and floods. Two months of almost continuous rain poured down on the Midwestern United States. Rivers swelled, putting close to 20,000 square miles (51,800 sq. km) under water. Similarly, four weeks of floods flowed through India, Nepal, and Bangladesh, as a result of the heaviest monsoon rains in decades. In Bangladesh alone, waters covered more than 18,500 square miles (47,915 sq. km), one-third of the country's total area.

The consequences of these floods, however, were far from similar for the two regions. While the U.S. floods claimed only about 45 lives, some 4,200 people died in South Asia. In the United States, the floods forced some 74,000 people to leave their homes. In contrast, the South Asian floods left more than 7 million people homeless. Yet while the floods had much higher human costs in South Asia, economic costs were higher in the United States. Estimates of damage in South Asia were in the millions of dollars, but cost estimates for the U.S. floods were more than 12 billion dollars.

Why did the South Asian floods cause so many more deaths, and leave so many more people homeless, than the U.S. floods did? Why did the U.S. floods cost so much more money? Several factors account for these differences. In the United States, an extensive system of dams, levees, and reservoirs helped control the waters and saved lives. No such system exists in South Asia. Dams and levees are expensive to build, and South Asia does not have the funds available for such projects. Furthermore, the building of major dams to control South Asia's rivers would require the consent of several countries, each with its own interests to protect.

Communication and transportation systems also played a role. South Asian residents did not receive the advance warnings that were available to their U.S. counterparts, nor did they have the benefit of evacuation plans. And when the floods did hit, local building materials became a factor. While in the United States most homes are built of materials such as brick or concrete, many structures in South Asia are made from mud and other substances that are easily washed away.

Another factor was population density. Much of the flooding in the United States took place on thinly populated farmland, so fewer people were affected. In South Asia, on the other hand, the river plains are some of the region's most densely settled areas. Differences in population density and land usage also explains, in part, the differences in economic costs. The U.S. farmlands drowned by the floods were mostly commercial farms worth hundreds of thousands of dollars. In contrast, much of the land flooded in South Asia was made up of small subsistence farms, whose primary worth was only to the families that lived on and worked them.

As you can see from this example, the effects of natural disasters are not the same the world over. Levels of economic development, political concerns, and a variety of other human factors all play a role in determining how people will respond to environmental challenges. In short, people have different ways of adapting to and changing their environments. Human geography and physical geography are firmly linked.

PRACTICING THE SKILL

Research the effects of a different type of natural disaster—hurricane, drought, earthquake—on two or more regions that have experienced that type of disaster over the past decade. How did the event affect each region? What factors were responsible for any differences? Did ways in which people had changed their environments play a role in how the disaster affected them? Did either region's response to the disaster involve further change to the environment? Write a summary of your findings and present it to the class.

BUILDING YOUR PORTFOLIO

Individually or in a group, complete one of the following projects to show your understanding of the geography concepts involved.

A *Cities of the Subcontinent*

Select any city of South Asia. Cities you might select include (but are not limited to) New Delhi, Bombay, Calcutta, and Madras, in India; Islamabad, Karachi, and Lahore, in Pakistan; Kathmandu, in Nepal; Dhaka, in Bangladesh; and Colombo, in Sri Lanka. Imagine that you are a government official in the city you have chosen. Your job is to write a brochure profiling the city's history, population, economy, and anything else you think is significant about the city. To prepare your brochure, you will need to consider the following questions.

1. What are the city's origins? How did location and physical features play a role in the city's origins? How has the city grown over time, and why has it grown? Is it still growing today? For your brochure, draw a map of the city itself or of its location in the region. Include a caption explaining the city's origins and pattern of growth.
2. How large is the city's population? What is the economic and cultural make-up of its citizens? Do different groups live in different areas of the city? A chart might help your brochure present information on the city's population.
3. What major economic activities take place in the city? What forms of entertainment are available? What are the city's main sights? You might use illustrations to show this information in your brochure.

Organize your materials into the brochure. Present the brochure to other government officials (the rest of your class). Conclude with a few statements discussing challenges facing the city and predicting the city's future over the next 20 years.

B *A Year in Rural India*

Imagine that you are a farmer in rural India. You live on the outskirts of a village on a small piece of land that your family has worked for decades. The village has changed little in your lifetime.

One year, however, a large factory is built in your village by a company based in Bombay. The factory employs a large number of workers, all of whom have moved to your community. How does the arrival of this factory and its workers affect your life? Write a series of monthly journal entries discussing the ways in which your life changes. Your journal entries might address the following questions.

1. How does the introduction of the factory affect your village and its physical environment? What resources does the factory need? How does the factory receive supplies and send its products to market? Where do the factory workers live?
2. How does the introduction of the factory and the arrival of the factory workers affect your life economically? Particularly consider the relationship between the newly increased population and the amounts of food, other goods, and money available.
3. How does the introduction of the factory and its workers change your belief systems, if at all? Particularly consider beliefs regarding tradition, caste, and so on. Also consider matters of literacy and the status of women.

Share your journal entries with other villagers (the rest of your class). Include a discussion of how you have responded to the changes in your life caused by the introduction of the factory.

UNIT 11

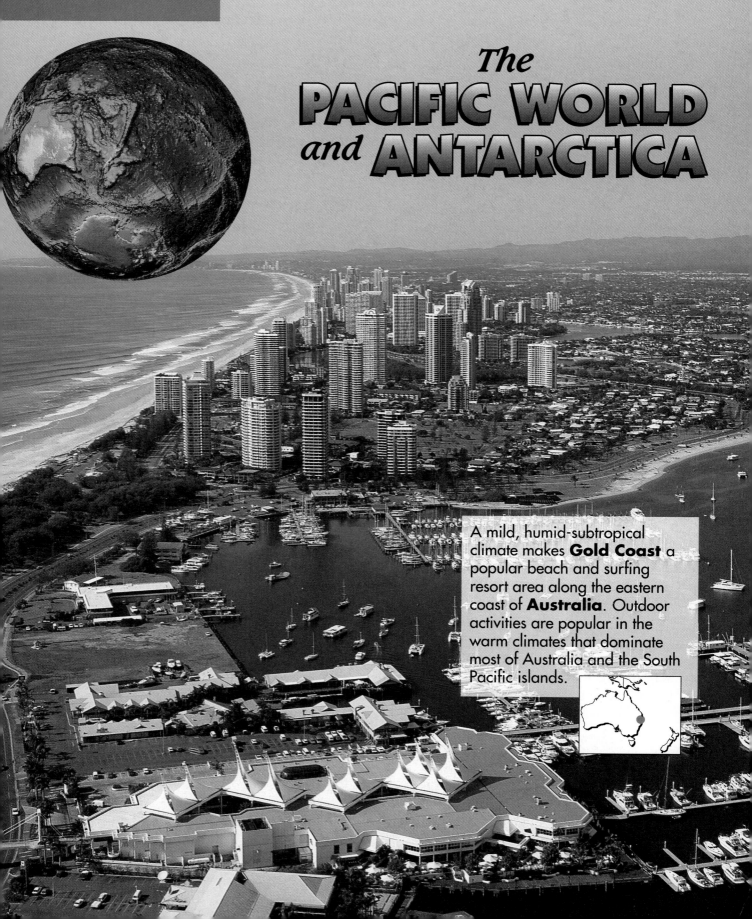

The PACIFIC WORLD *and* ANTARCTICA

A mild, humid-subtropical climate makes **Gold Coast** a popular beach and surfing resort area along the eastern coast of **Australia**. Outdoor activities are popular in the warm climates that dominate most of Australia and the South Pacific islands.

UNIT 11 CHAPTERS

Sheep graze in the beautiful countryside south of **Auckland, New Zealand**. New Zealand is one of the world's major wool producers.

Aboriginal bark-painting of the Big Dipper from **Australia**

Clam-shell pendants from the **Solomon Islands**

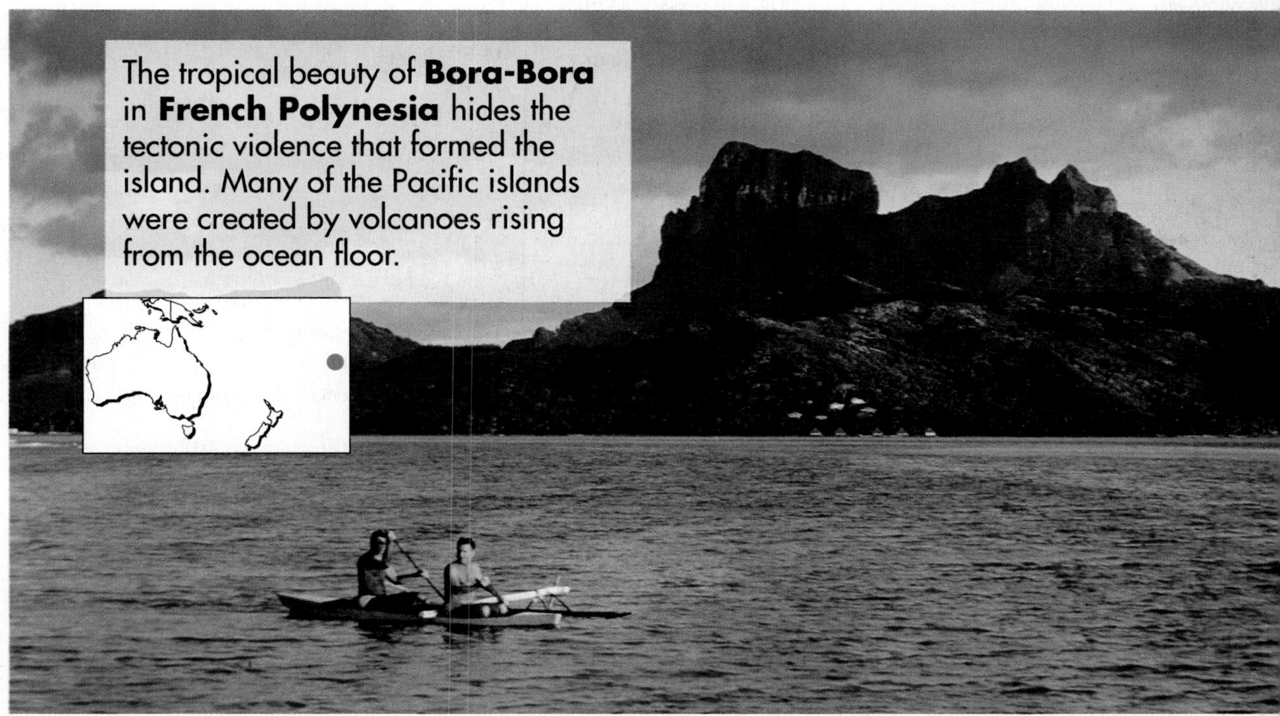

The tropical beauty of **Bora-Bora** in **French Polynesia** hides the tectonic violence that formed the island. Many of the Pacific islands were created by volcanoes rising from the ocean floor.

One World, Many Voices

LITERATURE

Maori stick dance, **New Zealand**

Wiremu Kingi Kerekere (1923–), of Maori heritage, is a poet and singer. Maori peoples are descendants of the original inhabitants of New Zealand. One Maori cultural tradition is the waiata, or action song, which is performed in a dance form. The poem below is a waiata. When Kerekere was only 13 years of age, he performed some of his original compositions on radio broadcasts. In recent years, Kerekere's work has been performed on television and by many theater, dance, and opera groups in New Zealand.

Taku Waiata

Horahia mai ki au nga taonga tupuna,
Nga kowhaiwhai, nga whakairo, nga
pakiwaitara, nga waiata Maori tuturu
o te ao tawhito,
Tuituia mai ki nga tikanga o toku ao o
tenei ra,
Te ao tawhito, te ao hou, Maori, Pakeha,
Homai nga tohu tika, i runga i te aroha.
Tohatohaina tenei kupu ki te ao katoa
Puta ake i te po, ka ao, ka ao, ka marama.

Give to us the treasured arts of our
ancestors,
The carvings, the designs and patterns,
the stories and the traditional chants.
Let us of this generation, try to combine
them with what talents we have in this
changing world of today,
Give to us your greatest gift, 'aroha', love
in its many connotations,
It will bring peace, goodwill, friendship
among peoples,
And will take us from the world of darkness
and ignorance, into the
light of knowledge and understanding.

Maori wood carving

Interpreting Literature

1. What does the author admire about his Maori ancestors?
2. What does this poem say about how the past and the present are connected?

FOR THE RECORD

J. H. Kerr was one of the many European settlers who came to Australia in the mid-1800s. Settlers such as Kerr from the British Isles confronted a very different climate from what they knew in their home countries. Australia's dry climate and heat sparked intense bushfires that could rage across large areas within a very short time. In the passage below, Kerr describes a particularly destructive fire that engulfed Victoria in southern Australia on February 6, 1851, later known as Black Thursday.

"I drew aside a curtain to peep at the weather, and found the glass of the window panes so heated that I was glad quickly to withdraw my hand. . . . The wind was still rising as the day wore on, moaning and whistling through the streets and hollows in fiery suffocating blasts, that seemed to issue from a furnace heated seven times. Clouds of dust and smoke obscured the sun, while the air was dense with dead leaves and other charred matter, whirled on high by the wind. . . . It was evident even from the town that there must be extensive bush-fires blazing not far distant, but the full extent of the damage which had been done was not suspected at the time. The storm attained its great height at about two o'clock p.m., when it became a perfect hurricane. Some idea may be formed of its fury, from the fact that singed leaves and other light matter were blown as far as to the adjoining colony of Tasmania, and fell in showers on the decks of ships in Bass's Straits. In the afternoon, as is usual after these burning gales, the wind suddenly veered round to the south, blowing back the dust and smoke once more, till at length its fury was exhausted and it sank to rest as night closed. Those who have not witnessed the violence of tempests in tropical climates cannot realize the force of those wild tumults of nature, in which it often seems as though 'Heaven and earth must pass away.' "

Analyzing Primary Sources

1. How did the author first realize that a fire was burning somewhere in Victoria?
2. What details in this passage indicate that the fire covered a large area?

CHAPTER 49

Australia

Aboriginal art from **Australia**

"The Great Barrier Reef swarms with life. The tides surging through the coral heads charge the water with oxygen and the tropical sun warms it and fills it with light. All the main kinds of sea animals seem to flourish here. . . . Purple eyes peer out from beneath shells; black sea urchins swivel their spines as they slowly perambulate [stroll] on needle tip; starfish of an intense blue spangle the sand; and patterned rosettes unfurl from holes in the smooth surface of the coral."

David Attenborough

GEOGRAPHY DICTIONARY

- *artesian well*
- *biogeography*
- *endemic species*
- *marsupial*
- *alien species*

Settled by Great Britain in 1788 as a prison colony, Australia today is a modern and prosperous democratic nation. It is the sixth largest country in the world, with a land area about the same size as the contiguous United States. Because it lies south of the equator, Australia is sometimes called the "Land Down Under."

The nation is made up of six states and two federal territories. The states include New South Wales, Victoria, Queensland, South Australia, Western Australia, and the island state of Tasmania. The two federal territories are the Northern Territory and the tiny Australian Capital Territory, in which Canberra, the national capital, is located. Australia also governs several small islands in the South Pacific and Indian oceans. The Australian population is concentrated in a few large coastal cities. The rest of the continent, known as the Outback, is dry, flat, and almost uninhabited.

Great Barrier Reef off **Australia**

MAPPING THE REGION

THE PACIFIC WORLD AND ANTARCTICA

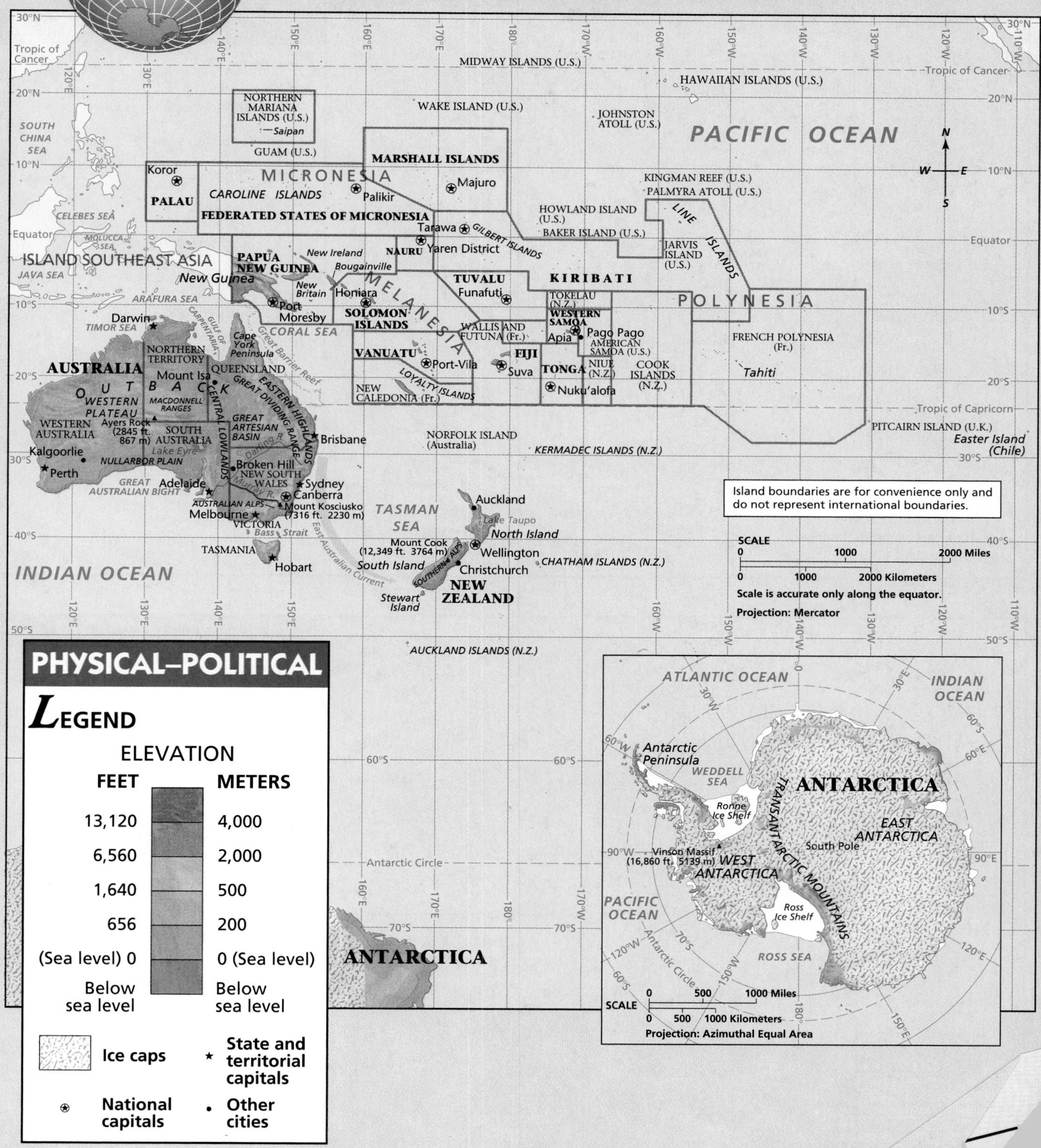

Tropic of Cancer
PACIFIC OCEAN
INDIAN OCEAN
MIDWAY ISLANDS (U.S.)
HAWAIIAN ISLANDS (U.S.)
WAKE ISLAND (U.S.)
JOHNSTON ATOLL (U.S.)
NORTHERN MARIANA ISLANDS (U.S.)
Saipan
GUAM (U.S.)
MARSHALL ISLANDS
Majuro
MICRONESIA
Koror
PALAU
CAROLINE ISLANDS
Palikir
FEDERATED STATES OF MICRONESIA
KINGMAN REEF (U.S.)
PALMYRA ATOLL (U.S.)
HOWLAND ISLAND (U.S.)
BAKER ISLAND (U.S.)
JARVIS ISLAND (U.S.)
LINE ISLANDS
Tarawa
GILBERT ISLANDS
NAURU
Yaren District
KIRIBATI
Equator
SOUTH CHINA SEA
CELEBES SEA
MOLUCCA SEA
JAVA SEA
ISLAND SOUTHEAST ASIA
PAPUA NEW GUINEA
New Guinea
New Ireland
Bougainville
New Britain
Honiara
Port Moresby
ARAFURA SEA
TIMOR SEA
Darwin
GULF OF CARPENTARIA
CORAL SEA
MELANESIA
SOLOMON ISLANDS
TUVALU
Funafuti
TOKELAU (N.Z.)
POLYNESIA
WESTERN SAMOA
Apia
Pago Pago
AMERICAN SAMOA (U.S.)
WALLIS AND FUTUNA (Fr.)
VANUATU
Port-Vila
FIJI
Suva
TONGA
Nuku'alofa
NIUE (N.Z.)
COOK ISLANDS (N.Z.)
FRENCH POLYNESIA (Fr.)
Tahiti
NEW CALEDONIA (Fr.)
LOYALTY ISLANDS
Tropic of Capricorn
PITCAIRN ISLAND (U.K.)
Easter Island (Chile)
AUSTRALIA
NORTHERN TERRITORY
Cape York Peninsula
Great Barrier Reef
QUEENSLAND
Mount Isa
OUTBACK
WESTERN PLATEAU
MACDONNELL RANGES
GREAT DIVIDING RANGE
EASTERN HIGHLANDS
CENTRAL LOWLANDS
GREAT ARTESIAN BASIN
WESTERN AUSTRALIA
Ayers Rock (2845 ft. 867 m)
SOUTH AUSTRALIA
Lake Eyre
Brisbane
Kalgoorlie
Perth
NULLARBOR PLAIN
Darling R.
Broken Hill
NEW SOUTH WALES
Sydney
Canberra
GREAT AUSTRALIAN BIGHT
Adelaide
Murray R.
AUSTRALIAN ALPS
Mount Kosciusko (7316 ft. 2230 m)
Melbourne
VICTORIA
Bass Strait
TASMANIA
Hobart
NORFOLK ISLAND (Australia)
KERMADEC ISLANDS (N.Z.)
TASMAN SEA
East Australian Current
Auckland
Lake Taupo
North Island
Mount Cook (12,349 ft. 3764 m)
Wellington
South Island
SOUTHERN ALPS
Christchurch
CHATHAM ISLANDS (N.Z.)
NEW ZEALAND
Stewart Island
AUCKLAND ISLANDS (N.Z.)
Island boundaries are for convenience only and do not represent international boundaries.
SCALE
0 1000 2000 Miles
0 1000 2000 Kilometers
Scale is accurate only along the equator.
Projection: Mercator
Antarctic Circle
ANTARCTICA
PHYSICAL–POLITICAL
LEGEND
ELEVATION
FEET
METERS
13,120 4,000
6,560 2,000
1,640 500
656 200
(Sea level) 0 0 (Sea level)
Below sea level
Ice caps
State and territorial capitals
National capitals
Other cities
ATLANTIC OCEAN
INDIAN OCEAN
Antarctic Peninsula
WEDDELL SEA
ANTARCTICA
Ronne Ice Shelf
TRANSANTARCTIC MOUNTAINS
EAST ANTARCTICA
South Pole
Vinson Massif (16,860 ft. 5139 m)
WEST ANTARCTICA
PACIFIC OCEAN
Ross Ice Shelf
ROSS SEA
SCALE
0 500 1000 Miles
0 500 1000 Kilometers
Projection: Azimuthal Equal Area

MAPPING THE REGION

THE PACIFIC WORLD AND ANTARCTICA

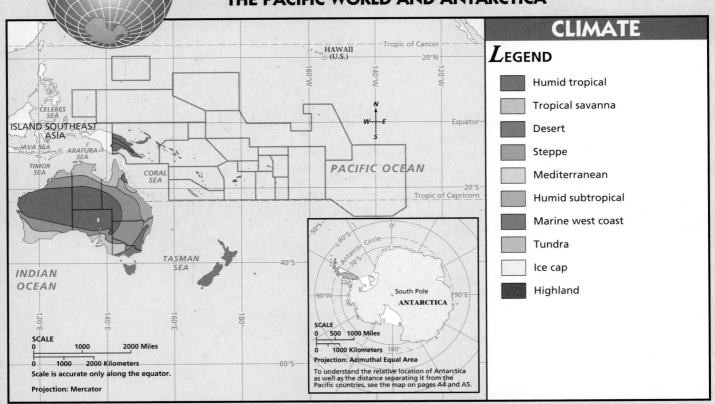

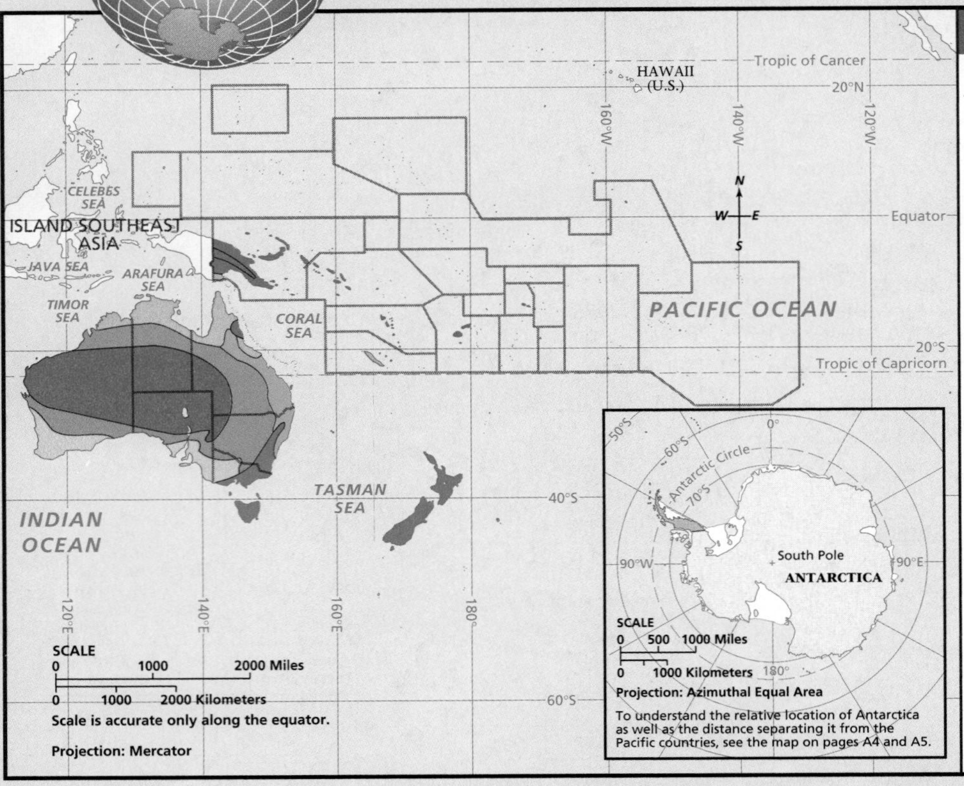

CLIMATE

HAWAII (U.S.)
Tropic of Cancer
20°N
160°W
140°W
120°W
Equator
N
W
E
S
CELEBES SEA
ISLAND SOUTHEAST ASIA
JAVA SEA
ARAFURA SEA
TIMOR SEA
CORAL SEA
PACIFIC OCEAN
20°S
Tropic of Capricorn
TASMAN SEA
INDIAN OCEAN
40°S
120°E
140°E
160°E
180°
60°S
SCALE
0 1000 2000 Miles
0 1000 2000 Kilometers
Scale is accurate only along the equator.
Projection: Mercator
50°S
60°S
Antarctic Circle
70°S
0°
South Pole
ANTARCTICA
90°W
90°E
180°
SCALE
0 500 1000 Miles
0 1000 Kilometers
Projection: Azimuthal Equal Area
To understand the relative location of Antarctica as well as the distance separating it from the Pacific countries, see the map on pages A4 and A5.

LEGEND

- Humid tropical
- Tropical savanna
- Desert
- Steppe
- Mediterranean
- Humid subtropical
- Marine west coast
- Tundra
- Ice cap
- Highland

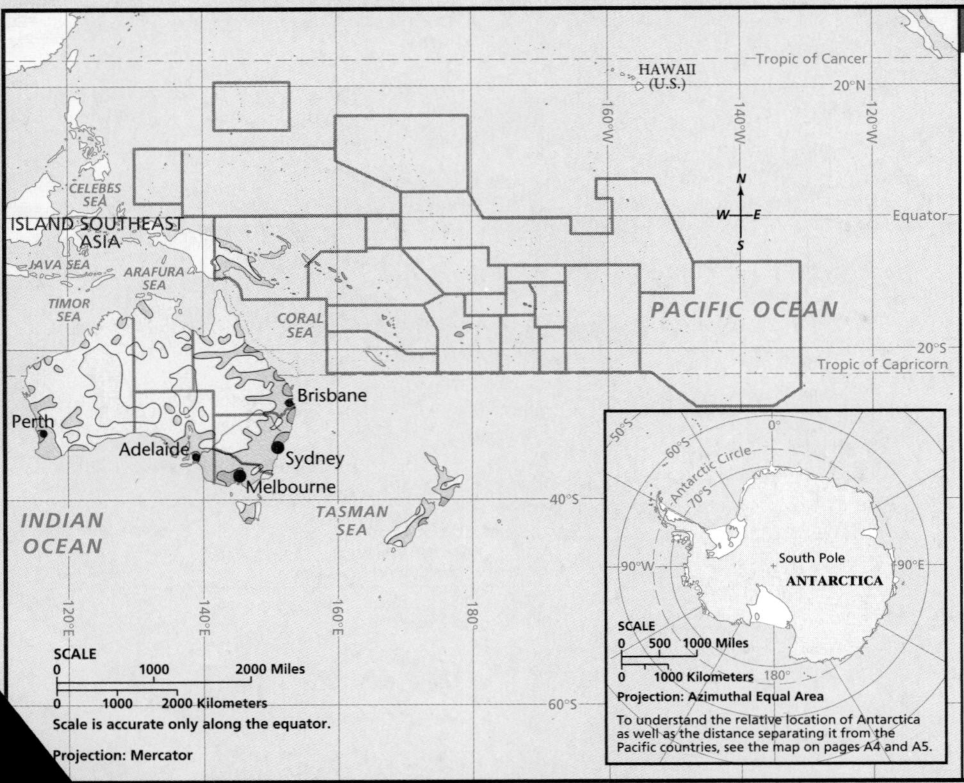

POPULATION

HAWAII (U.S.)
Tropic of Cancer
20°N
160°W
140°W
120°W
Equator
N
W
E
S
CELEBES SEA
ISLAND SOUTHEAST ASIA
JAVA SEA
ARAFURA SEA
TIMOR SEA
CORAL SEA
PACIFIC OCEAN
20°S
Tropic of Capricorn
Brisbane
Perth
Adelaide
Sydney
Melbourne
TASMAN SEA
INDIAN OCEAN
40°S
120°E
140°E
160°E
180°
60°S
SCALE
0 1000 2000 Miles
0 1000 2000 Kilometers
Scale is accurate only along the equator.
Projection: Mercator
50°S
60°S
Antarctic Circle
70°S
0°
South Pole
ANTARCTICA
90°W
90°E
180°
SCALE
0 500 1000 Miles
0 1000 Kilometers
Projection: Azimuthal Equal Area
To understand the relative location of Antarctica as well as the distance separating it from the Pacific countries, see the map on pages A4 and A5.

LEGEND

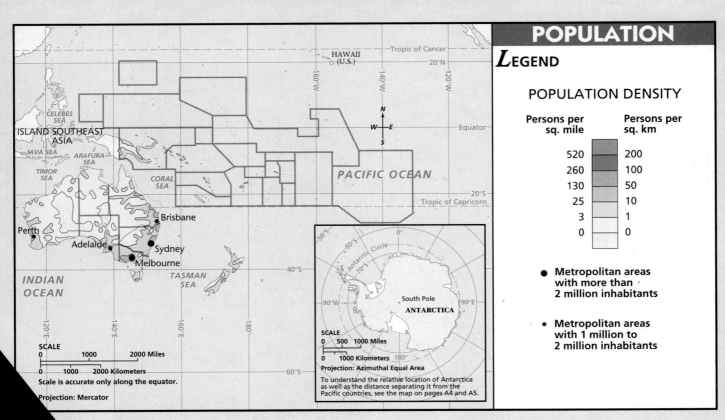

POPULATION DENSITY

Persons per sq. mile	Persons per sq. km
520	200
260	100
130	50
25	10
3	1
0	0

- ● Metropolitan areas with more than 2 million inhabitants
- • Metropolitan areas with 1 million to 2 million inhabitants

MAPPING THE REGION

THE PACIFIC WORLD AND ANTARCTICA

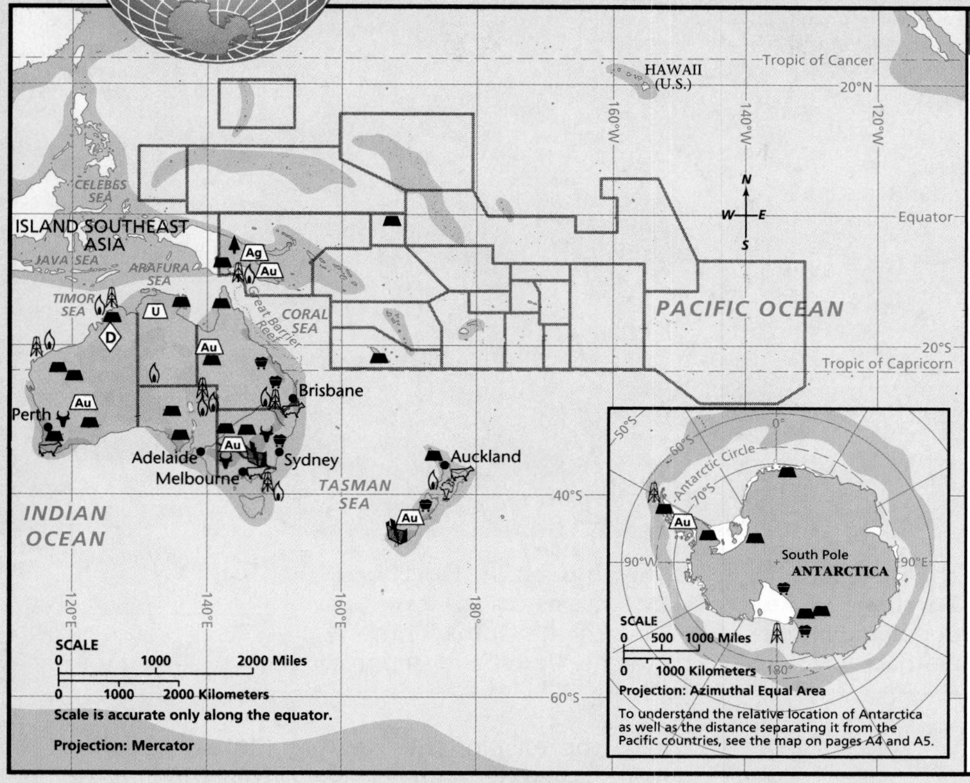

ECONOMY

LEGEND

- Livestock raising
- Commercial farming
- Dairying
- Subsistence farming
- Manufacturing and trade
- Commercial fishing
- Forests
- Limited economic activity

- Coal
- Natural gas
- Oil
- Hydroelectric power
- Geothermal power
- Gold
- Silver
- Diamonds
- Uranium
- Other minerals
- Timber

HAWAII (U.S.)
Tropic of Cancer
20°N
160°W
140°W
120°W
Equator
PACIFIC OCEAN
20°S
Tropic of Capricorn
CELEBES SEA
ISLAND SOUTHEAST ASIA
JAVA SEA
ARAFURA SEA
TIMOR SEA
CORAL SEA
Great Barrier Reef
Brisbane
Perth
Adelaide
Melbourne
Sydney
Auckland
TASMAN SEA
INDIAN OCEAN
40°S
120°E
140°E
160°E
180°
60°S

SCALE
0 1000 2000 Miles
0 1000 2000 Kilometers
Scale is accurate only along the equator.
Projection: Mercator

50°S
60°S
70°S
Antarctic Circle
0°
90°W
90°E
South Pole
ANTARCTICA
180°
SCALE
0 500 1000 Miles
0 1000 Kilometers
Projection: Azimuthal Equal Area

To understand the relative location of Antarctica as well as the distance separating it from the Pacific countries, see the map on pages A4 and A5.

CLIMATE

◆ Most of Australia is generally warm, dry, and sunny. Most precipitation occurs in regions along the coasts.

◆ Most of the Pacific islands have tropical climates.

◆ Because of its oceanic and middle-latitude location, New Zealand is dominated by a marine-west-coast climate.

? Which two climates cover most of Australia?

? Which climate types are dominant in Papua New Guinea?

Compare this map to the physical–political map on page 601.

? In which climate region are Perth and Adelaide located?

POPULATION

◆ Australia's most populous region is in the southeast.

◆ Most of New Zealand's population lives on North Island.

◆ Except for scientists at research stations, Antarctica has no permanent human population.

? Which Australian metropolitan areas have populations of more than 2 million?

Compare this map to the climate map on page 602.

? Which climate regions in Australia have the densest populations?

Compare this map to the economic map above.

? What economic activities are found on New Zealand's most populous island?

ECONOMY

◆ Papua New Guinea has an abundance of natural resources, including minerals, which are the country's dominant exports.

◆ Because of its extensive dry climates, the vast majority of Australia's land is not used for farming.

◆ Although believed to be rich in natural resources, Antarctica is not open to mining.

? Name the manufacturing and trade centers shown in Australia.

? Where in Australia will you find commercial farming?

? Which natural resources can be found in Antarctica?

Size comparison of Australia to the contiguous United States

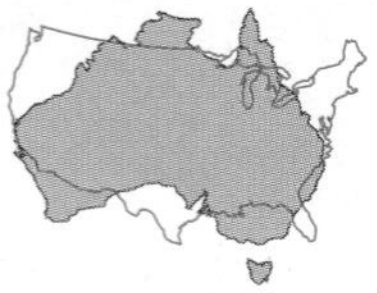

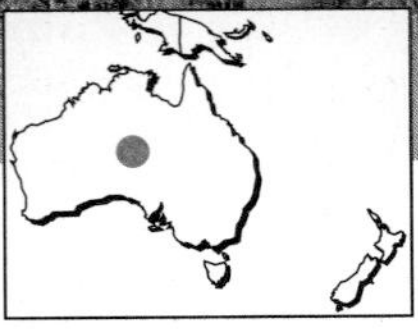

The Finke River winds its way through the **Northern Territory** of **Australia**. Australians call the less-populated interior of the country the Outback. The Outback is dominated by the dry desert and steppe climates.

PHYSICAL GEOGRAPHY

FOCUS

- *What are the landforms and climates of Australia? What water resources are found in Australia?*
- *How is Australia's animal, plant, and ocean life significant?*

REGION Landforms Australia is the world's smallest, flattest, and lowest continent. It can be divided into three major landform regions: the Western Plateau, the Central Lowlands, and the Eastern Highlands. (See the map on page 601.)

The Western Plateau covers more than one-half of the continent, including Western Australia, the Northern Territory, and most of South Australia. The treeless Nullarbor Plain, the flattest landform region on Earth, stretches along the southern edge of the plateau. Ayers Rock, a huge rock formation, lies just south of the Macdonnell Ranges in the western part of the region.

To the east of the Western Plateau are the Central Lowlands, which extend from the Gulf of Carpentaria in the north to the Great Australian Bight in the south. They include interior Queensland and New South Wales and adjoining areas of Victoria and South Australia. The area around dry Lake Eyre in South Australia is below sea level. The lake basin fills with water only about once every 10 years.

Australia's highest elevations are found in the Great Dividing Range of the Eastern Highlands. This range stretches from Cape York Peninsula in the north, along the country's east coast, to Victoria in the south. The mountains appear again on the island state of Tasmania. Mount Kosciusko (kahz-ee-ES-koe) is Australia's highest peak at 7,316 feet (2,230 m).

Climate Some geographers describe Australia as a desert with green edges. Indeed, most of the continent receives little rain, mainly for two reasons. First, the stable air of the subtropical high-pressure system prevents rain from reaching many areas. Second, the Great Dividing Range rises in the path of the southeast trade winds.

The range acts as a barrier blocking moist Pacific winds from reaching the leeward rain-shadow side of the range and the interior of the continent. Instead, moisture falls on the windward eastern slopes, forming a narrow band of humid climate regions on the east coast of the continent.

As a result, about half of Australia has a desert climate. Desert covers much of the continent's interior and stretches to the coast of Western Australia. Forming a semiarid rim around the desert is a steppe climate, dominated by scrub bushes.

On the northern coast of the Northern Territory and the Cape York Peninsula of Queensland is a tropical-savanna climate. The monsoon influences this hot region, causing distinct wet and dry seasons. A half year of drought and several months of flooding are typical.

A small area of humid-tropical climate and rain forests lies in the far northeast corner of coastal Queensland. Meanwhile, most of coastal Queensland and New South Wales has a humid-subtropical climate. Farther south, the cooler, marine-west-coast climate prevails. Here, winter storms pushed by the westerly winds bring rain and sometimes snow to the mountain areas. The marine-west-coast climate extends along coastal Victoria, northward into the mountains of New South Wales, and southward over all of the island of Tasmania. The windward western side of Tasmania supports lush temperate forests.

A Mediterranean climate bathes the southwest coast of Western Australia near Perth and the coast of South Australia near Adelaide. Winters here are mild and rainy, and summers are long, dry, and sunny.

Water Resources Almost all of Australia is subject to drought. These long dry spells can lead to severe water shortages, brushfires, and dust storms. Severe tropical cyclones (hurricanes) also occasionally threaten the northern and northeastern coasts.

Australia has only one major river system, the Murray-Darling. It flows from the Great Dividing Range westward across the dry, steppe country of southeastern Australia. In contrast, the smaller rivers along the eastern coast flow into the sea within a short distance of their origins. Large water projects have tried to move water through tunnels from the wet eastern slopes to the dry interior slopes.

Fortunately, groundwater is abundant in Australia. The largest source of underground well water lies in the Great Artesian Basin of interior Queensland. Here, **artesian wells** dot the area. Artesian wells are those in which water rises toward the surface without being pumped. This water originates as rain falling on the wet mountains to the east. The water enters an aquifer and flows westward under the desert. Much of this well water, however, is of poor quality and can be used only to water sheep.

FACTS IN BRIEF Australia

COUNTRY POPULATION (1994)	LIFE EXPECTANCY (1994)	LITERACY RATE	PER CAPITA GDP (1993)
AUSTRALIA 18,077,419	74, male 81, female	99% (1992)	$19,100
UNITED STATES 260,713,585	73, male 79, female	97% (1991)	$24,700

Sources: *The World Factbook 1994,* Central Intelligence Agency; *The World Almanac and Book of Facts 1995*

The life expectancy rates of Australians are slightly higher than those for residents of the United States.

Biogeography The study of the geographic distribution of plants and animals is called **biogeography**. Australia's island location has helped create a biogeography here that is much different from that in other parts of the world. Because of Australia's isolation from other continents, many **endemic species** developed. Endemic species are plants or animals that originate only in a particular geographic region. Most of Australia's plants and animals are found nowhere else on Earth.

Australia is especially well known for its **marsupials**, such as the kangaroo and the koala. Marsupials are mammals that carry their young in pouches. There are almost 50 varieties of kangaroos in Australia, from small mouselike ones to some that are larger than humans. The koala, on the other hand, is a tree-dwelling vegetarian known for its furry, bearlike appearance. Australia also is home to the platypus and spiny anteater, the only egg-laying mammals in the world.

The endemic plants of Australia are unusual in that they are dominated by a single plant variety, the eucalyptus (yoo-kuh-LIP-tus). Australia has more than 500 species of eucalyptus, and more than 90 percent of its trees are some variety of this plant.

Many plants and animals in Australia, however, are **alien species**. Alien species are those introduced to a region by humans. The Aborigines, the original inhabitants of Australia, brought the dingo, a dog from Asia. Europeans later introduced sheep, rabbits, foxes, water buffalo, birds, and hundreds of plants. Due to a lack of native predators, the number of rabbits in Australia exploded. They overran the continent, destroying crops and grasslands. In 1950, a disease introduced by humans killed most of the rabbits. The rabbit population, however, is increasing once again.

The koala lives in forested areas of eastern **Australia** and is protected by law. Eucalyptus leaves are its favorite food.

Managing Australia's unique wildlife is a source of concern and controversy to the country's people. Some Australian animals, such as the koala, may become endangered, mainly due to loss of habitat. Other animals, especially kangaroos, exist in great numbers and are hunted to protect sheep-grazing lands. Some oppose killing any of the animals.

The Great Barrier Reef

In the Coral Sea, off the northeastern coast of Queensland, lies the Great Barrier Reef. The world's largest coral reef, it stretches more than 1,200 miles (1,931 km). The reef first formed along the edges of the Australian continental shelf. These shallow waters, swept by the warm East Australian Current, are ideal for the formation of coral reefs. Then, in the years since the last ice age, the coral continued to grow upward with rising sea levels. Eventually, numerous coral sand islands built up on the reef.

Except for the tropical rain forests, no ecosystem on Earth is as complex as that of the Great Barrier Reef. It is home to the world's greatest variety of ocean life, including fish, shellfish, and birds. Many people are concerned that tourism and offshore oil drilling could damage the reef. To protect it, the government has made most of the Great Barrier Reef a national park.

1 SECTION REVIEW

1. Why do some geographers describe Australia as a desert with green edges?
2. What are two of Australia's endemic animals? What is significant about the Great Barrier Reef?
3. **Critical Thinking** **Explain** how problems can arise when the population of a certain kind of animal becomes very large. Use two examples from the section.

2 ECONOMIC GEOGRAPHY

FOCUS

- *What agricultural and mineral resources does Australia have?*
- *How have Australia's industry and trade been influenced by global forces?*

Agriculture

Australia produces about one-third of the world's wool supply. The country exports nearly all its wool, mainly to Japan and China. In the wetter eastern portions of Australia, sheep are raised for mutton and lamb, which are meat products. Southwest Asia imports much of this meat as well as many live sheep.

Australia is also the leading beef exporter in the world. It exports most of its beef to the United States and Japan. Beef cattle and sheep are usually raised on large ranches, called stations, in the nation's dry interior. Most dairy farms are located in wetter regions, such as Victoria and Tasmania, and near large urban areas.

Because of its dry climates, only about 6 percent of Australia's land is used for farming. Despite the limited amount of arable land and good water, Australia is self-sufficient in food and is a major food exporter as well. Its small population, efficient farmers, and the many different climate regions contribute to this success.

Wheat, the nation's major crop, covers more than half of Australia's cropland. The wheat belt stretches across the plains west of the Great Dividing Range in Victoria and New South Wales. Some wheat is grown, however, in all of the mainland Australian states. The country is one of the world's largest wheat exporters.

Workers pick grapes near **Adelaide, South Australia,** where a Mediterranean climate supports many vineyards.

Australia's other major export crops are sugar, rice, and cotton, which are grown mainly in coastal New South Wales and Queensland. Also from the Queensland coast come bananas, pineapples, and other tropical fruits. Citrus fruits, olives, and wine grapes grow well in the humid-subtropical and Mediterranean climate regions. Crops such as corn, apples, peaches, and pears grow in the marine-west-coast climate region on the southeastern coast and in Tasmania.

Mineral Resources Australia's Outback is a storehouse of mineral wealth. In fact, Australia is a world leader in iron-ore, bauxite, and coal exports and in diamond and opal processing. The country also has vast uranium reserves and deposits of nickel, gold, lead, zinc, and copper.

Australia produces about 70 percent of its needed oil. The country's major oil fields lie in the Bass Strait, between Tasmania and Victoria, on the offshore shelf areas of Western Australia, and in southern Queensland. Huge natural-gas reserves also have been discovered on the continental shelf off the northwest coast. Most of the country's coal mining is in eastern Australia.

Industries Although it has a small consumer population and a small labor force, Australia is an industrialized nation. Many industries, such as iron, steel, and food processing, are based on rich natural resources. Most of Australia's industries, however, are now in decline due to foreign competition. As a result, service industries are becoming the country's major employer. For example, Australia has a large transportation industry to maintain its vast network of roads and operate its railroads and airlines. The tourism and entertainment industries also are growing rapidly.

Asian-Pacific Connections With the formation of the European Community, Australia lost some of its close trade ties with the United Kingdom and other European countries. Since then, Australia has looked north to its Asian neighbors for international trade. Asian countries make up Australia's largest market. Japan in particular is a major customer for Australia's crops, meat, and minerals. The United States also is an important trade partner.

2 SECTION REVIEW

1. List three of Australia's most important agricultural exports and three of its major mineral resources.
2. How have foreign competition and the formation of the European Community affected Australia's industry and trade?
3. **Critical Thinking** **Determine** what factors have contributed to Australia's economic success.

3 POPULATION GEOGRAPHY

FOCUS

- *How is population distributed in Australia? What groups of people have settled in Australia?*
- *What environmental challenges does Australia face?*

Urban Geography Although Australians are often portrayed as rugged Outback settlers, more than 85 percent of all Australians live in urban areas. Each of Australia's six states is dominated by one large nodal capital city and seaport, with suburbs and transportation networks fanning out

Text continues on page 610.

GINZA
HOLLYWOOD
CITIES OF THE WORLD

Sydney

No longer a lonely British outpost, Sydney and its harbor open Australia to the Pacific Ocean and world trade.

Unlike other major cities of the world, Sydney did not begin as a bustling port, an industrial or transportation center, or a mining town. Instead, the British established Sydney as a prison colony in 1788. Australia's then-isolated location seemed an ideal place to house the empire's convicts.

For their survival, the first prisoners and their guards depended on supplies shipped from Great Britain. Soon, the small settlement on the shore of Sydney Cove became an important port for the entire continent and for the South Pacific. It also served as a base for explorers of Australia's interior.

Modern Sydney

In the past two centuries, Sydney has grown from a small, isolated prison colony to a city with a population of more than 3.5 million. The city covers 670 square miles (1,735 sq. km) and is Australia's largest urban area and leading seaport. Its residents are called Sydneysiders.

Sydney's growth is due in large part to its excellent harbor, which is one of the deepest in the world. The two main bays of the harbor area are Port Jackson and Middle Harbor. Dozens of fingerlike peninsulas creep into the harbor area, giving the impression that the city is built on a series of islands. The communities on each peninsula have their own distinct character.

Sydney grew rapidly in its early years. The colony first spread inland from Sydney Cove, then west and east along the harbor. Later, settlements sprang up across the harbor on the north shore. The original plan of the city was a regular grid pattern. As the city spread, however, this pattern changed.

The main roads into the interior of Australia followed the open areas between the hills, making travel easy for animal-drawn vehicles. When suburbs later grew around these roads, they followed the same winding pattern. The irregular coastline of the harbor's bays also adds to Sydney's unusual layout.

Also unusual is the fact that a high percentage of people in the city own their own house and garden. When Sydney first developed, large areas of land were inexpensive, and many people could afford to own large amounts of land. Land prices have risen dramatically, however. Today, a shortage of space, especially in the central business district, has resulted in many towering apartment complexes and high-rise buildings springing up.

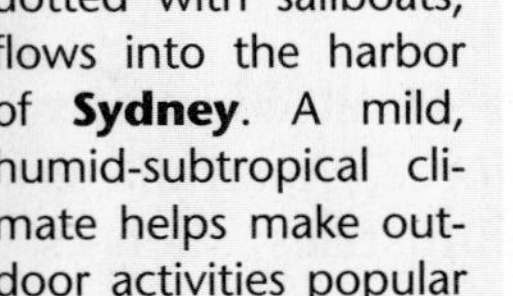

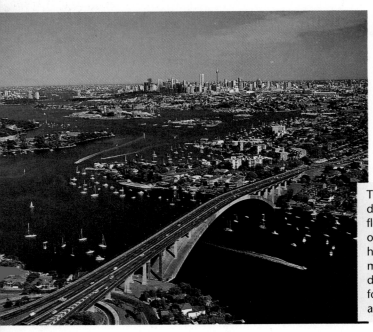

The Parramatta River, dotted with sailboats, flows into the harbor of **Sydney**. A mild, humid-subtropical climate helps make outdoor activities popular for Sydney residents and tourists.

Living Down Under

Growth brings many challenges to a city the size of Sydney. The irregular layout of the streets and the large

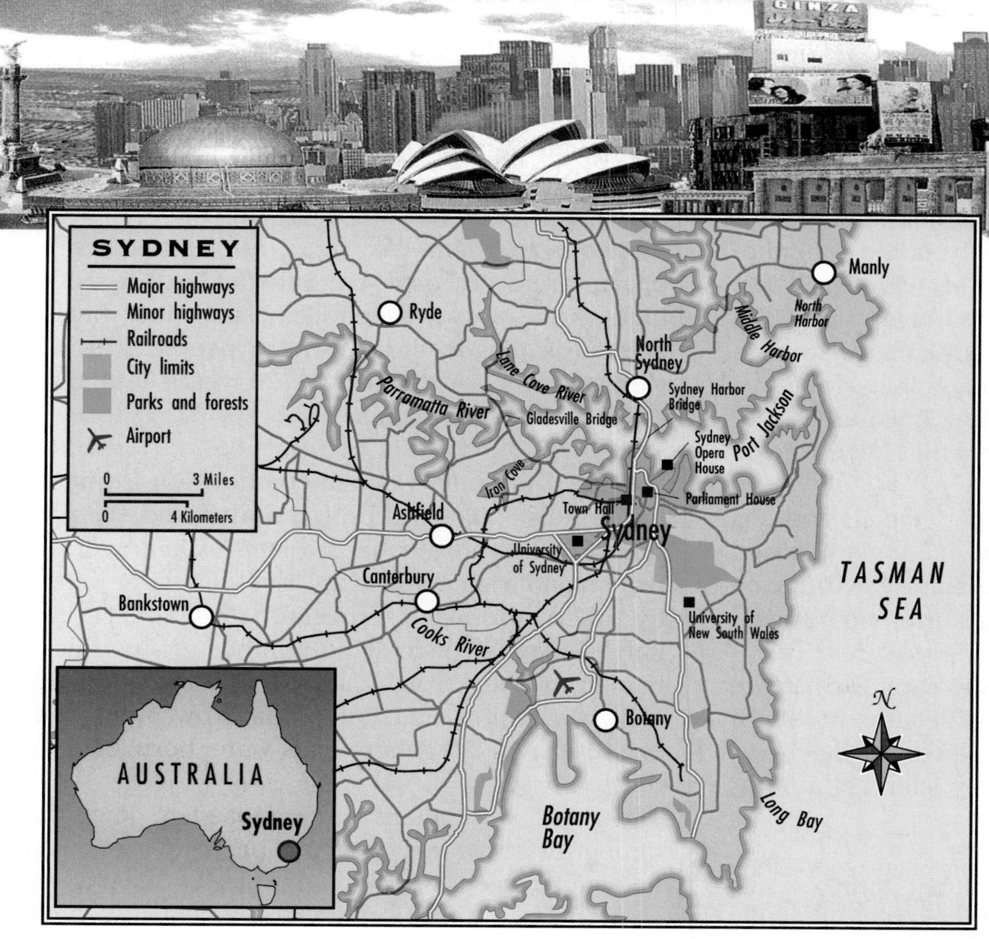
SYDNEY
Major highways
Minor highways
Railroads
City limits
Parks and forests
Airport
0 3 Miles
0 4 Kilometers
Manly
North Harbor
Middle Harbor
Ryde
North Sydney
Lane Cove River
Parramatta River
Gladesville Bridge
Sydney Harbor Bridge
Sydney Opera House
Port Jackson
Iron Cove
Parliament House
Town Hall
Ashfield
Sydney
University of Sydney
Canterbury
TASMAN SEA
Bankstown
Cooks River
University of New South Wales
Botany
Botany Bay
Long Bay
AUSTRALIA
Sydney
N

The metropolitan area of **Sydney** is divided by the Parramatta River and the harbor. Port Jackson and Middle Harbor provide Sydney's busy shipping industry access to the Tasman Sea and the Pacific Ocean beyond.

The Sydney Opera House, a famous city landmark, is located near the Sydney Harbor Bridge. The Australians use the British spelling of *harbour.*

number of commuters to the city's center each day have resulted in growing traffic congestion. People who live on the mainly residential north shore must drive to work over the Harbor Bridge to the commercial areas of the city in the south and west. Ferries ease some of this congestion, but automobiles remain residents' favorite form of transportation.

Sydney, host of the 2000 Olympics, is filled with opportunities for entertainment and recreation. City dwellers enjoy the Opera House; sports such as tennis, golf, and horse racing; and all water activities, especially sailing. The coastline provides many sandy beaches for swimming and surfing. These attractions, combined with the city's excellent harbor, business opportunities, and location on the Pacific Rim make Sydney an increasingly popular world city.

The famous Sydney Opera House, completed in 1973, overlooks the city's harbor. The Opera House, with its distinctive white concrete shells that resemble sails, hosts concerts, opera, and theater.

into the hinterland. In some cases, the hinterland is an entire state.

The southeastern corner of Australia is the country's heartland, containing the nation's three largest cities and most of its population. Sydney, Australia's oldest and largest city, is the capital of the most populous state, New South Wales. (See "Cities of the World" on pages 608–609.) Melbourne, the capital of Victoria, is Australia's second largest city. It is a busy seaport and a commercial and industrial center. Located inland, about halfway between Sydney and Melbourne, is Canberra, the nation's capital.

Brisbane, the capital of Queensland, is Australia's third largest city. Its economy is based mainly on agriculture and tourism. Near Brisbane are the Gold Coast, known as "surfer's paradise," and the Great Barrier Reef.

The continent's other nodal capital cities are Perth in Western Australia, Adelaide in South Australia, and Darwin in the Northern Territory. Tasmania, the smallest and only island state, has Hobart as its capital city.

The Aborigines The first inhabitants of Australia were the Aborigines. Originally a nomadic hunting-and-gathering people, they migrated to Australia from Southeast Asia about 50,000 years ago. When the first European settlers arrived during the 1700s, about 300,000 Aborigines lived in Australia. Diseases brought by Europeans greatly reduced this number. Today, Aborigines make up only about one percent of the population, and their nomadic way of life has disappeared.

An Australian Aborigine artist produces a bark painting. Aboriginal art forms, including cave paintings, have been practiced for centuries.

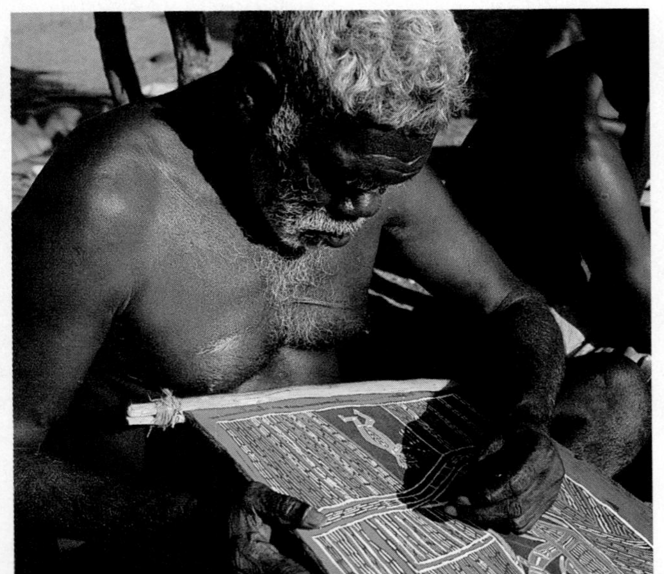

In recent years, the Aborigines have gained back some of their rights that were lost when Europeans settled the continent. They now have full citizenship, and the federal government has returned some of the land in the Northern Territory to them. Yet most Aborigines live in urban areas in poverty. Substandard housing, high crime rates, alcoholism, and low health standards plague the Aborigine people. The life expectancy of Aborigines is 20 years less than that of other Australians.

Issues Australia's ethnic composition is changing rapidly. Like the United States and Canada, Australia has always been a land of immigration. British and Irish colonial settlers were later joined by other Europeans, especially from Greece and Italy. Today, an increasing number of Australia's immigrants are from Asia, particularly Vietnam, Malaysia, China, and Cambodia. Now more than one-fourth of all Australians were born outside the country.

Some Australians are opposed to immigration growth, which over the past 40 years has helped double Australia's population as well as change its composition. Others, however, believe that these new citizens contribute to the nation's economic growth and diversity.

Rapid population growth and economic development, however, have led to environmental concerns among Australians. Although Australia's cities generally are clean, smog hangs over many urban centers. In addition, mining, damming of rivers, and forest destruction all threaten Australia's environment. Salinity and wind erosion of topsoil have become serious problems as well, especially during drought years.

3 SECTION REVIEW

1. What is the most densely populated part of Australia? How is Australia's ethnic composition changing?
2. How have population growth and economic development affected Australia's environment?
3. **Critical Thinking** **Compare** the experiences of the Aborigines in Australia with the experiences of Native Americans in the Western Hemisphere.

CHAPTER 49 CHECK

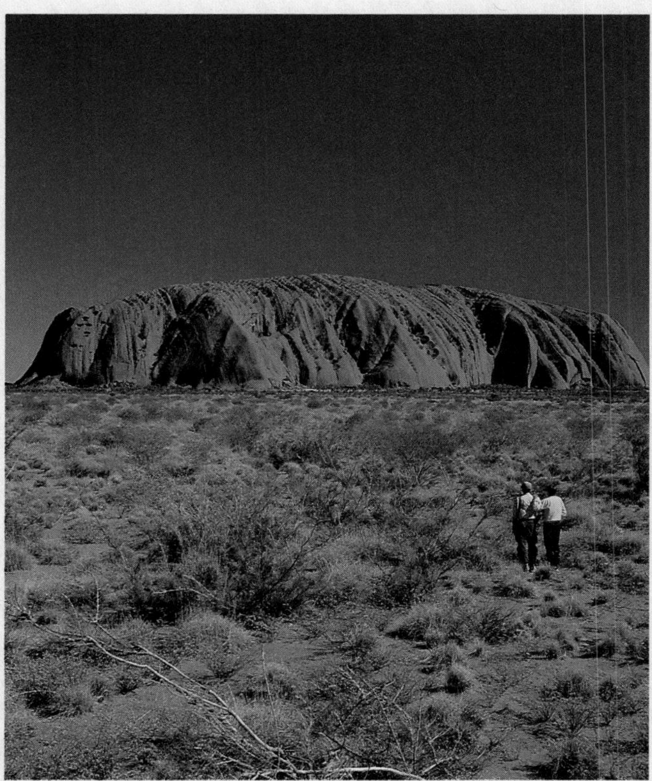

Ayers Rock, **Northern Territory, Australia**

Reviewing the Main Ideas

1. Australia is the world's smallest continent. It can be divided into three major landform regions. Although Australia has many climate types, most of the country is warm, sunny, and subject to drought. Because of Australia's isolation from other continents, many of its animals and plants are unique.
2. Australia is rich in agricultural and mineral resources. It also is an industrialized nation, although some industries are now in decline. The countries of Asia are Australia's major trade partners.
3. More than 85 percent of all Australians live in urban areas, particularly in the southeastern corner of the country. The Aborigines were Australia's first inhabitants but today make up only about one percent of the population. Many of Australia's immigrants now come from Asia. Population growth and economic development have presented some environmental challenges.

Building a Vocabulary

1. What is an artesian well? In what part of Australia are artesian wells found, and what is the source of their water?
2. Define *biogeography*. Why is Australia's biogeography distinct?
3. What is an endemic species? Give an example of a plant that is an endemic species in Australia.
4. What distinguishes marsupials from other mammals? Name two marsupials that live in Australia.

Recalling and Reviewing

1. What is the Great Dividing Range, and what effect does it have on Australia's climates?
2. What are Australia's major industries?
3. What are the country's three largest cities?

Thinking Critically

1. Australia was originally settled as a prison colony by Great Britain in 1788. Why might the British have chosen Australia for a prison colony?
2. What does Australia's agriculture have in common with that of Argentina and of the Great Plains region of the United States?

Learning About Your Local Geography

Cooperative Project

Contact a local wildlife organization to see what plants and animals are endemic species in your area and whether any of them are endangered. Make a poster showing the local wildlife and explaining why it is important to protect endangered species.

Using the Five Themes of Geography

Write a paragraph discussing the effects of **location** and **human-environment interaction** on Australia's biogeographical **place** characteristics.

CHAPTER 50

The Pacific Islands and Antarctica

GEOGRAPHY DICTIONARY

Oceania
atoll
Exclusive Economic Zone (EEZ)
polar desert
katabatic wind

"The North Island [of New Zealand] is all juts and spits and gulfs and bays. A protuberance [bulge] reaches north like a strand of taffy pulled away from the bulk of the island; no spot on . . . this peninsula is more than twenty miles from the sea. . . . All is fairly flat here for a hundred miles to the south of Auckland—then the country rises to volcanically active elevations of between three and nine thousand feet, and they dominate the rest of the island. This country is prone to convulse, rumble, boil, hiss, gurgle, seethe, tremble, and blow up."

John Gunther

Covering more than one-third of the earth's surface, the Pacific Ocean is the largest geographic feature on Earth. It contains more than 25,000 islands, more than are in all the other oceans and seas combined. Because most of the Pacific region is water, it is sometimes called Oceania.

Story board from **Papua New Guinea**

Some of the Pacific islands are little more than piles of jagged rock or coral reefs. Others cover thousands of square miles. Likewise, some islands are heavily populated, while others remain uninhabited. The first peoples arrived in the Pacific thousands of years ago from Southeast Asia. They were experienced navigators and eventually settled even the most isolated island groups. Today, modern influences have greatly changed the islanders' traditional ways of life.

South of the Pacific, and the rest of the world, lies the ice-covered continent of Antarctica. Its only human inhabitants are researchers studying this vast, frozen land.

Northern **Papua New Guinea**

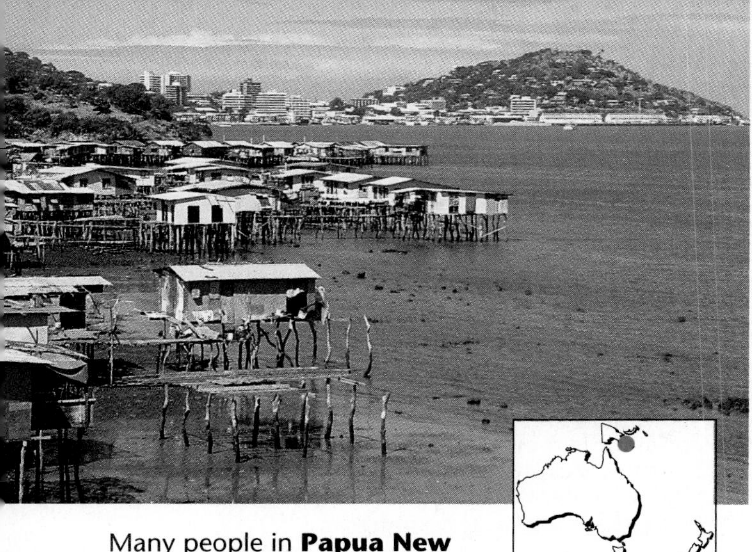

Many people in **Papua New Guinea** build their homes on stilts, like these houses near **Port Moresby**. The elevated houses are cooler and are better protected from rising water.

1 ISLANDS OF THE PACIFIC

FOCUS

- *What is the physical and historical geography of the Pacific islands?*
- *What are the major islands and island groups of the Pacific? What are their characteristics?*

REGION Physical Geography The islands of the Pacific can be divided into three island regions: Melanesia, Micronesia, and Polynesia. All of these groups contain high islands and low islands. The high islands have fresh water, good soils, and forest resources, and can thus support large human populations. Most high islands were formed by volcanoes rising from the deep sea. Tahiti, Hawaii, and Guam are examples of volcanic high islands. In the South Pacific, however, lie some large high islands, formed not from ocean volcanoes, but from sections of ancient continents. Examples of high continental islands are New Guinea, New Caledonia, and New Zealand.

The low islands, most of which are just above sea level, are made of coral. Because these islands have few resources, limited fresh water, thin soils, and few trees, they support only small human populations. Many small coral islands form **atolls**. An atoll is a ring of small islands built up on a coral reef surrounding a shallow lagoon.

Most of the Pacific islands lie in the tropics and have a humid-tropical climate, with rainfall all year. Some have a tropical-savanna climate, with most rainfall occurring during the Southern Hemisphere's summer.

MOVEMENT Historical Geography The first humans reached the Pacific islands about 50,000 years ago. Probably the first to arrive were the people who settled the large South Pacific islands of Melanesia. Thousands of years later, the Polynesian culture developed around the

FACTS IN BRIEF The Pacific Islands

COUNTRY POPULATION (1994)	LIFE EXPECTANCY (1994)	LITERACY RATE (1990)	PER CAPITA GDP (1993)
FIJI 764,382	63, male 68, female	85%	$4,000
KIRIBATI 77,853	53, male 56, female	90% (1985)	$525 (1990)
MARSHALL ISLANDS 54,031	62, male 65, female	86%	$1,500 (1992)
MICRONESIA 120,347	66, male 70, female	90% (1991)	$1,500 (1989) (GNP)
NAURU 10,019	64, male 69, female	99% (1988)	$10,000 (1989) (GNP)
NEW ZEALAND 3,388,737	73, male 80, female	99% (1991)	$15,700
PALAU 16,366	69, male 73, female	92% (1980)	$2,260 (1986)
PAPUA NEW GUINEA 4,196,806	56, male 57, female	52%	$2,000
SOLOMON ISLANDS 385,811	68, male 73, female	60% (1989)	$2,500 (1991)
TONGA 104,778	66, male 70, female	99% (1988)	$2,000
TUVALU 9,831	62, male 64, female	96%	$700 (1990) (GNP)
VANUATU 169,776	58, male 61, female	90%	$1,050 (1990)
WESTERN SAMOA 204,447	66, male 70, female	90% (1989)	$2,000 (1992)
UNITED STATES 260,713,585	73, male 79, female	97% (1991)	$24,700

Sources: *The World Factbook 1994,* Central Intelligence Agency; *The World Almanac and Book of Facts 1995*

Which is the most populous country of the region?

COUNTRIES OF MICRONESIA AND POLYNESIA

FEDERATED STATES OF MICRONESIA
This island country has chosen a Compact of Free Association with the United States. This means the country gets U.S. aid in exchange for allowing U.S. military bases on its land.

KIRIBATI
The capital of Kiribati is Tarawa, site of some of the fiercest fighting during World War II.

PALAU
Two-thirds of Palau's more than 16,000 people live in the capital, Koror. Palau was a U.S. trust territory until 1994.

TUVALU
The maximum elevation of any of the Polynesian islands of Tuvalu is 15 feet (4.6 m). Poor soil quality forces islanders to plant crops in trenches that reach down to the water table, as much as 12 feet (3.7 m) underground.

WESTERN SAMOA
Independent Western Samoa has a more traditional island culture than American Samoa, a U.S. territory. Copra and tropical crops are major exports of this Polynesian country.

MARSHALL ISLANDS
Two atolls of the Marshall Islands were sites for nuclear testing until 1958. Contamination from these tests left many islands of this Micronesian country uninhabitable for decades.

TONGA
Tonga, part of Polynesia, is a constitutional monarchy. In the 1200's, Tonga's king ruled areas as far away as Hawaii.

NAURU
The economy of this Micronesian country is based on phosphate mining for fertilizer.

islands of Tonga and Samoa. The Polynesians eventually spread all across the Pacific. Later, probably about 3,000 years ago, the Micronesians settled the islands of the North Pacific.

The first European explorers arrived in the region in the sixteenth century. Then in the late eighteenth century, Captain James Cook led major expeditions to all the main island regions of the Pacific. Soon after, the European powers, particularly Great Britain, France, Germany, and Spain, claimed lands in this part of the world. Only Tonga, a Polynesian kingdom, escaped colonial status. By the early twentieth century, Spanish and German claims in the Pacific were replaced by Japanese and U.S. claims.

During the colonial period, European missionaries, sailors, plantation owners, and miners heavily influenced the island cultures. For example, due to the missionary activity during the 1800s, most islanders today are Christians. European practices also changed the population makeup of the region because their plantations attracted many workers from India and China.

In this century, World War II had a tremendous impact on the region. Many islands were devastated first by invading Japanese forces and then by U.S., Australian, and New Zealand forces when they recaptured the islands toward the end of the war.

Melanesia The large islands of Melanesia stretch from New Guinea to Fiji. (See the map on page 601.) Melanesia includes a number of independent island nations, including Vanuatu, the Solomon Islands, and the Fiji Islands. New Caledonia belongs to France.

The world's second largest island and the most populated in the Pacific is New Guinea. Its western half, called Irian Jaya, is part of Indonesia. Its eastern half is the Pacific island nation of Papua New Guinea, which received independence from Australia in 1975.

The center of the island of New Guinea has a long east-west mountain range with peaks rising more than 14,000 feet (4,267 m). Most of the nearly 4 million people in Papua New Guinea are subsistence farmers who live in these central highlands. The rest of the population lives in the rain forests and along the rivers and coastline. Due to the geographic isolation of the many groups, more than 700 languages are spoken on the island. Many people also know Pidgin, a trade language based on English.

As for resources, the islands hold copper, gold, silver, and newly discovered oil deposits. In fact, minerals are now the region's dominant export and are helping fuel Papua New Guinea's rapidly growing economy.

Micronesia and Polynesia The more than 2,000 islands of Micronesia stretch mainly across the North Pacific Ocean. This region of mostly small coral islands is scattered over an area larger than the United States. Polynesia, however, is the largest of the three island regions. It forms a huge triangle across the Pacific, with corners at New Zealand, Hawaii, and Easter Island. (See the map on page 601.)

Many Micronesian and Polynesian islands were once colonies of other countries. Then after World War II, some island groups became trust territories. These territories later were able to choose their political future. Other islands remained territories of foreign countries. For example, France controls Tahiti, a popular tourist destination in Polynesia. U.S. island territories include Guam, Midway, and Wake Island. Some islands, such as the Northern Mariana Islands, have chosen to be a commonwealth of the United States, making the islanders U.S. citizens. To learn more about independent countries of Micronesia and Polynesia, see the chart on page 614.

Problems in Paradise Although outsiders often picture the Pacific islands as a paradise, the region faces challenging issues. For example, some of the islands have severe trade deficits, with their imports 10 times greater than their exports. Although tourism offers economic help for some islands, others must depend on the larger nations of the Pacific Rim, such as Australia, New Zealand, Japan, and the United States, for support.

For many islands, offshore resources of fish and deep-sea minerals promise future economic development. The island countries claim a 200-mile (322-km) **Exclusive Economic Zone (EEZ)** around their islands. They control all marine resources within that zone. Foreign nations that wish to fish or mine in the EEZ must have permission and pay fees to the controlling nation or island. Most of the world's nations claim EEZs off their coastlines.

In addition to the economy, the islanders are concerned about the effects of rapid acculturation. Many believe that tourism, television, processed foods, alcohol, mopeds, and other outside influences have had a negative impact on island cultures. Many hope to preserve the islands' traditional ways of life, distinct cultures, and beautiful environments.

SECTION REVIEW

1. How do the high islands differ physically from the low islands?
2. What are the three Pacific island regions? Which is the largest?
3. **Critical Thinking** **Decide** whether the Pacific islands form a culturally and politically unified region. Explain.

NEW ZEALAND

FOCUS

- *What are the physical features of New Zealand?*
- *What is New Zealand's economic geography? its human geography?*

LOCATION **Physical Geography** The island nation of New Zealand lies 1,000 miles (1,609 km) southeast of Australia. The Cook Strait separates the country's two large continental

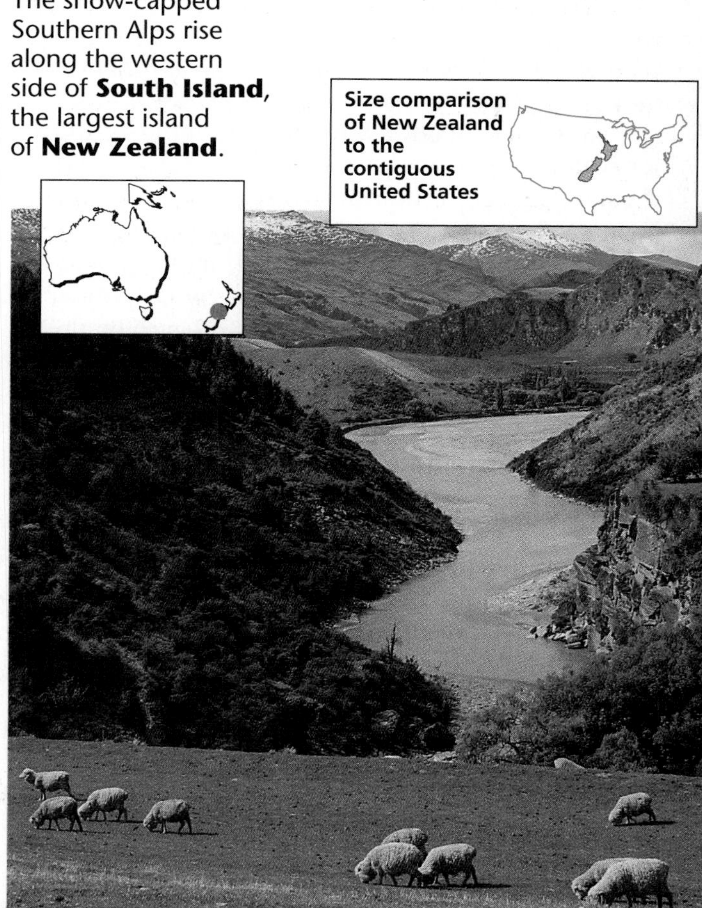
Size comparison of New Zealand to the contiguous United States

The snow-capped Southern Alps rise along the western side of **South Island**, the largest island of **New Zealand**.

NEWS FROM AROUND THE WORLD: New Zealand

New Zealand Teenagers Feel Global Influences

Thousands of miles and 11 time zones separate New Zealand and Great Britain. Yet, as a member of the British Commonwealth, New Zealand—particularly its schools—is steeped in British tradition. Morning tea is served promptly at 10 a.m. each school day to teachers and administrators, while students entertain themselves nearby. For 15 to 30 minutes, teachers share news, enjoy tea or coffee, and gear up for the day ahead.

Also borrowed from British tradition are school uniforms, like those shown above. New Zealanders believe that uniforms encourage students to take education seriously. They also maintain that uniform clothing lessens the distinction between rich and poor. For the most part, uniforms are worn proudly, and students are discouraged from any behavior that might "disgrace the school uniform."

Although most of the population is of British descent, New Zealand is a nation of diverse peoples. In recent years, increased immigration to New Zealand has meant a blending of world peoples and a sharing of new traditions. So that students of all backgrounds feel welcome, a class might be greeted each morning in several languages: "namaste" in Hindi, "magandáng umága" in Tagalog, "zǎo shàng hǎo" in Chinese, "kia ora" in Maori, "talofa" in Samoan, or "good morning" in English.

islands, North Island and South Island. (See the map on page 601.)

Mountains and hills cover almost three-fourths of New Zealand. On South Island, the spectacular glaciated Southern Alps rise from the Tasman Sea. Mount Cook, New Zealand's highest peak, is found here. North Island, meanwhile, has four volcanic peaks separated by a volcanic plateau. Most of the rest of the island is rolling hills and plains. New Zealand's location on the Pacific Ring of Fire makes the nation prone to earthquakes and volcanic eruptions. Geothermal activity on North Island causes hot springs, geysers, and steam vents.

Because of its oceanic, middle-latitude location, New Zealand has a mild marine-west-coast climate. The warmest area is on the northern peninsula of North Island. The coolest area is Stewart Island in the far south.

New Zealand also lies in the path of the westerly winds. As the westerlies rise over the Southern Alps, rain and snow fall on the windward side of the range. Some areas on the west coast receive up to 320 inches (813 cm) of rain per year. Dry air descending on the leeward side of the mountains results in some areas receiving only 14 inches (36 cm) of rain per year.

As in Australia, the biogeography of New Zealand is distinct. The country's endemic animals include several species of bats and flightless birds. The most famous flightless bird is the kiwi (KEE-wee), the country's national symbol. The many alien species, including deer, sheep, and opossums, have greatly altered the ecology of the islands.

MOVEMENT **Historical Geography** The first people to reach New Zealand were the Moa hunters, who arrived about 1,000 years ago. They were named after the huge, flightless birds they hunted for food. The Moas eventually died out, perhaps because they hunted the moa birds to extinction.

The Maoris (MAU-reez), a Polynesian people, settled the islands sometime between the tenth and fifteenth centuries. Although they built an advanced civilization, fierce wars broke out among various Maori groups.

The first European to sight New Zealand was the Dutch explorer Abel Tasman in 1642. Then in 1769, Captain James Cook claimed the islands for Great Britain. The first British settlements were established in the 1840s. Although the British settlers and the Maoris signed a treaty in 1840, fighting between them continued until the early 1870s. Today, the Maoris form about 10 percent of the population.

New Zealand is a parliamentary democracy and, like Australia, is a member of the British Commonwealth of Nations. It has had self-government since 1852.

Economic Geography New Zealand is a major exporter of lamb, beef, butter, cheese, and wool. Its mild climate also allows a great variety of crops, including grains, vegetables, and fruit. New Zealand grows enough food to supply most of the nation's needs. Its leading export crops are apples and kiwifruit.

The country must import oil, but it has large quantities of coal and offshore natural-gas deposits. Hydroelectric power and geothermal energy, however, supply more than 80 percent of the country's electricity. New Zealand is not rich in mineral resources, but the country does have some iron and gold deposits.

New Zealand's industries include tourism, processed foods, minerals, and wood and paper products. The country also produces steel, aluminum, boats, and clothing. Most manufactured goods, however, must be imported.

Before the formation of the European Union, New Zealand's exporting was largely dependent on British markets. Today, Japan, Australia, and the United States, as well as the EU, are its major trade partners. New Zealand seeks to maintain good relations with all countries because of its dependence on foreign trade.

Population About 75 percent of New Zealand's population lives on North Island, where most industry, agriculture, and the largest cities are found. In fact, more than 85 percent of New Zealanders live in urban areas. The city of Auckland has the nation's largest population, most industry, and main seaport. Wellington, the capital, is New Zealand's second largest city.

New Zealanders, sometimes called "Kiwis," generally enjoy a high standard of living. In return for high taxes, the government provides an extensive social-welfare system.

SECTION REVIEW

1. What climate type dominates New Zealand?
2. What are New Zealand's major exports? Who are the Maoris?
3. **Critical Thinking** **Explain** why most of New Zealand's population, industry, and agriculture are located on North Island.

ANTARCTICA

FOCUS

- *What is the physical geography of Antarctica?*
- *What human activity takes place in Antarctica?*

LOCATION **The Frozen Continent** At the southernmost part of the world lies Antarctica, Earth's coldest, driest, highest, and windiest continent. It remains the planet's last frontier, a place where harsh physical geography reigns over human activity.

Antarctica's total land area is about 5.5 million square miles (14.3 million sq. km), or about 10 percent of the world's land. Yet less than 5 percent of the continent is free of ice. For six months of the year, Antarctica is hidden from the sun and remains in total darkness. During its short summer, the sun never sets.

The Antarctic ice sheet contains more than 90 percent of the world's ice and averages over one mile (1.6 km) in depth. In some areas, the ice buries entire mountain ranges. The weight of the ice sheet causes it to flow slowly off the continent. When it reaches the sea, huge icebergs then break away and drift northward into the ocean. Some of these icebergs are larger than the state of Rhode Island.

The towering peaks of the Transantarctic Mountains divide the continent into East

Text continues on page 620.

Preserving Antarctica

In 1912, a search party combing the frozen lands of Antarctica found the body of explorer Robert Falcon Scott. Scott had written in his diary, "Great God! This is an awful place." His dream of being the first person to reach the South Pole had been thwarted. Scott's expedition reached the pole only to find the tent and flag of Norwegian explorer Roald Amundsen, who had arrived five weeks earlier than Scott.

Scientists believe that understanding Antarctica is an important key to understanding our planet.

Nearly a century later, Antarctica's modern-day explorers are research scientists. For the most part, Antarctica has become a successful experiment in international scientific research and cooperation.

A Scientific Laboratory

A wasteland to some, Antarctica is a scientific laboratory to many. The continent is of central importance to understanding our planet. For example, the icy waters bordering Antarctica move north, cooling warmer waters. This movement helps explain ocean currents, clouds, and weather patterns. The world's climate is also affected by Antarctic sea ice. The ice acts as a shield, keeping the earth cool by reflecting rather than absorbing the sun's heat energy.

The continent's ice, meanwhile, gives clues to the past. Buried within the layers of ice are gas bubbles and pollutants that are a "record" of Earth's air. Scientists have been able to compare, for example, traces of atmospheric gas trapped in Antarctic ice with atmospheric gas of today. From this comparison, they have learned that the use of fossil fuels has raised the amount of carbon dioxide in the air to the highest levels in human history.

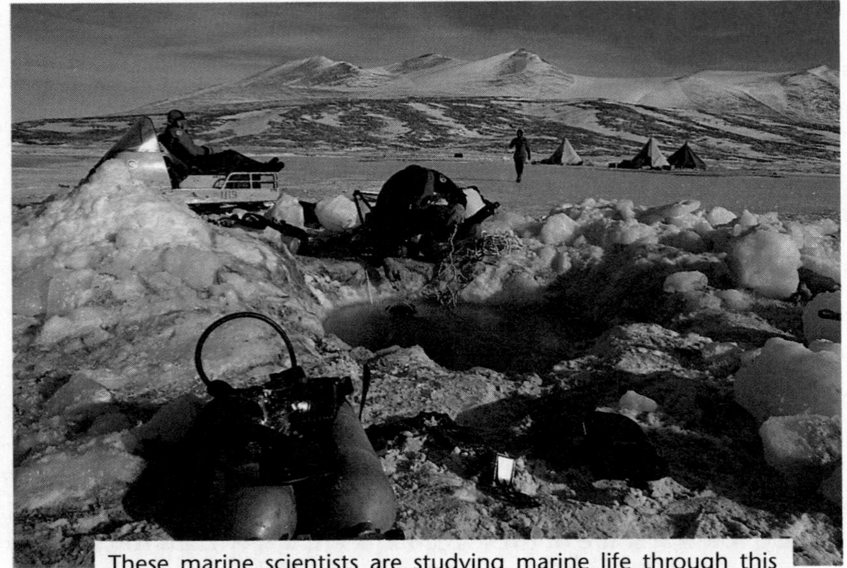

These marine scientists are studying marine life through this six-foot hole cut through the ice around **Antarctica**. A rich variety of marine life inhabits the water under Antarctica's ice.

Eyes on Antarctica

Antarctica is not owned by any one nation. Although a few countries claim sectors of Antarctica, none of these claims is recognized. In 1959, the Antarctic Treaty established Antarctica as "a continent for science and peace." Signed by 12 nations and later agreed to by 26 additional nations, the treaty banned military activity in the region and made Antarctica a nuclear-free zone. It also promoted free access to scientific research on the continent and banned further territorial claims by nations.

The Antarctic Treaty, however, did not cover mining rights on the continent. So when Antarctica's mineral riches were discovered, some nations began viewing the continent as a source of untapped wealth. Geographers, scientists, and environmentalists also took notice. They feared that if mining took place on Antarctica, the continent's environment would suffer. In particular, if a practical, inexpensive method of extracting the offshore oil were found, Antarctica's coastline and rich marine life would be threatened by oil spills.

At the same time, evidence of environmental neglect at the continent's research stations appeared. Environmentalists voiced concern about scientists' careless disposal of trash and sewage. A U.S. Coast Guard captain who worked on icebreakers described pollution at McMurdo, a U.S. scientific base. "Trash was just rolled down the hill. . . . One of the jobs of the icebreakers was to break up the ice where the trash was and push it out to sea." Concern for the Antarctic waters increased when, in 1989, an Argentine supply ship ran aground along Antarctica's rocky shore, spilling 170,000 gallons of oil.

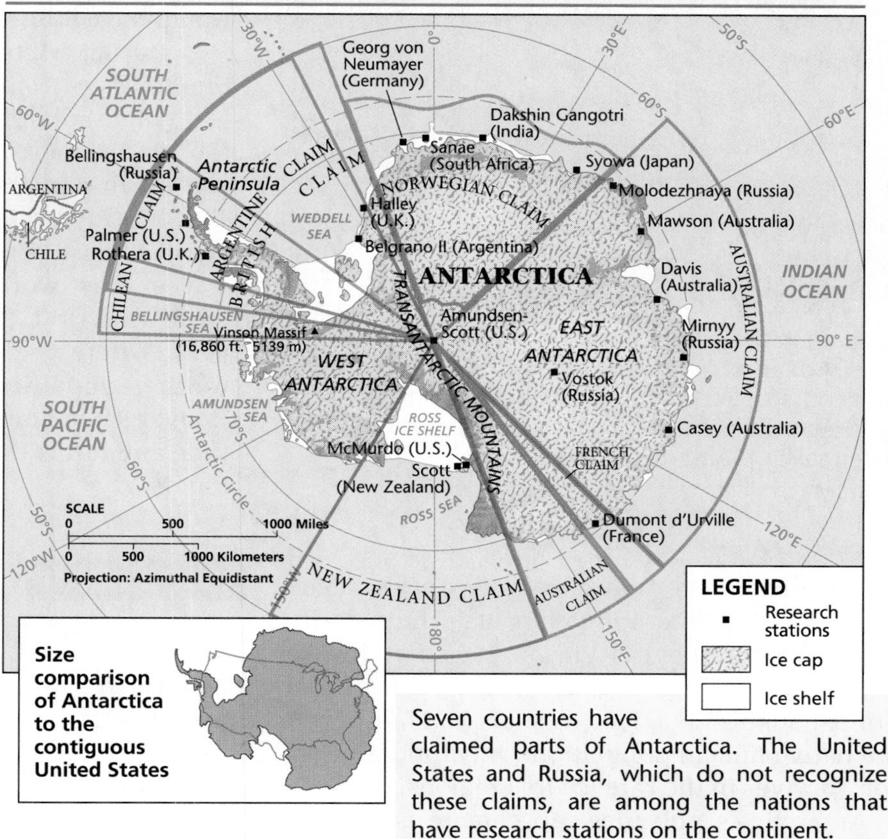

Antarctica
SOUTH ATLANTIC OCEAN
Georg von Neumayer (Germany)
Dakshin Gangotri (India)
Sanae (South Africa)
Syowa (Japan)
Molodezhnaya (Russia)
Mawson (Australia)
Bellingshausen (Russia)
Antarctic Peninsula
ARGENTINA
CHILE
Palmer (U.S.)
Rothera (U.K.)
WEDDELL SEA
Halley (U.K.)
Belgrano II (Argentina)
NORWEGIAN CLAIM
BRITISH CLAIM
ARGENTINE CLAIM
CHILEAN CLAIM
ANTARCTICA
TRANSANTARCTIC MOUNTAINS
Davis (Australia)
INDIAN OCEAN
AUSTRALIAN CLAIM
BELLINGSHAUSEN SEA
Vinson Massif (16,860 ft. 5139 m)
Amundsen-Scott (U.S.)
EAST ANTARCTICA
Mirnyy (Russia)
WEST ANTARCTICA
Vostok (Russia)
SOUTH PACIFIC OCEAN
AMUNDSEN SEA
ROSS ICE SHELF
Casey (Australia)
Antarctic Circle
McMurdo (U.S.)
Scott (New Zealand)
FRENCH CLAIM
ROSS SEA
Dumont d'Urville (France)
SCALE
0 500 1000 Miles
0 500 1000 Kilometers
Projection: Azimuthal Equidistant
NEW ZEALAND CLAIM
AUSTRALIAN CLAIM
LEGEND
Research stations
Ice cap
Ice shelf
Size comparison of Antarctica to the contiguous United States

Seven countries have claimed parts of Antarctica. The United States and Russia, which do not recognize these claims, are among the nations that have research stations on the continent.

Protecting Antarctica

In 1991, 39 nations abiding by the original Antarctic Treaty signed a new agreement called the Madrid Protocol. Unlike the original treaty of nearly 40 years ago, the new agreement is designed to protect the continent. It forbids most activities on Antarctica that do not have a scientific purpose. It also bans mining and drilling on the continent by the signing nations and sets limits on tourism. Pollution concerns are addressed specifically.

The treaty will remain in effect for 50 years. At that time, the agreement can be changed if two-thirds of the signing nations agree. Then, it becomes the responsibility of another generation to preserve Antarctica.

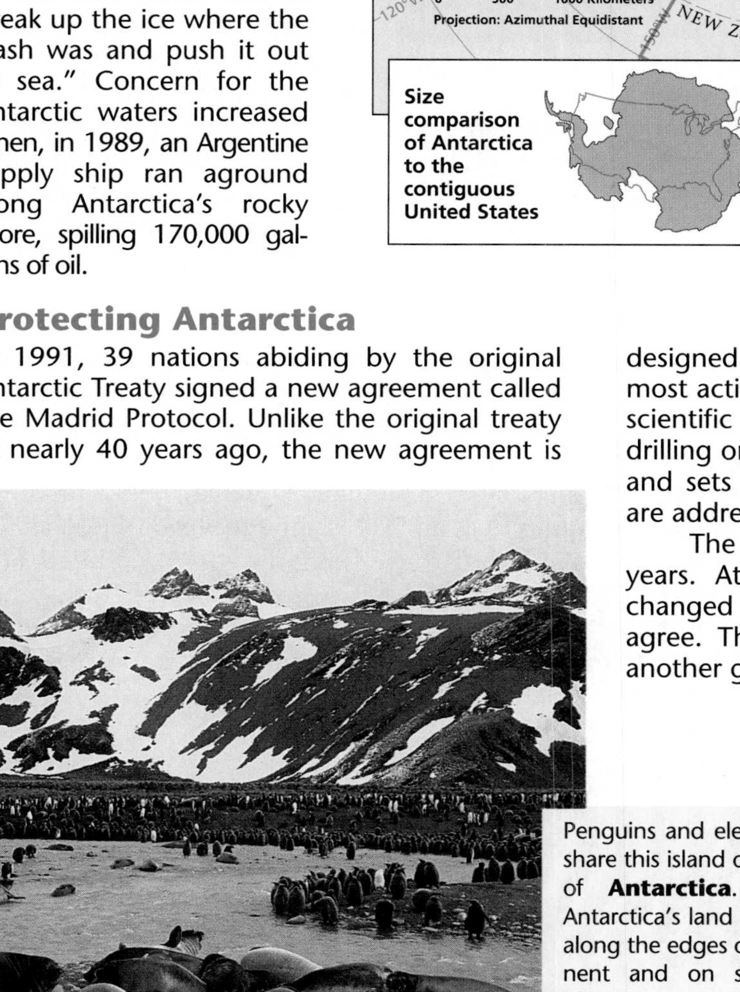
Penguins and elephant seals share this island off the coast of **Antarctica**. Most of Antarctica's land animals live along the edges of the continent and on surrounding islands.

Your Turn

Imagine you are helping develop the Madrid Protocol. **Write** a short description of the provisions you want included in the treaty and why you think they are important.

An iceberg in the bitter cold climate of **Antarctica** is a haven for these penguins. Except for researchers who temporarily live on the continent, only penguins and other wildlife make up Antarctica's population.

Antarctica and West Antarctica, a few dry coastal valleys, and the Antarctic Peninsula. The highest peak on the continent is Vinson Massif at 16,860 feet (5,139 m). Except for a few insects, this frozen continent supports no land animals and has never had a native human population. Only tundra vegetation survives in the rare ice-free regions.

Antarctica's polar ice-cap climate is severe. Because of its high average elevation and because more than 95 percent of the land is ice covered, Antarctica is much colder than even the Arctic region. For most of the year, temperatures remain bitterly cold, dropping to -120°F (-84°C) or lower. The Russian research station in East Antarctica recorded the world's record low temperature of -128.6°F (-89.2°C). Most of Antarctica is a **polar desert** and receives less precipitation than the Sahara in Africa. Here, however, there is almost no evaporation and melting. As a result, ice has built up over thousands of years.

Air flowing off this icy land makes Antarctica the windiest place on Earth. **Katabatic winds** gusting more than 100 miles per hour (161 kph) race along Antarctica's coast. A katabatic wind is a rapid flow of dense cold air down the slopes of high mountains or ice caps.

Antarctic Waters The "Southern Ocean" isolates Antarctica from the other continents. The Southern Ocean is really the far southern reaches of the Atlantic, Pacific, and Indian oceans. It has a rich and distinct ecosystem, with a large population of penguins, seals, and whales. This ocean life depends on a small shrimplike animal that forms the basis of the Antarctic food chain. Without these tiny shelled creatures, the Antarctic waters would have little or no ocean life.

During winter in the Southern Hemisphere, the Southern Ocean freezes over with about five feet (1.5 m) of drifting sea ice. During the summer, the sea ice melts, and research and supply ships can reach some sections of the Antarctic coast. The stormy waters around Antarctica are the world's most severe, with high winds and huge waves. The world's largest ocean current, the West Wind Drift, flows clockwise around the continent. Where the cold Antarctic waters meet the warmer middle-latitude waters is the Antarctic Convergence Zone. This is at about 50° south latitude and forms the true boundary of the Antarctic region.

INTERACTION **Research** British explorer James Cook was the first person known to reach Antarctic waters. His sighting of icebergs in 1772 confirmed that a vast continent existed. It was not until 1911, however, that the first recorded human expedition reached the South Pole. Today, the only people found in Antarctica are researchers working in scientific research stations, which are located at various points on the ice cap.

The Antarctic environment holds clues to the ice ages, global warming, the ozone layer, meteorites (found in the ice), and marine ecosystems. Indeed, research here has provided new and fascinating information on our planet and solar system. Minerals have been found on Antarctica, and oil is thought to be there as well. Mining of these minerals is prohibited. (See "Planet Watch" on pages 618–619.) Researchers, however, continue to mine Antarctica for answers to questions about the planet's past, present, and future.

SECTION REVIEW

1. Why is Antarctica called "the frozen continent"?
2. In what activity are most of the people in Antarctica involved?
3. **Critical Thinking** **Determine** whether the Antarctic Convergence Zone is an Exclusive Economic Zone. Explain your reasoning.

CHAPTER 50 CHECK

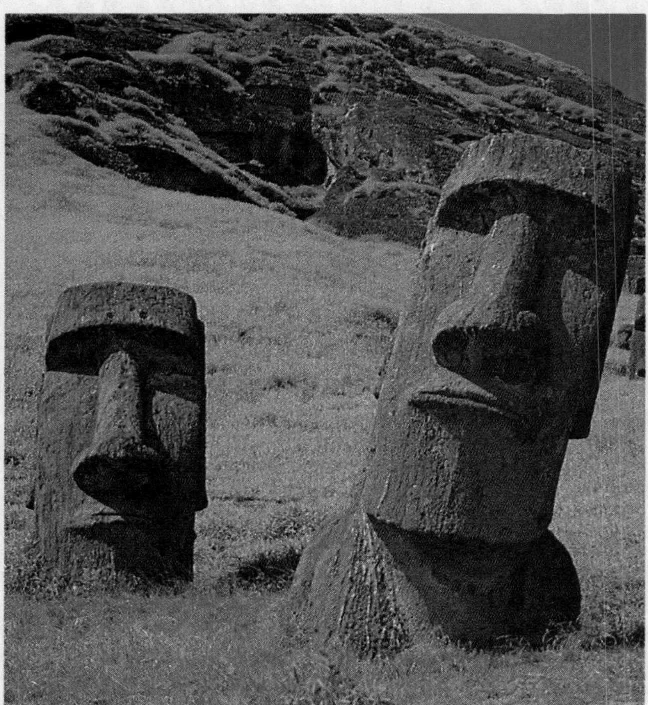
Carved hundreds of years ago, some 600 stone statues called *moai* (MOE-ie) still watch over isolated **Easter Island**.

Reviewing the Main Ideas

1. There are more than 25,000 islands in the Pacific Ocean. Some are small and uninhabited, while others are large and populated. The islands are divided into three regions: Melanesia, Micronesia, and Polynesia. All the island cultures have been influenced by Asian and European explorers and settlers.
2. New Zealand is made up of two large continental islands: North Island and South Island. New Zealand exports many agricultural products but must import oil and most manufactured goods. Auckland is New Zealand's largest city.
3. Ice covers almost all of Antarctica. The Southern Ocean supports life, but the only human inhabitants of the continent are scientific researchers.

Building a Vocabulary

1. Why is the Pacific region sometimes called Oceania?
2. Define *atoll*. Are the soils of atolls agriculturally productive?
3. What is the term for a rapid flow of dense cold air down high mountain slopes or ice caps?
4. Is a polar desert dry? Why or why not?

Recalling and Reviewing

1. How does New Zealand's climate differ from the climates of most of the other Pacific islands? What accounts for the difference?
2. What countries are the major trading partners of the Pacific islands and New Zealand?
3. What cultural challenges do the Pacific islands face?
4. Describe Antarctica's biogeography.

Thinking Critically

1. Both cultural diffusion and cultural isolation characterize the Pacific islands. Explain how both are possible in the same region and how geographic factors are involved.
2. Compare and contrast the economic geography of New Zealand and Australia.

Learning About Your Local Geography

Cooperative Project

World War II had a great impact on the Pacific islands. To what extent did World War II affect your own community? Interview an individual who has lived in your community long enough to remember the war and its local effects. After the interview, write a summary of what you have learned.

Using the Five Themes of Geography

What forms of **human-environment interaction** are currently possible in Antarctica, given its physical **place** characteristics? Make a list of adaptations humans might make to enable increased **movement** to and settlement in the region. Do you think that modification of the environment is possible? In your opinion, would such modification be desirable?

SKILLS

Developing Global Awareness

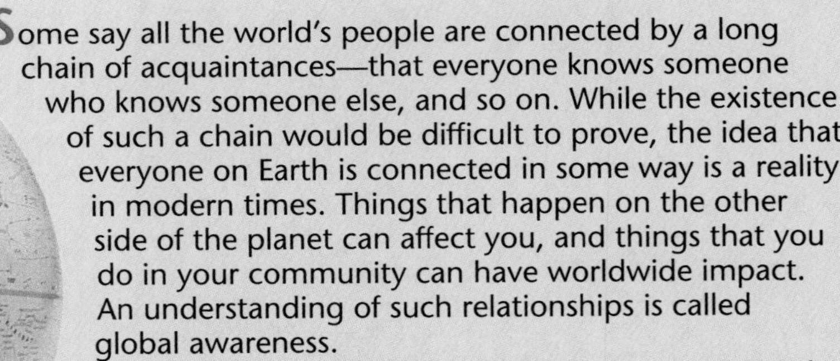

Some say all the world's people are connected by a long chain of acquaintances—that everyone knows someone who knows someone else, and so on. While the existence of such a chain would be difficult to prove, the idea that everyone on Earth is connected in some way is a reality in modern times. Things that happen on the other side of the planet can affect you, and things that you do in your community can have worldwide impact. An understanding of such relationships is called global awareness.

You might think that the Pacific islands, spread throughout the vast Pacific Ocean, have somehow managed to escape global importance. But nothing could be further from the truth. Despite the water that physically separates the islands from each other and from the continents, they are connected to the rest of the world. Many people from Europe, Asia, and the Americas have settled on the islands, and the societies there are a blend of cultures. Also, while the islands depend on income from tourism, they must adapt to tourists' expectations that the islands are a sort of paradise. This often means further cultural change.

Furthermore, due to their location between Asia and the Americas, the islands have global strategic value. Several of the islands have been used as military bases by larger nations. Some Pacific islands were the scene of intense fighting between the United States and Japan during World War II. And because of the islands' relative isolation and small populations, some of them have been used as sites for weapons testing. As a result of this testing, certain islands are now uninhabitable.

As you read or watch the news, think about the global meaning of the events being discussed. How is the United States involved politically, economically, militarily, environmentally, or culturally in events that take place throughout the world? How are other countries involved in U.S. affairs? How do events in the opposite hemisphere affect you directly or indirectly, and how do your actions affect people in the opposite hemisphere or even just on the opposite side of the country? With the pace of communication growing ever faster, the world's regions become linked in more ways, and almost everything we do can have a global effect.

PRACTICING THE SKILL

1. Analyze the global significance of a news event you have read or heard about. What countries are involved directly? What countries are involved indirectly? What effects will the outcomes of the event have on the rest of the world? Write a paragraph answering these questions.
2. Keep a log of your activities during one day. Identify from your log activities that have global impact, however minor. Make notes next to those selected entries, explaining their global implications.

BUILDING YOUR PORTFOLIO

Individually or in a group, complete one of the following projects to show your understanding of the geography concepts involved.

A *Biogeography*

As you know, biogeography is the study of the geographic distribution of plants and animals. From this unit or from outside sources, identify and select any animal or plant species found in Australia, the Pacific islands, Antarctica, the Pacific Ocean, or the Southern Ocean. Once you have made your selection, use library resources to research the animal or plant's geographic origins, its unique or unusual features, its role in and interaction with its environment, and its interaction with humans. Design a poster to display your findings. You will need to consider the following questions as you prepare your poster.

1. How did the animal or plant come to its present location? Is it endemic to the area, or was it introduced from another region? What unique or unusual features does the species have? How do these features relate to geography? Draw some pictures of the species for your poster. Next to your illustrations, write about the geographic origins and unusual features of the plant or animal.
2. How does the plant or animal interact with its environment, including humans? Where is the plant or animal in the food chain? Do many other species depend on it for survival, or is it near the top of the food chain? Has the plant or animal's presence changed the environment? Do humans hunt or use the plant or animal? If so, for what purpose? Do any human activities endanger the plant or animal? Create diagrams, graphic organizers, or flowcharts showing the plant or animal's patterns of interaction with its environment and with humans. Place these materials on your poster.

Present your poster to the rest of your class.

B *Antarctica and Outer Space*

You have read about how Antarctica is not a country or territory but is peacefully shared by many different countries. Antarctica is viewed by some as a model of a much larger mostly unexplored area—outer space. Using library resources, collect information on the Antarctic Treaty. Then, using what you have learned about the Antarctic Treaty, write a treaty outlining how Earth's nations might cooperate in their efforts to explore outer space, use its resources, and learn about any life-forms discovered in space in the future. To complete this project, you will need to do the following.

1. Make an outline of the provisions of the Antarctic Treaty. How is the treaty divided? What issues does the treaty address? How does the treaty provide for potential problems in dealing with Antarctica?
2. Based on your outline of the Antarctic Treaty, write a treaty for how nations should approach the exploration and use of outer space. Should space be divided into territories, each belonging to a country, or should it be shared by many countries, as Antarctica is? If valuable resources are discovered in space, what rules will govern their use? Do the environments of other planets and of space itself need protection? What policy should govern encounters with life-forms in space, should such encounters occur?

Present your outline of the Antarctic Treaty and your own treaty for outer space to your class. Discuss the extent to which the two treaties are similar, and explain why there are differences.

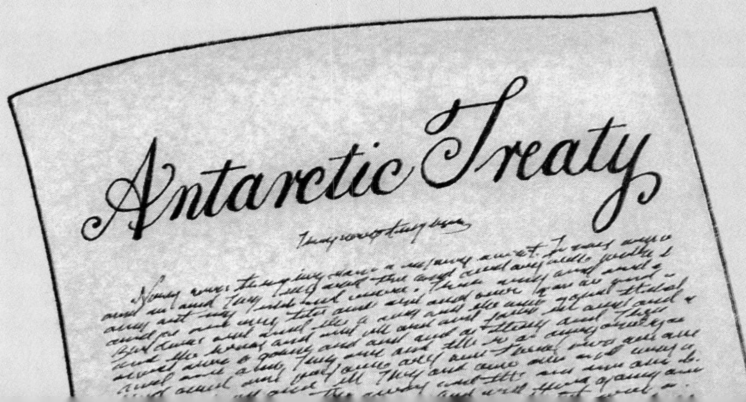
Antarctic Treaty

APPENDIX

Metric Conversion Table

	If you have	multiply by	to get
LENGTH	miles	1.609	kilometers
	kilometers	.62	miles
	feet	.3048	meters
	meters	39.37, then divide by 12	feet
	inches	2.54	centimeters
	centimeters	.39	inches
WEIGHT	pounds	.454	kilograms
	kilograms	2.2046	pounds
AREA	acres	.405	hectares
	hectares	2.47	acres
	square miles	2.59	square kilometers
	square kilometers	.3861	square miles
CAPACITY	quarts	.946	liters
	liters	1.057	quarts
TEMPERATURE	degrees Fahrenheit	subtract 32, then multiply by $5/9$	degrees Celsius
	degrees Celsius	multiply by $9/5$, then add 32	degrees Fahrenheit

Country ★ Capital	Population (1994)	Land Area (1994)	Principal Languages
THE UNITED STATES AND CANADA			
United States ★ Washington, D.C.	260,713,585	3,618,765 sq. mi. 9,372,610 sq. km	English, Spanish, Native American languages, other languages of the world
Canada ★ Ottawa	28,113,997	3,851,788 sq. mi. 9,976,140 sq. km	English, French
MIDDLE AND SOUTH AMERICA			
Antigua and Barbuda ★ St. John's	64,762	170 sq. mi. 440 sq. km	English, local dialects
Argentina ★ Buenos Aires	33,912,994	1,068,296 sq. mi. 2,766,890 sq. km	Spanish, English, Italian, German, French

Country ★ Capital	Population (1994)	Land Area (1994)	Principal Languages
Bahamas ★ Nassau	273,055	5,382 sq. mi. 13,940 sq. km	English, Creole
Barbados ★ Bridgetown	255,827	166 sq. mi. 430 sq. km	English
Belize ★ Belmopan	208,949	8,865 sq. mi. 22,960 sq. km	English, Spanish, Maya, Garifuna (Carib)
Bolivia ★ La Paz ★ Sucre	7,719,445	424,162 sq. mi. 1,098,580 sq. km	Spanish, Quechua, Aymara,
Brazil ★ Brasília	158,739,257	3,286,470 sq. mi. 8,511,965 sq. km	Portuguese, Spanish, English, French
Chile ★ Santiago	13,950,557	292,258 sq. mi. 756,950 sq. km	Spanish
Colombia ★ Bogotá	35,577,556	439,733 sq. mi. 1,138,910 sq. km	Spanish
Costa Rica ★ San José	3,342,154	19,730 sq. mi. 51,100 sq. km	Spanish, English
Cuba ★ Havana	11,064,344	42,803 sq. mi. 110,860 sq. km	Spanish
Dominica ★ Roseau	87,696	290 sq. mi. 750 sq. km	English, French patois
Dominican Republic ★ Santo Domingo	7,826,075	18,815 sq. mi. 48,730 sq. km	Spanish
Ecuador ★ Quito	10,677,067	109,483 sq. mi. 283,560 sq. km	Spanish, Quechua and other Native American languages
El Salvador ★ San Salvador	5,552,511	8,124 sq. mi. 21,040 sq. km	Spanish, Nahuatl
Grenada ★ St. George's	94,109	131 sq. mi. 340 sq. km	English, French patois
Guatemala ★ Guatemala City	10,721,387	42,042 sq. mi. 108,890 sq. km	Spanish, Native American dialects
Guyana ★ Georgetown	729,425	83,000 sq. mi. 214,970 sq. km	English, Native American dialects
Haiti ★ Port-au-Prince	6,491,450	10,714 sq. mi. 27,750 sq. km	French, Creole

Country ★ Capital	Population (1994)	Land Area (1994)	Principal Languages
Honduras ★ Tegucigalpa	5,314,794	43,278 sq. mi. 112,090 sq. km	Spanish, Native American dialects
Jamaica ★ Kingston	2,555,064	4,243 sq. mi. 10,990 sq. km	English, Creole
Mexico ★ Mexico City	92,202,199	761,602 sq. mi. 1,972,550 sq. km	Spanish, Mayan dialects
Nicaragua ★ Managua	4,096,689	49,998 sq. mi. 129,494 sq. km	Spanish, English, Native American languages
Panama ★ Panama City	2,630,000	30,193 sq. mi. 78,200 sq. km	Spanish, English
Paraguay ★ Asunción	5,213,772	157,046 sq. mi. 406,750 sq. km	Spanish, Guaraní
Peru ★ Lima	23,650,671	496,223 sq. mi. 1,285,220 sq. km	Spanish, Quechua, Aymara
St. Kitts and Nevis ★ Basseterre	40,671	104 sq. mi. 269 sq. km	English
St. Lucia ★ Castries	145,090	239 sq. mi. 620 sq. km	English, French patois
St. Vincent and the Grenadines ★ Kingstown	115,437	131 sq. mi. 340 sq. km	English, French patois
Suriname ★ Paramaribo	422,840	63,039 sq. mi. 163,270 sq. km	Dutch, English, Surinamese, Hindi, Javanese
Trinidad and Tobago ★ Port-of-Spain	1,328,282	1,981 sq. mi. 5,130 sq. km	English, Hindi, French, Spanish
Uruguay ★ Montevideo	3,198,910	68,039 sq. mi. 176,220 sq. km	Spanish
Venezuela ★ Caracas	20,562,405	352,143 sq. mi. 912,050 sq. km	Spanish, Native American dialects

EUROPE

Country ★ Capital	Population (1994)	Land Area (1994)	Principal Languages
Albania ★ Tiranë	3,374,085	11,100 sq. mi. 28,750 sq. km	Albanian, Greek
Andorra ★ Andorra la Vella	63,930	174 sq. mi. 450 sq. km	Catalan, French, Castilian

Country ★ Capital	Population (1994)	Land Area (1994)	Principal Languages
Austria ★ Vienna	7,954,974	32,374 sq. mi. 83,850 sq. km	German
Belgium ★ Brussels	10,062,836	11,780 sq. mi. 30,510 sq. km	Flemish (Dutch), French, German
Bosnia and Herzegovina ★ Sarajevo	4,651,485	19,781 sq. mi. 51,233 sq. km	Serbo-Croatian
Bulgaria ★ Sofia	8,799,986	42,822 sq. mi. 110,910 sq. km	Bulgarian, Turkish
Croatia ★ Zagreb	4,697,614	21,829 sq. mi. 56,538 sq. km	Serbo-Croatian
Czech Republic ★ Prague	10,408,280	30,387 sq. mi. 78,703 sq. km	Czech, Slovak
Denmark ★ Copenhagen	5,187,821	16,629 sq. mi. 43,070 sq. km	Danish, Faroese, Greenlandic, German
Estonia ★ Tallinn	1,616,882	17,413 sq. mi. 45,100 sq. km	Estonian, Latvian, Lithuanian, Russian
Finland ★ Helsinki	5,068,931	130,127 sq. mi. 337,030 sq. km	Finnish, Swedish, Lapp, Russian
France ★ Paris	57,840,445	211,208 sq. mi. 547,030 sq. km	French, regional dialects and languages
Germany ★ Berlin	81,087,506	137,803 sq. mi. 356,910 sq. km	German
Greece ★ Athens	10,564,630	50,942 sq. mi. 131,940 sq. km	Greek, English, French
Hungary ★ Budapest	10,319,113	35,919 sq. mi. 93,030 sq. km	Hungarian
Iceland ★ Reykjavik	263,599	39,768 sq. mi. 103,000 sq. km	Icelandic
Ireland ★ Dublin	3,539,296	27,135 sq. mi. 70,280 sq. km	English, Irish (Gaelic)
Italy ★ Rome	58,138,394	116,305 sq. mi. 301,230 sq. km	Italian, German, French, Slovene
Latvia ★ Riga	2,749,211	24,749 sq. mi. 64,100 sq. km	Lettish, Lithuanian, Russian

Country ★ Capital	Population (1994)	Land Area (1994)	Principal Languages
Liechtenstein ★ Vaduz	30,281	62 sq. mi. 160 sq. km	German, Alemannic dialect
Lithuania ★ Vilnius	3,848,389	25,174 sq. mi. 65,200 sq. km	Lithuanian, Polish, Russian
Luxembourg ★ Luxembourg	401,900	998 sq. mi. 2,586 sq. km	Luxembourgisch, German, French, English
Macedonia ★ Skopje	2,213,785	9,781 sq. mi. 25,333 sq. km	Macedonian, Albanian, Turkish, Serbo-Croatian
Malta ★ Valletta	366,767	124 sq. mi. 320 sq. km	Maltese, English
Monaco ★ Monaco	31,278	.73 sq. mi. 1.9 sq. km	French, English, Italian, Monegasque
Netherlands ★ Amsterdam ★ The Hague	15,367,928	14,413 sq. mi. 37,330 sq. km	Dutch
Norway ★ Oslo	4,314,604	125,181 sq. mi. 324,220 sq. km	Norwegian, Lapp, Finnish
Poland ★ Warsaw	38,654,561	120,726 sq. mi. 312,680 sq. km	Polish
Portugal ★ Lisbon	10,524,210	35,552 sq. mi. 92,080 sq. km	Portuguese
Romania ★ Bucharest	23,181,415	91,699 sq. mi. 237,500 sq. km	Romanian, Hungarian, German
San Marino ★ San Marino	24,091	23 sq. mi. 60 sq. km	Italian
Slovakia ★ Bratislava	5,403,505	18,859 sq. mi. 48,845 sq. km	Slovak, Hungarian
Slovenia ★ Ljubljana	1,972,227	7,836 sq. mi. 20,296 sq. km	Slovenian, Serbo-Croatian
Spain ★ Madrid	39,302,665	194,884 sq. mi. 504,750 sq. km	Castilian Spanish, Catalan, Galician, Basque
Sweden ★ Stockholm	8,778,461	173,731 sq. mi. 449,964 sq. km	Swedish, Lapp, Finnish
Switzerland ★ Bern	7,040,119	15,942 sq. mi. 41,290 sq. km	German, French, Italian, Romansh

Country ★ Capital	Population (1994)	Land Area (1994)	Principal Languages
United Kingdom ★ London	58,135,110	94,525 sq. mi. 244,820 sq. km	English, Welsh, Scottish Gaelic
Vatican City ★ Vatican City	821	0.169 sq. mi. 0.44 sq. km	Italian, Latin, other languages of the world
Yugoslavia (Serbia and Montenegro) ★ Belgrade	10,759,897	39,517 sq. mi. 102,350 sq. km	Serbo-Croatian, Albanian
RUSSIA AND NORTHERN EURASIA			
Armenia ★ Yerevan	3,521,517	11,506 sq. mi. 29,800 sq. km	Armenian, Russian
Azerbaijan ★ Baki	7,684,456	33,436 sq. mi. 86,600 sq. km	Azeri, Russian, Armenian
Belarus ★ Minsk	10,404,862	80,154 sq. mi. 207,600 sq. km	Belorussian, Russian
Georgia ★ T'bilisi	5,681,025	26,911 sq. mi. 69,700 sq. km	Georgian, Russian, Armenian, Azeri
Kazakhstan ★ Almaty	17,267,554	1,049,150 sq. mi. 2,717,300 sq. km	Kazakh, Russian
Kyrgyzstan ★ Bishkek	4,698,108	76,641 sq. mi. 198,500 sq. km	Kyrgyz, Russian
Moldova ★ Chişinău	4,473,033	13,012 sq. mi. 33,700 sq. km	Moldovan, Romanian, Russian, Turkish dialect
Russia ★ Moscow	149,608,953	6,592,735 sq. mi. 17,075,200 sq. km	Russian
Tajikistan ★ Dushanbe	5,995,469	55,251 sq. mi. 143,100 sq. km	Tajik, Russian
Turkmenistan ★ Ashgabat	3,995,122	188,455 sq. mi. 488,100 sq. km	Turkmen, Russian, Uzbek
Ukraine ★ Kyyiv	51,846,958	233,089 sq. mi. 603,700 sq. km	Ukrainian, Russian, Romanian, Polish, Hungarian
Uzbekistan ★ Toshkent	22,608,866	172,741 sq. mi. 447,400 sq. km	Uzbek, Russian, Tajik
SOUTHWEST ASIA			
Afghanistan ★ Kabul	16,903,400	250,000 sq. mi. 647,500 sq. km	Pashtu, Afghan Persian, Uzbek, Turkmen

Country ★ Capital	Population (1994)	Land Area (1994)	Principal Languages
Bahrain ★ Manama	585,683	239 sq. mi. 620 sq. km	Arabic, English, Farsi, Urdu
Cyprus ★ Nicosia	730,084	3,571 sq. mi. 9,250 sq. km	Greek, Turkish, English
Iran ★ Tehran	65,615,474	636,293 sq. mi. 1,648,000 sq. km	Persian and Persian dialects, Turkic dialects, Kurdish, Luri, Baloch, Arabic, Turkish
Iraq ★ Baghdad	19,889,666	168,753 sq. mi. 437,072 sq. km	Arabic, Kurdish, Assyrian, Armenian
Israel ★ Jerusalem	5,050,850	8,019 sq. mi. 20,770 sq. km	Hebrew, Arabic, English
Jordan ★ Amman	3,961,194	34,445 sq. mi. 89,213 sq. km	Arabic, English
Kuwait ★ Kuwait City	1,819,322	6,880 sq. mi. 17,820 sq. km	Arabic, English
Lebanon ★ Beirut	3,620,395	4,015 sq. mi. 10,400 sq. km	Arabic, French, Armenian, English
Oman ★ Muscat	1,701,470	82,031 sq. mi. 212,460 sq. km	Arabic, English, Balochi, Urdu, Indian dialects
Qatar ★ Doha	512,779	4,447 sq. mi. 11,000 sq. km	Arabic, English
Saudi Arabia ★ Riyadh	18,196,783	756,981 sq. mi. 1,960,582 sq. km	Arabic
Syria ★ Damascus	14,886,672	71,498 sq. mi. 185,180 sq. km	Arabic, Kurdish, Armenian, Aramaic, Circassian, French
Turkey ★ Ankara	62,153,898	301,382 sq. mi. 780,580 sq. km	Turkish, Kurdish, Arabic
United Arab Emirates ★ Abu Dhabi	2,791,141	29,182 sq. mi. 75,581 sq. km	Arabic, Persian, English, Hindi, Urdu
Yemen ★ Sanaa	11,105,202	203,849 sq. mi. 527,970 sq. km	Arabic

AFRICA

Country ★ Capital	Population (1994)	Land Area (1994)	Principal Languages
Algeria ★ Algiers	27,895,068	919,590 sq. mi. 2,381,740 sq. km	Arabic, French, Berber dialects

Country ★ Capital	Population (1994)	Land Area (1994)	Principal Languages
Angola ★ Luanda	9,803,576	481,351 sq. mi. 1,246,700 sq. km	Portuguese, Bantu, other African languages
Benin ★ Porto-Novo	5,341,710	43,483 sq. mi. 112,620 sq. km	French, Fon, Yoruba, other local languages
Botswana ★ Gaborone	1,359,352	231,803 sq. mi. 600,370 sq. km	English, Setswana
Burkina Faso ★ Ouagadougou	10,134,661	105,869 sq. mi. 274,200 sq. km	French, local languages
Burundi ★ Bujumbura	6,124,747	10,745 sq. mi. 27,830 sq. km	Kirundi, French, Swahili
Cameroon ★ Yaoundé	13,132,191	183,567 sq. mi. 475,440 sq. km	English, French, African languages
Cape Verde ★ Praia	423,120	1,556 sq. mi. 4,030 sq. km	Portuguese, Crioulo
Central African Republic ★ Bangui	3,142,182	240,533 sq. mi. 622,980 sq. km	French, Sangho, Arabic, Hunsa, Swahili
Chad ★ N'Djamena	5,466,771	495,752 sq. mi. 1,284,000 sq. km	French, Arabic, Sara, Sango, other dialects
Comoros ★ Moroni	530,136	838 sq. mi. 2,170 sq. km	Arabic, French, Comoran
Congo ★ Brazzaville	2,446,902	132,046 sq. mi. 342,000 sq. km	French, Lingala, Kikongo, other African languages
Côte d'Ivoire ★ Abidjan	14,295,501	124,502 sq. mi. 322,460 sq. km	French, Dioula, native dialects
Djibouti ★ Djibouti	412,599	8,494 sq. mi. 22,000 sq. km	French, Arabic, Somali, Afar
Egypt ★ Cairo	60,765,028	386,660 sq. mi. 1,001,450 sq. km	Arabic, English, French
Equatorial Guinea ★ Malabo	409,550	10,830 sq. mi. 28,050 sq. km	Spanish, pidgin English, Fang, Bubi, Ibo
Eritrea ★ Asmara	3,782,543	46,842 sq. mi. 121,320 sq. km	Tigre and Kunama, Tigre, Nora Bana, Arabic, other dialects
Ethiopia ★ Addis Ababa	54,927,108	435,184 sq. mi. 1,127,127 sq. km	Amharic, Tigrinya, Orominga, Guaraginga, Somali, Arabic, English

Country ★ Capital	Population (1994)	Land Area (1994)	Principal Languages
Gabon ★ Libreville	1,139,006	103,347 sq. mi. 267,670 sq. km	French, Fang, Myene, Bateke, Bapounou/Eschira, Bandjabi
Gambia ★ Banjul	959,300	4,363 sq. mi. 11,300 sq. km	English, Mandinka, Wolof, Fula, other local dialects
Ghana ★ Accra	17,225,185	92,100 sq. mi. 238,540 sq. km	English, Akan, Moshi-Dagomba, Ewe, Ga
Guinea ★ Conakry	6,391,536	94,927 sq. mi. 245,860 sq. km	French, local languages
Guinea-Bissau ★ Bissau	1,098,231	13,946 sq. mi. 36,120 sq. km	Portuguese, Criolo, African languages
Kenya ★ Nairobi	28,240,658	224,961 sq. mi. 582,650 sq. km	English, Swahili, local languages
Lesotho ★ Maseru	1,944,493	11,718 sq. mi. 30,350 sq. km	Sesotho, English, Zulu, Xhosa
Liberia ★ Monrovia	2,972,766	43,000 sq. mi. 111,370 sq. km	English, Niger-Congo languages
Libya ★ Tripoli	5,057,392	679,358 sq. mi. 1,759,540 sq. km	Arabic, Italian, English
Madagascar ★ Antananarivo	13,427,758	226,656 sq. mi. 587,040 sq. km	French, Malagasy
Malawi ★ Lilongwe	9,732,409	45,745 sq. mi. 118,480 sq. km	English, Chichewa, other regional languages
Mali ★ Bamako	9,112,950	478,764 sq. mi. 1,240,000 sq. km	French, Bambara, African languages
Mauritania ★ Nouakchott	2,192,777	397,953 sq. mi. 1,030,700 sq. km	Hasaniya Arabic, Pular, Soninke, Wolof
Mauritius ★ Port Louis	1,116,923	718 sq. mi. 1,860 sq. km	English, Creole, French, Hindi, Urdu, Hakka, Bojpoori
Morocco ★ Rabat	28,558,635	172,413 sq. mi. 446,550 sq. km	Arabic, Berber dialects, French
Mozambique ★ Maputo	17,346,280	309,494 sq. mi. 801,590 sq. km	Portuguese, many local dialects
Namibia ★ Windhoek	1,595,567	318,694 sq. mi. 825,418 sq. km	English, Afrikaans, German, local languages

Country ★ Capital	Population (1994)	Land Area (1994)	Principal Languages
Niger ★ Niamey	8,971,605	489,189 sq. mi. 1,267,000 sq. km	French, Hausa, Djerma
Nigeria ★ Abuja	98,091,097	356,668 sq. mi. 923,770 sq. km	English, Hausa, Yoruba, Ibo, Fulani
Rwanda ★ Kigali	8,373,963	10,170 sq. mi. 26,340 sq. km	Kinyarwanda, French, Kiswahili
São Tomé and Príncipe ★ São Tomé	136,780	371 sq. mi. 960 sq. km	Portuguese
Senegal ★ Dakar	8,730,508	75,749 sq. mi. 196,190 sq. km	French, Wolof, Pulaar, Diola, Mandingo
Seychelles ★ Victoria	72,113	176 sq. mi. 455 sq. km	English, French, Creole
Sierra Leone ★ Freetown	4,630,037	27,699 sq. mi. 71,740 sq. km	English, Mende, Temne, Krio
Somalia ★ Mogadishu	6,666,873	246,201 sq. mi. 637,660 sq. km	Somali, Arabic, Italian, English
South Africa ★ Pretoria	43,930,631	471,008 sq. mi. 1,219,912 sq. km	Afrikaans, English, Ndebele, Pedi, Sotho, Swati, Tsonga, Tswana, Venda, Xhosa, Zulu
Sudan ★ Khartoum	29,419,798	967,493 sq. mi. 2,505,810 sq. km	Arabic, Nubian, Ta Bedawie, various other dialects, English
Swaziland ★ Mbabane	936,369	6,703 sq. mi. 17,360 sq. km	English, siSwati
Tanzania ★ Dar es Salaam	27,985,660	364,899 sq. mi. 945,090 sq. km	Swahili, English, local languages
Togo ★ Lomé	4,255,090	21,927 sq. mi. 56,790 sq. km	French, Ewe, Mina, Dagomba, Kabye
Tunisia ★ Tūnis	8,726,562	63,170 sq. mi. 163,610 sq. km	Arabic, French
Uganda ★ Kampala	19,121,934	91,135 sq. mi. 236,040 sq. km	English, Luganda, Swahili, other Bantu and Nilotic languages
Zaire ★ Kinshasa	42,684,091	905,563 sq. mi. 2,345,410 sq. km	French, Lingala, Swahili, Kingwana, Kikongo, Tshiluba

Country ★ Capital	Population (1994)	Land Area (1994)	Principal Languages
Zambia ★ Lusaka	9,188,190	290,583 sq. mi. 752,610 sq. km	English, local languages
Zimbabwe ★ Harare	10,975,078	150,803 sq. mi. 390,580 sq. km	English, Shona, Sindebele

EAST AND SOUTHEAST ASIA

Country ★ Capital	Population (1994)	Land Area (1994)	Principal Languages
Brunei ★ Bandar Seri Begawan	284,653	2,228 sq. mi. 5,770 sq. km	Malay, English, Chinese
Burma (Myanmar) ★ Rangoon (Yangon)	44,277,014	261,969 sq. mi. 678,500 sq. km	Burmese, local languages
Cambodia ★ Phnom Penh	10,264,628	69,900 sq. mi. 181,040 sq. km	Khmer, French
China ★ Beijing	1,190,431,106	3,705,386 sq. mi. 9,596,960 sq. km	Mandarin Chinese, other Chinese dialects
Indonesia ★ Jakarta	200,409,741	741,096 sq. mi. 1,919,440 sq. km	Bahasa Indonesia, English, Dutch, Javanese and other local dialects
Japan ★ Tokyo	125,106,937	145,882 sq. mi. 377,835 sq. km	Japanese
Laos ★ Vientiane	4,701,654	91,428 sq. mi. 236,800 sq. km	Lao, French, English
Malaysia ★ Kuala Lumpur	19,283,157	127,316 sq. mi. 329,750 sq. km	Malay, English, Chinese dialects, Tamil, local dialects
Mongolia ★ Ulaanbaatar	2,429,762	604,247 sq. mi. 1,565,000 sq. km	Khalkha Mongol, Russian, Chinese, Turkic languages
North Korea ★ P'yŏngyang	23,066,573	46,540 sq. mi. 120,540 sq. km	Korean
Philippines ★ Manila	69,808,930	115,830 sq. mi. 300,000 sq. km	Pilipino, English
Singapore ★ Singapore	2,859,142	244 sq. mi. 633 sq. km	Chinese, Malay, Tamil, English
South Korea ★ Seoul	45,082,880	38,023 sq. mi. 98,480 sq. km	Korean, English
Taiwan ★ T'aipei	21,298,930	13,892 sq. mi. 35,980 sq. km	Mandarin Chinese, Taiwanese (Min), Hakka dialects

Country ★ Capital	Population (1994)	Land Area (1994)	Principal Languages
Thailand ★ Bangkok	59,510,471	198,455 sq. mi. 514,000 sq. km	Thai, English, regional dialects
Vietnam ★ Hanoi	73,103,898	127,243 sq. mi. 329,560 sq. km	Vietnamese, French, Chinese, English, Khmer, local languages

SOUTH ASIA

Country ★ Capital	Population (1994)	Land Area (1994)	Principal Languages
Bangladesh ★ Dhaka	125,149,469	55,598 sq. mi. 144,000 sq. km	Bangla, English
Bhutan ★ Thimphu	716,380	18,147 sq. mi. 47,000 sq. km	Dzongkha, Tibetan dialects, Nepalese dialects
India ★ New Delhi	919,903,056	1,269,338 sq. mi. 3,287,590 sq. km	Hindi, English, Hindustani, local languages
Maldives ★ Male	252,077	116 sq. mi. 300 sq. km	Divehi, English
Nepal ★ Kathmandu	21,041,527	54,363 sq. mi. 140,800 sq. km	Nepali, local dialects
Pakistan ★ Islamabad	128,855,965	310,401 sq. mi. 803,940 sq. km	English, Punjabi, Sindhi, Pashtu, Balochi, Urdu
Sri Lanka ★ Colombo	18,129,850	25,332 sq. mi. 65,610 sq. km	Sinhala, Tamil, English

THE PACIFIC WORLD

Country ★ Capital	Population (1994)	Land Area (1994)	Principal Languages
Australia ★ Canberra	18,077,419	2,967,893 sq. mi. 7,686,850 sq. km	English, local languages
Fiji ★ Suva	764,382	7,054 sq. mi. 18,270 sq. km	English, Fijian, Hindustani
Kiribati ★ Tarawa	77,853	277 sq. mi. 717 sq. km	English, Gilbertese
Marshall Islands ★ Majuro	54,031	70 sq. mi. 181 sq. km	English, Marshallese dialects, Japanese
Micronesia, Federated States of ★ Palikir	120,347	271 sq. mi. 702 sq. km	English, Trukese, Pohnpeian, Yapese, Kosrean
Nauru ★ Yaren District	10,019	8 sq. mi. 21 sq. km	Nauruan, English

Country ★ Capital	Population (1994)	Land Area (1994)	Principal Languages
New Zealand ★ Wellington	3,388,737	103,737 sq. mi. 268,680 sq. km	English, Maori
Palau ★ Koror	16,366	177 sq. mi. 458 sq. km	English, Palauan, Japanese, Sonsorolese, other local languages
Papua New Guinea ★ Port Moresby	4,196,806	178,259 sq. mi. 461,690 sq. km	local languages, English, pidgin English, Motu
Solomon Islands ★ Honiara	385,811	10,985 sq. mi. 28,450 sq. km	local languages, Melanesian pidgin, English
Tonga ★ Nuku'alofa	104,778	289 sq. mi. 748 sq. km	Tongan, English
Tuvalu ★ Funafuti	9,831	10 sq. mi. 26 sq. km	Tuvaluan, English
Vanuatu ★ Port-Vila	169,776	5,699 sq. mi. 14,760 sq. km	English, French, pidgin
Western Samoa ★ Apia	204,447	1,104 sq. mi. 2,860 sq. km	Samoan (Polynesian), English

THE UNITED STATES

State	Capital	Date entered Union
Alabama	Montgomery	1819
Alaska	Juneau	1959
Arizona	Phoenix	1912
Arkansas	Little Rock	1836
California	Sacramento	1850
Colorado	Denver	1876
Connecticut	Hartford	1788
Delaware	Dover	1787
Florida	Tallahassee	1845
Georgia	Atlanta	1788
Hawaii	Honolulu	1959
Idaho	Boise	1890
Illinois	Springfield	1818
Indiana	Indianapolis	1816
Iowa	Des Moines	1846
Kansas	Topeka	1861
Kentucky	Frankfort	1792
Louisiana	Baton Rouge	1812
Maine	Augusta	1820
Maryland	Annapolis	1788
Massachusetts	Boston	1788
Michigan	Lansing	1837
Minnesota	St. Paul	1858
Mississippi	Jackson	1817
Missouri	Jefferson City	1821

State	Capital	Date entered Union
Montana	Helena	1889
Nebraska	Lincoln	1867
Nevada	Carson City	1864
New Hampshire	Concord	1788
New Jersey	Trenton	1787
New Mexico	Santa Fe	1912
New York	Albany	1788
North Carolina	Raleigh	1789
North Dakota	Bismarck	1889
Ohio	Columbus	1803
Oklahoma	Oklahoma City	1907
Oregon	Salem	1859
Pennsylvania	Harrisburg	1787
Rhode Island	Providence	1790
South Carolina	Columbia	1788
South Dakota	Pierre	1889
Tennessee	Nashville	1796
Texas	Austin	1845
Utah	Salt Lake City	1896
Vermont	Montpelier	1791
Virginia	Richmond	1788
Washington	Olympia	1889
West Virginia	Charleston	1863
Wisconsin	Madison	1848
Wyoming	Cheyenne	1890

GLOSSARY

Pronunciation Key

This key will help you understand the pronunciations of unfamiliar words given in the textbook.

Key Word	Example	Key Word	Example
map	Africa (AF-ri-kuh)	book	Cooktown (KOOHCK-town)
day	Bombay (bahm-BAY)	student	Peru (puh-ROO)
date	Maine (MANE)	sun	Kentucky (ken-TUHCK-ee)
air	Canary (kuh-NAIR-ee)	deep	Armenia (ahr-MEE-nee-uh)
father	Baja (BAH-HAH)	tip	Britain (BRIT-en)
law	Bengal (ben-GAWL)	pie	Thailand (TIE-land)
wet	Genoa (JEHN-uh-wuh)	tide	Palestine (PAL-uh-stine)
boy	Detroit (di-TROIT)	top	Bosporus (BAHS-puh-ruhs)
now	Moscow (MAHS-kow)	toe	Ohio (oe-HIE-oe)
moon	Rangoon (ran-GOON)	note	Roanoke (ROE-uh-noke)

A

abdicate to resign or give up power, *329*

absolute location the exact spot on Earth where something is found, often stated in latitude and longitude, *3*

abyssal plain a large, flat area on the ocean floor, *47*

acculturation a process by which one culture changes a great deal through its meeting with another culture, *69*

acid rain a type of polluted rain, produced when acids from smokestacks combine with water vapor, that can harm lakes, forests, and human health, *92*

agribusiness farming on large modern farms that often are owned by big corporations, *160*

air pressure the measurement of the force exerted by air, *20*

alien species a species of plant or animal that has been introduced to a region by humans from a different region, *606*

alliance an agreement between countries to support one another against enemies, *265*

alluvial fan a fan-shaped deposit of sediment located at the base of a mountain, *49*

altiplano the broad highland plain between the eastern and western ridges of the Andes in Peru and Bolivia, *247*

animism a religious belief that honors the memories of ancestors and holds that waters, animals, trees, mountains, and other objects in nature have spirits, *427*

annex to formally join, as a territory, to one's own country or territory, *521*

Antarctic Circle the line of latitude, located $66\frac{1}{2}°$ south of the equator, south of which all areas have 24 hours of daylight during the December solstice and 24 hours of darkness during the June solstice, *15*

anthracite coal hard coal, *125*

apartheid the South African government policy of separation of races; the policy began to disappear in the late 1980s, *479*

aquaculture farming in seas and ponds, *500*

aqueduct a canal for carrying water, *155*

aquifer a rock layer through which groundwater flows easily, *39*

arable suitable for growing crops, *294*

archipelago a large group of islands, *540*

Arctic Circle the line of latitude, located $66\frac{1}{2}°$ north of the equator, north of which all areas have 24 hours of darkness during the December solstice and 24 hours of daylight during the June solstice, *15*

artesian well a well in which water rises toward the surface without being pumped, *605*

atlas an organized collection of maps in one book, *S4*

atmosphere the layer of gases that surrounds the earth, *16*

atoll a ring of small islands built up on a coral reef surrounding a shallow lagoon, *613*

autarky economic self-sufficiency, *330*

autonomy the right to self-government, *300*

B

badlands lands eroded by wind and water into small gullies stripped of vegetation and soil, *151*

balance of power a political stability that results when members of a group, such as countries of a region, have equal levels of strength, *265*

balance of trade the relationship between the value of a country's exports and its imports, *304*

barrier island a long, narrow, sandy island separated from the mainland by a shallow lagoon or wetland, *133*

basin a low area of land, generally surrounded by mountains, *113*

basin irrigation the type of irrigation in which farmers divide their land into shallow basins by building walls of earth to trap floodwaters, *433*

bayou a small, sluggish stream that cuts through the low and swampy land of a delta, *135*

Bedouin a migrating herder of Southwest Asia, *384*

bilingual having the ability to speak two languages, *138*

biogeography the study of the geographic distribution of plants and animals, *605*

biome a plant and animal community that covers a very large land area, *58*

biosphere the part of the earth in which all of the planet's plant and animal life exists, *16*

biotechnology the application of biology to industrial processes, *123*

birthrate the number of births per 1,000 people in a given year, *83*

bituminous coal soft coal, *125*

bog soft ground that is soaked with water, *276*
borough an administrative unit of a city, *127*
break-of-bulk center a place where goods are moved from one mode of transportation to another, *127*
Buddhism a religion, practiced primarily in Asia, whose followers seek enlightenment through meditation, *503*
buffer state a small country that separates two larger, competing countries, *234*

C

cacao a small tree on which the cocoa bean grows, *198*
caldera a large depression formed after a major eruption and collapse of a volcanic mountain, *164*
calico a fine fabric woven from cotton and printed with colorful patterns, *567*
calypso a type of music that originated in Trinidad and Tobago, *219*
canton a largely self-governing state within a country, as in Switzerland, *291*
capital a source of wealth used to make more wealth, *438*
capitalism an economic system, based on free enterprise, in which resources, industries, and businesses are owned by private individuals, *81*
cardamom a spice popular in many of the foods of Southwest Asia, *215*
cartel a group of business organizations that agree to limit supplies of a product and thereby control prices, *249*
cartography the branch of geography that studies maps and mapmaking, *8*
cash crop a crop produced primarily for direct sale in a market, rather than for other purposes such as consumption by the farmer, *204*
cassava a crop whose fleshy roots are nutritious, *443*
caste system the social system developed in ancient India which divides society into major classes, or castes, according to occupation, *578*
cease-fire an agreement to halt active hostility, *587*
center pivot irrigation the type of irrigation that uses sprinkler systems mounted on huge rotating wheels, *153*
channel a natural or artificial course for running water; to straighten a river by carving a deep lane for ships, *143*
chinook a strong, dry wind that blows from the Rocky Mountains onto the Great Plains, *152*
Christianity the religion founded on the teachings of Jesus Christ, *380*
city-state a self-governing city and its surrounding area, *286*
clan a group of people descended from a common ancestral family, *364*
client state a country that is politically, economically, or militarily dependent on a more powerful country, *331*
climate weather conditions in an area averaged over a long period of time, *18*
climax community the stable plant group that is best suited to use the resources of a particular area and forms the last stage of plant succession, *55*
coalition government a government in which several political parties join together to run the country, *304*
coke coking coal, used in blast furnaces to purify iron ore for making steel, *125*
command economy an economy in which the government determines wages, the kinds and amounts of goods produced, and the prices of goods, *81*
commercial agriculture the type of agriculture in which people grow crops for sale, usually on large farms using modern technology, *72*
commonwealth (1) a self-governing political unit, *223* (2) an association of self-governing states with similar backgrounds and united by a common loyalty, *275*
commune a large cooperative group, *509*
communism an economic and political system in which the government owns or controls almost all of the facilities and resources necessary to make products, *82*
compass rose a directional indicator that has arrows pointing north, south, east, and west, *S6*
complementary region a region particularly suited to certain activities, such as industry or mining, so that when combined with a region of different strengths, such as agriculture, the regions benefit each other, *312*
condensation the process by which water vapor changes from a gas into liquid droplets, *25*
confederation a group of states joined together for a common purpose, *291*
Confucianism a religion or code of ethics, practiced primarily in East Asia, that centers around family, social relationships, and duty, *503*
coniferous forest a middle-latitude forest whose plants remain green year-round, *58*
constitutional monarchy a country that has a reigning king or queen (a monarch), as well as a parliament that serves as the lawmaking branch of government, *273*
contiguous connecting or bordering; the 48 contiguous U.S. states border each other as a single unit, *113*
continent one of seven large landmasses on Earth, *S2*
Continental Divide the crest of the Rocky Mountains that divides the rivers of North America into those that flow east and those that flow west, *154*
continental shelf the part of the sea floor that slopes gently down from the continents; the most shallow part of the ocean, *39*
cooperative an organization owned by and run for the mutual benefit of its members, *222*
copra the dried meat of the coconut, from which coconut oil is obtained, *485*
coral reef a ridge found in warm, tropical water close to shore, made of a rocky limestone material formed by the skeletons of tiny marine animals, *56*
cork a bark stripped from the trunk of the Mediterranean oak tree, *299*
Corn Belt the area of the Midwestern United States, stretching mainly from Ohio to Iowa, in which raising corn is a major economic activity, *143*
cosmopolitan having many foreign influences, *290*
Cossack one of a group of people who controlled the northern Eurasian steppe region beginning in the 1500s, *353*
cottage industry an industry in which workers make small consumer items in homes and small workshops, *577*
coup an overthrow of an existing government by another political or military group, *248*
crop rotation a method of preserving soil fertility in which different types of crops are grown on the same piece of land over time, *87*
culture the features of a society's way of life, passed down from generation to generation through teaching, example, and imitation, *67*
culture hearth a region where many important new ideas and developments originate, *72*
culture region an area with many shared culture traits, *67*
culture trait an activity or behavior repeatedly practiced by people in a given culture, *67*
czar the title for the rulers of the Russian Empire (pre-Soviet era); literally, "emperor," *329*

D

Dairy Belt the area of the Midwestern United States, located north of the Corn Belt, in which dairy farming is a major economic activity, *144*
death rate the number of deaths per 1,000 people in a given year, *83*
deciduous forest a middle-latitude forest whose plants lose their leaves during a certain season each year, *58*
defoliant a chemical that makes plants' leaves drop off, usually killing the plants, *535*
deforestation the clearing or destruction of forests, *88*
degree a unit of measurement of temperature or of distance on the earth's surface, *S3*
delta the landform at the mouth of a river created by deposits of sediment, *49*
demilitarized zone (DMZ) a buffer zone into which military forces may not enter, *528*
democratic government a government in which everyone in the society has a voice, *75*
demography the study that emphasizes statistics to look at human population distribution, population density, and trends in population, *82*
deport to send out of a country, *175*
desalinization a process by which saltwater is changed into fresh water, *89*
desertification a process by which desertlike conditions are created by a loss of plant cover and soil due to human activity, *87*
developed country a wealthy country that has a good educational system, widely available health care, and many manufacturing and service industries, *80*
developing country a poorer country in which many people live by subsistence farming and in which there are few manufacturing and service industries, *81*
dialect a regional variety of a major language, *300*
diffusion a process by which an innovation or other culture trait spreads from one culture region to another, *68*
dike a wall built to keep water out of a low-lying area, *290*
directional indicator an element of a map that shows which directions are north, south, east, and west, *S6*
diversify to produce a variety of crops or goods rather than just one, *135*
doldrums calm areas centered along the equator, *21*
domestication the growing of a plant or taming of an animal by a people for their own use, *72*
dominion a territory or sphere of influence, *175*
double cropping an agricultural practice, enabled by a long growing season, in which two crops are grown in one year on the same land, *511*
dredge to dig or clean out a river to make it deeper, *143*
drip irrigation a process by which pipes placed near plant roots drip only as much water as the plants need, *382*
drought a long, dry period with little or no rain, *4*
dryland agriculture agriculture that depends on rainfall rather than irrigation to water crops, *361*
dynasty a government ruled by a family whose power lasts through several generations, *507*

E

earthquake the shock waves, or vibrations, caused by movement along a fault, *45*
earth system the interactions of elements on and around the earth, *16*
economic association an organization formed to break down trade barriers among member nations, *267*
ecosystem an interdependent community of plants and animals combined with its physical environment, *54*
edge city a suburb with large employment and commercial centers, *116*
ejido a communal farm, in Mexico, on which farmers work land together or work individual plots, *205*
elevation height on the earth's surface above or below sea level, *26*
El Niño a weather pattern, involving a warming of part of the Pacific Ocean, that occurs every five to ten years and causes a decrease in fish and an increase in flooding in some areas, *248*
embargo a legal restriction against the movement of freight, *480*
emigrant someone who moves out of a country, *84*
endemic species a plant or animal that originates only in a particular region, *605*
entrepot an intermediate port used to store goods before they are shipped to their final destinations, *544*
equator the imaginary line of latitude that lies halfway between the North and South poles and circles the globe, *S2*
equinox one of two times each year when the earth's polar axis is not pointed toward or away from the sun but is at a 90° angle to the sun, *15*
erg a large area covered by sand, *431*
erosion the wearing away of land by water, wind, waves, ice, and other agents, *42*
escarpment a steep slope capped by a plateau, *421*
estuary a semi-enclosed coastal body of water where seawater and fresh water mix, *37*
ethnic group a population that shares a common cultural background, *68*
ethnocentrism a form of cultural bias in which people use their own culture and experiences as a point of reference from which to make statements about other cultures, *552*
Eurasia the world's largest landmass, made up of the continents of Europe and Asia, *258*
European Union (EU) an economic association, made up of several European countries, whose ultimate goal is for Europe to be a unified economic region, *267*
evaporation the process by which water is changed from a liquid to a gas, *25*
evapotranspiration a measure of the total loss of water from the land to the atmosphere through evaporation and transpiration, *37*
exclave a part of a country that is separated from the rest of the country and surrounded by foreign territory, *473*
Exclusive Economic Zone (EEZ) an area off a country's coast, whose resources are claimed and controlled by that country, *615*
exile a person who has been forced to leave his or her country, *480*
exotic river a river that flows from a humid environment across a desert into a sea, *378*
export economy an economy in which most manufactured products are made for export, *527*
extinct no longer in existence, *60*

F

Fall Line the line in the United States along which the Coastal Plain mees the Piedmont, marked by waterfalls and rapids indicating the elevation change, *111*
fallow unused, as cropland, *400*
famine an extreme shortage of food, *266*

fault a feature of the earth's crust created by the breaking and movement of rock layers below the surface, *45*
favela a slum area near a Brazilian city, *237*
fjord a narrow, deep inlet of the sea between high, rocky cliffs created by glaciers, *277*
floodplain a landform of level ground built by sediment deposited by a river or stream, *49*
foehn a warm, dry wind in Europe that blows down the slopes of the Alps, *294*
fold a feature of the earth's crust created by the bending of rock layers, *45*
food chain a series of organisms in which energy is passed along through living things, *53*
foreign aid loans or gifts of money from one country to another, *81*
fossil fuel a fuel that may have been formed from the remains of prehistoric plants and animals, *93*
fossil water groundwater that is not replaced by rainfall, *379*
free enterprise an economic principle under which prices are determined mostly through competition and people are free to choose what and when to sell and buy, *81*
free port a port in which unloaded goods are subject to few or no taxes, *438*
front a zone formed by the meeting of two air masses with very different temperatures and amounts of moisture, *21*
fundamentalism a movement that stresses the strict following of basic traditional principles, *75*
furrow a narrow trench made by a plow, *379*

G

geothermal energy a renewable energy source generated by the escape of underground heat to the surface, *95*
geyser a hot spring that shoots hot water and steam into the air, *280*
ghetto a part of a city where one or more minority groups are concentrated because of economic pressure and social discrimination, *116*
ghost town a town that once flourished but is deserted today, usually because of the depletion of a nearby resource such as gold or silver, *154*
glacier a thick mass of ice that moves slowly across Earth's surface, *43*
glasnost the policy of open discussion promoted in the Soviet Union shortly before its breakup, *331*
globe a scale model of Earth, especially useful for looking at the entire earth or at large areas of its surface, *S2*
granite a speckled, hard, crystalline rock formed deep in the earth's crust, *122*
great-circle route the shortest route between any two points on the earth, *S5*
greenhouse effect (1) the warming of the earth created by the trapping of the sun's energy in Earth's atmosphere, *19* (2) the warming of the earth caused by the buildup of carbon dioxide in the lower atmosphere, possibly as a result of human industrial activity, *93*
Green Revolution the name given to the widespread introduction of new high-yield grains, *566*
grid the pattern of lines that circle the earth in east-west and north-south directions, *S2*
gross domestic product (GDP) the value of goods and services produced within a country in a year, not including income earned outside that country, *80*
gross national product (GNP) the value of goods and services produced by a country in a year, both inside and outside the country, *80*
groundwater fresh water found beneath the earth's surface in tiny spaces between soil and rock grains, *39*
growth pole a metropolitan area given official help to strengthen its economic development, *238*
guerrilla a person who fights as a member of an armed band and takes part in irregular warfare, such as harassment and raids, *229*
Gulf Stream an Atlantic Ocean current that moves warm water north along the U.S. East Coast, *114*
gum arabic a product that comes from the sap of acacia trees and is used in some medicines and candies, *455*

H

habitation fog a fog caused by fumes and smoke from cities, *336*
hacienda a land system, introduced by the Spanish to Middle and South America, in which large family-owned estates (haciendas) are worked by peasants in return for small plots on which to grow their own crops, *198*
headwaters the first and smallest streams to form from the runoff of precipitation flowing down the slopes of hills and mountains, *37*
heavy industry industry that focuses on manufacturing based on metals such as steel, *342*
hemisphere half of the earth, such as the Northern or Southern hemispheres or the Eastern and Western hemispheres, *S3*
Hinduism the major religion of India and Nepal, *564*
hinterland a region beyond the center of an area, *176*
Holocaust the name given to the persecution and killing of millions of Jews by Nazi Germany during World War II, *390*
homogeneous of the same kind, *355*
hot spot an area where molten material from the earth's interior rises through the crust, *169*
human-environment interaction one of the Five Themes of Geography; deals with how people adapt to and change their environment, *5*
humidity the amount of water vapor in the air, *25*
humus an important ingredient of soil, made of decayed plant or animal matter, *55*
hurricane a tropical storm that brings violent winds, heavy rain, and high seas; also called a typhoon, *25*
hydroelectricity electricity produced by harnessing the power of running water, *35*
hydrologic cycle the circulation of water among parts of the hydrosphere, *36*
hydrosphere the physical system made up of all of Earth's water, *16*

I

icebreaker a ship that can break up the ice of frozen waterways, *264*
immigrant someone who moves into a country, *84*
imperialism the policy of gaining control over territory outside a nation, *265*
indigenous native to a certain area, *188*
indigo a plant used to make a blue dye, *196*
industrialization the process in which manufacturing based on machine power becomes widespread in an area, *35*
inflation the rise in prices caused by a decrease in the value of a nation's currency, *212*
infrastructure facilities and structures, such as roads,

bridges, and power plants, that are needed to build industries and move goods in and out of a country, *80*
innovation a new idea that is accepted into a culture, *68*
intensive agriculture agriculture that uses much human labor, *532*
irrigation the watering of land through pipes, ditches, or canals, *35*
Islam the religion founded by Muhammad, *380*
island a landmass smaller than a continent and completely surrounded by water, *S2*
isthmus a narrow neck of land that acts as a bridge to connect two larger bodies of land, *188*

J

Judaism the religion of the Jews, which appeared in Southwest Asia between 3500 B.C. and 600 B.C., *380*
jute a plant fiber used to produce coarse fabrics, such as burlap and twine, *565*

K

kampong a traditional village in island Southeast Asia; a slum area around Jakarta, Indonesia, *546*
katabatic wind a rapid flow of dense cold air down the slopes of high mountains or ice caps, *620*
kibbutz an Israeli collective farm, *390*

L

labor intensive requiring a large human work force, *198*
landform a shape on the earth's surface, *46*
landlocked completely surrounded by land, without access to an ocean, *242*
land reclamation a method used after mining to prevent erosion and return the land to its original contours, *153*
land reform the changing of a land ownership system; usually the breaking up of large landholdings to allow more people to own their own land, *200*
Landsat a satellite that collects information about Earth's surface and the condition of Earth's environment, *23*
latifundia a land system, introduced by Europeans to South America, in which tenant farmers work part of an estate in exchange for food; the estates themselves, *250*
latitude a measure of distance north or south of the equator, *S2*
lava melted liquid rock, or magma, from within the earth that spills out on the earth's surface, *45*
leaching the process by which nutrients are washed down out of the topsoil by rainfall, *56*
legacy something received from a previous era, *311*
legend an element of a map that explains what the symbols on the map represent; also called a key, *S6*
levee a dirt ridge along a river that hinders flooding, *135*
light industry industry that focuses on producing lightweight goods, such as clothing, *342*
lingua franca a common language spoken and understood by the majority of people in an area, although other languages may be spoken at home, *445*
literacy the ability to read or write, *80*
lithosphere the solid surface of the planet that forms the continents and the ocean floor, *16*
llanos the plains of southern Colombia and Venezuela, *193*
location one of the Five Themes of Geography; deals with where places and regions are located; defined in terms of absolute and relative location, *3*
loch a long, deep lake carved by a glacier, *273*
lock a part of a waterway enclosed by gates, used to raise or lower boats from one water level to another, *135*
loess dust-sized soil particles deposited by wind, *287*
longitude a measure of distance east or west of the prime meridian, *S2*

M

map a flat diagram of all or part of the earth's surface, *S4*
map projection one of a variety of ways of presenting a round earth on a flat map, *S4*
maquiladora a foreign company's factory in Mexico, usually near the U.S. border, in which products are assembled for export, *208*
market economy an economy in which consumers determine what is to be bought or sold by buying or not buying certain goods and services, *81*
marsupial a mammal, such as a kangaroo, that carries its young in a pouch, *605*
martial law military rule, *515*
Megalopolis the giant urban area that stretches along the U.S. eastern seaboard, *120*
mental map a map that we create and see in our minds to help make sense of the world, *S9*
merengue the national music and dance of the Dominican Republic, *219*
meridian a line of longitude, *S3*
Meseta the rocky, treeless plateau that covers much of Spain, *297*
mestizo a person, usually from Middle or South America, with both Native American and European ancestors, *196*
meteorology the field of geography that specializes in weather and weather forecasting, *8*
microstate a very small country, *258*
mid-ocean ridge a chain of mountains on the ocean floor formed when molten rock rises from below the seafloor to fill openings created by plate movement, *47*
millet a grain crop that is resistant to drought, *443*
minaret the tower of a mosque, *395*
minifundio a small farm in Middle or South America formed when a large estate is broken up, *250*
mistral a strong, cool wind in Europe that blows from the Alps toward the Mediterranean coast, following the valley of the Rhone River, *284*
mixed forest an area where deciduous and coniferous forests blend, *59*
model a plan or pattern used to represent something, such as a city, *182*
monopoly a business that has no competitors and thus can control an industry's supply of goods and prices, *212*
monotheism the belief in one god, *381*
monsoon the wind that blows from the same direction for months at a definite season of the year, *27*
moraine a ridge of rocks, gravel, and sand deposited along the margins of a glacier or ice sheet, *122*
mosque an Islamic place of worship, *360*
movement one of the Five Themes of Geography; deals with the movement of people, goods, and ideas, *6*
mulatto a person, usually from Middle or South America, with both African and European ancestors, *196*
multicultural region an area in which people of many cultures have settled, *161*
multilingual having the ability to speak three or more languages, *267*
multinational company a business with activities in many countries, *81*

muskeg a forested swamp that is frozen except during the summer, *179*
Muslim a person who follows Islam, *380*

N

nationalism feelings of pride in one's country, *74*
nationalize to take over by the government, such as to assume government control of an industry, *274*
natural boundary a physical feature, such as a mountain range or a river, that determines a political border, *262*
navigable able to be sailed by large ships, such as a wide and deep river, *263*
neutral choosing not to take sides in international conflicts, *279*
newly industrialized country (NIC) a nation that is undergoing or has recently undergone rapid industrialization and economic growth, *501*
newsprint an inexpensive paper used mainly for printing newspapers, *174*
nodal region a central area distinguished by the movement or activities that take place in and around it, *101*
nonrenewable resource a resource, such as oil or gold, that cannot be replaced by natural processes or is replaced extremely slowly, *86*

O

oasis a site in the desert where water can be found in springs or wells or under dry riverbeds, *359*
ocean the largest part of the earth's water surface, often divided into four smaller oceans, *S2*
Oceania the ocean region containing the Pacific islands, *612*
orographic effect an effect of mountains on climate that produces moisture on the windward side and dryness on the leeward side, *26*
ozone a gas formed from an interaction between oxygen and sunlight; the ozone layer is a region in the earth's upper atmosphere that protects life beneath it by filtering out dangerous ultraviolet solar radiation, *92*

P

paddy (padi) a water-covered field usually used for rice cultivation, *500*
pampas the wide grassy plains of Argentina and neighboring countries in South America, *193*
panhandle a narrow arm of land attached to a larger region, *165*
parallel a line of latitude, *S2*
páramo the elevation zone of the Andes just above the tree line, *228*
parliament a lawmaking body, *175*
partition to divide an area, *585*
pastoralism an agricultural way of life that involves raising livestock and usually includes some farming, *443*
peat a substance made of decayed vegetable matter, usually mosses, that can be used for fuel, *276*
peninsula a landform bordered by water on three sides, *122*
perennial irrigation the type of irrigation in which dams store floodwater and release it year-round to irrigate crops, *433*
perestroika the policy of economic reform and restructuring promoted in the U.S.S.R. just before its breakup, *331*
perimeter outer boundary, *405*
periodic market an open-air trading market held regularly at a crossroads or in a town, *466*
permafrost water below a tundra surface that remains frozen throughout the year, *32*
petrochemical a chemical produced from petroleum; some chemical fertilizers, pesticides, food additives, explosives, and medicines are petrochemicals, *93*
photochemical smog a type of air pollution produced by the interaction of exhaust gases with sunlight, *92*
photosynthesis the process by which a plant converts sunlight into chemical energy, *53*
pictogram a written language character that is a simple picture of the object or idea it represents, *502*
piedmont an area of plains and low hills at or near the foot of a mountain region; in the United States, the Piedmont is a rolling plateau region inland from the Coastal Plain, east and south of the Appalachian Mountains, *111*
pilgrimage a journey, often for religious purposes, *404*
place one of the Five Themes of Geography; deals with the physical and human characteristics of a place, *4*
plain a nearly flat area of land, *46*
planet a spherical object that orbits around a star, *11*
plantain a type of banana, *223*
plantation a large farm that concentrates on one major crop, such as cotton or coffee, *115*
plant community a group of plants that live and grow together, *53*
plant succession the process by which one group of plants replaces another in a community, *55*
plateau an elevated flatland, *46*
plate tectonics theory that the earth's crust is divided into several rigid, slow-moving plates *46*
plaza a public square in the center of a town that serves as a marketplace and gathering place, *298*
polar desert a high-latitude region that receives very little precipitation, such as Antarctica, *620*
polar region a cold area near the North or South pole, *14*
polar wind the type of wind that dominates the high latitudes and brings cold conditions into the middle latitudes, usually blowing from the east, *21*
polder a lowland area that has been drained of water, *290*
portage a low land area across which boats and their cargoes can be carried between bodies of water, *175*
potash a mineral that is an important raw material for manufacturing fertilizers, *178*
prairie a region of tall grassland and fertile soil, *59*
precipitation condensed droplets of water vapor that fall as rain, snow, sleet, or hail, *25*
prevailing wind a wind that usually blows from the same direction year-round, *21*
primary industry an economic activity, such as agriculture, forestry, or mining, that makes direct use of natural resources or raw materials, *79*
primate city a city that ranks first in a nation in terms of population and economy, *284*
prime meridian the imaginary line of longitude that runs through Greenwich, England, from the North Pole to the South Pole, *S3*
privatization the selling of government-owned businesses or lands to private owners, *212*
puppet government a government that is controlled by outside forces, *509*

Q

quaternary industry an economic activity involving the movement and processing of information, 79

quota a limit on the amount of a particular good that can be imported, *76*
Qur'an (Koran) the book of Islam's sacred writings, *381*

R

rain-shadow effect the forming of a desert on the leeward side of mountains due to a lack of moisture, *27*
reforestation a process by which forests are renewed through the planting of seeds or young trees, *87*
refugee a person who flees his or her own country to another, often for economic or political reasons, *223*
reg a gravel-covered plain from which wind has blown the sand and dust away, *432*
reggae a type of music that originated in Jamaica, *219*
region one of the Five Themes of Geography, *7*; an area with more common characteristics than differences, *100*
regionalism the political and emotional support for one's region before support for one's country, *180*
regional specialization a plan in which each country of a region specializes in products it makes or grows best, *311*
relative location the position of a place in relation to other places, *3*
relief the difference in elevation between the top and bottom of a landform, *49*
remote sensor a device attached to a satellite that collects information about Earth, *22*
Renaissance the era from the 1300s to the 1600s, during which a renewed interest in learning spread throughout Europe; literally, "rebirth," *301*
renewable resource a resource, such as solar energy, soil, water, or forests, that can be replaced by the earth's natural processes, *86*
reunification the process of reuniting, or rejoining, into one unit, *266*
revolution the orbit of an object in space; the orbit of the earth around the sun, which takes one Earth year, *13*
rift a long, deep valley usually with mountainous slopes on both sides, *421*
rock weathering the process that breaks up rocks and causes them to decay, *43*
rotation the act of turning around an axis; one complete spin of Earth on its axis, which takes 24 hours, *12*

S

Sahel a region of bush savannas and grasslands along the southern edge of Africa's Sahara, *422*
sanction an economic or political measure, such as an embargo, used by one or more nations to force another nation to cease an illegal or immoral act, *480*
sanctuary an area in which endangered species are protected, *572*
sand dune a hill of wind-deposited sand, *44*
satellite a body that orbits a larger body, *11*
savanna a tropical grassland with scattered trees and shrubs, *59*
scale an element of a map that indicates distances between points on the map, *S6*
seaboard a land area near the ocean, *115*
secondary industry an economic activity that takes goods made in primary industries and processes them into products that are useful to consumers, *79*
second-growth forest a forest consisting of trees covering an area after the original forest has been removed, *123*
secularism the indifference to religion, *381*
sediment small particles of mud, sand, or gravel created by rock weathering, *43*
selvas the dense rain forest vegetation of South America's Amazon basin, *245*
serf a peasant bound to the land, *340*
shantytown a poor settlement, often near a city, of small, makeshift shelters, *199*
shelterbelt a row of trees along a field that blocks the wind and protects the soil, *152*
shifting agriculture a type of agriculture in which farmers move to new areas every few years, *532*
Shinto a Japanese religion whose followers believe that gods inhabit natural objects and that these gods should be worshiped in community rituals, *504*
Sikhism a religion formed from the blending of Muslim and Hindu teachings, *565*
silt sediment coarser than clay but finer than sand, *432*
sisal a strong fiber used to make rope and twine, *458*
site the place where something is located, *3*
situation the position of a place in relation to other places, *3*
slash-and-burn farming the type of agriculture in which forests are cut and burned to clear land for planting, *231*
smelter a factory that processes mineral ores, *342*
socialism an economic system in which the government owns and controls the means of producing goods, *278*
soil exhaustion the loss in soil nutrients that results from always planting the same crop in a particular area, *236*
soil horizon a distinct layer of soil formed over a long period of time, *55*
soil salinization salt buildup in the soil often caused by evaporation of irrigation water, *87*
solar system the sun and the nine planets and all other objects that revolve around the sun, *11*
solstice one of two times each year when the earth's polar axis points at its greatest angle toward or away from the sun, *14*
sorghum a grain crop that is resistant to drought, *443*
sovereignty authority to rule, *550*
soviet the local governing council of each republic of the Soviet Union during the Soviet era, *330*
spatial interaction the movement of people, goods, and ideas, *6*
Special Economic Zone (SEZ) an area designed to attract foreign companies and investment, *512*
staple a main food crop in an area, *443*
steppe short-grass vegetation; a steppe climate is often found between deserts and more humid climates, *30*
stereotype a generalization about a group of people, *552*
storm surge a rise in tide level that occurs as a storm approaches the coast, washing huge waves of seawater over the land, *591*
strip mining the type of mining in which soil and rock are stripped away by large machines to get at coal or other minerals buried just beneath the surface, *153*
subcontinent a large landmass that is smaller than a continent, *558*
subduction the process in which a tectonic plate collides with and dives under another tectonic plate, *47*
subregion a region that is part of a larger region, *100*
subsidy financial support given by a government to an individual or industry within a country, *521*
subsistence agriculture the type of agriculture in which people grow food on small farms mostly for their own use, *72*

sultanate a country that is ruled by a Muslim monarch (sultan), *406*
superpower a huge, powerful nation, *331*
suq a marketplace in parts of Southwest Asia and North Africa, *395*
sustained yield use a method of forest preservation in which people can take products from a forest without destroying the trees, *88*
Swahili a language used in East Africa that developed as a trade language, *454*

T

taiga the huge forest region, located in the Northern Hemisphere just south of the tundra, of tall, evergreen needle-leaf trees, *337*
tannin a substance, obtained from the quebracho tree, used in preparing leather, *240*
Taoism a religion practiced mainly in East Asia whose main idea is to live a simple life close to nature, *503*
tariff a tax placed on an import or export, *76*
telecommunications electronically transmitted communication, *80*
temperature the measurement of heat, *19*
tenant farmer a farmer who rents a plot of land on which to grow crops, *135*
terrorism the use of violence as a means of political force, *248*
tertiary industry an economic activity that provides services to primary and secondary industries and to consumers; also called a service industry, *79*
textile any of a number of products made from cloth, *121*
theocracy a country governed by religious law, *403*
tierra caliente the elevation zone of the Andes that is closest to sea level, *227*
tierra fría the elevation zone of the Andes just below the tree line, *228*
tierra helada the highest elevation zone, permanently covered with snow, of the Andes, *228*
tierra templada the elevation zone of the Andes, just above the *tierra caliente,* which has mild climates, *228*
topographic relating to surface features; a topographic map shows a region's surface features, *8*
toponym a place-name, *156*
tornado a small, twisting, highly destructive storm that usually forms along fronts in the middle latitudes, *26*
totalitarian government a government in which only one or a few people decide what is best for everyone in the society, *75*
township and range system the system, started by the U.S. government in the late 1780s, by which the land northwest of the Ohio River was divided into sections and offered to people who agreed to farm them, *143*
trade deficit the gap that occurs when the value of a nation's imports exceeds that of its exports, *118*
trade surplus the economic condition in which a country exports more than it imports, *522*
trade wind an east-to-west wind that blows from the subtropical high-pressure zone toward the equatorial low-pressure zone, *21*
transition zone an area that contains characteristics from more than one region as a result of these regions overlapping with each other, *101*
transpiration the process by which plants give off water vapor through their leaves, *37*
tree line the elevation, on mountain slopes, above which trees cannot grow, *154*
trench a long, deep valley on the ocean floor caused by subduction, *47*
tributary a stream or river that flows into a larger stream or river, *37*
tropical cyclone a hurricane-like storm that usually forms in low latitudes, *564*
Tropic of Cancer the line of latitude, located $23\frac{1}{2}°$ north of the equator, struck by the sun's most direct rays of energy during the June solstice, *15*
Tropic of Capricorn the line of latitude, located $23\frac{1}{2}°$ south of the equator, struck by the sun's most direct rays of energy during the December solstice, *15*
tropics warm areas in the low latitudes, *14*
truck farm a farm whose products are trucked to nearby city markets, *125*
trust territory a region that is placed under the control of another nation until it can govern itself, *472*
tsunami a large, potentially destructive sea wave caused by tectonic activity, *520*
typhoon a tropical storm that brings violent winds, heavy rain, and high seas; also called a hurricane, *25*

U

uniform region an area distinguished by one or more common characteristics, *101*
uninhabitable unable to support human life and settlements, *279*
urban agglomeration a densely inhabited and contiguous region surrounding a central city, *523*
urbanization a growth in the proportion of people living in towns and cities, *72*

V

veld a term used for a vegetation and landform region of South Africa, *477*
volcano the opening in the earth's crust through which lava flows; the surface feature created by the lava, *45*
voodoo a religion, common in Haiti, whose followers believe that spirits of good and evil play an important part in daily life, *223*

W

water table the top of an underground zone that is saturated with groundwater, *39*
watershed an area of land drained by a river and its tributaries, *37*
weather the condition of the atmosphere at a given place and time, *18*
westerly a west-to-east wind that dominates the middle latitudes, blowing from the subtropical high-pressure zone to the subpolar low-pressure zone, *21*
wetlands productive land areas that are flooded for at least part of the year, *38*
Wheat Belt the area of the United States, located in the Great Plains, in which wheat farming is a major economic activity, *152*
work ethic a belief that work is good in and of itself, *522*

Z

Zionism a movement, started in Europe in the 1800s, that believed a Jewish state should be created in Palestine and Jews from around the world should move there, *390*
zonal running in east-west bands of latitude, or zones; often used to describe climate patterns, *441*

INDEX

(def.) indicates definition g indicates graphic m indicates map p indicates photograph

A

B

C

D

E

F

G

H

I

J

K

L

M

N

O

P

T

U

V

W

X

Y

Z

ACKNOWLEDGMENTS

For permission to reprint copyrighted material, grateful acknowledgment is made to the following sources:

Samuel Allen: "Souffles" by Birago Diop, translated by Samuel Allen from *Poems from Africa,* selected by Samuel Allen. Copyright © 1973 by Samuel Allen. **Américas Magazine:** From "Stepping Lively" by Robert Fried from *Américas,* vol. 38, no. 1, January-February 1986. Copyright © 1986 by Américas. *Américas* is a bimonthly magazine published by the General Secretariat of the Organization of American States in Spanish and English. **Ruskin Bond:** From "The Tenacity of Mountain Water" by Ruskin Bond from *The Christian Science Monitor,* vol. 84, no. 132, June 3, 1992. Copyright © 1992 by Ruskin Bond. **The Boston Globe Newspaper Company:** From "Huck Finn Meets the Solar Car" by Ellen Goodman from *The Washington Post,* June 6, 1989. Copyright © 1989 by The Boston Globe Newspaper Company. **The Christian Science Publishing Society:** From "World Population" by George D. Moffett III from *The Christian Science Monitor,* vol. 84, no. 157, July 8, 1992. Copyright © 1992 by The Christian Science Publishing Society. All rights reserved. From "Costs of a One-Way Culture Flow" by Howard LaFranchi from *The Christian Science Monitor,* vol. 85, no. 42, January 27, 1993. Copyright © 1993 by The Christian Science Publishing Society. All rights reserved. **Doubleday, a division of Bantam Doubleday Dell Publishing Group, Inc.:** From *When Heaven and Earth Changed Places* by Le Ly Hayslip. Copyright © 1989 by Le Ly Hayslip and Charles Jay Wurts. **Dutton Signet, a division of Penguin Books USA Inc.:** From *Lost City of the Incas* by Hiram Bingham. Copyright 1948 by Hiram Bingham; copyright renewed © 1976 by Alfred Bingham. **Farrar, Straus and Giroux, Inc.:** From "Nemesio Antúnez" from *Passions and Impressions* by Pablo Neruda, translated by Margaret Sayers Peden, edited by Matilde Neruda and Miguel Otero Silva. Translation copyright © 1980, 1981, 1983 by Farrar, Straus and Giroux, Inc. From "A Journey Along the Oka" from *Stories and Prose Poems* by Alexander Solzhenitsyn, translated by Michael Glenny. Translation copyright © 1970, 1971 by Michael Glenny. **Hakluyt Society:** From *Ying-Yai Sheng-Lan: 'The Overall Survey of the Ocean's Shores'* by Ma Huan, translated from the Chinese text and edited by Feng Ch'eng-Chün, with introduction, notes and appendices by J.V.G. Mills. Copyright © 1970 by Cambridge University Press. **Harcourt Brace & Company:** From *The Gentle Tasaday: A Stone Age People in the Philippine Rain Forest* by John Nance. Copyright © 1975 by John Nance. **HarperCollins Publishers, Inc.:** From "Cedars of Lebanon" from *Strange Lands and Friendly People* by William O. Douglas. Copyright 1951 by William O. Douglas. From *John Gunther's Inside Australia,* compiled and edited by William H. Forbis. Copyright © 1972 by Jane Perry Gunther. From *My Africa* by Mbonu Ojike. Copyright 1946 by The John Day Company. From *Here Is New York* by E. B. White. Copyright 1949 by The Curtis Publishing Company. **William Heinemann, a division of Reed Publishing (NZ) Ltd.:** "Taku Waiata" by Wiremu Kingi Kerekere from *Into the World of Light: An Anthology of Maori Writing,* edited by Witi Ihimaera and D. S. Long. Copyright © 1982 by Witi Ihimaera and D. S. Long. **Houghton Mifflin Co.:** From "The Niger Described" from *The Strong Brown God* by Sanche de Gramont. Copyright © 1975 by Sanche de Gramont. All rights reserved. From "The Expreso del Sol to Bogotá" from *The Old Patagonian Express: By Train through the Americas* by Paul Theroux. Copyright © 1979 by Cape Cod Scriveners Company. **IMG–Julian Bach Literary Agency, Inc.:** From *Spain* by Jan Morris. Copyright © 1979 by Jan Morris. **Kegan Paul International:** From "Scattered Voices" by Khalil al-Fuzay, translated by Abdaf Soueif and Thomas G. Ezzy from *The Literature of Modern Arabia: An Anthology,* edited by Salma Khadra Jayyusi. Copyright © 1988 by King Saud University. **Liveright Publishing Corporation:** From *The Travels of Marco Polo (The Venetian),* revised and edited by Manuel Komroff. Copyright 1926 by Boni & Liveright, Inc.; copyright renewed 1953 by Manuel Komroff; copyright 1933 by Horace Liveright, Inc. **William Morrow & Company, Inc.:** From *The Walk West* by Peter and Barbara Jenkins. Copyright © 1981 by Peter and Barbara Jenkins. **National Council of Educational Research and Training, New Delhi:** From *Everest: Where the Snow Melts,* edited by Prabhakar Dwivedi. Copyright © 1987 by the National Council of Educational Research and Training. **National Geographic Society:** From "Egypt's Desert of Promise" by Farouk El-Baz from *National Geographic,* vol. 161, no. 2, February 1982. Copyright © 1982 by National Geographic Society. From "Monsoon's: Life Breath of Half the World" by Priit J. Vesilind from *National Geographic,* December 1984. Copyright © 1984 by National Geographic Society. From "Montreal: Spirited Heart of French Canada" from *National Geographic,* March 1991. Copyright © 1991 by National Geographic Society. From "India's Wildlife Dilemma" by Geoffrey G. Ward from *National Geographic,* May 1992. Copyright © 1992 by National Geographic Society. **National Wildlife Federation:** From "Kuna Indians: Building a Bright Future" by Norman Myers from *International Wildlife,* vol. 17, no. 4, July/August 1987. Copyright © 1987 by the National Wildlife Federation. From "Healing the Ravaged Land" by Maryanne Vollers from *International Wildlife,* vol. 18, no. 1, January/February 1988. Copyright © 1988 by the National Wildlife Federation. **The New York Review of Books:** From "The Pearl of Siberia" by Peter Matthiessen from *The New York Review of Books,* February 14, 1991. Copyright © 1991 by The New York Review of Books. **The New Yorker Magazine, Inc.:** Quote by Jamaica Kincaid from "The Talk of the Town" from *The New Yorker,* October 17, 1977. Copyright © 1977 by The New Yorker Magazine, Inc. **W. W. Norton & Company, Inc.:** From *Nature's Metropolis: Chicago and the Great West* by William Cronon. Copyright © 1991 by William Cronon. **The Office of Tibet:** From a speech, "Statement of Shared Vision" by the Dalai Lama given at Oxford University, 1988. **Oil & Gas Journal:** Data from "Worldwide Look at Reserves and Production" from *Oil & Gas Journal,* December 1992. Copyright © 1992 by Oil & Gas Journal. **Makoto Ōoka, c/o Graphic-sha Publishing Co., Ltd.:** From "The Japanese and Mt. Fuji" by Makoto Ōoka from *Mt. Fuji,* photographs by Yukio Ohyama. **The Putnam Publishing Group:** From "The Start of a New Life" from *My Life* by Golda Meir. Copyright © 1975 by Golda Meir. **Random House UK Limited:** From Chapter 2 from *The Flame Trees of Thika* by Elspeth Huxley. Copyright © 1959 by Elspeth Huxley. Published by Chattos & Windus. **Wendy Rose:** From "Loo-Wit" from *Bone Dance: New and Selected Poems 1967–1992* by Wendy Rose (University of Arizona Press, 1994). Copyright © 1985 by Wendy Rose. First appeared in *The Halfbreed Chronicles & Other Poems* by Wendy Rose (West End Press 1985/1993). **Scholastic Inc.:** From "A Teen Summit" from *Scholastic Update,* vol. 124, no. 9, January 24, 1992. Copyright © 1992 by Scholastic Inc. **Science News, the weekly newsmagazine of science:** From "Disarming Farming's Chemical Warriors" by Christopher Vaughan from *Science News,* August 20, 1988. Copyright © 1988 by Science Service, Inc. **Scribner's, an imprint of Simon & Schuster, Inc.:** From Introduction by Lewis Gannett from *Cry the Beloved Country* by Alan Paton. Copyright 1948 and renewed © 1976 by Alan Paton. **Patroclos Stavrou, Literary Agent and Representative of Mrs. Helen N. Kazantzakis:** From *England: A Travel Journal* by Nikos Kazantzakis. English translation copyright © 1965 by Simon and Schuster, Inc. **United Nations:** Statistics from *World Urbanization Prospects 1990.* (United Nations publication, Sales No. E.91.XIII.11). Copyright © 1991 by United Nations. Statistics from *The Sex and Age Distributions of World Populations: The 1994 Revision* (United Nations publication, Sales No. E.95.XIII.2). Copyright © 1995 by United Nations. **University of Chicago Press:** From *The English Teacher* by R. K. Narayan. Copyright 1945 by Rasipuram Krishnaswamier Narayanaswami. **Viking Penguin, a division of Penguin Books USA Inc.:** From "At Home" by Anton Checkhov from *The Portable Checkhov,* edited by Avrahm Yarmolinsky. Copyright 1947, © 1968 by The Viking Press, Inc. From *India: A Million Mutinies Now* by V. S. Naipaul. Copyright © 1990 by V. S. Naipaul. From *Don Quixote* by Miguel de Cervantes Saavedra, translated by Samuel Putnam. Translation copyright 1949 by The Viking Press, Inc. **The Washington Post:** From "MTV Age Dawning in India" by Steve Coll from *The Washington Post,* March 5, 1992. Copyright © 1992 by The Washington Post.

REFERENCES:
From *The Heated Debate: Greenhouse Predictions Versus Climate Reality* by Robert C. Balling, Jr. Copyright © 1992 by Pacific Research Institute for Public Policy. Published by Pacific Research Institute for Public Policy. From *Beyond the Limits* by Donella H. Meadows, Dennis L. Meadows, and Jorgen Randers. Copyright © 1992 by Donella H. Meadows, Dennis L. Meadows, and Jorgen Randers. Published by Chelsea Green Publishing Company. From "Let's Be Sensible on Global Warming" by Andrew Revkin from *The Christian Science Monitor* vol. 84, no. 151, June 30, 1992. Copyright © 1992 by The Christian Science Publishing Society. Published by The Christian Science Publishing Society. Quote by Julian Simon from *Are World Population Trends a Problem?* edited by Ben Wattenberg and Karl Zinsmeister. Copyright © 1985 by the American Enterprise Institute for Public Policy Research. Published by the American Enterprise Institute for Public Policy Research. From *Wild Animals in Central India* by A. A. Dunbar Brander. Copyright 1923 by Edward Arnold. Published by Edward Arnold.